The Electrical Engineering Handbook
Third Edition

Electronics, Power Electronics, Optoelectronics, Microwaves, Electromagnetics, and Radar

The Electrical Engineering Handbook Series

Series Editor
Richard C. Dorf
University of California, Davis

Titles Included in the Series

The Handbook of Ad Hoc Wireless Networks, Mohammad Ilyas
The Avionics Handbook, Cary R. Spitzer
The Biomedical Engineering Handbook, Third Edition, Joseph D. Bronzino
The Circuits and Filters Handbook, Second Edition, Wai-Kai Chen
The Communications Handbook, Second Edition, Jerry Gibson
The Computer Engineering Handbook, Vojin G. Oklobdzija
The Control Handbook, William S. Levine
The CRC Handbook of Engineering Tables, Richard C. Dorf
The Digital Signal Processing Handbook, Vijay K. Madisetti and Douglas Williams
The Electrical Engineering Handbook, Third Edition, Richard C. Dorf
The Electric Power Engineering Handbook, Leo L. Grigsby
The Electronics Handbook, Second Edition, Jerry C. Whitaker
The Engineering Handbook, Third Edition, Richard C. Dorf
The Handbook of Formulas and Tables for Signal Processing, Alexander D. Poularikas
The Handbook of Nanoscience, Engineering, and Technology, William A. Goddard, III, Donald W. Brenner, Sergey E. Lyshevski, and Gerald J. Iafrate
The Handbook of Optical Communication Networks, Mohammad Ilyas and Hussein T. Mouftah
The Industrial Electronics Handbook, J. David Irwin
The Measurement, Instrumentation, and Sensors Handbook, John G. Webster
The Mechanical Systems Design Handbook, Osita D.I. Nwokah and Yidirim Hurmuzlu
The Mechatronics Handbook, Robert H. Bishop
The Mobile Communications Handbook, Second Edition, Jerry D. Gibson
The Ocean Engineering Handbook, Ferial El-Hawary
The RF and Microwave Handbook, Mike Golio
The Technology Management Handbook, Richard C. Dorf
The Transforms and Applications Handbook, Second Edition, Alexander D. Poularikas
The VLSI Handbook, Wai-Kai Chen

The Electrical Engineering Handbook
Third Edition

Edited by
Richard C. Dorf

Circuits, Signals, and Speech and Image Processing

Electronics, Power Electronics, Optoelectronics, Microwaves, Electromagnetics, and Radar

Sensors, Nanoscience, Biomedical Engineering, and Instruments

Broadcasting and Optical Communication Technology

Computers, Software Engineering, and Digital Devices

Systems, Controls, Embedded Systems, Energy, and Machines

The Electrical Engineering Handbook
Third Edition

Electronics, Power Electronics, Optoelectronics, Microwaves, Electromagnetics, and Radar

Edited by
Richard C. Dorf
University of California
Davis, California, U.S.A.

A CRC title, part of the Taylor & Francis imprint, a member of the Taylor & Francis Group, the academic division of T&F Informa plc.

Published in 2006 by
CRC Press
Taylor & Francis Group
6000 Broken Sound Parkway NW, Suite 300
Boca Raton, FL 33487-2742

Printed in the United States of America on acid-free paper
10 9 8 7 6 5 4 3 2 1

International Standard Book Number-10: 0-8493-7339-5 (Hardcover)
International Standard Book Number-13: 978-0-8493-7339-8 (Hardcover)
Library of Congress Card Number 2005054344

Library of Congress Cataloging-in-Publication Data

Electronics, power electronics, optoelectronics, microwaves, electromagnetics, and radar / edited by Richard C. Dorf.
p. cm.
Includes bibliographical references and index.
ISBN 0-8493-7339-5 (alk. paper)
1. Electric engineering. 2. Electronics. I. Dorf, Richard C. II. Title.

TK145.E435 2005
621.381--dc22 2005054344

Visit the Taylor & Francis Web site at
http://www.taylorandfrancis.com

and the CRC Press Web site at
http://www.crcpress.com

Preface

Purpose

The purpose of *The Electrical Engineering Handbook, 3rd Edition* is to provide a ready reference for the practicing engineer in industry, government, and academia, as well as aid students of engineering. The third edition has a new look and comprises six volumes including:

Circuits, Signals, and Speech and Image Processing
Electronics, Power Electronics, Optoelectronics, Microwaves, Electromagnetics, and Radar
Sensors, Nanoscience, Biomedical Engineering, and Instruments
Broadcasting and Optical Communication Technology
Computers, Software Engineering, and Digital Devices
Systems, Controls, Embedded Systems, Energy, and Machines

Each volume is edited by Richard C. Dorf, and is a comprehensive format that encompasses the many aspects of electrical engineering with articles from internationally recognized contributors. The goal is to provide the most up-to-date information in the classical fields of circuits, signal processing, electronics, electromagnetic fields, energy devices, systems, and electrical effects and devices, while covering the emerging fields of communications, nanotechnology, biometrics, digital devices, computer engineering, systems, and biomedical engineering. In addition, a complete compendium of information regarding physical, chemical, and materials data, as well as widely inclusive information on mathematics is included in each volume. Many articles from this volume and the other five volumes have been completely revised or updated to fit the needs of today and many new chapters have been added.

The purpose of this volume (*Electronics, Power Electronics, Optoelectronics, Microwaves, Electromagnetics, and Radar)* is to provide a ready reference to subjects in the fields of electronics, integrated circuits, power electronics, optoelectronics, electromagnetics, light waves, and radar. We also include a section on electrical effects and devices. Here we provide the basic information for understanding these fields. We also provide information about the emerging fields of microlithography and power electronics.

Organization

The information is organized into four sections. The first three sections encompass 29 chapters and the last section summarizes the applicable mathematics, symbols, and physical constants.

Most articles include three important and useful categories: defining terms, references, and further information. *Defining terms* are key definitions and the first occurrence of each term defined is indicated in boldface in the text. The definitions of these terms are summarized as a list at the end of each chapter or article. The *references* provide a list of useful books and articles for follow-up reading. Finally, *further information* provides some general and useful sources of additional information on the topic.

Locating Your Topic

Numerous avenues of access to information are provided. A complete table of contents is presented at the front of the book. In addition, an individual table of contents precedes each section. Finally, each chapter begins with its own table of contents. The reader should look over these tables of contents to become familiar with the structure, organization, and content of the book. For example, see Section II: Electromagnetics, then Chapter 17: Antennas, and then Chapter 17.2: Aperture. This tree-and-branch table of contents enables the reader to move up the tree to locate information on the topic of interest.

Two indexes have been compiled to provide multiple means of accessing information: subject index and index of contributing authors. The subject index can also be used to locate key definitions. The page on which the definition appears for each key (defining) term is clearly identified in the subject index.

The Electrical Engineering Handbook, 3rd Edition is designed to provide answers to most inquiries and direct the inquirer to further sources and references. We hope that this handbook will be referred to often and that informational requirements will be satisfied effectively.

Acknowledgments

This handbook is testimony to the dedication of the Board of Advisors, the publishers, and my editorial associates. I particularly wish to acknowledge at Taylor & Francis Nora Konopka, Publisher; Helena Redshaw, Editorial Project Development Manager; and Susan Fox, Project Editor. Finally, I am indebted to the support of Elizabeth Spangenberger, Editorial Assistant.

Richard C. Dorf
Editor-in-Chief

Editor-in-Chief

Richard C. Dorf, Professor of Electrical and Computer Engineering at the University of California, Davis, teaches graduate and undergraduate courses in electrical engineering in the fields of circuits and control systems. He earned a Ph.D. in electrical engineering from the U.S. Naval Postgraduate School, an M.S. from the University of Colorado, and a B.S. from Clarkson University. Highly concerned with the discipline of electrical engineering and its wide value to social and economic needs, he has written and lectured internationally on the contributions and advances in electrical engineering.

Professor Dorf has extensive experience with education and industry and is professionally active in the fields of robotics, automation, electric circuits, and communications. He has served as a visiting professor at the University of Edinburgh, Scotland; the Massachusetts Institute of Technology; Stanford University; and the University of California, Berkeley.

Professor Dorf is a Fellow of The Institute of Electrical and Electronics Engineers and a Fellow of the American Society for Engineering Education. Dr. Dorf is widely known to the profession for his *Modern Control Systems, 10th Edition* (Addison-Wesley, 2004) and *The International Encyclopedia of Robotics* (Wiley, 1988). Dr. Dorf is also the co-author of *Circuits, Devices and Systems* (with Ralph Smith), *5th Edition* (Wiley, 1992), and *Electric Circuits, 7th Edition* (Wiley, 2006). He is also the author of *Technology Ventures* (McGraw-Hill, 2005) and *The Engineering Handbook, 2nd Edition* (CRC Press, 2005).

Advisory Board

Contributors

Samuel O. Agbo
California Polytechnic State University
San Luis Obispo, California

Ayse E. Amac
Illinois Institute of Technology
Chicago, Illinois

John Okyere Attia
Prairie View A&M University
Prairie View, Texas

John E. Ayers
University of Connecticut
Storrs, Connecticut

Inder J. Bahl
M/A-COM, Inc.
Roanoke, Virginia

R. Bartnikas
Institut de Recherche d'Hydro-Québec
Varennes, Quebec, Canada

Geoffrey Bate
Consultant in Information Storage Technology
Los Altos Hills, California

R.A. Becker
Integrated Optical Circuit Consultants
Cupertino, California

Melvin L. Belcher
Georgia Tech Research Institute
Atlanta, Georgia

Ashoka K.S. Bhat
University of Victoria
Victoria, Canada

Imran A. Bhutta
RFPP
Voorhees, New Jersey

Glenn R. Blackwell
Purdue University
West Lafayette, Indiana

Bimal K. Bose
University of Tennessee
Knoxville, Tennessee

Joe E. Brewer
University of Florida
Gainesville, Florida

John R. Brews
The University of Arizona
Tucson, Arizona

Gordon L. Carpenter
California State University
Long Beach, California

John Choma, Jr.
University of Southern California
Los Angeles, California

Richard C. Compton
DV Wireless Group
Sunnyvale, California

Kevin A. Delin
Jet Propulsion Laboratory
Pasadena, California

Kenneth Demarest
University of Kansas
Lawrence, Kansas

Allen Dewey
Duke University
Durham, North Carolina

Mitra Dutta
University of Illinois
Chicago, Illinois

Alexander C. Ehrlich
U.S. Naval Research Laboratory
Washington, District of Columbia

Aicha Elshabini-Riad
Virginia Polytechnic Institute and State University
Blacksburg, Virginia

Ali Emadi
Illinois Institute of Technology
Chicago, Illinois

Halit Eren
Curtin University of Technology
Bentley, Australia

K.F. Etzold
IBM T.J. Watson Research Center
Yorktown Heights, New York

J. Patrick Fitch
Lawrence Livermore National Laboratory
Livermore, California

Susan A.R. Garrod
Purdue University
West Lafayette, Indiana

Boris Gelmont
University of Virginia
Charlottesville, Virginia

Gennady Sh. Gildenblat
The Pennsylvania State University
University Park, Pennsylvania

Jeff Hecht
Laser Focus World
Auburndale, Massachusetts

Leland H. Hemming
McDonnell Douglas Helicopter Systems
Mesa, Arizona

Ken Kaiser
Kettering University
Flint, Michigan

Gerd Keiser
Photonics Comm Solutions, Inc.
Newton, Massachusettes

E.J. Kennedy
University of Tennessee
Knoxville, Tennessee

Nicholas J. Kolias
Raytheon Company
Andover, Massachusetts

Mark H. Kryder
Carnegie Mellon University
Pittsburgh, Pennsylvania

Fang Lin Luo
Nanyang Technological University
Singapore

Marc J. Madou
University of California
Irvine, California

Andrew Marshall
Texas Instruments Incorporated
Dallas, Texas

Sudip. K. Mazumder
University of Illinois
Chicago, Illinois

Michael S. Mazzola
Mississippi State University
Mississippi State, Mississippi

Miran Milkovic
Analog Technology Consultants
Alpharetta, Georgia

Mark B. Moffett
Antion Corporation
North Kingstown, Rhode Island

James E. Morris
Portland State University
Portland, Oregon

Wayne Needham
Intel Corporation
Chandler, Arizona

Sudarshan Rao Nelatury
The Pennsylvania State University
Erie, Pennsylvania

Josh T. Nessmith
Georgia Tech Research Institute
Atlanta, Georgia

Robert E. Newnham
The Pennsylvania State University
University Park, Pennsylvania

Terry P. Orlando
Massachusetts Institute of Technology
Cambridge, Massachusetts

Benjamin Y. Park
University of California
Irvine, California

Harold G. Parks
The University of Arizona
Tucson, Arizona

Christian Piguet
CSEM Centre Suisse d'Electronique
et de Microtechnique
and LAP-EPFL
Neuchâtel, Switzerland

Samuel O. Piper
Georgia Tech Research Institute
Atlanta, Georgia

S. Rajaram
Lucent Technologies
Piscataway, New Jersey

Kaushik Rajashekara
Delphi Energy & Engineering
Management System
Indianapolis, Indiana

Banmali S. Rawat
University of Nevada
Reno, Nevada

Ian D. Robertson
University of Leeds
Leeds, United Kingdom

Peter H. Rogers
Georgia Institute of Technology
Atlanta, Georgia

Matthew N.O. Sadiku
Prairie View A&M University
Prairie View, Texas

Tirthajyoti Sarkar
University of Illinois
Chicago, Illinois

Charles H. Sherman
Image Acoustics
North Kingstown, Rhode Island

Sidney Soclof
California State University
Los Angeles, California

J.W. Steadman
University of Wyoming
Laramie, Wyoming

Michael B. Steer
North Carolina State University
Raleigh, North Carolina

F.W. Stephenson
Virginia Polytechnic Institute and
State University
Blacksburg, Virginia

Harvey J. Stiegler
Texas Instruments Incorporated
Dallas, Texas

V. Sundar
Dentsply Prosthetics – Ceramco
Burlington, New Jersey

Ronald J. Tallarida
Temple University
Philadelphia, Pennsylvania

Moncef B. Tayahi
University of Nevada
Reno, Nevada

S.K. Tewksbury
Stevens Institute of Technology
Hoboken, New Jersey

Charles W. Therrien
Naval Postgraduate School
Monterey, California

Spyros Tragoudas
Southern Illinois University
Carbondale, Illinois

Robert J. Trew
North Carolina State University
Raleigh, North Carolina

Martin A. Uman
University of Florida
Gainesville, Florida

John V. Wait
University of Arizona (Retired)
Tucson, Arizona

Laurence S. Watkins
Lucent Technologies
Princeton, New Jersey

Joseph Watson
University of Wales
Swansea, United Kingdom

B.M. Wilamowski
University of Wyoming
Laramie, Wyoming

James C. Wiltse
Georgia Tech Research Institute
Atlanta, Georgia

Bert Wong
Western Australia Telecommunications Research Institute
Crawley, Australia

Hong Ye
Nanyang Technological University
Singapore

R. Yimnirun
Chiang Mai University
Chiang Mai, Thailand

Rabih Zaouk
University of California
Irvine, California

Mehdi R. Zargham
Southern Illinois University
Carbondale, Illinois

Contents

SECTION I Electronics

SECTION II Electromagnetics

SECTION III Electrical Effects and Devices

SECTION IV Mathematics, Symbols, and Physical Constants

Indexes

I

Electronics

1

Semiconductors

Gennady Sh. Gildenblat
The Pennsylvania State University

Boris Gelmont
University of Virginia

Miran Milkovic
Analog Technology Consultants

Aicha Elshabini-Riad
Virginia Polytechnic Institute and State University

F.W. Stephenson
Virginia Polytechnic Institute and State University

Imran A. Bhutta
RFPP

1.1 Physical Properties

Gennady Sh. Gildenblat and Boris Gelmont

Electronic applications of semiconductors are based on our ability to vary their properties on a very small scale. In conventional semiconductor devices, one can easily alter charge carrier concentrations, fields, and current densities over distances of 0.1 to 10 μm. Even smaller characteristic lengths of 10 to 100 nm are feasible in materials with an engineered band structure. This section reviews the essential physics underlying modern semiconductor technology.

Energy Bands

In crystalline semiconductors atoms are arranged in periodic arrays known as crystalline lattices. The lattice structure of silicon is shown in Figure 1.1. Germanium and diamond have the same structure but with different interatomic distances. As a consequence of this periodic arrangement, the allowed energy levels of electrons are grouped into **energy bands,** as shown in Figure 1.2. The probability that an electron will occupy an allowed quantum state with energy E is

$$f = [1 + \exp(E - F)/k_B T]^{-1} \tag{1.1}$$

Here, $k_B = 1/11{,}606$ eV/K denotes the Boltzmann constant, T is the absolute temperature, and F is a parameter known as the Fermi level. If the energy $E > F + 3k_BT$, then $f(E) < 0.05$ and these states are mostly empty.

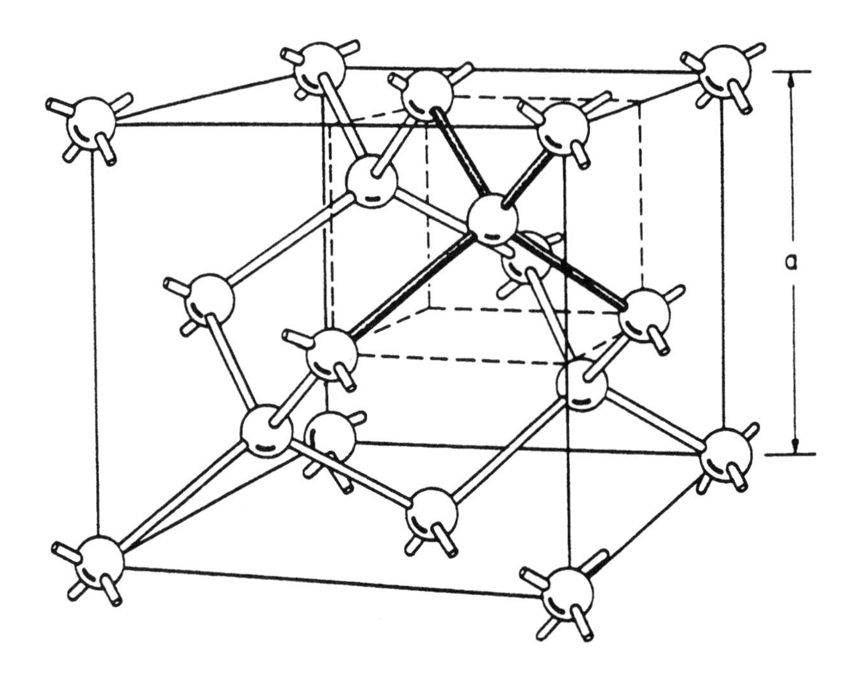

FIGURE 1.1 Crystalline lattice of silicon, $a = 5.43$ Å at 300°C. (*Source:* S.M. Sze, *Semiconductor Devices: Physics and Technology*, New York: John Wiley & Sons, 1985, p. 5. With permission.)

Similarly, the states with $E < F - 3k_BT$ are mostly occupied by electrons. In a typical metal (Figure 1.2(a)), the energy level $E = F$ is allowed, and only one energy band is partially filled. (In metals like aluminum, the partially filled band in Figure 1.2(a) may actually represent a combination of several overlapping bands.) The remaining energy bands are either completely filled or totally empty. Obviously, the empty energy bands do not contribute to the charge transfer. It is a fundamental result of solid-state physics that energy bands that are completely filled also do not contribute. What happens is that in the filled bands the average velocity of electrons is equal to zero. In semiconductors (and insulators) the Fermi level falls within a forbidden **energy gap** so that two of the energy bands are partially filled by electrons and may give rise to electron current. The upper partially filled band is called the **conduction band** while the lower is known as the **valence band**. The number of electrons in the conduction band of a semiconductor is relatively small and can be easily changed by adding impurities. In metals, the number of free carriers is large and is not sensitive to doping.

A more detailed description of energy bands in a crystalline semiconductor is based on the Bloch theorem, which states that an electron wave function has the form (Bloch wave):

$$\mathbf{\Psi}_{b\mathbf{k}} = u_{b\mathbf{k}}(\mathrm{r}) \ \exp(i\,\mathbf{kr}) \tag{1.2}$$

where **r** is the radius vector of electron, the modulating function $u_{b\mathbf{k}}(\mathbf{r})$ has the periodicity of the lattice, and the quantum state is characterized by wave vector **k** and the band number b. Physically, Equation (1.2) means that an electron wave propagates through a periodic lattice without attenuation. For each energy band one can consider the dispersion law $E = E_b(\mathbf{k})$. Since (see Figure 1.2(b)) in the conduction band only the states with energies close to the bottom, E_c, are occupied, it suffices to consider the $E(\mathbf{k})$ dependence near E_c. The simplified band diagrams of Si and GaAs are shown in Figure 1.3.

Electrons and Holes

The concentration of electrons in the valence band can be controlled by introducing impurity atoms. For example, the substitutional doping of Si with As results in a local energy level with an energy about $\Delta W_d \approx 45$ meV below the conduction band edge, E_c (Figure 1.2(b)). At room temperature this impurity center

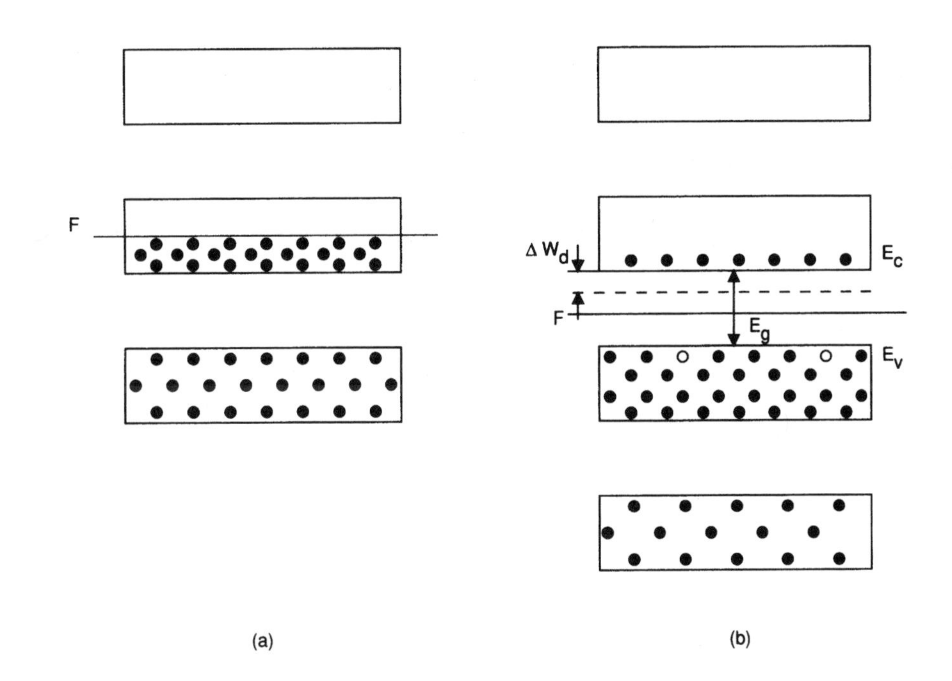

FIGURE 1.2 Band diagrams of metal (a) and semiconductor (b); ●, electron; ○, missing electron (hole).

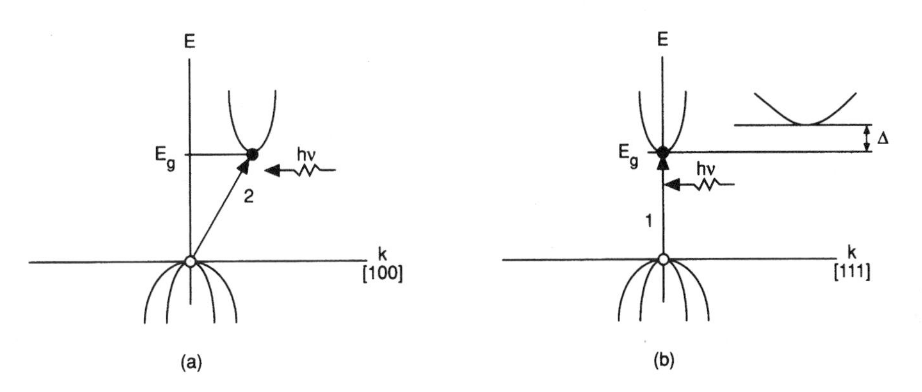

FIGURE 1.3 Simplified $E(\mathbf{k})$ dependence for Si (a) and GaAs (b). At room temperature $E_g(\text{Si}) = 1.12$ eV, E_g (GaAs) $=$ 1.43 eV, and $\Delta = 0.31$ eV; (1) and (2) indicate direct and indirect band-to-band transitions.

is readily ionized, and (in the absence of other impurities) the concentration of electrons is close to the concentration of As atoms. Impurities of this type are known as donors.

While considering the contribution $\mathbf{j}_p$ of the predominantly filled valence band to the current density, it is convenient to concentrate on the few missing electrons. This is achieved as follows: let $\mathbf{v}(\mathbf{k})$ be the velocity of electron described by the wave function (1.2). Then

$$\mathbf{j}_p = -q \sum_{\substack{\text{filled} \\ \text{states}}} \mathbf{v}(\mathbf{k}) = -q \left[\sum_{\text{all states}} \mathbf{v}(\mathbf{k}) - \sum_{\substack{\text{empty} \\ \text{states}}} \mathbf{v}(\mathbf{k}) \right] = q \sum_{\substack{\text{empty} \\ \text{states}}} \mathbf{v}(\mathbf{k}) \tag{1.3}$$

Here, we have noted again that a completely filled band does not contribute to the current density. The picture emerging from Equation (1.3) is that of particles (known as **holes**) with the charge $+q$ and velocities corresponding to those of missing electrons. The concentration of holes in the valence band is controlled by

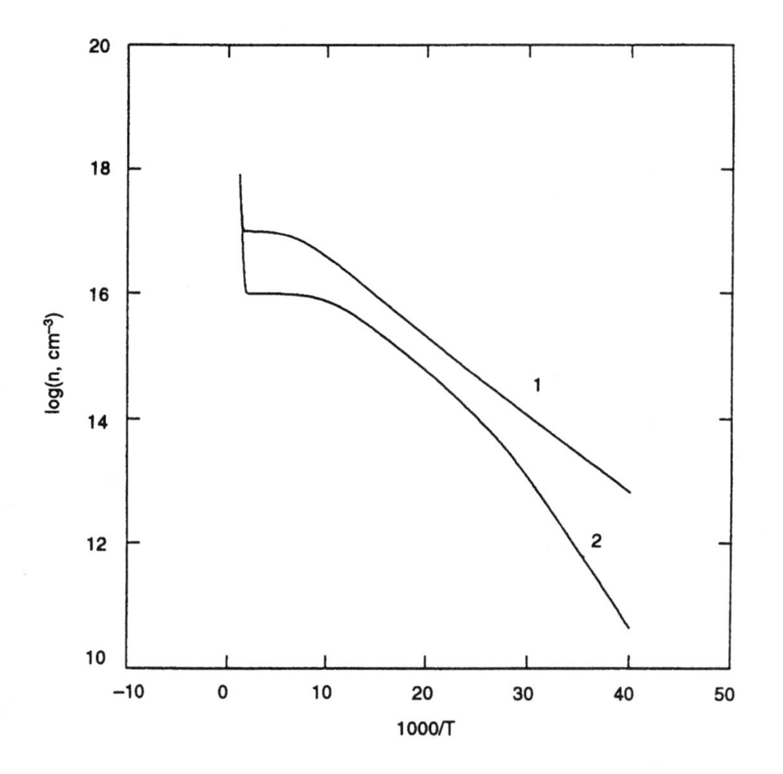

FIGURE 1.4 The inverse temperature dependence of electron concentration in Si; 1: $N_d = 10^{17}\text{cm}^{-3}$, $N_a = 0$; 2: $N_d = 10^{16}\ \text{cm}^{-3}$, $N_a = 10^{14}\,\text{cm}^{-3}$.

adding acceptor-type impurities (such as boron in silicon), which form local energy levels close to the top of the valence band. At room temperature these energy levels are occupied by electrons that come from the valence band and leave the holes behind. Assuming that the Fermi level is removed from both E_c and E_v by at least $3k_BT$ (a nondegenerate semiconductor), the concentrations of electrons and holes are given by

$$n = N_c \ \exp[(F - E_c)/k_B T] \tag{1.4}$$

and

$$p = N_v \ \exp[(E_v - F)/k_B T] \tag{1.5}$$

where $N_c = 2(2m_n^* \pi k_B T)^{3/2}/h^3$ and $N_v - 2(2m_p^* \pi k_B T)^{3/2}/h^3$ are the effective densities of states in the conduction and valence bands, respectively, h is Plank constant, and the effective masses m_n^* and m_p^* depend on the details of the band structure (Pierret, 1987).

In a nondegenerate semiconductor, $np = N_c N_v \ \exp(-E_g/k_B T) \triangleq n_i^2$ is independent of the doping level. The neutrality condition can be used to show that in an n-type (n > p) semiconductor at or below room temperature:

$$n(n + N_a)(N_d + N_a - n)^{-1} = (N_c/2) \ \exp(-\Delta W_d/k_B T) \tag{1.6}$$

where N_d and N_a denote the concentrations of **donors** and **acceptors**, respectively.

Corresponding temperature dependence is shown for silicon in Figure 1.4. Around room temperature $n = N_d - N_a$, while at low temperatures n is an exponential function of temperature with the activation energy $\Delta W_d/2$ for $n > N_a$ and ΔW_d for $n < N_a$. The reduction of n compared with the net impurity concentration $N_d - N_a$ is known as a freeze-out effect. This effect does not take place in the heavily doped semiconductors.

For temperatures $T > T_i = (E_g/2k_B)/\ln[\sqrt{N_c N_v}/(N_d - N_a)]$ the electron concentration $n \approx n_i >> N_d - N_a$ is no longer dependent on the doping level (Figure 1.4). In this so-called intrinsic regime electrons come directly from the valence band. A loss of technological control over n and p makes this regime unattractive for electronic applications. Since $T_i \propto E_g$ the transition to the intrinsic region can be delayed by using widegap semiconductors. Both silicon carbide (several types of SiC with different lattice structures are available with $E_g = 2.2$–2.86 eV) and diamond ($E_g = 5.5$ eV) have been used to fabricate diodes and transistors operating in the 300 to 700°C temperature range.

Transport Properties

In a semiconductor the motion of an electron is affected by frequent collisions with **phonons** (quanta of lattice vibrations), impurities, and crystal imperfections. In weak uniform electric fields, $\mathscr{E}$, the carrier drift velocity, $\mathbf{v}_d$, is determined by the balance of the electric and collision forces:

$$m_n^* \mathbf{v}_d/\tau = -q\mathscr{E} \tag{1.7}$$

where τ is the momentum relaxation time. Consequently $\mathbf{v}_d = -\mu_n \mathscr{E}$, where $\mu_n = q\tau/m_n^*$ is the electron mobility. For an n-type semiconductor with uniform electron density, n, the current density $\mathbf{j}_n = -qn\mathbf{v}_d$ and we obtain Ohm's law $j_n = \sigma\mathscr{E}$ with the conductivity $\sigma = qn\mu_n$. The momentum relaxation time can be approximately expressed as

$$1/\tau = 1/\tau_{ii} + 1/\tau_{ni} + 1/\tau_{ac} + 1/\tau_{npo} + 1/\tau_{po} + 1/\tau_{pe} + \ldots \tag{1.8}$$

where τ_{ii}, τ_{ni}, τ_{ac}, τ_{npo}, τ_{po}, τ_{pe} are the relaxation times due to ionized impurity, neutral impurity, acoustic phonon, nonpolar optical, polar optical, and piezoelectric scattering, respectively.

In the presence of concentration gradients, electron current density is given by the drift-diffusion equation:

$$\mathbf{j}n = qn\mu_n\mathscr{E} + qD_n\nabla n \tag{1.9}$$

where the diffusion coefficient D_n is related to mobility by the Einstein relation $D_n = (k_B T/q)\mu_n$.

A similar equation can be written for holes and the total current density is $\mathbf{j} = \mathbf{j}_n + \mathbf{j}_p$. The right-hand side of Equation (1.9) may contain additional terms corresponding to temperature gradient and compositional nonuniformity of the material (Wolfe et al., 1989).

In sufficiently strong electric fields the drift velocity is no longer proportional to the electric field. Typical velocity–field dependencies for several semiconductors are shown in Figure 1.5. In GaAs $v_d(\mathscr{E})$ dependence is

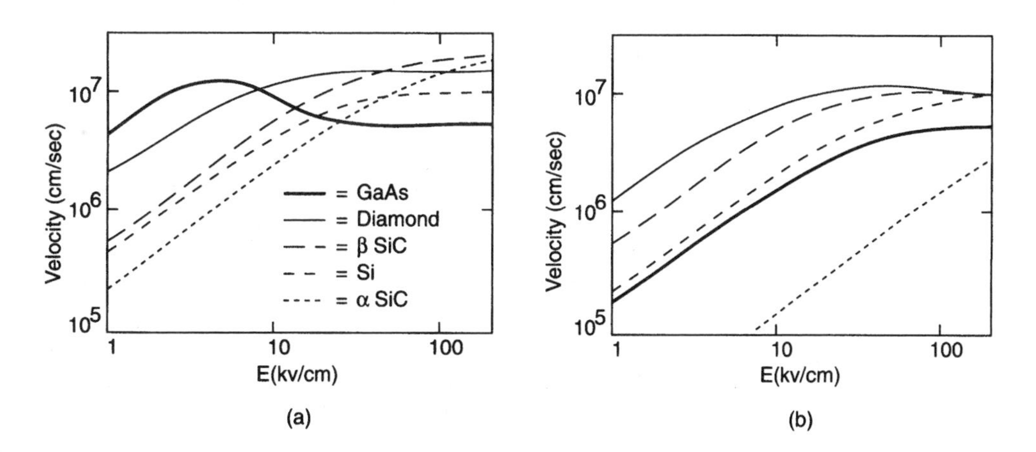

FIGURE 1.5 Electron (a) and hole (b) drift velocity versus electric field dependence for several semiconductors at $N_d = 10^{17}$ cm^{-3}. (*Source:* R.J. Trew, J.-B. Yan, and L.M. Mack, *Proc. IEEE*, vol. 79, no. 5, p. 602, May 1991. © 1991 IEEE.)

not monotonic, which results in negative differential conductivity. Physically, this effect is related to the transfer of electrons from the conduction band to a secondary valley (see Figure 1.3).

The limiting value $\mathbf{v}_s$ of the drift velocity in a strong electric field is known as the saturation velocity and is usually within the 10^7 to $3{\cdot}10^7$ cm/sec range. As semiconductor device dimensions are scaled down to the submicrometer range, $\mathbf{v}_s$ becomes an important parameter that determines the upper limits of device performance. The curves shown in Figure 1.5 were obtained for uniform semiconductors under steady-state conditions. Strictly speaking, this is not the case with actual semiconductor devices, where velocity can "overshoot" the value shown in Figure 1.5. This effect is important for Si devices shorter than 0.1 μm (0.25 μm for GaAs devices) (Shur, 1990; Ferry, 1991). In such extreme cases the drift-diffusion equation (1.9) is no longer adequate, and the analysis is based on the Boltzmann transport equation:

$$\frac{\partial f}{\partial t} + \mathrm{v}\nabla f + q\mathscr{E}\,\nabla_{\mathbf{p}} f = \left(\frac{\partial f}{\partial t}\right)_{\mathrm{coll}} \tag{1.10}$$

Here, f denotes the distribution function (number of electrons per unit volume of the phase space, i.e., $f = dn/d^3r d^3p$), $\mathbf{v}$ is electron velocity, $\mathbf{p}$ is momentum, and $(\partial f/\partial t)_{\mathrm{coll}}$ is the "collision integral" describing the change of f caused by collision processes described earlier. For the purpose of semiconductor modeling, Equation (1.10) can be solved directly using various numerical techniques, including the method of moments (hydrodynamic modeling) or Monte-Carlo approach. The drift-diffusion equation (1.9) follows from Equation (1.10) as a special case. For even shorter devices quantum effects become important and device modeling may involve quantum transport theory (Ferry, 1991).

Hall Effect

In a uniform magnetic field electrons move along circular orbits in a plane normal to the magnetic field $\mathbf{B}$ with the angular (cyclotron) frequency $\omega_c = q\mathbf{B}/m_n^*$. For a uniform semiconductor the current density satisfies the equation:

$$\mathbf{j} = \sigma(\mathscr{E} + R_H[\mathbf{jB}]) \tag{1.11}$$

In the usual weak-field limit $\omega_c\tau << 1$ the Hall coefficient $R_H = -r/nq$ and the Hall factor r depend on the dominating scattering mode. It varies between $3\pi/8 \approx 1.18$ (acoustic phonon scattering) and $315\pi/518 \approx 1.93$ (ionized impurity scattering).

The Hall coefficient can be measured as $R_H = V_y d/I_x B$ using the test structure shown in Figure 1.6. In this expression V_y is the Hall voltage corresponding to $I_y = 0$ and d denotes the film thickness.

Combining the results of the Hall and conductivity measurements one can extract the carrier concentration type (the signs of V_y are opposite for n-type and p-type semiconductors) and Hall mobility $\mu_H = r\mu$:

$$\mu_H = -R_H\sigma, \quad n = -r/qR_H \tag{1.12}$$

Measurements of this type are routinely used to extract concentration and mobility in doped semiconductors. The weak-field Hall effect is also used for the purpose of magnetic field measurements.

In strong magnetic fields $\omega_c\tau >> 1$ and on the average an electron completes several circular orbits without a collision. Instead of the conventional $E_b(\mathbf{k})$ dependence, the allowed electron energy levels in the magnetic field are given by ($\hbar = h/2\pi$; $s = 0, 1, 2, \ldots$):

$$E_s = \hbar\omega_c\,(s + 1/2) + \hbar^2 k_z^2/2m_n^* \tag{1.13}$$

The first term in Equation (1.13) describes the so-called Landau levels, while the second corresponds to the kinetic energy of motion along the magnetic field $\mathbf{B} = \mathbf{B}_z$. In a pseudo-two-dimensional system like the channel of a field-effect transistor the second term in Equation (1.13) does not appear, since the motion of

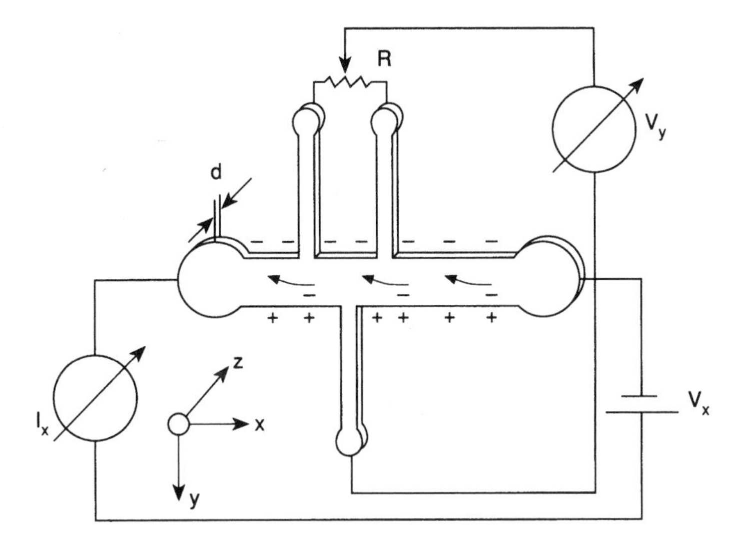

FIGURE 1.6 Experimental setup for Hall effect measurements in a long two-dimensional sample. The Hall angle is determined by a setting of the rheostat that renders $\mathbf{j}_y = 0$. Magnetic field $\mathbf{B} = \mathbf{B}_z$. (*Source:* K.W. Böer, *Surveys of Semiconductor Physics,* New York: Chapman & Hall, 1990, p. 760. With permission.)

electrons occurs in the plane perpendicular to the magnetic field.[1] In such a structure the electron density of states (number of allowed quantum states per unit energy interval) is peaked at the Landau level. Since $\omega_c \propto \mathbf{B}$, the positions of these peaks relative to the Fermi level are controlled by the magnetic field.

The most striking consequence of this phenomenon is the quantum Hall effect, which manifests itself as a stepwise change of the Hall resistance $\rho_{xy} = V_y/I_x$ as a function of magnetic field (see Figure 1.7). At low temperature (required to establish the condition $\tau << \omega_c^{-1}$) it can be shown (von Klitzing, 1986) that

$$\rho_{xy} = h/sq^2 \tag{1.14}$$

where s is the number of the highest occupied Landau level. Accordingly, when the increased magnetic field pushes the sth Landau level above the Fermi level, ρ_{xy} changes from h/sq^2 to $h/(s-1)q^2$. This stepwise change of ρ_{xy} is seen in Figure 1.7. Localized states produced by crystal defects determine the shape of the $\rho_{xy}(\mathbf{B})$ dependence between the plateaus given by Equation (1.14). They are also responsible for the disappearance of $\rho_{xx} = V_x/I_x$ between the transition points (see Figure 1.7). The quantized Hall resistance ρ_{xy} is expressed in terms of fundamental constants and can be used as a resistance standard that permits one to measure an electrical resistance with better accuracy than any wire resistor standard. In an ultraquantum magnetic field, i.e., when only the lowest Landau level is occupied, plateaus of the Hall resistance are also observed at fractional s (the fractional quantum Hall effect). These plateaus are related to the Coulomb interaction of electrons.

Electrical Breakdown

In sufficiently strong electric fields a measurable fraction of electrons (or holes) acquires sufficient energy to break the valence bond. Such an event (called impact ionization) results in the creation of an electron–hole pair by the energetic electron. Both the primary and secondary electrons as well as the hole are accelerated by the electric field and may participate in further acts of impact ionization. Usually, the impact ionization is balanced by recombination processes. If the applied voltage is high enough, however, the process of electron

[1]To simplify the matter we do not discuss surface subbands, which is justified as long as only the lowest of them is occupied.

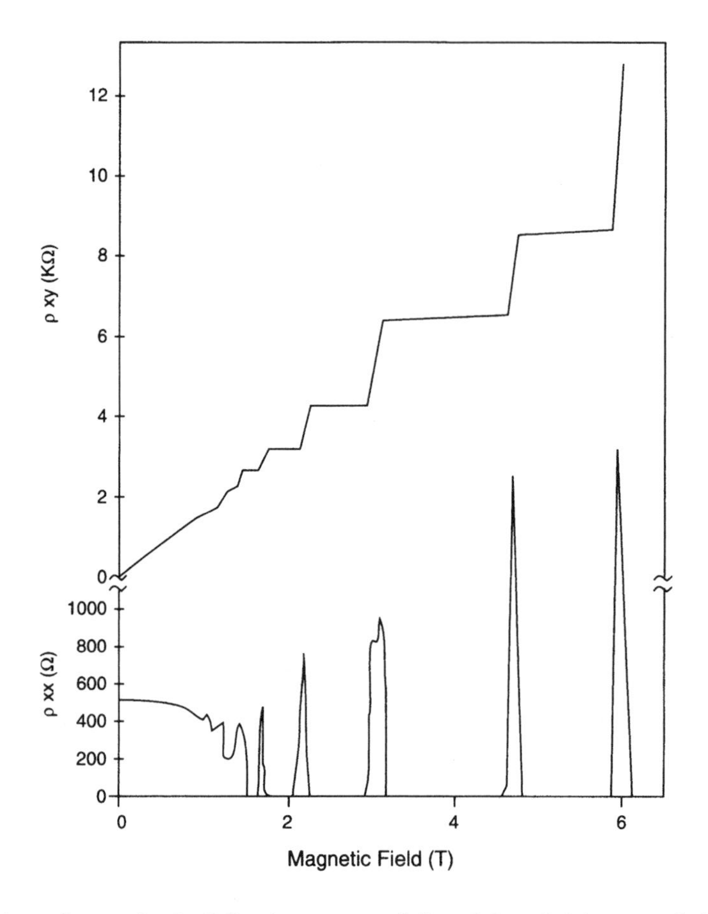

FIGURE 1.7 Experimental curves for the Hall resistance $\rho_{xy} = \mathscr{E}_y/j_x$ and the resistivity $\rho_{xx} = \mathscr{E}x/j_x$ of a heterostructure as a function of the magnetic field at a fixed carrier density. (*Source:* K. von Klitzing, *Rev. Modern Phys.*, vol. 58, no. 3, p. 525, 1986. With permission.)

TABLE 1.1 Impact Ionization Threshold Energy (eV)

Semiconductor	Si	Ge	GaAs	GaP	InSb
Energy gap, E_g	1.1	0.7	1.4	2.3	0.2
E_{th}, electron-initiated	1.18	0.76	1.7	2.6	0.2
E_{th}, hole-initiated	1.71	0.88	1.4	2.3	0.2

multiplication leads to avalanche breakdown. The threshold energy E_{th} (the minimum electron energy required to produce an electron–hole pair) is determined by energy and momentum conservation laws. The latter usually results in $E_{th} < E_g$, as shown in Table 1.1.

The field dependence of the impact ionization is usually described by the impact ionization coefficient α_i, defined as the average number of electron–hole pairs created by a charge carrier per unit distance traveled. A simple analytical expression for α_i (Okuto and Crowell, 1972) can be written as

$$\alpha_i = (\lambda/x)\ \exp\left(a - \sqrt{a^2 + x^2}\right) \tag{1.15}$$

where $x = q\mathscr{E}\lambda/E_{th}$, $a = 0.217\ (E_{th}/E_{opt})^{1.14}$, λ is the carrier mean free path, and E_{opt} is the optical phonon energy ($E_{opt} = 0.063$ eV for Si at 300°C).

An alternative breakdown mechanism is tunneling breakdown, which occurs in highly doped semiconductors when electrons may tunnel from occupied states in the valence band into the empty states of the conduction band.

Optical Properties and Recombination Processes

If the energy of an incident **photon** $\hbar\omega > E_g$, then the energy conservation law permits a direct band-to-band transition, as indicated in Figure 1.2(b). Because the photon's momentum is negligible compared to that of an electron or hole, the electron's momentum $\hbar\mathbf{k}$ does not change in a direct transition. Consequently, direct transitions are possible only in direct-gap semiconductors where the conduction band minimum and the valence band maximum occur at the same **k**. The same is true for the reverse transition, where the electron is transferred from the conduction to the valence band and a photon is emitted. Direct-gap semiconductors (e.g., GaAs) are widely used in optoelectronics.

In indirect-band materials (e.g., Si, see Figure 1.3(a)), a band-to-band transition requires a change of momentum that cannot be accomplished by absorption or emission of a photon. Indirect band-to-band transitions require the emission or absorption of a phonon and are much less probable than direct transitions.

For $\hbar\omega < E_g$ (i.e., for $\lambda > \lambda_c = 1.24$ μm/E_g [eV] – cutoff wavelength) band-to-band transitions do not occur, but light can be absorbed by a variety of the so-called subgap processes. These processes include the absorption by free carriers, formation of excitons (bound electron–hole pairs whose formation requires less energy than the creation of a free electron and a free hole), transitions involving localized states (e.g., from an acceptor state to the conduction band), and phonon absorption. Both band-to-band and subgap processes may be responsible for the increase of the free charge carriers concentration. The resulting reduction of the resistivity of illuminated semiconductors is called *photoconductivity* and is used in photodetectors.

In a strong magnetic field ($\omega_c\tau >> 1$) the absorption of microwave radiation is peaked at $\omega = \omega_c$. At this frequency the photon energy is equal to the distance between two Landau levels, i.e., $\hbar\omega = E_{s+1} - E_s$ with reference to Equation (1.13). This effect, known as cyclotron resonance, is used to measure the effective masses of charge carriers in semiconductors (in a simplest case of isotropic $E(\mathbf{k})$ dependence, $m_n^* = qB/\omega_c$).

In indirect-gap materials like silicon, the generation and annihilation (or recombination) of electron–hole pairs is often a two-step process. First, an electron (or a hole) is trapped in a localized state (called a recombination center) with the energy near the center of the energy gap. In a second step, the electron (or hole) is transferred to the valence (conduction) band. The net rate of recombination per unit volume per unit time is given by the Shockley–Read–Hall theory as

$$R = \frac{np - n_i^2}{\tau_n(p + p_1) + \tau_p(n + n_1)} \tag{1.16}$$

where τ_n, τ_p, p_1, and n_1 are parameters depending on the concentration and the physical nature of recombination centers and temperature. Note that the sign of R indicates the tendency of a semiconductor toward equilibrium (where $np = n_p^2$, and $R = 0$). For example, in the depleted region $np < n_i^2$ and $R < 0$, so that charge carriers are generated.

Shockley–Read–Hall recombination is the dominating recombination mechanism in moderately doped silicon. Other recombination mechanisms (e.g., Auger) become important in heavily doped semiconductors (Wolfe et al., 1989; Shur, 1990; Ferry, 1991).

The recombination processes are fundamental for semiconductor device theory, where they are usually modeled using the continuity equation:

$$\frac{\partial n}{\partial t} = \text{div}\frac{j_n}{q} - R \tag{1.17}$$

Nanostructure Engineering

Epitaxial growth techniques, especially molecular beam epitaxy and metal–organic chemical vapor deposition, allow monolayer control in the chemical composition process. Both single thin layers and superlattices can be obtained by such methods. The electronic properties of these structures are of interest for potential device applications. In a single quantum well, electrons are bound in the confining well potential. For example, in a rectangular quantum well of width b and infinite walls, the allowed energy levels are

$$E_s(\mathbf{k}) = \pi^2 s^2 \hbar^2/(2m_n^* b^2) + \hbar^2 k^2/(2m_n^*), \qquad s = 1, 2, 3, \ldots \tag{1.18}$$

where $\mathbf{k}$ is the electron wave vector parallel to the plane of the semiconductor layer. The charge carriers in quantum wells exhibit confined particle behavior. Since $E_s \propto b^{-2}$, well structures can be grown with distance between energy levels equal to a desired photon energy. Furthermore, the photoluminescence intensity is enhanced because of carrier confinement. These properties are advantageous in fabrication of lasers and photodetectors.

If a quantum well is placed between two thin barriers, the tunneling probability is greatly enhanced when the energy level in the quantum well coincides with the Fermi energy (resonant tunneling). The distance between this "resonant" energy level and the Fermi level is controlled by the applied voltage. Consequently, the current peaks at the voltage corresponding to the resonant tunneling condition. The resulting negative differential resistance effect has been used to fabricate microwave generators operating at both room and cryogenic temperatures.

Two kinds of superlattices are possible: compositional and doping. Compositional superlattices are made of alternating layers of semiconductors with different energy gaps. Doping superlattices consist of alternating n- and p-type layers of the same semiconductor. The potential is modulated by electric fields arising from the charged dopants. Compositional superlattices can be grown as lattice matched or as strained layers. The latter are used for modification of the band structure, which depends on the lattice constant to produce desirable properties.

In superlattices energy levels of individual quantum wells are split into minibands as a result of electron tunneling through the wide-bandgap layers. This occurs if the electron mean free path is larger than the superlattice period. In such structures the electron motion perpendicular to the layer is quantized. In a one-dimensional tight binding approximation the miniband can be described as

$$E(\mathbf{k}) = E_0[1 - \cos(\mathbf{k}a)] \tag{1.19}$$

where a is the superlattice period and E_0 is the half-width of the energy band. The electron group velocity:

$$v = \hbar^{-1} \partial E(\mathbf{k})/\partial \mathbf{k} = (E_0 a/\hbar) \quad \sin(\mathbf{k}a) \tag{1.20}$$

is a decreasing function of $\mathbf{k}$ (and hence of energy) for $\mathbf{k} > \pi/2a$. The higher energy states with $\mathbf{k} > \pi/2a$ may become occupied if the electrons are heated by the external field. As a result, a negative differential resistance can be achieved at high electric fields. The weak-field mobility in a superlattice may exceed that of the bulk material because of the separation of dopants if only barriers are doped. In such modulated structures, the increased spatial separation between electrons and holes is also responsible for a strong increase in recombination lifetimes.

Disordered Semiconductors

Both amorphous and heavily doped semiconductors are finding increasing applications in semiconductor technology. The electronic processes in these materials have specific features arising from the lack of long-range order.

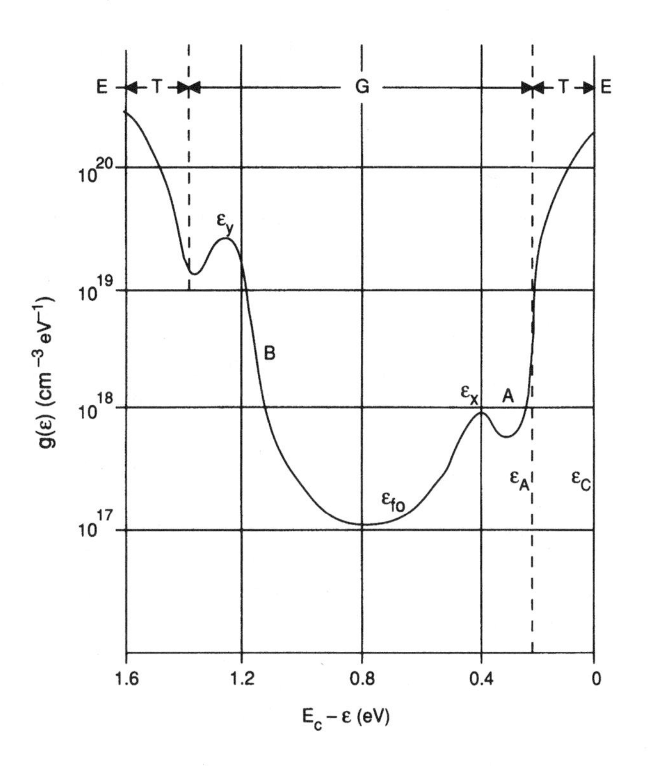

FIGURE 1.8 Experimentally determined density of states for a-Si. *A* and *B* are acceptor-like and donor-like states, respectively. The arrow marks the position of the Fermi level ε_{fo} in undoped hydrogenated a-Si. The energy spectrum is divided into extended states E, band-tail states T, and gap states G. (*Source:* M.H. Brodsky, Ed., *Amorphous Semiconductors,* 2nd ed., Berlin: Springer-Verlag, 1985. With permission.)

Amorphous semiconductors do not have a crystalline lattice, and their properties are determined by the arrangement of the nearest neighboring atoms. Even so, experimental data show that the forbidden energy band concept can be applied to characterize their electrical properties. However, the disordered nature of these materials results in a large number of localized quantum states with energies within the energy gap. The localized states in the upper and lower half of the gap behave like acceptors and donors, respectively. As an example, consider the density of states in hydrogenated amorphous silicon (a-Si) shown in Figure 1.8. The distribution of the localized states is not symmetrical with respect to the middle of the energy gap. In particular, the undoped hydrogenated amorphous silicon is an n-type semiconductor.

Usually amorphous semiconductors are not sensitive to the presence of impurity atoms, which saturate all their chemical bonds in the flexible network of the host atoms. (Compare this with a situation in crystalline silicon where an arsenic impurity can form only four chemical bonds with the host lattice, leaving the fifth responsible for the formation of the donor state.) Consequently, the doping of amorphous semiconductors is difficult to accomplish. However, in hydrogenated a-Si (which can be prepared by the glow discharge decomposition of silane), the density of the localized states is considerably reduced and the conductivity of this material can be controlled by doping. As in crystalline semiconductors, the charge carrier concentration in hydrogenated a-Si can also be affected by light and strong field effects. The a-Si is used in applications that require deposition of thin-film semiconductors over large areas (xerography, solar cells, thin-film transistors [TFT] for liquid-crystal displays). The a-Si device performance degrades with time under electric stress (TFTs) or under illumination (Staebler–Wronski effect) because of the creation of new localized states.

An impurity band in crystalline semiconductors is another example of a disordered system. Indeed, the impurity atoms are randomly distributed within the host lattice. For lightly doped semiconductors at room temperature, the random potential associated with charged impurities can usually be ignored. As the doping level increases, however, a single energy level of a donor or an acceptor is transformed into an energy band with a width determined by impurity concentrations. Unless the degree of compensation is unusually high, this reduces the activation energy compared to lightly doped semiconductors. The activation energy is further reduced by the overlap of the wave functions associated with the individual donor or acceptor states.

For sufficiently heavy doping, i.e., for $N_d > N_{dc} = (0.2/a_B)^3$, the ionization energy is reduced to zero, and the transition to metal-type conductivity (the Anderson–Mott transition) takes place. In this expression the effective electron Bohr radius $a_B = \hbar\sqrt{2m_n^* E_i}$, where E_i is the ionization energy of the donor state. For silicon, $N_{dc} \approx 3.8 \cdot 10^{18}\ \text{cm}^{-3}$. This effect explains the absence of freeze-out in heavily doped semiconductors.

Defining Terms

Conduction/valence band: The upper/lower of the two partially filled bands in a semiconductor.

Donors/acceptors: Impurities that can be used to increase the concentration of electrons/holes in a semiconductor.

Energy band: Continuous interval of energy levels that are allowed in the periodic potential field of the crystalline lattice.

Energy gap: The width of the energy interval between the top of the valence band and the bottom of the conduction band.

Hole: Fictitious positive charge representing the motion of electrons in the valence band of a semiconductor; the number of holes equals the number of unoccupied quantum states in the valence band.

Phonon: Quantum of lattice vibration.

Photon: Quantum of electromagnetic radiation.

References

D.K. Ferry, *Semiconductors,* New York: Macmillan, 1991.

Y. Okuto and C.R. Crowell, *Phys. Rev.*, vol. B6, p. 3076, 1972.

R.F. Pierret, *Advanced Semiconductor Fundamentals,* Reading, Mass.: Addison-Wesley, 1987.

M. Shur, *Physics of Semiconductor Devices,* Englewood Cliffs, NJ: Prentice-Hall, 1990.

K. von Klitzing, *Rev. Modern Phys.*, vol. 58, p. 519, 1986.

C.M. Wolfe, N. Holonyak, and G.E. Stilman, *Physical Properties of Semiconductors,* Englewood Cliffs, NJ: Prentice-Hall, 1989.

Further Information

Engineering aspects of semiconductor physics are often discussed in the *IEEE Transistors on Electron Devices, Journal of Applied Physics,* and *Solid-State Electronics.*

1.2 Diodes

Miran Milkovic

Diodes are the most widely used devices in low- and high-speed electronic circuits and in rectifiers and power supplies. Other applications are in voltage regulators, detectors, and demodulators. Rectifier diodes are capable of conducting several hundred amperes in the forward direction and less than 1 μA in the reverse direction. Zener diodes are ordinary diodes operated in the Zener or avalanche region and are used as voltage regulators. Varactor diodes are ordinary diodes used in reverse biasing as voltage-dependent capacitors. Tunnel diodes and quantum well devices have a negative differential resistance and are capable of operating in the upper gigahertz region. Photodiodes are ordinary diodes operated in the reverse direction. They are sensitive to light and are used as light sensors. Solar cells are diodes which convert light energy into electrical energy. Schottky diodes, also known as metal-semiconductor diodes, are extremely fast because they are **majority carrier** devices.

pn-Junction Diode

A pn-diode is a semiconductor device having a p-region, a n-region, and a junction between the regions. Modern planar semiconductor pn-junction diodes are fabricated by **diffusion** or implantation of impurities into a semiconductor. An n-type semiconductor has a relatively large density of free electrons to conduct electric current, and the p-type semiconductor has a relatively large concentration of "free" holes to conduct electric current. The pn-junction is formed during the fabrication process. There is a large concentration of holes in the p-semiconductor and a large concentration of electrons in the n-semiconductor. Because of their large concentration gradients, holes and electrons start to diffuse across the junction. As holes move across the junction, negative immobile charges (**acceptors**) are uncovered on the p side, and positive immobile charges (**donors**) are uncovered on the n side due to the movement of electrons across the junction. When sufficient numbers of the immobile charges on both sides of the junction are uncovered, a potential energy barrier voltage V_0 is created by the uncovered acceptors and donors. This **barrier voltage** prevents further diffusion of holes and electrons across the junction. The charge distribution of acceptors and donors establishes an opposing electric field, E, which at equilibrium prevents a further diffusion of carriers across the junction. This equilibrium can be regarded as the flow of two equal and opposite currents across the junction, such that the net current across the junction is equal to zero. Thus, one component represents the diffusion of carriers across the junction and the other component represents the **drift** of carriers across the junction due to the electric field E in the junction. The barrier voltage V_0 is, according to the **Boltzmann relation** (Grove, 1967; Foustad, 1994):

$$V_0 = V_T \ln[p_p/p_n] \tag{1.21}$$

In this equation, p_p is the concentration of holes in the p-material and p_n is the concentration of holes in the n-material. V_T is the thermal voltage. $V_T = 26\,\mathrm{mV}$ at room temperature (300 K). With

$$p_p \approx N_A \text{ and } p_n \approx \frac{n_i^2}{N_D}$$

where n_i is the intrinsic concentration, the barrier voltage V_0 becomes approximately (Sze, 1985; Fonstad, 1994):

$$V_0 = V_T \ln[N_A N_D/n_i^2] \tag{1.22}$$

Here, N_A denotes the concentration of immobile acceptors on the p side of the junction and N_D is the concentration of immobile donors on the n side of the junction. A depletion layer of immobile acceptors and donors causes an electric field E across the junction. For silicon, V_0 is at room temperature $T = 300°$ K, typically $V_0 = 0.67$ V for an abrupt junction with $N_A = 10^{17}$ at/cm^3 and $N_D = 10^{15}$ at/cm^3. The depletion layer width is typically about 4 μm, and the electric field E is about 60 kV/cm. Note the magnitude of the electric field across the junction.

pn-Junction with Applied Voltage

If the externally applied voltage V_D to the diode is opposite to the barrier voltage V_0, then p_p in the Boltzmann relation in Equation (1.21) is altered to

$$p_p = p_n \exp(V_0 - V_D)/V_T \tag{1.23}$$

This implies that the effective barrier voltage is reduced and the diffusion of carriers across the junction, is increased. Accordingly the concentration of diffusing holes into the n-material is at $x = 0$:

$$p_n(x = 0) = p_n \exp V_D/V_T \tag{1.24}$$

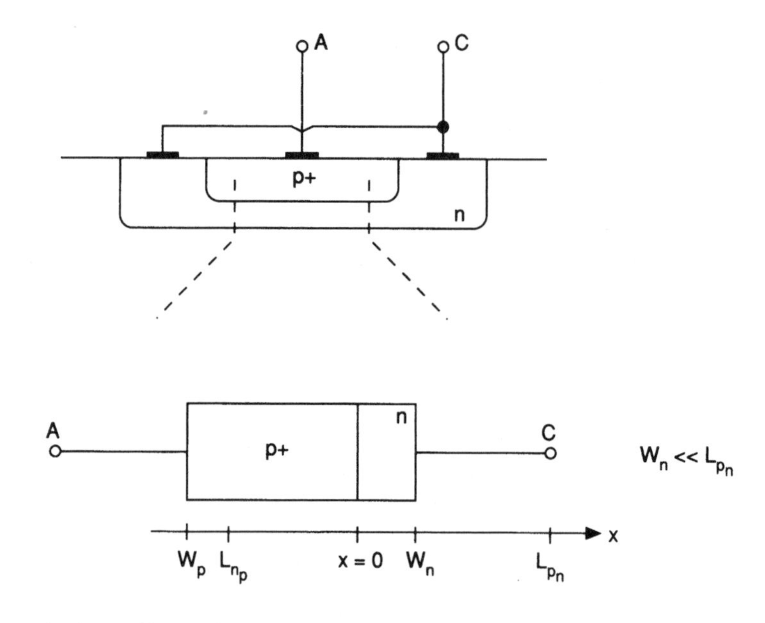

FIGURE 1.9 Planar diodes are fabricated in planar technology. Most modern diodes are unsymmetrical; thus $W_n << L_{pn}$. The p-type region is more highly doped than the n region.

and accordingly the concentration of electrons:

$$n_n(x = 0) = n_n \exp V_D/V_T \tag{1.25}$$

Most modern planar diodes are unsymmetrical. Figure 1.9 shows a pn-diode with the n region W_n much shorter than the diffusion length L_{pn} of holes in the n-semiconductor region. This results in a linear **concentration gradient** of injected diffusing holes in the n region given by

$$\mathrm{d}p/\mathrm{d}x = -(p_n \exp V_D/V_T - p_n)/W_n \tag{1.26}$$

The diffusion gradient is negative since the concentration of positive holes decreases with distance due to the hole–electron recombinations. The equation for the hole diffusion current is

$$I_p = -qA_jD_p\,\mathrm{d}p/\mathrm{d}x \tag{1.27}$$

where A_j is the junction area, D_p is the **diffusion constant** for holes, and q is the elementary charge.

By combining of above equations we obtain

$$I_p = (qA_jD_pp_n/W_n)\,(\exp V_D/V_T - 1) \tag{1.28}$$

In the p-semiconductor we assume that $L_{np} << W_p$; then

$$\mathrm{d}n/\mathrm{d}x = n_p \exp(V_D/V_T - 1) \tag{1.29}$$

By substituting this into the electron diffusion equation:

$$I_n = qA_jD_n\mathrm{d}n/\mathrm{d}x \tag{1.30}$$

we obtain

$$I_n = (qA_j D_n n_p)/L_{np}(\exp\ V_D/V_T - 1) \tag{1.31}$$

Thus, the total junction diffusion current is

$$I_D = I_p + I_n = \{qA_j D_p p_n/W_n + qA_j D_n n_p/L_{np}\}\ (\exp V_D/V_T - 1) \tag{1.32}$$

Since the recombination of the injected carriers establishes a diffusion gradient, this in turn yields a flow of current proportional to the slope. For $|-V_D| >> V_T$, i.e., $V_D = -0.1$ V:

$$I_S = (qA_i D_p p_n/W_n + qA_i D_n n_p/L_{np}) \tag{1.33}$$

Here, I_S denotes the **reverse saturation current**. In practical junctions, the p region is usually much more heavily doped than the n region; thus $n_p << p_n$. Also, since $W_n << L_{np}$ in Equation (1.33), we obtain

$$I_S = qA_j D_p p_n/W_n = qA_j D_p n_i^2/W_n N_D \tag{1.34}$$

The reverse saturation current in short diodes is mainly determined by the diffusion constant D_p and the width W_n of the n region, by intrinsic concentration n_i, by the doping concentration N_D in the n region, and by the diode area A_j. (In reality, I_S is also slightly dependent on the reverse voltage (Phillips, 1962).)

If V_D is made positive, the exponential term in Equation (1.32) rapidly becomes larger than one; thus:

$$I_D = I_S \exp V_D/V_T \tag{1.35}$$

where I_D is the diode forward current and I_S is the reverse saturation current.

Another mechanism predominates the reverse current I_S in silicon. Because of the recombination centers in the depletion region, a generation-recombination hole–electron current I_G is generated in the depletion region (Phillips, 1962; Sze, 1985):

$$I_G = KqA_j eX_d \tag{1.36a}$$

Here, e is the generation rate unit volume, A_j is the junction area, q is the elementary charge, X_d is the depletion layer thickness, and K is a dimensional constant. I_G is proportional to the thickness X_d of the depletion layer and to the junction area A_j. Since X_d increases with the square root of the reverse voltage, I_G increases accordingly, yielding a slight slope in the reverse I–V characteristic. The forward I–V characteristic of the practical diode is only slightly affected (slope $m = 2$) at very small forward currents ($I_D = 1$ nA to 1 μA). In practical diodes $n \approx 1$ at small to medium currents ($I_D = 1\ \mu$A to 10 mA). At large currents ($I_D > 10$ mA), $m = 1$ to 2 due to the high current effects (Phillips, 1962) and due to the series bulk resistance of the diode.

The reverse current I_R in silicon is voltage dependent. The predominant effect is the voltage dependence of the generation-recombination current I_G and to a smaller extent the voltage dependence of I_S.

The total reverse current of the diode is thus equal to

$$I_R = I_G + I_S \tag{1.36b}$$

Forward-Biased Diode

For most practical applications:

$$I_D = I_S\ \exp V_D/mV_T \tag{1.37}$$

where I_S is the reverse saturation current (about 10^{-14} A for a small-signal diode); $V_T = kT/q$ is the thermal voltage equal to 26 mV at room temperature; k = Boltzmann's constant, $1.38 \cdot 10^{-23}$ J/K; T is the absolute temperature in kelvin; q is the elementary charge $1.602 \cdot 10^{-19}$ C; m is the **ideality factor**, $m = 1$ for medium currents, $m = 2$ for very small and very large currents; I_S is part of the total reverse current I_R of the diode $I_R = I_S + I_G$; and I_S is the reverse saturation current and I_G is the generation-recombination current, also called diode leakage current because I_G is not a part of the carrier diffusion process in the diode. I_D is exponentially related to V_D in Figure 1.10.

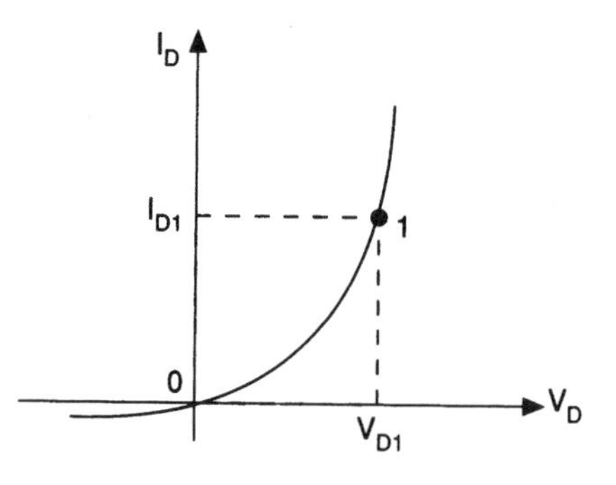

FIGURE 1.10 I_D versus V_D of a diode.

Temperature Dependence of V_D

Equation (1.37) solved for V_D yields

$$V_D = mV_T \ln(I_D/I_S) \tag{1.38}$$

at constant current I_D, the diode voltage V_D is temperature dependent because V_T and I_S are temperature dependent. Assume $m = 1$. The reverse saturation current I_S from Equation (1.34) is

$$I_S = qA_j n_i^2 D_p / W_n N_D = B_1 n_i^2 D_p = B_2 n_i^2 \mu_p$$

where $D_p = V_T \mu_p$. $\mu_p = B_3 T^{-n}$ and for n_i^2:

$$n_i^2 = B_4 T^{\gamma} \exp(-V_{G0}/V_T) \tag{1.39}$$

where $\gamma = 4 - n$, and V_{G0} is the extrapolated **bandgap energy** (Gray and Meyer, 1993). With Equation (1.39) into Equation (1.38), the derivative $\mathrm{d}V_D/\mathrm{d}T$ for I_D = const yields

$$\mathrm{d}V_D/\mathrm{d}T = (V_D - V_{G0})/T - \gamma k/q \tag{1.40}$$

At room temperature ($T = 300$ K), and $V_D = 0.65$ V, $V_{G0} = 1.2$ V, $\gamma = 3$, $V_T = 26$ mV, and $\mathbf{k}/q = 86\ \mu$V/degree, one gets $\mathrm{d}V_D/\mathrm{d}T \approx -2.1$ mV/degree. The **temperature coefficient** TC of V_D is thus:

$$\mathrm{TC} = \mathrm{d}V_D/V_D\,\mathrm{d}T = 1/T - V_{G0}/V_D T - \gamma k/qV_D \tag{1.41}$$

For the above case TC $\approx -0.32\%$/degree. In practical applications it is more convenient to use the expression:

$$V_D(\delta_2) = V_D(\delta_1) - \mathrm{TC}(\delta_2 - \delta_1) \tag{1.42}$$

where δ_1 and δ_2 are temperatures in degrees Celsius. For TC $= -0.32\%$/degree and $V_D = 0.65$ V at $\delta_1 = 27$°C, $V_D = 0.618$ V at $\delta_2 = 37$°C. Both $\mathrm{d}V_D/\mathrm{d}T$ and TC are I_D dependent. At higher I_D, both $\mathrm{d}V_D/\mathrm{d}T$ and TC are smaller than at a lower I_D, as shown in Figure 1.11.

I_D–V_D Characteristic

From the I_D–V_D characteristic of the diode one can find for $m = 1$:

$$I_{D1} = I_S \exp(V_{D1}/V_T) \quad \text{and} \quad I_{D2} = I_S \exp(V_{D2}/V_T) \tag{1.43}$$

Thus, the ratio of currents is

$$I_{D2}/I_{D1} = \exp(V_{D2} - V_{D1})/V_T \tag{1.44}$$

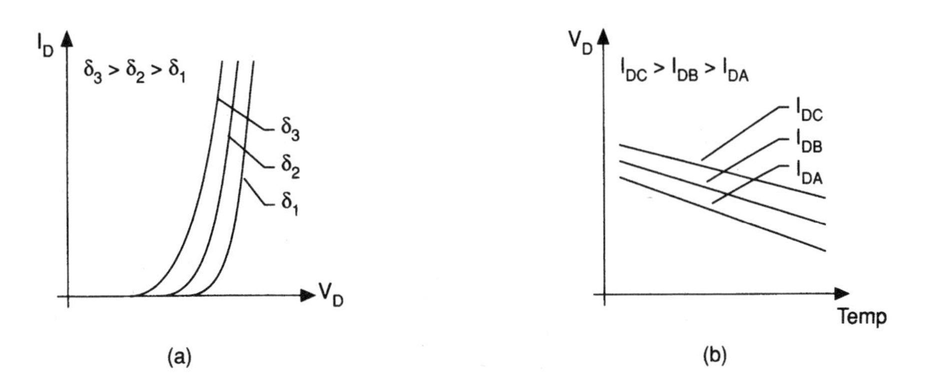

FIGURE 1.11 (a) I_D versus V_D of a diode at three different temperatures $\delta_3 < \delta_2 < \delta_1$; (b) $V_D = f(\text{temp})$, $I_{DC} < I_{DB} < I_{DA}$.

or the difference voltage:

$$V_{D2} - V_{D1} = V_T \ln(I_{D2}/I_{D1}) \tag{1.45}$$

in terms of base 10 logarithm:

$$V_{D2} - V_{D1} = V_T\, 2.3 \log(I_{D2}/I_{D1}) \tag{1.46}$$

For $(I_{D2}/I_{D1}) = 10$ (one decade), $V_{D2} - V_{D1} = {\sim}60$ mV, or $V_{D2} - V_{D1} = 17.4$ mV for $(I_{D2}/I_{D1}) = 2$. In a typical example, $m = 1$, $V_D = 0.67$ V at $I_D = 100\ \mu$A. At $I_D = 200\ \mu$A, $V_D = 0.67$ V + 17.4 mV = 0.687 V.

DC and Large-Signal Model

The diode equation in Equation (1.37) is widely utilized in diode circuit design. I_S and m can sometimes be found from the data book or they can be determined from measured I_D and V_D. From two measurements of I_D and V_D, for example, $I_D = 0.2$ mA at $V_D = 0.670$ V and $I_D = 10$ mA at $V_D = 0.772$ V, one can find $m = 1.012$ and $I_S = 1.78 \cdot 10^{-15}$ A for the particular diode. A practical application of the large-signal diode model is shown in Figure 1.13. Here, the current I_D through the series resistor R and a diode D is to be found:

$$I_D = (V_{CC} - V_D)/R \tag{1.47}$$

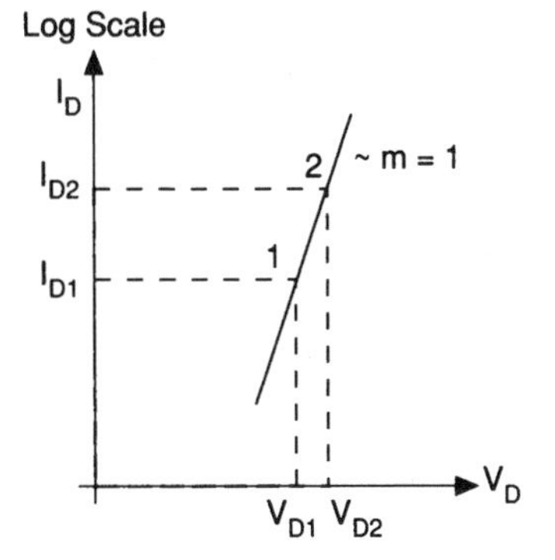

FIGURE 1.12 I_D versus V_D of a diode on a semi-logarithmic plot.

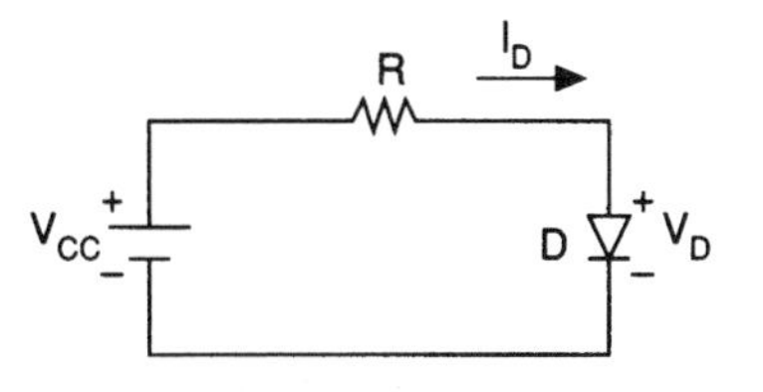

FIGURE 1.13 Diode–resistor biasing circuit.

The equation is implicit and cannot be solved for I_D since V_D is a function of I_D. Here, V_D and I_D are determined by using iteration. By assuming $V_D = V_{D0} = 0.6$ V (cut-in voltage), the first iteration yields

$$I_D(1) = (5\ \text{V} - 0.6\ \text{V})/1\ \text{k}\Omega = 4.4\ \text{mA}$$

Next, the first iteration voltage $V_D(1)$ is calculated (by using m and I_S above and $I_{D1} = 4.4$ mA); thus:

$$V_D(1) = mV_T[\ln\ I_D(1)/I_S] = 1.012 \times 26\ \text{mV}\ \ln(4.4\ \text{mA}/1.78 \cdot 10^{-15}\text{A}) = 0.751\ \text{V}$$

From the second iteration $I_D(2) = [V_{CC} - V_D(1)]/R = 4.25$ mA and thus $V_D(2) = 0.75$ V. The third iteration yields $I_D(3) = 4.25$ mA, and $V_D(3) = 0.75$ V. These are the actual values of I_D and V_D for the above example, since the second and the third iterations are almost equal.

Graphical analysis (in Figure 1.14) is another way to analyze the circuit in Figure 1.13. Here, the load line R is drawn with the diode I–V characteristic, where $V_{CC} = V_D + I_D R$. This type of analysis is illustrative but not well suited for a numerical analysis.

High Forward Current Effects

In the pn-junction diode analysis it was assumed that the density of injected carriers from the p region into the n region is small compared to the density of majority carriers in that region. Thus, all of the forward voltage V_D appears across the junction. Therefore, the injected carriers move only because of the diffusion. At high forward currents this is not the case anymore. When the voltage drop across the bulk resistance becomes comparable with the voltage across the junction, the effective applied voltage is reduced (Phillips, 1962). Because of the electric field created by the voltage drop in the bulk (neutral) regions, the current is not only a diffusion current anymore. The drift current due to the voltage drop across the bulk region opposes the diffusion current. The net effect is that, first, the current becomes proportional to twice the diffusion constant, second, the high-level current becomes independent of resistivity, and, third, the magnitude of the exponent is reduced by a factor of two in Equation (1.37). The effect of high forward current on the I–V characteristic is shown in Figure 1.15. In all practical designs, $m \approx 2$ at $I_D \geq 20$ mA in small-signal silicon diodes.

Large-Signal Piecewise Linear Model

Piecewise linear model of a diode is a very useful tool for quick circuit design containing diodes. Here, the diode is represented by asymptotes and not by the exponential I–V curve. The simplest piecewise linear model is shown in Figure 1.16(a). Here, D_i is an ideal diode with $V_D = 0$ at $I_D \geq 0$, in series with V_{D0}, where V_{D0} is the diode cut-in or threshold voltage. The current in the diode will start to flow at $V_D \geq V_{D0}$.

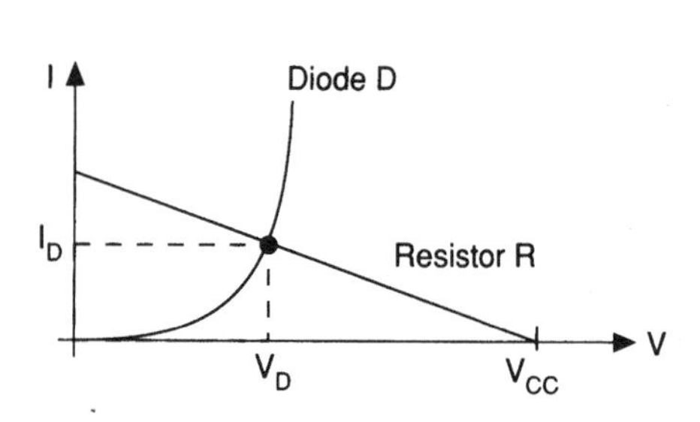

FIGURE 1.14 Graphical analysis of a diode–resistor circuit.

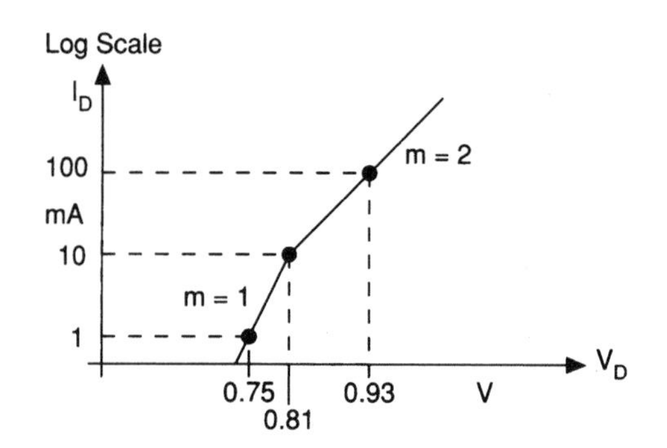

FIGURE 1.15 I_D versus V_D of a diode at low and high forward currents.

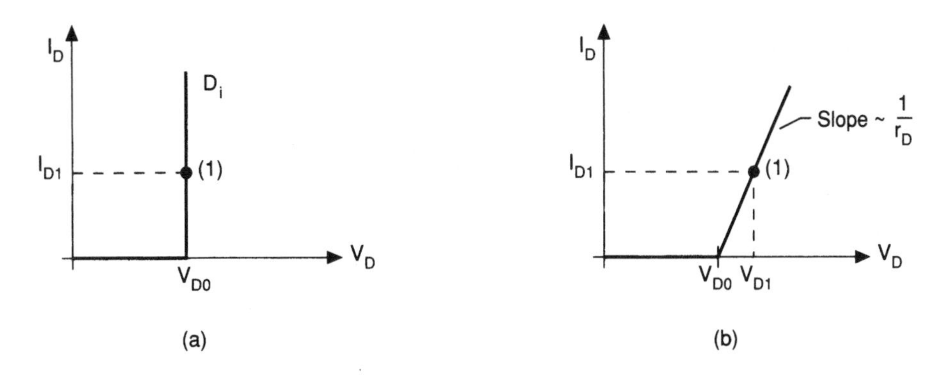

FIGURE 1.16 (a) Simplified piecewise linear model of a diode; (b) improved piecewise linear model of a diode. The diode cut-in voltage V_{D0} is defined as the voltage V_D at a very small current I_D typically at about 1 nA. For silicon diodes this voltage is typically $V_{D0} = 0.6$ V.

An improved model is shown in Figure 1.16(b), where V_{D0} is again the diode voltage at a very small current I_{D0}, r_D is the extrapolated diode resistance, and I_{D1} is the diode current in operating point 1. Thus, the diode voltage is

$$V_{D1} = V_{D0} + I_{D1} r_D \tag{1.48}$$

where V_{D1} is the diode voltage at I_{D1}. V_{D0} for silicon is about 0.60 V. r_D is estimated from the fact that V_D in a real diode is changing per decade of current by m 2.3 V_T. Thus, V_D changes about 60 mV for a decade change of current I_D at $m = 1$. Thus, in a 0.1 to 10 mA current change, V_D changes about 120 mV, which corresponds to $anr_D \approx 120\,\text{mV}/10\,\text{mA} = 12\ \Omega$.

The foregoing method is an approximation; however, it is quite practical for first-hand calculations. To compare this with the above iterative approach let us assume $m = 1$, $V_{D0} = 0.60$ V, $r_D = 12\,\Omega$, $V_{CC} = 5$ V, $R = 1\,\text{k}\Omega$. The current $I_{D1} = [V_{CC} - V_{D0}]/(R + r_D) = 4.34$ mA compared with $I_{D1} = 4.25$ mA in the iterative approach.

Small-Signal Incremental Model

In the small-signal **incremental model**, the diode is represented by linear elements. In small-signal (incremental) analysis, the diode voltage signals are assumed to be about $V_T/2$ or less, thus much smaller than the dc voltage V_D across the diode. In the forward-biased diode, three elements are of practical interest: **incremental resistance** (or small-signal or differential resistance) r_d, the **diffusion capacitance** C_d, and the **junction capacitance** C_j.

Incremental Resistance, r_d

For small signals the diode represents a small-signal resistance (often called incremental or differential resistance) r_d in the operating point (I_D, V_D) where

$$r_d = dV_D/dI_D = mV_T/I_S \exp(V_D/mV_T) = mV_T/I_D \tag{1.49}$$

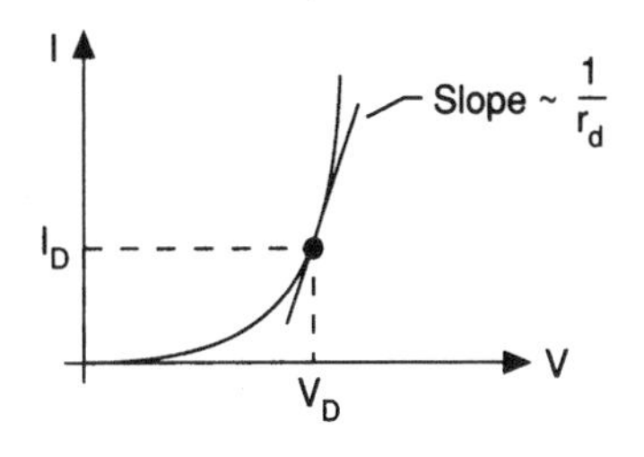

FIGURE 1.17 Small-signal incremental resistance r_d of a diode.

In Figure 1.17, r_d is shown as the tangent in the dc operating point (V_D, I_D). Note that r_d is independent of the geometry of the device and inversely proportional to the diode dc current. Thus, for $I_D = 1$ mA, $m = 1$ and $V_T = 26$ mV, the incremental resistance is $r_d = 26\ \Omega$.

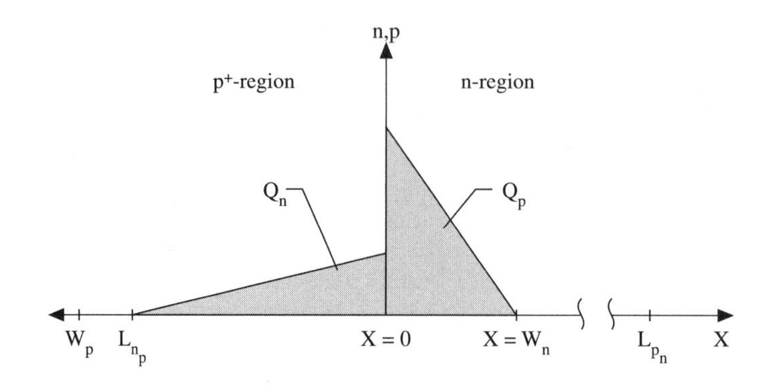

FIGURE 1.18 Minority carrier charge injection in a diode.

Diffusion Capacitance, C_d

C_d is associated with the injection of holes and electrons in the forward-biased diode. In steady state, holes and electrons are injected across the junction. Hole and electron currents flow due to the diffusion gradients on both sides of the junction in Figure 1.18. In a short diode, holes are traveling a distance $W_n << L_{pn}$. For injected holes, and since $W_n << L_{pn}$:

$$I_p = dq_p/dt = dq_p v/dx \tag{1.50}$$

where v is the average carrier velocity, D_p is the diffusion constant for holes and W_n is the travel distance of holes. By integrating of Equation (1.50) one gets

$$I_p \int_0^{W_n} dx = v \int_0^{Q_p} dq_p$$

and the charge Q_p of holes becomes

$$Q_p = I_p W_n / v = I_p \tau_p \tag{1.51}$$

$\tau_p = W_n/v$ is the transit time holes travel the distance W_n. Similarly, for electron charge Q_n, since $W_p >> L_{np}$:

$$Q_n = I_n L_{np}/v = I_n \tau_n \tag{1.52}$$

Thus the total diffusion charge Q_d is

$$Q_d = Q_p + Q_n \tag{1.53}$$

and the total transit time is

$$\tau_F = \tau_p + \tau_n \tag{1.54}$$

and with $I_p + I_n = I_D = I_S \exp V_D/mV_T$ and Equation (1.51), Equation (1.52), and Equation (1.54) one gets

$$Q_d = \tau_F I_S \exp V_D/mV_T = \tau_F I_D \tag{1.55}$$

The total diffusion capacitance is

$$C_d = C_p + C_n = dQ_d/dV_D = Q_d/mV_T \tag{1.56}$$

and from Equation (1.55) and Equation (1.56):

$$C_d = I_D \tau_F / m V_T \tag{1.57}$$

C_d is thus directly proportional to I_D and to the carrier transit time τ_F. For an unsymmetrical diode with $W_n << L_{pn}$ and $N_A >> N_D$ (Gray and Meyer, 1984):

$$\tau_F \approx W_n^2 / 2D_p \tag{1.58}$$

τ_F is usually given in data books or it can be measured.

For $W_n = 6\,\mu$ and $D_p = 14\ \text{cm}^2/\text{s}$, $\tau_F \approx 13$ ns, $I_D = 1$ mA, $V_T = 26$ mV, and $m = 1$, the diffusion capacitance is $C_d = 500$ pF.

Depletion Capacitance, C_j

The depletion region is always present in a pn-diode. Because of the immobile ions in the depletion region, the junction acts as a voltage-dependent plate capacitor C_j (Horenstein, 1990; Gray and Meyer, 1993):

$$C_j = C_{j0} / \sqrt{V_0 - V_D} \tag{1.59}$$

V_D is the diode voltage (positive value for forward biasing, negative value for reverse biasing), and C_{j0} is the zero bias depletion capacitance; A_j is the junction diode area:

$$C_{j0} = K A_j \tag{1.60}$$

K is a proportionality constant dependent on diode doping, and A_j is the diode area. C_j is voltage dependent. As V_D increases, C_j increases in a forward-biased diode in Figure 1.19. For $V_0 = 0.7$ V, $V_D = -10$ V and $C_{j0} = 3$ pF, the diode depletion capacitance is $C_j = 0.75$ pF. In Figure 1.20 the small-signal model of the diode is shown. The total small-signal time constant τ_d is thus (by neglecting the bulk series diode resistance R_{BB}):

$$\tau_d = r_d(C_d + C_j) = r_d C_d + r_d C_j = \tau_F + r_d C_j \tag{1.61}$$

τ_d is thus current dependent. At small I_D the $r_d C_j$ product is predominant. For high-speed operation $r_d C_j$ must be kept much smaller than τ_F. This is achieved by a large operating current I_D. The diode behaves to a first approximation as a frequency-dependent element. In the reverse operation, the diode behaves as a high ohmic resistor $R_p \approx V_R / I_G$ in parallel with the capacitor C_j. In forward small-signal operation, the diode behaves as a resistor r_d in parallel with the capacitors C_j and C_d (R_p is neglected). Thus, the diode is in a first approximation, a low-pass network.

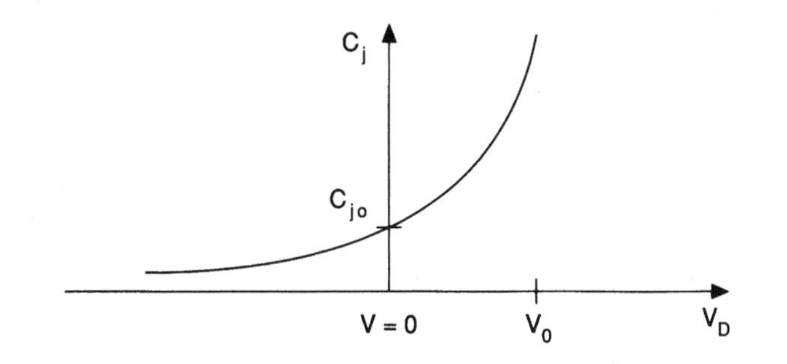

FIGURE 1.19 Depletion capacitance C_j of a diode versus diode voltage V_R.

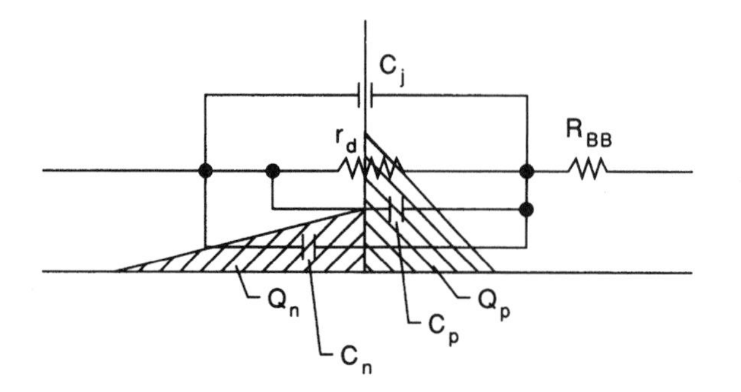

FIGURE 1.20 Simplified small-signal model of a diode.

Large-Signal Switching Behavior of a pn-Diode

When a forward-biased diode is switched from the forward into the reverse direction, the stored charge Q_d of **minority carriers** must first be removed. The charge of minority carriers in the forward-biased unsymmetrical diode is from Equation (1.55) and Equation (1.58):

$$Q_d = I_D \tau_F = I_D W_n^2 / 2D_p \tag{1.62}$$

where $W_n << L_{pn}$ is assumed. τ_F is minimized by making W_n very small. Very low-lifetime τ_F is required for high-speed diodes. **Carrier lifetime** τ_F is reduced by adding a large concentration of recombination centers into the junction. This is common practice in the fabrication of high-speed computer diodes (Phillips, 1962). The charge Q_d is stored mainly in the n region in the form of a concentration gradient of holes in Figure 1.21(a). The diode is turned off by moving the switch from position (a) into position (b) (Figure 1.21(a)). The removal of carriers is done in three time intervals. During the time interval t_1, also called the recovery phase, a constant reverse current $|I_R| = V_R/R$ flows in the diode. During the time interval $t_2 - t_1$ the charge in the diode is reduced by about 1/2 of the original charge. During the third interval $t_3 - t_2$, the residual charge is removed.

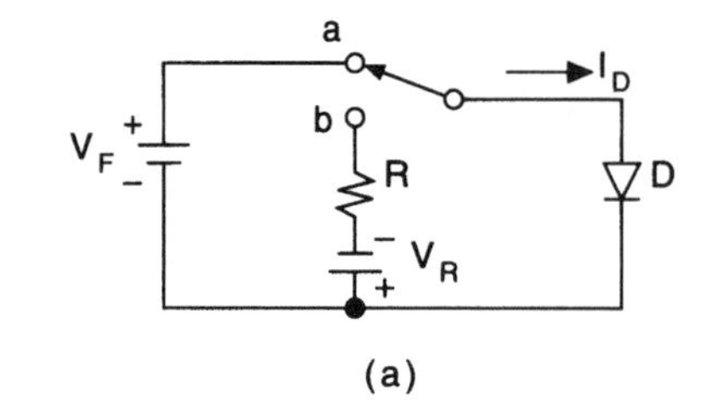

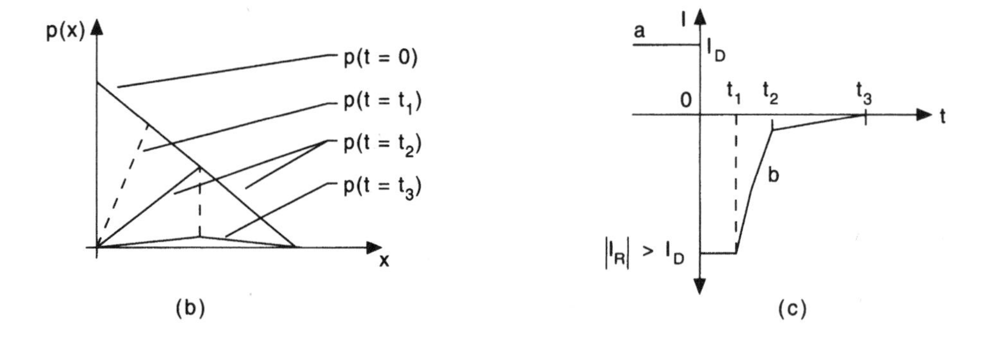

FIGURE 1.21 (a) Diode is switched from forward into reverse direction; (b) concentration of holes in the n region; (c) diode turns off in three time intervals.

If during the interval t_1, $|I_R| >> I_D$, then Q_d is reduced only by flow of reverse diffusion current; no holes arrive at the metal contact (Gugenbuehl et al., 1962), and

$$t_1 \approx \tau_F (I_D / |I_R|)^2 \tag{1.63}$$

During time interval $t_2 - t_1$, when $|I_R| = I_D$, in Figure 1.21(b):

$$t_2 - t_1 \approx \tau_F I_D / |I_R| \tag{1.64}$$

The residual charge is removed during the time $t_3 - t_2 \approx 0.5\,\tau_F$.

Diode Reverse Breakdown

Avalanche breakdown occurs in a reverse-biased plane junction when the critical electric field E_{crt} at the junction within the depletion region reaches about $3 \cdot 10^5$ V/cm for junction doping densities of about 10^{15} to 10^{16} at/cm^3 (Gray and Meyer, 1984). At this electric field E_{crt}, the minority carriers traveling (as reverse current) in the depletion region acquire sufficient energy to create new hole–electron pairs in collision with atoms. These energetic pairs are able to create new pairs, etc. This process is called the avalanche process and leads to a sudden increase of the reverse current I_R in a diode. The current is then limited only by the external circuitry. The avalanche current is not destructive as long as the local junction temperature does not create local hot spots, i.e., melting of material at the junction. Figure 1.22 shows a typical *I–V* characteristic for a junction diode in the avalanche breakdown. The effect of breakdown is seen by the large increase of the reverse current I_R when V_R reaches – *BV*. Here, *BV* is the actual breakdown voltage. It was found that $I_{RA} = M\, I_R$, where I_{RA} is the avalanche reverse current at *BV*, *M* is the multiplication factor, and I_R is the reverse current not in the breakdown region. *M* is defined as

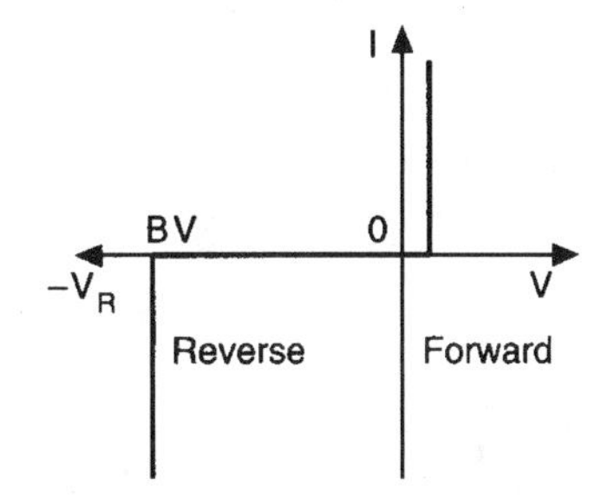

FIGURE 1.22 Reverse breakdown voltage of a diode at $-V_R = \text{BV}$.

$$M = 1/[1 - V_R/\text{BV}]^n \tag{1.65}$$

where $n = 3$ to 6. As $V_R = \text{BV}$, $M \to \infty$ and $I_{RA} \to \infty$. The above BV is valid for a strictly plane junction without any curvature. However, in a real planar diode as shown in Figure 1.9, the p-diffusion has a curvature with a finite radius x_j. If the diode is doped unsymmetrically, thus $\sigma_p >> \sigma_n$, then the depletion area penetrates mostly into the n region. Because of the finite radius, the breakdown occurs at the radius x_j, rather than in a plane junction (Grove, 1967). The breakdown voltage is significantly reduced due to the curvature. In very shallow planar diodes, the avalanche breakdown voltage BV can be much smaller than 10 V.

Zener and Avalanche Diodes

Zener diodes (ZD) and avalanche diodes are pn-diodes specially built to operate in reverse breakdown. They operate in the reverse direction; however, their operating mechanism is different. In a Zener diode the hole–electron pairs are generated by the electric field by direct transition of carriers from valence band into the conductance band. In an avalanche diode, the hole–electron pairs are generated by impact ionization due to high-energy holes and electrons.

Avalanche and Zener diodes are extensively used as voltage regulators and as overvoltage protection devices. T_C of Zener diodes is negative at $V_Z \leq 3.5$ to 4.5 V and is equal to zero at about $V_Z \approx 5$ V. T_C of a Zener diode operating above 5 V is in general positive. Above 10 V the pn-diodes operate as avalanche diodes with a strong positive temperature coefficient. The T_C of a Zener diode is more predictable than that of the avalanche diode. Temperature-compensated Zener diodes utilize the positive T_C of a 7-V Zener diode, which is compensated with a series-connected forward-biased diode with a negative T_C. The disadvantage of Zener diodes is a relatively large electronic noise.

Varactor Diodes

The varactor diode is an ordinary pn-diode that uses the voltage-dependent variable capacitance of the diode. The varactor diode is widely used as a voltage-dependent capacitor in electronically tuned radio receivers and in TV.

Tunnel Diodes

The tunnel diode is an ordinary pn-junction diode with very heavy doped n and p regions. Because the junction is very thin, a tunnel effect takes place. An electron can tunnel through the thin depletion layer from the conduction band of the n region directly into the valence band of the p region. Tunnel diodes create a negative differential resistance in the forward direction due to the tunnel effect. Tunnel diodes are used as mixers, oscillators, amplifiers, and detectors. They operate at very high frequencies in the gigahertz bands.

Photodiodes and Solar Cells

Photodiodes are ordinary pn-diodes that generate hole–electron pairs when exposed to light. A photocurrent flows across the junction, if the diode is reverse biased. Silicon pn-junctions are used to sense light at near-infrared and visible spectra around 0.9 μm. Other materials are used for different spectra.

Solar cells utilize the pn-junction to convert light energy into electrical energy. Hole–electron pairs are generated in the semiconductor material by light photons. The carriers are separated by the high electric field in the depletion region across the pn-junction.The electric field forces the holes into the p region and the electrons into the n region. This displacement of mobile charges creates a voltage difference between the two semiconductor regions. Electric power is generated in an external load connected between the terminals to the p and n regions. The conversion efficiency is relatively low, around 10 to 12%. With the use of new materials, an efficiency of about 30% has been reported. Efficiency up to 45% was achieved by using monochromatic light.

Schottky Barrier Diode

The Schottky barrier diode is a metal-semiconductor diode. Majority carriers carry the electric current. No minority carrier injection takes place. When the diode is forward biased, carriers are injected into the metal, where they reside as majority carriers at an energy level that is higher than the Fermi level in metals. The *I–V* characteristic is similar to conventional diodes. The barrier voltage is small, about 0.2 V for silicon. Since no minority carrier charge exists, the Schottky barrier diodes are very fast. They are used in high-speed electronic circuitry.

Defining Terms

Acceptor: Ionized, negative-charged immobile dopant atom (ion) in a p-type semiconductor after the release of a hole.

Avalanche breakdown: In the reverse-biased diode, hole–electron pairs are generated in the depletion region by ionization, thus by the lattice collision with energetic electrons and holes.

Bandgap energy: Energy difference between the conduction band and the valence band in a semiconductor.

Barrier voltage: A voltage which develops across the junction due to uncovered immobile ions on both sides of the junction. Ions are uncovered due to the diffusion of mobile carriers across the junction.

Boltzmann relation: Relates the density of particles in one region to that in an adjacent region, with the potential energy between both regions.

Carrier lifetime: Time an injected minority carrier travels before its recombination with a majority carrier.

Concentration gradient: Difference in carrier concentration.

Diffusion: Movement of free carriers in a semiconductor caused by the difference in carrier densities (concentration gradient). Also movement of dopands during fabrication of diffused diodes.

Diffusion capacitance: Change in charge of injected carriers corresponding to change in forward bias voltage in a diode.

Diffusion constant: Product of the thermal voltage and the mobility in a semiconductor.

Donor: Ionized, positive-charged immobile dopant atom (ion) in an n-type semiconductor after the release of an electron.

Drift: Movement of free carriers in a semiconductor due to the electric field.

Ideality factor: The factor determining the deviation from the ideal diode characteristic $m = 1$. At small and large currents $m \approx 2$.

Incremental model: Small-signal differential (incremental) semiconductor diode equivalent RC circuit of a diode, biased in a dc operating point.

Incremental resistance: Small-signal differential (incremental) resistance of a diode, biased in a dc operating point.

Junction capacitance: Change in charge of immobile ions in the depletion region of a diode corresponding to a change in reverse bias voltage on a diode.

Majority carriers: Holes are in majority in a p-type semiconductor; electrons are in majority in an n-type semiconductor.

Minority carriers: Electrons in a p-type semiconductor are in minority; holes are in majority. Similarly, holes are in minority in an n-type semiconductor and electrons are in majority.

Reverse breakdown: At the reverse breakdown voltage the diode can conduct a large current in the reverse direction.

Reverse generation-recombination current: Part of the reverse current in a diode caused by the generation of hole–electron pairs in the depletion region. This current is voltage dependent because the depletion region width is voltage dependent.

Reverse saturation current: Part of the reverse current in a diode which is caused by diffusion of minority carriers from the neutral regions to the depletion region. This current is almost independent of the reverse voltage.

Temperature coefficient: Relative variation $\Delta X/X$ of a value X over a temperature range, divided by the difference in temperature ΔT.

Zener breakdown: In the reverse-biased diode, hole–electron pairs are generated by a large electric field in the depletion region.

References

C.G. Fonstad, *Microelectronic Devices and Circuits,* New York: McGraw-Hill, 1994.

P.R. Gray and R.G. Meyer, *Analysis and Design of Analog Integrated Circuits,* New York: John Wiley & Sons, 1993.

A.S. Grove, *Physics and Technology of Semiconductor Devices*, New York: John Wiley & Sons, 1967.

W. Gugenbuehl, M.J.O. Strutt, and W. Wunderlin, *Semiconductor Elements,* Basel: Birkhauser Verlag, 1962.

M.N. Horenstein, *Microelectronic Circuits and Devices,* Englewood Cliffs, NJ: Prentice-Hall, 1990.

A.B. Phillips, *Transistor Engineering,* New York: McGraw-Hill, 1962.

S.M. Sze, *Semiconductor Devices, Physics, and Technology,* New York: John Wiley & Sons, 1985.

Further Information

A good classical introduction to diodes is found in P.E. Gray and C.L. Searle, *Electronic Principles,* New York: Wiley, 1969. Other sources include S. Soclof, *Applications of Analog Integrated Circuits,* Englewood Cliffs, NJ: Prentice-Hall, 1985 and E.J. Angelo, Jr., *Electronics: BJT's, FET's and Micro-Circuits,* New York: McGraw-Hill, 1969.

1.3 Electrical Equivalent Circuit Models and Device Simulators for Semiconductor Devices

Aicha Elshabini-Riad, F.W. Stephenson, and Imran A. Bhutta

In the past 15 years, the electronics industry has seen a tremendous surge in the development of new semiconductor materials, novel devices, and circuits. For the designer to bring these circuits or devices to the market in a timely fashion, he or she must have design tools capable of predicting the device behavior in a variety of circuit configurations and environmental conditions. Equivalent circuit models and semiconductor device simulators represent such design tools.

Overview of Equivalent Circuit Models

Circuit analysis is an important tool in circuit design. It saves considerable time at the circuit design stage by providing the designer with a tool for predicting the circuit behavior without actually processing the circuit.

An electronic circuit usually contains active devices, in addition to passive components. While the current and voltage behavior of passive devices is defined by simple relationships, the equivalent relationships in active devices are quite complicated in nature. Therefore, in order to analyze an active circuit, the devices are replaced by equivalent circuit models that give the same output characteristics as the active device itself. These models are made up of passive elements, voltage sources, and current sources. Equivalent circuit models provide the designer with reasonably accurate values for frequencies below 1 GHz for bipolar junction transistors (BJTs), and their use is quite popular in circuit analysis software. Some field-effect transistor (FET) models are accurate up to 10 GHz. As the analysis frequency increases, however, so does the model complexity. Since the equivalent circuit models are based on some fundamental equations describing the device behavior, they can also be used to predict the characteristics of the device itself.

When performing circuit analysis, two important factors that must be taken into account are the speed and accuracy of computation. Sometimes, the computation speed can be considerably improved by simplifying the equivalent circuit model, without significant loss in computation accuracy. For this reason, there are a number of equivalent circuit models, depending on the device application and related conditions. Equivalent circuit models have been developed for diodes, BJTs, and FETs. In this overview, the equivalent circuit models for BJT and FET devices are presented.

Most of the equivalent circuits for BJTs are based on the Ebers–Moll model (Ebers and Moll, 1954) or the Gummel–Poon model (Gummel and Poon, 1970). The original Ebers–Moll model was a large signal, nonlinear dc model for BJTs. Since then, a number of improvements have been incorporated to make the model more accurate for various applications. In addition, an accurate model has been introduced by Gummel and Poon (1970).

There are three main types of equivalent circuit models, depending on the device signal strength. On this basis, the models can be classified as follows:

1. Large-signal equivalent circuit model
2. Small-signal equivalent circuit model
3. DC equivalent circuit model

Use of the large-signal or small-signal model depends on the magnitude of the driving source. In applications where the driving currents or the driving voltages have large amplitudes, large-signal models are used. In circuits where the signal does not deviate much from the dc biasing point, small-signal models are more suitable. For dc conditions and very-low-frequency applications, dc equivalent circuit models are used. For dc and very-low-frequency analysis, the circuit element values can be assumed to be lumped, whereas in high-frequency analysis, incremental element values give much more precise results.

Large-Signal Equivalent Circuit Model

Depending on the frequency of operation, large-signal equivalent circuit models can be further classified as (1) high-frequency large-signal equivalent circuit model and (2) low-frequency large-signal equivalent circuit model.

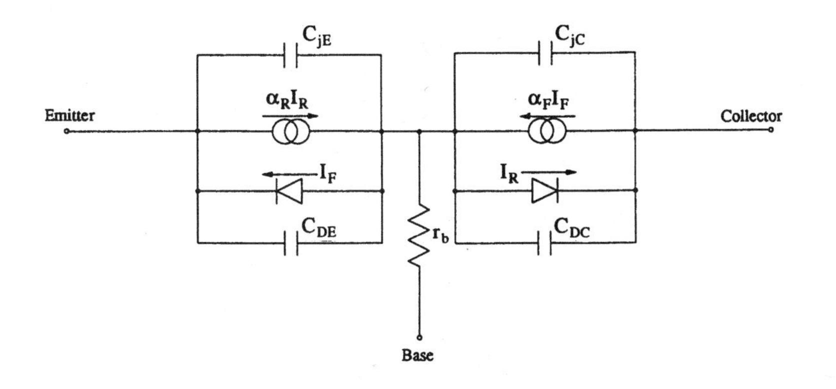

FIGURE 1.23 High-frequency large-signal equivalent circuit model of an *npn* BJT.

High-Frequency Large-Signal Equivalent Circuit Model of a BJT. In this context, high-frequency denotes frequencies above 10 kHz. In the equivalent circuit model, the transistor is assumed to be composed of two back-to-back diodes. Two current-dependent current sources are added to model the current flowing through the reverse-biased base–collector junction and the forward-biased base–emitter junction. Two junction capacitances, C_E and C_{jC}, model the fixed charges in the emitter–base space charge region and base–collector space charge region, respectively. Two diffusion capacitances, C_{DE} and C_{DC}, model the corresponding charge associated with mobile carriers, while the base resistance, r_b, represents the voltage drop in the base region. All the above circuit elements are very strong functions of operating frequency, signal strength, and bias voltage.

The high-frequency large-signal equivalent circuit model of an *npn* BJT is shown in Figure 1.23, where the capacitances C_{jE}, C_{jC}, C_{DE}, C_{DC} are defined as follows:

$$C_{jE}(V_{B'E'}) = \frac{C_{jEO}}{\left(1 - \frac{V_{B'E'}}{\phi_E}\right)^{m_E}} \tag{1.66}$$

$$C_{jC}(V_{B'C'}) = \frac{C_{jCO}}{\left(1 - \frac{V_{B'C'}}{\phi_C}\right)^{m_C}} \tag{1.67}$$

$$C_{DE} = \frac{\tau_F I_{CC}}{V_{B'E'}} \tag{1.68}$$

and

$$C_{DC} = \frac{\tau_R I_{EC}}{V_{B'E'}} \tag{1.69}$$

In these equations, $V_{B'E'}$ is the internal base–emitter voltage, C_{jEO} is the base–emitter junction capacitance at $V_{B'E'} = 0$, ϕ_E is the base–emitter barrier potential, and m_E is the base–emitter capacitance gradient factor. Similarly, $V_{B'C'}$ is the internal base–collector voltage, C_{jCO} is the base–collector junction capacitance at $V_{B'C'} = 0$, ϕ_C is the base–collector barrier potential, and m_C is the base–collector capacitance gradient factor. I_{CC} and I_{EC} denote the collector and emitter reference currents, respectively, while τ_F is the total forward transit time, and τ_R is the total reverse transit time. α_R and α_F are the large-signal reverse and forward current gains of a common base transistor, respectively.

This circuit can be made linear by replacing the forward-biased base–emitter diode with a low-value resistor, r_π, while the reverse-biased base–collector diode is replaced with a high-value resistor, r_μ. The junction and diffusion capacitors are lumped together to form C_π and C_μ, while the two current sources are lumped

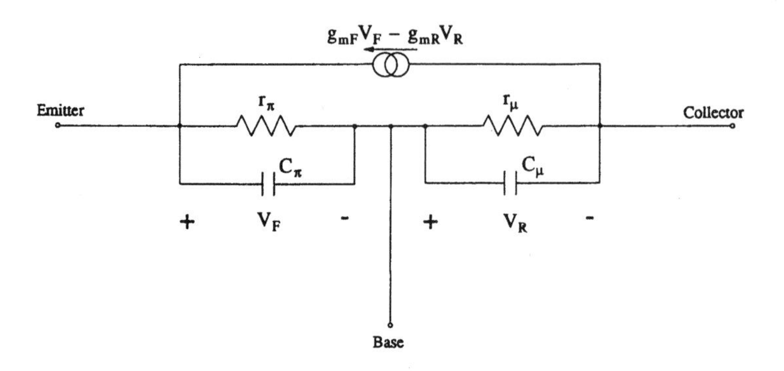

FIGURE 1.24 High-frequency large-signal equivalent circuit model (linear) of an *npn* BJT.

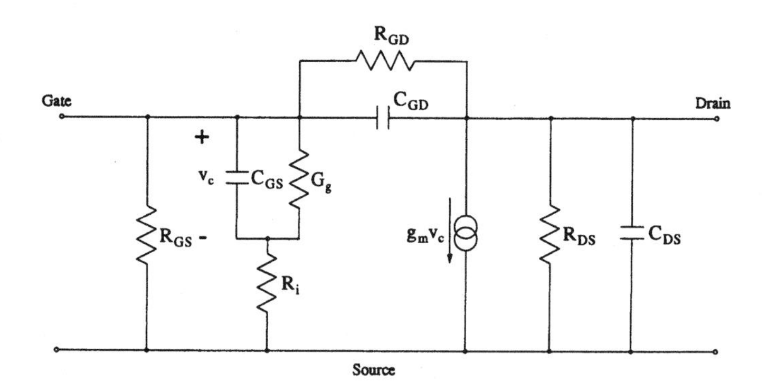

FIGURE 1.25 High-frequency large-signal equivalent circuit model of a FET.

into one source ($g_{\mathrm{mF}}V_{\mathrm{F}} - g_{\mathrm{mR}}V_{\mathrm{R}}$), where g_{mF} and g_{mR} are the transistor forward and reverse transconductances, respectively. V_{F} and V_{R} are the voltages across the forward- and reverse-biased diodes, represented by r_π and r_μ, respectively. r_π is typically about 3 k**Ω**, while r_μ is more than a few megohms, and C_π is about 120 pF. The linear circuit representation is illustrated in Figure 1.24.

The Gummel–Poon representation is very similar to the high-frequency large-signal linear circuit model of Figure 1.24. However, the terms describing the elements are different and a little more involved.

High-Frequency Large-Signal Equivalent Circuit Model of a FET. In the high-frequency large-signal equivalent circuit model of a FET, the fixed charge stored between the gate and the source and between the gate and the drain is modeled by the gate-to-source and the gate-to-drain capacitances, C_{GS} and C_{GD}, respectively. The **mobile charges** between the drain and the source are modeled by the drain-to-source capacitance, C_{DS}. The voltage drop through the active channel is modeled by the drain-to-source resistance, R_{DS}. The current through the channel is modeled by a voltage-controlled current source. For large signals, the gate is sometimes driven into the forward region, and thus the conductance through the gate is modeled by the gate conductance, G_{g}. The conductance from the gate to the drain and from the gate to the source is modeled by the gate-to-drain and gate-to-source resistances, R_{GD} and R_{GS}, respectively. A variable resistor, R_{i}, is added to model the gate charging time such that the time constant given by $R_{\mathrm{i}}C_{\mathrm{GS}}$ holds the following relationship:

$$R_i C_{\mathrm{GS}} = \text{constant} \tag{1.70}$$

For MOSFETs, typical element values are: C_{GS} and C_{GD} are in the range of 1 to 10 pF, C_{DS} is in the range of 0.1 to 1 pF, R_{DS} is in the range of 1 to 50 k**Ω**, R_{GD} is more than 10^{14} **Ω**, R_{GS} is more than 10^{10} **Ω**, and g_{m} is in the range of 0.1 to 20 mA/V.

Figure 1.25 illustrates the high-frequency large-signal equivalent model of a FET.

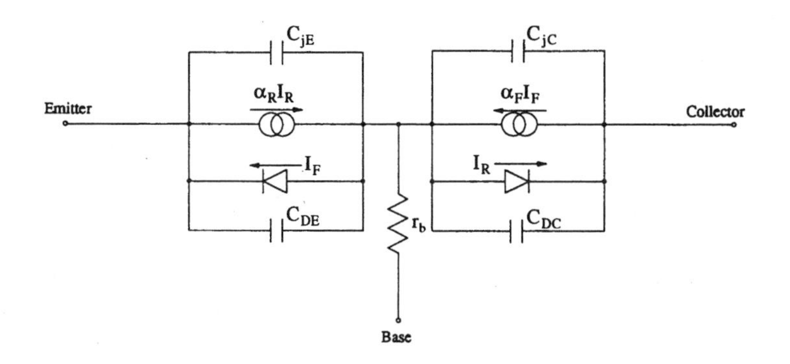

FIGURE 1.26 Low-frequency large-signal equivalent circuit model of an *npn* BJT.

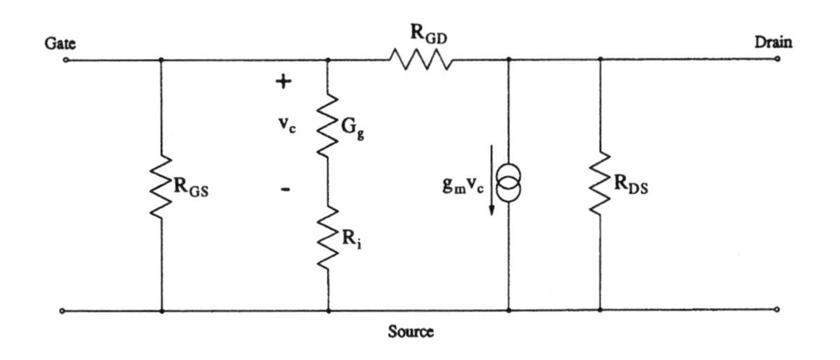

FIGURE 1.27 Low-frequency large-signal equivalent circuit model of a FET.

Low-Frequency Large-Signal Equivalent Circuit Model of a BJT. In this case, low frequency denotes frequencies below 10 kHz. The low-frequency large-signal equivalent circuit model of a BJT is based on its dc characteristics. Whereas at high frequencies one has to take incremental values to obtain accurate analysis, at low frequencies, the average of these incremental values yields the same level of accuracy in the analysis. Therefore, in low-frequency analysis, the circuit elements of the high-frequency model are replaced by their average values. The low-frequency large-signal equivalent circuit model is shown in Figure 1.26.

Low-Frequency Large-Signal Equivalent Circuit Model of a FET. Because of their high reactance values, the gate-to-source, gate-to-drain, and drain-to-source capacitances can be assumed to be open circuits at low frequencies. Therefore, the low-frequency large-signal model is similar to the high-frequency large-signal model, except that it has no capacitances. The resulting circuit describing low-frequency operation is shown in Figure 1.27.

Small-Signal Equivalent Circuit Model

In a small-signal equivalent circuit model, the signal variations around the dc-bias operating point are very small. Just as for the large-signal model, there are two types of small-signal models, depending upon the operating frequency: (1) the high-frequency small-signal equivalent circuit model and (2) the low-frequency small-signal equivalent circuit model.

High-Frequency Small-Signal Equivalent Circuit Model of a BJT. The high-frequency small-signal equivalent circuit model of a BJT is quite similar to its high-frequency large-signal equivalent circuit model. In the small-signal model, however, in addition to the base resistance r_b, the emitter and collector resistances, r_e and r_c, respectively, are added to the circuit. The emitter resistance is usually very small because of high emitter doping used to obtain better emitter injection efficiency. Therefore, whereas at large signal strengths the effect of r_e is overshadowed by the base resistance, at small signal strengths this emitter resistance cannot

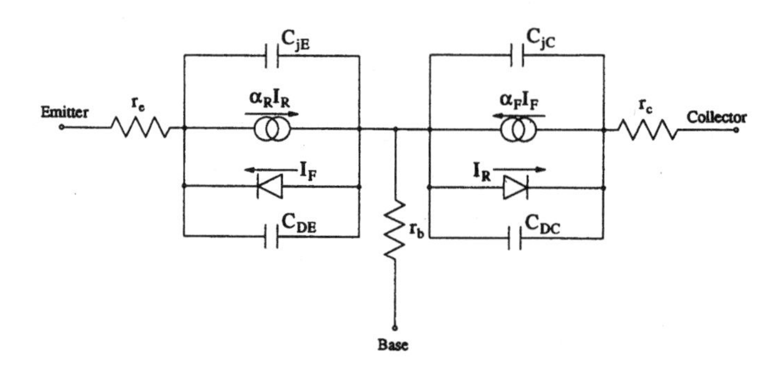

FIGURE 1.28 High-frequency small-signal equivalent circuit model of an *npn* BJT.

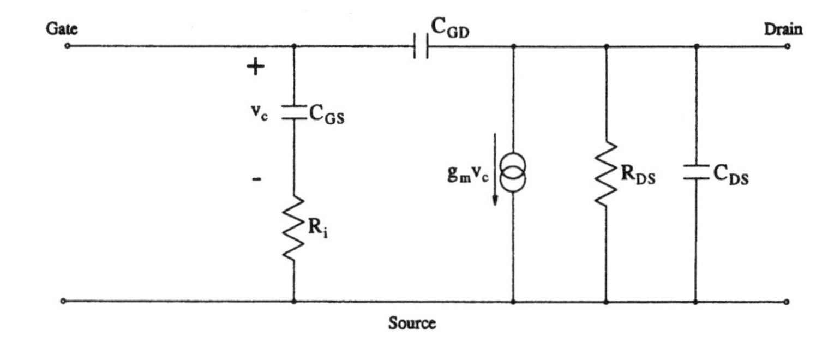

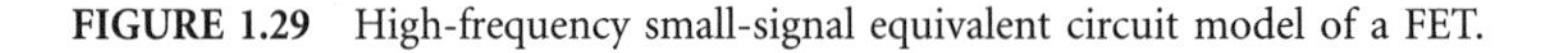

FIGURE 1.29 High-frequency small-signal equivalent circuit model of a FET.

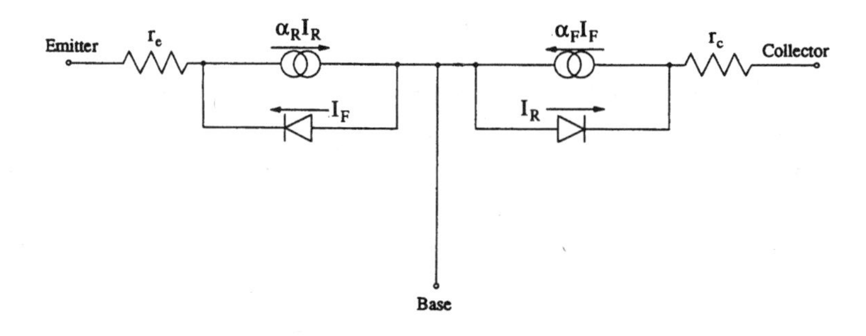

FIGURE 1.30 Low-frequency small-signal equivalent circuit model of an *npn* BJT.

be neglected. The collector resistance becomes important in the linear region, where the collector–emitter voltage is low. The high-frequency small-signal equivalent circuit model is shown in Figure 1.28.

High-Frequency Small-Signal Equivalent Circuit Model of a FET. For small-signal operations, the signal strength is not large enough to forward bias the gate-to-semiconductor diode; hence, no current will flow from the gate to either the drain or the source. Therefore, the gate-to-source and gate-to-drain series resistances, R_{GS} and R_{GD}, can be neglected. Also, since there will be no current flow from the gate to the channel, the gate conductance, G_g, can also be neglected. Figure 1.29 illustrates the high-frequency small-signal equivalent circuit model of a FET.

Low-Frequency Small-Signal Equivalent Circuit Model of a BJT. As in the low-frequency large-signal model, the junction capacitances, C_{jC} and C_{jE}, and the diffusion capacitances, C_{DE} and C_{DC}, can be neglected. Furthermore, the base resistance, r_b, can also be neglected, because the voltage drop across the base is not significant and the variations in the base width caused by changes in the collector–base voltage are also very small. The low-frequency small-signal equivalent circuit model is shown in Figure 1.30.

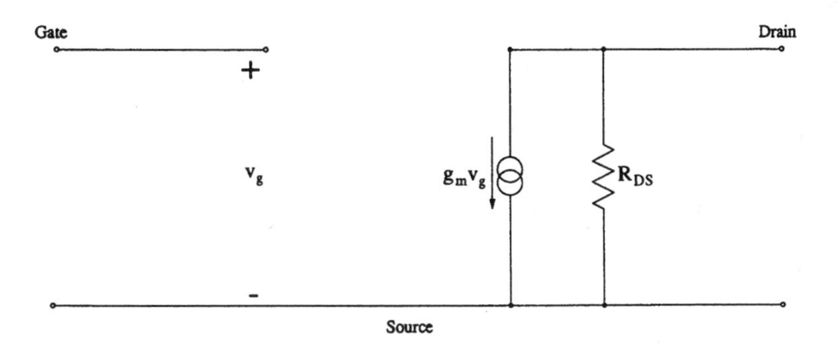

FIGURE 1.31 Low-frequency small-signal equivalent circuit model of a FET.

Low-Frequency Small-Signal Equivalent Circuit Model of a FET. Because the reactances associated with all the capacitances are very high, one can neglect the capacitances for low-frequency analysis. The gate conductance as well as the gate-to-source and gate-to-drain resistances can also be neglected in small-signal operation. The resulting low-frequency equivalent circuit model of a FET is shown in Figure 1.31.

DC Equivalent Circuit Model

DC Equivalent Circuit Model of a BJT. The dc equivalent circuit model of a BJT is based on the original Ebers–Moll model. Such models are used when the transistor is operated at dc or in applications where the operating frequency is below 1 kHz.

There are two versions of the dc equivalent circuit model—the *injection version* and the *transport version.* The difference between the two versions lies in the choice of the reference current. In the *injection version,* the reference currents are I_F and I_R, the forward- and reverse-biased diode currents, respectively. In the *transport version,* the reference currents are the collector transport current, I_{CC}, and the emitter transport current, I_{CE}. These currents are of the form:

$$I_F = I_{ES}\left[\exp\left(\frac{qV_{BE}}{kT}\right) - 1\right] \tag{1.71}$$

$$I_R = I_{CS}\left[\exp\left(\frac{qV_{BC}}{kT}\right) - 1\right] \tag{1.72}$$

$$I_{CC} = I_S\left[\exp\left(\frac{qV_{BE}}{kT}\right) - 1\right] \tag{1.73}$$

and

$$I_{EC} = I_S\left[\exp\left(\frac{qV_{BC}}{kT}\right) - 1\right] \tag{1.74}$$

In these equations, I_{ES} and I_{CS} are the base–emitter saturation current and the base–collector saturation current, respectively. I_S denotes the saturation current.

In most computer simulations, the *transport version* is usually preferred because of the following conditions:

1. I_{CC} and I_{EC} are ideal over many decades.
2. I_S can specify both reference currents at any given voltage.

The dc equivalent circuit model of a BJT is shown in Figure 1.32.

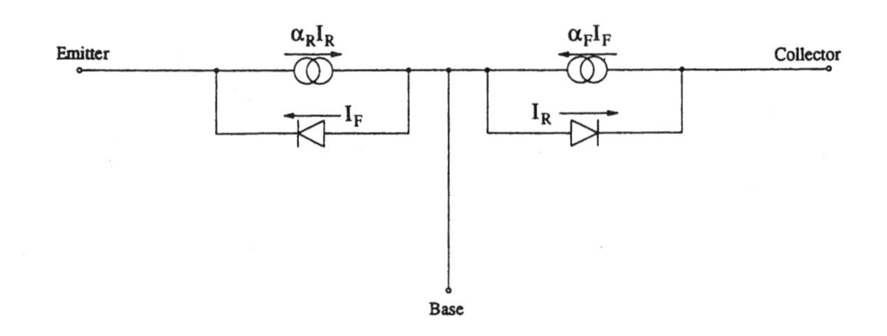

FIGURE 1.32 DC equivalent circuit model (injection version) of an *npn* BJT.

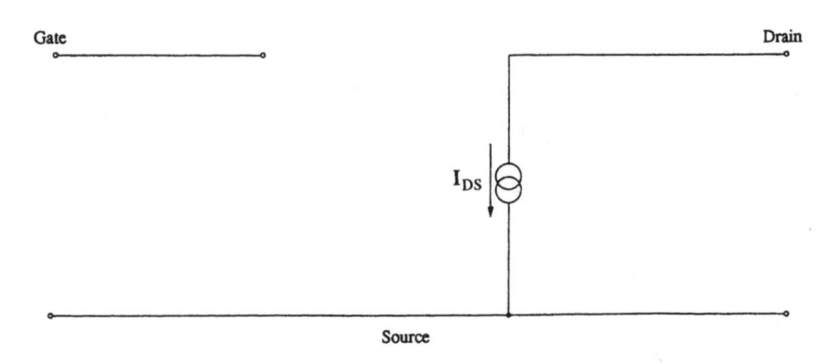

FIGURE 1.33 DC equivalent circuit model of a FET.

DC Equivalent Circuit Model of a FET. In the dc equivalent circuit model of a FET, the gate is considered isolated because the gate–semiconductor interface is formed as a reverse-biased diode and therefore is open circuited. All capacitances are also assumed to represent open circuits. R_{GS}, R_{GD}, and R_{DS} are neglected because there is no conductance through the gate and, because this is a dc analysis, there are no charging effects associated with the gate. The dc equivalent circuit of a FET is illustrated in Figure 1.33.

Commercially Available Packages

A number of circuit analysis software packages are commercially available, one of the most widely used being SPICE. In this package, the BJT models are a combination of the Gummel–Poon and the modified Ebers–Moll models. Figure 1.34 shows a common emitter transistor circuit and a SPICE input file containing the transistor model. Some other available packages are SLIC, SINC, SITCAP, and Saber.

Equivalent circuit models are basically used to replace the semiconductor device in an electronic circuit. These models are developed from an understanding of the device's current and voltage behavior for novel devices where internal device operation is not well understood. For such situations, the designer has another tool available, the semiconductor device simulator.

Overview of Semiconductor Device Simulators

Device simulators are based on the physics of semiconductor devices. The input to the simulator takes the form of information about the device under consideration such as material type, device, dimensions, doping concentrations, and operating conditions. Based on this information, the device simulator computes the electric field inside the device and thus predicts carrier concentrations in the different regions of the device. Device simulators can also predict transient behavior, including quantities such as current–voltage characteristics and frequency bandwidth. The three basic approaches to device simulation are (1) the classical approach, (2) the semiclassical approach, and (3) the quantum mechanical approach.

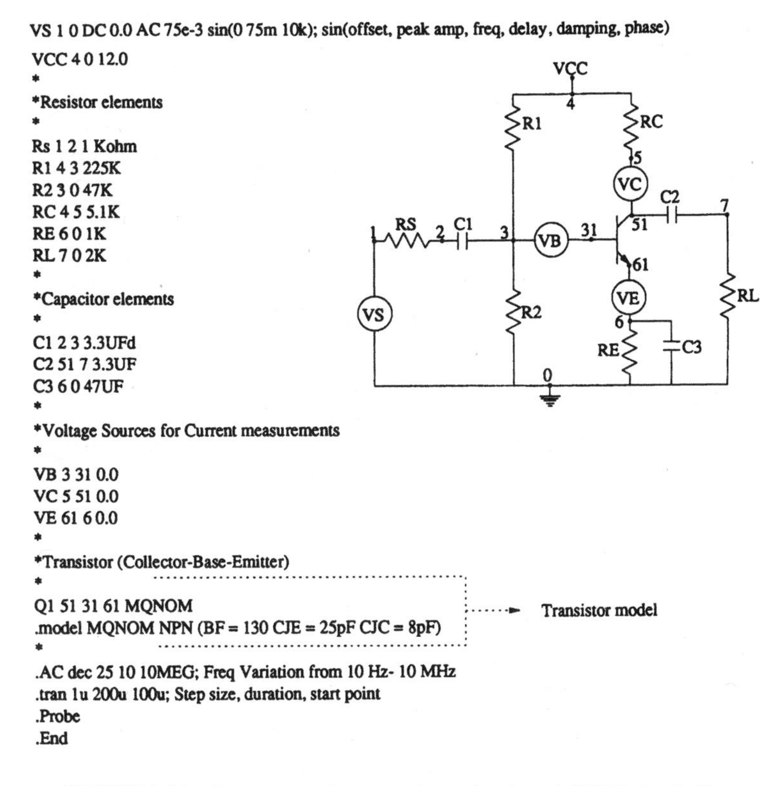

FIGURE 1.34 Common emitter transistor circuit and SPICE circuit file.

Device Simulators Based on the Classical Approach

The *classical approach* is based on the solution of Poisson's equation and the current continuity equations. The current consists of the drift and the diffusion current components.

Assumptions. The equations for the classical approach can be obtained by making the following approximations to the Boltzmann transport equation:

1. Carrier temperature is the same throughout the device and is assumed to be equal to the lattice temperature.
2. Quasi steady-state conditions exist.
3. Carrier **mean free path** must be smaller than the distance over which the **quasi-Fermi level** is changing by kT/q.
4. The impurity concentration is constant or varies very slowly along the mean free path of the carrier.
5. The energy band is parabolic.
6. The influence of the boundary conditions is negligible.

For general purposes, even with these assumptions and limitations, the models based on the classical approach give fairly accurate results. The model assumes that the driving force for the carriers is the quasi-Fermi potential gradient, which is also dependent upon the electric field value. Therefore, in some simulators, the quasi-Fermi level distributions are computed and the carrier distribution is estimated from this information.

Equations to be Solved. With the assumption of a quasi-steady-state condition, the operating wavelength is much larger than the device dimensions. Hence, Maxwell's equations can be reduced to the more familiar Poisson's equation:

$$\nabla^2\psi = -\frac{\rho}{\epsilon} \tag{1.75}$$

and, for a nonhomogeneous medium:

$$\nabla \cdot \epsilon(\nabla \psi) = -\rho \tag{1.76}$$

where ψ denotes the potential of the region under simulation, ϵ denotes the permittivity, and ρ denotes the charge enclosed by this region.

Also from Maxwell's equations, one can determine the current continuity equations for a homogeneous medium as:

$$\nabla \cdot J_{\mathrm{n}} - q\left(\frac{\partial n}{\partial t}\right) = +qU \tag{1.77}$$

where

$$J_{\mathrm{n}} - q\mu_{\mathrm{n}}E + qD_{\mathrm{n}}\nabla \cdot n \tag{1.78}$$

and

$$\nabla \cdot J_{\mathrm{p}} + q\left(\frac{\partial p}{\partial t}\right) = -\,qU \tag{1.79}$$

where

$$J_{\mathrm{p}} = q\mu_{\mathrm{p}}pE - qD_{\mathrm{p}}\nabla \cdot p \tag{1.80}$$

For nonhomogeneous media, the electric field term in the current expressions is modified to account for the nonuniform **density of states** and the bandgap variation (Lundstrom and Schuelke, 1983).

In the classical approach, the objective is to calculate the potential and the carrier distribution inside the device. Poisson's equation is solved to yield the potential distribution inside the device from which the electric field can be approximated. The electric field distribution is then used in the current continuity equations to obtain the carrier distribution and the current densities. The diffusion coefficients and carrier mobilities are usually field as well as spatially dependent.

The generation-recombination term U is usually specified by the Shockley–Read–Hall relationship (Yoshi et al., 1982):

$$Rn = \frac{pn - n_{\mathrm{ie}}^2}{\tau_{\mathrm{p}}(n + n_t) + \tau_{\mathrm{n}}(p + p_t)} \tag{1.81}$$

where p and n are the hole and electron concentrations, respectively, n_{ie} is the effective intrinsic carrier density, τ_{p} and τ_{n} are the hole and electron lifetimes, and p_t and n_t are the hole and electron trap densities, respectively.

The electron and hole mobilities are usually specified by the Scharfetter–Gummel empirical formula, as

$$\mu = \mu_0\left[1 + \frac{N}{(N/a) + b} + \frac{(E/c)^2}{(E/c) + d} + (E/e)^2\right]^{-1/2} \tag{1.82}$$

where N is the total ionized impurity concentration, E is the electric field, and a, b, c, d, and e are defined constants (Scharfetter and Gummel, 1969) that have different values for electrons and holes.

Boundary Conditions. Boundary conditions have a large effect on the final solution, and their specific choice is a very important issue. For ohmic contacts, infinite recombination velocities and space charge neutrality conditions are assumed. Therefore, for a p-type material, the ohmic boundary conditions take the form:

$$\psi = V_{\text{appl}} + \frac{kT}{q}\ln\left(\frac{n_{\text{ie}}}{p}\right) \tag{1.83}$$

$$p = \left[\left(\frac{N_{\text{D}}^{+} - N_{\text{A}}^{-}}{2}\right)^{2} + n_{\text{ie}}^{2}\right]^{1/2} - \left(\frac{N_{\text{D}}^{+} - N_{\text{A}}^{-}}{2}\right) \tag{1.84}$$

and

$$n = \frac{n_{\text{ie}}^{2}}{p} \tag{1.85}$$

where V_{appl} is the applied voltage, k is Boltzmann's constant, and $N_{\text{D}}^{+}$and N_{A}^{-} are the donor and acceptor ionized impurity concentrations, respectively.

For **Schottky contacts**, the boundary conditions take the form:

$$\psi = V_{\text{appl}} + \frac{E_{\text{G}}}{2} - \phi_{\text{B}} \tag{1.86}$$

and

$$n = n_{\text{ie}} \exp\left(\frac{(E_{\text{G}}/2) - \phi_{\text{B}}}{kT/q}\right) \tag{1.87}$$

where E_{G} is the semiconductor bandgap and ϕ_{B} is the barrier potential. For other boundaries with no current flow across them, the boundary conditions are of the form:

$$\frac{\partial \psi}{\partial n} = \frac{\partial \varphi|_{\text{n}}}{\partial n} = \frac{\partial \varphi_{\text{p}}}{\partial p} = 0 \tag{1.88}$$

where ϕ_{n} and ϕ_{p} are the electron and hole quasi-Fermi levels, respectively.

For field-effect devices, the potential under the gate may be obtained either by setting the gradient of the potential near the semiconductor–oxide interface equal to the gradient of potential inside the oxide (Kasai et al., 1982), or by solving Laplace's equation in the oxide layer, or by assuming a Dirichlet boundary condition at the oxide–gate interface and determining the potential at the semiconductor–oxide interface as:

$$\epsilon_{\text{Si}} \left.\frac{\partial \psi}{\partial y}\right|_{\text{Si}} = \epsilon_{\text{Ox}} \frac{\psi_{\text{G}} - \psi_{\text{S}}^{*}(x, z)}{T(z)} \tag{1.89}$$

where ϵ_{Si} and ϵ_{Ox} are the permittivities of silicon and the oxide, respectively, ψ_{G} is the potential at the top of the gate, $\psi_{\text{S}}^{*}(x, z)$ is the potential of the gate near the interface, and $T(z)$ is the thickness of the gate metal.

Solution Methods. Two of the most popular methods of solving the above equations are finite difference method (FDM) and finite element method (FEM).

In FDM, the region under simulation is divided into rectangular or triangular areas for two-dimensional cases or into cubic or tetrahedron volumes in three-dimensional cases. Each corner or vertex is considered as a node. The differential equations are modified using finite difference approximations, and a set of equations is constructed in matrix form. The finite difference equations are solved iteratively at only these nodes. The most commonly used solvers are Gauss–Seidel/Jacobi (G-S/J) techniques or Newton's technique (NT) (Banks et al., 1983). FDM has the disadvantage of requiring more nodes than the FEM for the same structure. A new variation of FDM, namely the finite boxes scheme (Franz et al., 1983), however, overcomes this problem by enabling local area refinement. The advantage of FDM is that its computational memory requirement is less than that required for FEM because of the band structure of the matrix.

In FEM, the region under simulation is divided into triangular and quadrilateral regions in two dimensions or into tetrahedra in three dimensions. The regions are placed to have the maximum number of vertices in areas where there is expected to be a large variation of composition or a large variation in the solution. The equations in FEM are modified by multiplying them with some shape function and integrating over the simulated region. In triangular meshes, the shape function is dependent on the area of the triangle and the spatial location of the node. The value of the spatial function is between 0 and 1. The solution at one node is the sum of all the solutions, resulting from the nearby nodes, multiplied by their respective shape functions. The number of nodes required to simulate a region is less than that in FDM; however, the memory requirement is greater.

Device Simulators Based on the Semiclassical Approach

The *semiclassical* approach is based upon the Boltzmann transport equation (BTE) (Engl, 1986) which can be written as

$$\frac{\mathrm{d}f}{\mathrm{d}t} = \frac{\partial f}{\partial t} + v \cdot \nabla_{\mathrm{r}} \pm \frac{q}{(h/2\pi)} E \cdot \nabla_{\mathrm{k}} f = \left(\frac{\partial f}{\partial t}\right)_{\mathrm{coll}} \tag{1.90}$$

where f represents the carrier distribution in the volume under consideration at any time t, v is the group velocity, E is the electric field, and q and h are the electronic charge and Planck's constant, respectively.

BTE is a simplified form of the Liouville–Von Neumann equation for the density matrix. In this approach, the free flight between two consecutive collisions of the carrier is considered to be under the influence of the electric field, whereas different scattering mechanisms determine how and when the carrier will undergo a collision.

Assumptions. The assumptions for the semiclassical model can be summarized as follows:

1. Carrier-to-carrier interactions are considered to be very weak.
2. Particles cannot gain energy from the electric field during collision.
3. Scattering probability is independent of the electric field.
4. Magnetic field effects are neglected.
5. No electron-to-electron interaction occurs in the collision term.
6. Electric field varies very slowly, i.e., electric field is considered constant for a wave packet describing the particle's motion.
7. The electron and hole gas is not degenerate.
8. Band theory and effective-mass theorems apply to the semiconductor.

Equations to Be Solved. As a starting point, Poisson's equation is solved to obtain the electric field inside the device. Using the Monte-Carlo technique (MCT), the BTE is solved to obtain the carrier distribution function, f. In the MCT, the path of one or more carriers, under the influence of external forces, is followed, and from this information the carrier distribution function is determined. BTE can also be solved by the momentum and energy balance equations.

The carrier distribution function gives the carrier concentrations in the different regions of the device and can also be used to obtain the electron and hole currents, using the following expressions:

$$J_{\mathrm{n}} = -q \int_{k} v f(r, k, t) d^3 k \tag{1.91}$$

and

$$J_p = +q \int_k v f(r,k,t) d^3k \tag{1.92}$$

Device Simulators Based on the Quantum Mechanical Approach

The *quantum mechanical approach* is based on the solution of the Schrodinger wave equation (SWE), which, in its time-independent form, can be represented as

$$\frac{(h/2\pi)^2}{2m} \nabla^2 \varphi_n + (E_n + qV)\varphi_n = 0 \tag{1.93}$$

where φ_n is the wave function corresponding to the subband n whose minimum energy is E_n, V is the potential of the region, m is the particle mass, and h and q are Planck's constant and the electronic charge, respectively.

Equations to Be Solved. In this approach, the potential distribution inside the device is calculated using Poisson's equation. This potential distribution is then used in the SWE to yield the electron wave vector, which in turn is used to calculate the carrier distribution, using the following expression:

$$n = \sum_n N_n |\varphi_n|^2 \tag{1.94}$$

where n is the electron concentration and N_n is the concentration of the subband n.

This carrier concentration is again used in Poisson's equation, and new values of φ_n, E_n, and n are calculated. This process is repeated until a self-consistent solution is obtained. The final wave vector is invoked to determine the scattering matrix, after which MCT is used to yield the carrier distribution and current densities.

Commercially Available Device Simulation Packages

The classical approach is the most commonly used procedure since it is the easiest to implement and, in most cases, the fastest technique. Simulators based on the classical approach are available in two-dimensional forms like FEDAS, HESPER, PISCES-II, PISCES-2B, MINIMOS, and BAMBI, or three-dimensional forms like TRANAL, SIERRA, FIELDAY, DAVINCI, and CADDETH.

Large-dimension a devices, where the carriers travel far from the boundaries, can be simulated based on a one-dimensional approach. Most currently used devices, however, do not fit into this category, and therefore one has to resort to either two- or three-dimensional simulators.

FEDAS (Field Effect Device Analysis System) is a two-dimensional device simulator that simulates MOSFETs, JFETs, and MESFETs by considering only those carriers that form the channel. The Poisson equation is solved everywhere except in the oxide region. Instead of carrying the potential calculation within the oxide region, the potential at the semiconductor–oxide interface is calculated by assuming a mixed boundary condition. FEDAS uses FDM to solve the set of linear equations. A three-dimensional variation of FEDAS is available for the simulation of small geometry MOSFETs.

HESPER (HEterostructure device Simulation Program to Estimate the performance Rigorously) is a two-dimensional device simulator that can be used to simulate heterostructure photodiodes, HBTs, and HEMTs. The simulation starts with the solution of Poisson's equation in which the electron and hole concentrations are described as functions of the composition (composition dependent). The recombination rate is given by the Shockley–Read–Hall relationship. Lifetimes of both types of carriers are assumed to be equal in this model.

PISCES-2B is a two-dimensional device simulator for simulation of diodes, BJTs, MOSFETs, JFETs, and MESFETs. Besides steady-state analysis, transient and ac small-signal analysis can also be performed.

Conclusion

The decision to use an equivalent circuit model or a device simulator depends upon the designer and the required accuracy of prediction. To save computational time, one should use as simple a model as accuracy will allow. At this time, however, the trend is toward developing quantum mechanical models that are more accurate, and with faster computers available, the computational time for these simulators has been considerably reduced.

Defining Terms

Density of states: The total number of charged carrier states per unit volume.

Fermi levels: The energy level at which there is a 50% probability of finding a charged carrier.

Mean free path: The distance traveled by the charged carrier between two collisions.

Mobile charge: The charge due to the free electrons and holes.

Quasi-Fermi levels: Energy levels that specify the carrier concentration inside a semiconductor under nonequilibrium conditions.

Schottky contact: A metal-to-semiconductor contact where, in order to align the Fermi levels on both sides of the junction, the energy band forms a barrier in the majority carrier path.

References

R.E. Banks, D.J. Rose, and W. Fitchner, "Numerical methods for semiconductor device simulation," *IEEE Trans. Electron Devices,* vol. ED-30, no. 9, pp. 1031–1041, 1983.

J.J. Ebers and J.L. Moll, "Large signal behavior of junction transistors," *Proc. IRE*, vol. 42, pp. 1761–1772, Dec. 1954.

W.L. Engl, *Process and Device Modeling*, Amsterdam: North-Holland, 1986.

A.F. Franz, G.A. Franz, S. Selberherr, C. Ringhofer, and P. Markowich, "Finite boxes—A generalization of the finite-difference method suitable for semiconductor device simulation," *IEEE Trans. Electron Devices*, vol. ED-30, no. 9, pp. 1070–1082, 1983.

H.K. Gummel and H.C. Poon, "An integral charge control model of bipolar transistors," *Bell Syst. Tech. J.*, vol. 49, pp. 827–852, May–June 1970.

R. Kasai, K. Yokoyama, A. Yoshii, and T. Sudo, "Threshold-voltage analysis of short- and narrow-channel MOSFETs by three-dimensional computer simulation," *IEEE Trans. Electron Devices,* vol. ED-21, no. 5, pp. 870–876, 1982.

M.S. Lundstrom and R.J. Schuelke, "Numerical analysis of heterostructure semiconductor devices," *IEEE Trans. Electron Devices*, vol. ED-30, no. 9, pp. 1151–1159, 1983.

D.L. Scharfetter and H.K. Gummel, "Large-signal analysis of a silicon read diode oscillator," *IEEE Trans. Electron Devices*, vol. ED-16, no. 1, pp. 64–77, 1969.

A. Yoshii, H. Kitazawa, M. Tomzawa, S. Horiguchi, and T. Sudo, "A three dimensional analysis of semiconductor devices," *IEEE Trans. Electron Devices*, vol. ED-29, no. 2, pp. 184–189, 1982.

Further Information

Further information about semiconductor device simulation and equivalent circuit modeling, as well as about the different software packages available, can be found in the following articles and books:

C.M. Snowden, *Semiconductor Device Modeling*, London: Peter Peregrinus Ltd., 1988.

C.M. Snowden, *Introduction to Semiconductor Device Modeling,* Teaneck, NJ: World Scientific, 1986.

W.L. Engl, *Process and Device Modeling*, Amsterdam: North-Holland, 1986.

J.-H. Chern, J.T. Maeda, L.A. Arledge, Jr., and P. Yang, "SIERRA: A 3-D device simulator for reliability modeling," *IEEE Trans. Computer-Aided Design,* vol. CAD-8, no. 5, pp. 516–527, 1989.

T. Toyabe, H. Masuda, Y. Aoki, H. Shukuri, and T. Hagiwara, "Three-dimensional device simulator CADDETH with highly convergent matrix solution algorithms," *IEEE Trans. Electron Devices*, vol. ED-32, no. 10, pp. 2038–2044, 1985.

PISCES-2B and DAVINCI are softwares developed by TMA Inc., Palo Alto, Calif. 94301.

Hewlett-Packard's first product, the model 200A audio oscillator (preproduction version). William Hewlett and David Packard built an audio oscillator in 1938, from which the famous firm grew. Courtesy of Hewlett-Packard Company.)

2

Semiconductor Manufacturing

Harold G. Parks
The University of Arizona

Wayne Needham
Intel Corporation

S. Rajaram
Lucent Technologies

Benjamin Y. Park
University of California

Rabih Zaouk
University of California

Marc J. Madou
University of California

2.1 Processes

Harold G. Parks

Integrated circuit (IC) fabrication consists of a sequence of processing steps referred to as unit-step processes that lead to the devices present on today's microchips. These unit-step processes provide the methodology for introducing and transporting dopants to change the conductivity of the semiconductor substrate, growing thermal oxides for inter- and intra-level isolation, depositing insulating and conducting films, and patterning and etching the various layers in formation of the IC. Many of these unit steps have remained essentially the same since discrete component processing. However, others have originated and grown with the evolution of integrated circuits from large-scale integration (LSI) with thousands of components per chip in the late 1970s through to the ultra-large-scale integration (ULSI) era of today, with billions of devices per chip. As technology continues to evolve into the nanometer regime, further modification of current unit-step processes will be required. In this section, the unit-step processes for silicon IC processing as they exist today are presented, with an eye toward the future. How they are combined to form the actual IC process will be discussed in a later section. Because of space limitations, only silicon processes will be discussed. This is not a major limitation, as many of the same steps are used to process other types of semiconductors, and perhaps more than 98% of all ICs today and in the near future are and will be made of silicon. Furthermore, only the highlights of the unit steps can be presented in this space, but ample references are provided for a more thorough presentation. The referenced processing textbooks provide a detailed discussion of all processes.

Thermal Oxidation

Silicon dioxide (SiO_2) layers are important in integrated-circuit technology for surface passivation as both a diffusion barrier and a surface dielectric. The fact that silicon readily forms a high-quality, dense, natural oxide is the major reason it is dominant in integrated-circuit technology today. If a silicon wafer is exposed to air, it will grow a thin (≈ 20 Å) oxide layer in a relatively short time period. To achieve the thickness required for SiO_2 layers used in integrated-circuit technology (1.2 nm to 2 μm), alternative steps are needed. Thermal oxidation is an extension of natural oxide growth at an elevated temperature (800 to 1200°C). The temperature is usually a compromise — it must be high enough to grow the oxide in a reasonable time period and as low as possible to minimize crystal damage and the unwanted diffusion of dopants already in the wafer.

Oxidation Process

The two most common oxidizing environments are dry and wet. As the name implies, dry oxides are grown in dry O_2 gas following the reaction:

$$\mathrm{Si} + \mathrm{O_2} \rightarrow \mathrm{SiO_2} \tag{2.1}$$

Wet oxides were originally grown by bubbling dry oxygen gas through water at 95°C. Most "wet" oxides today are produced by the pyrogenic reaction of H_2 and O_2 gas to form steam and are called steam oxidations. In either case, the reaction is essentially the same at the wafer:

$$\mathrm{Si} + \mathrm{H_2O} \rightarrow \mathrm{SiO_2} + 2\mathrm{H_2} \tag{2.2}$$

The oxidation process can be modeled as shown in Figure 2.1. The position X_0 represents the Si/SiO_2 interface, which is a moving boundary. The volume density of oxidizing species in the bulk gas, N_G, is depleted at the oxide surface, N_S, due to an amount, N_0, being incorporated in the oxide layer. The oxidizing species then diffuses across the growing oxide layer, where it reacts with silicon at the moving interface to form SiO_2. F_G represents the flux of oxidant transported by diffusion from the bulk gas to the oxide surface. The oxidizing species that enters the SiO_2 diffuses across the growing SiO_2 layer with a flux, F_{ox}. A reaction takes place at the Si/SiO_2 interface, consuming some or all of the oxidizing species, as represented by the flux, F_I.

In a steady state, these three flux terms are equal and can be used to solve for the concentrations N_I and N_0 in terms of the reaction rate and diffusion coefficient of the oxidizing species. This in turn specifies the flux terms, which can be used in the solution of the differential equation:

$$\frac{dx}{dt} = \frac{F}{N_{ox}} \tag{2.3}$$

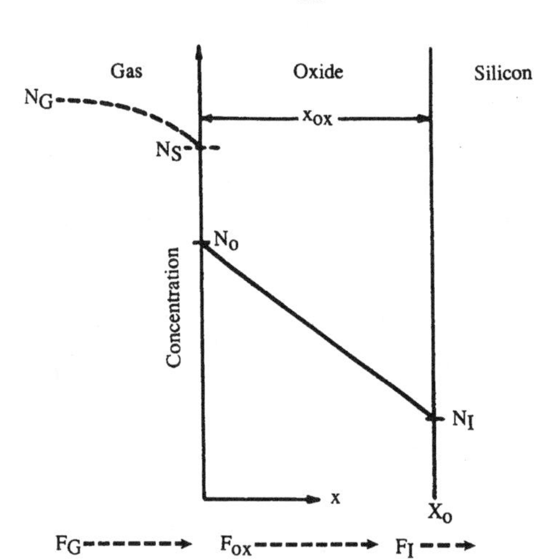

FIGURE 2.1 Model of the oxidation process.

for the oxide growth, x. In this equation, N_{ox} is the number of oxidant molecules per unit volume of oxide. An excellent derivation of the growth equation is given in Jaeger (1988, 2002). The result can be represented by

$$x_{ox} = \frac{A}{2}\left[\sqrt{1 + \frac{4B}{A^2}(t+\tau)} - 1\right] \tag{2.4}$$

where x_{ox} is the oxide thickness, B is the parabolic rate constant, B/A is the linear rate constant, t is the oxidation time, and τ is the initial oxide thickness.

Referring to Equation (2.4), we see there are two regimes of oxide growth. For thin oxides or short time periods such as the initial phase of the oxidation process, the equation reduces to

$$x_{ox} = \frac{B}{A}(t+\tau) \tag{2.5}$$

and growth is a linear function of time, limited by the surface reaction at the Si/SiO_2 interface. For thicker oxides and longer time periods, the reaction is limited by the diffusion of the oxidizing species across the growing oxide layer. The limiting form of Equation (2.4) is

$$x_{ox} = \sqrt{Bt} \tag{2.6}$$

Oxidation Rate Dependencies

Typical oxidation curves showing oxide thickness as a function of time, with temperature as a parameter for wet and dry oxidation of <100> silicon, are shown in Figure 2.2. This type of curve is qualitatively similar for all oxidations. The oxidation rates are highly temperature dependent, as both the linear- and parabolic-rate constants show an Arrhenius relationship with temperature. The linear rate is dominated by the temperature dependence of the interfacial growth reaction and the parabolic rate is dominated by the temperature dependence of the diffusion coefficient of the oxidizing species in SiO_2.

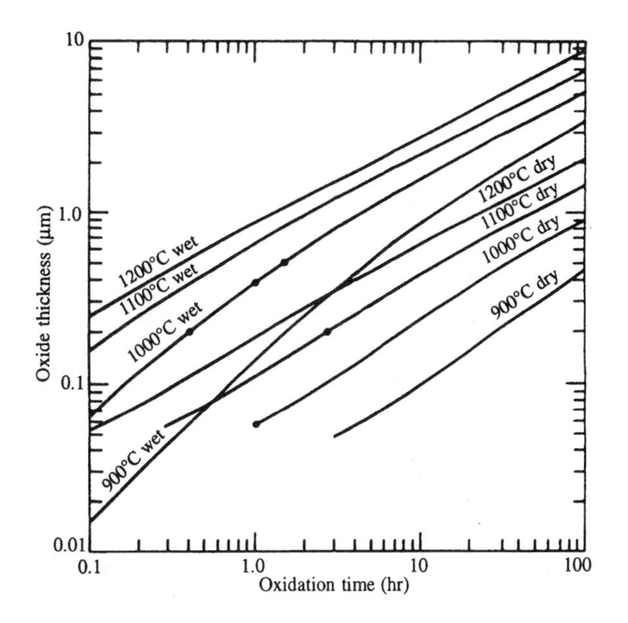

FIGURE 2.2 Thermal silicon-dioxide growth on <100> silicon for wet and dry oxides. (*Source:* R. C. Jaeger, *Introduction to Microelectronic Fabrication, Vol. 5 in the Modular Series on Solid State Devices*, G.W. Neudek and R.F. Pierret, Eds., Reading, MA: Addison-Wesley, 1988, p. 35. With permission.)

Wet oxides grow faster than dry oxides. Both the linear- and parabolic-rate constants are proportional to the equilibrium concentration of the oxidant in the oxide. The solubility of H_2O in SiO_2 is greater than that of O_2. Hence, the oxidation rate is enhanced for wet oxides.

Dry oxides with chlorine addition grow at a faster rate than pure dry oxides because the chlorine species and oxygen react to form water vapor as a by-product (Plummer et al., 2000).

The oxidation rate depends on substrate orientation (Jaeger, 1988, 2002). The rate is related to the surface atom density of the substrate — the higher the density, the faster the oxidation process. The oxidation rate also depends on pressure. The linear and parabolic rates are dependent on the equilibrium concentration of the oxidizing species in the SiO_2, which is directly proportional to the partial pressure of the oxidant in the ambient.

Oxide growth rate shows a doping dependence for heavily doped substrates ($>10^{20}$ cm^{-3}). Boron increases the parabolic-rate constant and phosphorus enhances the linear-rate constant (Wolf and Tauber, 1986).

Oxide Characteristics

Dry oxides grow more slowly than wet oxides, resulting in higher density, higher breakdown field strengths, and more controlled growth, thus making them ideal for metal-oxide semiconductor (MOS) gate dielectrics.

Wet oxidation is used for forming thick oxides for field isolation and for masking implants and diffusions. The slight degradation in oxide density is more than compensated for by the thickness in these applications.

<100> substrates have fewer dangling bonds at the surface, which results in lower fixed oxide charge and interface traps and therefore higher quality MOS devices.

Conventional dopants (B, P, As, and Sb) diffuse slowly in both wet and dry oxides and hence these oxides provide a good barrier for masking diffusions in integrated circuit fabrication.

High-pressure steam oxidations provide a means of growing relatively thick oxides in a reasonable time and at low temperatures, avoiding dopant diffusion. Conversely, low-pressure oxidations show promise for forming controlled, ultra-thin gate growth for ULSI technologies.

Chlorine added to gate oxides (Sze, 1988) has been shown to reduce mobile ions, reduce oxide defects, increase breakdown voltage, reduce fixed oxide charge and interface traps, reduce oxygen-induced stacking faults, and increase the lifetime of substrate minority carriers. Chlorine is introduced into dry oxidations, in less than 5% concentrations, as anhydrous HCl gas or by either trichloroethylene (TCE) or trichloroethane (TCA).

Dopant Segregation and Redistribution

Since silicon is consumed during oxidation, the dopant in the substrate will be redistributed due to segregation (Wolf and Tauber, 1986). The boundary condition across the Si/SiO_2 interface is that the chemical potential of the dopant is the same on both sides. This results in the definition of a segregation coefficient, m, as the ratio of the equilibrium concentration of dopant in Si to the equilibrium concentration of dopant in SiO_2. Depending on the value of m (whether less than or greater than 1) and the diffusion properties of the dopant in SiO_2, various redistributions are possible. For example, $m \approx 0.3$ for boron. It is a slow diffuser in SiO_2, so it tends to deplete from the Si surface and accumulate in the oxide at the Si/SiO_2 interface. Phosphorus, however, has $m \approx 10$. It is also a slow diffuser in SiO_2 and tends to pile up in the Si at the Si/SiO_2 interface. Antimony and arsenic behave similarly to phosphorus.

Oxidation Technology

Thermal oxidation of silicon is achieved by placing wafers in a quartz furnace tube. The tube is surrounded by a three-zone resistance heater and has provisions for the controlled flow of an inert gas, such as nitrogen, and the oxidant. The wafer zone has a flat temperature profile to within a fraction of a degree and can handle up to 50 parallel stacked wafers. The horizontal furnace was the mainstay of the semiconductor industry for decades; however, vertical furnaces are commonly used for processing wafers with diameters of 150 mm or larger. Modern furnaces are computer-controlled and programmable. Wafers are usually loaded in an inert environment, ramped to the correct temperature, and switched to the oxidant for a programmed time. When oxidation is complete, the gas is switched back to the inert gas and the temperature is ramped down to the unload temperature. All these complications in the process are to minimize thermal stress damage to the wafers; the procedures can vary considerably.

Oxidation has been, and is certain to remain, an integral part of silicon processing technology. However, as device dimensions shrink, process requirements shift to a need for lower temperatures to control unwanted impurity diffusion. A noted example is the move to low-temperature, low-pressure oxidations for thin-gate oxides. These are typically effected by oxygen dilution in a 10%:90% O_2:N_2 environment at a temperature of $\approx 800°C$. Another technology shift brought on by the shrinking technology is the use of oxynitride gate dielectrics to prevent boron diffusion through thin-gate oxides. The films are generally prepared by exposing a thermally grown SiO_2 film to an N_2O or NO ambient. The resulting oxynitride, SiO_xN_y, usually varies in composition throughout the film, with nitrogen-rich regions near the Si/dielectric interface and at the top of the dielectric at the SiO_2/ambient interface. The oxynitride interface causes rugged gate dielectrics that are more resistant to bond breaking during hot carrier stressing, more resistant to boron penetration, and have the additional advantage of a higher gate dielectric than does SiO_2.

Detailed discussions of the equipment and procedures can be found in Plummer et al. (2000); Campbell (2001); Jaeger (2002).

Diffusion

Diffusion is the traditional way dopants were introduced into silicon wafers to create junctions and control the resistivity of layers. Ion implantation has now superseded diffusion for this purpose. However, the principles and concepts of diffusion theory remain important, since they describe the movement and transport of dopants and impurities during the high-temperature processing steps of integrated circuit manufacture.

Diffusion Mechanism

Consider a silicon wafer with a high concentration of surface impurity. At any temperature, there are a certain number of vacancies in the Si lattice. If the wafer is subjected to an elevated temperature, the number of vacancies in the silicon will increase and the impurity will enter the wafer, moving from the high surface concentration to be redistributed in the bulk. The redistribution mechanism is diffusion. Depending on the impurity type, it will either be substitutional or interstitial (Ghandhi, 1982).

For substitutional diffusion, the impurity atom substitutes for a silicon atom at a vacancy lattice site and then moves into the wafer by hopping from lattice site to lattice site via the vacancies. Clearly, the hopping can be in a random direction; however, since the impurity is present initially in high concentration on the surface only, there is a net flow of the impurity from the surface into the bulk.

In the case of interstitial diffusion, the impurity diffuses by squeezing between the lattice atoms and taking residence in the interstitial space between lattice sites. Since this mechanism does not require the presence of a vacancy, it proceeds much faster than substitutional diffusion. Conventional dopants such as B, P, As, and Sb primarily diffuse by the substitutional method. This is beneficial because the diffusion process is much slower and can therefore be controlled more easily in the manufacturing process. In heavily doped regions, sustitutional impurities diffuse by a variety of mechanisms, which are further influenced by higher-order vacancy effects that cause the diffusion coefficient to vary with impurity level. In general, such diffusions cannot be treated analytically and require numerical solutions of the diffusion equations. Many of the undesired impurities such as Fe, Cu, and other heavy metals diffuse by the interstitial mechanism. Therefore, the process is extremely fast; this is beneficial because, at the temperatures used and during the fabrication processes, the unwanted metals can diffuse completely through the Si wafer. Gettering creates trapping sites on the back surface of the wafer for these impurities that would otherwise remain in the silicon and cause adverse device effects.

Simple diffusions, regardless of the diffusion mechanism, can be formalized mathematically in the same way by introducing a constant diffusion coefficient, D (cm^2/sec), that accounts for the diffusion rate. The diffusion constants follow an Arrhenius behavior according to the equation:

$$D = D_0 \exp\left[-\frac{E_A}{kT}\right] \tag{2.7}$$

where D_0 is the prefactor, E_A is the activation energy, k is Boltzmann's constant, and T is the absolute temperature. Conventional silicon dopants (substitutional diffusers) have diffusion coefficients on the order of 10^{-14} to 10^{-12} cm^2/sec at 1100°C, whereas heavy metal interstitial diffusers (Fe, Au, and Cu) have diffusion coefficients of 10^{-6} to 10^{-5} cm^2/sec at this temperature.

The diffusion process can be described using Fick's laws. Fick's first law states that the flux of impurity, F, crossing any plane is related to the impurity distribution, $N(x,t)$ per cm^3, by

$$F = D\frac{\partial N}{\partial x} \tag{2.8}$$

in the one-dimensional case. Fick's second law states that the time rate of change of the particle density is in turn related to the divergence of the particle flux:

$$\frac{\partial N}{\partial x} = \frac{\partial F}{\partial x} \tag{2.9}$$

Combining these two equations gives

$$\frac{\partial N}{\partial x} = \frac{\partial}{\partial x}\left(D\frac{\partial N}{\partial x}\right) = D\frac{\partial^2 N}{\partial x^2} \tag{2.10}$$

in the case of a constant diffusion coefficient, as is often assumed. This partial differential equation can be solved by the separation of variables or by Laplace transform techniques for specified boundary conditions.

For a constant source diffusion, the impurity concentration at the surface of the wafer is held constant throughout the diffusion process. The solution of Equation (2.10) under these boundary conditions, assuming a semi-infinite wafer, results in a complementary error function diffusion profile:

$$N_{(x,t)} = N_0 erfc\left(\frac{x}{2\sqrt{Dt}}\right) \tag{2.11}$$

Here, N_0 is the impurity concentration at the surface of the wafer, x is the distance into the wafer, and t is the diffusion time. As time progresses, the impurity profile penetrates deeper into the wafer while maintaining a constant surface concentration. The total number of impurity atoms/cm^2 in the wafer is the dose, Q, and it continually increases with time:

$$Q = \int_0^\infty N_{(x,t)}\mathrm{d}x = 2N_0\sqrt{\frac{Dt}{\pi}} \tag{2.12}$$

For a limited source diffusion, an impulse of impurity of dose Q is assumed to be deposited on the wafer surface. The solution of Equation (2.10) under these boundary conditions, assuming a semi-infinite wafer with no loss of impurity, results in a Gaussian diffusion profile:

$$N_{(x,t)} = \frac{Q}{\sqrt{\pi Dt}}\exp\left[-\left(\frac{x}{2\sqrt{Dt}}\right)^2\right] \tag{2.13}$$

In this case, as time progresses, the impurity penetrates more deeply into the wafer, and the surface concentration falls so as to maintain a constant dose in the wafer.

Practical Diffusions

Most real diffusions follow a two-step procedure, where the dopant is applied to the wafer with a short, constant-source diffusion, then driven in with a limited source diffusion. The reason for this is that in trying to control the dose, a constant-source diffusion must be performed at the solid solubility limit of the impurity in the Si, which is of the order of 10^{20} for most dopants. If only a constant-source diffusion were done, this would result in only very high surface concentrations. Therefore, to achieve lower concentrations, a short constant-source diffusion is done first, achieving a controlled dose of impurities in a near-surface layer. This diffusion is known as the predeposition, or predep, step. The source is then removed and the dose is diffused into the wafer, simulating a limited-source diffusion in the subsequent drive-in step. In modern IC processing, the predep step has been replaced by ion implantation, providing more accurate control of the placement and dose of impurities in the wafer.

If the *Dt* product for the drive-in step is much greater than the *Dt* product for the predep step, the resulting profile is very close to Gaussian. In this case the dose can be calculated by Equation (2.12) for the predep time and diffusion coefficient. In the case of ion implantation, the dose is determined by the implant conditions in a very thin layer and the resulting drive-in ensures a near Gaussian profile. This dose is then used in the limited source Equation (2.13) to describe the final profile based on the time and diffusion coefficient for the drive-in. If these (*Dt*) criteria are not met, an integral solution exists for the evaluation of the resulting profiles (Ghandhi, 1968).

Further Profile Considerations

A wafer typically goes through many temperature cycles during fabrication, potentially altering the impurity profile. The effects of many thermal cycles that take place at different times and temperatures are accounted for by calculating a total *Dt* product for the diffusion that is equal to the sum of the individual process *Dt* products:

$$(Dt)_{\text{tot}} = \sum_i D_i t_i \tag{2.14}$$

Here, D_i and t_i are the diffusion coefficient and time, respectively, that pertain to the *i*th process step.

Many diffusions are used to form junctions by diffusing an impurity opposite in type to the substrate. At the metallurgical junction, x_j, the impurity diffusion profile, has the same concentration as the substrate. For a junction with a surface concentration N_0 and substrate doping N_B, the metallurgical junction for a Gaussian profile is

$$x_j = 2\sqrt{Dt \ln\left(\frac{N_0}{N_B}\right)} \tag{2.15}$$

and for a complementary error function profile is

$$x_j = 2\sqrt{Dt}\; erfc^{-1}\left(\frac{N_B}{N_0}\right) \tag{2.16}$$

Modern processes use ion implantation as the predep and are required to produce shallow junctions with the drive-in diffusion. In doing this, it was discovered that impurities diffused considerably more than predicted by simple diffusion theory using the *Dt* product calculated for the annealing process. It was found that the ion implant introduced damage to the silicon lattice (cf. next section), enhancing the diffusion coefficient by a factor of 5 to 10 times. The effect is transient; the annealing removes the damage and thus the enhancement in diffusion coefficient. Nevertheless, the effect can significantly modify the profile, especially in the tail region, causing significant change in the junction depth. Because of the transient nature of the process, it has become known as transient enhanced diffusion (TED) (Jaeger, 2002).

So far, we have considered only vertical diffusion. In practical IC fabrication, usually only small regions are affected by the diffusion, by using an oxide mask and making a cut in it where specific diffusion is to occur, but we should also be concerned with lateral diffusion of the dopant so as not to affect adjacent devices. Two-dimensional numerical solutions exist for solving this problem (Jaeger, 1988); however, a useful rule of thumb is that the lateral junction is $y_j = 0.8x_j$.

Another parameter of interest is the sheet resistance of the diffused layer. This has been numerically evaluated for various profiles and presented as general-purpose graphs known as Irvin's curves. For a given profile type, such as n-type Gaussian, Irvin's curves plot surface dopant concentration vs. the product of sheet resistance and junction depth, with substrate doping as a parameter. Thus, given a calculated diffusion profile, one could estimate the sheet resistivity for the diffused layer. Alternatively, given the measured junction depth and sheet resistance, one could estimate the surface concentration for a given profile and substrate doping. Most processing books (e.g., Jaeger, 1988) contain Irvin's curves.

Ion Implantation

Diffusion places severe limits on device design, such as hard-to-control, low-dose diffusions, no tailored profiles, and appreciable lateral diffusion at mask edges. Ion implantation overcomes all these drawbacks and is an alternative approach to diffusion used in virtually all production-doping applications today. Although many different elements can be implanted, IC manufacture is primarily interested in the conventional silicon shallow dopants: B, P, As, and Sb.

Ion Implant Technology

A schematic drawing of an ion implanter is shown in Figure 2.3. The ion source operates at relatively high voltage ($\approx$ 20 to 25 kV) and for conventional dopants is usually a gaseous type that extracts the ions from a plasma. The ions are mass-separated with a 90° analyzer magnet that directs the selected species through a resolving aperture focused and accelerated to the desired implant energy. At the other end of the implanter is the target chamber, where the wafer is placed in the beam path. The beam line following the final accelerator and the target chamber is held at or near ground potential for safety reasons. After the final acceleration, the beam is bent slightly off-axis to trap neutrals and is scanned asynchronously over the wafer in the X and Y

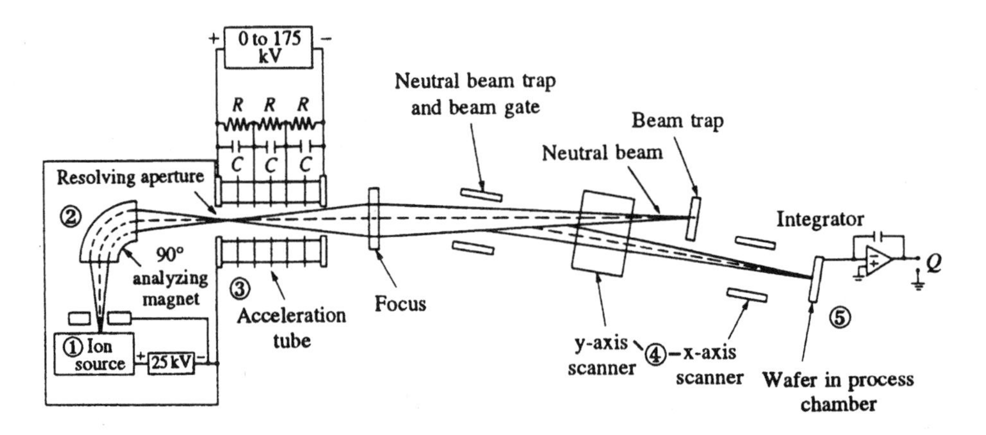

FIGURE 2.3 Schematic drawing of an ion implanter. (*Source:* R.C. Jaeger, *Introduction to Microelectronic Fabrication*, Vol. 5 in the Modular Series on Solid State Devices, G.W. Neudek and R.F. Pierret, Eds., Reading, MA: Addison-Wesley, 1988, p. 90. With permission.)

directions to maintain dose uniformity. This is often accompanied by rotation and sometimes also by translation of the target wafer.

The implant parameters of interest are the ion species, implant energy, and dose. The ion species can consist of singly ionized elements, doubly ionized elements, or ionized molecules. The molecular species can form shallow junctions with light ions, such as B, using BF_2^+. The beam energy is

$$E = nqV \tag{2.17}$$

where n represents the ionization state (1 for singly and 2 for doubly ionized species), q is the electronic charge, and V is the total acceleration potential (source + acceleration tube) seen by the beam. The dose, Q, from the implanter is:

$$Q = \int_0^{t_1} \frac{I}{nqA} \mathrm{d}t \tag{2.18}$$

where I is the beam current in amperes, A is the wafer area in cm^2, t_1 is the implant time in sec, and n is the ionization state.

Ion Implant Profiles

Ions impinge on the surface of the wafer at a certain energy and give up that energy in a series of electronic and nuclear interactions with the target atoms before coming to rest. As a result, the ions do not travel in a straight line but follow a zigzag path resulting in a statistical distribution of final placement. To first order, the ion distribution can be described with a Gaussian distribution:

$$N_x = N_\mathrm{p} \exp\left[-\frac{(x - R_\mathrm{p})^2}{2(\Delta R_\mathrm{p})^2} \right] \tag{2.19}$$

R_p is the projected range, which is the average depth of an implanted ion. The peak concentration, N_p, occurs at R_p and the ions are distributed about the peak with a standard deviation ΔR_p known as the straggle. Curves for projected range and straggle, taken from Lindhard–Scharff–Schiott (LSS) theory (Gibbons et al., 1975), are shown in Figure 2.4 and Figure 2.5, respectively, for the conventional dopants.

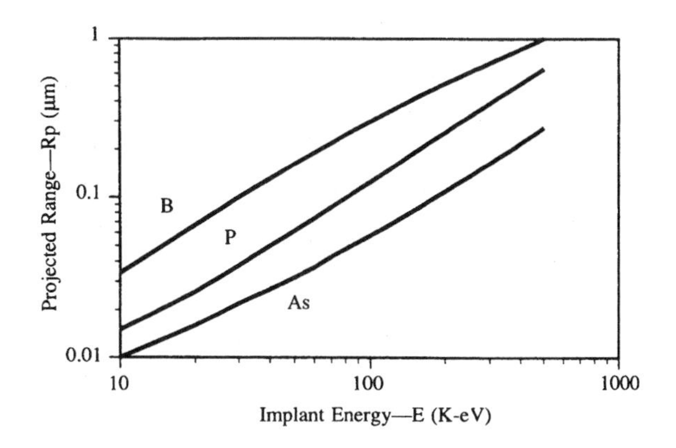

FIGURE 2.4 Projected range for B, P, and As based on LSS calculations.

The area under the implanted distribution represents the dose as given by

$$Q = \int_0^{\infty} N_{(x)} \mathrm{d}x = \sqrt{2\pi} N_\mathrm{p} \Delta R_\mathrm{p} \tag{2.20}$$

which can be related to the implant conditions by Equation (2.18). Implant doses can range from 10^{10} to 10^{18} per cm^2 and can be controlled within a few percentage points.

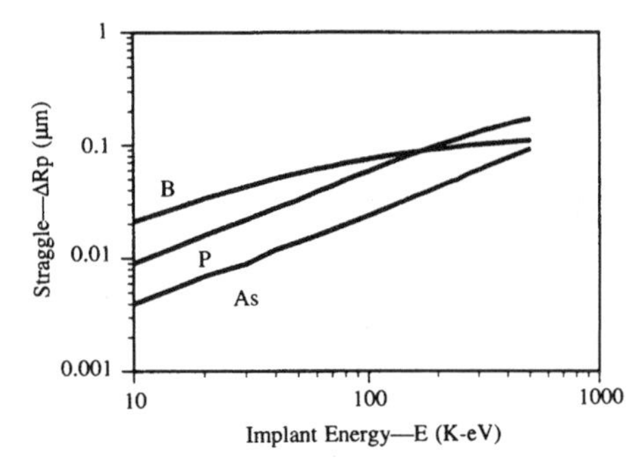

FIGURE 2.5 Implant straggle for B, P, and As based on LSS calculations.

The mathematical representation of the implant profile just presented pertains to an amorphous substrate. Silicon wafers are crystalline and therefore present the opportunity for the ions to travel much deeper into the substrate by a process known as channeling. The regular arrangement of atoms in the crystalline lattice leaves large amounts of open space that appear as channels in the bulk when viewed from the major orientation directions, i.e., <110>, <100>, and <111>. Practical implants are usually done through a thin oxide, with the wafers tilted off-normal by a small angle (typically 7°) and rotated by 30° to make the surface atoms appear more randomly oriented. Implants with these conditions agree well with the projected range curves of Figure 2.4, indicating that the wafers do appear amorphous.

Actual implant profiles deviate from the simple Gaussian profiles described in the previous paragraphs. Light ions tend to backscatter from target atoms and fill in the distribution on the surface side of the peak. Heavy atoms tend to forward-scatter from the target atoms and fill in the profile on the substrate side of the peak. This behavior has been modeled with distributions such as the Pearson Type-IV distribution (Jaeger, 1988). However, for implant energies below 200 keV and first-order calculations, the Gaussian model will more than suffice.

Masking and Junction Formation

It is usually desirable to implant species only in selected areas of the wafer to alter or create device properties. Hence, the implant must be masked. This is done by putting a thick layer of silicon dioxide, silicon nitride, or photoresist on the wafer and patterning and opening the layer where the implant is desired. To prevent significant alteration of the substrate doping in the mask regions. the implant concentration at the Si/mask interface, X_0, must be less than 1/10 of the substrate doping, N_B. Under these conditions, Equation (2.19) can be solved for the required mask thickness as

$$X_0 = R_\mathrm{p} + \Delta R_\mathrm{p} \sqrt{2 \ln\left(\frac{10 N_\mathrm{p}}{N_\mathrm{B}}\right)} \tag{2.21}$$

This implies that the range and straggle are known for the mask material being used. These are available in the literature (Gibbons et al., 1975) but can also be reasonably approximated by making the calculations for Si. SiO_2 is assumed to have the same stopping power as Si and thus would have the same mask thickness. Silicon nitride has more stopping power than SiO_2 and therefore requires only 85% of calculated mask thickness, whereas photoresist is less effective for stopping the ions and requires 1.8 times the equivalent Si thickness.

Analogous to the mask calculations is junction formation. Here, the metallurgical junction, x_j, occurs when the opposite-type implanted dopant profile is equal to the substrate doping, N_B. Solving Equation (2.19) for these conditions gives the junction depth, X_J, as

$$X_\mathrm{j} = R_\mathrm{p} \pm \Delta R_\mathrm{p} \sqrt{2 \ln\left(\frac{N_\mathrm{p}}{N_\mathrm{B}}\right)} \tag{2.22}$$

Note that both roots may be applicable, depending on the depth of the implant.

Lattice Damage and Annealing

During ion implantation, the impinging atoms can displace Si atoms in the lattice, causing damage to the crystal. For high implant doses, the damage can be severe enough to make the implanted region amorphous. Typically, light ions, or light doses of heavy ions, will cause primary crystalline defects (interstitials and vacancies), whereas medium-to-heavy doses of heavy ions will cause amorphous layers. Implant damage can be removed by annealing the wafers in an inert gas at 800 to 1000°C for approximately 30 min. Annealing cycles at high temperatures can cause appreciable diffusion, a consideration especially in the newer technologies where shallow junctions are required. Rapid thermal annealing (RTA) can be successfully applied to prevent undesirable diffusion in these cases (Campbell, 2001). Annealing cycles of from 950 to 1050°C, ranging from a few minutes down to only a few seconds, have been used to keep the Dt product manageable in modern proceses.

Deposition

During IC fabrication, thin films of dielectrics (SiO_2, Si_3N_4, etc.), polysilicon, and metal conductors are deposited on the wafer surface to form devices and circuits. The techniques used to form these thin films are physical vapor deposition (PVD), chemical vapor deposition (CVD), and epitaxy, a special case of CVD.

Physical Vapor Deposition

Vacuum evaporation and sputtering are the two methods of physical vapor deposition used in fabricating integrated circuits. Both processes are carried out in a vacuum to prevent substrate contamination and to provide a reasonable mean free path for the deposited material. In the early days of integrated circuits, aluminum was used exclusively for IC metallization and evaporation was used for its deposition. As IC technology matured, the need for metal alloys, alternative metals, and various insulating thin films stimulated the development and acceptance of sputter deposition as the PVD method of choice.

When the temperature is raised high enough to melt a solid, some atoms have enough internal energy to break the surface and escape into the surrounding atmosphere. These evaporated atoms strike the wafer and condense into a thin film. Typical film thicknesses used in the IC industry are in the few thousand angstroms to the 1 μm range. Heat is provided by resistance heating, electron beam heating, or rf inductive heating. Discussions of these techniques and their historical significance can be found in most processing books (e.g., Sze, 1983; Wolf and Tauber, 1986; Jaeger, 2002). The most commonly used system employs a focused electron beam scanned over a crucible of metal, as illustrated in Figure 2.6(a). Contamination levels can be quite low because only electrons come into contact with the melted metal. A high-intensity, electron beam, typically 15 keV, bent through a 270° angle to shield the wafers from filament contamination, provides the heating for evaporation of the metal. The wafers are mounted above the source on a planetary substrate holder that rotates around the source during the deposition to ensure uniform step coverage. For the planetary substrate holder, shown schematically in Figure 2.6(b), the growth rate is independent of substrate position. The relatively large size of the crucible provides an ample supply of deposition source material. The deposition rate is controlled by changing the current and energy of the electron beam. Dual beams with dual targets can be used to co-evaporate composite films. Device degradation due to x-ray radiation, generated by the electron-beam system, is of great concern in MOS processing. Because of this, sputtering has replaced e-beam evaporation in many process lines.

Sputtering is achieved by bombarding a target surface with energetic ions that dislodge target atoms from the surface by direct momentum transfer. Under the proper conditions, the sputtered atoms are transported to the wafer surface, where they deposit to form a thin film. Sputter deposition takes place in a chamber that is evacuated and then backfilled with an inert gas at roughly 10 mtorr pressure. A glow discharge within the gas between two electrodes, one of which is the target, creates a plasma that provides the source of ions for the sputter process. Metals can be sputtered in a simple, dc parallel plate reactor with the most negative electrode being the target and the most positive electrode (usually ground) being the substrate holder. If the dc voltage is replaced by an rf voltage, insulators as well as conductors can be sputter-deposited. Magnetron systems

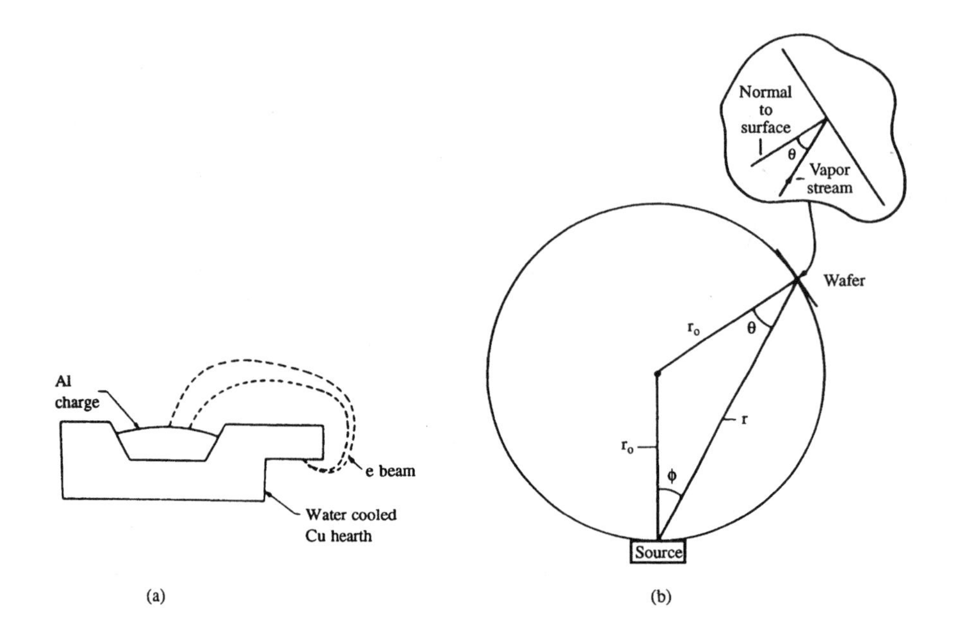

FIGURE 2.6 (a) Electron beam evaporation source. (b) Geometry for a planetary holder. (*Source:* R.C. Jaeger, *Introduction to Microelectronic Fabrication*, Vol. 5 in the Modular Series on Solid State Devices, G.W. Neudek and R.F. Pierret, Eds., Reading, MA: Addison-Wesley, 1988, pp. 110, 112. With permission.)

incorporate a magnetic field that enhances the efficiency of the discharge by trapping secondary electrons in the plasma and lowers substrate heating by reducing the energy of electrons reaching the wafer. Circular magnetrons (S-guns) virtually eliminate substrate heating by electron bombardment due to a ring-shaped cathode/anode combination, where the substrate is a non-participating system electrode (Sze, 1983). Besides insulators and conductors, sputtering can be used to deposit alloy films with the same composition as an alloy target.

Chemical Vapor Deposition

Chemical vapor deposition (CVD) is a method of forming thin films on a substrate in which energy is supplied for a gas-phase reaction. The energy can be supplied by heat, plasma excitation, or optical excitation. Since the reaction can take place close to the substrate, CVD can be performed at atmospheric pressure (i.e., low mean free path) or at low pressure. Relatively high temperatures can be used, resulting in excellent conformal step coverage; or relatively low temperatures can be used to passivate a low melting temperature film such as aluminum. Substrates can be amorphous or single crystalline. The epitaxial growth of Si is simply CVD on a single crystalline substrate, resulting in a single crystal layer. Generally, unless the qualifying adjective, epitaxial, is used, the depositions are assumed to result in amorphous or polycrystalline films. Typically all CVD depositions show a growth rate versus temperature dependence as illustrated in Figure 2.7. At low temperatures, the deposition is limited by surface reaction rates and shows an Arrhenius-type behavior. At high temperatures the rate is dominated by mass transfer of the

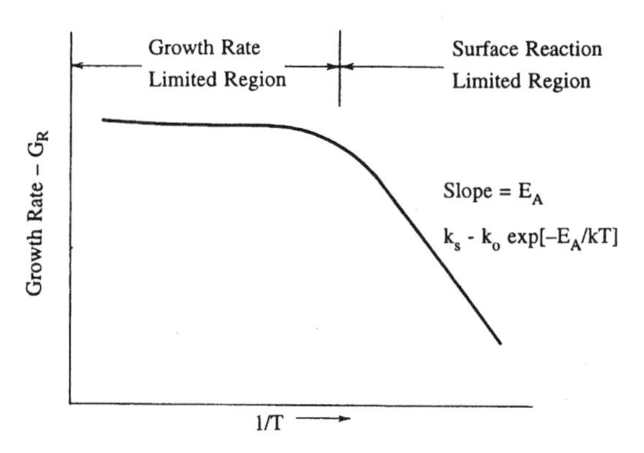

FIGURE 2.7 Typical chemical vapor deposition growth rate versus reciprocal temperature characteristics.

reactant and the growth rate is essentially temperature independent. The transition temperature, where the reaction switches from reaction rate to mass-flow dominated, is dependent on other factors such as pressure, reactant species, and energy source.

Several types of films can be deposited by CVD. Insulators and dielectrics (such as SiO_2 and Si_3N_4), doped glasses (PSG, BPSG), and nonstoichiometric dielectrics as well as semiconductors (such as Si, Ge, GaAs, and GaP) can be deposited. Conductors of pure metals, such as Al, Ni, Au, Pt, Ti, etc., as well as silicides such as WSi_2 and $MoSi_2$ can also be deposited.

CVD systems come in a multitude of designs and configurations and are selected for their compatibility with the method, energy source, and temperature range being used. Reaction chambers can be quartz, similar to diffusion or oxidation furnaces, or stainless steel. Wafer holders also depend on the type of reaction and can be graphite, quartz, or stainless steel. Cold-wall systems employ direct heating of the substrate or wafer holder by induction or radiation. As such, the reaction takes place right at the wafer surface and is usually cleaner because the film does not build up on the chamber walls. In a hot-wall deposition system, the reaction takes place in the gas stream and the reaction product is deposited on every surface in the system, including the walls. Such unwanted depositions build up and flake off these surfaces in time. Without proper cleaning and maintenance procedures, they become a source of contamination.

Atmospheric-pressure deposition systems were first used in the deposition of epitaxial Si in bipolar processes (such as buried layer formation). Early systems were horizontal, rf induction-heated systems using graphite substrate holders. Atmospheric pressure systems rely on flow dynamics in the chamber to produce uniform films. Since the reactant species is being depleted from the gas stream, it is imperative that the system design (flow pattern and rate) ensures that all wafers receive the same amount of deposit. Because of this, most atmospheric depositions are carried out using a horizontal or near-horizontal wafer holder.

Low-pressure CVD depositions are performed at pressures in the 0.5 to 1 torr range and most often in a horizontal hot-wall reactor. Wafers are held vertically and the system is operated in the reaction rate limited regime. A multi-zone furnace (typically three) is used to allow the temperature to be increased along the wafer holder to compensate for gas depletion by deposition along the flow path. LPCVD is typically used to deposit SiO_2 ($\approx 925°C$), Si_3N_4 ($\approx 850°C$), and polycrystalline Si ($\approx 630°C$). Because of the higher temperatures, excellent conformal step coverage can be obtained.

Plasma-enhanced CVD uses a plasma to supply the required reaction energy. Typically, rf energy is used to produce a glow discharge so that the gas constituents are in a highly reactive state. Because of the ability of the plasma to impart high energy to the reaction at low temperatures, these depositions maintain many of the excellent features of LPCVD, such as step coverage. They have low-temperature attributes, such as reduced wafer warpage, less impurity diffusion, and less film stress. Plasma-enhanced deposition results in nonstoichiometric, highly hydrogenated films such as SiO–H, SiN–H, and amorphous Si–H.

Lithography and Pattern Transfer

In the production of integrated circuits, thin films are fabricated in the Si wafer and either grown or deposited on the surface. Each of these layers has a functionality as either an active part of a device, a barrier or mask, or an inter- or intra-layer isolation. To perform its intended function, each of these layers must be located in specific regions on the wafer surface. This is accomplished either by patterning and etching a thin film to serve as a mask to form the desired function by a blanket process or by forming the desired layer and then patterning and etching it to provide the device function in local areas.

Wafer Patterning

Lithography is a transfer process where the pattern on a mask is replicated in a radiation-sensitive layer on the wafer surface. Typically, this has been accomplished with UV light as the radiation source and photoresist, or resist in conventional terminology, which is a UV-sensitive polymer as the mask layer. The wafers, along with the layer to be patterned, are cleaned and prebaked (400 to 800°C for 20 to 30 min) to drive off moisture and promote resist adhesion. Many processes also add an adhesion promoter such as hexamethyldisilazane (HMDS) at this point, which removes unwanted surface radicals that prevent adhesion. Then, a few drops of resist are deposited on a wafer that is spinning at a slow rate to produce a uniform coating. The spin speed is increased to enhance drying. The wafer with resist is softbaked at 80 to 90°C for 10 to 30 min to drive off the remaining solvents. The wafers are then put in an exposure system, and the mask pattern to be transferred is aligned to any existing wafer patterns. Present-day exposure systems are step and repeat cameras, where the mask is a 5× enlargement of a single chip pattern. Earlier systems used a 1× mask for the whole wafer and were of the contact, proximity or 1× projection variety (Anner, 1990). The resist is exposed through the mask to UV radiation that changes its structure, depending on whether the resist type is positive or negative. Negative resist becomes polymerized (or cross-linked) in areas exposed to radiation, whereas in a positive resist the polymer bonds are scissioned upon exposure. The resist is not affected in regions where the mask is opaque. After full wafer exposure, the resist is developed such that the unpolymerized regions are selectively dissolved in an appropriate solvent. The polymerized portion of the resist remains intact on the wafer surface, replicating the mask's opaque features in a positive resist, and the opposite for a negative resist.

As the minimum feature size approaches the wavelength of light used in optical exposure systems, resolution is lost due to diffraction limitations. Development alternatives are continually being developed to overcome these limitations and extend the life of optical lithography. The simplest method is to use a shorter wavelength. Initially, IC lithography was done with a Hg-rare gas discharge lamp as the source using the UV g-line ($\lambda = 436$ nm) emission. In submicron technology, the i-line ($\lambda = 365$ nm) was brought into use. At the deep submicron node, 0.25 μm, the krypton fluoride (KrF) excimer laser deep UV (DUV) source ($\lambda = 248$ nm) was introduced. The argon fluoride (ArF) source ($\lambda = 193$ nm) was introduced into mainstream processing around the 100 nm technology node and the F2 excimer laser ($\lambda = 157$ nm) is projected for the 70 nm tecnology node (Braun, 2003). With the shift to DUV sources, the resist also had to change as the conventional Novolac resin, DNQ, resists are opaque at these wavelengths. DUV resists are based on the chemically amplified resists that provide further contrast and resolution enhancements to the technology (Plummer et al., 2002). Further extensions of this technique introduce optical proximity correction and optical phase shifting in the photomask itself, extending the diffraction limit even further (Campbell, 2001; Plummer et al., 2002; Mauer, 2003). These techniques are eventually limited, somewhere in the range of the 50 nm technology node (Braun, 2003), and new-generation lithography (NGL) techniques, such as electron beam projection lithography (EPL), or x-ray systems will have to replace the optical systems for the finest dimension patterning. These techniques are capable of wavelengths much shorter than conventional deep UV or extreme ultraviolet lithography (EUV) and thus have much higher resolution capabilities. They are analogous in operation to optical systems but use resists specifically tailored for sensitivity in their appropriate wavelength range. Mix-and-match lithography for different feature sizes is currently used by some companies and is expected to become more prevalent in the future, to keep thoughput up and equipment costs down as device dimensions continue to scale (Hand, 2003).

Pattern Transfer

After the resist pattern is formed, it is transferred to the surface layer of the wafer. Sometimes, this is an invisible transfer, such as for ion implantation, but more often it involves a physical transfer of the pattern by etching the surface layer, using the resist as a mask. This results in the desired structure or produces a more etch-resistant mask for further pattern transfer operations.

Historically, the most common etch processes used wet chemicals. During wet etching, wafers with resist (or a resist transferred mask) are immersed in a temperature-controlled etchant for a fixed time period. The etch rate is dependent on the strength of the etchant, temperature, and material being etched. Such chemical etches are isotopic, which means the vertical and lateral etch rates are the same. Thus, the thicker the layer being etched, the greater the undercutting of the mask pattern. Most wet etches are stopped by an underlying etch-stop layer that is impervious to the etchant used to remove the top layer. To ensure that the layer is totally removed, an over-etch is allowed, exacerbating the undercutting.

Modern high-density, small-feature processes require anisotropic etch processes; this requirement has driven the development of dry etching techniques. Plasma etching is a dry-etch technique that uses an rf plasma to generate chemically active etchants that form volatile etch species with the substrate. Typically, chlorine or fluorine compounds, most notably CCl_4 and CF_4, have been tailored for etching polysilicon, SiO_2,

Si_3N_4, and metals. The etch rate can be significantly enhanced by adding 5 to 10% O_2 to the etch gas. However, this increases erosion of resist masks.

Early plasma etch systems were barrel reactors that used a perforated metal cylinder to confine the plasma in a region outside the wafers. In such reactors, the etch species are electrically neutral, so the reaction is entirely chemical and just as isotropic as wet chemical etch. A similar result occurs for a planar parallel plate rf reactor where the wafer is placed on the grounded electrode. However, some anisotropic etching is achieved in this configuration because ions can reach the wafer. This is further enhanced in reactive ion etching (RIE) where the wafer is placed on the rf electrode in a planar parallel plate reactor. Here, the ions experience a considerable acceleration to the wafer by the dc potential developed between plasma and cathode, resulting in anisotropic etching. The etch processes require significant characterization and development through optimization of pressure, gas flow rate, gas mixture, and power to produce the desired etch rate, anisotropy, selectivity, uniformity, and resist erosion. Nevertheless, RIE has been the workhorse of etch processes in modern ULSI processes. A less-popular version uses a beam of reactive species for the etch process and is called reactive ion beam etching (RIBE). A new type of plasma etch system that is becoming more popular is the high density plasma (HDP) etch system (Plummer et al., 2002). These systems produce a plasma density much higher than standard RIE systems (10^{11} to 10^{12} ions/cm^3), without a large sheath bias by using a non-capacitively coupled source such as electron cyclotron resonance (ECR) or inductively coupled plasma (ICP) sources. This allows lower pressures to be used in the system while still achieving high ion fluxes and etch rates. A lower pressure means fewer gas-phase collisions in the sheath. Thus, a more-directed etching is provided for more anisotropic etching, as needed with the small dimensions in modern processes.

Finally, it should be noted that near-total anisotropic dry etching can be achieved with ion etching. This is done with an inactive species (e.g., Ar ions) either in a beam or with a parallel plate sputtering system with the wafer on the rf electrode. This results in an etch process that is entirely physical through momentum transfer. The etch rate is primarily controlled by the sputtering efficiency for various materials and thus does not differ significantly from material to material. Hence, such processes suffer from poor selectivity. Resist or mask erosion is also a problem with these techniques. Because of these limitations, virtually the only application of pure ion etching in VLSI/ULSI is for sputter-cleaning of wafers before deposition.

New Processes

Rapid thermal processing (RTP) and chemical mechanical processing (CMP) are two new unit-step processes that have come into existence with the evolution to smaller and smaller dimensions in process technology and are worthy of note here.

As devices shrink in lateral dimensions, an attendant shrinkage in vertical dimensions is mandated to maintain proper device performance. This implies minimization of the diffusion and redistribution of dopants in thermal processes and thinner films (such as gate oxides, nitrides, and silicides). Low-temperature processing might seem to be a viable approach to solve these problems, as in the case of low-pressure oxidations. However, some processes, such as implant annealing, are not effective at low temperatures. Some dopants cannot be fully activated and certain types of implant damage cannot be annealed out unless high temperatures are used. The only possibility is to use high temperatures for short times. Standard furnace processing is not suited to short time-processing steps due to wafer warpage, which results from temperature gradients because the wafers are heated from the edges inward. RTP describes a family of single-wafer hot processes developed to minimize the thermal budget by reducing the time at a particular temperature in addition to, or instead of, reducing the temperature.

Rapid thermal annealing (RTA) systems were the first of the RTP processes developed to achieve the desired implant activation and damage annealing. High-intensity lamps are used to rapidly heat the wafer to the desired annealing temperature (e.g., 950 to 1050°C) in a very short time (minutes down to a few seconds). Extremely fast temperature ramp-up and ramp-down rates of 50°C/sec or more can be achieved. Using RTA, the effective *Dt* product can be kept very small. RTA represents only one form of RTP. Similar systems are used to grow very thin gate oxides in a process known as rapid thermal oxidation (RTO) and nitride layers by rapid thermal nitridation (RTN). RTP processes can be used for chemical vapor deposition (RTCVD) in the formation of silicides and for epitaxial growth.

CMP was introduced to fabrication processes in the early 1990s to achieve highly planar topologies. This has allowed optical lithography to go far beyond original expectations through the use of higher numerical aperture lenses, with their subsequent loss in depth of focus. In a CMP system, the wafer is mounted on a carrier and is brought into contact with a polishing pad mounted on a rotating platen. A liquid slurry is continuously dispensed onto the surface of the polishing pad. A combination of the vertical force between the wafer and the pad along with the chemical action of the slurry, permits a process capable of polishing the surface of the wafer to a highly planar state.

Defining Terms

Chemical mechanical polishing: A polishing mechanism using both mechanical abrasion and chemical interactions to polish a wafer surface to a highly planar state.

Chemical vapor deposition: A process in which insulating or conducting films are deposited on a substrate by use of reactant gases and an energy source, producing a gas-phase chemical reaction. The energy source may be thermal, optical, or plasma in nature.

Deposition: An operation in which a film is placed on a wafer surface, usually without a chemical reaction with the underlying layer.

Diffusion: A high-temperature process in which impurities on or in a wafer are redistributed within the silicon. If the impurities are desired dopants, this technique is often used to form specific device structures. If the impurities are undesired contaminants, diffusion often results in undesired device degradation.

Dry etching: Processes that use gas-phase reactants, inert or active ionic species, or a mixture of these to remove unprotected substrate layers by chemical processes, physical processes, or a mixture of these.

Ion implantation: A high-energy process, usually greater than 10 keV, that injects ionized species into a semiconductor substrate. It is often done to introduce dopants for device fabrication into silicon with boron, phosphorus, or arsenic ions.

Lithography: A patterning process in which a mask pattern is transferred by a radiation source to a radiation-sensitive coating that covers the substrate.

Physical vapor deposition: A process in which a conductive or insulating film is deposited on a wafer surface without the assistance of a chemical reaction. Examples are vacuum evaporation and sputtering.

Rapid thermal processing: A family of single-wafer hot processes developed to minimize the thermal budget of various processes by reducing the time at temperature in addition to, or instead of, reducing the temperature.

Thermal oxidation of silicon: A high-temperature chemical reaction, typically greater than 800°C, in which the silicon of the wafer surface reacts with oxygen or water vapor to form silicon dioxide.

Wet etching: A process that uses liquid chemical reactions with unprotected regions of a wafer to remove specific layers of the substrate.

References

G.E. Anner, *Planar Processing Primer*, New York: Van Nostrand Reinhold, 1990.

A.E. Braun, "Resist technology concentrates on 248 nm," *Semicond. Int.*, vol. 26. no. 2, p. 58, 2003.

S.A. Campbell, *The Science and Engineering of Microelectronic Fabrication*, New York: Oxford University Press, 2001.

C.Y. Chang and S.M. Sze, *ULSI Technology*, New York: McGraw-Hill, 1995.

B. Ciciani, *Manufacturing Yield of VLSI*, Piscataway, NJ: IEEE Press, 1995.

S.K. Ghandhi, *VLSI Fabrication Principles — Silicon and Gallium Arsenide*, New York: Wiley, 1982.

J.F. Gibbons, W.S. Johnson, and S.M. Mylroie, *Projected Range Statistics, Semiconductors and Related Materials*, Vol. 2, 2nd ed., Stroudsburg, PA: Dowden, Hutchinson, & Ross, 1975.

A. Hand, "Mix and match lithography tackles tighter requirements," *Semicond. Int.*, vol. 26. no.2, p. 52, 2003.
IEEE Symposium on Semiconductor Manufacturing, Piscataway, NJ: IEEE Press, 1995.
R.C. Jaeger, *Introduction to Microelectronic Fabrication*, Vol. 5 in the Modular Series on Solid State Devices, G.W. Neudek and R.F. Pierret, Eds., Reading, MA: Addison-Wesley, 1988.
R.C. Jaeger, *Introduction to Microelectronic Fabrication*, Vol. 5 in the Modular Series on Solid State Devices, G.W. Neudek and R.F. Pierret, Eds., Upper Saddle River, NJ: Prentice-Hall, 2002.
W. Maurer, "Application of Phase-Shift Masks," *Semicond. Int.*, Vol. 26. no. 10, p. 69, 2003.
J.D. Plummer, M.D. Deal, and P.B. Griffin, *Silicon VLSI Technology, Fundamentals, Practice, and Modeling*, Upper Saddle River, NJ: Prentice-Hall, 2002.
S.M. Sze, Ed., *VLSI Technology*, 2nd ed., New York: McGraw-Hill, 1988.
P. Van Zant, *Microchip Fabrication*, 3rd ed., New York: McGraw-Hill, 1996.
S. Wolf and R.N. Tauber, *Silicon Processing for the VLSI ERA*, Vol. 1, *Process Technology*, Sunset Beach, CA: Lattice Press, 1986.

Further Information

The references in this section have been chosen to provide more detail than is possible in the limited space for this section. Specifically, the referenced processing textbooks provide detailed discussion of all the presented unit-step processes. Further details and more recent process developments can be found in several magazines or journals, such as *Semiconductor International, Solid State Technology, IEEE Transactions on Semiconductor Manufacturing, IEEE Transactions on Electron Devices, IEEE Electron Device Letters, Journal of Applied Physics*, and the *Journal of the Electrochemical Society.*

2.2 Testing[1]

Wayne Needham

The function of testing a semiconductor device is twofold. First is design debug, to understand the failing section of the device, identify areas for changes, and verify correct modes of operation. The second major area is to simply separate good devices from bad devices in a production testing environment. Data collection and information analysis are equally important in both types of test, but for different reasons. The first case is obvious; debug requires understanding of the part. In the second case, data collected from a test program may be used for **yield** enhancement. This is done by finding marginal areas of the device or in the fabrication process and then making improvements to raise yields and therefore lower cost.

In this section, we will look at methods of testing that can be used for data collection, analysis, and debug of a new device. No discussion of testing would be useful if the test strategy is not thought out clearly before design implementation. Design for test (access for control and observation) is a requirement for successful debug and testing of any semiconductor device.

The basis for all testing of complex integrated circuits is a comparison of known good patterns to the response of a device under test (DUT). The simulation of the devices is done with input stimuli, and those same input stimuli (**vectors**) are presented on the DUT. Comparisons are made cycle by cycle with an option to ignore certain pins, times, or patterns. If the device response and the expected response are not in agreement, the devices are usually considered defective.

This section will cover common techniques for testing. Details of generation of test programs, simulation of devices, and tester restrictions are not covered here.

[1]Portions reproduced from W.M. Needham, *Designer's Guid to Testable ASIC Devices*, New York: Van Nostrand Reinhold, 1991. With permission.

Built-In Self-Test

Self-testing (**built-in self-test** or BIST) is essentially the implementation of logic built into the circuitry to do testing without the use of the **tester** for pattern generation and comparison purposes. A tester is still needed to categorize failures and to separate good from bad units. In this case, the test system supplies clocks to the device and determines pass/fail from the outputs of the device. The sequential elements are run with a known data pattern, and a signature is generated. The signature can be a simple go or no-go signal presented on one pin of the part, or the signal may be a polynomial generated during testing. This polynomial has some significance to actual states of the part. Figure 2.8 shows a typical self-testing technique implemented on a large block of random logic, inside a device. The number of unique inputs to the logic block should be limited to 20 bits so the total vectors are less than 1 million. This keeps test time to less than 1 sec in most cases.

Self-test capability can be implemented on virtually any size block. It is important to understand the tradeoffs between the extra logic added to implement self-testing and the inherent logic of the block to be tested. For instance, adding self-testing capability to a RAM requires adding counters and read, write, and multiplexor circuitry to allow access to the part. The access is needed not only by the self-test circuitry, but by the circuitry that would normally access the RAM.

When implementing self-testing on blocks such as RAMs and ROMs, it is worthwhile to note the typical failure mode of semiconductor devices. Single-bit defects can easily be detected using self-testing techniques. Single-point defects in the manufacturing process can show up as a single transistor failure in a RAM or ROM, or they may be somewhat more complex. If a single-point defect happens to be in the decoder section or in a row or column within the RAM, a full section of the device may be nonfunctional.

The problem with this failure mode is that RAMs and ROMs are typically laid out in a square or rectangular array. They are usually decoded in powers of 2, such as a 32 * 64 or a 256 * 512 array. If the self-testing circuitry is an 8-bit-wide counter or linear feedback shift register, there may be problems. There are 256 possible combinations within the states of the counter, and this may be a multiple of the row or columns. Notice that 256 possible rows or multiples of 256 rows in the array and 256 states in the counter make for a potential error-masking combination. If the right type of failure modes occur and a full row is nonfunctional, it may be masked. The implementation of the counter or shift register must be done with full-column or row failure modes in mind, or it can easily mask failures. This masking of the failure gives a false result that the device is passing the test.

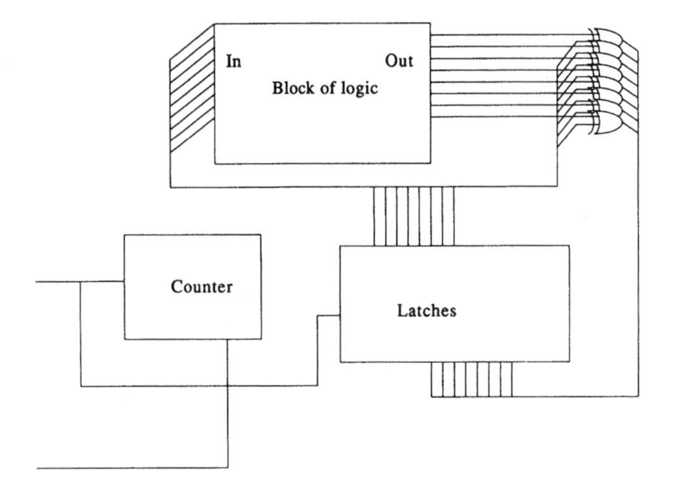

FIGURE 2.8 Example of self-test in circuit. (*Source:* W.M. Needham, *Designer's Guide to Testable ASIC Devices*, New York: Van Nostrand Reinhold, 1991, p. 88. With permission.)

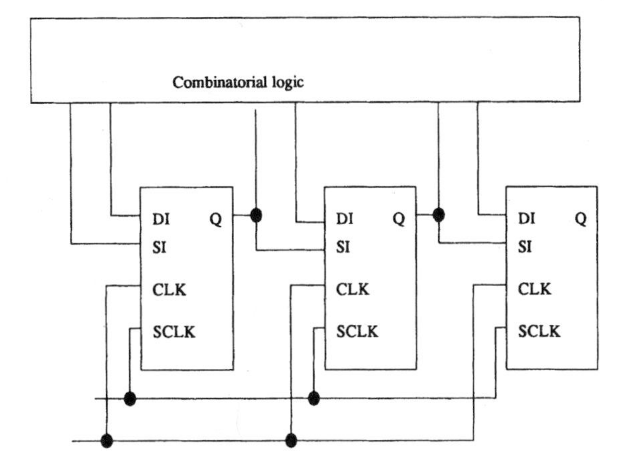

FIGURE 2.9 Scan test implementation. (*Source:* W.M. Needham, *Designer's Guide to Testable ASIC Devices*, New York: Van Nostrand Reinhold, 1991, p. 94. With permission.)

Scan

Scan is a test technique that ties together some or all the latches in the part to form a path through the part to shift data. Figure 2.9 shows the implementation of a scan latch in a circuit. Note that the latch has a dual mode of operation. In the normal mode, the latches act like normal flip-flops for data storage. In the scan mode, the latches act like a shift register connecting one element to another. This is a basic implementation of a shift register or scan latch in a block of logic. Data can be shifted in via the shift pin or can be clocked in from the data pin. Clock line is the normal system clock and SCLK is the shift clock for scan operations.

Data for testing are shifted in on the serial data in pins of the device. Patterns for the exercise of the combinatorial logic could be generated by truth table or by random generation of patterns. These patterns are then clocked a single time to store the results of the combinatorial logic. The latches now contain the results of the combinatorial logic operations. Testing of the logic becomes quite easy, as the sequential depth into the part is of no significance to the designer. Once the patterns are latched, the same serial technique is used to shift them out for comparison purposes. At the time of outward shift, new patterns are shifted in. Figure 2.10 shows this pattern shift for a circuit using scan. This is the actual implementation of scan in a small group of logic, including the truth table associated with it and one state of testing.

Direct Access Testing

Direct access is a method whereby one gains access to a device logic block by bringing signals from the block to the outside world via multiplexors. Data are then forced into the block directly and the outputs are directly measured. This is one of the simplest methods to check devices for logic functionality. This particular method would supplement previous testing methods by allowing the user to impose data patterns directly on large blocks of logic. The same feature holds true for output observation. In the direct access scheme, access of a test mode would force certain logic blocks via multiplexors to have access to the outside pins of the part. One could then drive the data patterns to the input pins, compare output pins of the part, and measure the access, status, and logical functionality of the block directly.

Figure 2.11 shows implementation of direct access test techniques on a block of logic. During normal operation, the logic block B has inputs driven by the logic block A, and outputs are connected to the logic block C. In the test mode, the two multiplexors are switched so that the input and output pins of the device can control and observe the logic block B directly.

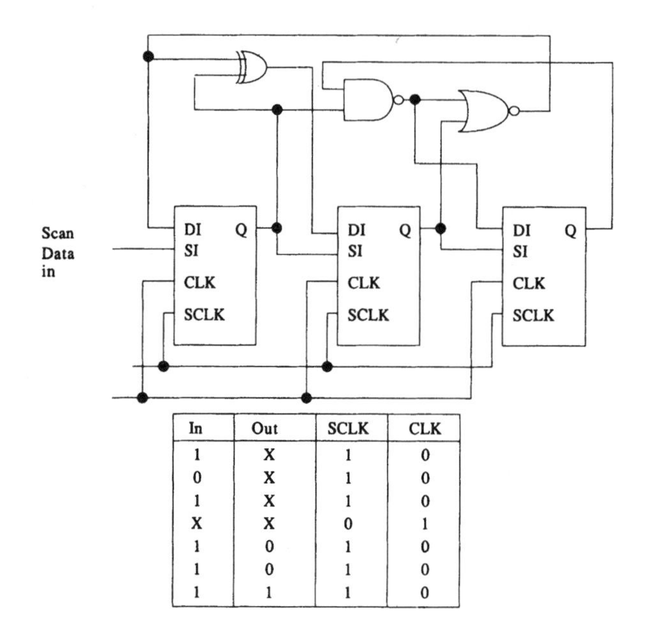

In	Out	SCLK	CLK
1	X	1	0
0	X	1	0
1	X	1	0
X	X	0	1
1	0	1	0
1	0	1	0
1	1	1	0

FIGURE 2.10 Scan test example and patterns. (*Source:* W.M. Needham, *Designer's Guide to Testable ASIC Devices*, New York: Van Nostrand Reinhold, 1991, p. 95. With permission.)

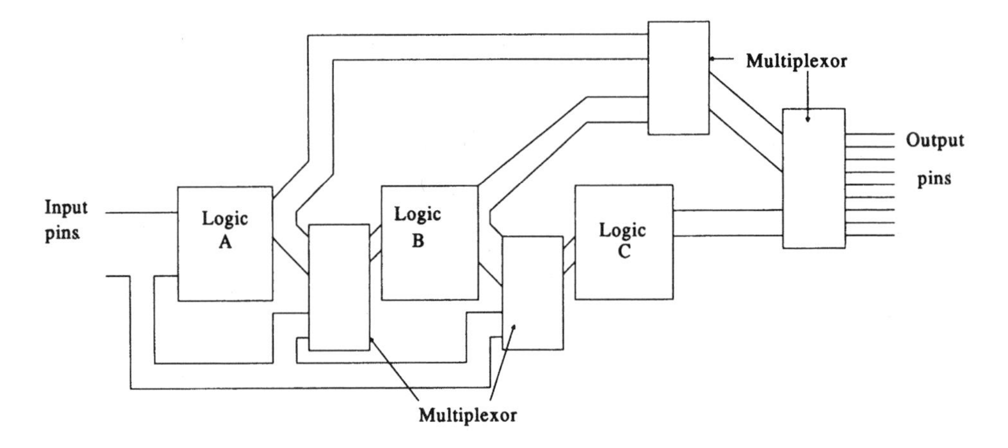

FIGURE 2.11 Direct access implementation. (*Source:* W.M. Needham, *Designer's Guide to Testable ASIC Devices*, New York: Van Nostrand Reinhold, 1991, p. 104. With permission.)

Joint Test Action Group (JTAG)

When implemented in a device, Joint Test Action Group (**JTAG**), or IEEE 1149.1, allows rapid and accurate measurement of the direct connections from one device to another on a PC board. This specification defines a Test Access Port (TAP) for internal and external IC testing. Figure 2.12 shows the TAP use in a small device, thus allowing accurate measurement and detection of solder connections and bridging on a PC board. This technique allows the shifting of data through the input and output pins of the part to ensure correct connections on the PC board. The four required pins are shown at the bottom of the drawing, and all the I/O ports are modified to include 1149.1 latches. The internal logic of the part does not need to change in order to implement 1149.1 boundary scan capability.

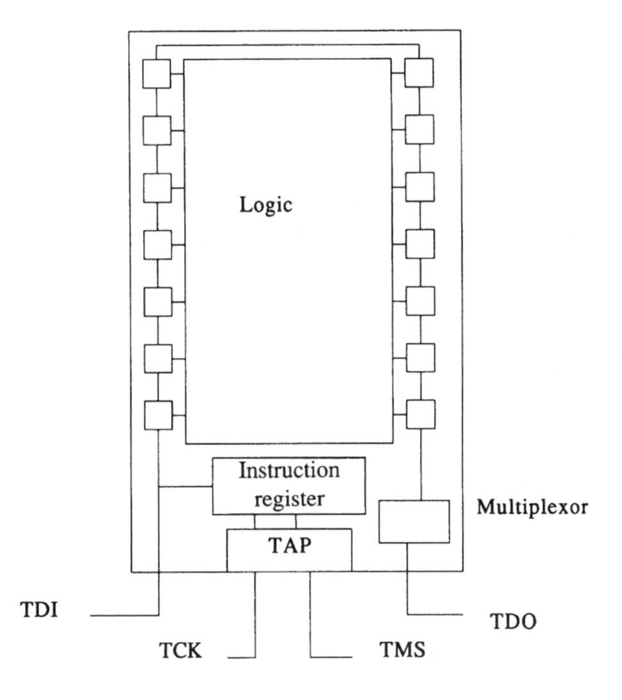

FIGURE 2.12 TAP circuit implementation.

JTAG test capability, or IEEE P1149.1 standard Test Access Port and boundary scan capability, allows the designer to implement features that enhance testing. Both PC board test capability and the internal logic verification of the device can be facilitated. For systems where remaining components on the boards are implemented with the 1149.1 scan approach, adding a TAP to a device is a wise step to take. Scan, BIST, and direct access can all be controlled by the 1149.1 test access port.

Pattern Generation for Functional Test Using Unit Delay

During simulation of the function portion of the device, patterns are captured and stored for exercise of the system. Figure 2.13 shows typical patterns for each block and the width of the data for test. The patterns check the functionality of the logic and ensure that the logic implemented performs the desired function in the device. The patterns can be generated by several methods if they were not generated by self-test. First is exercising of the system by use of code such as assemblers, macros, and high-level inputs that ensure the operation. Usually this is done by comparison of a high-level model to the actual logic. A second method of pattern generation is random number generation; in this case random numbers of ones and zeros are impressed on the logic and the results compared to a high-level model. Finally, coding of ones and zeros for logic checking is an alternative, but this is typically prohibitive in today's technology where devices may contain millions of logic gates.

Pattern Generation for Timing

After the functional patterns are complete, specific tests for timing may be generated. Timing tests are test procedures to verify the correction operation to the timing specification of the device. Typical timing tests include outputs relative to the clocks, propagation delays, set-up and hold times, access times, minimum and maximum speed of operation, rise and fall times, and others. These tests are captured in simulation and used in the test system to verify performance of the device. Table 2.1 shows the relationship between time-based simulation and cycle-based test files.

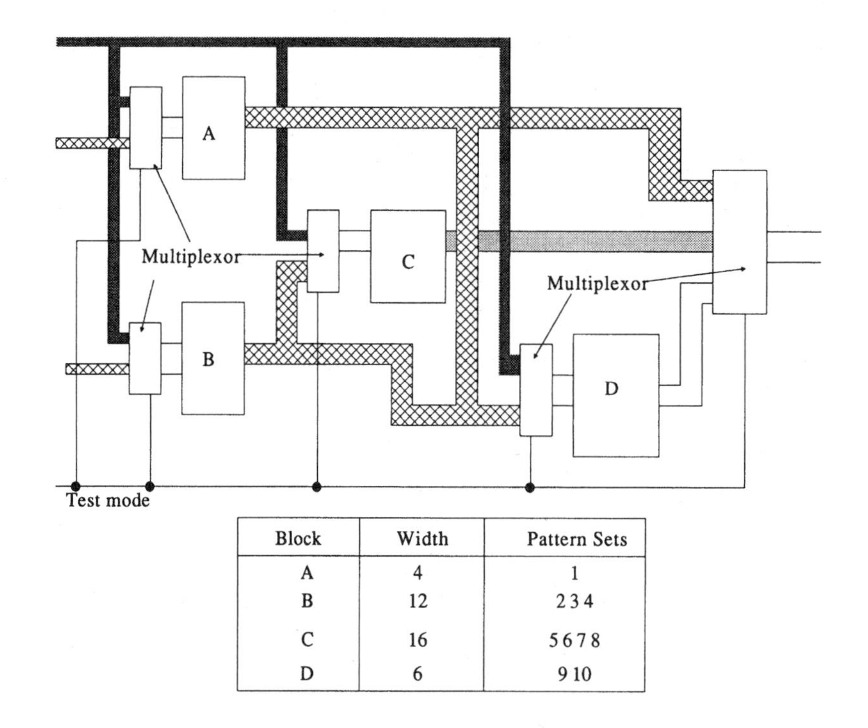

Block	Width	Pattern Sets
A	4	1
B	12	2 3 4
C	16	5 6 7 8
D	6	9 10

FIGURE 2.13 Typical patterns for function test. (*Source:* W.M. Needham, *Designer's Guide to Testable ASIC Devices*, New York: Van Nostrand Reinhold, 1991, p. 129. With permission.)

TABLE 2.1 Typical Devices Simulation and Test Files

		Pin Number																					
Time (nsec)		1	2	3	4	5	6	7	8	9	10	11	12	13	14	15	16	17	18	19	20		
31		1	0	1	1	1	0	1	1	1	1	1	1	1	1	0	1	1	1	1	1		
52		1	0	1	1	1	0	1	1	0	1	1	1	1	1	0	1	1	0	1	0		
97		1	0	1	1	1	0	1	1	1	1	1	1	1	1	0	1	1	1	0	0		
101		1	0	1	1	1	0	1	1	1	1	1	1	1	1	0	1	1	1	0	0		
156		1	0	1	1	1	0	1	1	1	1	1	1	1	1	0	1	1	1	0	0		
207		1	0	1	1	1	1	0	1	1	0	1	1	0	0	0	0	0	0	0	0		
229		1	0	1	1	1	0	1	1	0	1	1	1	1	1	0	0	0	0	1	1		
Vector	Time	Data																					
1	0	1	0	1	1	1	0	1	1	1	1	1	1	1	1	0	1	1	1	1	1	1	
2	50	1	0	1	1	1	0	1	1	0	1	1	1	1	1	0	1	1	0	1	1	1	N=NRZ
3	100	1	0	1	1	1	0	1	1	1	1	1	1	1	1	0	1	1	1	0	0	1	R=RZ
4	150	1	0	1	1	1	0	1	1	1	1	1	1	1	1	0	1	1	1	0	0	1	1=R1
5	200	1	0	1	1	1	1	0	1	1	0	1	1	0	0	0	0	0	0	0	0	1	
Format		N	N	N	N	N	R	1	1	1	1	N	N	1	1	R	N	N	1	N	N	N	

Temperature, Voltage, and Processing Effects

Figure 2.14 shows the impact on speed of process variations, and as a result of voltage or temperature variations. It is important to simulate with the total variation over the entire process, temperature, and voltage range to ensure testability. Remember that if the system design of the logic was not done with some

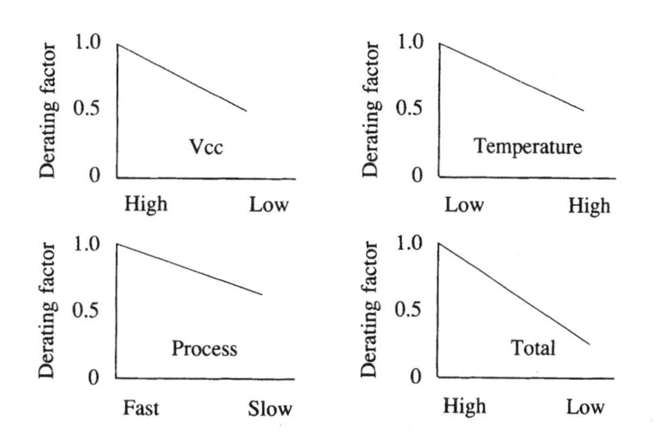

FIGURE 2.14 Effects of voltage, process, and temperature on test. (*Source:* W.M. Needham, *Designer's Guide to Testable ASIC Devices*, New York: Van Nostrand Reinhold, 1991, p. 172. With permission.)

guard-banding, there may be no margin. If the library used for the design did not include some amount of margin, there may be a need to add guardbanding by optimizing logic for speed or choosing faster gates. There must be a delta placed between the testing parameters used for initial production testing and final quality assurance (QA) test of the devices. This will ensure that they are electrically stable and manufacturable.

It is very important to review the critical paths, timing parameters, and the early simulations versus the timing parameters in the final simulation. The effect of parasitic capacitance and resistance will show up now. This is the time to look back to see whether the performance of the device will meet all the system specifications.

If the design meets all the device specifications and all the system specifications, now is the time to sign off and accept this simulation. If the simulations do not match the system requirements, this is the time to fix them once again.

Fault Grading

Fault grading is a measure of test coverage of a given set of stimuli and response for a given circuit. This figure of merit is used to ensure that the device is fully testable and measurable. Once the test patterns are developed for the device, artificial faults are put in via simulation to ensure that the part is testable and observable.

Figure 2.15 shows a single stuck-at-one (S@1) faults injected into logic circuitry with data patterns for checking the functionality. Notice in the truth table that faults on one gate show up on an output pin of the part or test observation point. Gaining high fault coverage is a very desirable method for ensuring that the device will function correctly in the system. Semiconductor failure modes include not only stuck-at-zeros and -ones but several other types, including delay faults, bridging faults, and leakage faults.

It is also important to mention that although fault grading using single stuck-at-zero and single S@1 faults covers many aspects of semiconductor failures, it does not cover all of them. For example, bridging faults are one of the more common types of failures within semiconductor devices; they may be caught with stuck-at-fault graded patterns if the patterns are done correctly.

Test Program Flow

Data patterns that were generated during simulation are then converted into a functional **test program.** Table 2.2 shows a typical test flow used by a semiconductor vendor for exercising the device. It describes the basic test done, along with the portions of the specification of the device tested in each step. The flow is from

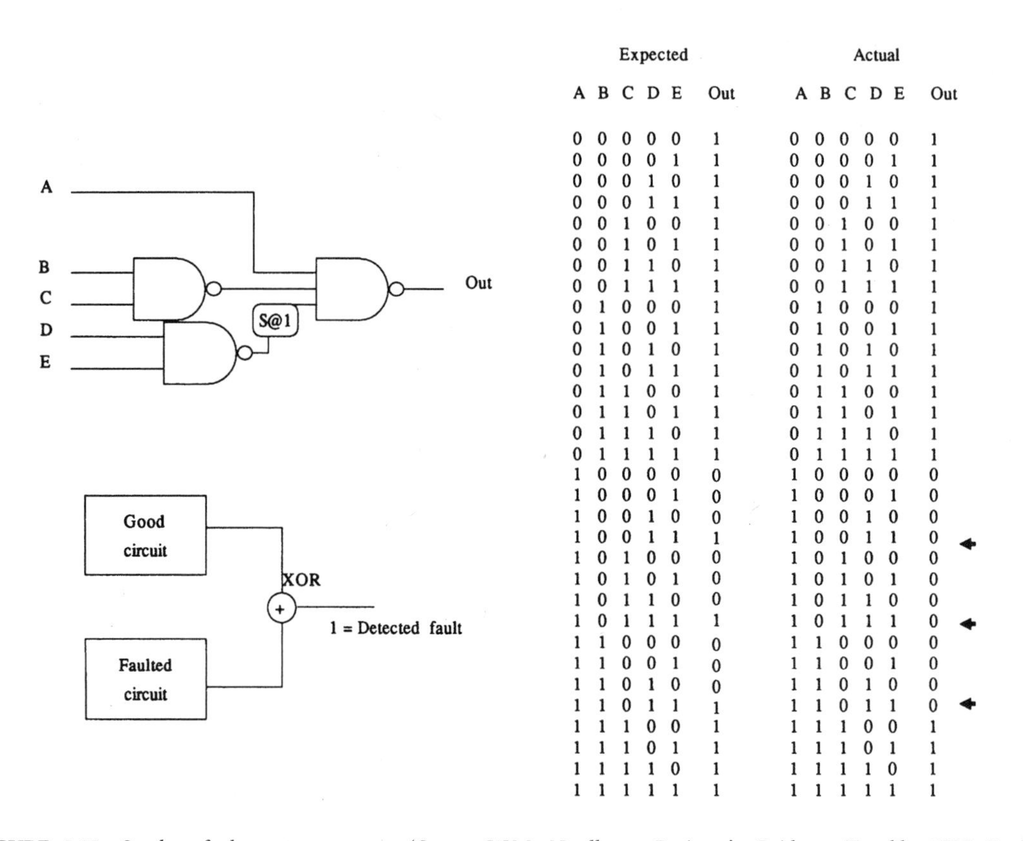

Expected A	B	C	D	E	Out	Actual A	B	C	D	E	Out	
0	0	0	0	0	1	0	0	0	0	0	1	
0	0	0	0	1	1	0	0	0	0	1	1	
0	0	0	1	0	1	0	0	0	1	0	1	
0	0	0	1	1	1	0	0	0	1	1	1	
0	0	1	0	0	1	0	0	1	0	0	1	
0	0	1	0	1	1	0	0	1	0	1	1	
0	0	1	1	0	1	0	0	1	1	0	1	
0	0	1	1	1	1	0	0	1	1	1	1	
0	1	0	0	0	1	0	1	0	0	0	1	
0	1	0	0	1	1	0	1	0	0	1	1	
0	1	0	1	0	1	0	1	0	1	0	1	
0	1	0	1	1	1	0	1	0	1	1	1	
0	1	1	0	0	1	0	1	1	0	0	1	
0	1	1	0	1	1	0	1	1	0	1	1	
0	1	1	1	0	1	0	1	1	1	0	1	
0	1	1	1	1	1	0	1	1	1	1	1	
1	0	0	0	0	0	1	0	0	0	0	0	
1	0	0	0	1	0	1	0	0	0	1	0	
1	0	0	1	0	0	1	0	0	1	0	0	
1	0	0	1	1	1	1	0	0	1	1	0	←
1	0	1	0	0	0	1	0	1	0	0	0	
1	0	1	0	1	0	1	0	1	0	1	0	
1	0	1	1	0	0	1	0	1	1	0	0	
1	0	1	1	1	1	1	0	1	1	1	0	←
1	1	0	0	0	0	1	1	0	0	0	0	
1	1	0	0	1	0	1	1	0	0	1	0	
1	1	0	1	0	0	1	1	0	1	0	0	
1	1	0	1	1	1	1	1	0	1	1	0	←
1	1	1	0	0	1	1	1	1	0	0	1	
1	1	1	0	1	1	1	1	1	0	1	1	
1	1	1	1	0	1	1	1	1	1	0	1	
1	1	1	1	1	1	1	1	1	1	1	1	

FIGURE 2.15 Stuck-at-faults test sequences. (*Source:* W.M. Needham, *Designer's Guide to Testable ASIC Devices*, New York: Van Nostrand Reinhold, 1991, p. 152. With permission.)

TABLE 2.2 Typical Test Flow

Test	Function
Shorts	Checks if the adjacent pins are shorted
Open	Checks to see if the p–n junction exits (pad protection device)
Basic function test	Checks functionality of the part, uses most vectors, loose timing, and nominal voltages
dc spec test	Checks the inputs and outputs compared to spec for dc levels
ac spec and margin test	Checks the vectors with timing set to spec; checks at minimum and maximum voltage

Source: W.M. Needham, *Designer's Guide to Testable ASIC Devices*, New York: Van Nostrand Reinhold, 1991, p. 181. With permission.

loose testing of the device to detailed checking of specifications; this basic flow is both efficient and helpful in debugging failures encountered during test. Table 2.3 shows the approximate test time for a 100-pin device as executed on a verification system. Notice that a large number of tests can be done in a very short time. The test times are approximate and will vary based on the type of system used, by the type of test executed, length of patterns, and pin count.

The cost of testing is very much related to test time. In production, package handlers, wafer probe equipment, people, and perhaps a computer network are all needed to run production tests. When comparing test time on a small system to a large system, total cost must also be compared. In this case, total cost is the cost of the test equipment along with the needed support equipment. Table 2.4 shows the same test time for the same device on a large production test system. Again, the test time is approximate and will vary depending on options, program flow, and test details. The biggest reason for the difference in test time from a verification system to a production system is due to pattern reload time.

TABLE 2.3 Test Execution Time on a Verification Tester

Test	Number of Test	Type of Test	Time to Execute
Opens	100	Parametric	1 sec
Shorts	100	Parametric	1 sec
Basic function test	40,000	Vector pattern	0.04 sec*
ac spec test	300–500	Parametric-functional	3–5 sec
ac spec and margin test	100,000	Vector pattern	0.1 sec*

*Execution time only; vector reload time may be between 10 and 500 sec.
Source: W.M. Needham, *Designer's Guide to Testable ASIC Devices,* New York: Van Nostrand Reinhold, 1991, p. 182. With permission.

TABLE 2.4 Test Execution Time on a Production Tester

Test	Number of Test	Type of Test	Time to Execute
Opens/shorts	2	Parametric	0.05 sec
Basic function test	40,000	Vector pattern	0.04 sec*
ac spec test	300–500	Parametric-functional	0.5 sec
ac spec and margin test	100,000	Vector pattern	0.75 sec*

*Execution time only; vector reload time may be between 10 and 500 sec.
Source: W.M. Needham, *Designer's Guide to Testable ASIC Devices,* New York: Van Nostrand Reinhold, 1991, p. 182. With permission.

TABLE 2.5 Test Data Sources

Test	Source of Data
Opens	Process description and net list for I/O pads
Shorts	Process description and net list for I/O pads
Basic function test	Simulations done for test
	Internal library for cores selected by the user
dc spec test	Net list for types of pads
	Automatic place and route data for placement
	Simulation data for vector setup sequences
ac spec and margin test	Simulations done for test and performance
	Internal library for cores selected by the user

Source: W.M. Needham, *Designer's Guide to Testable ASIC Devices,* New York: Van Nostrand Reinhold, 1991, p. 183. With permission.

Regardless of the techniques that were used for the generation of the data patterns, it will be necessary to include those patterns into a test program. This is similar to that shown in Table 2.5. This is the same flow with the addition of information from the net list and vector simulations. Vectors are used for functional tests and some portions of I/O testing. Net list description and selection of input pads and power connections are used in the parametric tests. Although the flow may vary from one manufacturer to another, most of them perform essentially the same kinds of tests.

Even if the device uses special testing techniques such as JTAG, BIST, scan, or direct access, the flow is always the same. It is basically always done with a setup sequence or a preconditioning of the part, and a formal measurement of the data on the device under test. The testing may be on a cycle-by-cycle basis or as a pass or fail conclusion at the end of a long routine. There may be a burst of data to set up the part and a burst of data to measure it. Finally, information is interpreted by the program to categorize or bin the device as a pass or fail rating.

Defining Terms

Built-in self-test (BIST): A design process where logic is added to the part to perform testing. Advantages include easy testing of buried logic. The biggest negative is patterns are set in hardware and the design must be changed to improve test coverage.

DUT: Device under test.

Fault coverage: A metric of test pattern coverage. Calculated by measuring faults tested, divided by total possible faults in the circuits.

JTAG: A specification from the IEEE 1149.1 that defines a test access port (TAP).

Scan: A method of connection of latches within an integrated circuit that allows (a) patterns to be shifted in, (b) combinatorial logic check to be exercised, and (c) patterns to be shifted out and compared to known simulation results.

T_0: A timing reference point used in simulation and on the tester. This point T_0 is where all timing is referenced.

Test program: A software routine consisting of patterns, flow information, voltage and timing control, and decision processes to make certain the DUT is correct.

Tester: A piece of equipment used to verify the device, often called automate test equipment (ATE).

Vector: A series of ones and zeros that describe the input and output states of the device.

Yield: A metric of good devices after test, divided by total tested.

References

J.M. Acken, *Deriving Accurate Fault Models,* Palo Alto, CA: Stanford University, Computer Systems Laboratory, 1989.

V.D. Agrawal and S.C. Seth, "Fault coverage requirements in production testing of LSI circuits," *IEEE Journal of Solid-State Circuits,* vol. SC-17, no. 1, pp. 57–61, February 1982.

E.J. McCluskey, *Logic Design Principles: With Emphasis on Testable Semicustom Circuits,* Englewood Cliffs, NJ: Prentice-Hall, 1986.

W.M. Needham, *Designer's Guide to Testable ASIC Devices,* New York: Van Nostrand Reinhold, 1991.

F. Tsui, *LSI/VLSI Testability Design,* New York: McGraw-Hill, 1987.

Further Information

IEEE Design and Test Computers is a bi-monthly publication focusing on testing digital circuitry.

Proceedings of the IEEE Test Conference—This conference occurs each year in the late summer. The conference is the major focal point for test equipment vendors, test technology, and a forum for advanced test papers.

Design Automation Conference—This conference is held annually in the late spring. The focus is the design process, but many of the papers and vendors at the conference have programs for test design. Vendor material is available from design and test equipment vendors.

2.3 Electrical Characterization of Interconnections

S. Rajaram

Semiconductor technology provides electronic system designers with high-speed integrated circuit (IC) devices that operate at switching speeds approaching 1 nsec and lower. For the past 25 years, the performance and cost of digital electronic systems have improved continuously because of the technological advances in manufacturing physically smaller devices. With the reduction in the feature sizes of the gates and cells that make up these devices, there has also been a tremendous increase in the density of IC integration. These physically smaller devices with their intrinsic faster switching speeds and lower power per gate have given rise

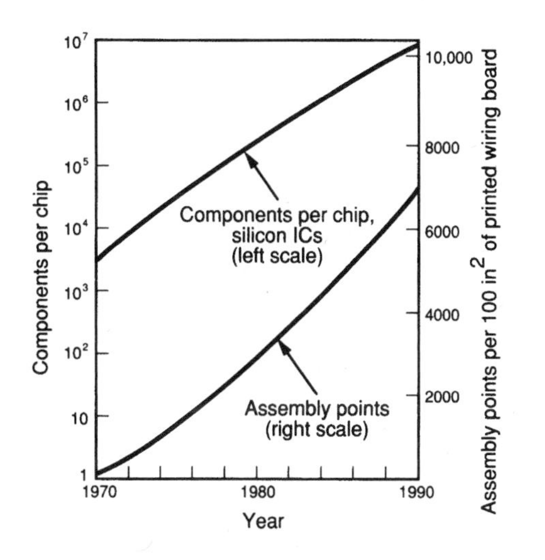

FIGURE 2.16 The history of silicon integrated circuit scale of integration and its effect on interconnection density.

to the very large-scale integration (VLSI) technology. We therefore have a situation today in which IC technology includes both high density and high speed. The unprecedented gains in the increasing scales of integration in IC technology are shown in Figure 2.16.

For leading-edge technology, the maximum number of components (transistors) per chip increases by a factor of 100 per decade. It is tempting to believe that these increases may be extended indefinitely, leading to ever smaller devices with increasingly larger electronic functionality. However, the limiting factor for such spectacular gains in IC technology is the electrical performance capability of its associated **interconnections and packaging (I&P)**. Thus, interconnections and packaging is the bottleneck to IC performance. The signal originates from an IC device referred to as the driver and is received by another IC device called the receiver. The path between the driver and the receiver is the I&P medium. The various elements of I&P are shown in Figure 2.17.

The elements that make up I&P are the IC package (including its wirebonds inside the package and the leads or pins external to the package), printed circuit board (PCB) with conductor (copper) traces, electrical PCB connector, backplane or motherboard, and external cables. The driver and receiver may be on the same PCB or on different PCBs within the same equipment or in different equipment. I&P should be designed so as to cause minimal signal degradation from the driver to the receiver. Thus, there are several interfaces within I&P, and signal degradation occurs at every interface. The elements of I&P between the driver and the receiver resemble the mechanical links in a chain.

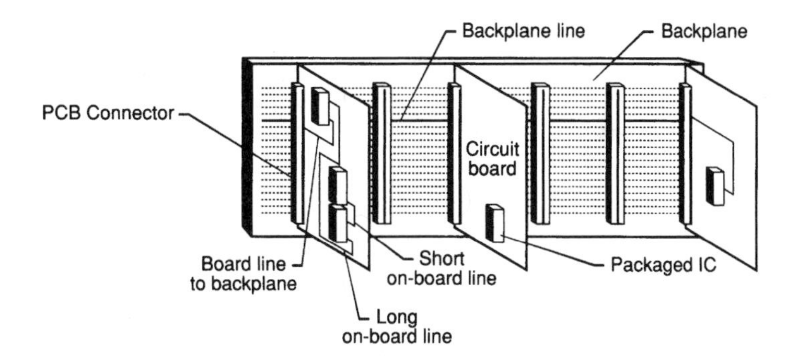

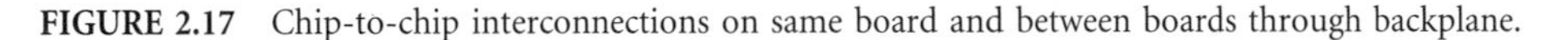
FIGURE 2.17 Chip-to-chip interconnections on same board and between boards through backplane.

The key role of I&P is to enable electronic system designers to fully exploit the advancements taking place in device and manufacturing technologies related to ICs. I&P technology is dictated by three main external forces. These are silicon IC technology, automated systems for assembly and testing, and the fundamental architecture of electronic systems (Hoover et al., 1987). We have already seen the 100-fold increase in component density at the IC level. Most of the electronic functions are built into these dense ICs. However, the system functionality is contained in several ICs that must "talk" to each other. The medium connecting various ICs in a system is generically referred to as I&P. Therefore, in order to attain the performance capability of the IC, it is necessary to have an I&P medium to match its electrical performance. The electrical parameters of the I&P must be tailored to meet IC performance. This becomes the crux of any high-speed electronic system design. To optimize IC performance, therefore, it is also necessary to minimize the size of all interconnections. However, to optimize manufacturing costs, I&P must be large enough to attain reliable and high-yield assembly operations, test access, repair, and maintenance. Thus, these requirements imply automated assembly and test equipment for high-quality manufacture of electronic systems with optimal cost. Finally, I&P is also dictated by system architecture requirements.

Historically, telecommunications systems have had system requirements dictating the high density of interconnection and input/output (I/O) signals at moderately low speeds and low interconnection density or low I/O at moderately higher speeds. However, today the trend is toward high I/O and high speed, which increases the interconnection density on I&P elements. These have caused a fundamental change in the IC package and manufacturing and assembly technologies associated with electronic systems. Electrical performance requirements of the I&P have forced the change from dual inline package (DIP) ICs to surface mount (SM). The SM ICs have much smaller package outlines and leads than the corresponding DIPs. While the DIPs are soldered into holes drilled through the PCB, SM components are soldered onto pads on the surface of the PCB. While the pitch on the leads of a DIP is 100 mil (0.100 in.), the pitch on SM IC package leads is typically 25 to 50 mil. I&P technology has therefore evolved rapidly as the driving forces behind it have advanced at a rapid pace. It is fairly accurate to state that the ultimate quality and reliability of many complex electronic systems are determined primarily by the reliability of the I&P of the system. The stringent electrical and density requirements impose severe constraints and challenges on mechanical design and manufacturing capability. Most failure mechanisms today are related to interconnections in the system.

Interconnection Metrics

In Figure 2.16, we observed the tremendous increase in the IC scale of integration. We have also made the statement that the increased IC component density has led to increased interconnection density. It is important to understand the relationship between the scale of component integration inside the IC package and the interconnection density requirements outside the IC package. Figure 2.16 also shows the evolution of the number of pins interconnected (assembly points) on a PCB within an area of 100 in.^{2}, and the trend is obvious. This data is derived from CAD packages used within Bell Labs for PCB layout. IC circuits are usually described in terms of number of gates (1 gate $\approx$ 4 transistors). The relationship between the number of gates of an IC (G) to the number of signal I/O pins in the package (P) is given empirically by Rent's rule (Hoover et al., 1987). For random logic, Rent's rule states that $P = KG^{\alpha}$. The value of K ranges from 3 to 6 and α typically ranges from 0.4 to 0.55. For a functionally complete chip, $K \approx 7$ and $\alpha \approx 0.2$ (Moresco, 1990). The empirical Rent's rule has been found to apply remarkably well over a large range of circuit sizes, as shown in Figure 2.18.

This simple empirical relationship is important in understanding the fundamental trade-offs between interconnection costs, reliability, and increased scales of integration. Let us assume that a certain electrical function requires the use of 100,000 gates. Let us partition the circuitry into ICs with 100, 1000, and 10,000 gates per chip (device). Let us apply Rent's rule by taking $K = 4$ and $\alpha = 0.5$, and examine the I/O pins, and therefore the interconnection requirements for the three circuit partition choices, as shown in Table 2.6 (Hoover et al., 1987).

The most important observation to be made from Table 2.6 is that increasing the IC scale of integration by a factor of 100 moves 90% of the external interconnections into silicon. The density of the interconnection at

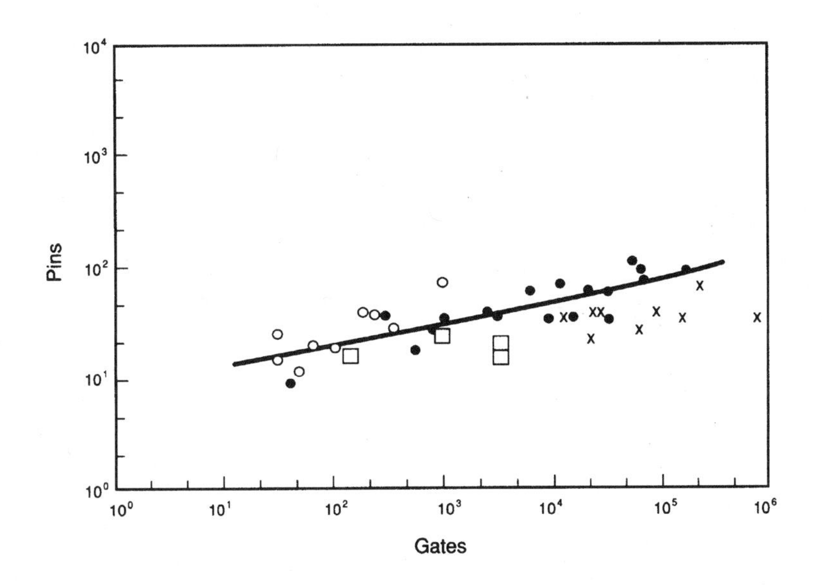

FIGURE 2.18 Rent's rule for LSI and VLSI chips. (●) Silicon microprocessors and random logic, (x) silicon memories, (o) GaAs functional chips, and (□) GaAs memory.

the silicon increases, but the overall cost of interconnections for the system decreases since the relative costs per interconnection in silicon are 1 to 2 orders of magnitude cheaper than those on PCBs and hybrid integrated circuits (HICs). This is clearly shown in Figure 2.19.

TABLE 2.6 Distribution of Interconnections for 100,000-Gate Random Logic Circuit

	Gates per IC		
Interconnections	100	1,000	10,000
On silicon	160,000	187,400	196,000
On PCB	40,000	12,600	4,000

This is the driving force for integrating as much functionality in silicon, with reduced external interconnections. However, the problem is compounded by the fact that many such dense ICs are now placed on a board, and these have to be interconnected on a dense multilayer board. This has led to the requirement for fine-line PCB technology with increasing numbers of layers. The dense traces on the board lead to high I/O (pin count) connectors requiring as many as 400 I/Os on an 8-in. board edge. Thus, there is an increasing density of interconnection at every level, both internal and external to silicon, as shown in Table 2.7.

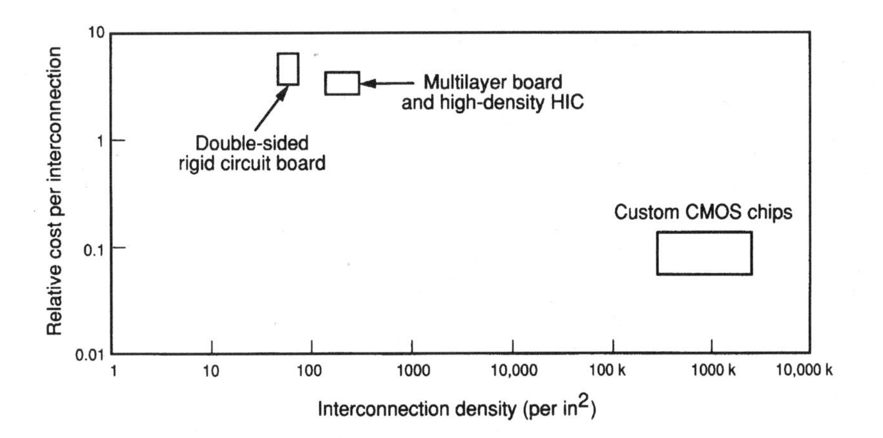

FIGURE 2.19 Interconnections cost trade-off.

TABLE 2.7 Interconnection Limits for Telecommunications Systems

Packaging Technologies	1970	1978	1986	1990s
PCB area, in.^{2}	50	100	200	200
Number of layers in PCB	2	6	8	12–14
Line width, mil	25	7	6	4
Number of device terminal connections on PCB	100	2,000	9,000	16,000
External I/Os	80	300	600	>800
Logic gates	10^2	10^4	10^5	10^6

TABLE 2.8 Effect of Increasing Scale of Integration on System Reliability

	Gates per IC		
	100	1,000	10,000
Number of ICs	1,000	100	10
IC FIT count, total	50,000	5,000	500
Number of external interconnections	40,000	12,600	4,000
Interconnection FIT count, total	4,000	750	200
System FIT count	54,000	5,750	700

Another important aspect of increasing the scale of integration at the IC level is the improved overall reliability of the system. Practical experience indicates that the Failure unIT (FIT) count of an IC is approximately constant for a given IC technology and manufacturer and is almost independent of the number of gates per chip. A FIT is defined as one failure in 10^9 h. Therefore, system FIT count due to devices should decrease inversely in proportion to scale of integration. Once again, let us consider the example of partitioning 100,000 gates into devices with 100, 1000, and 10,000 gates per IC. Thus, from 100 gates per IC, we have increased the scale of integration by a factor of 10 and 100 as we go to 1000 and 10,000 gates per chip, respectively. The FIT count for an IC is taken as 50. The total system FIT count is the sum total of the IC and interconnections FIT count. These are shown in Table 2.8, where the FIT counts for interconnections are those appropriate for the PCB types for the levels of integration based on experience at AT&T (Hoover et al., 1987).

The important message is that there is a tremendous improvement in system reliability (FIT count) with increasing scales of integration in silicon. It is important to remember that the example shown in Table 2.8 is an ideal situation. In any practical situation, the circuit is partitioned into many devices with different levels of integration, with a few VLSI, but 50 or more small scale integration devices. Typically, the total system FIT count, including devices and interconnections, may be in the range of 3000 to 5000. The remarkable attribute of IC technology is that as the scale of integration increases, the total system cost decreases, while the overall system reliability improves considerably. It is this remarkable attribute that has led to a continuous decrease in the cost of electronics products with increased functionality with time. However, the miniaturization and the consequent increased scales of integration and interconnections density at the packaging level together with increasing switching speeds have resulted in a whole range of interesting electrical problems. Although these problems appear to be new, solutions to many of them are available from extensive work that has been well documented in the microwave area. In the sections that follow we will discuss the significance of the important electrical parameters and techniques to evaluate them.

Interconnection Electrical Parameters

The interconnection medium is the basic path for transmitting the pulse from the driver to the receiver. As the speed of the circuits goes beyond 10 MHz, it is necessary to use high-frequency techniques developed by RF engineers. The fundamental electrical parameters of the circuit such as inductance L and capacitance C behave as lumped elements at low frequencies and as distributed parameters at high frequencies where transmission line techniques must be employed. The transition from lumped element to transmission line behavior depends upon the risetime of the pulse T_r and on the total delay in the pulse transmission through the interconnection T_d. In the lumped element mode, the inductance and capacitance appear to the pulse to be concentrated at a point. However, in the transmission line mode, inductance and capacitance appear to be uniformly distributed throughout the interconnection; and as far as the pulse is concerned, the medium is infinite in length, and all the characteristics of wave propagation must be taken into consideration. Interestingly, similar techniques may be used for electrical characterization of all of the interconnection elements such as wirebonds, package leads, PCBs, connectors, and cables. The basic pulse transmission parameters of interest are propagation delay (T_d), characteristic impedance (Z_0), reflection coefficient (Γ), **crosstalk** (X), and **risetime degradation** (T_{dr}).

Propagation Delay (T_d)

For a pulse being transmitted through a medium of length l and wave velocity v, delay $T_d = l/v$. The speed of pulse transmission depends upon the dielectric constant of the material ε_r and is given by $v = c_0/\sqrt{\varepsilon_r}$, where c_0 is the speed of light in air $= 3 \times 10^8$ m/sec. Thus, **propagation delay** is proportional to length and also the square root of the dielectric constant. In order to reduce delay and reduce machine cycle time and increase speed, it is necessary to reduce the dielectric constant of the material. There is tremendous interest in developing low dielectric constant materials for this reason.

Characteristic Impedance (Z_0)

In general, a transmission line is any structure that propagates an electromagnetic wave from one point to another. However, in the world of microwave, RF, or high-speed digital designs, the use of the term *transmission line* is far more restrictive. In order for a structure to be a transmission line, the electrical length of the (transmission) line must be much larger than the wavelength at the frequency of interest. For an interconnection medium that behaves as a transmission line as shown in Figure 2.20, the **characteristic impedance** is defined as

$$Z_0 = \sqrt{\frac{R + j\omega L}{G + j\omega C}} \quad (\Omega) \tag{2.23}$$

where R = series resistance (Ω/m), L = series inductance (H/m), G = shunt conductance ($\Omega^{-1}\text{m}^{-1}$), C = shunt capacitance (F/m), and $\omega = 2\pi f$ = radian frequency. In general, Z_0 is complex, and the distributed per unit length quantities R, L, G, and C must be determined from the material and structural characteristics of the transmission line. For most applications, $R << L$, $G << C$, and

$$Z_0 = \sqrt{\frac{L}{C}} \tag{2.24}$$

The speed of wave propagation through the transmission line is given by $v = 1/\sqrt{LC}$. Therefore, we can also define $Z_0 = 1/vC = vL$.

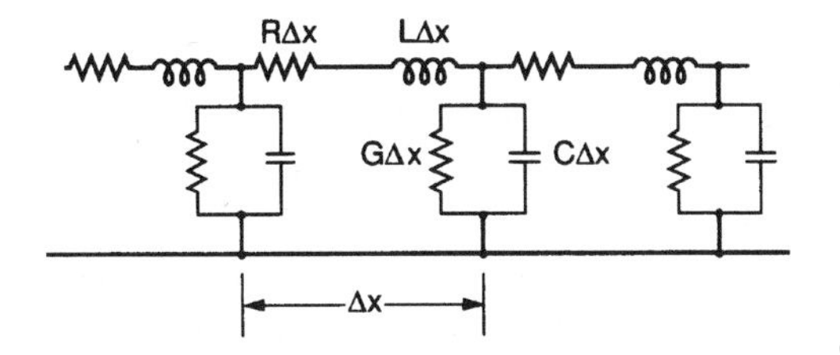

FIGURE 2.20 Representation of a short section of transmission line.

A digital pulse propagating in a circuit consists of voltages and currents of different frequencies (ac components). A digital pulse is therefore an electromagnetic wave of many frequencies propagating down the transmission line. In any circuit with transients (ac components), the current and voltage are not in phase because of inductance and capacitance. In a pure inductor, voltage leads current by 90°, and in a pure capacitor, voltage lags behind current by 90°. The total energy (W_T) of the electromagnetic wave is made up of the magnetic energy (W_m) and the electric energy (W_e). Magnetic energy is stored in inductance and is given by $W_m = 1/2LI^2$, where I is the current. Similarly, electric energy is stored in its capacitance and is given by $1/2CV^2$. Therefore, the total energy is

$$W_T = W_m + W_e = \frac{1}{2}LI^2 + \frac{1}{2}CV^2 \tag{2.25}$$

In an alternating field, the total energy is continually being swapped between the magnetic and electrical elements, one at the expense of the other. Because of the phase relationships discussed earlier, magnetic energy is a maximum when electric energy is 0 and vice versa. Since it is the same stored energy that appears alternately as magnetic and electrical energy, we can write

$$W_T = \frac{1}{2}LI_{max}^2 = \frac{1}{2}CV_{max}^2 \tag{2.26}$$

Therefore, $Z_0 = V_{max}/I_{max} = \sqrt{L/C}$.

Therefore, characteristic impedance Z_0 gives the relationship between the maximum voltage and maximum current in a transmission line and has the units of impedance (Ω). Thus, it is important to note that the current required to drive a transmission line is determined by its characteristic impedance. Z_0 is really the impedance (resistance) to energy transfer associated with electromagnetic wave propagation. In fact, characteristic impedance is not unique to electromagnetic waves alone. Characteristic impedance Z_0 is an important parameter associated with propagation of waves in a medium. Some examples are given below:

- Electromagnetic waves: $Z_0 = \sqrt{L/C} = vL = 1/vC$
- Transverse vibrating string: $Z_0 = \lambda_0 v = \sqrt{\lambda_0 T_0}$, where λ_0 = mass/unit length and T_0 = force
- Longitudinal waves in a rod: $Z_0 = \rho_0 v = \sqrt{\rho_0 E}$, where ρ_0 = mass/unit volume and E = Young's modulus
- Plane acoustic waves: $Z_0 = \rho_0 v = \sqrt{B\rho_0}$, where ρ_0 = density and B = bulk modulus

Note that Z_0 is a unique characteristic of material properties and geometry alone.

The performance of an interconnect that behaves as a transmission line is measured in terms of the efficiency of energy (information) transfer from input to output, with minimal loss and dispersion effects. At high speeds, the interconnect must maintain a uniform Z_0 along the length of the signal. Any mismatch in characteristic impedance across interconnect interfaces will cause reflection of the signal at the interface, which can cause errors in digital circuits. Reflections are part of the losses in interconnect and lead to loss of information. These reflections are sent back to the signal source. Therefore, multiple reflections from interfaces can distort the input signal. The magnitude of reflection is defined by the reflection coefficient Γ, given by

$$\Gamma = \frac{Z_L - Z_0}{Z_L + Z_0} \tag{2.27}$$

where Z_L is the load impedance. Note that when the load and impedances are matched, $Z_0 = Z_L$, reflection coefficient $\Gamma = 0$, and the wave (energy, information) is transmitted without loss. For an open circuit, $Z_L = \infty$, and $K = 1$. Thus, the entire pulse is reflected back to the source. For a short circuit, $Z_L = 0$, and $K = -1$. Here, the pulse is reflected back to the source with the same amplitude, but with the sign reversed. For practical purposes, Z_0 of PCBs may be considered to be independent of frequency up to nearly 1 GHz. In other words, PCBs may be considered lossless transmission lines for impedance calculations. Beyond 1 GHz, the skin effect in conductors, and dielectric and dispersion losses, cause signal degradation. Z_0 is a parameter associated with PCBs and cables but not with connectors at lower speeds, because of the definition of transmission line. The length of an electrical path through a connector is approximately 1 in., which is very small compared to the wavelength at lower frequencies. However, as speeds increase, the connector also has a Z_0 associated with it. Otherwise, a connector acts as a lumped L and C in a transmission path with its associated losses such as reflection and degradation. Z_0 values for rectangular and circular transmission lines with and without ground planes applicable for cables are given in Everitt (1970). We shall now look at evaluation of Z_0 for typical PCB structures.

***Z_0* of PCB Structures.** The two commonly used PCB structures in electronic circuits are the microstrip and stripline designs, shown in Figure 2.21 and Figure 2.22. In a pure microstrip, there is only one ground plane below the conductor. The space below the conductor is filled with a dielectric material, and above the conductor it is air. In most applications, however, the conductor traces are protected with a solder mask coating of 2 to 3 mil thickness above it, as in Figure 2.23. This solder mask has the effect of reducing the Z_0 of

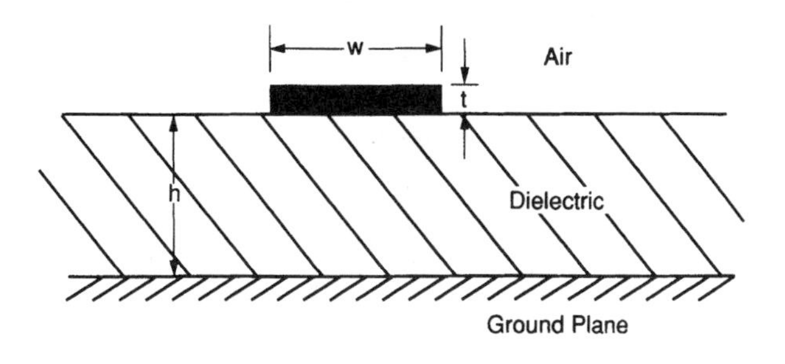

FIGURE 2.21 Classical microstrip.

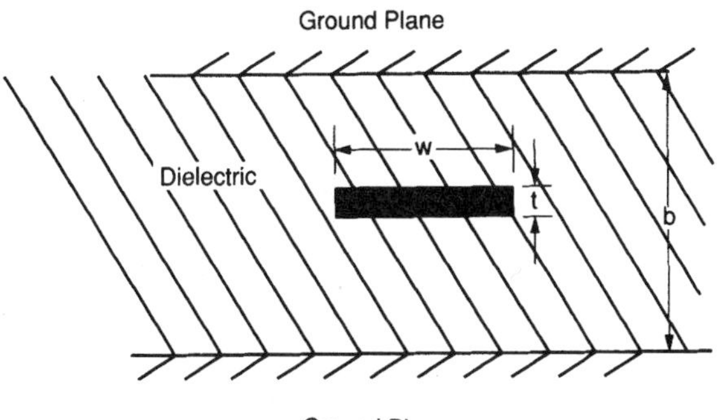

FIGURE 2.22 Classical stripline.

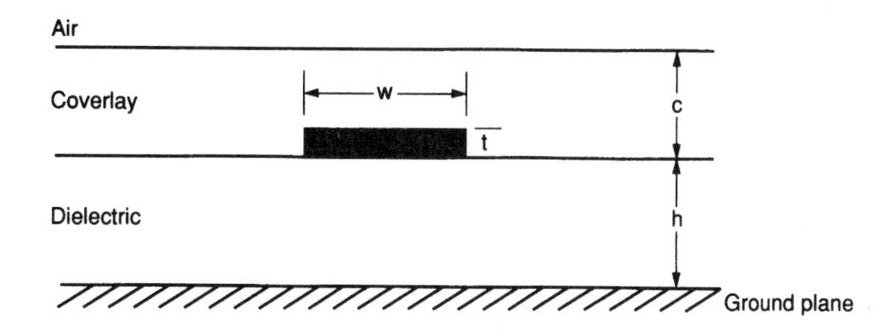

FIGURE 2.23 Covered microstrip.

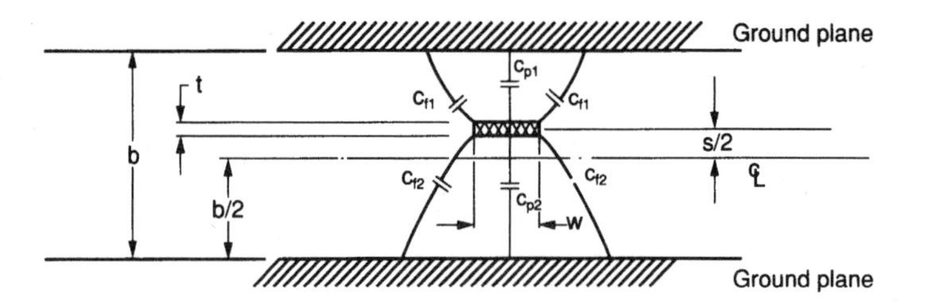

FIGURE 2.24 Asymmetric stripline.

the microstrip. In a pure microstrip, the electromagnetic wave is transmitted partially in the dielectric and partially in air. Therefore, the effective dielectric constant is a weighted average between that of air and the dielectric material. In a stripline design, the conductor is placed symmetrically between two ground planes and the space filled with dielectric material. Thus, in a stripline, the wave is completely transmitted in the dielectric. A variation of the stripline design is the asymmetric stripline where the conductor is closer to one ground plane than the other, as in Figure 2.24.

For a microstrip transmission line structure, the characteristic impedance Z_0 is given by Kaupp (1967):

$$Z_0 = \frac{60}{\sqrt{0.475\,\varepsilon_r + 0.67}}\ln\left[\frac{4h}{0.67(0.8w+t)}\right]$$

$$= \frac{87}{\sqrt{\varepsilon_r + 1.41}}\ln\left[\frac{5.98h}{(0.8w+t)}\right]\Omega \tag{2.28}$$

The effective dielectric constant $\varepsilon_{r\,\mathrm{eff}} = \sqrt{0.475\varepsilon_r + 0.67}$. Experimental measurements with fiberglass epoxy boards have shown that Equation (2.28) predicts Z_0 accurately for most practical applications.

For stripline configuration, characteristic impedance is given by Howe (1974).

For $w/b \le 0.35$:

$$Z_0 = \frac{60}{\sqrt{\varepsilon_r}}\ln\left[\frac{4b}{\pi d}\right]\Omega \tag{2.29a}$$

$$d = \frac{w}{2}\left[1 + \frac{t}{\pi w}\left(1 + \ln\frac{4\pi w}{t} + 0.51\pi\left(\frac{t}{w}\right)^2\right)\right] \tag{2.29b}$$

For $w/b \ge 0.35$:

$$Z_0 = \frac{94.15}{\sqrt{\varepsilon_r}}\frac{1}{\dfrac{w}{b\left(1-\frac{t}{b}\right)} + \dfrac{C_f'}{\varepsilon}}\Omega \tag{2.29c}$$

$$\frac{C_f'}{\varepsilon} = \frac{1}{\pi}\left\{\frac{2}{1-\frac{t}{b}}\ln\left(\frac{1}{1-\frac{t}{b}}+1\right)\right\} - \frac{1}{\pi}\left\{\left(\frac{1}{1-\frac{t}{b}}-1\right)\ln\left(\frac{1}{\left(1-\frac{t}{b}\right)^2}-1\right)\right\} \tag{2.29d}$$

where C_f'/ε is the ratio of the static fringing capacitance per unit length between conductors to the permittivity (in the same units) of the dielectric material. This ratio is independent of the dielectric constant. $\varepsilon = \varepsilon_r\varepsilon_0$, where ε_0 is the permittivity of air 8.854*1.0e-12 F/m. For the asymmetric stripline, where the conductor is closer to one ground plane than the other:

$$Z_0 = \frac{376.7}{\sqrt{\varepsilon_r \frac{C}{\varepsilon}}} \tag{2.30a}$$

Here, C/ε is the ratio of the static capacitance per unit length between the conductors to the permittivity of the dielectric medium and is once again independent of the dielectric constant. Capacitance is made up of both the parallel plate capacitances C_p and fringing capacitances C_f. Subscripts 1 and 2 refer to the two ground planes, which are different distances from the line,

$$\frac{C}{\varepsilon} = \frac{C_{p1}}{\varepsilon} + \frac{C_{p2}}{\varepsilon} + 2\frac{C_{f1}}{\varepsilon} + 2\frac{C_{f2}}{\varepsilon} \tag{2.30b}$$

$$\frac{C_{p1}}{\varepsilon} = \frac{\frac{2w}{b-s}}{1-\frac{t}{b-s}} \text{ and } \frac{C_{p2}}{\varepsilon} = \frac{\frac{2w}{b+s}}{1-\frac{t}{b+s}} \tag{2.30c}$$

$$\frac{C_{f1}}{\varepsilon} = \frac{1}{\pi}\left\{\frac{2}{1-\frac{t}{b-s}}\ln\left(\frac{1}{1-\frac{t}{b-s}}+1\right)\right\} - \frac{1}{\pi}\left\{\left(\frac{1}{1-\frac{t}{b-s}}-1\right)\ln\left(\frac{1}{\left(1-\frac{t}{b-s}\right)^2}-1\right)\right\} \tag{2.30d}$$

$$\frac{C_{f2}}{\varepsilon} = \frac{1}{\pi}\left\{\frac{2}{1-\frac{t}{b+s}}\ln\left(\frac{1}{1-\frac{t}{b+s}}+1\right)\right\} - \frac{1}{\pi}\left\{\left(\frac{1}{1-\frac{t}{b+s}}-1\right)\ln\left(\frac{1}{\left(1-\frac{t}{b+s}\right)^2}-1\right)\right\} \tag{2.30e}$$

The formulas given above are for Z_0 of a single (isolated) line. As we shall see later, the presence of adjacent lines alters the value of Z_0. The appropriate line widths and the dielectric thicknesses required for Z_0 in the range of 50 to 75 Ω in glass epoxy FR4 boards with $\varepsilon_r = 4.2$ are shown in Table 2.9 and Table 2.10 for

TABLE 2.9 Microstrip in Glass Epoxy, $\varepsilon_r = 4.2$, $t = 1.2$ mil

w (mil)	s (mil)	h (mil)	w/h	s/h	Z_0 (Ω)	X_{max} (%)
8	8	5	1.6	1.6	50.3	4.1
8	8	10	0.8	0.8	75.8	9.5
6	6	7.5	0.8	0.8	73.9	9.7

TABLE 2.10 Stripline in Glass Epoxy, $\varepsilon_r = 4.2$, $t = 1.2$ mil

w (mil)	s (mil)	b (mil)	w/b	s/b	Z_0 (Ω)	X_{max} (%)
8	8	22	0.363	0.363	50.5	6.8
8	8	40	0.2	0.2	67.7	14.6
6	6	18.0	0.33	0.33	51.2	7.8
6	6	35	0.17	0.17	70.6	16.9

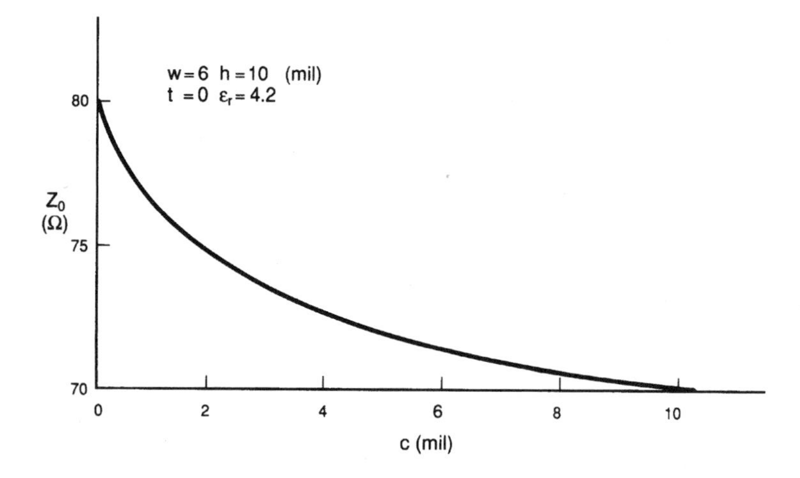

FIGURE 2.25 Z_0 reduction due to cover layer.

microstrip and stripline. Also shown is the maximum crosstalk noise X_{max} between two conductors in percent, for a given spacing *s* between conductors in the same signal layer as in Figure 2.27 and Figure 2.38b.

An important observation from Table 2.9 and Table 2.10 is that the thickness of the dielectric for a microstrip is considerably less than that of the stripline for the same Z_0. Thus, microstrip structures lend themselves to much thinner PCBs, and this is an advantage from a manufacturing point of view, since thicker boards are more expensive to make. In practical PCBs, the traces are protected by a coat of solder mask (cover coat). This cover coat is usually a dry film or is screen printed and has an $\varepsilon_r \approx 4$. Thus, we never have a classical microstrip with air at the top, but rather a covered microstrip. The effect of the cover layer is to increase the effective dielectric constant. Consequently, the capacitance of the line increases and Z_0 decreases. The reduction in Z_0 with cover layer thickness is shown in Figure 2.25.

In practical situations, the thickness of the cover layer is $\approx$3 to 4 mil, which can lead to a reduction in Z_0 of $\approx$10%. The covered microstrip can be considered to be a three dielectric layer problem, and the appropriate line parameters may be evaluated by techniques given in Das and Prasad (1984). For most practical designs today, the line width is typically 6 or 8 mil. There are some high-speed designs where 4-mil lines have been used, where the technology is migrating due to the demand for high circuit and interconnection density. There are some innovative design techniques that enable the user to attain high interconnection (circuit) density without increasing the board thickness by a large amount. We noted earlier that the microstrip structure lends itself to thinner dielectrics. However, the stripline provides a pure transverse electromagnetic (EM) wave and the protection of two ground planes. We can combine the advantages of both of these structures by using the asymmetric stripline.

To design a stripline with $Z_0 = 50\ \Omega$ in FR4 material ($\varepsilon_r \approx 4.2$) for a line width of 8 mil, we require a dielectric thickness of 22 mil. Now consider a classical stripline with $w = 6$ mil and $b = 18$ mil. In FR4, $Z_0 \approx 51\ \Omega$. If we now start moving one of the ground planes away from the line, Z_0 increases up to a certain distance, beyond which the line is not influenced by the presence of this ground plane. From Figure 2.26 we see that when b_2 is greater than 20 to 25 mil, the influence of this ground plane on Z_0 is negligible. We can therefore move the ground plane sufficiently far so as to have another signal plane to obtain an asymmetrical stripline as shown in Figure 2.27.

Now for $w = 6$ mil and $b = 22$ mil, we obtain two signal layers, while with 8 mil and $b = 22$ mil, we can only get one signal layer. Thus, by going to an asymmetrical stripline design, we can almost double the interconnection (circuit) density. There is the potential therefore to reduce an eight-layer multilayer board to four layers. The penalty that we pay is in increased crosstalk, but that can be addressed by wider interline spacing, as will be shown later.

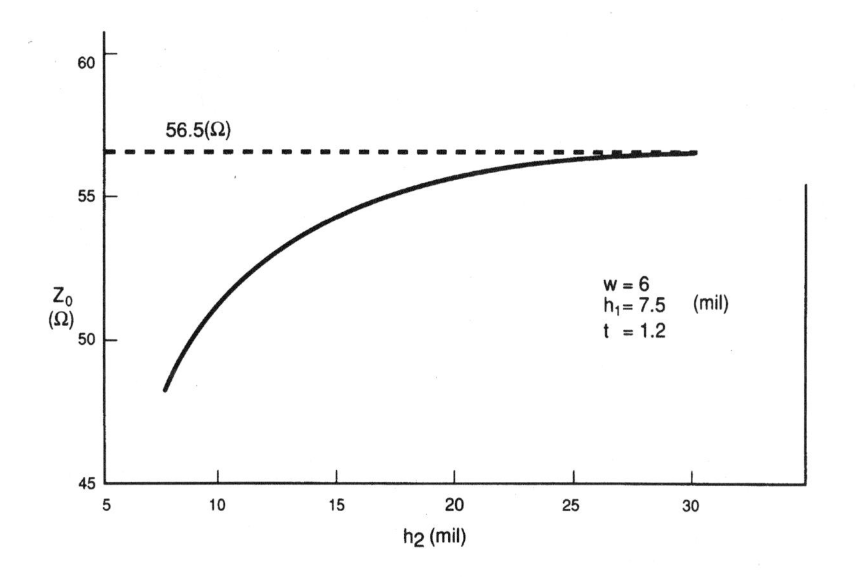

FIGURE 2.26 Increase in Z_0 by moving second groundplate.

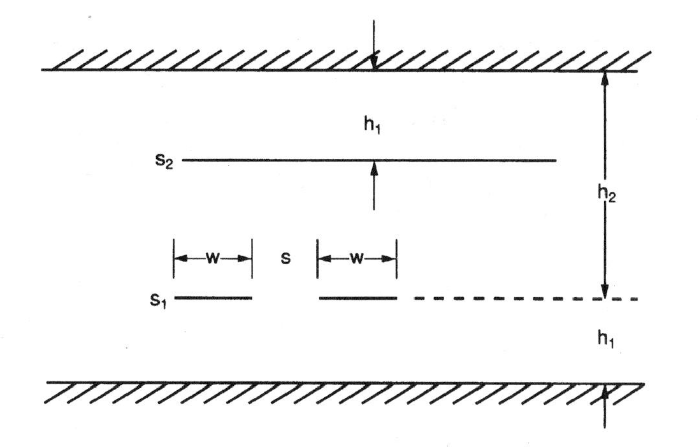

FIGURE 2.27 Two signal layers S_1 and S_2 in asymmetric stripline. S_1 and S_2 are orthogonal lines.

Effect of Manufacturing Tolerances on Z_0. For high-speed systems, designers require a tight control on impedance of boards. However, there is variation in Z_0 because of manufacturing tolerances associated with line w, t, dielectric thickness h or b, and with ε_r. A common design requirement is that the variation in Z_0 not exceed 10%. In a manufacturing environment, the variation in the relevant parameters is random in nature. We may therefore express the variation in Z_0 as

$$\Delta Z_0 \approx \sqrt{\left(\frac{\partial Z_0}{\partial w}\right)^2 \Delta w^2 + \left(\frac{\partial Z_0}{\partial h}\right)^2 \Delta h^2 + \left(\frac{\partial Z_0}{\partial t}\right)^2 \Delta t^2 + \left(\frac{\partial Z_0}{\partial \varepsilon_r}\right)^2 \Delta \varepsilon_r^2} \tag{2.31}$$

The partial derivatives may be evaluated from the appropriate expressions given in Equation (2.28), Equation (2.29), and Equation (2.30). They are shown in Table 2.11 for $Z_0 = 50$ and 75 Ω for classical microstrip and stripline.

TABLE 2.11 Variation in Z_0 due to Parameter Changes

Design	Z_0 (Ω)	w (mil)	$\partial Z_0/\partial h$ or $\partial Z_0/\partial b$ (Ω/mil)	$\partial Z_0/\partial w$ (Ω/mil)	$\partial Z_0/\partial \varepsilon r$ (Ω)
Microstrip	50	8	7.5	−3.8	−4.6
Microstrip	50	6	9.5	−4.8	−4.6
Microstrip	75	8	3.8	−3.8	−6.9
Microstrip	75	6	4.8	−4.8	−6.9
Stripline	50	8	1.3	−2.7	−6.2
Stripline	50	6	1.6	−3.4	−6.2
Stripline	50	4	2.1	−4.6	−6.2

However, the worst-case tolerances or the limits of the tolerance are defined by a Taylor series as

$$\Delta Z_0 \approx \left(\frac{\partial Z_0}{\partial w}\right)\Delta w + \left(\frac{\partial Z_0}{\partial h}\right)\Delta h + \left(\frac{\partial Z_0}{\partial t}\right)\Delta t + \left(\frac{\partial Z_0}{\partial \varepsilon_r}\right)\Delta \varepsilon_r \tag{2.32}$$

These tolerance limits are shown in Figure 2.28 and Figure 2.29 for a microstrip of $Z_0 = 75\ \Omega$ and for a stripline of 50 Ω. The line width is 6 mil ($\approx 150\ \mu$m) and the values chosen for Z_0 are of practical interest. The limits of variation in Z_0 are shown for a change in dielectric constant $\Delta\varepsilon_r = 0.1$, which is very realistic for variation in material properties. Also, dielectric constant for FR4 is not a constant, but varies with frequency as shown in Figure 2.30, taken from measurements made by S. Mumby at Bell Laboratories.

We observe that ε_r decreases with frequency, and that in the range of 100 Hz to 1 GHz, ε_r decreases from 4.8 to 4, a change of $\approx 20\%$. From the equations for Z_0, we observe that it scales approximately as $1/\sqrt{\varepsilon_r}$. Thus, a change of 20% in ε_r will result in $\approx 14\%$ change in Z_0 as we go from 100 Hz to 1 GHz. This is just to point out that ε_r is not a constant but does vary slightly with frequency. In Figure 2.29 and Figure 2.30, it is clear that in addition to a change of 0.1 in ε_r, any variation in w (Δw) and dielectric thickness (Δh or Δb) moves us up along the contours to increase $\Delta Z_0/Z_0$. The tolerance limits are shown for Δh or Δb of 12.5, 25, and 50 μm

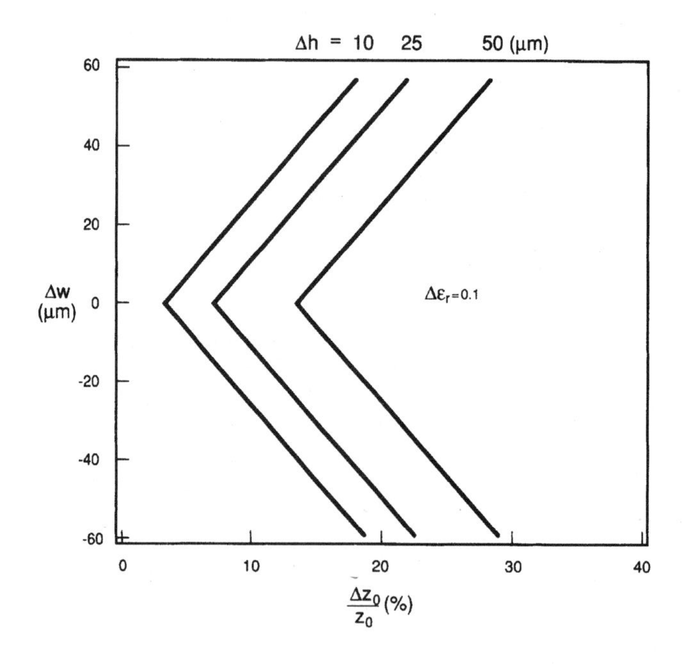

FIGURE 2.28 Tolerance limits for microstrip line. $Z_0 = 75\ \Omega$, $w = 6$ mil, $t = 1.4$ mil, $\varepsilon_r = 4.2$ (1 mil = 25 μm).

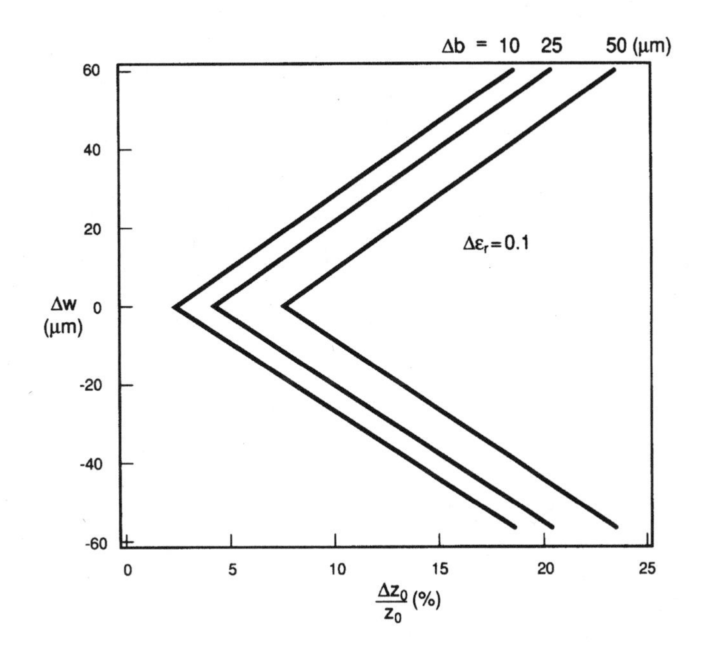

FIGURE 2.29 Tolerance limits for stripline. $Z_0 = 50\ \Omega$, $w = 6$ mil, $t = 1.4$ mil, $\varepsilon_r = 4.2$.

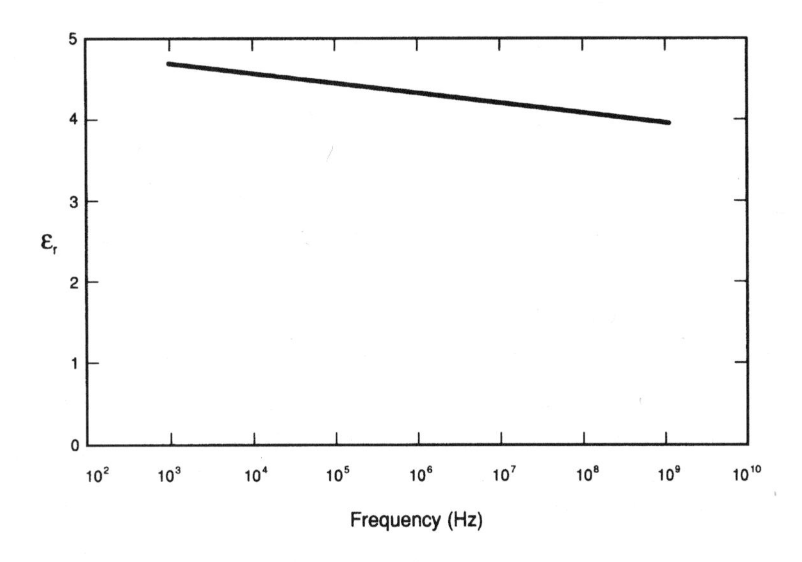

FIGURE 2.30 Variation of ε_r with frequency for FR4.

(0.5, 1, and 2 mil). In order to limit the Z_0 tolerance to 10%, we need tight control over the manufacturing process. Designers should be aware of the manufacturing tolerances of PCB manufacturing to determine the ability to meet their controlled impedance requirements. These aspects are discussed in Ritz (1988).

Crosstalk (X)

Crosstalk may be defined as noise that occurs on idle lines due to interactions with stray EM fields that originate from active (pulsed) lines. This interaction is shown pictorially in Figure 2.31 and Figure 2.32.

There are two components to crosstalk. They are inductive crosstalk due to mutual inductance, L_m, and capacitive crosstalk due to mutual capacitance, C_m. Inductive crosstalk is proportional to $L_m\ dI/dt$, and

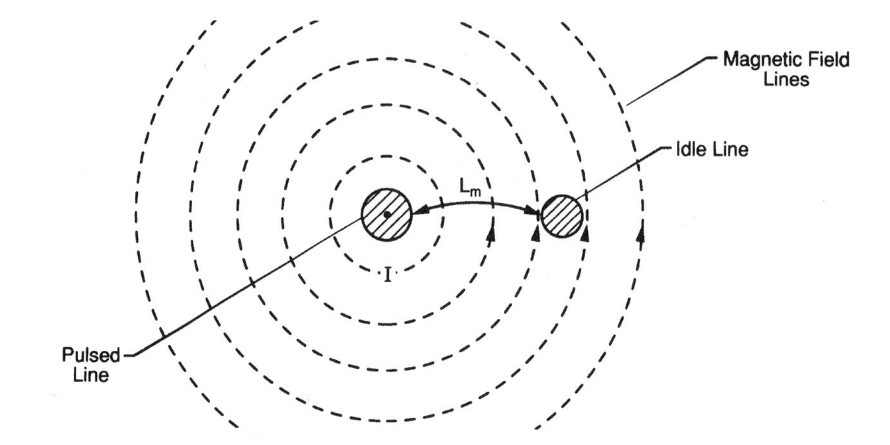

FIGURE 2.31 Schematic of inductive crosstalk coupling.

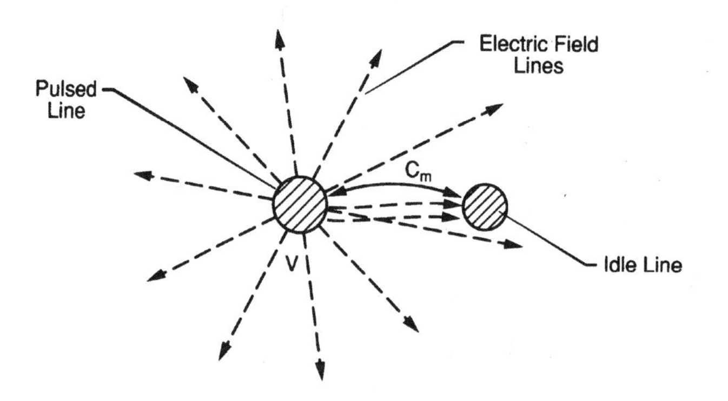

FIGURE 2.32 Schematic of capacitive crosstalk coupling.

capacitive crosstalk is proportional to C_m dV/dt. In any conductor in which we have a transient current propagating perpendicular to the plane of the paper, circular magnetic field lines are generated in space as shown in Figure 2.31. Any conductor that interacts with this magnetic field has crosstalk noise imposed on it due to mutual inductive coupling L_m between the pulsed and the idle lines. Similarly, a transient voltage pulse in the conductor generates radial electric fields emanating from the conductor as shown in Figure 2.32. An idle line that interacts with this electric field has crosstalk noise imposed on it due to mutual capacitance coupling between the pulsed and idle lines. Thus, to predict crosstalk, we must be able to evaluate the mutual coupling coefficients L_m and C_m.

Consider a pulsed line and an idle line that are terminated in their characteristic impedances as shown in Figure 2.33.

If we generate a digital pulse on the active pulsed line, then the crosstalk noise on the idle line is given by

$$V_{NE} = \frac{1}{2}\left(L_m \frac{dI}{dt} + Z_0 C_m \frac{dV}{dt}\right) \text{V/m} \tag{2.33}$$

$$V_{FE} = \frac{1}{2}\left(Z_0 C_m \frac{dV}{dt} - L_m \frac{dI}{dt}\right) \text{V/m} \tag{2.34}$$

The section of the line near the source is called the "near end" (NE), and the end of the line away from the source is called the "far end" (FE). Note that the NE crosstalk noise V_{NE} is the sum of the inductive and

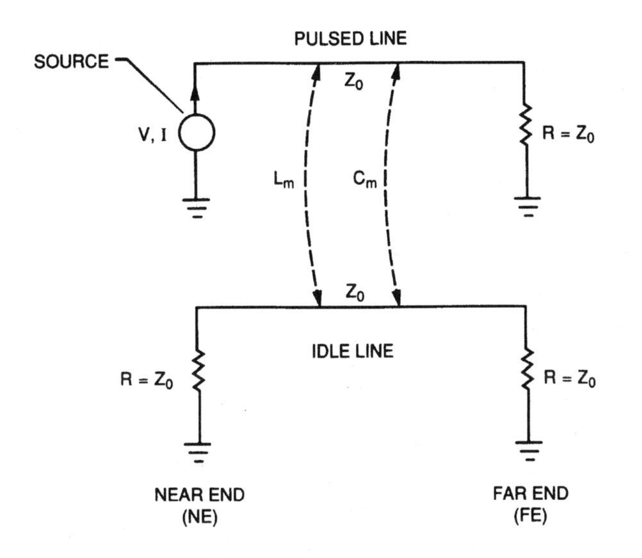

FIGURE 2.33 Crosstalk model showing near end and far end.

capacitive components, while the FE crosstalk noise V_{FE} is the difference between the capacitive and inductive components. The crosstalk noise given by Equation (2.33) and Equation (2.34) is per unit length of the line. Therefore, to evaluate the total crosstalk, we must multiply V_{NE} and V_{FE} by the length of coupling. This is generally true for PCBs and cables where L_{m} and C_{m} are expressed in nH/m and pF/m. For a pure transverse EM (TEM) wave propagation, $Z_0 = \sqrt{L_{\mathrm{m}}/C_{\mathrm{m}}}$. Therefore, the inductive and capacitive crosstalk noise components are equal, and $V_{\mathrm{FE}} = 0$. However, in most practical applications, we do not have a pure TEM wave, and the two components differ by a small amount. Therefore, in connectors and PCBs where the length of coupling ranges from 0.5 to 1.0 in. for connectors, and about 10 to 20 in. for PCBs and backplanes, $V_{\mathrm{FE}} \approx 0$. However, in cables that may run for several meters between equipment, the total FE crosstalk V_{FE} can become quite large since we are multiplying a small difference between capacitive and inductive crosstalk by a large length. Thus, for connectors and PCBs, we are mainly interested in NE crosstalk noise V_{NE}, but for cables, we are interested in both V_{NE} and V_{FE}.

From Equation (2.33) and Equation (2.34) it is obvious that V depends on L_{m}, C_{m}, Z_0, the rate of change of current and voltage, $\mathrm{d}I/\mathrm{d}t$ and $\mathrm{d}V/\mathrm{d}t$, respectively, and the length of coupling. What is of interest is to know the shape and duration of both V_{NE} and V_{FE}. We shall see that the answer to this question depends upon the relative magnitudes of the propagation delay in the line T_{d} (which is proportional to length) and the risetime of the pulse T_{r} in digital systems, or the frequency of the pulse for analog systems. The sign of the inductive and capacitive noise components may be easily understood by looking at the simple mechanical analogy of the spring–mass system. The inductive noise given by $L_{\mathrm{m}}\,\mathrm{d}I/\mathrm{d}t$ has the effect of inserting a voltage source V_{L} in the idle line as shown in Figure 2.34.

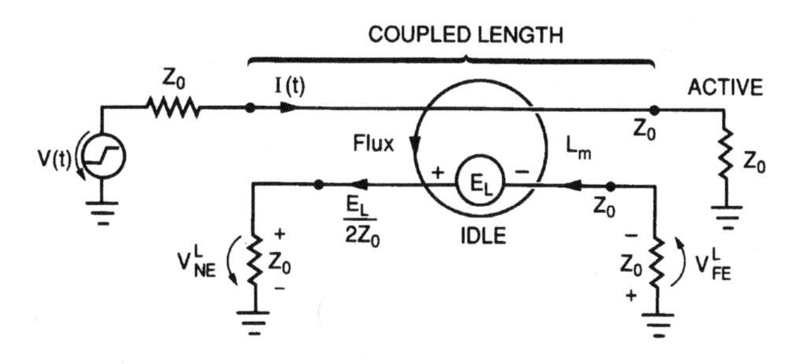

FIGURE 2.34 Inductive noise on idle line.

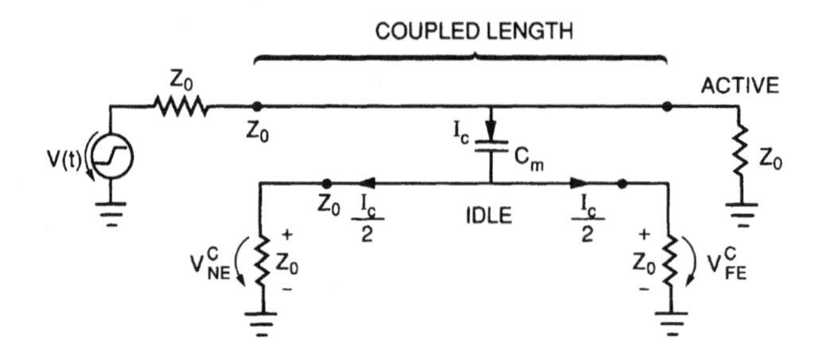

FIGURE 2.35 Capacitive noise on idle line.

However, inductance has the effect of opposing the source current pulse. The source creates a clockwise loop of current (action). Inductance behaves like the mechanical spring, which produces a reaction $-kx$ to the displacement x (action). So, as a reaction to the clockwise loop of current in the source line (action), mutual inductance induces a counterclockwise loop of current in the idle line (reaction) to oppose the source pulse. This therefore determines the polarity of the voltage source on the idle line as shown in Figure 2.34. It is this voltage source that generates the counterclockwise loop of current. Therefore, we observe from Figure 2.34 that since the current flows from the near end to ground, it must be positive with respect to ground. Similarly, since the direction of current is from ground to far end, it must be negative with respect to ground. Thus, we may observe from Equation (2.33) and Equation (2.34) that the inductive coupling term is positive at the near end and negative at the far end. The current produced by mutual inductance is $E_L/2Z_0$. Therefore, the inductive coupling noise at the near end and far end is given by $V_{NE}^{L} = -V_{FE}^{L} = E_{L/2}$, where $E_L = L_m dI/dt$. For the source line, we can write $dI/dt \approx V_s / Z_0 T_r$. Here, V_s is the magnitude of the source pulse, and T_r is the risetime of the pulse.

Capacitive coupling has the effect of inserting a mutual capacitor C_m between the source and idle lines as shown in Figure 2.35. The current that passes through C_m divides evenly in the idle line, with half going to ground through the near end and the other half going to ground through the far end. Thus, the capacitive crosstalk voltages at the near end and the far end are given by $V_{NE}{}^{C} = V_{FE}{}^{C} = I_C Z_0/2$. The coupled capacitive current $I_C = C_m\, dV/dt \approx C_m V_s/T_r$. Therefore, mutual capacitance may be considered to be like mass (inertia) in the mechanical spring–mass analogy. Mass (inertia) tends to keep going in the direction of displacement. Similarly, mutual capacitance induces the same noise in both the near and far end as can be seen in Equation (2.33) and Equation (2.34), and the sign of the noise is the same as dV/dt (inertia). The total crosstalk noise (X) is due to inductive (X_L) and capacitive (X_C) noise and is usually expressed as a fraction of the input source pulse V_s as $X = V/V_s$. Therefore, for a coupling length of L, we can write crosstalk noise as

$$X_{NE} = \frac{V_{NE}}{V_s = \frac{1}{2}\left(Z_0 C_m + \frac{L_m}{Z_0}\right)\frac{L}{T_r}} \tag{2.35}$$

$$X_{FE} = \frac{V_{FE}}{V_s} = \frac{1}{2}\left(Z_0 C_m - \frac{L_m}{Z_0}\right)\frac{L}{T_r} \tag{2.36}$$

However, we know that length $L = vT_d$. Therefore,

$$X_{NE} = \frac{V_{NE}}{V_s} = \frac{v}{4}\left(Z_0 C_m + \frac{L_m}{Z_0}\right)\frac{2T_d}{T_r} = K_{NE}\left(\frac{2T_d}{T_r}\right) \tag{2.37}$$

$$X_{FE} = \frac{V_{FE}}{V_s} = \frac{1}{2}\left(Z_0 C_m - \frac{L_m}{Z_0}\right)\frac{L}{T_r} = K_{FE}\left(\frac{L}{T_r}\right) \tag{2.38}$$

$$K_{\mathrm{NE}} = \frac{v}{4}\left(Z_0 C_{\mathrm{m}} + \frac{L_{\mathrm{m}}}{Z_0}\right) \tag{2.39}$$

$$K_{\mathrm{FE}} = \frac{1}{2}\left(Z_0 C_{\mathrm{m}} - \frac{L_{\mathrm{m}}}{Z_0}\right) \tag{2.40}$$

The expressions derived here are for noise in digital systems. It can be shown analytically and experimentally (Rainal, 1979) that near-end noise, X_{NE}, increases with length of coupling, reaches a maximum value, and saturates there. Any increase in coupling length beyond a critical length will not increase the value of X_{NE}. This critical coupling length for maximum crosstalk $X_{\max}$ depends upon the delay and signal risetime and is given by

$$X_{\mathrm{NE}} = K_{\mathrm{NE}} = X_{\max} \quad \text{for } \frac{2T_{\mathrm{d}}}{T_{\mathrm{r}}} \geqslant 1 \quad \text{(maximum crosstalk)} \tag{2.41}$$

$$X_{\mathrm{NE}} = K_{\mathrm{NE}}\left(\frac{2T_{\mathrm{d}}}{T_{\mathrm{r}}}\right) \quad \text{for } \frac{2T_{\mathrm{d}}}{T_{\mathrm{r}}} < 1 \tag{2.42}$$

Equation (2.41) is the condition for transmission line behavior, that is, $2T_{\mathrm{d}}/T_{\mathrm{r}} \geq 1$. Equation (2.42) is the limit for lumped parameter analysis, that is, $2T_{\mathrm{d}}/T_{\mathrm{r}} < 1$. Note that in the transmission line condition, X_{NE} is independent of risetime. For the transmission line condition (long coupling length as in cables and backplanes), the shape of the near-end noise is trapezoidal, while for lumped parameter condition (as in connectors), the shape of the near-end noise is triangular. These are clear in Figure 2.36a and b.

Also, we observe from Figure 2.36a that X_{NE} reaches a maximum value and saturates there. It is instructive to determine the lengths of lines for which transmission line characteristics are valid and when maximum crosstalk occurs. If we consider FR4 glass epoxy boards for which $\varepsilon_{\mathrm{r}} \approx 4$, the wave speed v is $\approx 1.5 \times 10^8$ m/sec $\approx$ 6 in./nsec, or the delay is 1/6 nsec/in. The lengths for which $2T_{\mathrm{d}} \approx T_{\mathrm{r}}$ are shown in Table 2.12 for FR4 boards.

Note that for a risetime of 1 nsec, it only takes 3 in. of coupling for maximum crosstalk to occur. Thus, while the near-end noise X_{NE} has a maximum limit, far-end noise X_{FE} given by Equation (2.38) increases linearly with length. This is the reason that X_{FE} is important for cables running over long distances. This is obvious from Figure 2.36a, where near-end and far-end noise are almost of the same magnitude. Another very important observation to be made from Figure 2.36a is that while the active pulse (source) leaves the source line at time T_{d}, the near-end noise has a width of $2T_{\mathrm{d}}$. Thus, the crosstalk noise stays around for an extra time delay T_{d}, even though the active pulse causing the noise has already left the source line. However, observe that far-end noise occurs as a single pulse at time T_{d}. The reason for this phenomenon is that near-end coupling occurs the moment the pulse enters the active line. As the source pulse travels on the active line, the coupled noise on the idle line is sent back continuously to the near-end from every point of coupling between the two lines. Therefore, there is a delay in the noise as it travels from a point of coupling back to the near-end. As the coupled noise travels back to the near-end, the last portion of the noise that coupled at the end of the active line (far end) has to travel back all the way to the near-end. It took the active pulse a time T_{d} to reach the end of the line, and it takes another T_{d} for the coupled noise to travel back to the near end. Thus, the near-end noise is continuously being sent back for a duration of time $2T_{\mathrm{d}}$, even though the active pulse has left the line at T_{d}. This is why near-end crosstalk is also referred to as backward crosstalk. On the other hand, far-end noise travels forward with the active pulse and appears as a single noise pulse at T_{d} just as the active pulse leaves the line. For this reason, far-end crosstalk is also referred to as forward crosstalk.

For a periodic signal, the crosstalk formulas are given in Rainal (1979) as

$$X_{\mathrm{NE}} = 2K_{\mathrm{NE}}\sin(2\pi f T_{\mathrm{d}}) \tag{2.43}$$

$$X_{\mathrm{FE}} = K_{\mathrm{FE}}(2\pi f L) \tag{2.44}$$

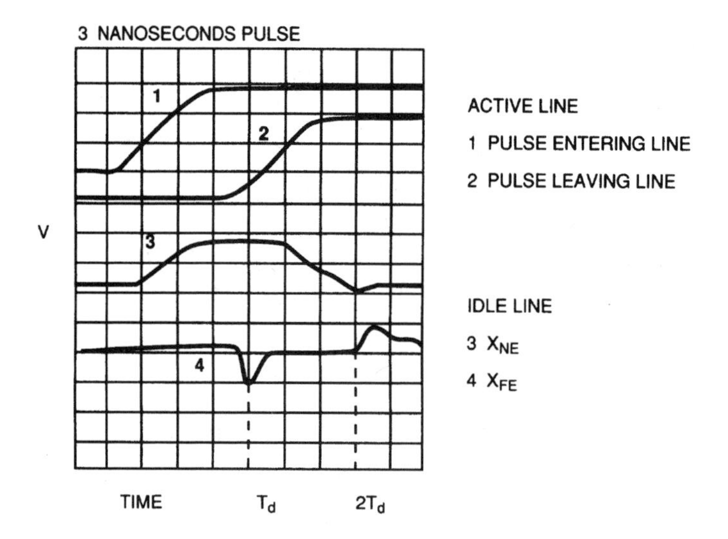

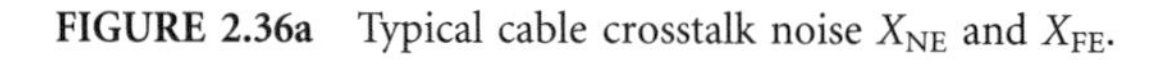

FIGURE 2.36a Typical cable crosstalk noise X_{NE} and X_{FE}.

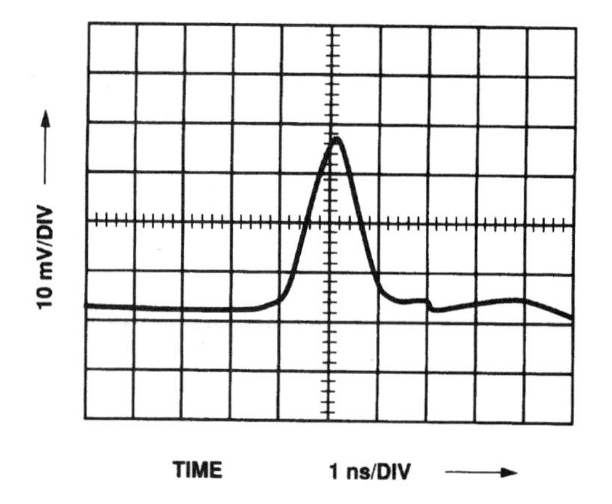

FIGURE 2.36b Typical connector near-end crosstalk X_{NE}.

TABLE 2.12 Lengths for Maximum Crosstalk in FR4 Boards

T_r (nsec)	$T_d = T_r/2$ (nsec)	L (in.)
0.2	0.1	0.5
0.5	0.25	1.5
1.0	0.5	3
2.0	1.0	6
5.0	2.5	15
10	5	30

Note that the maximum value of near-end crosstalk for a periodic signal given by Equation (2.43) is twice that obtained for a digital pulse. Also, the magnitude of X_{NE} depends upon both the length of coupling (delay T_d), as well as frequency. Since X_{NE} is sinusoidal, we can obtain greater crosstalk with smaller coupling length and vice versa. Once again, far-end crosstalk increases linearly with coupling length, and in addition also increases linearly with frequency. However, note that for both analog and digital signals, crosstalk noise X_{NE} and X_{FE}

depend upon the mutual coupling coefficients L_m and C_m. Thus, the key to reducing crosstalk is to reduce both L_m and C_m. The formulas given so far are for crosstalk due to a single source. If there are n sources pulsing simultaneously, then the worst-case crosstalk on an idle line is given by

$$X_{TOTAL} = X_1 + X_2 + \cdots + X_n \tag{2.45a}$$

For most practical applications, the pulses may never be synchronized, and a more realistic estimate of crosstalk may be given by

$$X_{TOTAL} = \sqrt{X_1^2 + X_2^2 + \cdots + X_n^2} \tag{2.45b}$$

Normally, X_{NE} and X_{FE} are measured experimentally using a time-domain reflectometer (TDR). From these measured values, we may evaluate the capacitive crosstalk X_C and inductive crosstalk X_L as

$$X_C = \frac{1}{2}(X_{NE} + X_{FE}) \tag{2.46a}$$

$$X_L = \frac{1}{2}(X_{NE} - X_{FE}) \tag{2.46b}$$

From these, we can evaluate the mutual coefficients L_m and C_m, which enable us to scale the results for crosstalk noise for different values of T_p, T_d, L and T_r, or f. L_m and C_m may also be evaluated analytically for many practical situations, and these will now be considered.

Expressions for L_m and C_m. In order to analytically evaluate L_m and C_m, we use the coupled lines techniques developed by microwave engineers. For this, we consider the two coupled lines to be in the odd and even mode configurations as shown in Figure 2.37a and b for a microstrip design.

In the odd mode configuration, the potential on one line is the negative potential of the other line, while in the even mode, both lines have the same potential imposed upon them. Let C_a be the capacitance of an isolated line to ground. Then, the odd and even mode capacitances, inductances, and characteristic impedances are

$$C_{odd} = C_a + 2C_m, \quad L_{odd} = L_a - L_m, \quad Z_{0odd} = \sqrt{\frac{L_{odd}}{C_{odd}}} \tag{2.47a}$$

$$C_{even} = C_a, \quad L_{even} = L_a + L_m, \quad Z_{0even} = \sqrt{\frac{L_{even}}{C_{even}}} \tag{2.47b}$$

The characteristic impedance for a balanced or differential line Z_{0b} is given by

$$Z_{0b} = 2Z_{0odd} \tag{2.47c}$$

Expressions for C_{odd} and C_{even}, and Z_{0odd} and Z_{0even} are given in many references (such as Davis, 1990). From these, we can then evaluate L_{odd} and L_{even}, and we may express the mutual coefficients as

$$C_m = \frac{C_{odd} - C_{even}}{2}, \quad L_m = \frac{L_{even} - L_{odd}}{2} \tag{2.47d}$$

Mathematically, the odd and even mode impedances are the minimum and maximum values of impedances for coupled lines. Observe that this change in impedance is due to the mutual coefficients. This has

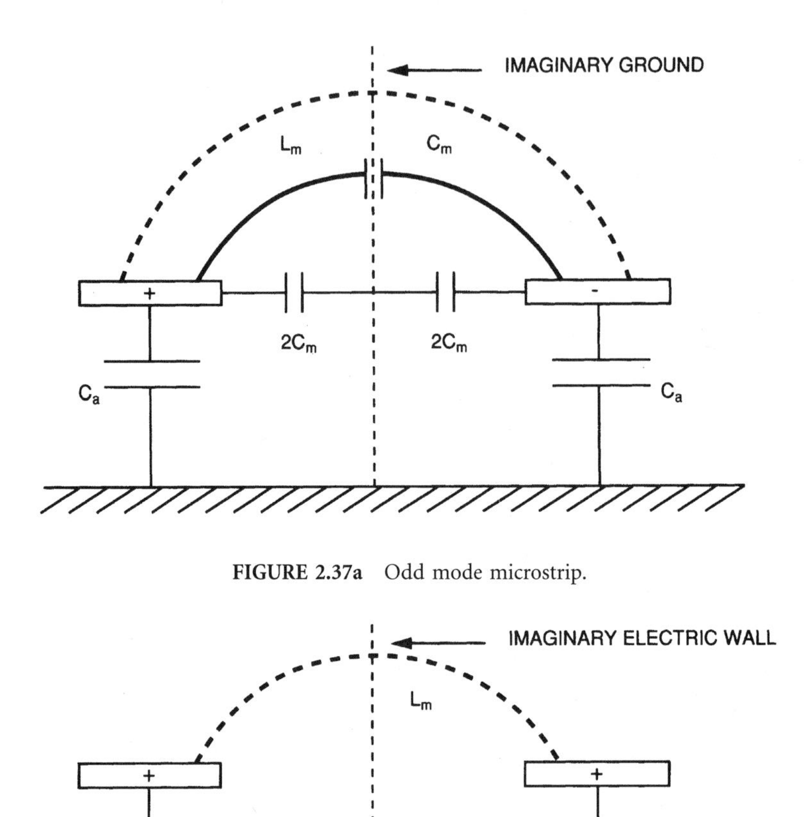

FIGURE 2.37a Odd mode microstrip.

FIGURE 2.37b Even mode microstrip.

important implications in high-speed and high-density designs where many lines are closely spaced. The very presence of an idle line in proximity alters Z_0 of the line. Consider a completely isolated line (alone) with its parameters C_a, L_a, and Z_a as in Figure 2.38a. Next consider one idle neighbor line as in Figure 2.38b.

The presence of an idle line sets up a series capacitive path from active line to ground through mutual coupling to idle line. Thus, with one idle neighbor line, capacitance increases from C_a to C_{1i} given by

$$C_{1i} = C_a + \frac{C_a C_m}{C_a + C_m} \approx C_a + C_m \text{ for } \frac{C_m}{C_a} << 1 \tag{2.48}$$

Thus, the presence of one idle neighbor increases the capacitance by $\approx C_m$. Similarly, the presence of two idle neighbors increases capacitance by $\approx 2C_m$. However, in both cases, inductance remains the same as the isolated line, L_a. Therefore, the impedances for one and two idle neighbors are given respectively as

$$Z_{0i1} \approx \sqrt{\frac{L_a}{C_a + C_m}}, \quad Z_{0i2} \approx \sqrt{\frac{L_a}{C_a + 2C_m}} \tag{2.49}$$

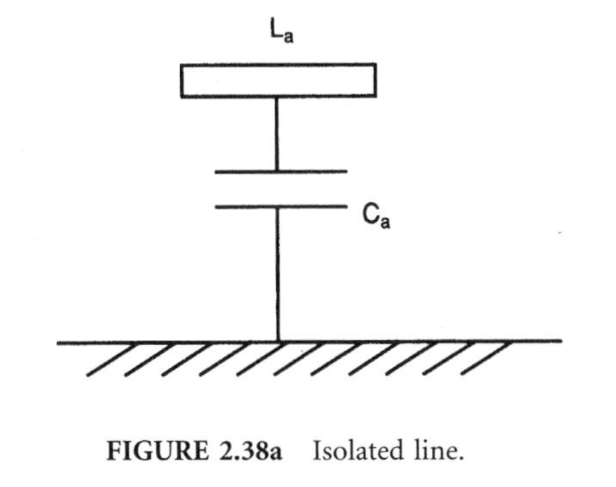

FIGURE 2.38a Isolated line.

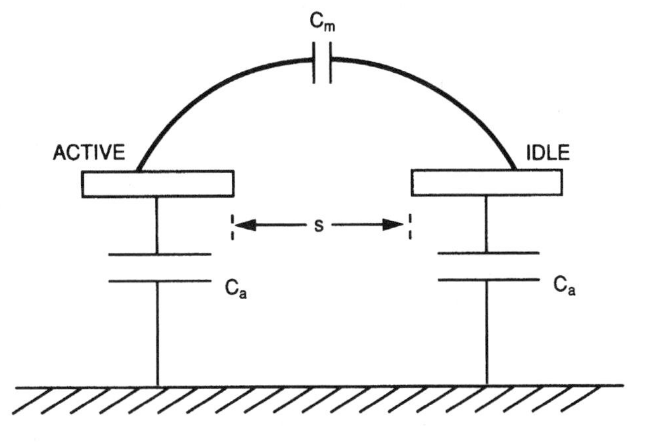

FIGURE 2.38b Line with one idle neighbor.

Therefore, since the presence of idle lines increases capacitance, there is a reduction in characteristic impedance, which may be important for controlled impedance lines. The decrease in impedance due to one and two idle neighbor lines on a 50-Ω stripline design in FR4 ($\varepsilon_r = 4.2$, and $t = 1.2$ mil) is shown in Figure 2.39a for different line width and space (w/s in mil/mil) combinations. In Figure 2.39b are shown the maximum crosstalk for the same w/s combinations. Observe that the shape of the lines in both figures is identical. The larger the crosstalk, the greater will be the change in impedance. Thus, crosstalk and impedance are related, and when we design for low crosstalk we are automatically designing for good impedance control. Also, from Figure 2.39b we observe that to reduce crosstalk we must increase the space s between lines. As we go from a 6/6 design to a 6/10 design, maximum crosstalk reduces from 7.5 to 3.5%, a reduction of 4%. This has relevance to Figure 2.27 where we showed that the asymmetric stripline design lends itself to higher interconnection density. We also noted that the penalty to be paid for moving one ground plane farther away is increased crosstalk. However, by increasing s, we can reduce crosstalk and reduce this penalty. As shown in Figure 2.40, as we change from a w/s design of 6/6 to 6/8, we get an almost uniform reduction of 2% in maximum crosstalk. Thus, we can adjust the spacing to achieve crosstalk immunity. We should always keep in mind that in addition to manufacturing tolerances, the presence of adjacent lines in dense boards has the effect of altering Z_0.

Noise in Connectors, IC Leads, and Wirebonds. Unlike boards, backplanes, and cables, the lengths of the signal paths in connectors, IC pins, and wirebonds are very short so that delay is small compared to risetime. Therefore, they behave like lumped parameters. Therefore, near-end crosstalk noise is important. We also noted that in a pure TEM wave, both inductive and capacitive noise contributions are almost equal, which is appropriate for boards and cables. However, in connectors, pins, and wirebonds, the predominant noise contribution is inductive in nature. Fortunately, inductive noise can be evaluated from some simple analytical

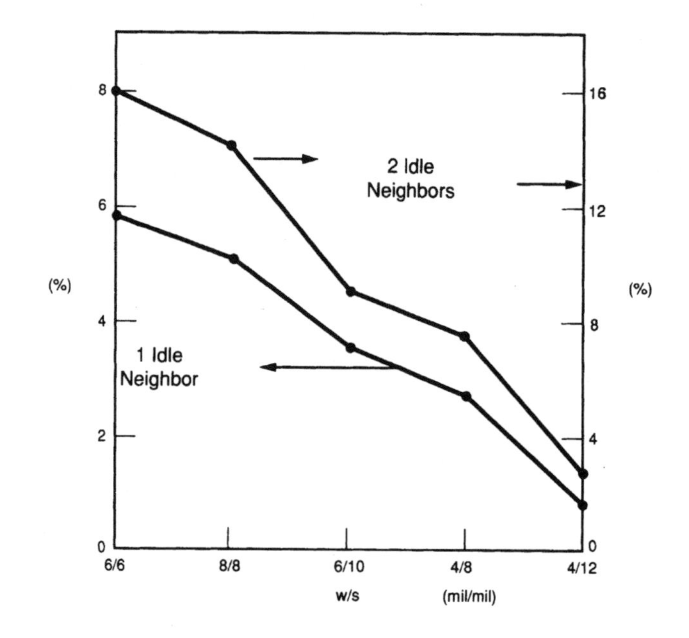

FIGURE 2.39a Reduction in Z_0 due to presence of adjacent lines. Stripline nominal $Z_0 = 50\ \Omega$, $\varepsilon_r = 4.2$, $t = 1.2$ mil.

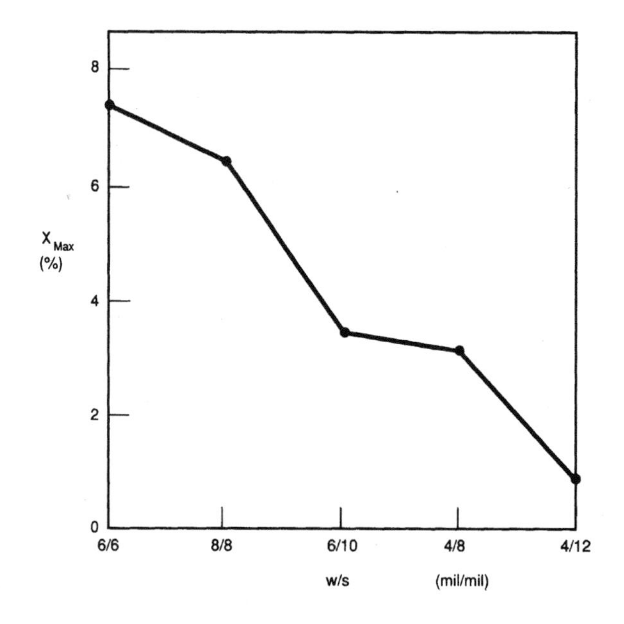

FIGURE 2.39b Maximum crosstalk for various *w*/*s* combinations. Stripline nominal $Z_0 = 50\ \Omega$, $\varepsilon_r = 4.2$, $t = 1.2$ mil.

expressions that have been verified experimentally. However, to evaluate capacitive noise, we have to resort to detailed field solutions of the electromagnetic equations. This can be very complicated. However, from measurements in connectors, we have determined that the maximum capacitive noise contribution is ≈30% of X_{NE}. In a practical situation, this may be ≈15 to 20%, which may be added to the inductive noise. Inductive noise itself is made up of two components: noise due to self-inductance of the pins or wires and noise due to mutual inductance between pins or wires. The mutual inductance, L_m (nH), between two wires of

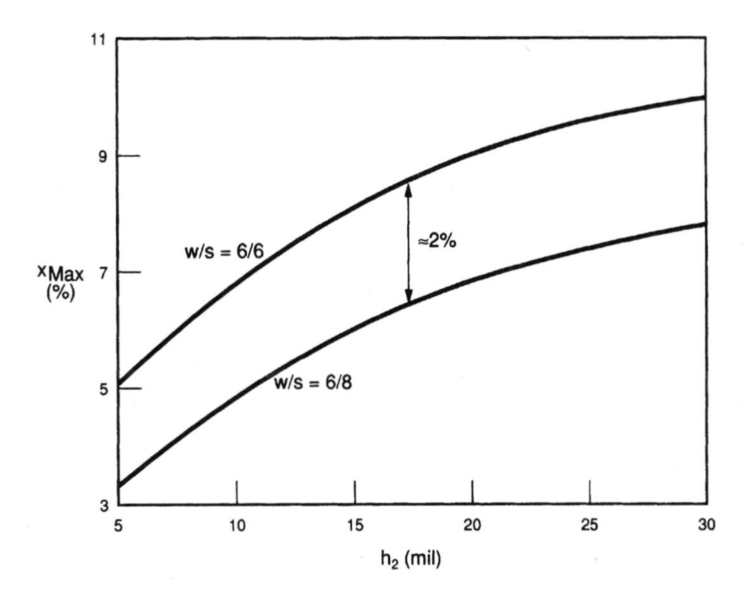

FIGURE 2.40 Asymmetric stripline crosstalk reduction. $w=6$ mil, $t=1.2$ mil, $h_1=7.5$ mil.

length l (in.) and center line separation of d (in.) is given by Rainal (1984):

$$L_{\mathrm{m}} \approx 5l\left[\ln\left(\frac{l}{d}+\sqrt{1+(l/d)^2}\right)-\sqrt{1+(d/l)^2}+d/l\right]\mathrm{nH} \tag{2.50}$$

Similarly, the self-inductance L_s (nH) of a pin of radius ρ (in.) and length l (in.) is given by

$$L_{\mathrm{s}} = L_{\mathrm{m}}|_{d=\rho} + \frac{5l}{4}\ \mathrm{nH}$$

Thus, total self-inductance is the sum of the contributions from both external and internal magnetic fields. For a rectangular conductor of perimeter p (in.), self-inductance is given by

$$L_{\mathrm{s}} \approx 5l\left[\ln\left(\frac{4l}{p}\right)+\frac{1}{2}\right]\mathrm{nH} \tag{2.52}$$

Thus, inductive noise due to self-inductance can be significant even in the presence of mutual inductance. It is this phenomenon that gives rise to ground noise in connector pins and wire-bonds in IC packages and may be as high as 100 mV or more (Rainal, 1984). This becomes an important issue in the allocation of ground return leads or wires during the signal layout of connectors and ICs. The layout of a connector or IC leads (wirebonds) consists of arrays or clusters of conductors. Of these, a certain number are allocated to signals and the remaining to ground. In connectors, we are mainly interested in X_{NE} on an idle signal pin, when one or more of the adjacent signal lines are pulsed. If an array of n conductors are active, then the total noise on the idle pin due to mutual coupling is given by

$$V_{\mathrm{m}} = \sum_{i=1}^{i=n} L_{\mathrm{m}i}(\mathrm{d}I/\mathrm{d}t)_i \tag{2.53}$$

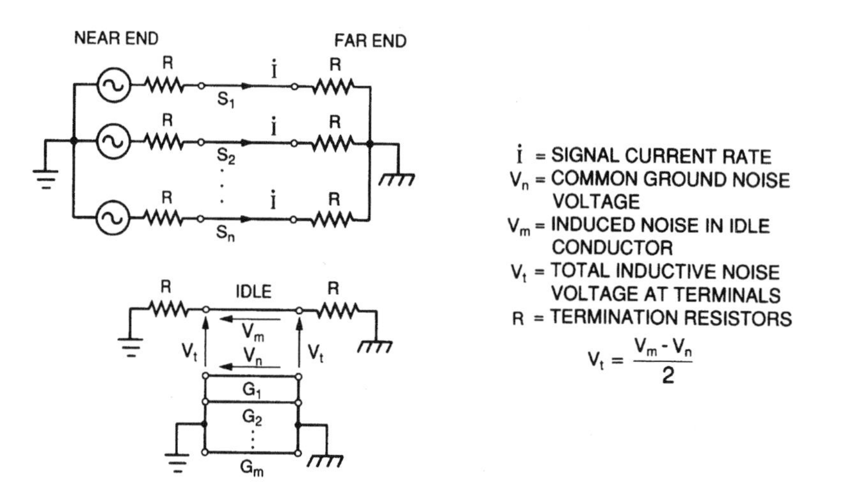

FIGURE 2.41 Inductive noise model for an array of signal (S_i), ground (G_i), and idle conductors.

Noise on the ground pins is the sum of self and mutual inductances and is given by

$$V_{\mathrm{n}} = -\sum_{i=1}^{i=n} (L_{\mathrm{s}} - L_{\mathrm{mi}})(\mathrm{d}I/\mathrm{d}t)_i \tag{2.54}$$

The return currents in the ground pins or leads will vary according to the inductive field around it. Since the ground has shifted, the near-end inductive crosstalk on the idle pin is given by

$$V_{\mathrm{t}} = \frac{V_{\mathrm{m}} - V_{\mathrm{n}}}{2} = V_{\mathrm{NE}}^{\mathrm{L}} \tag{2.55}$$

The model for noise evaluation in connectors and wirebonds is shown in Figure 2.41.

Note that ground noise becomes an important part of inductive noise. This phenomenon, also called "ground bounce," can be a significant problem when 16 or 32 signal bits are switched simultaneously in a chip. For high-speed logic designs, ground noise must be kept to a minimum. This requires as many ground return leads as possible. In addition, ground and signal leads must be closely alternated in a checkerboard pattern. Keeping all the signals clustered in one region and all the grounds clustered in another region will cause significant ground noise problems (Rainal, 1984).

Risetime Degradation

When a pulse is passed through an interconnection element, there is an increase in the risetime of the pulse, and this slowing down of the wave is referred to as risetime degradation. If the input risetime is T_{ri} and the output risetime is T_{ro}, then risetime degradation T_{rd} is defined as

$$T_{\mathrm{rd}} = \sqrt{T_{\mathrm{ro}}^2 - T_{\mathrm{ri}}^2} \tag{2.56}$$

Thus, if we have a step input, the output risetime is T_{rd}. If there are n interconnection elements, each with risetime degradation T_{rd1}, T_{rd2}, etc., then the risetime at output of n interconnection elements is

$$T_{\mathrm{ro}}^2 = T_{\mathrm{ri}}^2 + T_{\mathrm{rd1}}^2 + T_{\mathrm{rd2}}^2 + \ldots T_{\mathrm{rdn}}^2 \tag{2.57}$$

This relationship is important because risetime degradation is not just the sum, but the square root of the sum of squares. Thus, if T_{rd1} is 0.2 and T_{rd2} is 0.5, it is almost six times (0.25/0.04) more important than T_{rd1}. Therefore, attention should be paid to those elements that are most important. In general, risetime degradation occurs due to resistive, inductive, and capacitive effects of circuits. In PCBs with long traces as transmission lines, T_{rd} is due to a combination of dc (IR loss) and skin effect. At very low frequencies, the current fills the entire cross-section of the conductor. As we go to higher frequencies, the current is concentrated in a very thin layer (skin) around the perimeter of the conductor. This thin layer is called skin depth, $\delta = 1/\sqrt{\pi f \sigma \mu}$, where σ is the conductivity, μ the permeability, and f the frequency. For copper, $\delta = 2.09$ μm at 1 GHz. Skin effect adds a series resistance to the line, and for a rectangular conductor, the first-order approximation is

$$R_s = \frac{1}{[\sigma \times \text{perimeter} \times \delta]} \tag{23.58}$$

Skin effect slows down the signal and also lowers the magnitude of the pulse. For a copper conductor with $w = 4$ mil and $t = 2$ mil, for a risetime of 0.5 nsec, skin effect becomes critical when the length of the trace approaches 40 in. (Chang, 1988). Therefore, for most practical board designs today, the lossless line is a good approximation.

Risetime degradation is important because it is related to the **bandwidth** (BW) of the interconnection medium given as

$$\text{BW} = \frac{0.35}{T_{rd}} \tag{2.59}$$

The concept of bandwidth itself comes from low-pass RC filter theory. We know that the classical low-pass RC filter passes low frequencies readily, but attenuates high frequencies. For this RC filter, the bandwidth is given by $f_2 = 1/2(\pi RC)$ and is the frequency at which the gain of the filter falls to 3 dB (70%) of the low-frequency value. It is like a cutoff frequency and is a qualitative measure of the transfer of energy through the filter. If we put a step input to the filter, the output from the filter is shown in Figure 2.42.

From theory, we can show that $f_2 = 0.35/t_r$. Typically, good design requires that the interconnection loss be limited to ≈ 1 dB. For connector characterization we pass a very fast pulse (almost a step input) and observe the output pulse leaving the connector and its bandwidth expressed according to Equation (2.59).

The low-pass filter concept is very useful because most interconnection elements such as connectors, short PCB traces, and IC package pins behave as low-pass filters. Since their lengths are relatively small compared to pulse risetime, they act as lumped parameters and discontinuities along the transmission path.

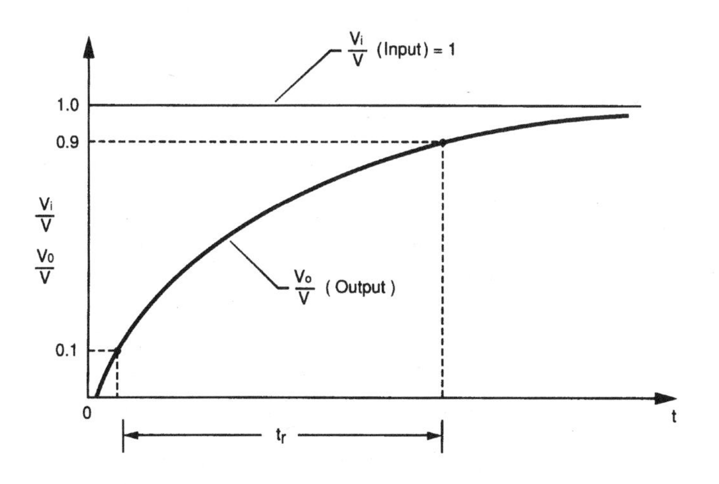

FIGURE 2.42 Low-pass RC circuit response to step input.

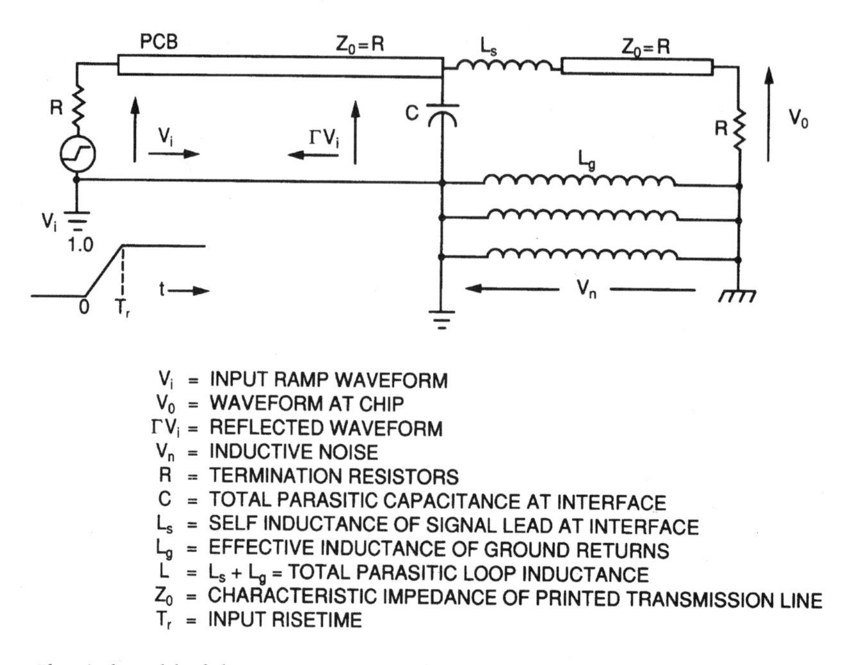

FIGURE 2.43 Electrical model of the transmission path from a printed wiring board to a matched high-speed chip. (*Source:* A.J. Rainal, "Performance limits of electrical interconnections to a high-speed chip," *IEEE Trans. CHMT,* vol. 11, no. 3, pp. 260–266, Sept. 1988. © 1988 IEEE.)

The degradation of the signal is mainly due to capacitance and inductance of the pins and leads. The appropriate model for such an interconnection discontinuity at a PCB IC package interface is shown in Figure 2.43 (Rainal, 1988).

This generic model is valid for any interface, including connector and wirebonds or short PCB traces which act as lumped elements. The interface is dominated by both parasitic capacitance and inductance of the signal and ground leads. As the signal enters the interface, part of the energy is absorbed, part of it is reflected due to the discontinuity, and the remaining is transmitted. This combination of events causes both an attenuation of the magnitude of the pulse, as well as degradation of the risetime. Interestingly, all of the parameters of interest, such as reflection, magnitude, and T_{rd}, may be predicted from two nondimensional parameters. These are

$$\alpha = \frac{L}{R^2C} = \frac{L/R}{RC}, \quad \beta = \frac{T_r}{RC} \tag{2.60}$$

α is the ratio of inductive to capacitive time constants, β is a nondimensional risetime, and R is the terminating resistor $= Z_0$. The results for various combinations of α and β are given in Figure (2.44), Figure (2.45), and Figure (2.46) for the practical application of a high-speed chip package.

Other cases may be obtained from equations given in Rainal (1988). Typical values of L for PCB connectors vary from 15 to 25 nH, and the capacitance varies from 1.5 to 2 pF. If the connector is placed on a 75-Ω line through which we pass a pulse with a risetime T_r of 1 nsec, then for $L = 20$ nH and $C = 2$ pF, $\alpha \approx 1.8$, and $\beta \approx 6.7$. Therefore, from Figure 2.44, the maximum reflection coefficient from the connector is 0.1 (10%). Also, from Figure 2.45, for a 1-V input pulse, the output from the connector will only be 0.75 V (75%). Thus, this connector configuration may not be suitable for high-speed designs as we would like the reflection coefficient to be $< 5\%$ and the output voltage at least 80%. In addition, there is also the degradation of the risetime. Note that this degradation of signal is just at the connector alone. In any interconnection medium, there may be many interfaces at which signal distortion also occurs. The goal is to minimize distortion at every interconnecting interface. Observe that the minimum reflection occurs for $\alpha = 1$, when $L/R = RC$, or

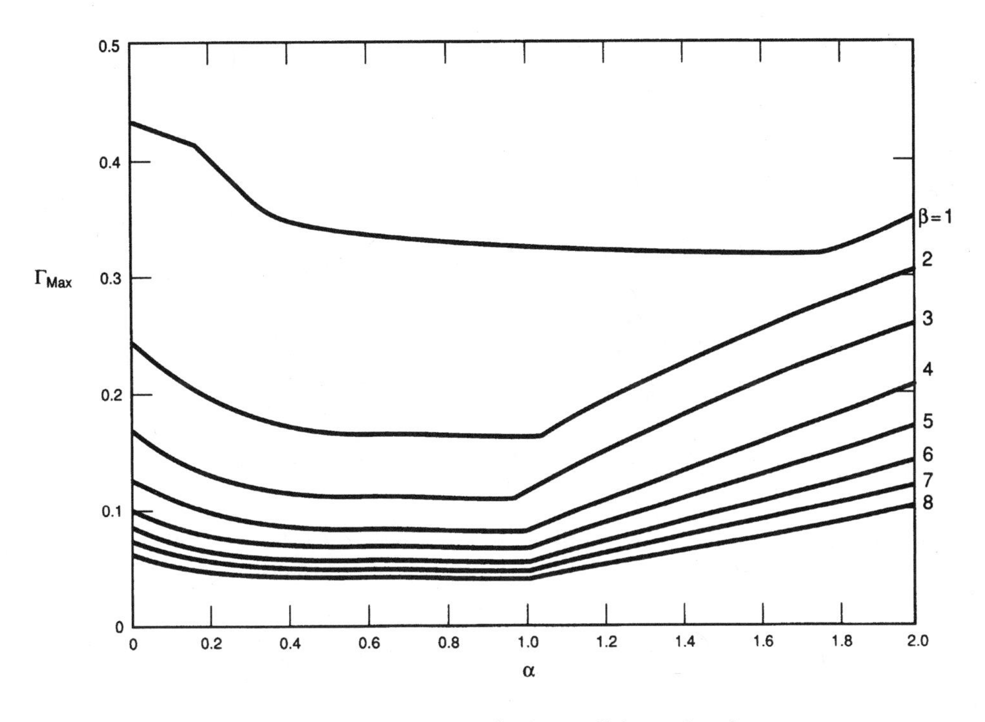

FIGURE 2.44 Maximum reflection coefficient at interface.

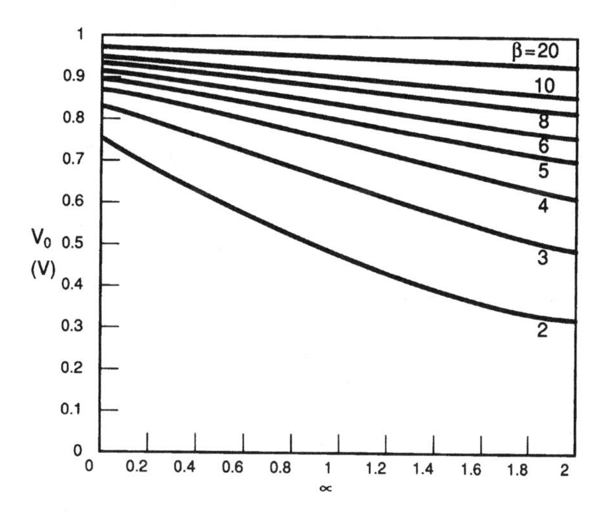

FIGURE 2.45 For a 1-V ramp input, V_0 denotes the output response at the interface for $t = \beta RC$. (*Source:* A.J. Rainal, "Performance limits of electrical interconnections to a high-speed chip," *IEEE Trans. CHMT,* vol. 11, no. 3, pp. 260–266, Sept. 1988. © 1988 IEEE.)

$R = \sqrt{L/C} = Z_0$, when the inductive and capacitive time constants are equal. This may be considered as a "matched" discontinuity. For $\alpha > 1$, the interface may be considered inductive in nature, while it is capacitive for $\alpha < 1$. Thus, the reflection will be positive in one case and negative in the other. The α–β theory given above also implies some important considerations. From Figure 2.45, we observe that β must be large to have a high value for the output voltage. β may be made large by reducing C, but when we reduce C, we increase α, and we observe that both the output voltage falls as α increases, and at the same time the reflection coefficient

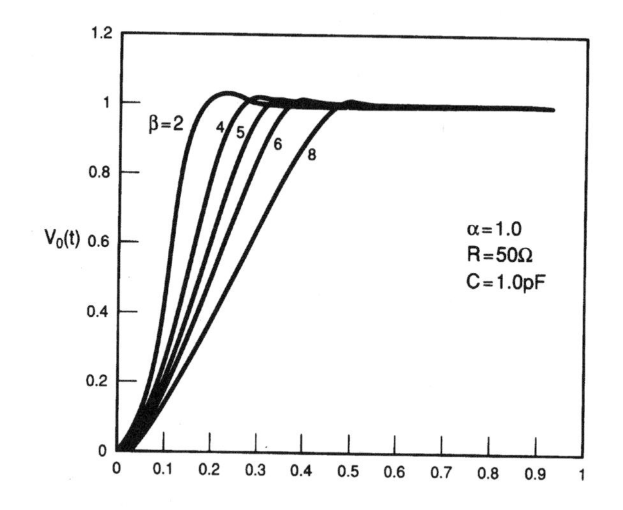

FIGURE 2.46 Output waveform at interface for a 1-V ramp input. (*Source:* A.J. Rainal, "Performance limits of electrical interconnections to a high-speed chip," *IEEE Trans. CHMT,* vol. 11, no. 3, pp. 260–266, Sept. 1988. © 1988 IEEE.)

also increases. Therefore, just reducing C results in signal degradation. It is not possible to maintain signal integrity by tuning either L or C alone. We need to specify joint bounds for L and C to minimize signal degradation. These are discussed in Rainal (1988). This simple electrical model is a very practical tool for evaluating interconnection elements for signal integrity.

Conclusions

This section is an overview of electrical characterization of interconnections and packaging (I&P) of electronic systems. It is shown that I&P technology is mainly driven by advances in silicon IC technology, such as VLSI where there has been a hundred-fold increase in component density per decade. Such a rapid increase in silicon integration has also led to larger connections outside of the IC package, where the pin count is dictated by the empirical Rent's rule. Integration of such dense ICs in turn has led to multilayer PCBs with increased track density and layer count. The signal traces from these dense multilayer boards terminate on dense PCB connectors with 300 to 400 I/Os on an 8-in. height. The trend in electronics packaging is toward both high speed and high density. I&P technology has therefore evolved at a rapid pace along with the driving forces behind it.

The elements that make up I&P are the IC package with the wirebonds inside and the pins outside, PCB with the conductor traces, and bias between the layers, PCB connector, motherboard with pins and conductor traces, and cabling between boards and/or equipment. The elements of I&P are analogous to a mechanical chain, with each element forming a link in the chain. The mechanical analogy may be extended further by noting that the I&P chain is only as strong as the weakest link. As the signal travels from a driver to a receiver, it passes through several interfaces between elements. There is signal degradation at every interface. I&P should be designed so as to minimize the degradation at each interface. A high-performance IC may be degraded due to improper interconnections. Today, I&P is the bottleneck to designing high-performance electronic systems. Reliability of the system depends critically upon the integrity of the interconnection scheme.

The important electrical parameters characterizing I&P are delay, characteristic impedance, crosstalk, and risetime degradation. The article outlines the significance of each of these parameters and techniques to evaluate them. The electrical transmission path from driver to receiver is a very complicated one. It is extremely difficult to model the problem electrically from end to end. The best approach is to electrically characterize each interface (element) individually for signal integrity. This breaks the problem down to simple parts. The goal is to address those elements that are the cause of severe signal degradation and in the process optimize the entire interconnection chain. Breaking the problem into individual elements identifies the important links. If the entire problem is solved from end to end, it is often difficult to isolate those elements that are the major cause of degradation. Fortunately, most of the electrical parameters can be evaluated using simple analytical techniques that are readily available in literature. These have been used and verified experimentally at Bell Labs in the design of many systems. In the design of high-speed PCBs, designers should also consider the effect of manufacturing tolerances on electrical parameters, such as characteristic impedance. Because of the increased speed and density requirements of many systems, careful electrical characterization and optimization of I&P are the keys to system performance.

Defining Terms

Bandwidth (BW): Related to risetime degradation and the frequency at which the energy transfer through I&P falls to –3 dB below the low-frequency value. It is only a qualitative measure to compare different options, and the concept comes from classical RC low-pass filter theory.

Characteristic impedance (Z_0): The impedance (resistance) to energy transfer associated with wave propagation in a line that is much larger than the wavelength. It gives the relationship between the maximum voltage and maximum current in a line.

Crosstalk (X): Noise that appears on a line due to interactions with stray electromagnetic fields that originate from adjacent pulsed lines.

Interconnections and packaging (I&P): Elements of the electrical signal transmission path from the driver chip to the receiver chip. Various elements that make up I&P are chip wirebonds and package pins, circuit boards, connectors, motherboards, and cables.

Propagation delay (T_d): Time required by a signal to travel from source to receiver.

Risetime degradation (T_{rd}): A measure of the slowing down of the pulse as it passes through an I&P element. It includes both the increase in risetime of the pulse, as well as loss in amplitude.

References

C.S. Chang, "Electrical design of signal lines for multilayer printed circuit boards," *IBM J. Res. Develop.*, vol. 32, no. 5, pp. 647–657, Sept. 1988.

B.N. Das and K.V.S.V.R. Prasad, "A generalized formulation of electromagnetically coupled striplines," *IEEE Trans. Microwave Theory and Techniques*, vol. MTT-32, no. 11, pp. 1427–1433, Nov. 1984.

W.A. Davis, *Microwave Semiconductor Circuit Design*, New York: Van Nostrand Reinhold, 1990.

W.L. Everitt (Ed.), *Physical Design of Electronic Systems, Volume 1, Design Technology*, Englewood Cliffs, NJ: Prentice-Hall, 1970, pp. 362–363.

C.W. Hoover, W.L. Harrod, and M.I. Cohen, "The technology of interconnection," *AT&T Technical Journal*, vol. 66, issue 4, pp. 2–12, July/August 1987.

H. Howe, *Stripline Circuit Design*, Burlington, MA: Microwave Associates, 1974.

H.R. Kaupp, "Characteristics of microstrip transmission lines," *IEEE Trans. Electronic Computers*, vol. EC-16, no. 2, pp. 185–193, April 1967.

L.L. Moresco, "Electronic system packaging: The search for manufacturing the optimum in a sea of constraints," *IEEE Trans. CHMT*, vol. 13, no. 3, pp. 494–508, Sept. 1990.

A.J. Rainal, "Transmission properties of various styles of printed wiring boards," *Bell System Tech. Journal*, vol. 58, no. 5, pp. 995–1025, May–June 1979.

A.J. Rainal, "Computing inductive noise of chip packages," *AT&T Bell Laboratories Tech. Journal*, vol. 63, no. 1, pp. 177–195, 1984.

A.J. Rainal, "Performance limits of electrical interconnections to a high-speed chip," *IEEE Trans. CHMT*, vol. 11, no. 3, pp. 260–266, Sept. 1988.

K. Ritz, "Manufacturing tolerances for high-speed PCBs," *Circuit World*, vol. 15, no. 1, pp. 54–56, 1988.

Further Information

The issue of the *AT&T Technical Journal* in Hoover et al. (1987) focuses on the technology of electronic interconnections. It is an excellent source of information on important areas of interconnections and packaging. It begins with an overview and includes areas such as systems integration and architecture, materials and media, computer-aided design (CAD) tools, reliability evaluation, and standardized systems packaging techniques. The book *Microelectronics Packaging Handbook,* edited by Rao R. Tummala and Eugene J. Rymaszewski and published by Van Nostrand Reinhold, is a standard reference for packaging engineers, covering all major areas in detail. Most of the work on electrical characterization was done by researchers in the microwave area. These studies have been extensively published in the *IEEE Transactions on Microwave Theory and Techniques,* a publication of the IEEE Microwave Theory and Techniques Society. Parameters for electrical package design and evaluation have also been published in the *IBM Journal of Research and Development.* A source for the general areas of components, connector technologies, and manufacturing aspects related to packaging is the *IEEE Transactions on Components, Hybrids, and Manufacturing Technology.*

2.4 Microlithography for Microfabrication

Benjamin Y. Park, Rabih Zaouk, and Marc J. Madou

Introduction

The advent of photolithography has shaped the latter part of the twentieth century, bringing about the integrated circuit (IC) revolution. Almost all electronic devices used today have one or more ICs inside. More advanced lithography led to smaller and smaller transistors, which translated into faster and more efficient computing machines. Photolithography also powered the advent of microelectromechanical systems (MEMS), which are now starting to be used in more and more diverse commercial products, from mechanical to biomedical devices, and are helping to change the way people perceive the applicability of IC technology. This chapter examines the basics of the photolithography process. It also addresses current and future lithography trends toward further decreasing the minimum feature size, allowing the fabrication of high-aspect-ratio structures and facilitating the introduction of new processes and materials. These are all contributing factors to the continued expansion of application domains for IC technology.

Photolithography

In this section, a succinct overview of the photolithography process is presented. The reader is referred to Madou's *Fundamentals of Microfabrication* [1] for a more detailed description.

History of Lithography

The word *lithography* (Greek for the words "stone [lithos]" and "to write [graphein]") refers to a process invented in 1796 by Aloys Senefelder. Senefelder discovered that stone (in his case, Bavarian limestone), when properly inked and treated with chemicals, could transfer a carved image onto paper [2].

By the early 1960s, methods were devised whereby a photoetching process produced large numbers of transistors on a thin slice of silicon (Si). At that time, pattern resolution was no better than 5 μm [3]. Today, photolithography, x-ray lithography, and charged-particle lithography all achieve submicron printing accuracy. The 64-bit PowerPC 970FX chip, available from IBM Corp. and introduced in 2004, has a minimum feature size of 90 nm, a core speed of 2 GHz, and 58 million transistors. It also incorporates advanced Si technologies such as SOI and strained silicon [4].

Photolithography Overview

The most widely used form of lithography is photolithography. In the IC industry, pattern transfer from masks onto thin photosensitive films is accomplished almost exclusively by photolithography. In this process, a pattern is transferred to a photosensitive polymer (a photoresist) by exposure to a light source through an optical mask. An optical mask usually consists of opaque patterns (usually chrome or iron oxide) on a transparent support (usually quartz) that is used to define features on a wafer. The photoresist pattern is then transferred to the underlying substrate by subtractive (etching) or additive (deposition) techniques. The combination of the accurate alignment of a successive set of photomasks and the exposure of these successive patterns leads to complex, multilayered ICs or micromachines. Photolithography has matured rapidly in the ability to resolve ever smaller features. For the IC industry, this continued improvement in resolution has impeded the adoption of alternative, higher-resolution lithography techniques, such as x-ray lithography and other next-generation lithography (NGL) techniques. Research in high-aspect-ratio resist features, driven by the field of MEMS, is also being actively pursued, as opposed to the essentially two-dimensional processes used in traditional IC manufacturing.

Photolithography and pattern transfer involve a set of process steps that are summarized in Figure 2.47. As an example, an oxidized Si wafer and a negative photoresist are used to transfer a pattern from a mask to a layer of silicon dioxide. An oxidized wafer (A) is coated with a 1-μm thick negative photoresist layer (B). After exposure (C), the wafer is rinsed in a developing solution or sprayed with a spray developer, to remove the unexposed areas of photoresist and leave a pattern of bare and photoresist-coated oxide on the wafer surface (D). The resulting photoresist pattern is the negative image of the photomask pattern. In a typical next step after development, the wafer is placed in a solution of HF or a mixture of HF and NH_4F that attacks the oxide at a much faster rate than the photoresist or the underlying silicon (E). The photoresist prevents the oxide underneath from being attacked. Once the exposed oxide has been etched away, the remaining photoresist can be stripped off with a solution that only attacks the photoresist such as a strong acid, for instance, H_2SO_4, or an acid–oxidant combination such as Piranha (H_2SO_4:H_2O_2) (F). Other liquid strippers include organic solvent strippers and alkaline strippers (with or without oxidants). The oxidized Si wafer with the etched windows in the oxide (F) now awaits further processing. Recently, photoresists have been increasingly used in applications where the resist is a permanent part of the final device rather than just a sacrificial layer for patterning the substrate.

Wafer Preparation

Wafer Cleaning. All lithography processes used in manufacturing take place inside a semiconductor clean room, a specially constructed enclosed area with yellow lighting. The clean room is environmentally controlled with respect to airborne particulates, temperature ($\pm 0.1°F$), air pressure, humidity (from 0.5 to 5% relative humidity [RH]), and vibration. Airborne contaminants are especially detrimental to the fabrication process yield. The number of usable devices out of the total number of manufactured devices is crucial to the economic viability of the final product. Physical contaminants such as dust particles can hinder the lithography process by preventing light from exposing the photoresist or by disturbing the surface uniformity of a coated photoresist. Chemical contaminants can react with various materials used in the lithography process to create unwanted effects.

Silicon is the material of choice for today's microfabrication, but there is a trend toward different substrate materials, especially in BioMEMS. A variety of cleaning methods can be used to prepare a wafer for the lithography process. Usually, new wafers are already cleaned before shipping and are kept in a contamination-free container. If cleaning of the wafer is needed, a variety of methods (HF dip, RCA1, RCA2, Piranha clean, etc.) can remove different types of contaminants. The presence of water or water vapor compromises the adhesion between the photoresist and the wafer. Before the photoresist is applied, a dehydration bake is performed to remove water from the wafer surface. Adhesion can be further promoted by applying an adhesion promoter (hexamethyldisilizane [HMDS]) or by roughening the wafer surface by plasma etching. This is referred to as wafer priming.

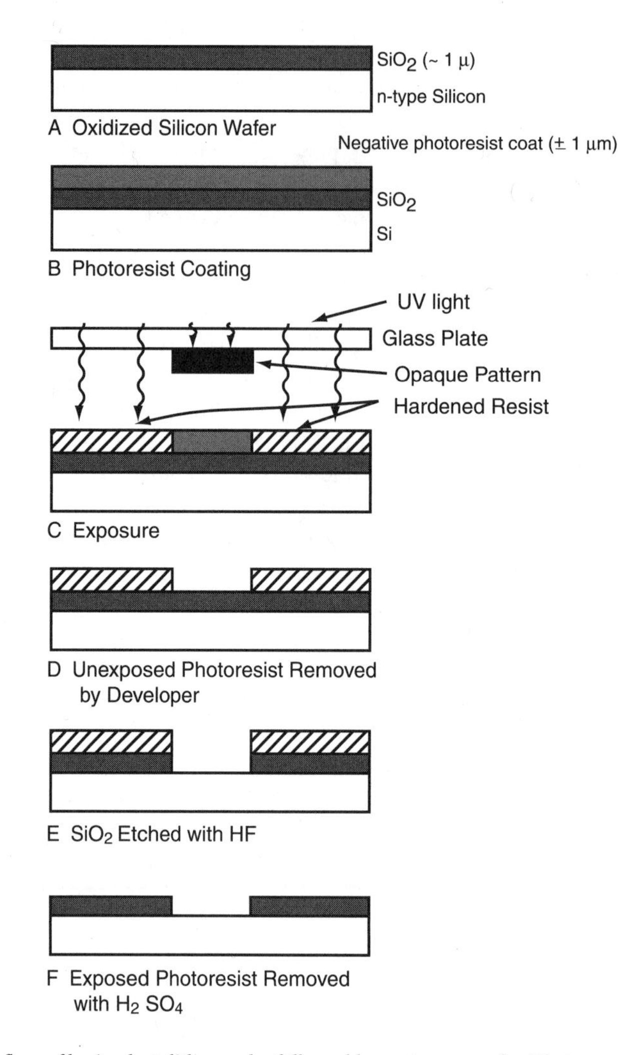

FIGURE 2.47 Process flow of basic photolithography followed by pattern transfer. The example uses an oxidized Si wafer and a negative photoresist system. The process steps include photoresist coating, exposure, development, oxide etching, resist stripping, and oxide etching. Steps A to F are explained in detail in the text.

Oxide Growth. In many cases, an oxide layer is desired as a mask for subsequent processes (for example, an etch or an implant process) or as an insulating layer. This is usually done by heating a silicon wafer to between 900 and 1150°C in a dry or humidified oxygen stream in a tube furnace.

Resist Spinning and Soft Bake

As the first step in the lithography process itself, a thin layer of an organic polymer, a photoresist sensitive to ultraviolet radiation, is deposited on the oxide surface (Figure 2.47B). The liquid photoresist is dispensed onto a wafer held by a vacuum chuck in a resist spinner. The wafer is then spun in one or more steps at precisely controlled speeds. The spin speed (between 1500 and 8000 rpm) allows the formation of a uniform film. At these speeds, the centrifugal force causes the liquid to flow to the edges, where it builds up until expelled when the surface tension is exceeded. The resulting polymer thickness, T, is a function of spin speed, solution concentration and molecular weight (measured by intrinsic viscosity). The empirical expression for T is given by

$$T = \frac{KC^{\beta}\eta^{\gamma}}{\omega^{\alpha}} \tag{2.61}$$

where K = overall calibration constant, C = polymer concentration in g/100 ml solution, η = intrinsic viscosity, and ω = rotations per minute (rpm).

Once the various exponential factors (α, β, and γ) have been determined, Equation (2.61) can be used to predict the film thickness for various molecular weights and solution concentrations of a given polymer and solvent system [5].

The spinning process is of primary importance to the effectiveness of pattern transfer. The quality of the resist coating determines the density of defects transferred to the device being made. The resist-film uniformity across a single substrate and from substrate to substrate must be $\pm$ 5 nm (for a 1.5-μm film, this is $\pm$ 0.3%) to ensure reproducible line widths and development times in subsequent steps.

After spin coating, the resist still contains up to 15% solvent and may have built-in stresses. The wafers are therefore soft-baked (prebaked) at 75 to 100°C to remove solvents and stress, and to promote adhesion of the resist layer to the wafer. The optimization of this prebaking step may substantially increase device yield.

Exposure and Postexposure Treatment

Pattern transfer onto a photoresist is done by shining light through the mask (Figure 2.47C). In photolithography, wavelengths of the light source used for exposure of the resist-coated wafer range from the very short wavelengths of extreme ultraviolet light (EUV) (10 to 14 nm) to longer wavelengths such as deep ultraviolet (DUV) (150 to 300 nm) and near-ultraviolet (UV) (350 to 500 nm) light. In near-UV, one typically uses the g-line (435 nm) or i-line (365 nm) of a mercury lamp. For a projection lithography system (with lenses), the theoretical resolution limit is given by the Rayleigh expression [1,6]:

$$R = \frac{0.61\lambda}{2\text{NA}} \tag{2.62}$$

where R = minimum achievable resolution, λ = wavelength of exposure source, and NA = numerical aperture:

$$\text{NA} = n\sin\theta \tag{2.63}$$

where n = index of refraction of the ambient and $\sin\theta$ = the angular half-aperture of the lens.

In general, the smallest feature that can be printed using projection lithography is roughly equal to the wavelength of the exposure source. For a KrF excimer laser, 248 nm would be the expected minimum feature size. Resolution enhancement techniques (RET), discussed below, are used to go beyond the conventional Rayleigh diffraction limit. Smaller features require shorter wavelengths but as the wavelength becomes shorter, it becomes difficult to manufacture lenses that can transmit the light efficiently due to absorption. Thus, the projection system must be circumvented completely (as in the case of x-ray lithography), or reflective optics must be used (EUV lithography). At short wavelengths, there is also a lack of reliable light sources of sufficient quality and power output. Next-generation lithography (NGL) schemes will be discussed below.

The action of light on a photoresist either increases or decreases the resist solubility, depending on whether a positive or negative photoresist is applied. Thus, for a positive-tone photoresist, the opaque pattern on the mask will determine the features remaining in the resist layer after development (Figure 2.48). Conversely, after development of a negative photoresist, the clear pattern of the mask determines the remaining photoresist features (Figure 2.48). The profile of the photoresist side walls (Figure 2.49) is critical to many applications, such as the patterning of hard-to-etch metals (lift-off) and mold fabrication. Figure 2.50 illustrates the use of a lift-off profile. The resist wall profile can be controlled by adjusting resist tone, exposure dose, developer strength, development time, and by other means.

Post exposure treatment is often desired because the reactions initiated during exposure might not have been completed. To halt the reactions or to induce new ones, several postexposure treatments can be

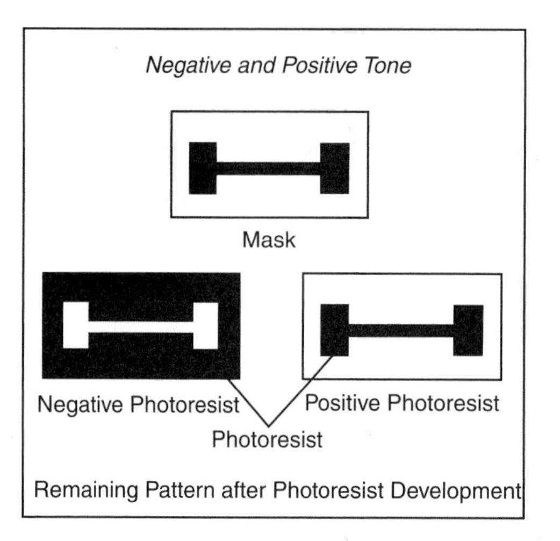

FIGURE 2.48 The figure shows resulting patterns after exposure and development of a positive- and negative-tone photoresist. The opaque image on the mask is transferred as is onto the positive photoresist. The image is reversed in the case of a negative photoresist.

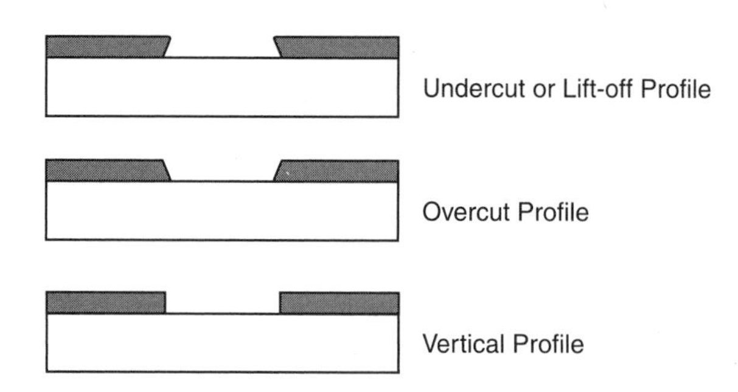

FIGURE 2.49 The three different photoresist profiles. The undercut (lift-off) profile is used mainly in the patterning of metals in the process lift-off (see Figure 2.50). The overcut profile is normally obtained from a positive-tone photoresist. The vertical profile achieves the best pattern fidelity but is relatively difficult to obtain.

used: postexposure baking, flood exposure with other types of radiation, treatment with reactive gas, and vacuum treatment.

Development, Descumming, and Postbaking

During the development process, selective dissolving of resist takes place (Figure 2.47D). Development can be done using a liquid (wet development), a gas, or a plasma (dry development). Positive resists are typically developed in aqueous alkaline solutions (such as tetramethyl ammonium hydroxide [TMAH]), and negative resists in organic solvents. Dry development methods have some advantages over wet methods, including less swelling of the resist and environmental friendliness. These dry methods are becoming more prevalent.

Unwanted, residual photoresist sometimes remains after development. Descumming is a procedure for removing unwanted photoresist with a mild plasma treatment. In this process, highly energetic oxygen ions bombard the sample surface and physically etch away the unwanted photoresist.

Postbaking or hard baking removes residual solvents and anneals the film to promote interfacial adhesion of the resist that has been weakened either by developer penetration along the resist/substrate interface or by

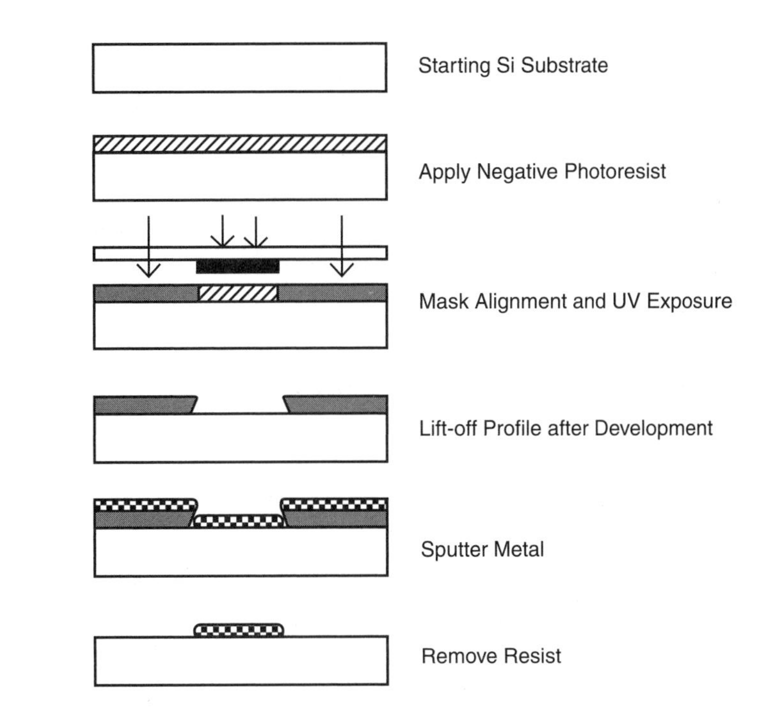

FIGURE 2.50 Example of the lift-off sequence using negative resist as a sacrificial layer. This method is used in cases where the metal is difficult to etch (e.g., gold or platinum).

swelling of the resist (mainly for negative resists). Hard baking also improves the hardness of the film. Improved hardness increases the resistance of the resist to subsequent etching and deposition steps. Postbaking is usually done at higher temperatures (120°C) and for longer time periods (e.g., 20 min) than soft baking or prebaking.

Pattern Transfer

In cases where the photoresist is a permanent part of the final device (microfluidics, C-MEMS, etc.), further processing may not be needed. However, in most cases, the sacrificial photoresist pattern is used as a mask for etching (subtractive) or deposition (additive) of or on the underlying substrate (Figure 2.47E shows a subtractive process). In a subtractive process, the resist acts as a protective barrier to the etching agent, which can be a liquid solution, a gas, or a plasma. After pattern transfer, the resist can be removed for further processing. Similarly, pattern transfer can involve a deposition technique: chemical vapor deposition (CVD) or e-beam evaporation.

Current and Future Trends

The electronics and photonics industries are interested in developing new nanolithography methods in order to continue the long-term trend of building ever smaller, faster, and less expensive devices (Moore's law). Conventional projection lithography is based on the 4× reduction of reticle features and is limited by optical diffraction, where the smallest feature size achievable is approximately the order of the wavelength of the radiation source [1]. For wavelengths shorter than 157 nm, refractive optics are not suitable because of the strong absorption of shorter wavelengths by the optics elements. Much more expensive and demanding reflective optics are needed when shorter-wavelength light sources are used.

There is no clear consensus as yet on what the favored next generation lithography process for IC manufacturing will be, although extreme ultra-violet (EUV) lithography seems to be the leading contender. On a research level, there are many technologies that are competing. EUV lithography, x-ray lithography, e-beam and ion-beam lithography, soft lithography, and proximal probe lithography are all potential contenders in this crowded arena. These lithography methods, all of them, significantly different from current lithography processes, are called NGL systems.

EUV Lithography

The most logical extension of the lithography used today is to move to shorter wavelengths. EUV systems use 13.4-nm radiation in a vacuum environment with an all-reflective optical exposure station to fabricate devices and structures with sub-70-nm dimensions. Some major challenges involve the fabrication of highly reflective, multilayer coatings for the optics and the synthesis of new resists to work in this specific wavelength regime [7–9]. In 1997, a U.S. consortium called the Extreme Ultraviolet Limited Liability Company (EUV LLC) was created to spearhead research to bring EUV lithography to market. The consortium includes Intel Corp., Motorola Corp., Advanced Micro Devices Corp., IBM, Infineon Technologies AG, Micron Technology Inc., and the Sandia and Lawrence Livermore National Laboratories. The work so far has led to a prototype EUV lithography chip-making machine, but there are still many obstacles to overcome to achieve sufficient yield in a manufacturing setting [10,11].

Resolution Enhancement Technologies (RET)

RET are not part of NGL technology, but they have enabled the semiconductor industry to continue miniaturization using current photolithography technologies. Resolution enhancement technologies such as phase shift masks (PSM)[11] (Figure 2.51), off-axis illumination (OAI) [11], optical proximity correction (OPC)[11] (Figure 2.52), and liquid immersion lithography (Figure 2.53)[6,12,13] and interferometric lithography [14] all enable the writing of photoresist features with dimensions smaller than the wavelength of the exposing light source. Although RET is extensively used in current IC manufacturing, the added manpower, extra computing and databases, software and technology licensing is expensive [15]. RET technologies have delayed the transition to NGLs, but because of cost and other limitations, the transition to NGL systems is inevitable.

Serial Lithography Schemes (Proximal Probe, E-beam, and Ion Beam)

Using atomic force microscopy (AFM) [16,17], scanning tunneling microscopy (STM) [18,19], near-field scanning optical microscopy (NSOM) [20,21], apertureless near-field scanning optical microscopy (ANSOM) [22], electron-beam lithography (EBL), ion beam lithography (IBL), "dip-pen" nanolithography [23], and other sequential writing schemes, smaller features can be defined than with photolithography. However, these technologies are too expensive to implement, and because of their serial nature, they are also too slow to be practical. Arrays of these serial-writing tools have been investigated, but in the case of the proximal probe approach (AFM, STM, NSOM, ANSOM), tip stability is an issue that needs to be resolved, as demonstrated by

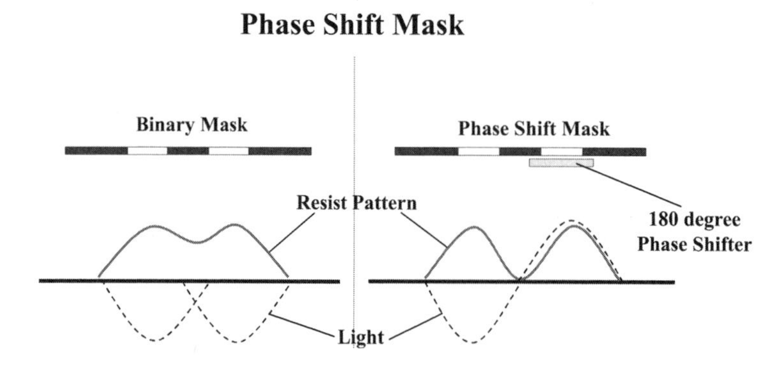

FIGURE 2.51 Illustration of resolution enhancement by the use of a phase shift mask. Opposite phases of light interfere destructively and cancel each other to enhance feature contrast.

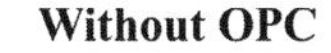

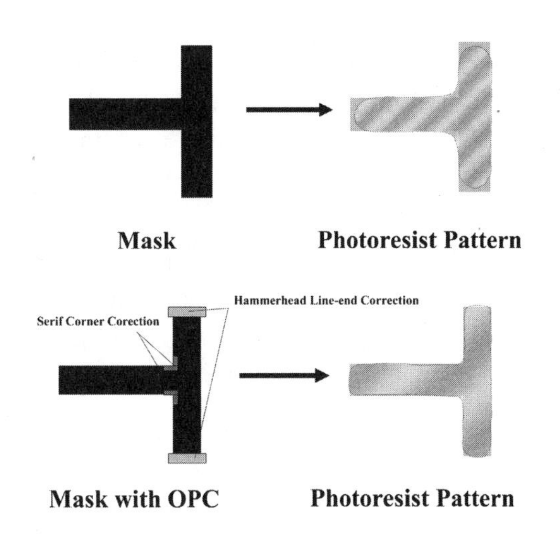

FIGURE 2.52 Illustration of resolution enhancement using optical proximity correction. Hammerhead correction marks are line ends to compensate for the lack of exposure, and serif-corner correction features (more transmissive) are added to compensate for the unneeded exposure within the inner corners.

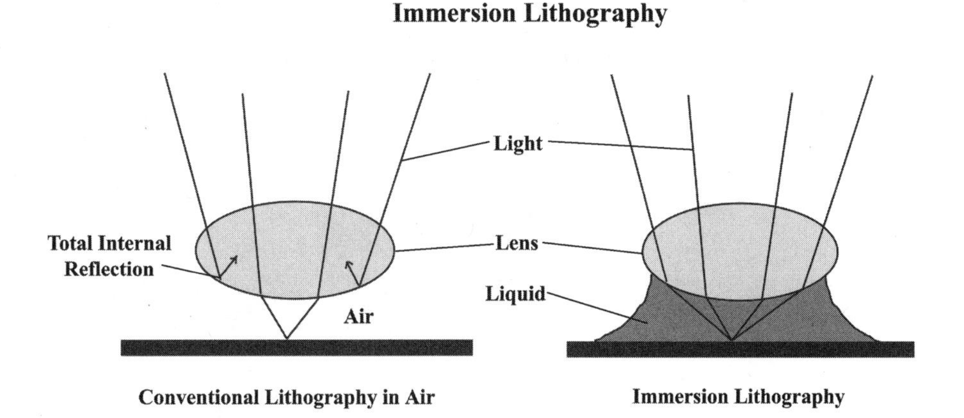

FIGURE 2.53 Illustration of resolution enhancement using immersion lithography. Using liquid (water) as the separating medium between the focusing lens and the photoresist permits the use of higher numerical apertures, with less degradation in the depth of focus as compared to having air as the separating medium. From Equation (2.62) and Equation (2.63), an n-fold increase in index of refraction results in an n-fold increase in resolution.

the work on IBM's AFM arrays (the so-called millipede [24,25]). EBL is commonly used in the prototyping of devices with fine features and in making masks for production projection systems, but the technology is considered too slow to be used in a manufacturing setting.

SCALPEL

Systems that use electrons in a flood-exposure mode instead of in a fine beam have also been developed. In this method, an electron source is used as a "light" source for projection lithography. The lenses in a conventional projection lithography system are replaced by magnetic 'lenses.' The method, called scattering with angular limitation projection electron-beam lithography (SCALPEL)[26–30] has many advantages over conventional lithography, including smaller features and better depth-of-focus, but the price of mask fabrication and of the equipment is a major disadvantage.

Imprint Lithography

Imprint lithography techniques all incorporate an imprint step, where the topography of a template defines patterns created on a substrate. The three main approaches to imprint lithography are soft lithography [33–35], nanoimprint lithography (NIL) [31], and step-and-flash imprint lithography (SFIL) [31] (Figure 2.54). Imprinting processes are economical due to the simplicity of the equipment and the potential for high throughput. As the rate of improvements in optical lithography decelerates and as the costs of manufacturing continue to escalate, there is increased interest in printing and molding as alternative processes for microfabrication. As imprint lithography depends on another lithography technique (perhaps an NGL) to make the master, it does not really qualify as an NGL itself. Resolutions of sub-10 nm have been achieved using imprint techniques [32]. In most cases, imprint lithography resolution is only limited by the resolution of the template-fabrication process.

Soft lithography[33–35] refers to a series of methods that use a patterned elastomer as a stamp, mold, or mask to generate micropatterns and microstructures instead of using a rigid photomask. These methods include replica molding (REM), microcontact printing (μCP), micromolding in capillaries (MIMIC), and micro-transfer molding (μTM) [36]. The major advantage of soft lithography is its short turn-around time.

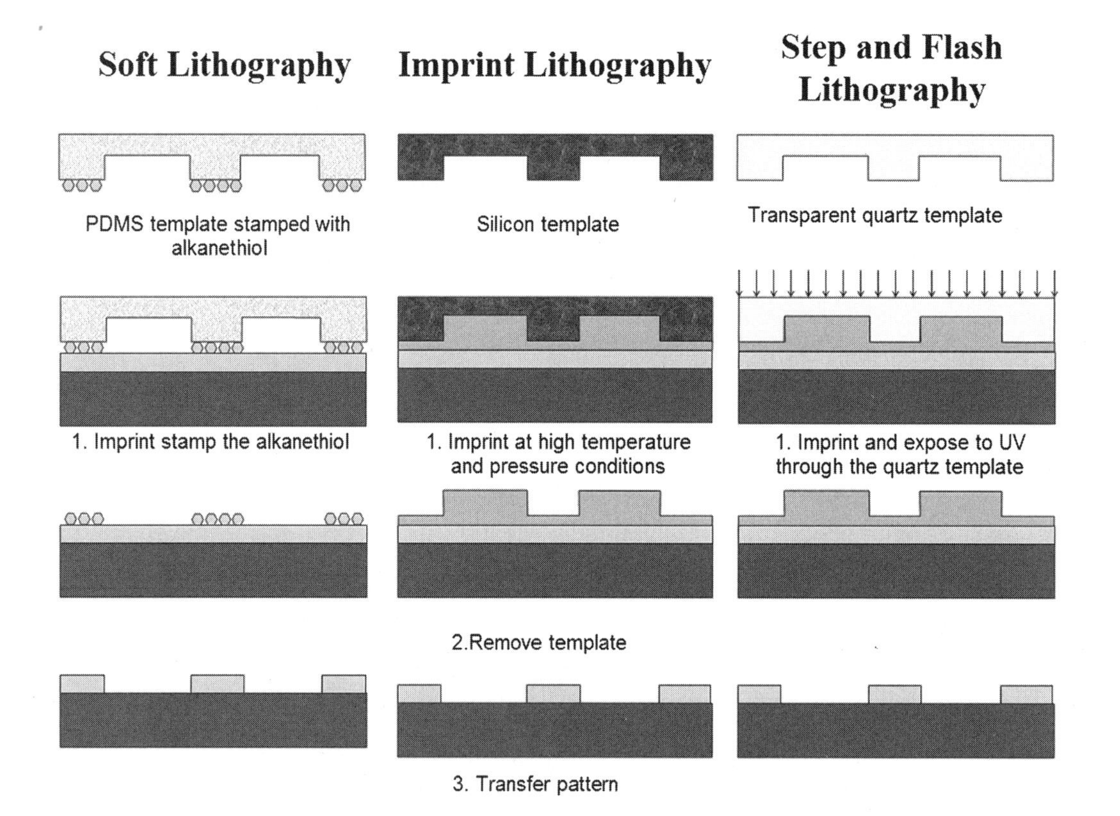

FIGURE 2.54 Illustration of imprint lithography methods: soft lithography or microcontact printing (μCP), nanoimprint lithography (NIL), and step-and-flash imprint lithography (SFIL). All these techniques can result in features only as good as those used in the imprint template. While the fidelity of soft lithography is compromised by the use of a flexible stamp (PDMS), NIL and SFIL faithfully duplicate the finest features of the template.

It is possible to go from design to production of replicated structures in less than 24 hours. The method is low in cost and, unlike photolithography, is applicable to almost all polymers and, thus, many materials that can be prepared from polymeric precursors. Since soft materials are used, deformation of the stamp/mold, low reproducibility (due to distortion), and defects (yield) are problems that prevent this technology from being a viable manufacturing technique [33]. The process constitutes a very handy research tool but most likely will not reach the commercial manufacturing stage.

NIL [31,50] and SFIL [31,37] are techniques that use hard molds instead of soft molds. In NIL, a template made of a hard material (usually Ni or Si) is pressed against a polymer layer. High temperature and pressure conditions mold and harden the polymer layer. In SFIL, a hard-but-transparent template is used to mold a photoresist. The photoresist is exposed to light through the template to harden the resist. Although more progress is needed for widespread adoption of NIL and SFIL, sub-100-nm feature-size devices can be reliably fabricated at a reasonable throughput. It has been demonstrated that the mold templates do not deteriorate, even after 1500 imprints with sub-100-nm feature sizes [38].

MEMS

MEMS, or the science of miniaturization, refers to a class of devices that have at least one of their dimensions in the micrometer range. While IC devices only use electrical components (transistors, diodes, capacitors, etc.), MEMS devices take advantage of a wide range of other equipment, from mechanical to biological (BioMEMS). The materials and fabrication methods used in MEMS are much more varied than those used for IC fabrication (where one deals principally with silicon, oxides, and metals patterned using photolithography). In contrast to the IC industry, where the devices are carefully packaged and protected from the environment, MEMS devices, such as pressure or glucose sensors, often must have surfaces directly exposed to the sensing environment. Examples of successfully commercialized mechanical devices are accelerometers, gyroscopes, tilt meters, membrane-pressure sensors, micromirrors, optical MEMS switches, and inkjet-print heads. Commercial BioMEMS devices include glucose sensors and DNA arrays. Because of the irreversible chemical reactions involved and contamination considerations, BioMEMS devices tend to be disposable.

Nonplanar Lithography

In many MEMS and BioMEMS applications, nonplanar geometries are required. True 3D structures (with actual 3D curves) can be built using stereolithography [39–42] and holographic lithography [43,44]. Stereolithography has a very short turn-around time and is relatively inexpensive. It is an alternative to the LIGA process (see next section) for prototyping three-dimensional structures when dimensional accuracy is not critical.

Lithographie Galvanoformung und Abformtechnik (LIGA, the German acronym for x-ray lithography [x-ray lithographie], electrodeposition [galvanoformung], and molding [abformtechnik]) [45,46] is a process that was devised to create very precise and high-aspect-ratio structures. The process, illustrated in Figure 2.55, involves a thick layer of x-ray resist (from microns to centimeters), high-energy x-ray radiation exposure and development to arrive at a three-dimensional resist structure. Subsequent metal deposition fills the resist mold with a metal and, after resist removal, a freestanding metal structure results. The metal shape may be a final product or serve as a mold insert for precision plastic molding. The lack of x-ray sources and the high cost of LIGA masks has led many researchers toward development of LIGA-like processes that result in 3D, high-aspect-ratio structures. One of these methods is the use of SU-8 negative photoresist (MicroChem, Newton, MA). SU-8 is a chemically amplified negative photoresist with high transparency. This allows light to penetrate through thick layers of photoresist. Because thick layers of photoresist can be exposed with near-vertical sidewall profiles, SU-8 photoresist is commonly called the "poor man's LIGA." SU-8 structures have been used as molds for microfluidic applications. In these applications, PDMS is poured onto the patterned SU-8 mold, then the PDMS is cured, removed, and pressed or bonded onto a flat substrate to create microchannels. Figure 2.56 shows an SEM photo of a PDMS microchannel for a microfluidic system.

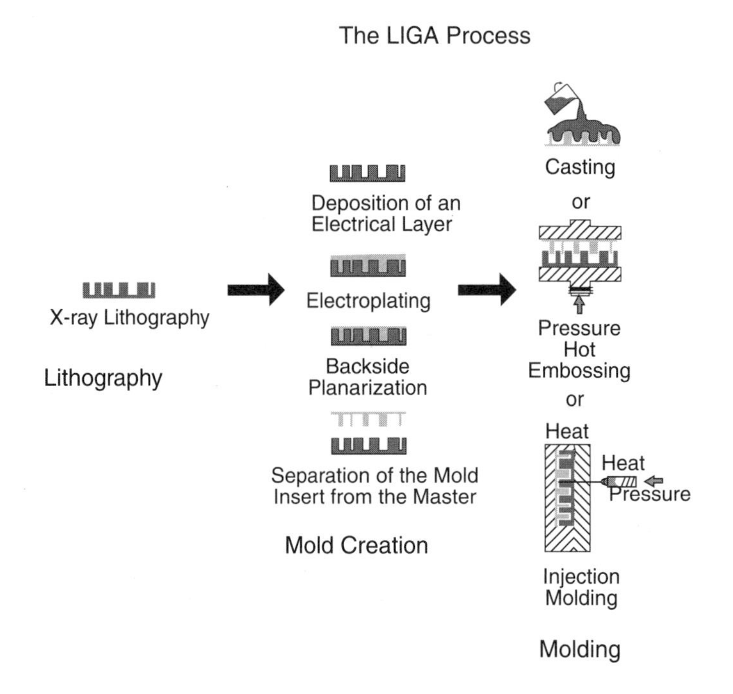

FIGURE 2.55 Details of the LIGA process. X-rays from a synchrotron are used to create a PMMA template that can have extremely high-aspect ratios with vertical side profiles. A metal mold is created from this PMMA template through electroplating. The metal mold can then be used in plastic molding machines to create parts that have fine structures with extremely high vertical walls.

FIGURE 2.56 SEM photo of a PDMS microfluidic channel. This photo shows a PDMS microchannel used for hybridization of DNA on a microfluidic platform. (Courtesy of Guangyao Jia, Madou Research Group, UCI.)

Nontraditional Materials

Different materials that are being used or investigated for use in MEMS processes include polymers, ceramics, nitinol (shape memory alloys), biomaterials, and carbon. The Madou research group at UCI [47] has been pursuing research in carbon MEMS (C-MEMS) derived from pyrolysed photoresist. Patterned SU-8 photoresist is converted into carbon electrodes by subjecting the photoresist to high temperatures in an inert environment in a process called pyrolysis. C-MEMS electrodes can be easily shaped into complex 3D geometries that were previously difficult or expensive to fabricate using conventional carbon-electrode

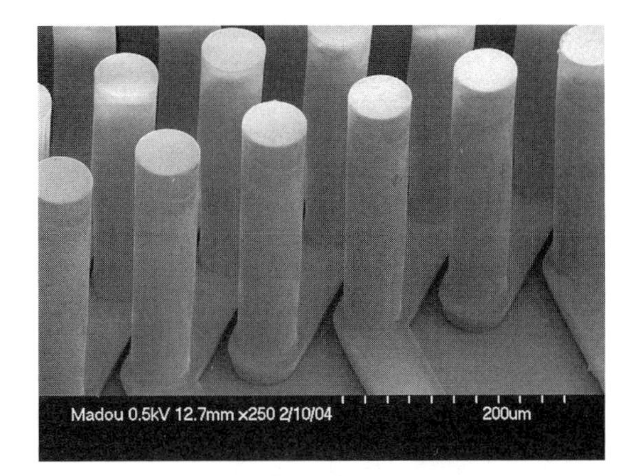

FIGURE 2.57 SEM photo of a high-aspect-ratio SU-8 negative photoresist structure. This SEM photo was taken before the structure was pyrolysed into carbon. The resulting electrodes and interconnects were used for battery research. (Courtesy of Chunlei Wang, Madou Research Group, UCI.)

FIGURE 2.58 SEM photo of carbon (C-MEMS) posts obtained by pyrolysing SU-8 negative photoresist. (Courtesy of Chunlei Wang, Madou Research Group, UCI.)

fabrication methods. The electrochemical properties of carbon make it an excellent electrode material [48]. High-aspect-ratio carbon structures (Figure 2.58) can be obtained from SU-8 photoresist structures (Figure 2.57). These high-aspect-ratio carbon structures are being evaluated for their potential use in batteries, and chemical and bio-sensing applications.

Summary and Conclusion

In this section, we introduced various microlithography techniques, as well as some fundamentals of microfabrication. The application domain of lithography has expanded beyond the IC world to encompass the fabrication of mechanical and biological devices. The fact that some DNA arrays are now made through lithography illustrates this point. With the advent of MEMS, different materials and high-aspect-ratio fabrication techniques have been introduced. Photolithography is continuously merging with traditional manufacturing techniques such as molding and electroplating (including LIGA and imprint lithography). The field of microfabrication is so vast that it would be impossible to offer a comprehensive review within the space limitations of this section. The reader is referred to textbooks [1,49] and recent reviews for a more thorough investigation.

References

1. M.J. Madou, *Fundamentals of Microfabrication — The Science of Miniaturization*, 2nd ed., Boca Raton, FL: CRC Press, 2002.
2. Compton, *Compton's Interactive Encyclopedia (Interactive Multimedia): Computer Data and Program*, Carlsbad: Compton's New Media, 1994.
3. T.W. Harris, *Chemical Milling*, Oxford: Clarendon Press, 1976.
4. IBM PowerPC Website: http://www.ibm.com/powerpc.
5. L.F. Thompson et al., Eds., "Resist processing," in *Introduction to Microlithography*, Washington, DC: American Chemical Society, 1994, pp. 1–17.
6. M. Switkes et al., "Immersion lithography: beyond the 65 nm node with optics," *Microlith. World*, p. 4, May 2003.
7. C.W. Gwyn and P.J. Silverman, "EUV lithography: transition from research to commercialization," in *Photomask Meeting*, Yokohama, Japan, April 2003.
8. E.J. Lerner, "Next generation lithography," *Ind. Phys.*, June edition, 1999.
9. H.J. Levinson, "Overview of lithography: challenges and metrologies," *Adv. Micro Devices*, April 2003.
10. J. Cobb et al., "EUV photoresist performance results from the VNL and the EUV LLC," in *Proc. SPIE Emerg. Lithographic Technol.*, vol. 4688, March 2002.
11. P. Naulleau et al., "Sub-70 nm extreme ultraviolet lithography at the advanced light source static microfield exposure station using the engineering test stand set-2 optic," *J. Vac. Sci. Technol. B*, vol. 20, no. 6, 2002, pp. 2829–2833.
12. F.M. Schellenberg, "Resolution enhancement technology: the past, the present, and extensions for the future," in *Proc. SPIE Opt. Microlith. XVII*, vol. 5377, 2004, 1–20.
13. M. Switkes and M. Rothschild, "Immersion lithography at 157 nm," *J. Vac. Sci. Technol. B*, vol. 19, no. 6, 2353 pp., 2001.
14. M. Switkes and M. Rothschild, "Resolution enhancement of 157 nm lithography by liquid immersion," *J. Microlith., Microfab., Microsyst.* vol. 1, p. 225, 2002.
15. S.H. Zaidi and S.R.J. Brueck, "Multiple-exposure interferometric lithography," *J. Vac. Sci. Technol. B*, vol. 11, 1993, pp. 658–666.
16. M.E. Mason, "The real cost of RETs. (Feature article)," *Microlith. World*, vol. 12, no. i2, p8(4), 2003.
17. D.M. Schaefer et al., "Fabrication of two-dimensional arrays of nanometer-size clusters with the atomic force microscope," *Appl. Phys. Lett.*, vol. 66, p. 1012, 1995.
18. T. Junno et al., "Controlled manipulation of nanoparticles with an atomic force microscope," *Appl. Phys. Lett.*, vol. 66, p. 3627, 1995.
19. J. Jersch and K. Dickmann, "Nanostructure fabrication using laser field enhancement in the near field of a scanning tunneling microscope tip," *Appl. Phys. Lett.*, 68, 868–870, 1996.
20. M.C. McCord and R.F.W. Pease, "Lithography with the scanning tunneling microscope," *J. Vac. Sci. Technol. B*, vol. 4, p. 86, 1986.
21. E. Betzig and J.K. Trautman, "Single molecules observed by near-field scanning optical microscopy," *Science*, vol. 257, p. 189, 1992.
22. E. Betzig et al., "Near-field magneto-optics and high density data storage," *Appl. Phys. Lett.*, vol. 61, p. 142–144, 1992.
23. A. Tarun et al., "Apertureless optical near-field fabrication using an atomic force microscope on photoresists," *Appl. Phys. Lett.*, vol. 80, 2002, pp. 3400–3402.
24. R.D. Piner et al., "Dip-Pen Nanolithography," *Science*, pp. 661–663, January 29, 1999.
25. P. Vettiger et al., "Ultrahigh density, high-data-rate NEMS-based AFM data storage system," *J. Microelectron. Eng.*, 46, 11–17, 1999.
26. P. Vettiger et al., "The 'Millipede' — more than one thousand tips for future AFM data storage," *IBM J. Res. Dev.*, vol. 44, no. 3, 2000.
27. J.M. Gibson and S.D. Berger, "New approach to projection-electron lithography with demonstrated 0.1 μm linewidth," *Appl. Phys. Lett.*, 57, 153, 1990.

28. S.D. Berger et al., "The SCALPEL system," *Proc. SPIE*, 2322, 434, 1994.
29. L.R. Harriott et al., "The SCALPEL proof of concept system," *Microelectron. Eng.*, 35, 477, 1997.
30. W.K. Waskiewicz et al., "SCALPEL proof-of-concept system: preliminary lithography results," *Proc. SPIE*, 3048, 255, 1997.
31. L.R. Harriott, "Scattering with angular limitation projection electron beam lithography for suboptical lithography," *J. Vac. Sci. Technol B*, vol. 15, no. 6, pp. 2130–2135, 1997.
32. Resnick et al., "Imprint lithography for IC fabrication," *J. Vac. Sci. Technol. B*, vol. 21, no. 6, 2003.
33. P.M. St. John et al., "Diffraction-based cell detection using a microcontact printed antibody grating," *Anal. Chem.*, 70, 1998, 1108–1111.
34. Y. Xia and G.M. Whitesides, "Soft lithography," *Annu. Rev. Mater. Sci.*, 28, 153–184. 1998A.
35. A. Kumar and G.M. Whitesides, "Features of gold having micrometer to centimeter dimensions can be formed through a combination of stamping with an elastomeric stamp and an alkanethiol 'ink' followed by chemical etching," *Appl. Phys. Lett.*, 63, 2002–2004, 1993.
36. A. Kumar et al., "Patterned self-assembled monolayers and meso-scale phenomena," *Acc. Chem. Res.*, 28, 219, 1995.
37. Y. Xia et al., "Unconventional methods for fabricating and patterning nanostructures," *Chem. Rev.*, 99, 1999, 1823–1848.
38. M. Colburn et al., "Step and flash imprint lithography: a new approach to high resolution patterning," *Proc. SPIE*, vol. 3676, no. I, p. 379, 1999.
39. F. Xu, "Development of imprint materials for the step and flash imprint lithography process," *Emerg. Lithographic Technol. VIII, Proc. SPIE*, 5375, 2004, 232–241.
40. K. Ikuta and K. Hirowatari, "Real three-dimensional micro fabrication using stereo lithography," in *Proc. IEEE Micro Electro Mech. Syst. (MEMS'93)*. Ft. Lauderdale, FL, 1993, pp. 42–47.
41. K. Ikuta, K. Hirowatari, and T. Ogata, "Three-dimensional integrated fluid systems (MIFS) fabricated by stereo lithography," In *IEEE Int. Workshop Micro Electro Mech. Syst. (MEMS '94)*, Oiso, Japan, 1994, pp. 1–6.
42. T. Takagi and N. Nakajima, "Photoforming applied to fine machining," In *Proc. IEEE Micro Electro Mech. Syst. (MEMS '93)*, Fort Lauderdale, FL, 1993, pp. 173–178.
43. T. Takagi and N. Nakajima, "Architecture combination by micro photoforming process," In *IEEE Int. Workshop Micro Electro Mech. Syst. (MEMS '94)*, Oiso, Japan, 1994, pp. 211–216.
44. J. Brook and R. Dandliker, "Submicrometer holographic photolithography," *Solid State Technol.*, 32, 91–94, 1989.
45. B. Omar et al., "Advances in holographic lithography," *Solid State Technol.*, 89–94, September 1991.
46. W. Ehrfeld, "The LIGA process for microsystems," in *Proc. Micro Syst. Technol. '90*, Berlin, 1990, pp. 521–528.
47. H. Lehr and M. Schmidt, *The LIGA Technique: Commercial Brochure*, Mainz-Hechtsheim: IMM Institut fur Mikrotechnnik GmbH, 1995.
48. Madou Research Group Homepage: http://www.biomems.net.
49. N.E. Hebert et al., "Performance of pyrolyzed photoresist carbon films in a microchip capillary electrophoresis device with sinusoidal voltammetric detection," *Anal. Chem.*, vol. 75, no. 16, 4265–4271, 2003.
50. M. Gad-el-Hak, *The MEMS Handbook*, Boca Raton, FL: CRC Press, 2002.
51. S.Y. Chou, P.R. Krauss, and P.J. Renstrom, "Nanoimprint lithography," *J. Vac. Sci. Technol. B*, 14, 4129, 1996.

3

Transistors

Sidney Soclof
California State University

Joseph Watson
University of Wales

John R. Brews
The University of Arizona

Harvey J. Stiegler
Texas Instruments Incorporated

James E. Morris
Portland State University

3.1 Junction Field-Effect Transistors

Sidney Soclof

A junction field-effect transistor (JFET) is a type of transistor in which the current flow through the device between the drain and source electrodes is controlled by the voltage applied to the gate electrode. A simple physical model of the JFET is shown in Figure 3.1. In this JFET an n-type conducting channel exists between drain and source. The gate is a p^+ region that surrounds the n-type channel. The gate-to-channel pn-junction is normally kept reverse-biased. As the reverse bias voltage between gate and channel increases, the depletion region width increases, as shown in Figure 3.2. The depletion region extends mostly into the n-type channel because of the heavy doping on the p^+ side. The depletion region is depleted of mobile charge carriers and thus cannot contribute to the conduction of current between drain and source. Thus, as the gate voltage increases, the cross-sectional areas of the n-type channel available for current flow decreases. This reduces the current flow between drain and source. As the gate voltage increases, the channel gets further constricted and the current flow gets smaller. Finally, when the depletion regions meet in the middle of the channel, as shown in Figure 3.3, the channel is pinched off in its entirety between source and drain. At this point the current flow between drain and source is reduced to essentially zero. This voltage is called the **pinch-off voltage,** V_P. The pinch-off voltage is also represented by V_{GS} (off) as being the gate-to-source voltage that turns the

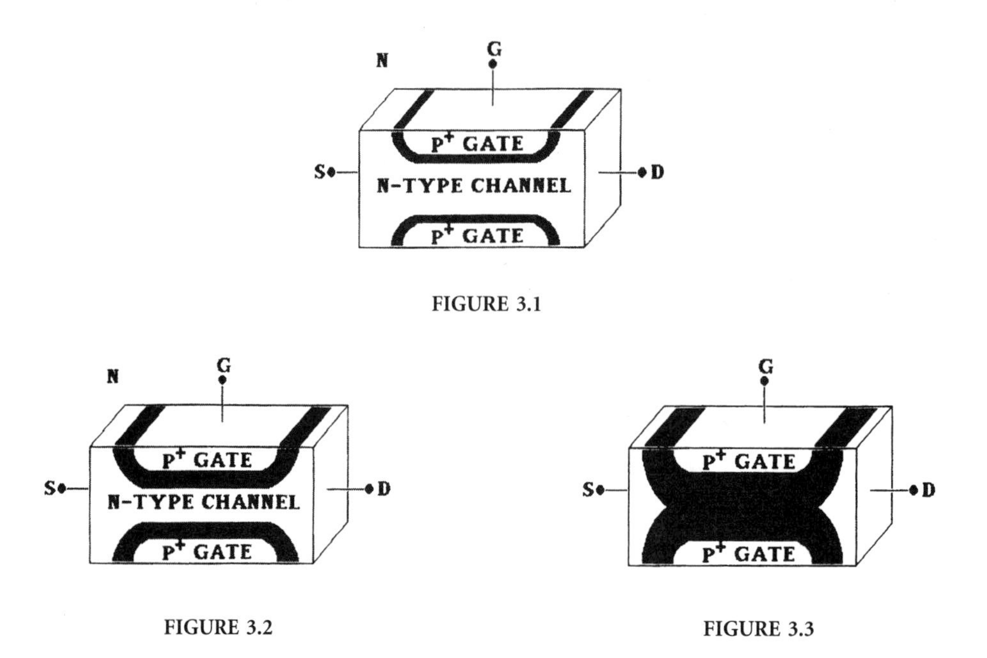

FIGURE 3.1

FIGURE 3.2

FIGURE 3.3

drain-to-source current I_{DS} off. We have been considering here an n-channel JFET. The complementary device is the p-channel JFET that has an n^+ gate region surrounding a p-type channel. The operation of a p-channel JFET is the same as for an n-channel device, except the algebraic signs of all dc voltages and currents are reversed.

We have been considering the case for V_{DS} small compared to the pinch-off voltage such that the channel is essentially uniform from drain to source, as shown in Figure 3.4(a). Now let us see what happens as V_{DS} increases. As an example let us assume an n-channel JFET with a pinch-off voltage of $V_P = -4$ V. We will see what happens for the case of $V_{GS} = 0$ as V_{DS} increases. In Figure 3.4(a) the situation is shown for the case of $V_{DS} = 0$ in which the JFET is fully "on" and there is a uniform channel from source to drain. This is at point A on the I_{DS} vs. V_{DS} curve of Figure 3.5. The drain-to-source conductance is at its maximum value of g_{ds} (on), and the drain-to-source resistance is correspondingly at its minimum value of r_{ds} (on). Now let us consider the case of $V_{DS} = +1$ V, as shown in Figure 3.4(b). The gate-to-channel bias voltage at the source end is still $V_{GS} = 0$. The gate-to-channel bias voltage at the drain end is $V_{GD} = V_{GS} - V_{DS} = -1$ V, so the depletion region will be wider at the drain end of the channel than at the source end. The channel will thus be narrower at the drain end than at the source end, and this will result in a decrease in the channel conductance g_{ds} and, correspondingly, an increase in the channel resistance r_{ds}. So the slope of the I_{DS} vs. V_{DS} curve that corresponds to the channel conductance will be smaller at $V_{DS} = 1$ V than it was at $V_{DS} = 0$, as shown at point B on the I_{DS} vs. V_{DS} curve of Figure 3.5.

In Figure 3.4(c) the situation for $V_{DS} = +2$ V is shown. The gate-to-channel bias voltage at the source end is still $V_{GS} = 0$, but the gate-to-channel bias voltage at the drain end is now $V_{GD} = V_{GS} - V_{DS} = -2$ V, so the depletion region will now be substantially wider at the drain end of the channel than at the source end. This leads to a further constriction of the channel at the drain end, and this will again result in a decrease in the channel conductance g_{ds} and, correspondingly, an increase in the channel resistance r_{ds}. So the slope of the I_{DS} vs. V_{DS} curve will be smaller at $V_{DS} = 2$ V than it was at $V_{DS} = 1$ V, as shown at point C on the I_{DS} vs. V_{DS} curve of Figure 3.5.

In Figure 3.4(d) the situation for $V_{DS} = +3$ V is shown, and this corresponds to point D on the I_{DS} vs. V_{DS} curve of Figure 3.5.

When $V_{DS} = +4$ V, the gate-to-channel bias voltage will be $V_{GD} = V_{GS} - V_{DS} = 0 - 4\text{ V} = -4\text{ V} = V_P$. As a result the channel is now pinched off at the drain end but is still wide open at the source end since $V_{GS} = 0$, as

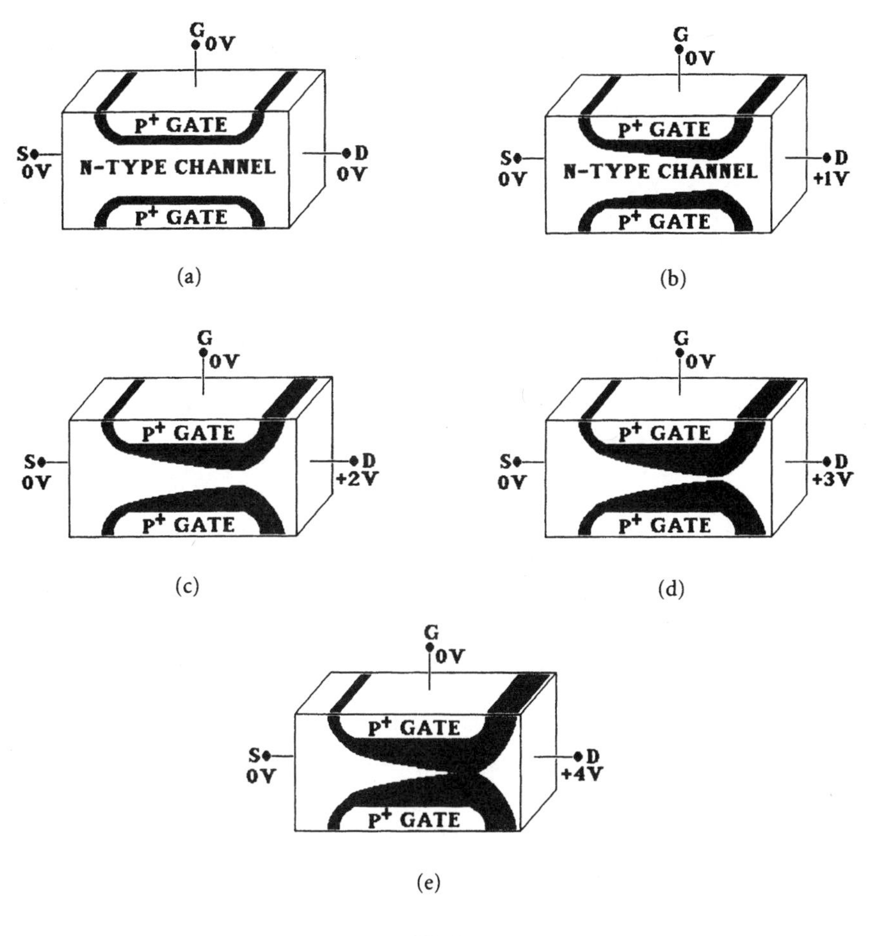

FIGURE 3.4

shown in Figure 3.4(e). It is very important to note that the channel is pinched off just for a very short distance at the drain end so that the drain-to-source current I_{DS} can still continue to flow. This is not at all the same situation as for the case of $V_{GS} = V_P$, where the channel is pinched off in its entirety, all the way from source to drain. When this happens, it is like having a big block of insulator the entire distance between source and drain, and I_{DS} is reduced to essentially zero. The situation for $V_{DS} = +4\ \text{V} = -V_P$ is shown at point E on the I_{DS} vs. V_{DS} curve of Figure 3.5.

For $V_{DS} > +4$ V, the current essentially saturates and does not increase much with further increases in V_{DS}. As V_{DS} increases above +4 V, the pinched-off region at the drain end of the channel gets wider, which increases r_{ds}. This increase in r_{ds} essentially counterbalances the increase in V_{DS} such that I_{DS} does not increase much. This region of the I_{DS} vs. V_{DS} curve in which the channel is pinched off at the drain end is called the **active region** and is also known as the *saturated region.* It is called the active region because when the JFET is to be used as an amplifier, it should be biased and operated in this region. The saturated value of drain current up in the active region for the case of $V_{GS} = 0$ is called the **drain saturation current**, I_{DSS} (the third subscript S refers to I_{DS} under

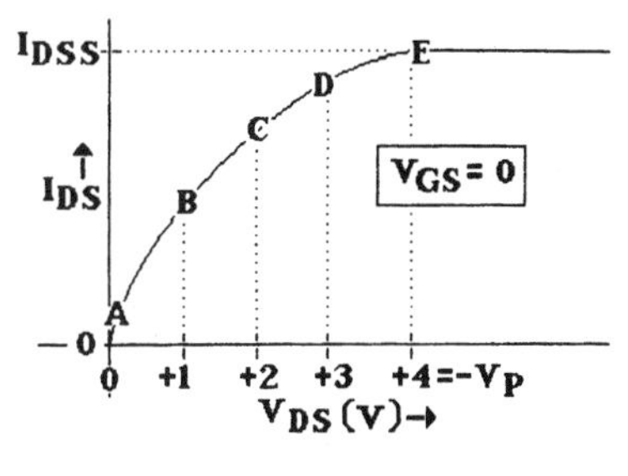

FIGURE 3.5

the condition of the gate *shorted* to the source). Since there is not really a true saturation of current in the active region, I_{DSS} is usually specified at some value of V_{DS}. For most JFETs, the values of I_{DSS} fall in the range of 1 to 30 mA.

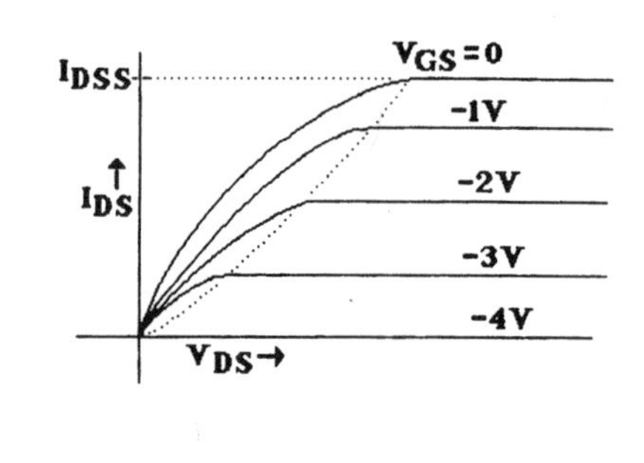

FIGURE 3.6

The region below the active region where $V_{DS} < +4$ V $= -V_P$ has several names. It is called the **nonsaturated region**, the **triode region**, and the **ohmic region**. The term *triode region* apparently originates from the similarity of the shape of the curves to that of the vacuum tube triode. The term *ohmic region* is due to the variation of I_{DS} with V_{DS} as in Ohm's law, although this variation is nonlinear except for the region of V_{DS} that is small compared to the pinch-off voltage where I_{DS} will have an approximately linear variation with V_{DS}.

The upper limit of the active region is marked by the onset of the breakdown of the gate-to-channel pn-junction. This will occur at the drain end at a voltage designated as BV_{DG}, or BV_{DS}, since $V_{GS} = 0$. This breakdown voltage is generally in the 30- to 150-V range for most JFETs.

So far we have looked at the I_{DS} vs. V_{DS} curve only for the case of $V_{GS} = 0$. In Figure 3.6 a family of curves of I_{DS} vs. V_{DS} for various constant values of V_{GS} is presented. This is called the *drain characteristics,* also known as the *output characteristics,* since the output side of the JFET is usually the drain side. In the active region where I_{DS} is relatively independent of V_{DS}, a simple approximate equation relating I_{DS} to V_{GS} is the square-law *transfer equation* as given by $I_{DS} = I_{DSS}[1 - (V_{GS}/V_P)]^2$. When $V_{GS} = 0$, $I_{DS} = I_{DSS}$ as expected, and as $V_{GS} \rightarrow V_P$, $I_{DS} \rightarrow 0$. The lower boundary of the active region is controlled by the condition that the channel be pinched off at the drain end. To meet this condition the basic requirement is that the gate-to-channel bias voltage at the drain end of the channel, V_{GD}, be greater than the pinch-off voltage V_P. For the example under consideration with $V_P = -4$ V, this means that $V_{GD} = V_{GS} - V_{DS}$ must be more negative than -4 V. Therefore, $V_{DS} - V_{GS} \geqslant +4$ V. Thus, for $V_{GS} = 0$, the active region will begin at $V_{DS} = +4$ V. When $V_{GS} = -1$ V, the active region will begin at $V_{DS} = +3$ V, for now $V_{GD} = -4$ V. When $V_{GS} = -2$ V, the active region begins at $V_{DS} = +2$ V, and when $V_{GS} = -3$ V, the active region begins at $V_{DS} = +1$ V. The dotted line in Figure 3.6 marks the boundary between the nonsaturated and active regions.

The upper boundary of the active region is marked by the onset of the avalanche breakdown of the gate-to-channel pn-junction. When $V_{GS} = 0$, this occurs at $V_{DS} = BV_{DS} = BV_{DG}$. Since $V_{DG} = V_{DS} - V_{GS}$ and breakdown occurs when $V_{DG} = BV_{DG}$, as V_{GS} increases the breakdown voltage decreases, as given by $BV_{DG} = BV_{DS} - V_{GS}$. Thus, $BV_{DS} = BV_{DG} + V_{GS}$. For example, if the gate-to-channel breakdown voltage is 50 V, the V_{DS} breakdown voltage will start off at 50 V when $V_{GS} = 0$ but decrease to 46 V when $V_{GS} = -4$ V.

In the nonsaturated region I_{DS} is a function of both V_{GS} and V_{DS}, and in the lower portion of the nonsaturated region where V_{DS} is small compared to V_P, I_{DS} becomes an approximately linear function of V_{DS}. This linear portion of the nonsaturated is called the *voltage-variable resistance* (VVR) region, for in this region the JFET acts like a linear resistance element between source and drain. The resistance is variable in that it is controlled by the gate voltage. This region and VVR application will be discussed in a later section. The JFET can also be operated in this region as a switch, and this will also be discussed in a later section.

JFET Biasing

Voltage Source Biasing

Now we will consider the biasing of JFETs for operation in the active region. The simplest biasing method is shown in Figure 3.7, in which a voltage source V_{GG} is used to provide the quiescent gate-to-source bias voltage V_{GSQ}. In the active region the transfer equation for the JFET has been given as $I_{DS} = I_{DSS}[1 - (V_{GS}/V_P)]^2$, so for a quiescent drain current of I_{DSQ}, the corresponding gate voltage will be given by $V_{GSQ} = V_P(1 - \sqrt{I_{DSQ}/I_{DSS}})$. For a Q point in the middle of the active region, we have that $I_{DSQ} = I_{DSS}/2$, so $V_{GSQ} = V_P(1 - \sqrt{1/2}) = 0.293\ V_P$.

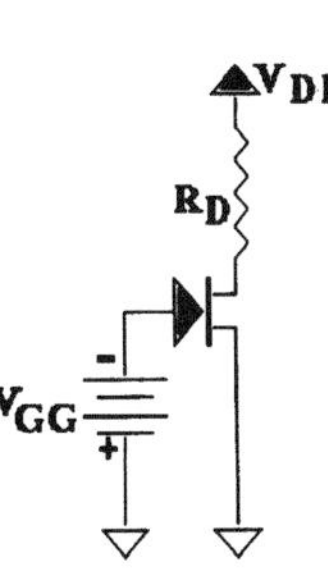

FIGURE 3.7 Voltage source biasing.

The voltage source method of biasing has several major drawbacks. Since V_P will have the opposite polarity of the drain supply voltage V_{DD}, the gate bias voltage will require a second power supply. For the case of an n-channel JFET, V_{DD} will come from a positive supply voltage and V_{GG} must come from a separate negative power supply voltage or battery. A second, and perhaps more serious, problem is the "open-loop" nature of this biasing method. The JFET parameters of I_{DDS} and V_P will exhibit very substantial unit-to-unit variations, often by as much as a 2:1 factor. There is also a significant temperature dependence of I_{DDS} and V_P. These variations will lead to major shifts in the position of the Q point and the resulting distortion of the signal. A much better biasing method is shown in Figure 3.8.

Self-Biasing

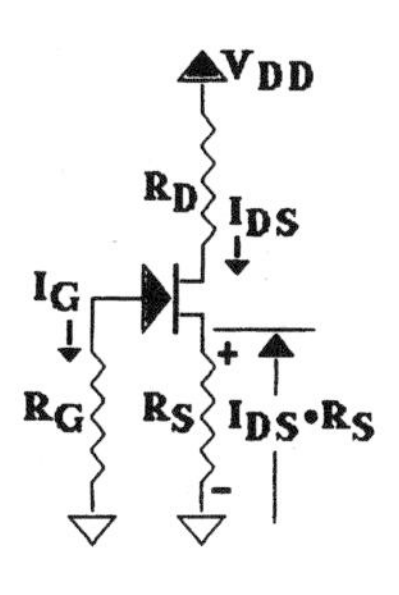

FIGURE 3.8 Self-biasing.

The biasing circuit of Figure 3.8 is called a *self-biasing* circuit in that the gate-to-source voltage is derived from the voltage drop produced by the flow of drain current through the source biasing resistor R_S. It is a closed-loop system in that variations in the JFET parameters can be partially compensated for by the biasing circuit. The gate resistor R_G is used to provide a dc return path for the gate leakage current and is generally up in the megohm range.

The voltage drop across R_S is given by $V_S = I_{DS} \cdot R_S$. The voltage drop across the gate resistor R_G is $V_G = I_G \cdot R_G$. Since I_G is usually in the low nanoampere or even picoampere range, as long as R_G is not extremely large, the voltage drop across R_G can be neglected, so $V_G \cong 0$. Thus, we have that $V_{GS} = V_G - V_S \cong -V_S = -I_{DS} \cdot R_S$. For example, if $I_{DSS} = 10$ mA and $V_P = -4$ V, and for a Q point in the middle of the active region with $I_{DSQ} = I_{DSS}/2 = 5$ mA, we have that $V_{GSQ} = 0.293 V_P = -1.17$ V. Therefore, the required value for the source biasing resistor is given by $R_S = -V_{GS}/I_{DSQ} = 1.17$ V/5 mA $= 234\ \Omega$. This produces a more stable quiescent point than voltage source biasing, and no separate negative power supply is required.

The closed-loop nature of this biasing circuit can be seen by noting that if changes in the JFET parameters were to cause I_{DS} to increase, the voltage drop across R_S would also increase. This will produce an increase in V_{GS} (in the negative direction for an n-channel JFET), which will act to reduce the increase in I_{DS}. Thus, the net increase in I_{DS} will be less due to the feedback voltage drop produced by the flow of I_{DS} through R_S. The same basic action would, of course, occur for changes in the JFET parameters that would cause I_{DS} to decrease.

Bias Stability

Now let us examine the stability of the Q point. We will start again with the basic transfer equation as given by $I_{DS} = I_{DSS}[1 - (V_{GS}/V_P)]^2$. From this equation the change in the drain current, ΔI_{DS}, due to changes in I_{DSS}, V_{GS}, and V_P can be written as

$$\Delta I_{DS} = g_m \Delta V_{GS} - g_m \frac{V_{GS}}{V_P} \Delta V_P + \frac{I_{DS}}{I_{DSS}} \Delta I_{DSS}$$

Since $V_{GS} = -I_{DS} \cdot R_S$, $\Delta V_{GS} = -R_S \cdot \Delta I_{DS}$, we obtain that

$$\Delta I_{DS} = -g_m R_S \Delta I_{DS} - g_m \frac{V_{GS}}{V_P} \Delta V_P + \frac{I_{DS}}{I_{DSS}} \Delta I_{DSS}$$

Collecting terms in ΔI_{DS} on the left side gives

$$\Delta I_{DS}(1 + g_m R_S) = -g_m \frac{V_{GS}}{V_P} \Delta V_P + \frac{I_{DS}}{I_{DSS}} \Delta I_{DSS}$$

Now solving this for ΔI_{DS} yields

$$\Delta I_{DS} = \frac{-g_m (V_{GS}/V_P) \Delta V_p + \frac{I_{DS}}{I_{DSS}} \Delta I_{DSS}}{1 + g_m R_S}$$

From this we see that the shift in the quiescent drain current, ΔI_{DS}, is reduced by the presence of R_S by a factor of $1 + g_m R_S$.

If $I_{DS} = I_{DSS}/2$, then:

$$g_m = \frac{2\sqrt{I_{DS} \cdot I_{DSS}}}{-V_P} = \frac{2\sqrt{I_{DS} \cdot 2I_{DS}}}{-V_P} = \frac{2\sqrt{2} I_{DS}}{-V_P}$$

Since $V_{GS} = 0.293 V_P$, the source biasing resistor will be $R_S = -V_{GS}/I_{DS} = -0.293\ V_P/I_{DS}$. Thus:

$$g_m R_S = \frac{2\sqrt{2} I_{DS}}{-V_P} \times \frac{-0.293 V_P}{I_{DS}} = 2\sqrt{2} \times 0.293 = 0.83$$

so $1 + g_{m \cdot S} = 1.83$. Thus the sensitivity of I_{DS} due to changes in V_P and I_{DSS} is reduced by a factor of 1.83.

The equation for ΔI_{DS} can now be written in the following form for the fractional change in I_{DS}:

$$\frac{\Delta I_{DS}}{I_{DS}} = \frac{-0.83(\Delta V_P/V_P) + 1.41(\Delta I_{DSS}/I_{DSS})}{1.83}$$

so $\Delta I_{DS}/I_{DS} = -0.45\ (\Delta V_P/V_P) + 0.77\ (\Delta I_{DSS}/I_{DSS})$, and thus a 10% change in V_P will result in approximately a 4.5% change in I_{DS}, and a 10% change in I_{DSS} will result in an 8% change in I_{DS}. Thus, although the situation is improved with the self-biasing circuit using R_S, there will still be a substantial variation in the quiescent current with changes in the JFET parameters.

A further improvement in bias stability can be obtained by the use of the biasing methods of Figure 3.9 and Figure 3.10. In Figure 3.9 a gate bias voltage V_{GG} is obtained from the V_{DD} supply voltage by means of the $R_{G1}-R_{G2}$ voltage divider. The gate-to-source voltage is now $V_{GS} = V_G - V_S = V_{GG} - I_{DS}R_S$. So now for R_S we have $R_S = (V_{GG} - V_{GS})/I_{DS}$. Since V_{GS} is of opposite polarity to V_{GG}, this will result in a larger value for R_S than before. This in turn will result in a larger value for the $g_m R_S$ product and hence improved bias stability. If we continue with the preceding examples and now let $V_{GG} = V_{DD}/2 = +10$ V, we have that $R_S = (V_{GG} - V_{GS})/I_{DS} = [+10\text{V} - (-1.17\text{V})]/5$ mA $= 2.234$ kΩ, as compared to $R_S = 234\ \Omega$ that was obtained before. For g_m we have $g_m = 2\sqrt{I_{DS} \cdot I_{DSS}}/(-V_P) = 3.54$ mS, so $g_m R_S = 3.54$ mS$\cdot 2.234$ k$\Omega = 7.90$. Since $1 + g_m R_S = 8.90$, we now have an improvement by a factor of 8.9 over the open-loop voltage source biasing and by a factor of 4.9 over the self-biasing method without the V_{GG} biasing of the gate.

Another biasing method that can lead to similar results is the method shown in Figure 3.10. In this method the bottom end of the source biasing resistor goes to a negative supply voltage V_{SS} instead of to ground. The gate-to-source bias voltage is now given by $V_{GS} = V_G - V_S = 0 - (I_{DS} \cdot R_S + V_{SS})$ so that for R_S we now have $R_S = (-V_{GS} - V_{SS})/I_{DS}$. If $V_{SS} = -10$ V, and as before $I_{DS} = 5$ mA and $V_{GS} = -1.17$ V, we have $R_S = 11.7$ V/5 mA $= 2.34$ kΩ, and thus $g_m R_S = 7.9$ as in the preceding example. So this method does indeed lead to results similar to that for the R_S and V_{GG} combination biasing. With either of these two methods the change in I_{DS} due to a 10% change in V_P will be only 0.9%, and the change in I_{DS} due to a 10% change in I_{DSS} will be only 1.6%.

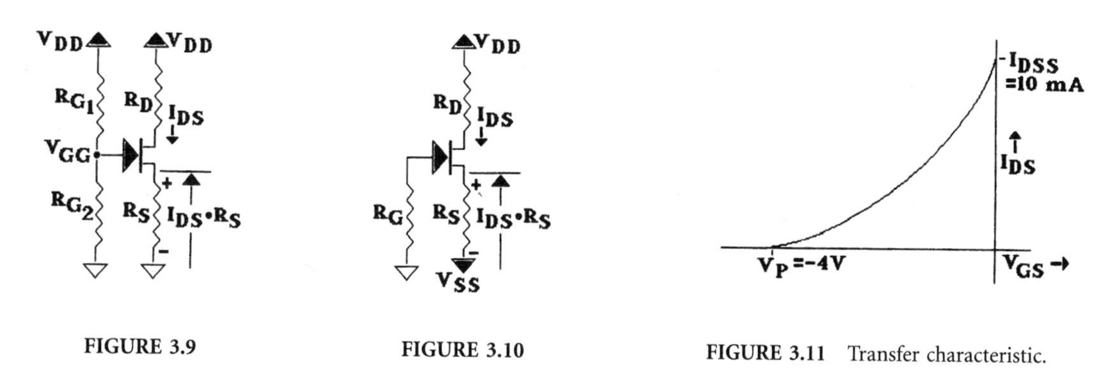

FIGURE 3.9

FIGURE 3.10

FIGURE 3.11 Transfer characteristic.

The biasing circuits under consideration here can be applied directly to the common-source (CS) amplifier configuration, and can also be used for the common-drain (CD), or source-follower, and common-gate (CG) JFET configurations.

Transfer Characteristics

Transfer Equation

Now we will consider the *transfer characteristics* of the JFET, which is a graph of the output current I_{DS} vs. the input voltage V_{GS} in the active region. In Figure 3.11 a transfer characteristic curve for a JFET with $V_P = -4$ V and $I_{DSS} = +10$ mA is given. This is approximately a square-law relationship as given by $I_{DS} = I_{DSS}\,[1 - (V_{GS}/V_P)]^2$. This equation is not valid for V_{GS} beyond V_P (i.e., $V_{GS} < V_P$), for in this region the channel is pinched off and $I_{DS} \cong 0$.

At $V_{GS} = 0$, $I_{DS} = I_{DSS}$. This equation and the corresponding transfer curve can actually be extended up to the point where $V_{GS} \cong +0.5$ V. In the region where $0 < V_{GS} < +0.5$ V, the gate-to-channel pn-junction is *forward-biased* and the depletion region width is reduced below the width under zero bias conditions. This reduction in the depletion region width leads to a corresponding expansion of the conducting channel and thus an increase in I_{DS} above I_{DSS}. As long as the gate-to-channel forward bias voltage is less than about 0.5 V, the pn-junction will be essentially "off" and very little gate current will flow. If V_{GS} is increased much above +0.5 V, however, the gate-to-channel pn-junction will turn "on" and there will be a substantial flow of gate voltage I_G. This gate current will load down the signal source and produce a voltage drop across the signal source resistance, as shown in Figure 3.12. This voltage drop can cause V_{GS} to be much smaller than the signal source voltage V_{in}. As V_{in} increases, V_{GS} will ultimately level off at a forward bias voltage of about +0.7 V, and the signal source will lose control over V_{GS}, and hence over I_{DS}. This can result in severe distortion of the input signal in the form of clipping, and thus this situation should be avoided. Thus, although it is possible to increase I_{DS} above I_{DSS} by allowing the gate-to-channel junction to become forward-biased by a small amount ($\leq$0.5 V), the possible benefits are generally far outweighed by the risk of signal distortion. Therefore, JFETs are almost always operated with the gate-to-channel pn-junction reverse-biased.

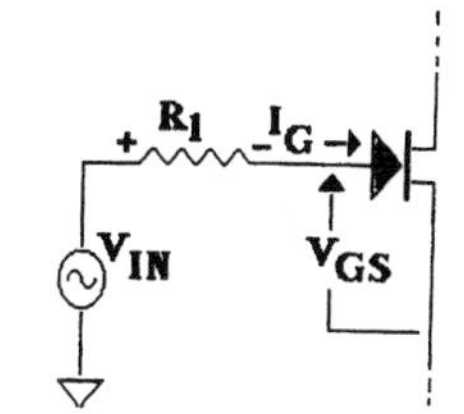

FIGURE 3.12 Effect of forward bias on V_{GS}.

Transfer Conductance

The slope of the transfer curve, dI_{DS}/dV_{GS}, is the dynamic forward transfer conductance, or mutual transfer conductance, g_m. We see that g_m starts off at zero when $V_{GS} = V_P$ and increases as I_{DS} increases, reaching a maximum when $I_{DS} = I_{DSS}$. Since $I_{DS} = I_{DSS}[1-(V_{GS}/V_P)]^2$, g_m can be obtained as

$$g_m = \frac{dI_{DS}}{dV_{GS}} = 2I_{DSS}\frac{\left(1 - \frac{V_{GS}}{V_P}\right)}{-V_P}$$

Since

$$1 - \left(\frac{V_{GS}}{V_P}\right) = \sqrt{\frac{I_{DS}}{I_{DSS}}}$$

we have that

$$g_m = 2I_{DSS}\frac{\sqrt{I_{DS}/I_{DSS}}}{-V_P} = 2\frac{\sqrt{I_{DS}\cdot I_{DSS}}}{-V_P}$$

The maximum value of g_m is obtained when $V_{GS} = 0$ ($I_{DS} = I_{DSS}$) and is given by g_m ($V_{GS} = 0$) $= g_{m0}$ $= 2I_{DS}/(-V_P)$.

Small-Signal AC Voltage Gain

Let us consider the CS amplifier circuit of Figure 3.13. The input ac signal is applied between gate and source, and the output ac voltage is taken between drain and source. Thus, the source electrode of this triode device is common to input and output, hence the designation of this JFET configuration as a CS amplifier.

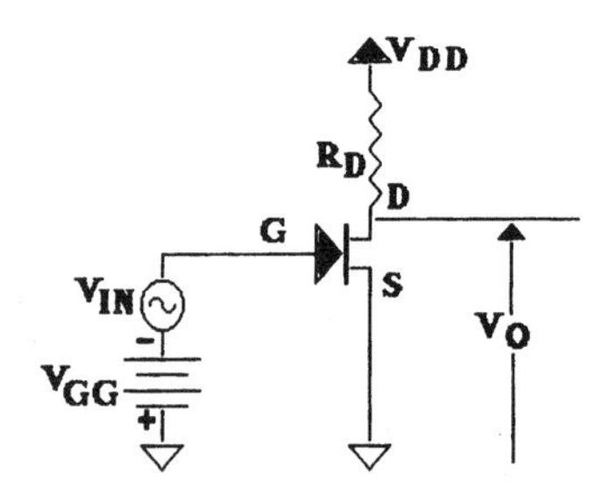

FIGURE 3.13 Common-source amplifier.

A good choice of the dc operating point or quiescent point (Q point) for an amplifier is in the middle of the active region at $I_{DS} = I_{DSS}/2$. This allows for the maximum symmetrical drain current swing, from the quiescent level of $I_{DSQ} = I_{DSS}/2$, down to a minimum of $I_{DS} \cong 0$, and up to a maximum of $I_{DS} = I_{DSS}$. This choice for the Q point is also a good one from the standpoint of allowing for an adequate safety margin for the location of the actual Q point due to the inevitable variations in device and component characteristics and values. This safety margin should keep the Q point well away from the extreme limits of the active region, and thus ensure operation of the JFET in the active region under most conditions. If $I_{DSS} = +10$ mA, then a good choice for the Q point would thus be around +5.0 mA. If $V_P = -4$ V, then:

$$g_m \frac{2\sqrt{I_{DS}\cdot I_{DSS}}}{-V_P} = \frac{2\sqrt{5\,\mathrm{mA}\cdot 10\mathrm{mA}}}{4V} = 3.54\ \mathrm{mA/V} = 3.54\ \mathrm{mS}$$

If a small ac signal voltage v_{GS} is superimposed on the dc gate bias voltage V_{GS}, only a small segment of the transfer characteristic adjacent to the Q point will be traversed, as shown in Figure 3.14. This small segment will be close to a straight line, and as a result the ac drain current i_{ds} will have a waveform close to that of the ac voltage applied to the gate. The ratio of i_{ds} to v_{GS} will be the slope of the transfer curve as given by $i_{ds}/v_{GS} \cong dI_{DS}/dV_{GS} = g_m$. Thus, $i_{ds} \cong g_m v_{GS}$. If the net load driven by the drain of the JFET is the drain load resistor R_D, as shown in Figure 3.13, then the ac drain current i_{ds} will produce an ac drain voltage of $v_{ds} = -i_{ds}\cdot R_D$. Since $i_{ds} = g_m v_{GS}$, this becomes $v_{ds} = -g_m v_{GS}\cdot R_D$. The ac small-signal voltage gain from gate to drain thus becomes $A_V = v_O/v_{in} = v_{ds}/v_{GS} = -g_m\cdot R_D$. The negative sign indicates signal inversion as is the case for a CS amplifier.

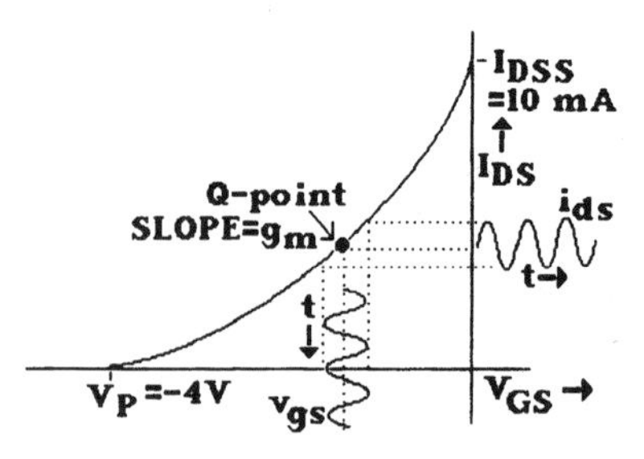

FIGURE 3.14 Transfer characteristic.

If the dc drain supply voltage is $V_{DD} = +20$ V, a quiescent drain-to-source voltage of $V_{DSQ} = V_{DD}/2 = +10$ V will result in the JFET being biased in the middle of the active region. Since $I_{DSQ} = +5$ mA in the example under consideration, the voltage drop across the drain load resistor R_D is 10 V. Thus, $R_D = 10\ \text{V}/5\ \text{mA} = 2\ \text{k}\Omega$. The ac small-signal voltage gain A_V thus becomes $A_V = -g_m \cdot R_D = -3.54\ \text{mS} \cdot 2\ \text{k}\Omega = -7.07$. Note that the voltage gain is relatively modest as compared to the much larger voltage gains that can be obtained in a bipolar-junction transistor (BJT) common-emitter amplifier. This is due to the lower transfer conductance of both JFETs and MOSFETs (metal-oxide semiconductor field-effect transistors) as compared to BJTs. For a BJT the transfer conductance is given by $g_m = I_C/V_T$, where I_C is the quiescent collector current and $V_T = kT/q \cong 25$ mV is the thermal voltage. At $I_C = 5$ mA, $g_m = 5\ \text{mA}/25\ \text{mV} = 200$ mS, as compared to only 3.5 mS for the JFET in this example. With a net load of 2 kΩ, the BJT voltage gain will be -400 as compared to the JFET voltage gain of only 7.1. Thus, FETs do have the disadvantage of a much lower transfer conductance, and therefore voltage gain, than BJTs operating under similar quiescent current levels, but they do have the major advantage of a much higher input impedance and a much lower input current. In the case of a JFET the input signal is applied to the *reverse-biased* gate-to-channel pn-junction and thus sees a very high impedance. In the case of a common-emitter BJT amplifier, the input signal is applied to the *forward-biased* base-emitter junction, and the input impedance is given approximately by $r_{in} = r_{BE} \cong 1.5 \cdot \beta \cdot V_T/I_C$. If $I_C = 5$ mA and $\beta = 200$, for example, then $r_{in} \cong 1500\ \Omega$. This moderate input resistance value of 1.5 kΩ is certainly no problem if the signal source resistance is less than around 100 Ω. However, if the source resistance is above 1 kΩ, then there will be a substantial signal loss in the coupling of the signal from the signal source to the base of the transistor. If the source resistance is in the range of above 100 kΩ, and certainly if it is above 1 MΩ, then there will be severe signal attenuation due to the BJT input impedance, and the FET amplifier will probably offer a greater overall voltage gain. Indeed, when high-impedance signal sources are encountered, a multistage amplifier with an FET input stage followed by cascaded BJT stages is often used.

JFET Output Resistance

Dynamic Drain-to-Source Conductance

For the JFET in the active region, the drain current I_{DS} is a strong function of the gate-to-source voltage V_{GS} but is relatively independent of the drain-to-source voltage V_{DS}. The transfer equation has previously been stated as $I_{DS} = I_{DSS}\,[1-(V_{GS}/V_P)]^2$.

The drain current will, however, increase slowly with increasing V_{DS}. To take this dependence of I_{DS} on V_{DS} into account, the transfer equation can be modified to give

$$I_{DS} = I_{DSS}\left(1 - \frac{V_{GS}}{V_P}\right)^2\left(1 + \frac{V_{DS}}{V_A}\right)$$

where V_A is a constant called the *early voltage* and is a parameter of the transistor with units of volts. The early voltage V_A is generally in the range of 30 to 300 V for most JFETs. The variation of the drain current with drain voltage is the result of the *channel-length modulation effect* in which the channel length decreases as the drain voltage increases. This decrease in the channel length results in an increase in the drain current. In BJTs a similar effect is the *base-width modulation effect.*

The *dynamic drain-to-source conductance* is defined as $g_{ds}=dI_{DS}/dV_{DS}$ and can be obtained from the modified transfer equation $I_{DS} = I_{DSS}\,[1-(V_{GS}/V_P)]^2\,[1+V_{DS}/V_A]$ as simply $g_{ds} = I_{DS}/V_A$. The reciprocal of g_{ds} is *dynamic drain-to-source resistance* r_{ds}, so $r_{ds} = 1/g_{ds} = V_A/I_{DS}$. If, for example, V_A=100 V, we have that $r_{ds} = 100\ \text{V}/I_{DS}$. At $I_{DS} = 1$ mA, $r_{ds} = 100\ \text{V}/1\ \text{mA} = 100\ \text{k}\Omega$, and at $I_{DS} = 10$ mA, $r_{ds} = 10\ \text{k}\Omega$.

Equivalent Circuit Model of CS Amplifier Stage

A small-signal equivalent circuit model of a CS FET amplifier stage is shown in Figure 3.15. The ac small-signal voltage gain is given by $A_V = -\,g_m \cdot R_{net}$, where $R_{net} = [r_{ds} \| R_D \| R_L]$ is the net load driven by the drain for the FET and includes the dynamic drain-to-source resistance r_{ds}. Since r_{ds} is generally much larger than $[R_D \| R_L]$, it will usually be the case that $R_{net} \cong [R_D \| R_L]$, and r_{ds} can be neglected. There are, however, some cases in which r_{ds} must be taken into account. This is especially true for the case in which an active load is used, as shown in Figure 3.16. For this case $R_{net} = [r_{ds}1 \| r_{ds}2 \| R_L]$, and r_{ds} can be a limiting factor in determining the voltage gain.

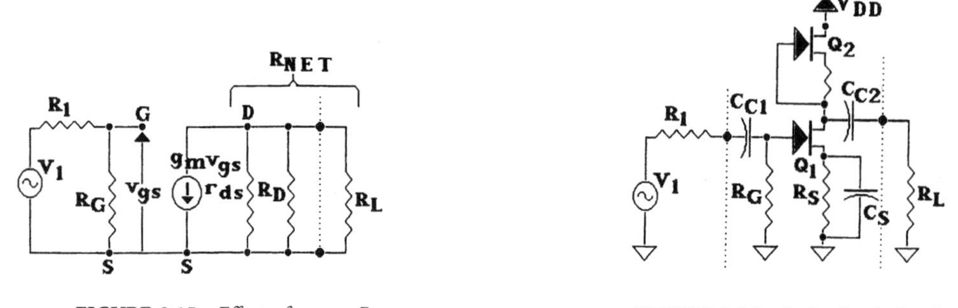

FIGURE 3.15 Effect of r_{ds} on R_{net}.

FIGURE 3.16 Active load circuit.

Consider an example for the active load circuit of Figure 3.16 for the case of identical JFETs with the same quiescent current. Assume that $R_L >> r_{ds}$ so that $R_{net} \cong [r_{ds}1 \| r_{ds}2] = V_A/(2I_{DSQ})$. Let $I_{DSQ} = I_{DSS}/2$, so $g_m = -2\sqrt{I_{DSS} \cdot I_{DSQ}}/(-V_P) = 2\sqrt{2}I_{DSQ}/(-V_P)$. The voltage gain is

$$A_V = -g_m \cdot R_{net} = \frac{2\sqrt{2}I_{DSQ}}{V_P} \times \frac{V_A}{2I_{SDQ}} = \sqrt{2}\frac{V_A}{V_P}$$

If $V_A = 100$ V and $V_P = -2$ V, we obtain $A_V = -70$, so we see that with active loads relatively large voltage gains can be obtained with FETs.

Another circuit in which the dynamic drain-to-source resistance r_{ds} is important is the constant-current source or current regulator diode. In this case the current regulation is directly proportional to the dynamic drain-to-source resistance.

Source Follower

Source-Follower Voltage Gain

We will now consider the CD JFET configuration, which is also known as the source follower. A basic CD circuit is shown in Figure 3.17. The input signal is supplied to the gate of the JFET. The output is taken from the source of the JFET, and the drain is connected directly to the V_{DD} supply voltage, which is ac ground.

For the JFET in the active region we have that $i_{ds} = g_m v_{GS}$. For this CD circuit we also have that $v_{GS} = v_G - v_S$ and $v_S = i_{ds}R_{net}$, where $R_{net} = [R_S \| R_L]$ is the net load resistance driven by the transistor. Since $v_{GS} = i_{ds}/g_m$, we have that $i_{ds}/g_m = v_G - i_{ds}R_{net}$. Collecting terms in i_{ds} on the left side yields $i_{ds}\,[(1/g_m) + R_{net}] = v_G$, so:

$$i_{ds} = \frac{v_G}{(1/g_m) + R_{net}} = \frac{g_m v_G}{1 + g_m R_{net}}$$

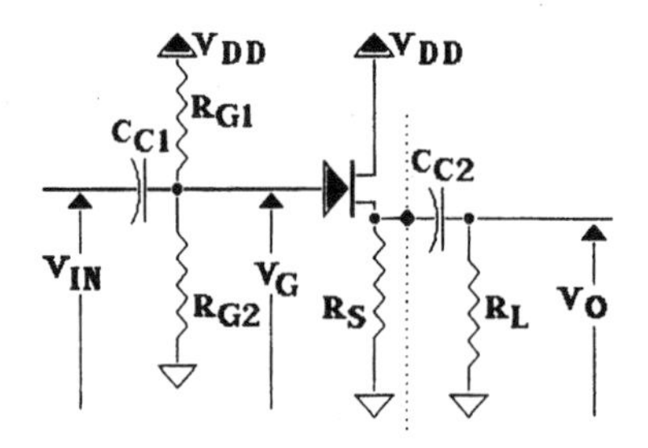

FIGURE 3.17 Source follower.

The output voltage is

$$v_o = v_s = i_{ds}R_{net} = \frac{g_m R_{net} v_G}{1 + g_m R_{net}}$$

and thus the ac small-signal voltage gain is

$$A_V = \frac{v_o}{v_G} = \frac{g_m R_{net}}{1 + g_m R_{net}}$$

Upon dividing through by g_m this can be rewritten as

$$A_V = \frac{R_{net}}{(1/g_m) + R_{net}}$$

From this we see that the voltage gain will be positive, and thus the source follower is a noninverting amplifier. We also note that A_V will always be less than unity, although for the usual case of $R_{net} >> 1/g_m$, the voltage gain will be close to unity.

The source follower can be represented as an amplifier with an open-circuit (i.e., no load) voltage transfer ratio of unity and an output resistance of $r_O = 1/g_m$. The equation for A_V can be expressed as $A_V = R_{net}/(R_{net} + r_O)$, which is the voltage division ratio of the $r_O = R_{net}$ circuit.

Source-Follower Examples

Let us consider an example of a JFET with $I_{DSS} = 10$ mA and $V_P = -4$ V. Let $V_{DD} = +20$ V and $I_{DSQ} = I_{DSS}/2 = 5$ mA. For $I_{DS} = I_{DSS}/2$ the value of V_{GS} is -1.17 V. To bias the JFET in the middle of the active region, we will let $V_{GQ} = V_{DD}/2 = +10$ V, so $V_{SQ} = V_{GQ} - V_{GS} = +10\text{ V} - (-1.17\text{ V}) = +11.17$ V. Thus $R_S = V_{SQ}/I_{DSQ} = 11.17\text{ V}/5\text{ mA} = 2.23\text{ k}\Omega$.

The transfer conductance at $I_{DS} = 5$ mA is 3.54 mS so that $r_O = 1/g_m = 283\ \Omega$. Since $g_m R_S = 7.9$, good bias stability will be obtained. If $R_L >> R_S$, then $A_V \cong R_S/(r_O + R_S) = 2.23\text{ k}\Omega/(283\ \Omega + 2.23\text{ k}\Omega) = 0.887$. If $R_L = 1\text{ k}\Omega$, then $R_{net} = 690\ \Omega$, and A_V drops to 0.709, and if $R_L = 300\ \Omega$, $R_{net} = 264\ \Omega$ and A_V is down to 0.483. A BJT emitter-follower circuit has the same equations for the voltage gain as the FET source follower. For the BJT case, $r_O = 1/g_m = VT/I_C$, where VT = thermal voltage $= kT/q \cong 25$ mV and I_C is the quiescent collector current. For $I_C = 5$ mA, we get $r_O \cong 25\text{ mV}/5\text{ mA} = 5\Omega$ as compared to $r_O = 283\ \Omega$ for the JFET case at the same quiescent current level. So the emitter follower does have a major advantage over the source follower since it has a much lower output resistance r_O and can thus drive very small load resistances with a voltage gain close to unity. For example, with $R_L = 100\ \Omega$, we get $A_V \cong 0.26$ for the source follower as compared to $A_V \cong 0.95$ for the emitter follower.

The FET source follower does, however, offer substantial advantages over the emitter follower of a much higher input resistance and a much lower input current. For the case in which a very high-impedance source, up in the megohm range, is to be coupled to a low-impedance load down in the range of 100 Ω or less, a good combination to consider is that of a cascaded FET source follower followed by a BJT emitter follower. This combination offers the very high input resistance of the source follower and the very low output resistance of the emitter follower.

For the source-follower circuit under consideration the input resistance will be $R_{in} = [R_{G1} \| R_{G2}] = 10\text{ M}\Omega$. If the JFET gate current is specified as 1 nA (max), and for good bias stability the change in gate voltage due to the gate current should not exceed $|V_P|/10 = 0.4$ V, the maximum allowable value for $[R_{G1} \| R_{G2}]$ is given by $IG \cdot [R_{G1} \| R_{G2}] < 0.4$ V. Thus $[R_{G1} \| R_{G2}] < 0.4\text{ V}/1\text{ nA} = 0.4\text{ G}\Omega = 400\text{ M}\Omega$. Therefore, R_{G1} and R_{G2} can each be allowed to be as large as 800 MΩ, and very large values for R_{in} can thus be obtained. At higher frequencies the input capacitance C_{in} must be considered, and C_{in} will ultimately limit the input impedance of the circuit. Since the input capacitance of the FET will be comparable to that of the BJT, the advantage of the FET source follower over the BJT emitter follower from the standpoint of input impedance will be obtained only at relatively low frequencies.

Source-Follower Frequency Response

The input capacitance of the source follower is given by $C_{in} = C_{GD} + (1-A_V)\, C_{GS}$. Since A_V is close to unity, C_{in} will be approximately given by $C_{in} \cong C_{GD}$. The source-follower input capacitance can, however, be reduced below C_{GD} by a bootstrapping circuit in which the drain voltage is made to follow the gate voltage. Let us consider a representative example in which $C_{GD} = 5$ pF, and let the signal-source output resistance be $R_1 = 100$ kΩ. The input circuit is in the form of a simple RC low-pass network. The RC time constant is

$$\tau = [R \| R_{G1} \| R_{G2}] \cdot C_{in} \cong R_1 \cdot C_{GD}$$

Thus $\tau \cong 100$ kΩ·5 pF = 500 nsec = 0.5 μsec. The corresponding 3-dB or half-power frequency is $f_H = 1/(2\,\pi\tau) = 318$ kHz. If R_1=1 MΩ, the 3-dB frequency will be down to about 30 kHz. Thus, we see indeed the limitation on the frequency response that is due to the input capacitance.

Frequency and Time-Domain Response

Small-Signal CS Model for High-Frequency Response

We will now consider the frequency- and time-domain response of the JFET CS amplifier. In Figure 3.18 an ac representation of a CS amplifier is shown, the dc biasing not being shown. In Figure 3.19 the JFET small-signal ac equivalent circuit model is shown, including the junction capacitances C_{GS} and C_{GD}. The gate-to-drain capacitance C_{GD} is a feedback capacitance in that it is connected between output (drain) and input (gate). Using Miller's theorem for shunt feedback this feedback capacitance can be transformed into an equivalent input capacitance $C_{GD'} = (1-A_V)C_{GD}$ and an equivalent output capacitance $C_{GD''} = (1-1/A_V)C_{GD}$, as shown in Figure 3.20. The net input capacitance is now $C_{in} = C_{GS} + (1-A_V)C_{GD}$ and the net output capacitance is $C_O = (1-\ 1/A_V)C_{GD} + C_L$. Since the voltage gain A_V is given by $A_V = -g_m R_{net}$, where R_{net} represents the net load resistance, the equations for C_{in} and C_O can be written approximately as $C_{in} = C_{GS} + (1 + g_m R_{net})C_{GD}$ and $C_O = [1 + 1/(g_m R_{net})]C_{GD} + C_L$. Since usually $A_V = g_m R_{net} >> 1$, C_O can be written as $C_O \cong C_{GD} + C_L$. Note that the voltage gain given by $A_V = -g_m R_{net}$ is not valid in the higher frequency, where A_V will decrease with increasing frequency. Therefore the expressions for C_{in} and C_O will not be exact but will still be a useful approximation for the determination of the frequency- and time-domain responses. We also note that the contribution of C_{GD} to the input capacitance is increased by the Miller effect factor of $1+ g_m R_{net}$.

The circuit in Figure 3.21 is in the form of two cascaded RC low-pass networks. The RC time constant on the input side is $\tau_1 = [R_1 \|\ RG] \cdot C_{in} \cong R_1 \cdot C_{in}$, where R_1 is the signal-source resistance. The RC time

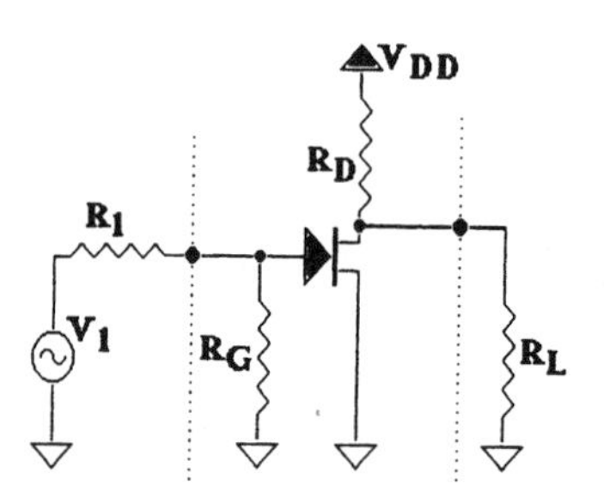

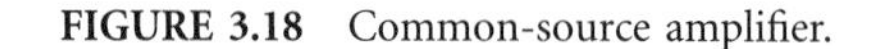
FIGURE 3.18 Common-source amplifier.

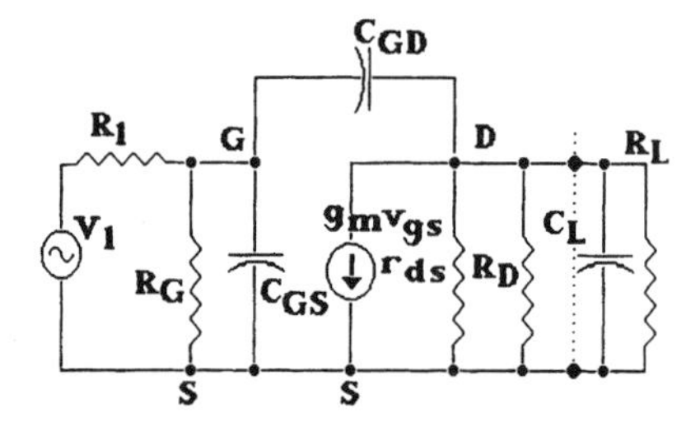

FIGURE 3.19 AC small-signal model.

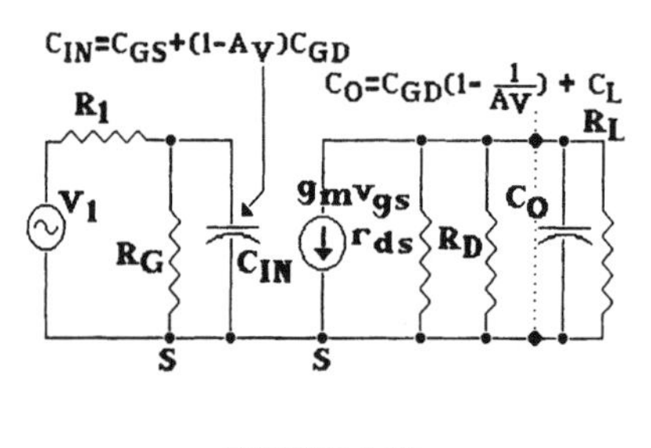

FIGURE 3.20

constant on the output side is given by $\tau_2 = R_{net} \cdot C_O$. The corresponding breakpoint frequencies are

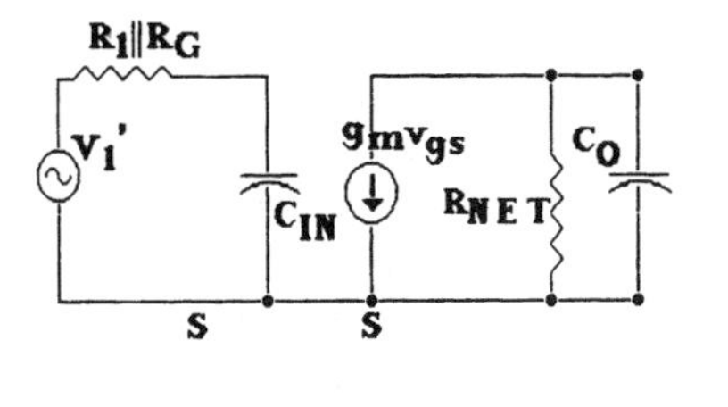

FIGURE 3.21

$$f_1 = \frac{1}{2\pi\tau_1} = \frac{1}{2\pi R_1 \cdot C_{in}}$$

and

$$f_2 = \frac{1}{2\pi\tau_2} = \frac{1}{2\pi R_{net} \cdot C_o}$$

The 3-dB or half-power frequency of this amplifier stage will be a function of f_1 and f_2. If these two breakpoint frequencies are separated by at least a decade (i.e., 10:1 ratio), the 3-dB frequency will be approximately equal to the lower of the two breakpoint frequencies. If the breakpoint frequencies are not well separated, then the 3-dB frequency can be obtained from the following approximate relationship: $(1/f_{3dB})^2 \cong (1/f_1)^2 + (1/f_2)^2$. The time-domain response as expressed in terms of the 10 to 90% rise time is related to the frequency-domain response by the approximate relationship that $t_{rise} \cong 0.35/f_{3dB}$.

We will now consider a representative example. We will let $C_{GS} = 10$ pF and $C_{GD} = 5$ pF. We will assume that the net load driven by the drain of the transistors is $R_{net} = 2$ kΩ and $C_L = 10$ pF. The signal-source resistance $R_1 = 100$ Ω. The JFET will have $I_{DSS} = 10$ mA, $I_{DSQ} = I_{DSS}/2 = 5$ mA, and $V_P = -4$ V, so $g_m = 3.535$ mS. Thus the midfrequency gain is $A_V = -g_m R_{net} = -3.535 \text{ mS} \cdot 2 \text{ k}\Omega = -7.07$. Therefore, we have that:

$$C_{in} \cong C_{GS} + (1 + g_m R_{net}) C_{GD} = 10\,\text{pF} + 8.07.5\,\text{pF}$$

and

$$C_o \cong C_{GD} + C_L = 15\,\text{pF}$$

Thus, $\tau_1 = R_1 \cdot C_{in} = 100\ \Omega \cdot 50.4$ pF $= 5040$ ps $= 5.04$ nsec, and $\tau_2 = R_{net} \cdot C_O = 2\ \text{k}\Omega \cdot 15$ pF $= 30$ nsec. The corresponding breakpoint frequencies are $f_1 = 1/(2\pi \cdot 5.04 \text{ nsec}) = 31.6$ MHz and $f_2 = 1/(2\pi \cdot 30 \text{ ns}) = 5.3$ MHz. The 3-dB frequency of the amplifier can be obtained from $(1/f_3\text{dB})^2 \cong (1/f_1)^2 + (1/f_2)^2 = (1/31.6 \text{ MHz})^2 + (1/5.3 \text{ MHz})^2$, which gives $f_3\text{dB} \cong 5.2$ MHz. The 10 to 90% rise time can be obtained from $t_{rise} \cong 0.35/f_3\text{dB} = 0.35/5.2 \text{ MHz} = 67$ nsec.

In the preceding example the dominant time constant is the output circuit time constant of $\tau_2 = 30$ nsec due to the combination of load resistance and output capacitance. If we now consider a signal-source resistance of 1 kΩ, the input circuit time constant will be $\tau_1 = R_1 \cdot C_{in} = 1000\ \Omega \cdot 50.4$ pF $= 50.4$ nsec. The corresponding breakpoint frequencies are $f_1 = 1/(2\pi \cdot 50.4 \text{ nsec}) = 3.16$ MHz and $f_2 = 1/(2\pi \cdot 30 \text{ nsec}) = 5.3$ MHz. The 3-dB frequency is now $f_3\text{dB} \cong 2.7$ MHz, and the rise time is $t_{rise} \cong 129$ nsec. If R_1 is further increased to 10 kΩ, we obtain $\tau_1 = R_1 \cdot C_{in} = 10\ \text{k}\Omega \cdot 50.4$ pF $= 504$ nsec, giving breakpoint frequencies of $f_1 = 1/(2\pi \cdot 504 \text{ nsec}) = 316$ kHz and $f_2 = 1/(2\pi \cdot 30 \text{ nsec}) = 5.3$ MHz. Now τ_1 is clearly the dominant time constant, the 3-dB frequency is now down to $f_3\text{dB} \cong f_1 = 316$ kHz, and the rise time is up to $t_{rise} \cong 1.1\ \mu$sec. Finally, for the case of $R_1 = 1$ MΩ, the 3-dB frequency will be only 3.16 kHz and the rise time will be 111 μsec.

Use of Source Follower for Impedance Transformation

We see that large values of signal-source resistance can seriously limit the amplifier bandwidth and increase the rise time. In these cases, the use of an impedance transforming circuit such as an FET source follower or a BJT emitter follower can be very useful. Let us consider the use of a source follower as shown in Figure 3.22. We will assume that both FETs are identical to the one in the preceding examples and are biased at $I_{DSQ} = 5$ mA. The source follower Q_1 will have an input capacitance of $C_{in} = C_{GD} + (1 - A_{V1})\, C_{GS} \cong C_{GD} = 5$ pF, since A_V will be very close to unity for a source follower that is driving a CS amplifier. The source-follower output resistance will be $r_O = 1/g_m = 1/3.535$ mS $= 283\ \Omega$. Let us again consider the case of $R_1 = 1\ \text{M}\Omega$. The time constant due to the combination of R_1 and the input capacitance of the source follower is $t_{SF} = 1\ \text{M}\Omega \cdot 5$ pf $= 5\ \mu$sec. The time constant due to the combination of the source-follower output resistance r_O and the input capacitance of the CS stage is $\tau_1 = r_O \cdot C_{in} = 283\ \Omega \cdot 50.4$ pF $=$ 14 nsec, and the time constant of the output circuit is $\tau_2 = 30$ nsec, as before. The breakpoint frequencies are $f_{SF} = 31.8$ - kHz, $f_1 = 11$ MHz, and $f_2 = 5.3$ MHz. The 3-dB frequency of the system is now $f_3\text{dB} \cong f_{SF} = 31.8$ kHz, and the rise time is $t_{rise} \cong 11$ μsec. The use of the source follower thus results in an improvement by a factor of 10:1 over the preceding circuit.

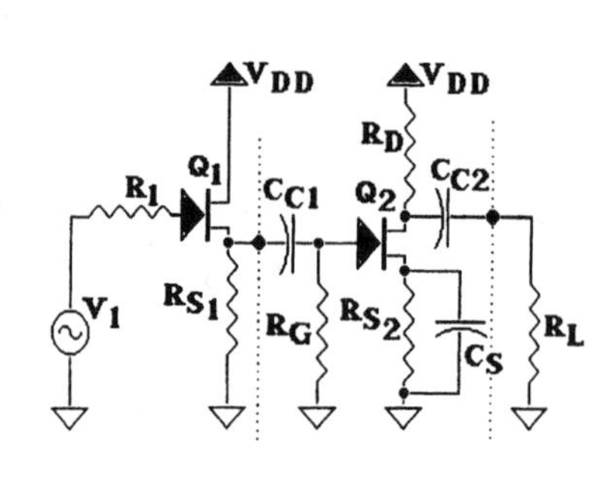

FIGURE 3.22

Voltage-Variable Resistor

Operation of a JFET as a Voltage-Variable Resistor

We will now consider the operation of a JFET as a voltage-variable resistor (VVR). A JFET can be used as a VVR in which the drain-to-source resistance r_{ds} of the JFET can be varied by variation of V_{GS}. For values of $V_{DS} << V_P$ the I_{DS} vs. V_{DS} characteristics are approximately linear, so the JFET looks like a resistor, the resistance value of which can be varied by the gate voltage as shown in Figure 3.23.

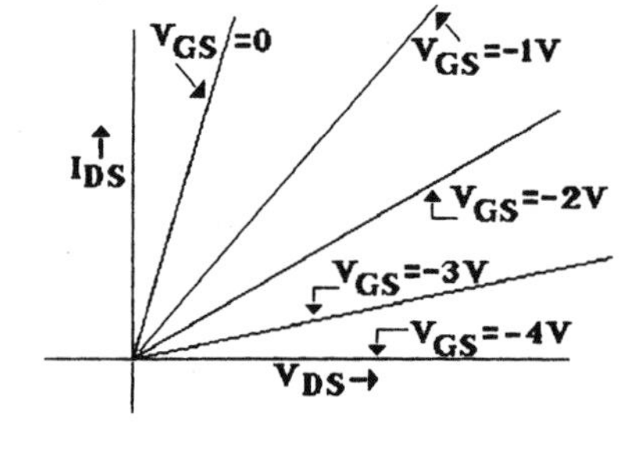

FIGURE 3.23

The channel conductance in the region where $V_{DS} << V_P$ is given by $g_{ds} = A\sigma/L = WH\sigma/L$, where the channel height H is given by $H = H_0 - 2W_D$. In this equation, W_D is the depletion region width and H_0 is the value of H as $W_D \rightarrow 0$. The depletion region width is given by, $W_D = K\sqrt{V_J} = K\sqrt{V_{GS} + \phi}$ where K is a constant, V_J is the junction voltage, and ϕ is the pn-junction contact potential (typically around 0.8 to 1.0 V). As V_{GS} increases, W_D increases and the channel height H decreases, as given by $H = H_0 - 2K\sqrt{V_{GS} + \phi}$. When $V_{GS} = V_P$, the channel is completely pinched off, so $H = 0$ and thus $2K\sqrt{V_p + \phi} = H_0$. Therefore, $2K = H_0/\sqrt{V_p + \phi}$, and thus:

$$H = H_0 - H_0 \frac{\sqrt{V_{GS} + \phi}}{\sqrt{V_p + \phi}} = H_0 \left(1 - \frac{\sqrt{V_{GS} + \phi}}{\sqrt{V_p + \phi}} \right)$$

For g_{ds} we now have:

$$g_{ds} = \frac{\sigma WH}{L} = \sigma \frac{WH_0}{L} \left(1 - \frac{\sqrt{V_{GS} + \phi}}{\sqrt{V_p + \phi}} \right)$$

When $V_{GS} = 0$, the channel is fully open or "on," and:

$$g_{ds} = g_{ds}(\text{on}) = \sigma \frac{WH_0}{L} \left(1 - \frac{\sqrt{\phi}}{\sqrt{V_P + \phi}} \right)$$

The drain-to-source conductance can now be expressed as

$$g_{ds} = g_{ds}(\text{on})\frac{1-(\sqrt{V_{GS}+\phi}/\sqrt{V_P+\phi})}{1-(\sqrt{\phi}/\sqrt{V_P+\phi})}$$

The reciprocal quantity is the drain-to-source resistance r_{ds} as given by $r_{ds} = 1/g_{ds}$, and r_{ds} (on) $= 1/g_{ds}$ (on), so:

$$r_{ds} = r_{ds}(\text{on})\frac{1-(\sqrt{\phi}/\sqrt{V_P+\phi})}{1-(\sqrt{V_{GS}+\phi}/\sqrt{V_P+\phi})}$$

As $V_{GS} \rightarrow 0$, $r_{ds} \rightarrow r_{ds}$ (on), and as $V_{GS} \rightarrow V_P$, $r_{ds} \rightarrow \infty$. This latter condition corresponds to the channel being pinched off in its entirety all the way from source to drain. This is like having a big block of insulator (i.e., the depletion region) between source and drain. When $V_{GS} = 0$, r_{ds} is reduced to its minimum value of r_{ds} (on), which for most JFETs is in the 20- to 400-Ω range. At the other extreme, when $V_{GS} > V_P$, the drain-to-source current I_{DS} is reduced to a very small value, generally down into the low nanoampere or even picoampere range. The corresponding value of r_{ds} is not really infinite but is very large, generally well up into the gigaohm (1000 MΩ) range. Thus by variation of V_{GS}, the drain-to-source resistance can be varied over a very wide range. As long as the gate-to-channel junction is reverse-biased, the gate current will be very small, generally down in the low nanoampere or even picoampere range, so the gate as a control electrode draws very little current. Since V_P is generally in the 2- to 5-V range for most JFETs, the V_{DS} values required to operate the JFET in the VVR range are generally <0.1 V. In Figure 3.23 the VVR region of the JFET I_{DS} vs. V_{DS} characteristics is shown.

VVR Applications

Applications of VVRs include automatic gain control (AGC) circuits, electronic attenuators, electronically variable filters, and oscillator amplitude control circuits.

When using a JFET as a VVR, it is necessary to limit V_{DS} to values that are small compared to V_P to maintain good linearity. In addition, V_{GS} should preferably not exceed 0.8 V_P for good linearity, control, and stability. This limitation corresponds to an r_{ds} resistance ratio of about 10:1. As V_{GS} approaches V_P, a small change in V_P can produce a large change in r_{ds}. Thus unit-to-unit variations in V_P as well as changes in V_P with temperature can result in large changes in r_{ds} as V_{GS} approaches V_P.

The drain-to-source resistance r_{ds} will have a temperature coefficient (TC) due to two causes: (1) the variation of the channel resistivity with temperature and (2) the temperature variation of V_P. The TC of the channel resistivity is positive, whereas the TC of V_P is negative due to the negative TC of the contact potential *f*. The positive TC of the channel resistivity will contribute to a positive TC of r_{ds}. The negative TC of V_P will contribute to a negative TC of r_{ds}. At small values of V_{GS}, the dominant contribution to the TC is the positive TC of the channel resistivity, so r_{ds} will have a positive TC. As V_{GS} gets larger, the negative TC contribution of V_P becomes increasingly important, and there will be a value of V_{GS} at which the net TC of r_{ds} is zero, and above this value of V_{GS} the TC will be negative. The TC of r_{ds}(on) is typically +0.3%/°C for *n*-channel JFETs and +0.7%/°C for p-channel JFETs. For example, for a typical JFET with an r_{ds} (on) = 500 Ω at 25°C and $V_P = 2.6$ V, the zero TC point will occur at $V_{GS} = 2.0$ V. Any JFET can be used as a VVR, although there are JFETs that are specifically made for this application.

A simple example of a VVR application is the electronic gain control circuit of Figure 3.24. The voltage gain is given by $A_V = 1 + (\text{RF}/r_{ds})$. If, for example, RF = 19 kΩ and $r_{ds}(\text{on}) = 1$ kΩ, then the maximum gain will be $A_V\text{max} = 1 + [\text{RF}/r_{ds}\text{ (on)}] = 20$. As V_{GS} approaches V_P, r_{ds} will increase and become very large such that $r_{ds} >> \text{RF}$, so that A_V will decrease to a minimum

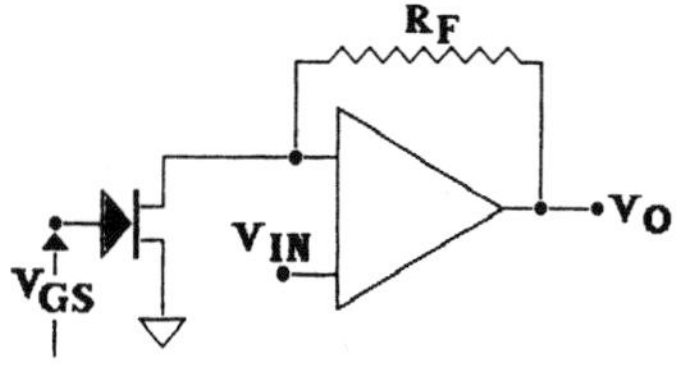

FIGURE 3.24 Electronic gain control.

value of close to unity. Thus, the gain can be varied over a 20:1 ratio. Note that $V_{DS} \cong$ Vin, so to minimize distortion the input signal amplitude should be small compared to V_P.

Defining Terms

Active region: The region of JFET operation in which the channel is pinched off at the drain end but still open at the source end such that the drain-to-source current I_{DS} approximately saturates. The condition for this is that $|V_{GS}| < |V_P|$ and $|V_{DS}| > |V_P|$. The active region is also known as the saturated region.

Ohmic, nonsaturated, or triode region: The three terms all refer to the region of JFET operation in which a conducting channel exists all the way between source and drain. In this region the drain current varies with both V_{GS} and V_{DS}.

Drain saturation current, I_{DSS}: The drain-to-source current flow through the JFET under the conditions that $V_{GS} = 0$ and $|V_{DS}| > |V_P|$ such that the JFET is operating in the active or saturated region.

Pinch-off voltage, V_P: The voltage that when applied across the gate-to-channel pn-junction will cause the conducting channel between drain and source to become pinched off. This is also represented as V_{GS} (off).

References

R. Mauro, *Engineering Electronics*, Englewood Cliffs, NJ: Prentice-Hall, 1989, pp. 199–260.

J. Millman and A. Grabel, *Microelectronics*, 2nd ed., New York: McGraw-Hill, 1987, pp. 133–167, 425–429.

F.H. Mitchell, Jr. and F.H. Mitchell, Sr., *Introduction to Electronics Design*, 2nd ed., Englewood Cliffs, NJ: Prentice-Hall, 1992, pp. 275–328.

C.J. Savant, M.S. Roden, and G.L. Carpenter, *Electronic Design*, 2nd ed., Menlo Park, CA: Benjamin-Cummings, 1991, pp. 171–208.

A.S. Sedra and K.C. Smith, *Microelectronic Circuits*, 3rd ed., Philadelphia, PA: Saunders, 1991, pp. 322–361.

3.2 Bipolar Transistors

Joseph Watson

Modern amplifiers abound in the form of *integrated circuits* (ICs), which contain transistors, diodes, and other structures diffused into single-crystal *dice*. As an introduction to these ICs, it is convenient to examine single-transistor amplifiers, which in fact are also widely used in their own right as *discrete* circuits—and indeed much more complicated discrete signal-conditioning circuits are frequently found following sensors of various sorts.

There are two basic forms of transistor, the *bipolar* family and the *field-effect* family, and both appear in ICs. They differ in their modes of operation but may be incorporated into circuits in quite similar ways. To understand elementary circuits, there is no need to become too familiar with the physics of transistors, but some basic facts about their electrical properties must be known.

Consider the bipolar transistor, of which there are two types, *npn* and *pnp*. Electrically, they differ only in terms of current direction and voltage polarity. Figure 3.25(a) illustrates the idealized structure of an *npn* transistor, and diagram (b) implies that it corresponds to a pair of diodes with three leads. This representation does *not* convey sufficient information about the actual operation of the transistor, but it does make the point that the flow of conventional current (positive to negative) is easy from the *base* to the *emitter*, since it passes through a *forward-biased diode*, but difficult from the *collector* to the *base*, because flow is prevented by a *reverse-biased diode.*

Figure 3.25(c) gives the standard symbol for the *npn* transistor, and diagram (d) defines the direction of current flow and the voltage polarities observed when the device is in operation. Finally, diagram (e) shows that for the *pnp* transistor, all these directions are reversed and the polarities are inverted.

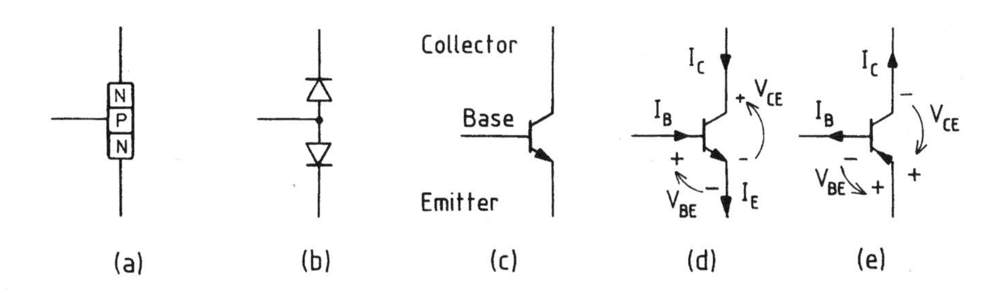

FIGURE 3.25 The bipolar transistor. (a) to (d) *npn* transistor; (e) *pnp* transistor.

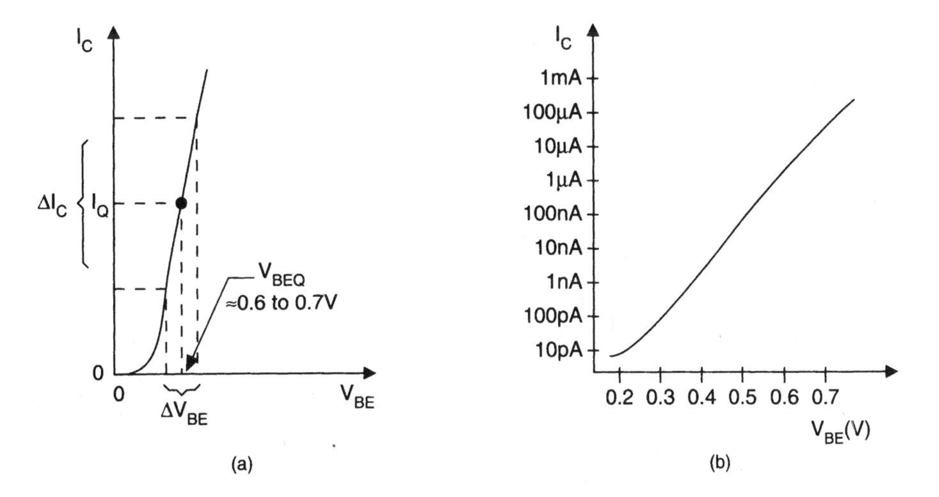

FIGURE 3.26 The transconductance curve for a transistor on (a) linear and (b) logarithmic axes.

For a transistor, there is a main current flow between the collector and the emitter, and a very much smaller current flow between the base and the emitter. So, the following relations may be written:

$$I_E = I_C + I_B \tag{3.1}$$

(Note that the arrow on the transistor symbol defines the emitter and the direction of current flow—*out* for the *npn* device, and *in* for the *pnp*.) Also

$$I_C/I_B = h_{FE} \tag{3.2}$$

Here, h_{FE} is called the *dc common-emitter current gain*, and because $I_C >> I_B$ then h_{FE} is large, typically 50 to 300. The implication of this may be seen immediately: if the small current I_B can be used to control the large current I_C, then the transistor may obviously be used as a current amplifier. (This is why Figure 3.25(b) is inadequate—it completely neglects this all-important current-gain property of the transistor.) Furthermore, if a load resistance is connected into the collector circuit, it will become a voltage amplifier as well.

Unfortunately, h_{FE} is an ill-defined quantity and varies not only from transistor to transistor, but also changes with temperature. The relationship between the base-emitter voltage V_{BE} and the collector current is much better defined and closely follows an exponential law over at least eight decades. This relationship is shown in both linear and logarithmic form in Figure 3.26. Because the output current I_C is dependent upon the input voltage V_{BE}, the plot must be a transfer conductance or *transconductance* characteristic. The relevant law is

$$I_C = I_{ES}(e^{(q/kT)V_{BE}} - 1) \tag{3.3}$$

Here, I_{ES} is an extremely small leakage current internal to the transistor, q is the electronic charge, k is Boltzmann's constant, and T is the absolute temperature in Kelvin. Usually, kT/q is called V_T and is about 26 mV at a room temperature of 25°C. This implies that for any value of V_{BE} over about 100 mV, then $\exp(V_{BE}/V_T) \gg 1$, and for all normal operating conditions, Equation (3.3) reduces to

$$I_C = I_{ES}e^{V_{BE}/V_T} \quad \text{for } V_{BE} > 100\,\text{mV} \tag{3.4}$$

The term "normal operating conditions" is easily interpreted from Figure 3.26(a), which shows that when V_{BE} has reached about 0.6 to 0.7 V, any small fluctuations in its value cause major fluctuations in I_C. This situation is illustrated by the dashed lines enclosing ΔV_{BE} and ΔI_C, and it implies that to use the transistor as an amplifier, working values of V_{BE} and I_C must be established, after which signals may be regarded as fluctuations around these values.

Under these *quiescent, operating,* or *working* conditions:

$$I_C = I_Q \quad \text{and} \quad V_{CE} = V_Q$$

and methods of defining these quiescent or operating conditions are called *biasing*.

Biasing the Bipolar Transistor

A fairly obvious way to bias the transistor is to first establish a constant voltage V_B using a potential divider $R1$ and $R2$ as shown in the **biasing circuit** of Figure 3.27. Here:

$$V_B \simeq \frac{V_{CC}R2}{R1 + R2}$$

if I_B is very small compared with the current through $R2$, which is usual. If it is not, this fact must be taken into account.

This voltage will be much greater than V_{BE} if a realistic power supply is used along with realistic values of $R1$ and $R2$. Hence, when the transistor is connected into the circuit, an emitter resistor must also be included so that:

$$V_{BE} = V_B - I_E R_E \tag{3.5}$$

Now consider what happens when the power supply is connected. As V_B appears, a current I_B flows into the base and produces a much larger current $I_C = h_{FE}I_B$ in the collector. These currents add in the emitter to give

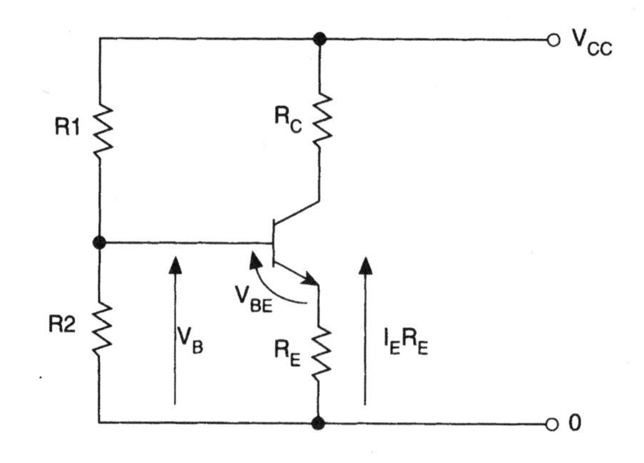

FIGURE 3.27 A transistor biasing circuit.

$$I_E = I_B + h_{FE}I_B = (1 + h_{FE})I_B \simeq h_{FE}I_B \tag{3.6}$$

Clearly, I_E will build up until a fixed or quiescent value of base-emitter voltage V_{BEQ} appears. Should I_E try to build up further, V_{BE} will fall according to Equation (3.5) and, hence, so will I_E. Conversely, should I_E not build up enough, V_{BE} will increase until it does so.

This is actually a case of current-derived negative feedback, and it successfully holds the collector current near the quiescent value I_Q. Furthermore, it does so despite different transistors with different values of h_{FE} being used and despite temperature variations. Actually, V_{BE} itself falls with temperature at about –2.2 mV/°C for constant I_C, and the circuit will also compensate for this. The degree of success of the negative feedback in holding I_Q constant is called the *bias stability.*

This is one example of a **common-emitter** (CE) circuit, so-called because the emitter is the common terminal for both base and collector currents. The behavior of the transistor in such a circuit may be illustrated by superimposing a *load line* on the *output characteristics* of the transistor, as shown in Figure 3.28.

If the collector current I_C is plotted against the collector-to-emitter voltage V_{CE}, a family of curves for various fixed values of V_{BE} or I_B results, as in Figure 3.28. These curves show that as V_{CE} increases, I_C rises very rapidly and then turns over as it is limited by I_B. In the CE circuit, if I_B were reduced to zero, then I_C would also be zero (apart from a small leakage current I_{CEO}). Hence there would be no voltage drop in either R_C or R_E, and practically all of V_{CC} would appear across the transistor. That is, under *cut-off* conditions:

$$V_{CE} \longrightarrow V_{CC} \quad \text{for } I_B = 0 \tag{3.7}$$

Conversely, if I_B were large, I_C would be very large, almost all of V_{CC} would be dropped across $R_{C+}R_E$ and

$$I_C \longrightarrow \frac{V_{CC}}{R_C + R_E} \quad \text{for large } I_B \tag{3.8}$$

Actually, because the initial rise in I_C for the transistor is not quite vertical, there is always a small *saturation voltage* V_{CES} across the transistor under these conditions, where V_{CES} means the voltage across the transistor in

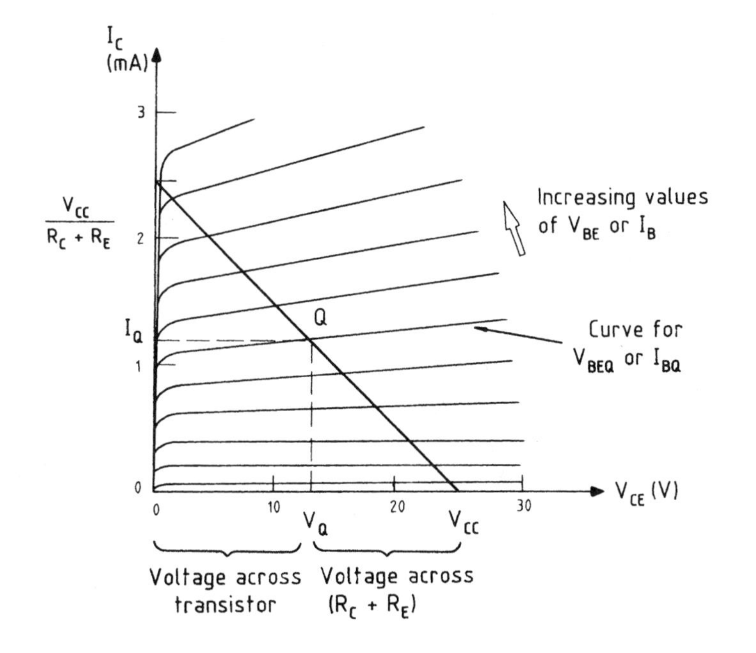

FIGURE 3.28 The load-line diagram.

the common-emitter mode when saturated. In this saturated condition, $V_{CES} \simeq 0.3\,V$ for small silicon transistors. Both these conditions are shown in Figure 3.28.

From the circuit of Figure 3.27:

$$V_{CE} = V_{CC} - I_C(R_C + R_E) \tag{3.9a}$$

which may be rewritten as

$$I_C = -V_{CE}/(R_C + R_E) + V_{CC}/(R_C + R_E) \tag{3.9b}$$

This is the straight-line equation to the *dc load-line* (compare $y = mx+c$), showing that its slope is $-1/(R_C + R_E)$ and that it crosses the I_C axis at $V_{CC}/(R_C + R_E)$ as expected. The actual position of a point is determined by where this load line crosses the output characteristic in use, that is, by what value of V_{BE} or I_B is chosen. For example, the quiescent point for the transistor is where the load line crosses the output curve defined by $V_{BE} = V_{BEQ}$ (or $I_B = I_{BQ}$) to give $V_{CE} = V_Q$ and $I_C = I_Q$.

Note that because the transistor is *nonohmic* (that is, it does not obey Ohm's law), the voltage across it may only be determined by using the (ohmic) voltage drop across the resistors R_C and R_E according to Equation (3.9). At the quiescent point this is

$$V_Q = V_{CC} - I_Q(R_C + R_E)$$

A design example will illustrate typical values involved with a small-transistor CE stage.

Example 1

A transistor is to be biased at a collector current of 1 mA when a 12-V power supply is applied. Using the circuit of Figure 3.27, determine the values of $R1$, $R2$, and R_E if 3.4 V is to be dropped across R_E and if the current through $R2$ is to be 10 I_{BQ}. Assume that for the transistor used, $V_{BEQ} = 0.6$ V and $h_{FE} = 100$.

Solution. In this circuit $I_Q = 1\text{ mA} \simeq I_E$ (because $I_B << lI_C$). Hence,

$$R_E = \frac{V_{R_E}}{I_Q} = \frac{3.4}{1} = 3.4\,\text{k}\Omega$$

Also, $V_B = V_{RE}+V_{BE} = 3.4+0.6 = 4$ V. This gives:

$$R2 = \frac{V_B}{10I_{BQ}}$$

where $I_{BQ} = I_Q/h_{FE} = 1/100 = 0.01$ mA, so:

$$R2 = \frac{4}{10 \times 0.01} = 40\,\text{k}\Omega$$

Now $V_{R1} = V_{CC} - V_B = 12 - 4 = 8$ V, and the current through $R1$ is 10 $I_{BQ}+I_{BQ} = 11\ I_{BQ}$, so:

$$R1 = \frac{V_{R1}}{I_{R1}} = \frac{8}{11 \times 0.01} = 72.7\,\text{k}\Omega$$

In the above design example, the base current I_{BQ} has been included in the current passing through $R1$. Had this not been done, $R1$ would have worked out at 80 kΩ. Usually, this difference is not very important because *discrete* (or individual) resistors are available only in a series of nominal values, and each of these is subject to a *tolerance*, including 10, 5, 2, and 1%.

In the present case, the following (5%) values could reasonably be chosen:

$$R_E = 3.3\,\text{k}\Omega \quad R1 = 75\,\text{k}\Omega \quad R2 = 39\,\text{k}\Omega$$

All this means that I_Q cannot be predetermined very accurately, but the circuit nevertheless settles down to a value close to the chosen one, and, most importantly, stays there almost irrespective of the transistor used and the ambient temperature encountered.

Having biased the transistor into an operating condition, it is possible to consider *small-signal operation.*

Small-Signal Operation

In the biasing circuit of Figure 3.27, the collector resistor R_C had no discernible function because it is simply the load resistor across which the signal output voltage is developed. However, it was included because it also drops a voltage due to the bias current flowing through it. This means that its value must not be so large that it robs the transistor of adequate operating voltage; that is, it must not be responsible for moving the operating point too far to the left in Figure 3.28.

If the chosen bias current and voltage are I_Q and V_Q, then small signals are actually only fluctuations in these bias (or average) values that can be separated from them using coupling capacitors.

To inject an input signal to the base, causing V_{BE} and I_B to fluctuate by v_{be} and i_b, a signal source must be connected between the base and the common or zero line (also usually called ground or earth whether it is actually connected to ground or not!). However, most signal sources present a resistive path through themselves, which would shunt $R2$ and so change, or even destroy, the bias conditions. Hence, a coupling capacitor C_c must be included, as shown in Figure 3.29, in series with a signal source represented by a Thévenin equivalent.

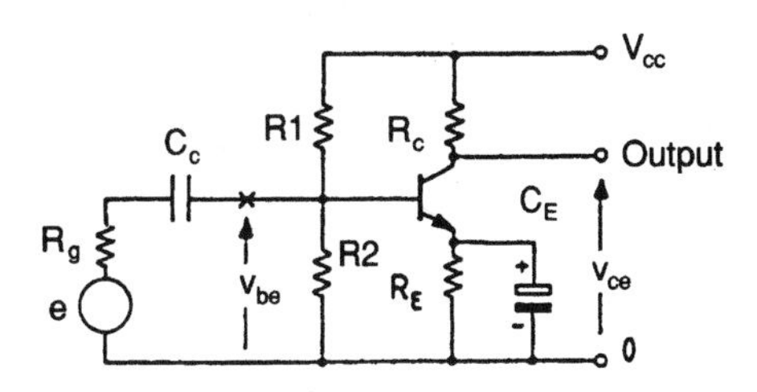

FIGURE 3.29 A complete common-emitter stage.

The emitter resistor R_E was included for biasing reasons (although there are other bias circuits that omit it), but for signal amplification purposes it must be shunted by a high-value capacitor C_E so that the signal current can flow down to ground without producing a signal voltage drop leading to negative feedback (as did the bias current). The value of C_E must be much greater than is apparent at first sight, and this point will be developed later; for the present, it will be assumed that it is large enough to constitute a short circuit at all the signal frequencies of interest. So, for ac signals R_E is short-circuited and only R_C acts as a load. This implies that a *signal* or *ac load line* comes into operation with a slope of $-1/R_C$, as shown in Figure 3.30.

The ways in which the small-signal quantities fluctuate may now be examined. If v_{be} goes positive, this actually means that V_{BE} increases a little. This in turn implies that I_C increases by an amount i_c, so the voltage drop in R_C increases by v_{ce}. Keeping in mind that the top of R_C is held at a constant voltage, this means that the voltage at the bottom of R_C must fall by v_{ce}. This very important point shows that because v_{ce} falls as v_{be} rises, there is 180° phase shift through the stage. That is, the CE stage is an *inverting voltage amplifier.* However, because i_c increases into the collector as i_b increases into the base, it is also a *noninverting current amplifier.*

Now consider the amount by which v_{ce} changes with v_{be}, which is the *terminal voltage gain* of the stage. In Figure 3.26, the slope of the transconductance curve at any point defines by how much I_C changes with a fluctuation in V_{BE}. That is, it gives the ratio i_c/v_{be} at any operating point Q. Equation (3.4) is

$$I_c = I_{ES}e^{V_{BE}/V_T}$$

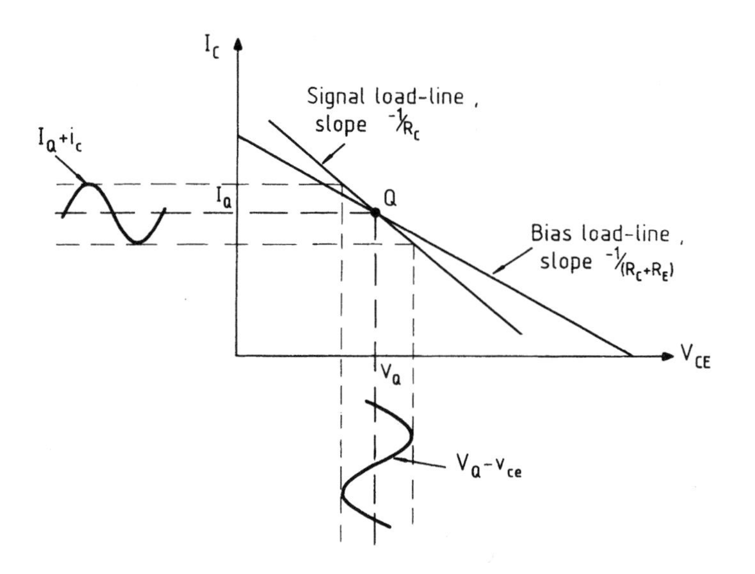

FIGURE 3.30 The signal or ac load line.

so that

$$\frac{dI_C}{dV_{BE}} = \frac{1}{V_T} I_{ES} e^{V_{BE}/V_T}$$

or

$$\frac{i_c}{v_{be}} = \frac{I_C}{V_T} = g_m = \text{the transconductance} \tag{3.10}$$

Now the signal output voltage is

$$v_{ce} \simeq -i_c R_C$$

(Here, the approximation sign is because the collector-emitter path within the transistor does present a large resistance r_{ce} through which a very small part of i_c flows.)

The terminal voltage gain is therefore:

$$A_v = \frac{v_{ce}}{v_{be}} \simeq \frac{-i_c R_C}{v_{be}} = -g_m R_C \tag{3.11}$$

where the negative sign implies signal inversion.

In practice, $V_T \simeq 26\,\text{mV}$ at room temperature, as has been mentioned, and this leads to a very simple numerical approximation. From Equation (3.10) and using $I_C = I_Q$:

$$g_m = \frac{I_Q}{V_T} \simeq \frac{I_Q}{0.026} \simeq 39 I_Q \quad \text{mA/V}$$

if I_Q is in milliamps and at room temperature. This shows that irrespective of the transistor used, the transconductance may be approximated knowing only the quiescent collector current.

The magnitude and phase relationships between v_{ce} and i_c can easily be seen by including them on the signal load-line diagram as shown in Figure 3.30, where the output characteristics of the transistor have been omitted for clarity. Sinusoidal output signals have been inserted, and either may be obtained from the other by following the signal load-line locus.

Now consider the small-signal current gain. Because the value of h_{FE} is not quite linear on the I_C/I_B graph, its slope too must be used for small-signal work. However, the departure from linearity is not great over normal working conditions, and the small-signal value h_{fe} is usually quite close to that of h_{FE}. Hence:

$$A_i = \frac{i_c}{i_b} \simeq h_{fe} \tag{3.12}$$

The small-signal or incremental input resistance to the base itself (to the right of point X in Figure 3.29) may now be found:

$$R_{in} = \frac{v_{be}}{i_b} = \frac{v_{be}}{i_c}\frac{i_c}{i_b} \simeq \frac{h_{fe}}{g_m} \tag{3.13}$$

Three of the four main (midfrequency) parameters for the CE stage have now been derived, all from a rather primitive understanding of the transistor itself. The fourth, R_{out}, is the dynamic, incremental, or small-signal resistance of the transistor from collector to emitter, which is the slope of the output characteristic at the working point r_{ce}. Being associated with a reverse-biased (CB) junction, this is high—typically about 0.5 MΩ—so that the transistor acts as a current source feeding a comparatively low load resistance R_C. Summarizing, at mid-frequency:

$$A_i \simeq h_{fe} \quad A_v \simeq -g_m R_C \quad R_{in} \simeq \frac{h_{fe}}{g_m} \quad R_{out} \simeq r_{ce}$$

Example 2

Using the biasing values for $R1$, $R2$, and R_E already obtained in Example 1, calculate the value of R_C to give a terminal voltage gain of –150. Then determine the input resistance R_{in} if h_{fe} for the transistor is 10% higher than h_{FE}.

Solution. Because $I_Q = 1$ mA, $g_m \simeq 39 \times 1 = 39\,\text{mA/V}$. Hence, $A_v \simeq -g_m R_C$ or $-150 \simeq -39\,R_C$, giving

$$R_C = 150/39 \simeq 3.9\,\text{k}\Omega$$

(*Note:* This value *must* be checked to determine that it is reasonable insofar as biasing is concerned. In this case, it will drop $I_Q R_C = 1\text{–}3.9 = 3.9$ V. Because $V_{RE} = 3.4$ V, this leaves 12 – 3.9 – 3.4 = 4.7 V across the transistor, which is reasonable.)

Finally:

$$R_{in} \simeq \frac{h_{fe}}{g_m} = \frac{110}{39} = 2.8\,\text{k}\Omega$$

A Small-Signal Equivalent Circuit

The conclusions reached above regarding the performance of the bipolar transistor are sufficient for the development of a basic equivalent circuit, or model, relevant *only* to small-signal operation. Taking the operating CE amplifier, this may be done by first "looking into" the base, shown as b in Figure 3.31. Between this point and the actual active part of the base region b′, it is reasonable to suppose that the intervening

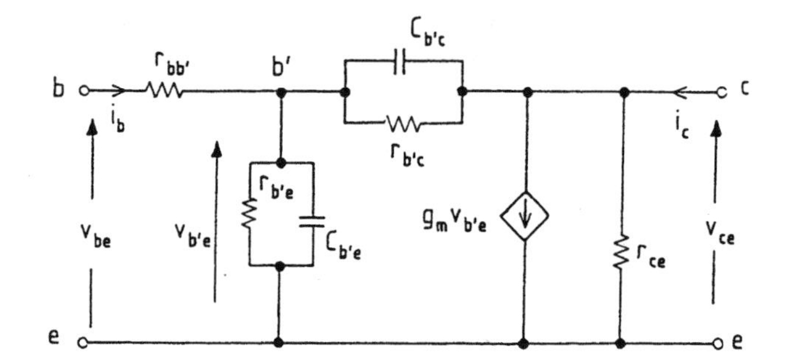

FIGURE 3.31 The hybrid-π small-signal transistor equivalent circuit or model.

(inactive) semiconductor material will present a small resistance $r_{bb'}$. This is called the *base spreading resistance*, and it is also shown in Figure 3.31.

From b′ to the emitter e, there will be a dynamic or incremental resistance given by

$$r_{b'e} = \frac{v_{b'e}}{i_b} = \frac{v_{b'e}}{i_c}\frac{i_c}{i_b} = \frac{h_{fe}}{g_m} \tag{3.14}$$

so that the full resistance from the base to the emitter must be

$$R_{in} = r_{bb'} + r_{b'e} = r_{bb'} + \frac{h_{fe}}{g_m} \simeq \frac{h_{fe}}{g_m} \tag{3.15}$$

because $r_{bb'}$ is only about 10 to 100 Ω, which is small compared with $r_{b'e}$, this being several kilohms (as shown by the last example). It will now be understood why Equation (3.13) gave $R_{in} \simeq h_{fe}/g_m$.

The reverse-biased junction that exists from b' to the collector ensures that the associated dynamic resistance $r_{b'c}$ will be very large indeed, which is fortunate; otherwise, signal feedback from the output to the input would modify the gain characteristics of the amplifier. Typically, $r_{b'c}$ will be some tens of megohms.

However, because of transistor action, the dynamic resistance from collector to emitter, r_{ce}, will be smaller than $r_{b'c}$ and will typically be below a megohm. This "transistor action" may be represented by a current source from collector to emitter that is dependent upon either i_b or $v_{b'e}$. That is, it will be either $h_{fe}i_b$ or $g_m v_{b'e}$. The latter leads to the well-known hybrid-π model, and it is this that is shown in Figure 3.31.

Where junctions or interfaces of any sort exist, there will always be distributed capacitances associated with them; and to make these easy to handle analytically, they may be "lumped" into single capacitances. In the present context, two lumped capacitances have been incorporated into the hybrid-π model, $C_{b'e}$ from base to emitter and $C_{b'c}$ from base to collector, respectively. These now complete the model, and it will be appreciated that they make it possible to analyze high-frequency performance. Typically, $C_{b'e}$ will be a few picofarads and will always be larger than $C_{b'c}$.

Figure 3.31 is the hybrid-π small-signal, dynamic, or incremental model for a bipolar transistor; and when external components are added and simplifications made, it makes possible the determination of the performance of an amplifier using that transistor not only at midfrequencies but also at high and low frequencies.

Low-Frequency Performance

In Figure 3.32 both a source and a load have been added to the hybrid-π equivalent circuit to model the complete CE stage of Figure 3.29. Here, both $C_{b'e}$ and $C_{b'c}$ have been omitted because they are too small to

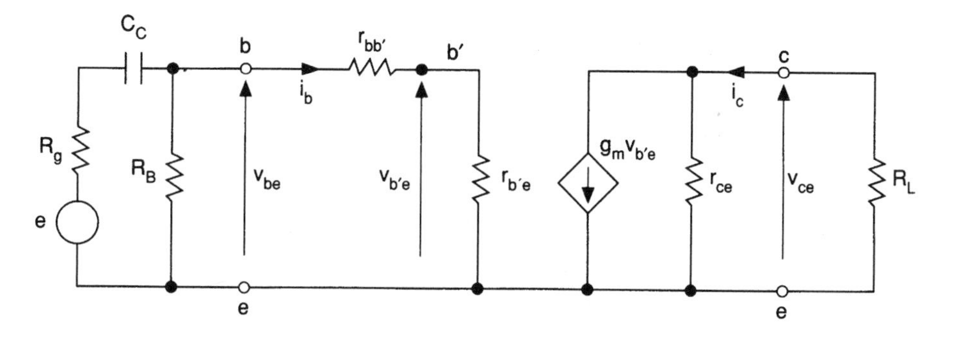

FIGURE 3.32 The loaded hybrid-π model for low frequencies.

affect the low-frequency performance, as has $r_{b'c}$ because it is large and so neither loads the source significantly compared to $r_{bb'}+r_{b'e}$ nor applies much feedback.

The signal source has been represented by a Thévenin equivalent that applies a signal via a coupling capacitor C_c. Note that this signal source has been returned to the emitter, which implies that the emitter resistor bypass capacitor C_E has been treated as a short circuit at all signal frequencies for the purposes of this analysis.

Because the top of biasing resistor $R1$ (Figure 3.29) is taken to ground via the power supply insofar as the signal is concerned, it appears in parallel with $R2$, and the emitter is also grounded to the signal via C_E. That is, a composite biasing resistance to ground R_B appears:

$$R_B = \frac{R1 \cdot R2}{R1 + R2}$$

Finally, the collector load is taken to ground via the power supply and hence to the emitter via C_E.

Figure 3.32 shows that v_{be} is amplified independently of frequency, so the terminal voltage gain A_v may easily be determined:

$$A_v = \frac{v_{ce}}{v_{be}} = \frac{-i_c R_L}{v_{be}}$$

Now:

$$v_{be} = \frac{v_{b'e}(r_{bb'} + r_{b'e})}{r_{b'e}} \simeq v_{b'e} \quad \text{and} \quad i_c = \frac{g_m v_{b'e} r_{ce}}{r_{ce} + R_L} \simeq g_m v_{b'e}$$

because $r_{b'e} >> r_{bb'}$ and $r_{ce} >> R_L$. So:

$$A_v \simeq -g_m R_L$$

which is as expected.

The model shows that v_{be} is amplified independently of frequency because there are no capacitances to its right, so an analysis of low-frequency response devolves down to determining v_{be} in terms of e. Here, part of e will appear across the capacitive reactance X_{Cc}, and the remainder is v_{be}. So, to make the concept of reactance valid, a sinusoidal signal E must be postulated, giving a sinusoidal value for $v_{be} = V_{be}$.

At mid-frequencies, where the reactance of C_c is small, the signal input voltage is

$$V_{be}(f_m) = \frac{E \cdot R_{BP}}{R_g + R_{BP}} \quad (3.16)$$

where $R_{BP} = R_B.R_{in}/(R_B+R_{in})$ and $R_{in} = r_{bb'}+r_{b'e}$ as before.

At low frequencies, where the reactance of C_c is significant:

$$V_{be}(f_{low}) = \frac{E \cdot R_{BP}}{\sqrt{(R_g + R_{BP})^2 + X_{C_c}^2}} \quad (3.17)$$

Dividing Equation (3.16) by Equation (3.17) gives

$$\frac{V_{be}(f_m)}{V_{be}(f_{low})} = \frac{\sqrt{(R_g + R_{BP})^2 + X_{C_c}^2}}{R_g + R_{BP}}$$

There will be a frequency f_L at which $|X_{Cc}| = R_g + R_{BP}$ given by

$$\frac{1}{2\pi f_L C_c} = R_g + R_{BP} \quad \text{or} \quad f_L = \frac{1}{2\pi C_c(R_g + R_{BP})} \quad (3.18)$$

At this frequency, $V_{be}(f_m)/V_{be}(f_L) = \sqrt{2}$ or $V_{be}(f_L)$ is 3 dB lower than $V_{be}(f_m)$.

Example 3

Using the circuit components of the previous examples along with a signal source having an internal resistance of $R_g = 5\,\text{k}\Omega$, find the value of a coupling capacitor that will define a low-frequency –3 dB point at 42 Hz.

Solution. Using Equation (3.18):

$$C_c = \frac{1}{2\pi(R_g + R_{BP})f_L}$$

where $R_{BP} = R1|R2\ _{in} = 75|39|2.8 = 2.5\,\text{k}\Omega$. That is:

$$C = \frac{10^6}{2\pi(5000 + 2500)(42)} \simeq 0.5\,\mu\text{F}$$

Since a single RC time constant is involved, the voltage gain of the CE stage will appear to fall at 6 dB/octave as the frequency is reduced because more and more of the signal is dropped across C_c. However, even if C_E is very large, it too will contribute to a fall in gain as it allows more and more of the output signal to be dropped across the $R_E||X_{CE}$ combination, this being applied also to the input loop, resulting in negative feedback. So, at very low frequencies, the gain roll-off will tend to 12 dB/octave. The question therefore arises of how large C_E should be, and this can be conveniently answered by considering a second basic form of transistor connection as follows.

The Emitter-Follower or Common-Collector (CC) Circuit

Suppose that R_C is short-circuited in the circuit of Figure 3.29. This will not affect the biasing because the collector voltage may take any value (the output characteristic is nearly horizontal, as seen in Figure 3.28).

However, the small-signal output voltage ceases to exist because there is now no load resistor across which it can be developed, although the output current i_c will continue to flow as before.

If now C_E is removed, i_c flows entirely through R_E and develops a voltage that can be observed at the emitter $i_e R_E$ ($\simeq i_c R_E$). Consider the magnitude of this voltage. Figure 3.26(a) shows that for a normally operating transistor, the signal component of the base-emitter voltage ΔV_{BE} (or v_{be}) is very small indeed, whereas the constant component needed for biasing is normally about 0.6 to 0.7 V. That is, $v_{be} \ll V_{BE}$. This implies that the emitter voltage must always follow the base voltage but at a dc level about 0.6 to 0.7 V below it. So, if an output signal is taken from the emitter, it is almost the same as the input signal at the base. In other words, *the voltage gain of an* ***emitter follower*** *is almost unity.*

If this is the case, what is the use of the emitter follower? The answer is that because the signal *current gain* is unchanged at $i_e/i_b = (h_{fe} + 1) \simeq h_{fe}$, then the power gain must also be about h_{fe}. This means in turn that the output resistance must be the resistance "looking into" the transistor from the emitter, divided by h_{fe}. If the parallel combination of R_g and the bias resistors is R_G, then:

$$R_{out(CC)} = \frac{R_G + r_{bb'} + r_{b'e}}{h_{fe}} \tag{3.19}$$

where $R_G = R_g \| R1 \| R2 (\text{or } R_g B)$.

If a voltage generator with zero internal resistance ($R_g = 0$) were applied to the input, then this would become

$$R_{out(CC)} = \frac{r_{bb'} + r_{b'e}}{h_{fe}}$$

and if $r_{b'e} >> r_{bb'}$ (which is usual), then:

$$R_{out(CC)} \simeq \frac{r_{b'e}}{h_{fe}} = \frac{1}{g_m} \tag{3.20}$$

Consider the numerical implications of this: if $I_C = 1$ mA, then $g_m \simeq 39\,\text{mA/V}$ (at room temperature), so $1/g_m \simeq 26\,\Omega$, which is a very low output resistance indeed. In fact, although it appears in parallel with R_E, it is unlikely that R_E will make any significant contribution because it is usually hundreds or thousands of ohms.

Example 4

Using the same bias resistors as for the CE examples, find the output resistance at the emitter of a CC stage.

Solution. The parallel resistances to the left of the base are

$$R_G = R_g \| R1 \| R2 = 5 \| 75 \| 39 \approx 4.2\,\text{k}\Omega$$

Using Equation (3.19):

$$R_{out} \approx \frac{R_G + r_{b'e}}{h_{fe}} = \frac{R_G}{h_{fe}} + \frac{1}{g_m} \quad (\text{neglecting } r_{bb'})$$

where $g_m \approx 39 I_C$, $I_Q = 1$ mA, and $h_{fe} = 110$, so:

$$R_{out(CC)} \approx \frac{4200}{110} + \frac{1000}{39} \approx 63.8\,\Omega$$

From values like this, it is clear that the output of an emitter follower can be thought of as a good practical dependent voltage source of very low internal resistance.

The converse is also true: the input at the base presents a high resistance. This is simply because whereas much the same signal voltage appears at the base as at the emitter, the base signal current i_b is smaller than the emitter signal current i_e by a factor of $(h_{fe}+1) \simeq h_{fe}$. Hence, the apparent resistance at the base must be at least $h_{fe}R_E$. To this must be added $r_{bb'}+r_{b'e}$ so that:

$$R_{in(CC)} \simeq r_{bb'} + r_{b'e} + h_{fe}R_E \tag{3.21a}$$

Now h_{fe} is rarely less than about 100, so $h_{fe}R_E$ is usually predominant and

$$R_{in(CC)} \simeq h_{fe}R_E \tag{3.21b}$$

The emitter-follower circuit is therefore a *buffer stage* because it can accept a signal at a high resistance level without significant attenuation and reproduce it at a low resistance level and with *no phase shift* (except at high frequencies).

In this configuration, the unbypassed emitter resistor R_E is obviously in series with the input circuit as well as the output circuit. Hence, it is actually a feedback resistor and so may be given the alternative symbol R_F, as in Figure 3.33. Because all the output signal voltage is fed back in series with the input, this represents 100% voltage-derived series negative feedback.

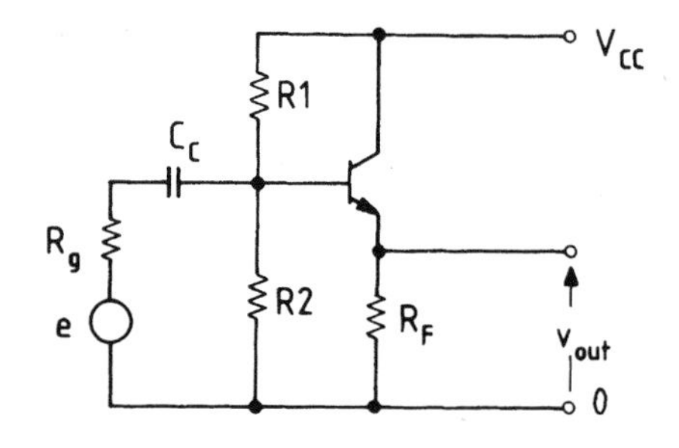

FIGURE 3.33 The emitter follower (or CC stage).

The hybrid-π model for the bipolar transistor may now be inserted into the emitter-follower circuit of Figure 3.33, resulting in Figure 3.34, from which the four midfrequency parameters may be obtained. As an example of the procedures involved, consider the derivation of the voltage gain expression.

Summing signal currents at the emitter:

$$v_{out}\left(\frac{1}{R_F}+\frac{1}{r_{ce}}\right) = v_{b'e}\left(\frac{1}{r_{b'e}}+g_m\right)$$

Now $1/r_{ce} << 1/R_F$ and thus may be neglected, and $v_{b'e} = v_{in} - v_{out}$, so:

$$v_{out}\left(\frac{1}{R_F}\right) = (v_{in}-v_{out})\left(\frac{1}{r_{b'e}}+g_m\right)$$

or

$$v_{out}\left(\frac{1}{R_F}+\frac{1}{r_{b'e}}+g_m\right) = v_{in}\left(\frac{1}{r_{b'e}}+g_m\right)$$

giving

$$\begin{aligned} A_{v(CC)} &= \frac{v_{out}}{v_{in}} = \frac{1/r_{b'e}+g_m}{1/r_{b'e}+g_m+1/R_F} = \frac{1+g_m r_{b'e}}{1+g_m r_{b'e}+r_{b'e}/R_F} \\ &\simeq \frac{g_m r_{b'e}}{g_m r_{b'e}+r_{b'e}/R_F} = \frac{g_m R_F}{g_m R_F+1} \end{aligned} \tag{3.22}$$

which is a little less than unity as expected.

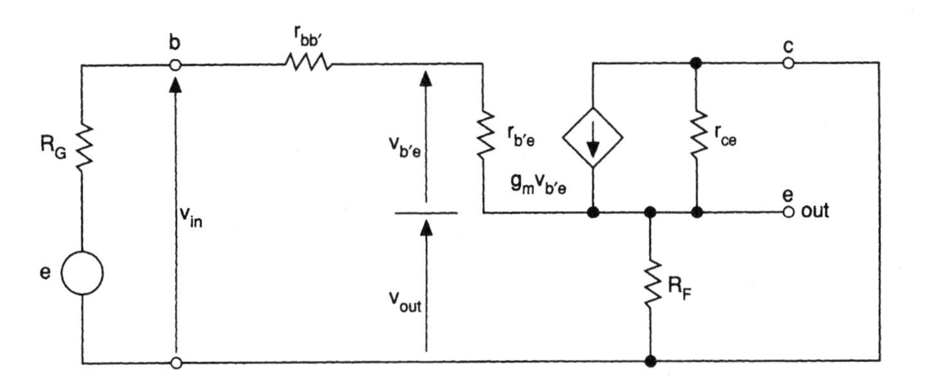

FIGURE 3.34 An emitter-follower equivalent circuit for low frequencies.

Similar derivations based on the equivalent circuit of Figure 3.34 result in the other three basic midband operating parameters for the emitter follower, and all may be listed:

$$A_{i(\mathrm{CC})} \simeq h_{\mathrm{fe}} \quad A_{v(\mathrm{CC})} \rightarrow +1$$

$$R_{\mathrm{in(CC)}} \simeq r_{\mathrm{bb'}} + r_{\mathrm{b'e}} + h_{\mathrm{fe}} R_{\mathrm{F}} \simeq h_{\mathrm{fe}} R_{\mathrm{F}}$$

and

$$R_{\mathrm{out(CC)}} \simeq \frac{R_{\mathrm{G}} + r_{\mathrm{bb'}} + r_{\mathrm{b'e}}}{h_{\mathrm{fe}}} \| R_{\mathrm{F}} \simeq \frac{R_{\mathrm{G}} + r_{\mathrm{bb'}} + r_{\mathrm{b'e}}}{h_{\mathrm{fe}}} \simeq \frac{1}{g_{\mathrm{m}}} \quad \text{if } R_{\mathrm{g}} \rightarrow 0 \text{ and } r_{\mathrm{bb'}} << r_{\mathrm{b'e}}$$

The Common-Emitter Bypass Capacitor C_E

In a CE circuit such as that of Figure 3.29, suppose C_c is large so that the low-frequency –3-dB point f_L is defined only by the parallel combination of the resistance at the emitter and C_E. It will now be seen why the emitter-follower work is relevant: the resistance appearing at the emitter of the CE stage is the same as the output resistance of the emitter-follower stage, and this will now appear in parallel with R_E. If this parallel resistance is renamed R_{emitter}, then, neglecting $r_{bb'}$:

$$R_{\mathrm{emitter}} = R_{\mathrm{out(CC)}} \| R_{\mathrm{E}} \simeq \frac{R_{\mathrm{G}} + r_{\mathrm{b'e}}}{h_{\mathrm{fe}}} \| R_{\mathrm{E}} \simeq \frac{R_{\mathrm{G}} + r_{\mathrm{b'e}}}{h_{\mathrm{fe}}}$$

and if C_E were to define f_L, then:

$$f_{\mathrm{L}} = \frac{1}{2\pi R_{\mathrm{emitter}} C_{\mathrm{E}}} \tag{3.23a}$$

For design purposes, C_E can be extracted for any given value of f_L:

$$C_{\mathrm{E}} = \frac{1}{2\pi R_{\mathrm{emitter}} f_{\mathrm{L}}} \tag{3.23b}$$

Example 5

In Example 4, let C_c be large so that only C_E defines f_L at 42 Hz, and find the value of C_E.

Solution. In the emitter-follower example, where $R_g = 5\ \text{k}\Omega$, $R_{out(CC)}$ was found to be 63.8 Ω, and this is the same as $R_{emitter}$ in the present case. Therefore:

$$C_E = \frac{10^6}{2\pi 63.8 \times 42} \simeq 60\ \mu\text{F}$$

This is the value of C_E that would define f_L if C_c were large. However, if C_E is to act as a short circuit at this frequency, thus allowing C_c to define f_L, then its value would have to be one or two orders of magnitude greater, that is, 600 to 6000 μF.

Summarizing, three possibilities exist:

1. If C_E is very large, C_c defines f_L and a 6-dB/octave roll-off results.
2. If C_c is large, C_E defines f_L and again a 6-dB/octave roll-off results.
3. If both C_c and C_E act together, a 12-dB/octave roll-off results.

In point of fact, at frequencies much less than f_L, both conditions (1) and (2) eventually produce 12-dB/octave roll-offs as the alternate "large" capacitors come into play at very low frequencies; but since the amplifier will not still have a useful gain at such frequencies, this is of little importance.

High-Frequency Response

Unlike the low-frequency response situation, the high-frequency response is governed by the small distributed capacitances inside the transistor structure, and these have been lumped together in the hybrid-π model of Figure 3.31 as $C_{b'e}$ and $C_{b'c}$. At high frequencies, $r_{b'c}$ may be neglected in comparison with the reactance of $C_{b'c}$, so the model may be simplified as in Figure 3.35(a). From this it will be seen that $C_{b'c}$ is a capacitance

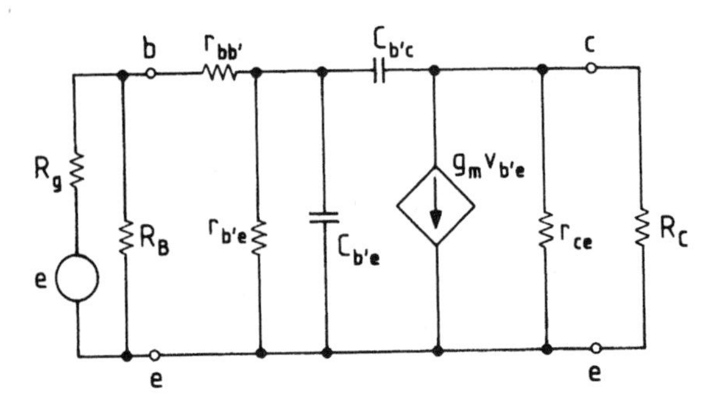

(a)

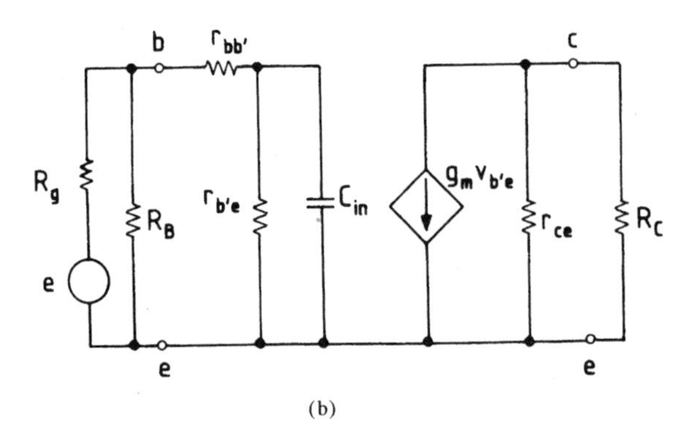

(b)

FIGURE 3.35 (a) The high-frequency hybrid-π model and (b) its simplification.

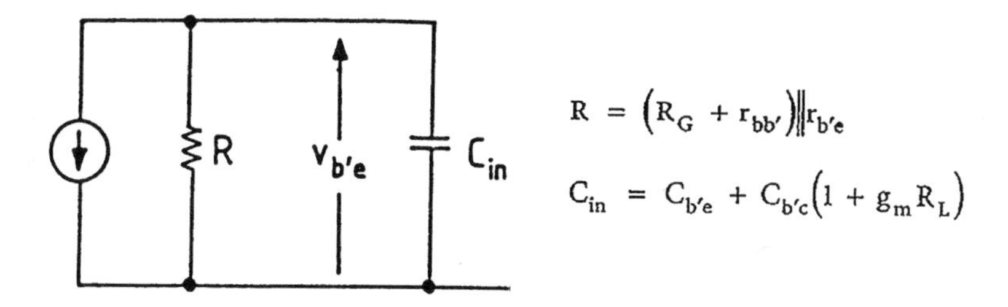

FIGURE 3.36 Simplification of the input part of the high-frequency hybrid-π model.

that appears from the output to the input so that it may be converted by the Miller effect into a capacitance at the input of value:

$$C_{b'c}(1 - A_v) = C_{b'c}(1 + g_m R_C)$$

This will now add to $C_{b'e}$ to give C_{in}:

$$C_{in} = C_{b'e} + C_{b'c}(1 + g_m R_C) \tag{3.24}$$

This simplification is shown in Figure 3.35(b), where C_{in} is seen to be shunted by the input parts of the model. These input parts may be reduced by sequential use of Thévenin–Norton transformations to result in Figure 3.36, which is a simple parallel RC circuit driven by a current source. The actual value of this current source is immaterial—what matters is that the input signal to be amplified, $v_{b'e}$, will be progressively reduced as the frequency rises and the reactance of C_{in} falls.

Using a sinusoidal source, $V_{b'e}$ will be 3 dB down when $R = | X_{C_{in}} |$, which gives

$$R = \frac{1}{2\pi f_H C_{in}} \quad \text{or} \quad f_H = \frac{1}{2\pi R C_{in}} \tag{3.25}$$

where $R = (R_G + r_{bb'})\|r_{b'e}$ from the circuit reduction.

Complete Response

Now that both the low- and high-frequency roll-offs have been related to single time constants (except when C_c and C_E act together), it is clear that the complete frequency response will look like Figure 3.37, where the midband voltage gain is $A_v = -g_m R_C$.

Design Comments

The design of a simple single-transistor amplifier stage has now been covered in terms of both biasing and small-signal performance. These two concepts have been kept separate, but it will have been noticed that they are bridged by the transconductance, because $g_m = (q/kT)I_Q$ ($\simeq 39 I_Q$ at room temperature). That is, when I_Q has been determined, then the small-signal performance follows from expressions involving g_m.

In fact, once the quiescent voltage across the load resistor of a CE stage has been determined, the voltage gain follows from this, irrespective of the values of I_Q and R_C.

If the quiescent voltage at the collector is V_{out}, then in dc biasing terms:

$$V_{RC} = I_Q R_C = (V_{CC} - V_{out})$$

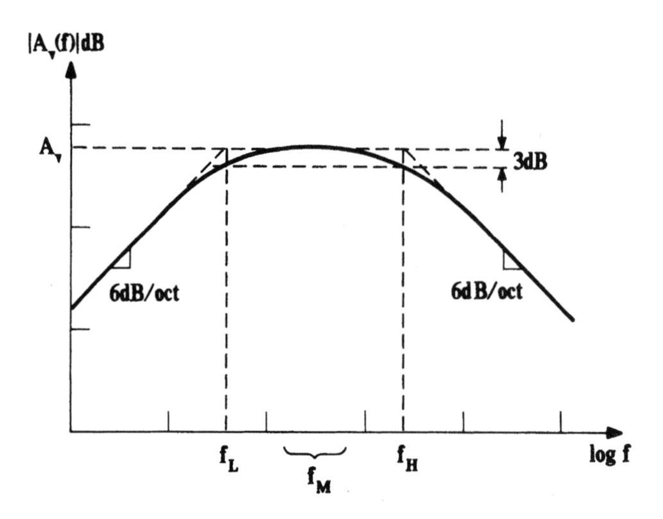

FIGURE 3.37 The complete frequency response.

and in small-signal terms:

$$A_v = -g_m R_C \cong -39 I_Q R_C \quad (\text{at } 25^\circ\text{C}) = -39(V_{CC} - V_{out})$$

Thus, g_m really does act as a bridge between the bias and the small-signal conditions for the bipolar transistor amplifier stage.

Unfortunately, however, there are serious problems with such a stage from a practical viewpoint. For example, it cannot amplify down to dc because of the existence of C_c; and if a larger gain is needed, the cascading of such stages will present problems of phase shift and hence feedback stability. Furthermore, it cannot be produced in IC form because of the incorporation of large capacitances and somewhat critical and high-valued resistors. This leads to a reevaluation of the basic tenets of circuit design, and these can be summed up as follows: circuit design using *discrete* components is largely concerned with voltage drops across resistors (as has been seen), but the design of ICs depends extensively on *currents* and *current sources and sinks.*

Integrated Circuits

Monolithic ICs are fabricated on single chips of silicon or *dice* (the singular being *die*). This means that the active and passive structures on the chips are manufactured all at the same time, so it is easy to ensure that a large number of such structures are identical, or bear some fixed ratio to one another, but it is more difficult to establish precise values for such sets of structures. For example, a set of transistors may all exhibit almost the same values of h_{FE}, but the actual numerical value of h_{FE} may be subject to wider tolerances. Similarly, many pairs of resistors may bear a ratio n:1 to each other, but the actual values of these resistors are more difficult to define. So, in IC design, it is very desirable to exploit the close similarity of devices (or close ratios) rather than depend upon their having predictable absolute values. This approach has led to two ubiquitous circuit configurations, both of which depend upon device similarity: the **long-tailed pair or difference amplifier** (often called the *differential amplifier*) and the **current mirror.** This section will treat both, and the former is best introduced by considering the **degenerate common-emitter** stage.

The Degenerate Common-Emitter Stage

Consider two CE stages that are identical in every respect but which have no emitter resistor bypass capacitors, as shown in Figure 3.38. Also, notice that in these diagrams, two power supply rails have been used, a positive one at V_{CC}^{+} and a negative one at V_{CC}^{-}. The reason for this latter, negative, rail is that the bases may be

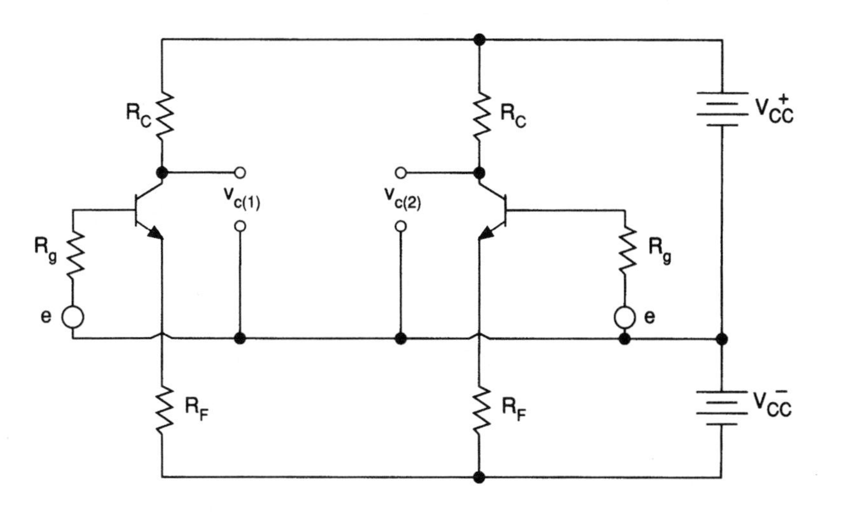

FIGURE 3.38 Two degenerate CE stages.

operated via signal sources referred to a common line or ground. (If, for example, V_{CC}^{+} and V_{CC}^{-} are obtained from batteries as shown, then the common line is simply the junction of the two batteries, as is also shown.) The absence of capacitors now means that amplification down to dc is possible.

It is now very easy to find the quiescent collector currents I_Q because from a dc bias point of view the bases are connected to ground via resistances R_g, which will be taken as having low values so that they drop negligibly small voltages. Hence:

$$I_Q = \frac{|V_{CC}^{-}| - V_{BE}}{R_F} \tag{3.26}$$

(For example, if industry-standard supplies of ± 15 V are used, $V_{BE} = 0.6$ V, and for $R_F = 15$ kΩ, then $I_Q = (15 - 0.6)/15 = 0.96 \simeq 1$ mA.)

Now suppose that identical signals e are applied. At each collector, this will result in an output signal voltage v_c, where $v_c = -i_c R_C$. Also, at each emitter, the output signal voltage will be $v_e = i_e R_F \simeq i_c R_F$. That is:

$$\frac{v_c}{v_e} \simeq \frac{-i_c R_c}{i_c R_F} = -\frac{R_C}{R_F}$$

If the voltage gain from base to collector of a degenerate CE stage is $A_{v(dCE)}$ and the voltage gain from base to emitter is simply the emitter-follower gain $A_{v(CC)}$, then:

$$v_c = A_{v(dCE)}e \text{ and } v_e = A_{v(CC)}e$$

giving

$$\frac{A_{v(dCE)}}{A_{v(CC)}} \simeq -\frac{R_C}{R_F}$$

Now $A_{v(CC)}$ is known from Equation (3.22) so that:

$$A_{v(dCE)} \simeq -A_{v(CC)}\frac{R_C}{R_F} = -\frac{g_m R_C}{1 + g_m R_F} \simeq -\frac{R_C}{R_F} \tag{3.27}$$

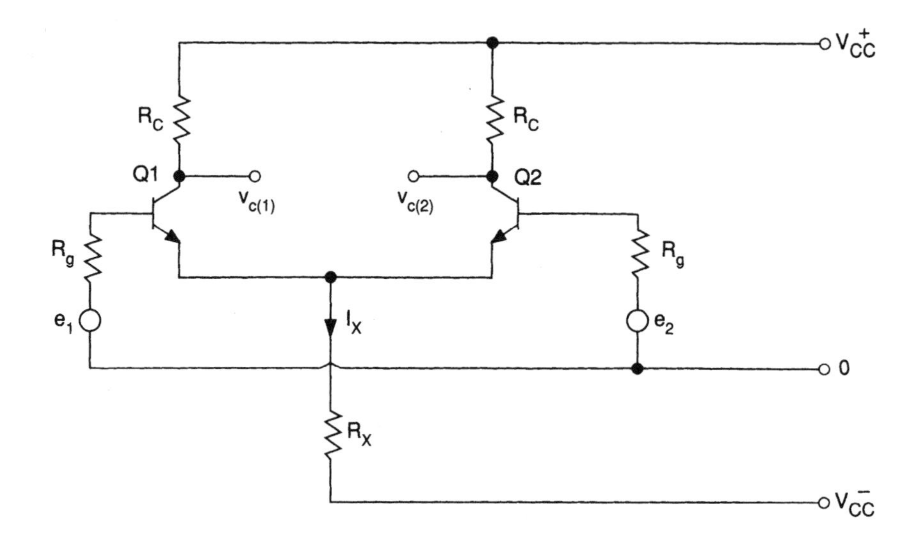

FIGURE 3.39 The difference amplifier.

Note that the input resistance to each base is as for the emitter-follower stage:

$$R_{\mathrm{in(dCE)}} = r_{\mathrm{bb'}} + r_{\mathrm{b'e}} + h_{\mathrm{fe}}R_{\mathrm{F}} \simeq h_{\mathrm{fe}}R_{\mathrm{F}}$$

Now consider what happens if the emitters are connected together as in Figure 3.39, where the two resistors R_{F} have now become R_{X}, where $R_{\mathrm{X}} = \frac{1}{2}R_{\mathrm{F}}$.

The two quiescent emitter currents now combine to give $I_{\mathrm{X}} = 2I_{\mathrm{E}} \simeq 2I_{\mathrm{C}}$, and otherwise the circuit currents and voltages remain undisturbed. So, if the two input signals are identical, then the two output signals will also be identical. This circuit is now called a *difference amplifier*, and the reason will become obvious as soon as the two input signals differ.

The Difference Amplifier

In Figure 3.39, if $e_1 = e_2$, these are called common-mode input signals, $e_{\mathrm{in(CM)}}$, and they will be amplified by $-R_{\mathrm{C}}/R_{\mathrm{F}}$ as for the degenerate CE stage. However, if $e_1 \neq e_2$, then $e_1 - e_2 = e_{\mathrm{in}}$, the difference input signal. The following definitions now apply:

$$\frac{e_1 + e_2}{2} = e_{\mathrm{in(CM)}} \quad \text{the common mode component}$$

and

$$\frac{\pm(e_1 - e_2)}{2} = e_{\mathrm{in(diff)}} \quad \text{the difference component, or } \frac{1}{2}e_{\mathrm{in}}$$

Hence, $e_1 = e_{\mathrm{in(CM)}} + e_{\mathrm{in(diff)}}$ and $e_2 = e_{\mathrm{in(CM)}} - e_{\mathrm{in(diff)}}$.

Consider the progress of a signal current driven by $e_1 - e_2$ and entering the base of $Q1$. It will first pass through R_{g}, then into the resistance R_{in} at the base of $Q1$, and will arrive at the emitter of $Q2$. Here, if R_{X} is large, most of this signal current will pass into the resistance presented by the $Q2$ emitter and eventually out of the $Q2$ base via another R_{g} to ground. The total series resistance is therefore:

$$R_{\mathrm{g}} + R_{\mathrm{in}} = R_{\mathrm{g}} + r_{\mathrm{bb'}} + r_{\mathrm{b'e}} + h_{\mathrm{fe}}R_{\mathrm{emitter(2)}}$$

However:

$$R_{emitter(2)} = \frac{R_g + r_{bb'} + r_{b'e}}{h_{fe}}$$

so

$$R_g + R_{in} = 2(R_g + r_{bb'} + r_{b'e})$$

which is the resistance between the two signal sources. Hence:

$$i_{b(1)} = -i_{b(2)} = \frac{e_1 - e_2}{2(R_g + r_{bb'} + r_{b'e})}$$

giving

$$v_{c(1)} = -v_{c(2)} = \frac{h_{fe}R_C(e_1 - e_2)}{2(R_g + r_{bb'} + r_{b'e})}$$

so that the overall difference voltage gain to each collector is

$$A_{ov} = \frac{v_c}{e_1 - e_2} = \frac{\pm h_{fe}R_C}{2(R_g + r_{bb'} + r_{b'e})} \tag{3.28a}$$

If the voltage gain with the input signal measured between the actual bases is needed, R_g can be removed to give

$$A_v = \frac{\pm h_{fe}R_C}{2(r_{bb'} + r_{b'e})} \tag{3.28b}$$

Finally, if the output signal is measured between the collectors (which will be twice that at each collector because they are in antiphase), the difference-in-to-difference-out voltage gain will be

$$A_{v(diff)} = \frac{h_{fe}R_C}{r_{bb'} + r_{b'e}} \simeq \frac{h_{fe}R_C}{r_{b'e}} = g_m R_C \tag{3.28c}$$

which is the same as for a single CE stage.

Note that this is considerably larger than the gain for a common-mode input signal; that is, the difference stage amplifies difference signals well but largely rejects common-mode signals. This common-mode rejection property is very useful, for often, small signals appear across leads, both of which may contain identical electrical noise. So, the difference stage tends to reject the noise while still amplifying the signal. Furthermore, the difference stage has the advantage that it needs no coupling or bypass capacitors and so will amplify frequencies down to zero (dc). Also, it is very stable biaswise and lends itself perfectly to realization on a monolithic IC.

To make the above derivation valid, the long-tail resistance R_X should be as large as possible so that most of the signal current enters the emitter of Q2. However, R_X must also carry the quiescent current, which would produce a very high quiescent voltage drop and so require a very high value of V_{CC}. To overcome this, another transistor structure can be used within a configuration known as a current mirror.

The Current Mirror

The two transistors in Figure 3.40 are assumed to be identical, and $Q1$ has its base and collector connected so that it acts simply as a diode (formed by the base-emitter junction). The current through it is therefore:

$$I = \frac{V_{CC} - V_{BE}}{R} \tag{3.29}$$

The voltage drop V_{BE} so produced is applied to $Q2$ as shown so that it is forced to carry the same collector current I; that is, it mirrors the current in $Q1$.

The transistor $Q2$ is now a device that carries a dc $I_{C(2)} = I$ but presents a large incremental resistance r_{ce} at its collector. This is exactly what is required by the difference amplifier pair, so it may be used in place of R_X.

The Difference Stage with Current Mirror Biasing

Figure 3.41 shows a complete difference stage complete with a current mirror substituting for the long-tail resistor R_X, where the emitter quiescent currents combine to give I_X:

$$I_x = \frac{V_{cc^+} + |V_{cc^-}| - V_{BE(3)}}{R} = I \tag{3.30}$$

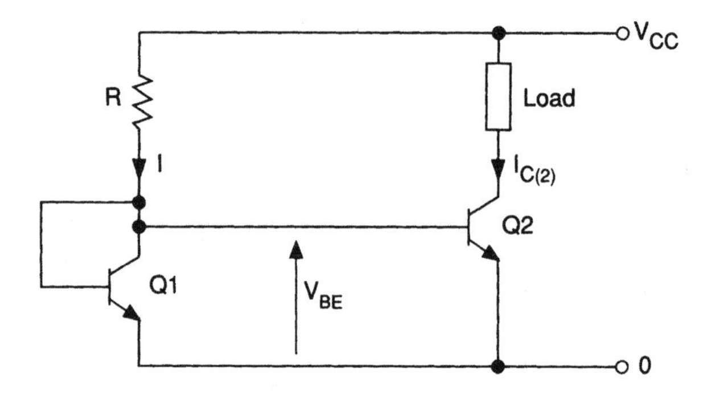

FIGURE 3.40 The current mirror.

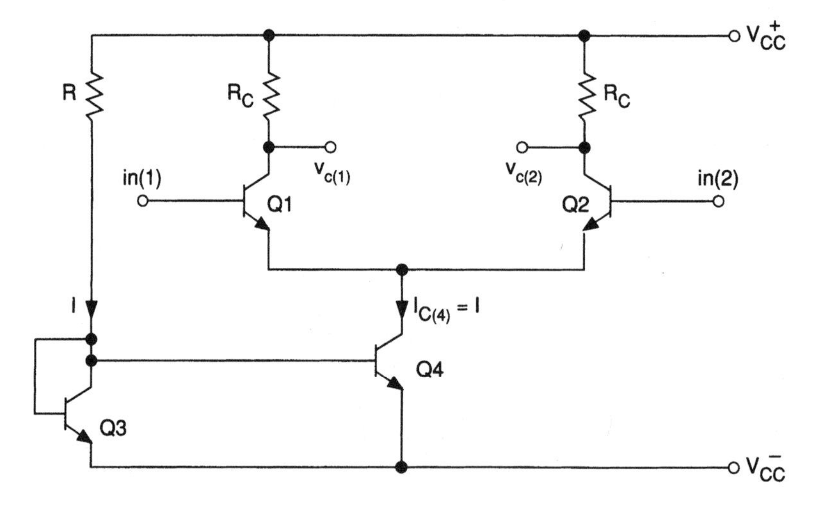

FIGURE 3.41 Current mirror biasing.

This quiescent or bias current is very stable because the change in $V_{BE(3)}$ due to temperature variations is exactly matched by that required by Q4 to produce the same current. The difference gain will be as discussed above, but the common-mode gain will be extremely low because of the high incremental resistance r_{ce} presented by the long-tail transistor.

The Current Mirror as a Load

A second current mirror can be used as a load for the difference amplifier, as shown in Figure 3.42. This must utilize *pnp* transistor structures so that the Q6 collector loads the Q2 collector with a large incremental resistance $r_{ce}(6)$, making for an extremely high voltage gain. Furthermore, Q5 and Q6 combine the signal output currents of both Q1 and Q2 to perform a double-ended-to-single-ended conversion as follows. Taking signal currents:

$$i_{out} = i_c(6) - i_c(2)$$

However:

$$i_c(6) = i_c(5)$$

by current mirror action, and

$$i_c(5) = i_c(1)$$

so

$$i_c(6) = i_c(1)$$

Also:

$$i_c(2) = -i_c(1)$$

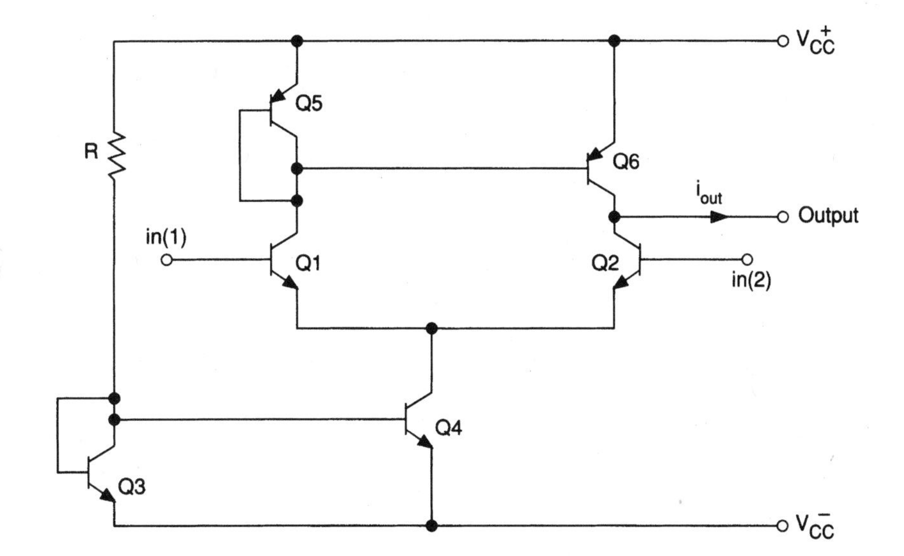

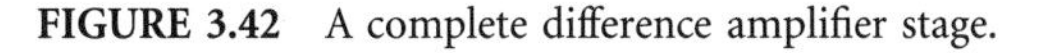
FIGURE 3.42 A complete difference amplifier stage.

by difference amplifier action, so:

$$i_{out} = i_c(1) + i_c(1) = 2i_c(1) \tag{3.31}$$

Thus, both sides of the long-tailed pair are used to provide an output current that may then be applied to further stages to form a complete amplifier. Also, because no capacitors and only one resistor are needed, it is an easy circuit for monolithic integration on a single die.

Summary

It has been shown how a limited knowledge of bipolar operation can lead to properly biased amplifier stages using discrete transistors. An equivalent circuit—the hybrid-π model—was then derived, again from limited information, which made possible the analysis of such stages, and some purely practical design results were favorably compared with its predictions. Finally, the tenets of this equivalent circuit were used to evaluate the performance of the difference amplifier and current mirror circuits, which are the cornerstones of modern electronic circuit design in a very wide variety of its manifestations. These circuits are, in fact, the classic transconductance and translinear elements that are ubiquitous in modern IC signal conditioning and function networks.

It should be recognized that there are many models other than the one introduced here, from the simple but very common h-parameter version to complex and comprehensive versions developed for computer-aided design (CAD) methods. However, the present elementary approach has been from a design rather than an analytical direction, for it is obvious that powerful modern computer-oriented methods such as the SPICE variants become useful only when a basic circuit configuration has been established, and at the time of writing, this is still the province of the human designer.

Defining Terms

Biasing circuit: A circuit that holds a transistor in an operating condition ready to receive signals.

Common emitter: A basic transistor amplifier stage whose emitter is common to both input and output loops. It amplifies voltage, current, and hence power.

Current mirror: An arrangement of two (or more) transistors such that a defined current passing into one is mirrored in another at a high resistance level.

Degenerate common emitter: A combination of the common-emitter and emitter-follower stages with a very well-defined gain.

Difference amplifier or long-tailed pair: An arrangement of two transistors that amplifies difference signals but rejects common-mode signals. It is often called a differential pair.

Emitter follower or common collector: A basic transistor amplifier stage whose collector is common to both input and output loops. Its voltage gain is near unity, but it amplifies current and hence power. It is a high-input resistance, low-output resistance, or buffer, circuit.

Further Information

The following list of recent textbooks covers topics mainly related to analog circuitry containing both integrated and discrete semiconductor devices.

G.M. Glasford, *Analog Electronic Circuits*, Englewood Cliffs, NJ: Prentice-Hall, 1986.

P.R. Gray and R.G. Meyer, *Analysis and Design of Analog Integrated Circuits*, 2nd ed., New York: Wiley, 1984.

J. Keown, *PSPICE and Circuit Analysis*, New York: Macmillan, 1991.

R.B. Northrop, *Analog Electronic Circuits*, Reading, MA: Addison-Wesley, 1990.

A.S. Sedra and K.C. Smith, *Microelectronic Circuits*, 3rd ed., Philadelphia, PA: Saunders, 1991.

T. Schubert and E. Kim, *Active and Non-Linear Electronics*, New York: Wiley, 1996.

J. Watson, *Analog and Switching Circuit Design*, New York: Wiley, 1989.

3.3 The Metal-Oxide Semiconductor Field-Effect Transistor (MOSFET)

John R. Brews (revised by Harvey J. Stiegler)

The *MOSFET* is a transistor that uses a control electrode, the **gate**, to capacitively modulate the conductance of a surface **channel** joining two end contacts, the **source**, and the **drain**. The gate is separated from the semiconductor **body** underlying the gate by a thin *gate insulator*, usually silicon dioxide. The surface channel is formed at the interface between the semiconductor body and the gate insulator (see Figure 3.43).

The MOSFET can be understood by contrasting with other field-effect devices, like the *JFET* or junction field-effect transistor, and the *MESFET* or metal semiconductor field-effect transistor (Hollis and Murphy, 1990). These other transistors modulate the conductance of a *majority-carrier* path between two *ohmic* contacts by capacitive control of its cross section. (Majority carriers are those in greatest abundance in a field-free semiconductor, electrons in n-type material and holes in p-type material.) This modulation of the cross section can take place at any point along the length of the channel, so the gate electrode can be positioned anywhere and need not extend the entire length of the channel.

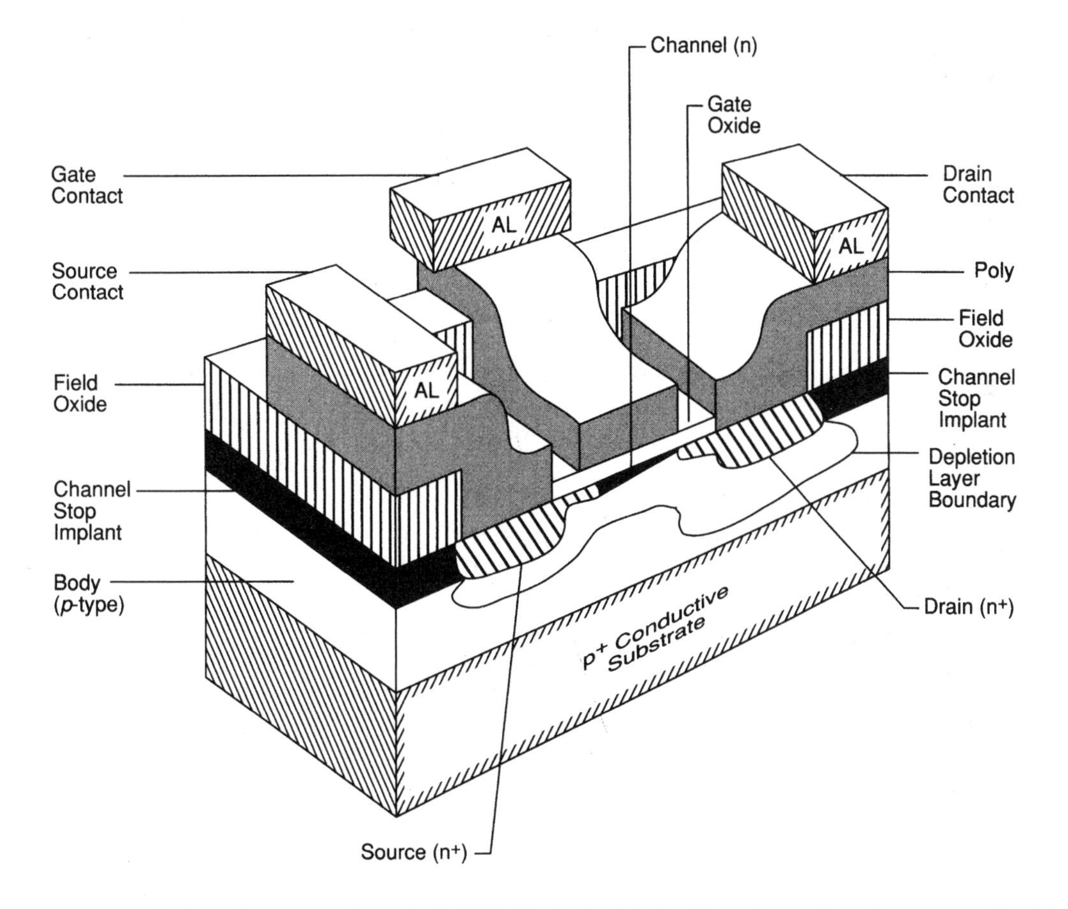

FIGURE 3.43 A high-performance *n*-channel MOSFET. The device is isolated from its neighbors by a surrounding thick *field oxide* under which is a heavily doped *channel stop implant* intended to suppress accidental channel formation that could couple the device to its neighbors. The *drain contacts* are placed over the field oxide to reduce the capacitance to the body, a parasitic that slows response times. These structural details are described later. (*Source:* after Brews, 1990.)

Analogous to these field-effect devices is the *buried-channel*, *depletion-mode*, or *normally-on* MOSFET, which contains a surface layer of the same doping type as the source and drain (opposite type to the semiconductor body of the device). As a result, it has a built-in or normally-on channel from source to drain with a conductance that is reduced when the gate depletes the majority carriers.

In contrast, the true MOSFET is an *enhancement-mode* or *normally-off* device. The device is normally off because the body forms pn-junctions with both the source and the drain, so no majority-carrier current can flow between them. Instead, *minority-carrier* current can flow provided minority carriers are available. As discussed later, for gate biases that are sufficiently attractive, above **threshold**, minority carriers are drawn into a surface channel, forming a conducting path from source to drain. The gate and channel then form two sides of a capacitor separated by the gate insulator. As additional attractive charges are placed on the gate side, the channel side of the capacitor draws a balancing charge of minority carriers from the source and the drain. The more charges on the gate, the more populated the channel, and the larger the conductance. Because the gate *creates* the channel, to ensure electrical continuity, the gate must extend over the entire length of the separation between source and drain.

The MOSFET channel is created by attraction to the gate and relies upon the insulating layer between the channel and the gate to prevent leakage of minority carriers to the gate. As a result, MOSFETs can be made only in material systems that provide very good gate insulators, and the best-known system is the silicon–silicon dioxide combination. This requirement for a good gate insulator is not so important for JFETs and MESFETs, where the role of the gate is to *push away* majority carriers rather than to attract minority carriers. Thus, in Ga–As systems where good insulators are incompatible with other device or fabrication requirements, MESFETs are used.

A more recent development in Ga–As systems is the heterostructure field-effect transistor or HFET (Pearton and Shaw, 1990), made up of layers of varying compositions of Al, Ga, and As or In, Ga, P, and As. These devices may be made using, for example, molecular beam epitaxy or by organometallic vapor phase epitaxy. HFETs include a variety of structures, the best known of which is the modulation-doped FET, or MODFET. HFETs are field-effect devices, not MOSFETs, because the gate simply modulates the carrier density in a preexistent channel between ohmic contacts. The channel is formed spontaneously, regardless of the quality of the gate insulator as a condition of equilibrium between the layers, just as a depletion layer is formed in a pn junction. The resulting channel is created very near to the gate electrode, resulting in gate control as effective as in a MOSFET.

The silicon-based MOSFET has been successful primarily because the silicon–silicon dioxide system provides a stable interface with low trap densities, and because the oxide is impermeable to many environmental contaminants, has a high breakdown strength, and is easy to grow uniformly and reproducibly (Nicollian and Brews, 1982). These attributes allow easy fabrication using lithographic processes, resulting in integrated circuits (ICs), with very small devices, very large device counts, and very high reliability at low cost. Because the importance of the MOSFET lies in this relationship to high-density manufacture, an emphasis of this section is to describe the issues involved in continuing miniaturization.

An additional advantage of the MOSFET is that it can be made using either electrons or holes as channel carriers. Using both types of devices in so-called complementary MOS (CMOS) technology allows circuits that draw essentially no *dc* power if current paths include at least one series connection of both types of device because, in steady state, only one or the other type conducts, not both at once. Of course, in exercising the circuit, power is drawn during switching of the devices. This flexibility in choosing *n*- or *p*-channel devices has enabled large circuits to be made that use low power levels. Hence, complex systems can be manufactured without expensive packaging or cooling requirements.

Current–Voltage Characteristics

The derivation of the current–voltage characteristics of the MOSFET can be found in several sources (Brews, 1981; Annaratone, 1986; Pierret, 1990). Here a qualitative discussion is provided.

Strong-Inversion Characteristics

In Figure 3.44 the source-drain current I_D is plotted vs. drain-to-source voltage V_D (the *I–V* curves for the MOSFET). At low V_D the current increases approximately linearly with increased V_D, behaving like a simple resistor with a resistance that is controlled by the gate voltage V_G. As the gate voltage is made more attractive for channel carriers, the channel becomes stronger, more carriers are contained in the channel, and its resistance R_{ch} drops. Hence, at larger V_G the current is larger.

At large V_D the curves flatten out, and the current is less sensitive to drain bias. The MOSFET is said to be in *saturation*. There are different reasons for this behavior, depending upon the field along the channel caused by the drain voltage. If the source-drain separation is short, where "short" is technology dependent but generally denotes a distance in the submicrometer regime, the usual drain voltage is sufficient to create fields along the channel of more than a few $\times 10^4$ V/cm. In this case the carrier energy is sufficient for carriers to lose energy by causing vibrations of the silicon atoms composing the crystal (optical phonon emission). Consequently, the carrier velocity does not increase much with increased field, saturating at a value $v_{sat} \approx 10^7$ cm/sec in silicon MOSFETs. Because the carriers do not move faster with increased V_D, the current also saturates.

For longer devices the current–voltage curves saturate for a different reason. Consider the potential along the insulator-channel interface, the surface potential. Whatever the surface potential is at the source end of the channel, it varies from the source end to a value larger at the drain end by V_D because the drain potential is V_D higher than the source. The gate, on the other hand, is at the same potential everywhere. Thus, the difference in potential between the gate and the source is larger than that between the gate and the drain. Correspondingly, the oxide field at the source is larger than that at the drain, and as a result less charge can be supported at the drain. This reduction in attractive power of the gate reduces the number of carriers in the

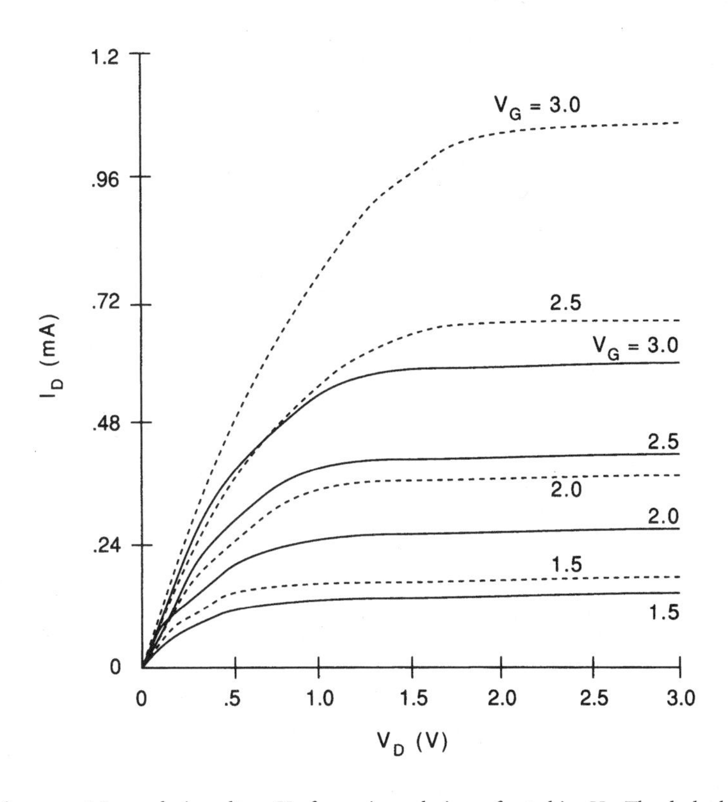

FIGURE 3.44 Drain current I_D vs. drain voltage V_D for various choices of gate bias V_G. The dashed-line curves are for a long-channel device for which the current in saturation increases quadratically with gate bias. The solid-line curves are for a *short-channel* device that is approaching *velocity saturation* and thus exhibits a more linear increase in saturation current with gate bias, as discussed in the text.

channel at the drain end, increasing channel resistance. In short, we have $I_D \approx V_D/R_{ch}$, but the channel resistance $R_{ch} = R_{ch}(V_D)$ is increasing with V_D. As a result, the current–voltage curves do not continue along the initial straight line, but bend over and saturate.

Another difference between the current–voltage curves for short devices and those for long devices is the dependence on gate voltage. For long devices, the current level in saturation, $I_{D,sat}$, increases quadratically with gate bias. The reason is that the number of carriers in the channel is proportional to $V_G - V_{TH}$ (where V_{TH} is the *threshold voltage*), as is discussed later, the channel resistance $R_{ch} \propto 1/(V_G - V_{TH})$, and the drain bias in saturation is approximately V_G. Thus, $I_{D,sat} = V_D/R_{ch} \propto (V_G - V_{TH})^2$, and we have quadratic dependence. When the carrier velocity is saturated, however, the dependence of the current on drain bias is suppressed because the speed of the carriers is fixed at υ_{sat}, and $I_{D,sat} \propto \upsilon_{sat}/R_{ch} \propto (V_G - V_{TH})\upsilon_{sat}$, a linear gate-voltage dependence. As a result, the current available from a short device is not as large as would be expected if we assumed it behaved like a long device.

Subthreshold Characteristics

Quite different current–voltage behavior is seen in **subthreshold**, that is, for gate biases so low that the channel is in *weak inversion*. In this case the number of carriers in the channel is so small that their charge does not affect the potential, and channel carriers simply must adapt to the potential set up by the electrodes and the dopant ions. Likewise, in subthreshold any flow of current is so small that it causes no potential drop along the interface, which becomes an equipotential region.

As there is no lateral field to move the channel carriers, they move by diffusion only, driven by a gradient in carrier density set up because the drain is effective in reducing the carrier density at the drain end of the channel. In the subthreshold the current is then independent of drain bias once this bias exceeds a few tens of millivolts, enough to reduce the carrier density at the drain end of the channel to near zero.

In short devices, however, the source and drain are close enough together to begin to share control of the potential with the gate. If this effect is too strong, a drain-voltage dependence of the subthreshold characteristic then occurs, which is undesirable, because it increases the MOSFET off current and can cause a drain-bias dependent threshold voltage.

Although for a well-designed device there is no drain-voltage dependence in subthreshold, gate-bias dependence is exponential. The surface is lowered in energy relative to the semiconductor body by the action of the gate. If this *surface potential* is ϕ_S below that of the body, the carrier density is enhanced by a Boltzmann factor $\exp(q\phi_S/kT)$ relative to the body concentration, where kT/q = the thermal voltage $\approx$ 25 mV at 290 K. As ϕ_S is roughly proportional to V_G, this exponential dependence on ϕ_S leads to an exponential dependence upon V_G for the carrier density and, hence, for the current in subthreshold.

Important Device Parameters

A number of MOSFET parameters are important to the performance of a MOSFET. In this subsection some of these parameters are discussed, particularly from the viewpoint of digital ICs.

Threshold Voltage

The threshold voltage is vaguely defined as the gate voltage V_{TH} at which the channel begins to form. At this voltage devices begin to switch from "off" to "on," and circuits depend on a voltage swing that straddles this value. Thus, threshold voltage helps in deciding the necessary supply voltage for circuit operation, and also it helps in determining the leakage or "off" current that flows when the device is in the off state.

Threshold voltage is controlled by oxide thickness d and by body doping. To control the body doping, ion implantation is used so that the dopant-ion density is not simply a uniform extension of the bulk, background level N_B ions/unit volume but has superimposed upon it an implanted-ion density. To estimate the threshold voltage, we need a picture of what happens in the semiconductor under the gate as the gate voltage is changed from its off level toward threshold.

If we imagine changing the gate bias from its off condition toward threshold, at first the result is to repel majority carriers, forming a surface *depletion layer* (refer to Figure 3.43). In the depletion layer there are almost no carriers present, but there are dopant ions. In n-type material these dopant ions are positive donor impurities that cannot move under fields because they are locked in the silicon lattice, where they have been deliberately introduced to replace silicon atoms. In p-type material these dopant ions are negative acceptors. Thus, each charge added to the gate electrode to bring the gate voltage closer to threshold causes an increase in the depletion-layer width sufficient to balance the gate charge by an equal but opposite charge of dopant ions in the silicon depletion layer.

This expansion of the depletion layer continues to balance the addition of gate charge until threshold is reached. Then this charge response changes: above threshold any additional gate charge is balanced by an increasingly strong inversion layer or channel. The border between a depletion-layer and an inversion-layer response, threshold, should occur when:

$$\frac{\mathrm{d}qN_{\mathrm{inv}}}{\mathrm{d}\phi_{\mathrm{S}}} = \frac{\mathrm{d}Q_{\mathrm{D}}}{\mathrm{d}\phi_{\mathrm{S}}} \tag{3.32}$$

where $\mathrm{d}\phi_{\mathrm{S}}$ is the small change in surface potential that corresponds to our incremental change in gate charge, qN_{inv} is the inversion-layer charge/unit area, and Q_{D} is the depletion-layer charge/unit area. According to Equation (3.32), the two types of response are equal at threshold, so one is larger than the other on either side of this condition. To be more quantitative, the rate of increase in qN_{inv} is exponential; that is, its rate of change is proportional to qN_{inv}, so as qN_{inv} increases, so does the left side of Equation (3.32). However, Q_{D} has a square-root dependence on ϕ_{S}, which means its rate of change becomes smaller as Q_{D} increases. Thus, as surface potential is increased, the left side of Equation (3.32) increases proportional to qN_{inv} until, at threshold, Equation (3.32) is satisfied. Then, beyond threshold, the exponential increase in qN_{inv} with ϕ_{S} swamps Q_{D}, making change in qN_{inv} the dominant response. Likewise, below threshold, the exponential decrease in qN_{inv} with decreasing ϕ_{S} makes qN_{inv} negligible and change in Q_{D} becomes the dominant response. The abruptness of this change in behavior is the reason for the term *threshold* to describe MOSFET switching.

To use Equation (3.32) to find a formula for threshold voltage, we need expressions for N_{inv} and Q_{D}. Assuming the interface is held at a lower energy than the bulk due to the charge on the gate, the minority-carrier density at the interface is larger than in the bulk semiconductor, even below threshold. Below threshold and even up to the threshold of Equation (3.32), the number of charges in the channel/unit area N_{inv} is given for n channel devices approximately by (Brews, 1981):

$$N_{\mathrm{inv}} \approx d_{\mathrm{INV}} \frac{n_{\mathrm{i}}^2}{N_{\mathrm{B}}} e^{q(\phi_{\mathrm{S}} - V_{\mathrm{S}})/kT} \tag{3.33}$$

where the various symbols are defined as follows: n_{i} = intrinsic carrier density/unit volume $\approx 10^{10}/\mathrm{cm}^3$ in silicon at 290 K and V_{S} = body reverse bias, if any. The first factor, d_{INV}, is an effective depth of minority carriers from the interface given by

$$d_{\mathrm{INV}} = \frac{\varepsilon_{\mathrm{S}} kT/q}{Q_{\mathrm{D}}} \tag{3.34}$$

where Q_{D} = depletion-layer charge/unit area due to charged dopant ions in the region where there are no carriers and ε_{S} is the dielectric permittivity of the semiconductor.

Equation (3.33) expresses the net minority-carrier density/unit area as the product of the bulk minority-carrier density/unit volume, $n_{\mathrm{i}}^2/N_{\mathrm{B}}$, with the depth of the minority-carrier distribution d_{INV} multiplied in turn by the customary Boltzmann factor $\exp[q(\phi_{\mathrm{S}} - V_{\mathrm{S}})/kT]$ expressing the enhancement of the interface density over the bulk due to lower energy at the interface. The depth d_{INV} is related to the carrier distribution near the interface using the approximation (valid in *weak inversion*) that the minority-carrier density decays exponentially with distance from the oxide-silicon surface. In this approximation, d_{INV} is the *centroid* of the minority-carrier density. For example, for a uniform bulk doping of 10^{16} dopant ions/cm^3 at 290 K, using

Equation (3.33) and the surface potential at threshold from Equation (3.38) below ($\phi_{TH} = 0.69$ V), there are $Q_D/q = 3\times10^{11}$charges/cm^2 in the depletion layer at threshold. This Q_D corresponds to a $d_{INV} = 5.4$ nm and a carrier density at threshold of $N_{inv} = 5.4\times10^9$charges/cm^2.

The next step in using the definition of threshold, Equation (3.32), is to introduce the depletion-layer charge/unit area Q_D. For the ion-implanted case, Q_D is made up of two terms (Brews, 1981):

$$Q_D = qN_BL_B\{2[q\phi_{TH}/(kT) - m_1 - 1]\}^{1/2} + qD_I \tag{3.35}$$

where the first term is Q_B, the depletion-layer charge from bulk dopant atoms in the depletion layer with a width that has been reduced by the first moment of the implant, namely, m_1 given in terms of the centroid of the implant x_C by

$$m_1 = \frac{D_I x_C}{N_B L_B^2} \tag{3.36}$$

The second term is the additional charge due to the implanted-ion density within the depletion layer, D_I ions per unit area. The Debye length L_B is defined as

$$L_B^2 \equiv \frac{kT}{q}\frac{\varepsilon_S}{qN_B} \tag{3.37}$$

where ε_S is the dielectric permittivity of the semiconductor. The Debye length is a measure of how deeply a variation of surface potential penetrates into the body when $D_I = 0$ and the depletion layer is of zero width.

Approximating qN_{inv} by Equation (3.33) and Q_D by Equation (3.35), Equation (3.32) determines the surface potential at threshold, ϕ_{TH}, to be

$$\phi_{TH} = 2\frac{kT}{q}\ln\frac{N_B}{n_i} + \frac{kT}{q}\ln\left(1 + \frac{qD_I}{Q_B}\right) \tag{3.38}$$

where the new symbols are defined as follows: Q_B = depletion-layer charge/unit area due to bulk body dopant N_B in the depletion layer, and qD_I = depletion-layer charge/unit area due to implanted ions in the depletion layer between the inversion-layer edge and the depletion-layer edge. Because even a small increase in ϕ_S above ϕ_{TH} causes a large increase in qN_{inv}, which can balance a rather large change in gate charge or gate voltage, ϕ_S does not increase much as $V_G - V_{TH}$ increases. Nonetheless, in strong inversion, $N_{inv} \approx 10^{12}$charges/cm^2, so in strong inversion ϕ_S will be about $10kT/q$ larger than ϕ_{TH}.

Equation (3.38) indicates for uniform doping (no implant, $D_I = 0$) that threshold occurs approximately for $\phi_S = \phi_{TH} = 2(kT/q)\ln(N_B/n_i) \equiv 2\phi_B$, but for the nonuniformly doped case a larger surface potential is needed, assuming the case of a normal implant where D_I is positive, increasing the dopant density. The implant increases the required surface potential because the field at the surface is larger, narrowing the inversion layer and reducing the channel strength for $\phi_S = 2\phi_B$. Hence, a somewhat larger surface potential is needed to increase qN_{inv} to the point that Equation (3.32) is satisfied. Equation (3.38) would not apply if a significant fraction of the implant were confined to lie within the inversion layer itself. However, no realistic implant can be confined within a distance comparable to an inversion-layer thickness (a few tens of nanometers), so Equation (3.38) covers practical cases.

With the surface potential ϕ_{TH} known, the potential on the gate at threshold Φ_{TH} can be found if we know the oxide field F_{ox} by simply adding the potential drop across the semiconductor to that across the oxide. That is, $\Phi_{TH} = \phi_{TH} + F_{ox}d$, with d = oxide thickness and F_{ox} is given by Gauss's law as

$$\varepsilon_{ox}F_{ox} = Q_D \tag{3.39}$$

There are two more complications in finding the threshold voltage. First, the *gate voltage* V_{TH} usually differs from the gate potential Φ_{TH} at threshold because of a work-function difference between the body and the gate material. This difference causes a spontaneous charge exchange between the two materials as soon as the MOSFET is placed in a circuit allowing charge transfer to occur. Thus, even before any *voltage* is applied to the device, a *potential difference* exists between the gate and the body due to spontaneous charge transfer. The second complication affecting threshold voltage is the existence of charges in the insulator and at the insulator–semiconductor interface. These nonideal contributions to the overall charge balance are due to traps and fixed charges incorporated during the device processing.

Ordinarily, interface-trap charge is negligible ($<10^{10}/cm^2$ in silicon MOSFETs), and the other nonideal effects upon threshold voltage are accounted for by introducing the *flatband voltage* V_{FB}, which corrects the gate bias for these contributions. Then, using Equation (3.39) with $F_{ox} = (V_{TH} - V_{FB} - \phi_{TH})/d$ we find:

$$V_{TH} = V_{FB} + \phi_{TH} + Q_D \frac{d}{\varepsilon_{ox}} \tag{3.40}$$

which determines V_{TH} even for the nonuniformly doped case, using Equation (3.38) for ϕ_{TH} and Q_D at threshold from Equation (3.35). If interface-trap charge/unit area is not negligible, then terms in the interface-trap charge/unit area Q_{IT} must be added to Q_D in Equation (3.40).

From Equation (3.35) and Equation (3.38), the threshold voltage depends upon the implanted dopant-ion profile only through two parameters, the net charge introduced by the implant in the region between the inversion layer and the depletion-layer edge qD_I, and the centroid of this portion of the implanted charge x_C. As a result, a variety of implants can result in the same threshold, ranging from the extreme of a δ-function spike implant of dose D_I/unit area located at the centroid x_C, to a box-type rectangular distribution with the same dose and centroid, namely, a rectangular distribution of width $x_W = 2x_C$ and volume density D_I/x_W (of course, x_W must be no larger than the depletion-layer width at threshold for this equivalence to hold true, and x_C must not lie within the inversion layer). This weak dependence on the details of the profile leaves flexibility to satisfy other requirements, such as control of off current.

As already said, for gate biases $V_G > V_{TH}$ any gate charge above the threshold value is balanced mainly by inversion-layer charge. Thus, the additional oxide field, given by $(V_G - V_{TH})/d$, is related by Gauss's law to the inversion-layer carrier density approximately by

$$\varepsilon_{ox} \frac{V_G - V_{TH}}{d} \approx qN_{inv} \tag{3.41}$$

which shows that channel strength above threshold is proportional to $V_G - V_{TH}$, an approximation often used in this section. Thus, the switch in balancing gate charge from the depletion layer to the inversion layer causes N_{inv} to switch from an exponential gate-voltage dependence in subthreshold to a linear dependence above threshold.

For circuit analysis, Equation (3.41) is a convenient *definition* of V_{TH} because it fits current–voltage curves. If this definition is chosen instead of the charge-balance definition of Equation (3.32), then Equation (3.32) and Equation (3.38) result in an *approximation* to ϕ_{TH}.

Driving Ability and $I_{D,sat}$

The driving ability of the MOSFET is proportional to the current it can provide at a given gate bias. One might anticipate that the larger this current, the faster the circuit. Here, this current is used to find some response times governing MOSFET circuits.

MOSFET current is dependent upon the carrier density in the channel or upon $V_G - V_{TH}$ (see Equation (3.41)). For a long-channel device, driving ability depends also on channel length. The shorter the channel length, L, the greater the driving ability, because the channel resistance is directly proportional to the channel length. Although it is an oversimplification, let us suppose that the MOSFET is primarily in saturation during the driving of its load. This simplification will allow a clear discussion of the issues involved in making faster MOSFETs without complicated mathematics. Assuming the MOSFET to be saturated over most of the switching period, driving ability is proportional to current in saturation, or to

$$I_{D,sat} = \frac{\varepsilon_{ox} Z \mu}{2dL} (V_G - V_{TH})^2 \tag{3.42}$$

where the factor of two results from the saturating behavior of the *I–V* curves at large drain biases and Z is the width of the channel normal to the direction of current flow. Evidently, for long devices driving ability is quadratic in $V_G - V_{TH}$ and inversely proportional to d.

The result of Equation (3.42) holds for long devices. For short-channel devices, as explained for Figure 3.44, the larger fields exerted by the drain electrode cause *velocity saturation* and, as a result, $I_{D,sat}$ is given roughly by (Einspruch and Gildenblat, 1989)

$$I_{D,sat} \approx \frac{\varepsilon_{ox} Z \upsilon_{sat}}{d} \frac{(V_G - V_{TH})^2}{V_G - V_{TH} + F_{sat} L} \tag{3.43}$$

where υ_{sat} is the carrier saturation velocity, about 10^7 cm/sec for silicon at 290 K, F_{sat} is the field at which velocity saturation sets in, about 5×10^4 V/cm for electrons and not well established as $\gtrsim 10^5$ V/cm for holes in silicon MOSFETs. For Equation (3.43) to agree with Equation (3.42) at long L, we need $\mu \approx 2\upsilon_{sat}/F_{sat} \approx 400\ \text{cm}^2/(\text{V}\cdot\text{s})$ for electrons in silicon MOSFETs, which is only roughly correct. Nonetheless, we can see that for devices in the submicron channel length regime, $I_{D,sat}$ tends to become independent of channel length L and becomes more linear with $V_G - V_{TH}$, and less quadratic (see Figure 3.44). Equation (3.43) shows that velocity saturation is significant when $V_G/L \lesssim F_{sat}$ for example, when $L \lesssim 0.5\ \mu\text{m}$ if $V_G - V_{TH} = 2.5$ V.

To relate $I_{D,sat}$ to a gate response time τ_G, consider one MOSFET driving an identical MOSFET as load capacitance. Then the current from Equation (3.43) charges this capacitance to a voltage V_G in a gate response time τ_G given by (Shoji, 1988):

$$\tau_G = \frac{C_G V_G}{I_{D,sat}} = \frac{L}{\upsilon_{sat}} \left(1 + \frac{C_{par}}{C_{ox}}\right) \frac{V_G(V_G - V_{TH} + F_{sat} L)}{(V_G - V_{TH})^2} \tag{3.44}$$

where C_G is the MOSFET gate capacitance $C_G = C_{ox} + C_{par}$, with $C_{ox} = \varepsilon_{ox} ZL/d$ the MOSFET oxide capacitance, and C_{par} the parasitic component of the gate capacitance (Chen, 1990). The parasitic capacitance C_{par} is due to overlap of the gate electrode over the source and drain, and to fringing-field and channel-edge capacitances. For short-channel lengths, C_{par} is a significant part of C_G. Typically, $V_{TH} \approx V_G/4$, so:

$$\tau_G = \left(\frac{L}{\upsilon_{sat}}\right)\left(1 + \frac{C_{par}}{C_{ox}}\right)\left(1.3 + 1.8 \frac{F_{sat} L}{V_G}\right) \tag{3.45}$$

Thus, on an intrinsic level, the gate response time is closely related to the transit time of an electron from source to drain, which is L/υ_{sat} in velocity saturation. At shorter L, a linear reduction in delay with L is predicted, while for longer devices the improvement can be quadratic in L, depending upon how V_G is scaled as L is reduced.

The gate response time is not the only delay in device switching, because the drain-body pn-junction also must charge or discharge for the MOSFET to change state (Shoji, 1988). Hence, we must also consider a drain response time τ_D. Following Equation (3.44), we suppose that the drain capacitance C_D is charged by the supply voltage through a MOSFET in saturation so that:

$$\tau_D = \frac{C_D V_G}{I_{D,sat}} = \frac{C_D}{C_G} \tau_G \tag{3.46}$$

Equation (3.46) suggests that τ_D will show a similar improvement to τ_G as L is reduced, provided that C_D/C_G does not increase as L is reduced. However, $C_{ox} \propto L/d$, and the major component of C_{par}, namely, the overlap capacitance contribution, leads to $C_{par} \propto L_{ovlp}/d$ where L_{ovlp} is roughly three times the length of overlap of the gate over the source or drain (Chen, 1990). Then $C_G \propto (L + L_{ovlp})/d$ and, to keep the C_D/C_G ratio from increasing as L is reduced, either C_D or oxide-thickness d must be reduced along with L.

Clever design can reduce C_D. For example, various *raised-drain* designs reduce the drain-to-body capacitance by separating much of the drain area from the body using a thick oxide layer. The contribution to drain capacitance stemming from the sidewall depletion-layer width next to the channel region is more difficult to handle because the sidewall depletion layer is deliberately reduced during miniaturization to avoid *short-channel* effects, that is, drain influence upon the channel in competition with gate control. As a result this sidewall contribution to the drain capacitance tends to increase with miniaturization unless junction depth can be shrunk.

Equation (3.45) and Equation (3.46) predict reduction of response times by reduction in channel length L. Decreasing oxide thickness leads to no improvement in τ_G, but Equation (3.46) shows a possibility of improvement in τ_D because C_D is independent of d while C_G increases as d decreases. The *ring oscillator*, a closed loop of an odd number of inverters, is a test circuit whose performance depends primarily on τ_G and τ_D. Gate delay/stage for ring oscillators is found to be near 12 ps/stage at 0.1 μm channel length, and 60 ps/stage at 0.5 μm channel length.

For circuits, interconnection capacitances and fan-out (multiple MOSFET loads) will increase response times beyond the device response time, even when parasitics are taken into account. Thus, we are led to consider interconnection delay, τ_{INT}. Although a lumped model suggests, as with Equation (3.46), that $\tau_{INT} \approx (C_{INT}/C_G)\tau_G$, the length of interconnections requires a *distributed* model. Interconnection delay is then:

$$\tau_{INT} = \frac{R_{INT}C_{INT}}{2} + R_{INT}C_G + \left(1 + \frac{C_{INT}}{C_G}\right)\tau_G \tag{3.47}$$

where the new symbols are R_{INT} = interconnection resistance, C_{INT} = interconnection capacitance, and we have assumed that the interconnection joins a MOSFET driver in saturation to a MOSFET load C_G. For small R_{INT}, τ_{INT} is dominated by the last term, which resembles Equation (3.44) and Equation (3.46). However, unlike the ratio C_D/C_G in Equation (3.46), it is difficult to reduce or even maintain the ratio C_{INT}/C_G in Equation (3.47) as L is reduced. Recall that $C_G \propto Z(L+L_{ovlp})/d$. Reduction of L therefore tends to increase C_{INT}/C_G, especially because interconnect cross sections cannot be reduced without impractical increases in R_{INT}. What is worse, along with reduction in L, chip sizes usually increase, making line lengths longer, increasing R_{INT} even at constant cross section. As a result, interconnection delay becomes a major problem as L is reduced. The obvious way to keep C_{INT}/C_G under control is to increase the device width Z so that $C_G \propto Z(L+L_{ovlp})/d$ remains constant as L is reduced. A better way is to cascade drivers of increasing Z (Shoji, 1988; Chen, 1990). Either solution requires extra area, however, reducing the packing density that is a major objective in decreasing L in the first place. An alternative is to reduce the oxide thickness d.

Transconductance

Another important device parameter is the small-signal transconductance g_m (Haznedar, 1991; Sedra and Smith, 1991;), which determines the amount of output current swing at the drain that results from a given input voltage variation at the gate, that is, the small-signal gain:

$$g_m = \left.\frac{\partial I_D}{\partial V_G}\right|_{V_D=\text{const}} \tag{3.48}$$

Using the chain rule of differentiation, the transconductance in saturation can be related to the small-signal *transition* or *unity-gain frequency*, which determines at how high a frequency ω the small-signal current gain $|i_{out}/i_{in}| = g_m/(\omega C_G)$ drops to unity. Using the chain rule:

$$g_{\mathrm{m}} = \frac{\partial I_{\mathrm{D,sat}}}{\partial Q_{\mathrm{G}}} \frac{\partial Q_{\mathrm{G}}}{\partial V_{\mathrm{G}}} = \omega_T C_{\mathrm{G}} \tag{3.49}$$

where C_{G} is the oxide capacitance of the device, $C_{\mathrm{G}} = \partial Q_{\mathrm{G}}/\partial V_{\mathrm{G}}|V_{\mathrm{D}}$ with Q_{G} = the charge on the gate electrode. The frequency ω_{T} is a measure of the small-signal, high-frequency speed of the device, neglecting parasitic resistances. Using Equation (3.43) in Equation (3.49) we find that the transition frequency also is related to the transit time L/v_{sat} of Equation (3.45), so that the digital and small-signal circuit speeds are related to this parameter.

Output Resistance and Drain Conductance

For small-signal circuits the output resistance r_{o} of the MOSFET (Sedra and Smith, 1991) is important in limiting the gain of amplifiers. This resistance is related to the small-signal drain conductance g_{D} in saturation by

$$r_{\mathrm{o}} = \frac{1}{g_{\mathrm{D}}} = \left.\frac{\partial V_{\mathrm{D}}}{\partial I_{\mathrm{D,sat}}}\right|_{V_{\mathrm{G}}=\mathrm{const}} \tag{3.50}$$

If the MOSFET is used alone as a simple amplifier with a load line set by a resistor R_{L}, the gain becomes

$$\left|\frac{v_{\mathrm{o}}}{v_{\mathrm{in}}}\right| = g_{\mathrm{m}} \frac{R_{\mathrm{L}} r_{\mathrm{o}}}{R_{\mathrm{L}} + r_{\mathrm{o}}} \leqslant g_{\mathrm{m}} R_{\mathrm{L}} \tag{3.51}$$

showing how gain is reduced if r_{o} is reduced to a value approaching R_{L}.

As devices are miniaturized, r_{o} is decreased, g_{D} increased, due to several factors. At moderate drain biases, the main factor is channel-length modulation, the reduction of the channel length with increasing drain voltage that results when the depletion region around the drain expands toward the source, causing L to become drain-bias dependent. At larger drain biases, a second factor is drain control of the inversion-layer charge density, which can compete with gate control in short devices. This is the same mechanism discussed later in the context of subthreshold behavior. At rather high drain bias, carrier multiplication further lowers r_{o}.

In a digital inverter, a lower r_{o} widens the voltage swing needed to cause a transition in output voltage. This widening increases power loss due to current spiking during the transition and reduces noise margins (Annaratone, 1986). It is not, however, a first-order concern in device miniaturization for digital applications. Because small-signal circuits are more sensitive to r_{o} than digital circuits, MOSFETs designed for small-signal applications are not usually made as small as those for digital applications.

Limitations upon Miniaturization

A major factor in the success of the MOSFET has been its compatibility with processing useful down to very small dimensions. In this section some of the limits that must be considered in miniaturization are outlined (Brews, 1990).

It should be noted that scaling of MOSFET technology has continued at an aggressive pace for approximately three decades and is projected to continue well into the future. Today channel lengths (source-to-drain spacings) of 50 nm are manufacturable and further reduction to less than 15 nm is projected by 2015. At each generation the limitations discussed in this section, as well as others, have been addressed with innovative engineering solutions. These areas remain the subject of active research and the reader is referred to the current literature, and the latest edition of *International Technology Roadmap for Semiconductors* for the most current information (Semiconductor Industry Association, 2003).

Subthreshold Control

When a MOSFET is in the "off" condition, that is, when the MOSFET is in *subthreshold*, the off current drawn with the drain at supply voltage must not be too large in order to avoid power consumption and discharge of ostensibly isolated nodes (Shoji, 1988). In small devices, however, the source and drain are closely spaced, so there exists a danger of direct interaction of the drain with the source, rather than an interaction mediated by the gate and channel. In an extreme case, the drain may draw current directly from the source, even though the gate is off (*punchthrough*). A less extreme but also undesirable case occurs when the drain and gate jointly control the carrier density in the channel (*drain-induced barrier lowering*, or drain control of threshold voltage). In such a case, the on–off behavior of the MOSFET is not controlled by the gate alone and switching can occur over a range of gate voltages dependent on the drain voltage. Reliable circuit design under these circumstances is very complicated and testing for design errors is prohibitive. Hence, in designing MOSFETs, a drain-bias independent subthreshold behavior is necessary.

A measure of the range of influence of the source and drain is the depletion-layer width of the associated pn-junctions. The depletion layer of such a junction is the region in which all carriers have been depleted, or pushed away, due to the potential drop across the junction. This potential drop includes the applied bias across the junction and a spontaneous *built-in* potential drop induced by spontaneous charge exchange when p and n regions are brought into contact. The depletion-layer width W of an abrupt junction is related to potential drop V and dopant-ion concentration/unit volume N by:

$$W = \left(\frac{2\varepsilon_S V}{qN}\right)^{1/2} \tag{3.52}$$

To avoid subthreshold problems, a commonly used rule of thumb is to make sure that the channel length is longer than a minimum length L_{min} related to the junction depth r_j, the oxide thickness d, and the depletion-layer widths W_S and W_D of the source and drain, respectively, by (Brews, 1990):

$$L_{min} = A[r_j d(W_S + W_D)^2]^{1/3} \tag{3.53}$$

where the empirical constant $A = 0.88\ \text{nm}^{-1/3}$ if r_j, W_S, and W_D are in micrometers and d is in nanometers.

Equation (3.53) shows that smaller devices require shallower junctions (smaller r_j), thinner oxides (smaller d), or smaller depletion-layer widths (smaller voltage levels or heavier doping). These requirements introduce side effects that are difficult to control. For example, if the oxide is made thinner while voltages are not reduced proportionately, then oxide fields increase, requiring better oxides. If junction depths are reduced, better control of processing is required and the junction resistance is increased due to smaller cross-sections. To control this resistance, various *self-aligned contact* schemes have been developed to bring the source and drain contacts closer to the gate (Einspruch and Gildenblat, 1989; Brews, 1990), reducing the resistance of these connections. If depletion-layer widths are reduced by increasing the dopant-ion density the *driving ability* of the MOSFET suffers because the threshold voltage increases. That is, Q_D increases in Equation (3.40), reducing $V_G - V_{TH}$. Thus, for devices that are not velocity-saturated, that is, devices where $V_G/L \lesssim F_{sat}$, increasing V_{TH} results in slower circuits.

As secondary consequences of increasing dopant-ion density, channel conductance is further reduced due to the combined effects of increased scattering of electrons from the dopant atoms and increased oxide fields that pin carriers in the inversion layer closer to the insulator–semiconductor interface, increasing scattering at the interface. These effects also reduce driving ability, although for shorter devices they are important only in the linear region (that is, below saturation), assuming that mobility μ is more strongly affected than saturation velocity υ_{sat}.

Hot-Electron Effects

Another limit upon how small a MOSFET can be made is a direct result of the larger fields in small devices. Let us digress to consider why proportionately larger voltages, and thus larger fields, are used in smaller devices.

First, according to Equation (3.45), τ_G is shortened if voltages are increased, at least so long as $V_G/L \lesssim F_{sat}$ 5×10^4V/cm. If τ_G is shortened this way, then so are τ_D and τ_{INT} (Equation (3.46) and Equation (3.47)). Thus, faster response is gained by increasing voltages into the velocity saturation region. Second, the control of fabrication tolerances for smaller devices has not improved proportionately as L has shrunk, so there is a larger percentage variation in device parameters with smaller devices. Thus, disproportionately larger voltages are needed to ensure that all devices operate in the circuit to overcome this increased fabrication control "noise." Thus, to increase speed and to cope with fabrication control variations, fields go up in smaller devices.

As a result of these larger fields along the channel direction, a small fraction of the channel carriers have enough energy to enter the insulating layer near the drain. In silicon-based *p*-channel MOSFETs, energetic holes can become trapped in the oxide, leading to a positive oxide charge near the drain that reduces the strength of the channel, degrading device behavior. In *n*-channel MOSFETs, energetic electrons entering the oxide create interface traps and oxide wear-out, eventually leading to gate-to-drain shorts (Pimbley et al., 1989).

To cope with these problems "drain-engineering" has been tried, the most common solution being the *lightly doped drain* (Einspruch and Gildenblat, 1989; Pimbley et al., 1989; Chen, 1990). In this design, a lightly doped extension of the drain is inserted between the channel and the drain proper. To keep the field moderate and reduce any peaks in the field, the lightly doped drain extension is designed to spread the drain-to-channel voltage drop as evenly as possible. The aim is to smooth out the field at a value close to F_{sat} so that energetic carriers are kept to a minimum. The expense of this solution is an increase in drain resistance and a decreased gain. To increase packing density, this lightly doped drain extension can be stacked vertically alongside the gate, rather than laterally under the gate, to control the overall device area.

Thin Oxides

According to Equation (3.53), thinner oxides allow shorter devices and therefore higher packing densities for devices. In addition, driving ability is increased, shortening response times for capacitive loads, and output resistance and transconductance are increased. There are some basic limitations upon how thin the oxide can be made. For instance, there is a maximum oxide field that the insulator can withstand. It is thought that the intrinsic breakdown voltage of SiO_2 is of the order of 10^7 V/cm, a field that can support $\approx 2 \times 10^{13}$ charges/cm^2, a large enough value to make this field limitation secondary. Of more serious concern are factors such as defect density and precise control of oxide thickness.

Some improvement has been made in gate dielectric properties by adding nitrogen to silicon dioxide to produce a nitrided silicon dioxide. This has the effect of improving breakdown and leakage characteristics, reducing defect density, and somewhat increasing the dielectric constant of the gate insulator. Looking forward, more exotic "high-κ" materials are being considered for use as gate dielectrics which, by virtue of higher dielectric constants, offer the promise of performance equivalent to thinner silicon dioxide with physically thicker films.

Dopant-Ion Control

As devices are made smaller, the precise positioning of dopant inside the device is critical. At high temperatures during processing, dopant ions can move. For example, source and drain dopants can enter the channel region, causing position dependence of threshold voltage. Similar problems occur in isolation structures that separate one device from another (Einspruch and Gildenblat, 1989; Pimbley et al., 1989; Wolf, 1995).

To control these thermal effects, process sequences are carefully designed to limit high-temperature steps. This design effort is shortened and improved by the use of computer modeling of the processes. Dopant-ion movement is complex, however, and its theory is made more difficult by the growing trend to use *rapid thermal processing* that involves short-time heat treatments. As a result, dopant response is not steady state, but transient. Computer models of transient response are primitive, forcing further advances in small-device design to be more empirical.

As device dimensions are further reduced, the possibility exists that the smallest of devices may have only of the order of a few hundred or fewer dopant ions within the active channel region. This means that a variation of dopant ions that is small in absolute number may imply a significant difference in doping density between otherwise identical devices. Under such conditions control of the statistical variation of device parameters is extremely challenging.

Other Limitations

Besides limitations directly related to the MOSFET, there are some broader difficulties in using MOSFETs of smaller dimension in chips involving even greater numbers of devices. Already mentioned is the increased delay due to interconnections that are lengthening due to increasing chip area and increasing complexity of connection. The capacitive loading of MOSFETs that must drive signals down these lines can slow circuit response, requiring extra circuitry to compensate. Another limitation is the need to isolate devices from each other (Einspruch and Gildenblat, 1989; Pimbley et al., 1989; Brews, 1990; Chen, 1990; Wolf, 1995), so their actions remain uncoupled by parasitics. One method to address device isolation at decreasing dimensions is *shallow-trench isolation*, which has been developed to replace local oxidation isolation methods. As isolation structures are reduced in size to increase device densities, new parasitics are discovered. A developing solution to this problem is the manufacture of circuits on insulating substrates, silicon-on-insulator technology (Colinge, 1991).

Defining Terms

Channel: The conducting region in a MOSFET between source and drain. In an *enhancement*-mode (or normally off) MOSFET, the channel is an inversion layer formed by attraction of minority carriers toward the gate. These carriers form a thin conducting layer that is prevented from reaching the gate by a thin *gate-oxide* isulating layer when the gate bias exceeds *threshold*. In a *buried-channel*, or *depletion-mode* (or normally on) MOSFET, the channel is present even at zero gate bias, and the gate serves to increase the channel resistance when its bias is nonzero. Thus, this device is based on majority-carrier modulation, like a MESFET.

Gate: The control electrode of a MOSFET. The voltage on the gate capacitively modulates the resistance of the connecting channel between the source and drain.

Source, drain: The two output contacts of a MOSFET, usually formed as pn-junctions with the *substrate* or *body* of the device.

Strong inversion: The range of gate biases corresponding to the "on" condition of the MOSFET. At a fixed gate bias in this region, for low drain-to-source biases the MOSFET behaves as a simple gate-controlled resistor. At larger drain biases, the channel resistance can increase with drain bias, even to the point that the current *saturates*, or becomes independent of drain bias.

Substrate or body: The portion of the MOSFET that lies between the *source* and *drain*, and under the *gate*. The gate is separated from the body by a thin *gate insulator*, usually silicon dioxide. The gate modulates the conductivity of the body, providing a gate-controlled resistance between the source and drain. The body is sometimes dc-biased to adjust overall circuit operation. In some circuits the body voltage can swing up and down as a result of input signals, leading to "body-effect" or "back-gate bias" effects that must be controlled for reliable circuit response.

Subthreshold: The range of gate biases corresponding to the "off" condition of the MOSFET. In this regime the MOSFET is not perfectly "off" but conducts a leakage current that must be controlled to avoid circuit errors and power consumption.

Threshold: The gate bias of a MOSFET that marks the boundary between "on" and "off" conditions.

References

The following references are not to the original sources of the ideas discussed in this section, but have been chosen to be generally useful to the reader.

M. Annaratone, *Digital CMOS Circuit Design*, Boston, Mass.: Kluwer Academic, 1986.

J.R. Brews, "Physics of the MOS transistor" in *Applied Solid State Science*, Suppl. 2A, D. Kahng, Ed., New York: Academic, 1981.

J.R. Brews, "The submicron MOSFET" in *High-Speed Semiconductor Devices*, S. M. Sze, Ed., New York: Wiley, 1990, pp. 139–210.

J.Y. Chen, *CMOS Devices and Technology for VLSI*, Englewood Cliffs, N.J.: Prentice-Hall, 1990.

J.-P. Colinge, *Silicon-on-Insulator Technology: Materials to VLSI*, Boston, Mass.: Kluwer Academic, 1991.

N.G. Einspruch and G. Sh. Gildenblat, Eds., "VLSI Microstructure Science", vol. 18, *Advanced MOS Device Physics*, New York: Academic, 1989.

P.R. Gray, P.J. Hurst, S.H. Lewis, and R.G. Meyer, *Analysis and Design of Analog Integrated Circuits*, 4th ed., New York: Wiley, 2001.

H. Haznedar, *Digital Microelectronics*, Redwood City, Calif.: Benjamin-Cummings, 1991.

Chenming Hu, et al., *BSIM4.4.0 MOSFET Model – User's Manual*, University of California, Berkeley: Department of Electrical Engineering and Computer Sciences, 2004. Available at: http://www-device.eecs.berkeley.edu/~bsim3/bsim4.html.

M.A. Hollis and R.A. Murphy, "Homogeneous field-effect transistors," in *High-Speed Semiconductor Devices*, S.M. Sze, Ed., New York: Wiley, 1990, pp. 211–282.

M. Lundstrom, *Fundamentals of Carrier Transport*, Cambridge: Cambridge University Press, 2000.

N.R. Malik, *Electronic Circuits: Analysis, Simulation, and Design*, Englewood Cliffs, N.J.: Prentice-Hall, 1995.

E.H. Nicollian and J.R. Brews, *MOS Physics and Technology*, New York: Wiley, 1982, Chapter 1.

S.J. Pearton and N.J. Shaw, "Heterostructure field-effect transistors" in *High-Speed Semiconductor Devices*, S.M. Sze, Ed., New York: Wiley, 1990, pp. 283–334.

R.F. Pierret, *Modular Series on Solid State Devices, Field Effect Devices*, 2nd ed., vol. 4, Reading, Mass.: Addison-Wesley, 1990.

J.M. Pimbley, M. Ghezzo, H.G. Parks, and D.M. Brown, *VLSI Electronics Microstructure Science, Advanced CMOS Process Technology*, vol. 19, N. G. Einspruch, Ed., New York: Academic, 1989.

S.S. Sedra and K.C. Smith, *Microelectronic Circuits*, 3rd ed., Philadelphia, Pa.: Saunders, 1991.

Semiconductor Industry Association, *International Technology Roadmap for Semiconductors*, 2003 Edition, 2003.

B.G. Streetman and S. Banerjee, *Solid State Electronic Devices*, Upper Saddle River, N.J.: Prentice Hall, 2000.

M. Shoji, *CMOS Digital Circuit Technology*, Englewood Cliffs, N.J.: Prentice-Hall, 1988.

Y. Tsividis, *Operation and Modeling of the Mos Transistor*, 2nd ed., Oxford: Oxford University Press, 2003.

S. Wolf, *Silicon Processing for the VLSI era: volume 3 — the Submicron MOSFET*, Sunset Beach, Calif.: Lattice Press, 1995.

Further Information

The references given in this section have been chosen to provide more detail than is possible to provide in the limited space of this section. In particular, Annaratone (1986) and Shoji (1988) provide much more detail about device and circuit behavior. Gray et al (2001) provides extensive detail on relevant parameters and the use of MOSFETs in analog circuits. Chen (1990), Pimbley et al. (1989), and Wolf (1995) provide many technological details of processing and its device impact. Semiconductor Industry Association (2003) contains many details regarding the latest technology advances, challenges, and future roadmap. Haznedar (1991), Sedra and Smith (1991), and Malik (1995) provide much information about circuits. Brews (1981), Pierret (1990), and Streetman and Banerjee (2000) provide good discussions of the derivation of the device current-voltage curves and device behavior in all bias regions. Lundstrom (2000) discusses device operation from a different perspective and provides many advanced details on the properties of carrier transport in semiconductor devices. Readers interested in MOSFET models used in circuit simulation programs may wish to consult Hu (2004).

3.4 Single-Electron Transistors

James E. Morris

Moore's Law has been used for many years to predict the eventual demise of complementary metal-oxide-semiconductor (CMOS) technology [1] as shrinking devices either become too small to function or too small to make. As MOS gate thicknesses shrink, electron tunneling and leakage currents due to **electron tunneling** increase, forcing the development of high dielectric constant materials to restore gate thickness. The end of functional optical lithography [2] was expected when device dimensions reached optical wavelengths, but was averted by high-tech tricks like phase contrast masking. The industry has proved most inventive in delaying the end, but eventually a new technology is still expected to take over. Single-electron devices, conceptually similar to MOS devices, provide one possible solution. The key requirements for the adoption of any of the competing successor technologies will be size, speed, and price, with competitive pricing requiring parallel manufacturing techniques.

One of the most exciting recent developments in microelectronics has been the development of single-electron transistor (SET) logic devices from the basic concept of the two-terminal Coulomb block/blockade (CB). The CB consists of a small (metal) **quantum dot** separated from source and drain electrodes by tunneling junctions. The SET structure adds a gate electrode. In the CB, current cannot flow until the metal dot of radius r is charged, which requires energy:

$$\Delta E \geqslant q^2/4\pi\varepsilon r \tag{3.54}$$

in a medium of dielectric constant ε, where q is the electronic charge. At absolute zero, the applied field provides the electrostatic energy, i.e., $V \geqslant \Delta E/q$ for current to flow. The fundamental SET idea is that the charge can be placed on the island by a third, gate-control electrode, permitting the flow of low voltage "on" current. This section develops the basic principle of operation of the SET in terms of electron tunneling and electrostatic charging of quantum dot capacitances, the CB, and the "electron box." SET operation is considered in terms of stability and thermal effects, and the latter's effect on practical device design. The section concludes with a discussion of fabrication issues.

Electron Tunneling

CB and SET operation is based on electron tunneling between the quantum-dot island and the electrodes. Figure 3.45(a) contrasts the normal **thermionic emission** (TE) process over the potential barrier through the conduction band of the insulator with electron tunneling (ET) through it. The tunneling transition rate depends upon the product of the probability $F_1(E)$ that electron states at the electron energy level E are occupied in the source electrode and the probability $[1 - F_2(E)]$ that the receiving electrode's energy states are empty, i.e., the net electron transfer rate $T_{1\rightarrow 2} - T_{2\rightarrow 1}$ is proportional to:

$$\begin{aligned} &D_{12}(E_1)F_1(E_1)[1 - F_2(E_2)] - D_{21}(E_2)F_2(E_2)[1 - F_1(E_1)] \\ &= D_{12}(E_1)F_1(E_1) - D_{21}(E_2)F_2(E_2) \end{aligned} \tag{3.55}$$

where $E_2 = E_1 + qV$ at bias V, and D_{i2} is the tunneling probability from 1 to 2. For tunneling, the transition probability is exponentially dependent on the barrier width, s, and height, Φ_0 (which equals the **work function** in vacuum). For a rectangular barrier (e.g., in Figure 3.45(a)) the zero-field probability is [3]:

$$D_{12} = \left[\cosh^2 \zeta s + \frac{1}{4}\frac{(1 - 2E/\Phi_0)^2}{(E/\Phi_0)(1 - E/\Phi_0)} \sinh^2 \zeta s \right]^{-1} \tag{3.56}$$

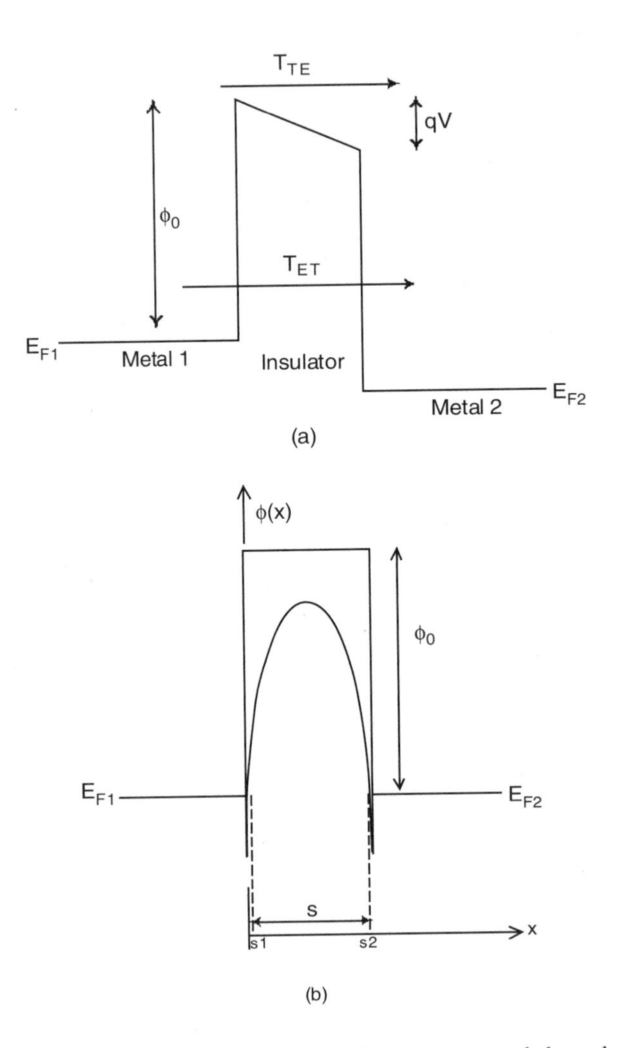

FIGURE 3.45 Electron energy diagrams of the potential barrier between two metal electrodes separated by an insulator (or vacuum) with applied bias V. (a) Thermionic emission *over* the barrier (e.g., through the conduction band of the insulator) requires high energy; tunneling *through* the barrier is possible at lower energies near the Fermi Level, E_F (where there are more electrons), if the barrier is sufficiently thin. (b) Electron image effects in the electrodes convert the rectangular barrier to a parabolic function.

where $\zeta = (2\pi/h)[2m(\Phi_0 - E)]^{1/2}$, h is Planck's constant, and m is the electron mass, which reduces to

$$D_{12} \approx \frac{1}{4}\exp[-(4\pi/h)\{2m(\Phi_0 - E)\}^{1/2}s] \tag{3.57}$$

if $\Phi_0 \sim 2E$ and $\zeta s >> 0$. The more realistic barrier shape is parabolic (Figure 3.45(b)) due to **image charge effects** [4], and the **WKB approximation** gives the similar result [3]:

$$D_{12} = \exp[(-4\pi/h)(2m\Phi_{\text{eff}})^{1/2}s]$$

with $\Phi_{\text{eff}}(E) = s^{-1}\int_{s_1}^{s_2}(\Phi(x) - E)\text{d}x$. The exponential dependence on gap width s means that electron tunneling is only observable for s~nanometers (nm).

Electrostatic Charging Energy

The metal–insulator–metal system described above also forms a capacitor, and the transfer of an electron by tunneling changes the system energy by

$$\Delta E = q^2/2C \tag{3.58}$$

where $C = \varepsilon A/s$ for a parallel plate capacitor of plate area A. If one of these "plates" is actually a co-planar electrode, and the other a small sphere of radius r, the charging energy and capacitance can be estimated as the classical results for a spherically symmetric system, i.e., (more completely than in Equation (3.54)), [5]:

$$\Delta E = (q^2/4\pi\varepsilon)(r^{-1} - (r+s)^{-1}) \tag{3.59}$$

or

$$C = 2\pi\varepsilon r(1 + r/s) \tag{3.60}$$

This energy becomes significant, i.e., $\Delta E >> kT$, where T is the absolute temperature, as the island dimension decreases. ΔE is plotted as a function of island size in Figure 3.46, for the cases of $\varepsilon_r = 1, 2, 4$ and $s = 1, 2, 5$ nm, and ∞. The thermal energy is also indicated for $T = 4, 77$, and $300\ K$.

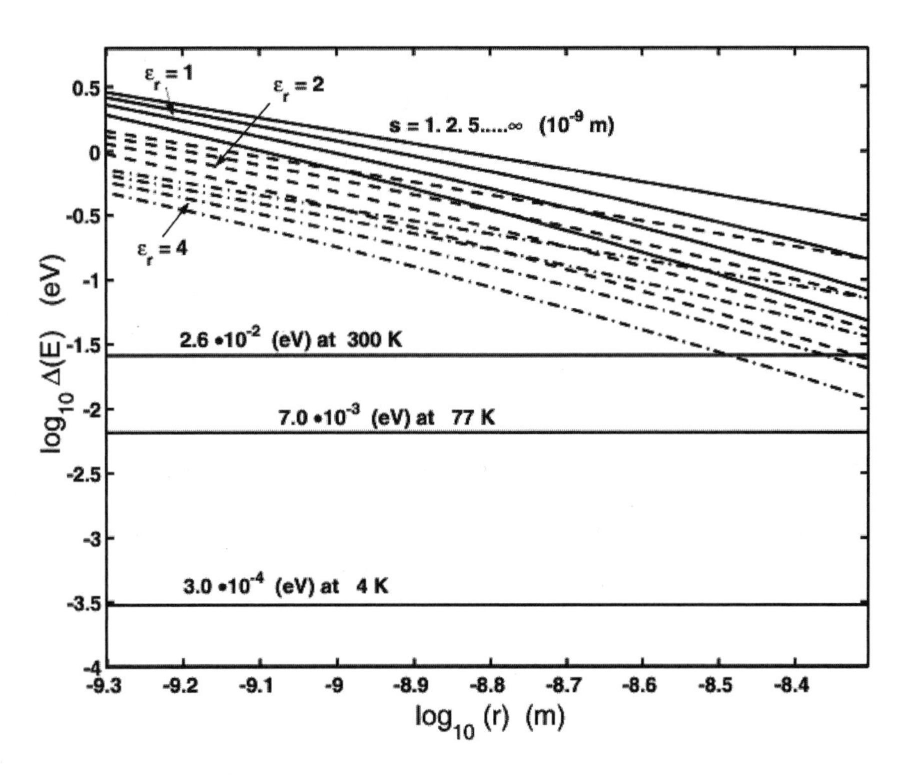

FIGURE 3.46 Electrostatic charging energy ΔE as a function of island radius, r, gap width, $s = 1, 2, 5$, and ∞ nm, and relative dielectric constant $\varepsilon_r = 1, 2$, and 4. (Dielectric constant $\varepsilon = \varepsilon_r \varepsilon_0$ where ε_0 is the dielectric constant of free space.) The three clusters of lines correspond to ε_r values, with $\varepsilon_r = 1$ at the top, and within each cluster, the four lines correspond to s values, with $s = 1$ nm at the bottom.

The number N of an array of N_∞ identical islands that are randomly charged by thermal energy at T is given by the Boltzmann distribution

$$N = N_\infty \exp(-\Delta E/kT) \tag{3.61}$$

Conversely, an individual island would be randomly charged a similar proportion of the time. In practice the islands will not be spherical, and ΔE requires further modification [6].

Heisenberg's uncertainty principle requires that

$$\Delta E \cdot \Delta t \geqslant h \tag{3.62}$$

For $\Delta E = q^2/2C$ and $\Delta t = R_t C$, where R_t is the effective tunneling resistance, this gives:

$$R_t \geqslant 2h/q^2 \tag{3.63}$$

i.e., the tunneling resistance must be greater than the "quantum resistance" of around 52 kΩ in order for the charge to be considered as localized on the quantum dot [7]. (Lower values permit "co-tunneling," i.e., an electron tunnels off the island at the same time as one tunnels on, maintaining charge neutrality of the island.)

Coulomb Block

The charging of the small island requires energy, and at absolute zero, with electrons confined to the Fermi level and below, electron tunneling to the island cannot occur (Figure 3.47) until voltage $V_{ds} \geqslant 2(\Delta E/2q)$, as shown in Figure 3.48(b). For island charge $Q = Q_s - Q_d$:

$$V_s = (C_d V_{ds} + Q)/(C_s + C_d) \tag{3.64}$$

$$V_d = (C_s V_{ds} - Q)/(C_s + C_d) \tag{3.65}$$

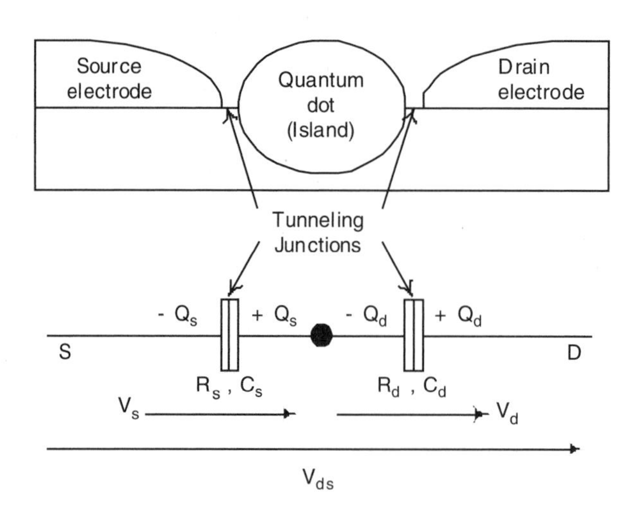

FIGURE 3.47 Coulomb block: physical structure and equivalent circuit, with tunnel junction symbols.

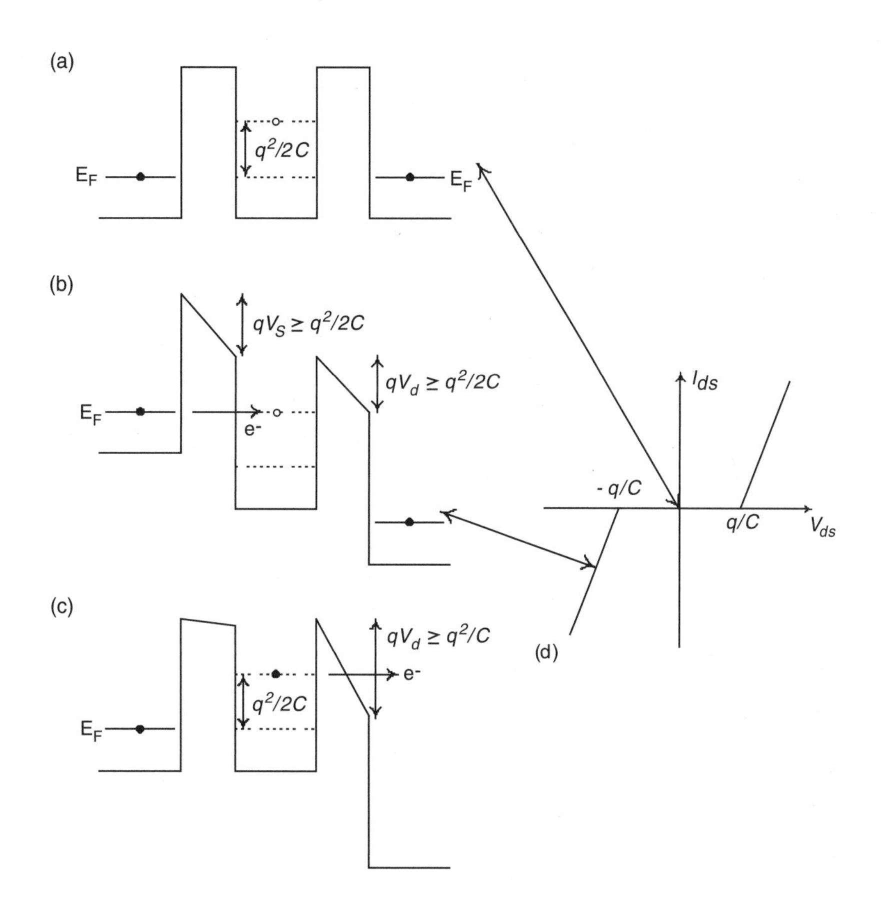

FIGURE 3.48 Electron energy diagram for the double tunnel junction Coulomb block. (a) Zero bias, showing charging energy ΔE relative to the barrier height. (b) An electron can tunnel to the island when $qV > \Delta E$. (Assumes equal gap widths.) (c) With the island charged, E_F is raised ΔE, and the electron tunnels on with high probability. (d) CB characteristic at absolute zero, showing zero current below $2\Delta E/q$.

and for identical tunnel junctions $C_s = C_d = C/2$, so the bias is equally distributed for $Q = 0$. Once an excess electron is resident on the island, it raises the Fermi level by ΔE, and it can clearly tunnel on to the second electrode without energy constraint, (Figure 3.48(c)), i.e., a current flows. The low T current–voltage (I–V) characteristic of the CB is shown in Figure 3.48(d).

At finite temperatures, random thermal charging effects will lead to finite subthreshold currents. As V_{ds} increases, ΔE is reduced by the field, and the current increases as the thermal charging probability increases.

Single-Electron Box

The single-electron box (Figure 3.49) looks similar to the CB, but one of the gaps is sufficiently thick to preclude the possibility of tunneling, i.e., there is one tunnel junction and one capacitor to the structure. The application of sufficient voltage again induces charging of the island when the tunnel junction voltage, V_{XS}, exceeds $\Delta E/q$, i.e., when V_{gs} exceeds $(1 + C_s/C_g)\Delta E/q$ (Figure 3.50(a)). For $C_g << C_s$, almost all V_{gs} appears across the capacitor, and the electron transfer takes place when $V_{gs} \geqslant q^2/2C \approx q^2/2C_g$. However, unlike the CB case, once the island is charged, the electron cannot tunnel on through the capacitor to discharge it. The "stability" diagram for multiple-electron charging is shown in Figure 3.50(b).

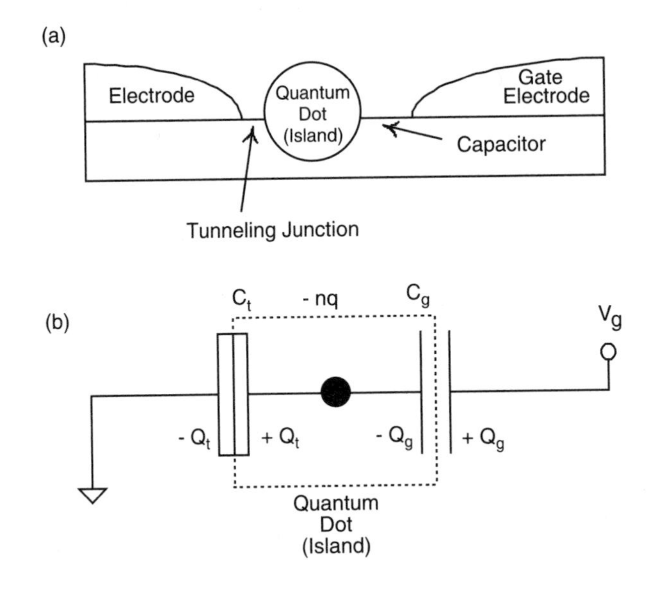

FIGURE 3.49 Single-electron box: physical structure and equivalent circuit [7].

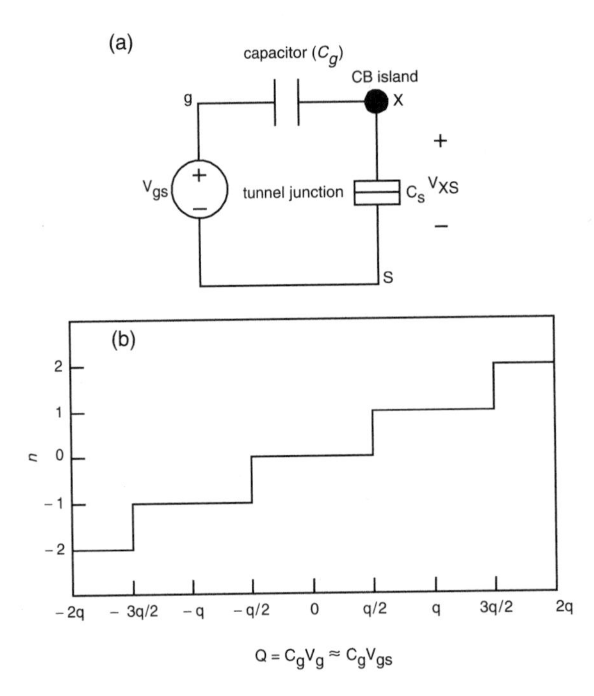

FIGURE 3.50 Single-electron box: (a) circuit for analysis and (b) stability diagram.

Single-Electron Transistor (SET)

The SET combines the CB source to drain "channel" with the single-electron box gate structure (Figure 3.51). Here:

$$V_s = (C_d V_{ds} + C_g V_{gs} + Q_{net})/(C_d + C_s + C_g) \tag{3.66}$$

$$V_d = ((C_s + C_g)V_{ds} - C_g V_{gs} - Q_{net})/(C_d + C_s + C_g) \tag{3.67}$$

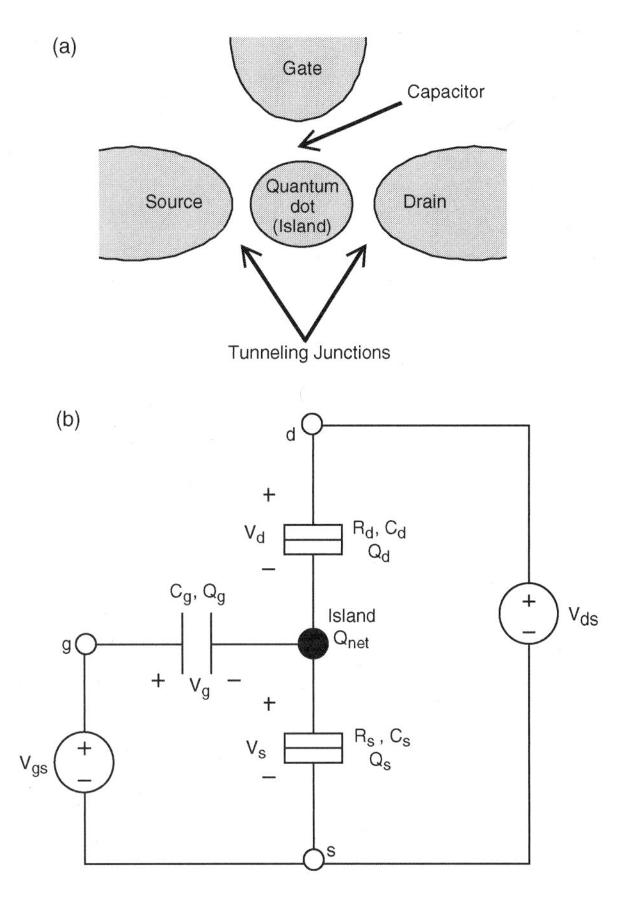

FIGURE 3.51 Single-electron transistor: (a) physical structure and (b) circuit for analysis.

Current can flow if V_s or V_d exceeds $Q/2C$ where $C = C_d + C_s + C_g$ for any $Q = nq$. This gives rise to the stability diagram of Figure 3.52, where current can flow for conditions defined as outside the n-periodic diamond shapes in Figure 3.52(a), as shown in Figure 3.52(b) for the application of finite V_{ds} [8].

Thermal Effects

An example of a basic SET structure is shown in Figure 3.53. The source and drain electrodes are shown at high magnification, where random islands are seen as white dots. The lower magnification picture shows the gate configuration. The 77 K characteristics shown in Figure 3.54 are actually for a different but very similar device. The thermal charging effects are evident, as is the turn-on threshold somewhere between $0.5\ \mathrm{V} < V_t < 1.0\ \mathrm{V}$. To minimize thermal charging effects, the charging capacitance needs to go down, either by achieving smaller island sizes, or by using a series combination of islands (so $C_{series} = C_{islands}/n$ for n islands), or both.

The series chain of islands has been used by Yano et al. in the fabrication of room-temperature memory circuits [10]. The islands are established on silicon crystals in a polycrystalline film thinned to widen grain boundaries. In similar 2D island arrays of discontinuous metal thin films (DTF), it has been shown that dc (i.e., on-state) conduction requires the establishment of space charge distribution, specifically with negatively charged islands at the positive electrode [11]. This space distribution takes finite time to establish, and the simple SET concept may require a high-frequency refresh mode of operation.

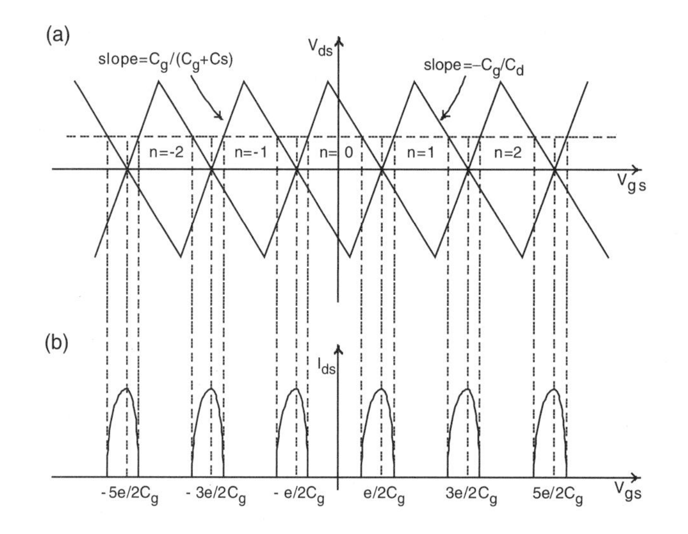

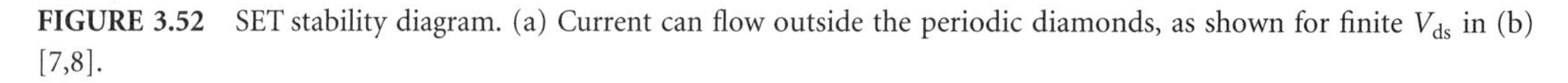
FIGURE 3.52 SET stability diagram. (a) Current can flow outside the periodic diamonds, as shown for finite V_{ds} in (b) [7,8].

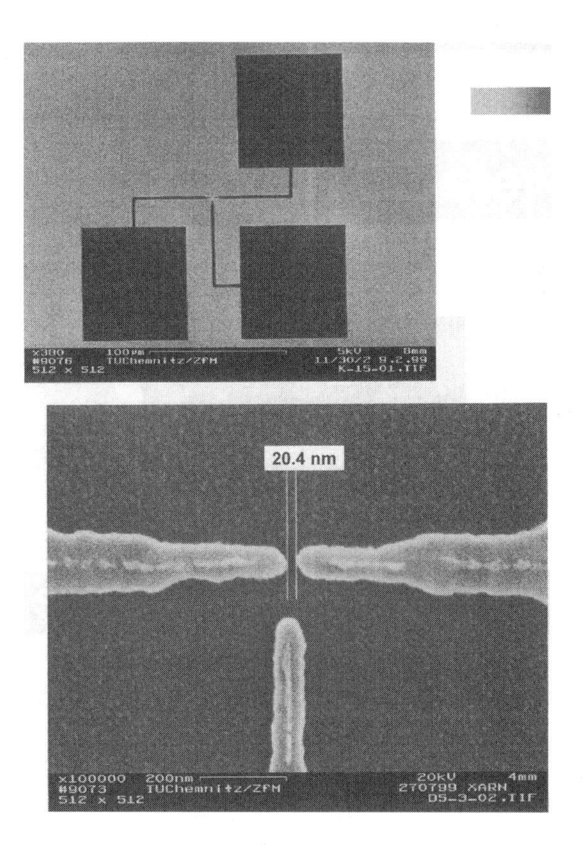

FIGURE 3.53 Electron micrographs of the SET structure [9].

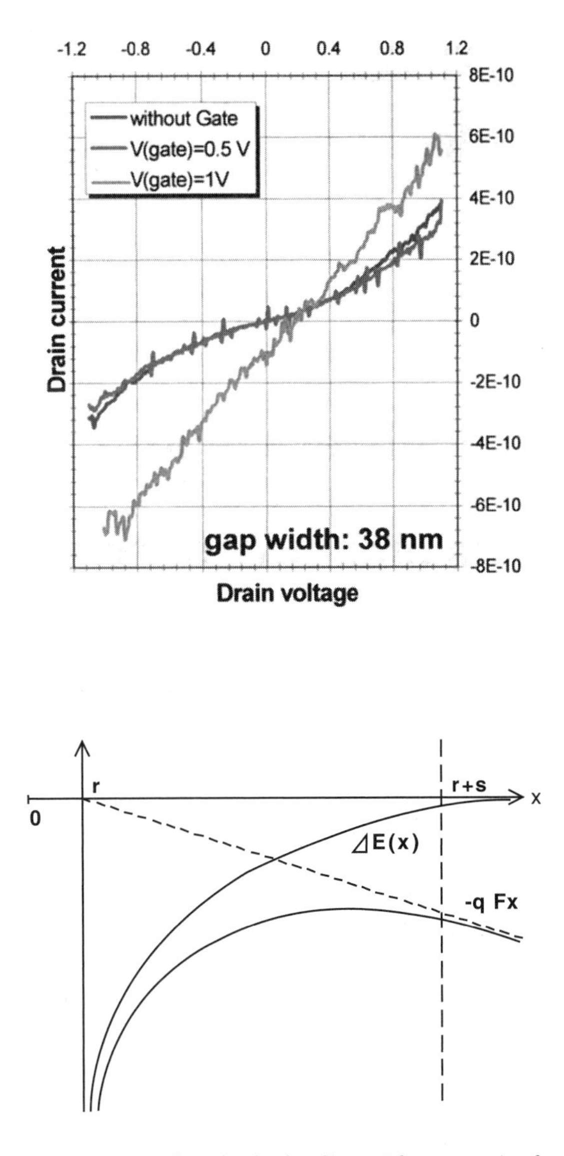

FIGURE 3.55 Electrostatic charging energy for island of radius r. The composite function $\Delta E(x) - qFx$ develops a maximum at high fields, where $\Delta E(x) = q^2/4\pi\varepsilon x$.

DTF electrical conductance may be expressed as

$$\sigma = \sigma_0 \exp(-\Delta E/kT), \tag{3.68}$$

where ΔE is also the electrostatic charging energy of the metal islands. At finite T, the thermal energy and applied field, F, both contribute to island charging, and the form of ΔE needs further consideration. In a treatment that exactly parallels **Schottky effect** theory (Figure 3.55) [12]:

$$\Delta E = (q^2/4\pi\varepsilon)(r^{-1} - (r + s)^{-1}) - (2r + s)qF \tag{3.69}$$

at low electric field

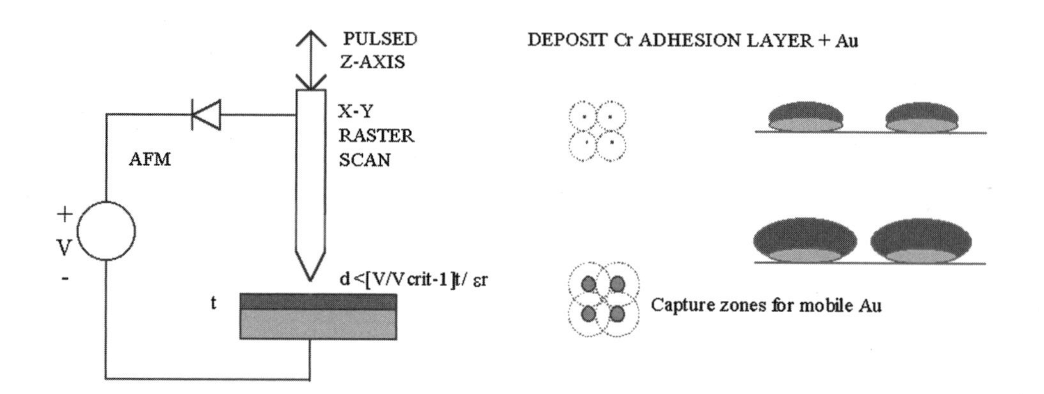

FIGURE 3.56 (a) AFM-piezoelectric drive system and (b) deposition model [23].

$$F < (q/4\pi\varepsilon)(r+s)^{-2} \tag{3.70}$$

and

$$\Delta E = (q^2/4\pi\varepsilon)r^{-1} + (r/s)(2r+s)qF - (q^2/4\pi\varepsilon s)^{1/2}((2r+s)qF)^{1/2} \tag{3.71}$$

at high field. $\delta E \rightarrow 0$ at

$$qV = (1 + s/r)\Delta E_{V=0} \tag{3.72}$$

which is the standard CB result. The nonlinear subthreshold CB characteristics of $kT \cdot \log(\sigma/\sigma_0)$ should follow the predicted variation of ΔE with F.

Fabrication

The basic concept has evolved in many different directions. One approach has been to adopt silicon processing technology, and this has led to a number of modified configurations [13] that look little like the conceptual device described here and shown above (Figure 3.53). Here, gold islands have been deposited upon pre-deposited, etched gold electrodes on SiO_2. The electrode **capture distance** provides some consistency to structures obtained with this approach, but an alternative approach to defining regular stable structures in an atomic force microscope (AFM) is under development (Figure 3.56). In this approach, Cr ions are deposited on the insulating substrate by field emission from the AFT tip. The Cr forms stable bonding nuclei for the subsequent deposition of gold (or other noble metal) islands and electrodes. The AFM can also be used to define SET patterns by field oxidation of metal thin films [14–16]. The AFM technique is clearly unsuitable for wide area parallel device fabrication, but may be useful to define initial structures for subsequent imprint processing of polymers. In all cases, the real challenge is to develop interconnect technologies on a scale to match the SET dimensions.

Circuits

One of the simplest circuits is the inverter shown in Figure 3.57, which clearly shows some conceptual similarity to the CMOS inverter. An estimate of the high-frequency performance of SET circuits can be obtained from the RC product of the minimum tunneling resistance (the quantum resistance of, say, 50 KΩ (Equation (3.63)), and the SET capacitance [17] of around 0.1 aF from Equation (3.60) for $r = s = 1$ nm, $\varepsilon_r = 1$. These figures suggest an upper limit of 30 THz. With such high potential for device performance, the limiting factor for circuit speed will be the interconnect technology (and background charge effects which have been ignored in this idealized introduction).

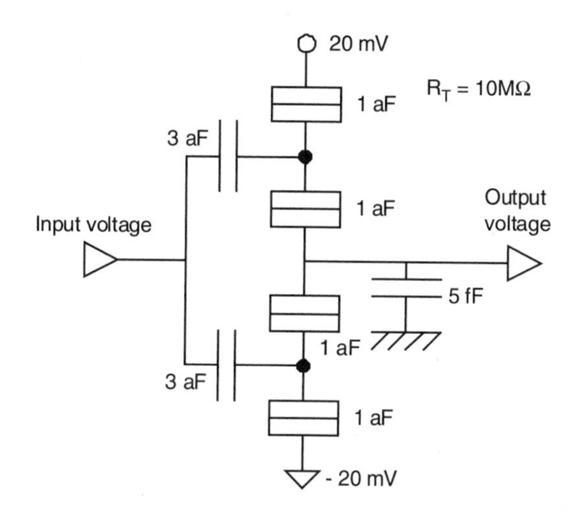

FIGURE 3.57 Simple SET inverter [7].

Defining Terms

Moore's law: Notes that integrated circuit (IC) component density doubles about every 18 months. However, the title is used generically for a wide variety of related linear semi-log plots of IC parameters vs. time, e.g., device size, number of transistors per chip, etc.

Electron tunneling: A quantum-mechanical effect which enables electrons to pass through a sufficiently thin insulating layer separating two conductors. It comes from the wave nature of the electron in a crystal lattice. Where $E < \Phi_0$, the solution to the wave equation becomes an exponentially decaying function which may be nonzero on the other side of a very thin barrier. When the electron is positioned in the barrier, image charges in the electrodes, and images of the images, exert forces on the electron, much as for the Schottky effect referred to above. The net effect of the **image charge effect** is to reduce the barrier height, and convert the rectangular barrier to a parabolic shape. With the barrier slope changing sufficiently with the parabolic shape, the **WKB approximation** can be employed to calculate a more realistic electron transmission probability than for the rectangular barrier.

Quantum dots: Nanoparticles of metal or semiconductor which display quantum(-mechanical) effects, such as quantized electron energy levels, due to the small size.

The electrons in a metal with energy greater than the **work function** of the material may be emitted from the surface by **thermionic emission**. The **Schottky effect** lowers the effective work function by the application of an electric field.

When an atom is adsorbed on a substrate, it dissipates thermal energy by moving about the surface. In this process, it may be captured by a preexisting condensate island, electrode, etc., if it was deposited within the characteristic **capture distance** of such an entity.

Acknowledgment

Jeahuck Lee's assistance with manuscript preparation is gratefully recognized.

References

1. G.L. Moss, "IC Logic Family Operation and Characteristics," in *The Electrical Engineering Handbook*, Volume 5, 3rd ed., Boca Raton, FL: CRC Press, 2006.
2. H.G. Parks, "Processes", in *The Electrical Engineering Handbook*, Volume 2, Boca Raton, FL: CRC Press, 2006, Chapter 2.1.

3. A.A. Sokolov, Y.M. Loskutov, and I.M. Ternov, *Quantum Mechanics*, New York: Holt, Rinehart & Winston, 1966.
4. J.G. Simmons, *J. Appl. Phys.*, 35, 2472–2481, 1964.
5. C.A. Neugebauer and M.B. Webb, *J. Appl. Phys.*, 33, 74–82, 1962.
6. J.E. Morris, *Thin Solid Films*, 11, 259–272, 1972.
7. K. Uchida, "Single-electron devices for logic applications," in *Nanoelectronics and Information Technology: Advanced Electronic Materials and Novel Devices*, New York: Wiley-VCH, 2003, pp. 425–443, Chapter 16.
8. P. Hadley, "Single-electron tunneling devices", in *Superconductivity in Networks and Mesoscopic Structures*, TMR Euroschool, 1997.
9. J.E. Morris, F. Wu, C. Radehaus, M. Hietschold, A. Henning, K. Hofmann, and A. Kiesow, *Proc. 7th Int. Conf. Solid State IC Technol.*, Beijing, 2004.
10. K. Yano, T. Ishii, T. Hashimoto, T. Kobayashi, F. Marai, and K. Seki, *IEEE Trans. Electron Devices*, 41, 1628, 1994.
11. J.E. Morris and F. Wu, *Thin Solid Films*, 317, 178, 1998.
12. J.E. Morris, *J. Appl. Phys.*, 39, 6107–6109, 1968.
13. Y. Takahashi, Y. Ono, A, Fujiwara and H. Inokawa, *J. Phys.: Condensed Matter*, 14, 995–1033, 2002.
14. K. Matsumoto, *Proc. IEEE*, vol. 85, no. 4, pp. 612–628, 1997.
15. J. Servat, P. Gorastiza, F. Sanz, F. Perez-Murano, N. Barniol, G. Abagal and X. Aymerich, *J. Vac. Sci. Technol. A*, vol. 14, no. 3, p. 1208, 1996.
16. F. Perez-Murano, G. Abagal, N. Barniol, X. Aymerich, J. Servat, P. Gorastiza and F. Sanz, *J. Appl. Phys.*, 78, 6797, 1995.
17. J.R. Tucker, *J. Appl. Phys.*, 72, 4399–4413, 1992.
18. D.K. Ferry, J.B. Barker, and C. Jacoboni, Eds., *Granular Nanoelectronics*, New York: Plenum, 1991.
19. K.K. Likharev, *Proc. IEEE*, 87, 606, 1999.
20. S. Goodnick and J. Bird, *IEEE Trans. Nanotechnol.*, 2, 368, 2003.
21. H. Ahmed and K. Nakazato, *Microelectron. Eng.*, 32, 297, 1996.
22. J.E. Morris and T.J. Coutts, *Thin Solid Films*, 47, 1–66, 1977.
23. J.E. Morris, *Vacuum*, 5, 107–113, 1998.

Further Reading

There are now a great many books on the broad field of Nanotechnology, and SETs appear in many of these, but a good place to start is with Ref. [7] or [18]. Other review articles found useful in the preparation of this chapter include Refs. [19, 20]. Some alternative device configurations can be found in Refs. [7, 13, 19, 21] and circuits in Refs. [13, 21], with a range of future application proposals in Ref. [19]. DTF background can be found in Refs. [22, 23]. For new developments, there will be a rapidly expanding range of magazines and journals covering or touching upon nanotechnology in general. The *IEEE Transactions on Nanotechnology* seems to carry a couple of SET papers in each issue, and *Nanotechnology* (Institute of Physics) is also proving fruitful.

4

Integrated Circuits

Joe E. Brewer
University of Florida

Mehdi R. Zargham
Southern Illinois University

Spyros Tragoudas
Southern Illinois University

S.K. Tewksbury
Stevens Institute of Technology

Christian Piguet
CSEM Centre Suisse d'Electronique et de Microtechnique and LAP-EPFL

4.1 Integrated Circuit Technology

Joe E. Brewer

Integrated circuit technology, the cornerstone of the modern electronics industry, is subject to rapid change. Electronic engineers, especially those engaged in research and development, can benefit from an understanding of the structure and pattern of growth of this technology.

Technology Perspective

A solid-state integrated circuit (IC) is a group of interconnected circuit elements formed on or within a continuous substrate. While an integrated circuit may be based on many different material systems, silicon is by far the dominant material. For quite a number of years, about 98% of contemporary electronic devices were based on silicon technology, and about 85% of those silicon integrated circuits were complementary metal-oxide semiconductor (CMOS) devices. In recent years, the use of nonsilicon integrated circuits such as Ga–As, and silicon variants such as Si–Ge, has increased because of proliferation of communications-oriented applications. It is also expected that the nature of the classical CMOS-type circuits will be changing to incorporate new materials and geometric structures.

From an economic standpoint the most important metric for an IC is the "level of functional integration." The accepted way to measure functional integration is by component count such as the number of transistors per chip or the number of bits per chip, but other measures such as million instructions per second (MIPS) or million floating point operations per second (MFLOPS) can also be used. Since the invention of the IC by Jack Kilby in 1958, the level of integration has steadily increased. The pleasant result is that cost and physical size per function reduce continuously, and we enjoy a flow of new affordable information processing products that pervade all aspects of our day-to-day lives.

The approximate historical rate of increase was a doubling of functional content per chip every 18 months. The related resulting historical experience has been a cost reduction per function of ~25% per year combined with a growth in the market for integrated circuits of ~17% per year. Because of increased difficulty in scaling devices these trends are slowing and it is expected that functional content per chip will increase at about half the historical rate.

For engineers who work with products that use semiconductor devices, the challenge is to anticipate and make use of these enhanced capabilities in a timely manner. It is not an overstatement to say that survival in the marketplace depends on rapid "design-in" and deployment.

For engineers who work in the semiconductor industry, or its myriad supporting industries, the challenge is to maintain this relentless exponential growth. The entire industry is marching to a drum beat. The cost of technology development and the investment in plant and equipment has risen to billions of dollars. Companies that lag behind face serious loss of market share, and possibly dire economic consequences.

Technology Generations

The concept of a technology generation emerged from analysis of historical records, and was outlined by Gordon Moore in the 1960s and captured as Moore's law. This was an observation that circuit complexity per chip was doubling at approximately equal time intervals. For quite a while the simple rule of thumb was that succeeding generations would support a four times increase in circuit complexity, and new generations emerge on an approximate 3-year interval. The associated observations are that linear dimensions of device features change by a factor of 0.7, and the economically viable die area grows by a factor of 1.6. While technology generations are still being defined as having a 0.7 reduction in linear dimensions, succeeding generations no longer support a 4× complexity increase.

Minimum feature size stated in nanometers is the term used most frequently to label a technology generation. "Feature" refers to a geometric object in the mask set such as a line width or a gate length. The "minimum feature" is the smallest dimension that can be reliably used to form the entity. Examples of features might be the width of a metal or polysilicon line or the space between those lines. To establish a numeric measure of the capability of a given technology it is common practice to use the half pitch of critical metal lines. The pitch is the sum of the width of a line and the width of the space to an adjacent line. The "half" pitch is, of course, one half of that sum. The minimum feature used to define a technology generation (also referred to as a technology node) is the half pitch of metal lines in a dynamic random access memory (DRAM) array. In recent years the metal one level half pitch for a microprocessor unit (MPU) has rivaled the aggressiveness of the DRAM half pitch, and it is sometimes also used to specify the node.

Figure 4.1 displays the technology evolution sequence. In the diagram, succeeding generations are numbered using the current generation as the "zero" reference. Because this material was written in 2004, the "zero" generation is the 90-nm minimum feature size technology that began volume production in 2004.

generation 0	generation 1	generation 2	generation 3	generation 4	generation 5
2004	2007	2010	2013	2016	2019
90 nm	65 nm	45 nm	32 nm	22 nm	15 nm
production	development	research	research	research	research

FIGURE 4.1 Semiconductor technology generation time sequence.

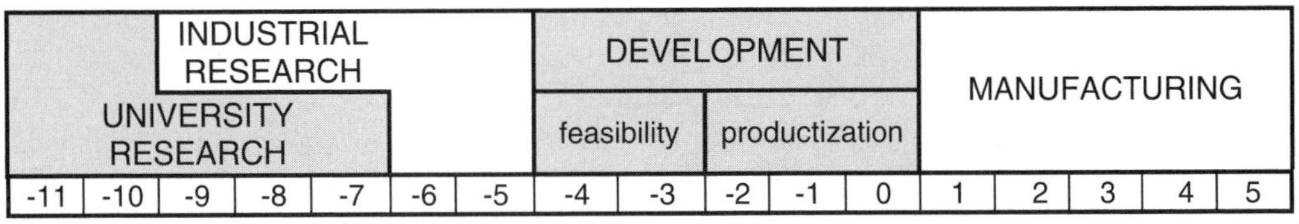

UNIVERSITY RESEARCH			INDUSTRIAL RESEARCH				DEVELOPMENT: feasibility		DEVELOPMENT: productization			MANUFACTURING				
-11	-10	-9	-8	-7	-6	-5	-4	-3	-2	-1	0	1	2	3	4	5

FIGURE 4.2 Life cycle of a semiconductor technology generation.

04	05	06	07	08	09	10	11	12	13	14	15	16	17	18	19	20
90 nm	1	2	3	4	5											
-3	-2	-1	65 nm	1	2	3	4	5								
-6	-5	-4	-3	-2	-1	45 nm	1	2	3	4	5					
-9	-8	-7	-6	-5	-4	-3	-2	-1	32 nm	1	2	3	4	5		
	-11	-10	-9	-8	-7	-6	-5	-4	-3	-2	-1	22 nm	1	2	3	4
				-11	-10	-9	-8	-7	-6	-5	-4	-3	-2	-1	15 nm	1

FIGURE 4.3 Time overlap of semiconductor technology generations.

An individual device generation has been observed to have a reasonably well-defined life cycle that covers about 17 years. The first year of volume manufacture is the reference point for a generation, but its lifetime actually extends further in both directions. As shown in Figure 4.2, one can think of the stages of maturity as ranging over a linear scale that measures years to production in both the plus and minus directions.

The 17-year life cycle of a single generation, with new generations being introduced at 3-year intervals, means that at any given time up to six generations are being worked on. This tends to blur the significance of research news and company announcements unless the reader is sensitive to the technology overlap in time.

To visualize this situation, consider Figure 4.3. The top row lists calendar years. The second row shows how the life cycle of the 90-nm generation relates to the calendar. The third row shows the life cycle of the 65-nm generation versus the calendar. Looking down any column corresponding to a specific calendar year, one can see which generations are active and identify their respective life-cycle year.

One should not interpret the 17-year life cycle as meaning that no work is being performed that is relevant to a generation before the 17-year period begins. For example, many organizations are currently (2004) conducting experiments directed at transistors with gate lengths smaller than 15 nm. This author's interpretation is that when basic research efforts have explored technology boundaries, the conditions are ripe for a specific generation to begin to coalesce as a unique entity. When a body of research begins to seek compatible materials and processes to enable design and production at the target feature size, the generation life cycle begins. This is a rather diffused activity at first, and it becomes more focused as the cycle proceeds.

International Technology Roadmap for Semiconductors

The recurring great expense of the research and development necessary to establish the new capabilities and physical plant to produce devices at exponentially reduced feature sizes at 3-year intervals is more than a single manufacturer (or country for that matter) can afford. In order to manage at least some of these costs, manufacturers around the world have banded together to formulate a common roadmap that identifies the technical requirements for five or six technology generations—about a 15-year time horizon.

The International Technology Roadmap for Semiconductors (ITRS) is sponsored by industry groups from Europe, Japan, Korea, Taiwan, and the United States.

Development of the ITRS is an ongoing activity that involves almost 1000 technical experts, who treat every aspect of the complex device geometry and materials definition, fabrication processes and tooling, circuit design and design tooling, test and assembly, and all of the supporting disciplines. Current practice is to prepare a new version of the roadmap during odd-numbered years, and prepare only critical updates during even-numbered years. In 2003 a complete roadmap was released and is available to the public over the Internet at http://public.itrs.net.

TABLE 4.1 Overall Roadmap Technology Characteristics

Year of First DRAM Shipment Minimum Feature (nm)	2004 90	2007 65	2010 45	2013 32	2016 22
Memory					
Bits/chip (DRAM)	4G	16G	32G	64G	128G
Bits/cm^2 (DRAM)	0.77G	2.22G	5.19G	10.37G	24.89
Logic (High-Volume Microprocessor)					
Transistors per chip	193M	386M	773M	1546M	3902M
Transistors/cm^2	138M	276M	552M	1104M	2209M
Logic (Low-Volume ASIC)					
Maximum transistors per chip	1020M	2041M	4081M	8163M	16326M
Usable transistors/cm^2 (auto layout)	178M	357M	714M	1427M	2854M
Number of Chip I/Os					
MPU maximum total pads	3072	3072	3840	4224	4416
ASIC maximum total pads	3600	4400	4800	5400	6000
Number of Package Pins/Balls					
Microprocessor/controller	1600	2140	2782	3616	5426
ASIC (high performance)	3000	4000	4009	5335	8450
Chip Frequency (GHz)					
On-chip local clock	4.2	9.3	15	23	40
Chip-to-board speed, high performance	2.0	4.9	9.5	19	36
Chip Size at Production (mm^2)					
DRAM	140	140	140	140	140
Microprocessor	310	310	310	310	310
ASIC	572	572	572	572	572
Fabrication Process					
On-chip wiring levels—maximum	10–14	11–15	12–16	12–16	14–18
Electrical defect density—MPU (faults/m^2)	1395	1395	1395	1395	1395
Mask count—MPU	31	33	35	35	39
Wafer diameter—bulk/epitaxial/SOI (mm)	300	300	300	450	450
Power Supply Voltage (V)					
Desktop	1.2	1.1	1.0	0.9	0.8
Battery	0.9	0.8	0.7	0.6	0.5
Maximum Power					
High performance with heatsink (W)	158	189	218	251	288
Logic without heatsink (W)	84	104	120	138	158
Battery (W)	2.2	2.5	2.8	3.0	3.0

The starting assumption of the ITRS is that a 0.7× reduction of the minimum feature size will continue to define technology nodes. The overall roadmap is comprised of many individual roadmaps that address defined critical areas of semiconductor research, development, engineering, and manufacturing. In each area, needs and potential solutions for each technology generation are reviewed. Of course, this process is more definitive for the early generations because knowledge is more complete and the range of alternatives is restricted.

Table 4.1 presents some highlights of the 90 to 22 nm technology nodes extracted from various tables within the 2003 ITRS. This brief summary shows trends across the generations, but the reader should consult the actual ITRS for full explanations of the basis for the individual parameters.

References

Current and archival versions of the *International Technology Roadmap for Semiconductors* (ITRS) are available at http://public.itrs.net.

Information concerning the integrated circuit life cycle can be found in Larrabee, G.B. and Chatterjee, P., "DRAM manufacturing in the 90s—part 1: the history lesson, and part 2: the roadmap," *Semiconductor International*, pp. 84–92, May 1991.

4.2 Layout, Placement, and Routing

Mehdi R. Zargham and Spyros Tragoudas

Very large scale integrated (VLSI) electronics present a challenge, not only to those involved in the development of fabrication technology, but also to computer scientists, computer engineers, and electrical engineers. The ways in which digital systems are structured, the procedures used to design them, the trade-offs between hardware and software, and the design of computational algorithms will all be greatly affected by the coming changes in integrated electronics.

A VLSI chip can today contain hundreds of millions of transistors. One of the main factors contributing to this increase is the effort that has been invested in the development of computer-aided design (CAD) systems for VLSI design. The VLSI CAD systems are able to simplify the design process by hiding the low-level circuit theory and device physics details from the designer, and allowing him or her to concentrate on the functionality of the design and on ways of optimizing it.

A VLSI CAD system supports descriptions of hardware at many levels of abstraction, such as system, subsystem, register, gate, circuit, and layout levels. It allows designers to design a hardware device at an abstract level and progressively work down to the layout level. A layout is a complete geometric representation (a set of rectangles) from which the latest fabrication technologies directly produce reliable working chips. A VLSI CAD system also supports verification, synthesis, and testing of the design. Using a CAD system, the designer can make sure that all of the parts work before actually implementing the design.

A variety of VLSI CAD systems are commercially available that perform all or some of the levels of abstraction of design. Most of these systems support a *layout editor* for designing a circuit **layout**. A layout editor is software that provides commands for drawing lines and boxes, copying objects, moving objects, erasing unwanted objects, and so on. The output of such an editor is a design file that describes the layout. Usually, the design file is represented in a standard format, called Caltech Intermediate Form (CIF), which is accepted by the fabrication industry.

What Is Layout?

For a specific circuit, a layout specifies the position and dimension of the different layers of materials as they would be laid on the silicon wafer. However, the layout description is only a symbolic representation, which simplifies the description of the actual fabrication process. For example, the layout representation does not explicitly indicate the thickness of the layers, thickness of oxide coating, amount of ionization in the transistors channels, etc., but these factors are implicitly understood in the fabrication process. Some of the main layers used in any layout description are n-diffusion, p-diffusion, poly, metal-1, and metal-2. Each of these layers is represented by a polygon of a particular color or pattern. As an example, Figure 4.4 presents a specific pattern for each layer that will be used throughout the rest of this section.

As is shown in Figure 4.5, an n-diffusion layer crossing a poly layer implies an nMOS transistor, and a p-diffusion crossing poly implies a pMOS transistor.

Note that the widths of diffusion and poly are represented with a scalable parameter called *lambda*. These measurements, referred to as *design rules*, are introduced to prevent errors on the chip, such as preventing thin lines from opening (disconnecting) and short-circuiting.

Implementing the design rules based on lambda makes the design process independent of the fabrication process. This allows the design to be rescaled as the fabrication process improves.

Metal layers are used as wires for connections between the components. This is because metal has the lowest propagation delay compared to the other layers. However, sometimes a poly layer is also used for short wires in order to reduce the complexity of the wire routing. Any wire can cross another wire without being electrically affected as long as they are in different layers. Two different layers can be electrically connected together using *contacts*. The fabrication process of the contacts depends on the types of layers that are to be connected. Therefore, a layout editor supports different types of contacts by using different patterns.

From the circuit layout, the actual chip is fabricated. Based on the layers in the layout, various layers of materials, one on top of the others, are laid down on a silicon wafer. Typically, the process of laying down each of these materials involves several steps, such as masking, oxide coating, lithography, and etching (Mead and Conway, 1980). For example, as shown in Figure 4.6(a), for fabricating an nMOS transistor, first, two masks, one for poly and one for n-diffusion, are obtained from the circuit layout. Next, the n-diffusion mask is used

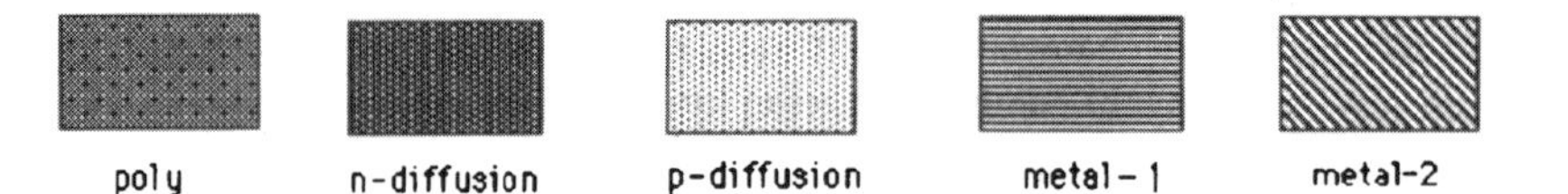

FIGURE 4.4 Different layers.

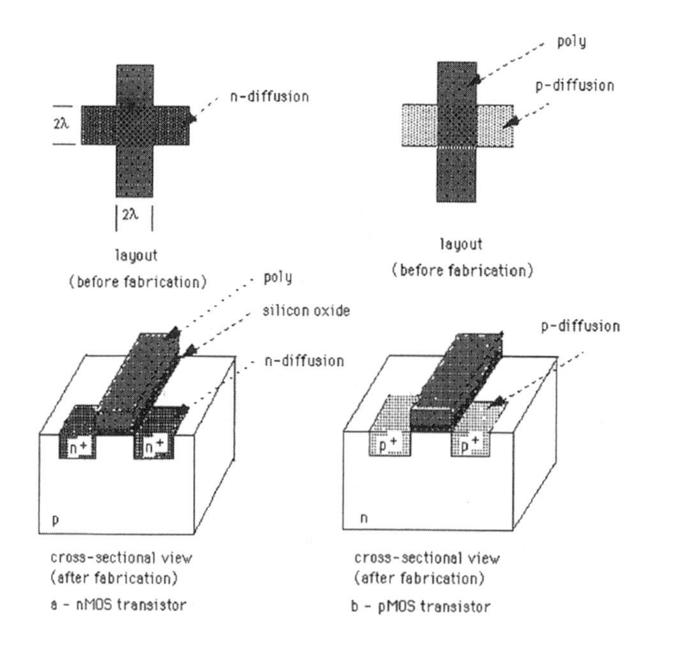

FIGURE 4.5 Layout and fabrication of MOS transistors.

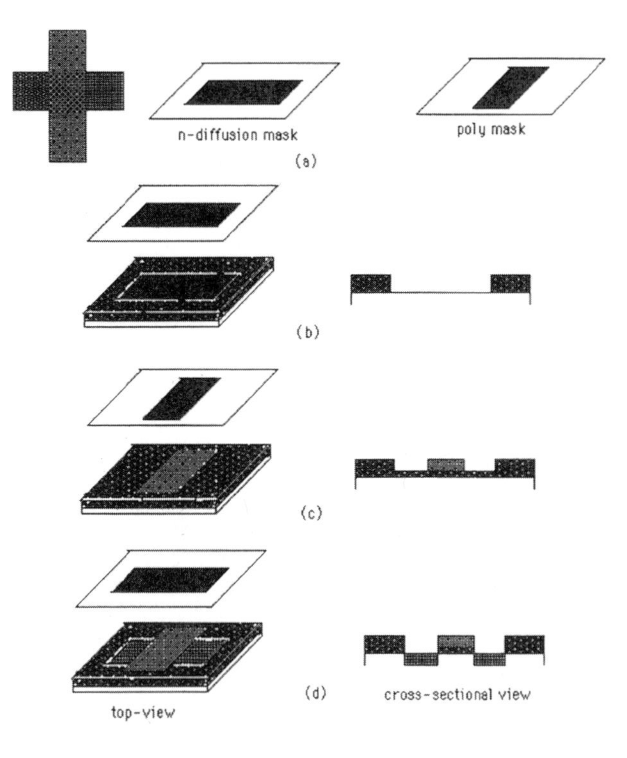

FIGURE 4.6 Fabrication steps for an nMOS transistor.

to create a layer of silicon oxide on the wafer (see Figure 4.6(b)). The wafer will be covered with a thin layer of oxide in places where the transistors are supposed to be placed as opposed to a thick layer in other places. The poly mask is used to place a layer of polysilicon on top of the oxide layer to define the gate terminals of the transistor (see Figure 4.6(c)). Finally, the *n*-diffusion regions are made to form the source and drain terminals of the transistor (see Figure 4.6(d)).

To better illustrate the concept of layout design, the design of an inverter in CMOS technology is shown in Figure 4.7. An inverter produces an output voltage that is the logical inverse of its input. Considering the circuit diagram of Figure 4.7(a), when the input is one, the lower nMOS is on, but the upper pMOS is off. Thus, the output becomes zero by becoming connected to the ground through the nMOS. However, if the input is zero, the pMOS is on and the nMOS is off, so the output must find a charge-up path through the pMOS to the supply and therefore becomes one. Figure 4.7(b) represents a layout for such an inverter. As can be seen from this figure, the problem of a layout design is essentially reduced to drawing and painting a set of polygons. Layout editors provide commands for drawing such polygons. The commands are usually entered at the keyboard or with a mouse and, in some menu-driven packages, can be selected as options from a pull-down menu.

In addition to the drawing commands, often a layout system provides tools for minimizing the overall area of the layout (i.e., size of the chip). Today, a VLSI chip consists of many individual cells, with each one laid out separately. A cell can be an inverter, a NAND gate, a multiplier, a memory unit, etc. The designer can make the layout of a cell and then store it in a file called the *cell library.* Later, each time the designer wants to design a circuit that requires the stored cell, he or she simply copies the layout from the cell library. A layout may consist of many cells. Most of the layout systems provide routines, called **floor planning**, **placement**, and **routing** routines for placing the cells and then interconnecting them with wires in such a way to minimize the layout area. As an example, Figure 4.8 presents the placement of three cells. The area between the cells is used for routing. We can also route above placed cells. This is called over the cell (OTC) routing. The entire routing surface is divided into a set of rectangular routing areas called channels. The sides of each channel consist of a set of terminals. A wire that connects the terminals with the same ID is called a net. The router finds a location

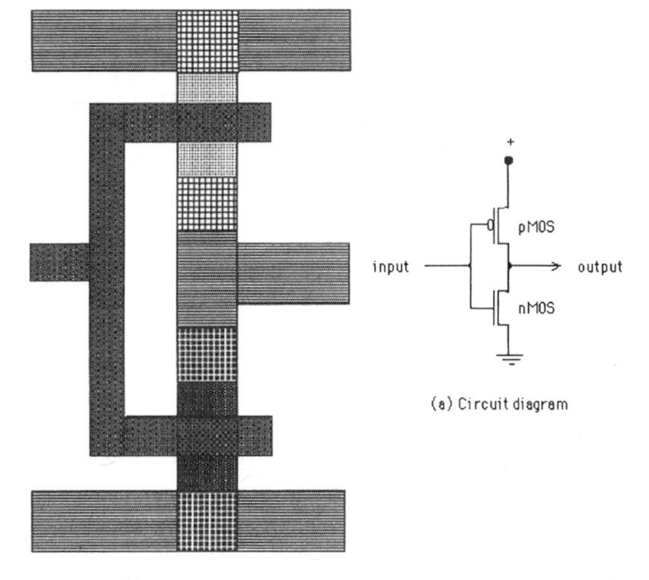

FIGURE 4.7 An inverter.

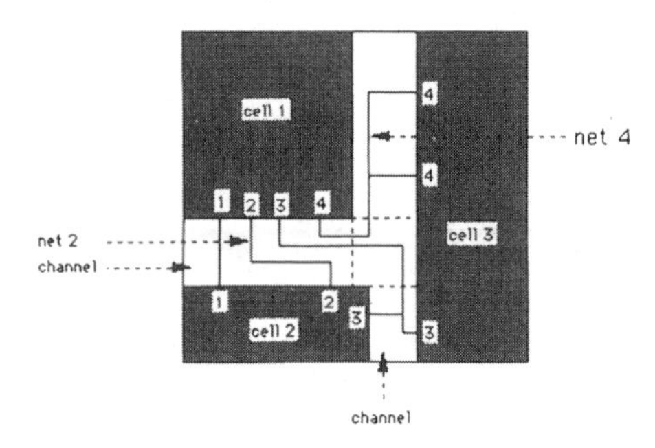

FIGURE 4.8 Placement and routing.

for the wire segments of each net within the channel. The following sections classify various types of placement and routing techniques and provide an overview of the main steps of some of these techniques.

Floor Planning Techniques

The floor planning problem in computer-aided design of integrated circuits is similar to that in architecture and the goal is to find a location for each cell based on proximity (layout adjacency) criteria to other cells. We consider rectangular floor plans whose boundaries are rectangles. It is desirable to obtain a floor plan that minimizes the overall area of the layout.

An important goal in floor planning is the cell sizing problem where the goal is to determine the dimensions of variable cells whose area is invariant. All cells are assumed to be rectangular, and in the cell sizing problem the goal is to determine the width and height of each cell subject to predetermined upper and lower bounds on their ratio, and to their product being equal to its area, so that the final floor plan has optimal area.

One of the early approaches in floor planning is the hierarchical, where recursive bipartition or partition into more than two parts is recursively employed and a floor plan tree is constructed. The tree simply reflects the hierarchical construction of the floor plan. Figure 4.9 shows a hierarchical floor plan and its associated tree. The partitioning problem and related algorithms are discussed extensively later in this section.

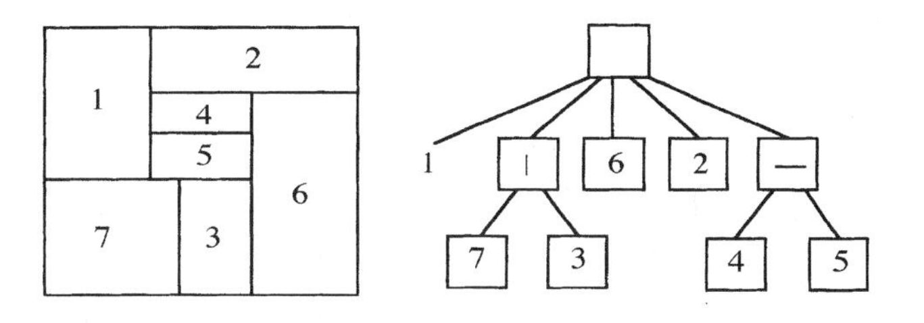

FIGURE 4.9 A hierarchical floor plan and its associated tree. The root node has degree 5. The internal node labeled with "|" indicates a vertical slicing. The internal node labeled with "—" indicates a horizontal slicing.

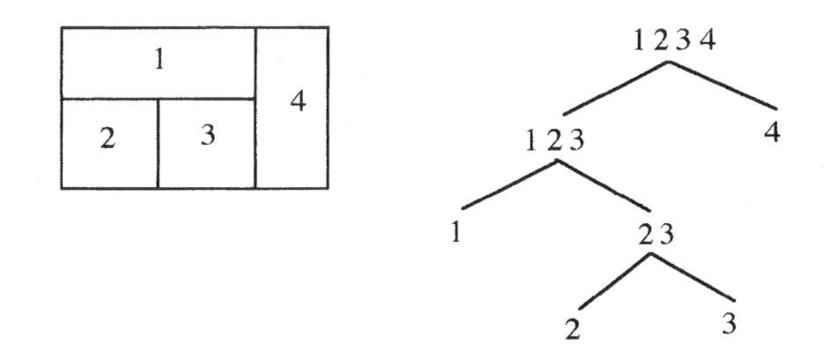

FIGURE 4.10 A sliceable floor plan and its associated binary tree.

Many early hierarchical floor planning tools insist that the floor plan be sliceable. A sliceable floor plan is recursively defined as follows: (a) a cell or (b) a floor plan that can be bipartitioned into two sliceable floor plans with either a horizontal or vertical line. Figure 4.10 shows a sliceable floor plan whose tree is binary.

Many tools that produce sliceable floor plans are still in use because of their simplicity. In particular, many problems arising in sliceable floor planning are solvable optimally in polynomial time (Sarrafzadeh and Wong, 1996). Unfortunately, sliceable floor plans are rarely optimal (in terms of their areas), and they often result in layouts with very difficult routing phases. (Routing is discussed later in this section.) Figure 4.11 shows a compact floor plan that is not sliceable.

Hierarchical tools that produce nonsliceable floor plans have also been proposed (Sarrafzadeh and Wong, 1996). The major problem in the development of such tools is that we are often facing problems that are intractable and thus we have to rely on heuristics in order to obtain fast solutions. For example, the cell sizing problem can be tackled optimally in sliceable floor plans (Otten, 1983; Stockmeyer, 1983) but the problem is intractable for general nonsliceable floor plans.

A second approach to floor planning is the rectangular dual graph. The idea here is to use duality arguments and express the cell adjacency constraints in terms of a graph, and then use an algorithm to translate the graph into a rectangular floor plan. A rectangular dual graph of a rectangular floor plan is a planar graph $G = (V,E)$, where V is the set of cells and E is the set of edges, and an edge (C_1,C_2) is in E if and only if cells C_1 and C_2 are adjacent in the floor plan. See Figure 4.12 for a rectangular floor plan and its rectangular dual graph G.

Let us assume that the floor plan does not contain cross junctions. Figure 4.13 shows a cross junction. This restriction does not significantly increase the area of a floor plan because, as Figure 4.13 shows, a cross junction can be replaced by two T junctions by simply adding a short edge e.

It has been shown that in the absence of cross junctions, the dual graph is planar triangulated (PT) and every T-junction corresponds to a triangulated face of the dual PT graph. Unfortunately, not all PT graphs have a rectangular floor plan. For example, in the graph of Figure 4.14 we cannot satisfy the adjacency requirements of edges (a,b), (b,c), and (c,a) at the same time. Note that the later edges form a cycle of length 3 that is not a face. It has been shown that a PT graph has a rectangular floor plan if and only if it does not contain such cycles of length 3. Moreover, a linear time algorithm to obtain such a floor plan has been presented (Sarrafzadeh and Wong, 1996). The rectangular dual graph approach is a new method for floor planning, and many floor planning problems, such as the sizing problem, have not been tackled yet.

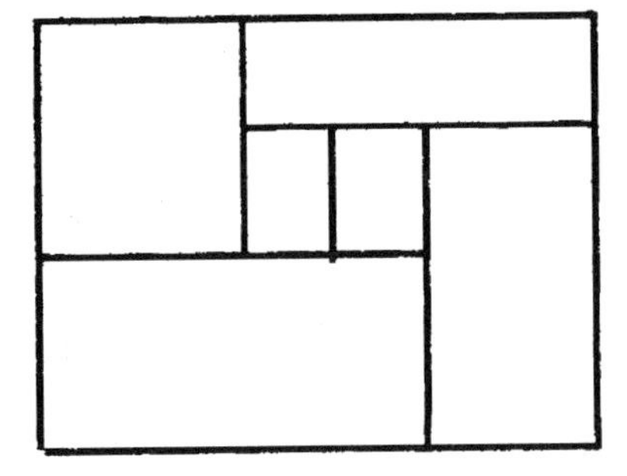

FIGURE 4.11 A compact layout that is not sliceable.

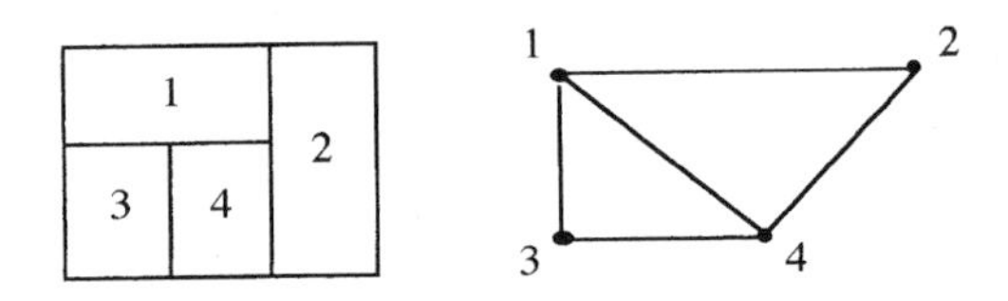

FIGURE 4.12 A rectangular floor plan and its associated dual planar graph.

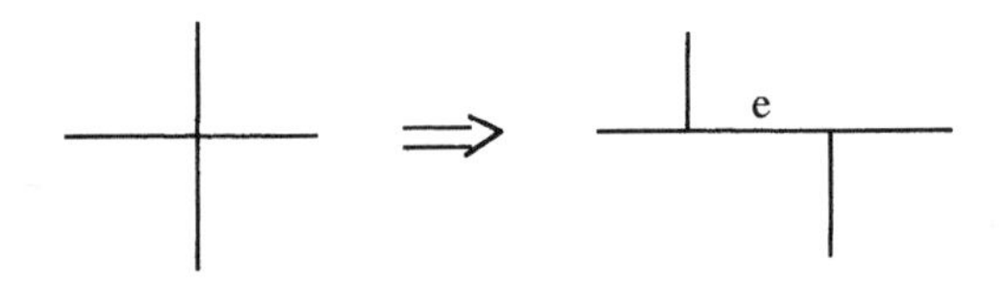

FIGURE 4.13 A cross junction can be replaced by two T junctions.

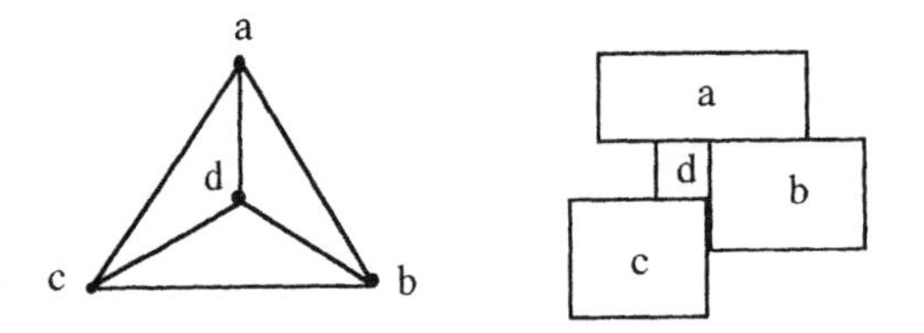

FIGURE 4.14 For a cycle of size 3 that is not a face, we cannot satisfy all constraints.

Rectangular floor plans can be obtained using simulated annealing and genetic algorithms. Both techniques are used to solve general optimization problems for which the solution space is not well understood. The approaches are easy to implement, but the algorithms have many parameters that require empirical adjustments, and the results are usually unpredictable.

A final approach to floor planning, which unfortunately requires substantial computational resources and results to an intractable problem, is to formulate the problem as a mixed-integer linear programming (LP). Consider the following definitions:

W_i, H_i, R_i: width, height, and area of cell C_i
X_i, Y_i: coordinates of lower left corner of cell C_i
X, Y: the width and height of the final floor plan
A_i, B_i: lower and upper bound for the ratio W_i/H_i of cell C_i
P_{ij}, Q_{ij}: variables that take zero/one values for each pair of cells C_i and C_j

The goal is to find X_i, Y_i, W_i, and H_i for each cell so that all constraints are satisfied and XY is minimized. The latter is a nonlinear constraint. However, we can fix the width W and minimize the height of the floor plan as follows:

$$\begin{aligned} &\min Y \\ &X_i + W_i \leq W \\ &Y \geq Y_i + H_i \end{aligned}$$

The complete mixed-integer LP formulation is (Sutanthavibul et al., 1991):

$$\begin{aligned} &\min Y \\ &X_i, Y_i, W_i \geq 0 \\ &P_{ij}, Q_{ij} = 0 \text{ or } 1 \\ &X_i + W_i \leq W \\ &Y \geq Y_i + H_i \\ &X_i + W_i \leq X_j + W(P_{ij} + Q_{ij}) \\ &X_j + W_j \leq X_i + W(1 - P_{ij} + Q_{ij}) \\ &Y_i + H_i \leq Y_j + H(1 + P_{ij} - Q_{ij}) \\ &Y_j + H_j \leq Y_i + H(2 - P_{ij} - Q_{ij}) \end{aligned}$$

When H_i appears in the above equations, it must be replaced (using first-order approximation techniques) by $H_i = D_i W_i + E_i$, where D_i and E_i are defined below:

$$\begin{aligned} &W_{\min} = (R_i A_i)^{1/2} \\ &W_{\max} = (R_i B_i)^{1/2} \\ &H_{\min} = (R_i / B_i)^{1/2} \\ &H_{\max} = (R_i / A_i)^{1/2} \\ &D_{\mathrm{i}} = H_{\max} - H_{\min}) / (W_{\min} - W_{\max}) \\ &E_i = H_{\max} - D_i W_{\min} \end{aligned}$$

The unknown variables are X_i, Y_i, W_i, P_{ij}, and Q_{ij}. All other variables are known. The equations can then be fed into an LP solver to find a minimum cost solution for the unknowns.

Placement Techniques

Placement is a restricted version of floor planning where all cells have fixed dimension. The objective of a placement routine is to determine an optimal position on the chip for a set of cells in a way that the total occupied area and total estimated length of connections are minimized. Given that the main cause of delay in a chip is the length of the connections, providing shorter connections becomes an important objective in placing a set of cells. The placement should be such that no cells overlap and enough space is left to complete all the connections.

All exact methods known for determining an optimal solution require a computing effort that increases exponentially with number of cells. To overcome this problem, many heuristics have been proposed (Preas and Lorenzetti, 1988). There are basically three strategies of heuristics for solving the placement problem, namely, *constructive*, *partitioning*, and *iterative* methods. Constructive methods create placement in an incremental manner where a complete placement is only available when the method terminates. They often start by placing a *seed* (a seed can be a single cell or a group of cells) on the chip and then continuously placing other cells based on some heuristics such as size of cells, connectivity between the cells, design condition for connection lengths, or size of chip. This process continues until all the cells are placed on the chip. Partitioning methods divide the cells into two or more partitions so that the number of connections that cross the partition boundaries is minimized. The process of dividing is continued until the number of cells per partition becomes less than a certain small number. Iterative methods seek to improve an initial placement by repeatedly modifying it. Improvement might be made by transforming one cell to a new position or switching positions of two or more cells. After a change is made to the current placement configuration based on some cost function, a decision is made to see whether to accept the new configuration. This process continues until an optimal (in most cases a near optimal) solution is obtained. Often the constructive methods are used to create initial placement on which an iterative method subsequently improves.

Constructive Method

In most of the constructive methods, at each step an unplaced cell is selected and then located in the proper area. There are different strategies for selecting a cell from the collection of unplaced cells (Wimer and Koren, 1988). One strategy is to select the cell that is most strongly connected to already placed cells. For each unplaced cell, we find the total of its connections to all of the already placed cells. Then we select the unplaced cell that has the maximum number of connections. As an example consider the cells in Figure 4.15. Assume that cells c_1 and c_2 are already placed on the chip. In Figure 4.16 we see that cell c_5 has been selected as the next cell to be placed. This is because cell c_5 has the largest number of connections (i.e., three) to cells c_1 and c_2.

The foregoing strategy does not consider area as a factor and thus results in fragmentation of the available free area; this may make it difficult to place some of the large unplaced cells later. This problem can be overcome, however, by considering the product of the number of connections and the area of the cell as a criterion

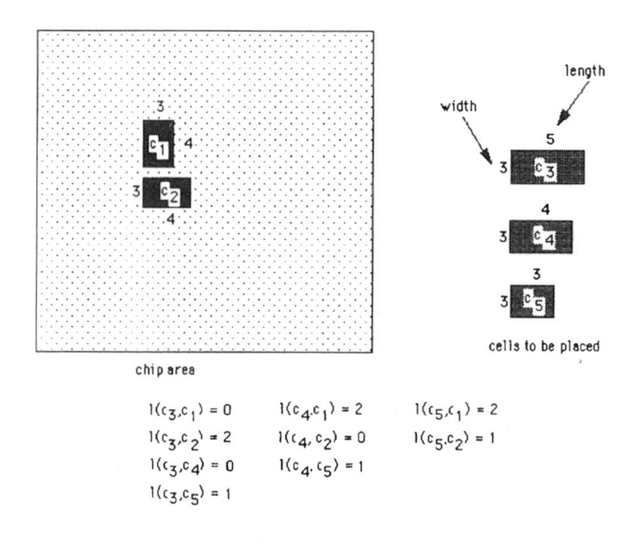

FIGURE 4.15 Initial configuration.

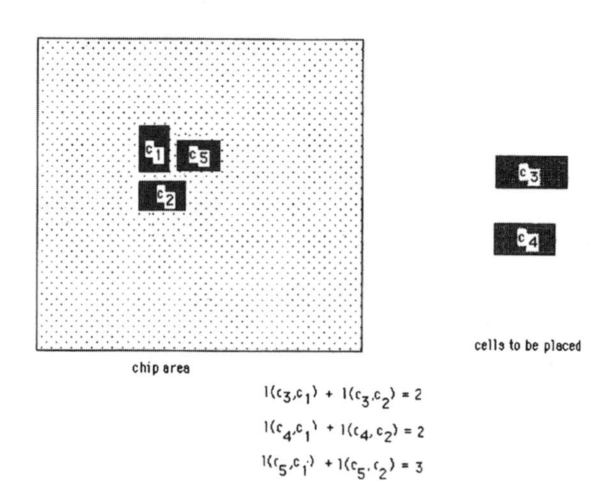

FIGURE 4.16 Selection based on the number of connections.

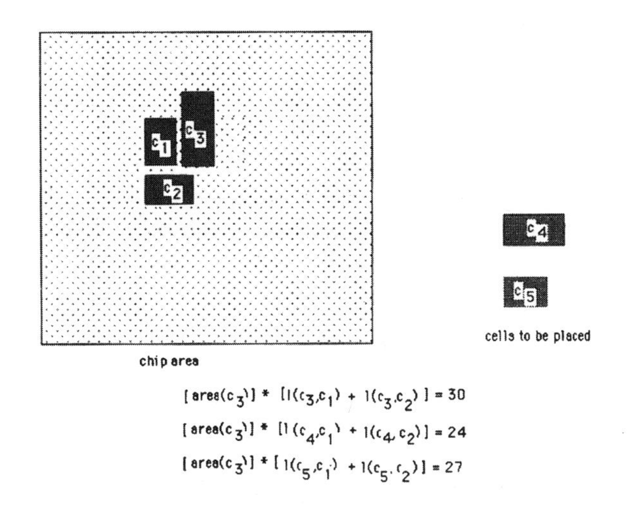

FIGURE 4.17 Selection based on the number of connections and area.

for the selection. Figure 4.17 presents an example of such a strategy. Cell c_3 is selected as the next choice since the product of its area and its connections to c_1 and c_2 combine to associate with the maximum value.

Partitioning Method

The approaches for the partitioning method can be classified as quadratic and sliced bisection. In both approaches the layout is divided into two sub-areas, A and B, each having a size within a predefined range. Each cell is assigned to one of these sub-areas. This assignment is such that the number of interconnections between the two sub-areas is minimal. For example, Figure 4.18 presents successive steps for the quadratic and sliced-bisection methods. As shown in Figure 4.18(a), in the first step of the quadratic method the layout area is divided into two almost equal parts; in the second step the layout is further divided into four almost equal parts in the opposite direction. This process continues until each sub-area contains only one cell. Similar to the quadratic method, the sliced bisection also divides the layout area into several sub-areas.

The sliced-bisection method has two phases. In the first phase, the layout area is iteratively divided into a certain number of almost equal sub-areas in the same direction. In this way, we end up with a set of slices (see Figure 4.18(b)). Similarly, the second phase divides the area into a certain number of sub-areas; however, the slicing is done in the opposite direction.

Several heuristics have been proposed for each of the preceding partitioning methods. Here, for example, we emphasize the work of Fiduccia and Mattheyses (1982), which uses the quadratic method. For simplicity, their algorithm is only explained for one step of this method. Initially the set of cells is randomly divided into two sets, A and B. Each set represents a sub-area of the layout and has size equal to the area it represents. A cell is selected from one of these sets to be moved to the other set. The selection of the cell depends on three criteria.

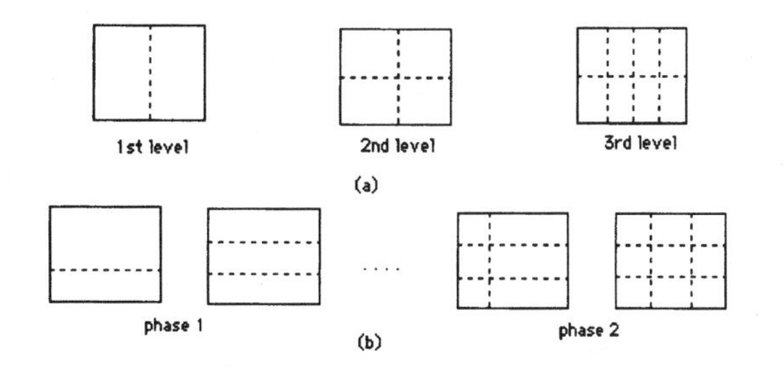

FIGURE 4.18 Partitioning.

The cell should be free, i.e., the cell must have a minimum gain among the gains of all other free cells. A cell c has gain g if the number of interconnections between the cells of the two sets decreases by g units when c is moved from its current set to the other. Finally, the selected set's move should not violate a predefined balancing criterion that guarantees that the sizes of the two sets are almost equal. After moving the selected cell from its current set to the complementary set, it is no longer free. A new partition, which corresponds to the new instance of the two sets A and B, is created. The cost of a partition is defined as the number of interconnections between cells in the two sets of the partition. The Fiduccia–Mattheyses algorithm keeps track of the best partition encountered so far, i.e., the partition with the minimum cost. The algorithm will move the selected cell to the complementary set even if its gain is negative. In this case, the new partition is worse than the previous one, but the move can eventually lead to better partitions.

This process of generating new partitions and keeping track of the best encountered partition is repeated until no free cells remain. At this point in the process, which is called a pass, the algorithm returns the best partition and terminates. To obtain better partitions, the algorithm can be modified such that more passes occur. This can easily be done by selecting the best partition of a pass as the initial partition of the next pass. In this partition, however, all cells are free. The modified algorithm terminates whenever a new pass returns a partition that is no better than the partition returned by the previous pass. This way, the number of passes generated will never be more than the number of the interconnections in the circuit and the algorithm always terminates.

The balancing criterion in the Fiduccia–Mattheyses algorithm is easily maintained when the cells have uniform areas. The only way a pass can be implemented to satisfy the criterion is to start with a random initial partition in which the two sets differ by one cell, each time select the cell of maximum gain from the larger sized set, move the cell and generate a new partition, and repeat until no free cells remain.

In the example of Figure 4.19, the areas of the cells are nonuniform. However, the assigned cell areas ensure that the previously described operation occurs so that the balancing criterion is satisfied. (The cell areas are omitted in this figure, but they correspond to the ones given in Figure 4.15.) Figure 4.19 illustrates a pass of the Fiduccia–Mattheyses algorithm. During this pass, six different partitions are generated, i.e., the initial partition and five additional ones. Note that according to the description of the pass, the number of additional partitions equals the number of cells in the circuit.

Each partition consists of the cells of the circuit (colored according to the set to which they belong and labeled with an integer), the gain value associated with each free cell in the set from which the selected cell will be moved (this value can be a negative number), and the nets (labeled with letters). In the figure, a cell that is no longer free is distinguished by a circle placed inside the rectangle that represents that cell.

The initial partition has cost 5 since nets a, b, h, g, f connect the cells in the two sets. Then the algorithm selects cell 1, which has the maximum gain. The new partition has cost 3 (nets e, g, f), and cell 1 is no longer free. The final partition has no free cells. The best partition in this pass has cost 3.

Iterative Method

Many iterative techniques have been proposed. Here, we emphasize one of these techniques called simulated annealing. Simulated annealing, as proposed by Kirkpatrick et al. (1983), makes the connection between statistical mechanics and combinatorial optimization problems. The main advantage with simulated annealing is its hill-climbing ability, which allows it to back out of inferior local solutions and find better solutions.

Sechen (1990) has applied simulated annealing to the placement problem and has obtained good solutions. The method basically involves the following steps:

BEGIN

1. Find an initial configuration by placing the cells randomly. Set the initial temperature T and the maximum number of iterations.
2. Calculate the cost of the initial configuration.
   A general form of the cost function may be: Cost $= c_1 \times$ Area of layout $+ c_2 \times$ Total interconnection length, where c_1 and c_2 are tuning factors.
3. While (stopping criterion is not satisfied)
   {587
   a. For (maximum number of iteration)

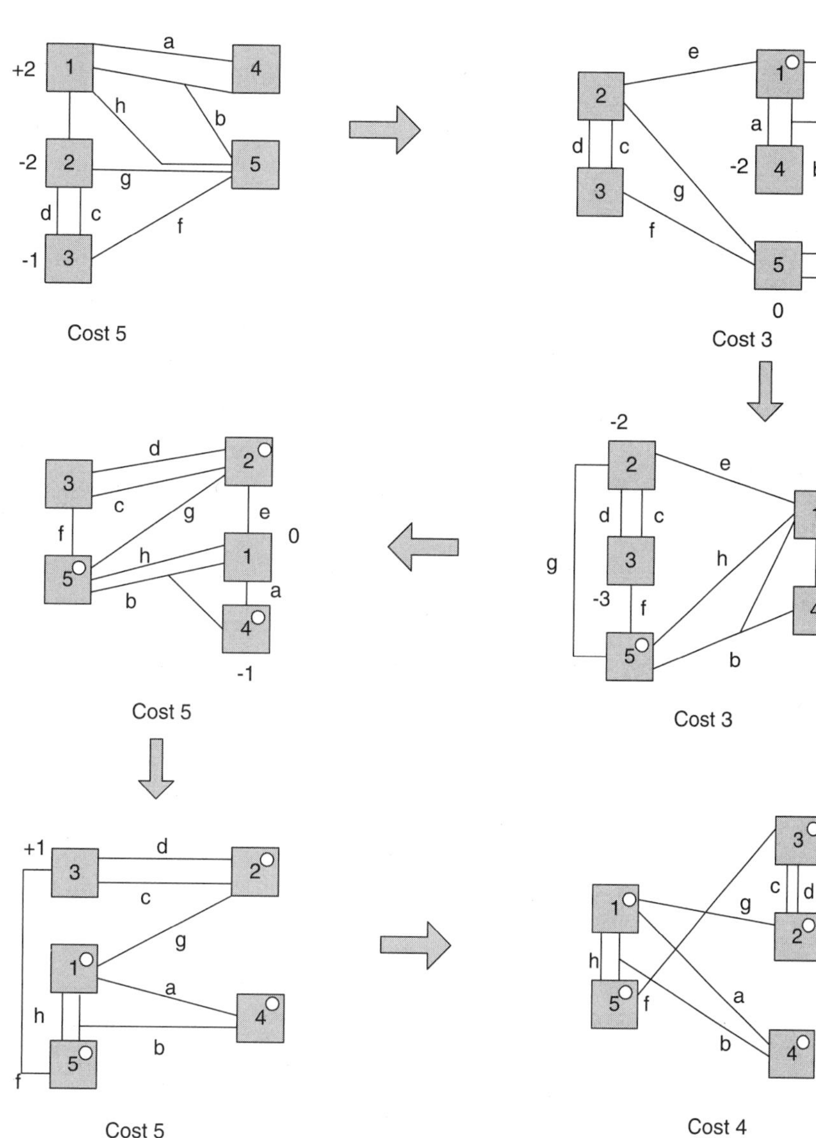

FIGURE 4.19 Illustration of a pass.

{

1. Transform the old configuration into a new configuration.
   This transformation can be in the form of exchange of positions of two randomly selected cells or change of position of a randomly selected cell.
2. Calculate the cost of the new configuration.
3. If (new cost $>$ old cost) accept the iteration, else check if the new iteration could be accepted with the probability: e (|new cost − old cost|/T). There are also other options for the probability function.

}

b. Update the temperature.

}

END

The parameter T is called *temperature*; it is initially set to a very large value so that the probability of accepting "uphill" moves is very close to one, that it is slowly decreasing toward zero, according to a rule called the cooling schedule. Usually, the new reduced temperature is calculated as follows:

$$\text{new temperature} = (\text{Cooling rate}) \times (\text{old temperature})$$

Using a faster cooling rate can result in getting stuck at local minima; however, a cooling rate that is too slow can pass over the possible global minima. In general, the cooling rate is taken from approximately 0.80 to 0.95.

Usually, the stopping criterion for the while-loop is implemented by recording the cost function's value at the end of each temperature stage of the annealing process. The stopping criterion is satisfied when the cost function's value has not changed for a number of consecutive stages.

Though simulated annealing is not the ultimate solution to placement problems, it gives very good results compared to the other popular techniques. The long execution time of this algorithm is its major disadvantage. Although a great deal of research has been done in improving this technique, substantial improvements have not been achieved.

Routing Techniques

Given a collection of cells placed on a chip, the routing problem is to connect the terminals (or ports) of these cells for a specific design requirement. The routing problem is often divided into subproblems: global, area, detailed, and specialized routing.

The global router considers the overall routing region in order to distribute the nets over the channels based on their capacities while keeping the length of each net as short as possible. For every channel that a net passes through, the net's id is placed on the sides of the channel. Once the terminals of each channel are determined, the detailed router connects all the terminals with the same id by a set of wire segments. (A wire segment is a piece of material described by a layer, two end-points, and a width.)

Area routers can be used for the general routing problem. The general purpose routers impose very few constraints on the routing problem and operate on a single connection at a time. Since these routers work on the entire design in a serial fashion, the size of the problems they can attempt is limited. They are typically used for specialized routing problems or small routing regions. These routers can do a good job of modeling the contention of nets for the routing resources and therefore can be viewed as routing the nets in parallel.

The specialized router is designed to solve a specific problem such as routing of the wires for the clock signals, and power and ground wires. Wires for clock signals must satisfy several timing constraints. Power and ground wires also require special attention for two reasons: (1) they are usually routed in one layer in order to reduce the parasitic capacitance of contacts, and (2) they are usually wider than other wires (signal and data) since they carry more current.

Global Routing

Global routers generate a nonfully specified route for each net. They assign a list of channels to each net without specifying the actual geometric layout for each wire. The global routing phase typically consists of three phases. The first phase partitions the routing space into channels and is called the channel definition phase. Channels include areas between placed blocks or above them. The second phase is called region assignment and identifies the list of channels each net is going through. The last phase is called pin assignment and assigns for each net a pin for each channel boundary it crosses.

At the end of the global routing phase the length of each net can be estimated. If some net fails to meet its timing specifications, it needs to be ripped up and re-routed. Area routers can be used for rerouting of ripped-up nets or the global routing phase is repeated.

A typical objective in the global routing problem is to minimize the sum of the estimated lengths of the globally routed nets. In addition, the number of nets crossing each channel boundary should not violate a pre-specified capacity for the boundary. This problem is NP hard even for a single multiterminal net and is known as the Steiner tree problem. In high-performance circuits, the objective is modified so that the maximum diameter of selected Steiner trees is minimized. The diameter of a Steiner tree is defined as the maximum length of a path between any two pins on the Steiner tree. This is also an intractable problem.

The global routing problem is typically studied as a graph problem. The most general and accurate model for global routing is the channel intersection model. For a given layout, each vertex of the graph represents a channel intersection. Two vertices are adjacent if there exists a channel between them. Often channel intersection graphs are extended so that the pins on the nets are included in the vertices of the graph. That way, connections between the pins can be considered.

Area Routing

Algorithms for this problem are categorized as maze routing, line probing, and shortest path-based. Since the problem of routing a multiterminal net is difficult, each multiterminal net is decomposed into several two-terminal nets. Then each resulting two terminal net is routed using either: (a) maze routing algorithms that use grid graphs; (b) line probing algorithms where the search of the minimum connection is guided by line segments instead of grid nodes; or (c) modifications of Dijkstra-based shortest path algorithms for general graphs. This subsection focuses on maze routing algorithms.

A maze router is a shortest-path algorithm on grid graphs. It finds the shortest rectilinear path by propagating a wavefront from a source point toward a destination point (Lee, 1969). Considering the routing surface as a rectangular array of cells, the algorithm starts by marking the source cell as visited. In successive steps, it visits all the unvisited neighbors of visited cells. This continues until the destination cell is visited. For example, consider the cell configuration given in Figure 4.20.

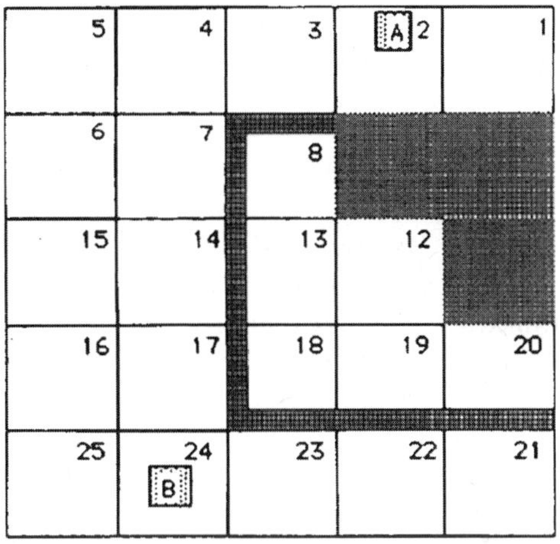

FIGURE 4.20 Initial configuration.

We would like to find a minimal-crossing path from source cell A (cell 2) to destination cell B (cell 24). (The minimal-crossing path is defined as a path that crosses over the fewest number of existing paths.) The algorithm begins by assigning the start cell 2 to a list, denoted as L, i.e., L consists of the single entry {2}. For each entry in L, its immediate neighbors (which are not blocked) will be added to an auxiliary list L′. (The auxiliary cell list L′ is provided for momentary storage of names of cells.) Therefore, list L′ contains entries {1,3}. To these cells a chain coordinate and a weight are assigned, as denoted in Figure 4.21. For example, in cell 3, we have the pair (0, →), meaning that the chain coordinate is toward the right and the cell weight is zero. The weight for a cell represents the number of wires that should be crossed in order to reach that cell from the source cell. The cells with minimum weight in list L′ are appended to list L. Thus, cells 1 and 3 are appended to list L. Moreover, cell 2 is erased from list L. Appending the immediate neighbors of the cells in L to L′, we find that list L′ now contains entries 4 and 8. Note that cell 8 has a weight of 1; this is because a wire must be crossed in order to reach this cell. Again the cells with minimum weight in list L′ are appended to list L, and cell 3 and cell 1 are erased from L. Now L contains entry {4}. The above procedure is repeated until it reaches the final cell B.

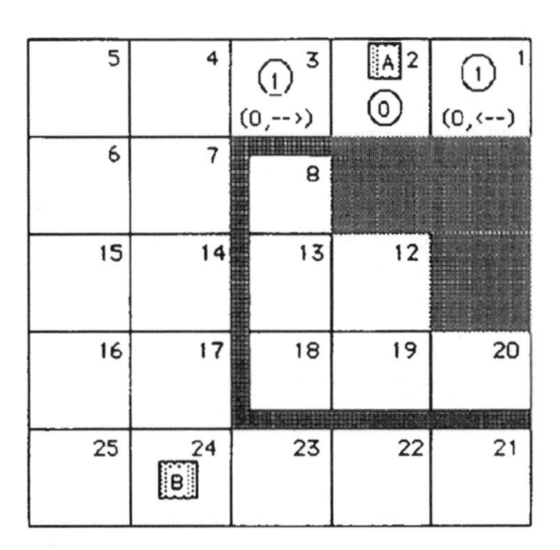

FIGURE 4.21 First step.

Then a solution is found by tracing the chain coordinated from cell B to cell A as shown in Figure 4.22.

The importance of Lee's algorithm is that it always finds the shortest path between two points— the source and the target. Since it routes one net at a time, however, there is a possibility of having some nets unrouted at the end of the routing process. The other weak points of this technique are the requirements of a large memory space and long execution time. For this reason, the maze router is often used as a side router for the routing of critical nets and/or routing of leftover unrouted nets.

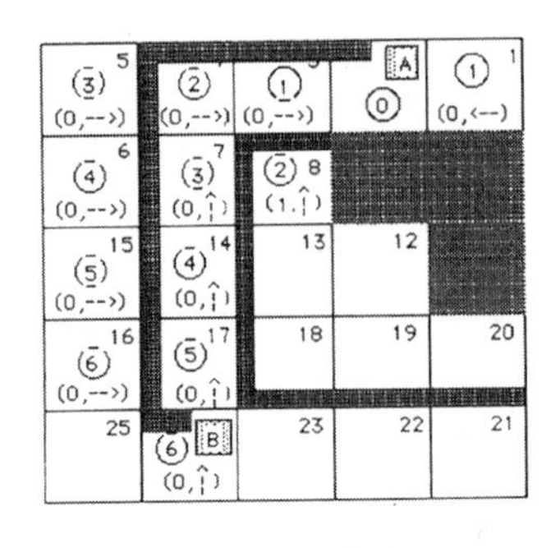

FIGURE 4.22 Final step.

Significant reductions in the space requirements are obtained using the A* search method (Sherwani, 1999), also known as Hadlock's minimum detour algorithm. The length of the path is obtained by adding on the invariant Manhattan distance twice the number of path detour points, which is called the path's detour number. A vertex on a path is called a detour point if it is directed as we move from the source to the target along the path. Thus, the path length is minimized by minimizing its detour number. Lee's algorithm is modified so that the wavefront is guided by detour numbers rather than distances from the source.

Detailed Routing

Detailed routers typically distinguish between channels that only have pins on two opposite sides as opposed to channels that have pins on three or more. Provided that no routing regions are available above blocks, the former type, called a simple channel, is defined by the region between placed blocks, whereas the latter type is defined by their intersection areas. The latter type is also called a 2D switchbox. For regions where routing over placed cells is possible, 3D switchboxes are defined. A 3D switchbox is a channel with pins on six sides (Sherwani, 1999). Typically, simple channels are routed first, since they expand, then the 2D switchboxes and finally the 3D switchboxes.

To reduce the complexity of the problem, this type of router often uses a rectangular grid on which trunks (horizontal wire segments) and branches (vertical wire segments) are typically placed on different layers. This is known as the Manhattan model. However, in a non-Manhattan model, the assignment of a layer to a vertical or horizontal direction is not enforced. Given the freedom of direction, assignment to layers reduces the channel width and vias in many cases; the latter model usually produces a better result than the former.

Two or more layers are used for detailed routing. In the following, we assume the Manhattan model, and two layers for routing. One layer accommodates trunks and the other branches. We consider only simple channels whose two boundaries are horizontal lines. The goal is to connect all the net endpoints so that the number of horizontal lines that contain a net segment is minimized. These horizontal lines are called tracks. This amounts to minimizing the total area required for the simple channel. This is also called the width of the simple channel. In the literature, many different techniques have been proposed for simple channel routers. In general, these techniques can be grouped into four different approaches: (1) algorithms (such as left-edge, greedy, hierarchical); (2) expert systems; (3) neural networks; and (4) genetic algorithms (Lengauer, 1990; Zobrist, 1994; and Sarrafzadeh, 1996).

Many algorithmic approaches resemble the operation of the left-edge algorithm, which was proposed for printed circuit board (PCB) routing. In PCB routing there is enough room to shift the position of the pins on the channel boundaries so that are no channel boundary pins on the same vertical line.

The left-edge algorithm first sorts the nets according to their leftmost endpoint. See also Figure 4.23(a). The nets are examined in their sorted order. A net is allocated to a new track only if it cannot be assigned to an existing track due to overlapping with the horizontal portion of a net that has already been assigned in the track. Figure 4.23(b) shows the routed channel by the left-edge algorithm, which uses four tracks, i.e., the channel width is 4. Notice that each net has only one trunk.

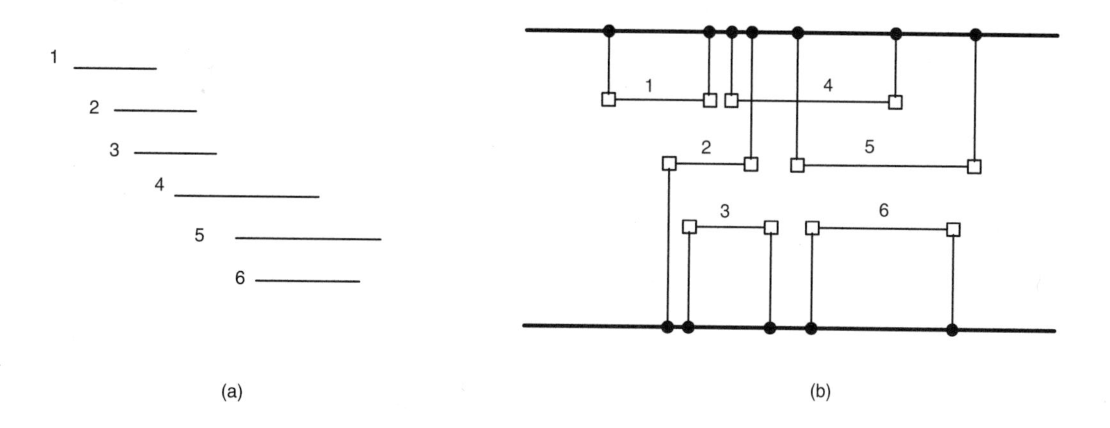

FIGURE 4.23 Simple channel routing with the left-edge routing algorithm.

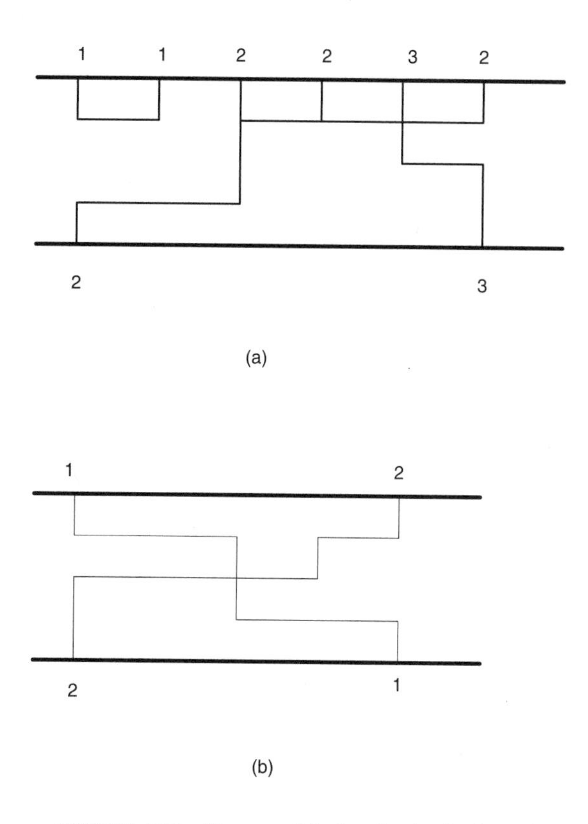

FIGURE 4.24 The use of doglegs in simple channel routing.

The left-edge algorithm guarantees minimum channel width if no channel boundary pins are on the same vertical line. However, the two-layer simple channel routing problem is intractable in the more general case where several pairs of channel boundary pins share vertical lines. In this case, more complex routing algorithms apply. Typically, a net is routed with more than one trunk. This is also called routing using doglegs.

Doglegs are useful in reducing the channel width. The channel of Figure 4.24(a) has two tracks but without doglegs, three tracks are needed. Some channel instances are infeasible to route without doglegs. This case is illustrated in Figure 4.24(b).

Defining Terms

Floor planning: An approximate position for each cell so that the total area is minimized.

Layout: Specifies the position and dimension of the different layers of materials as they would be laid on the silicon wafer.

Placement: An optimal position on the chip for a set of cells with fixed dimensions so that the total occupied area and the total estimated length of connections are minimized.

Routing: Given a collection of cells placed on a chip, the routing routine connects the terminals of these cells for a specific design requirement.

References

C.M. Fiduccia and R.M. Mattheyses, "A linear-time heuristic for improving network partitions," in *Proc. 19th Ann. Des. Autom. Conf.*, (July), pp. 175–181, 1982.

S. Kirkpatrick, C.D. Gelatt, and M.P. Vecchi, "Optimization by simulated annealing," *Science*, vol. 220, no. 4598, pp. 671–680, 1983.

C.Y. Lee, "An algorithm for path connections and its application," *IRE Trans. Electron. Comput.*, (Sept.), pp. 346–365, 1969.

T. Lengauer, *Combinatorial Algorithms for Integrated Circuit Layout*, New York: Wiley, 1990.

C.A. Mead and L.A. Conway, *Introduction to VLSI Systems*, Reading, MA: Addison-Wesley, 1980.

R.H.J.M. Otten, "Efficient floor plan optimization," *Int. J. Comput. Des.*, pp. 499–503, IEEE/ACM, 1983.

B. Preas and M. Lorenzetti, *Physical Design Automation of VLSI Systems*, Menlo Park, CA: Benjamin/Cummings, 1988.

M. Sarrafzadeh and C.K. Wong, *An Introduction to VLSI Physical Design*, New York: McGraw-Hill, 1996.

C. Sechen, "Chip-planning, placement and global routing of macro-cell integrated circuits using simulated annealing," *Int. J. Comp. Aided VLSI Des.*, 2, 127–158, 1990.

N.A. Sherwani, *Algorithms for VLSI Physical Design Automation*, 3rd ed., Dordecht, Germany: Kluwer, 1999.

L. Stockmeyer, "Optimal orientation of cells in slicing floor plan designs," *Inform. Control*, vol. 57, no. 2, pp. 91–101, 1983.

S. Sutanthavibul, E. Shargowitz, and J.B. Rosen, "An analytical approach to floor plan design and optimization," *IEEE Trans. Comput. Aided-Des.*, vol. 10, no. 6, pp. 761–769, 1991.

S. Wimer and I. Koren, "Analysis of strategies for constructive general block placement," *IEEE Trans. Comput. Aided Des.*, vol. 7, no. 3, pp. 371–377, 1988.

G.W. Zobrist, Ed., *Routing, Placement, and Partitioning*, Norwood, NJ: Ablex Publishing, 1994.

Further Information

Other recommended layout design publications include Weste and Eshraghian, *Principles of CMOS VLSI Design: A Systems Perspective*, Reading, MA: Addison-Wesley, 1988; and the book by B. Preas and M. Lorenzetti, *Physical Design Automation of VLSI Systems*, Menlo Park, CA: Benjamin/Cummings, 1988. The first book describes the design and analysis of a layout. The second book describes different techniques for development of CAD systems.

Another source is *IEEE Transactions on Computer-Aided Design of Integrated Circuits and Systems*, which is published monthly by the Institute of Electrical and Electronics Engineers.

4.3 Application-Specific Integrated Circuits

S.K. Tewksbury

Introduction

Very large scale integration (VLSI), complementary metal-oxide semiconductor (CMOS) integrated circuit (IC) technologies have evolved from simple digital circuits consisting of basic logic functions a few decades ago, to today's extraordinarily complex and sophisticated digital circuits consisting of hundreds of millions of transistors (Weste and Eshraghian, 1993; Wolf, 1994; Chandrakasan and Broderson, 1995; Baker et al., 1998; Chen, 1999; Uyemura, 1999). The routine placement of such a vast number of transistors on a thumbnail-sized substrate has enabled a vast number of technological advances, ranging from personal computers through sophisticated consumer products to new technologies appearing almost everywhere. These integrated circuits are the primary representation of microelectronics, electronic circuits/systems created using micron dimension devices. Already, dimensions have decreased below the micron level and we are entering the new era of nanoelectronics with dimensions in the nanometer range (1 nanometer equals 1/1000th of a micron, a small scale that can be visualized by recognizing that the diameter of a human hair is about 30 to 40 microns). This progression is continuing, following Moore's "law", which states that the number of transistors per IC doubles every 18 months. Table 4.2 summarizes the information provided in the 1994 version (National Technology Roadmap for Semiconductors, 1994) of the industry roadmap, providing a look both forward to the future generations and backward to prior generations. The most recent report is the 2003 version (International Technology Roadmap for Semiconductors, 2003).

Accompanying the evolution to these highly complex ICs has been an evolution of the tools and principles supporting correct and efficient design of such sophisticated circuits. These tools and principles capture the lessons of over three decades of experience obtained during earlier generations of the IC technologies. Although design of a contemporary VLSI circuit remains a significant challenge, computer-aided design (CAD) and electronics design automation (EDA) software tools (Rubin 1987; Hill and Peterson 1993; Sherwani, 1993; Banerjee, 1994; Trimberger, 2002) substantially simplify the effort involved.

Perhaps the most familiar example of the rapidly advancing VLSI technologies is the personal computer (PC), which today provides the casual user with computing power exceeding that of supercomputers of the recent past. The microprocessor (e.g., Pentium, PowerPC, etc.) is a general-purpose IC whose specific function can be customized to an application through the use of programs running on the PC. The flexibility seen in microprocessors is the result of a design intended to efficiently perform several basic functions in a flexible environment in which those functions can be manipulated by the user to perform the desired application. However, that flexibility is obtained at the expense of performance. If the function to be performed by an IC

TABLE 4.2 Prediction of VLSI Evolution (Technology Roadmap)

Year	1995	1998	2001	2004	2007	2010
Feature size (microns)	0.35	0.25	0.18	0.13	0.10	0.07
DRAM bits/chip	64M	256M	1G	4G	16G	64G
ASIC gates/chip	5M	14M	26M	50M	210M	430M
Chip size (ASIC) (mm^2)	450	660	750	900	1100	1400
Max. number of wiring levels	4–5	5	5–6	6	6–7	7–8
On-chip speed (MHz)	300	450	600	800	1000	1100
Chip-to-board speed (MHz)	150	200	250	300	375	475
Desktop supply voltage (V)	3.3	2.5	1.8	1.5	1.2	0.9
Maximum power (W), heatsink	80	100	120	140	160	180
Maximum power (W), no heatsink	5	7	10	10	10	10
Power (W), battery systems	2.5	2.5	3.0	3.5	4.0	4.5
Number of I/O connections	900	1350	2000	2600	3600	4800

Adapted from *The National Technology Roadmap for Semiconductors,* Semiconductor Industry Association, San Jose, CA, 1994.

can be made specific (e.g., analyzing images for content), then significant performance advantages normally result by designing an IC specifically for that function (and no others) (Parhi, 1998). Representative performance metrics include power dissipation (e.g., for battery-operated products) (Sanchez-Sinencio, 1999), computation rates (e.g., for handling image analysis) and other factors. In applications for which the performance advantages are considerable, application-specific integrated circuits (ASICs) provide an important alternative to general-purpose integrated circuits.

The traditional distinction between general-purpose and application-specific integrated circuits is, however, increasingly blurred as a result of the substantial complexity available in a contemporary VLSI circuit. For example, it is possible to integrate a microprocessor that was a full IC a few years ago with application-specific circuitry to create an integrated, application-specific system on a single IC. This leads to the concept of "system on a chip" (SoC) (Wolf, 2002) in which a single IC is essentially the entire digital system. For the purposes of this section, we define "application-specific" as meaning that the IC is designed to provide substantial performance benefits compared to the performance achievable with general-purpose digital components.

Primary Steps of VLSI ASIC Design

The VLSI IC design process consists of a sequence of well-defined steps (Preas and Lorenzetti, 1988; Hill et al., 1989; DeMicheli, 1994a,b) related to the definition of the functions to be designed; organization of the circuit blocks implementing these logic functions within the area of the IC; verification and simulation at several stages of design (e.g., behavioral simulation, gate-level simulation, circuit simulation [White and Sangiovanni-Vincentelli, 1987; Lee et al., 1993]); routing of physical interconnections among the functional blocks; and final detailed placement and transistor-level layout of the VLSI circuit. The design process starts with a hierarchical top-down planning (Gajski, 2002), functionally specifying one of the blocks comprising the IC, representing that block in terms of simpler blocks and proceeding to more detailed levels until the transistor-level description has been completed. During this process, information gained as one moves toward the most detailed level may lead to changes in the assumptions used at higher levels of the design. In this case, "back-annotation" is used to update the representations of the higher-level functions, perhaps requiring a retraversal of the design steps to the most detailed level.

Representative steps in this process are illustrated generally in Figure 4.25(a), where details of feedback loops have been suppressed. An illustrative example (Lipman, 1995) of a design process including the specific feedback loops is shown in Figure 4.25(b). The general steps shown in Figure 4.25(a) are as follows:

A. *Behavioral Specification of Function:* The behavioral specification is essentially a generalized description of what the function will do, without detailed regard for the manner in which the function is constructed. High-level description languages (HDLs) such as VHDL (Armstrong, 1989; Lipsett et al., 1990; Mazor and Langstraat, 1992; Bhaskar, 1999; Rustin, 2001) and Verilog (Thomas and Moorby, 1991) provide a framework in which the behavior of a function can be specified using a programming language (VHDL based on the ADA programming language and Verilog based on the C++ programming language). These HDLs support system descriptions that are strongly tied to the physical realization of the function. The behavioral description relates generally to a data sheet for a circuit function. Also provided are structural descriptions (describing how the function is constructed from simpler subfunctions) and dataflow descriptions (describing how signals flow through the subfunctions). Figure 4.26(a) illustrates this specification of an overall function in terms of subfunctions (A[s], B[s], ..., E[s]) as well as expansion of one of these subfunctions (C[s]) in terms of still simpler functions (c1, c2, c3, ..., c6).

B. *Verification of Function's Behavior:* It is important to verify that the design descriptions (behavioral, structural, and others) truly provide the function sought by the designer. VHDL and Verilog are specialized forms of their underlying programming languages (ADA and C++, respectively) developed to explicitly support simulation of the outputs of represented functions, given inputs to the functions. The designer can then apply an appropriate set of test inputs to the design function and compare the

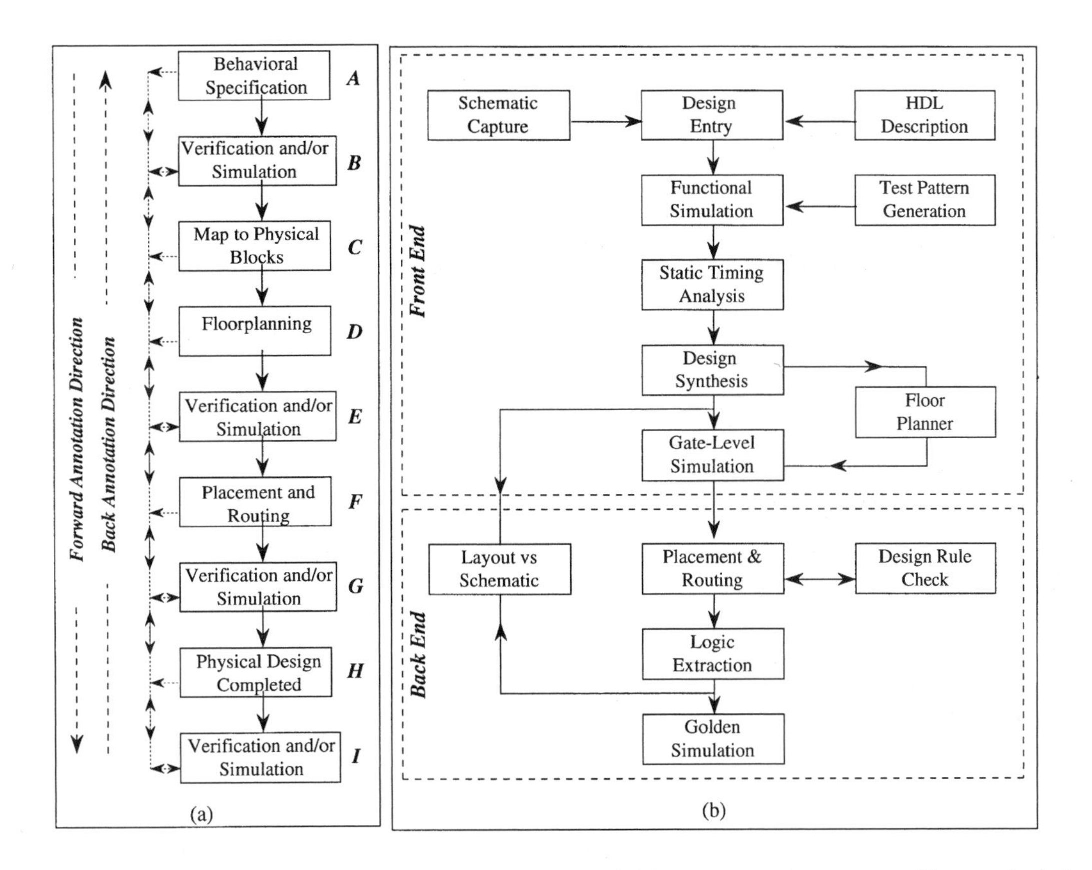

FIGURE 4.25 Representative VLSI design sequences. (a) Simplified but representative sequence; (b) example design approach (Lipman, 1995).

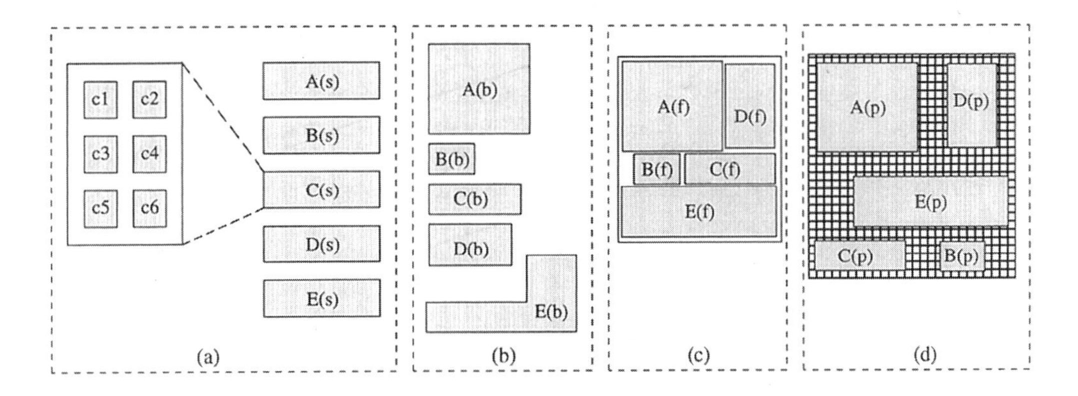

FIGURE 4.26 Circuit stages of design. (a) Initial specifications (e.g., HDL, schematic, etc.) of ASIC function in terms of functions, with the next lower level description of function C illustrated. (b) Estimated size of physical blocks implementing functions. (c) Floorplanning to organize blocks on an IC. (d) Placement and routing of interconnections among blocks.

function's output with the expected outputs. This verification is repeated for all levels of the hierarchical design, ranging from the simplest low-level functions up through the overall function of the IC. These HDLs also provide means of including realistic signal properties such as delays through circuits and along interconnections, thereby allowing evaluations of signal timing effects.

C. *Mapping of Logical Functions into Physical Blocks*: Next, the logical functions, e.g., A(s), B(s), . . ., E(s) in Figure 4.26(a) are converted into physical circuit blocks A(b), B(b), . . ., E(b) in Figure 4.26(b). Each physical block represents a logical function as a set of interconnected gates. Although details of the physical layout are not known at this point, the estimated area and aspect ratio (ratio of height to width) of each circuit are needed to organize these blocks within the area of the IC.
D. *Floor Planning:* Next, the individual circuit blocks are compactly arranged to fit within the minimum area (to allow the greatest amount of logic to be placed in the IC area). Floor planning establishes this organization, as illustrated in Figure 4.26(c). At this stage, routing of interconnections among blocks has not been performed, perhaps requiring modifications to the floor plan to provide space when these interconnections are added. During floor planning, the design of a logic function in terms of a block function can be modified to achieve a shape that better matches the available IC area. For example, the block E(b) in Figure 4.26(b) has been redesigned to provide a different geometric shape for block E(f) in the floor plan in Figure 4.26(c).
E. *Verification/Simulation of Function Performance:* Given the floor plan, it is possible to estimate the average length of interconnections among blocks (with the actual length not known until after interconnection routing is completed in Step F below). Signal timing throughout the IC is estimated, allowing verification that the various circuit blocks interact within timing margins.
F. *Placement and Routing:* When an acceptable floor plan has been established, the next step is to complete the routing of interconnections among the various blocks of the design. As interconnections are routed, the overall IC area may need to expand to provide the area needed for the wiring, with the possibility that the starting floorplan (which ignored interconnections) is not optimal. Placement (Shahookar and Mazumder, 1991) considers various rearrangements of the circuit blocks, without changing their internal design but allowing rotations, reflections, etc. For example, the arrangement of some of the blocks in Figure 4.26(c) has been changed in the arrangement in Figure 4.26(d).
G. *Verification/Simulation of Performance:* Following placement and routing, the detailed layout of the circuit blocks and interconnections has been established and more accurate simulations of signal timing and circuit behavior can be performed, verifying that the circuit behaves as desired with the interblock interconnections in place.
H. *Physical Design at Transistor/Cell Level:* Each block in the discussion above may itself consist of subblocks of logic functions. The same sequence of steps listed above can be applied to place those subblocks, route their interconnections, and simulate performance. This process continues at successive levels of detail in the circuit design, eventually reaching the point at which the elements being placed are individual transistors and the interconnections being routed are those among the transistors. Upon completion of this detailed level, the set of physical masks can be specified, translating the transistor-level design into the actual physical structures within the layers of the silicon device fabrication. Upon completion of the mask designs, the details of the physical circuit dimensions are available.
I. *Verification/Simulation of Performance:* Before fabricating the masks and proceeding with manufacture of the ASIC circuit, a final verification of the ASIC is normally performed. Figure 4.25(b) represents this step as "golden simulation," a process based on detailed and accurate simulation tools, tuned to the process of the foundry and providing the final verification of the performance of the ASIC.

Increasing Impact of Interconnection Delays on Design

In earlier generations of VLSI technology (with larger transistors and wider interconnection lines/spacings), delays through low-level logic gates greatly dominated delays along interconnection lines. Under these conditions, it was often possible to move blocks of circuitry in the steps discussed above without excessive concern regarding the lengths of connections among the blocks since the interconnection delays were of secondary concern. However, as feature sizes have decreased, delays through logic gates have also decreased. For technologies with feature sizes less than about 0.5 microns, interconnection delays became larger than logic delays.

Figure 4.27 illustrates these delay issues on technology scaling to smaller feature sizes. A logic function *F* in a previous-generation technology requires a smaller physical area and has a higher speed in a later scaled

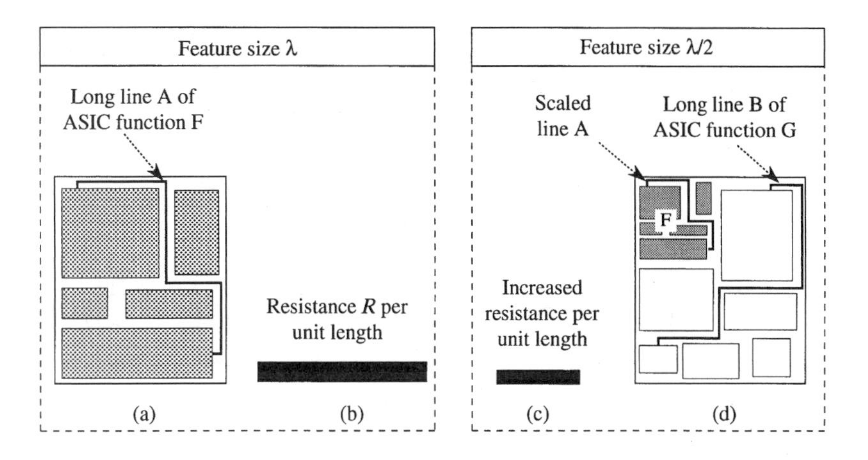

FIGURE 4.27 Interconnect lengths under scaling of feature size. (a) Initial VLSI ASIC function *F* with line A extending across IC. (b) Interconnection cross-section with resistance R^* per unit length. (c) Interconnection cross-section, in scaled technology, with increased resistance per unit length. (d) VLSI ASIC function G in scaled technology containing function *F* but reduced in size (including interconnection A) and containing a long line B extending across IC.

technology (i.e., a technology with feature size decreased). Although intrablock line lengths decrease (relaxing the impact within the block of the higher resistance R^* per unit length and the higher capacitance C^* per unit length of the interconnect line, leading to a larger R^*C^* delay), the interblock lines continue to have lengths proportional to the overall IC size. The result is a larger R^*C^* delay on the interblock lines.

When interconnect delays dominate logic delays, movements of logic blocks in the design steps in the previous section become more complicated because rearranging the logic blocks causes interconnection line lengths to change, leading to substantial changes in the timing of signals throughout the rearranged circuit. The distinction made here can best be understood by considering the two limiting cases:

- Case 1: Logic delays are nonzero and interconnect delays are zero (roughly representing earlier generations of VLSI circuits). In this limiting case, the speed performance of a VLSI circuit does not depend on where the various blocks are placed on the IC.
- Case 2: Interconnect delays are nonzero and logic delays are zero (roughly representing current generations of VLSI circuits). In this limiting case, the speed performance of a VLSI circuit depends completely and critically on the detailed placement of the various blocks.

This challenge can be relaxed in part through the use of architectural approaches that allow long interconnect delays among blocks of the VLSI circuit. However, the sensitivity to placement and routing also requires that designs evolve through an iterative process as illustrated by the back- and forward-annotation arrows in Figure 4.25(a). Initial estimates of delays in Step B need to be refined through back-annotation of interconnection delay parameters obtained after floorplanning and/or after placement, and routing to reflect the actual interconnection characteristics, perhaps requiring changes in the initial specification of the desired function in terms of logical and physical blocks. This iterative process moving between the logical design and the physical design of an ASIC has been problematic since often the logical design is performed by the company developing the ASIC, whereas the physical design is performed by the company (e.g., the "foundry") fabricating the ASIC. CAD tools are important vehicles for coordination of the interface between the designer and the foundry.

General Transistor Level Design of CMOS Circuits

The previous subsection emphasized the CAD tools and general design steps involved in designing an ASIC. A top-down approach was described, with the designer addressing successively more detailed portions of the overall design through a hierarchical organization of the overall description of the function. However,

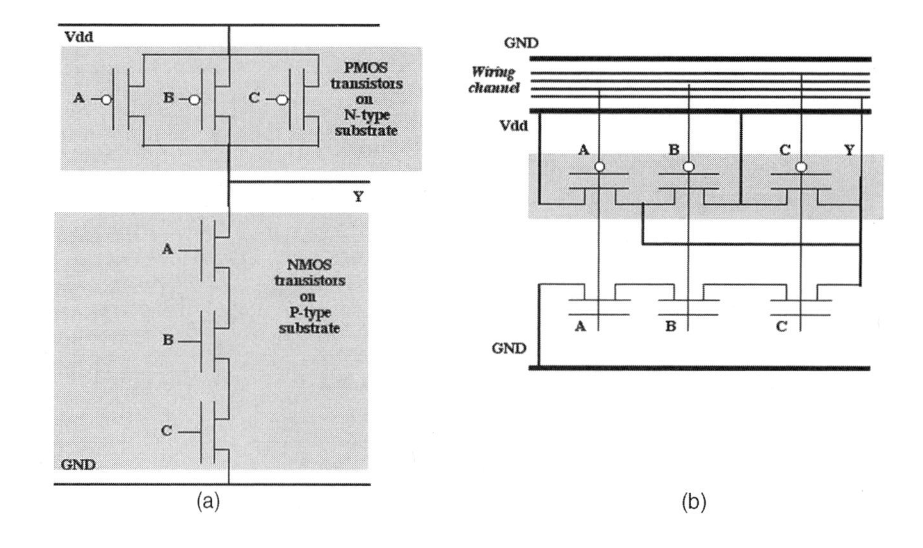

FIGURE 4.28 Transistor representation of three-input NAND gate. (a) Transistor representation without regard to layout. (b) Transistor representation using parallel rows of PMOS and NMOS transistors, with interconnections connected from wiring channel.

the design process also presumes a considerable understanding of the bottom-up principles through which the overall IC function will eventually appear as a fully detailed specification of all the transistor and interconnection structures throughout the overall IC (Dillinger, 1988; Weste and Eshraghian, 1993; Kang and Leblebici, 1996; Rabaey, 1996; Baker et al., 1998).

Figure 4.28 illustrates the transistor-level description of a simple three-input NAND gate. VLSI ASIC logic cells are dominated by this general structure where the PMOS transistors (used to create the pull-up section) are connected to the supply voltage Vdd and the NMOS transistors (used to create the pull-down section) are connected to the ground return GND. When the logic function generates a logic "one" output, the pull-up section is shorted through its PMOS transistors to Vdd while the pull-down section is open (no connection to GND). For a logic "zero" output, the pull-down section is shorted to GND whereas the pull-up section is open. Since the logic output is either "zero" or "one," only one of the sections (pull-down or pull-up) is shorted and the other section is open, with no DC flowing directly from Vdd to GND through the logic circuitry. This leads to a low DC power dissipation, a factor that has driven the dominance of CMOS for VLSI circuits.

The PMOS transistors used in the pull-up section are fabricated with *p*-type source and drain regions on *n*-type substrates. The NMOS transistors used in the pull-down section, however, are fabricated with *n*-type source and drain regions on *p*-type substrates. Since a given silicon wafer is either *n*-type or *p*-type, a deep, opposite doping-type region must be placed in the silicon substrate for those transistors needing a substrate of the opposite type. In Figure 4.28, a *p*-type substrate (supporting the NMOS transistors) is assumed. The shaded region illustrates the area requiring the deep doping to create an *n*-type well in which the PMOS transistors can be fabricated. As suggested in this figure, the *n*-type well can extend across a significant length, covering the area required for a multiplicity of PMOS transistors. For classical CMOS logic cells, the same set of input signals is applied to both the pull-up and the pull-down transistors. As shown in Figure 4.28(b), proper sequencing of the PMOS and NMOS transistors allows the external connection (A, B, etc.) to extend straight down (on polysilicon, passing under the Vdd metal interconnection) to contact both the PMOS and corresponding NMOS transistor. Algorithms to determine the optimum sequence of transistors evolved in early CAD tools.

Each logic cell must be connected to power (Vdd) and ground (GND), requiring that external power and ground connections from the edge of the IC be routed (on continuous metal to avoid resistive voltage drops) to each logic cell on the IC (Zhu, 2004). Early IC technologies provided a single level of metalization on which to route power and ground, leading to the interdigitated layout illustrated in Figure 4.29(a). Given this layout

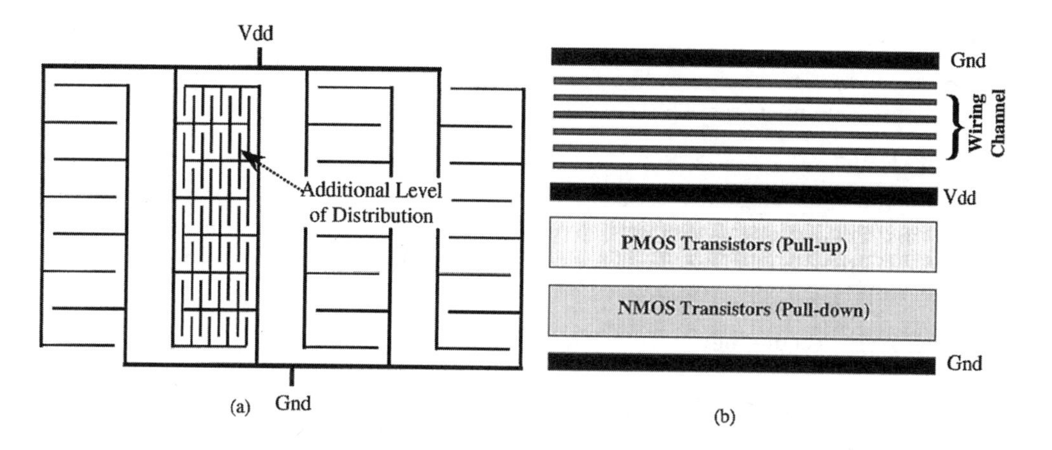

FIGURE 4.29 Power and ground distribution (interdigitated lines) with rows of logic cells and rows of wiring channels. (a) Overall power distribution and organization of logic cells and wiring channels. (b) Local region of power distribution network.

approach, channels of pull-up sections and channels of pull-down sections were placed between the power and ground lines (Figure 4.29(b)). If all logic cells had the same height (regardless of the number of transistors in the pull-up and in the pull-down sections), the layout of the circuit becomes much simpler and areas for logic cells can be more easily estimated. This can be achieved by organizing the PMOS and the NMOS transistors in single rows, as illustrated in Figure 4.28(b). Under these conditions, logic cells can be abutted against one another as shown in Figure 4.30 while retaining the convenience of straight power and ground lines. To interconnect transistors within a logic cell, area between the PMOS and NMOS transistors is provided for intra-cell transistor interconnections. This allows each cell to be identified by an enclosing box (dotted lines in Figure 4.30) and each such box labeled (e.g., by the name of the logic function). Connections to and from the enclosing box are inter-cell interconnections, which are completed in the inter-cell wiring channel in Figure 4.30.

After completing the detailed transistor-level designs of the logic cell, that cell can be used as a "block" to build a higher-level logic function (e.g., a logic function comprised of several basic logic gates and flip-flops). At this higher level of description, the details within the box are discarded and only the named box appears. The designer at this stage does not "see" the internal wiring within the logic cells and therefore cannot run interconnection wires across named boxes without being in danger of shorting to internal "hidden" interconnections. This restriction has been relaxed by current VLSI technologies using multiple layers of metalization. For example, if all internal interconnections within a box are restricted to layer one of metal, then higher layers of metal can pass over the named box without danger of shorting. However, the problem of not "seeing" internal details of boxes reappears at the next higher level of design and special "boxes" (switch boxes) containing interconnections only are often used as a means of routing signals in one inter-cell wiring channel to a wiring channel above or below that channel.

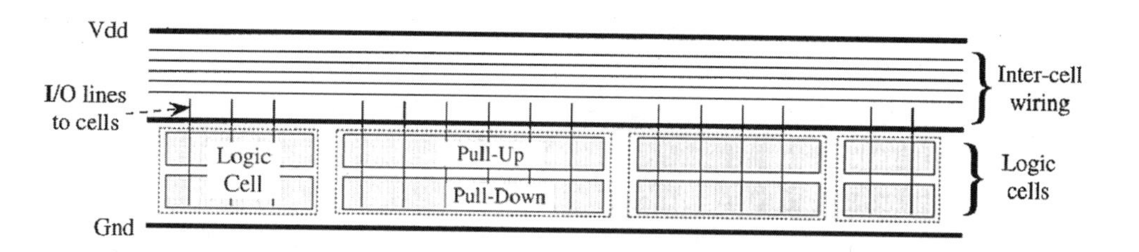

FIGURE 4.30 Cell-based logic design, with cells organized between power and ground lines and with intercell wiring in channels above (and/or below also) the cell row.

ASIC Technologies

Drawing on the discussion above regarding transistor-level design, the primary ASIC technologies (full-custom, gate arrays, field-programmable gate arrays, etc.) can be easily summarized.

Full Custom Design

In *full custom design,* custom logic cells are designed for the specific low-level functions desired and the logic cells are selected, placed, and interconnected to optimally create the desired IC logic function. As illustrated in Figure 4.31(a), all elements of the overall design are generated in detail according to the function desired. The design can also incorporate nonstandard, low-level logic circuitry as desired.

Standard Cell ASIC Technology

With a sufficiently rich set of predesigned, low-level logic cells, it is possible to design the entire VLSI circuit using only that set of cells. The *standard cell design* (Heinbuch, 1987; *SCMOS Standard Cell Library,* 1989) makes use of libraries of standard cell layouts which are used as low-level "blocks" to create the overall design. The overall design proceeds as in full custom design (and is represented by Figure 4.32(a)), but uses someone else's low-level cell designs.

Gate Array ASIC Technology

The *gate array design* (Hollis, 1987) is based on partially prefabricated (up to but not including the final metalization layer) wafers populated with low-level logic cells and stored until the final metalization is completed. Since they are prefabricated, a decision has been made regarding the number and placement of each of the low-level logic cell functions as well as the inter-cell interconnections. From the perspective of the VLSI designer, low-level logic cells have been defined and placed already, without regard for the overall VLSI function to be performed. The ASIC designer maps the desired logic function onto this array of logic cells, using those needed and ignoring those not needed (these will be "wasted" logic cells). Figure 4.33 illustrates this step.

The gate array technology shares the costs of masks for all layers except the metalization layer among all ASIC customers, exploiting high-volume production of these gate array wafers. The ASIC customer incurs the cost of the metalization mask and the fabrication cost for that last metalization step. There is clearly wasted circuitry (and performance lost, since interconnections will tend to be longer than in full custom or standard cell designs).

CMOS Custom Circuits with Megacell Elements

As the complexity of VLSI ICs has increased, it has become possible to include standard, higher-level functions (e.g., microprocessors, DSPs, PCI interfaces, MPED coders, RAM arrays, etc.) within a custom ASIC.

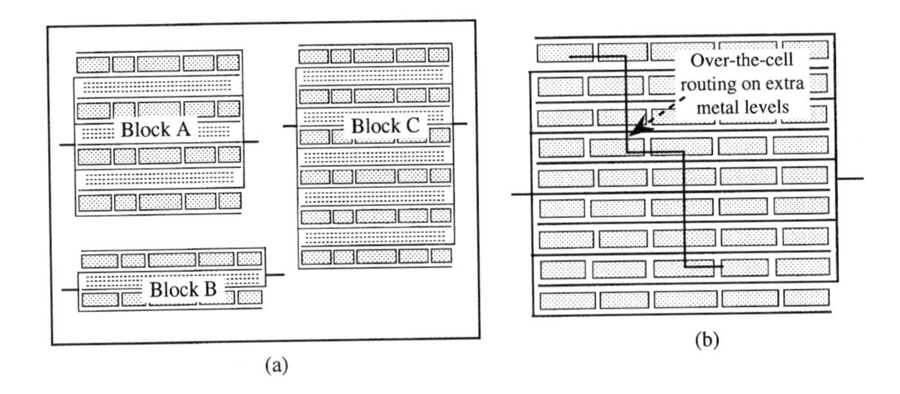

FIGURE 4.31 (a) Construction of larger blocks from cells with wiring channels between rows of cells within the block. (b) Over-the-cell routing on upper metal levels which are not used within the cells.

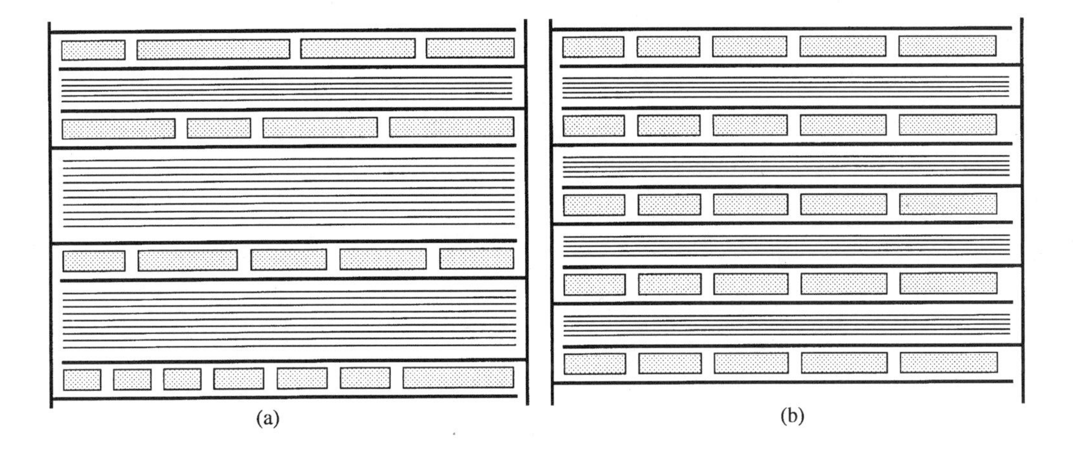

FIGURE 4.32 (a) Full custom layout (using custom cells) or standard cell ASIC layout (using library cells). (b) Gate array layout with fixed width channels (in prefabricated, up to metalization, wafers).

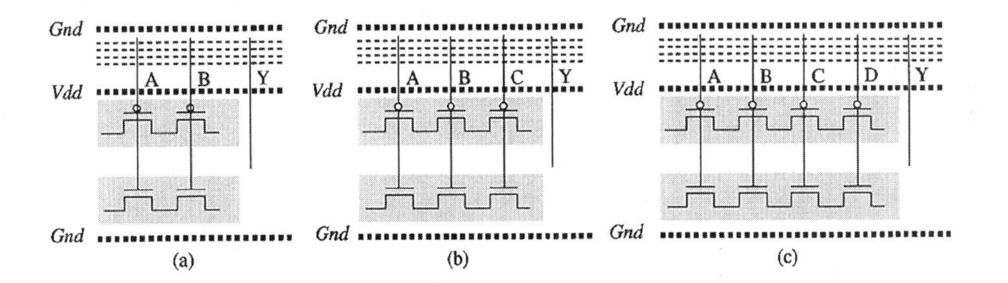

FIGURE 4.33 Representative gate array cells. (a) Two-input gate, (b) three-input gate, and (c) four-input gate. The dashed lines represent the power and ground lines, as well as interconnections in the wiring channel, which are placed on the IC during customization.

For example, an earlier generation microprocessor might offer a desired functionality and would occupy only a small portion of the area of the custom VLSI IC. Including such a microprocessor has the advantage of allowing the custom designer to focus on other parts of the design and also provides users with a standard microprocessor instruction set and software development tools. Such large cells are called *megacells* and are essentially detailed, mask-level designs (or other usable function descriptions) provided by the "owner" of the higher-level function.

Programmable Logic: Use of General Purpose Logic Cells

The *programmable gate array* (*PGA*), like the gate array discussed above, places fixed cells in a two-dimensional array organization on the IC and the application user designs an application function from this array. However, in contrast to the simple logic functions seen in standard gate array technology, the cells of the PGA are complex logic functions (often including flip-flops) whose function can be specified by the user. In this sense, the user is "designing" the individual cells of the array. In addition to specifying the function to be performed by each of the generalized logic cells, the user specifies the interconnections among the logic cells and the overall input/output connections of the IC.

One version of this technology requires a final fabrication step to embed the user's design in the IC. The flexibility of the generalized logic cells typically leads to a more efficient design (logic per unit area, speed, etc.) than standard gate array approaches. In addition, rather than designing masks to create the interconnections,

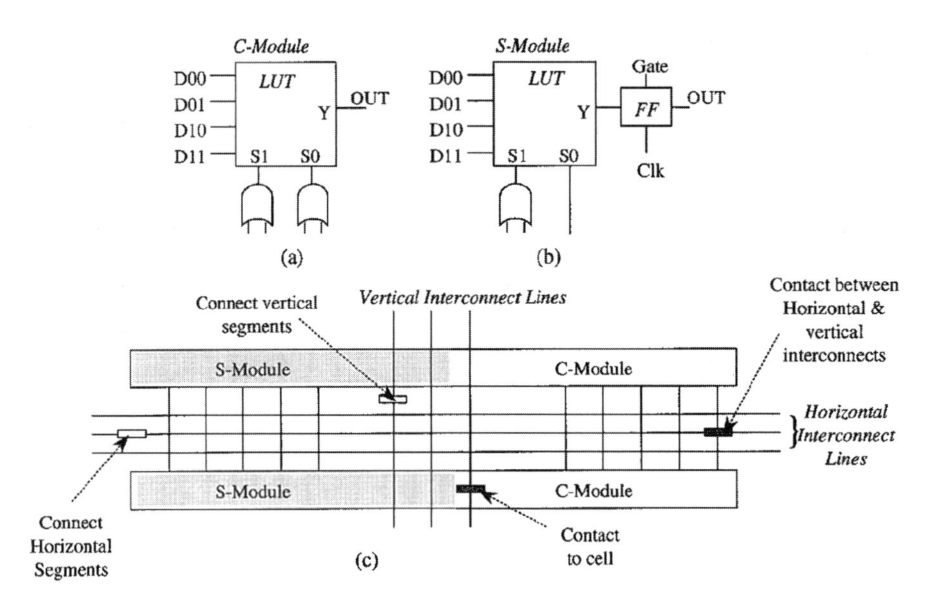

FIGURE 4.34 Example of FPGA elements (Actel FPGA family [Actel, 1995]). Combinational cells (a) and sequential cells (b). (c) Programmable wiring organization.

normally closed "antifuses" are included to allow one-time programming of the logic cells and their interconnectionsby opening the fuses.

The second version, the *field-programmable gate array* (FPGA),[1] is completely fabricated and packaged when obtained by the user. "Field programmability" refers to the ability of a user to program the functionality of the IC by loading data information into the FPGA, in much the same manner that one can program the information stored in memory by loading ("writing") that information into the memory IC. Rather than using antifuses that must be opened during a final fabrication step, the FPGA provides electronically alterable fuses that can be written once or electronic switches whose control signals are stored in local storage elements (after being "written") in the FPGA. Storage of programming information on flip-flops within the FPGA leads to volatile operation (the stored FPGA program information is lost when power is turned off), whereas storage of the information in EPROM (write once memory) or EEPROM (multiple write memory) provides nonvolatile operation.

Figure 4.34 shows an example of cells and programmable interconnections for an earlier simple but representative FPGA technology (Actel, 1995). The array of cells is constructed from two types of cell, which alternate along the logic cell rows of the FPGA. The combinational logic cell — the "C-module" in Figure 4.34(a) — provides a ROM-based lookup table (LUT) capable of realizing any complex logic function with four inputs and two control signals. The sequential cell — the "S module" in Figure 4.34(b) — adds a flip-flop to the C module, allowing efficient realization of sequential circuits. The interconnection approach illustrated in Figure 4.34(c) is based on (1) short vertical interconnections directly connecting adjacent modules, (2) long vertical interconnections extending across the IC height, (3) long horizontal interconnections extending across the IC width, (4) points at which the long vertical interconnections can be connected to cells, and (5) points at which the long vertical and horizontal lines can be used for general routing. The long vertical and horizontal lines are broken into segments, with programmable links between successive segments. The programmer can then connect a set of adjacent line segments to create the desired

[1]The term "field-programmable gate array" is copyrighted by Xilinx Corporation (Xilinx, 2004), a major producer of programmable logic circuits. However, like the term "Xerox" which has come to be used generically, "field-programmable gate array" (FPGA) has come to be used generically for this style of VLSI circuit.

interconnection line between nonlocal modules. In addition, programmable connection points allow the programmer to make transitions between the long vertical and long horizontal lines. By connecting various control inputs to a module of the FPGA to either Vdd or GND, the module can be "programmed" to perform one of its possible functions. The basic array of modules is complemented by additional driver and other circuitry around the perimeter of the FPGA for interfacing to the "external world."

Different FPGA manufacturers have developed FPGA families with differing basic cells, seeking to provide the most useful functionality for generation of the overall FPGA custom function.

The ability of the user to completely change the function performed by the FPGA simply by loading a different set of programming information has led to a number of interesting architectural opportunities for systems designers. For example, microprocessor applications in which the actual hardware of the microprocessor and the instruction set of the microprocessor can be changed during execution of a software program to provide an overall architecture optimized for the specific operations being performed. FPGAs do not provide the level of performance seen in full custom VLSI designs and are substantially more expensive than large volume production of the full custom designs. Software tools have been developed to allow easy translation of an FPGA design into a higher performance (and lower cost) PGA or other VLSI technologies.

Interconnection Performance Modeling

Accurate estimation of signal timing throughout a VLSI circuit is increasingly important in contemporary VLSI ASICs (in which interconnection delays often dominate the speed performance) and becomes increasingly important as technologies are scaled to smaller feature sizes. High clock rates impose tighter timing margins, requiring more accurate modeling of the delays on both the global clock signal and other signals.

The increasing importance of interconnection delays (Gradinski, 2000; Hall et al., 2000) relative to gate delays can be seen in the history of VLSI technologies. In the earlier 1-micron VLSI technologies, typical gate delays were about six times the average interconnection delays. For 0.3-micron technologies, the decreasing gate delays led to average interconnection delays being about six times greater than typical gate delays. Accurate estimation of signal delays early in the design process is increasingly difficult since the designer does not have, at that point in the design process, accurate estimates of interconnection lengths and nearby lines causing crosstalk noise. Only much later in the design when the blocks have been placed and interconnections routed do these lengths and interline couplings become well known. As the design proceeds and the interconnection details become better understood, parameters related to signal timing can be fed back (back-annotated) to the earlier design steps, allowing modifications in those steps to achieve the required performance.

In earlier VLSI technologies, a linear delay model was adequate, representing the overall delay T from the signal input to one cell (cell A in Figure 4.35) to the input of the connected cell (cell B in Figure 4.35). Such a delay model has the general form $\tau = \tau(0) + k_1 \cdot C(\text{out}) + k_2 \cdot \tau(s)$, where τ_0 is the intrinsic (internal) delay of the cell with no output loading, $C(\text{out})$ is the capacitive load seen by the output driver of the cell, $\tau(s)$ is the (no load) **rise/fall time** of the cell's output signal, and the parameters k_1 and k_2 are constants (perhaps geometry dependent). In the case of deep submicron CMOS technologies, the overall delay must be divided

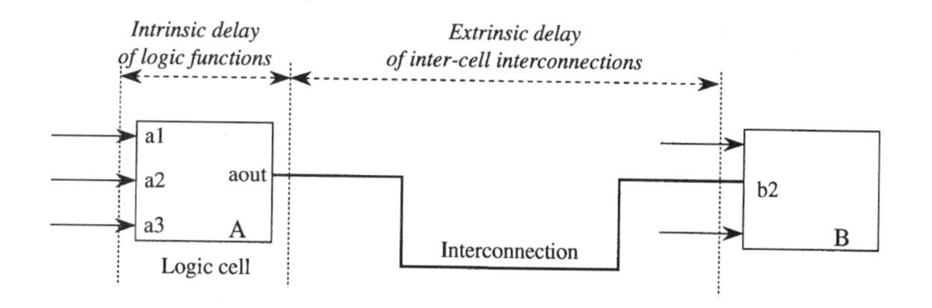

FIGURE 4.35 Logic cell delay (intrinsic delay) and intercell interconnection (extrinsic delay).

into the **intrinsic delay** of the cell and the **extrinsic delay** of the interconnect—each delay having substantially more complex models than the linear model.

Factors impacting the intrinsic delay of cells include the following, with the input and output signals referring to logic cell A in Figure 4.35:

1. Different input states during an input signal's transition may lead to different delays through the cell.
2. Starting from the time when the input starts to change, slower transition times lead to longer delays before the threshold voltage is reached, leading to longer delays to the output transition.
3. Once the input passes the threshold voltage, a slower changing input may lead to a longer delay to the output transition.
4. The zero-to-one change in an input may cause a delay different than a one-to-zero change.

These are merely representative examples of the more complex behavior seen in the logic cells as the feature size decreases.

The models used for interconnections (Tewksbury, 1994; Hall et al., 2000) have also changed, reflecting the changing interconnection parameters and increasing clock rates. The four primary models are as follows:

Lumped RC Model: If the rise/fall times of the signal are substantially greater than the round-trip propagation delay of the signal, then the voltage and current are approximately constant across the length of the interconnection. As a result, the interconnection can be modeled using a single lumped resistance and a single lumped capacitance leading to an RC delay model (the resistance of the driver and the capacitance of the load must be included to obtain the total RC time constant).

Distributed RC Model: If the interconnection line is sufficiently long, the rise/fall time of the signal will be less than the round-trip propagation delay of the signal. In this case, the voltages and currents at any instant in time are not constant along the length of the interconnection. In this case, the long line can be divided into smaller segments, each sufficiently small to allow use of a lumped RC model for the segment. The result is a series of lumped RC sections and a delay resulting from this series.

Distributed RLC Model: As the rise/fall times become shorter, the relative contributions of capacitance and inductance change. The impedance associated with the line capacitance is inversely proportional to frequency and decreases as the signal's frequencies increase. The impedance associated with the line inductance (negligible in the cases above) is proportional to frequency and increases as the frequencies increase. At signal frequencies sufficiently high that the inductive impedance is not negligible, the distributed RC model must be replaced by a distributed RLC model.

Transmission Line Model: At still higher signal frequencies, the representation of the interconnection as a series of discrete segments fails and a differential model of the interconnection is required. This leads to the traditional transmission line model with characteristic impedance, reflections, and other standard transmission line effects. The delay is represented by a signal "propagating" along the interconnection line, arriving at different points along the line at different times. Termination of the interconnection in the characteristic impedance of the transmission line is needed to avoid problems associated with signal reflections at the ends of the line.

Given the wide range of interconnection lengths found in a typical IC (with very short interconnects connecting adjacent logic cells and with very long interconnections extending across the entire IC), all four models above are relevant (the transmission line model arising, however, only for very high signal frequencies). In all cases, knowledge of the capacitance C^* and inductance L^* per unit length along the lines is essential. To model crosstalk effects, the coupling capacitance per unit length to nearby lines must be known. With multiple metal layers now used for interconnections, this calculation of the coupling capacitance has become more difficult, particularly since the value of the coupling capacitance depends on the routing and placement of nearby interconnections.

The discussion above has highlighted signal interconnections. However, the voltages appearing on the power and ground interconnections are not constants, but instead reflect the resistive losses and changing currents on those power and ground interconnections. Overall, *signal integrity* management is a critical element of the design process for very high speed VLSI.

Clock Distribution

The challenge of managing signal timing across an entire IC containing millions of gates is substantially relaxed by the use of *synchronous* designs. In such designs, signals are regularly retimed using clocked flip-flops, synchronizing the times when signals change to the transition time of a clock signal. In this manner, the complexities of signal timing can generally be restricted to local regions of an IC. Today's ICs use a vast number of retiming points (often in the form of data registers) for this purpose. To be generally effective (and to avoid extreme complexities in adjusting signal delays throughout an IC), these synchronous designs employ a common clock signal distributed in such a manner that its transition times are the same throughout the IC. *Clock skew* is the maximum difference between the times of clock transitions at any two flip-flops in the overall IC and typically has a value substantially smaller than the clock period. For example, part of the clock period of a high speed VLSI circuit is consumed by the rise/fall times of the signals appearing at the inputs to the flip-flops. Another part of the clock period is consumed by the required time that the signal at the input to a flip-flop must remain constant before (setup time) and following (hold time) the clock transition at the flip-flop. As a result, to achieve maximum speed, the clock skew must generally be less than about 20% of the clock period. For a 500-MHz clock (with clock period equal to 2 nsec), this represents a clock skew of only about 0.2 nsec.

The distance over which the clock signal can travel along an interconnection line before incurring a delay greater than the clock skew defines *isochronous* regions within the IC. If the common external clock signal can be delivered to each such region with zero clock skew, then clock routing within the isochronous region is not critical. Figure 4.36(a) illustrates the H-tree approach, providing clock paths from the clock input point to all points in the circuit along paths of equal length (and ideally, therefore, delivering clocks to each of the terminal points with zero clock skew). In a real circuit, precisely zero clock skew is not achieved since different network segments encounter different environments of data lines coupled electrically to the clock line segment. In addition, the net load seen by a clock signal once distributed to an isochronous region can differ among those regions.

In Figure 4.36(a), a single buffer drives the entire H-tree network, requiring a large area buffer and wide clock lines toward the point where the H-tree is connected to the clock input. Such a large buffer can account for up to 30% or more of the total VLSI circuit power dissipation. Figure 4.36(b) illustrates a distributed buffer approach, with a given buffer only having to drive those clock line segments to the next level of buffers. In this case, the buffers can be smaller and the clock lines narrower.

The constraint on clock timing is a bound on clock skew, not a requirement for zero clock skew. In Figure 4.36(c), the clock network uses multiple buffers but allows different path lengths consistent with

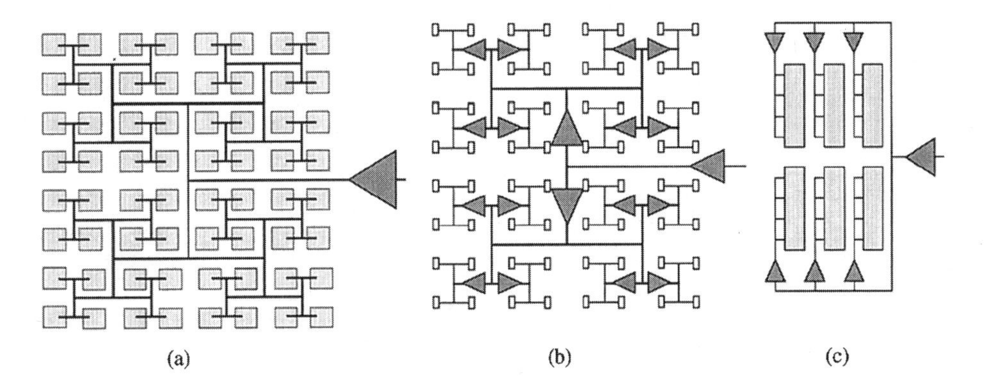

FIGURE 4.36 H-tree clock distribution. (a) Single driver layout. (b) Distributed driver layout. (c) Clock distribution with unequal line lengths but within skew tolerances.

clock skew margins. For tight margins, an H-tree can be used to deliver clock pulses to local regions in which distribution proceeds using a different buffered approach such as that in Figure 4.36(c).

Other approaches for clock distribution are gaining in importance as clock rates increase but sizes of ICs remain constant. For example, a lower frequency clock can be distributed across the IC, with locally distributed phase-locked loops (PLLs) used to multiply the low clock rate to the desired high clock rate. In addition to multiplying the clock rate, the PLL can also adjust the phase of the high rate clock. Using this approach, clocks confined to regions of an IC can be synchronous, whereas different regions do not have tightly synchronized clocks. Architectural level techniques support the connection of the *asynchronous regions*, which then can be regarded as "communicating" with one another rather than being "interconnected."

Power Distribution

Present-day VLSI circuits can be broadly separated into two groups: one for performance-oriented applications and the other for portable (battery-operated) applications. Power dissipations in the first group can be considerable, with contemporary VLSI circuits consuming 40 to 80 watts, corresponding to currents in the 10 to 40 amp range. Voltage and ground lines must be properly sized to prevent the peak current density from exceeding the level at which the metal interconnection will be physically "blown out," leading to a catastrophic failure of the circuit. Even if total power dissipation is modest, current densities appearing on power and ground lines can be large due to the small cross-sectional area of those lines. Another serious limitation is imposed by *electromigration.* The flow of current (moving electrons) creates a pressure on the metal ions of an interconnection, causing a slow migration (electromigration) of the metal atoms in the direction of the current flow. The problem is particularly important in power and ground lines where the current direction (pressure direction) is always in the same direction (i.e., signal lines tend to have currents flowing in alternating directions, with less net pressure in any given direction). At points in the aluminum interconnection where discontinuities appear (e.g., due to the grain structure of the aluminum), the flow becomes nonuniform and voids in the interconnection line gradually develop. There is a threshold level (about 1 mamp/micron) below which electromigration is greatly reduced and circuit designs seek to operate below this threshold. In addition, copper can be added to the aluminum (filling the intergrain spaces) to reduce the creation of voids due to electromigration. This issue may be relaxed with the trend toward use of copper metalization as a replacement for aluminum due to its lower resistivity.

A quite different issue in power distribution concerns *ground bounce* (or *simultaneous switching noise*). The problem arises due to the inductance associated with connecting the input/output pads of an IC to the IC package. The voltage drop ΔV_{L} across this inductance $L_{\mathrm{I/O}}$ due to a changing current $I_{\mathrm{L}}(t)$ through the inductance is given by $\Delta V_{\mathrm{L}} = L_{\mathrm{I/O}} \cdot [\mathrm{d}I_{\mathrm{L}}/\mathrm{d}t] \approx L_{\mathrm{I/O}} \cdot I_{\mathrm{L,0}}/\tau_{\mathrm{trans}}$ where $L_{\mathrm{L,0}}$ is the average current through the I/O and τ_{trans} is the average transition time of the current through the I/O. This voltage drop appears on the signal I/O pins, adding a degree of noise to the signal. However, the current flowing out of (into) the I/O pin is delivered from (into) the power (ground) connection to the IC. If M output signals change simultaneously, then the corresponding voltage swing across the inductance at the power connection to the IC is M times larger than that seen on the signal. With contemporary ICs having hundreds of I/O pins, this multiplier factor is large. Furthermore, with the increasing speeds of ICs, the transition times of signals at the I/O pins is decreasing. All these factors lead to the potential for very substantial voltage swings on the IC side of the power connection. To reduce these swings, much lower inductance is required for the power (and for the ground) connection. This is achieved by using a large number of power input pins (placing their respective inductances in parallel). It is not unusual for the number of power and ground connections to be comparable to the number of signal interconnections in today's VLSI circuits.

Analog and Mixed Signal ASICs

The discussion above has focused on digital ASICs not because digital ICs are the most common form of application-specific circuit but rather because the underlying issues related to the VLSI design are easier to describe. Ideally, digital circuits involve signals at only two levels: Vdd corresponding to logic value "one" and GND corresponding to logic value "zero." Analog circuits are profoundly more complex since the signals are continuously varying signals and the behavior and performance of an analog circuit depend critically on this continuous variation. Digital circuits can be reduced, at their lowest level, to simple logic gates and flip-flops. Analog circuits, however, have far more diverse and complex lowest-level functions. Digital circuits are basically switches and are simplified considerably by this basis. Analog circuits typically seek to obtain linear circuit functions with nonlinear devices (or exploit nonlinear devices to create specialized nonlinear analog circuit functions). Digital circuit technologies providing user programmable logic cells are straightforward approaches. Generic high-performance analog circuits based on interconnections of user programmable analog circuit cells are far less straightforward. For all these reasons and others, analog circuits tend to be custom ICs almost by their nature. To provide an environment supporting the efficient and correct custom design of analog circuits, a distinct set of CAD tools (Rutenbar et al., 2002) have emerged, along with a rich history of designs for standard analog circuit functions such as amplifiers, voltage dividers, current mirrors, etc. Similar to the hardware description languages (HDLs) widely used for design of digital circuits, analog hardware description languages (AHDLs) are advancing rapidly to support design of complex analog circuits.

Much of the history of integrated circuits has separated the design of digital and analog circuits, relegating each to separate ICs. However, there have been a rapidly increasing number of applications in which the co-integration of digital and analog circuits is favored. Such **mixed-signal ICs** are particularly important examples of application-specific ICs, with both the analog and digital sections customized to perform the overall "system" function. Such mixed signal circuits have been a part of the IC family for some time (Geiger et al., 1990; Comer, 1994; Ismail and Fiez, 1994; Laker and Sansen, 1994; Sanchez-Sinencio and Andreou, 1999; Handkiewicz, 2002; Tsividis, 2002). General-purpose mixed-signal ICs are seen in the example of microcontrollers (e.g., Park and Barrett, 2002), combining a basic microprocessor with memory and analog peripheral components. Special-purpose mixed-signal ICs are seen, for example, in contemporary wireless transceiver designs (Abidi et al., 1999; Leung, 2002), combining the analog circuitry to handle the higher frequency operations and digital circuitry to handle baseband operations. These mixed-signal ICs are also seen in today's automobiles, providing an interface between sensors monitoring the automobile and the control associated with such sensors (e.g., engine controls).

Summary

For over four decades, microelectronics technologies have been evolving, starting with primitive digital logic functions and evolving to the extraordinary capabilities available in present-day VLSI ICs containing complete systems or subsystems and vast amounts of memory on a single IC. ASIC technologies (including the EDA/CAD tools that guide the designer to a correctly operating circuit) provide the innovative user with opportunities to create ICs with performance and functionalities not readily available using general-purpose ICs, often making the difference between a successful product and a product that can not be differentiated from others. VLSI technologies have advanced to the point that entire systems are implemented on a single IC (the SoC approach), providing applications with miniaturized electronics with substantial capabilities. This trend toward entire systems on a single IC will drive the interest in ASICs in the future. In a real sense, the customization achieved by custom designing a printed circuit board to hold a selected set of commercial ICs has migrated to the IC level.

Defining Terms

ASIC: Application-specific integrated circuit — an integrated circuit designed for a special applications.

CAD: Computer-aided design — software programs that assist the design of electronic, mechanical, and other components and systems.

CMOS: Complementary metal-oxide semiconductor transistor circuit composed of PMOS and NMOS transistors.

EDA: Electronics design automation — software programs that automate various steps in the design of electronics components and systems.

Extrinsic delay: Also called *point-to-point delay,* the delay from the transition of the output of a logic cell to the transition at the input to another logic cell.

HDL: Hardware description language — a "software language" used to describe the function performed by a circuit and the structure of the circuit.

IC: Integrated circuit — a (normally silicon) substrate in which electronic devices and interconnections have been fabricated.

Intrinsic delay: Also called *pin-to-pin delay,* the delay between the transition of an input to a logic cell to the transition at the output of that logic cell.

Mixed-signal ICs: Integrated circuits including circuitry performing digital logic functions as well as circuitry performing analog circuit functions.

NMOS transistor: A metal-oxide semiconductor transistor that is in the on state when the voltage input is high (forming an n-type channel) and in the off state when the voltage input is low.

PMOS transistor: A metal-oxide semiconductor transistor that is in the on state when the voltage input is low (forming a p-type channel) and in the off state when the voltage input is high.

Rise/(fall) time: The time required for a signal (normally voltage) to change from a low (high) value to a high (low) value.

Vdd: The supply voltage used to drive logic within an IC.

VLSI: Very large scale integration — microelectronic integrated circuits containing a large number (presently tens of millions) of transistors and their associated interconnections to realize a complex electronic function.

Wiring channel: A region extending between the power and ground lines on an IC and dedicated for placement of interconnections among the cells.

References

A.A. Abidi, P.R. Gray, and R.G. Meyer (Eds.), *Integrated Circuits for Wireless Communications,* Piscataway, NJ: IEEE Press, 1999.

Actel. *Actel FPGA Data Book and Design Guide*, Sunnyvale, CA: Actel Corporation, 1995.

J.R. Armstrong, *Chip-Level Modeling with VHDL*, Englewood Cliffs, NJ: Prentice-Hall, 1989.

R.J. Baker, H.W. Li, and D.E. Boyce, *CMOS Circuit Design: Layout and Simulation,* Piscataway, NJ: IEEE Press, 1998.

P. Banerjee, *Parallel Algorithms for VLSI Computer-Aided Design,* Englewood Cliffs, NJ: Prentice-Hall, 1994.

J. Bhasker, *VHDL Primer,* Upper Saddle River, NJ: Prentice Hall, 1999.

R. Camposano and W. Wolf (Eds.), *High Level VLSI Synthesis,* Norwell, MA: Kluwer, 1995.

A. Chandrakasan and R. Broderson, *Low Power Digital CMOS Design,* Norwell, MA: Kluwer, 1995.

W.-K. Chen, *The VLSI Handbook,* Boca Raton, FL: CRC Press, 1999.

D.T. Comer, *Introduction to Mixed Signal VLSI,* Hirespire, PA: Array Publishing Co., 1994.

G. De Micheli, *Synthesis of Digital Circuits,* New York: McGraw-Hill, 1994a.

G. De Micheli, *Synthesis and Optimization of Digital Circuits,* New York: McGraw-Hill, 1994b.

T.E. Dillinger, *VLSI Engineering,* Englewood Cliffs, NJ: Prentice-Hall, 1988.

D. Gajski, *High-Level Synthesis: Introduction to Chip and System Design,* Norwell, MA: Kluwer, 2002.

R.L. Geiger, P.E. Allen, and N.R. Strader, *VLSI Design Techniques for Analog and Digital Circuits,* New York: McGraw-Hill, 1990.

H. Gradinski, *Interconnects in VLSI Design,* Norwell, MA: Kluwer, 2000.

S.H. Hall, G.W. Hall, and J.A. McCall, *High-Speed Digital System Design: A Handbook of Interconnect Theory and Design Practice,* New York: Wiley, 2000.

A. Handkiewicz, *Mixed-Signal Systems: A Guide to CMOS Circuit Design,* Hoboken, NJ: Wiley, 2002.

D.V. Heinbuch, *CMOS Cell Library,* New York: Addison-Wesley, 1987.

F.J. Hill and G.R. Peterson, *Computer Aided Design with Emphasis on VLSI,* New York: Wiley, 1993.

D. Hill, D. Shugard, J. Fishburn, and K. Keutzer, *Algorithms and Techniques for VLSI Layout Synthesis,* Norwell, MA: Kluwer, 1989.

E.E. Hollis, *Design of VLSI Gate Array ICs,* Englewood Cliffs, NJ: Prentice-Hall, 1987.

M. Ismail and T. Fiez, *Analog VLSI: Signal and Information Processing,* New York: McGraw Hill, 1994.

N. Jha and S. Kundu, *Testing and Reliable Design of CMOS Circuits,* Norwell, MA: Kluwer, 1990.

S.-M. Kang and Y. Leblebici, *CMOS Digital Integrated Circuits: Analysis and Design,* New York: McGraw-Hill, 1996.

K.R. Laker and W.M.C. Sansen, *Design of Analog Integrated Circuits and Systems,* New York: McGraw-Hill, 1994.

K. Lee, M. Shur, T.A. Fjeldly, and Y. Ytterdal, *Semiconductor Device Modeling,* Englewood Cliffs, NJ: Prentice-Hall, 1993.

B. Leung, *VLSI for Wireless Communications,* Upper Saddle River, NJ: Prentice Hall, 2002.

J. Lipman, "EDA tools put it together," *Electronics Design News (EDN),* pp. 81–92, October 26, 1995.

R. Lipsett, C. Schaefer, and C. Ussery, *VHDL: Hardware Description and Design,* Norwell, MA: Kluwer, 1990.

S. Mazor and P. Langstraat, *A Guide to VHDL,* Norwell, MA: Kluwer, 1992.

The National Technology Roadmap for Semiconductors, San Jose, CA: Semiconductor Industry Association, 1994.

The International Technology Roadmap for Semiconductors, http://public.itrs.net/, International Technology Roadmap for Semiconductors, 2003.

K.K. Parhi, *VLSI Digital Signal Processing Systems: Design and Implementation,* Hoboken, NJ: Wiley, 1998.

D.J. Park and S.F. Barrett, *68HC12 Microcontroller: Theory and Practice,* Upper Saddle River, NJ: Prentice Hall, 2002.

K.P. Parker, *The Boundary-Scan Handbook,* Norwell, MA: Kluwer, 1992.

B. Preas and M. Lorenzetti, *Physical Design Automation of VLSI Systems,* Menlo Park, CA: Benjamin-Cummings, 1988.

J.M. Rabaey, *Digital Integrated Circuits: A Design Perspective,* Englewood Cliffs, NJ: Prentice-Hall, 1996.

S. Rubin, *Computer Aides for VLSI Design,* New York: Addison-Wesley, 1987.

R.A. Rutenbar, G.G.E. Gielen, and B.A. Antao (Eds.), *Computer-Aided Design of Analog Integrated Circuits and Systems,* Hoboken, NJ: Wiley, 2002.

A. Rustin, *VHDL for Logic Synthesis,* New York: Wiley, 2001.

E. Sanchez-Sinencio and A.G. Andreou, *Low-Voltage/Low-Power Integrated Circuits and Systems,* Pisacataway, NJ: IEEE Press, 1999.

SCMOS Standard Cell Library, Center for Integrated Systems, Mississippi State University, 1989.

K. Shahookar and P. Mazumder, "VLSI placement techniques," *ACM Comput. Surv.,* vol. 23, no. 2, pp. 143–220, 1991.

N.A. Sherwani, *Algorithms for VLSI Design Automation,* Norwell, MA: Kluwer, 1993.

N.A. Sherwani, S. Bhingarde, and A. Panyam, *Routing in the Third Dimension: From VLSI Chips to MCNs,* Piscataway, NJ: IEEE Press, 1995.

M.J.S. Smith, *Application-Specific Integrated Circuits,* Reading, MA: Addison-Wesley, 1997.

S, Tewksbury (Ed.), *Microelectronic Systems Interconnections: Performance and Modeling,* Piscataway, NJ: IEEE Press, 1994.

The International Technology Roadmap for Semiconductors http://public.itrs.net/, International Technology Roadmap for Semiconductors, 2003.

The National Technology Roadmap for Semiconductors, San Jose, CA: Semiconductor Industry Association, 1994.
D.E. Thomas and P. Moorby, *The Verilog Hardware Description Language*, Norwell, MA: Kluwer, 1991.
S. Trimberger (Ed.), *An Introduction to CAD for VLSI*, Norwell, MA: Kluwer, 2002.
Y. Tsividis, *Mixed Analog-Digital VLSI Devices and Technology*, River Edge, NJ: World Scientific, 2002.
J.P. Uyemura, *CMOS Logic Circuit Design*, Boston, MA: Kluwer, 1999.
N.H.E. Weste and K. Eshraghian, *Principles of CMOS VLSI Design*, New York: Addison-Wesley, 1993.
J. White and A. Sangiovanni-Vincentelli, *Relaxation Methods for Simulation of VLSI Circuits*, Norwell, MA: Kluwer, 1987.
W. Wolf, *Modern VLSI Design: System-on-Chip Design*, Upper Saddle River, NJ: Prentice-Hall, 2002.
Xilinx Corporation, *http://www.xilinx.com.*
Q.K. Zhu, *Power Distribution Network Design for VLSI*, Hoboken, NJ: Wiley, 2004.

4.4 Low-Power Design Techniques

Christian Piguet

Introduction

Low-power design of integrated circuits has to be addressed at all levels of design, i.e., from systems, architectures, logic, and physical levels. It is shown that most of the power can be saved at the highest levels, mainly by modifying or simplifying the specifications. Less power can be saved at low levels, but reducing power consumption of standard cell libraries and memories can be applied to all applications, contrary to power savings at high levels, which is strongly application dependent.

Large Power Reduction at High Level

If one considers the past history of microelectronics, the extrapolated power consumption for high-end microprocessors would be about 2000 watts and 3000 amperes in 2010.[1] Clearly, this is impossible, and a set of low-power techniques, as described in this chapter, have to be applied to get more reasonable power consumption.

The most important observation regarding low-power techniques is that most of the power can be saved at the highest design levels. At the system level, low-power design techniques such as partition, activity reduction, reduction of the number of steps, simplicity, data representation, and locality (cache or distributed memory instead of a centralized memory) have to be chosen. These choices are, however, strongly application dependent.[2] At the lowest design levels, for instance, for a low-power library, only a moderate factor (about two) in power reduction can be reached.

Power Consumption of Digital Circuits

Power consumption of digital circuits results in three components, i.e., dynamic, short-circuit, and static power. Although static power is larger and larger in deep submicron technologies, the first part of this section focuses on the reduction of dynamic power. The dynamic power consumption of a digital circuit is given by $P = a \times f \times C \times \text{Vdd}^2$ in which a is the circuit activity (number of switching gates over the total number of gates), f is the frequency of switching, C is the capacitance of the gate or of the module (extracted from the layout), and Vdd is the supply voltage. Energy can be considered as well. The energy is the power × delay product, and using the expression of the delay $= 1/f$, the energy becomes energy $= a \times C \times \text{Vdd}^2$.

In the design process, one can consider different architectures for a specified function. It is therefore necessary to perform a speed/power comparison in order to select the "best" architecture. Such a comparison is not so easy, as it depends on a large number of design parameters. One has to consider a basic operation (or an instruction) in a logic block or a microprocessor. Such an operation (instruction) is generally executed in clocks per operation (CPO) (clocks per instruction [CPI]) clock periods or steps. A step is executed at the frequency f in a clock period $T = 1/f$. The energy per operation, which is independent of the frequency, can therefore be computed as energy/operation = CPO $\times a \times C \times$ Vdd^2.

Some Basic Rules

There are some basic rules that can be proposed to reduce power consumption at system levels:

Reduction of the number N of operations to execute a given task.
A too high sequencing always consumes more than the same functions executed in parallel.
Consequently, parallel architectures provide better CPO or CPI, as well as pipelined and RISC architectures.
The lowest Vdd for the specific application has to be chosen.
The goal is to design a chip that just fits the speed requirements.[3]

Large Power Reduction at High Level

As mentioned before, a large part of the power can be saved at high level. Factors of 10 to 100 or more are possible. However, it means that the resulting system could be quite different, with less functionality or less programmability. The choice among various systems is strongly application dependent. One has to think about systems and low power, to ask good questions to the customers, and to get reasonable answers. This will be illustrated by some examples. Power estimation at high level is a very useful tool to check what will be the estimated total power consumption.

Processor vs. Random Logic

It should be noted that a microcontroller-based implementation results in a very high sequencing. This is due to the microcontroller structure, which is based on the reuse of the same operators and registers (ALU, accumulators). For instance, only one step ($N = 1$) is necessary to update a hardware counter (Figure 4.37). For its software counterpart, the number of steps is much higher while executing several instructions with many clock cycles in sequence. This simple example shows that the number of steps executed for the same task can be very different depending on the architecture. For a wristwatch, 2000 instructions were necessary to update the time vs. $N = 1$ step for a random logic circuit.

The conclusion is that a microprocessor is a stupid idea to save power, as it provides too high sequencing for a given task. However, if they are preferred today, it is due to their flexibility.

Number of Clock Cycles per Instruction

An example of too high sequencing is provided by complex instruction set computers (CISC) micro-controllers. Their architectures come from the first Von Neumann microprocessor architectures with a single

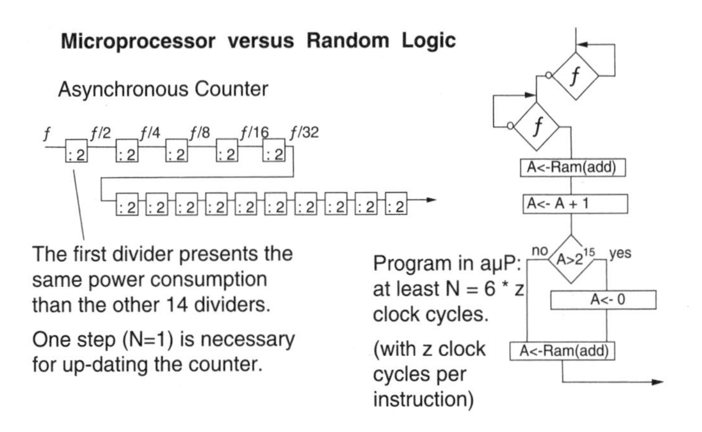

FIGURE 4.37 Hardware counter vs. microprocessor.

unified memory. Furthermore, this memory is organized in bytes, resulting in many memory accesses through a single 8-bit bus, as most of the instructions are coded with 2 or 3 bytes. It basically explains why the number of CPI is very high.

As an example, Table 4.3 shows a comparison of the number of instructions used to program a short routine (synchrone transmission, shifting out 8-bit data and clock). It can be seen that the number of instructions in the program, the number of bytes, as well as the number of executed instructions are quite similar for these microcontrollers. However, regarding low power, Table 4.3 shows the number of executed clocks for this routine. One can see that the resulting CPI is quite large for the first CISC microcontrollers in the list. The last two in the list are more reduced instruction set computers (RISC)-like and do present a reduced CPI, mainly the CoolRISC with a single clock cycle per instruction.[4] As the energy per clock cycle is roughly the same, it means that the energy per instruction is proportional to CPI shown in Table 4.3, demonstrating that too high sequencing impacts highly on power consumption.

Processor Types

The choice of a processor is very important regarding power consumption. First, the data width of the processor has to be the same as the processed data. It results in a substantially increased sequencing to manage, for instance, 16-bit data on 8-bit microcontrollers. For a 16-bit multiply, 30 instructions are required (add-shift algorithm) on a 16-bit processor, while 127 instructions are required on an 8-bit machine (double precision). A better architecture is to have a 16×16-bit parallel-parallel multiplier with only one instruction to execute a multiplication.

Another issue is to choose the right processor for the right task. For control tasks, digital signal processing DSP (digital signal procesing) processors are largely inefficient. But conversely, microcontrollers for DSP tasks are also quite inefficient! For instance, to perform a JPEG compression on an 8-bit microcontroller requires about 10 million executed instructions for a 256×256 image (CoolRISC, 10 MHz, 10 MIPS, 1 sec per image). It is quite inefficient. Factor 100 in energy reduction can be achieved with JPEG-dedicated hardware.

Figure 4.38 shows probably the best architecture to save power. For many applications, there is some control that is performed by a microcontroller (the best machine to perform control). However, in most applications, there are also other tasks to execute, which could be a DSP task, convolution, JPEG, image compression, or another task. The best architecture is consequently to use a specific machine (co-processor) to execute these tasks, i.e., one co-processor per task so that these tasks are executed by the smallest and the most energy-efficient machines.

Memories

Memory organization is very important to reduce the energy per memory access. Generally, memories consume most of the power. So it is immediately obvious that memories have to be designed hierarchically. No memory technology can simultaneously maximize speed and capacity at lowest cost and power. Data for immediate use are stored in expensive registers, in cache memories, and less used data are stored in large memories.

For each application, the choice of the memory architecture is very important. One has to think of hierarchical, parallel, interleaved, and cache memories, sometimes several levels of cache, to try to find the best trade-off. The application algorithm has to be analyzed from the data point of view to find the best way to organize the data arrays and how to access these structured data.

TABLE 4.3 Number of Executed Clock Cycles per Instruction in Some Microcontrollers

Microcontroller	Instructions in the Routine	Bits in the Routine	Executed Instructions	Executed Clock Cycles	CPI
ST62xx	12	152	60	2704	45
COP800	12	120	60	2000	33
8048	8	112	35	1125	32
Z86Cxx	8	168	35	692	20
68HC05	11	160	59	904	15
PIC16C5x	11	132	59	300	5
CoolRISC	10	180	58	58	1

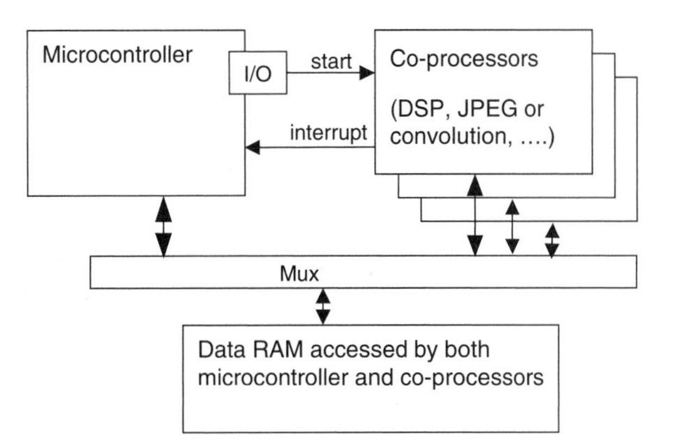

FIGURE 4.38 Microcontroller and co-processors.

If a cache memory is used, it is possible, for instance, to minimize the number of cache misses while using adequate programming as well as good data organization in the data memory. For instance, in inner loops of a program manipulating structured data, it is not equivalent to write: (1) **do *i*** then **do *j*** or (2) **do *j*** then **do *I***, depending on how the data are located in the data memory.

The Energy-Flexibility Gap

Figure 4.39 shows that the flexibility,[5] i.e., the use of a general-purpose processor or of a specialized DSP processor, has a large impact on the energy required to perform a given task compared to the execution of the same given task on dedicated hardware.

Pipelined and Parallel Digital Circuits

As a too high sequencing is not favorable for reducing power, many techniques such as pipelining and parallelism have been proposed at architecture levels to reduce dynamic power consumption.

Pipelining

Pipelining is a very effective technique to increase the throughput without generating a too large overhead that could compromise the power consumption reduction. Figure 4.40 shows an execution unit composed of several logic blocks connected in series. By inserting registers between the logic blocks (Figure 4.41), the delay

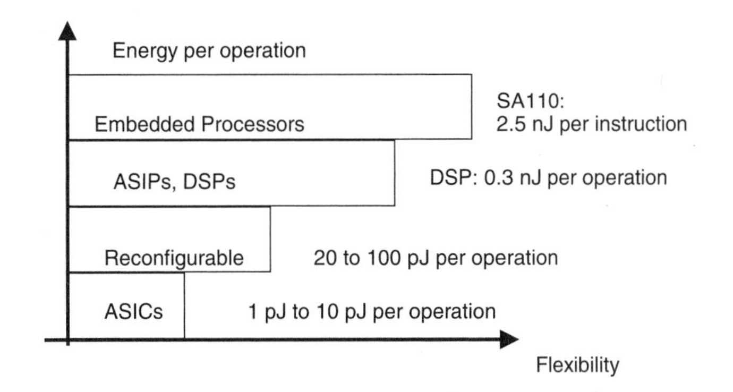

FIGURE 4.39 Energy-flexibility gap.

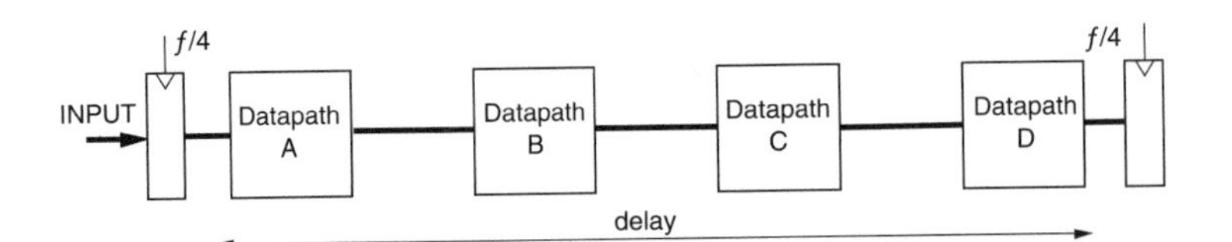

FIGURE 4.40 Large execution unit.

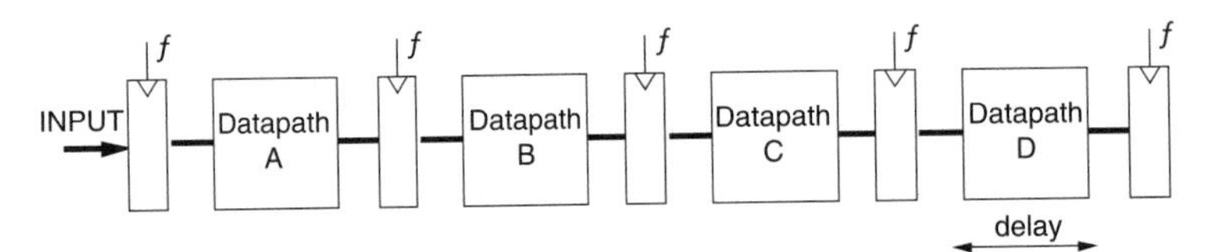

FIGURE 4.41 Pipelined execution unit.

for each logic block is reduced to a quarter of the total delay of the previous execution unit. Therefore the frequency and the throughput can be increased by a factor of 4 with a small overhead and a small power consumption increase.

The pipelining technique is largely used in microprocessors. One has to take into account the latency, i.e., the data output is issued after four delays (Figure 4.41). Pipelining can also be used to significantly reduce the power consumption if the throughput is kept constant in both the execution units described, respectively, in Figure 4.40 and Figure 4.41. In the latter, frequency can be reduced by a factor of 4 while maintaining the same throughput, and consequently the power consumption, proportional to frequency, is similarly reduced by a factor of 4.

Parallelization

Circuit parallelization has been proposed to maintain, at a reduced Vdd, the throughput of logic modules that are placed on the critical path.[6–8] It can be achieved with M parallel units clocked at f/M. Results are provided at the nominal frequency f through an output multiplexer controlled by $f/2$ (Figure 4.42). Each unit can compute its result in a time slot M times longer, and can therefore be supplied at a reduced supply voltage. If the units are data paths or processors,[7] the latter have to be duplicated, resulting in an M times area and switched capacitance increase. Applying a dynamic power formula, one can write $P = M \times C \times f/M \times \mathrm{Vdd}^2 = \mathrm{C} \times f \times \mathrm{Vdd}^2$. Power consumption reduction is only achieved by the reduction of Vdd. One can see on the timing diagram in Figure 4.42 that the output multiplexer is controlled at $f/2$. The operation or access of unit 2 is started before the completion of the operation of unit 1. Therefore, M successive computations do not have to be dependent on each other.

However, some parallelized logic modules do not require M unit duplication. It is the case, for instance, for memories[8] in which each unit contains $1/M$ data or instructions, resulting in the same total number of bits to store the information (Figure 4.43). In such a case, the dynamic power is $P = C \times f/M \times \mathrm{Vdd}^2$. Consequently, power could be saved even if Vdd is not reduced. However, some overhead has to be considered, such as the address registers duplication and the output multiplexer (Figure 4.43).

Figure 4.44 shows a parallelized shift register. Such a concept has been proposed for CCD serial memories.[9,10] The input is successively provided to the upper or to the lower half shift register at a reduced frequency, while the output multiplexer restores the output at the frequency f. There is no latency due to the fact that the combinatorial circuit of the state machine "shift register" is only constituted by wires, with no associated delay. The total number of D flip-flops is the same as in the nonparallelized shift register.[8] In this case also, power is saved without reducing Vdd, but it is still possible to save more power by reducing Vdd as each part of the shift registers is clocked at $f/2$.

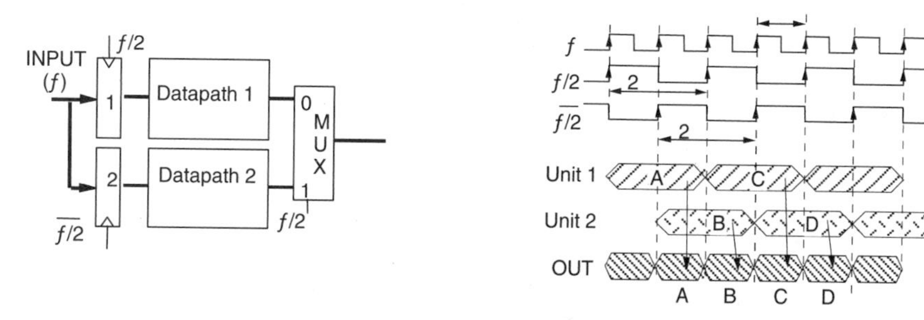

FIGURE 4.42 Data path parallelization.

Table 4.4 shows the simulation results of an 800-bit shift register working at 6 GHz and 1.8 volt in a 0.18 μm technology. It has been parallelized from two to eight parts. Power consumption reduction at the same given supply voltage is quite interesting.

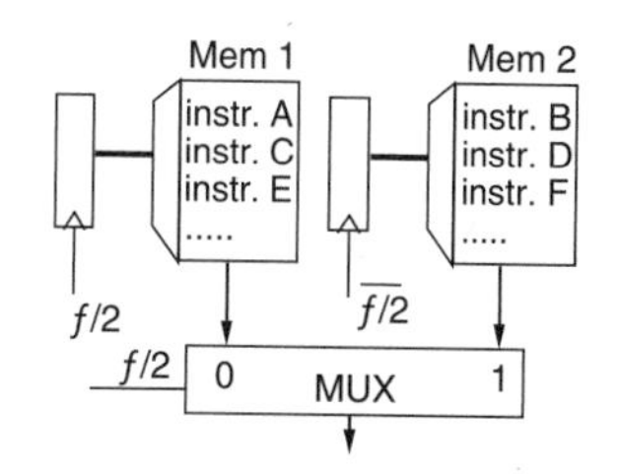

FIGURE 4.43 Memory parallelization.

Latch-Based and Gated Clocked Circuits

At the architecture level, a major issue regarding low power is the clocking scheme. In deep submicron technologies, as wire delays are larger than gate delays, the distribution of the clock through the complete chip is the main problem for synchronous architectures. Several low-power techniques have been proposed, such as asynchronous architectures that work without any clock.[11] Other techniques that still keep synchronous architectures while drastically reducing power consumption are presented in the following.

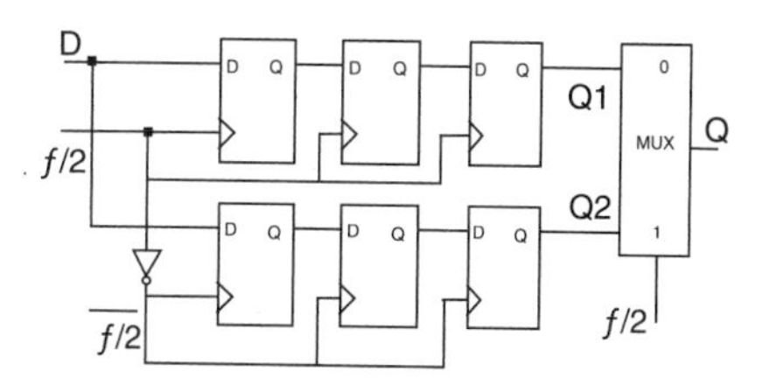

FIGURE 4.44. Parallelized shift register.

Latch-Based Design

Digital circuits are today described in VHDL and are automatically synthesized with a logic synthesizer. Most VHDL descriptions are based on a single clock that is connected to master-slave flip-flops. This clocking scheme implies the generation of a single clock tree that is more and more difficult to design to obtain a very small clock skew, i.e., a time difference between the arrival of the clock at all flip-flops, which is as small as possible.

Another clocking scheme called "latch-based" can be used with many advantages. Figure 4.45 shows this concept, consisting of latch-based registers clocked by two different nonoverlapping clocks Ø1 and Ø2 generated by two different clock trees. This scheme has been chosen to be more robust to clock skew, flip-flop failures, and timing problems at very low voltage. The clock skew between various Ø1 and Ø2 pulses has to be

TABLE 4.4 Power Consumption Reduction in Parallelized Shift Registers

	Cells	Power Consumption (mW)
No//	800	75
2-//	851	54
4-//	850	32
8-//	840	15

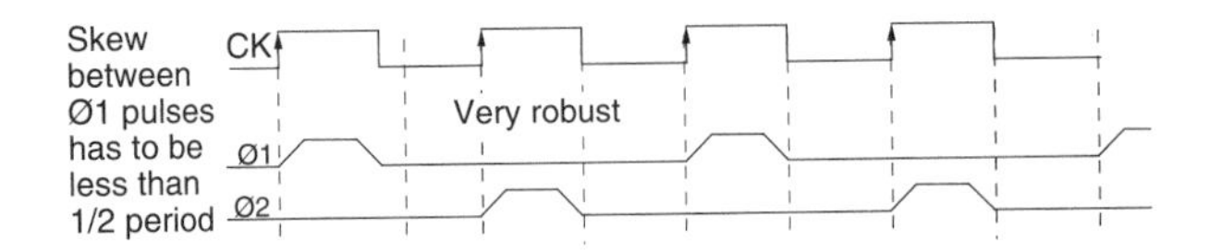

FIGURE 4.45 Two phased clocks for a latch-based clocking scheme.

shorter than half a period of the master clock CK, which is not too difficult to achieve. Because of the nonoverlapping of the clocks and the additional time barrier caused by having two latches instead of one flip-flop, latch-based designs support greater clock skew before failing than a similar flip-flop-based design, each targeting the same throughput.

This allows the synthesizer and router to use smaller clock buffers and to simplify the clock tree generation, which will reduce the power consumption of the clock tree. Therefore, the clock skew becomes relevant only when its value is close to the nonoverlapping of the clocks. Furthermore, if the chip has clock skew problems at the targeted frequency after integration, it is possible in a latch-based design to reduce the clock frequency. The result is that the clock skew problem will disappear, allowing the designer to test the chip functionality and eventually detect other bugs or validate the design functionality. This can reduce the number of test integrations needed to validate the chip. With a flip-flop-based design, when a clock skew problem appears, the complete circuit must be rerouted and integrated again.

Using latches can also reduce the MOS of a design. For instance, a microcontroller register bank has 16×32 bit registers, i.e., 512 flip-flops with about 13,000 MOS. With latches, the master part of the registers can be common for all the registers, which gives 544 latches, or about 6500 MOS. In this example, the register bank transistor count is reduced by a factor of 2.

Gated Clocks

The gated clock technique is extensively used in the design of low-power circuits.[12,13] It is based on the fact that a register need not be clocked when its next state is identical to its previous state. Some logic is used to detect if such registers have to be clocked or not. ALUs, for instance, can be designed with input and control registers that are loaded (or clocked) only when an ALU operation has to be executed. During the execution of another instruction (branch, load/store), these registers are not clocked; and thus no transitions occur in the ALU (Figure 4.46). A similar mechanism is used for pipeline stages if an instruction execution is finished at a given pipeline stage. The clock of the next pipeline stages is thus gated as the next pipelines have nothing to do and therefore no transitions occur in these last stages of the pipeline.

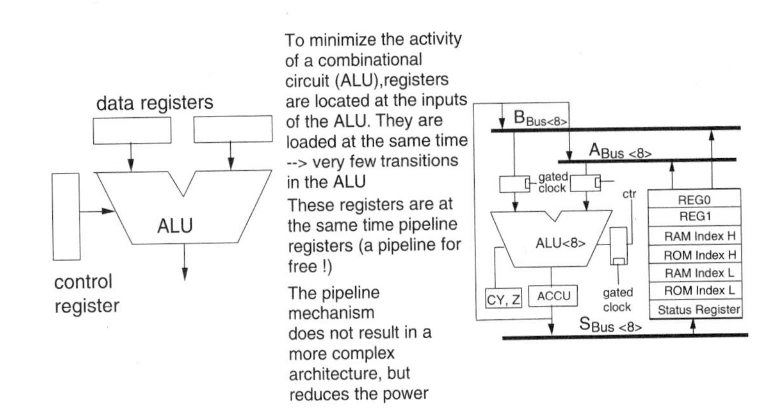

FIGURE 4.46 Gated clock applied to an ALU.

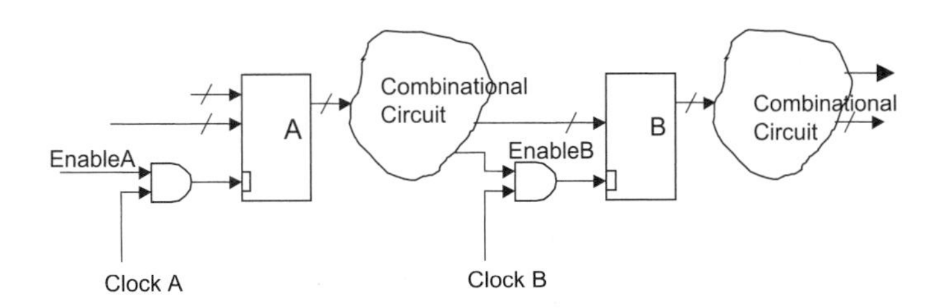

FIGURE 4.47 Clock gating in a latch-based clocking scheme.

Gated Clock with Latch-Based Designs

The latch-based design also allows a very natural and safe clock gating methodology. Figure 4.47 shows a simple and safe way of generating enable signals for clock gating. This method gives glitch-free clock signals without the addition of memory elements, as is needed with flip-flop clock gating. Logic synthesizers handle the latch-based design methodology very nicely. However, generally, clock gating combined with latch-based design cannot be inserted automatically by a logic synthesizer. The designer has to write the description of the clock gating in his VHDL code. The most critical problem is to prevent the synthesizer from optimizing the clock gating AND gate with the rest of the combinational logic. To ensure a glitch-free clock, this AND gate has to be placed as shown in Figure 4.47. This can easily be done manually by the designer by placing these AND gates in a separate level of hierarchy of the design or placing a "don't touch" attribute on them. Forcing a "don't touch" on these gates presents the drawback that this part of the clock tree will not be optimized for speed or clock buffering.

Low-Power Standard Cell Libraries

At the logic level, digital logic design is performed today by using standard cell libraries and place and route CAD tools. As mentioned earlier, power savings at this level are limited. However, even if it is only a factor of 2, it is not application dependent, and all chips designed with a low-power library will present reduced power consumption. Many different logic styles or logic families have been and continue to be proposed for general purpose and specialized standard cell libraries. Low power is even more important than speed and silicon area, but it is increasingly difficult to achieve such a power reduction in very deep submicron technologies as well as for specialized libraries for self-timed or cryptographic applications.

Static CMOS Logic

Static CMOS is the older and still most used logic family. It is still considered the simplest and most robust logic style.[14] Each CMOS gate is constructed with two dual N-ch and P-ch networks connected, respectively, between Vss and Vdd and the gate output. Any logic Boolean function can be designed by connecting CMOS transistors in series and/or parallel in the two N-ch and P-ch networks, provided that inputs are available in true and complemented forms.

z \ ab	00	01	11	10
c = 0	1	1	0	1
c = 1	0	1	1	1

$$z = (\bar{a}\,\bar{b}\,c + a\,\bar{b}\,c)\{0\} + (a\,b + a\,\bar{c} + b\,c)$$

$$zN = \bar{a}\,\bar{b}\,c + a\,\bar{b}\,c$$

$$zP = a\,\bar{b} + \bar{a}\,c + b\,c$$

FIGURE 4.48 CMOS gate synthesized by the "separated simplification" method.

The design of a CMOS gate is generally performed by synthesizing the N-ch network by taking the "zero" cubes in the Karnaugh map of the Boolean function. The P-ch network is then derived as the dual network by connecting in series (parallel) the transistors that are in parallel (series) in the N-ch network. Figure 4.48 shows the synthesis of a Boolean function given by a Karnaugh map, in which the *z* symmetrical equation contains two terms, the first one with the "zero" cubes indicated by {0} and the second term with the "one" cubes indicated with {1}. The MOS structure of the N-ch and P-ch networks are then designed as zN and zP, by taking the first term {0} as such and by inverting each letter in the second term {1}, as P-ch transistors are conducting when they have a "zero" on their gate. In the zN and zP expressions, AND operators mean a serial connection of transistors while OR operators mean a parallel connection. This method has been introduced in order to be capable of having the two N-ch and P-ch expressions as sums of products. As described in the next paragraph, it results in the so-called "branch-based" logic style that provides some advantages with respect to layout regularity, better performances in speed and power, and a better testability.[15,16]

Branch-Based Logic

In branch-based logic,[15] logic cells are designed exclusively with branches composed of transistors in series connected between a supply line and the gate output (Figure 4.49). The number of MOS in series is limited to three for speed performances. The main advantage of such an implementation is the layout density. For instance, the symbolic layout of the nonbranch-based P-ch network (Figure 4.49) contains two supplementary contacts with two drain parasitic capacitances that can be removed in the more compact branch-based implementation. The symbolic layout of Figure 4.49 comes from the logical equation $S = (B+C)\ (A \times C\text{bar} +\ A\text{bar} \times D)$. If implemented as such, the P-ch network is shown in the top of Figure 4.49. If a Karnaugh map is designed, the minimal number of blocks of "one" that are necessary is three, resulting in the branch-based implementation with six transistors shown in the bottom of Figure 4.49. It can be seen that the latter is a very regular layout consisting of three branches. It provides no diffusion interruption, a common drain for two branches, a minimal number of contacts, and few metal connections. This is not the case, for instance, for the implementation shown at the top of Figure 4.49, where a product of sum has to be implemented with a supplementary wire. This branch-based technique, first introduced to reduce parasitic capacitances for achieving low power,[14] is also beneficial for high-speed logic such as fast adders in SOI technology.[16]

An obvious drawback of this technique is the possible increase in the number of transistors when realizing a MOS network of complex CMOS gates with a sum of products (a transistor controlled by the same input in two parallel branches is repeated). However, this problem is not too serious. A standard cell library does not contain a large number of complex gates. The most used cells are simple gates, flip-flops, latches, and multiplexers. In a 200 cell library, some cells contain one supplementary transistor (XOR, AOI, OAI, latches with reset, *D* flip-flops, frequency dividers) and a few cells contain two supplementary transistors (latches and flip-flops with set/reset).

Transmission Gates

Transmission gate-based design has been largely used in the past, as MOS transistors are very good switches. Transmission gates use two complementary transistors, as a single N-ch pass transistor (for instance,

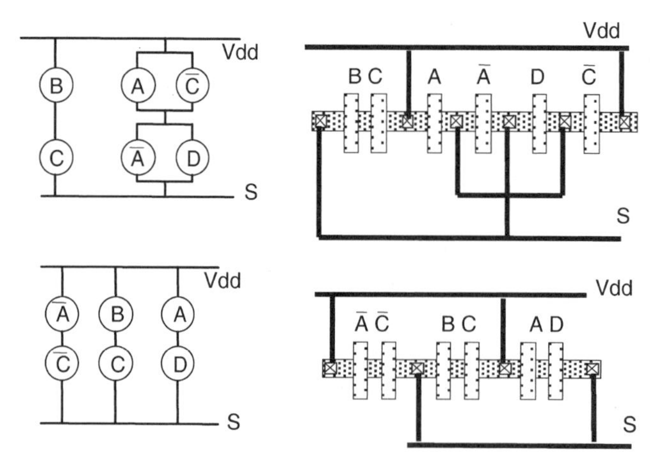

FIGURE 4.49 Branch-based layout.

presenting $V_{\text{gate-source}} = 0$ when it has to conduct Vdd from its source to its drain) only reaches Vdd-VT at its output. In the same situation, the complementary P-ch transistor has a full $V_{\text{gate-source}} = \text{Vdd}$ and provides Vdd at the output of the transmission gate. Rules of thumb are often used for the design of these transmission gate-based cells. It is shown here that the same basic methodology introduced for 'branch-based' logic can be applied to transmission gates or pass-transistor circuits.

A transmission gate, controlled by an input variable, connects or not another input variable to the gate output. This means that some inputs are connected to transistor sources and not just to the gates of transistors. Compared to the branch-based style, for which sources of transistors or branches are always connected to Vss {0} or to Vdd {1}, in transmission gate designs, sources of transistors, or branches can also be connected to input variables. Thus, some cubes in the Karnaugh map are not only 'zero' cubes indicated by {0} or 'one' cubes by {1}, but also some cubes identified by {input}. The content of the cubes is not 'zero' or 'one', but is identical to a given input variable (or the complemented input). As a result, some cubes containing 'zero' and 'one' can be chosen, provided that the arrangement of 'zero' and 'one' are identical to a given input.

Figure 4.50 provides an example for which one cube is a conventional cube with {0} and for which two other cubes are transmission cubes containing zero and one. From Figure 4.50, it can be seen that the top cube content is identical to the input variable d, while the other cube content is identical to the complement of the input variable *d*. As such, the designer can write the symmetrical equation of output *x* by a first term representing the 'zero' cube {0} and by two other terms for which the only difference is that they refer to input variables, i.e., {*d*} and {*d*-bar}. The N-ch and P-ch networks are then derived using the same rules, i.e., the terms with {0} and {input} are selected without any change for the N-ch network *x*N, while the terms with {1} and {input} are selected for the P-ch network *x*P by inverting each letter in the expression of the cubes. The transmission cubes give a contribution in both N-ch and P-ch networks, as the designer wants to obtain a transmission gate-based circuit. If only N pass transistors are required, only the contribution of the transmission cubes in the N-ch network is necessary. The example of Figure 4.50 shows that both cube types, {0} or {1}, and {input} can be simultaneously synthesized. The resulting circuit will therefore contain some branches connected to Vss (and Vdd) and other branches connected to input variables.

The symbolic schematic of the resulting circuit (Figure 4.50), designed in such a way that branches are highlighted, shows, for instance, that two branches are connected to input *d*, i.e., one P-ch branch controlled by *c* and *b* bar and one N-ch branch controlled by *c* bar and *b*. These two branches do implement two transmission gates connected in series.

Transmission gate-based circuits do have a smaller number of transistors compared to static CMOS logic. However, in terms of layout density, depending on the layout style, the cell area is often very similar. Some

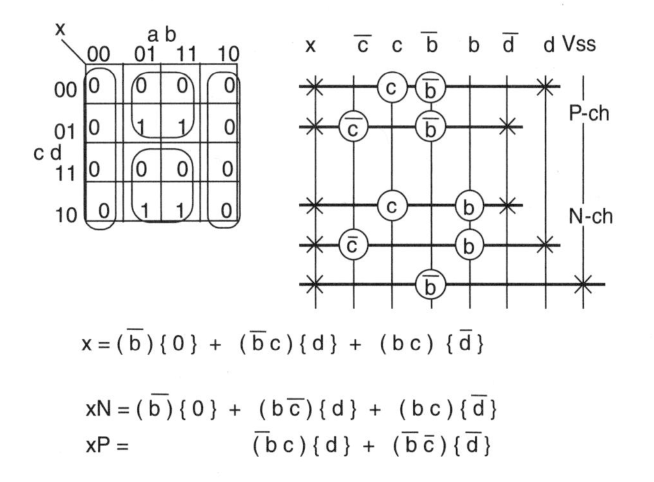

FIGURE 4.50 Synthesis of a complex CMOS gate with transmission gates.

circuits are advantageously designed as transmission gate-based design, such as XOR gates and adders (based on XOR gates), but other basic cells in a library can be designed in static CMOS logic without any penalty.

N Pass Logic

Before CMOS became the mainstream technology, N-MOS logic was extensively used with a depleted transistor as the load device. N-MOS pass transistor logic was also used for many cells, resulting in a very low transistor count. However, in N-MOS logic, the gate output produces a Vdd-VT voltage. Although it was not a problem many years ago with Vdd at 5.0 V, it is a major drawback today with supply voltages close to or below 1.0 V. In order to keep the transistor count low, N pass logic can be used by having an output keeper or a restoring transistor as shown in Figure 4.51 for an XOR gate. This logic style is called single pass transistor logic (SPL). The keeper device is used to force full Vdd at the output when the inverted output is zero. This logic style seems to be interesting only for multiplexers and XOR gates, explaining why it is generally benchmarked for adders and multipliers. This SPL could compete with static CMOS at high Vdd, but not at low Vdd, where SPL is slower and consumes much more than static CMOS.

Figure 4.51 shows another similar logic style called complementary pass transistor logic (CPL) and shows a 2:1 multiplexer. This CPL style is dual rail logic, as both true and complemented outputs are provided. A comparison (Table 4.5) with static CMOS shows some advantages in speed at high Vdd, but not better performances at low Vdd,[17] the transistor count being similar to static CMOS. There are many other logic styles inspired by SPL and CPL, such as dual rail differential cascode voltage switch logic (DCVSL),[18] for which the two N-ch networks are not designed as N-pass logic but as conventional N-ch networks. Two cross-coupled P-ch MOS are used as load devices, similar to two P-ch MOS of the CPL logic shown in Figure 4.51. It is a ratioed logic, as the N-ch networks have to fight against the P-ch devices.

Standard Cell Libraries

Standard cell libraries often provide a huge number of cells, up to 300 or even 500. A new approach is proposed, which is based on a limited set of standard cells.[20] The number of functions for the new library has been reduced to 22 and the number of layouts to 92. It can be seen that the ratio between the number of layouts and the number of functions is larger (92/22 = 4.2 instead of 220/60 = 3.6 for the previous library). This means that the number of cell and buffer drives is larger. For speed and power optimization achieved by the logic synthesizer, the increased ratio of layouts to functions, as shown above, is beneficial.

It seems obvious that the logic synthesizer could do a better job if the number of cells in the library was large. With a larger choice, it should be possible to provide a better solution. However, this is not the case. Experiments in Table 4.6 and Table 4.7 show that the delay of some operators is significantly reduced with the new library, resulting in a very small increase in silicon area (Table 4.6) and that the silicon area is reduced at the same speed with the new library (Table 4.7). These results show that the logic synthesizer is more efficient

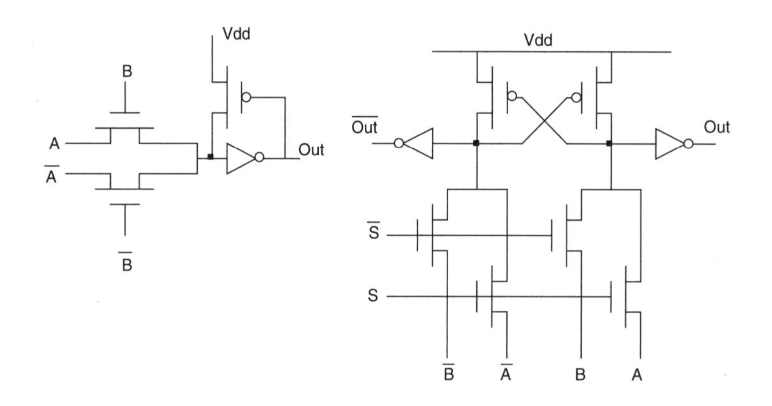

FIGURE 4.51 SPL XOR gate and CPL (dual-rail) 2:1 multiplexer.

TABLE 4.5 Comparison between Static CMOS and CPL for a Full Adder [19]

Logic Family	Delay (nsec)		Power (mW)		Power × Delay	
	3.3 V	1.5 V	3.3 V	1.5 V	3.3 V	1.5 V
CMOS	1.89	7.88	32.9	6.4	1.00	1.00
CPL	1.39	8.33	34.1	6.0	0.76	0.99

TABLE 4.6 Delay Comparison (Synthesis for Maximum Speed, 0.5 μm Process)

	Old Library Delay (nsec)	μm^2	New Library Delay (nsec)	μm^2
32-bit multiplier	16.4	907 K	12.1	999 K
Floating-point adder	27.7	510 K	21.1	548 K
CoolRISC ALU [21]	10.8	140 K	7.7	170 K

TABLE 4.7 Silicon Area Comparison (Synthesis for a Given Delay, 0.5 μm Process)

	Old Library Delay (nsec)	μm^2	New Library Delay (nsec)	μm^2
32-bit multiplier	17.1	868 K	17.0	830 K
Floating-point adder	28.1	484 K	28.0	472 K
CoolRISC ALU [21]	11.0	139 K	11.0	118 K

because it has a limited set of well-chosen cells and cell sizing adapted to the considered logic synthesizer. With significantly fewer cells than conventional libraries, the synthesizer is not lost in some optimization loops due to a too large choice of cells.

The design of the library has been based on keeping only very fast cells, i.e., to remove all the cells with three P-ch transistors in series and to have a very limited number of cells with two P-ch transistors in series. The number of cell layouts for the same function has been increased. However, it is not a simple increase from, for instance, sizing D1 (small transistors), D2, and D3 (medium-sized transistors) to D1, D2, D3, D4, and D5 (very large transistors). The cell sizing performed takes into account how the synthesizer uses the considered cells. The third consideration is based on buffer insertion, i.e., the combination of a given cell and of a buffer to replace complex gates.

Such a strategy must be checked through many experiments. The choice of the 22 functions was performed with a large number of experiments with and without a specific cell, and then the decision was made to either insert this cell or not in the library. Similar experiments were performed with various sizing and buffering of the cells. At the end, only 22 functions and 92 layouts were kept in the new library.

Furthermore, as the number of layouts is drastically reduced, it takes less time to design a new library for a more advanced process. Substantial time can also be saved for the library characterization, which is known to be often the most time-consuming activity in library design. Reducing the number of layouts from 220 to 92 is a significant advantage. The reduction of the number of cells implies removal of complex gates from the library, forcing the logic synthesizer to decompose complex gates, which, as described, is beneficial in terms of speed.

It will also be a crucial point in future libraries for which more versions of the same function will be required while considering static power problems. The same function could be realized, for instance, with low or high VT for double-VT technologies, or with several cells such as a generic cell with typical VT, a low-power cell with high VT and a fast cell with low VT.

Logic Styles for Specific Applications: Self-Timed Design

The design of cell libraries is largely considered to be independent of the applications, i.e., any library can be used for any application. This assumption no longer holds today, as certain specific applications require special cell libraries. It is the case, for instance, for self-timed design, but also for other applications such as adiabatic circuits, cryptographic applications, and fault tolerant logic. Self-timed logic, i.e., digital circuits without any master clock or asynchronous logic, has been introduced to solve the problem of the clock tree synthesis, which is proved to be more and more difficult, and power consuming.[11,22]

Several asynchronous techniques have been proposed at the block and/or cell levels. They are based on handshaking, i.e., a local control of the data shifted in a pipeline. This control logic is largely based on C Muller elements, not generally proposed by conventional libraries. For asynchronous design, it would be beneficial to have this C gate as a library cell. Another logic style that is used in self-timed architectures is dynamic DCVSL dual-rail logic. During precharged phase, both outputs are "11," an invalid state. After evaluation, the valid state is reached ("01" or "10"), indicating that the operation is completed. This signal is consequently used to start the next operation (Request) and to acknowledge the previous pipeline stage (Acknowledge). In a manner similar to global clocks, in which rising edges are used for synchronization, rising and falling edges of these 'Request' and 'Acknowledge' signals are used in self-timed logic. A static logic family, called "event logic," can also be designed while using these signal edges for which these edges are the only events of interest.[11] There are two different protocols, i.e., the two phases (for which a rising edge has the same significance as a falling edge) and the four phases protocol (for which only the rising edge is taken into account while the falling edge means only reset).

Library Cells for Cryptographic Applications

ASICs for smart cards can be attacked by differential power attacks (DPA) that consist of tracing the power consumption and identifying operations that are data dependent after removal of the power consumption that is data independent. In this way, secret keys could be obtained. Consequently, DPA-resistant circuits will be of crucial importance in the future. DPA was demonstrated in 1999.[23] It has been shown that it is possible to examine the power consumed by the circuit when processing data or executing instructions. By analyzing the variation in power consumption and the data processed, an attacker can discover the secure information being processed and the keys hidden in the circuit. The attacker can analyze a single power trace (SPA) or can perform a statistical analysis of many collected power traces (DPA). These power traces will provide an average power trace that represents the data-independent power trace. By having a given power trace with a given hidden key, the attacker can subtract the average power trace and determine the difference that represents only the data-dependent power. By comparing the difference to the simulated power traces, the attacker can quite easily deduce what is the secure key of the considered circuit.

The DPA attack is based on the fact that executed operations or instructions present power consumption that is data dependent. A very general goal is to find circuit implementations that are not sensitive to operands regarding the power. Some circuit techniques, such as current steering logic or dual-rail DCVSL logic, are known to be less data dependent. Self-timed logic does not have a global clock, so there is not a global timing signal for use as a reference and the analysis and correlation of power traces of power consumption are more difficult.[24] It is difficult to determine when an operation or instruction starts and stops. However, as power issues are crucial for smart cards, these logic families do not meet this requirement. Furthermore, recent research has revealed weaknesses in the basic DCVSL scheme and suggested improvements.[25] In dual-rail DCVSL logic, parasitic capacitances are different in the two N-ch networks implementing the dual Boolean functions and, therefore, the power consumption too. A new SABL logic family is introduced in Ref. [25], based on the fact that the output will charge the same capacitance for each clock, even if the output transition is 1–1 or 0–0. This SABL logic is based on the differential StrongArm SAFF Flip-Flop; it consumes, however, twice the power of static CMOS.

Leakage Reduction at Architecture Level

Many techniques have been proposed to reduce the leakage power, such as multi-VT technologies, gated Vdd, DTMOS, and static and dynamic SATS.[26] These techniques are effective at the cost of more or less complex circuits or technologies. Another technique already proposed is weak inversion logic for which transistors work in the weak inversion regime. Despite the use of these techniques, reducing static power in very deep submicron technologies has to be addressed at all design levels, including systems and architecture levels. This section addresses this problem at the architecture level.

Activity, Logic Depth, and Number of Gates

By analyzing the ratio of dynamic over static power, it is observable that circuits presenting a low or very low activity will present a too large static power compared to dynamic power. Performing a logic function with a very small activity can be considered far from the optimum, as the circuit is idle most of the time. In other words, if a transistor or a logic gate is not switching for a very long period, it is not very efficient, as the ratio between the switching time (related to dynamic power) and the idle time (for which the transistor or the logic gate is leaky) is very small. It is obviously the case when the circuit is in sleep mode, as no dynamic power is present. Hence, the only consumed power is the static one. However, in that case, nothing can be done at the architecture level and only circuit techniques[26] can be used to reduce leakage.

The presented design methodology aims at searching for an optimum in which one has better use of switching transistors or gates in the reference period of time, i.e., an increased activity in such a way that dynamic power is not too small compared to static power. Obviously, this relative increase in activity has to be understood as useful and not, for instance, by suppressing gated clocks or increasing glitches.

The conventional definition of activity is the factor a in the dynamic power formula $P = a \times f \times C \times \mathrm{Vdd}^2$. It means that a is the ratio between the number of switching gates in a clock period over the total number of gates. Combinational circuits generally present activities around 1 to 5%. One has also to consider idle gates when they are connected in series. If there are 20 gates in series for a given pipeline stage or logic block (logic depth [LD] = 20), these 20 gates have to switch in series in a clock period. So only 1/20 of the clock period is used for the switching of a given gate; the rest of the time the considered gate is only a leaky gate. In running modes, with more and more leaky transistors, it seems that it would be better to avoid designing very inactive gates and therefore to search for an optimum ratio of dynamic power over static power. Leakage energy could be considered roughly proportional to the number of gates and to the duration of the clock period. It is not the case of the dynamic energy which is only proportional to the number of switching gates. To have a better balance of dynamic versus static energy, it could be mandatory to decrease significantly the total number of gates, resulting in an increase in global activity.

Optimal Total Power

This optimum of the total power (dynamic+static) is roughly obtained with similar amounts of static and dynamic power.[27,28] To reduce the total power consumption, an attractive goal could be to have fewer but more active transistors or gates to perform the same logic function. If a given logic function requires 10,000 gates for its implementation and presents an activity factor of 1%, this means that, on average, 100 gates are switching in a clock period. If the same logic function could be implemented with only 1000 gates, keeping the same number of switching gates (100), the activity will be 10%, with the same dynamic power but with a leakage reduced by a factor of 10 due to the reduction in the total number of gates.[29]

Figure 4.52 (left) shows such a theoretical situation with the same dynamic power and a significantly reduced leakage power for the architecture containing fewer gates. This is, however, a naive situation as nothing is said about the speed of the two architectures. The architecture with the smaller number of gates could be slower and would therefore require a larger Vdd to achieve the same speed, impacting dynamic and static power.

Figure 4.52 (right) shows a practical situation comparing a 16×16 multiplier automatically synthesized with a logic synthesizer according to two different architectures. The architecture A is realized with four RCA multipliers working in parallel (3250 gates) and a second architecture B is implemented with two Wallace multipliers working in parallel (1960 gates). Both present 208 transitions for executing a 16×16 multiply (3250 gates $\times$ 6.4% $=$ 1960 gates $\times$ 10.6% $=$ 208 transitions), showing that various architectures with a quite different number of gates can achieve their function with the same number of transitions. The architecture B (second bar, at same Vdd and VT) with fewer gates than A presents a larger activity, a smaller leakage, but also a smaller dynamic power (less sized transistors, due to improved delay slack). However, at the optimum of the total power consumption, this architecture B can be supplied with smaller Vdd and VT

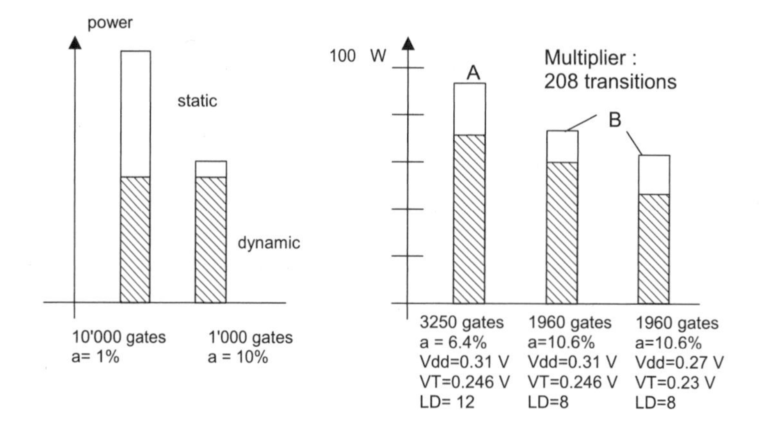

FIGURE 4.52 Various architectures with the same number of transitions for a given function (white rectangle: static; filled rectangle: dynamic).

(third bar) to meet the same speed performances as A, with better results in total power consumption. This shows, however, that the speed performances are a very important parameter in such comparisons.

The goal of designing architectures with more global activity is not only to reduce leakage power with same dynamic power (Figure 4.52, left), but also to find an optimum of the total power, as shown in Figure 4.52 (right). This optimum has to be searched for the same speed constraints; and looking at first and last bars of Figure 4.52 (right) architecture B with fewer gates and more activity is better in terms of optimal total power, however, with a similar ratio of dynamic over static power of, respectively, 3.2 for A and 2.5 for B. This example shows very clearly that the reduction in the total number of gates and the increase in the activity could result in the same or even smaller dynamic power and optimal total power. It is the paradigm shift, not increasing the number of switching gates, but reducing the total number of gates. In this respect, the resulting activity although increased, can be considered as useful activity. However, as already pointed out, the reduction in the total number of gates and the increase in the activity cannot be a goal *per se*, as some architectures for a given logic function with significantly fewer gates could be extremely slow. So one has to dramatically increase the supply voltage and decrease the VT to satisfy the speed constraints, resulting in a much larger total power. It could be the case, for instance, for a sequential multiplier using add-shift mechanism compared to a parallel multiplier.

Design Methodologies Based on Optimal Total Power

In Ref. [30], it is shown that at optimal total power, the ratio I_{on}/I_{off} of a single device in a given technology is proportional to $K_1 \times (\text{LD}/a)$, i.e., proportional to architectural parameters such as logic depth (LD) and activity (a). K_1 is not fixed as shown in Ref. [31] and Ref. [32], but goes from three to six depending on the architecture. Dynamic over static power is also equal to K_1, i.e., dynamic power is three to six times larger than static power at the optimum of the total power. It turns out that I_{on}/I_{off} is quite low in very deep submicron technologies, showing that leakage-tolerant architectures do have to present small LD and large activities. So pipelined architectures, reducing LD proportionally to the number of stages with the same activity (Figure 4.41), have to be preferred to parallel architectures (Figure 4.42) that reduce LD and activity. In other words, parallel architectures do reduce LD, but the transistor count is also largely increased, impacting too much on the number of inactive gates and therefore the leakage.

Obviously, what is searched for is a design methodology starting from a given logic function architecture and generating a new architecture in which the number of cells is reduced, the number of transitions for this given logic function is constant, the activity is increased, but LD is also constant (or even smaller). If LD is increased, as is generally the case by increasing sequencing to reduce the gate count, then the speed constraints have to be satisfied by increasing Vdd and reducing VT, resulting in a larger optimal total power. Figure 4.52 shows such an example, i.e., architecture B (Wallace parallelized twice) is much better than architecture A (RCA parallelized four times), as the former (B) has fewer gates, the same number of transitions, increased activity, and furthermore even a smaller LD.

References

1. S. Borkar, "Technology trends and design challenges for microprocessor design," in *ESSCIRC'98*, September 22–24, The Hague, Netherlands, 1998, pp. 7–8.
2. E.A. Vittoz, "Design of low-voltage low-power ICs," *Symposium on Low Voltage Low Power Silicon ICs, ESSDERC'93*, September 13–16, Grenoble, 1993.
3. V. von Kaenel, P. Macken, and M. Degrauwe, "A voltage reduction technique for battery-operated systems," *IEEE J. Solid-State Circ.*, vol. 25, no. 5, pp. 1136–1140, 1990.
4. C. Piguet, J.-M. Masgonty, C. Arm, S. Durand, T. Schneider, F. Rampogna, C. Scarnera, C. Iseli, J.-P. Bardyn, R. Pache, and E. Dijkstra, "Low-power design of 8-bit embedded CoolRISC micro-controller cores," *IEEE JSSC*, vol. 32, no. 7, pp. 1067–1078, 1997.
5. J.M. Rabay, "Managing power dissipation in the generation-after-next wireless systems," *FTFC'99*, Paris, France, June 1999.
6. A.P. Chandrakasan, S. Sheng, and R.W. Brodersen, "Low-power CMOS digital design," *IEEE J. Solid-State Circ.*, vol. 27, no. 4, pp. 473–484, 1992.
7. C. Piguet, J.-M. Masgonty, V. von Kaenel, and T. Schneider, "Logic design for low-voltage/low-power CMOS circuits," *International Symposium on Low Power Design*, Dana Point, CA, April 23–26, 1995.
8. T. Schneider, V. von Kaenel, and C. Piguet, "Low-voltage/low-power parallelized logic modules," in *Proc. PATMOS'95*, Paper S4.2, Oldenburg, Germany, pp. 147–160, October 4–6, 1995.
9. J.P. Hayes, *Computer Architecture and Organization*, New York: McGraw-Hill, 1978, p. 382.
10. G. Panigrahi, "The implications of electronic serial memories," *Computer*, pp. 18–25, July 1977.
11. J. Sparsoe and S. Furber, *Principles of Asynchronous Circuit Design*. Dordecht: Kluwer, 2001.
12. L. Benini, P. Siegel, and G. DeMicheli, "Saving power by synthesizing gated clocks for sequential circuits," *IEEE Des. Test Comput.*, vol. 11, no. 4, pp. 32–41, 1994
13. C. Piguet, "Low-power and low-voltage CMOS digital design," *Elsevier Microelectron. Eng.* vol. 39, pp. 179–208, 1997.
14. T.G. Noll and E. De Man, "Pushing the performances limits due to power dissipation of future ULSI chips," *IEEE International Symposium on Circuits and Systems, ISCAS'92*, San Diego, May 1992, pp. 1652–1655.
15. J.-M. Masgonty and C. Piguet, "Technology- and power-supply-independent cell library" *IEEE CICC'91*, San Diego, CA, May 12–15, 1991.
16. A. Nève et al., "Design of a branch-based 64-bit carry-select adder in 0.18 μm partially-depleted SOI CMOS," in *Proc. ISLPED'02*, Monterey, CA, August 12–14, 2002, pp. 108–111.
17. S. Nikolaidis and A. Chatzigeorgiou, "Circuit-level low-power design," in *Designing CMOS Circuits for Low-Power*, D. Soudris, C. Piguet, and C. Goutis, Eds., Dordecht: Kluwer Academic Press, 2002.
18. L.G. Heller and W.R. Griffin, "Cascode voltage switch logic: a differential CMOS Logic Family," in *Proc. ISSCC*, San Francisco, CA, 1984, pp. 16–17.
19. R. Zimmermann and W. Fichtner, "Low-power logic styles: CMOS versus pass-transistor logic," *IEEE JSSC*, vol. 37, no. 7, pp. 1079–1090, 1997.
20. J.-M. Masgonty, S. Cserveny, C. Arm, P.-D. Pfister, and C. Piguet, "Low-power low-voltage standard library cells with a limited number of cells," *PATMOS 2001*, Yverdon, Switzerland, September 26–28, 2001.
21. C. Arm, J.-M. Masgonty, and C. Piguet, "Double-latch clocking scheme for low-power I.P. cores," *PATMOS 2000*, Goettingen, Germany, September 13–15, 2000.
22. "Asynchronous circuits and systems," *Special Issue of IEEE Proceedings*, 1999.

23. P. Kocher, "Differential power analysis" *Advances in Cryptology – Crypto 99*, Springer LNCS, vol. 1666, pp. 388–397.
24. M. Renaudin and C. Piguet, *Asynchronous and Locally Synchronous Low-Power SoCs,* Münich, 2001, pp. 490–491.
25. C. Tiri, M. Akmal, and I. Verbauwhede, "A dynamic and differential CMOS logic with signal independent power consumption to withstand differential power analysis on smart cards," in *Proc. ESSCIRC 2002*, Florence, Italy, 2002, pp. 403–406.
26. M. Anis and M. Elmasry, *Multi-Threshold CMOS Digital Circuits*, Dordecht: Kluwer, 2003.
27. C. Heer, "Designing low-power circuits: an industrial point of view," *PATMOS 2001*, Yverdon, September 26–28, 2001.
28. C. Piguet, S. Cserveny, J.-F. Perotto, and J.-M. Masgonty, "Techniques de circuits et méthodes de conception pour réduire la consommation statique dans les technologies profondément submicroniques," in *Proc. FTFC'03*, Paris, May 15–16, 2003, pp. 21–29.
29. Piguet, C. and Narayanan, V., "Guest Editorial," *IEEE Trans. VLSI Syst.*, vol. 12, no. 2 and no. 3, 2004.
30. C. Piguet, C. Schuster, and J.-L. Nagel, "Optimizing architecture activity and logic depth for static and dynamic power reduction," in *Proc. 2nd Northeast Workshop Circ. Syst. – NewCAS'04*, June 20–23, 2004, Montréal, Canada.
31. R.W. Brodersen, M.A. Horowitz, D. Markovic, B. Nikolic, and V. Stojanovic, "Methods for true power minimization," in *Proc. Int. Conf. Comput. Aided Des.*, San Jose, CA, 2002, November, pp. 35–42.
32. K. Nose and T. Sakurai, "Optimization of Vdd and Vth for low-power and high-speed applications," *ASPDAC*, January 2000, pp. 469–474.

5

Surface Mount Technology

Glenn R. Blackwell
Purdue University

5.1 Introduction

This chapter on surface mount technology (SMT) will familiarize the reader with the process steps in a successful SMT design. Being successful with the implementation of SMT means the engineers involved must commit to the principles of concurrent engineering. It also means that a continuing commitment to a quality technique is necessary, whether that is Taguchi, TQM, SPC, DOE, another technique, or a combination of several quality techniques, lest you too have quality problems with SMT (Figure 5.1).

5.2 Definition and Considerations

SMT is a collection of scientific and engineering methods needed to design, build, and test products made with electronic components that mount to the surface of the printed circuit board (PCB) without holes for leads (Higgins, 1991). This definition notes the breadth of topics necessary to understand SMT, and also clearly says that the successful implementation of SMT will require the use of concurrent engineering (Shina, 1991; Classon, 1993). Concurrent engineering means that a team of design, manufacturing, test, and marketing people will concern themselves with board layout, parts and parts placement issues, soldering, cleaning, test, rework, and packaging before any product is made. The careful control of all these issues improves both yield and reliability of the final product. In fact, SMT cannot be reasonably implemented without the use of concurrent engineering and/or the principles contained in design for manufacturability (DFM) and design for testability (DFT), and therefore any facility that has not embraced these principles should do so if implementation of SMT is its goal.

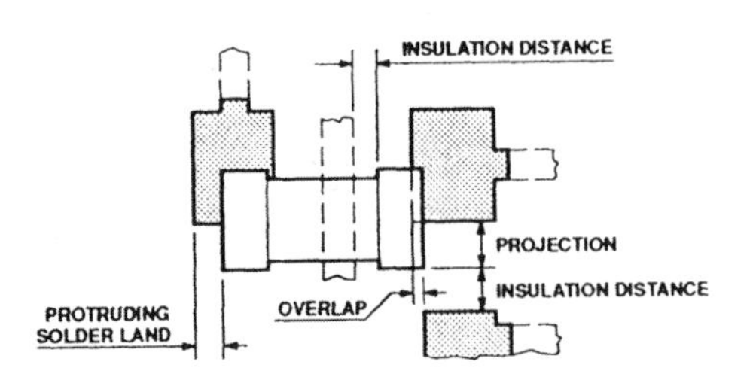

FIGURE 5.1 Placement misalignment of an SMT chip resistor. (*Source*: Phillips Semiconductors, *Surface Mount Process and Application Notes*, Sunnyvale, CA: Phillips Semiconductors, 1991. With permission.)

Considerations in the Implementation of SMT

Main reasons to consider implementation of SMT include:

- Reduction in circuit board size
- Reduction in circuit board weight
- Reduction in number of layers in the circuit board
- Reduction in trace lengths on the circuit board, with correspondingly shorter signal transit times and potentially higher-speed operation

Not all these reductions may occur in any given product redesign from **through-hole technology** (THT) to SMT. Variations in considerations for manufacturing and testing will affect the reductions possible.

Most companies that have not converted to SMT are considering doing so. All is of course not golden in SMT Land. During the assembly of a **through-hole** board, either the component leads go through the holes or they do not, and the component placement machines can typically detect the difference in force involved. During SMT board assembly, the placement machine does not have such direct feedback, and the accuracy of final soldered placement becomes a stochastic (probability-based) process dependent on such items as component pad design, accuracy of the PCB artwork, and fabrication which affects the accuracy of pad location, accuracy of solder paste deposition location and deposition volume, accuracy of adhesive deposition location and volume if adhesive is used, accuracy of placement machine vision systems, variations in component sizes from the assumed sizes, and thermal issues in the solder reflow process. In THT tests, there is a through-hole at every potential test point, making it easy to align a bed-of-nails tester. In SMT designs, there are not holes corresponding to every device lead. DFT principles must be followed to create testable products.

The design team must also consider form, fit, and function, time-to-market, existing capabilities, testing, rework capabilities, and the cost and time to characterize a new process when deciding on a change of technologies.

5.3 SMT Design, Assembly, and Test Overview

- Circuit design (not covered in this chapter)
- Substrate (typically PCB) design
- Thermal design considerations
- Bare PCB fabrication and tests (not covered in this chapter)
- Application of adhesive, if necessary
- Application of solder paste
- Placement of components in solder paste
- Reflowing of solder paste
- Cleaning, if necessary
- Testing of populated PCB (not covered in this chapter)

Any work in SMT must follow standards as set forth by the IPC. These standards cover all aspects of printed circuit design and assembly, including:

- Laminates
- Reinforcements
- Circuit board and circuit assembly acceptance

- Assembly
- Assembly support
- Assembly materials
- Components
- Cleaning and cleanliness
- Solderability
- Other relevant topics

Once circuit design is complete, substrate design and fabrication, most commonly of a PCB, enters the process. Generally, PCB assembly configurations using surface mount devices (SMDs) are classified as shown in Figure 5.2. While the IPC has further subclassifications of SMT assembly types, these three basic definitions serve for most assemblies.

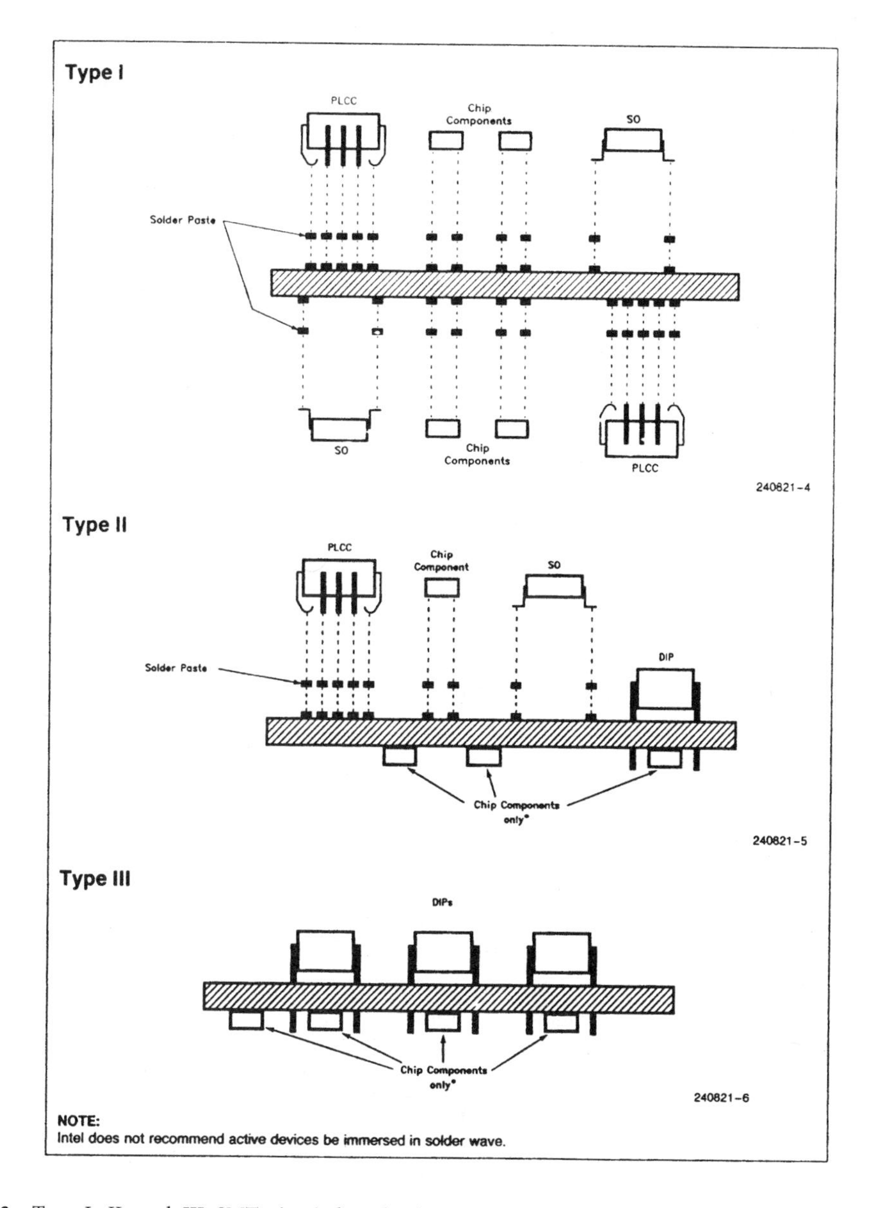

FIGURE 5.2 Type I, II, and III SMT circuit boards. (*Source*: Intel Corporation, *Packaging*, Santa Clara, CA: Intel Corporation, 2000. With permission.)

- **Type I**—only SMDs are used, typically on both sides of the board. No through-hole components are used. Top and bottom may contain both large and small active and passive SMDs. This type of board uses reflow soldering only.
- **Type II**—a double-sided board, with SMDs on both sides. The top side may have all sizes of active and passive SMDs, as well as through-hole components, while the bottom side carries passive SMDs and small active components such as transistors. This type of board requires both reflow and wave soldering, and will require placement of bottom-side SMDs in adhesive.
- **Type III**—the top side has only through-hole components, which may be active and/or passive, while the bottom side has passive and small active SMDs. This type of board uses wave soldering only, and also requires placement of the bottom-side SMDs in adhesive. It should be noted that with the ongoing increase in usage of various techniques to place IC dice directly on circuit boards, Type III in some articles means a mix of packaged SMT ICs and bare die on the same board.

A Type I bare board will first have solder paste applied to the component pads on the board. Once solder paste has been deposited, active and passive parts are placed in the paste. For prototype and low-volume lines, this can be done with manually guided XY tables using vacuum needles to hold the components, while in medium- and high-volume lines automated placement equipment is used. This equipment will pick parts from reels, sticks, or trays, then place the components at the appropriate pad locations on the board; hence the term "pick and place" equipment.

After all parts are placed in the solder paste, the entire assembly enters a reflow oven to raise the temperature of the assembly high enough to reflow the solder paste and create acceptable solder joints at the component lead/pad transitions. As discussed in more detail in Reflow International's *Reflow Technology Handbook*, reflow ovens most commonly use convection and IR heat sources to heat the assembly above the point of solder liquidus, which for 63/37 tin-lead eutectic solder is 183°C. Because of the much higher thermal conductivity of the solder paste compared to the IC body, reflow soldering temperatures are reached at the leads/pads before the IC chip itself reaches damaging temperatures. The board is inverted and the process repeated.

The move to lead-free solder will require higher reflow temperatures, with increasing concerns about potential damage to both substrates and components. Lead-free solder is discussed later in this chapter.

If mixed-technology Type II is being produced, the board will first go through the Type I reflow process, then be inverted, an adhesive will be dispensed at the centroid of each SMD, parts placed, the adhesive cured, the assembly rerighted, through-hole components mounted, and the circuit assembly will then be wave-soldered which will create acceptable solder joints for both the through-hole components and bottom-side SMDs.

A Type III board will first be inverted, adhesive dispensed, SMDs placed on the bottom side of the board, the adhesive cured, the board rerighted, through-hole components placed, and the entire assembly wave-soldered. It is imperative to note that only passive components and small active SMDs can be successfully bottom-side wave-soldered without considerable experience on the part of the design team and the board assembly facility. It must also be noted that successful wave soldering of SMDs requires a dual-wave machine with one turbulent wave and one laminar wave.

It is common for a manufacturer of through-hole boards to convert first to a Type II or Type III substrate design before going to an all-SMD Type I design. This is especially true if amortization of through-hole insertion and wave-soldering equipment is necessary. Many factors contribute to the reality that most boards are mixed-technology Type II or Type III boards. While most components are available in SMT packages, through-hole connectors are still commonly used for the additional strength the through-hole soldering process provides, and high-power devices such as three-terminal regulators are still commonly through-hole due to off-board heat-sinking demands. Both of these issues are addressed by manufacturers, and solutions exist which allow Type I boards with connectors and power devices.

With regard to Type II and Type III boards and the use of wave soldering, be certain to consult with all component manufacturers of any SMT components that will be placed on the wave-solder side of the assembly. For example, while some IC manufacturers allow their smaller ICs to be wave soldered, others do not recommend that any of their ICs be subjected to the wave-solder process.

Again, it is imperative that all members of the design, build, and test teams be involved from the design stage. Today's complex board designs mean that it is entirely possible to exceed the ability to adequately test a board if the test is not designed-in, or to robustly manufacture the board if in-line inspections and handling are not adequately considered. Robustness of both test and manufacturing are only assured with full involvement of all parties to overall board design and production.

It cannot be overemphasized that the speed with which packaging issues are moving requires anyone involved in SMT board or assembly issues to stay current and continue to learn about the processes. Subscribe to one or more of the industry-oriented journals noted in the "Further Information" section at the end of this chapter, obtain any IC industry references, and purchase several SMT reference books.

5.4 SMD Definitions

The new user of SMDs must rapidly learn the packaging sizes and types for SMDs. Resistors, capacitors, and most other passive devices come in two-terminal packages that have end-terminations designed to rest on substrate pads/lands (Figure 5.2 and Figure 5.3).

SMD ICs come in a wide variety of packages, from eight-pin small outline packages (SOLs) to 1000+ connection packages in a variety of sizes and lead configurations, as shown in Figure 5.3. The most common commercial packages currently include **plastic leaded chip carriers** (PLCCs), SOLs **quad flat packs** (QFPs), and plastic quad flat packs (PQFPs). Add in tape automated bonding (TAB), ball grid array (BGA), and other newer technologies, and the IC possibilities become overwhelming. Space prevents examples of all these technologies from being included here. The reader is referred to the standards of the Institute for Interconnecting and Packaging Electronic Circuits (IPC) to find the latest package standards, and to the proceedings of the most recent APEX, NEPCON, SMTAI, and other conferences for information on industry uses of the latest SMT packages. A good overview of package styles is found in packaging handbooks from various manufacturers, including Intel and National Semiconductor.

Each IC manufacturer's data books will have packaging information for their products. Shown in Figure 5.4 are two of the many SMT IC plastic packages.

The engineer should be familiar with the term "lead pitch," which means the center-to-center distance between IC leads. Pitch may be in thousandths of an inch, also known as mils, or in millimeters. Common pitches are 0.050 in. (50 mil pitch), 0.025 in. (25 mil pitch) frequently called "fine pitch," and 0.020 in. and smaller, frequently called "ultra-fine pitch." Metric equivalents are 1.27 mm, 0.635 mm, and 0.508 mm and smaller. Conversions from metric to inches are easily approximated if one remembers that 1 mm approximately equals 40 mils. Do remember that this approximation is fine for "back-of-the-envelope" designs, but for actual designs exact conversions must be performed.

For process control, design teams must consider the minimum and maximum package size variations allowed by their part suppliers, the moisture content of parts as-received, and the relative robustness of each

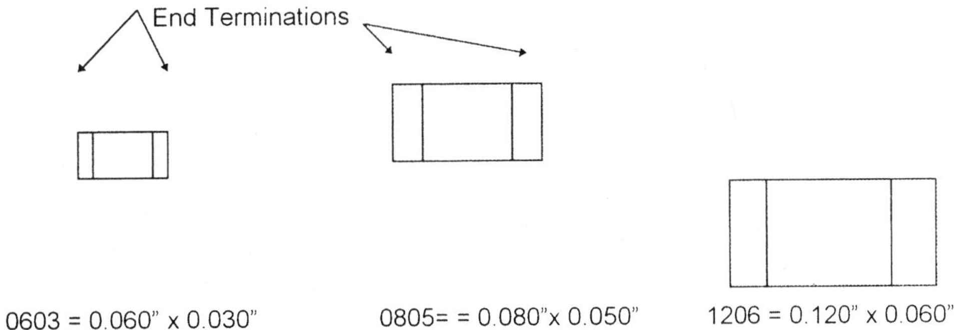

FIGURE 5.3 Example of passive component sizes (top view) (not to scale).

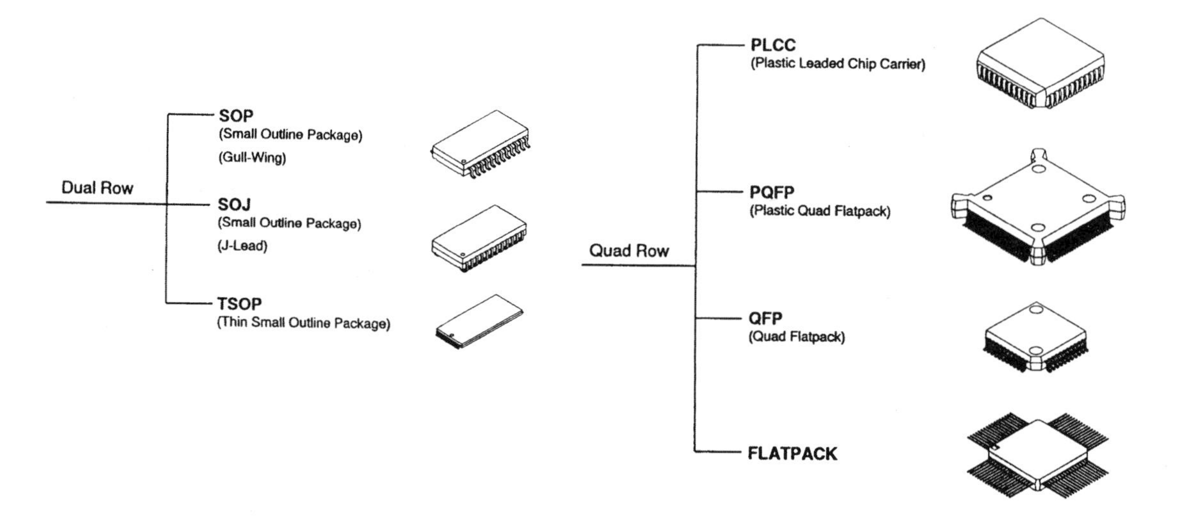

FIGURE 5.4 Examples of SMT plastic packages. (*Source*: Intel Corporation, *Packaging*, Santa Clara, CA: Intel Corporation, 1994. With permission.)

lead type. Incoming inspection should consist of both electrical and mechanical tests. Whether these are spot checks, lot checks, or no checks will depend on the relationship with the vendor.

5.5 Substrate Design Guidelines

As noted previously, substrate (typically PCB) design has an effect not only on board/component layout, but also on the actual manufacturing process. Incorrect land design or layout can negatively affect the placement process, the solder process, the test process, or any combination of the three. Substrate design must take into account the mix of SMDs that are available for use in manufacturing.

The considerations noted here as part of the design process are neither all-encompassing, nor in sufficient detail for a true SMT novice to adequately deal with all the issues involved in the process. They are intended to guide an engineer through the process, allowing the engineer to access more detailed information as necessary. General references are noted at the end of this chapter, and specific references will be noted as applicable. Although these guidelines are noted as "steps," they are not necessarily in an absolute order, and may require several iterations back-and-forth among the steps to result in a final satisfactory process and product.

After the circuit design (schematic capture) and analysis, Step 1 in the process is to determine whether all SMDs will be used in the final design making a Type I board, or whether a mix of SMDs and through-hole parts will be used, leading to a Type II or Type III board. This decision will be governed by some or all of the following considerations:

- Current parts stock
- Existence of current through-hole placement and wave solder equipment
- Amortization of current through-hole placement and solder equipment
- Existence of reflow soldering equipment or cost of new reflow soldering equipment
- Desired size of the final product
- Panelization of smaller Type I boards
- Thermal issues related to high power circuit sections on the board

It may be desirable to segment the board into areas based on function: RF, low power, high power, etc., using all SMDs where appropriate, and mixed-technology components as needed. Any RF designs will find Howard Johnson's book of value. Power and connector portions of the circuit may point to the use of

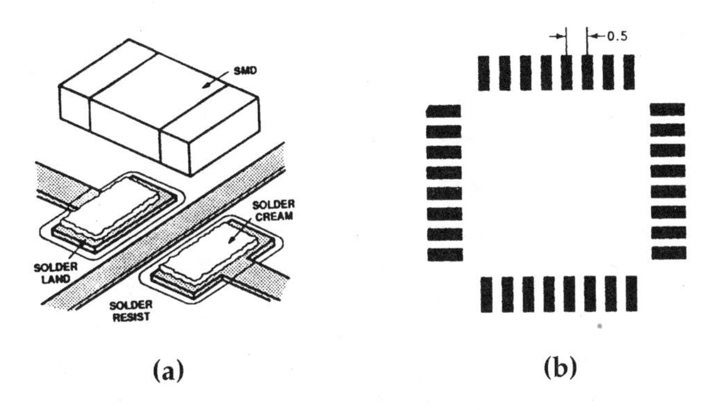

FIGURE 5.5 (a) Footprint land and resist. (*Source*: Phillips Semiconductors, *Surface Mount Process and Application Notes*, Sunnyvale, CA: Phillips Semiconductors, 1991. With permission.) (b) QFP footprint. (*Source*: Intel Corporation, *Packaging*, Santa Clara, Calif.: Intel Corporation, 1994. With permission.)

through-hole components, although as mentioned both these issues are being addressed by circuit board material and connector manufacturers. Using one solder technique (reflow or wave) simplifies processing and may outweigh other considerations.

Step 2 in the SMT process is to define all the footprints of the SMDs under consideration for use in the design. The footprint is the copper pattern or "land," on the circuit board upon which the SMD will be placed. Footprint examples are shown in Figure 5.5, and footprint recommendations are available from IC manufacturers and in the appropriate data books. They are also available in various ECAD packages used for the design process, or in several references that include an overview of the SMT process. However, thereader is seriously cautioned about using the general references for anything other than the most common passive and active packages. Even the position of pin 1 may be different among IC manufacturers of the "same" chip. The footprint definition may also include the position of the solder resist pattern surrounding the copper pattern. Footprint definition sizing will vary depending on whether reflow or wave solder process is used. Wave-solder footprints will require recognition of the direction of travel of the board through the wave to minimize solder shadowing in the final fillet, as well as requirements for solder thieves. The copper footprint must allow for the formation of an appropriate, inspectable solder fillet.

If done as part of the EDA process (electronic design automation, using appropriate electronic CAD software), the software will automatically assign copper directions to each component footprint, as well as appropriate coordinates and dimensions. These may need adjustment based on considerations related to wave soldering, test points, RF and/or power issues, and board production limitations. Allowing the software to select 5 mil traces when the board production facility to be used can only reliably do 10 mil traces would be inappropriate. Likewise, the solder resist patterns must be governed by the production capabilities.

Final footprint and trace decisions will:

- Allow for optimal solder fillet formation
- Minimize necessary trace and footprint area
- Allow for adequate test points
- Minimize board area, if appropriate
- Set minimum interpart clearances for placement and test equipment to safely access the board (Figure 5.6)
- Allow adequate distance between components for post-reflow operator inspections
- Allow room for adhesive dots on wave-soldered boards
- Minimize solder bridging

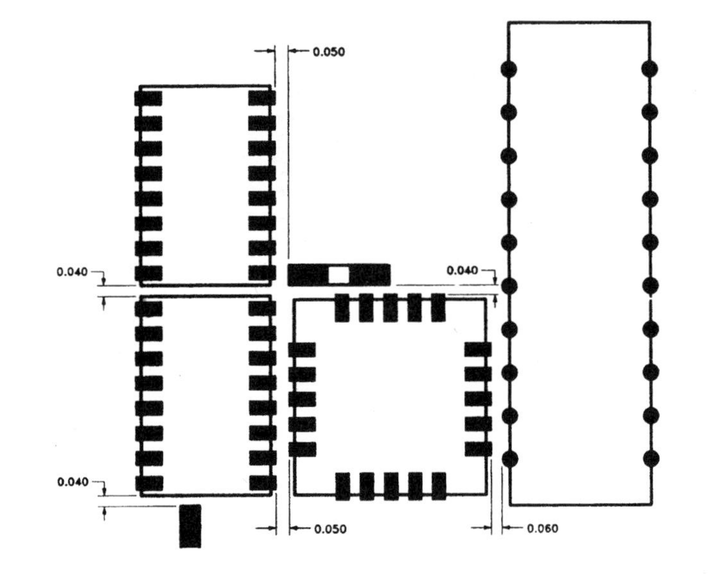

FIGURE 5.6 Minimum land-to-land clearance examples. (*Source*: Intel Corporation, *Packaging*, Santa Clara, CA: Intel Corporation, 1994. With permission.)

Decisions that will provide optimal footprints include a number of issues, including:

- Component dimension tolerances
- Board production capabilities, both artwork and physical tolerances across the board relative to a 0–0 fiducial
- How much artwork/board shrink or stretch is allowable
- Solder deposition volume consistencies with respect to fillet sizes
- Placement machine accuracies
- Test probe location controls and bed-of-nails grid pitch

Design teams should restrict wave-solder-side SMDs to passive components and transistors. While small SMT ICs can be successfully wave-soldered, this is inappropriate for an initial SMT design, and is not recommended by some IC manufacturers.

Certain design decisions may require a statistical computer program, if available, to the design team. The stochastic nature of the overall process suggests a statistical programmer will be of value.

5.6 Thermal Design Considerations

Thermal management issues remain major concerns in the successful design of an SMT board and product. Consideration must be taken of the variables affecting both board temperature and junction temperature of the IC. The reader is referred to other chapters in this Handbook for the basics of thermal management, and to Bar-Cohen and Kraus (1988) for a more detailed treatment on thermal issues affecting ICs and PCB design.

The design team must understand the basic heat transfer characteristics of most SMT IC packages (Capillo, 1993). Since the silicon chip of an SMD is equivalent to the chip in an identical-function DIP package, the smaller SMD package means the internal lead frame metal has a smaller mass than the lead frame in a DIP package. This lesser ability to conduct heat away from the chip is somewhat offset by the leadframe of many SMDs being constructed of copper, which has a lower thermal resistance than the Kovar and Alloy 42 materials commonly used for DIP packages. However, with less metal and shorter lead lengths to transfer heat to ambient air, more heat is typically transferred to the circuit board itself. Since all electronics components generate heat in use, and elevated temperatures negatively affect the reliability and failure rate of semiconductors, it is important that heat generated by SMDs be removed as efficiently as possible. The design team needs to have expertise with the variables related to thermal transfer:

- Junction temperature: T_j
- Thermal resistances: Θ_{jc}, Θ_{ca}, Θ_{cs}, Θ_{sa}
- Temperature sensitive parameter (TSP) method of determining Θs
- Power dissipation: P_D
- Thermal characteristics of substrate material

Power SMT packages have been developed to maximize heat transfer to the substrate. These include PLCCs with integral heat spreaders, the SOT-89 power transistor package, the DPAK power transistor package, and many others. Note that all of these devices are designed primarily for processing with the solder paste process, and some specifically recommend against their use with wave-solder applications. Heat sinks and heat pipes could also be considered for high-power ICs. Excellent references for thermal issues are the monthly publications *Electronics Cooling*, www.electoronics-cooling.com, and *Cooling Zone's* online magazine, www.coolingzone.com. Several board thermal analysis software packages are available and are highly recommended for boards that are expected to develop high thermal gradients. Their advertisements can be found on the two web sites just mentioned.

In the conduction process, heat is transferred from one element to another by direct physical contact between the elements. Ideally the material to which heat is being transferred should not be adversely affected by the transfer. As an example, the **glass-transition temperature** T_g of FR-4 is 125°C. Heat transferred to the board has little or no detrimental effect as long as the board temperature stays at least 50°C below T_g. Good heat sink materials exhibit high thermal conductivity, which is not a characteristic of fiberglass. Therefore, the traces must be depended on to provide the thermal transfer path (Choi et al., 1994; Brooks, 1998). Conductive heat transfer is also used in the transfer of heat from IC packages to heat sinks, which also requires use of thermal grease to fill all air gaps between the package and the "flat" surface of the sink.

The previous discussion of lead properties of course does not apply to leadless devices such as leadless ceramic chip carriers (LCCCs). Design teams using these and similar packages must understand the better heat transfer properties of the alumina used in ceramic packages, and must match TCEs between the LCCC and the substrate, since there are no leads to bend and absorb mismatches of expansion.

Since the heat transfer properties of the system depend on substrate material properties, it is necessary to understand several of the characteristics of the most common substrate material, FR-4 fiberglass. The glass-transition temperature has already been noted, and board designers must also understand that multilayer FR-4 boards do not expand identically in the *X*-, *Y*-, and *Z*-directions as temperature increases. Plated-through-holes will constrain *Z*-axis expansion in their immediate board areas, while non-through-hole areas will expand further in the *Z*-axis, particularly as the temperature approaches and exceeds T_g (Lee et al., 1984; Prasad, 1997). This unequal expansion can cause delamination of layers and plating fracture.

If the design team knows that there will be a need for higher abilities to dissipate heat and/or a need for higher glass transition temperatures and lower coefficients of thermal expansion (TCE) than FR-4 possesses, many other materials are available, as discussed in Hollomon (1995) and Prasad (1997).

Note in Table 5.1 that copper-clad Invar has both variable T_g and variable thermal conductivity, depending on the volume mix of copper and Invar in the substrate. Copper has a high TCE and Invar has a low TCE, so the TCE increases with the thickness of the copper layers. In addition to heat transfer considerations, board material decisions must also be based on the expected vibration, stress, and humidity in the application.

Convective heat transfer involves transfer due to the motion of molecules, typically airflow over a heat sink, and depends on the relative temperatures of the two media involved. It also depends on the velocity of air flow over the boundary layer of the heat sink. Convective heat transfer primarily occurs when forced air flow is provided across a substrate, and when convection effects are maximized through the use of heat sinks. The rules that designers are familiar with when designing THT heat-sink device designs also apply to SMT design.

The design team must consider whether passive conduction and convection will be adequate to cool a populated substrate, or whether forced-air cooling or liquid cooling will be needed. Passive conductive cooling is enhanced with thermal layers in the substrate, such as the previously mentioned copper/Invar.

TABLE 5.1

Substrate Material (Units)	T_g (°C)	TCE *X–Y* (PPM/°C)	TherCon (W/M°C)	MoistAbs (%)
FR-4 Epoxy glass	125	13–18	0.16	0.10
Polymide glass	250	12–16	0.35	0.35
Copper-clad invar (depends on resin)	5–7	160XY 15–20Z	NA	NA
Poly aramid fiber	250	3–8	0.15	1.65
Alumina/ceramic	NA	5–7	20–45	NA

T_g = glass-transition temperature
TCE = thermal coefficient of expansion, in this case in the *X–Y* plane
TherCon = thermal conductivity
MoistAbs = moisture absorption

There will also be designs that will rely on the traditional through-hole device with heat sink to maximize heat transfer. An example of this would be the typical three-terminal voltage regulator mounted on a heat sink or directly to a metal chassis for heat conduction, for which standard calculations apply (Lee et al., 1993).

Many specific examples of heat transfer may need to be considered in board design, and of course most examples involve both conductive and convective transfer. For example, the air gap between the bottom of a standard SMD and the board affects the thermal resistance from the case to ambient, Θ_{ca}. A wider gap will result in a higher resistance, due to poorer convective transfer, whereas filling the gap with a thermally conductive epoxy will lower the resistance by increasing conductive heat transfer (Blackwell, 2000). Thermal-modeling software is the best way to deal with these types of issues, due to the need for rigorous application of computational fluid dynamics (CFD).

5.7 Adhesives

In the surface mount assembly process, Type II and Type III boards will always require adhesive to mount the SMDs for passage through the solder wave. This is apparent when one envisions components on the bottom side of the substrate with no through-hole leads to hold them in place. Adhesives will stay in place after the soldering process and throughout the life of the substrate and the product, since there is no convenient means of adhesive removal once the solder process is complete. This means the adhesive used must meet a number of both physical and chemical characteristics that should be considered during the three phases of adhesive use in SMT production: pre-application properties relating to storage and dispensing issues, curing properties relating to time and temperature needed for cure, and post-curing properties relating to final strength, mechanical stability, and reworkability. Among these characteristics are:

- Electrically nonconductive
- Thermal coefficient of expansion similar to the substrate and the components
- Stable in both storage and after application, prior to curing
- Stable physical drop shape—retains drop height and fills *Z*-axis distance between the board and the bottom of the component; thixotropic with no adhesive migration
- Noncorrosive to substrate and component materials
- Chemically inert to flux, solder, and cleaning materials used in the process
- Curable as appropriate to the process: UV, oven, or air-cure
- Removable for rework and repair
- Once cured, unaffected by temperatures in the solder process
- Adhesive color, for easy identification by operators

One-part adhesives are easier to work with than two-part adhesives because an additional process step is not required. The user must verify that the adhesive has sufficient shelf life and pot life for the user's perceived process requirements. Both epoxy and acrylic adhesives are available as one- or two-part systems, and must be cured thermally. Generally, epoxy adhesives are cured by oven-heating, while acrylics may be formulated to be cured by long-wave UV light or heat.

Adhesive can be applied by screening techniques similar to solder paste screen application, by pin-transfer techniques, and by syringe deposition. Screen and pin-transfer techniques are suitable for high-volume production lines with few product changes over time. Syringe deposition using an X–Y table riding over the board with a volumetric pump and syringe tip is more suitable for lines with a varying product mix, prototype lines, and low-volume lines where the open containers of adhesive necessary in pin-transfer and screen techniques are avoided. Newer syringe systems are capable of handling high-volume lines. See Figure 5.8 for methods of adhesive deposition, which include stenciling and syringe deposition. If Type II or Type III assemblies are used, and thermal transfer between components and the substrate is a concern, the design team should consider thermally conductive adhesives.

Regardless of the type of assembly, the type of adhesive used, or the curing technique used, adhesive volume and height must be carefully controlled. Slump of adhesive after application is undesirable because the adhesive must stay high enough to solidly contact the bottom of the component, and must not spread and contaminate any pad associated with the component.

If adhesive dot height $= X$, substrate metal height $= Y$, and SMD termination thickness $= Z$, then $X > Y + Z$, allowing for all combinations of potential errors, e.g., end termination min and max thickness, adhesive dot min and max height, and substrate metal min and max height.

Typically, end termination thickness variations are available from the part manufacturer. Solder pad thickness variations are a result of the board manufacturing process, and will vary not only with the type of board metalization (standard etch vs. plated-through-hole) but also with the variations within each type. For adequate dot height, which will allow for some dot compression by the part, X should be between $1.5X$ and $2.5X$ of the total $Y + Z$, or just Z when dummy tracks are used. If adhesive dots are placed on masked areas of the board, mask thickness must also be considered.

A common variation on the above design is to place "dummy" copper pads under the center of the part. Since these pads are etched and plated at the same time as the actual solder pads, the variation in metal height Y is eliminated as an issue. Adhesive dots are placed on the dummy pads and $X > Z$ is the primary concern.

Adhesive dispensing quality issues are addressed by considerations of:

- Type of adhesive to be used
- Process-area ambient temperature and humidity
- Incoming quality control
- No voids in cured adhesive to prevent trapping of flux, dirt, etc.
- Volume control
- Location control

As shown in Figure 5.7, all combinations of termination, dot, and substrate height/thickness must be considered. Prasad (1997) has an excellent in-depth discussion of adhesives in SMT production.

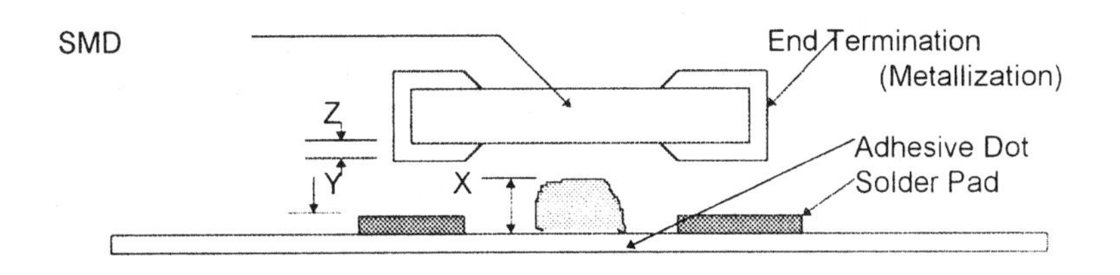

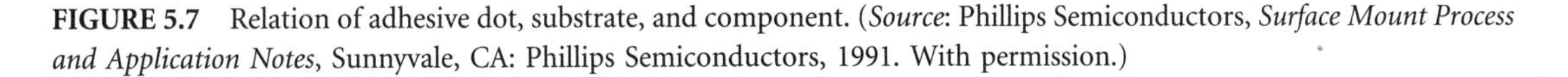

FIGURE 5.7 Relation of adhesive dot, substrate, and component. (*Source*: Phillips Semiconductors, *Surface Mount Process and Application Notes*, Sunnyvale, CA: Phillips Semiconductors, 1991. With permission.)

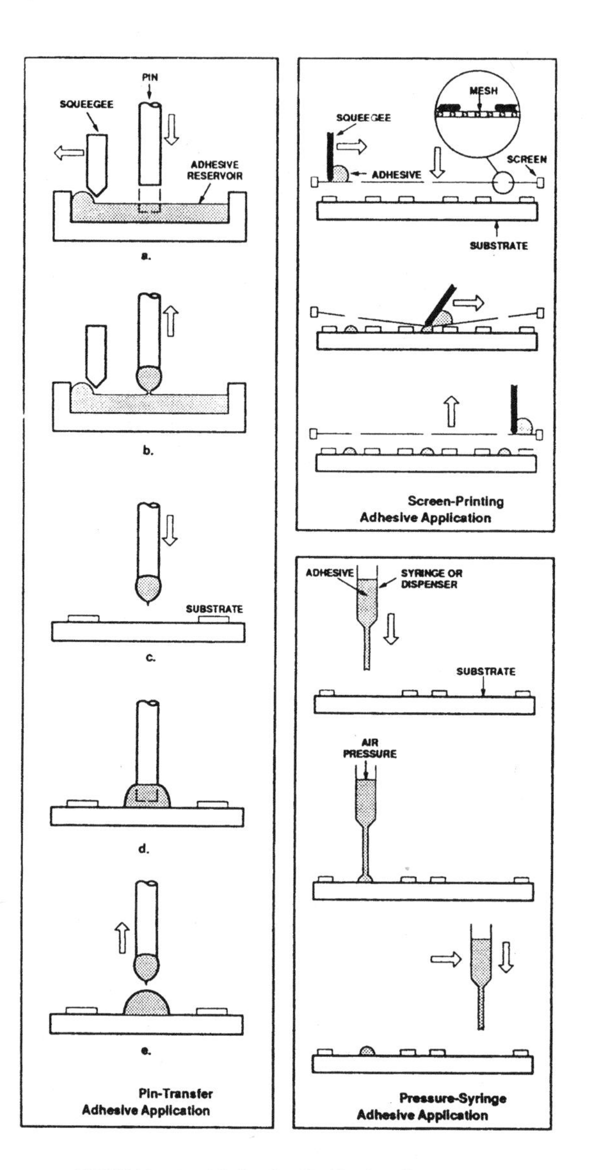

FIGURE 5.8 Methods of adhesive deposition.

5.8 Solder Paste and Joint Formation

Solder joint formation is the culmination of the entire process. Regardless of the quality of the design or any other single portion of the process, if high-quality reliable solder joints are not formed, the final product is not reliable. It is at this point that ppm levels take on their finest meaning. For a medium-size substrate (nominal 6 in. × 8 in.), with a medium density of components, a typical mix of active and passive parts on the topside and only passive and three- or four-terminal active parts on bottom side, there may be in excess of 1000 solder joints per board. If solder joints are manufactured at the 3 sigma level (99.73% good joints, 0.27% defect rate, or 2700 defects per one million joints), *there will be 2.7 defects per board*! At the 6 sigma level, of 3.4 ppm, there will be a defect on one board out of every 294 boards produced. If your anticipated production level is 1000 units per day, you will have 3.4 rejects based solely on solder joint problems, not counting other sources of defects.

Solder paste may be deposited by syringe, or by screen or stencil printing techniques. Stencil techniques are best for high-volume/speed production although they do require a specific stencil for each board design.

Syringe and screen techniques may be used for high-volume lines, and are also suited to mixed-product lines where only small volumes of a given board design are to have solder paste deposited. Syringe deposition is the only solder paste technique that can be used on boards that already have some components mounted. It is also well suited for prototype lines and for any use requires only software changes to develop a different deposition pattern.

Solder joint defects have many possible origins:

- Poor or inconsistent solder paste quality
- Inappropriate solder pad design/shape/size/trace connections
- Substrate artwork or production problems, e.g., mismatch of copper and mask, warped substrate
- Solder paste deposition problems, e.g., wrong volume or location
- Component lead problems, e.g., poor coplanarity or poor tinning of leads
- Placement errors, e.g., part rotation or *X–Y* offsets
- Reflow profile, e.g., preheat ramp too fast or too slow; wrong temperatures created on substrate
- Board handling problems, e.g., boards getting jostled prior to reflow

Once again, a complete discussion of all of the potential problems that can affect solder joint formation is beyond the scope of this chapter. Many references are available that address these issues. An excellent overview of solder joint formation theory is found in Lau and Pau (1996). Updated information on this and all SMT topics is available each year at conferences, such as APEX, SMTAI, and NEPCON.

While commonly used solder paste for both THT and SMT production contains 63/37 eutectic tin-lead solder, other metal formulations are available, including 62/36/2 Sn/Pb/Ag and 96/4 Sn/Ag (aka silver solder). The fluxes available include rosin mildly activated (RMA), water-soluble, and no-clean. The correct decision rests as much on the choice of flux as it does on the proper metal mixture. A solder paste supplier can best advise on solder pastes for specific needs.

Many studies have been done and are in process to determine an optimal no-lead replacement for lead-based solder in commercial electronic assemblies. The design should investigate the current status of these studies as well as the status of no-lead legislation as part of the decision-making process.

The major concern leading to the implementation of lead-free assembly stems not from the assembly process itself but from the leaching of heavy metals out of electronic assemblies that are discarded in landfills. Banning of electronics assemblies from landfills has already started, as Sioux Falls, South Dakota has banned electronics assemblies from its landfill. Other cities and states are considering similar legislation.

No-lead legislation in the European Community now exists, requiring the phased elimination of lead-bearing materials (as well as others) over the 2005–2010 timeframe. While there is no current U.S.A. federal legislation, designers and manufacturers who want to maintain an EU market must prepare for the transition. As Bradley et al. (2003) report, the National Electronics Manufacturing Initiative (NEMI) has now recommended that alloys of tin, silver, and copper be used in the lead-free process. This alloy mix appears to also be gaining acceptance in Europe and Japan. The major process drawback to this mix is that the reflow temperatures will rise approximately 35 to 40°C. This temperature rise will require careful consideration of the effects on both substrate materials and component survival. Apell discusses these and other lead-free reflow considerations.

To better understand solder joint formation, one must understand the make-up of solder paste used for SMT soldering. The solder paste consists of microscopic balls of solder, most commonly tin-lead with the accompanying oxide film, flux, and activator and thickener solvents as shown in Figure 5.9.

The fluxes are an integral part of the solder paste, and are discussed further in Section "Cleaning". RMA, water-soluble, and no-clean flux/pastes are available. An issue directly related to fluxes, cleaning, and fine-pitch components (25 mil pitch and less) is reflowing in an inert environment. Inert gas blanketing the oven markedly reduces the development of oxides in the elevated temperatures present. Oxide reduction needs are greater with the smaller metal balls in paste designed for fine-pitch parts because there is more surface area on which oxides can form. No-clean fluxes are not as active as other fluxes and therefore have a lesser ability to reduce the oxides formed on both the paste metal and substrate metalizations. Inerting the oven tends to solve these problems. However, it brings with it control issues that must be considered.

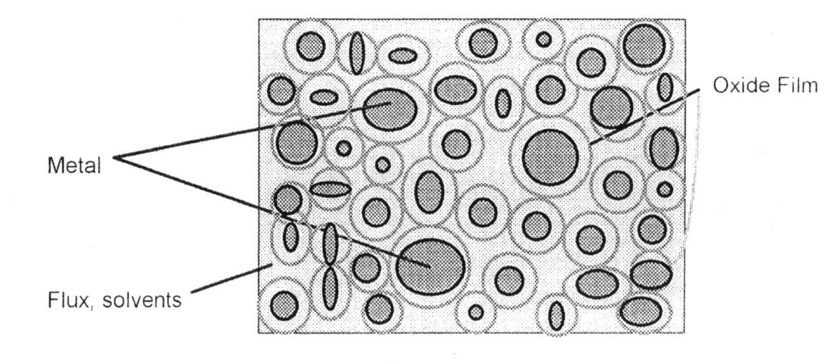

FIGURE 5.9 Make-up of SMT solder paste.

Regardless of the supplier, frequent solder paste tests are advisable, especially if the solder is stored for prolonged periods before use. As a minimum, viscosity, percent metal, and solder sphere formation should be tested (Capillo, 1990). Solder sphere formation is particularly important because acceptable particle sizes will vary depending on the pitch of the smallest-pitch part to be used, and the consistency of solder sphere formation will affect the quality of the final solder joint. Round solder spheres have the smallest surface area for a given volume. Therefore, they will have the least amount of oxide formation. Uneven distribution of sphere sizes within a given paste can lead to uneven heating during the reflow process, with the result that the unwanted solder balls will be expelled from the overall paste mass at a given pad/lead site. Fine-pitch paste has smaller ball sizes and consequently more surface area on which oxides can form.

It should be noted at this point that there are three distinctly different "solder balls" referred to in this chapter and in publications discussing SMT. The solder sphere test refers to the ability of a volume of solder to form a ball shape due to its inherent surface tension when reflowed (melted) on a nonwettable surface. This ball formation is dependent on minimum oxides on the microscopic metal balls that make up the paste—the second type of "solder ball." It is also dependent on the ability of the flux to reduce the oxides that are present, as well the ramp-up of temperature during the preheat and drying phases of the reflow oven profile. Too steep a time/temperature slope can cause rapid escape of entrapped volatile solvents, resulting in expulsion of small amounts of metal that will form undesirable "solder balls" of the third type, small metal balls scattered around the solder joints on the substrate itself rather than on the tinned metal of the joint. This third type of ball can also be formed by excess solder paste on the pad, and by misdeposition on nonwettable areas of the substrate.

The reader is referred to Lau and Pau (1996) for discussions of finite element modeling of solder joints, and detailed analytical studies of most aspects of basic joints and of joint failures. Various articles by Engelmaier et al. also address many solder joint reliability issues and their analytical analysis. These and other sources discuss in detail the quality issues that affect solder paste:

- Viscosity and its measurement
- Printability
- Open time
- Slump
- Metal content
- Particle/ball size in mesh
- Particle/ball size consistency
- Wetting
- Storage conditions

Note with regard to viscosity measurements, some paste manufacturers will prefer the spindle technique and some the spiral technique. To properly compare the paste manufacturer's readings with your tests, the same technique must be used.

5.9 Parts Inspection and Placement

Briefly, all parts must be inspected prior to use. Functional parts testing should be performed on the same basis as for through-hole devices. Each manufacturer of electronic assemblies is familiar with the various processes used on through-hole parts, and similar processes must be in place on SMDs. Problems with solderability of leads and lead planarity are two items that can lead to the largest number of defects in the finished product. Solderability is even more important with SMDs than with through-hole parts because all electrical and mechanical strength rests within the solder joint, there being no hole-with-lead to add mechanical strength.

Lead coplanarity is defined as follows. If a multilead part, e.g., IC, is placed on a planar surface, lack of ideal coplanarity exists if the lowest solderable part of any lead does not touch that surface. **Coplanarity** requirements vary depending on the pitch of the component leads and their shape, but generally out-of-plane measurements should not exceed 4 mils (0.004 in.) for 50 mil pitch devices, and 2 mils for 25 mil pitch devices. All SMDs undergo thermal shocking during the soldering process, particularly if the SMDs are to be wave-soldered (Type II or Type III boards), which means they will be immersed in the molten solder wave for 2 to 4 sec. Therefore, all plastic-packaged parts must be controlled for moisture content. If the parts have not been stored in a low-humidity environment (<25% RH), then the absorbed moisture will expand during the solder process and crack the package—a phenomenon know as "popcorning" because the crack is accompanied by a loud "pop" and the package expands due to the expansion of moisture, just like real popcorn. IC suppliers have strict recommendations on storage ambient humidity and temperature, and also on the baking procedures necessary to allow safe reflow soldering if those recommendations are not heeded. Follow them carefully. Typically, if storage RH is above 20%, baking must be considered prior to reflow.

Parts Placement

Proper parts placement not only places the parts within an acceptable window relative to the solder pad pattern on the substrate, but the placement machine will apply enough downward pressure on the part to force it halfway into the solder paste as well (Figure 5.10a and b). This assures both that the part will sit still when the board is moved, and that coplanarity offsets within limits will still result in an acceptable solder joint. The effects of coplanarity can be calculated mathematically by considering:

- the thickness of the solder paste deposit, T
- maximum coplanarity offsets among leads, C
- lead penetration in paste, P

Parts placement may be done manually for prototype or low-volume operations, although this author suggests the use of guided X–Y tables with vacuum part pickup for even the smallest operation. Manual

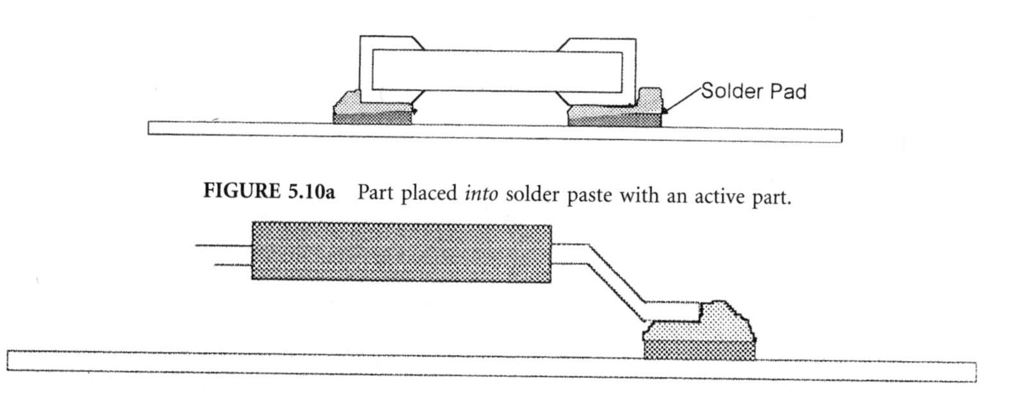

FIGURE 5.10a Part placed *into* solder paste with an active part.

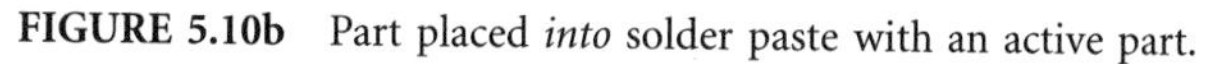

FIGURE 5.10b Part placed *into* solder paste with an active part.

placement of SMDs does not lend itself to repeatable work. For medium and high volume work, a multitude of machines are available (see Figure 5.11). Major manufacturers of medium and high volume SMT placement equipment include Assembleon (neé Philips) Fuji, Juki, Panasonic, Siplace, and Universal.

One good source for manufacturer's information on placement machines and most other equipment used in the various SMT production and testing phases is the annual *Directory of Suppliers to the Electronics Manufacturing Industry*, published by Electronic Packaging and Production. Among the elements to consider in the selection of placement equipment, whether fully automated or *X*–*Y*-vacuum assist tables, are:

- Volume of parts to be placed per hour
- Conveyorized links to existing equipment
- Packaging of components to be handled: tubes, reels, trays, bulk, etc.
- Ability to download placement information from CAD/CAM systems
- Ability to modify placement patterns by the operator
- Vision capability needed, for board fiducials and/or fine-pitch parts

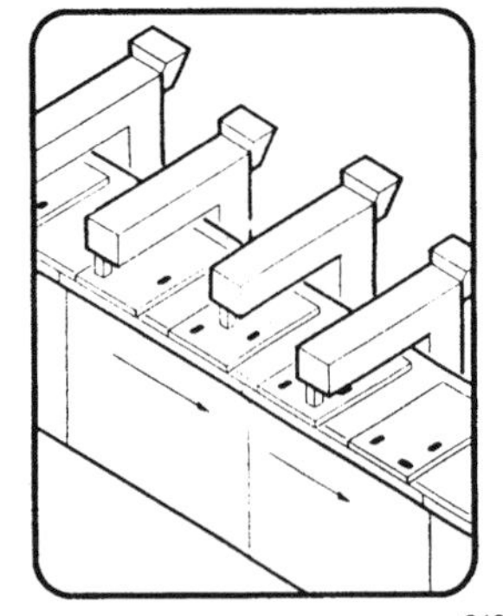

240821-15

- Moving board/fixed head
- Each head places one component
- 1.8 to 4.5 seconds/board

a. In-Line Placement Equipment

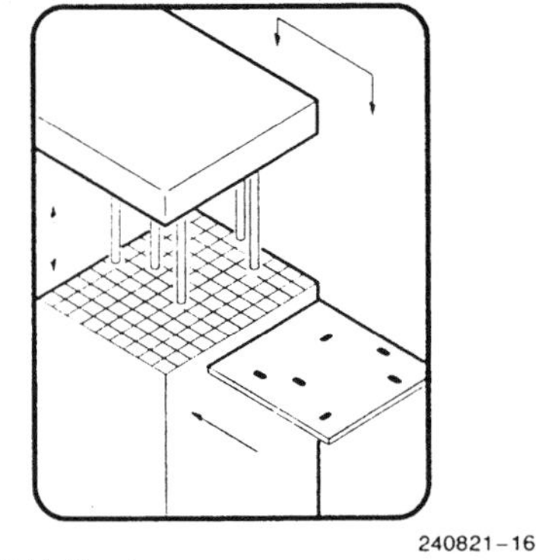

240821-16

- Fixed table/head
- All components placed simultaneously
- Seven to 10 seconds/board

b. Simultaneous Placement Equipment

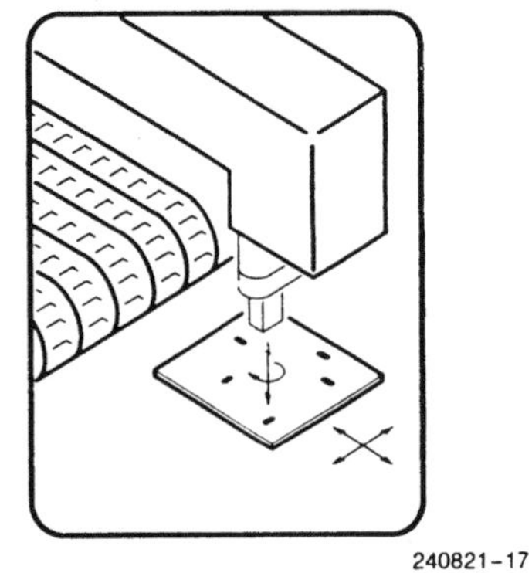

240821-17

- X-Y movement of table/head
- Components placed in succession individually
- 0.3 to 1.8 seconds/component

c. Sequential Placement Equipment

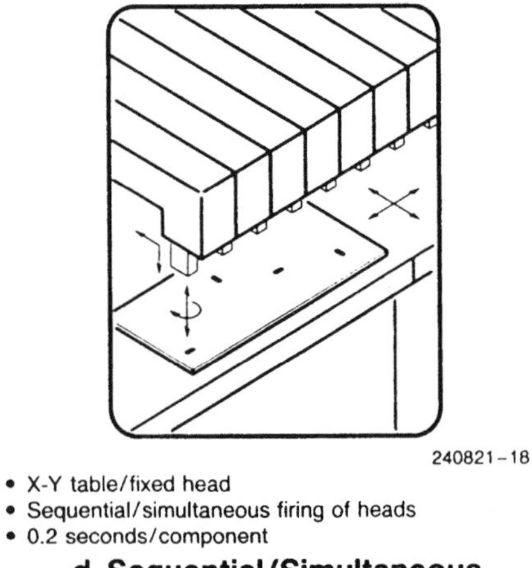

240821-18

- X-Y table/fixed head
- Sequential/simultaneous firing of heads
- 0.2 seconds/component

d. Sequential/Simultaneous Placement Equipment

FIGURE 5.11 Four major categories of placement equipment. (*Source*: Intel Corporation, *Packaging*, Santa Clara, CA: Intel Corporation, 1994. With permission.)

Checks of new placement equipment, or in-place equipment when problems occur, should include:

- *X*-, *Y*-, and *Z*-axis accuracy
- Placement pressure
- Vision system checks both downward and upward

It must be emphasized that placement accuracy checks cannot be made using standard circuit boards. Special glass plates with engraved measurement patterns and corresponding glass parts must be used because standard circuit boards and parts vary too widely to allow accurate *X*-, *Y*-, and *Q*-measurement.

5.10 Reflow Soldering

Once SMDs have been placed in solder paste, the assembly will be reflow soldered. This can be done in either batch-type ovens or conveyorized continuous-process ovens. The choice depends primarily on the board throughput per hour required. While many early ovens were of the vapor phase type, most ovens today use infrared (IR) heating, convection heating, or a combination of the two. In IR ovens, the absorbance of the paste, parts, glue, etc., as a function of color should be considered. Convection ovens tend to be more forgiving with variations in color and thermal masses on the substrates. This author does not recommend vapor-phase (condensation heating) ovens to new users of SMT. All ovens are zoned to provide a thermal profile necessary for successful SMD soldering. An example of an oven profile is shown in Figure 5.12, and the phases of reflow soldering reflected in that example include:

- Preheating: The substrate, components, and solder paste are preheated.
- Drying: Solvents evaporate from the solder paste. Flux activates, reduces oxides, and evaporates. Both low- and high-mass components have enough soak time to reach temperature equilibrium.
- Reflow: The solder paste temperature exceeds the liquidus point and reflows, wetting both the component leads and the board pads. Surface tension effects occur, minimizing wetted volume.
- Cooling: The solder paste cools below the liquidus point, forming acceptable (shiny and appropriate volume) solder joints.

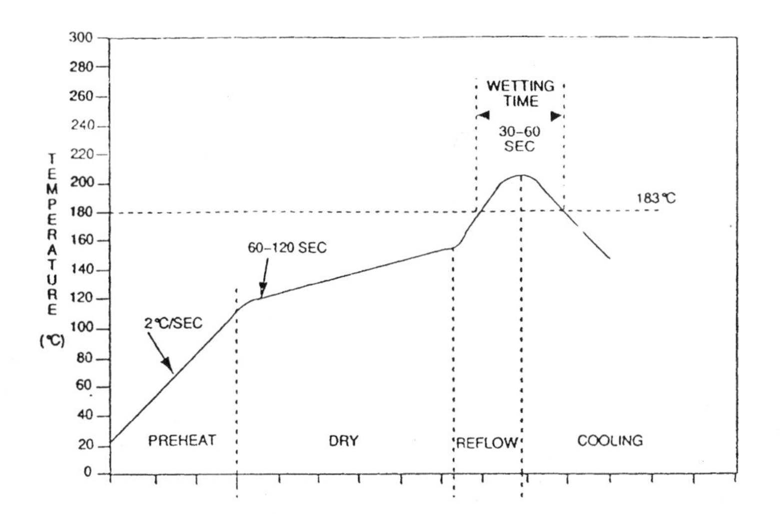

FIGURE 5.12 Typical thermal profile for SMT reflow soldering (Type I or II assemblies). (*Source*: Cox, N.R., *Reflow Technology Handbook*, Minneapolis, Minn.: Research, Inc., 1992. With permission.)

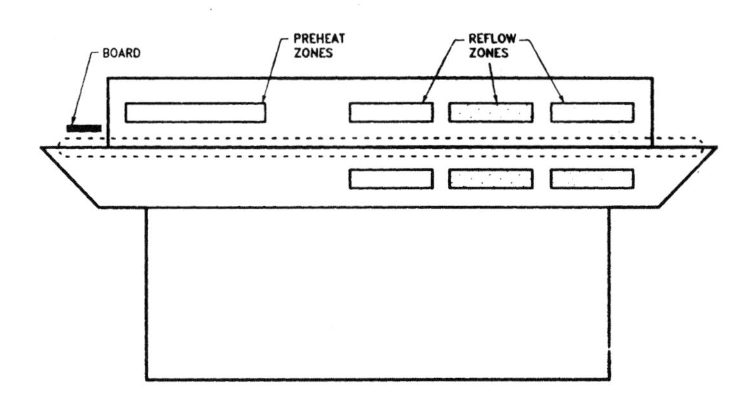

FIGURE 5.13 Conveyorized reflow oven showing zones that create the profile. (*Source*: Intel Corporation, *Packaging*, Santa Clara, CA: Intel Corporation, 1994. With permission.)

The setting of the reflow profile for an oven such as that shown in Figure 5.13 is not trivial. It will vary depending on whether the flux is RMA, water soluble, or no-clean, and it will also vary depending on both the mix of low- and high-thermal mass components, and on how those components are laid out on the board. The profile should exceed the liquidus temperature of the solder paste by 20 to 25°C. While final setting of the profile will depend on the actual quality of the solder joints formed in the oven, initial profile setting should rely heavily on information from the solder paste vendor, as well as the oven manufacturer. Remember that the profile shown is the profile to be developed on the substrate, and the actual control settings in various stages of the oven itself may be considerably different, depending on the thermal inertia of the product in the oven and the heating characteristics of the particular oven being used. This should be determined not by the oven settings but by instrumenting actual circuit boards with thermocouples and determining that the profiles at various locations on the circuit board meet the specifications necessary for good soldering.

Defects as a result of a poor thermal profile in the oven may include:

- Component thermal shock
- Solder splatter
- Solder balls formation
- Dewetted solder
- Cold or dull solder joints

It should be noted that many other problems might contribute to defective solder joint formation. One example would be placement misalignment which contributes to the formation of solder bridges, as shown in Figure 5.14.

Other problems that may contribute to defective solder joints include poor solder mask adhesion, and unequal solder land areas at opposite ends of passive parts, which creates unequal moments as the paste liquifies and develops surface tension. Wrong solder paste volumes, whether too much or too little, will create defects, as will board shake in placement machines and coplanarity problems in IC components. Many of these problems should be covered and compensated for during the design process and the qualification of SMT production equipment.

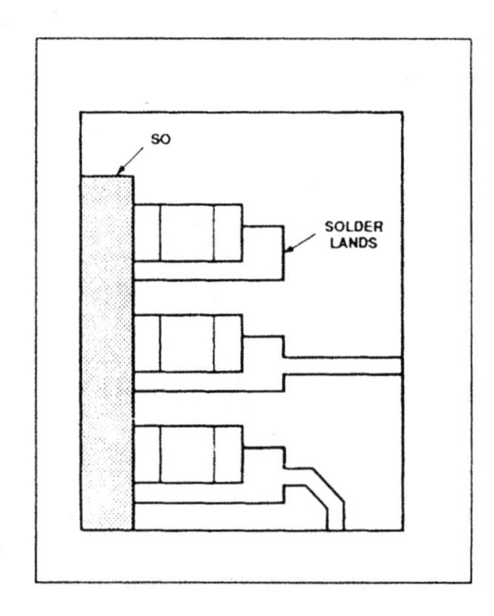

FIGURE 5.14 Solder bridge risk due to misalignment. (*Source*: Phillips Semiconductors, *Surface Mount Process and Application Notes*, Sunnyvale, CA: Phillips Semiconductors, 1991. With permission.)

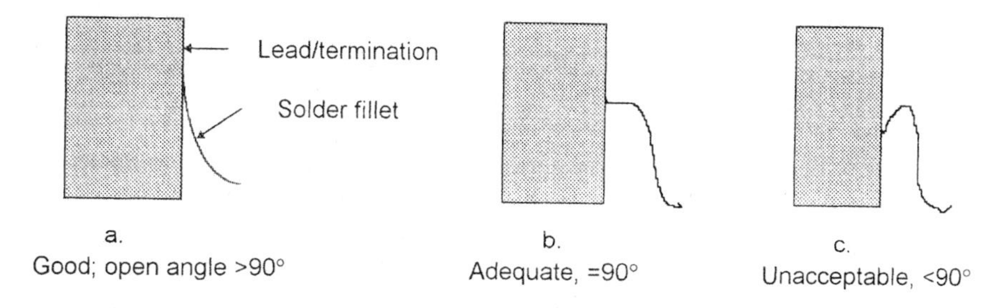

FIGURE 5.15 Solder joint inspection criteria.

Postreflow Inspection

Final analysis of the process is performed based on the quality of the solder joints formed in the reflow process. Whatever criteria may have been followed during the overall process, solder joint quality is the final determining factor of the correctness of the various process steps. As noted earlier, the quality level of solder joint production is a major factor in successful board assembly. A primary criterion is the indication of wetting at the junction of the reflowed solder and the part termination. This same criterion shown in Figure 5.15 applies to both through-hole and SMDs, with only the inspection location being different.

Note that criteria shown in Figure 5.15 are for any solderable surface, whether component or board, SMT or THT. Some lead surfaces are defined as not solderable, e.g., the cut and not-tinned end of an SO or QFP lead is not considered solderable. Parts manufacturers will define whether a given surface is designed to be solderable.

Presentation of criteria for all the various SMD package types and all the possible solder joint problems is beyond the scope of this chapter. The reader is directed to Hollomon (1995) and Prasad (1997) for an in-depth discussion of these issues.

5.11 Cleaning

Cleaning, like all other parts of the process, should be considered during the design phase. Cleaning requirements are determined largely by the flux in the solder paste, and should be determined before production is ever started. Design issues may affect the choice of flux, which determines cleaning. For example, low-clearance (close to the substrate) parts are difficult to clean under, and the use of no-clean flux may be suggested.

RMA is the old standard, and if cleaning is needed with RMA, either solvent-based cleaners or water with saponifiers must be used. (Saponifiers are alkaline materials that react with the rosin so that it becomes water soluble.) RMA tends to be nonactive at room temperatures, and may not need to be cleaned on commercial products designed for indoor use. The major limitation on RMA at this point is the need for chemicals in the cleaning process.

Water-soluble fluxes are designed to be cleaned with pure water. They remain active at room temperatures, and therefore *must* be cleaned when used. Beyond their activity, the other disadvantage to water-soluble fluxes is that their higher surface tension relative to solvent-based cleaners means there is more difficulty cleaning under low-clearance components.

No-clean fluxes are designed not to be cleaned, and this means they are of low activity (they do not reduce oxides as well as other types of flux) and when used they should *not* be cleaned. No-clean fluxes are designed so that after reflow they microscopically encapsulate themselves, sealing in any active components. If the substrate is subsequently cleaned, the encapsulants may be destroyed, leaving the possibility of active flux components remaining on the board.

The lead-free process may bring with it new flux compounds. These may require additional considerations in design, assembly, and cleaning processes.

5.12 Prototype Systems

Systems for all aspects of SMT assembly are available to support low-volume/prototype needs. These systems will typically have manual solder-paste deposition and parts placement systems, with these functions being assisted for the user. Syringe solder paste deposition may be as simple as a manual medical-type syringe dispenser that must be guided and squeezed freehand. More sophisticated systems will have the syringe mounted on an *X–Y* arm to carry the weight of the syringe, and will apply air pressure to the top of the syringe with a foot-pedal control, freeing the operator's arm to guide the syringe to the proper location on the substrate and perform the negative *Z*-axis maneuver that will bring the syringe tip into the proper location and height above the substrate. Dispensing is then accomplished by a timed air pressure burst applied to the top of the syringe under foot-pedal control. Paste volume is likewise determined by trial-and-error with the time/pressure relation and depends on the type and manufacturer of paste being dispensed.

Parts placement likewise may be as simple as tweezers and may progress to hand-held vacuum probes to allow easier handling of the components. As mentioned in Section "Parts Inspection and Placement", *X–Y* arm/tables are available, which have vacuum-pick nozzles to allow the operator to pick a part from a tray, reel, or stick, and move the part over the correct location on the substrate. The part is then moved down into the solder paste, the vacuum is turned off manually or automatically, and the nozzle is raised away from the substrate.

Soldering of prototype/low-volume boards may be carried out by contact soldering of each component, by a manually guided hot-air tool, or in a small batch or conveyorized oven. Each step up in soldering sophistication is, of course, accompanied by an increase in the investment required.

For manufacturers with large prototype requirements, it is possible to set up an entire line that would involve virtually no hardware changes from one board to another:

- ECAD design and analysis, producing Gerber files.
- CNC circuit board mill takes Gerber files and mills out two-sided boards.
- Software translation package generates solder-pad centroid information.
- Syringe solder paste deposition system takes a translated Gerber file and dispenses an appropriate amount at each pad centroid.
- Software translation package generates part centroid information.
- Parts placement equipment places parts based on translated part centroid information.
- Assembly is reflow soldered.

The only manual process in the above system is adjustment of the reflow profile based on the results of soldering an assembly. The final step in the process would be to test the finished prototype board. This system could also be used for very small-volume production runs, and all components as described are available. With a change from milled boards to etched boards, the system can be used as a flexible assembly system.

Defining Terms

Coefficient of thermal expansion (CTE, aka TCE): A measure of the ratio between the measure of a material and its expansion as temperature increases. May be different in *X*-, *Y*-, and *Z*-axes. Expressed in ppm/°C. A measure that allows comparison of materials that are to be joined.

Coplanarity: A simplified definition of planarity, which is difficult to measure. Coplanarity is the distance between the highest and lowest leads, and is easily measured by placing the IC on a flat surface such as a glass plate. The lowest leads will then rest on the plate, and the measured difference to the lead highest above the plate is the measurement of coplanarity.

Glass-transition temperature (T_g): Below T_g, a polymer substance, such as fiberglass, is relatively linear in its expansion/contraction due to temperature changes. Above T_g, the expansion rate increases dramatically and becomes nonlinear. The polymer will also lose its stability, i.e., an FR-4 board "droops" above T_g.

Gull wing: An SMD lead shape as shown in Figure 5.4 for SOPs and QFPs, and in Figure 5.11b. It is so called because it looks like a gull's wing in flight.

J-lead: An SMD lead shape as shown in the PLCC definition. It is so called because it is in the shape of the capital letter J.

Land: A metalized area intended for the placement of one termination of a component. Lands may be tinned with solder, or be bare copper in the case of SMOBC circuit board fabrication.

PLCC: Plastic leaded chip carrier. Shown in Figure 5.4, it is a common SMT IC package and is the only package that has the leads bent back under the IC itself.

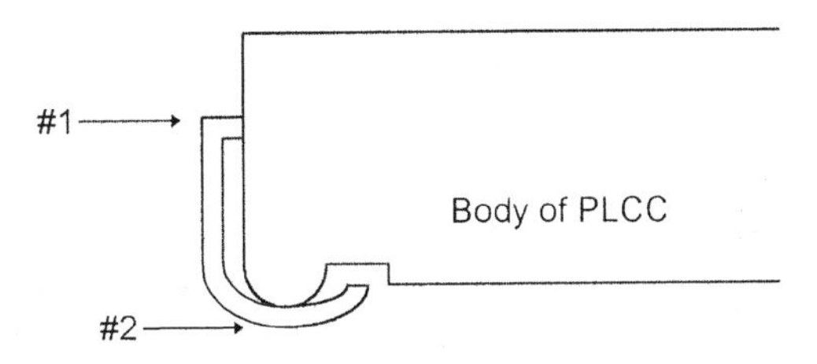

Planarity: Lying in the same plane. A plane is defined by the exit of the leads from the body of the IC (arrow #1 above). A second plane is defined as the average of the lowest point all leads are below the first plane (arrow #2 above). Nonplanarity is the maximum variation in mils or mm of any lead of an SMD from the lowest point plane.

Quad flat pack: Any flat pack IC package that has gull-wing leads on all four sides.

Through-hole: Also a plate-through-hole (PTH). A via that extends completely through a substrate and is solder-plated.

SMOBC: An acronym for "solder mask on bare copper," a circuit board construction technique that does not tin the copper traces with solder prior to the placement of the solder mask on the board.

Through-hole technology (THT): The technology of using leaded components that require holes through the substrate for their mounting (insertion) and soldering.

References

M.C. Apell, "Considerations for lead-free reflow soldering," *Circuits Assembly*, vol. 15, no. 3, 2004.

G.R. Blackwell, *The Electronics Packaging Handbook*, Boca Raton, Fla.: CRC/IEEE Publications, 2000.

E. Bradley, C. Handwerker, and J.E. Sohn, "NEMI report: a single lead-free alloy is recommended," *Surf. Mount Technol. Mag.*, vol. 17, no. 1, 2003.

D. Brooks, "Temperature rise in PCB traces," in *Proc. PCB Des. Conf.*, West, March 23–27 1998.

C. Capillo, *Surface Mount Technology, Materials, Processes and Equipment*, New York: McGraw-Hill, 1990, chap. 3.

C. Capillo, "Conduction heat transfer measurements for an array of surface mounted heated components," *Am. Soc. Mech. Eng., Heat Transfer Div., Proc. 1993 ASME Ann. Meet.*, 263, pp. 69–78, 1993.

C.Y. Choi, S.J. Kim, and A. Ortega, "Effects of substrate conductivity on convective cooling of electronic components," *J. Electron. Packaging*, vol. 116, no. 3, pp. 198–205, 1994.

F. Classon, *Surface Mount Technology for Concurrent Engineering and Manufacturing*, New York: McGraw-Hill, 1993.

N.R. Cox, *Reflow Technology Handbook*, Minneapolis, Minn.: Research International, 2000.

Flotherm, *Advanced Thermal Analysis of Packaged Electronic Systems,* Westborough, Mass.: Flomerics, Inc., 1995.

C. Higgins, Signetics Corp., Presentation, November 1991.

J.K. Hollomon, Jr., *Surface Mount Technology for PC Board Design*, Indianapolis, Ind.: Prompt Publishing, 1995.

J.G. Holmes, "Surface mount solution for power devices," *Surf. Mt. Tech.*, vol. 7, no. 9, pp. 18–20, 1993.

J.W. Hwang, *Solder Paste in Electronics Packaging*, New York: Van Nostrand Reinhold, 1989.

R.J. Klein-Wassink, *Soldering in Electronics*, 2nd ed., Port Erin, Isle of Man, UK.: Electrochemical Publications, 1998.

J.H. Lau and Y-H.Pau, *Solder Joint Reliability of BGA, CSP, Flip-Chip and Fine Pitch SMT Assemblies*, New York: Van Nostrand Reinhold, 1996.

"Leaded Surface Mount Technology," *Intel Packaging Handbook*, chap. 7, http://www.intel.com/design/packtech/packbook.htm.

T.Y. Lee, "Application of a CFD tool for system-level thermal simulation," *IEEE Trans. Compon. Packaging Manuf. Technol., Part A*, vol. 17, no. 4, pp. 564–571, 1994.

L.C. Lee, V.S. Darekar, and C.K. Lim, "Micromechanics of multilayer printed circuit board," *IBM J. Res. Dev.*, vol. 28, no. 6, 1984.

P.P. Marcoux, *Fine Pitch Surface Mount Technology*, New York: Van Nostrand Reinhold, 1992.

R.P. Prasad, *Surface Mount Technology Principles and Practice*, 2nd ed., New York: Van Nostrand Reinhold, 1997.

R. Rowland, *Applied Surface Mount Assembly*, New York: Van Nostrand Reinhold, 1993.

S.G. Shina, *Concurrent Engineering and Design for Manufacture of Electronic Products*, New York: Van Nostrand Reinhold, 1991.

Further Information

Specific journal references are available on any aspect of SMT. A search of the COMPENDEX Engineering Index 1997–present will show over 1000 references specifically to SMT topics.

Education/Training: a partial list of organizations that specialize in education and training directly related to issues in surface mount technology:

IPC, www.ipc.org, 2215 Sanders Rd., Northbrook, IL 60062–6135.

SMT Plus, Inc., www.smtplus.com, 14178 Pepperwood Dr., Penn Valley, CA 95946.

Surface Mount Technology Association (SMTA), www.smta.org, 5200 Wilson Rd, Ste. 100, Edina, MN 55424-1338. 612-920-7682.

6

Ideal and Practical Models

E.J. Kennedy
University of Tennessee

John V. Wait
University of Arizona (Retired)

The concept of the **operational amplifier** (usually referred to as an *op amp*) originated at the beginning of the Second World War with the use of vacuum tubes in dc amplifier designs developed by the George A. Philbrick Co. (some of the early history of operational amplifiers is found in Williams, 1991). The op amp was the basic building block for early electronic servomechanisms, for synthesizers, and in particular for analog computers used to solve differential equations. With the advent of the first monolithic integrated-circuit (IC) op amp in 1965 (the μA709, designed by the late Bob Widlar, then with Fairchild Semiconductor), the availability of op amps was no longer a factor, while within a few years the cost of these devices (which had been as high as $200 each) rapidly plummeted to close to that of individual discrete transistors.

Although the digital computer has now largely supplanted the analog computer in mathematically intensive applications, the use of inexpensive op amps in instrumentation applications, in pulse shaping, in filtering, and in signal processing applications in general has continued to grow. There are currently many commercial manufacturers whose main products are high-quality op amps. This competitiveness has ensured a marketplace featuring a wide range of relatively inexpensive devices suitable for use by electronic engineers, physicists, chemists, biologists, and almost any discipline that requires obtaining quantitative analog data from instrumented experiments.

Most op amp circuits can be analyzed, at least for first-order calculations, by considering the op amp to be an "ideal" device. For more quantitative information, however, and particularly when frequency response and dc offsets are important, one must refer to a more "practical" model that includes the internal limitations of the device. If the op amp is characterized by a really complete model, the resulting circuit may be quite complex, leading to rather laborious calculations. Fortunately, however, computer analysis using the program **SPICE** significantly reduces the problem to one of a simple input specification to the computer. Today, nearly all the op amp manufacturers provide SPICE models for their line of devices, with excellent correlation obtained between the computer simulation and the actual measured results.

6.1 The Ideal Op Amp

An **ideal operational amplifier** is a dc-coupled amplifier having two inputs and normally one output (although in a few infrequent cases there may be a differential output). The inputs are designated as

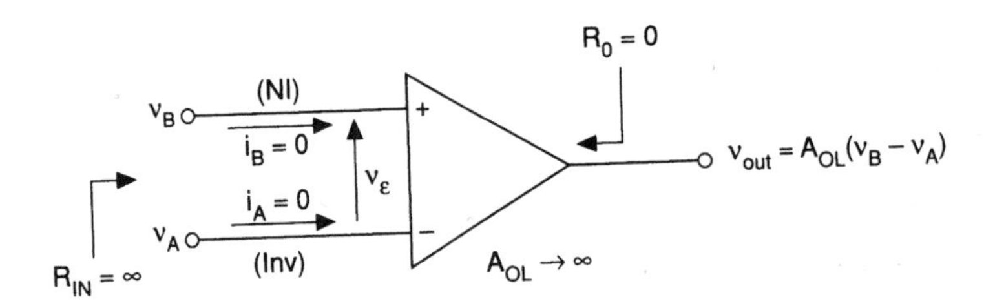

FIGURE 6.1 Configuration for an ideal op amp.

noninverting (designated + or NI) and inverting (designated – or Inv.). The amplified signal is the *differential signal*, v_ε, between the two inputs, so that the output voltage as indicated in Figure 6.1 is

$$v_{out} = A_{OL}(v_B - v_A) \tag{6.1}$$

The general characteristics of an ideal op amp can be summarized as follows:

1. The open-loop gain A_{OL} is infinite. Or, since the output signal v_{out} is finite, then the differential input signal v_ε must approach zero.
2. The input resistance R_{IN} is infinite, while the output resistance R_O is zero.
3. The amplifier has zero current at the input (i_A and i_B in Figure 6.1 are zero), but the op amp can either sink or source an infinite current at the output.
4. The op amp is not sensitive to a common signal on both inputs (i.e., $v_A = v_B$); thus, the output voltage change due to a common input signal will be zero. This common signal is referred to as a common-mode signal, and manufacturers specify this effect by an op amp's *common-mode rejection ratio* (CMRR), which relates the ratio of the open-loop gain (A_{OL}) of the op amp to the common-mode gain (A_{CM}). Hence, for an ideal op amp CMRR $= \infty$.
5. A somewhat analogous specification to the CMRR is the *power-supply rejection ratio* (PSRR), which relates the ratio of a power supply voltage change to an equivalent input voltage change produced by the change in the power supply. Because an ideal op amp can operate with any power supply, without restriction, then for the ideal device PSRR $= \infty$.
6. The gain of the op amp is not a function of frequency. This implies an infinite bandwidth.

Although the foregoing requirements for an ideal op amp appear to be impossible to achieve practically, modern devices can quite closely approximate many of these conditions. An op amp with a field-effect transistor (FET) on the input would certainly not have zero input current and infinite input resistance, but a current of <10 pA and an $R_{IN} = 10^{12}\ \Omega$ is obtainable and is a reasonable approximation to the ideal conditions. Further, although a CMRR and PSRR of infinity are not possible, there are several commercial op amps available with values of 140 dB (i.e., a ratio of 10^7). Open-loop gains of several precision op amps now have reached values of $>10^7$, although certainly not infinity. The two most difficult ideal conditions to approach are the ability to handle large output currents and the requirement of a gain independence with frequency.

Using the ideal model conditions it is quite simple to evaluate the two basic op amp circuit configurations, (1) the inverting amplifier and (2) the noninverting amplifier, as designated in Figure 6.2.

For the ideal inverting amplifier, since the open-loop gain is infinite and since the output voltage v_o is finite, then the input differential voltage (often referred to as the *error signal*) v_ε must approach zero, or the input current is

$$i_I = \frac{v_I - v_\varepsilon}{R_1} = \frac{v_I - 0}{R_1} \tag{6.2}$$

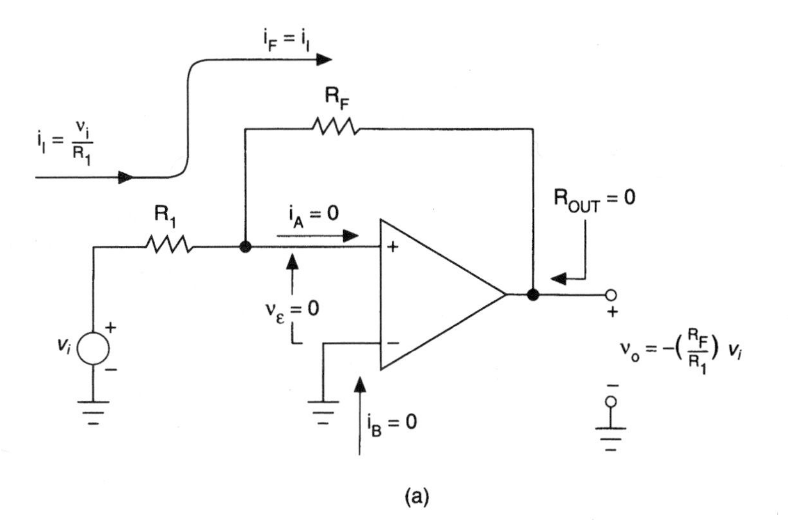

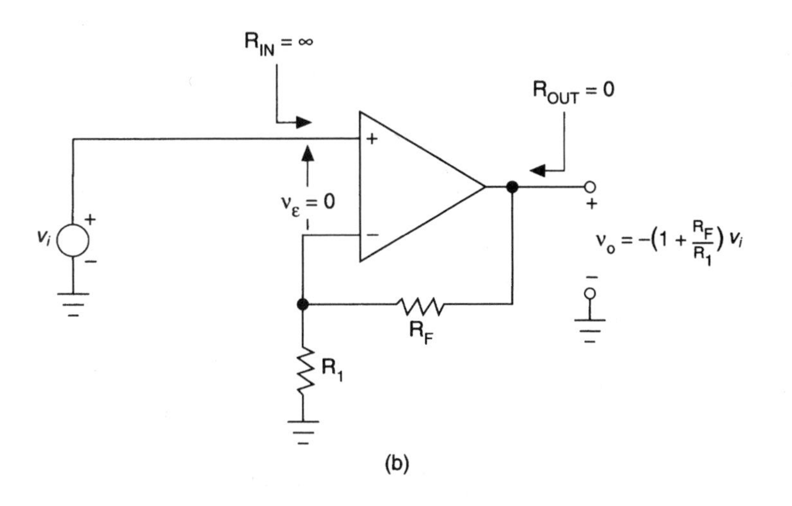

FIGURE 6.2 Illustration of (a) the inverting amplifier and (b) the noninverting amplifier. (*Source:* E.J. Kennedy, *Operational Amplifier Circuits, Theory and Applications,* New York: Holt, Rinehart and Winston, 1988, pp. 4, 6. With permission.)

The feedback current i_F must equal i_I, and the output voltage must then be due to the voltage drop across R_F, or

$$v_O = -i_F R_F + v_\varepsilon = -i_I R_F = -\left(\frac{R_F}{R_1}\right) v_I \tag{6.3}$$

The inverting connection thus has a voltage gain v_o/v_I of $-\ R_F/R_1$, an input resistance seen by v_I of R_1 ohms (from Equation (6.2)), and an output resistance of 0 Ω. By a similar analysis for the noninverting circuit of Figure 6.2(b), since v_ε is zero, then signal v_I must appear across resistor R_1, producing a current of v_I/R_1, which must flow through resistor R_F. Hence the output voltage is the sum of the voltage drops across R_F and R_1, or

$$v_O = R_F\left(\frac{v_1}{R_1}\right) + v_1 = \left(1 + \frac{R_F}{R_1}\right) v_1 \tag{6.4}$$

As opposed to the inverting connection, the input resistance seen by the source v_I is now equal to an infinite resistance, since R_{IN} for the ideal op amp is infinite.

6.2 Practical Op Amps

A nonideal op amp is characterized not only by finite open-loop gain, input and output resistance, finite currents, and frequency bandwidths, but also by various nonidealities due to the construction of the op amp circuit or external connections. A complete model for a practical op amp is illustrated in Figure 6.3. The nonideal effects of the PSRR and CMRR are represented by the input series voltage sources of ΔV_{supply}/PSRR and V_{CM}/CMRR, where ΔV_{supply} would be any total change of the two power supply voltages, V_{dc}^{+} and V_{dc}^{-}, from their nominal values, while V_{CM} is the voltage common to both inputs of the op amp. The open-loop gain of the op amp is no longer infinite but is modeled by a network of the output impedance Z_{out} (which may be merely a resistor but could also be a series R–L network) in series with a source $A(s)$, which includes all the open-loop poles and zeroes of the op amp as

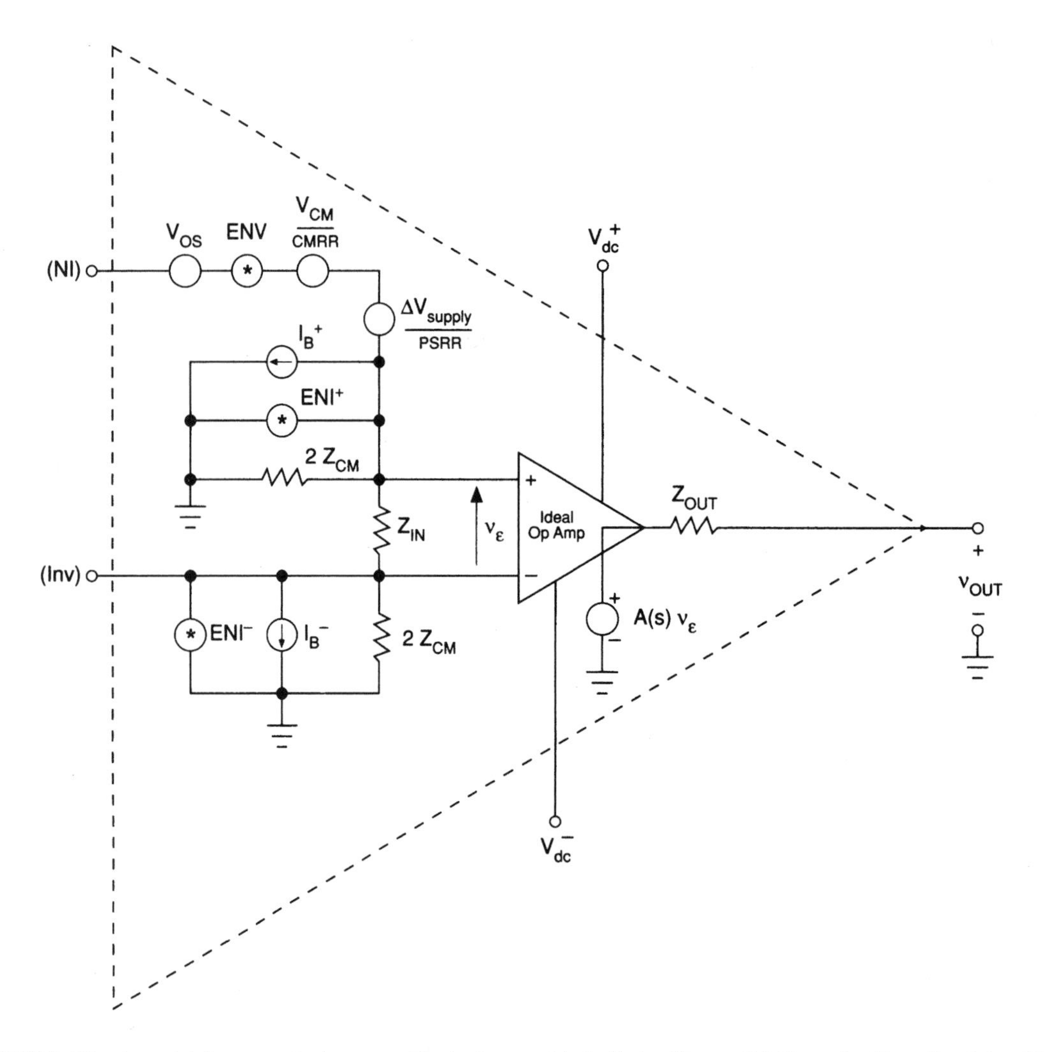

FIGURE 6.3 A model for a practical op amp illustrating nonideal effects. (*Source:* E.J. Kennedy, *Operational Amplifier Circuits, Theory and Applications*, New York: Holt, Rinehart and Winston, 1988, pp. 53, 126. With permission.)

$$A(s) = \frac{A_{OL}\left(1 + \frac{s}{\omega_{z_1}}\right)(1 + \cdots)}{\left(1 + \frac{s}{\omega_{p1}}\right)\left(1 + \frac{s}{\omega_{p2}}\right)(1 + \cdots)} \tag{6.5}$$

where A_{OL} is the finite dc open-loop gain, while poles are at frequencies ω_{p1}, ω_{p2},... and zeroes are at ω_{Z1}, etc. The differential input resistance is Z_{IN}, which is typically a resistance R_{IN} in parallel with a capacitor C_{IN}. Similarly, the common-mode input impedance Z_{CM} is established by placing an impedance $2Z_{CM}$ in parallel with each input terminal. Normally, Z_{CM} is best represented by a parallel resistance and capacitance of $2R_{CM}$ (which is $>> R_{IN}$) and $C_{CM}/2$. The dc bias currents at the input are represented by I_B^+ and I_B^- current sources that would equal the input base currents if a differential bipolar transistor were used as the input stage of the op amp, or the input gate currents if FETs were used. The fact that the two transistors of the input stage of the op amp may not be perfectly balanced is represented by an equivalent input offset voltage source, V_{OS}, in series with the input.

The smallest signal that can be amplified is always limited by the inherent random noise internal to the op amp itself. In Figure 6.3 the noise effects are represented by an *equivalent input voltage source* (ENV), which when multiplied by the gain of the op amp would equal the total output noise present if the inputs to the op amp were shorted. In a similar fashion, if the inputs to the op amp were open circuited, the total output noise would equal the sum of the noise due to the *equivalent input current sources* (ENI^+ and ENI^-), each multiplied by their respective current gain to the output. Because noise is a random variable, this summation must be accomplished in a squared fashion, i.e.:

$$E_O^2(\text{rms volt}^2/\text{Hz}) = (\text{ENV})^2 A_v^2 + (\text{ENI}^+)^2 A_{I1} + (\text{ENI}^-)^2 A_{I2}^2 \tag{6.6}$$

Typically, the correlation (C) between the ENV and ENI sources is low, so the assumption of $C \approx 0$ can be made.

For the basic circuits of Figure 6.2(a) or (b), if the signal source v_I is shorted then the output voltage due to the nonideal effects would be (using the model of Figure 6.3):

$$v_o = \left(V_{OS} + \frac{V_{CM}}{\text{CMRR}} + \frac{\Delta V_{\text{supply}}}{\text{PSRR}}\right)\left(1 + \frac{R_F}{R_1}\right) + I_B^- R_F \tag{6.7}$$

provided that the loop gain (also called loop transmission in many texts) is related by the inequality

$$\left(\frac{R_1}{R_1 + R_F}\right) A(s) >> 1 \tag{6.8}$$

Inherent in Equation (6.8) is the usual condition that $R_1 << Z_{IN}$ and Z_{CM}. If a resistor R_2 were in series with the noninverting input terminal, then a corresponding term must be added to the right hand side of Equation (6.7) of value $-I_B^+ R_2(R_1 + R_F)/R_1$. On manufacturers' data sheets the individual values of I_B^+ and I_B^- are not stated; instead the average input bias current and offset current are specified as

$$I_B = \frac{I_B^+ + I_B^-}{2}; \quad I_{\text{offset}} = |I_B^+ - I_B^-| \tag{6.9}$$

The output noise effects can be obtained using the model of Figure 6.3 along with the circuits of Figure 6.2 as

$$E_{\text{out}}^2(\text{rmsvolts}^2/\text{Hz}) = E_1^2\left(\frac{R_F}{R_1}\right)^2 + E_F^2 + (\text{ENV}^2 + E_2^2) \times \left(1 + \frac{R_F}{R_1}\right)^2 + (\text{ENI}^-)^2 R_F^2 + (\text{ENI}^+)^2 R_2^2\left(1 + \frac{R_F}{R_1}\right)^2 \tag{6.10}$$

where it is assumed that a resistor R_2 is also in series with the noninverting input of either Figure 6.2(a) or (b). The thermal noise (often called Johnson or Nyquist noise) due to the resistors R_1, R_2, and R_F is given by (in rms volt^2/Hz):

$$\begin{aligned} E_1^2 &= 4kT\ R_1 \\ E_2^2 &= 4kT\ R_2 \\ E_F^2 &= 4kT\ R_F \end{aligned} \tag{6.11}$$

where k is Boltzmann's constant and T is absolute temperature (Kelvin). To obtain the total output noise, one must multiply the E_{out}^2 expression of Equation (6.10) by the noise bandwidth of the circuit, which typically is equal to $\pi/2$ times the –3 dB signal bandwidth, for a single-pole response system (Kennedy, 1988).

6.3 SPICE Computer Models

The use of op amps can be considerably simplified by computer-aided analysis using the program SPICE. SPICE originated with the University of California, Berkeley, in 1975 (Nagel, 1975), although more recent user-friendly commercial versions are now available such as HSPICE, HPSPICE, IS-SPICE, PSPICE, and ZSPICE, to mention a few of those most widely used. A simple macromodel for a near-ideal op amp could be simply stated with the SPICE subcircuit file (* indicates a comment that is not processed by the file):

```
.SUBCKT IDEALOA 1 2 3
*A near-ideal op amp: (1) is noninv, (2) is inv, and (3) is output.
RIN 1 2 1E12
E1 (3,0) (1,2) 1E8
.ENDS IDEALOA
```
(6.12)

The circuit model for IDEALOA would appear as in Figure 6.4(a). A more complete model, but not including nonideal offset effects, could be constructed for the 741 op amp as the subcircuit file OA741, shown in Figure 6.4(b).

```
.SUBCKT OA741 1 2 6
*A linear model for the 741 op amp: (1) is noninv, (2) is inv, and
*(6) is output. RIN = 2MEG, AOL = 200,000, ROUT = 75 ohm,
*Dominant open-loop pole at 5 Hz, gain-bandwidth product
*is 1 MHz.
RIN 1 2 2MEG
E1 (3,0) (1,2)2E5
R1 3 4 100 K
C1 4 0 0.318UF; R1 × C1 = 5HZPOLE
E2 (5,0) (4,0) 1.0
ROUT 5 6 75
.ENDS OA741
```
(6.13)

The most widely used op amp macromodel that includes dc offset effects is the **Boyle model** (Boyle et al., 1974). Most op amp manufacturers use this model, usually with additions to add more poles (and perhaps zeroes). The various resistor and capacitor values, as well as transistor, and current and voltage generator, values are intimately related to the specifications of the op amp, as shown earlier in the nonideal model of

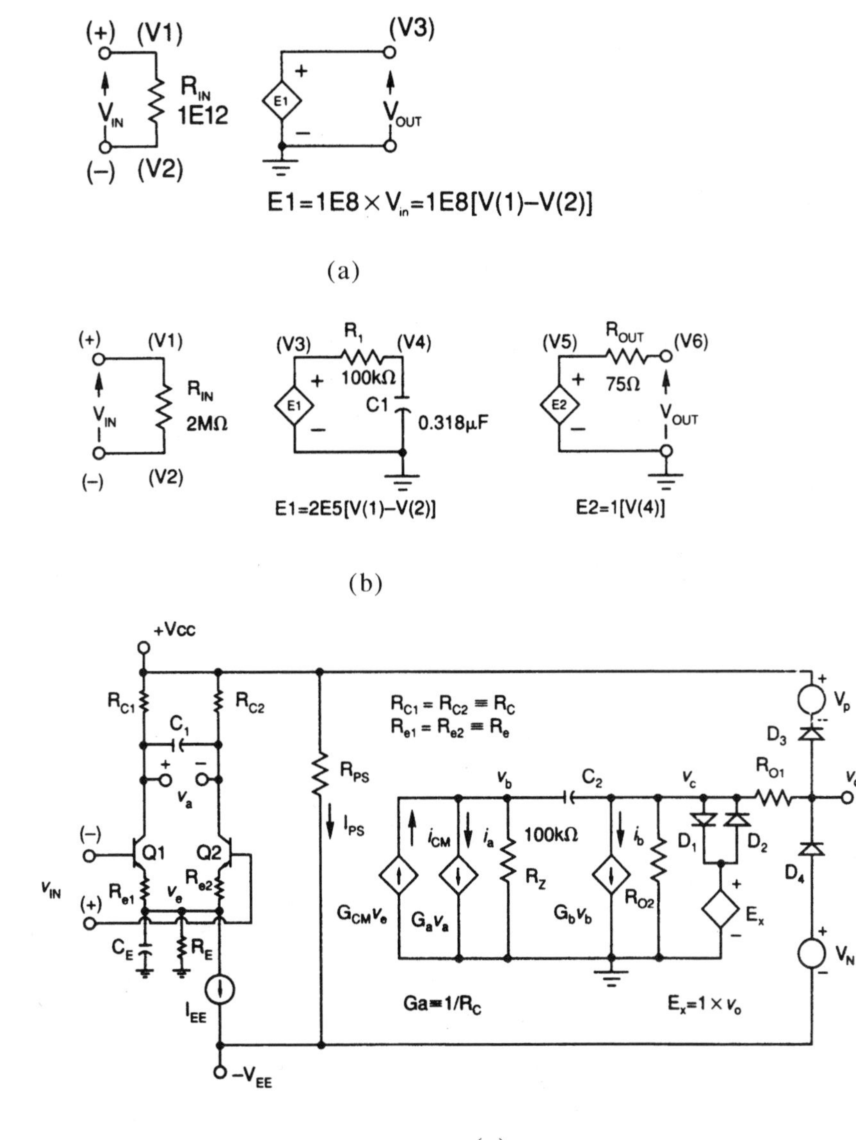

FIGURE 6.4 Some simple SPICE macromodels. (a) A near ideal op amp. (b) A linear model for a 741 op amp. (c) The *Boyle* macromodel. (*Source:* Reprinted, with permission, from J. Williams (Ed.), *Analog Circuit Design,* Stoneham, MA: Butterworth-Heinemann, 1991, p. 304.)

Figure 6.3. The appropriate equations are too involved to list here; instead, the interested reader is referred to the article by Boyle in the listed references. The Boyle model does not accurately model noise effects, nor does it fully model PSRR and CMRR effects.

A more circuits-oriented approach to modeling op amps can be obtained if the input transistors are removed and a model formed by using passive components along with both fixed and dependent voltage and current sources. Such a model is shown in Figure 6.5. This model not only includes all the basic nonideal effects of the op amp, allowing for multiple poles and zeroes, but can also accurately include ENV and ENI noise effects. The circuits-approach macromodel can also be easily adapted to current-feedback op amp designs, whose input impedance at the noninverting input is much greater than that at the inverting input

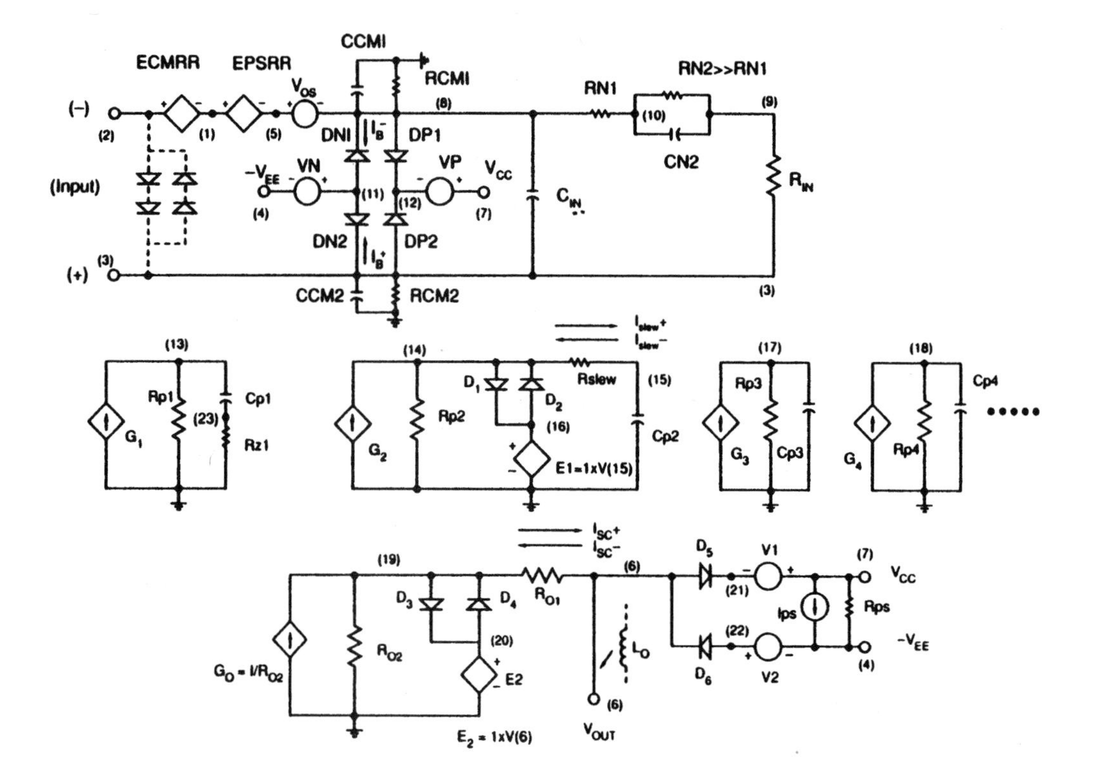

FIGURE 6.5 A SPICE circuits-approach macromodel. (*Source:* Reprinted, with permission, from J. Williams (Ed.), *Analog Circuit Design*, Stoneham, MA: Butterworth-Heinemann, 1991, p. 314.)

(see Williams, 1991). The interested reader is referred to the text edited by J. Williams, listed in the references, as well as the SPICE modeling book by Connelly and Choi (1992).

A comparison of the SPICE macromodels with actual manufacturer's data for the case of an LM318 op amp is demonstrated in Figure 6.6, for the open-loop gain versus frequency specification.

Defining Terms

Boyle macromodel: A SPICE computer model for an op amp. Developed by G.R. Boyle in 1974.

Equivalent noise current (ENI): A noise current source that is effectively in parallel with either the noninverting input terminal (ENI^+) or the inverting input terminal (ENI^-) and represents the total noise contributed by the op amp if either input terminal is open circuited.

Equivalent noise voltage (ENV): A noise voltage source that is effectively in series with either the inverting or noninverting input terminal of the op amp and represents the total noise contributed by the op amp if the inputs were shorted.

Ideal operational amplifier: An op amp having infinite gain from input to output, with infinite input resistance and zero output resistance and insensitive to the frequency of the signal. An ideal op amp is useful in first-order analysis of circuits.

Operational amplifier (op amp): A dc amplifier having both an inverting and noninverting input and normally one output, with a very large gain from input to output.

SPICE: A computer simulation program developed by the University of California, Berkeley, in 1975. Versions are available from several companies. The program is particularly advantageous for electronic circuit analysis, since dc, ac, transient, noise, and statistical analysis are possible.

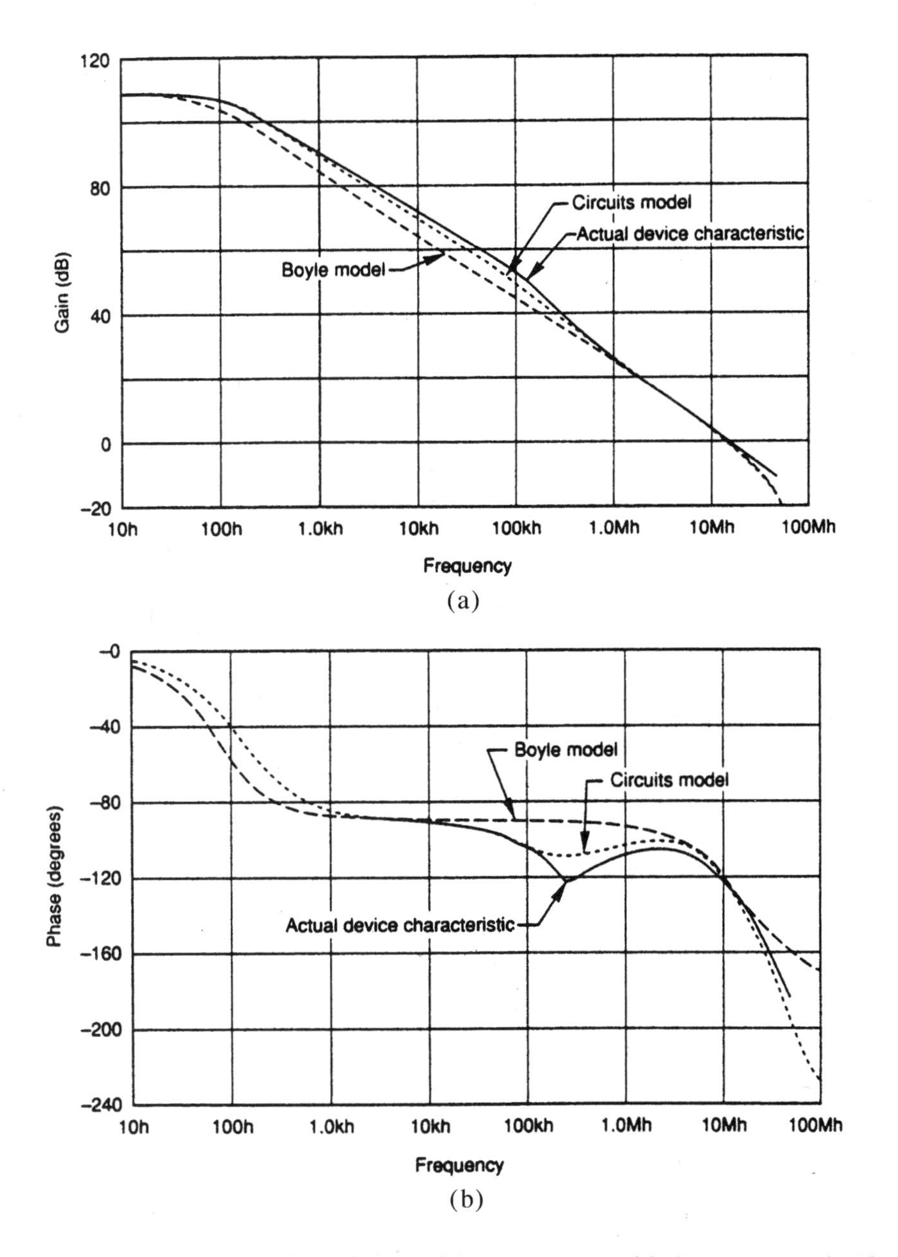

FIGURE 6.6 Comparison between manufacturer's data and the SPICE macromodels. (*Source:* Reprinted, with permission, from J. Williams (Ed.), *Analog Circuit Design*, Stoneham, MA: Butterworth-Heinemann, 1991, p. 319.)

References

G.R. Boyle et al., "Macromodeling of integrated circuit operational amplifiers," *IEEE J. S. S. Circuits,* pp. 353–363, 1974.

J.A. Connelly and P. Choi, *Macromodeling with SPICE,* Englewood Cliffs, NJ: Prentice-Hall, 1992.

E.J. Kennedy, *Operational Amplifier Circuits, Theory and Applications,* New York: Holt, Rinehart and Winston, 1988.

L.W. Nagel, *SPICE 2: A Computer Program to Simulate Semiconductor Circuits,* ERL-M520, University of California, Berkeley, 1975.

J. Williams (Ed.), *Analog Circuit Design,* Boston, Mass.: Butterworth-Heinemann, 1991.

7

Amplifiers

Gordon L. Carpenter
California State University
John Choma, Jr.
University of Southern California

7.1 Large Signal Analysis

Gordon L. Carpenter

Large signal amplifiers are usually confined to using bipolar transistors as their solid-state devices because of the large linear region of amplification required. One exception to this is the use of VMOS for large power outputs due to their ability to have a large linear region. There are three basic configurations of amplifiers: common emitter (CE) amplifiers, common base (CB) amplifiers, and common collection (CC) amplifiers. The basic configuration of each is shown in Figure 7.1.

In an amplifier system, the last stage of a voltage amplifier string has to be considered as a large signal amplifier, and generally EF amplifiers are used as large signal amplifiers. This then requires that the dc bias or dc operating point (quiescent point) be located near the center of the load line in order to get the maximum output voltage swing. Small signal analysis can be used to evaluate the amplifier for voltage gain, current gain, input impedance, and output impedance, all of which are discussed later.

DC Operating Point

Each transistor connected in a particular amplifier configuration has a set of characteristic curves, as shown in Figure 7.2.

When amplifiers are coupled together with capacitors, the configuration is as shown in Figure 7.3. The load resistor is really the input impedance of the next stage. To be able to evaluate this amplifier, a dc equivalent circuit needs to be developed as shown in Figure 7.4. This will result in the following dc bias equation:

$$I_{CQ} = \frac{V_{BB} - V_{BE}}{R_B/\text{beta} + R_E} \qquad \text{Assume } h_{FE} >> 1$$

where beta (h_{FE}) is the current gain of the transistor and V_{BE} is the conducting voltage across the base-emitter junction. This equation is the same for all amplifier configurations. Looking at Figure 7.3, the input circuit can be reduced to the dc circuit shown in Figure 7.4 using circuit analysis techniques, resulting in the following

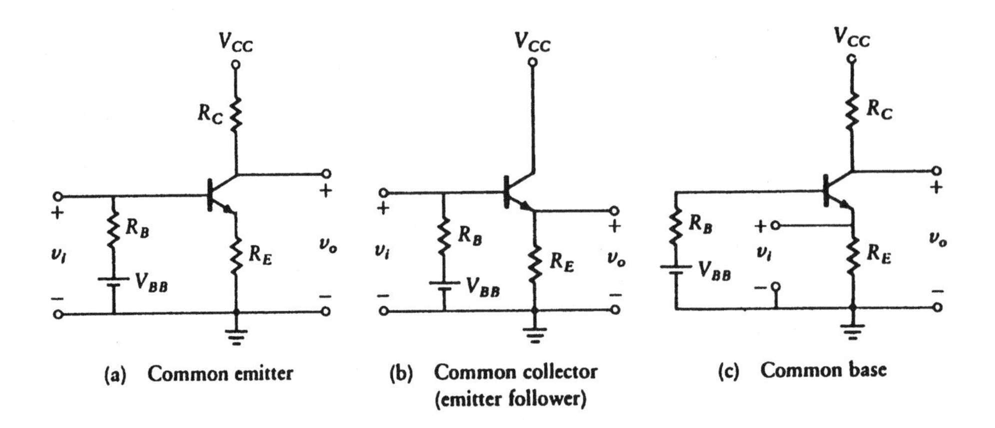

FIGURE 7.1 Amplifier circuits. (*Source:* C.J. Savant, M. Roden, and G. Carpenter, *Electronic Design, Circuits and Systems,* 2nd ed., Redwood City, CA: Benjamin-Cummings, 1991, p. 80. With permission.)

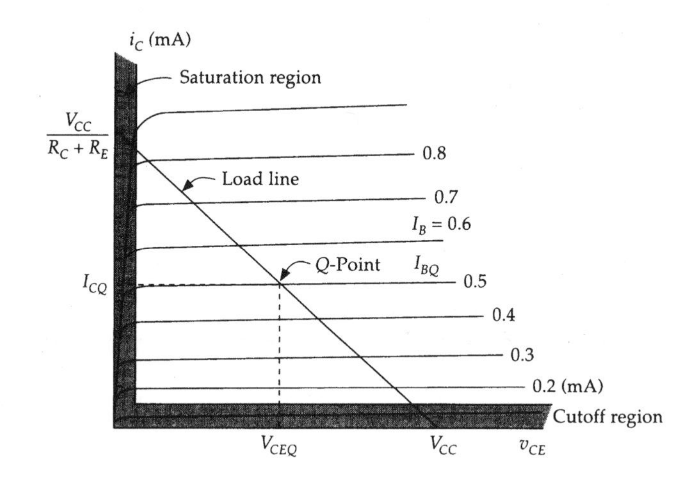

FIGURE 7.2 Transistor characteristic curves. (*Source:* C.J. Savant, M. Roden, and G. Carpenter, *Electronic Design, Circuits and Systems,* 2nd ed., Redwood City, CA: Benjamin-Cummings, 1991, p. 82. With permission.)

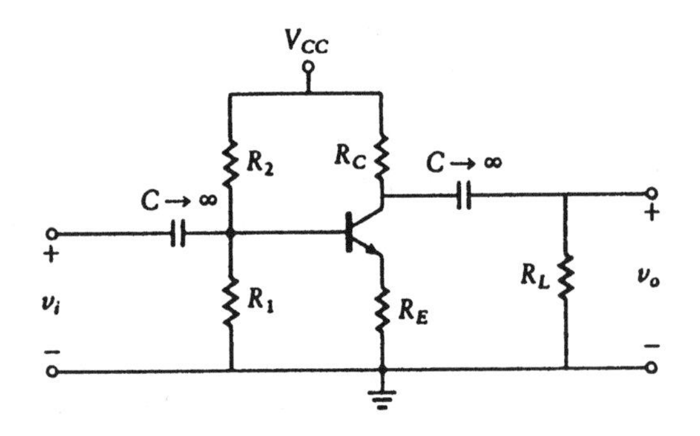

FIGURE 7.3 Amplifier circuit. (*Source:* C.J. Savant, M. Roden, and G. Carpenter, *Electronic Design, Circuits and Systems,* 2nd ed., Redwood City, CA: Benjamin-Cummings, 1991, p. 92. With permission.)

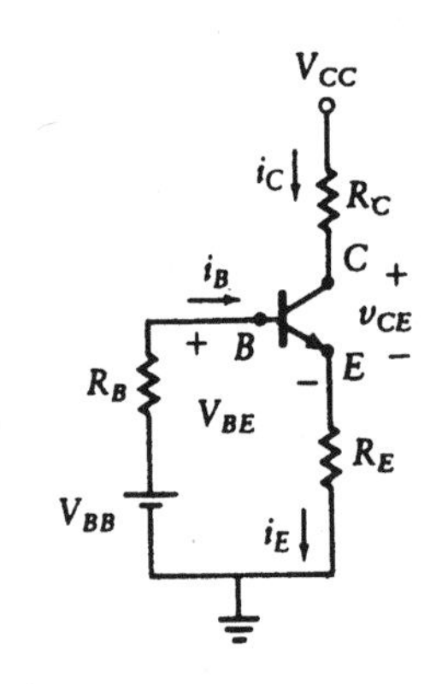

FIGURE 7.4 Amplifier equivalent circuit. (*Source:* C.J. Savant, M. Roden, and G. Carpenter, *Electronic Design, Circuits and Systems,* 2nd ed., Redwood City, CA: Benjamin-Cummings, 1991, p. 82. With permission.)

equations:

$$V_{BB} = V_{TH} = V_{CC}(R_1)/(R_1 + R_2)$$

$$R_B = R_{TH} = R_1//R_2$$

For this biasing system, the Thévenin equivalent resistance and the Thévenin equivalent voltage can be determined. For design with the biasing system shown in Figure 7.3, then

$$R_1 = R_B/(1 - V_{BB}/V_{CC})$$

$$R_2 = R_B(V_{CC}/V_{BB})$$

Graphical Approach

To understand the graphical approach, a clear understanding of the dc and ac load lines is necessary. The dc load line is based on the Kirchhoff's equation from the dc power source to ground (all capacitors open):

$$V_{CC} = v_{CE} + i_C R_{DC}$$

where R_{DC} is the sum of the resistors in the collector-emitter loop.

The ac load line is the loop, assuming the transistor is the ac source and the source voltage is zero, then

$$V'_{CC} = v_{ce} + i_C R_{ac}$$

where R_{ac} is the sum of series resistors in that loop with all the capacitors shorted. The load lines then can be constructed on the characteristic curves as shown in Figure 7.5. From this it can be seen that to get the maximum output voltage swing, the quiescent point, or Q point, should be located in the middle of the

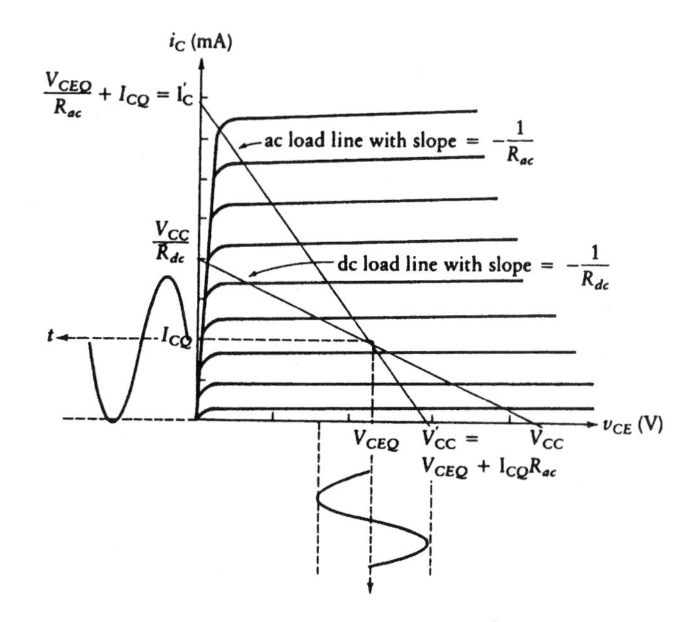

FIGURE 7.5 Load lines. (*Source:* C.J. Savant, M. Roden, and G. Carpenter, *Electronic Design, Circuits and Systems,* 2nd ed., Redwood City, CA: Benjamin-Cummings, 1991, p. 94. With permission.)

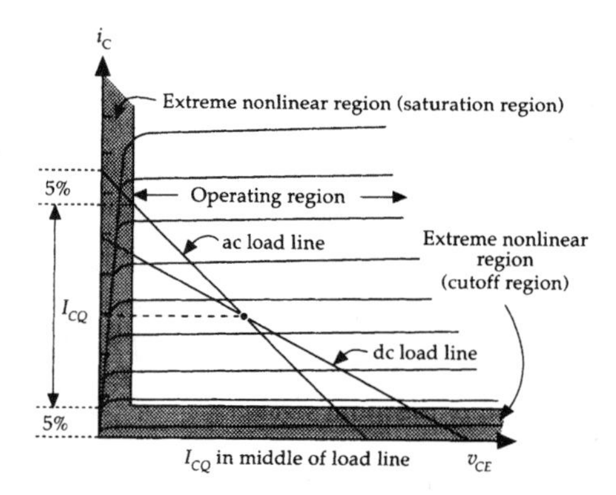

FIGURE 7.6 Q point in middle of load line. (*Source:* C.J. Savant, M. Roden, and G. Carpenter, *Electronic Design, Circuits and Systems,* 2nd ed., Redwood City, CA: Benjamin-Cummings, 1991, p. 135. With permission.)

ac load line. To place the Q point in the middle of the ac load line, I_{CQ} can be determined from the equation:

$$I_{CQ} = V_{CC}/(R_{DC} + R_{ac})$$

To minimize distortion caused by the cutoff and saturation regions, the top 5% and the bottom 5% are discarded. This then results in the equation (Figure 7.6):

$$V_o\,(\text{peak to peak}) = 2\,(0.9)\,I_{CQ}\,(R_C//R_L)$$

If, however, the Q point is not in the middle of the ac load line, the output voltage swing will be reduced. Below the middle of the ac load line (Figure 7.7(a)):

$$V_o\,(\text{peak to peak}) = 2\,(I_{CQ} - 0.05 I_{CMax})\,R_C//R_L$$

Above the middle of the ac load line (Figure 7.7(b)):

$$V_o\,(\text{peak to peak}) = 2\,(0.95 I_{CMax} - I_{CQ})\,R_C//R_L$$

These values allow the highest allowable input signal to be used to avoid any distortion by dividing the voltage gain of the amplifier into the maximum output voltage swing. The preceding equations are the same for the CB configuration. For the EF configurations, the R_C is changed to R_E in the equations.

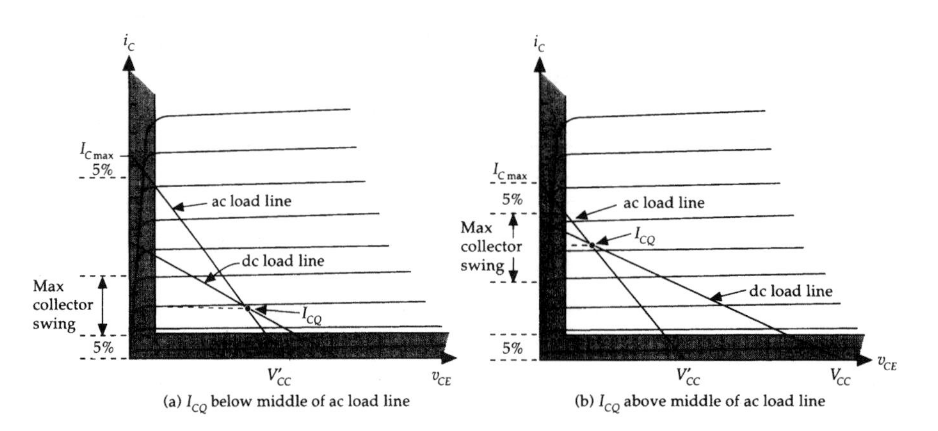

FIGURE 7.7 Reduced output voltage swing. (*Source:* C.J. Savant, M. Roden, and G. Carpenter, *Electronic Design, Circuits and Systems,* 2nd ed., Redwood City, CA: Benjamin-Cummings, 1991, p. 136. With permission.)

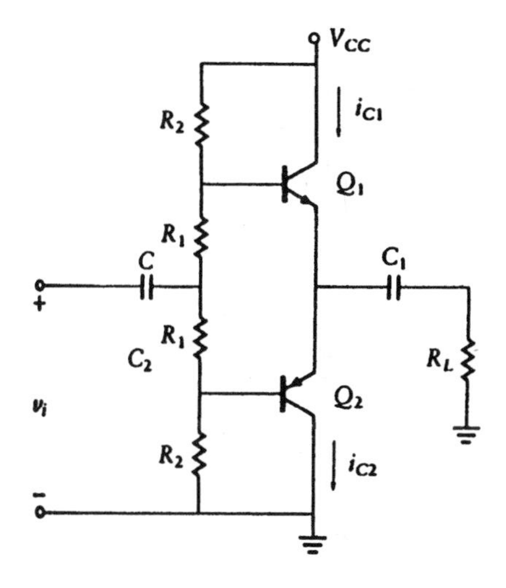

FIGURE 7.8 Complementary symmetry power amplifier. (*Source:* C.J. Savant, M. Roden, and G. Carpenter, *Electronic Design, Circuits and Systems,* 2nd ed., Redwood City, CA: Benjamin-Cummings, 1991, p. 248. With permission.)

Power Amplifiers

Emitter followers can be used as power amplifiers. Even though they have less than unity voltage gain they can provide high current gain. Using the standard linear EF amplifier for a maximum output voltage swing provides less than 25% efficiency (ratio of power in to power out). The dc current carrying the ac signal is where the loss of efficiency occurs. To avoid this power loss, the Q point is placed at I_{CQ} equal to zero, thus using the majority of the power for the output signal. This allows the efficiency to increase to as much as 70%. Full signal amplification requires one transistor to amplify the positive portion of the input signal and another transistor to amplify the negative portion of the input signal. In the past, this was referred to as push-pull operation. A better system is to use an NPN transistor for the positive part of the input signal and a PNP transistor for the negative part. This type of operation is referred to as Class B complementary symmetry operation (Figure 7.8).

In Figure 7.8, the dc voltage drop across R_1 provides the voltage to bias the transistor at cutoff. Because these are power transistors, the temperature will change based on the amount of power the transistor is absorbing. This means the base-emitter junction voltage will have to change to keep $I_{CQ} = 0$. To compensate for this change in temperature, the R_1 resistors are replaced with diodes or transistors connected as diodes with the same turn-on characteristics as the power transistors. This type of configuration is referred to as the complementary symmetry diode compensated (CSDC) amplifier and is shown in Figure 7.9. To avoid crossover distortion, small resistors can be placed in series with the diodes so that I_{CQ} can be raised slightly above zero to get increased amplification in the cutoff region. Another problem that needs to be addressed is the possibility of thermal runaway. This can be easily solved by placing small resistors in series with the emitters of the power transistors. For example, if the load is an 8-Ω speaker, the resistors should not be greater than 0.47 Ω to avoid output signal loss.

To design this type of amplifier, the dc current in the bias circuit must be large enough so that the diodes remain on during the entire input signal. This requires the dc diode current to be equal to or larger than the zero to peak current of the input signal, or

$$I_D \geqslant I_{ac}\,(0 \text{ to peak})$$

$$(V_{CC}/2 - V_{BE})/R_2 = I_B\,(0 \text{ to peak}) + V_L\,(0 \text{ to peak})/R_2$$

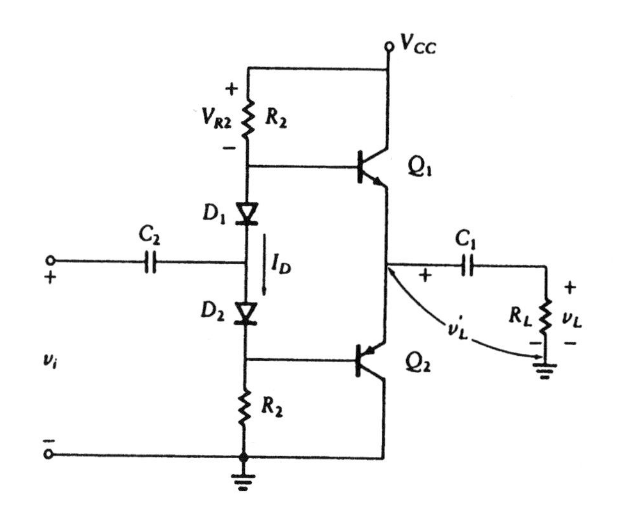

FIGURE 7.9 Complimentary symmetry diode compensated power amplifier. (*Source:* C.J. Savant, M. Roden, and G. Carpenter, *Electronic Design, Circuits and Systems*, 2nd ed., Redwood City, CA: Benjamin-Cummings, 1991, p. 251. With permission.)

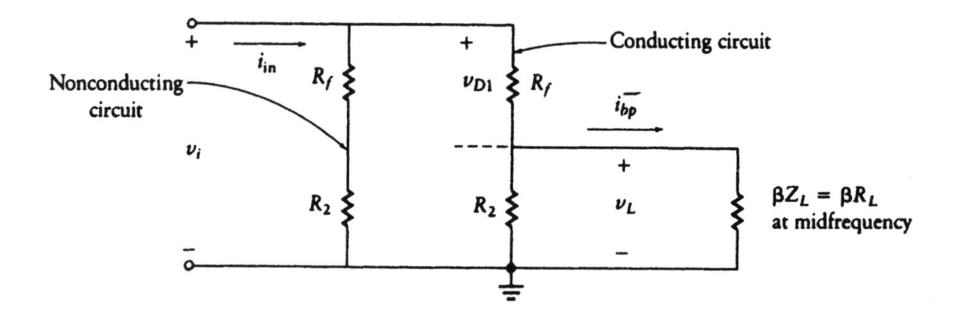

FIGURE 7.10 AC equivalent circuit of the CSDC amplifier. (*Source:* C.J. Savant, M. Roden, and G. Carpenter, *Electronic Design, Circuits and Systems*, 2nd ed., Redwood City, CA: Benjamin-Cummings, 1991, p. 255. With permission.)

When designing to a specific power, both I_B and V_L can be determined. This allows the selection of the value of R_2 and the equivalent circuit shown in Figure 7.10 can be developed. Using this equivalent circuit, both the input resistance and the current gain can be shown. R_f is the forward resistance of the diodes:

$$R_{in} = (R_f + R_2)//[R_f + (R_2//\text{Beta}\, R_L)]$$

$$P_0 = I_{Cmax} R_L/2$$

The power rating of the transistors to be used in this circuit should be greater than

$$P_{rating} = V_{CC}^2/(4\text{Pi}^2 R_L)$$

$$C_1 = 1/(2\text{Pi}\, f_{low} R_L)$$

$$C_2 = 10/[2\text{Pi}\, f_{low} R_{in} + R_i)]$$

where R_i is the output impedance of the previous stage and f_{low} is the desired low frequency cutoff of the amplifier.

References

P.R. Gray and R.G. Meyer, *Analysis and Design of Analog Integrated Circuits*, New York: Wiley, 1984.
J. Millman and A. Grabel, *Microelectronics*, New York: McGraw-Hill, 1987.
P.O. Neudorfer and M. Hassul, *Introduction to Circuit Analysis*, Needham Heights, MA: Allyn and Bacon, 1990.
C.J. Savant, M. Roden, and G. Carpenter, *Electronic Design, Circuits and Systems*, 2nd ed., Redwood City, CA: Benjamin-Cummings, 1991.
D.L. Schilling and C. Belove, *Electronic Circuits*, New York: McGraw-Hill, 1989.

7.2 Small Signal Analysis

John Choma, Jr.

This section introduces the reader to the analytical methodologies that underlie the design of small signal, analog bipolar junction transistor (BJT) amplifiers. Analog circuit and system design entails complementing basic circuit analysis skills with the art of architecting a circuit topology that produces acceptable input-to-output (*I*/*O*) electrical characteristics. Because design is not the inverse of analysis, analytically proficient engineers are not necessarily adept at design. However, circuit and system analyses that conduce an insightful understanding of meaningful topological structures arguably foster design creativity. Accordingly, this section focuses more on the problems of interpreting analytical results in terms of their circuit performance implications than it does on enhancing basic circuit analysis skills. Insightful interpretation breeds engineering understanding. In turn, such an understanding of the electrical properties of circuits promotes topological refinements and innovations that produce reliable and manufacturable, high performance electronic circuits and systems.

Hybrid-Pi Equivalent Circuit

In order for a BJT to function properly in linear amplifier applications, it must operate in the forward active region of its volt–ampere characteristic curves. Two conditions ensure BJT operation in the forward domain. First, the applied emitter-base terminal voltage must forward bias the intrinsic emitter-base junction diode at all times. Second, the instantaneous voltage established across the base-collector terminals of the transistor must preclude a forward biased intrinsic base-collector diode. The simultaneous satisfaction of these two conditions requires appropriate biasing subcircuits, and it imposes restrictions on the amplitudes of applied input signals (Clarke and Hess, 1978).

The most commonly used BJT equivalent circuit for investigating the dynamical responses to small input signals is the **hybrid-pi model** offered in Figure 7.11 (Sedra and Smith, 1987). In this model, R_b, R_c, and R_e, respectively, represent the *internal base, collector,* and *emitter resistances* of the considered BJT. Although these

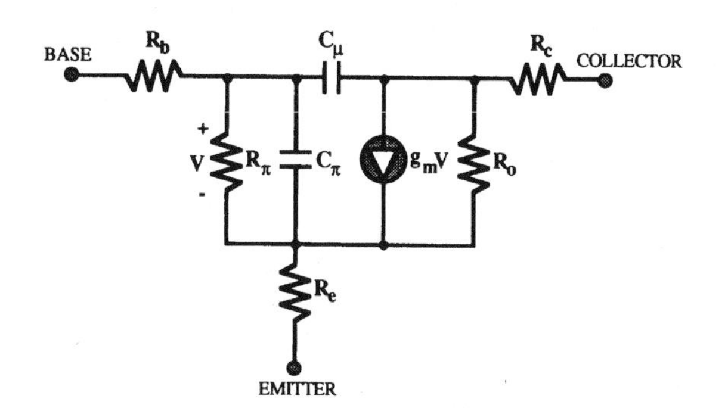

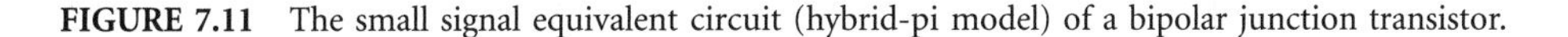
FIGURE 7.11 The small signal equivalent circuit (hybrid-pi model) of a bipolar junction transistor.

series resistances vary somewhat with quiescent operating point (de Graaf, 1969), they can be viewed as constants in first-order manual analyses.

The *emitter-base junction diffusion resistance*, R_π, is the small signal resistance of the emitter-base junction diode. It represents the inverse of the slope of the common emitter static input characteristic curves. Analytically, R_π is given by

$$R_\pi = \frac{h_{FE} N_F V_T}{I_{CQ}} \tag{7.1}$$

where h_{FE} is the *static common emitter current gain* of the BJT, N_F is the *emitter-base junction injection coefficient*, V_T is the *Boltzmann voltage corresponding to an absolute junction operating temperature of T*, and I_{CQ} is the *quiescent collector current.*

The expression for the resistance, R_o, which accounts for *conductivity modulation* in the neutral base, is

$$R_o = \frac{V'_{CEQ} + V_{AF}}{I_{CQ}\left(1 - \frac{I_{CQ}}{I_{KF}}\right)} \tag{7.2}$$

where V_{AF} is the *forward early voltage*, V'_{CEQ} is the quiescent voltage developed across the internal collector–emitter terminals, and I_{KF} symbolizes the *forward knee current.* The knee current is a measure of the onset of *high injection effects* (Gummel and Poon, 1970) in the base. In particular, a collector current numerically equal to I_{KF} implies that the forward biasing of the emitter-base junction promotes a net minority carrier charge injected into the base from the emitter that is equal to the background majority charge in the neutral base. The Early voltage is an inverse measure of the slope of the common emitter output characteristic curves.

The final low frequency parameter of the hybrid-pi model is the *forward transconductance*, g_m. This parameter, which is a measure of the forward small signal gain available at a quiescent operating point, is given by

$$g_m = \frac{I_{CQ}}{N_F V_T}\left(\frac{1 - \frac{I_{CQ}}{I_{KF}}}{1 + \frac{V_{CEQ'}}{V_{AF}}}\right) \tag{7.3}$$

Two capacitances, C_π and C_μ, are incorporated in the small signal model to provide a first-order approximation of steady-state transistor behavior at high signal frequencies. The capacitance, C_π, is the *net capacitance of the emitter-base junction diode* and is given by

$$C_\pi = \frac{C_{JE}}{\left(1 - \frac{V_E}{V_{JE} - 2V_T}\right)^{M_{JE}}} + \tau_f g_m \tag{7.4}$$

where the first term on the right-hand side represents the *depletion component* and the second term is the *diffusion component* of C_π. In Equation (7.4), τ_f is the average forward transit time of minority carriers in the field-neutral base, C_{JE} is the zero bias value of emitter-base junction depletion capacitance, V_{JE} is the built-in potential of the junction, V_E is the forward biasing voltage developed across the intrinsic emitter-base junction, and M_{JE} is the grading coefficient of the junction. The capacitance, C_μ, has only a depletion component, owing to the reverse (or at most zero) bias impressed across the internal base-collector junction. Accordingly, its analytical form is analogous to the first term on the right-hand side of

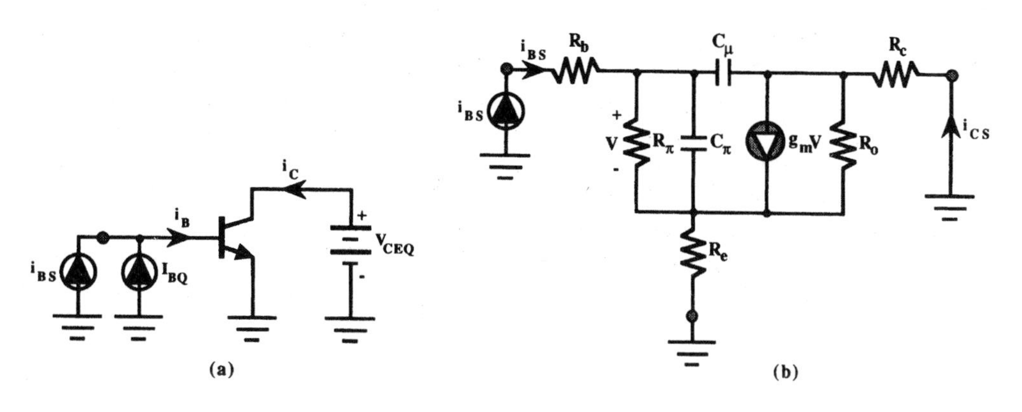

FIGURE 7.12 (a) Schematic diagram pertinent to the evaluation of the short circuit, common emitter, small signal current gain; (b) high-frequency small signal model of the circuit in part (a).

Equation (7.4). Specifically:

$$C_\mu = \frac{C_{JC}}{\left(1 - \dfrac{V_C}{V_{JC} - 2V_T}\right)^{M_{JC}}} \tag{7.5}$$

where the physical interpretation of C_{JC}, V_{JC}, and M_{JC} is analogous to C_{JE}, V_{JE}, and M_{JE}, respectively.

A commonly invoked figure-of-merit for assessing the high-speed, small signal performance attributes of a BJT is the *common emitter,* **short circuit gain-bandwidth product,** ω_T, which is given by

$$\omega_T = \frac{g_m}{C_\pi + C_\mu} \tag{7.6}$$

The significance of Equation (7.6) is best appreciated by studying the simple circuit diagram of Figure 7.12(a), which depicts the grounded emitter configuration of a BJT biased for linear operation at a quiescent base current of I_{BQ} and a quiescent collector-emitter voltage of V_{CEQ}. Note that the battery supplying V_{CEQ} grounds the collector for small signal conditions. The small signal model of the circuit at hand is resultantly seen to be the topology offered in Figure 7.12(b), where i_{BS} and i_{CS}, respectively, denote the signal components of the net instantaneous base current, i_B, and the net instantaneous collector current, i_C.

For negligibly small internal collector (R_c) and emitter (R_e) resistances, it can be shown that the *small signal, short circuit, high frequency common emitter current gain*, $\beta_{ac}(j\omega)$, is expressible as

$$\beta_{ac}(j\omega) \Delta \frac{i_{CS}}{i_{BS}} = \frac{\beta_{ac}\left(\dfrac{1 - j\omega C_\mu}{g_m}\right)}{1 + \dfrac{j\omega}{\omega_\beta}} \tag{7.7}$$

where β_{ac}, the low frequency value of $\beta_{ac}(j\omega)$, or simply the *low frequency beta*, is

$$\beta_{ac} = \beta_{ac}(0) = g_m R_\pi \tag{7.8}$$

and

$$\omega_\beta = \frac{1}{R_\pi(C_\pi + C_\mu)} \tag{7.9}$$

symbolizes the so-called *beta cutoff frequency* of the BJT. Because the frequency, g_m/C_μ, is typically much larger than ω_β, ω_β is the approximate **3-dB bandwidth** of $\beta_{ac}(j\omega)$; that is:

$$|\beta_{ac}(j\omega_\beta)| \cong \frac{\beta_{ac}}{\sqrt{2}} \tag{7.10}$$

It follows that the corresponding *gain-bandwidth product*, ω_T, is the product of β_{ac} and ω_β, which, recalling Equation (7.8), leads directly to the expression in Equation (7.6). Moreover, in the neighborhood of ω_T:

$$\beta_{ac}(j\omega) \cong \frac{\beta_{ac}\omega_\beta}{j\omega} = \frac{\omega_T}{j\omega} \tag{7.11}$$

which suggests that ω_T is the approximate frequency at which the magnitude of the small signal, short circuit, common emitter current gain degrades to unity.

Hybrid-Pi Equivalent Circuit of a Monolithic BJT

The conventional hyprid-pi model in Figure 7.11 generally fails to provide sufficiently accurate predictions of the high frequency response of monolithic diffused or implanted BJTs. One reason for this modeling inaccuracy is that the hybrid-pi equivalent circuit does not reflect the fact that monolithic transistors are often fabricated on lightly doped, noninsulating substrates that establish a distributed, large area, pn-junction with the collector region. Since the substrate-collector pn-junction is back biased in linear applications of a BJT, negligible static and low frequency signal currents flow from the collector to the substrate. At high frequencies, however, the depletion capacitance associated with the reverse-biased substrate-collector junction can cause significant susceptive loading of the collector port. In Figure 7.13, the lumped capacitance, C_{bb}, whose mathematical definition is similar to that of C_μ in Equation (7.5), provides a first-order account of this collector loading. Observe that this substrate capacitance appears in series with a substrate resistance, R_{bb}, which reflects the light doping nature of the substrate material. For monolithic transistors fabricated on insulating or semi-insulating substrates, R_{bb} is a very large resistance, thereby rendering C_{bb} unimportant with respect to the problem of predicting steady-state transistor responses at high signal frequencies.

A problem that is even more significant than parasitic substrate dynamics stems from the fact that the hybrid-pi equivalent circuit in Figure 7.11 is premised on a uniform transistor structure whose emitter-base and base-collector junction areas are identical. In a monolithic device, however, the effective base-collector junction area is much larger than that of the emitter-base junction because the base region is diffused or implanted into the collector (Glaser and Subak-Sharpe, 1977). The effect of such a geometry is twofold. First, the actual value of C_μ is larger than the value predicated on the physical considerations that surround a simplified uniform structure BJT. Second, C_μ is not a single lumped capacitance that is incident with only the intrinsic base-collector junction. Rather, the effective value of C_μ is distributed between the intrinsic collector and the entire base-collector junction interface. A first-order account of this capacitance distribution entails partitioning C_μ in Figure 7.11 into two capacitances, say $C_{\mu1}$ and $C_{\mu2}$, as indicated in Figure 7.13. In general, $C_{\mu2}$ is 3 to 5 times larger than $C_{\mu1}$. Whereas $C_{\mu1}$ is proportional to the emitter-base junction area, $C_{\mu2}$ is proportional to the net base-collector junction area, less the area of the emitter-base junction.

Just as $C_{\mu1}$ and $C_{\mu2}$ superimpose to yield the original C_μ in the simplified high frequency model of a BJT, the effective base resistances, R_{b1} and R_{b2}, sum to yield the original base resistance, R_b. The resistance, R_{b1}, is the *contact resistance* associated with the base lead and the inactive BJT base region. It is inversely proportional to the surface area of the base contact. However, R_{b2}, which is referred to as the *active base resistance*, is nominally an inverse function of emitter finger length. Because of submicron base widths and the relatively light average doping concentrations of active base regions, R_{b2} is significantly larger than R_{b1}.

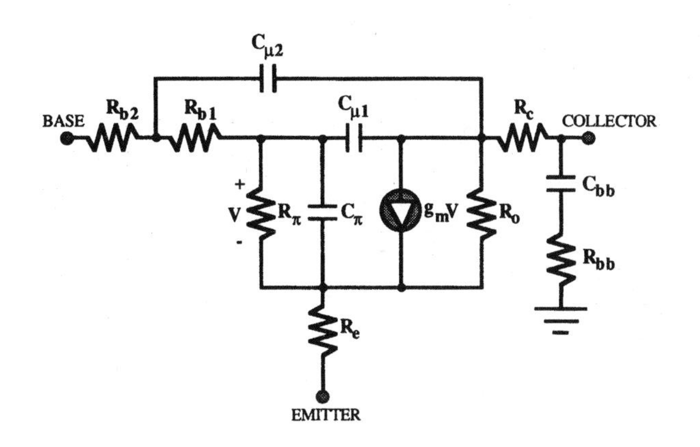

FIGURE 7.13 The hybrid-pi equivalent circuit of a monolithic bipolar junction transistor.

Common Emitter Amplifier

The most commonly used canonic cell of linear BJT amplifiers is the *common emitter amplifier*, whose basic circuit schematic diagram is depicted in Figure 7.14(a). In this diagram, R_{ST} is the Thévenin resistance of the applied signal source, V_{ST}, and R_{LT} is the effective, or Thévenin, load resistance driven by the amplifier. The signal source has zero average, or dc, value. Although requisite biasing is not shown in the figure, it is tacitly assumed that the transistor is biased for linear operation. Hence, the diagram at hand is actually the **ac schematic diagram**; that is, it delineates only the signal paths of the circuit. Note that in the common emitter orientation, the input signal is applied to the base of the transistor, while the resultant small signal voltage response, V_{OS}, is extracted at the transistor collector.

The hybrid-pi model of Figure 7.11 forms the basis for the small signal equivalent circuit of the common emitter cell, which is given in Figure 7.14(b). In this configuration, the capacitance, C_o, represents an effective output port capacitance that accounts for both substrate loading of the collector port (if the BJT is a monolithic device) and the net effective shunt capacitance associated with the load.

At low signal frequencies, the capacitors, C_π, C_μ, and C_o in the model of Figure 7.14(b), can be replaced by open circuits. A straightforward circuit analysis of the resultantly simplified equivalent circuit produces analytical expressions for the low frequency values of the small signal *voltage gain*, $A_{vCE} = V_{OS}/V_{ST}$; the *driving point input impedance*, Z_{inCE}; and the *driving point output impedance*, Z_{outCE}. Because the early resistance, R_o, is invariably much larger than the resistance sum $(R_c + R_e + R_{LT})$, the low frequency voltage gain of the

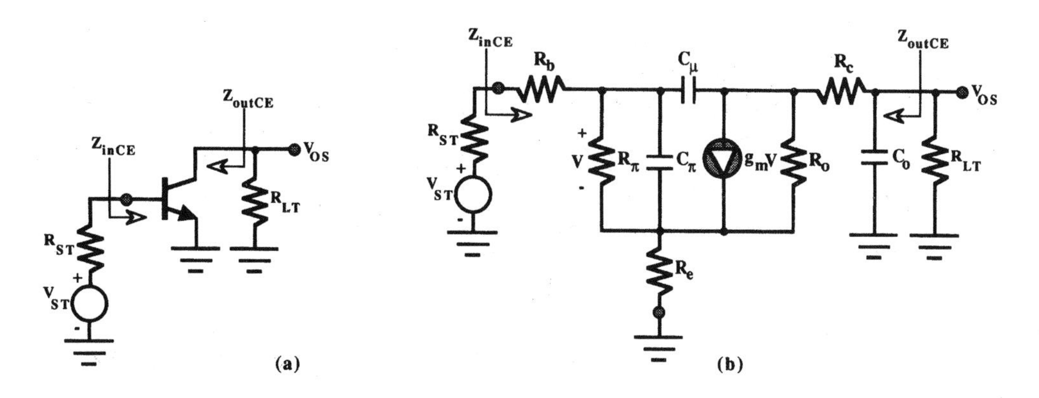

FIGURE 7.14 (a) AC schematic diagram of a common emitter amplifier; (b) modified small signal, high-frequency equivalent circuit of common emitter amplifier.

common emitter cell is expressible as

$$A_{vCE}(0) \cong -\left[\frac{\beta_{ac}R_{LT}}{R_{ST} + R_b + R_\pi + (\beta_{ac} + 1)R_e}\right] \tag{7.12}$$

For large R_o, conventional circuit analyses also produce a low frequency driving point input resistance of

$$R_{inCE} = z_{inCE}(0) \cong R_b + R_\pi + (\beta_{ac} + 1)R_e \tag{7.13}$$

and a low-frequency driving point output resistance of

$$R_{outCE} = Z_{outCE}(0) \cong \left(\frac{\beta_{ac}R_e}{R_e + R_b + R_\pi + R_{ST}} + 1\right)R_o \tag{7.14}$$

At high signal frequencies, the capacitors in the small signal equivalent circuit of Figure 7.14(b) produce a third-order voltage gain frequency response whose analytical formulation is algebraically cumbersome (Singhal and Vlach, 1977; Haley, 1988). However, because the poles produced by these capacitors are real, lie in the left half complex frequency plane, and generally have widely separated frequency values, the dominant pole approximation provides an adequate estimate of high frequency common emitter amplifier response in the usable passband of the amplifier. Accordingly, the high-frequency voltage gain, say $A_{vCE}(s)$, of the common emitter amplifier can be approximated as

$$A_{vCE}(s) \cong A_{vCE}(0)\left[\frac{1 + sT_{zCE}}{1 + sT_{pCE}}\right] \tag{7.15}$$

In this expression, T_{pCE} is of the form:

$$T_{pCE} = R_{C\pi}C_\pi + R_{C\mu}C_\mu + R_{Co}C_o \tag{7.16}$$

where $R_{C\pi}$, $R_{C\mu}$, and R_{Co}, respectively, represent the Thévenin resistances seen by the capacitors, C_π, C_μ, and C_o, under the conditions that (1) all capacitors are supplanted by open circuits, and (2) the independent signal generator, V_{ST}, is reduced to zero. Analogously, T_{zCE} is of the form:

$$T_{zCE} = R_{C\pi o}C_\pi + R_{C\mu o}C_\mu + R_{Coo}C_o \tag{7.17}$$

where $R_{C\pi o}$, $R_{C\mu o}$, and R_{Coo}, respectively, represent the Thévenin resistances seen by the capacitors, C_π, C_μ, and C_o, under the conditions that (1) all capacitors are supplanted by open circuits and (2) the output voltage response, V_{OS}, is constrained to zero while maintaining nonzero input signal source voltage. It can be shown that when R_o is very large and R_c is negligibly small:

$$R_{C\pi} = \frac{R_\pi || (R_{ST} + R_b + R_e)}{1 + \dfrac{\beta_{ac}R_e}{R_{ST} + R_b + R_\pi + R_e}} \tag{7.18}$$

$$R_{C\mu} = (R_{LT} + R_c) + \{(R_{ST} + R_b) || [R_\pi + (\beta_{ac} + 1)R_e]\}\left[1 + \frac{\beta_{ac}(R_{LT} + R_c)}{R_\pi + (\beta_{ac} + 1)R_e}\right] \tag{7.19}$$

and

$$R_{\mathrm{CO}} = R_{\mathrm{LT}} \tag{7.20}$$

Additionally, $R_{\mathrm{C}\pi\mathrm{o}} = RC_{\mathrm{oo}} = 0$, and

$$R_{\mathrm{C}\mu\mathrm{o}} = -\frac{R_\pi + (\beta_{\mathrm{ac}} + 1)R_{\mathrm{e}}}{\beta_{\mathrm{ac}}} \tag{7.21}$$

Once T_{pCE} and T_{zCE} are determined, the 3-dB voltage gain bandwidth, B_{CE}, of the common emitter amplifier can be estimated in accordance with

$$B_{\mathrm{CE}} \cong \frac{1}{T_{\mathrm{pCE}}\sqrt{1 - 2\left(\frac{T_{\mathrm{zCE}}}{T_{\mathrm{pCE}}}\right)^2}} \tag{7.22}$$

The high frequency behavior of both the driving point input and output impedances, $Z_{\mathrm{inCE}}(s)$ and $Z_{\mathrm{outCE}}(s)$, respectively, can be approximated by mathematical functions whose forms are analogous to the gain expression in Equation (7.15). In particular:

$$Z_{\mathrm{inCE}}(s) \cong R_{\mathrm{inCE}}\left[\frac{1 + sT_{\mathrm{zCE1}}}{1 + sT_{\mathrm{pCE1}}}\right] \tag{7.23}$$

and

$$Z_{\mathrm{outCE}}(s) \cong R_{\mathrm{outCE}}\left[\frac{1 + sT_{\mathrm{zCE1}}}{1 + sT_{\mathrm{pCE2}}}\right] \tag{7.24}$$

where R_{inCE} and R_{outCE} are defined by Equation (7.13) and Equation (7.14). The dominant time constants, T_{pCE1}, T_{zCE1}, T_{pCE2}, and T_{zCE2}, derive directly from Equation (7.16) and Equation (7.17) in accordance with (Choma and Witherspoon, 1990):

$$T_{\mathrm{pCE1}} = \lim_{R_{\mathrm{ST}}\to\infty} [T_{\mathrm{pCE}}] \tag{7.25}$$

$$T_{\mathrm{zCE1}} = \lim_{R_{\mathrm{ST}}\to 0} [T_{\mathrm{pCE}}] \tag{7.26}$$

$$T_{\mathrm{pCE2}} = \lim_{R_{\mathrm{LT}}\to\infty} [T_{\mathrm{pCE}}] \tag{7.27}$$

and

$$T_{\mathrm{zCE2}} = \lim_{R_{\mathrm{LT}}\to 0} [T_{\mathrm{pCE}}] \tag{7.28}$$

For reasonable values of transistor model parameters and terminating resistances, $T_{\mathrm{pCE1}} > T_{\mathrm{zCE1}}$, and $T_{\mathrm{pCE2}} > T_{\mathrm{zCE2}}$. It follows that both the input and output ports of a common emitter canonic cell are capacitive at high signal frequencies.

Design Considerations for the Common Emitter Amplifier

Equation (7.12) underscores a serious shortcoming of the canonical common emitter configuration. In particular, since the internal emitter resistance of a BJT is small, the low-frequency voltage gain is sensitive to the processing uncertainties that accompany the numerical value of the small signal beta. The problem can be rectified at the price of a diminished voltage gain magnitude by inserting an *emitter degeneration resistance*, R_{EE}, in series with the emitter lead, as shown in Figure 7.15(a). Since R_{EE} appears in series with the internal emitter resistance, R_e, as suggested in Figure 7.15(b), the impact of emitter degeneration can be assessed analytically by replacing R_e in Equation (7.12) to Equation (7.28) by the resistance sum $(R_e + R_{EE})$. For sufficiently large R_{EE}, such that:

$$R_e + R_{EE} \cong R_{EE} >> \frac{R_{ST} + R_b + R_\pi}{\beta_{ac} + 1} \tag{7.29}$$

the low-frequency voltage gain becomes

$$A_{vCE}(0) \cong -\frac{\alpha_{ac} R_{LT}}{R_{EE}} \tag{7.30}$$

where α_{ac}, which symbolizes the *small signal, short circuit, common base current gain*, or simply the ac *alpha*, of the transistor is given by

$$\alpha_{ac} = \frac{\beta_{ac}}{\beta_{ac} + 1} \tag{7.31}$$

Despite numerical uncertainties in β_{ac}, minimum values of β_{ac} are much larger than one, thereby rendering the voltage gain in Equation (7.30) almost completely independent of small signal BJT parameters.

A second effect of emitter degeneration is an increase in both the low-frequency driving point input and output resistances. This contention is confirmed by Equation (7.13), which shows that if R_o remains much larger than $(R_c + R_e + R_{EE} + R_{LT})$, a resistance in the amount of $(\beta_{ac}+1)R_{EE}$ is added to the input resistance established when the emitter of a common emitter amplifier is returned to signal ground. Likewise, Equation (7.14) verifies that emitter degeneration increases the low-frequency driving point output resistance. In fact, a very large value of R_{EE} produces an output resistance that approaches a limiting value of $(\beta_{ac} + 1)R_o$. It follows that a common emitter amplifier that exploits emitter degeneration behaves as a voltage-to-current converter at low signal frequencies. In particular, its high input resistance does not incur an

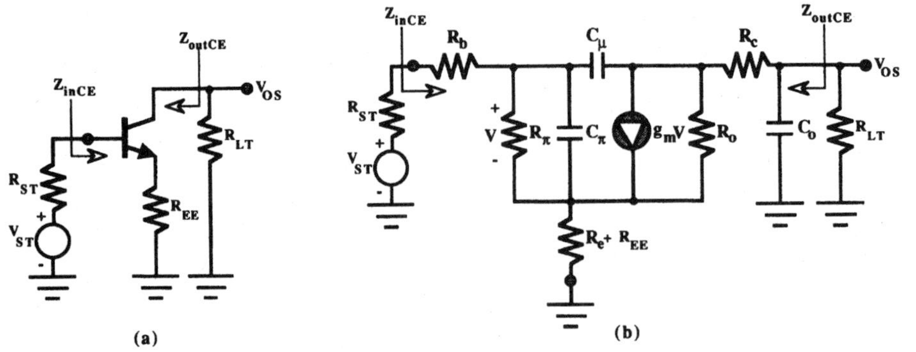

FIGURE 7.15 (a) ac schematic diagram of a common emitter amplifier using an emitter degeneration resistance; (b) small signal, high-frequency equivalent circuit of amplifier in part (a).

appreciable load on signal voltage sources that are characterized by even moderately large Thévenin resistances, while its large output resistance comprises an almost ideal current source at its output port.

A third effect of emitter degeneration is a decrease in the effective pole time constant, T_{pCE}, as well as an increase in the effective zero time constant, T_{zCE}, which can be confirmed by reinvestigating Equation (7.18) to Equation (7.21) for the case of R_e replaced by the resistance sum $(R_e + R_{EE})$. The use of an emitter degeneration resistance therefore promotes an increased 3-dB circuit bandwidth. Unfortunately, it also yields a diminished circuit gain-bandwidth product; that is, a given emitter degeneration resistance causes a degradation in the low frequency gain magnitude that is larger than the corresponding bandwidth increase promoted by this resistance. This deterioration of circuit gain-bandwidth product is a property of all *negative feedback circuits* (Choma, 1984).

For reasonable values of the emitter degeneration resistance, R_{EE}, the Thévenin time constant, $R_{C\mu}C_{\mu}$, is likely to be the dominant contribution to the effective first-order time constant, T_{pCE}, attributed to the poles of a common emitter amplifier. Hence, C_{μ} is the likely device capacitance that dominantly imposes an upper limit to the achievable 3-dB bandwidth of a common emitter cell. The reason for this substantial bandwidth sensitivity to C_{μ} is the so-called Miller multiplication factor, say M, which appears as the last bracketed term on the right-hand side of Equation (7.19), namely:

$$M = 1 + \frac{\beta_{ac}(R_{LT} + R_c)}{R_{\pi} + (\beta_{ac} + 1)R_e} \tag{7.32}$$

The Miller factor, M, which effectively multiplies C_{μ} in the expression for $R_{C\mu}C_{\mu}$, increases sharply with the load resistance, R_{LT}, and hence with the gain magnitude of the common emitter amplifier. Note that in the limit of a large emitter degeneration resistance (which adds directly to R_e), Equation (7.30) reduces Equation (7.32) to the factor:

$$M \cong 1 + |A_{vCE}(0)| \tag{7.33}$$

Common Base Amplifier

A second canonic cell of linear BJT amplifiers is the *common base amplifier*, whose ac circuit schematic diagram appears in Figure 7.16(a). In this diagram, R_{ST}, V_{ST}, R_{LT}, and V_{OS} retain the significance they respectively have in the previously considered common emitter configuration. Note that in the common base orientation, the input signal is applied to the base, while the resultant small signal voltage response is extracted at the collector of a transistor.

The relevant small signal model is shown in Figure 7.16(b). A straightforward application of Kirchhoff's circuit laws gives, for the case of large R_o, a low-frequency voltage gain, $A_{vCB}(0) = V_{OS}/V_{ST}$, of

$$A_{vCB}(0) \cong \frac{\alpha_{ac} R_{LT}}{R_{ST} + R_{inCB}} \tag{7.34}$$

where R_{inCB} is the low frequency value of the common base driving point input impedance:

$$R_{inCB} = Z_{inCB}(0) \cong R_e + \frac{R_b + R_{\pi}}{\beta_{ac} + 1} \tag{7.35}$$

Moreover, it can be shown that the low frequency driving point output resistance is

$$R_{outCB} = Z_{outCB}(0) \cong \left[\frac{\beta_{ac}(R_e + R_{ST})}{R_e + R_b + R_{\pi} + R_{ST}} + 1\right] R_o \tag{7.36}$$

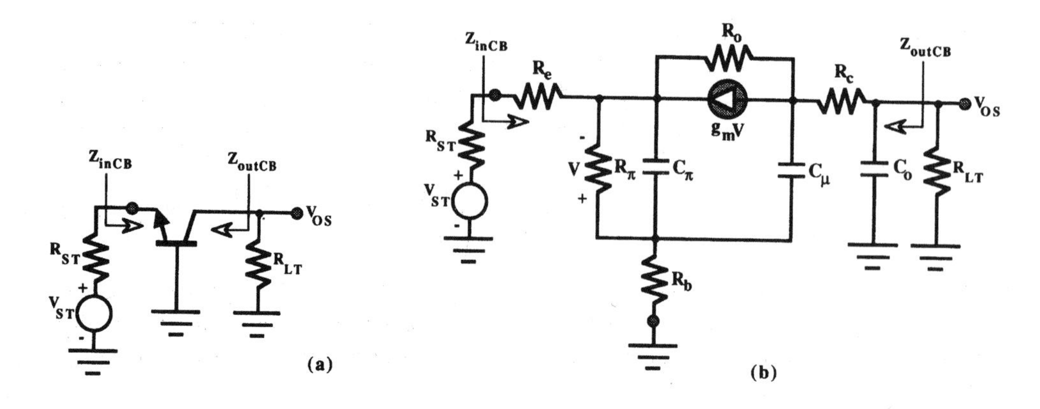

FIGURE 7.16 (a) ac schematic diagram of a common base amplifier; (b) small signal, high-frequency equivalent circuit of amplifier in part (a).

The preceding three equations underscore several operating characteristics that distinguish the common base amplifier from its common emitter counterpart. For example, Equation (7.35) suggests a low-frequency input resistance that is significantly smaller than that of a common emitter unit. To underscore this contention, consider the case of two identical transistors, one used in a common emitter amplifier and the other used in a common base configuration, that are biased at identical quiescent operating points. Under this circumstance, Equation (7.35) and Equation (7.13) combine to deliver:

$$R_{\text{inCB}} \cong \frac{R_{\text{inCE}}}{\beta_{\text{ac}} + 1} \tag{7.37}$$

which shows that the common base input resistance is a factor of $(\beta_{\text{ac}} + 1)$ times smaller than the input resistance of the common emitter cell. The resistance reflection factor, $(\beta_{\text{ac}} + 1)$, in Equation (7.37) represents the ratio of small signal emitter current to small signal base current. Accordingly, Equation (7.37) is self-evident when it is noted that the input resistance of a common base stage is referred to an input emitter current, whereas the input resistance of its common emitter counterpart is referred to an input base current.

A second difference between the common emitter and common base amplifiers is that the voltage gain of the latter displays no phase inversion between source and response voltages. Moreover, for the same load and source terminations and for identical transistors biased identically, the voltage gain of the common base cell is likely to be much smaller than that of the common emitter unit. This contention is verified by substituting Equation (7.37) into Equation (7.34) and using Equation (7.31), Equation (7.13), and Equation (7.12) to write:

$$A_{\text{vCB}}(0) \cong \frac{|A_{\text{vCE}}(0)|}{1 + \dfrac{\beta_{\text{ac}} R_{\text{ST}}}{R_{\text{ST}} + R_{\text{inCE}}}} \tag{7.38}$$

At high signal frequencies, the voltage gain, driving point input impedance, and driving point output impedance can be approximated by functions whose analytical forms mirror those of Equation (7.15), Equation (7.23), and Equation (7.24). Let T_{pCB} and T_{zCB} designate the time constants of the effective dominant pole and the effective dominant zero, respectively, of the common base cell. An analysis of the structure of Figure 7.16(b) resultantly produces, with R_{o} and R_{c} ignored:

$$T_{\text{pCB}} = R_{\text{C}\pi} C_{\pi} + R_{\text{C}\mu} C_{\mu} + R_{\text{Co}} C_{\text{o}} \tag{7.39}$$

where

$$R_{C\pi} = \frac{R_{\pi}||(R_{ST} + R_b + R_e)}{1 + \dfrac{\beta_{ac}(R_{ST} + R_e)}{R_{ST} + R_b + R_{\pi} + R_e}} \tag{7.40}$$

$$R_{C\mu} = R_b||[R_{\pi} + (\beta_{ac} + 1)(R_{ST} + R_e)] + R_{LT}\left[1 + \frac{\beta_{ac}R_b}{R_b + R_{\pi} + (\beta_{ac} + 1)(R_{ST} + R_e)}\right] \tag{7.41}$$

and R_{Co} remains given by Equation (7.20). Moreover:

$$T_{zCB} = \frac{R_b C_{\mu}}{\alpha_{ac}} \tag{7.42}$$

Design Considerations for the Common Base Amplifier

An adaptation of Equation (7.25) to Equation (7.28) to the common base stage confirms that the driving point input impedance is capacitive at high signal frequencies. However, $g_m R_b > 1$ renders a common base driving point input impedance that is inductive at high frequencies. This impedance property can be gainfully exploited to realize monolithic shunt peaked amplifiers in which the requisite circuit inductance is synthesized as the driving point input impedance of a common base stage (or the driving point output impedance of a common collector cell) (Grebene, 1984).

The common base stage is often used to broadband the common emitter amplifier by forming the *common emitter–common base cascode*, whose ac schematic diagram is given in Figure 7.17. The broadbanding afforded by the cascode structure stems from the fact that the effective low frequency load resistance, say R_{Le}, seen by the common emitter transistor, QE, is the small driving point input resistance of the common base amplifier, QB. This effective load resistance, as witnessed by C_{μ} of the common emitter transistor, is much smaller than the actual load resistance that terminates the output port of the amplifier, thereby decreasing the Miller multiplication of the C_{μ} in QE. If the time constant savings afforded by decreased Miller multiplication is larger than the sum of the additional time constants presented to the circuit by the common base transistor, an enhancement of common emitter bandwidth occurs. Note that such bandwidth enhancement is realized without compromising the common emitter gain–bandwidth product, since the voltage gain of the common emitter–common base unit is almost identical to that of the common emitter amplifier alone.

Common Collector Amplifier

The final canonic cell of linear BJT amplifiers is the *common collector amplifier*. The ac schematic diagram of this stage, which is often referred to as an *emitter follower*, is given in Figure 7.18(a). In emitter followers, the input signal is applied to the base, and the resultant small signal output voltage is extracted at the transistor emitter.

The small signal equivalent circuit corresponding to the amplifier in Figure 7.18(a) is shown in Figure 7.18(b). A straightforward circuit analysis gives, for the case of large R_o, a low-frequency voltage gain, $A_{vCC}(0) = V_{OS}/V_{ST}$, of:

$$A_{vCC}(0) \cong \frac{R_{LT}}{R_{LT} + R_{outCC}} \tag{7.43}$$

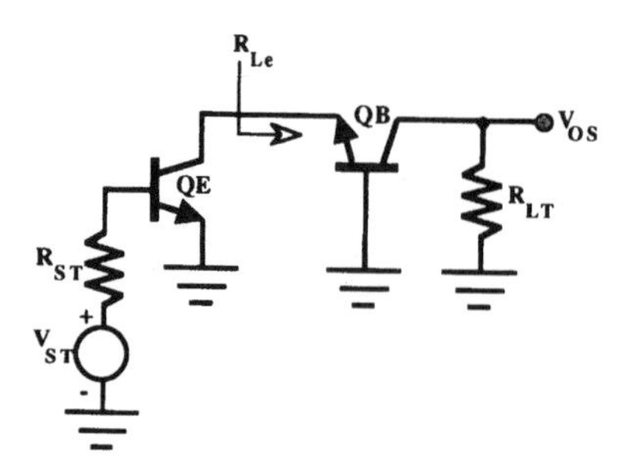

FIGURE 7.17 ac schematic diagram of a common emitter–common base cascode amplifier.

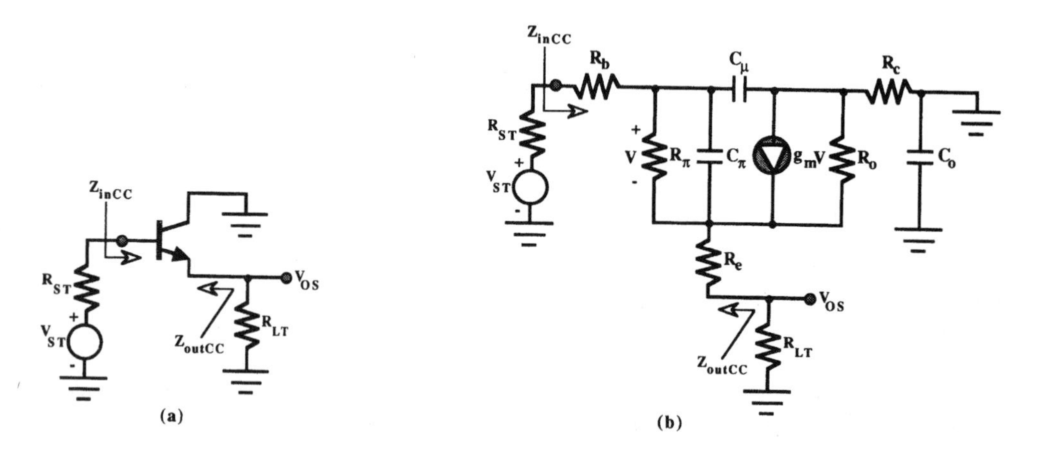

FIGURE 7.18 (a) ac schematic diagram of a common collector (emitter follower) amplifier; (b) small signal, high-frequency equivalent circuit of amplifier in part (a).

where R_{outCC} is the low frequency value of the driving point output impedance:

$$R_{\text{outCC}} = Z_{\text{outCC}}(0) \cong R_e + \frac{R_b + R_\pi + R_{\text{ST}}}{\beta_{\text{ac}} + 1} \tag{7.44}$$

The low-frequency driving point output resistance is

$$R_{\text{inCC}} = Z_{\text{inCC}}(0) \cong R_b + R_\pi + (\beta_{\text{ac}} + 1)(R_e + R_{\text{LT}}) \tag{7.45}$$

The facts that the voltage gain is less than one and is without phase inversion, the output resistance is small, and the input resistance is large make the emitter follower an excellent candidate for impedance buffering applications.

As in the cases of the common emitter and the common base amplifiers, the high-frequency voltage gain, driving point input resistance, and driving point output resistance can be approximated by functions having analytical forms that are similar to those of Equation (7.15), Equation (7.23), and Equation (7.24). Let T_{pCC} and T_{zCC} designate the time constants of the effective dominant pole and the effective dominant zero, respectively, of the emitter follower. Since the output port capacitance, C_o, appears across a short circuit, T_{pCC} is expressible as

$$T_{\text{pCC}} = R_{C\pi} C_\pi + R_{C\mu} C_\mu \tag{7.46}$$

With R_o ignored:

$$R_{C\pi} = \frac{R_\pi || (R_{ST} + R_b + R_{LT} + R_e)}{1 + \dfrac{\beta_{ac}(R_{LT} + R_e)}{R_{ST} + R_b + R_\pi + R_{LT} + R_e}} \tag{7.47}$$

and

$$\begin{aligned} R_{C\mu} &= (R_{ST} + R_b) || [R_\pi + (\beta_{ac} + 1)(R_{LT} + R_e) \\ &\quad + \left[1 + \frac{\beta_{ac}(R_{ST} + R_b)}{R_{ST} + R_b + R_\pi + (\beta_{ac} + 1)(R_{LT} + R_e)}\right] R_c \end{aligned} \tag{7.48}$$

The time constant of the effective dominant zero is

$$T_{zCC} = \frac{R_\pi C_\pi}{\beta_{ac} + 1} \tag{7.49}$$

Although the emitter follower possesses excellent wideband response characteristics, it should be noted in Equation (7.48) that the internal collector resistance, R_c, incurs some Miller multiplication of the base-collector junction capacitance, C_μ. For this reason, monolithic common collector amplifiers work best in broadband impedance buffering applications when they exploit transistors that have collector sinker diffusions and buried collector layers, which collectively serve to minimize the parasitic internal collector resistance.

Defining Terms

ac schematic diagram: A circuit schematic diagram, divorced of biasing subcircuits, that depicts only the dynamic signal flow paths of an electronic circuit.

Driving point impedance: The effective impedance presented at a port of a circuit under the condition that all other circuit ports are terminated in the resistances actually used in the design realization.

Hybrid-pi model: A two-pole linear circuit used to model the small signal responses of bipolar circuits and circuits fabricated in other device technologies.

Miller effect: The deterioration of the effective input impedance caused by the presence of feedback from the output port to the input port of a phase-inverting voltage amplifier.

Short circuit gain–bandwidth product: A measure of the frequency response capability of an electronic circuit. When applied to bipolar circuits, it is nominally the signal frequency at which the magnitude of the current gain degrades to one.

Three-decibel bandwidth: A measure of the frequency response capability of low-pass and bandpass electronic circuits. It is the range of signal frequencies over which the maximum gain of the circuit is constant to within a factor of the square root of two.

References

W.K. Chen, *Circuits and Filters Handbook*, Boca Raton, FL: CRC Press, 1995.

J. Choma, "A generalized bandwidth estimation theory for feedback amplifiers," *IEEE Transactions on Circuits and Systems*, vol. CAS-31, Oct. 1984.

J. Choma, and S. Witherspoon, "Computationally efficient estimation of frequency response and driving point impedances in wide-band analog amplifiers," *IEEE Transactions on Circuits and Systems*, vol. CAS-37, June 1990.

K.K. Clarke and D.T. Hess, *Communication Circuits: Analysis and Design,* Reading, MA: Addison-Wesley, 1978.

H.C. de Graaf, "Two New Methods for Determining the Collector Series Resistance in Bipolar Transistors With Lightly Doped Collectors," Phillips Research Report, 24, 1969.

A.B. Glaser, and G.E. Subak-Sharpe, *Integrated Circuit Engineering: Design, Fabrication, and Applications,* Reading, MA: Addison-Wesley, 1977.

A.B. Grebene, *Bipolar and MOS Analog Integrated Circuit Design,* New York: Wiley Interscience, 1984.

H.K. Gummel, and H.C. Poon, "An integral charge-control model of bipolar transistors," *Bell System Technical Journal*, 49, May–June 1970.

S.B. Haley, "The general eigenproblem: pole-zero computation," *Proc. IEEE*, 76, Feb. 1988.

J.D. Irwin, *Industrial Electronics Handbook,* Boca Raton, FL: CRC Press, 1997.

A.S. Sedra, and K.C. Smith, *Microelectronic Circuits,* 3rd ed., New York: Holt, Rinehart and Winston, 1991.

K. Singhal and J. Vlach, "Symbolic analysis of analog and digital circuits," *IEEE Transactions on Circuits and Systems*, vol. CAS-24, Nov. 1977.

Further Information

The *IEEE Journal of Solid-State Circuits* publishes state-of-the-art articles on all aspects of integrated electronic circuit design. The December issue of this journal focuses on analog electronics.

The *IEEE Transactions on Circuits and Systems* also publishes circuit design articles. Unlike the *IEEE Journal of Solid-State Circuits,* this journal addresses passive and active, discrete component circuits, as well as integrated circuits and systems, and it features theoretic research that underpins circuit design strategies.

The *Journal of Analog Integrated Circuits and Signal Processing* publishes design-oriented papers with emphasis on design methodologies and design results.

8

Active Filters

John E. Ayers
University of Connecticut

J.W. Steadman
University of Wyoming

B.M. Wilamowski
University of Wyoming

8.1 Synthesis of Low-Pass Forms

John E. Ayers

Introduction

Passive filters are frequency-selective circuits constructed using only passive components such as resistors, capacitors, and inductors. Other passive components such as distributed RC components, quartz crystals, and surface acoustic wave (SAW) devices are also used in special-purpose filters. However, *active filters* involve active components such as operational amplifiers (op amps) or discrete transistors. Compared to their passive counterparts, active filters are better suited to circuit board implementation because they avoid the need for inductors. *Switched capacitor* filters are best suited to monolithic integration because they avoid the need for large-valued passive components (resistors and capacitors).

This section describes basic filter theory, outlines the design of low-pass active filters, and introduces switched capacitor filter design. Other types of filters (high-pass, band-pass, and band-stop) may be designed by the translation of low-pass designs.

Transfer Functions of Filters

Filter design, whether active or passive, involves the determination of a suitable circuit topology and the computation of the circuit component values within the topology, such that the desired frequency response is obtained. The frequency response is commonly specified by the voltage transfer function $H(s)$, which is a rational function of the complex frequency variable s. For an nth-order filter, the denominator of $H(s)$ will be an nth-order polynomial in s and the numerator of $H(s)$ will be a polynomial in s of up to nth-order. For example, the transfer function of a second-order filter can be expressed as

$$H(s) = \frac{b_2 s^2 + b_1 s + b_0}{s^2 + a_1 s + a_0} \tag{8.1}$$

where s is the complex frequency and the a_i and b_j are constant coefficients. The roots of the denominator are the poles, and the roots of the numerator are the zeros. Therefore, the transfer function can also be written in the form:

$$H(s) = \frac{k \prod_{i=1}^{n} (s - z_i)}{\prod_{j=1}^{n} (s - p_i)} \tag{8.2}$$

where the z_i are the zeros of the transfer function, the p_i are the poles of the transfer function, and k is a constant. Filters with high selectivity (a steep transition between the passband and stopband) have complex conjugate pole-pairs. Often the frequency response information is displayed graphically using a Bode plot. The Bode plot involves a graph of the voltage gain in decibels ($20 \log_{10}|H(j\omega)|$) as a function of frequency, and the phase angle ($\angle H(j\omega)$) as a function of frequency. For most applications, however, the phase response is of secondary importance and the filter specifications may be given entirely in terms of the amplitude response.

Figure 8.1 shows an example of a frequency response specification for a low-pass filter. According to this specification, the passband gain (for frequencies less than ω_1) should be between $(1 - \delta_1)$ and 1. In the stopband (for frequencies greater than ω_2), the response should be less than δ_2. Also shown in Figure 8.1 is the response of a Butterworth filter which meets the given specifications.

Modern low-pass filter designs are usually based on one of these four classical types: *Butterworth*, *Chebyshev*, *Elliptic*, or *Bessel*. Each of these is optimum in a particular sense and therefore well-suited to certain applications. The *Butterworth* filter is optimum in the sense that its passband response is optimally flat, i.e., for an nth-order Butterworth filter the first $2n-1$ derivatives of the response are zero at zero frequency. Also, the response of the Butterworth filter is monotonic with no ripple. The *Chebyshev* filter is optimum in the sense that it gives the steepest transition from passband to stopband, for a given order n, *among all-pole filters.* However, it exhibits ripple in the passband. The *Elliptic* filter is optimum in the sense that it gives the steepest transition from passband to stopband, for a given order n. However, it introduces ripple in the stopband as well as the passband. The *Bessel* filter is optimum in the sense that it introduces a linear phase shift, and therefore a constant propagation delay for different frequency components. Infinitely many other filter designs are possible, but these four prototypes are a good starting point because of their special characteristics. Other types of responses, such as high-pass, band-pass, and band-stop, may be obtained from translated low-pass forms or their combinations.

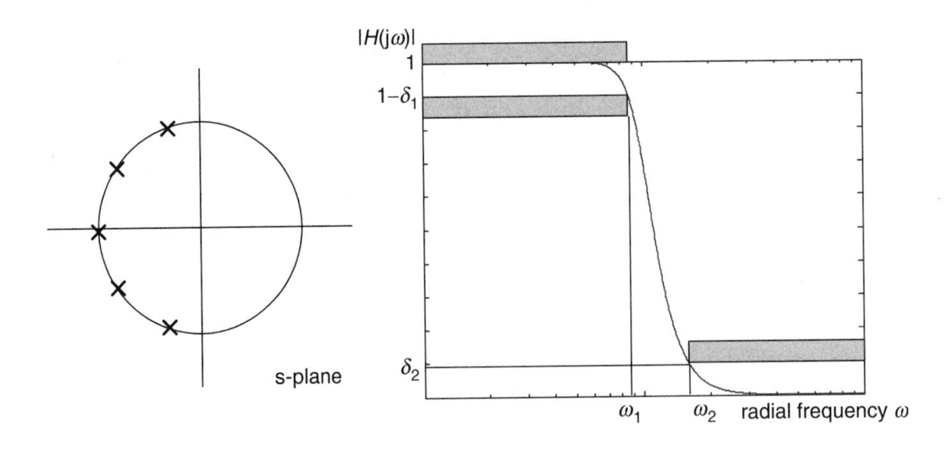

FIGURE 8.1 Butterworth filter: pole-zero plot and amplitude response for a fifth-order design.

TABLE 8.1 Butterworth Pole Positions

n	k	Real $-a$	Imaginary $\pm jb$
2	1.000	0.7071	0.07071
3	1.000	0.5000	0.8660
		1.0000	
4	1.000	0.3827	0.9239
		0.9239	0.3827
5	1.000	0.3090	0.9511
		0.8090	0.5878
		1.0000	
6	1.000	0.2588	0.9659
		0.7071	0.7071
		0.9659	0.2588
7	1.000	0.2225	0.9749
		0.6235	0.7818
		0.9010	0.4339
		1.0000	
8	1.000	0.1951	0.9808
		0.5556	0.8315
		0.8315	0.5556
		0.9808	0.1951
9	1.000	0.1737	0.9848
		0.5000	0.8660
		0.7660	0.6428
		0.9397	0.3420
		1.0000	
10	1.000	0.1564	0.9877
		0.4540	0.8910
		0.7071	0.7071
		0.8910	0.4540
		0.9877	0.1564

Butterworth Filter

The Butterworth filter is often employed because it has a maximally-flat passband response, with no ripple. The transfer function for an nth-order Butterworth filter is such that the amplitude response is given by

$$|H(j\omega)| = \frac{1}{\sqrt{1 + (\omega/\omega_0)^{2n}}} \tag{8.3}$$

where ω_0 is the -3 dB frequency for the filter.

The nth-order Butterworth filter has n poles which lie on a unit circle in the complex frequency plane. The Butterworth filter also has n zeros; however, because they all occur at $\omega =$ infinity, this filter is usually called an all-pole design. Figure 8.1 shows the pole-zero plot of a Butterworth filter and its amplitude response. Table 8.1 gives the pole locations for Butterworth filters up to the tenth order.

Chebyshev Filter

Compared to the Butterworth design, the Chebyshev[1] filter exhibits a steeper rolloff in the stopband. In fact, the Chebyshev filter exhibits the steepest rolloff available from a filter having only poles. However, it sacrifices monotonicity in the passband. The transfer function for an nth-order Chebyshev filter is devised so that the

[1]The Chebyshev filter also goes by the names "Chebyshev Type I," "Tchebyshev," and "Chebycheff."

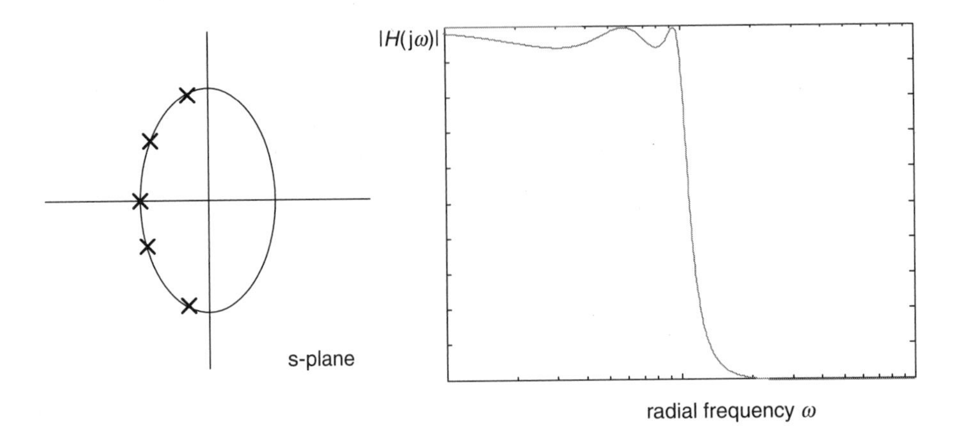

FIGURE 8.2 Chebyshev filter: pole-zero plot and amplitude response for a fifth-order design with 0.5 dB passband ripple.

amplitude response is given by

$$|H(j\omega)| = \frac{1}{\sqrt{1 + \left[\frac{1}{(1-\delta_1)^2} - 1\right] C_n^2(\omega/\omega_0)}} \tag{8.4}$$

where ω_0 is the frequency of the passband edge, δ_1 is the passband ripple, and $C_n(x)$ is the nth-order Chebyshev polynomial defined by

$$C_n(x) = \begin{cases} \cos(n\cos^{-1}x); & |x| \leqslant 1 \\ \cosh(n\cosh^{-1}x); & |x| \geqslant 1. \end{cases} \tag{8.5}$$

The nth-order Chebyshev filter has n poles which lie on an ellipse in the complex frequency plane. As with the Butterworth filter, the zeros all occur at ω = infinity and do not appear explicitly in the transfer function.

Figure 8.2 shows the pole-zero plot of a fifth-order Chebyshev filter and its amplitude response. Table 8.2 to Table 8.5 give the pole locations for Chebyshev filters up to tenth order, for the cases of 0.1 dB, 0.2 dB, 0.5 dB, and 1 dB passband ripple, respectively.

Bessel Filter

The Bessel filter achieves phase shift which increases linearly with frequency. This results in a uniform propagation delay for different frequency components, thus improving the transient response. This is achieved by pushing the poles out from the unit circle in the *s*-plane as shown in Figure 8.3. However, this results in a more gradual passband–stopband transition than even the Butterworth filter. Therefore, a higher order filter will be needed to achieve a given bandedge selectivity. Figure 8.3 shows the pole-zero plot of a Bessel filter and its amplitude response. Table 8.6 gives the pole locations for Bessel filters up to tenth order.

Elliptic Filter

The elliptic filter is optimum in the sense that it provides the steepest transition from passband to stopband, for a given order n. However, it exhibits ripple in the stopband as well as the passband. The transfer function

TABLE 8.2 Chebyshev Pole Positions (0.1 dB Ripple)

n	k	Real $-a$	Imaginary $\pm jb$
2	3.2761	1.1862	1.3809
3	1.6381	0.4847	1.2062
		0.9694	
4	0.8190	0.2642	1.1226
		0.6377	0.4650
5	0.4095	0.1665	1.0804
		0.4360	0.6677
		0.5389	
6	0.2048	0.1147	1.0565
		0.3133	0.7734
		0.4280	0.2831
7	0.1024	0.0838	1.0418
		0.2349	0.8355
		0.3395	0.4637
		0.3768	
8	0.0512	0.0640	1.0322
		0.1822	0.8750
		0.2727	0.5847
		0.3216	0.2053
9	0.0256	0.0504	1.0255
		0.1452	0.9018
		0.2225	0.6694
		0.2729	0.3562
		0.2905	
10	0.0128	0.0408	1.0207
		0.1184	0.9208
		0.1844	0.7307
		0.2323	0.4692
		0.2575	0.1617

for an nth-order Elliptic filter is such that the amplitude response is given by

$$|H(j\omega)| = \frac{1}{\sqrt{1 + \left[\frac{1}{(1-\delta_1)^2} - 1\right] E_n^2(\omega/\omega_0)}} \tag{8.6}$$

where ω_0 is the frequency of the passband edge, δ_1 is the passband ripple, and $C_n(x)$ is the nth-order Jacobian elliptic function defined by

$$E_n(x) = \int_0^x \frac{dy}{\sqrt{(1-y^2)(1-n^2y^2)}} \tag{8.7}$$

The nth-order elliptic filter has n poles which lie on an ellipse in the complex frequency plane and appear in conjugate pairs. However, the n zeros of the elliptic filter are spaced out on the $j\omega$ axis. Figure 8.4 shows the pole-zero plot of an elliptic filter and its amplitude response. Table 8.7 and Table 8.8 give the pole and zero locations for Bessel filters with 0.5 dB passband ripple and 20 dB stopband rejection (Table 8.7) and for 0.1 dB passband ripple and 40 dB stopband rejection (Table 8.8).

Passive Filter Synthesis

A passive low-pass filter may be designed using the LCR ladder structure shown in Figure 8.5.

The network shown is a fifth-order filter, but higher-order filters may be implemented by adding elements, while preserving the series inductor/parallel capacitor pattern. The drawback of this passive design is the need

TABLE 8.3 Chebyshev Pole Positions (0.2 dB Ripple)

n	k	Real $-a$	Imaginary $\pm jb$
2	2.3032	0.9635	1.1952
3	1.1516	0.4073	1.1170
		0.8146	
4	0.5758	0.2248	1.0715
		0.5427	0.4438
5	0.2879	0.1426	1.0474
		0.3733	0.6473
		0.4614	
6	0.1439	0.0985	1.0335
		0.2692	0.7566
		0.3677	0.2769
7	0.0720	0.0722	1.0249
		0.2022	0.8219
		0.2922	0.4561
		0.3243	
8	0.0360	0.0551	1.0192
		0.1570	0.8640
		0.2350	0.5773
		0.2772	0.2027
9	0.0180	0.0435	1.0153
		0.1253	0.8928
		0.1919	0.6627
		0.2355	0.3526
		0.2506	
10	0.0090	0.0352	1.0124
		0.1022	0.9133
		0.1592	0.7248
		0.2005	0.4653
		0.2223	0.1603

for inductors. Inductors present a number of practical problems, including their large series resistance, large physical size, and their coupling with electromagnetic fields. Furthermore, it is impractical to implement any but the very smallest values of inductance on integrated circuits. To avoid these problems, active filters incorporating only resistors and capacitors may be used.

Active Filter Synthesis

The design of a filter starts with required response, usually in the form of the amplitude response. The type of filter can be determined based on the detailed specifications. A maximally flat passband response dictates a Butterworth design. If a flat passband response is not required, sharper cutoff characteristics can be obtained in lower-order filters using Chebyshev or elliptic filters. If constant propagation delay is important, then a Bessel filter may be employed. Once the basic type of filter has been selected, the order of the filter is determined based on the steepness of the transition from passband to stopband.

An nth-order active filter can be synthesized by cascading second-order sections. If n is odd, an additional first-order section may be added. The process requires first the determination of the transfer function, or equivalently, the pole and zero positions in the s-plane. Computer tools such as MATLAB™ are used extensively to determine the required poles and zeros to obtain the desired filter response. Then individual second-order sections can be designed to provide the required second-order polynomials in the denominator and numerator (the required poles and zeros) of the transfer function. If the preceding design tables are used, then the pair of complex conjugate poles $-a \pm jb$ corresponds to a quadratic term in the denominator of the form:

$$s^2 - 2as + a^2 + b^2 = (s - a - jb)(s - a + jb) \tag{8.8}$$

TABLE 8.4 Chebyshev Pole Positions (0.5 dB Ripple)

n	k	Real $-a$	Imaginary $\pm\, jb$
2	1.4314	0.7128	1.0040
3	1.4314	0.3132	1.0219
		0.6265	
4	0.3578	0.1754	1.0163
		0.4233	0.4209
5	0.1789	0.1120	1.0116
		0.2931	0.6252
		0.3623	
6	0.0895	0.0777	1.0085
		0.2121	0.7382
		0.2898	0.2702
7	0.0447	0.0570	1.0064
		0.1597	0.8071
		0.2308	0.4479
		0.2562	
8	0.0224	0.0436	1.0050
		0.1242	0.8520
		0.1859	0.5693
		0.2193	0.1999
9	0.0112	0.0345	1.0040
		0.0992	0.8829
		0.1520	0.6553
		0.1864	0.3487
		0.1984	
10	0.0056	0.0279	1.0033
		0.0810	0.9051
		0.1261	0.7183
		0.1589	0.4612
		0.1761	0.1589

Two commonly used types of second-order sections are the *Sallen and Key* and the *state variable biquad*. However, the same design process can be used with other second-order circuit designs.

Sallen and Key Section

A Sallen and Key second-order filter section takes the form shown in Figure 8.6 (Sallen and Key, 1955). Its transfer function is given by

$$H(s) = \frac{K\dfrac{1}{C_1C_2R_1R_2}}{s^2 + s\left[\dfrac{1}{C_2R_2} + \dfrac{1}{C_2R_1} + \dfrac{1-K}{C_1R_1}\right] + \dfrac{1}{C_1C_2R_1R_2}} \tag{8.9}$$

This is an all-pole form since the numerator involves only a constant (there are no finite zeros). However, this is entirely adequate for the realization of Butterworth, Chebyshev, or Bessel filters.

The specification for a second-order section may be in terms of a coefficient form:

$$H(s) = \frac{k}{s^2 + a_1s + a_0} \tag{8.10}$$

TABLE 8.5 Chebyshev Pole Positions (1 dB Ripple)

n	k	Real $-a$	Imaginary $\pm\ jb$
2	0.9826	0.5489	0.8951
3	0.4913	0.2471	0.9660
		0.4942	
4	0.2457	0.1395	0.9834
		0.3369	0.4073
5	0.1228	0.0895	0.9901
		0.2342	0.6119
		0.2895	
6	0.0614	0.0622	0.9934
		0.1699	0.7272
		0.2321	0.2662
7	0.0307	0.0457	0.9953
		0.1281	0.7982
		0.1851	0.4429
		0.2054	
8	0.0154	0.0350	0.9965
		0.0997	0.8448
		0.1492	0.5644
		0.1760	0.1982
9	0.0077	0.277	0.9972
		0.0797	0.8769
		0.1221	0.6509
		0.1497	0.3463
		0.1593	
10	0.0038	0.0224	0.9978
		0.0650	0.9001
		0.1013	0.7143
		0.1277	0.4586
		0.1415	0.1580

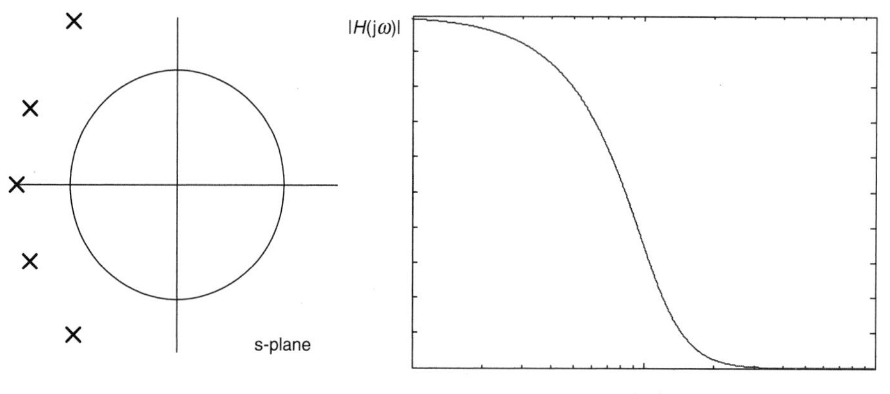

FIGURE 8.3 Bessel filter: pole-zero plot and amplitude response for a fifth-order design.

where

$$k = \frac{K}{C_1 C_2 R_1 R_2} \tag{8.11}$$

$$a_1 = \frac{1}{C_2 R_2} + \frac{1}{C_2 R_1} + \frac{1 - K}{C_1 R_1} \tag{8.12}$$

TABLE 8.6 Bessel Pole Positions

n	k	Real $-a$	Imaginary $\pm jb$
2	1.0000	1.1030	0.6368
3	1.0000	1.0509	1.0025
		1.3270	
4	1.0000	0.9877	1.2476
		1.3596	0.04071
5	1.0000	0.9606	1.4756
		1.3851	0.7201
		1.5069	
6	1.0000	1.3836	0.9727
		1.5735	0.3213
		0.9318	1.6640
7	1.0000	0.9104	1.8375
		1.3797	1.1923
		1.6130	0.5896
		1.6853	
8	1.0000	0.8955	2.0044
		1.3780	1.3926
		1.6419	0.8253
		1.7627	0.2737
9	1.0000	0.8788	2.1509
		1.3683	1.5685
		1.6532	1.0319
		1.8081	0.5126
		1.8575	
10	1.0000	0.9091	0.1140
		0.8688	0.3430
		0.7838	0.5759
		0.6418	0.8176
		0.4083	1.0813

and

$$a_0 = \frac{1}{C_1 C_2 R_1 R_2} \tag{8.13}$$

Alternatively, the transfer function for the second-order section may be given in the standard $Q - \omega_0$ second-order form:

$$H(s) = \frac{k}{s^2 + \frac{\omega_0}{Q}s + \omega_0^2} \tag{8.14}$$

where

$$k = \frac{K}{C_1 C_2 R_1 R_2} \tag{8.15}$$

$$\omega_0 = \frac{1}{\sqrt{C_1 C_2 R_1 R_2}} \tag{8.16}$$

and

$$Q = \frac{1}{\sqrt{\frac{C_1 R_1}{C_2 R_2}} + \sqrt{\frac{C_1 R_2}{C_2 R_1}} + (1 - K)\sqrt{\frac{C_2 R_2}{C_1 R_1}}} \tag{8.17}$$

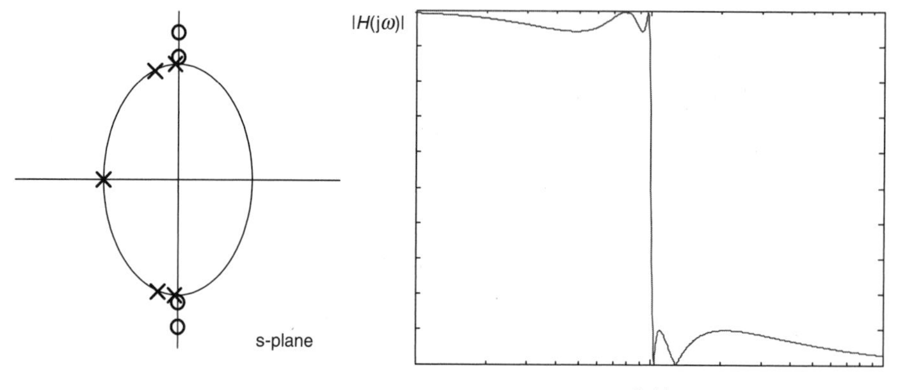

FIGURE 8.4 Elliptic filter: pole-zero plot and amplitude response for a fifth-order design.

TABLE 8.7 Elliptic Filter Pole and Zero Positions (0.5 dB passband ripple, 20 dB stopband rejection)

		Pole positions		Zero positions	
n	k	Real $-a$	Imaginary $\pm jb$	Real $-a$	Imaginary $\pm jb$
2	0.1000	0.6715	1.0539	0	3.8397
3	0.3642	0.2139	1.0500	0	1.5801
		0.7920			
4	0.1000	0.4950	0.6673	0	2.2268
		0.0726	1.0217	0	1.1743
5	0.3099	0.2107	0.9054	0	1.3131
		0.0252	1.0081	0	1.0582
		0.6809			
6	0.1000	0.4698	0.6320	0	2.1059
		0.0786	0.9710	0	1.1004
		0.0088	1.0029	0	1.0202
7	0.3039	0.2071	0.8883	0	1.2876
		0.0282	0.9904	0	1.0343
		0.0031	1.0010	0	1.0071
		0.6680			
8	0.1000	0.4667	0.6277	0	2.0917
		0.0783	0.9646	0	1.0928
		0.0100	0.9967	0	1.0120
		0.0011	1.0004	0	1.0025
9	0.3032	0.2067	0.8862	0	1.2846
		0.0282	0.9881	0	1.0318
		0.0035	0.9988	0	1.0042
		0.0004	1.0001	0	1.0009
		0.6664			
10	0.1000	0.4663	0.6272	0	2.0900
		0.0782	0.9638	0	1.0919
		0.0100	0.9959	0	1.0111
		0.0013	0.9996	0	1.0015
		0.0001	1.0000	0	1.0003

State-Variable Biquad Section

Sallen and Key sections are adequate for the synthesis of Butterworth, Chebyshev, or Bessel filters, which are all-pole filters. However, elliptic filters require the introduction of finite zeros as well as poles. Their design

TABLE 8.8 Elliptic Filter Pole and Zero Positions (0.1 dB Passband Ripple, 40 dB Stopband Rejection)

		Pole Positions		Zero Positions	
n	k	Real $-a$	Imaginary $\pm jb$	Real $-a$	Imaginary $\pm jb$
2	0.0100	0.5458	0.9001	0	9.9335
3	0.1034	0.4533	1.2116	0	4.0421
		1.0100			
4	0.0100	0.6731	0.5421	0	4.7249
		0.2124	1.1162	0	2.0661
5	0.0601	0.4119	0.7935	0	2.1728
		0.1067	1.0637	0	1.4692
		0.6706			
6	0.0100	0.5460	0.4273	0	3.6243
		0.2366	0.9062	0	1.5032
		0.0552	1.0346	0	1.2275
7	0.0527	0.3728	0.7016	0	1.8927
		0.1312	0.9558	0	1.2421
		0.2900	1.0187	0	1.1157
		0.5940			
8	0.0100	0.5119	0.3996	0	3.3834
		0.2268	0.8495	0	1.3972
		0.0714	0.9783	0	1.1226
		0.0154	1.0100	0	1.0602
9	0.0508	0.3605	0.6771	0	1.8248
		0.1296	0.9237	0	1.1946
		0.0385	0.9890	0	1.0637
		0.0082	1.0054	0	1.0318
		0.5734			

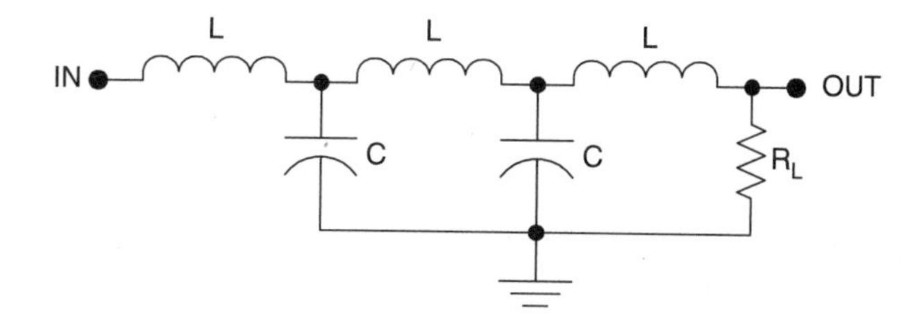

FIGURE 8.5 Fifth-order passive low-pass filter constructed using a LC ladder network.

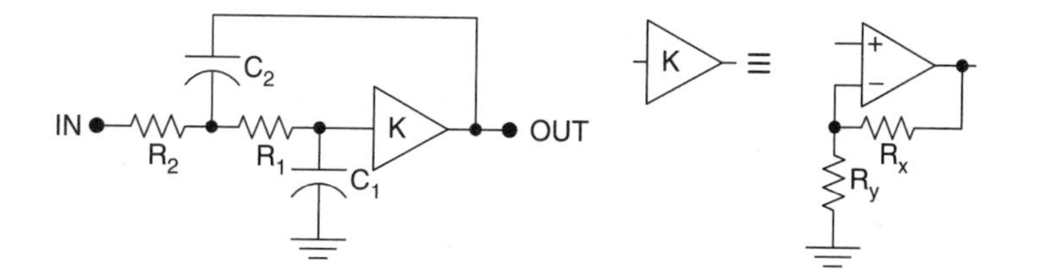

FIGURE 8.6 Sallen and Key second-order filter section.

thus dictates the use of biquadratic (biquad) sections. A biquad section is one for which the transfer function numerator and denominator are both quadratic functions of the complex frequency s.

For the state-variable biquad section shown in Figure 8.7 (Kerwin et al., 1967), the transfer function is given by

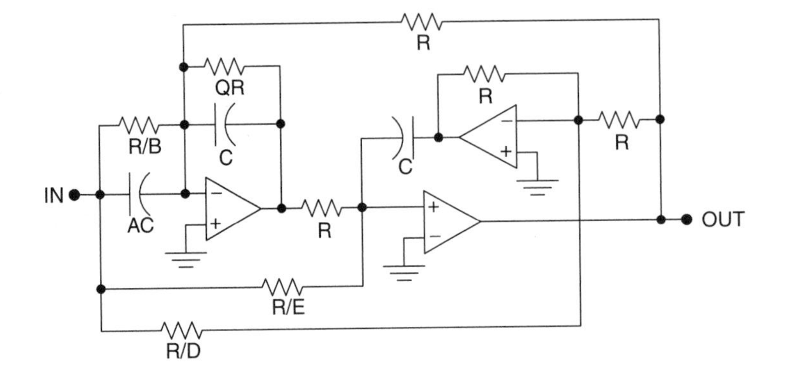

FIGURE 8.7 State-variable biquad second-order filter section.

$$H(s) = -\frac{As^2 + \omega_0(B - D)s + E\omega_0^2}{s^2 + \frac{\omega_0}{Q}s + \omega_0^2} \tag{8.18}$$

where

$$\omega_0 = \frac{1}{RC} \tag{8.19}$$

and A, B, D, E, F, R, C, and Q are defined in the figure.

Component Scaling

Once a prototype filter has been designed, the resistors and capacitors may be scaled to values that are practical for implementation. Typically, a prototype filter is designed using all 1 Ω resistors and with a cutoff frequency of 1 rad/sec. In practical active filters, the resistors should all be greater than 1 kΩ to avoid overloading the op amps and less than 1 MΩ to minimize electrical noise. If all resistors in the prototype are scaled up by a factor of 10^4 while all capacitors are scaled down by a factor of 10^{-4}, the frequency response will be left unchanged. A different scaling factor may be used if necessary, as long as the RC products are left unchanged.

Frequency Scaling

Typically, a prototype filter is designed with a cutoff frequency of 1 rad/sec. The cutoff frequency can be readily scaled up by simply scaling down all capacitors. For example, to scale the cutoff frequency up to 10 kHz, all capacitors would be scaled down by a factor of $1/(2\pi \times 10^4)$. However, it is prudent to keep all capacitors greater than 100 pF to minimize the effects of parasitic capacitances in the devices and circuit board.

Sensitivity and Stability

Once a filter has been completely designed, the circuit should be verified using a circuit simulation program such as SPICE. All component tolerances should be taken into account for such verification, and not just the nominal component values. Some filter circuits, such as higher-order Elliptic and Chebyshev filters, have poles that are very close to the $j\omega$ axis. It is especially important in these cases to make sure that the component tolerances will not cause one or more poles to appear in the right half of the s-plane, which would result in an unstable filter.

Fifth-Order Chebyshev Filter Design Example

Consider the design of a fifth-order Chebyshev filter with unity passband gain, 0.5 dB passband ripple, and a rolloff frequency of 10 kHz. From Table 8.4, the pole positions for the 1 rad/sec prototype are $-0.1120 \pm j1.0116$, $-0.2931 \pm j0.6252$, and -0.3623, with $k = 0.1789$. The required transfer function is therefore:

$$H(s) = \frac{0.1789}{(s^2 + 0.2240s + 1.0359)(s^2 + 0.5862s + 0.4768)(s + 0.3623)}$$
$$= \left(\frac{1.0359}{s^2 + 0.2240 + 1.0359}\right)\left(\frac{0.4768}{s^2 + 0.5862s + 0.4768}\right)\left(\frac{0.3623}{s + 0.3623}\right) \quad (8.20)$$

This transfer function can be realized by cascading two Sallen and Key sections and one first-order section as shown in Figure 8.8.

The first term on the left may be implemented using a Sallen and Key section with $K_1 = 1$. The following design equations are obtained:

$$\frac{1}{C_2R_2} + \frac{1}{C_2R_1} + \frac{1 - K_1}{C_1R_1} = 0.2240 \quad (8.21)$$

and

$$\frac{1}{C_1C_2R_1R_2} = 1.0359 \quad (8.22)$$

There are more unknowns than equations so multiple solutions exist. If we arbitrarily set $R_1 = R_2 = 1\ \Omega$, and solve Equation (8.21) and Equation (8.22) simultaneously for C_1 and C_2, we obtain $C_1 = 0.6512$ F and $C_2 = 3.6140$ F.

The second term may be implemented using a second Sallen and Key section with $K_2 = 1$. The design equations are

$$\frac{1}{C_4R_4} + \frac{1}{C_4R_3} + \frac{1 - K_2}{C_3R_3} = 0.5862 \quad (8.23)$$

and

$$\frac{1}{C_3C_4R_3R_4} = 0.4768 \quad (8.24)$$

We can arbitrarily set $R_3 = R_4 = 1\ \Omega$, and solve Equation (8.23) and Equation (8.24) simultaneously for C_3 and C_4 to obtain $C_3 = 0.325$ F and $C_4 = 4.913$ F.

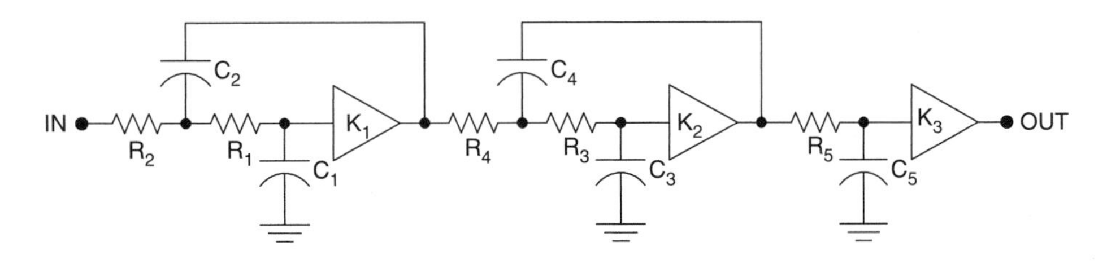

FIGURE 8.8 Realization of a fifth-order Chebyshev filter using Sallen and Key sections.

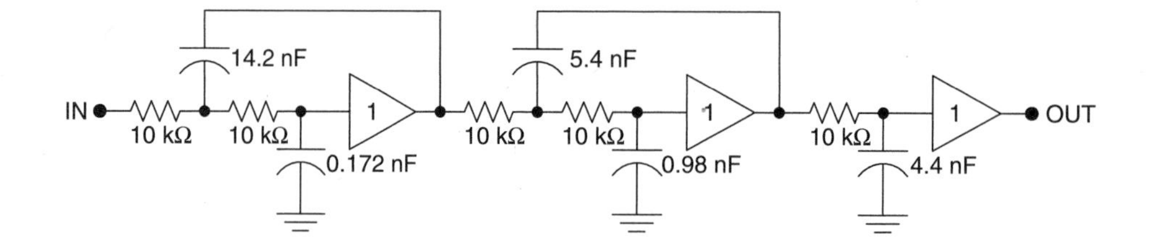

FIGURE 8.9 Fifth-order Chebyshev filter with 0.5 dB passband ripple and a rolloff frequency of 10 kHz, realized using Sallen and Key sections.

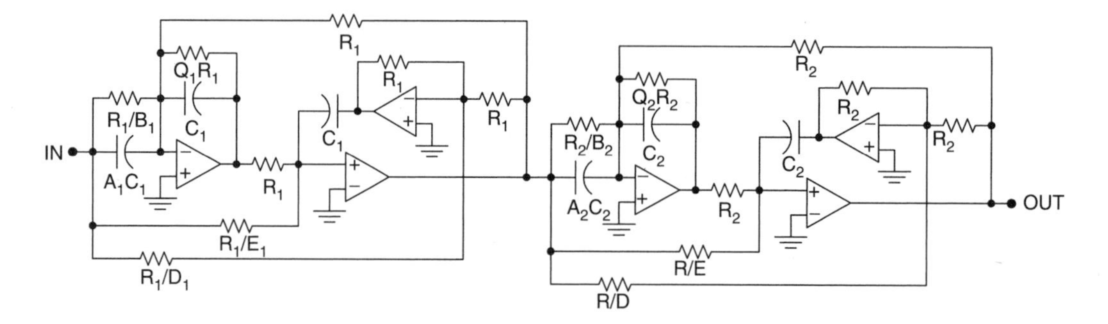

FIGURE 8.10 Realization of a fourth-order elliptic filter using state-variable biquad sections.

The third term of $H(s)$ may be implemented using a first-order section with $K_3 = 1$. The design equation is

$$\frac{1}{C_5 R_5} = 0.3623 \tag{8.25}$$

If we set $R_5 = 1\ \Omega$, we obtain $C_5 = 2.760$ F.

Now that the prototype filter has been designed, we can scale the components to practical values by multiplying all resistances by 10^4 and all capacitances by 10^{-4}. The resulting values are $R_1 = R_2 = R_3 = R_4 = R_5 = 10\ \text{k}\Omega$, $C_1 = 0.10812\text{E-}4$ F, $C_2 = 8.929\text{E-}4$ F, $C_3 = 0.6145\text{E-}4$ F, $C_4 = 3.419\text{E-}4$ F, and $C_5 = 2.760\text{E-}4$ F.

In order to scale the rolloff frequency from 1 rad/sec to 10 kHz, we can scale all capacitors by $1/(2\pi \times 10^4)$ while leaving the resistances unchanged. The resulting values are $R_1 = R_2 = R_3 = R_4 = R_5 = 10\ \text{k}\Omega$, $C_1 = 0.172$ nF, $C_2 = 14.2$ nF, $C_3 = 0.98$ nF, $C_4 = 5.4$ nF, and $C_5 = 4.4$ nF. The design for the fifth-order Chebyshev filter appears in Figure 8.9. When such a design is implemented using standard value components, it is necessary to verify the final design using a circuit analysis tool such as SPICE to ensure that the filter response is still within the specifications.

Fourth-Order Elliptic Filter Design Example

Consider the design of a fourth-order elliptic filter with unity passband gain, 0.5 dB passband ripple, 20 dB stopband rejection, and a rolloff frequency of 10 kHz. From Table 8.7, the pole positions for the 1 rad/sec prototype are $-0.4950 \pm j0.6673$ and $-0.0726 \pm j1.0217$, and the zero positions are at $\pm j1.1743$ and $\pm j2.2268$, and $k = 0.1000$. The required transfer function for the 1 rad/sec prototype is therefore

$$H(s) = \frac{0.1(s^2 + 1.379)(s^2 + 4.96)}{(s^2 + 0.990s + 0.690)(s^2 + 0.1452s + 1.049)} \tag{8.26}$$

The four poles and four zeros may be realized by cascading two biquad sections as shown in Figure 8.10.

If the section on the left is designed to have the transfer function:

$$H_1(s) = -\frac{0.1(s^2 + 1.379)}{(s^2 + 0.990s + 0.690)} \tag{8.27}$$

the following design equations result:

$$A_1 = 0.1 \tag{8.28}$$

$$B_1 - D_1 = 0 \tag{8.29}$$

$$E_1 = \frac{(0.1)(1.379)}{0.690} = 0.200 \tag{8.30}$$

$$\frac{1}{R_1 C_1} = 0.83 \text{ rad/sec} \tag{8.31}$$

and

$$Q_1 = \frac{0.990}{0.83} = 1.193 \tag{8.32}$$

Choosing $B_1 = D_1 = 1$ and $R_1 = 1\ \Omega$, we obtain $C_1 = 1.205$ F.

If the section on the right is designed to have the transfer function:

$$H_2(s) = -\frac{s^2 + 4.96}{(s^2 + 0.1452s + 1.049)} \tag{8.33}$$

the following design equations result:

$$A_2 = 1 \tag{8.34}$$

$$B_2 - D_2 = 0 \tag{8.35}$$

$$E_2 = \frac{4.96}{1.049} = 4.73 \tag{8.36}$$

$$\frac{1}{R_2 C_2} = 1.024 \text{ rad/sec} \tag{8.37}$$

and

$$Q_2 = \frac{0.1452}{0.1024} = 1.418 \tag{8.38}$$

Choosing $B_2 = D_2 = 1$ and $R_2 = 1\ \Omega$, we obtain $C_2 = 0.976$ F.

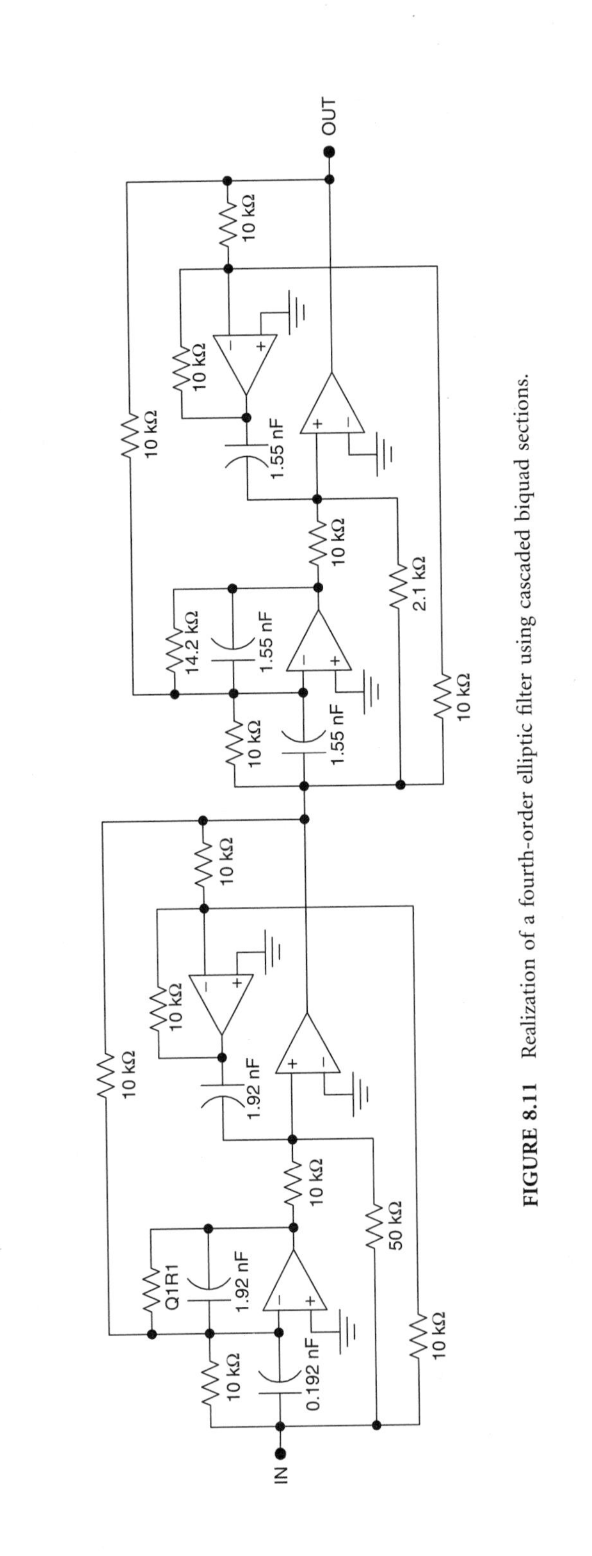

FIGURE 8.11 Realization of a fourth-order elliptic filter using cascaded biquad sections.

Now that the prototype filter has been designed, we can scale the components to practical values by multiplying all resistances by 10^4 and all capacitances by 10^{-4}. The resulting values are $A_1 = 0.1$, $B_1 = D_1 = 1$, $E_1 = 0.200$, $Q_1 = 1.193$, $R_1 = 10\ \text{k}\Omega$, $C_1 = 1.205\text{E-4 F}$, $A_2 = 1$, $B_2 = D_2 = 1$, $E_2 = 4.73$, $Q_2 = 1.418$, $R_2 = 10\ \text{k}\Omega$, and $C_2 = 0.976\text{E-4 F}$.

In order to scale the rolloff frequency from 1 rad/sec to 10 kHz, we can scale all capacitors by $1/(2\pi \times 10^4)$ while leaving the resistances unchanged. The resulting values are $A_1 = 0.1$, $B_1 = D_1 = 1$, $E_1 = 0.200$, $Q_1 = 1.193$, $R_1 = 10\ \text{k}\Omega$, $C_1 = 1.92$ nF, $A_2 = 1$, $B_2 = D_2 = 1$, $E_2 = 4.73$, $Q_2 = 1.418$, $R_2 = 10\ \text{k}\Omega$, and $C_2 = 1.55$ nF. The design for the fourth-order elliptic filter appears in Figure 8.11. Once standard value components have been chosen, the entire design should be verified using SPICE.

Switched Capacitor Filters

Active filters provide a great deal of design flexibility and use only standard op amps, capacitors, and resistors. However, the values of the resistors and capacitors in active filters are large enough to preclude their monolithic integration. Moreover, integrated capacitors are usually subject to tolerances as large as 30% due to process variations. Switched capacitor (SC) filters avoid these problems while providing unique capabilities (Taylor and Huang, 1997; Schaumann and van Valkenburg, 2001; Ghausi and Laker, 2003). First, resistors are not needed in SC filters. Second, the frequency response for an SC filter depends on ratios of capacitors, which can be controlled much more tightly than their absolute values. Third, the frequency response of an SC filter scales with the clock frequency, and can be changed as needed.

A basic SC filter is shown in Figure 8.12. The switch (implemented using metal oxide semiconductor field effect transistors, or MOSFETs) is toggled at a frequency f_{CLK}. Each time the switch toggles to the left-hand side, a charge of $C_1 V_{IN}$ is placed on the capacitor C_1. This charge is then transferred to the inverting input of the op amp when the switch toggles to the right-hand side. The average current delivered to the inverting input of the op amp is thus $V_{IN} C_1 f_{CLK}$, and so the switch–capacitor combination emulates a resistor of value $1/(C_1 f_{CLK})$. The transfer function for the SC filter shown in Figure 8.12 is therefore:

$$H(s) = -\frac{f_{CLK} C_1}{s C_2} \tag{8.39}$$

This is therefore a first-order low-pass filter with a single pole at the origin of the *s*-plane. The rolloff frequency of the filter is $(C_2/C_1)(f_{CLK}/2\pi)$, which depends on the ratio of the capacitors but not their absolute values. Also, the rolloff frequency can be changed by the alteration of the clock frequency.

The SC filter can be generalized to higher-order filters of various types. Second-order SC sections can be derived from the Sallen and Key and state variable biquad designs. All that is required is that each resistor be replaced by a capacitor and the associated MOSFET switch transistors.

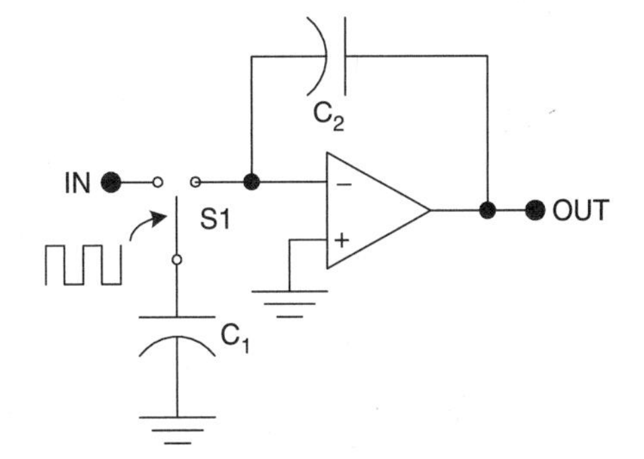

FIGURE 8.12 First-order switched capacitor filter.

Monolithic SC filters are readily available and typically require only a small number of external resistors and/or jumpers to configure their frequency response characteristics. The clock frequency is usually derived by an external quartz crystal oscillator and can be set very precisely as needed. A detailed description of SC filter theory is beyond the scope of this chapter, but can be found in Ghausi and Laker (2003) and Schaumann and Van Valkenburg (2001).

Defining Terms

Active filter: An electronic filter whose design includes one or more active devices.

Biquad: An active filter section whose transfer function comprises a ratio of second-order numerator and denominator polynomials in the frequency variable.

Electronic filter: An electronic circuit designed to transmit some range of signal frequencies while rejecting others. Phase and time-domain specifications may also occur.

MATLAB™: A software package produced by The MathWorks, Inc. and which is employed extensively in system and signal analysis. MATLAB has built-in filter design functions.

Sensitivity: A measure of the extent to which a given circuit performance measure is affected by a given component within the circuit.

SPICE: Simulation Program with Integrated Circuit Emphasis, a circuit simulation program available from many sources.

Switched capacitor filter: An active filter which uses capacitors and MOSFET switches in place of resistors.

References

P. Bowron and F.W. Stephenson, *Active Filters for Communications and Instrumentation*, New York: McGraw-Hill, 1979.

M. Ghausi and K. Laker, *Modern Filter Design: Active RC and Switched Capacitor*, Atlanta, GA: Noble Publishing, 2003.

W.J. Kerwin, L.P. Huelsman, and R.W. Newcomb, "State-variable synthesis for insensitive integrated circuit transfer functions," *IEEE J.*, vol. SC-2, pp. 87–92, 1967.

D. Lancaster, *Lancaster's Active Filter Cookbook*, 2nd ed., New York: Newnes, 1996.

P.R. Sallen and E.L. Key, "A practical method of designing RC active filters," *IRE Trans.*, vol. CT-2, pp. 74–85, 1955.

R. Schaumann and M. Van Valkenburg, *Design of Analog Filters*, New York: Oxford University Press, 2001.

J.T. Taylor and Q. Huang, *CRC Handbook of Electrical Filters*, Boca Raton, FL: CRC Press, 1997.

A. Waters, *Active Filter Design*, New York: Macmillan, 1991.

A. I. Zverev, *Handbook of Filter Synthesis*, New York: John Wiley, 1967.

Further Information

Tabulations of representative standard filter specification functions appear in Schaumann and van Valkenburg (2001), Bowron and Stephenson (1979), and Zverev (1967). The design process for analog filters is described with many practical details in Lancaster (1996). The theory of switched capacitor filters is covered in detail in Ghausi and Laker (2003). Good sources of journal and conference papers on filters include the *IEEE Transactions on Circuits and Systems*, the *IEEE International Symposium on Circuits and Systems* (*ISCAS*), and the *European Conference on Circuit Theory and Design* (*ECCTD*). Internet sources are too numerous to list here, and their number is rapidly growing. Some of the best practical design information relating to SC filters is posted on the Internet by manufacturers. These sources are easily found using any common Internet search engine.

8.2 Realization

J.W. Steadman and B.M. Wilamowski

After the appropriate low-pass form of a given **filter** has been synthesized, the designer must address the realization of the filter using **operational amplifiers.** If the required filter is not low-pass but high-pass, bandpass, or bandstop, transformation of the prototype function is also required (Budak, 1974; Van Valkenburg, 1982). While a detailed treatment of the various transformations is beyond the scope of this work, most of the filter designs encountered in practice can be accomplished using the techniques given here.

When the desired filter function has been determined, the corresponding electronic circuit must be designed. Many different circuits can be used to realize any given transfer function. For purposes of this Handbook, we present several of the most popular types of realizations. Much more detailed information on various circuit realizations and the advantages of each may be found in the literature, in particular Van Valkenburg (1982), Huelseman and Allen (1980), and Chen (1986). Generally the design trade-offs in making the choice of circuit to be used for the realization involve considerations of the number of elements required, the sensitivity of the circuit to changes in component values, and the ease of tuning the circuit to given specifications. Accordingly, limited information is included about these characteristics of the example circuits in this section.

Each of the circuits described here is commonly used in the realization of **active filters.** When implemented as shown and used in the appropriate gain and bandwidth specifications of the amplifiers, they will provide excellent performance. Computer-aided filter design programs are available which simplify the process of obtaining proper element values and simulation of the resulting circuits (Krobe et al., 1989; Wilamowski et al., 1992).

Transformation from Low-Pass to Other Filter Types

To obtain a high-pass, bandpass, or bandstop filter function from a low-pass prototype, one of two general methods can be used. In one of these, the circuit is realized and then individual circuit elements are replaced by other elements or subcircuits. This method is more useful in **passive filter** designs and is not discussed further here. In the other approach, the transfer function of the low-pass prototype is transformed into the required form for the desired filter. Then a circuit is chosen to realize the new filter function. We give a brief description of the transformation in this section, then give examples of circuit realizations in the following sections.

Low-Pass to High-Pass Transformation

Suppose the desired filter is, for example, a high-pass Butterworth. Begin with the low-pass Butterworth transfer function of the desired order and then *transform* each pole of the original function using the formula:

$$\frac{1}{S - S_j} \rightarrow \frac{Hs}{s - s_j} \tag{8.40}$$

which results in one complex pole and one zero at the origin for each pole in the original function. Similarly, each zero of the original function is transformed using the formula:

$$S - S_j \rightarrow \frac{s - s_j}{Hs} \tag{8.41}$$

which results in one zero on the imaginary axis and one pole at the origin. In both equations, the scaling factors used are

$$H = \frac{1}{S_j} \quad \text{and} \quad s_j = \frac{\omega_0}{S_j} \tag{8.42}$$

where ω_0 is the desired cut-off frequency in radians per second.

Low-Pass to Bandpass Transformation

Begin with the low-pass prototype function in factored, or *pole-zero*, form. Then each pole is transformed using the formula:

$$\frac{1}{S - S_j} \rightarrow \frac{Hs}{(s - s_1)(s - s_2)} \tag{8.43}$$

resulting in one zero at the origin and two conjugate poles. Each zero is transformed using the formula:

$$S - S_j \rightarrow \frac{(s - s_1)(s - s_2)}{Hs} \tag{8.44}$$

resulting in one pole at origin and two conjugate zeros. In Equation (8.43) and Equation (8.44):

$$H = -B;\ s_{1,2} = \omega_c\left(\alpha \pm \sqrt{\alpha^2 - 1}\right);\ \text{and}\ \alpha = \frac{BS_j}{2\omega_c} \tag{8.45}$$

where ω_c is the center frequency and B is the bandwidth of the bandpass function.

Low-Pass to Bandstop Transformation

Begin with the low-pass prototype function in factored, or pole-zero, form. Then each pole is transformed using the formula:

$$\frac{1}{S - S_j} \rightarrow \frac{H(s - s_1)(s - s_2)}{(s - s_3)(s - s_4)} \tag{8.46}$$

transforming each pole into two zeros on the imaginary axis and into two conjugate poles. Similarly, each zero is transformed into two poles on the imaginary axis and into two conjugate zeros using the formula:

$$S - S_j \rightarrow \frac{(s - s_3)(s - s_4)}{H(s - s_1)(s - s_2)} \tag{8.47}$$

where

$$H = \frac{1}{S_j};\ s_{1,2} = \pm j\omega_c;\ s_{3,4} = \omega_c\left(\beta \pm \sqrt{\beta^2 - 1}\right);\ \text{and}\ \beta = \frac{B}{2\omega_c S_j} \tag{8.48}$$

Once the desired transfer function has been obtained through obtaining the appropriate low-pass prototype and transformation, if necessary, to the associated high-pass, bandpass, or bandstop function, all that remains is to obtain a circuit and the element values to realize the transfer function.

Circuit Realizations

Various electronic circuits can be found to implement any given transfer function. Cascade filters and ladder filters are two of the basic approaches for obtaining a practical circuit. Cascade realizations are much easier to find and to tune, but ladder filters are less sensitive to element variations. In cascade realizations, the transfer function is simply factored into first- and second-order parts. Circuits are built for the individual parts and then cascaded to produce the overall filter. For simple to moderately complex filter designs, this is the most common method, and the remainder of this section is devoted to several examples of the circuits used to obtain the first- and second-order filters. For very high-order transfer functions, ladder filters should be considered, and further information can be obtained by consulting the literature.

In order to simplify the circuit synthesis procedure, very often ω_0 is assumed to be equal to one and then after a circuit is found, the values of all capacitances in the circuit are divided by ω_0. In general, the following magnitude and frequency transformations are allowed:

$$R_{\text{new}} = K_{\text{M}} R_{\text{old}} \text{ and } C_{\text{new}} = \frac{1}{K_{\text{F}} K_{\text{M}}} C_{\text{old}} \tag{8.49}$$

where K_{M} and K_{F} are magnitude and frequency scaling factors, respectively.

Cascade filter designs require the transfer function to be expressed as a product of first- and second-order terms. For each of these terms a practical circuit can be implemented. Examples of these circuits are presented in Figure 8.13 to Figure 8.23. In general the following first- and second-order terms can be distinguished:

(a) First-order low-pass:

$$T(s) = \frac{H\omega_0}{s + \omega_0}$$

Assumption: $r_1 = 1$

$$c_1 = \frac{1}{\omega_0} \qquad r_2 = |H|\omega_0$$

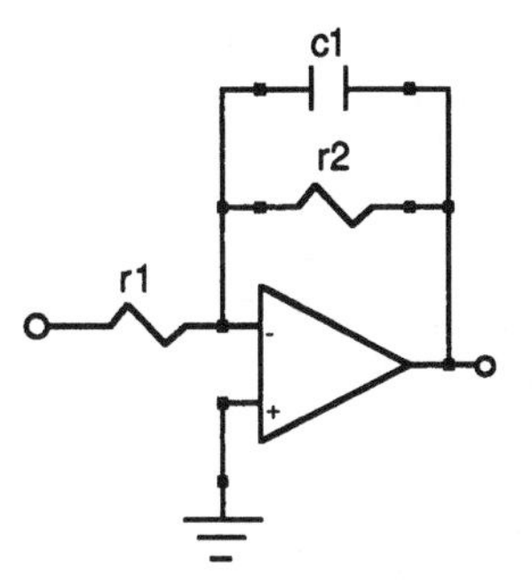

FIGURE 8.13 First-order low-pass filter.

This filter is inverting, i.e., H must be negative, and the scaling factors shown in Equation (8.49) should be used to obtain reasonable values for the components.

(b) First-order high-pass:

$$T(s) = \frac{Hs}{s + \omega_0}$$

Assumption: $r_1 = 1$

$$c_1 = \frac{1}{\omega_0} \qquad r_2 = |H|$$

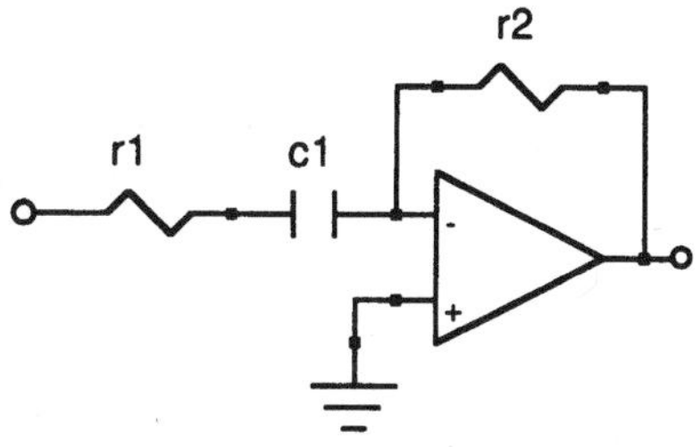

FIGURE 8.14 First-order high-pass filter.

This filter is inverting, i.e., H must be negative, and the scaling factors shown in Equation (8.49) should be used to obtain reasonable values for the components.

While several passive realizations of first-order filters are possible (low-pass, high-pass, and lead-lag), the active circuits shown here are inexpensive and avoid any loading of the other filter sections when the individual circuits are cascaded. Consequently, these circuits are preferred unless there is some reason to avoid the use of the additional operational amplifier. Note that a second-order filter can be realized using one operational amplifer as shown in the following paragraphs, so it is common practice to choose even-order transfer functions, thus avoiding the use of any first-order filters.

(c) There are several second-order low-pass circuits:

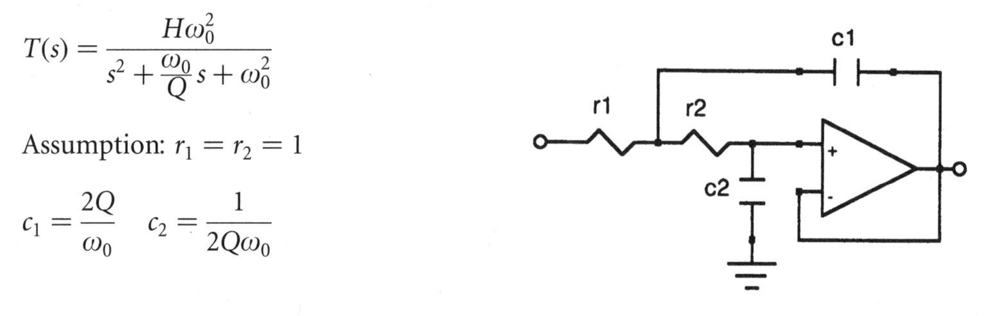

$$T(s) = \frac{H\omega_0^2}{s^2 + \frac{\omega_0}{Q}s + \omega_0^2}$$

Assumption: $r_1 = r_2 = 1$

$$c_1 = \frac{2Q}{\omega_0} \quad c_2 = \frac{1}{2Q\omega_0}$$

FIGURE 8.15 Second-order low-pass Sallen–Key filter.

This filter is noninverting and unity gain, i.e., H must be one, and the scaling factors shown in Equation (8.49) should be used to obtain reasonable element values. This is a very popular filter for realizing second-order functions because it uses a minimum number of components and since the operation amplifier is in the unity gain configuration it has very good bandwidth.

Another useful configuration for second-order low-pass filters uses the operational amplifier in its inverting "infinite gain" mode as shown in Figure 8.16.

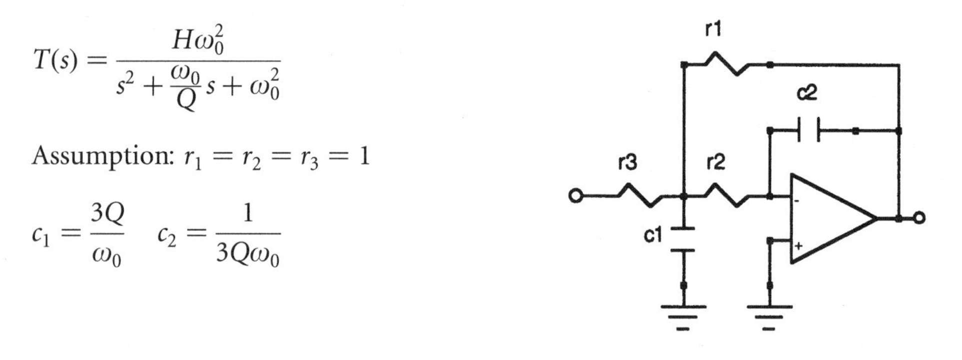

$$T(s) = \frac{H\omega_0^2}{s^2 + \frac{\omega_0}{Q}s + \omega_0^2}$$

Assumption: $r_1 = r_2 = r_3 = 1$

$$c_1 = \frac{3Q}{\omega_0} \quad c_2 = \frac{1}{3Q\omega_0}$$

FIGURE 8.16 Second-order low-pass filter using the inverting circuit.

This circuit has the advantage of relatively low sensitivity of ω_0 and Q to variations in component values. In this configuration the operational amplifier's gain–bandwidth product may become a limitation for high-Q and high-frequency applications (Budak, 1974). There are several other circuit configurations for low-pass filters. The references given at the end of the section will guide the designer to alternatives and the advantages of each.

(d) Second-order high-pass filters may be designed using circuits very much like those shown for the low-pass realizations. For example, the Sallen–Key low-pass filter is shown in Figure 8.17.

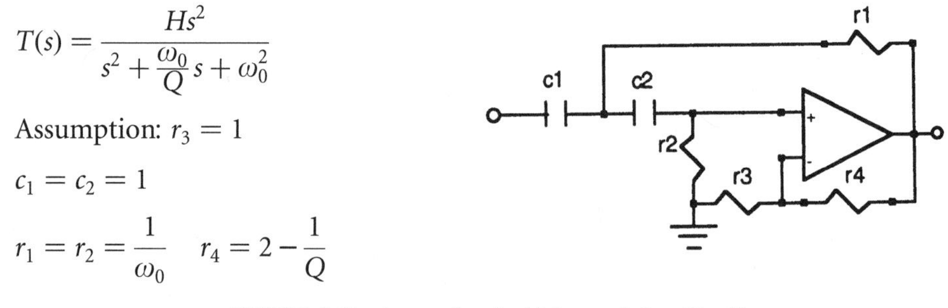

$$T(s) = \frac{Hs^2}{s^2 + \frac{\omega_0}{Q}s + \omega_0^2}$$

Assumption: $r_3 = 1$

$$c_1 = c_2 = 1$$

$$r_1 = r_2 = \frac{1}{\omega_0} \quad r_4 = 2 - \frac{1}{Q}$$

FIGURE 8.17 A second-order high-pass Sallen–Key filter.

As in the case of the low-pass Sallen–Key filter, this circuit is noninverting and requires very little gain from the operational amplifier. For low to moderate values of Q, the **sensitivity functions** are reasonable and the circuit performs well.

The inverting *infinite gain* high-pass circuit is shown in Figure 8.18 and is similar to the corresponding low-pass circuit:

$$T(s) = \frac{Hs^2}{s^2 + \frac{\omega_0}{Q}s + \omega_0^2}$$

Assumption: $r_1 = 1$

$$r_2 = 9Q^2 \quad c_1 = c_2 = c_3 = \frac{1}{3Q^2}$$

FIGURE 8.18 An inverting second-order high-pass circuit.

This circuit has relatively good sensitivity figures. The principal limitation occurs with high-Q filters since this requires a wide spread of resistor values.

Both low-pass and high-pass frequency response circuits can be achieved using three operational amplifier circuits. Such circuits have some sensitivity function and tuning advantages but require far more components. These circuits are used in the sections describing bandpass and bandstop filters. The designer wanting to use the three-operational-amplifier realization for low-pass or high-pass filters can easily do this using simple modifications of the circuits shown in the following sections.

(e) Second-order bandpass circuits may be realized using only one operational amplifier. The Sallen–Key filter shown in Figure 8.19 is one such circuit:

$$T(s) = \frac{H\frac{\omega_0}{Q}s}{s^2 + \frac{\omega_0}{Q}s + \omega_0^2}$$

Assumption: $c_1 = c_2 = 1;\ r_5 = 1$

$$r_2 = r_3 = \frac{\sqrt{2}}{\omega_0} \qquad r_1 = \frac{\frac{4Q}{\sqrt{2}} - 1}{H}$$

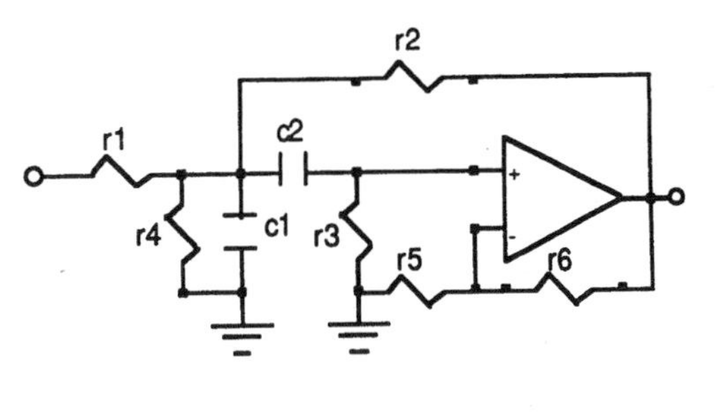

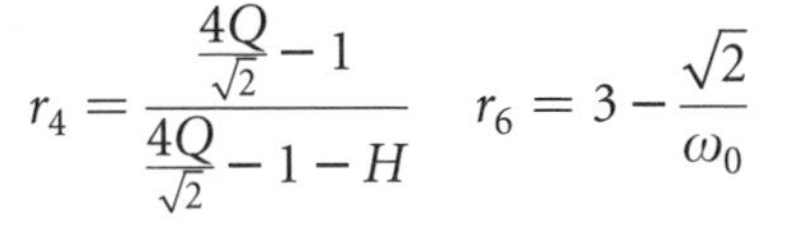

$$r_4 = \frac{\frac{4Q}{\sqrt{2}} - 1}{\frac{4Q}{\sqrt{2}} - 1 - H} \qquad r_6 = 3 - \frac{\sqrt{2}}{\omega_0}$$

FIGURE 8.19 A Sallen–Key bandpass filter.

This is a noninverting amplifier that works well for low- to moderate-Q filters and is easily tuned (Budak, 1974). For high-Q filters the sensitivity of Q to element values becomes high, and alternative circuits are recommended. One of these is the bandpass version of the inverting amplifier filter as shown in Figure 8.20.

This circuit has few components and relatively small sensitivity of ω_0 and Q to variations in element values. For high-Q circuits, the range of resistor values is quite large as r_1 and r_2 are much larger than r_3.

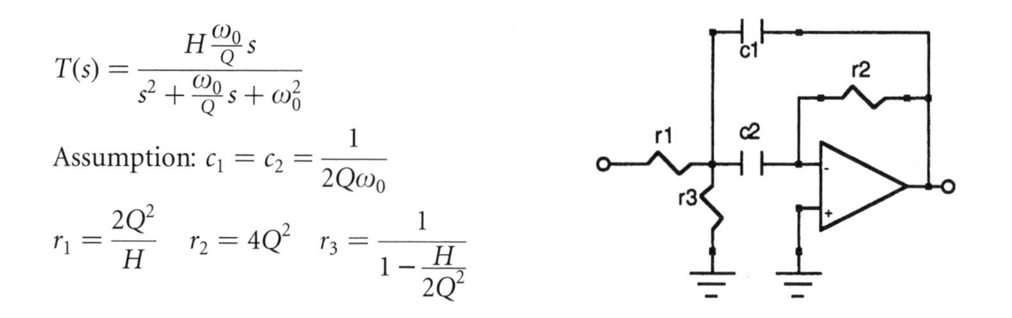

$$T(s) = \frac{H\frac{\omega_0}{Q}s}{s^2 + \frac{\omega_0}{Q}s + \omega_0^2}$$

$$\text{Assumption: } c_1 = c_2 = \frac{1}{2Q\omega_0}$$

$$r_1 = \frac{2Q^2}{H} \quad r_2 = 4Q^2 \quad r_3 = \frac{1}{1 - \frac{H}{2Q^2}}$$

FIGURE 8.20 The inverting amplifier bandpass filter.

When ease of tuning and small sensitivities are more important than the circuit complexity, the three-operational-amplifier circuit of Figure 8.21 may be used to implement the bandpass transfer function.

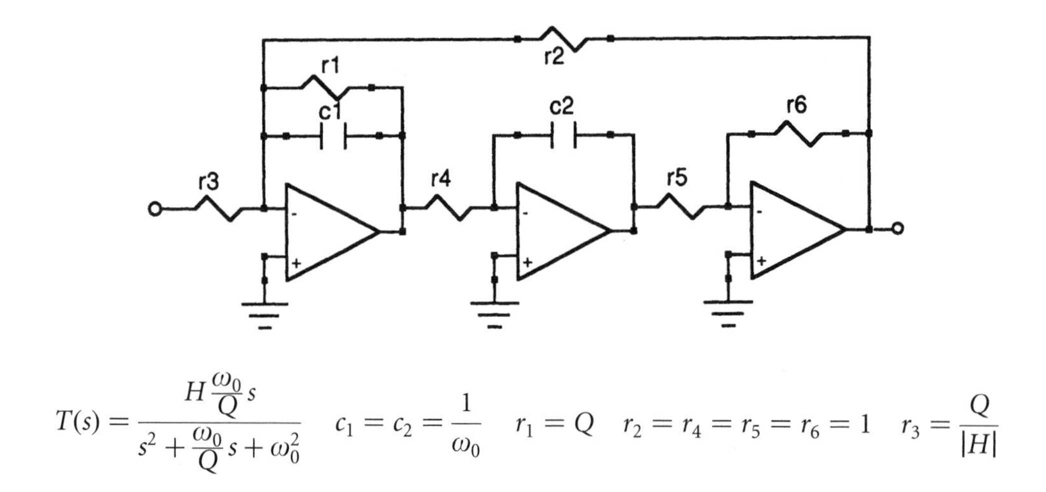

$$T(s) = \frac{H\frac{\omega_0}{Q}s}{s^2 + \frac{\omega_0}{Q}s + \omega_0^2} \quad c_1 = c_2 = \frac{1}{\omega_0} \quad r_1 = Q \quad r_2 = r_4 = r_5 = r_6 = 1 \quad r_3 = \frac{Q}{|H|}$$

FIGURE 8.21 The three-operational-amplifier bandpass filter.

The filter as shown in Figure 8.21 is inverting. For a noninverting realization, simply take the output from the middle amplifier rather than the right one. This same configuration can be used for a three-operational-amplifier low-pass filter by putting the input into the summing junction of the middle amplifier and taking the output from the left operational amplifier. Note that Q may be changed in this circuit by varying r_1 and that this will not alter ω_0. Similarly, ω_0 can be adjusted by varying c_1 or c_2 and this will not change Q. If only variable resistors are to be used, the filter can be tuned by setting ω_0 using any of the resistors other than r_1 and then setting Q using r_1.

(f) Second-order bandstop filters are very useful in rejecting unwanted signals such as line noise or carrier frequencies in instrumentation applications. Such filters are implemented with methods very similar to the bandpass filters just discussed. In most cases, the frequency of the zeros is to be the same as the frequency of the poles. For this application, the circuit shown in Figure 8.22 can be used.

$$T(s) = \frac{H(s^2 + \omega_z^2)}{s^2 + \frac{\omega_0}{Q} s + \omega_0^2}$$

Assumption: $c_1 = c_2 = 1$

$$r_1 = \frac{1}{2Q\omega_0} \quad r_3 = \frac{1}{Q\omega_0} \quad r_2 = r_4 = \frac{2Q}{\omega_0}$$

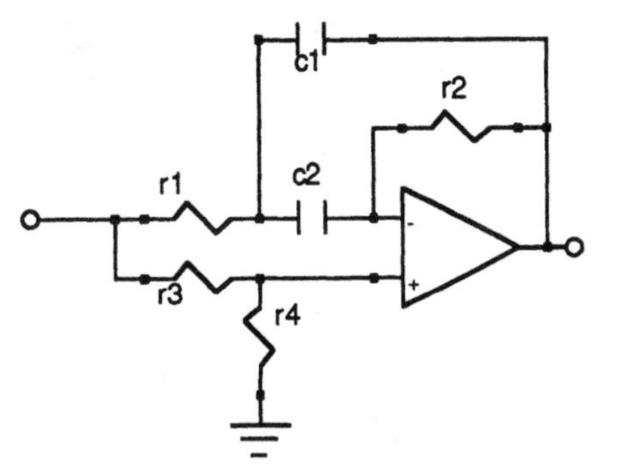

FIGURE 8.22 A single operational-amplifier bandstop filter.

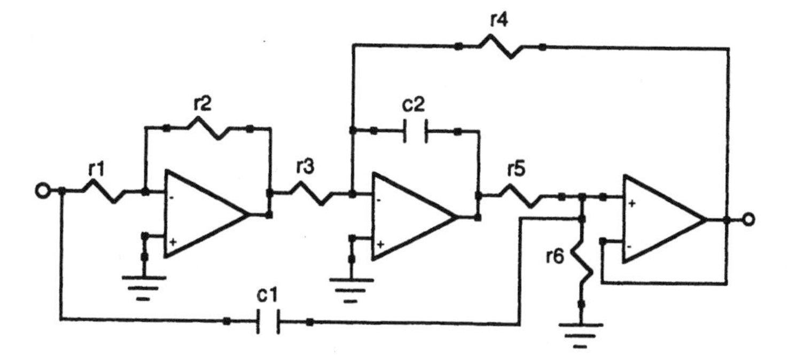

$$T(s) = \frac{H(s^2 + \omega_z^2)}{s^2 + \frac{\omega_0}{Q} s + \omega_0^2} \quad c_1 = c_2 = \frac{1}{\omega_0} \quad r_1 = 1 \quad r_2 = H \quad r_5 = r_6 = 2Q \quad r_3 = \frac{H\omega_0^2}{2Q\omega_z^2} \quad r_4 = \frac{1}{2Q}$$

FIGURE 8.23 A three-operational-amplifier bandstop filter.

The primary advantage of this circuit is that it requires a minimum number of components. For applications where no tuning is required and the Q is low, this circuit works very well. When the bandstop filter must be tuned, the three-operational-amplifier circuit is preferable.

The foregoing circuits provide a variety of useful first- and second-order filters. For higher-order filters, these sections are simply cascaded to realize the overall transfer function desired. Additional details about these circuits as well as other circuits used for active filters may be found in the references.

Defining Terms

Active filter: A filter circuit that uses active components, usually operational amplifiers.

Filter: A circuit that is designed to be frequency selective. That is, the circuit will emphasize or "pass" certain frequencies and attenuate or "stop" others.

Operational amplifier: A very high-gain differential amplifier used in active filter circuits and many other applications. These monolithic integrated circuits typically have such high gain, high input impedance, and low output impedance that they can be considered "ideal" when used in active filters.

Passive filter: A filter circuit that uses only passive components, i.e., resistors, inductors, and capacitors. These circuits are useful at higher frequencies and as prototypes for ladder filters that are active.

Sensitivity function: A measure of the fractional change in some circuit characteristic, such as center frequency, to variations in a circuit parameter, such as the value of a resistor. The sensitivity function is normally defined as the partial derivative of the desired circuit characteristic with respect to the element value and is usually evaluated at the nominal value of all elements.

References

A. Budak, *Passive and Active Network Analysis and Synthesis*, Boston, MA: Houghton Mifflin, 1974.

W.K. Chen, *Passive and Active Filters, Theory and Implementations*, New York: Wiley, 1986.

L.P. Huelseman and P.E. Allen, *Introduction to the Theory and Design of Active Filters*, New York: McGraw-Hill, 1980.

M.R. Krobe, J. Ramirez-Angulo, and E. Sanchez-Sinencio, "FIESTA—A filter educational synthesis teaching aid," *IEEE Trans. on Education*, vol. 12, no. 3, pp. 280–286, August 1989.

M.E. Van Valkenburg, *Analog Filter Design*, New York: Holt, Rinehart and Winston, 1982.

B.M. Wilamowski, S.F. Legowski, and J.W. Steadman, "Personal computer support for teaching analog filter analysis and design," *IEEE Trans. on Education*, vol. 35, no. 4, November 1992.

Further Information

The monthly journal *IEEE Transactions on Circuits and Systems* is one of the best sources of information on new active filter functions and associated circuits.

The British journal *Electronics Letters* also often publishes articles about active circuits.

The *IEEE Transactions on Education* has carried articles on innovative approaches to active filter synthesis as well as computer programs for assisting in the design of active filters.

9

Power Electronics

Andrew Marshall
Texas Instruments Incorporated

Kaushik Rajashekara
Delphi Energy & Engineering Management System

Ashoka K.S. Bhat
University of Victoria

Bimal K. Bose
University of Tennessee

Sudip. K. Mazumder
University of Illinois

Tirthajyoti Sarkar
University of Illinois

Mitra Dutta
University of Illinois

Michael S. Mazzola
Mississippi State University

Ayse E. Amac
Illinois Institute of Technology

Ali Emadi
Illinois Institute of Technology

9.1 Power Semiconductor Devices

Andrew Marshall

The era of semiconductor power devices began with the availability of thyristors in the late 1950s. Now, there are several types of power devices available for high-voltage, high-current, and high-power applications. These are used for power conversion and motor control systems. Particularly noteworthy are gate turn-off thyristors, Darlington transistors, power MOSFETs, and insulated-gate bipolar transistors (IGBTs) (Bird and King, 1984). This section reviews the basic characteristics of the most common of these.

Thyristor

The thyristor, also called a silicon-controlled rectifier (SCR), is a four-layer three-junction pnpn device. It has three terminals: anode, cathode, and gate (Figure 9.1), and is turned on by applying a short current pulse into

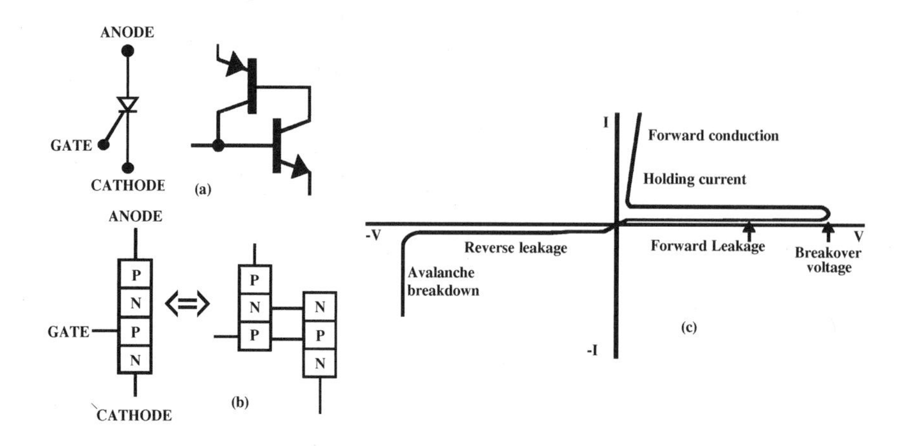

FIGURE 9.1 (a) Thyristor symbol and transistor equivalent; (b) physical junction arrangement; (c) volt–ampere characteristics.

the gate. Once the device turns on, the gate loses its control to turn off the device. Turn-off is achieved by applying a zero or reverse voltage across the anode and cathode.

Some thyristors include what is termed an amplifying gate. In this configuration a small auxiliary thyristor is placed in parallel with the power SCR. The two devices' anodes are common and the cathode of the auxiliary device is connected to the gate of the power device. This allows the power device to be switched on with substantially reduced current, at the cost of slightly higher voltage.

In cases where electrical isolation is required between the power supply and gate control, a light-activated SCR may be considered. In these applications a light pulse onto the thyristor triggers it. From an electrical schematic point of view this appears as a thyristor with its gate connected to its anode through a light-sensitive resistor or diode.

There are two major classifications of thyristor: converter grade and inverter grade. Converter-grade thyristors are slow (turn-off measured in tens of milliseconds) and are used in natural commutation (or phase-controlled) applications. Inverter-grade thyristors are used in forced commutation applications such as dc–dc choppers and dc–ac inverters, where faster turn-off is required. Inverter-grade thyristors may be turned off by forcing current to zero using an external commutation circuit. While providing improved switching times, this results in circuit losses and increased circuit complexity.

Thyristors have high di/dt and dv/dt capabilities. These are, respectively, the maximum permitted rate of change of current through a device and rate of change of voltage across the device beyond which failure may occur. Thyristors are available that can stand off up to around 6000 V, and with current ratings of several thousand amperes. The on-state forward voltage drop in thyristors (the voltage across the device when the anode is positive with respect to the cathode) is about 1.5 to 2 V, and this does not increase much even at very high currents. While the forward voltage determines the on-state power loss of the device at any given current, the switching power loss becomes dominant at high frequencies. Because of this, the maximum switching frequencies possible using thyristors are limited in comparison with other power devices considered in this section. In many applications protection against excessive switching rates is required to prevent thyristor turn-on without a gate pulse, which can occur as a result of capacitive coupling from anode to gate. In dc-to-ac conversion applications it is necessary to use an antiparallel diode of similar rating across each power thyristor to protect the thyristor from reverse damage.

Gate Turn-Off Thyristor (GTO)

The GTO is a power switching device that can be turned on by a short pulse of gate current and turned off by a reverse gate pulse. This reverse gate current amplitude is dependent on the anode current being turned off; hence, there is no need for an external commutation circuit to turn it off. Because turn-off is provided by bypassing carriers directly to the gate circuit, its turn-off time is short, providing higher-frequency operation than thyristors. The GTO symbol and turn-off characteristics are shown in Figure 9.2.

For reliable operation of GTOs, proper design of the gate turn-off circuit and the snubber circuit are critical. AGTO has a poor turn-off current gain of the order of 4 to 5. Thus, a 1000-A peak current GTO may require 200 A of reverse gate current. GTO devices also have an increased tendency to latch at temperatures above 125°C. GTOs are available with similar ratings to SCRs.

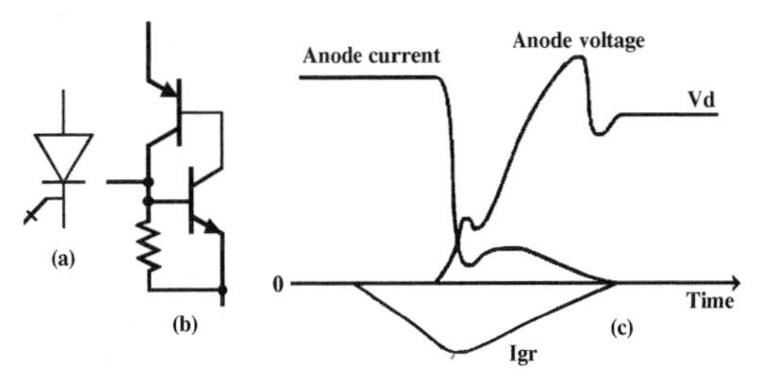

FIGURE 9.2 (a) GTO symbol; (b) transistor equivalent for the GTO; (c) turn-off characteristics. (*Source:* B.K. Bose, *Modern Power Electronics: Evaluation, Technology, and Applications*, p. 5. © 1992 IEEE).

Triac

The triac (Figure 9.3) is functionally a pair of thyristors connected in antiparallel. The result is a two-directional switch that can conduct current in both directions. It is turned on with a positive or negative voltage, and turns off when the current through the device drops below a threshold value. Because of the integration, the triac has poor gate current sensitivity at turn-on and a longer turn-off time than thyristors. Also triacs have a poor capability to withstand reapplied dv/dt (voltage appearing across the device during turn-off, due usually to inductive commutation). Triacs are mainly used in low-frequency phase control applications, such as in ac regulators for lighting, fan control, and in solid-state ac relays. Triacs are not available to the same ratings as thyristors, having typical ratings of up to about 1000 V and a few tens of amperes.

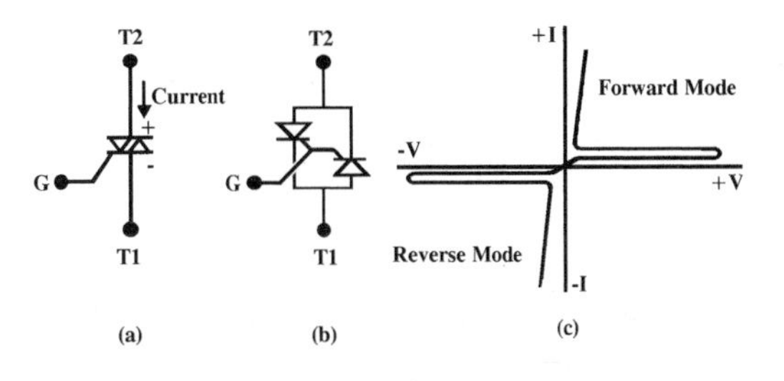

FIGURE 9.3 (a) Triac symbol; (b) thyristor equivalent circuit of Triac; (c) voltage–current characteristics.

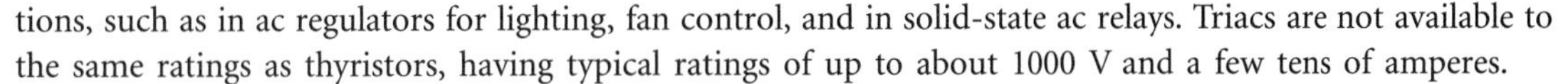

Reverse-Conducting Thyristor (RCT) and Asymmetrical Silicon-Controlled Rectifier (ASCR)

It is usual in power inverter applications to add an antiparallel diode between the thyristor anode and cathode for commutation/freewheeling purposes (Figure 9.4). In RCTs, the diode is integrated with a fast switching thyristor in a single silicon chip, and thus component count may be reduced. This integration improves the static and dynamic characteristics as well as the overall circuit performance. RCTs are usually designed for specific applications such as motor drivers. The antiparallel diode limits the reverse voltage across the thyristor (the voltage across the device when the anode is negative with respect to the cathode) to between 1 and 2 V. However, because of the reverse recovery behavior of the integrated diodes, the thyristor may see very high reapplied dv/dt when the diode recovers from its reverse voltage (Whitaker, 1998).

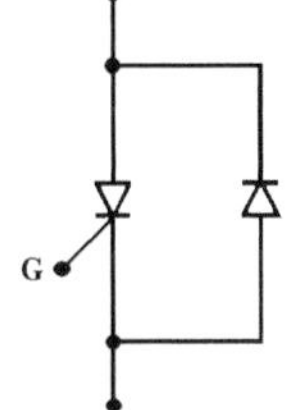

FIGURE 9.4 Schematic of the RCT showing how it is formed from the basic thyristor and diode.

The ASCR has a similar forward blocking capability as a power inverter-grade thyristor, but it has a limited reverse blocking capability of about 20 to 30 V. It has a reduced on-state voltage drop and faster turn-off rate than a thyristor of similar rating.

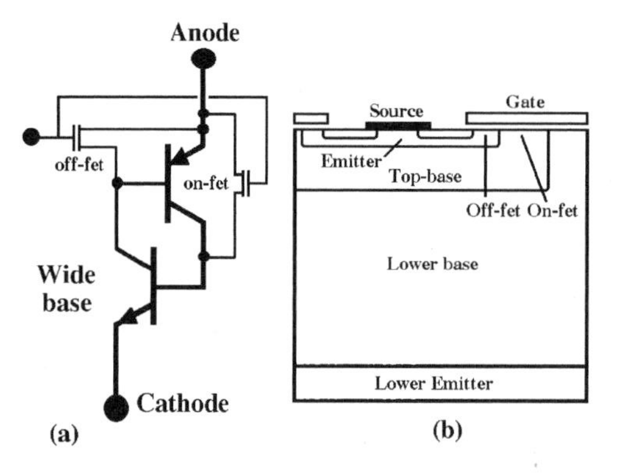

FIGURE 9.5 MOS controlled thyristor: (a) equivalent circuit and (b) device cross-section.

MOS-Controlled Thyristor (MCT)

The MCT combines the benefits of thyristor current density with MOS gated turn-on and turn-off. It is a rugged high-power, high-frequency component with low conduction drop when used in medium- to high-power applications. A cross-sectional structure of a *p*-type MCT with its circuit schematic is shown in Figure 9.5. Most MCTs are constructed of multiple parallel cells. The device is turned on by a negative voltage pulse at the gate with respect to the anode, and is turned off by a positive voltage pulse.

Compared to IGBTs, MCTs have a lower forward voltage drop and improved reverse bias safe operating area and switching speed. The MCT is capable of high current densities and blocking voltages in both directions. An MCT has high di/dt (of the order of 2500 A/μsec) and high dv/dt (of the order of 20,000 V/μsec) capability. It is often a good choice for applications such as motor drives, uninterrupted power supplies, and high-power active power line conditioners (Temple, 1989).

Power Transistors and Darlingtons

Power transistors are used in applications ranging up to several hundred kilowatts, at voltages up to a few kilovolts, and switching frequencies up to about 10 kHz. Power transistors used in power conversion applications are generally npn-type for reasons of efficiency, processing simplicity, and cost. The power transistor is turned on by supplying current to the base, which has to be maintained throughout the conduction period. It is turned off by removing the base drive and may be speeded up by making the base voltage slightly negative. This permits the more rapid removal of charge from the base region. The saturation voltage of the device is normally 0.5 to 2.5 V and increases with current; hence the on-state losses increase more than proportionately with current. In addition, power transistors can block only forward voltages, with reverse peak voltage rating of these devices only a few volts.

Because of their relatively long switching times, switching losses are significant, limiting the maximum operating frequency. During switching, the reverse-biased collector junction may show hot-spot breakdown (where current flows preferentially through one part of the device, causing it to heat up) due to exceeding the safe operating area (SOA). Bipolar devices have highly interdigitated emitter-base geometries to force more uniform current distribution and improve the SOA.

To reduce high base current requirements, Darlington configurations are commonly used. They are available as monolithic and multichip versions. The basic Darlington configuration (Figure 9.6) can considerably increase the current switched by the transistor for a given base drive, although the VCE(sat) for the Darlington is generally higher than that of a single transistor of similar rating with corresponding increase in on-state power loss.

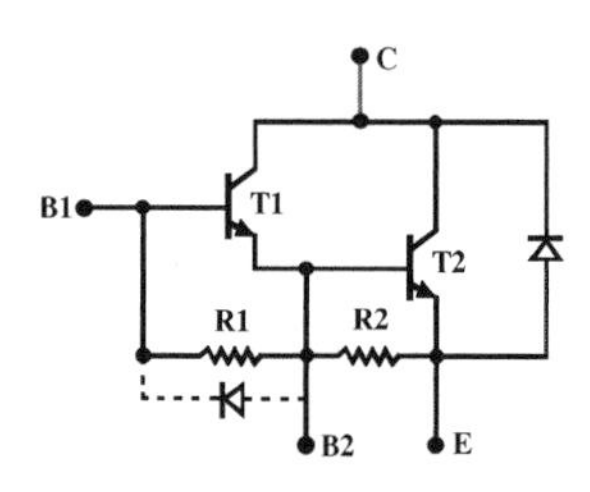

FIGURE 9.6 Schematic of Darlington, including turn-off and anti-parallel diode options.

Some Darlington devices use a high-gain T1 and low-gain T2 device to improve SOA. The improved SOA results from an early nondestructive BVcbo breakdown of T1, which turns on T2, into a robust FBSOA conduction state before RBSOA of T2 can occur. Faster turn-off is provided by the integration of a diode from what is often an internal B2 to external B1 node. Thus, for turn-off of a Darlington with integrated speed-up diode the base must be switched more than a Vd below ground to take advantage of the feature.

Power MOSFET

Various types of power MOSFETs are available with differences in internal geometry; examples include the VDMOS, HEXFET, SIPMOS, and TMOS. All power MOSFETs are voltage-driven rather than current-driven devices, unlike bipolar transistors. They are also the most common power device type used in power ICs, although most discrete power MOSFETs use the bottom of the wafer as the drain contact, whereas integrated circuit power MOSFETs bring all the device's terminals out to the top surface. MOSFETs are available with ratings up to about 1000 V and a few hundred watts.

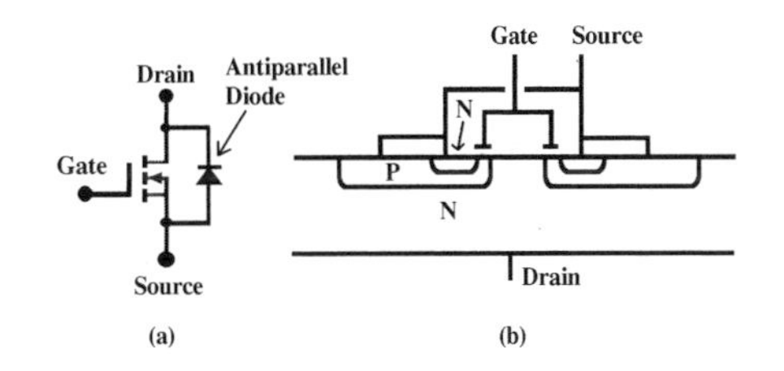

FIGURE 9.7 (a) Power MOSFET schematic, and (b) cross-section of vertical DMOS (VDMOS) device.

As with conventional MOSFETS, the gate of a power MOSFET is isolated electrically from the source by an insulating layer, often of silicon dioxide. The gate draws only a very small leakage current of the order of nanoamperes. Hence, the gate drive circuit is simple and dc power loss is minimal. Under switching conditions, however, capacitive losses occur as gate-to-source, and gate-to-drain capacitances have to be charged and discharged. The circuit symbol and silicon cross-section of a DMOS-type power MOSFET is shown in Figure 9.7.

Power MOSFETs are majority carrier devices and display no minority carrier storage time, leading to fast rise and fall times. They behave as resistive devices when turned on. At low currents a power MOSFET may have a lower conduction loss than a comparable bipolar device, but at higher currents, the conduction losses exceed those of bipolars. In addition, the Rds (on) increases with temperature, where bipolar performance may not necessarily increase with temperature, dependent upon current density.

An important feature of a power MOSFET is the absence of a secondary breakdown effect. As a result it has a more rugged switching performance than a bipolar device. In MOSFETs, Rds(on) increases with temperature, and thus current is automatically diverted away from hot spots, providing an added level of SOA stability. The drain body junction appears as an anti-parallel diode between source and drain. Thus, power MOSFETs will not support voltage in the reverse direction. Although this inverse diode is relatively fast, it is slow by comparison with the MOSFET. In high-performance commutating systems, where the diode may be forced to conduct, an external, fast recovery diode is often placed in parallel with the internal diode to achieve the circuit performance required.

If a MOSFET is operating within its specification range at all times, its chances of failing catastrophically are minimal. However, if its absolute maximum rating is exceeded, failure probability increases dramatically. Under actual operating conditions, a MOSFET may be subjected to transients—either externally from the power bus supplying the circuit or from the circuit itself due, for example, to inductive kicks going beyond the absolute maximum ratings. A combination of transistor and circuit design provide the ruggedness required for a MOSFET to operate in an environment of dynamic electrical stresses without activating any of the parasitic bipolar junction transistors inherent to most MOSFETs.

Insulated-Gate Bipolar Transistor (IGBT)

The IGBT has the high input impedance and high-speed characteristics of a MOSFET with the conductivity characteristic (low saturation voltage) of a bipolar transistor. The IGBT is turned on by applying a positive voltage between the gate and emitter and, as in the MOSFET, it is turned off by making the gate signal zero or slightly negative. The IGBT has a much lower voltage drop than a MOSFET of similar ratings. For a given IGBT, there is a critical value of collector current that will cause a large enough voltage drop to activate the thyristor. Hence, the peak allowable collector current that can flow without latch-up occurring is a characteristic which is specified in the datasheet, and must be adhered to in the circuit design. There is also a corresponding gate source voltage that permits this current to flow, which also should not be exceeded.

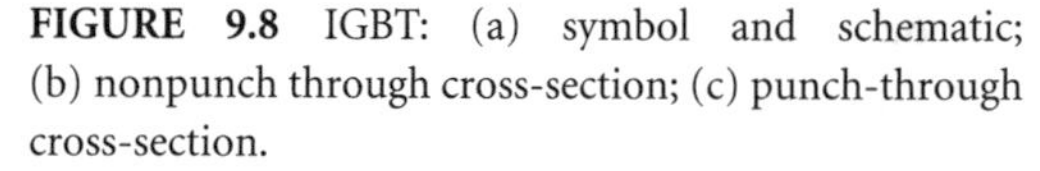

FIGURE 9.8 IGBT: (a) symbol and schematic; (b) nonpunch through cross-section; (c) punch-through cross-section.

Like the power MOSFET, the IGBT does not exhibit the secondary breakdown phenomenon common to bipolar transistors. However, care should be taken not to exceed the maximum power dissipation and specified maximum junction temperature of the device under all conditions for guaranteed reliable operation. The on-state voltage of the IGBT is heavily dependent on the gate voltage. To obtain a low on-state voltage, a sufficiently high gate voltage must be applied.

In general, IGBTs can be classified into punch-through (PT) and nonpunch-through (NPT) structures, as shown in Figure 9.8. In the PT IGBT, an N+ buffer layer is normally introduced between the P+ substrate and the N− epitaxial layer, so that the whole N− drift region is depleted when the device is blocking the off-state voltage, and the electrical field shape inside the N− drift region is close to a rectangular shape. Because a shorter N− region can be used in the PT IGBT, a better trade-off between the forward voltage drop and turn-off time can be achieved. PT IGBTs are available up to about 1200 V. High-voltage IGBTs normally use NPT processing, and are built on N− substrates which serve as the N− base drift region. NPT IGBTs of up to about 4 kV have been reported. NPT IGBTs are more robust than PT IGBTs, particularly under short-circuit conditions, but have a higher forward voltage drop.

The PT IGBTs should not normally be paralleled. Factors that inhibit current sharing of parallel-connected IGBTs are: (1) on-state current unbalance, caused by Vce(sat) distribution and circuit resistance, and (2) current unbalance at turn-on and turn-off, caused by the switching time differences between the parallel connected devices and wiring inductance. NPT IGBTs may be paralleled as they have a positive temperature coefficient, and thus evenly distribute current.

Alternative Materials

Most power semiconductors are fabricated on bulk silicon (where the starting material is a large wafer of silicon cut from a monocrystalline ingot). Alternative starting materials have a variety of advantages for certain power semiconductors, including silicon carbide, silicon-on-insulator, and III–V materials.

Development continues in silicon carbide power MOSFETs and thyristors. With their wider bandgap these permit operation to greater than 300°C, with promise of excellent switching characteristics and higher and more stable blocking voltages.

SOI has been used for a number of years for logic devices, giving SOI the advantage of being readily available and well understood (Marshall, Udrea). SOI has numerous advantages and drawbacks. It is readily integratable, without DC paths that can lead to latch-up. The most common SOI power device is the lateral DMOS (LDMOS) configuration, which is similar to the VDMOS, but the drain terminal is accessed from the top surface of the silicon (Figure 9.9). SOI power transistors have the advantage of high-temperature operation and no need for isolation in integrated systems; as a result, LDMOS devices are widely used in system-on-a-chip solutions.

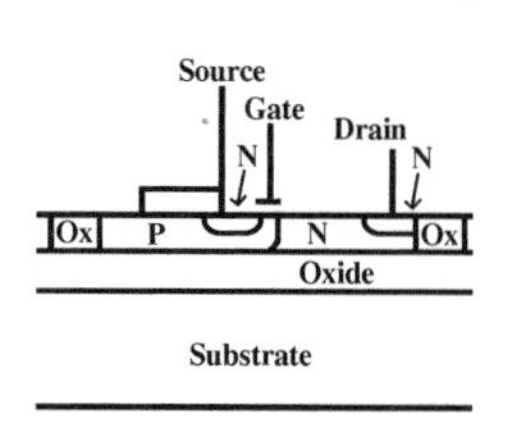

FIGURE 9.9 Lateral DMOS on SOI. Unlike conventional power MOS transistors, the drain contact is to the top of the silicon.

The most common III–V material for power devices is gallium nitride. In theory this material is capable of extremely high power, with power densities approximately five times that of silicon. However, the major difficulty has proved to be manufacturing low enough defect density GaN. Usually GaN is deposited as an epitaxial layer on SiC, sapphire, or silicon. The high electron mobility transistor (HEMT) is a popular III–V power transistor. The HEMT bypasses the need for a highly doped channel in conventional MOS devices, which slows the electron mobility. The structure is similar to a lateral MOS device, but the gate oxide is replaced with a heavily doped semiconductor region.

References

B.K. Bose, *Modern Power Electronics: Evaluation, Technology, and Applications*, New York: IEEE Press, 1992.

Harris Semiconductor, *User's Guide of MOS Controlled Thyristor.*

B.M. Bird and K.G. King, *An Introduction to Power Electronics*, New York: Wiley-Interscience, 1984.

A. Marshall and S. Natarajan, *SOI Design: Analog, Memory and Digital Techniques*, Dordrecht: Kluwer Academic Publishers, 2001.

T. Mimura et al., "A new field-effect transistor with selectively doped Gas/n–Alx Ga1–xAs heterojunctions," *Jpn J. Appl. Phys.*, 19, L225, 1980.

U.K. Mishra et al., "AlGaN/GaN HEMTs — an overview of device operation and applications," *Proc. IEEE*, vol. 90, no. 6, 2002, pp. 1022–1031.

V.A.K. Temple, "Advances in MOS controlled thyristor technology and capability," *Power Convers.*, 544–554, 1989 (October).

F. Udrea et al., "SOI power devices," *Electron. Commun. Eng. J.*, 27–40, 2000 (February).

J. Whitaker, *AC Power Systems Handbook*, 2nd ed., Boca Raton, FL: CRC Press, 1998.

9.2 Power Conversion

Kaushik Rajashekara

Power conversion deals with the process of converting electric power from one form to another. The power electronic apparatuses performing the power conversion are called *power converters.* Because they contain no moving parts, they are often referred to as *static* power converters. The power conversion is achieved using power semiconductor devices, which are used as switches. The power devices used are SCRs (silicon controlled rectifiers, or thyristors), triacs, power transistors, power MOSFETs, insulated gate bipolar transistors (IGBTs), and MCTs (MOS-controlled thyristors). The power converters are generally classified as:

1. ac–dc converters (phase-controlled converters)
2. direct ac–ac converters (cycloconverters)
3. dc–ac converters (inverters)
4. dc–dc converters (choppers, buck, and boost converters)

ac–dc Converters

The basic function of a **phase-controlled converter** is to convert an alternating voltage of variable amplitude and frequency to a variable dc voltage. The power devices used for this application are generally **SCRs**. The average value of the output voltage is controlled by varying the conduction time of the SCRs. The turn-on of the SCR is achieved by providing a gate pulse when it is forward-biased. The turn-off is achieved by the **commutation** of current from one device to another at the instant the incoming ac voltage has a higher instantaneous potential than that of the outgoing wave. Thus there is a natural tendency for current to be commutated from the outgoing to the incoming SCR, without the aid of any external commutation circuitry. This commutation process is often referred to as *natural commutation.*

A single-phase half-wave converter is shown in Figure 9.10. When the SCR is turned on at an angle α, full supply voltage (neglecting the SCR drop) is applied to the load. For a purely resistive load, during the positive

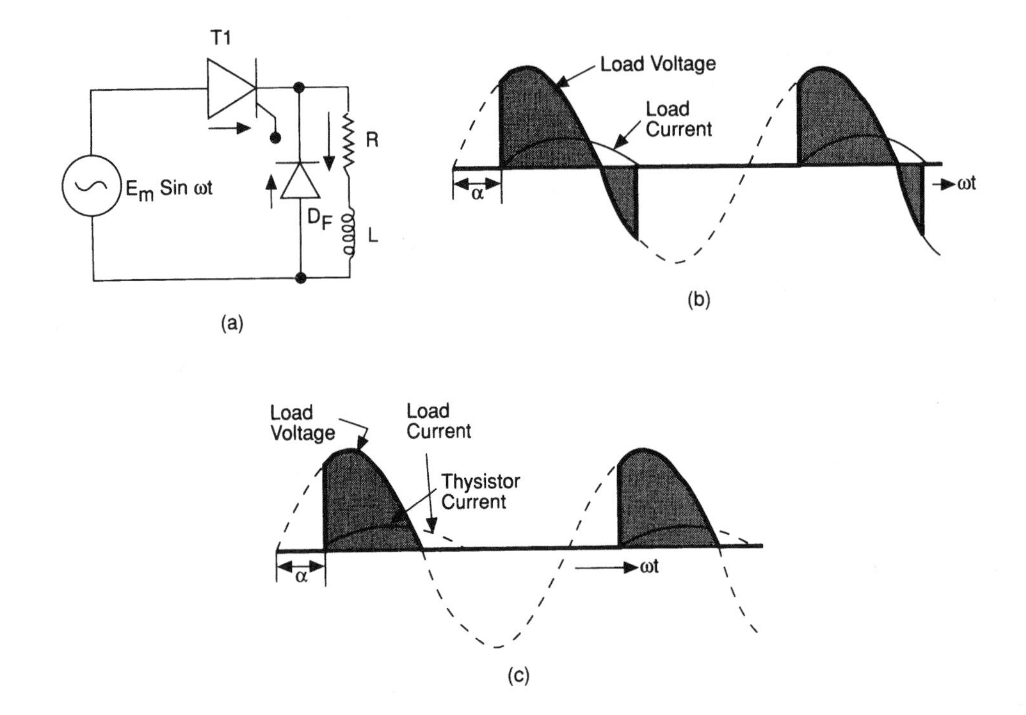

FIGURE 9.10 Single-phase half-wave converter with freewheeling diode. (a) Circuit diagram; (b) waveform for inductive load with no freewheeling diode; (c) waveform with freewheeling diode.

half cycle, the output voltage waveform follows the input ac voltage waveform. During the negative half cycle, the SCR is turned off. In the case of inductive load, the energy stored in the inductance causes the current to flow in the load circuit even after the reversal of the supply voltage, as shown in Figure 9.10(b). If there is no freewheeling diode D_F, the load current is discontinuous. A freewheeling diode is connected across the load to turn off the SCR as soon as the input voltage polarity reverses, as shown in Figure 9.11(c). When the SCR is off, the load current will freewheel through the diode. The power flows from the input to the load only when the SCR is conducting. If there is no freewheeling diode, during the negative portion of the supply voltage, SCR returns the energy stored in the load inductance to the supply. The freewheeling diode improves the input power factor.

The controlled full-wave dc output may be obtained by using either a center tap transformer (Figure 9.11) or by bridge configuration (Figure 9.12). The bridge configuration is often used when a transformer is undesirable and the magnitude of the supply voltage properly meets the load voltage requirements. The average output voltage of a single-phase full-wave converter for continuous current conduction is given by

$$v_{d\alpha} = 2\frac{E_m}{\pi}\cos\alpha$$

where E_m is the peak value of the input voltage and α is the firing angle. The output voltage of a single-phase bridge circuit is the same as that shown in Figure 9.11. Various configurations of the single-phase bridge circuit can be obtained if, instead of four SCRs, two diodes and two SCRs are used, with or without freewheeling diodes.

A three-phase full-wave converter consisting of six thyristor switches is shown in Figure 9.13(a). This is the most commonly used three-phase bridge configuration. Thyristors T_1, T_3, and T_5 are turned on during the positive half cycle of the voltages of the phases to which they are connected, and thyristors T_2, T_4, and T_6 are turned on during the negative half cycle of the phase voltages. The reference for the angle in each

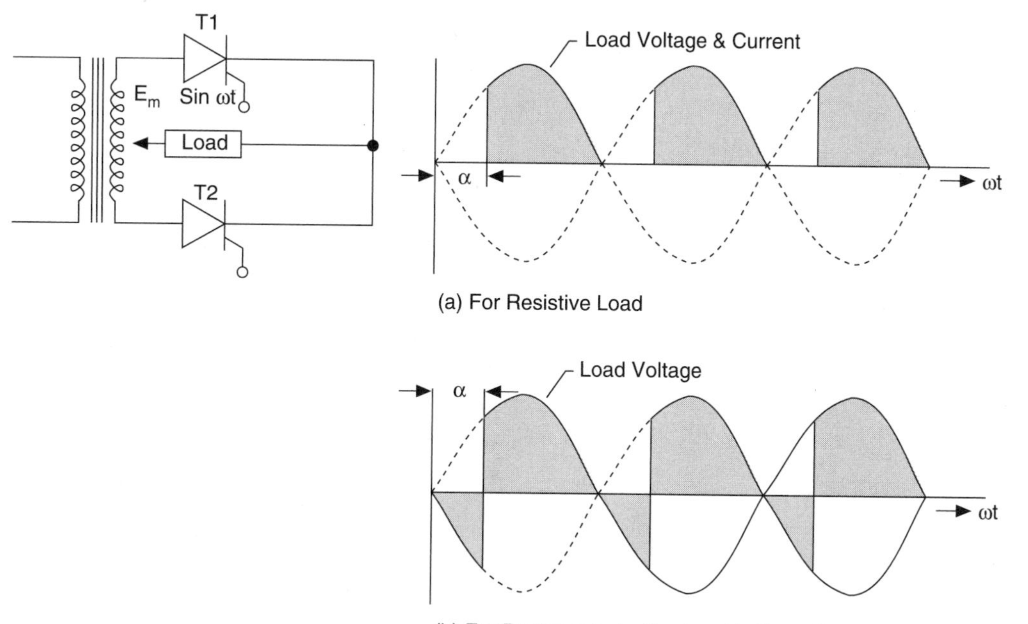

FIGURE 9.11 Single-phase full-wave converter with transformer.

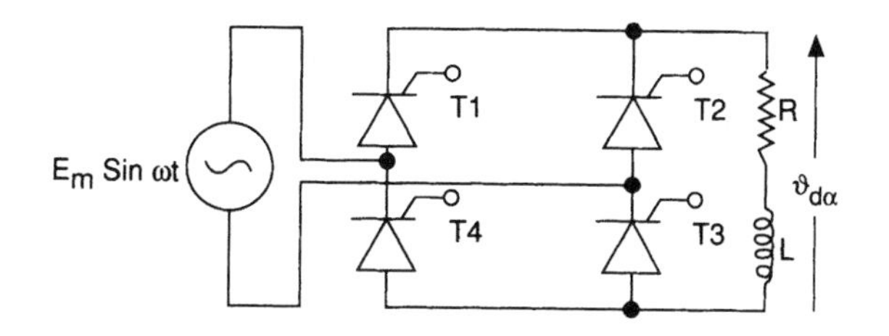

FIGURE 9.12. Single-phase bridge converter.

cycle is at the crossing points of the phase voltages. The ideal output voltage, output current, and input current waveforms are shown in Figure 9.13(b). The output dc voltage is controlled by varying the firing angle α. The average output voltage under continuous current conduction operation is given by

$$v_o = \frac{3\sqrt{3}}{\pi} E_m \cos\alpha$$

where E_m is the peak value of the phase voltage. At $\alpha = 90°$, the output voltage is zero. For $0 < \alpha < 90°$, v_o is positive and power flows from ac supply to the load. For $90° < \alpha < 180°$, v_o is negative and the converter operates in the inversion mode. If the load is a dc motor, the power can be transferred from the motor to the ac supply, a process known as *regeneration*.

In Figure 9.13(a), the top or bottom thyristors could be replaced by diodes. The resulting topology is called a *thyristor semiconverter*. With this configuration, the input power factor is improved, but the regeneration is not possible.

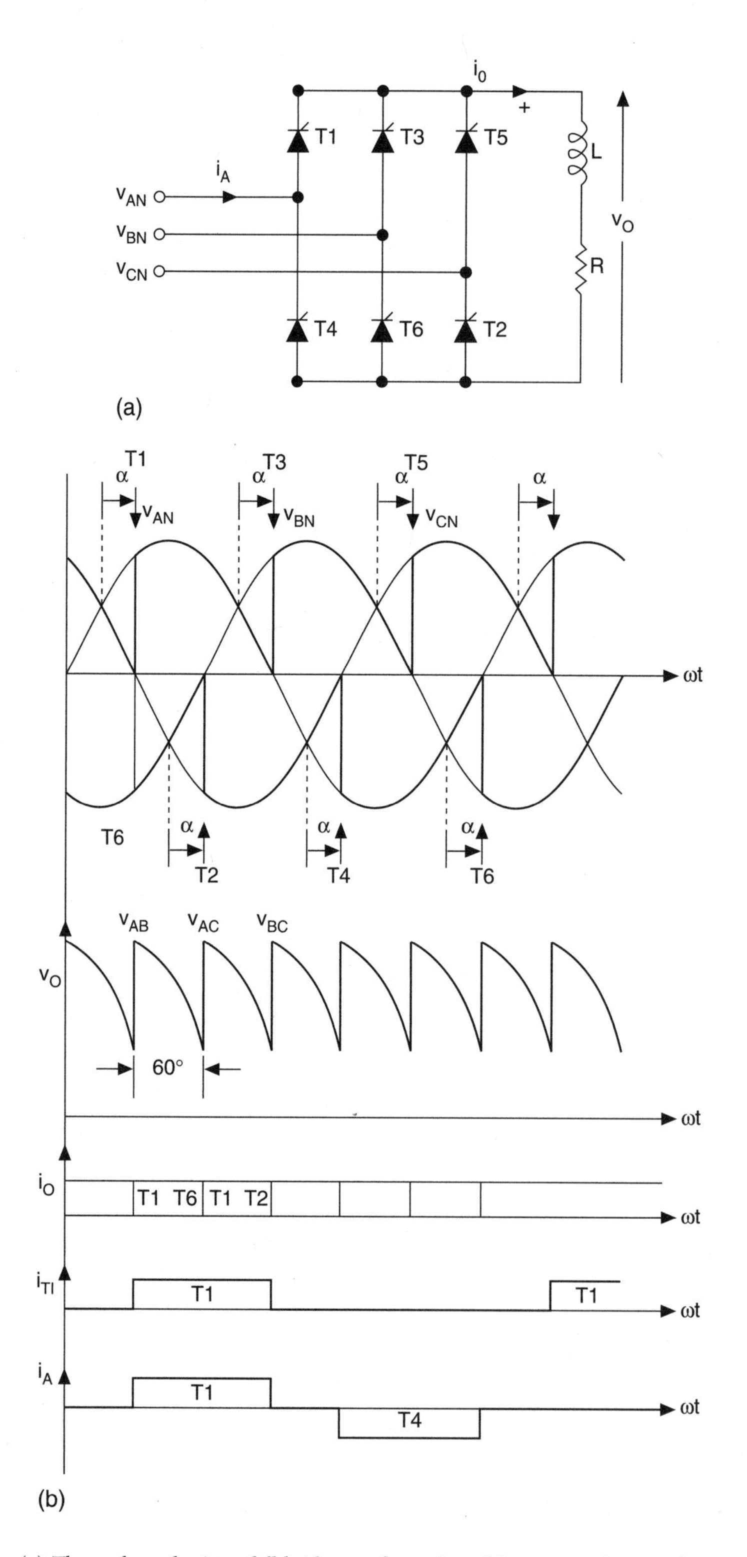

FIGURE 9.13 (a) Three-phase thyristor full bridge configuration; (b) output voltage and current waveforms.

Cycloconverters

Cycloconverters are direct ac-to-ac frequency changers. The term *direct conversion* means that the energy does not appear in any form other than the ac input or ac output. The output frequency is lower than the input frequency and is generally an integral multiple of the input frequency. A cycloconverter permits energy to be fed back into the utility network without any additional measures. Also, the phase sequence of the output voltage can be easily reversed by the control system. Cycloconverters have found applications in aircraft systems and industrial drives. These cycloconverters are suitable for synchronous and induction motor control. The operation of the cycloconverter is illustrated in Section "Converter Control of Machines" of this chapter.

dc-to-ac Converters

The dc-to-ac converters are generally called *inverters*. The ac supply is first converted to dc, which is then converted to a variable-voltage and variable-frequency power supply. This generally consists of a three-phase bridge connected to the ac power source, a dc link with a filter, and the three-phase inverter bridge connected to the load. In the case of battery-operated systems, there is no intermediate dc link. Inverters can be classified as voltage source inverters (VSIs) and current source inverters (CSIs). A voltage source inverter is fed by a stiff dc voltage, whereas a current source inverter is fed by a stiff current source. A voltage source can be converted to a current source by connecting a series inductance and then varying the voltage to obtain the desired current. A VSI can also be operated in current-controlled mode, and similarly a CSI can also be operated in the voltage-control mode. The inverters are used in variable frequency ac motor drives, uninterrupted power supplies, induction heating, static VAR compensators, etc.

Voltage Source Inverter

A three-phase voltage source inverter configuration is shown in Figure 9.14(a). The VSIs are controlled either in square-wave mode or in pulsewidth-modulated (PWM) mode. In square-wave mode, the frequency of the output voltage is controlled within the inverter, the devices being used to switch the output circuit between the plus and minus bus. Each device conducts for 180°, and each of the outputs is displaced 120° to generate a six-step waveform, as shown in Figure 9.14(b). The amplitude of the output voltage is controlled by varying the dc link voltage. This is done by varying the firing angle of the thyristors of the three-phase bridge converter at the input. The square-wave-type VSI is not suitable if the dc source is a battery. The six-step output voltage is rich in harmonics and thus needs heavy filtering.

In PWM inverters, the output voltage and frequency are controlled within the inverter by varying the width of the output pulses. Hence at the front end, instead of a phase-controlled thyristor converter, a diode bridge rectifier can be used. A very popular method of controlling the voltage and frequency is by sinusoidal pulsewidth modulation. In this method, a high-frequency triangle carrier wave is compared with a three-phase sinusoidal waveform, as shown in Figure 9.15. The power devices in each phase are switched on at the intersection of sine and triangle waves. The amplitude and frequency of the output voltage are varied, respectively, by varying the amplitude and frequency of the reference sine waves. The ratio of the amplitude of the sine wave to the amplitude of the carrier wave is called the *modulation index*.

The harmonic components in a PWM wave are easily filtered because they are shifted to a higher-frequency region. It is desirable to have a high ratio of carrier frequency to fundamental frequency to reduce the harmonics of lower-frequency components. There are several other PWM techniques mentioned in the literature. The most notable ones are selected harmonic elimination, hysteresis controller, and space vector PWM technique.

In inverters, if SCRs are used as power switching devices, an external forced commutation circuit has to be used to turn off the devices. Now, with the availability of IGBTs above 1000-A, 1000-V ratings, they are being used in applications up to 300-kW motor drives. Above this power rating, GTOs are generally used. Power Darlington transistors, which are available up to 800 A, 1200 V, could also be used for inverter applications.

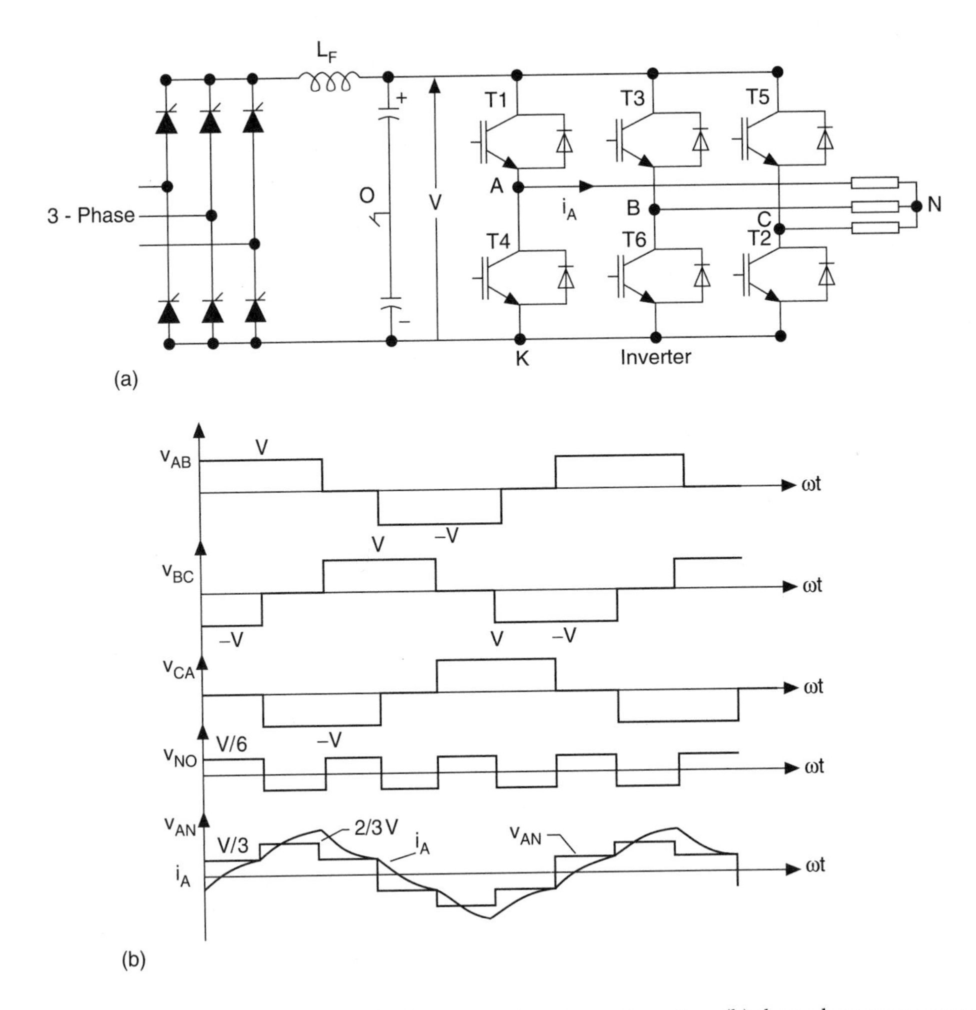

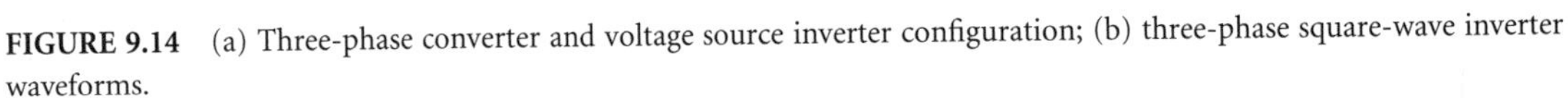

FIGURE 9.14 (a) Three-phase converter and voltage source inverter configuration; (b) three-phase square-wave inverter waveforms.

Current Source Inverter

Contrary to the voltage source inverter where the voltage of the dc link is imposed on the motor windings, in the current source inverter the current is imposed into the motor. Here the amplitude and phase angle of the motor voltage depend on the load conditions of the motor. The current source inverter is described in detail in Section "Converter Control of Machines".

Resonant-Link Inverters

The use of resonant switching techniques can be applied to inverter topologies to reduce the switching losses in the power devices. They also permit high switching frequency operation to reduce the size of the magnetic components in the inverter unit. In the resonant dc-link inverter shown in Figure 9.16, a resonant circuit is added at the inverter input to convert a fixed dc voltage to a pulsating dc voltage. This resonant circuit enables the devices to be turned on and turned off during the zero voltage interval. Zero voltage or zero current switching is often termed *soft switching.* Under soft switching, the switching losses in the power devices are almost eliminated. The electromagnetic interference (EMI) problem is less severe because resonant voltage pulses have lower dv/dt compared to those of hard-switched PWM inverters. Also, the machine insulation is less stretched because of lower dv/dt resonant voltage pulses. In Figure 9.16, all the inverter devices are turned

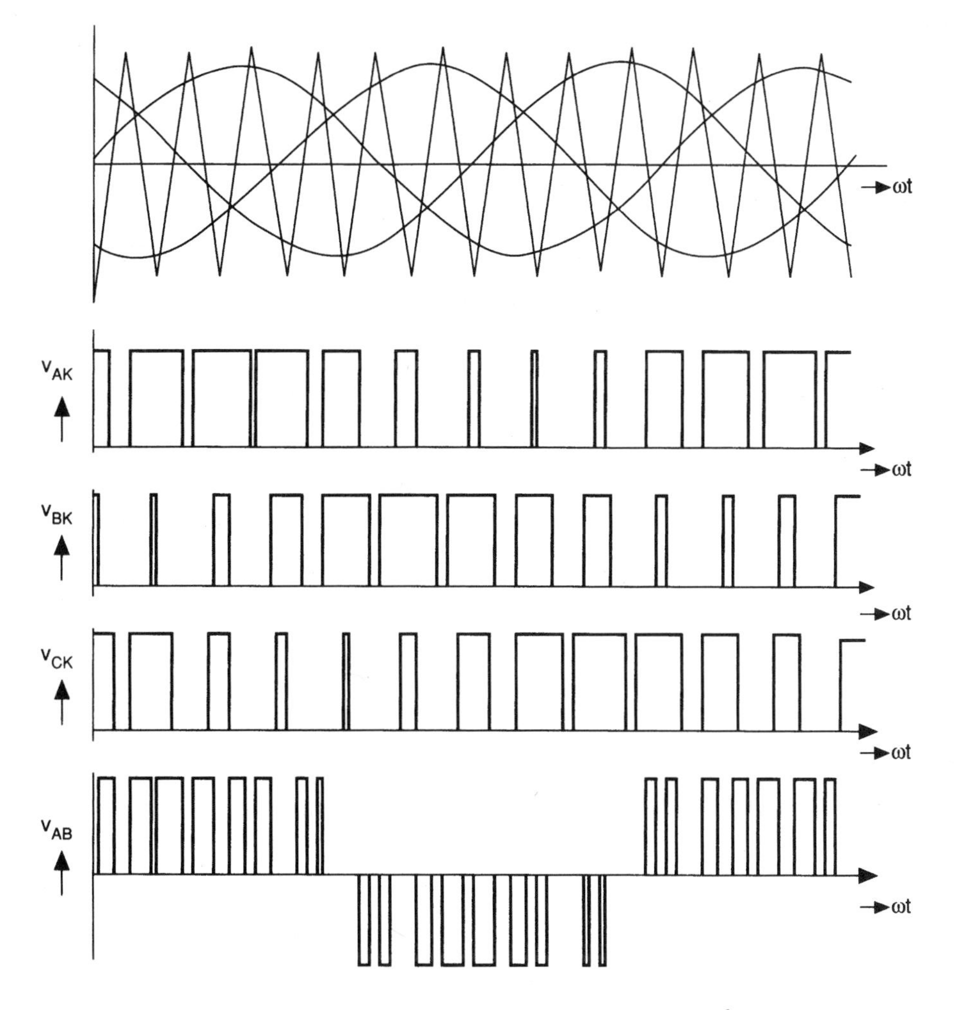

FIGURE 9.15 Three-phase sinusoidal PWM inverter waveforms.

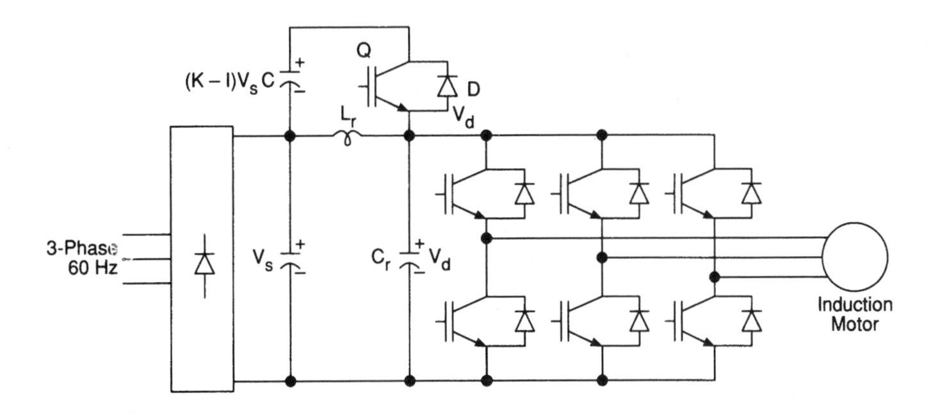

FIGURE 9.16 Resonant dc-link inverter system with active voltage clamping.

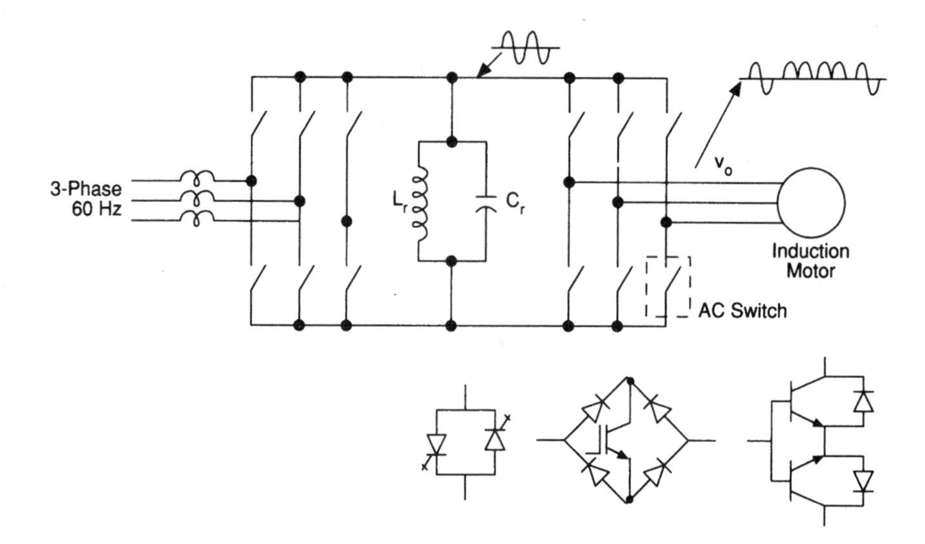

FIGURE 9.17 Resonant ac-link converter system showing configuration of ac switches.

on simultaneously to initiate a resonant cycle. The commutation from one device to another is initiated at the zero dc-link voltage. The inverter output voltage is formed by the integral numbers of quasi-sinusoidal pulses. The circuit consisting of devices Q, D, and the capacitor C acts as an active clamp to limit the dc voltage to about 1.4 times the diode rectifier voltage V_s.

There are several other topologies of resonant link inverters mentioned in the literature. There are also resonant link ac–ac converters based on bidirectional ac switches, as shown in Figure 9.17. These resonant link converters find applications in ac machine control and uninterrupted power supplies, induction heating, etc. The resonant link inverter technology is still in the development stage for industrial applications.

dc–dc Converters

dc–dc converters are used to convert unregulated dc voltage to regulated or variable dc voltage at the output. They are widely used in switch-mode dc power supplies and in dc motor drive applications. In dc motor control applications, they are called *chopper-controlled drives.* The input voltage source is usually a battery or derived from an ac power supply using a diode bridge rectifier. These converters are generally either hard-switched PWM types or soft-switched resonant-link types. There are several dc–dc converter topologies, the most common ones being buck converter, boost converter, and buck–boost converter, shown in Figure 9.18.

Buck Converter

A buck converter is also called a *step-down* converter. Its principle of operation is illustrated by referring to Figure 9.18(a). The IGBT acts as a high-frequency switch. The IGBT is repetitively closed for a time t_{on} and opened for a time t_{off}. During t_{on}, the supply terminals are connected to the load, and power flows from supply to the load. During t_{off}, load current flows through the freewheeling diode D_1, and the load voltage is ideally zero. The average output voltage is given by

$$V_{out} = DV_{in}$$

where D is the **duty cycle** of the switch and is given by $D = t_{on}/T$, where T is the time for one period. $1/T$ is the switching frequency of the power device IGBT.

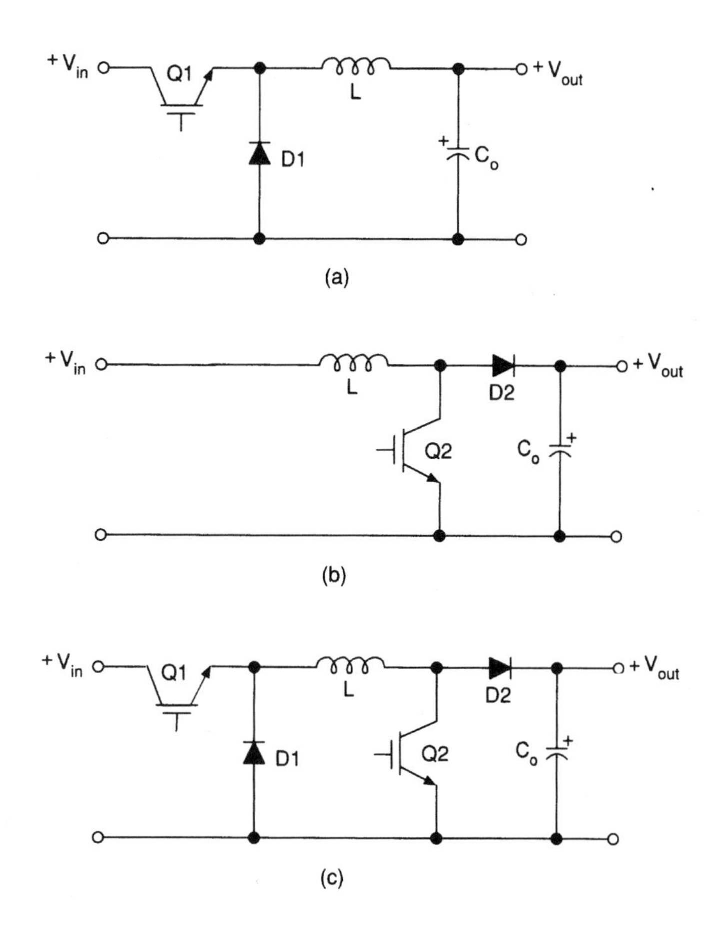

FIGURE 9.18 dc–dc converter configurations: (a) buck converter; (b) boost converter; (c) buck–boost converter.

Boost Converter

A boost converter is also called a *step-up* converter. Its principle of operation is illustrated by referring to Figure 9.18(b). This converter is used to produce higher voltage at the load than the supply voltage. When the power switch is on, the inductor is connected to the dc source and the energy from the supply is stored in it. When the device is off, the inductor current is forced to flow through the diode and the load. The induced voltage across the inductor is negative. The inductor adds to the source voltage to force the inductor current into the load. The output voltage is given by

$$V_{\text{out}} = \frac{V_{\text{in}}}{1 - D}$$

Thus for variation of D in the range $0 < D < 1$, the load voltage V_{out} will vary in the range $V_{\text{in}} < V_{\text{out}} < \infty$.

Buck–Boost Converter

A buck–boost converter can be obtained by the cascade connection of the buck and the boost converter. The steady-state output voltage V_{out} is given by

$$V_{\text{out}} = V_{\text{in}} \frac{D}{1 - D}$$

This allows the output voltage to be higher or lower than the input voltage, based on the duty cycle D. A typical buck–boost converter topology is shown in Figure 9.18(c). When the power device is turned on, the

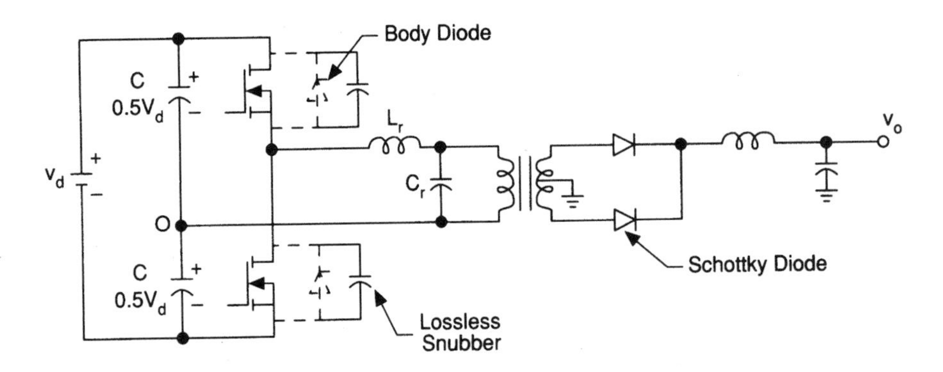

FIGURE 9.19 Resonant-link dc–dc converter.

input provides energy to the inductor and the diode is reverse biased. When the device is turned off, the energy stored in the inductor is transferred to the output. No energy is supplied by the input during this interval. In dc power supplies, the output capacitor is assumed to be very large, which results in a constant output voltage. In dc drive systems, the chopper is operated in step-down mode during motoring and in step-up mode during regeneration operation.

Resonant-Link dc–dc Converters

The use of resonant converter topologies would help to reduce the switching losses in dc–dc converters and enable the operation at switching frequencies in the megahertz range. By operating at high frequencies, the size of the power supplies could be reduced. There are several types of resonant converter topologies. The most popular configuration is shown in Figure 9.19. The dc power is converted to high-frequency alternating power using the MOSFET half-bridge inverter. The resonant capacitor voltage is transformer-coupled, rectified using the two Schottky diodes, and then filtered to get output dc voltage. The output voltage is regulated by control of the inverter switching frequency.

Instead of parallel loading as in Figure 9.19, the resonant circuit can be series-loaded; that is, the transformer in the output circuit can be placed in series with the tuned circuit. The series resonant circuit provides the short-circuit limiting feature.

There are other forms of resonant converter topologies mentioned in the literature such as quasi-resonant converters and multiresonant converters. These resonant converter topologies find applications in high-density power supplies.

Defining Terms

Commutation: Process of transferring the current from one power device to another.

Duty cycle: Ratio of the on-time of a switch to the switching period.

Full-wave control: Both the positive and negative half cycle of the waveforms are controlled.

IGBT: Insulated-gate bipolar transistor.

Phase-controlled converter: Converter in which the power devices are turned off at the natural crossing of zero voltage in ac-to-dc conversion applications.

SCR: Silicon-controlled rectifier.

References

B.K. Bose, *Modern Power Electronics*, New York: IEEE Press, 1992.

Motorola, *Linear/Switchmode Voltage Regulator Handbook*, 1989.

K.S. Rajashekara, H. Le-Huy, et al., "Resonant DC Link Inverter-Fed AC Machines Control," *IEEE Power Electronics Specialists Conference*, 1987, pp. 491–496.

P.C. Sen, *Thyristor DC Drives*, New York: John Wiley, 1981.
G. Venkataramanan and D. Divan, "Pulse Width Modulation with Resonant DC Link Converters," *IEEE IAS Annual Meeting*, 1990, pp. 984–990.

Further Information

B.K. Bose, *Power Electronics & AC Drives*, Englewood Cliffs, NJ: Prentice-Hall, 1986.
R. Hoft, *Semiconductor Power Electronics*, New York: Van Nostrand Reinhold, 1986.
B.R. Pelly, *Thyristor Phase Controlled Converters and Cycloconverters*, New York: Wiley-Interscience, 1971.
A.I. Pressman, *Switching and Linear Power Supply, Power Converter Design*, Carmel, IN: Hayden Book Company, 1977.
M.H. Rashid, *Power Electronics, Circuits, Devices and Applications*, Englewood Cliffs, NJ: Prentice-Hall, 1988.

9.3 Power Supplies

Ashoka K.S. Bhat

Introduction

Power supplies are used in many industrial and aerospace applications, and also in consumer products. Some of the requirements of power supplies are small size, light weight, low cost, and high power conversion efficiency. In addition to these, some power supplies require the following: electrical isolation between the source and load, low harmonic distortion for the input and output waveforms, and high power factor (PF) if the source is ac voltage. Some special power supplies require controlled direction of power flow.

Basically two types of power supplies are required: ac and dc. The output of dc power supplies is regulated or controllable dc, whereas the output for ac power supplies is ac. The input to these power supplies can be ac or dc.

DC Power Supplies

If ac source is used, then ac-to-dc converters explained, in Section "Power Conversion", can be used. In these converters, electrical isolation can only be provided by bulky line frequency transformers. The ac source can be rectified with a diode rectifier to get an uncontrolled dc, and then a dc-to-dc converter can be used to get a controlled dc output. Electrical isolation between the input source and the output load can be provided in the dc-to-dc converter using a high-frequency (HF) transformer. Such HF transformers have small size, light weight, and low cost compared to bulky line frequency transformers. Whether the input source is dc (e.g., battery) or ac, dc-to-dc converters form an important part of dc power supplies and they are explained in this subsection.

DC power supplies can be broadly classified as linear and switching power supplies.

A *linear power supply* is the oldest and simplest type. The output voltage is **regulated** by dropping the extra input voltage across a series transistor (therefore also referred to as a series regulator). They have very small output ripple, theoretically zero noise, large hold-up time (typically 1 to 2 msec), and fast response. Linear power supplies have the following disadvantages: very low efficiency, electrical isolation can only be on 60 Hz ac side, larger volume and weight, and in general, only a single output possible. However, they are still used in very small regulated power supplies and in some special applications (e.g., magnet power supplies). Three-terminal linear regulator integrated circuits (ICs) are readily available (e.g., μA7815 has +15 V, 1 A output), are easy to use, and have built-in load short-circuit protection.

Switching power supplies use power semiconductor switches in the on and off switching states, resulting in high efficiency, small size, and light weight. With the availability of fast switching devices, HF magnetics and capacitors, and high-speed control ICs, switching power supplies have become very popular. They can be further classified as **pulse-width-modulated (PWM) converters** and **resonant converters**, and they are explained below.

Pulse-Width-Modulated Converters

These converters employ square-wave pulse-width modulation to achieve voltage regulation. The average output voltage is varied by varying the duty cycle of the power semiconductor switch. The voltage waveform across the switch and at the output are square-wave in nature (refer to Figure 9.14(b)), and they generally result in higher switching losses when the switching frequency is increased. Also, the switching stresses are high with the generation of large electromagnetic interference (EMI), which is difficult to filter. However, these converters are easy to control, well understood, and have a wide load control range.

The methods of control of PWM converters are discussed next.

The Methods of Control. The PWM converters operate with a fixed frequency, variable duty cycle. Depending on the duty cycle, they can operate in either continuous current mode (CCM) or discontinuous current mode (DCM). If the current through the output inductor never reaches zero (refer to Figure 9.14) then the converter operates in CCM; otherwise DCM occurs.

The three possible control methods [1–5] are briefly explained below.

1. *Direct duty cycle control* is the simplest control method. A fixed frequency ramp is compared with the control voltage (Figure 9.20(a)) to obtain a variable duty cycle base drive signal for the transistor. This is the simplest method of control. Disadvantages with this method are: (a) no voltage feedforward provision to anticipate the effects of input voltage changes, slow response to sudden input changes, poor audio susceptibility, poor open-loop line regulation, higher loop gain required to achieve specifications; and (b) poor dynamic response.

2. *Voltage feedforward control.* In this case the ramp amplitude varies in direct proportion to the input voltage (Figure 9.20(b)). The open-loop regulation is very good, and the problems in 1(a) above are corrected.

3. *Current mode control.* In this method, a second inner control loop compares the peak inductor current with the control voltage which provides improved open-loop line regulation (Figure 9.20(c)). All of the problems of the direct duty cycle control method 1 above are corrected with this method. An additional advantage of this method is that a two-pole second-order filter is reduced to a single-pole (the filter capacitor) first-order filter, resulting in simpler compensation networks.

The above control methods can be used in all the PWM converter configurations explained below.

PWM converters can be classified as: (1) single-ended and (2) double-ended converters. These converters may or may not have a high-frequency transformer for isolation.

Nonisolated Single-Ended PWM Converters

The basic nonisolated single-ended converters are: (a) buck (step-down), (b) boost (step-up), (c) buck–boost (step up or down, also referred to as flyback), and (d) Cuk converters (Figure 9.21). The first three of these converters were discussed in Section "Power Conversion". The Cuk converter provides the advantage of nonpulsating input–output current ripple requiring smaller size external filters. Output voltage expression is the same as the buck–boost converter (refer to Section "Power Conversion") and can be less than or greater than the input voltage. There are many variations of the above basic nonisolated converters and most of them use high-frequency transformers for ohmic isolation between the input and the output. Some of them are discussed below.

Isolated Single-Ended Topologies

1. The flyback converter (Figure 9.22) is an isolated version of the buck–boost converter. In this converter (Figure 9.22), when the transistor is on, energy is stored in the coupled inductor (not a transformer) and this energy is transferred to the load when the switch is off.

   Some of the advantages of this converter are that the leakage inductance is in series with the output diode when current is delivered to the output and, therefore, no filter inductor is required; cross-regulation for multiple output converters is good; it is ideally suited for high-voltage output applications; and it has the lowest cost.

   Some of the disadvantages are that large output filter capacitors are required to smooth the pulsating output current; inductor size is large since air-gaps are to be provided; and due to stability reasons,

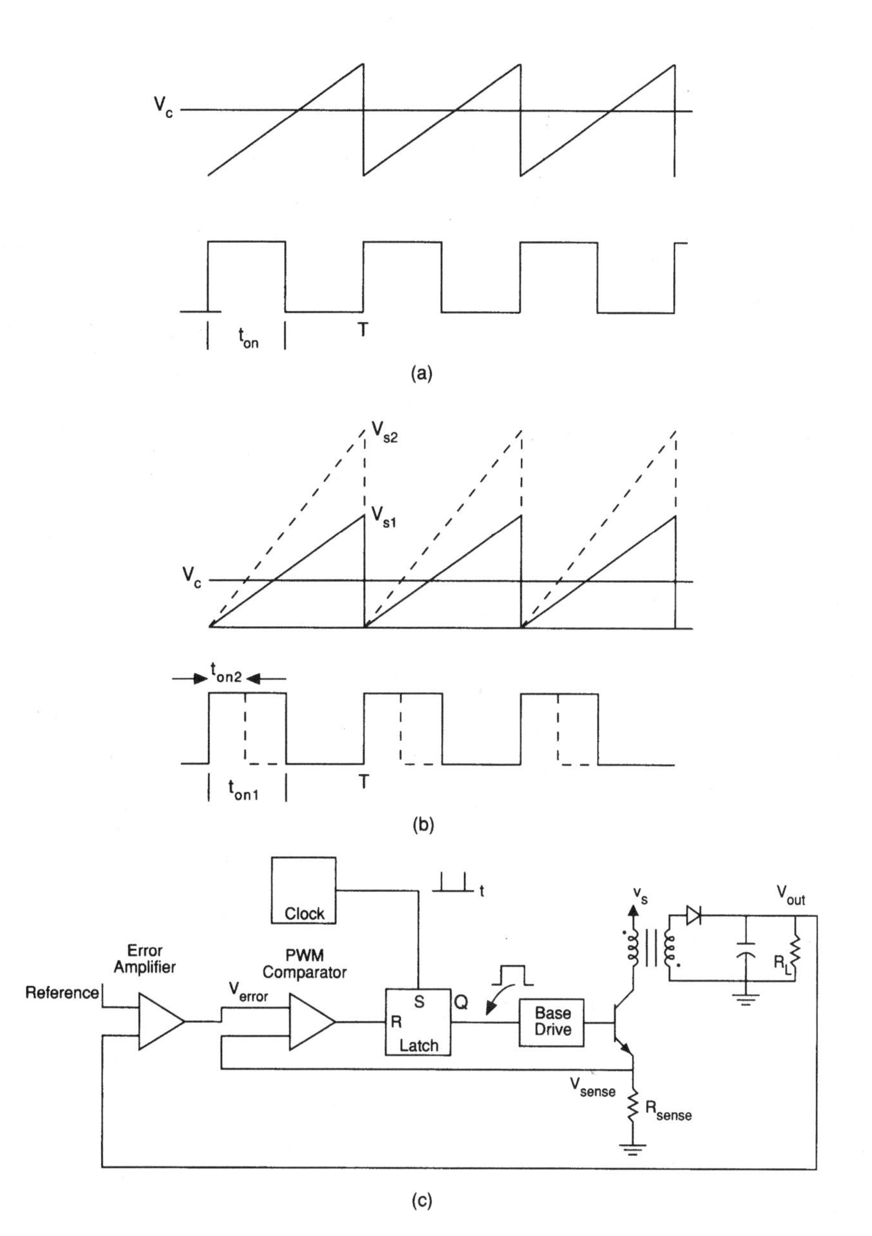

FIGURE 9.20 PWM converter control methods: (a) direct duty cycle control; (b) voltage feedforward control; (c) current mode control (illustrated for flyback converter).

flyback converters are usually operated in the DCM, which results in increased losses. To avoid the stability problem, flyback converters are operated with current mode control explained earlier. Flyback converters are used in the power range of 20 to 200 W.

2. The forward converter (Figure 9.23) is based on the buck converter. It is usually operated in the CCM to reduce the peak currents and does not have the stability problem of the flyback converter.

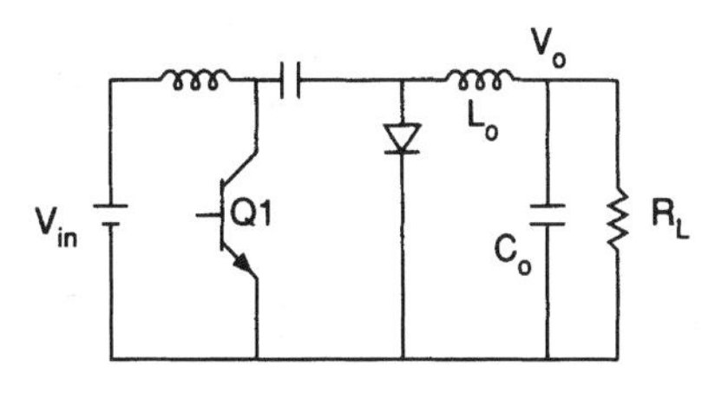

FIGURE 9.21 Nonisolated Ćuk converter.

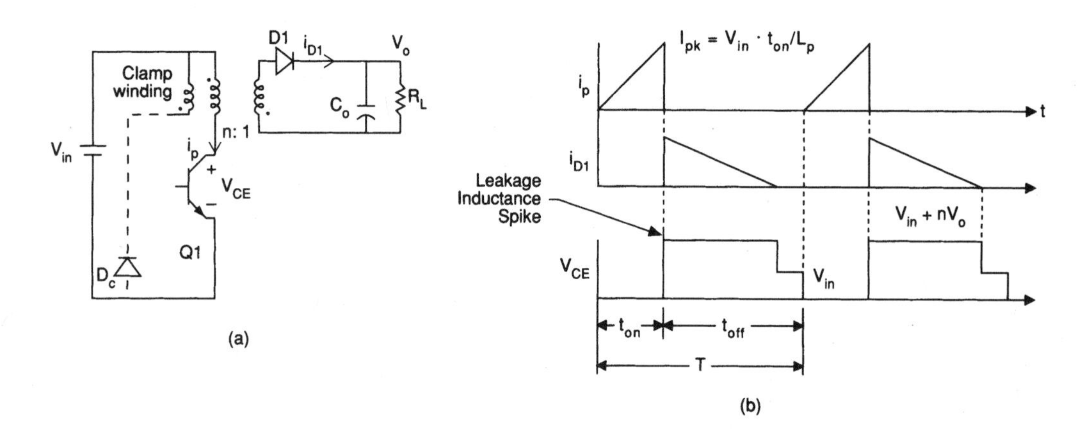

FIGURE 9.22 (a) Flyback converter. The clamp winding shown is optional and is used to clamp the transistor voltage stress to $V_{in} + nV_o$. (b) Flyback converter waveforms without the clamp winding. The leakage inductance spikes vanish with the clamp winding.

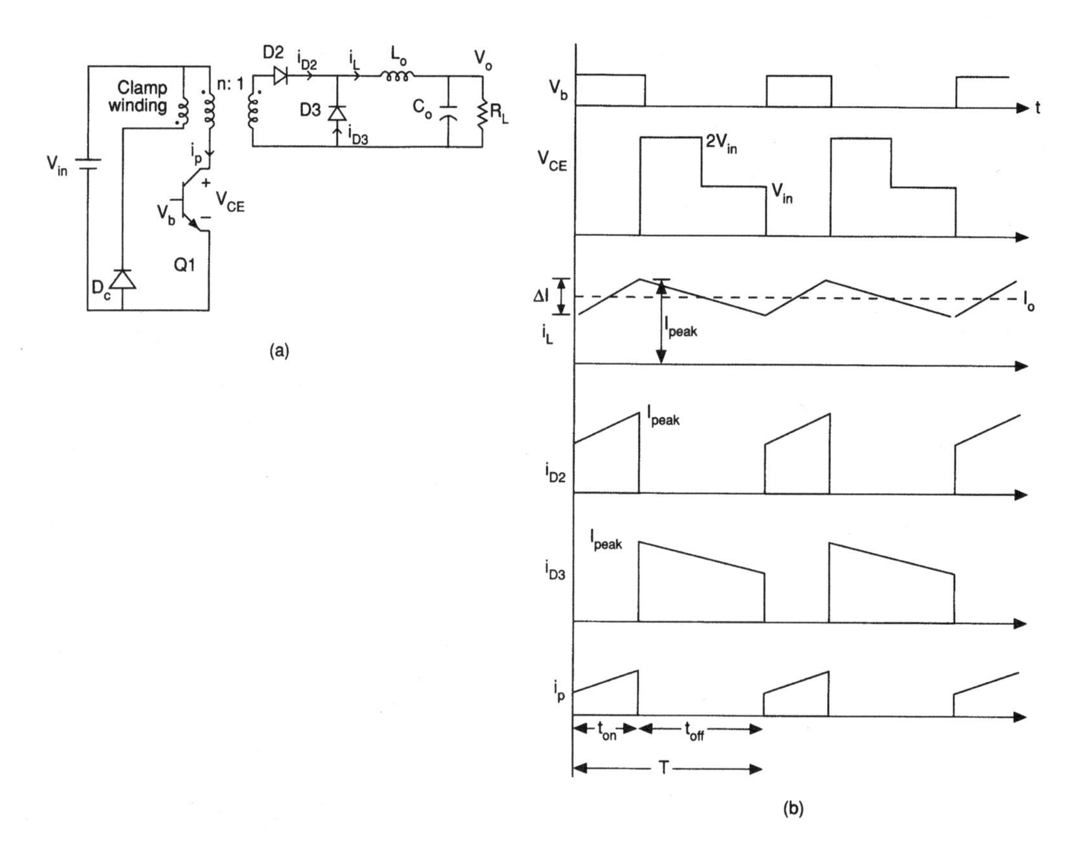

FIGURE 9.23 (a) Forward converter. The clamp winding shown is required for operation. (b) Forward converter waveforms.

The HF transformer transfers energy directly to the output with very small stored energy. The output capacitor size and peak current rating are smaller than they are for the flyback. Reset winding is required to remove the stored energy in the transformer. Maximum duty cycle is about 0.45 and limits the control range. This topology is used for power levels up to about 1 kW.

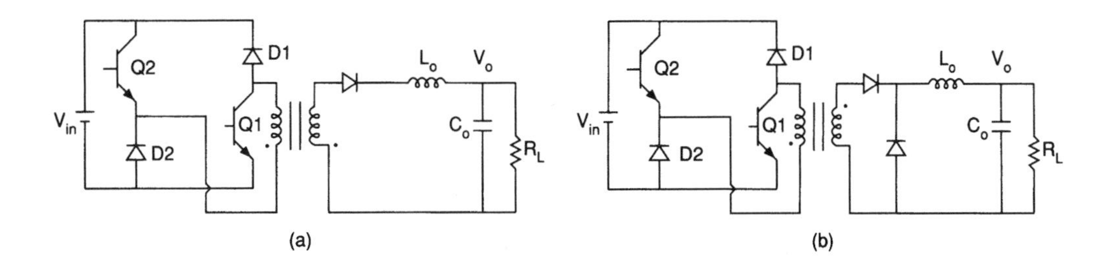

FIGURE 9.24 (a) Two-transistor single-ended flyback converter. (b) Two-transistor single-ended forward converter.

3. The flyback and forward converters explained above require the rating of power transistors to be much higher than the supply voltage. Two transistor flyback and forward converters shown in Figure 9.24 limit the voltage rating of transistors to the supply voltage.
4. The Sepic converter shown in Figure 9.25 is another isolated single-ended PWM converter.

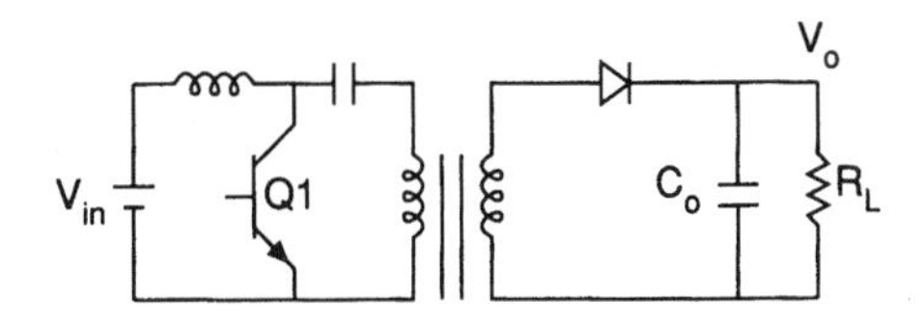

FIGURE 9.25 Sepic converter.

Double-Ended PWM Converters

Usually, for power levels above 300 W, double-ended converters are used. In double-ended converters, full-wave rectifiers are used and the output voltage ripple will have twice the switching frequency. Three important double-ended PWM converter configurations are push–pull (Figure 9.26), half-bridge (Figure 9.27), and full-bridge (Figure 9.28).

1. *The push–pull converter.* The duty ratio of each transistor in a push–pull converter (Figure 9.26) is less than 0.5. Two of the advantages are that the transformer flux swings fully, and thereby the size of the transformer is much smaller (typically half the size) than single-ended converters, and output ripple is twice the switching frequency of transistors, therefore needing smaller filters.

   Some of the disadvantages of this configuration are that the transistors must block twice the supply voltage, flux symmetry imbalance can cause transformer saturation with special control circuitry required to avoid this problem, and use of a center-tap transformer requires extra copper resulting in a higher volt–ampere (VA) rating.

   Current mode control (for the primary current) can be used to overcome the flux imbalance. This configuration is used in the 100 to 500 W output range.
2. *The half-bridge.* In the half-bridge configuration (Figure 9.27) a center-tapped dc source is created by two smoothing capacitors (C_{in}), and this configuration utilizes the transformer core efficiently. The voltage across each transistor is equal to the supply voltage (half of push–pull) and, therefore, suitable for high-voltage inputs. One salient feature of this configuration is that the input filter capacitors can be used to change between 110 and 220 V mains as selectable inputs to the supply.

   The disadvantage of this configuration is the requirement for large-size input filter capacitors. The half-bridge configuration is used for power levels on the order of 500 to 1000 W.
3. *The full-bridge.* The full-bridge configuration (Figure 9.28) requires only one smoothing capacitor, and for the same transistor type as that of half-bridge, output power can be doubled. It is usually used for power levels above 1 kW and the design is more costly due to the increased number of components (four transistors compared to two in push–pull and half-bridge converters).

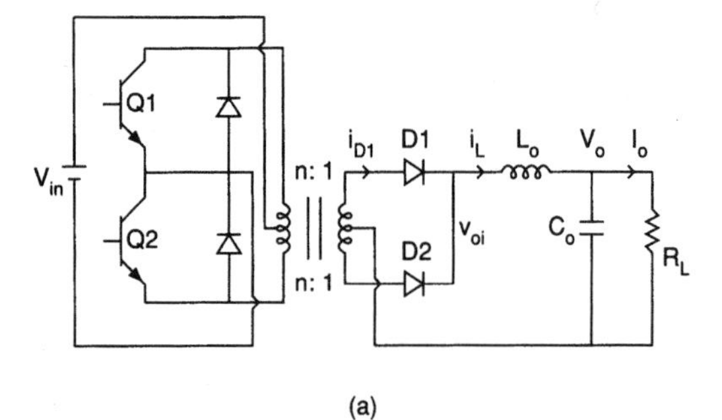

(a)

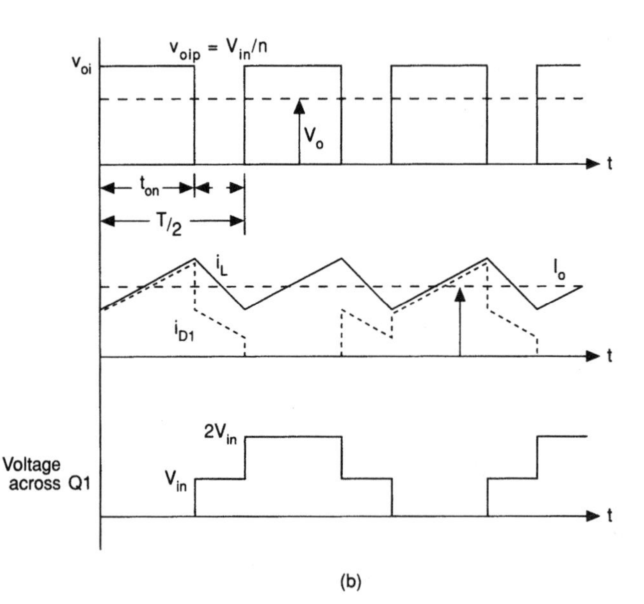

(b)

FIGURE 9.26 (a) Push-pull converter and (b) its operating waveforms.

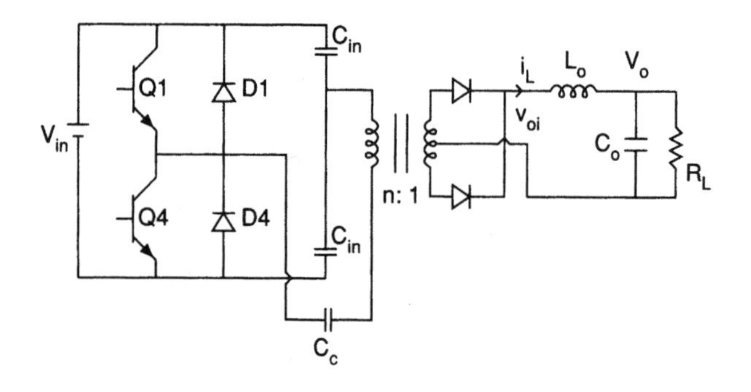

FIGURE 9.27 Half-bridge converter. Coupling capacitor C_c is used to avoid transformer saturation.

One of the salient features of a full-bridge converter is that, using proper control technique, it can be operated in zero-voltage switching (ZVS) mode [19,22], which results in negligible switching losses. However, at reduced load currents, the ZVS property is lost. Recently, there has been a lot of effort made to overcome this problem (for example Refs. [20,21]).

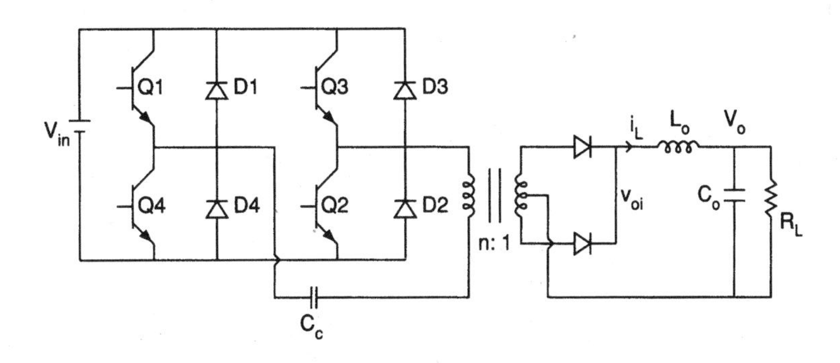

FIGURE 9.28 Full-bridge converter.

Resonant Power Supplies

Similar to the PWM converters, there are two types of resonant converters: single-ended and double-ended. Resonant converter configurations are obtained from the PWM configurations explained earlier by adding inductor-capacitor (LC) resonating elements to obtain sinusoidally varying voltage and/or current waveforms. This approach reduces the switching losses and the switch stresses during switching instants, enabling the converter to operate at high switching frequencies, resulting in reduced size, weight, and cost. Some other advantages of resonant converters are that leakage inductances of HF transformers and the junction capacitances of semiconductors can be used profitably in the resonant circuit, reducing EMI. The major disadvantage of resonant converters is increased peak current (or voltage) stress. To overcome this problem, resonant transition converter configurations have been proposed [14–18,20,21].

Single-Ended Resonant Converters

They are referred to as quasi-resonant converters (QRCs) since the voltage (or current) waveforms are quasi-sinusoidal in nature. The QRCs can operate with zero-current switching (ZCS) or ZVS, or both. All the QRC configurations can be generated by replacing the conventional switches by the resonant switches shown in Figure 9.29 and Figure 9.30. A number of configurations are realizable. Basic principles of ZCS and ZVS are explained briefly below.

1. *Zero-current switching QRCs* [7,8]. Figure 9.31(a) shows an example of a ZCS QR buck converter implemented using a ZC resonant switch. Depending on whether the resonant switch is a half-wave or full-wave type, the resonating current will be only half-wave sinusoidal (Figure 9.31(b)) or a full sine-wave

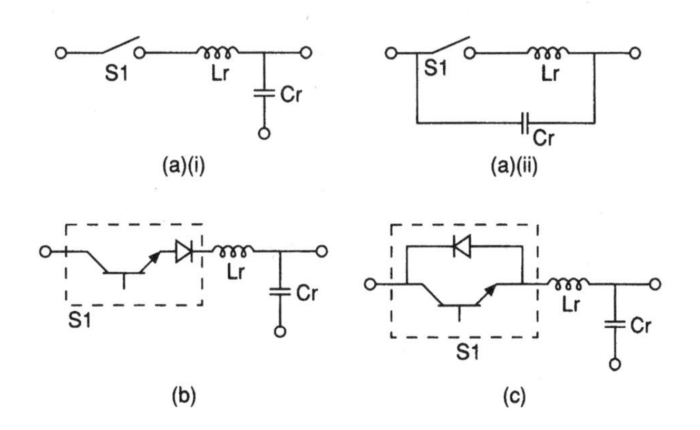

FIGURE 9.29 (a) Zero-current resonant switch: (i) L-type and (ii) M-type. (b) Half-wave configuration using L-type ZC resonant switch. (c) Full-wave configuration using L-type ZC resonant switch.

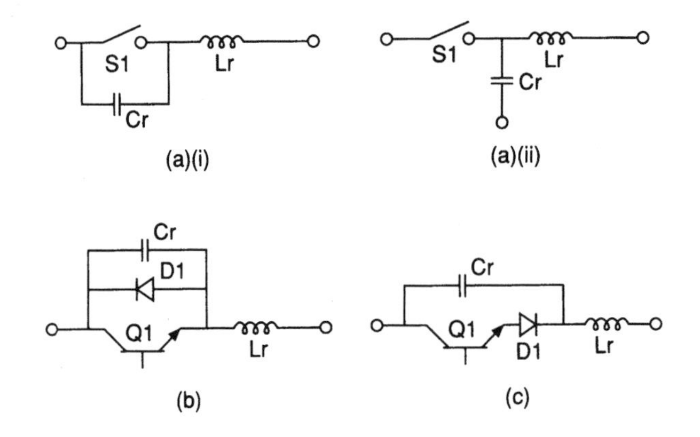

FIGURE 9.30 (a) Zero-voltage resonant switches. (b) Half-wave configuration using ZV resonant switch shown in (a)(i). (c) Full-wave configuration using ZV resonant switch shown in (a)(i).

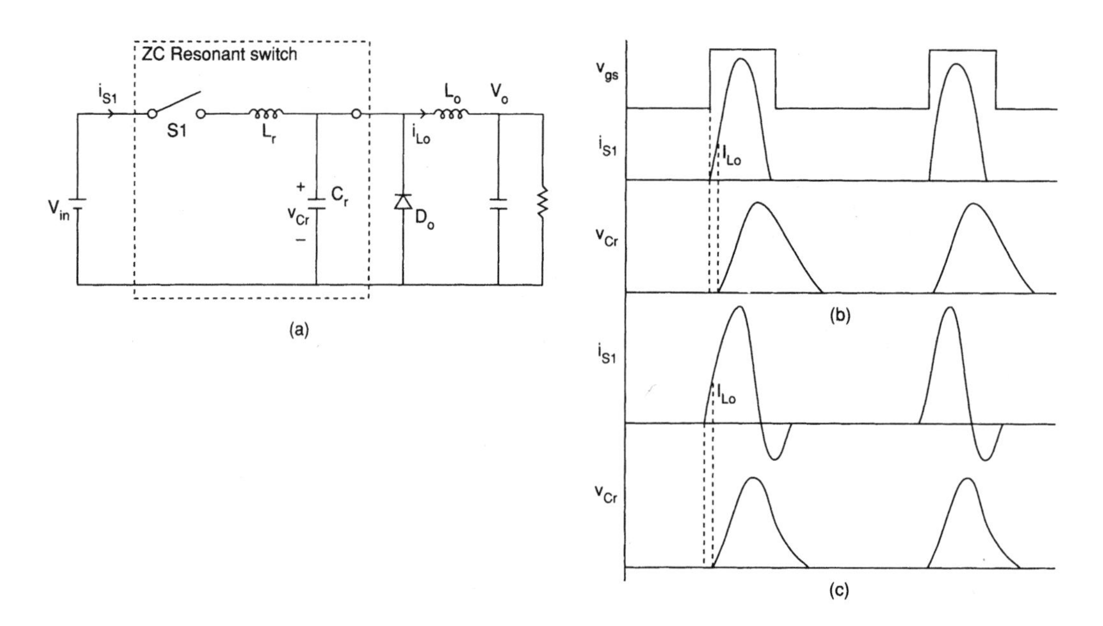

FIGURE 9.31 (a) Implementation of ZCS QR buck converter using L-type resonant switch. (b) Operating waveforms for half-wave mode. (c) Operating waveforms for full-wave mode.

(Figure 9.31(c)). The device currents are shaped sinusoidally and, therefore, the switching losses are almost negligible with low turn-on and turn-off stresses. ZCS QRCs can operate at frequencies on the order of 2 MHz. The major problems with this type of converter are high peak currents through the switch and capacitive turn-on losses.

2. *Zero-voltage switching QRCs* [7,9]. ZVS QRCs are duals of ZCS QRCs. The auxiliary LC elements are used to shape the switching device's voltage waveform at off time in order to create a zero-voltage condition for the device to turn on. Figure 9.32(a) shows an example of ZVS QR boost converter implemented using a ZV resonant switch. The circuit can operate in the half-wave (Figure 9.32(b)) or full-wave mode (Figure 9.32(c)) depending on whether a half-wave or full-wave ZV resonant switch is used, and the name comes from the capacitor voltage waveform. The full-wave mode ZVS circuit suffers from capacitive turn-on losses. The ZVS QRCs suffer from increased voltage stress on the switch. However, they can be operated at much higher frequencies compared to ZCS QRCs.

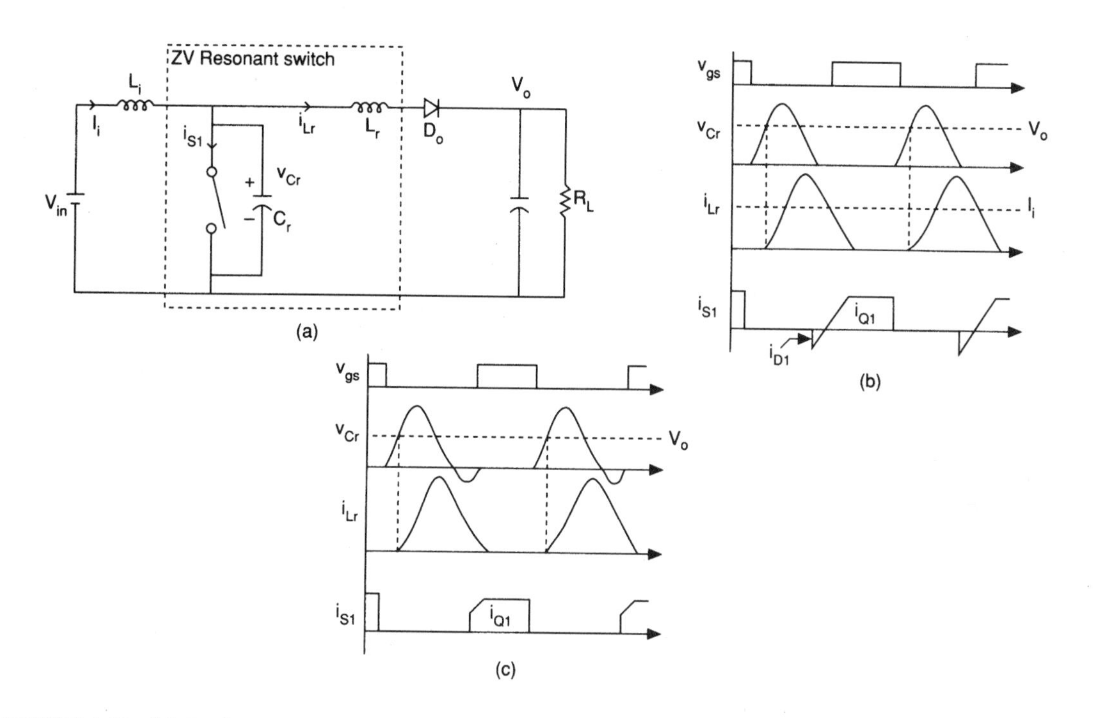

FIGURE 9.32 (a) Implementation of ZVS QR boost converter using resonant switch shown in Figure 9.30(a)(i). (b) Operating waveforms for half-wave mode. (c) Operating waveforms for full-wave mode.

Double-Ended Resonant Converters [7,10–13].

These converters use full-wave rectifiers at the output, and they are generally referred to as resonant converters. A number of resonant converter configurations are realizable by using different resonant tank circuits, and the three most popular configurations—the series resonant converter (SRC), the parallel resonant converter (PRC), and the series-parallel resonant converter (SPRC) (also called the LCC-type PRC)—are shown in Figure 9.33. Recently, the LCL-type resonant converter [13] has also become popular.

Series resonant converters (Figure 9.33(a)) have high efficiency from full load to part load. Transformer saturation is avoided due to the series blocking resonating capacitor. The major problems with the SRC are that it requires a very wide change in switching frequency to regulate the load voltage and the output filter capacitor must carry high ripple current (a major problem, especially in low-output voltage, high-output current applications).

Parallel resonant converters (Figure 9.33(b)) are suitable for low-output voltage, high-output current applications due to the use of filter inductance at the output with low ripple current requirements for the filter capacitor. The major disadvantage of the PRC is that the device currents do not decrease with the load current, resulting in reduced efficiency at reduced load currents.

The SPRC (Figure 9.33(c)) takes the desirable features of SRC and PRC.

Load voltage regulation in resonant converters for input supply variations and load changes is achieved by either varying the switching frequency or using fixed-frequency (variable pulse-width) control.

1. *Variable-frequency operation.* Depending on whether the switching frequency is below or above the natural resonance frequency (ω_r), the converter can operate in different operating modes as explained below.
   a. Below-resonance (leading PF) mode. When the switching frequency is below the natural resonance frequency, the converter operates in below resonance mode (Figure 9.34). The equivalent impedance across *AB* presents a leading PF so that natural turn-off of the switches is assured and any type of fast turn-off switch (including asymmetric SCRs) can be used. Depending on the instant of turn-on of switches S_1 and S_2, the converter can enter two modes of operation: continuous and discontinuous current modes. The steady-state operation in CCM (Figure 9.34(a)) is explained briefly as follows.

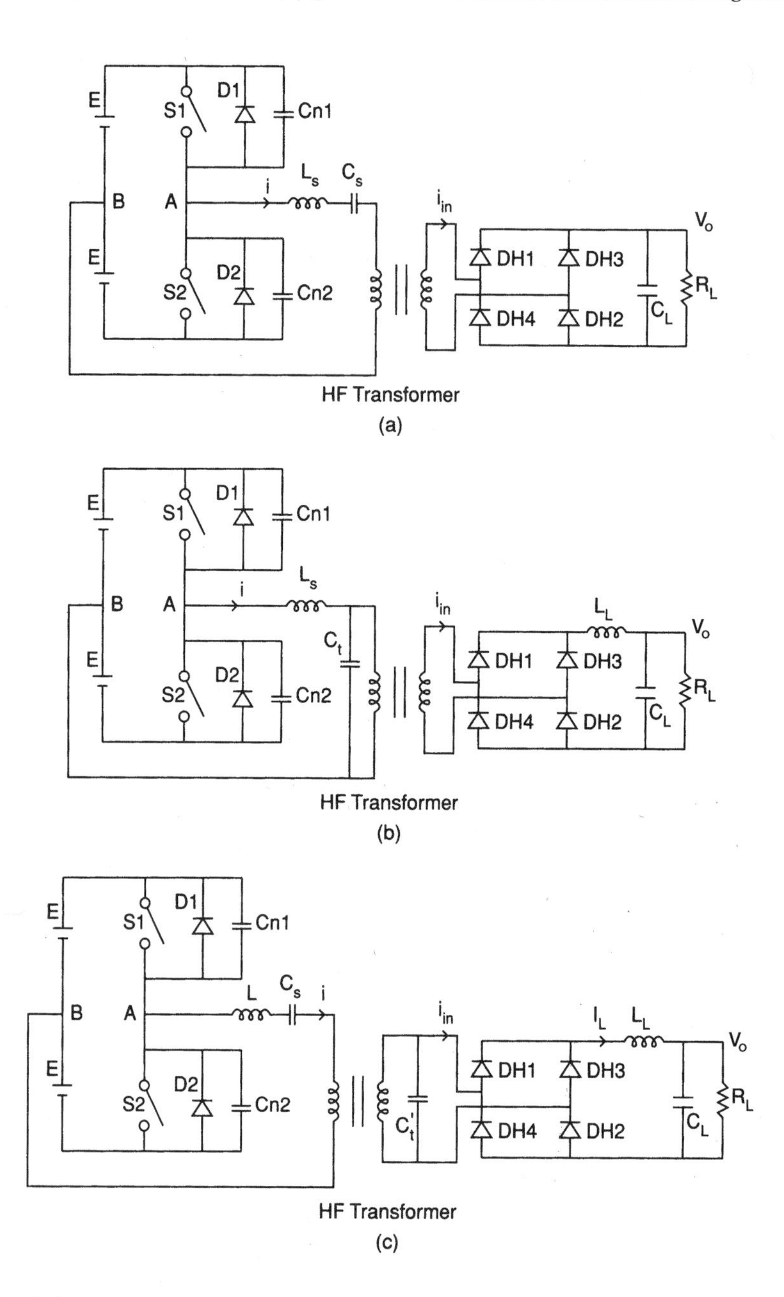

FIGURE 9.33 High-frequency resonant converter (half-bridge version) configurations suitable for operation above resonance. Cn1 and Cn2 are the snubbler capacitors. (*Note*: For operation below resonance, di/dt limiting inductors and RC snubbers are required. For operation above resonance, only capacitive snubbers are required as shown.) (a) Series resonant converter. Leakage inductances of the HF transformer can be part of resonant inductance. (b) Parallel resonant converter. (c) Series-parallel (or LCC-type) resonant converter with capacitor C_t placed on the secondary side of the HF transformer.

Assume that diode D2 was conducting and switch S1 is turned on. The current carried by D2 will be transferred to S1 almost instantaneously (except for a small time of recovery of D2 during which input supply is shorted through D2 and S1, and the current is limited by the di/dt limiting inductors). The current i then oscillates sinusoidally and goes to zero in the natural way. The current tries to reverse, and the path for this current is provided by the diode D1. Conduction of D1 feeds the reactive energy in the load and the tank circuit back to the supply. The on-state of D1 also provides a reverse voltage across S1,

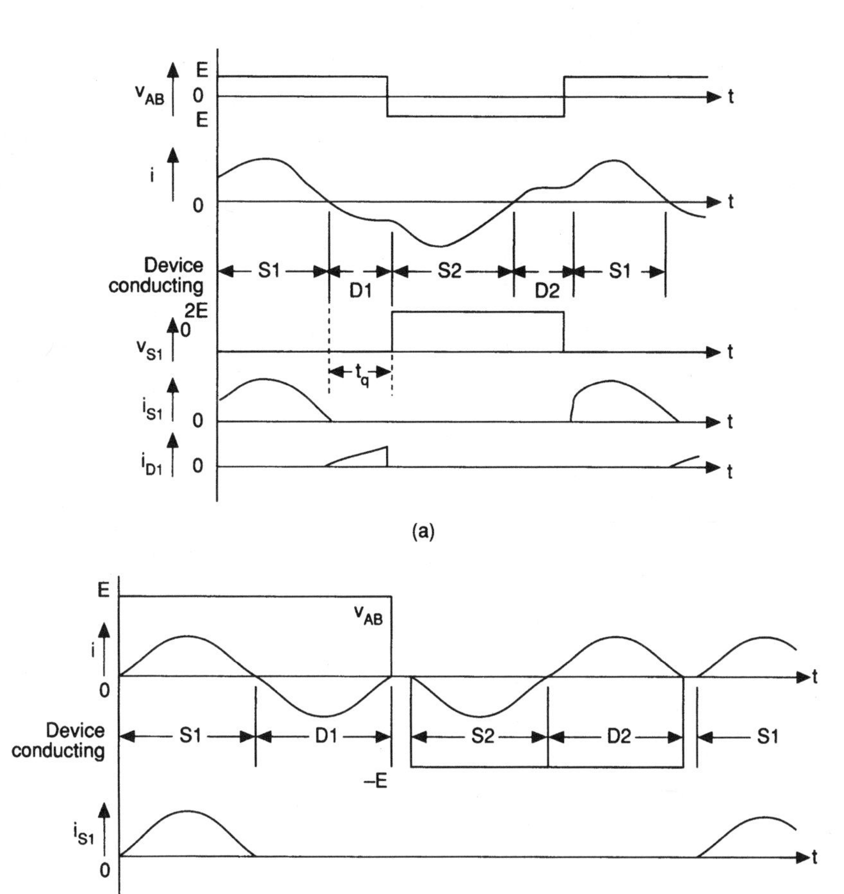

FIGURE 9.34 Typical waveforms at different points of a resonant converter operating below resonance in (a) continuous current mode and (b) discontinuous current mode.

allowing it to turn off. After providing a time equal to or greater than the turn-off time of S1, switch S2 can be turned on to initiate the second half cycle. The process is similar to the first half cycle, with the voltage across v_{AB} being of opposite polarity, and the functions of D1, S1 will be assumed by D2, S2. With this type of operation, the converter works in the continuous current mode as the switches are turned on before the currents in the diodes reach zero. If the switching on of S1 and S2 is delayed such that the currents through the previously conducting diodes reach zero, then there are zero current intervals and the inverter operates in the DCM (Figure 9.34(b)).

Load voltage regulation is achieved by decreasing the switching frequency below the rated value. Since the inverter output current i leads the inverter output voltage v_{AB}, this type of operation is also called a leading PF mode of operation. If transistors are used as the switching devices, then for operation in DCM, the pulse width can be kept constant while decreasing the switching frequency to avoid CCM operation. DCM operation has the advantages of negligible switching losses due to ZCS, lower di/dt and dv/dt stresses, and simple control circuitry. However, DCM operation results in higher switch peak currents.

From the waveforms shown in Figure 9.34, the following problems can be identified for operation in the below-resonance mode: requirement of di/dt inductors to limit the large turn-on switch currents, and a need for lossy RC-snubbers and fast recovery diodes. Since the switching frequency is decreased to control the load power, the HF transformer and magnetics must be designed for the lowest switching frequency, resulting in increased size of the converter.

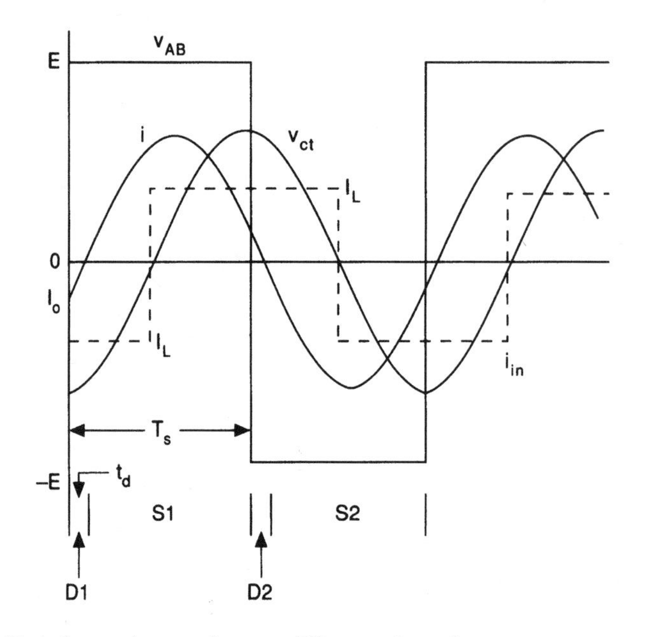

FIGURE 9.35 Typical operating waveforms at different points of an SPRC operating above resonance.

b. Above-resonance (lagging PF) mode. If switches capable of gate or base turn-off (e.g., MOSFETs, bipolar transistors) are used, then the converter can operate in the above-resonance mode (lagging PF mode). Figure 9.35 shows some typical operating waveforms for such type of operation, and it can be noticed that the current i lags the voltage v_{AB}. Since the switch takes current from its own diode across it at zero-current point, there is no need for di/dt limiting inductance, and a simple capacitive snubber can be used. In addition, the internal diodes of MOSFETs can be used due to the large turn-off time available for the diodes. Major problems with lagging PF mode of operation are that there are switch turn-off losses, and since the voltage regulation is achieved by increasing the switching frequency above the rated value, the magnetic losses increase and the design of a control circuit is difficult.

Exact analysis of resonant converters is complex due to the nonlinear loading on the resonant tanks. The rectifier-filter-load resistor block can be replaced by a square-wave voltage source (for SRC; Figure 9.33(a)) or a square-wave current source (for PRC and SPRC; Figure 9.33(b) and (c)). Using fundamental components of the waveforms, an approximate analysis [10,11] using a phasor circuit gives a reasonably good design approach. This analysis approach is illustrated next for the SPRC.

2. *Approximate analysis of SPRC.* Figure 9.36 shows the equivalent circuit at the output of the inverter and the phasor circuit used for the analysis. All the equations are normalized using the base quantities:

$$\text{base voltage}, V_B = E_{min}$$

$$\text{base impedance}, Z_B = R_L' = n^2 R_L$$

$$\text{base current}, I_B = V_B/Z_B$$

The converter gain (normalized output voltage in per unit [p.u.] referred to the primary side) can be derived as [10,11]:

$$M = 1/\left\{(\pi^2/8)^2\left[1 + (C_t/C_s)(1 - y_s^2)\right]^2 + Q_s^2[y_s - (1/y_s)]^2\right\}^{1/2} \quad \text{p.u.} \qquad (9.1)$$

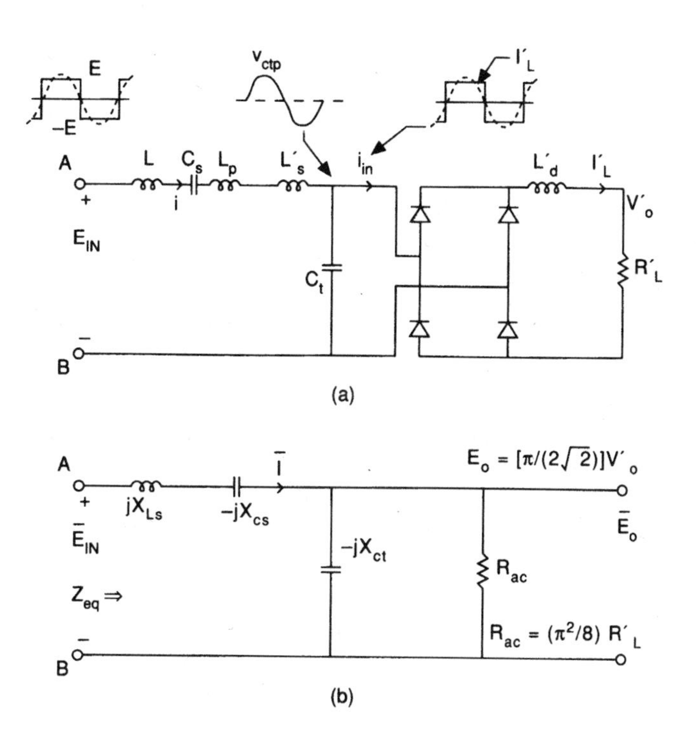

FIGURE 9.36 (a) Equivalent circuit for a SPRC at the output of the inverter terminals (across AB) of Figure 9.33(c), L_p and L'_s are the leakage inductance of the primary and primary referred leakage inductance of the secondary, respectively. (b) Phasor circuit model used for the analysis of the SPRC converter.

where

$$Q_s = (L_s/C_s)^{1/2}/R_L'; \quad L_s = L + L_p + L_p' \tag{9.2}$$

$$y_s = f_s/f_r \tag{9.3}$$

and

$$\begin{aligned} f_s &= \text{switching frequency} \\ f_r &= \text{series resonance frequency} \\ &= \omega_r/(2\pi) = 1/\left[2\pi(L_s C_s)^{1/2}\right] \end{aligned} \tag{9.4}$$

The equivalent impedance looking into the terminals AB is given by

$$Z_{eq} = [B_1 + jB_2]/B_3 \qquad \text{p.u.} \tag{9.5}$$

where

$$B_1 = (8/\pi^2)(C_s/C_t)^2(Q_s/y_s)^2 \tag{9.6}$$

$$B_2 = Q_s[y_s - (1/y_s)]\left[1 + (8/\pi^2)^2(C_s/C_t)^2(Q_s/y_s)^2\right] - (C_s/C_t)(Q_s/y_s) \tag{9.7}$$

$$B_3 = 1 + (8/\pi^2)^2 (C_s/C_t)^2 (Q_s/y_s)^2 \tag{9.8}$$

The peak inverter output (resonant inductor) current can be calculated using

$$I_p = 4/[\pi |Z_{eq}|] \text{ p.u.} \tag{9.9}$$

The same current flows through the switching devices.

The value of initial current I_0 is given by

$$I_0 = I_p \sin(-\phi) \quad \text{p.u.} \tag{9.10}$$

where $\phi = \tan^{-1}(B_2/B_1)$ rad. B_1 and B_2 are given by Equation (9.6) and Equation (9.7), respectively. If I_0 is negative, then forced commutation is necessary and the converter is operating in the lagging PF mode. The peak voltage across the capacitor C_t' (on the secondary side) is

$$V_{ctp} = (\pi/2) V_o \text{ V} \tag{9.11}$$

The peak voltage across C_s and the peak current through C_t' are given by

$$V_{csp} = (Q_s/y_s).I_p \quad \text{p.u.} \tag{9.12}$$

$$I_{ctp} = V_{ctp}/(X_{ctpu}.R_L) \quad \text{A} \tag{9.13}$$

$$X_{ctpu} = (C_s/C_t)(Q_s/y_s) \quad \text{p.u.} \tag{9.14}$$

The plot of converter gain versus the switching frequency ratio y_s, obtained using Equation 9.1 is shown for $C_s/C_t = 1$ in Figure 9.37, for the lagging PF mode of operation. If the ratio C_s/C_t increases, then the converter takes the characteristics of SRC and the load voltage regulation requires a very wide range in the

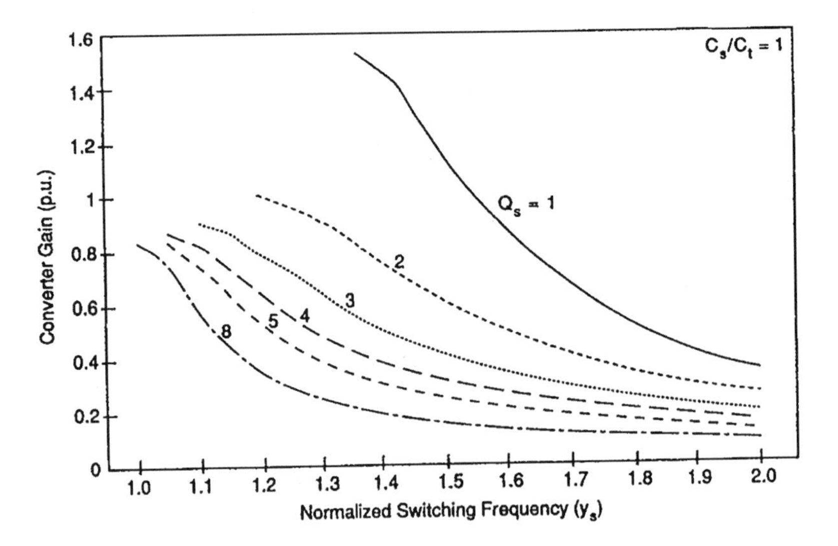

FIGURE 9.37 The converter gain M (p.u.) (normalized output voltage) versus normalized switching frequency y_s of SPRC operating above resonance for $C_s/C_t = 1$.

frequency change. Lower values of C_s/C_t take the characteristics of a PRC. Therefore, a compromised value of $C_s/C_t = 1$ is chosen.

It is possible to realize higher-order resonant converters with improved characteristics and many of them are presented in [10,13].

3. *Fixed-frequency operation.* In order to overcome some of the problems associated with the variable frequency control of resonant converters, they are operated with fixed frequency [7,12,13]. A number of configurations and control methods for fixed-frequency operation are available in the literature (Refs. [12,13] give a list of papers). One of the most popular methods of control is the phase-shift control (also called clamped-mode or PWM operation) method. Figure 9.38 illustrates the clamped-mode fixed-frequency operation of the SPRC. The load power control is achieved by changing the phase-shift angle ϕ between the gating signals to vary the pulse width of v_{AB}.

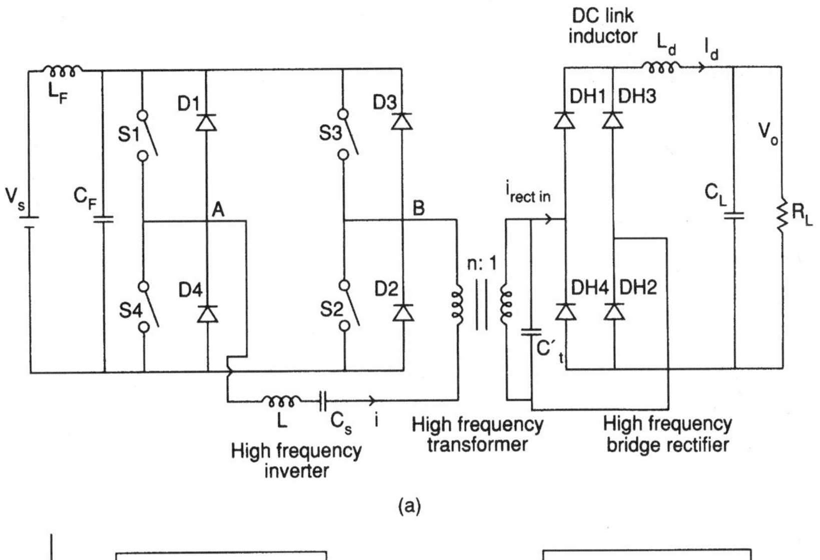

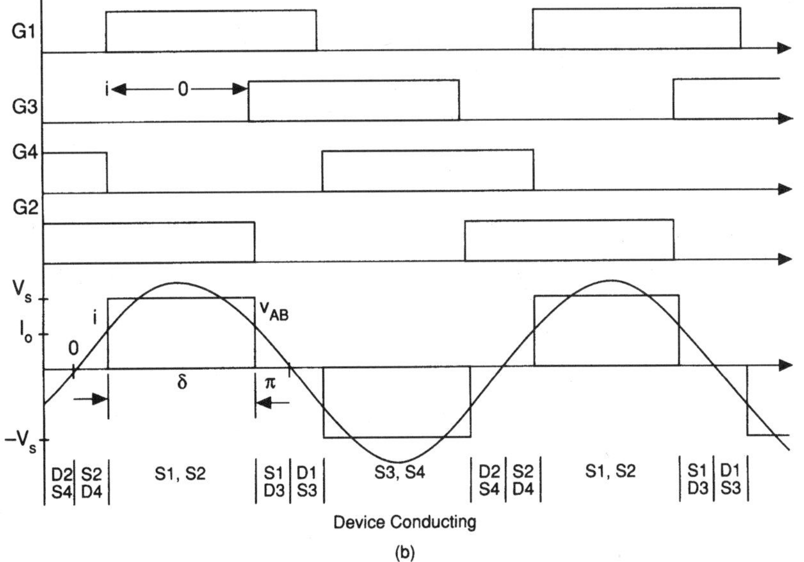

FIGURE 9.38 (a) Basic circuit diagram of series-parallel resonant converter suitable for fixed-frequency operation with PWM (clamped-mode) control. (b) Waveforms to illustrate the operation of fixed-frequency PWM series-parallel resonant converter working with a pulsewidth δ.

4. *Design example.* Design a 500 W output SPRC (half-bridge version) with secondary-side resonance (operation in lagging PF mode and variable-frequency control) with the following specifications:

$$\text{minimum input supply voltage} = 2E_{\text{min}}230 \text{ V}$$

$$\text{load voltage, } V_o = 48 \text{ V}$$

$$\text{switching frequency, } f_s = 100 \text{ kHz}$$

$$\text{maximum load current} = 10.42 \text{ A}$$

As explained in item 2, $C_s/C_t = 1$ is chosen. Using the constrains: (1) minimum kVA rating of tank circuit per kW output power, (2) minimum inverter output peak current, and (3) enough turn-off time for the switches, it can be shown [10] that $Q_s = 4$ and $y_s = 1.1$ satisfy the design constraints. From Figure 9.37, $M = 0.8$ p.u.
Average load voltage referred to primary-side of the HF transformer $= 0.8 \times 115$ V $= 92$ V. Therefore, the transformer turns ratio required $\cong 1.84$:

$$R_L' = n^2(V_o^2/P_o) = 15.6 \text{ ohms}$$

The values of L_s and C_s can be obtained by solving

$$(L_s/C_s)^{1/2} = 4 \times 15.6\ \Omega \quad \text{and} \quad \omega_r = 1/(L_s C_s)^{1/2} = 2\pi \times f_s/y_s$$

Solving the above equations gives $L_s = 109\ \mu$H and $C_s = 0.0281\ \mu$F. Leakage inductance $(L_p + L_s')$ of the HF transformer can be used as part of L_s. Typical value for a 100 kHz practical transformer (using Tokin Mn–Zn 2500B2 Ferrite, E–I type core) for this application is about 5 μH. Therefore, the external resonant inductance required is $L = 104\ \mu$H.

Since $C_s/C_t = 1$ is chosen, $C_t = 0.0281\ \mu$F. The actual value of C_t used on the secondary side of the HF transformer $= (1.84)^2 \times 0.0281 = 0.09514\ \mu$F. The resonating capacitors must be HF type (e.g., polypropylene) and must be capable of withstanding the voltage and current ratings obtained above (enough safety margin must be provided).

Using Equation (9.9) and Equation (9.11) through Equation (9.13):

$$\text{peak current through switches} = 7.6 \text{ A}$$

$$\text{peak voltage across } C_s,\ V_{csp} = 430 \text{ V}$$

$$\text{peak voltage (on secondary side) across } C_t',\ V_{ctp} = 76 \text{ V}$$

$$\text{peak current through the capacitor } C_t' \text{ (on secondary side)},\ I_{ctp} = 4.54 \text{ A}$$

A simple control circuit can be built using PWM IC SG3525 and TSC429 MOSFET driver ICs.

Resonant Transition Converters

There are two types of resonant transition converters, namely, zero-current transition (ZCT) [14] and zero-voltage transition (ZVT) [15–17] converters. These converters are realized from their PWM counterparts using LC resonant elements together with one (or more) auxiliary switch. The resonance circuit is activated using an auxiliary switch for a short duration either during turn-off transition (for ZCT) to achieve ZCS or during turn-on transition (for ZVT) to achieve ZVS. This family of converters operates in a similar way to PWM converters (square-wave voltage or current), retaining their advantages, while the main switch turns-off with zero-current in ZCT converter and turns-on with zero-voltage in ZVT converters. This concept can be used for single-ended [14–19] and double-ended converters [19–22].

With the development of digital ICs operating on low voltage (on the order of 3 V) supplies, the use of MOSFETs as *synchronous rectifiers* with very low voltage drop (~0.2 V) has become essential [4] to increase the efficiency of the power supply.

AC Power Supplies

Some applications of ac power supplies are ac motor drives, **uninterruptible power supply (UPS)** used as a standby ac source for critical loads (e.g., in hospitals, computers), and dc source-to-utility interface (either to meet peak power demands or to augment energy by connecting unconventional energy sources like photovoltaic arrays to the utility line). In ac induction motor drives, the ac main is rectified and filtered to obtain a smooth dc source, and then an inverter (single-phase version is shown in Figure 9.39) is used to obtain a variable-frequency, variable-voltage ac source. The sinusoidal pulse-width modulation technique described in Section "Power Conversion" can be used to obtain a sinusoidal output voltage. Some other methods used to get sinusoidal voltage output are [6] a number of phase-shifted inverter outputs summed in an output transformer to get a stepped waveform which approximates a sine wave, and the use of a bang-bang controller shown in Figure 9.39. All these methods use line-frequency (60 Hz) transformers for voltage translation and isolation purposes. To reduce the size, weight, and cost of such systems, one can use dc-to-dc converters (discussed earlier) as an intermediate stage. Figure 9.40 shows such a system in block schematic form. One can use an HF inverter circuit (discussed earlier) followed by a cycloconverter stage. The major

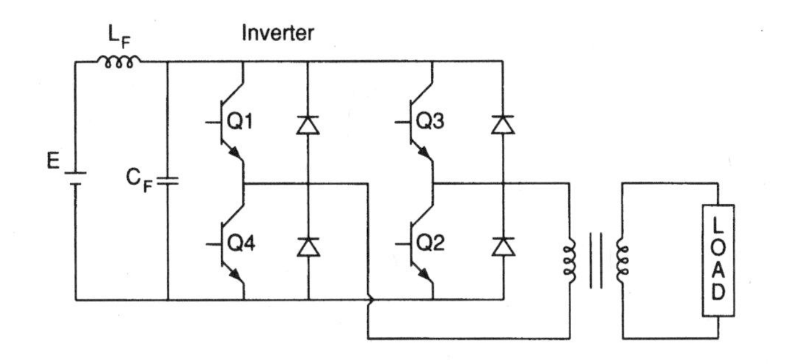

FIGURE 9.39 An inverter circuit to obtain variable-voltage, variable-frequency ac source. Using sinusoidal pulsewidth modulation control scheme, sine-wave ac output voltage can be obtained.

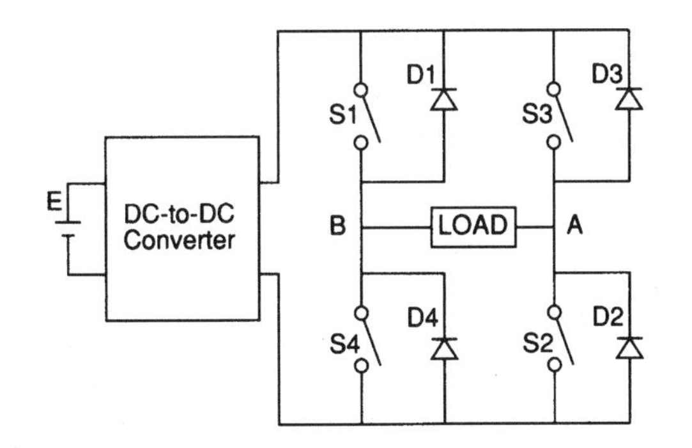

FIGURE 9.40 ac power supplies using HF switching (PWM or resonant) dc-to-dc converter as an input stage. HF transformer isolated dc-to-dc converters can be used to reduce the size and weight of the power supply. Sinusoidal voltage output can be obtained using the modulation in the output inverter stage or in the dc-to-dc converter.

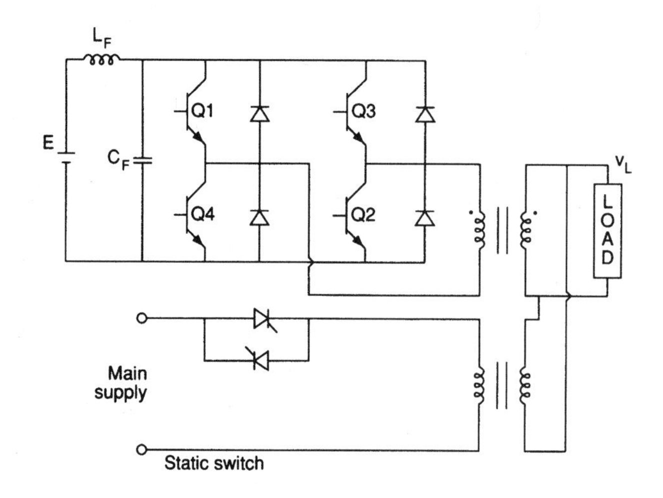

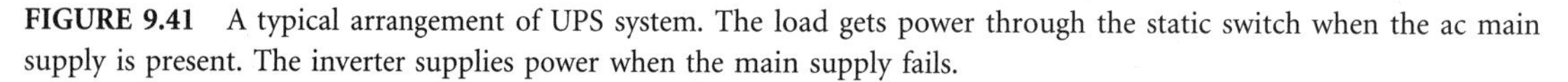

FIGURE 9.41 A typical arrangement of UPS system. The load gets power through the static switch when the ac main supply is present. The inverter supplies power when the main supply fails.

problem with these schemes is the reduction in efficiency due to the extra power stage. Figure 9.41 shows a typical UPS scheme. The battery shown has to be charged by a separate rectifier circuit.

ac-to-ac conversion can also be achieved using cycloconverters (e.g., Ref. [6]).

Special Power Supplies [6]

Using the inverters and cycloconverters, it is possible to realize bidirectional ac and dc power supplies. In these power supplies [6], power can flow in both directions, i.e., from input to output or from output to input. It is also possible to control the ac-to-dc converters to obtain sinusoidal line current with unity PF and low harmonic distortion at the ac source.

References

1. R. Severns and G. Bloom, *Modern Switching DC-to-DC Converters*, Princeton, NJ: Van Nostrand, 1988.
2. E.R. Hnatek, *Design of Solid-State Power Supplies*, 2nd ed., Princeton, NJ: Van Nostrand, 1981.
3. *Unitrode Switching Regulated Power Supply Design Seminar Manual*, Unitrode Corporation, 1984.
4. Motorola, *Linear/Switchmode Voltage Regulator Handbook*, 1989.
5. Philips Semiconductors, *Power Semiconductor Applications*, 1991.
6. M.H. Rashid, *Power Electronics: Circuits, Devices, and Applications*, Englewood Cliffs, NJ: Prentice-Hall, 1988.
7. K. Kit Sum, *Recent Developments in Resonant Power Conversion*, Intertech Communications Inc., 1988.
8. K.H. Liu, R. Oruganti, and F.C. Lee, "Resonant switches-topologies and characteristics," *IEEE Power Electron. Spec. Conf. Rec.*, 1985, pp. 106–116.
9. K.H. Liu and F.C. Lee, "Zero-voltage switching technique in DC/DC converters," *IEEE Power Electron. Spec. Conf. Rec.*, 1986, pp. 58–70.
10. A.K.S. Bhat, "A unified approach for the steady-state analysis of resonant converters," *IEEE Trans. Ind. Electron.*, vol. 38, no. 4, 1991, pp. 251–259.
11. R.L. Steigerwald, "A comparison of half-bridge resonant converter topologies," *IEEE Trans. Power Electron.*, vol. PE-3, no. 2, 1988, pp. 174–182.
12. A.K.S. Bhat, "Fixed frequency PWM series-parallel resonant converter," *IEEE Ind. Appl. Conf. Rec.*, pp. 1115–1121, 1988 (revised version appeared in *IEEE Trans. Ind. Appl.*, vol. 28, no. 5, 1992).
13. A.K.S. Bhat, "Analysis and design of a fixed-frequency LCL-type series-resonant converter with capacitive output filter," *IEE Proc. Circ. Devices Syst.*, vol. 144, no. 2, 1997, pp. 97–103.
14. G. Hua, E. Yang, and F.C. Lee, "Novel zero current transition PWM converters," *IEEE Power Electron. Spec. Conf. Rec.*, 1993, pp. 538–544.

15. G. Hua, C. Leu and F.C. Lee, "Novel zero voltage transition PWM converters," *IEEE Trans. Power Electron.*, vol. 9, no. 2, 1994, pp. 213–219.
16. R. Streit and D. Tollik, "High efficiency telecom rectifier using a novel soft-switched boost based input current shaper," *IEEE INTELEC Conf. Rec.*, 1991, pp. 720–726.
17. R. Gurunathan and A.K.S. Bhat, "A soft-switched boost converter for high-frequency Operation," *IEEE Power Electron. Spec. Conf. Rec.*, 1999, pp. 463–468.
18. I. Batarseh, *Power Electronic Circuits*, London: Wiley, 2004.
19. R.W. Erickson and D. Maksimovic, *Fundamentals of Power Electronics*, 2nd ed., Dordrecht, Germany: Kluwer, 2001.
20. J.G Cho, J.A Sabate, and F.C. Lee, " Novel full bridge zero voltage transition pwm dc–dc converter for high power applications," *IEEE Appl. Power Electron. Conf. Rec.*, 1994, pp. 143–149.
21. F.S. Hamdad and A.K.S. Bhat, "A novel pulse width control scheme for fixed-frequency zero-voltage-switching DC-to-DC PWM bridge converter," *IEEE Trans. Ind. Electron.*, vol. 48, no. 1, 2001, pp. 101–110.
22. N. Mohan, T.M. Undeland, and W.P. Robbins, *Power Electronics: Converters, Applications, and Design*, 2nd ed., New York: Wiley, 1995.

9.4 Converter Control of Machines

Bimal K. Bose

Converter-controlled electrical machine drives are very important in modern industrial applications. Some examples in the high-power range are metal rolling mills, cement mills, and gas line compressors. In the medium-power range are textile mills, paper mills, and subway car propulsion. Machine tools and computer peripherals are examples of converter-controlled electrical machine drive applications in the low-power range. The converter normally provides a variable-voltage dc power source for a dc motor drive and a variable-frequency, variable-voltage ac power source for an ac motor drive. The drive system efficiency is high because the converter operates in switching mode using power semiconductor devices. The primary control variable of the machine may be torque, speed, or position, or the converter can operate as a solid-state starter of the machine. The recent evolution of high-frequency power semiconductor devices and high-density and economical microelectronic chips, coupled with converter and control technology developments, is providing a tremendous boost in the applications of drives.

Converter Control of dc Machines

The speed of a dc motor can be controlled by controlling the dc voltage across its armature terminals. A phase-controlled thyristor converter can provide this dc voltage source. For a low-power drive, a single-phase bridge converter can be used, whereas for a high-power drive, a three-phase bridge circuit is preferred. The machine can be a permanent magnet or wound field type. The wound field type permits variation and reversal of field and is normally preferred in large power machines.

Phase-Controlled Converter dc Drive

Figure 9.42 shows a dc drive using a three-phase thyristor bridge converter. The converter rectifies line ac voltage to variable dc output voltage by controlling the firing angle of the thyristors. With rated field excitation, as the armature voltage is increased, the machine will develop speed in the forward direction until the rated, or base, speed is developed at full voltage when the firing angle is zero. The motor speed can be increased further by weakening the field excitation. Below the base speed, the machine is said to operate in constant torque region, whereas the field weakening mode is defined as the constant power region. At any operating speed, the field can be reversed and the converter firing angle can be controlled beyond 90° for **regenerative braking** mode operation of the drive. In this mode, the motor acts as a generator (with negative induced voltage) and the converter acts as an inverter so that the mechanical energy stored in the inertia is converted to electrical energy and pumped back to the source. Such **two-quadrant** operation gives improved efficiency if the drive accelerates and decelerates frequently. The speed of the machine can be controlled with precision by a feedback loop where the command

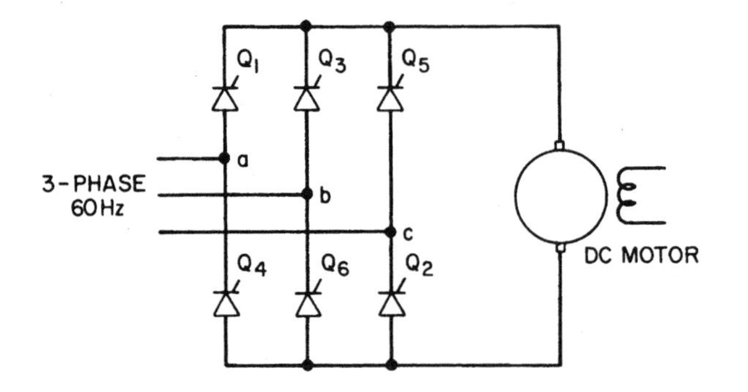

FIGURE 9.42 Three-phase thyristor bridge converter control of a dc machine.

speed is compared with the machine speed measured by a tachometer. The speed loop error generally generates the armature current command through a compensator. The current is then feedback controlled with the firing angle control in the inner loop. Since torque is proportional to armature current (with fixed field), a current loop provides direct torque control, and the drive can accelerate or decelerate with the rated torque. A second bridge converter can be connected in antiparallel so that the dual converter can control the machine speed in all the four quadrants (motoring and regeneration in forward and reverse speeds).

Pulsewidth Modulation Converter dc Machine Drive

Four-quadrant speed control of a dc drive is also possible using an H-bridge pulsewidth modulation (PWM) converter as shown in Figure 9.43. Such drives (using a permanent magnet dc motor) are popular in low-power applications, such as robotic and instrumentation drives. The dc source can be a battery or may be obtained from ac supply through a diode rectifier and filter. With PWM operation, the drive response is very fast and the armature current ripple is small, giving less harmonic heating and torque pulsation. Four-quadrant operation can be summarized as follows:

Quadrant 1: Forward motoring (buck or step-down converter mode)
- Q_1—on
- Q_3, Q_4—off
- Q_2—chopping
- Current freewheeling through D_3 and Q_1

Quadrant 2: Forward regeneration (boost or step-up converter mode)
- Q_1, Q_2, Q_3—off
- Q_4—chopping
- Current freewheeling through D_1 and D_2

Quadrant 3: Reverse motoring (buck converter mode)
- Q_3—on
- Q_1, Q_2—off

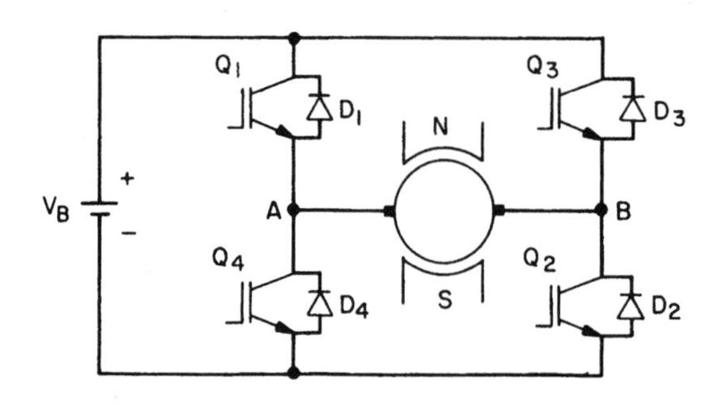

FIGURE 9.43 Four-quadrant dc motor drive using an H-bridge converter.

Q_4—chopping

Current freewheeling through D_1 and Q_3

Quadrant 4: Reverse regeneration (boost converter mode)

Q_1, Q_3, Q_4—off

Q_2—chopping

Current freewheeling through D_3 and D_4

Often a drive may need only a one- or two-quadrant mode of operation. In such a case, the converter topology can be simple. For example, in one-quadrant drive, only Q_2 chopping and D_3 freewheeling devices are required, and the terminal A is connected to the supply positive. Similarly, a two-quadrant drive will need only one leg of the bridge, where the upper device can be controlled for motoring mode and the lower device can be controlled for regeneration mode.

Converter Control of ac Machines

Although application of dc drives is quite common, disadvantages are that the machines are bulky and expensive, and the commutators and brushes require frequent maintenance. In fact, commutator sparking prevents machine application in an unclean environment, at high speed, and at high elevation. ac machines, particularly the cage-type induction motor, are favorable when compared with all the features of dc machines. Although converter system, control, and signal processing of ac drives is definitely complex, the evolution of ac drive technology in the past two decades has permitted more economical and higher performance ac drives. Consequently, ac drives are finding expanding applications, pushing dc drives toward obsolescence.

Voltage-Fed Inverter Induction Motor Drive

A simple and popular converter system for speed control of an induction motor is shown in Figure 9.44. The front-end diode rectifier converts 60 Hz ac to dc, which is then filtered to remove the ripple. The dc voltage is then converted to variable-frequency, variable-voltage output for the machine through a PWM bridge inverter. Among a number of PWM techniques, the sinusoidal PWM is common, and it is illustrated in Figure 9.45 for one phase only. The stator sinusoidal reference phase voltage signal is compared with a high-frequency carrier wave, and the comparator logic output controls switching of the upper and lower transistors in a phase leg. The phase voltage wave shown refers to the fictitious center tap of the filter capacitor. With the PWM technique, the fundamental voltage and frequency can be easily varied. The stator voltage wave contains high-frequency ripple, which is easily filtered by the machine leakage inductance. The voltage-to-frequency ratio is kept constant to provide constant airgap flux in the machine. The machine voltage-frequency relation, and the corresponding torque, stator current, and slip, are shown in Figure 9.46. Up to the base or rated frequency ω_b, the machine can develop constant torque. Then, the field flux weakens as the frequency is increased at

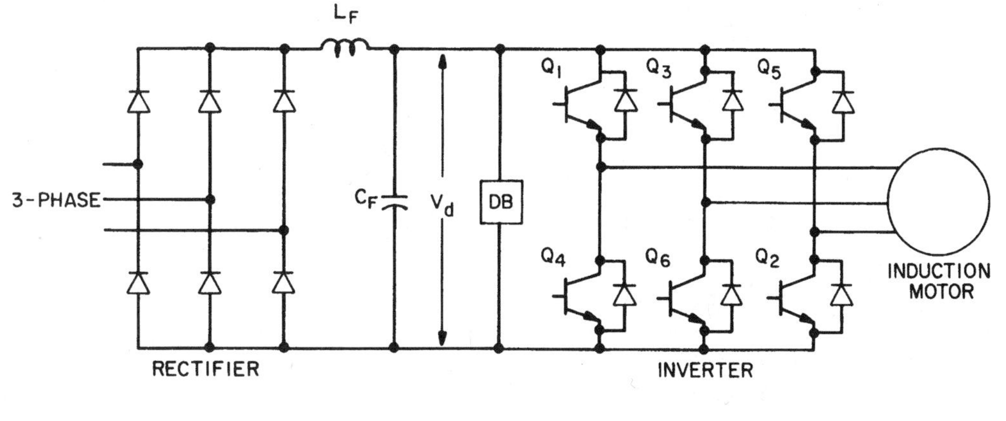

FIGURE 9.44 Diode rectifier PWM inverter control of an induction motor.

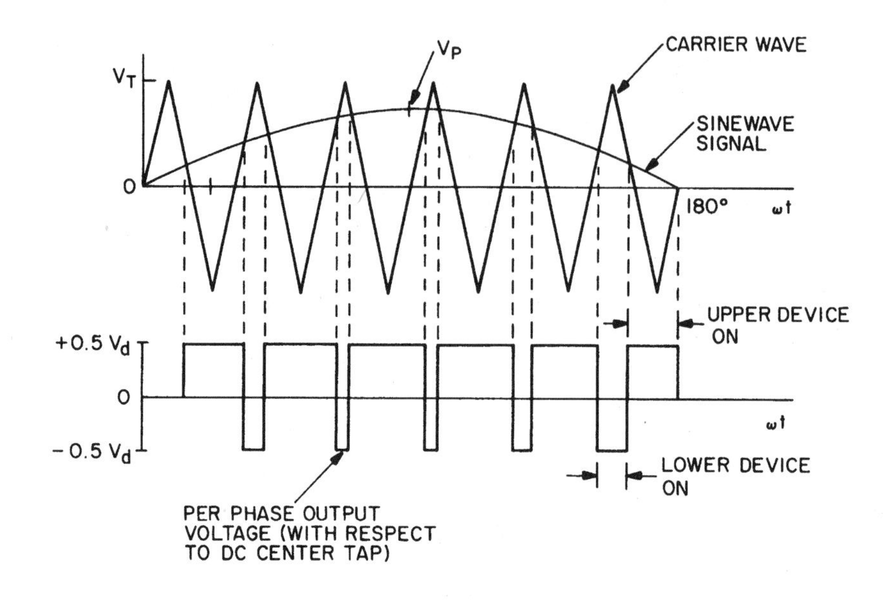

FIGURE 9.45 Sinusoidal pulse width modulation principle.

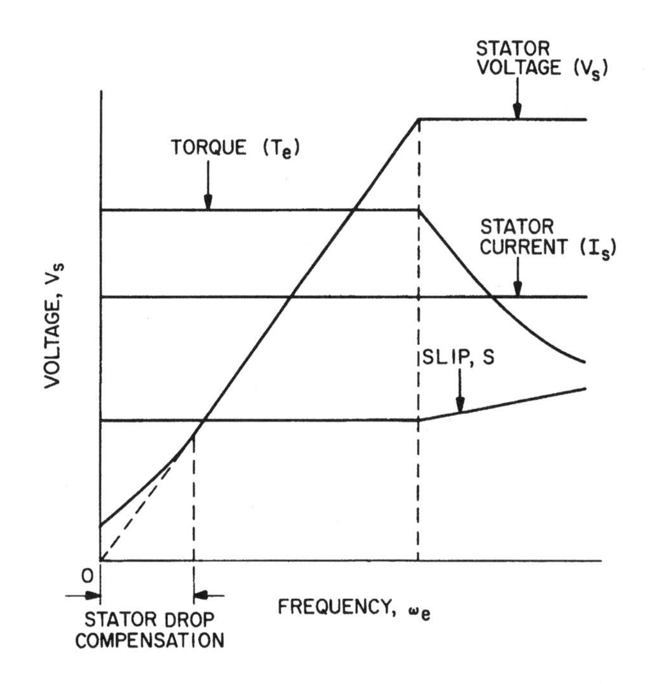

FIGURE 9.46 Voltage-frequency relation of an induction motor.

constant voltage. The speed of the machine can be controlled in a simple open-loop manner by controlling the frequency and maintaining the proportionality between the voltage and frequency. During acceleration, machine-developed torque should be limited so that the inverter current rating is not exceeded. By controlling the frequency, the operation can be extended in the field weakening region. If the supply frequency is controlled to be lower than the machine speed (equivalent frequency), the motor will act as a generator and the inverter will act as a rectifier, and energy from the motor will be pumped back to the dc link. The **dynamic brake** shown is nothing but a buck converter with resistive load that dissipates excess power to maintain the dc

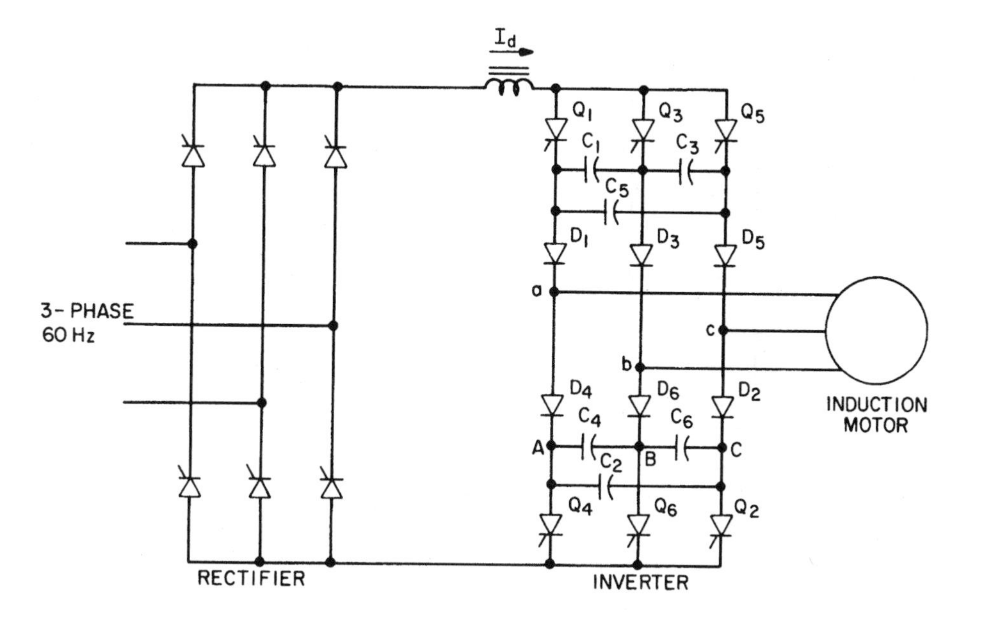

FIGURE 9.47 Force-commutated current-fed inverter control of an induction motor.

bus voltage constant. When the motor speed is reduced to zero, the phase sequence of the inverter can be reversed for speed reversal. Therefore, the machine speed can be easily controlled in all four quadrants.

Current-Fed Inverter Induction Motor Drive

The speed of a machine can be controlled by a current-fed inverter as shown in Figure 9.47. The front-end thyristor rectifier generates a variable dc current source in the dc link inductor. The dc current is then converted to six-step machine current wave through the inverter. The basic mode of operation of the inverter is the same as that of the rectifier, except that it is **force-commutated**, that is, the capacitors and series diodes help commutation of the thyristors. One advantage of the drive is that regenerative braking is easy because the rectifier and inverter can reverse their operation modes. Six-step machine current, however, causes large harmonic heating and torque pulsation, which may be quite harmful at low-speed operation. Another disadvantage is that the converter system cannot be controlled in open loop like a voltage-fed inverter.

Current-Fed PWM Inverter Induction Motor Drive

The force-commutated thyristor inverter in Figure 9.47 can be replaced by a **self-commutating** gate turn-off (GTO) thyristor PWM inverter as shown in Figure 9.48. The output capacitor bank shown has two functions: (1) it permits PWM switching of the GTO by diverting the load inductive current, and (2) it acts as a low-pass filter causing sinusoidal machine current. The second function improves machine efficiency and attenuates the irritating magnetic noise. Note that the fundamental machine current is controlled by the front-end rectifier, and the fixed PWM pattern is for controlling the harmonics only. The GTO is to be the reverse-blocking type. Such drives are popular in the multimegawatt power range. For lower power, an **insulated gate bipolar transistor** (**IGBT**) or transistor can be used with a series diode.

Cycloconverter Induction Motor Drive

A phase-controlled cycloconverter can be used for speed control of an ac machine (induction or synchronous type). Figure 9.49 shows a drive using a three-pulse half-wave or 18-thyristor cycloconverter. Each output phase group consists of positive and negative converter components which permit bidirectional current flow. The firing angle of each converter is sinusoidally modulated to generate the variable-frequency, variable-voltage output required for ac machine drive. Speed reversal and regenerative mode operation are easy.

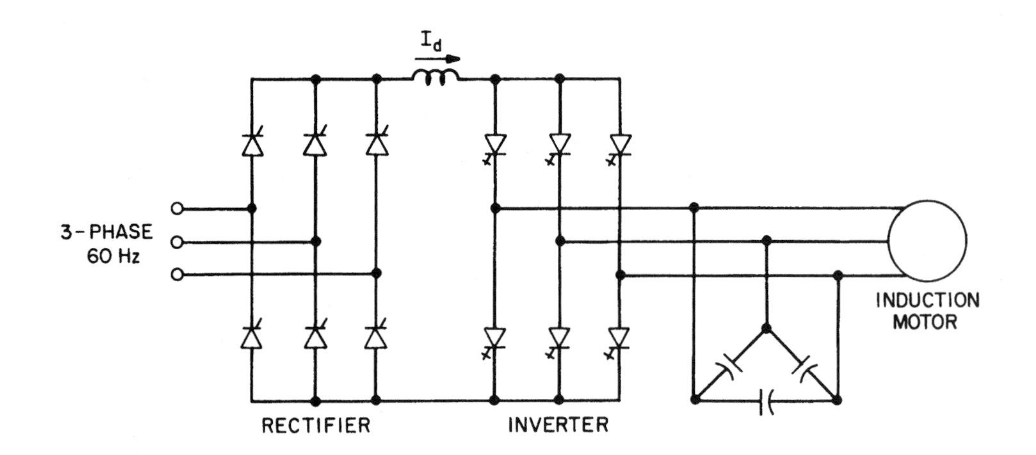

FIGURE 9.48 PWM current-fed inverter control of an induction motor.

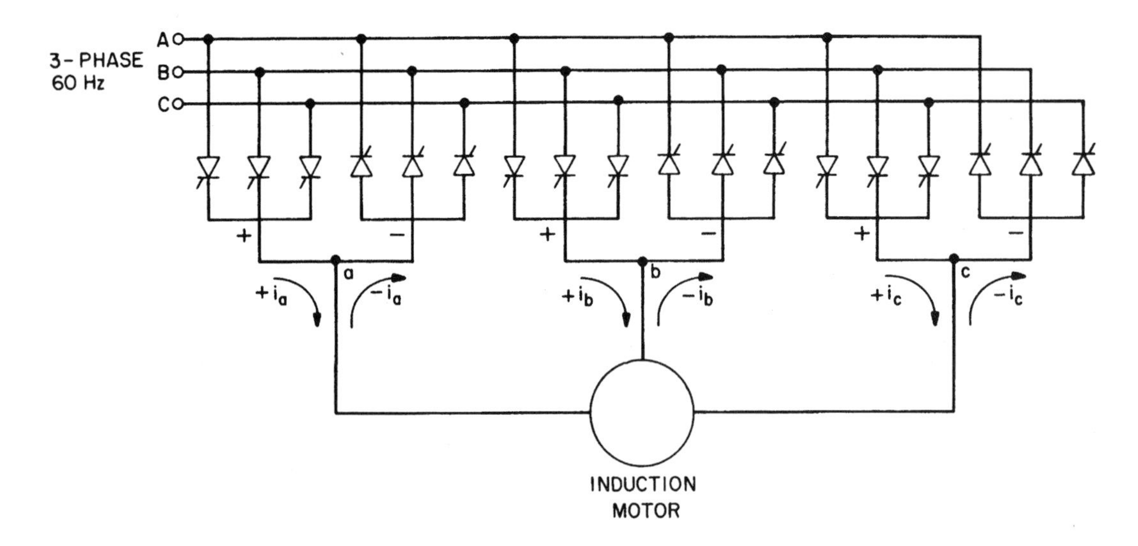

FIGURE 9.49 Cycloconverter control of an induction motor.

The cycloconverter can be operated in blocking or circulating current mode. In blocking mode, the positive or negative converter is enabled, depending on the polarity of the load current. In circulating current mode, the converter components are always enabled to permit circulating current through them. The circulating current reactor between the positive and negative converter prevents short circuits due to ripple voltage. The circulating current mode gives simple control and a higher range of output frequency with lower harmonic distortion.

Slip Power Recovery Drive of Induction Motor

In a cage-type induction motor, the rotor current at slip frequency reacting with the airgap flux develops the torque. The corresponding slip power is dissipated in the rotor resistance. In a wound rotor induction motor, the slip power can be controlled to control the torque and speed of a machine. Figure 9.50 shows a popular slip power-controlled drive, known as a static Kramer drive. The slip power is rectified to dc with a diode rectifier and is then pumped back to an ac line through a thyristor phase-controlled inverter. The method permits speed control in the subsynchronous speed range. It can be shown that the developed machine torque is proportional to the dc link current I_d and the voltage V_d varies directly with speed deviation from the

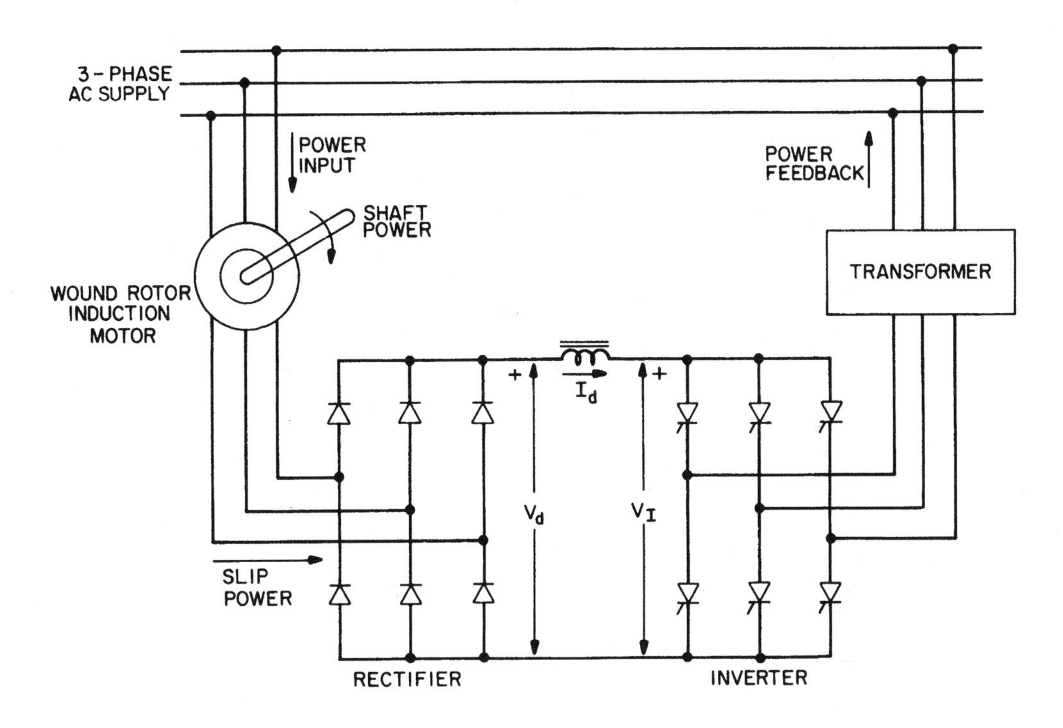

FIGURE 9.50 Slip power recovery control of a wound rotor induction motor.

synchronous speed. The current I_d is controlled by the firing angle of the inverter. Since V_d and V_I voltages balance at steady state, at synchronous speed the voltage V_d is zero and the firing angle is 90°. The firing angle increases as the speed falls, and at 50% synchronous speed the firing angle is near 180°. This is practically the lowest speed in static Kramer drive. The transformer steps down the inverter input voltage to get a 180° firing angle at lowest speed. The advantage of this drive is that the converter rating is low compared with the machine rating. Disadvantages are that the line power factor is low and the machine is expensive. For limited speed range applications, this drive has been popular.

Wound Field Synchronous Motor Drive

The speed of a wound field synchronous machine can be controlled by a current-fed converter scheme as shown in Figure 9.47, except that the forced-commutation elements can be removed. The machine is operated at leading power factor by overexcitation so that the inverter can be load commutated. Because of the simplicity of converter topology and control, such a drive is popular in the multimegawatt range.

Permanent Magnet Synchronous Motor Drive

Permanent magnet (PM) machine drives are quite popular in the low-power range. A PM machine can have sinusoidal or concentrated winding, giving the corresponding sinusoidal or trapezoidal induced stator voltage wave. Figure 9.51 shows the speed control system using a trapezoidal machine, and Figure 9.52 explains the wave forms. The power MOSFET inverter supplies variable-frequency, variable-magnitude six-step current wave to the stator. The inverter is self-controlled, that is, the firing pulses are generated by the machine position sensor through a decoder. It can be shown that such a drive has the features of dc drive and is normally defined as *brushless dc drive.* The speed control loop generates the dc current command, which is then controlled by the **hysteresis-band** method to construct the six-step phase current waves in correct phase relation with the induced voltage waves as shown in Figure 9.52. The drive can easily operate in four-quadrant mode.

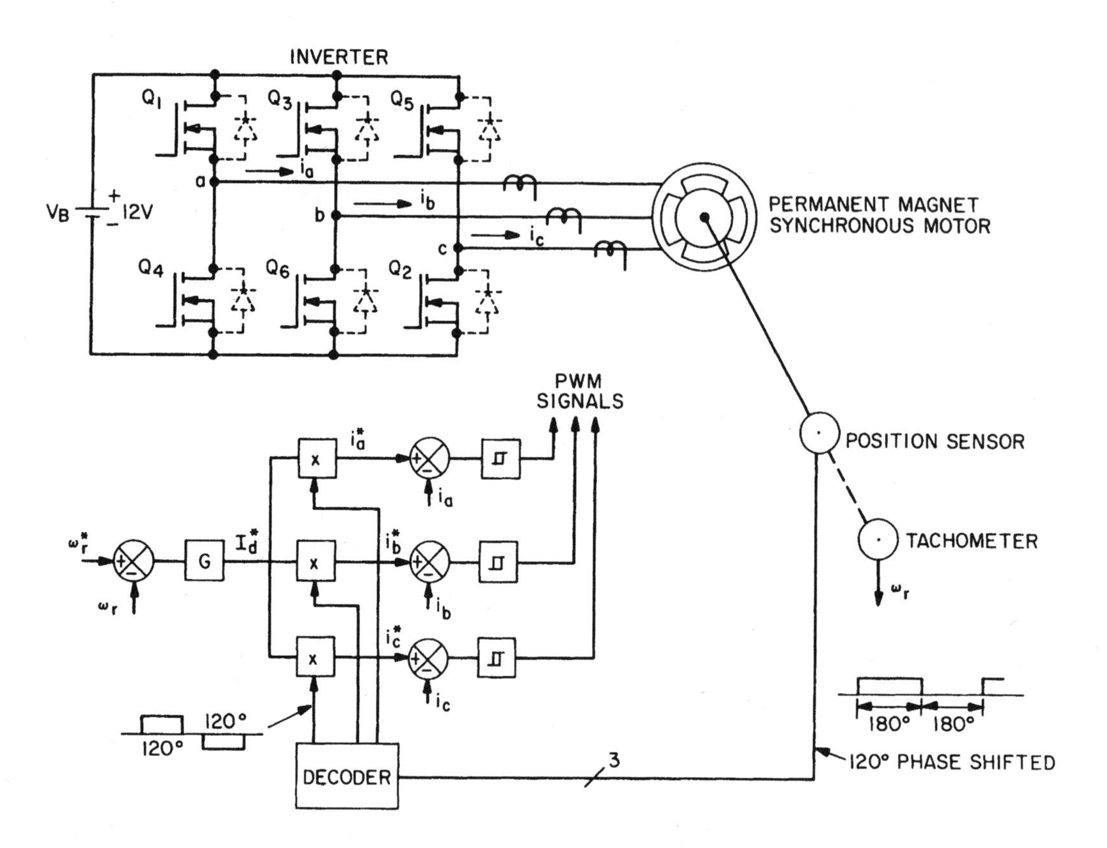

FIGURE 9.51 Permanent magnet synchronous motor control with PWM inverter.

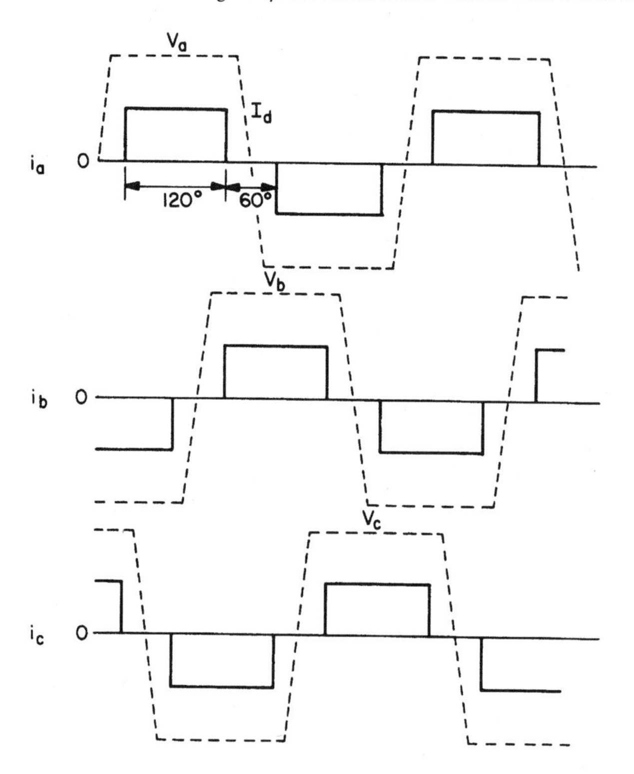

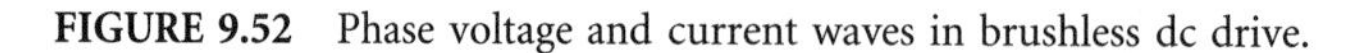
FIGURE 9.52 Phase voltage and current waves in brushless dc drive.

Defining Terms

Dynamic brake: The braking operation of a machine by extracting electrical energy and then dissipating it in a resistor.
Forced-commutation: Switching off a power semiconductor device by external circuit transient.
Four-quadrant: A drive that can operate as a motor as well as a generator in both directions.
Hysteresis-band: A method of controlling current where the instantaneous current can vary within a band.
Insulated gate bipolar transistor (IGBT): A device that combines the features of a power transistor and MOSFET.
Regenerative braking: The braking operation of a machine by converting its mechanical energy into electrical form and then pumping it back to the source.
Self-commutation: Switching of a power semiconductor device by its gate or base drive.
Two-quadrant: A drive that can operate as a motor as well as a generator in one direction.

References

B.K. Bose, *Power Electronics and AC Drives*, Englewood Cliffs, NJ: Prentice-Hall, 1986.
B.K. Bose, "Adjustable speed AC drives—A technology status review," *Proc. IEEE*, vol. 70, pp. 116–135, Feb. 1982.
B.K. Bose, *Modern Power Electronics*, New York: IEEE Press, 1992.
J.M.D. Murphy and F.G. Turnbull, *Power Electronic Control of AC Motors*, New York: Pergamon Press, 1988.
P.C. Sen, *Thyristor DC Drives*, New York: John Wiley, 1981.

9.5 Photoconductive Devices in Power Electronics

Sudip K. Mazumder, Tirthajyoti Sarkar, Mitra Dutta, and Michael S. Mazzola

Introduction

Photoconductive Devices

Semiconductor electronic devices operate by the controlled motion of charge carriers inside them. One has to induce some nonequilibrium in the charge carrier density to operate a semiconductor device. This can be done either by external electric field (*electrical injection*), optical pulse (*optical injection*), or by thermal energy (*thermionic emission*). Photoconductive devices fall within a special class of electronic devices where the transport or the initiation of transport of these charge carriers is done by optical energy.

Photoconductive devices operate on the principle of photogeneration. The phenomenon is shown in Figure 9.53, where an electron is transported from the valence band to the conduction band by the incidence of a photon. The photon energy ($h\nu$), where ν is the frequency of the light and h is the Planck's constant, must be greater than the bandgap energy (E_g) of the semiconductor, for photogeneration to occur. Transport of one electron from valence band to conduction band also implies the generation of a hole in the valence band. So photogeneration always occurs in electron–hole (E–H) pairs. These E–H pairs take part in current conduction in the device and because they are generated by light incidence, we call the device a photoconductive device.

Why Photoconductive Devices in Power Electronics?

Advancement in power electronics (PE) has always been dependent upon the advance in power semiconductor switches [1]. Starting from the early days of p-n-junction rectifier and power BJTs, and cruising through power MOSFET, thyristor, gate turn-off thyristor (GTO), we have now reached the age of advance devices like merged p-i-n/Schottky (MPS) rectifier or nonpunch-through insulated gate bipolar transistor (NPT-IGBT). They are all electrically operated, i.e., the control signal to open or close the switch in a power electronics system is electrical. This can be either the base drive current for a power BJT or the gate drive voltage for a power MOSFET or an IGBT. Photoconductive or optically triggered devices (OTD), when used in PE

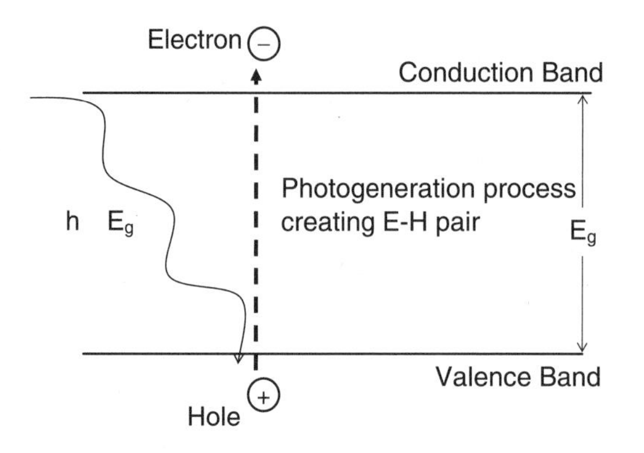

FIGURE 9.53 Photogeneration process by the incidence of an optical pulse having energy ($h\nu$) greater than the bandgap energy (E_g) of the semiconductor and the corresponding creation of the electron–hole (E–H) pair.

systems, have the potential to significantly increase the switching frequencies for all power levels and eliminate some of the major problems of electrically-triggered devices (ETD). In an ETD, the gate of the power device is fired electrically; while the triggering action in an OTD is provided by the photo-generation of E–H pairs, which may be direct band-to-band transition or between the band and the deep levels in the energy gap depending directly upon the energy of the photonic source [2]. Because of this unique characteristic, OTDs provide jitter-free operations, fast opening and closing times, high repetition rates, and fast recovery times, all of which enable high frequency of operation without sacrificing switching efficiency. In addition, OTDs are scalable for handling low, medium, and high power, and have low storage effects and inductance, which are useful for monolithic integration.

Advantages: Device Physics Point of View

Demand for ever-increasing switching frequency in power electronics has placed stress not only on the external control system, but also on the design of the semiconductor switches. The switches must be physically large to handle power, but at the same time, their large size places limitations on the ability to rapidly control the switch, especially during controlled turn-off. The physical ability to remove the charge, stored in the junctions of bipolar devices, becomes, in particular, a serious limitation as device size is increased. The alternative is to parallel devices in multi-chip modules, which has limitations both technically and economically. Since highly efficient devices are still mostly built from bipolar configurations (such as BJTs, GTOs, and IGBTs), this represents a fundamental limitation: High power or high switching frequency? In other words, today's power electronics systems can be either high power or high power density, but it is difficult to have both simultaneously.

The basic problem is scaling the device dimensions, which presents two fundamental physical problems:

- Limitations imposed by multi-dimensional field distributions
- Limitations imposed by multi-dimensional charge transport

With respect to the second limitation, the problem is how to inject or extract charge from large volumes to realize conventional gate control. The conventional approach is to create structures with large surface-area-to-volume ratios (or large perimeter-length-to-area ratios). A high-power gate-turn-off (GTO) thyristor with a convoluted gate structure is a common example. The trench structure used in power field-effect transistors (FET) is another. The basic problem is one of the boundary conditions. "Spreading out" the current allows a power device to be scaled for higher current ratings without exceeding the safe current density limit of the material. At some point, however, boundaries must intrude to permit gate or base control and to collect charge flowing to or from the external circuit. Nonuniform current density distributions eventually arise as the switching speed increases. This may be transient, but the effects can still be inconvenient or even catastrophic.

Current filamentation, thermal runaway, second breakdown, and snap back are all manifestations of the influence of this multi-dimensional charge transport in power semiconductor devices at the limits of switching speed.

Optical control of power semiconductor devices is an important technical approach for pushing these fundamental limits up. The optical injection of either minority or majority carriers over large areas is straightforward. For example, a device base or gate region can be irradiated with photons possessing at least the bandgap energy to inject E–H pairs in thin regions where charge injection initiates conduction. Absorption of photons with subbandgap energy (through trap-assisted optical excitation) also makes it possible to optically inject carriers into large volumes, such as the drift region. Both processes fundamentally eliminate the delay and the nonuniform current density distributions associated with the electrical transport of charge carriers to the gate or base control regions of large power devices.

However, there is another reason to consider optically activated power switches. The optical stimulation of deep trap levels can also remove mobile charge from the conduction and valence bands, thus providing a method for optically controlling switching a power semiconductor on and off. The physical mechanism of infrared quenching of photoconductivity has long been known, and will be discussed in the context of the bistable optically stimulated switch (BOSS) in the next section. Infrared quenching has been successfully applied to provide a purely optical control technique for command turn-on and turn-off of power switches [3].

Advantages: Circuit Point of View

On a circuit level, the simple difference between optically and electrically triggered systems results in some key additional advantages for the former, that are especially significant at high switching frequencies.

First, in an OTD there is complete electrical isolation between the gate driver and the power stage. As such, very high di/dt and dv/dt, which cause significant reliability problems in an ETD at a high switch frequency, have no impact on an OTD. Therefore, the basic architecture of the gate driver in an OTD is simple.

Second, for two- and higher-level electrically triggered switching converters, different designs of low- and high-side drivers are required; the latter is especially difficult to design for medium- and high-power applications. For an optically triggered converter, the designs of high- and low-side drivers remain the same.

Third, for firing OTDs (as opposed to that for ETDs), optically multiplexed triggering signals (as often encountered in modern optical fiber-based control systems) do not need to be converted back to electrical signals. This leads to simplicity of overall design, enhanced controlled bandwidth, enhanced reliability of the system, and monolithic integration.

Fourth, as the switching frequency of an ETD increases, parasitic oscillations may be induced in the driver circuit owing to coupling effects between the device capacitance and the parasitic inductance of the gate connection and also due to transmission-line effects. This may lead to failure of the gate driver; with an OTD such possibilities do not arise.

Fifth, an OTD based converter does not suffer from gate-driver failure due to short circuiting.

Clearly, photoconductive or optically triggered devices have the potential to address these problems faced with conventional ETDs. They are emerging devices for power electronic applications and much research is going on in this topic. In the next sections, some of the state-of-the-art photoconductive devices will be discussed along with their operational features.

Principal Photoconductive Devices in Power Electronics

All of the photoconductive devices work on the principle of photogeneration of excess carriers which disturbs the equilibrium of charge carriers inside semiconductor and results in current conduction; but some are particularly suitable for power electronic applications, which demand specific attributes from a semiconductor switch such as high reverse voltage blocking capability, large current density handling capacity, low on-state voltage drop, low opening and closing delays, and precise controllability. Good thermal stability, low contact and surface defect densities, and radiation hardness are additional requirements for higher reliability and for extreme applications (military or space applications). We discuss the important photoconductive devices and their respective pros and cons as a power semiconductor switch.

Photoconductive Switch

One of the main uses of optical control could be enabling a different scaling law for the design of high blocking-voltage power devices. The depletion region charge distribution fundamentally limits the blocking voltage in conventional devices. And, even then the fundamental limit is seldom achieved because surfaces and curvature in the doped regions produce electric field magnitudes in excess of the critical value at reduced voltage. The width W of the triangular field distribution in the one-dimensional solution to the Poisson equation describing the uniformly doped depletion region in a p^+-n-junction is given by:

$$W = \sqrt{\frac{2\varepsilon_s}{q}\frac{V_{bi} + V_B}{N_D}} \approx \sqrt{\frac{2\varepsilon_s}{q}\frac{V_B}{N_D}} = \frac{\varepsilon_s}{qN_D E_c} \tag{9.15}$$

where ε_s is the dielectric constant of the semiconductor, q is the elementary charge, N_D is the net doping concentration in the controlling (lightly doped) portion of the depletion region, and E_c is the critical electric field of the semiconductor. In simplifying Equation (9.15) we neglect the built-in potential $V_{bi} \sim 1$ V. Solving for the maximum blocking voltage V_B of the junction we obtain:

$$V_B = \frac{1}{2}\frac{\varepsilon_s}{qN_D}E_c^2 \tag{9.16}$$

This well-known result demonstrates that in a conventional power device the maximum blocking voltage is limited by the net doping of the depletion region. Practical limits on both N_D and the drift region thickness W limits blocking voltages to the order of 10 kV in silicon.

A conceptually different approach uses a neutral semiconductor with very low equilibrium conductivity σ between two ohmic contacts to block larger voltages. This approach, the heart of the photoconductive switch (PCS), which is discussed in the next section, typically uses a semiinsulating (SI) semiconductor such as gallium arsenide. SI GaAs is routinely grown with $\sigma < 10^{-8}/\Omega\text{cm}$. Recently, SI-silicon carbide (SiC) has become commercially available with more than 1000 times lower conductivity.

Provided that the diffusion length of the majority carriers is much less than the thickness of the SI region W, the blocking voltage can be estimated using Ohm's law

$$V_B = \frac{J}{\sigma}W \tag{9.17}$$

where J is the ohmic current density. To the extent that ohmic current–voltage characteristics (zeroth-order dimensional physics) apply, then V_B can be increased independently of doping simply by increasing W. Until instabilities, associated with multidimensional charge transport, intrude the only theoretical limits on V_B are that of $V_B/W < E_c$ and that J^2/σ does not exceed safe operating thermal limits. Under laboratory conditions, single PCSs, based on highly resistive silicon, have demonstrated $V_B > 100$ kV [1]. The problem of making such a highly resistive switch turn on is resolved by the optical injection of photoconductors that drift across the SI material.

There are two operating modes of PCSs that have been researched rather extensively. The first is the optically triggered PCS and the second is the optically sustained PCS. The device structure is typically the same (two ohmic contacts separated by a highly resistive semiconductor). The separate operating modes arise from the use of photoconductive gain γ defined as

$$\gamma = \frac{\tau_L}{t_d} \tag{9.18}$$

where τ_L is the carrier lifetime and t_d is the average time required by a carrier to drift across the photoconductive switch. If $\gamma > 1$ then the PCS can be optically triggered; if $\gamma < 1$ then the PCS must be optically sustained. PCSs with large photoconductive gain ($\gamma >> 1$) are used as efficient closing switches, because they can be triggered on with a short pulse of light after which they remain in conduction for durations of the order of τ_L. Regenerative conduction is supplied by the carrier injection from the ohmic contacts (as opposed to across a junction as in a thyristor). Since the PCS is intended for power applications, a relatively large gap between the electrodes is required, so that τ_L must be exceptionally long (0.1 to 1 msec) to observe large photoconductive gain. Bipolar conduction is possible in materials with inefficient direct recombination such as silicon, which is an indirect bandgap semiconductor available in high purity so that trap-assisted recombination is minimized. Semiconductors with efficient E–H recombination such as gallium arsenide can also exhibit large photoconductive gain via unipolar conduction. Here, a trap is introduced into the bandgap that efficiently traps only one of the carrier types (e.g., holes). The opposite carrier type becomes the majority carrier (e.g., electrons) provided that the majority carrier lifetime satisfies $\gamma > 1$. This approach has long been used in CdS photocells. An example of this type of photoconductor applied to a power switch is copper-doped GaAs, which will be treated in greater detail next.

There are advantages and disadvantages of the optically triggered PCS in power electronics. One advantage is the voltage scalability of a PCS by increasing the linear dimension; thus, a single PCS can be made to handle much larger voltages than a typical conventional power switch (i.e., thyristor or IGBT). Another advantage is that the regenerative conduction mechanism is not easily triggered by electrical means, making an optically triggered PCS virtually immune to spurious dv/dt triggering in conventional power electronics applications. Indeed, the dielectric strength of the PCS usually increases under transient voltages (i.e., with increasing dv/dt) [4]. The combination of fewer switches and optical triggering can simplify the design of a multi-valve converter. The main disadvantage is that it is difficult to design the switch for a useful blocking voltage and an acceptable conduction loss while achieving an acceptable average power from the triggering light source. One result is that there are no widely accepted, commercially available PCSs for power electronics applications.

Optically sustained PCSs ($\gamma < 1$) terminate the current pulse automatically with the removal of the sustaining light source (which is usually a laser pulse). These naturally opening switches interrupt current over a time scale given by τ_L. Hard switching of approximately 50 kV and 500 A with an impressive $di/dt = 25$ kA/μsec and $dv/dt = 2500$ kV/μsec has been demonstrated with a single gold-doped Si PCS [5]. However, for power electronics applications the laser energy required to sustain conduction against such an efficient free carrier loss mechanism is prohibitive. These devices appear limited to highly specialized pulsed power applications.

Bistable Optical Switch (BOSS)

There is another attractive physical mechanism available in optically activated power switches. The optical stimulation of deep trap levels can also remove mobile charge from the conduction and valence bands, thus providing a method for optically controlling the conduction state of power semiconductor switches both on and off. The observation of infrared quenching of photoconductivity in copper-doped GaAs was first reported in [6]. Infrared quenching in copper-doped semiinsulating GaAs has been applied to the PCS concept to produce the bistable optically stimulated switch (BOSS) [7]. BOSS is triggered on to produce majority carrier conduction with large photo-conductive gain by laser photons with wavelength of the order of 1 μm. Majority carrier lifetimes, significantly greater than 1 μsec, have been stimulated by nanosecond laser pulses [8]. Termination of the current pulse can be triggered at any time during the conduction period by stimulating infrared quenching with laser photons with wavelength about twice that required for turn on (i.e., $\lambda = 2$ μm). This process does not redirect the current flow, but instead interrupts it by optically stimulating fast free carrier recombination. Hard switching at 4.5 kV and 50 A ($di/dt = 650$ A/μsec and $dv/dt = 38$ kV/μsec) has been reported [9]. Less than 10 nsec rise and fall times and conduction periods greater than 1 μsec are possible with inexpensive GaAs. The conduction time can be arbitrarily extended by applying multiple $\lambda = 1$ μm conduction-stimulating laser pulses prior to a single $\lambda = 2\ \mu m$ conduction-extinguishing laser pulse. Special preparation of the GaAs has produced subnanosecond turn-off [10].

The BOSS, fabricated as a conventional PCS, shares the advantages and disadvantages of the optically triggered (closing) PCS described above without suffering, to the same degree, the inefficiency of the optically sustained (naturally opening) PCS. However, the need for two laser wavelengths increases complexity and cost, and the problem of designing an efficient cost-effective PCS for power electronics applications remains. One solution could be to include the optically active material that forms BOSS in the control region of a conventional power semiconductor switch, such as the base of an optically controlled BJT or the channel of an optically controlled MOSFET.

Light-Triggered Thyristor (LTT)/Optothyristor

A thyristor is a four-layer (p–n–p–n), regenerative, bistable semiconductor switch. It is used mostly for high-voltage dc (HVDC) transmission systems. Because of the very high voltage ratings of power thyristors and the stacking of tens of thyristors in each valve in the converter circuit (this is needed for very high current density), decoupling of the control circuit and the device itself is a very good option [4]. A light-triggered thyristor could be the most elegant device for many applications. In addition, an electrically triggered thyristor (ETT) valve requires many thyristor level electronic parts. An LTT valve eliminates most of these parts, resulting in higher reliability and more compact system design. Furthermore, a triggering signal is sent to the LTT in the form of light energy. This gives the LTT valve much better noise immunity than a conventional ETT valve, which suffers from electromagnetic noise problems.

Light triggering of the thyristors is achieved by localized illumination at a centered site of the device. There is an optical well and surrounding it there is a pilot thyristor and an amplifying thyristor structure. Photogenerated E–H pairs cause a lateral current flow from the center of the well, driven by the anode bias. This lateral current induces a voltage drop (~0.7 V) in the main cathode junction, which is sufficient to turn on the pilot thyristor. The pilot thyristor then turns on the larger amplifying thyristor. This, in turn, is able to drive the gate arms of the main device. The well is very sensitive and normally few μJ of optical energy is sufficient for taking the device into full conduction from a reverse blocking state. Sometimes, there can be more than one amplifying gate depending upon the current requirement of the device. Figure 9.54 shows a schematic of an LTT, with an optical gate, three amplifying gates, and an integrated breakover-cum-photodiode (BOD). This diode is for overvoltage protection and it turns on the thyristor by avalanche current if the reverse voltage exceeds the thyristor's safe limit.

Optothyristors are normally made of III–V compounds, e.g., GaAs, AlGaAs, InP, because of their higher temperature, radiation, and breakdown voltage handling capability compared to silicon. The direct bandgap nature of these materials allows optical efficiency to be higher and also facilitates the exploration of heterostructures [11].

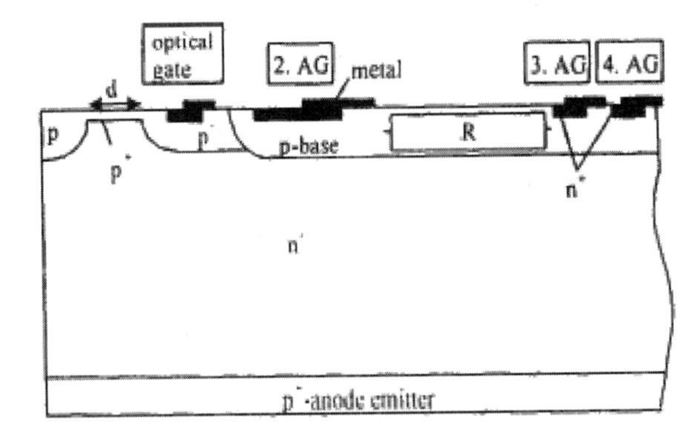

FIGURE 9.54 Schematic of an LTT, with an optical gate, three amplifying gates, and an integrated breakover-cum-photodiode (BOD). © IEEE, 2000, F.J. Niedernostheide, H.J. Schulze, J. Dorn, U. Kellner-Werdehausen, and D. Westerholt.

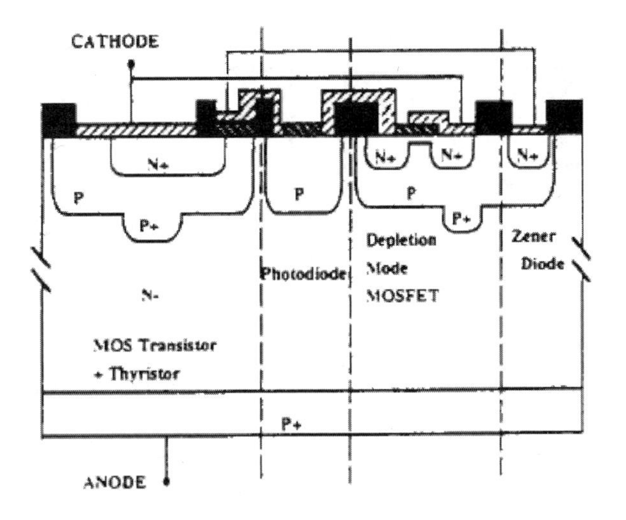

FIGURE 9.55 The cross-section schematic of the amplifying MOS-gated optothyristor which integrates a MOS-gated thyristor, photodiode, and breakover zener diode. © IEEE, 1993, J.L. Sanchez, R. Berriane, and J. Jalade.

Alhough successful fabrication and application of optothyristors have been demonstrated for high voltage, pulsed power systems [12–14], surface degradation, edge breakdown, and deep-level effects of the semiinsulating base layer can lead to early breakdown at a voltage much less than the theoretical value [12]. Like most power thyristors, LTTs are not suitable for fast switching (e.g., $dv/dt > 2\,kV/\mu sec$). As a result, the principal application of these devices remains the high-voltage utility power market, where the required switching speed is modest.

Another interesting approach has been reported in Ref. [15] to integrate the optical control detection circuitry with power device and to create a LTT with a MOS amplifying gate structure. Figure 9.55 shows the cross-section schematic of the device which integrates a MOS-gated thyristor, a photodiode, and a breakover zener diode.

Optically Triggered Static Induction Transistor

The static induction transistor (SIT) is a unipolar, vertical channel field-controlled semiconductor device [16]. The basic device structure is shown in Figure 9.56. It exhibits intrinsic turn-on and turn-off capability, and may be triggered electrically as well as optically. Some high-power systems, such as solid-state laser driver modules and microwave traveling tube amplifiers require high switching speeds (10 to 200 nsec) and high switched voltages (10 to 20 kV). This can only be achieved by series-coupled field-effect devices. Designing gate driver circuitry for MOSFETs poses problems due to high voltage floating potentials. Optically modulated series-coupled SITs are excellent choice because the ground-referenced low-power optical trigger sources are isolated from the high-voltage switch assembly through optical fibers, and they result in reliable, jitter-free operation [17]. Optically triggered SITs are also made from III–V materials to benefit from the advantage of high critical electric field and higher carrier mobility of these elements.

A transparent gate structure is used for optical absorption. Photogenerated electrons transit across the channel and are collected at the drain, whereas photogenerated holes drift to the gate junction. Based upon simple one-dimensional calculation, the relationship between the forward blocking voltage (V_B), the saturation (maximum) value of the device's blocking gain (G_M), and the optical parameters has been shown [17] to be

$$P_{opt}R_g = \frac{V_B h\nu}{qG_M(1-R)}\frac{A_{opt}}{L_g l} \tag{9.19}$$

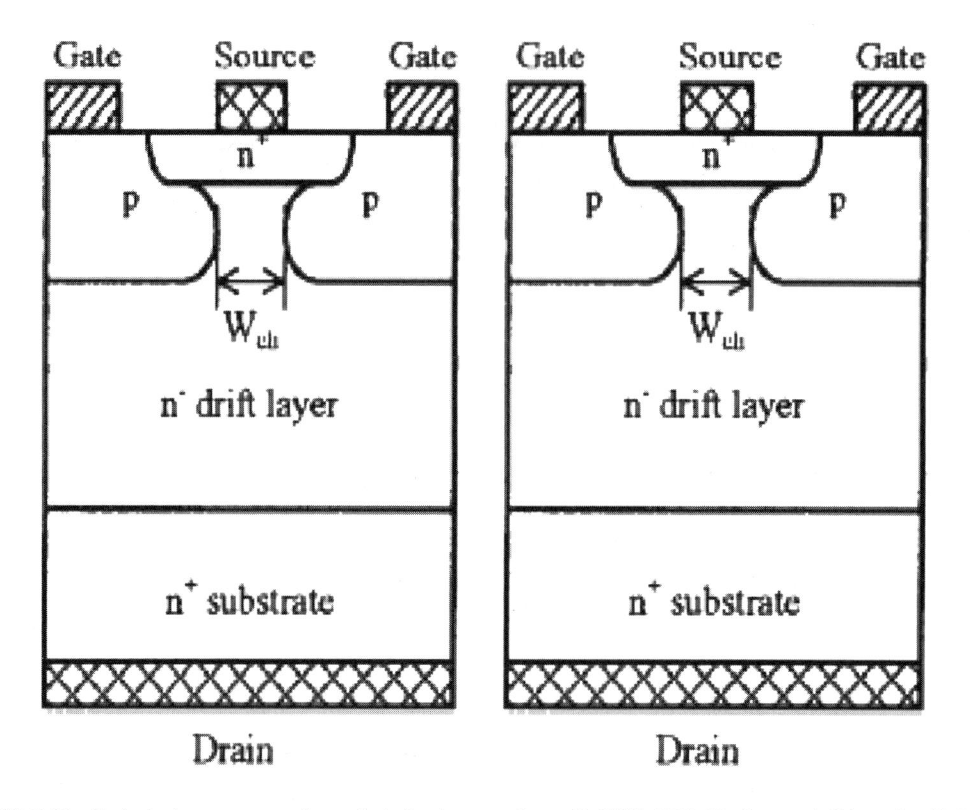

FIGURE 9.56 Basic device structure of a static induction transistor. © IEEE, 2002, H. Onose, A. Watanabe, T. Someya, and Y. Kobayashi.

Here, P_{opt} is the incident optical power, R_g is the external series gate resistance, A_{opt} is the optical window area, R is the surface optical reflectivity of the material, q is the charge of the electron, L_g is the length of the gate junction, and l is the distance of the generated charge boundary to the gate junction.

Optically Triggered MOSFET and IGBT

Sustained growth of power electronics will depend, among other factors, on the ability to develop power systems with higher power densities [18,19]. A widely accepted and viable approach for enhancing the power density is to increase the switching frequency of a power processor/system, because that lowers the values of parasitic components, i.e., capacitors and inductors, which takes up almost double the space of the power stage in a full-fledged system.

Among the power semiconductor devices, MOSFET is the preferred choice for high-frequency applications due to its unipolar conduction, which does not introduce any minority carrier storage delay, as in the case of a power BJT or thyristor. Also it has good dv/dt and di/dt handling capability and a negative thermal resistance coefficient, which makes paralleling of multiple devices much easier than in the case of a BJT or thyristor, by eliminating the chance of thermal runway. For moderate voltage and current levels (<600 V and ~20 to 30 A continuous current), MOSFET is the device of choice in the frequency range of >50 kHz. Applications include motor drives, switch-mode power supplies, solid-state ballasts, etc.

Interestingly, the optical switching of power MOSFETs has been rarely studied. Optical control of MOSFET or M–I–S (metal–insulator–semiconductor) structure, in general, has been reported [20,21], but primarily in the domain of low power optical communication or sensor applications. Recently, simulation studies on the modulation techniques for an optically controlled power MOSFET (OT-MOS) have been reported [22]. A vertical double-diffused MOS (DMOS) is controlled by a turn-on and a turn-off pulse and the hybrid device structure schematic, comprising an Si-based DMOS, a GaAs-based photodiode, and a GaAs-based photoconductive switch (PCS), is shown in Figure 9.57. The photodiode acts as a photogenerated current source, which, when exposed to the turn-on pulse, modulates the i-drift region resistance, conducts current (supplied by the reverse bias source), and charges the gate capacitance of the DMOS above the threshold voltage to turn on the OT-MOS. Again, during turn-off, another optical pulse falls on the PCS to modulate its conductivity, providing a short-circuit path through the PCS, and discharging the gate capacitance of the DMOS to turn it off. So the turn-on and -off time can be adjusted either by manipulating the absolute values, i.e., amplitudes of the photogenerated current source strength and PCS conductivity, or by varying the time-period of the triggering pulses, thus allowing the gate capacitance of the OT-MOS to charge or discharge for a varying time.

Consequently, there are possibilities of using both amplitude and pulse-width modulation techniques for triggering optical pulses. Figure 9.58 and Figure 9.59 show the switching result of the OT-MOS using these techniques. The circuit used for these simulations is a hard-switching arrangement comprising a 50 V drain voltage, 5 Ω external resistance, and 12 V photodiode reverse bias. Though only the modulation technique is reported [22], further work continues and the OT-MOS has also been simulated for sustained switching performance on a synchronous buck converter circuit. The results are shown in Figure 9.60(a) and (b), which indicate that the performance of the OT-MOS matches that of an electrically triggered MOSFET of the same voltage and current ratings, under identical circuit conditions. Analytical modeling and steady-state

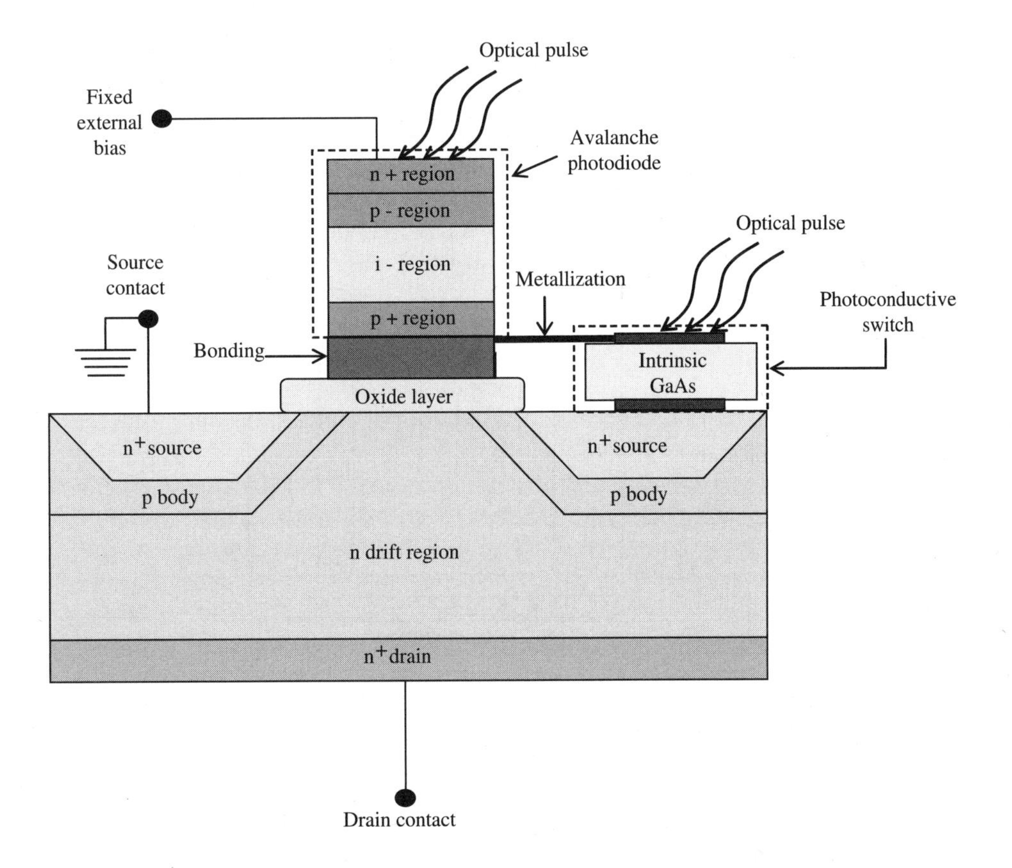

FIGURE 9.57 The novel optically triggered power MOSFET structure schematic, showing the triggering photodiode, MOSFET, and photoconductive switch.

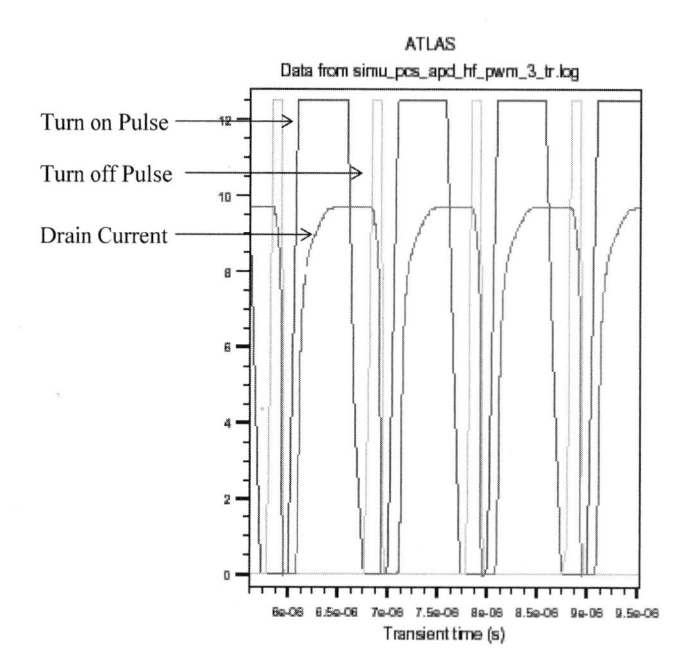

FIGURE 9.58 Switching behavior of OT-MOS using amplitude modulation at 1 MHz. 800 nm optical beam of 12.5 W/cm^2 intensity and pulse-widths of 0.8 μsec and 0.18 μsec, for the turn-on and off pulses. © IEEE, 2004, T. Sarkar and S.K. Mazumder.

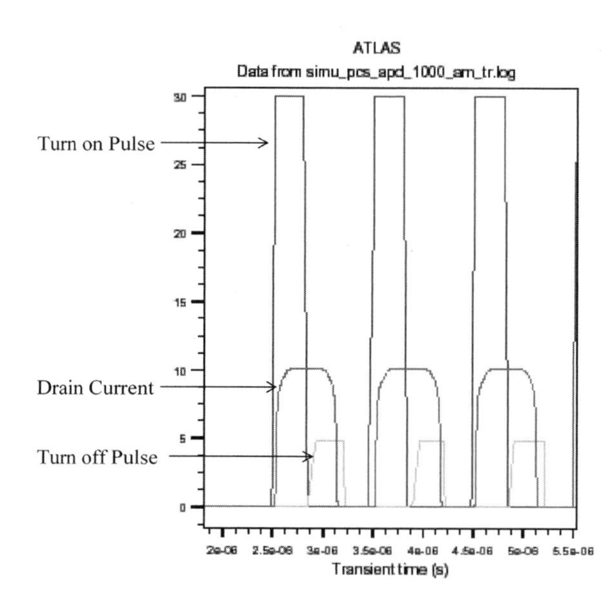

FIGURE 9.59 Switching behavior of OT-MOS using pulse-width modulation at 1 MHz; 800 nm optical beam of intensities 30.0 W/cm^2 and 4.9 W/cm^2, for the turn-on and -off pulses. © IEEE, 2004, T. Sarkar and S.K. Mazumder.

characterizations have also been done to reveal an interesting result, that the OT-MOS triggering pulses should have an optimized wavelength (or should lie within a small range of wavelengths) for fastest triggering of the device with minimum optical energy, and also that increasing the optical beam energy or equivalently decreasing the wavelength may actually lead to poor switching performance. The wavelength comparison

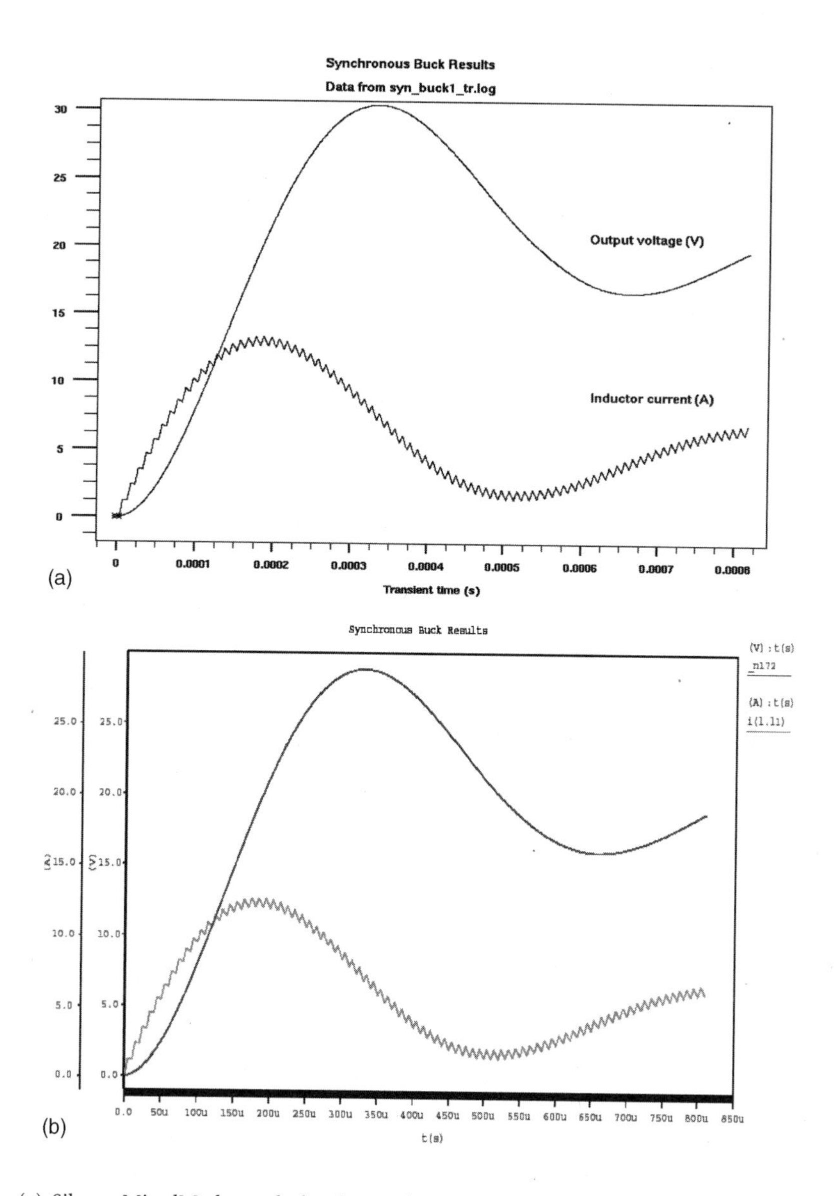

FIGURE 9.60 (a) Silvaco MixedMode result for the synchronous buck circuit using OT-MOS and (b) SaberDesigner result for the IRF460P are shown. Both of these results have been generated under the same circuit and parametric conditions. Here, we show the output voltage across the capacitor and the inductor current.

result in Figure 9.61 depicts that clearly the 800 nm wavelength turns out to be the best choice. The photodiode optically modulated resistance (R_{Photo}) is a function of photogenerated current (I_{Photo}) and it affects the OT-MOS switching performance by varying its turn-on time (τ_{on}). The equations for these quantities are given by

$$I_{\text{Photo}} = A \frac{q(1-R)\eta_{\text{i}}\left(\frac{I_{\text{opt}}\lambda}{hc}\right)\alpha}{\left(\frac{1}{\mu_{\text{a}}\tau_{\text{a}}E}-\alpha\right)}\left(\mathrm{e}^{-\alpha W_1}-\mathrm{e}^{-\alpha W_2}\right) - q\left(\frac{(1-R)\eta_{\text{i}}\left(\frac{I_{\text{opt}}\lambda}{hc}\right)\alpha L_{\text{a}}}{1+\alpha L_{\text{a}}}\mathrm{e}^{-\alpha W_2} + n_{\text{p}_0}\frac{D_{\text{a}}}{L_{\text{a}}}\right) \quad (9.20)$$

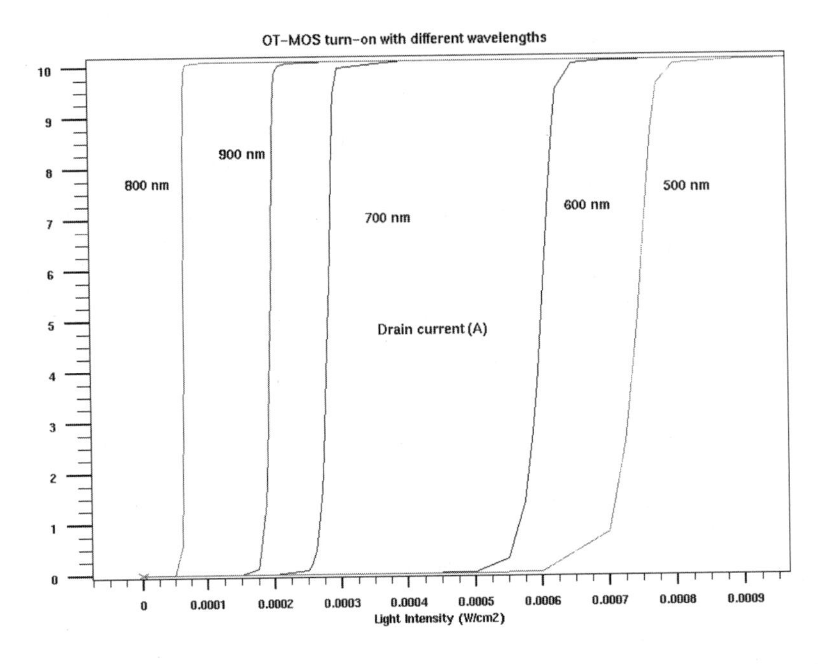

FIGURE 9.61 OT-MOS drain current vs. optical intensity for five different wavelengths: 500 nm, 600 nm, 700 nm, 800 nm, and 900 nm. The 800 nm shows sharpest turn-on and needs least optical power to initiate turn-on.

$$\tau_{\text{on}} = \frac{V_{\text{T}} C_{\text{in}}}{I_{\text{Gate}}} = \frac{V_{\text{T}}(C_{\text{GS}} + C_{\text{GD}}) R_{\text{photo}}}{V_{\text{Bias}}} \tag{9.21}$$

In Equation (9.20) and Equation (9.21), A is the photodiode electrode cross-section area, I_{opt} is the incident optical power density (optical power/area), η_{i} is the intrinsic quantum efficiency, λ is the wavelength, h and c denote the Planck's constant and the velocity of electromagnetic wave, respectively, α is the optical absorption coefficient for the photodiode material, μ_{a}, τ_{a}, D_{a}, and L_{a} denote the ambipolar mobility, lifetime, diffusion constant, and diffusion length, respectively, E is the electric field imposed by the external reverse bias, W_1 and W_2 are the p-i-n photodiode depletion width boundaries, and $n_{\text{p}0}$ is the minority carrier density in the p type region. V_{T} is the threshold voltage of the DMOS, C_{GS} and C_{GD} are the gate-source and gate-drain capacitances of the DMOS, respectively, V_{Bias} is the external reverse bias connected to the photodiode, and I_{Gate} is the gate charging current of the DMOS, which is equal to the bias supplied current, neglecting the small dark current and relatively smaller photogenerated current. These two equations relate the optical parameters, e.g., wavelength, illumination power density to device switching parameter τ_{on}.

The insulated-gate bipolar transistor (IGBT) is a relatively new power device, introduced in the early 1980s. It features higher current drive and voltage stand-off capability than MOSFET, but lower frequency of operation due to a bipolar conduction mode and higher on-state voltage drop. However, its operating frequency is much higher than that of the power BJT. Overall, it is the device of choice for relatively low frequency, high current density and >600 V applications, such as motor control, uninterrupted power supply (UPS), and welding equipment. A similar approach to trigger an IGBT with a direct optical pulse, incident in its channel region (the approach is schematically shown in Figure 9.62a), has been reported in Ref. [23]. It shows promising improvement in the on-state voltage drop of the optical IGBT because the channel width in this case depends upon the absorption depth for the particular semiconductor material–optical wavelength pair and it can be significantly higher than the intrinsic Debye length (a few nanometers), which is the default channel width for a MOSFET.

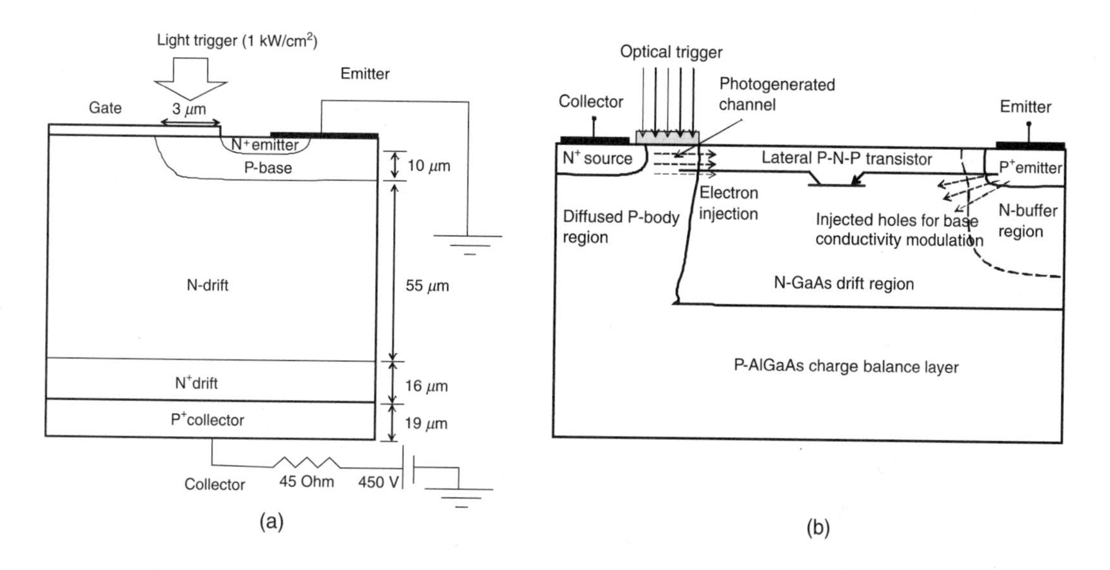

FIGURE 9.62 (a) Schematic of an optically triggered IGBT showing the light incidence in the channel region underneath the gate.© IEEE, 2001, T.S. Liao, P. Yu, and O. Zucker. (b) Structural schematic of the optically-gated bipolar transistor. © IEEE, 2005, S.K. Mazumder and T. Sarkar.

The on-state voltage drops of the conventional electrically triggered IGBT and the optical IGBT are given as follows [23]:

$$V_{\mathrm{on}}\big|_{\mathrm{electrical}} = \frac{2kT}{q}\ln\left[\frac{I_{\mathrm{c}}d}{2qWZD_{\mathrm{a}}n_{\mathrm{i}}F(d/l_{\mathrm{a}})(1-\alpha_{\mathrm{PNP}})}\right] + \frac{I_{\mathrm{c}}L_{\mathrm{ch}}}{\mu_{\mathrm{on}}C_{\mathrm{ox}}Z(V_{\mathrm{G}}-V_{\mathrm{T}})} \tag{9.22}$$

and

$$V_{\mathrm{on}}\big|_{\mathrm{optical}} = \frac{2kT}{q}\ln\left[\frac{Gd}{2qWZD_{\mathrm{a}}n_{\mathrm{i}}F(d/l_{\mathrm{a}})(1-\alpha_{\mathrm{PNP}})}\right] + GR_{\mathrm{s}} \tag{9.23}$$

In Equation (9.22) and Equation (9.23), k is the Boltzmann's constant, T is the absolute temperature, I_{c} is the collector current, V_{G} is the gate voltage, V_{T} is the threshold voltage, G is the photogeneration rate, d is the drift region thickness, W is the channel width, Z is the gate length, D_{a} is the ambipolar diffusion constant, L_{ch} is the channel length, μ_{on} is the on-state mobility, C_{ox} is the oxide capacitance, n_{i} is the intrinsic carrier density, α_{PNP} is the intrinsic BJT base transport factor, and the function F is given by

$$F\left(\frac{d}{L_{\mathrm{a}}}\right) = \frac{(d/L_{\mathrm{a}})\tanh(d/L_{\mathrm{a}})}{\sqrt{1-0.25\tanh^4(d/L_{\mathrm{a}})}}\exp\left(-\frac{qV_{\mathrm{m}}}{2kT}\right) \tag{9.24}$$

where V_{m} is the maximum blocking voltage.

From Equation (9.22), since I_{c} is controlled by the gate overdrive, i.e., $(V_{\mathrm{G}}-V_{\mathrm{T}})$, the on-state drop cannot be minimized much, but as R_{s} is small, and from Equation (9.23), the on-state drop in optical-IGBT is controlled by the photogeneration rate (G), which is a function of optical wavelength, absorption coefficient, and incident power, and can be made smaller. Though one report [23] treats only the improved current density and voltage drop and does not report switching results for the optical IGBT, nevertheless this approach is worth exploring and can lead to fully optically controlled IGBT based high-power converters.

Futher research on these types of devices has led to GaAs-AlGaAs based optically controllable lateral switching power devices [24]. Optically gated bipolar transistor (OGBT) (structural schematic has been shown in Figure 9.62b) features high device gain due to optically initiated base conductivity modulation. Furthermore, due to shorter carrier lifetime in GaAs compared to silicon, OBGT can be switched faster than conventional electrically triggered power bipolar transistors. Details of the structure and operation have been explained in Ref. [24].

Other Optically Controlled Power Devices

Power diodes have been in use as rectifiers in power electronics from its earliest days. Optically-controlled Si- and GaAs-based lateral and vertical p-i-n diodes have been reported [25,26]. Most of these devices are targeted for low-frequency but high power (in the range of 100 kW) applications, e.g., pulsed-power switching or microwave power generation. Figure 9.63 illustrates a simple p-i-n diode schematic with the incident optical beam and photogeneration in the intrinsic region. Normally the diode is reverse-biased and conducts very small dark current in its off-state. When the light falls on it through an optical window, excess carriers are generated, the conductivity increases significantly, and the diode conducts current as given by Equation (9.16). An excellent treatment on the steady-state characteristics and device geometry for an optically controlled p-i-n diode is given in Ref. [27].

The GaAs metal-semiconductor field effect transistor (MESFET) is the basic building block of high-speed monolithic microwave integrated circuits. Although MESFET itself is not an ideal switch for use in high-power PE converters (due to its low current carrying capability), for radio-frequency (RF) power applications optical control of MESFET can be a viable proposition. Figure 9.64 shows an illuminated MESFET structure. When the device is illuminated with light of photon energy greater than the bandgap energy, the illuminated channel region contributes optically generated gate photocurrent. Illuminated interelectrode epilayer photocurrent modulates the conductivity of the epilayer. In the episubstrate barrier region excess photogenerated carriers reduce the barrier height and the illuminated part of the substrate contributes to the substrate current. A comprehensive model of the optical switching of a GaAs MESFET is given in [28]. It can be inferred from the reported results that the MESFET is not a very fast optical switch and its rise and fall times lie in the range of the submicrosecond region. Recently, numerical simulation studies on vertical optically triggered power transistors (OTPT) have been reported in [29]. Vertical structure allows increased current carrying capability

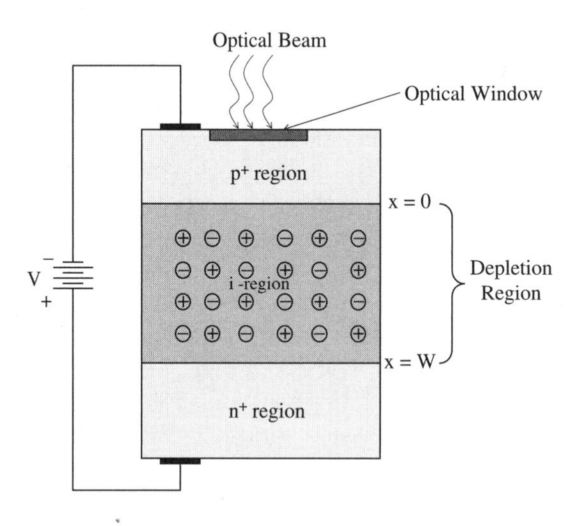

FIGURE 9.63 Schematic of a photodiode showing the p-i-n structure, optical window, and photogeneration in the intrinsic region.

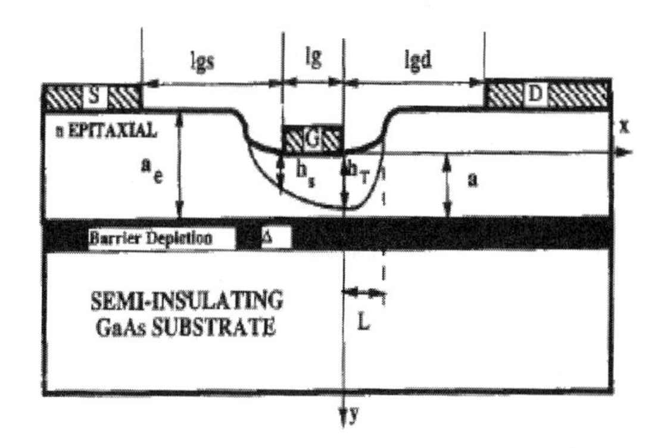

FIGURE 9.64 An illuminated MESFET structure. © IEEE, 1994, A. Madjar, A. Paolella, and P.R. Herczfeld.

maintaining fast optical control by direct photogeneration–recombination in the channel region of the device. OTPT does not have a gate oxide structure, thus making it possible to use GaAs–AlGaAs material system for fabrication (because in general, III-V compound semiconductors pose problems in growing high-quality insulator to realize MOSFET structures), which results in fast switching times due to lower carrier lifetime of GaAs compared to silicone.

Research Issues

Power-electronic converters need not only high switching frequencies but also high efficiencies. Thus, an optical switch must have fast opening and recovery (to achieve high repetition rates) and high switch-transfer efficiency. Conventionally, the applications of optical switching to ultra-low-power applications (such as high-speed data transmission) have focused primarily on speed and not on efficiency because in such applications speed is of the essence. However, some of the very high-power applications have primarily focused on efficiency and not on speed. With high switching frequency, an additional difficulty arises because completely opposite requirements exist between the conduction and transition (on- and off-phases) states of an optical switch. During the conduction state, high carrier mobility and long carrier lifetime (for high conduction efficiency) is highly desired; while during the on and off transition phases the switch must show a short carrier lifetime for fast closing and opening actions, and high open-state impedance.

While lowering the dynamic losses of an optically switched power converter is necessary at high switching frequencies, a reduction in conduction losses is equally important to increase efficiency. As such, an OTD should also have low on-resistance and forward voltage drop. OTDs must also have good voltage-standoff properties and the ability to minimize leakage current under dc voltage and long-pulse bias. The importance of voltage standoff increases with the higher operating voltage of a converter. To reduce leakage current under high biased conditions, which can lead to thermal runaway, an OTD should have very high impedance after it opens.

Irrespective of the power level, most power-electronic systems need to either regulate or track reference voltages and currents. This is achieved, in general, by controlling the power switches using constant or variable frequency pulse-width modulation, pulse-amplitude modulation, or other related modulation strategies under closed-loop (or open-loop) feedback operation. The success of each of these modulation strategies relies on the ability of the OTD to follow feedback control pulses accurately; the delay between the optical illumination and the onset of the switching should be minimal. Furthermore, the OTD for power-electronic applications must be jitter-free, have controllable conduction times, and a large dynamic range of operation.

Device life of an OTD is another important consideration for power-conversion systems. The power supply companies want operational life to be several million operations. However, most optical switches to date fail to hold off the voltage for that long. In addition, there are a number of possible effects that could cause device failure [11]. These include metal migration from the contacts, defects in the material synthesis, nonuniform (filamentary) current flow, and thermal effects. Additionally, to enhance the life of an OTD at high switching frequencies, adequate power dissipation is of concern, since experience has shown that the second most common source of failure after electrical breakdown is thermal. Hence, an OTD with superior thermal stability (that is, better thermal capacity and conductivity) is necessary. Another factor of prime importance is the temperature coefficient of the OTDs. For example, it has been shown [11] that optical switches comprised of materials with negative temperature coefficients when subjected to dc voltages are often susceptible to thermal runaway; needless to say, a significant number of power-electronic systems are powered by dc electrical sources.

A promising approach is to add true optical control to existing device topologies. One possibility is to incorporate the optically active materials used in the BOSS to form an optically controlled BJT or MOSFET hybrid. In particular this approach could eliminate the main two disadvantages of the PCS approach when applied as a controlled opening switch; namely the prohibitively high amount of light energy required to control the photoconductivity and scaling limits imposed by current instabilities observed during reestablishment of large blocking voltages. At the same time, devices can be scaled in size without imposing severe limits on switching speed caused by the problem of electrically removing stored charge in gate and base regions. Candidates for new, optically controlled conventional power devices include copper-doped GaAs and boron-doped SiC, incorporation of which is limited to the gate region of the FET or the base region of a BJT. The drift regions, relying on majority carrier injection from the much smaller optically controlled regions of the device, remain the same. The resulting hybrid device reduces the optically controlled volume by orders of magnitude compared to that of a conventional PCS, with a corresponding increase in control gain (given by the electrical-to-optical power ratio).

References

1. B.J. Baliga, "Trends in power discrete devices," *Proc. 10th Int. Symp. Power Semicond. Devices ICs*, 1998, pp. 5–10.
2. P. Bhattacharya, *Semiconductor Optoelectronic Devices*, Englewood Cliffs, NJ: Prentice-Hall, 1994.
3. M.S. Mazzola, Schoenbach, K.H., Lakdawala, V.K., and Roush, R.A., "Infrared quenching of conductivity at high electric fields in a bulk, copper-compensated, optically activated GaAs switch," *IEEE Trans. Electron. Devices*, vol. 37, no. 12, pp. 2499–2505, 1990.
4. G.M. Loubriel, M.W. O'Malley, F.J. Zutavern, B.B. McKenzie, W.R. Conley, and H.P. Hjalmarson, "High current photoconductive semiconductor switches," *18th IEEE Power Modulator Symp.*, 1988, pp. 312–317.
5. F.J. Zutavern, G.M. Loubriel, and M.W. O'Malley, "Recent developments in opening photoconductive semiconductor switches," *6th IEEE Pulsed Power Conf. Proc.*, 1987, pp. 577–580.
6. J. Blanc, R.H. Bube, and H.E. MacDonald, "Properties of high-resistivity gallium arsenide compensated with diffused copper," *J. Appl. Phys.*, 32, 1666–1679, 1961.
7. M.S. Mazzola, K.H. Schoenbach, V.K. Lakdawala, and S.T. Ko, "Nanosecond optical quenching of photoconductivity in a bulk GaAs switch," *Appl. Phys. Lett.*, 55, 2102–2104, 1989.
8. M.S. Mazzola, K.H. Schoenbach, V.K. Lakdawala, R. Germer, G.M. Loubriel, and F.J. Zutavern, "GaAs photoconductive closing switches with high dark resistance and microsecond conductivity decay," *Appl. Phys. Lett.*, 54, 742–744, 1989.
9. D.C. Stoudt, R.A. Roush, M.S. Mazzola, and S.F. Griffiths, "Investigation of a laser-controlled, copper-doped GaAs closing and opening switch for pulsed power applications," *8th IEEE Pulsed Power Conf. Proc.*, 1991, pp. 41–44.
10. D.C. Stoudt, R.P. Brinkmann, R.A. Roush, M.S. Mazzola, F.J. Zutavern, and G.M. Loubriel, "Subnanosecond high-power performance of a bistable optically controlled GaAs switch," *9th IEEE Pulsed Power Conf.*, 1993, pp. 72–75.

11. A. Rosen and F. Zutavern, *High Power Optically Activated Solid-State Switches*, Norwood, MA: Artech House, 1994.
12. J.H. Hur, P. Hadizad, S.R. Hummel, P.D. Dapkus, H.R. Fetterman, and M.A. Gundersen, "GaAs optothyristor for pulsed power applications," *IEEE Conf. Rec. Nineteenth Power Modulator Sym.*, 26–28 June, 1990, pp. 325–329.
13. J.H. Zhao, T. Burke, D. Larson, M. Weiner, A. Chin, J.M. Ballingall, and T. Yu, "Sensitive optical gating of reverse-biased AlGaAs/GaAs optothyristors for pulsed power switching applications," *IEEE Trans. Electron. Devices*, vol. 40, no. 4, 1993, pp. 817–823.
14. R.J. Lis, J.H. Zhao, L.D. Zhu, J. Illan, S. McAfee, T. Burke, M. Weiner, W.R. Buchwald, and K.A. Jones, "An LPE grown InP based optothyristor for power switching applications," *IEEE Trans. Electron. Devices*, vol. 41, no. 5, 1994, pp. 809–813.
15. J.-L. Sanchez, R. Berriane, J. Jalade, and J.P. Laur, "Functional integration of MOS and thyristor devices: a useful concept to create new light triggered integrated switches for power conversion," *Fifth Eur. Conf. Power Electron. Appl.*, 2, 1993 (September 13–16), pp. 5–9.
16. J. Nishizawa, T. Terasaki, and J. Shibata, "Field-effect transistor versus analog transistor (static induction transistor)," *IEEE Trans. Electron Devices*, vol. 22, no. 4, 1975, pp. 185–197.
17. P. Hadizad, J.H. Hur, H. Zhao, K. Kaviani, M.A. Gundersen, and H.R. Fetterman, "A high-voltage optoelectronic GaAs static induction transistor," *IEEE Electron Device Lett.*, vol. 14, no. 4, 1993, pp. 190–192.
18. F.C. Lee, D. Boroyevich, S.K. Mazumder, K. Xing, H. Dai, I. Milosavljevic, Z. Ye, T. Lipo, J. Vinod, G. Sinha, G. Oriti, J.B. Baliga, A. Ramamurthy, A.W. Kelley, and K. Armstrong, *Power Electronics Building Blocks and System Integration*, Blacksburg, VA: Virgin Power Electronics Center, 1997 (project report submitted to the Office of Naval Research).
19. J.B. Baliga, "Trends in power semiconductor devices," *IEEE Trans. Electron Devices*, vol. 43, no. 10, pp. 1717–1731, 1996.
20. T. Yamagata and K. Shimomura, "High responsivity in integrated optically controlled metal-oxide semiconductor field-effect transistor using directly bonded SiO_2–InP," *IEEE Photonics Tech Lett.*, 9, 1143–1145, 1997.
21. M. Madheswaran and P. Chakrabarti, "Intensity modulated photoeffects in InP–MIS capacitors," *IEE Proc. Optoelectron.*, 143, 1996, pp. 248–251.
22. T. Sarkar and S.K. Mazumder, "Amplitude, pulse-width, and wavelength modulation of a novel optically-triggered power DMOS," *35th IEEE Power Electron. Specialists Conf. PESC 2004*, Germany, June 2004, pp. 3004–3008.
23. T.S. Liao, P. Yu, and O. Zucker, "Analysis of high pulse power generation using novel excitation of IGBT," *Proc. 6th Int. Conf. Solid-State Integr.-Circ. Technol.*, 1, 2001, pp. 143–148.
24. S.K. Mazumder and T. Sarkar, "Device technologies for photonically-switched power-electronic systems," *IEEE Intl. Pulsed Power Conf.*, in press, 2005.
25. A. Rosen, P. Stabile, W. Janton, A. Gombar, P. Basile, J. Delmaster, and R. Hurwitz, "Laser-activated p-i-n diode switch for RF application," *IEEE Trans. Microwave Theory Tech.*, vol. 37, no. 8, 1989, pp. 1255–1257.
26. A. Rosen, P.J. Stabile, A.M. Gombar, W.M. Janton, A. Bahasadri, and P. Herczfeld, "100 kW DC biased, all semiconductor switch using Si P-I-N diodes and AlGaAs 2-D laser arrays," *IEEE Photonics Technol. Lett.*, vol. 1, no. 6, 1989, pp. 132–134.
27. J.W. Schwartzenberg, C.O. Nwankpa, R. Fischl, A. Rosen, D.B. Gilbert, and D. Richardson, "Evaluation of a PIN diode switch for power applications," *26th Ann. IEEE Power Electron. Specialists Conf.*, 1, 68–73, 1995.
28. A. Madjar, A. Paolella, and P.R. Herczfeld, "Modeling the optical switching of MESFET's considering the external and internal photovoltaic effects," *IEEE Trans. Microwave Theory Tech.*, vol. 42, no. 1, 1994, pp. 62–67.
29. T. Sarkar and S.K. Mazumder, "Steady-state and switching characteristics of an optically-triggered power transistor (OTPT)," Power Systems World Conference, 2004.

9.6 Nonlinear Control of Interactive Power-Electronics Systems

Sudip K. Mazumder

Introduction

Interactive power-electronics systems (IPNs), such as parallel dc–dc or parallel muniphase converters, are nonlinear hybrid dynamical systems [1,2]. The instability in such switching systems, owing to their discontinuity, can evolve on slow and fast scales. Conventional analyses of IPNs and their subsystems are based on averaged models, which ignore fast-scale instability and analyze the stability on a reduced-order manifold. As such, the validity of the averaged models varies with the switching frequency, even for the same topological structure. The prevalent procedure for analyzing the stability of IPNs and their subsystems is based on linearized averaged (small-signal) models that require a smooth averaged model. Yet there are systems (in active use) that yield a nonsmooth averaged model [2]. Even for systems for which a smooth averaged model is realizable, small-signal analyses of the nominal solution/orbit do not provide anything about three important characteristics [1–3]: region of attraction of the nominal solution, dependence of the converter dynamics on the initial conditions of the states, and the postinstability dynamics. As such, conventional linear controllers for IPNs, designed based on small-signal analyses, may be conservative and may not be robust and optimal. *Clearly, there is a need to design nonlinear controllers for such hybrid systems*, thereby achieving a wider stability margin, improved robustness against parametric variations, feedforward and feedback disturbances, switching nonlinearities, interactions, and enhanced performance.

Applications

Parallel dc–dc converter

Parallel dc–dc converters, as shown in Figure 9.65 are widely used in telecommunication power supplies [4–16]. They operate under closed-loop feedback control to regulate the bus voltage and enable load sharing [1, 4–16]. These closed-loop converters are inherently nonlinear systems. The major sources of nonlinearities are switching nonlinearity and interaction among the converter modules.[1] Yet, most conventional controllers for parallel dc–dc converters are linear. Recently, there have been many studies of the nonlinear control of standalone dc–dc converters [17–26], which have focused on variable-structure controllers (VSC) [27,28], Lyapunov-based controllers [29–33], feedback linearized and nonlinear H_∞ controllers [34–39], and fuzzy logic controllers [40–42]. However, there are few studies on the nonlinear control of parallel dc–dc converters where, unlike standalone converters, there is a strong interaction among the converter modules apart from the feedforward and feedback disturbances.

In Ref. [43], a fuzzy-logic compensator is proposed for the master–slave control of a parallel dc–dc converter. The controller uses a proportional-integral-derivative (PID) expert to derive the fuzzy inference rules; it shows improved robustness as compared to linear controllers. However, the control design is purely heuristic and the stability of the overall system has not been proven. In Reference [44], a VSC has been developed for a buck converter using interleaving. However, the interleaving scheme works only for three parallel modules. Besides, this paper gives no details regarding the existence and stability of the sliding manifolds.

In Ref. [45], Mazumder et al. have developed integral-variable-structure control (IVSC) schemes for N parallel dc–dc buck converters. The choice of a VSC is logical for power converters because the control and plant are both discontinuous. All of the nonlinear controllers mentioned earlier [21–26], which are not based on VSC, have completely relied on the smooth averaged models of the power converters. Therefore, control is valid only on a reduced-order manifold. The IVSC retains all of the properties of a VSC; that is, simplicity in

[1]The uniform distribution of power flow among the parallel connected converter modules is important for reasons of cost effectiveness and long-term reliability. Parallel modules are not identical because of finite tolerances in power stage and control parameters. As a result, the load current is not equally distributed among the modules, leading to excessive component stresses. Modules delivering higher currents will have a shorter lifetime and system reliability is degraded. Therefore, the control scheme should ensure equal distribution of power among the parallel connected converters.

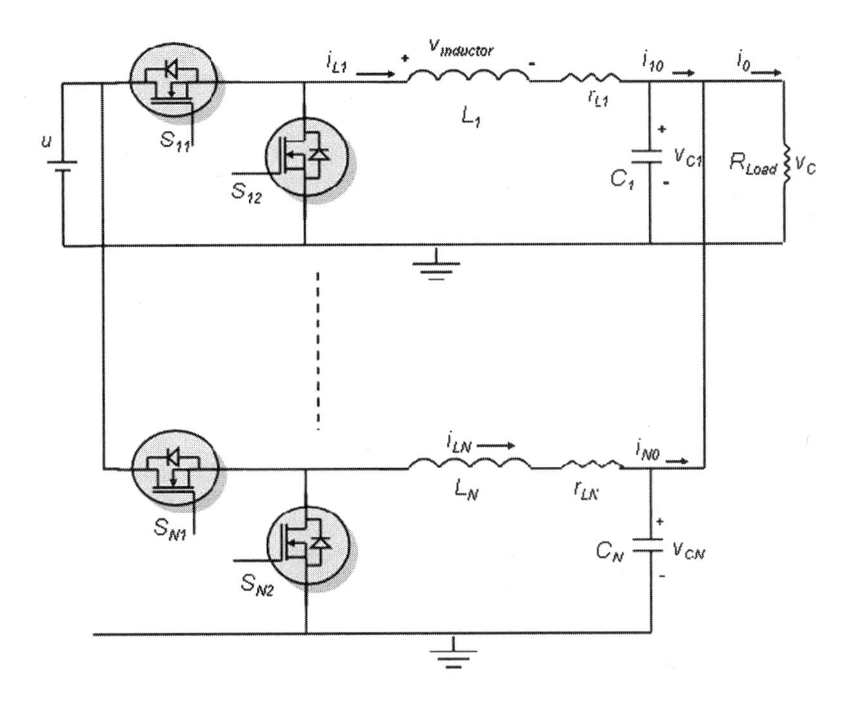

FIGURE 9.65 A parallel dc–dc converter.

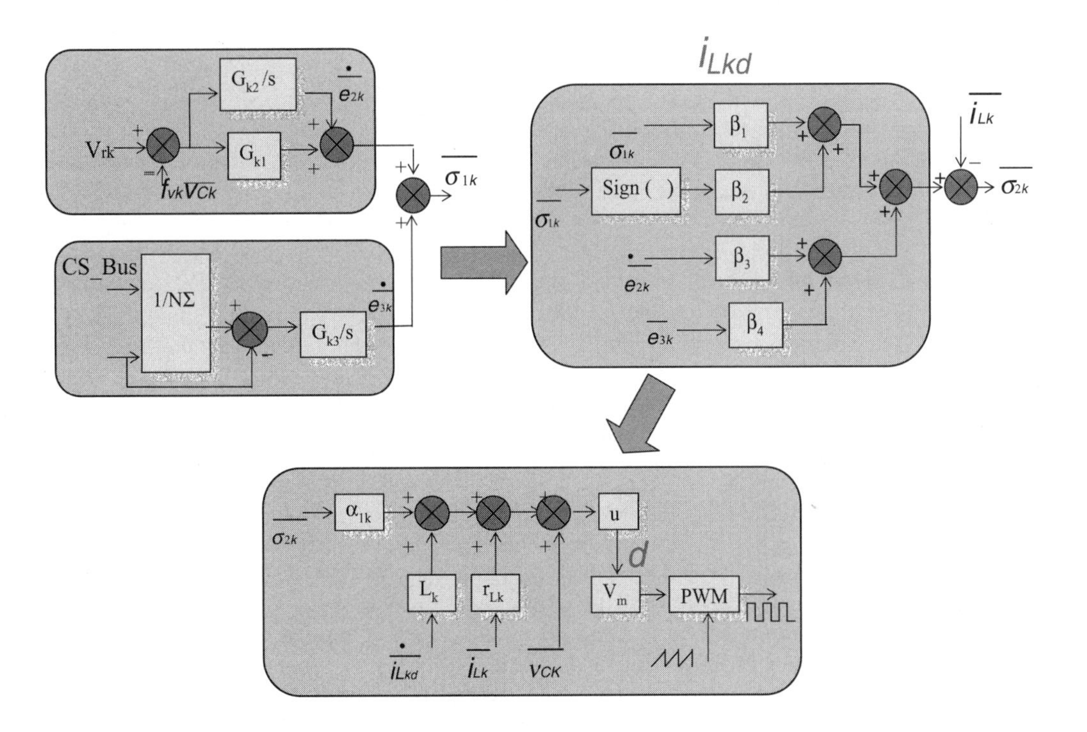

FIGURE 9.66 Methodology for duty-ratio ("d") and PWM signal generation using the nonlinear controller.

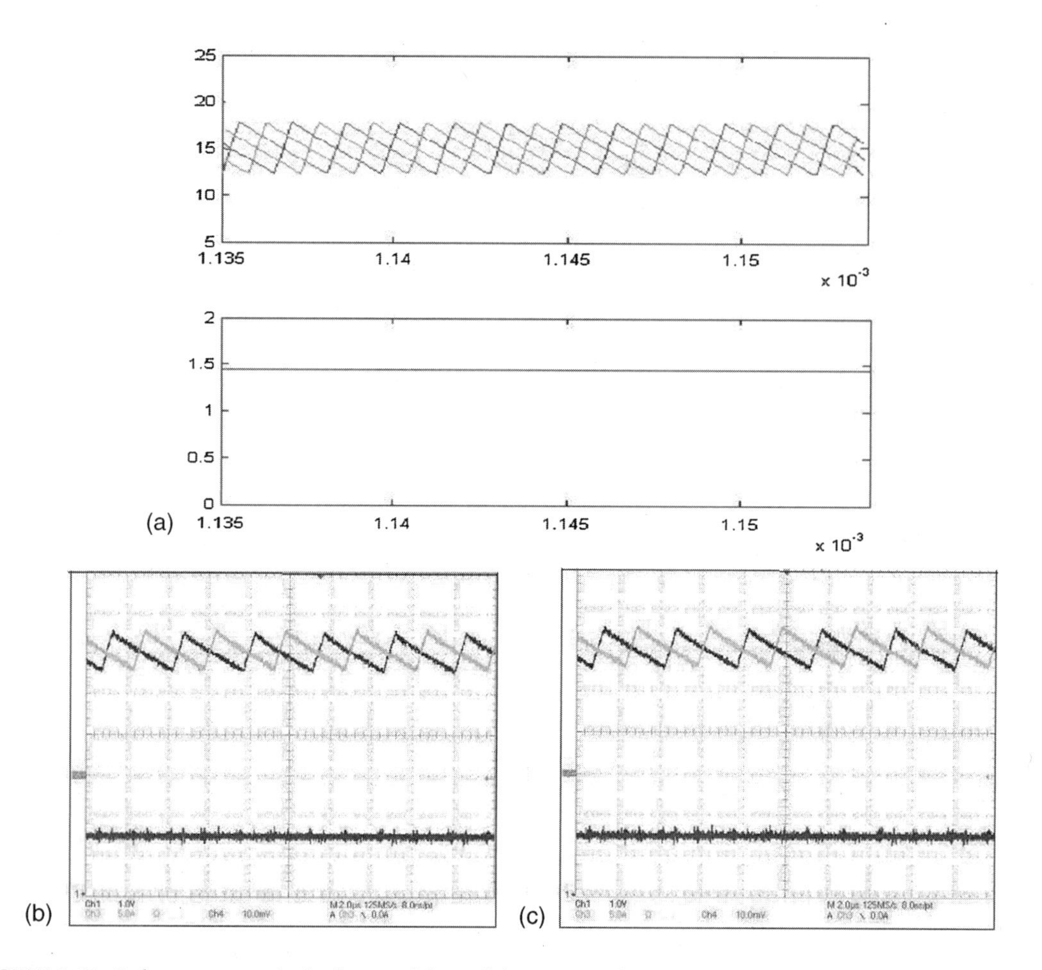

FIGURE 9.67 Inductor currents in the four modules and the output voltage: (a) simulated; (b) experimental (modules 1 and 3); and (c) experimental (modules 2 and 4).

design, good dynamic response, and robustness. In addition, the integral action of the IVSC eliminates the bus-voltage error and the error between the load currents of the converter modules under steady-state conditions. It also reduces the impact of very high-frequency dynamics due to parasitics on the closed-loop system.

Finally, when the error trajectories are inside the boundary layer[1] by modifying the control using the concepts of multiple-sliding-surface control (MSSC) [46,47] or the block-control principle [48,49], we are able to reject mismatched disturbances [50–53] and keep the steady-state switching frequency constant. This is achieved by calculating the duty ratio for each buck converter, as shown in Figure 9.66, "from a global stability" point of view (using Lyapunov's method). The error signals ensure that the output bus is regulated at a predetermined voltage reference while the load-sharing among the *N* modules is maintained under transient and steady-state conditions. The fundamental difference in computing the duty ratio for the nonlinear controller and that for a conventional linear controller yields superior transient performance of the buck converter without compromising its steady-state performance [45]. This is because inside the boundary layer, using the duty ratio, one can implement PWM control as well as interleaving [45].

[1]The limits of this boundary layer correspond to the maximum and minimum values of a ramp of switching frequency f_s. At the beginning of each switching cycle, the mode of operation is determined by whether the trajectories are within the boundaries or outside.

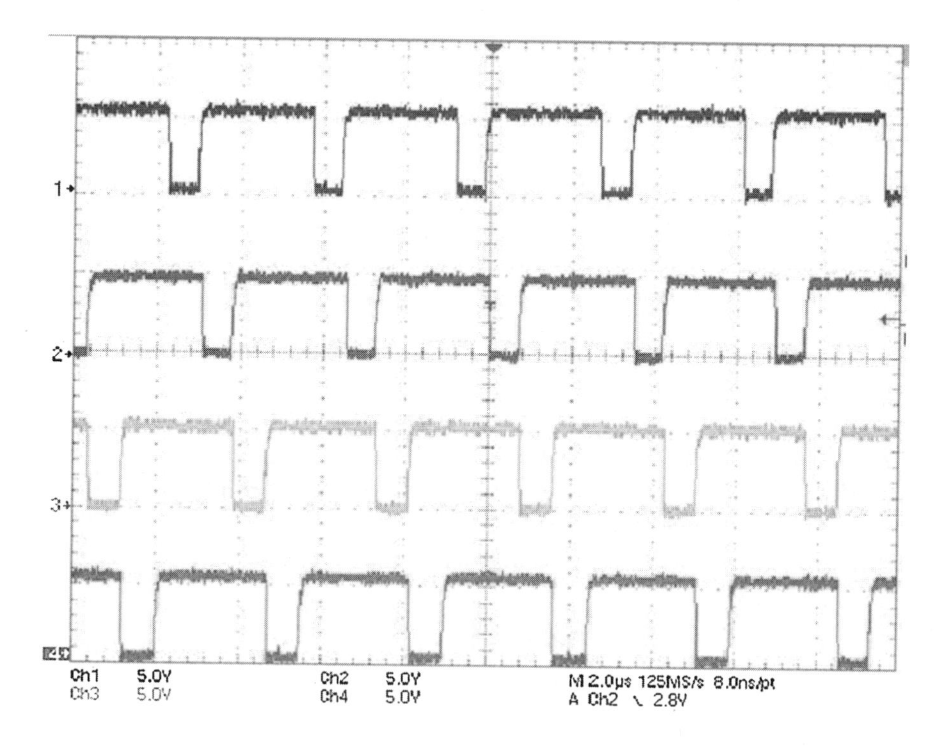

FIGURE 9.68 Gate signals of the four VRM modules showing the equal-phase-shifted operation.

The results of an application of the nonlinear controller, for a four-phase voltage-regulated module (VRM) for next-generation Intel processor, are shown in Figure 9.67 to Figure 9.69. The power-stage architecture of the VRM is the same as that shown in Figure 9.65. The VRM transient and steady-state specifications, which are "extremely stringent," are described in Table 9.1. Figure 9.67 shows the output voltage and the inductor currents in the four modules of the VRM, under steady-state conditions. The four inductor currents are interleaved, i.e., they differ in phase by 90°, which yields an output-ripple frequency four times that of the module switching frequency, thereby reducing the size of the output capacitor. The interleaving is also confirmed by the four equally-phase-shifted low-side gate-driver signals, as shown in Figure 9.68. Additionally, the VRM steady-state performance satisfies the stringent 2% tolerance. Finally, the transient performance of the VRM using the nonlinear controller is shown in Figure 9.69. The slew rate of the load transient is 50 A/μsec. The results show that, even though the slew rate of the load transient is extremely fast, the nonlinear controller satisfies the VRM specifications in Table 9.1 during both no-load to full-load as well as vice versa conditions.

Parallel Multiphase Converter

Applications of parallel multiphase power converters are on the rise [54–65], because they provide several advantages, including capability to handle high power, modularity, high reliability, less voltage or current ripple, and fast-dynamic response. Traditionally, a parallel multiphase converter either has a transformer at the ac side [57–59] or uses separate power supplies [56]. This approach, however, results in a bulky and expensive system because of the line-frequency transformer and the additional power supplies.

A recent approach to overcome these problems is to directly connect three-phase converters in parallel; one such system is shown in Figure 9.70. The parameters of the parallel three-phase boost rectifier (PTBR) are tabulated in Table 9.2. When two three-phase PWM modules are directly connected, circulating currents can exist in all of the phases [60,64,65], as shown in Figure 9.71(a) and (b). Several methods have been proposed to reduce the cross-current among the modules. Using a linear controller and space-vector modulation (SVM) schemes, which do not use the zero vectors, Xing et al. [60] have developed schemes for standardized

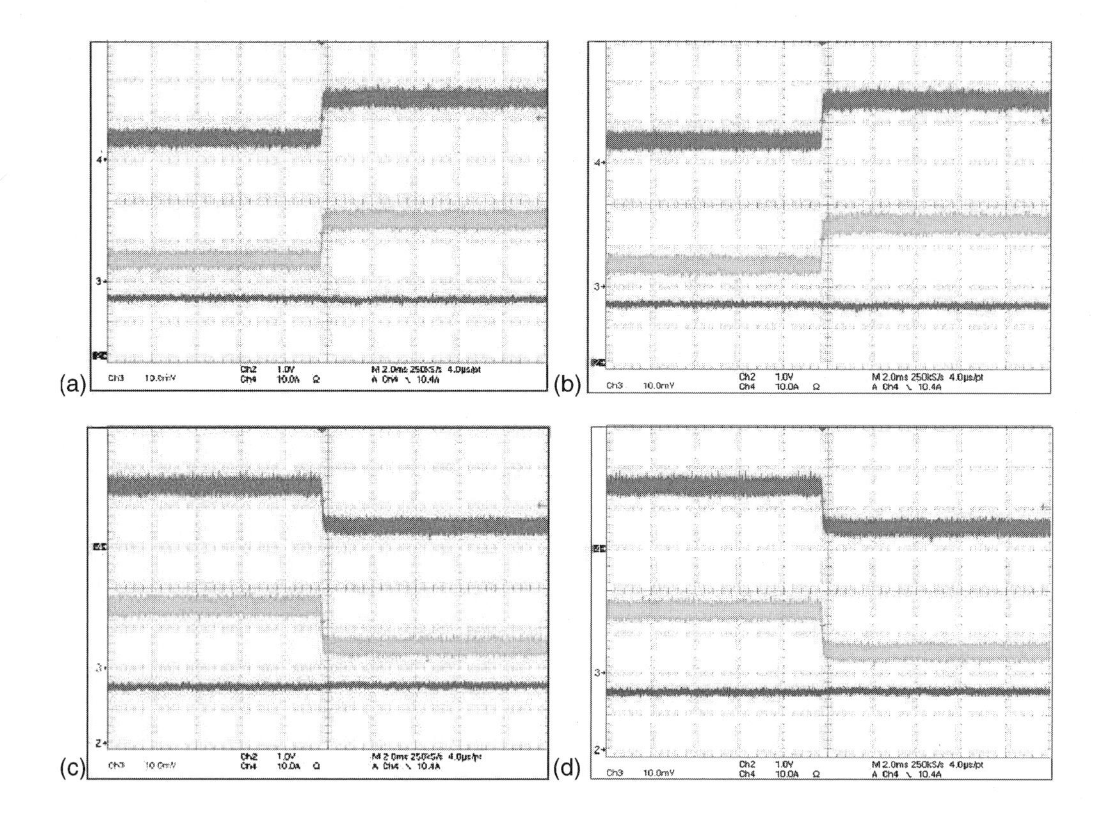

FIGURE 9.69 Experimental inductor currents and the output voltage during a load transient. Step-up load transient: (a) modules 1 and 3 and (b) modules 2 and 4. Step-down load transient: (c) modules 1 and 3 and (d) modules 2 and 4.

TABLE 9.1 Intel VRM 9.0 Design Guidelines

Electrical Specifications	Intel VRM 9.0 Design Guidelines
Output voltage	1.408 – 1.5 V (our nominal reference: 1.45 V)
Output current	60 A
No-load operation	Outputs ≤110% of the maximum value
Overshoot at turn-on/turn-off	Must be within 2% of the nominal output voltage set by VID code
Slew rate	50 A/μsec
Current sharing	Should be accurate within 10% of the rated output current, except during initial power-up and transient responses

three-phase modules to reduce the cross-current. The advantage of such schemes is that the communication between the modules is minimal. However, the transient response of the PTBR is not satisfactory and the magnitude of the zero-sequence current under steady-state conditions is not shown. Recently, Ye et al. [63] have proposed a linear control scheme,[1] which is simple and minimizes the zero-sequence current under steady-state conditions by simply varying the duration of the zero space-vector. The steady-state performance of the PTBR using CS_{LINEAR} is shown in Figure 9.71(c). However, if the system saturates, the control scheme will not work effectively, even under steady-state conditions. This is because, when the system saturates, the zero vector cannot be applied. Furthermore, the performance of the system under transient conditions has not been demonstrated [63].

Recently, three nonlinear control schemes were proposed by the author [64,65] to improve the transient performances of the PTBR as compared to those obtained using CS_{LINEAR}. The first two control schemes

[1]In this section, we will refer to this control proposed by Ye et al. [63] as CS_{LINEAR}.

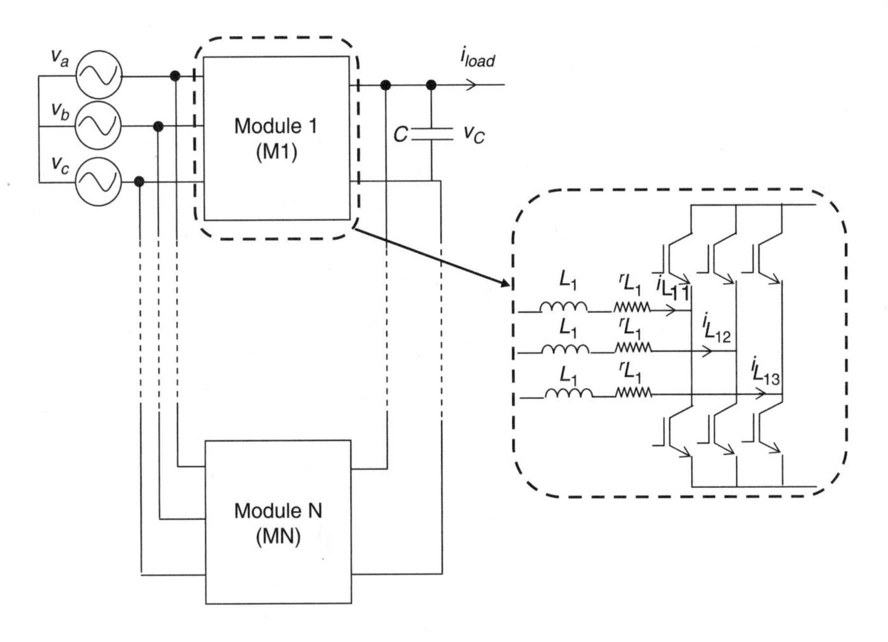

FIGURE 9.70 Schematic of a parallel three-phase boost rectifier (PTBR) with N modules.

TABLE 9.2 Nominal Parameters of the PTBR

Parameter	Nominal Values
$v_{ab} = v_{bc} = v_{ca} = v_n$	208 V (rms)
v_C (regulated)	400 V
Nominal switching frequency (=1/T)	32 kHz
$L_1 = L_2 = L_n$	500 μH
$^rL_1 = {}^rL_2 = {}^rL_n$	0.5 Ω
Bus capacitance (C)	1200 μF
Load resistance (R)	4 Ω
Power ratings of M1 and M2	20 kVA

(CS_{CONT1} and CS_{CONT2}) are developed in the continuous domain, whereas the third scheme ($CS_{DISCRETE}$) is developed in the discrete domain. The former control schemes stabilize the errors on the *dq*-axis sliding surfaces and rely on blocking the pure zero-sequence current path, the inductor size, and the switching frequency to bind the errors on the zero-axis sliding surfaces. The steady-state ripple of the PTBR obtained using CS_{CONT2} is slightly better than that obtained using CS_{CONT1}, because the former uses a hysteretic comparator, which has an inner and an outer hysteretic band. The steady-state ripple of the PTBR obtained with $CS_{DISCRETE}$ is better than that obtained with the other two proposed control schemes because the former combines SVM and nonlinear control, and stabilizes the zero-axis disturbance as well. Hence, the steady-state ripple has a constant frequency, and the deviation of the zero-axis current from its reference value (= 0) is minimized.

Figure 9.72 demonstrates the transient and steady-state performances of the PTBR using these three nonlinear control schemes. We also compare the performances of the three proposed controllers with the linear controller CS_{LINEAR}. We find that CS_{LINEAR} stabilizers the circulating current. However, its transient response is inferior to the proposed control schemes for even moderate feedforward and feedback disturbances. For even larger disturbances, the transient performance of the controller proposed by Ye et al. [63] suffers considerably.

Next, using Figure 9.73, we investigate the sharing of the line currents between M1 and M2, when the PTBR is subjected to a large disturbance in either the voltage (case 1) or the load (case 2). For case 1, we see that the best transient response is achieved using CS_{CONT1}; the response time is comparable to the other two proposed schemes. The recovery time of the PTBR obtained with CS_{LINEAR} is the longest. Moreover,

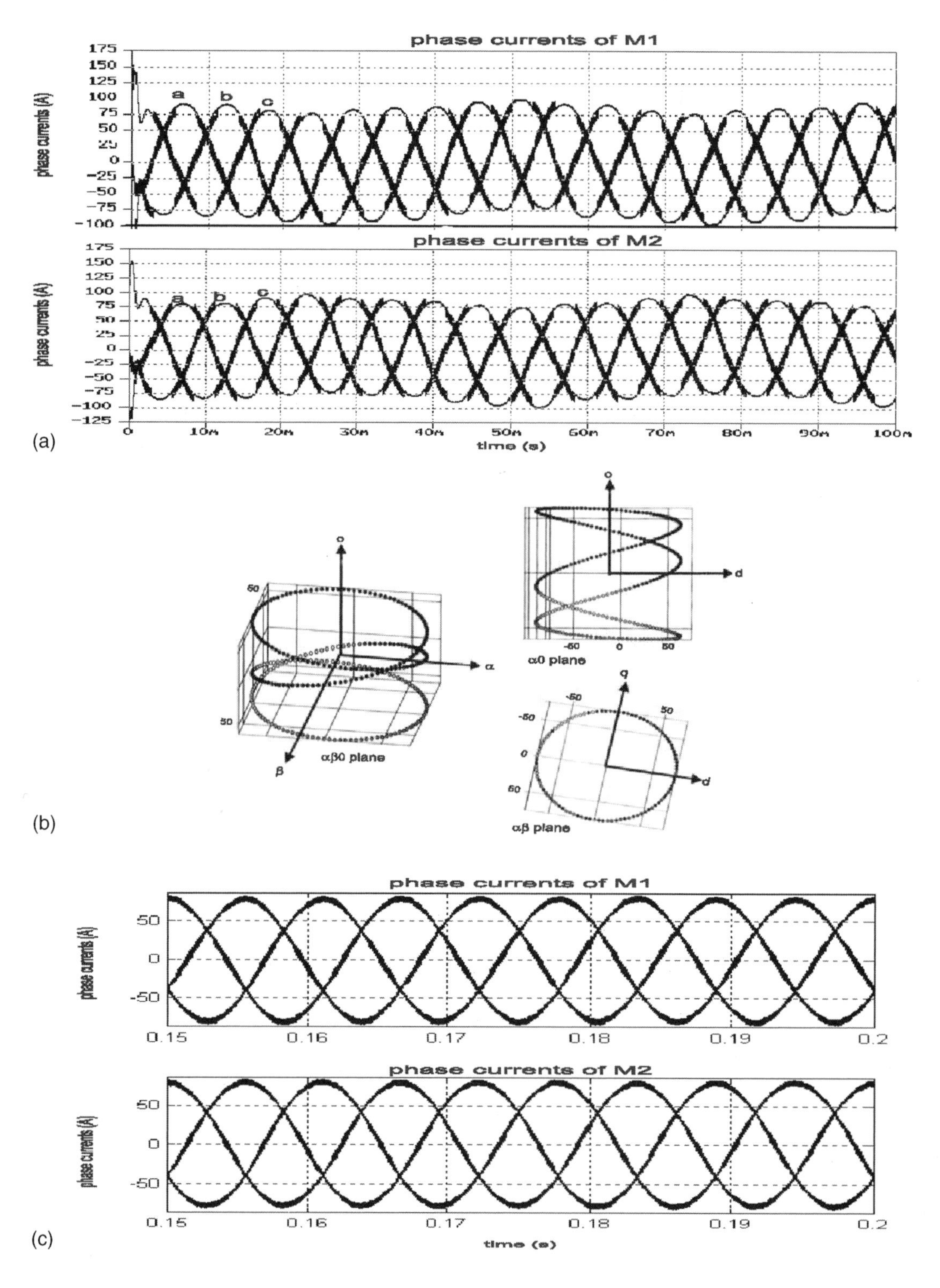

FIGURE 9.71 (a) The phase currents of M1 and M2, using a conventional *dq* controller, when the parameters of the modules are the same, except L_1 is 95% of L_2. The result shows the limitation of a conventional *dq* controller in ensuring even-load distribution when the two modules have parametric variations. (b) Three-dimensional view of the unbalanced phase currents of M1 in the αβo frame. It shows that a conventional *dq* controller can not see the zero-sequence current because it lies on a perpendicular axis. (c) The phase currents of M1 and M2 obtained using CS_{LINEAR} when the parameters of the modules are the same, except L_1 is 95% of L_2. By adding a zero-sequence controller, the effect of the overall unbalance as seen in (a) has been minimized.

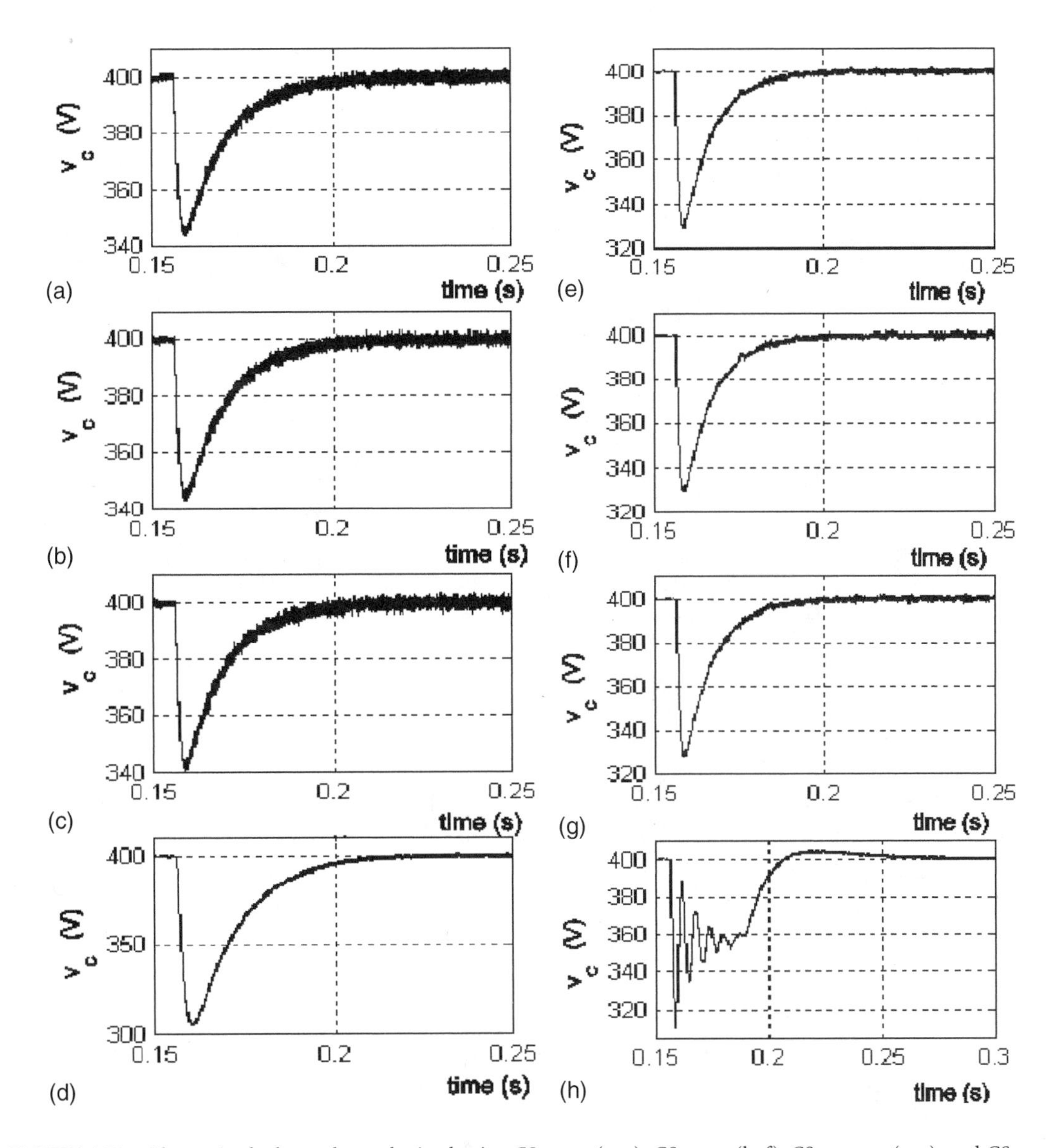

FIGURE 9.72 Change in the bus voltage obtained using CS_{CONT1} (a, e), CS_{CONT2} (b, f), $CS_{DISCRETE}$ (c, g), and CS_{LINEAR} (d, h) for case 1 (figures on the left) and case 2 (figures on the right). Case 1 corresponds to a large transient in the input voltage, while case 2 corresponds to a large transient in the load. For either case, the drop in the bus voltage is larger when using CS_{LINEAR}, even though it is implemented for a smaller variation (5%) L_1 as compared to the proposed control schemes (15%).

immediately after the change in the voltage, there is a undershoot and an overshoot in two of the phase currents, which are not evident in the responses obtained with the proposed control schemes. For case 2, among the three proposed control schemes, $CS_{DISCRETE}$ achieves the best compromise between the response time and current sharing. The recovery times of CS_{CONT1} and CS_{CONT2} are smaller than that of $CS_{DISCRETE}$. The response of the PTBR obtained with CS_{LINEAR} is significantly inferior to those obtained with the proposed control schemes, both in terms of the response time and current sharing.

Finally, in Figure 9.74, we show the impact of the proposed control schemes on the steady-state ripples of the phase currents (in the $\alpha\beta$ frame) and on the zero-axis current that circulates between the two modules. For all of these plots, we choose $L_1 = 0.85L_n$ and $L_2 = L_n$. All other parameters are kept the same as before.

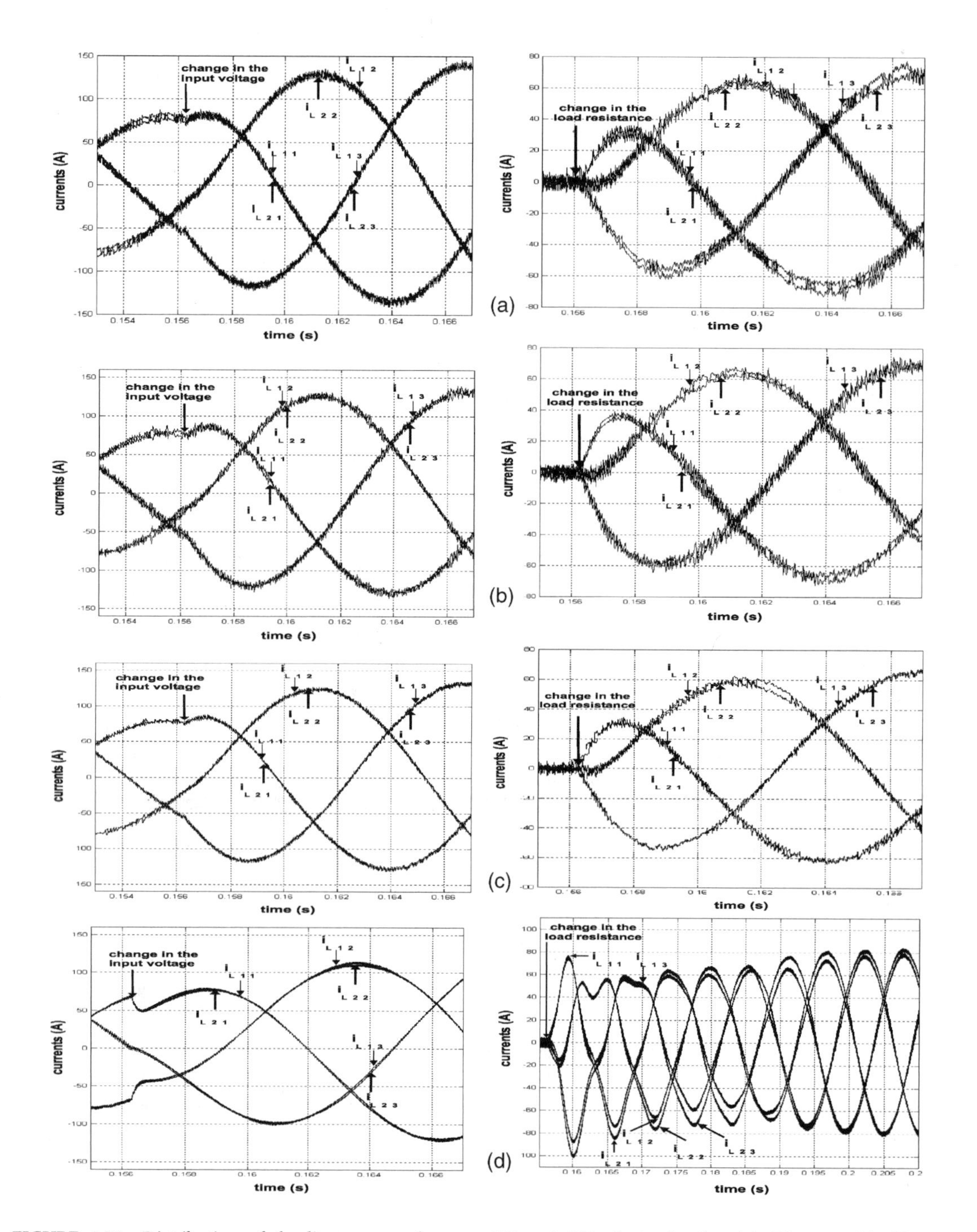

FIGURE 9.73 Distribution of the line currents between M1 and M2 obtained using (a) CS_{CONT1}, (b) CS_{CONT2}, (c) CS_{CONT3}, and (d) CS_{LINEAR} for case 1 (figures on the left) and case 2 (figures on the right). The proposed control schemes and CS_{LINEAR} operate with $L_1 = 85\%\ L_n$ and $L_1 = 95\%\ L_n$, respectively.

The steady-state ripple obtained with $CS_{DISCRETE}$ is better than those obtained using CS_{CONT1} and CS_{CONT2}. More importantly, the zero-axis current obtained with $CS_{DISCRETE}$ has a smaller magnitude compared to the previous cases. The steady-state results obtained using $CS_{DISCRETE}$ and CS_{LINEAR} are close. *Therefore, the nonlinear controller $CS_{DISCRETE}$ attains the best compromise between the dynamic- and steady-state performances.*

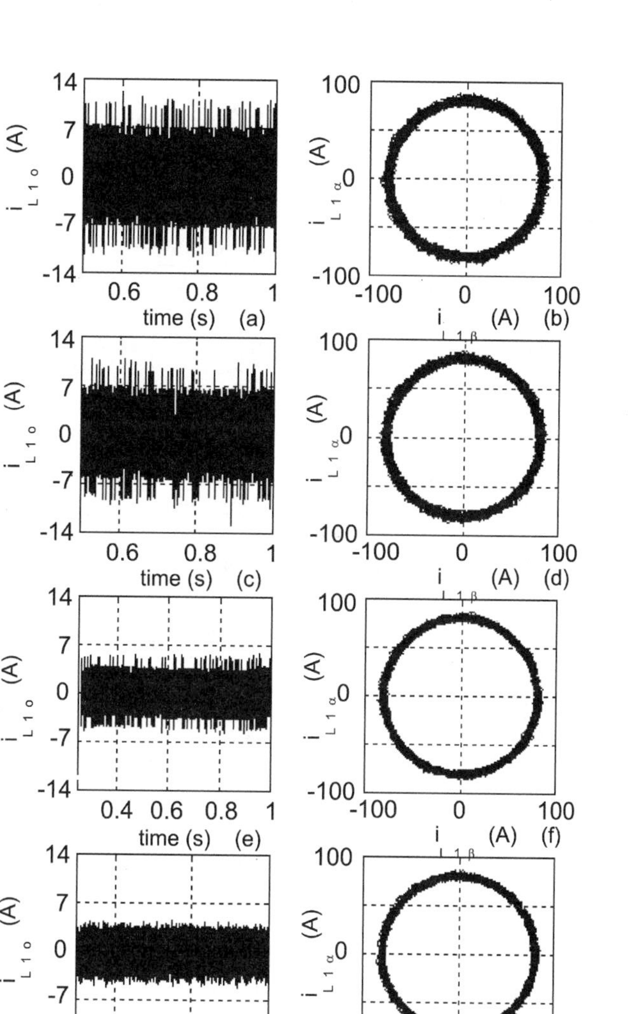

FIGURE 9.74 Steady-state currents on the zero axis and the $\alpha\beta$ axes for M1 obtained using CS_{CONT1}, (a) CS_{CONT2}, (c) $CS_{DISCRETE}$ (e and f), and CS_{LINEAR} (g and h). All of the cases have the same parametric variations. As such, the harmonic distortion and the zero-sequence current of $CS_{DISCRETE}$ and CS_{LINEAR} are close.

Research Issues

It is obvious from the above results that nonlinear controllers can and will play an increasingly effective role for the stabilization and performance optimization of progressively complex IPNs. However, to maximize the effectiveness of these nonlinear controllers, they need to be designed by treating an IPN model as a hybrid system and not as a smooth averaged model. This is a major challenge because the issues of existence of solutions and equilibrium stability for hybrids are difficult propositions to resolve. Additionally, to enhance the use of nonlinear controllers, their designs need to be systematic and not just intuitive. A further issue to be resolved is the need and procedure to quantify the improvement in performances of an IPN using a nonlinear controller with those that are obtained using a linear controller. Finally, strategies for distributed nonlinear control of IPNs needs to be developed, which take into account dynamic changes in the network and network processing delays leading to wider stability and power optimization.

References

1. S.K. Mazumder, A.H. Nayfeh, and D. Boroyevich, "Nonlinear analysis of parallel DC–DC converters, Special Issue on Nonlinear Controls and Dynamics," *J. Vib. Control*, Vol. 9, pp. 775–789, 2003.
2. S.K. Mazumder and A.H. Nayfeh, "A new approach to the stability analysis of boost power-factor-correction circuits, Special Issue on Nonlinear Controls and Dynamics," *J. Vib. Control*, 9, 749–773, 2003.
3. S.K. Mazumder, A.H. Nayfeh, and D. Boroyevich, "A theoretical and experimental investigation of the fast- and slow-scale instabilities of a DC–DC converter," *IEEE Trans. Power Electron.*, vol. 16, no. 2, pp. 201–216, 2001.
4. M. Jordan, "Load share IC simplifies power supply design," *High Freq. Power Convers. Conf.*, 1991, pp. 65–76.
5. M. Jordan, "UC3907 load share IC simplifies parallel power supply design," *Unitrode Application Note U-129*, 1991.
6. T. Kohama, T. Ninomiya, M. Shoyama, and F. Ihara, "Dynamic analysis of parallel-module converter system with current balance controllers," *IEEE Int. Telecommun. Energy Conf.*, 1994, pp. 190–195.
7. V.J. Thottuvelil and G.C. Verghese, "Stability analysis of paralleled DC–DC converters with active current sharing," *IEEE Power Electron. Spec. Conf.*, 1996, pp. 1080–1086.
8. K. Siri, C.Q. Lee, and T-F. Wu, "Current distribution control for parallel-connected converters," *IEEE Trans. Aerospace Electron. Syst.*, vol. 28, no.3, 1992, pp. 829–850.
9. Q. Chen, "Stability analysis of paralleled rectifier systems," *IEEE Int. Telecommun. Energy Conf.*, 1995, pp. 35–40.
10. J. Rajagopalan, K. Xing, Y. Guo, and F.C. Lee, "Modeling and dynamic analysis of paralleled DC–DC converters with master/slave current sharing control," *IEEE Appl. Power Electron. Conf.*, 1996, pp. 678–684.
11. Y. Panov, J. Rajagopalan, and F.C. Lee, "Analysis and design of *N* paralleled DC–DC converters with master-slave current sharing control," *IEEE Appl. Power Electron. Conf.*, 1996, pp. 678–684.
12. D.S. Garabandic and T.B. Petrovic, "Modeling parallel operating PWM DC–DC power supplies," *IEEE Trans. Ind. Electron.*, vol. 42, no. 5, 1995, pp. 545–550.
13. V.J. Thottuvelil and G.C. Verghese, "Analysis and control design of paralleled DC/DC converters with current sharing," *IEEE Trans. Power Electron.*, vol. 13, no. 4, 1998, pp. 635–644.
14. K. Siri, C.Q. Lee, and T-F. Wu, "Current distribution control for parallel connected converters: I," *IEEE Trans. Aerospace Electron. Syst.*, vol. 28, no. 3, 1992, pp. 829–840.
15. K. Siri, C. Q. Lee, and T-F. Wu, "Current distribution control for parallel connected converters: II," *IEEE Trans. Aerospace Electron. Syst.*, vol. 28, no. 3, 1992, pp. 841–851.
16. I. Batarseh, K. Siri, and H. Lee, "Investigation of the output droop characteristics of parallel connected DC–DC converters," *IEEE Power Electron. Spec. Conf.*, 1, 597–606.
17. H. Sira-Ramirez, G. Escobar, and R. Ortega, "On passivity-based sliding mode control of switched DC-to-DC power converters," *IEEE Conf. Decis. Control*, 3, 2525–2526, 1996.
18. M. Rios-Bolivar, A.S.I. Zinober, and H. Sira-Ramirez, "Dynamical sliding mode control via adaptive input-output linearization: a backstepping approach," *Lecture Notes in Control and Information Sciences: Robust Control via Variable Structure and Lyapunov Techniques*, New York: Springer-Verlag, 1996.
19. S.R. Sanders and G.C. Verghese, "Lyapunov-based control for switching power converters," *IEEE Power Electron. Spec. Conf.*, 1990, pp. 51–58.
20. M.I. Angulo-Nunez and H. Sira-Ramirez, "Flatness in the passivity based control of DC-to-DC power converters," *IEEE Conf. Decis. Control*, 4, 4115–4120, 1998.
21. A.M. Stankovic, D.J. Perreault, and K. Sato, "Synthesis of dissipative nonlinear controllers for series resonant DC–DC converters," *IEEE Trans. Power Electron.*, vol. 14, no. 4, 1999, pp. 673–682.
22. M. Rios-Bolivar, H. Sira-Ramirez, and A.S.I. Zinober, "Output tracking control via adaptive input-output linearization: a backstepping approach," *IEEE Conf. Decis. Control*, 2, 1579–1584, 1995.
23. A. Kugi and K. Schlacher, "Nonlinear H_∞-controller design for a DC–DC power converter," *IEEE Trans. Control Syst. Technol.*, vol. 7, no. 2, 1999, pp. 230–237.

24. P. Mattavelli, L. Rossetto, G. Spiazzi, and P. Tenti, "General-purpose fuzzy controller for DC/DC converters," *IEEE Power Electron. Spec. Conf.*, 1995, pp. 723–730.
25. F. Ueno, T. Inoue, I. Oota, and M. Sasaki, "Regulation of Cuk converters using fuzzy controllers," *IEEE Int. Telecommun. Energy Conf.*, 1991, pp. 261–267.
26. M. Scheffer, A. Bellini, R. Rovatti, A. Zafarana, and C. Diazzi, "A fuzzy analog controller for high performance microprocessor power supply," *Proc. EUFIT*, 1996.
27. V.I. Utkin, *Sliding modes in control optimization*, New York: Springer Verlag, 1992.
28. V.I. Utkin, J. Guldnerm, and J. Shi, *Sliding Mode Control in Electromechanical Systems*, London: Taylor & Francis, 1999.
29. P.V. Kokotovic, "The joy of feedback: nonlinear and adaptive," *IEEE Control Syst. Magazine*, vol. 12, no. 3, pp. 7–17, 1992.
30. K. Miroslav, I. Kanellakopoulos, and P.V. Kokotovic, *Nonlinear and Adaptive Control Design*, New York: Wiley, 1995.
31. R.A. Freeman and P.V. Kokotovic, *Robust Nonlinear Control Design: State-Space and Lyapunov Techniques*, Boston, MA: Birkhäuser, 1996.
32. S. Gutman, "Uncertain dynamical systems — a Lyapunov min–max approach," *IEEE Trans. Autom. Control*, 24, 437–443, 1979.
33. Z. Qu, *Robust Control of Nonlinear Uncertain Systems*, New York: John Wiley & Sons, 1998.
34. S. Sastry, *Nonlinear Systems: Analysis, Stability, and Control*, New York: Springer, 1999.
35. A. Isidori, *Nonlinear Control Systems*, New York: Springer, 1995.
36. R. Marino and P. Tomei, *Nonlinear Control Design: Geometric, Adaptive, and Robust*, Simon and Schuster International Group, 1995.
37. J. Ball and J.W. Helton, "Factorizaion of nonlinear systems: toward a theory for nonlinear H_∞ control," *IEEE Conf. Decis. Control*, 3, 2376–2381, 1988.
38. H.J. William, "Extending H_∞ control to nonlinear systems: control of nonlinear systems to achieve performance objectives," *Society for Industrial and Applied Mathematics*, 1999.
39. A.J. van der Schaft, *L2-gain and Passivity Techniques in Nonlinear Control*, New York: Springer, 1996.
40. M.M. Gupta and T. Yamakawa, Eds., *Fuzzy Computing: Theory, Hardware, and Applications*, Amsterdam: Elsevier Science, 1988.
41. M. Jamshidi, A. Titli, L. Zadeh, and S. Boverie, *Applications of Fuzzy Logic: Towards High Machine Intelligency Quotient Systems*, Englewood Cliffs, NJ: Prentice-Hall, 1997.
42. G. Langholz and A. Kandel, Eds., *Fuzzy Control Systems*, Boca Raton, FL: CRC Press, 1993.
43. B. Tomescu and H.F. VanLandingham, "Improved large-signal performance of paralleled DC–DC converters current sharing using fuzzy logic control," *IEEE Trans. Power Electron.*, vol. 14, no. 3, 1999, pp. 573–577.
44. M. Lopez, L.G. de Vicuña, M. Castilla, O. Lopez, and J. Majo, "Interleaving of parallel DC–DC converters using sliding mode control," *Proc. IEEE Ind. Electron. Soc.*, 2, 1055–1059, 1998.
45. S.K. Mazumder, A.H. Nayfeh, and D. Boroyevich, "Robust control of parallel DC–DC buck converters by combining integral-variable-structure and multiple-sliding-surface control schemes," *IEEE Trans. Power Electron.*, vol. 17, no. 3, 2002, pp. 428–437.
46. J.H. Green and K. Hedrick, "Nonlinear speed control of automative engines," *IEEE Am. Control Conf.*, 1990, pp. 2891–2897.
47. D. Swaroop, J.C. Gerdes, P.P. Yip, and J.K. Hedrick, "Dynamic surfce control of nonlinear system," *IEEE Am. Control Conf.*, 1990, pp. 3028–3034.
48. S.V. Drakunov, D.B. Izosimov, A.G. Lukajanov, V. Utkin, and V. I. Utkin, "Block control principle: part I," *Autom. Remot. Control*, vol. 51, no. 5, 1990, pp. 38–46.
49. S.V. Drakunov, D.B. Izosimov, A.G. Lukajanov, V. Utkin, and V.I. Utkin, "Block control principle: part II," *Autom. Remot. Control*, vol. 51, no. 6, 1990, pp. 20–31.
50. B.R. Barmish and G. Leitmann, "On ultimate boundedness control of uncertain systems in the absence of matching assumptions," *IEEE Trans. Autom. Control*, vol. 27, no. 1, 1982, pp. 153–158.

51. M. Corless and G. Leitmann, "Continuous state feedback guarantees uniform ultimate boundedness for uncertian dynamical systems," *IEEE Trans. Autom. Control*, vol. 26, 1981, pp. 1139–1144.
52. J.C. Gerdes, *Decoupled design of robust controllers for nonlinear systems: as motivated by and applied to coordinated throttle and brake control for automated highways*, Ph.D. dissertation, Department of Mechanical Engineering, University of California, Berkeley, CA, 1996.
53. A.A. Stotsky, J.K. Hedrick, and P.P. Yip, "The use of sliding modes to simplify the backstepping control method," *Appl. Math. Comput. Sci.*, vol. 8, no. 1, 1998, pp. 123–133.
54. S. Ogasawara, J. Takagaki, H. Akagi, and A. Nabae, "A novel control scheme of duplex current-controlled PWM inverters," *Proc. IEEE Ind. Appl. Soc.*, 1987, pp. 330–337.
55. I. Takahashi and M. Yamane, "Multiparallel asymmetrical cycloconverter having improved power factor and waveforms," *IEEE Trans. Ind. Appl.*, vol. 22, no. 6, 1986, pp. 1007–1016.
56. T. Kawabata and S. Higashino, "Parallel operation of voltage source inverters," *IEEE Trans. Ind. Appl.*, vol. 24, no. 2, 1988, pp. 281–287.
57. J.W. Dixon and B.T. Ooi, "Series and parallel operation of hysterisis current-controlled PWM rectifiers," *IEEE Trans. Ind. Appl.*, vol. 25, no. 4, 1989, pp. 644–651.
58. Y. Komatsuzaki, "Cross current control for parallel operating three-phase inverter," *IEEE Power Electron. Spec. Conf.*, 1994, pp. 943–950.
59. Z. Zhang and B. Ooi, "Multimodular current-source SPWM converters for superconducting a magnetic energy storage system," *IEEE Trans. Power Electron.*, vol. 8, no. 3, 1993, pp. 250–256.
60. K. Xing, S.K. Mazumder, Z. Ye, D. Borojevic, and F.C. Lee, "The circulating current in paralleled three-phase boost PFC rectifiers," *IEEE Power Electron. Specialists Conf.*, 1999, pp. 783–789.
61. Y. Sato and T. Kataoka, "Simplified control strategy to improve AC input-current waveform of parallel connected current-type PWM rectifiers," *Proc. Inst. Electr. Eng.*, 142, 246–254, 1995.
62. K. Matsui, "A pulse-width modulated inverter with parallel-connected transistors by using current sharing reactors," *Proc. Ind. Appl. Soc.*, 1985, pp. 1015–1019.
63. Z. Ye, D. Boroyevich, J.Y. Choi, and F.C. Lee, "Control of circulating current in parallel three-phase boost converters," *IEEE Appl. Power Electron. Conf.*, 2000.
64. S.K. Mazumder, *Nonlinear analysis and control of interactive power-electronics systems, First NSF CAREER Annual Report for 2003–2004*, February 2004.
65. S.K. Mazumder, "A novel discrete control strategy for independent stabilization of parallel three-phase boost converters by combining space-vector modulation with variable-structure control," *IEEE Trans. Power Electron.*, vol. 18, no. 4, 2002, pp. 1070–1083.
66. S.L. Kamisetty and S.K. Mazumder, "Hybrid nonlinear controller for multiphase VRM," IECON'03, the 29th Annual Conference of the IEEE, 2003, pp. 574–579.

9.7 Uninterruptible Power Supplies

Ayse E. Amac and Ali Emadi

Uninterruptible power supply (UPS) systems are presented in this section. Topologies, operation, and control principles of UPS systems are explained in detail. In addition, a brief description of conventional UPS systems, their disadvantages, scope for improvement, and advanced architectures in UPS research are presented.

Introduction

Uninterruptible power supply (UPS) systems are designed to provide reliable and high-quality continuous power to critical loads in the face of events on the utility supply. These events can range from overvoltage and undervoltage conditions to complete disruption of the mains. UPS systems ensure power without break for the load as operating along with the mains as well as suppress line transients and harmonic disturbances.

UPS systems are being applied for a wide variety of critical equipment, such as medical facilities, life support systems, financial transaction handlers, data storage and computer systems, telecommunications, industrial processing, and on-line management systems [1].

The objective of UPS systems is to provide sinusoidal input current with low total harmonic distortion (THD) and to realize power line conditioning that has sinusoidal output voltage and unity power factor. In addition, an ideal UPS should have seamless transition capability when a failure occurs, high reliability, and high efficiency. Furthermore, the UPS system should be low maintenance, low cost, and lightweight [2].

UPS systems are reviewed here in terms of classification, operation, and control. They are explained as static, rotary, and hybrid static/rotary systems in the section "Classification". Static UPS systems are defined in detail and in the section "Applications" distributed and centralized applications are presented. The section "Control Techniques" deals with suitable control techniques for these systems. Finally, the "Conclusion" summarizes the results obtained.

Classification

When the critical loads need to be supplied without interruption, a UPS system is generally employed. Both financial and industrial sectors need clean and reliable power supplies. Therefore, the UPS market grows increasingly and a wide range of UPS systems with various designs are now available in the market. There are three types of UPS systems in the literature: static, rotary, and hybrid static/rotary systems. These three categories are explained in this section.

Static UPS

Static UPS systems are available from 100 VA to 1 MVA. They are the most commonly used UPS systems from low-power personal computers and telecommunication systems, to medium-power medical systems, to high-power utility systems. They have high efficiency, high reliability, and low THD. Their main disadvantages are poor performance with nonlinear and unbalanced loads and high cost for achieving very high reliability. The main types of static UPS systems are off-line, on-line, and line-interactive topologies.

Off-Line UPS

Off-line topology is also called standby UPS or line-preferred UPS in the literature. These are widely used as a backup system for various electrical loads. Off-line UPS consists of an ac/dc converter, a battery bank, a dc/ac converter, and static and transfer switches as shown in Figure 9.75. There are three operating modes for off-line UPS systems: normal mode of operation, recovery mode, and backup mode.

When the ac line is normal, the transfer and the static switches are on. The ac line supplies the load through the switches. The dc/ac converter is connected in parallel to the load and stays on standby during the normal mode of operation. Since it is traditionally off in this mode, an off-line UPS does not usually correct the power factor. However, in the normal mode of operation, the dc/ac converter may be used as an active filter by

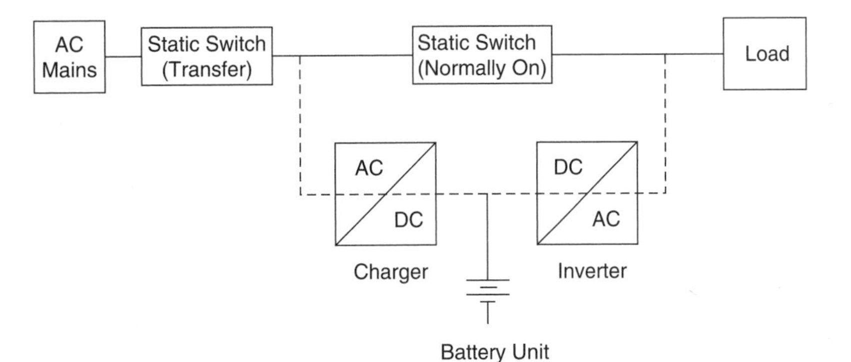

FIGURE 9.75 Block diagram of off-line UPS topology.

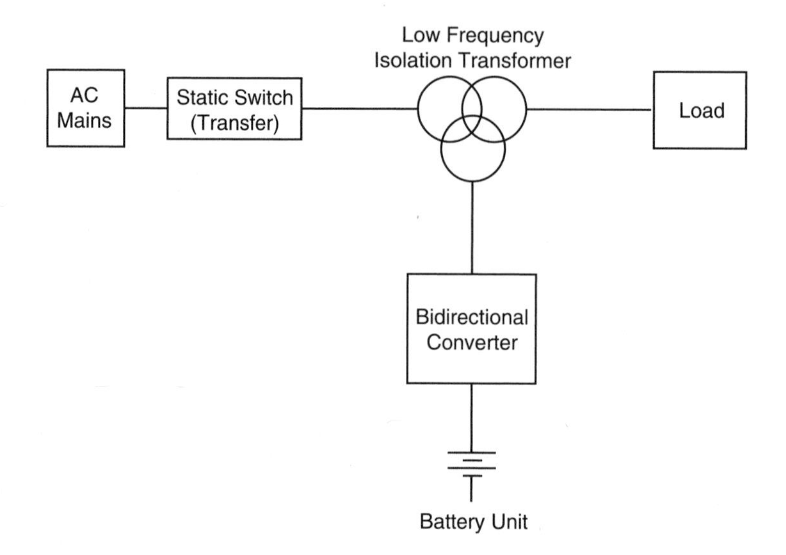

FIGURE 9.76 Block diagram of off-line UPS with low-frequency isolation transformer.

adding a simple control unit to the system. Therefore, reducing the line current harmonics and improving the load power factor can be achieved by the off-line UPS system. In addition, the ac/dc converter can be adjusted to decrease the harmonic content of the sinking current of the dc side. However, these improvements cause increased cost and complexity [3,4].

If necessary, the ac line also feeds the ac/dc converter as well as the load to charge the battery. This mode is called the recovery mode of operation. Since the ac/dc converter is used only as a charger, it does not need to meet the power demand of the load. Therefore, it is rated at low power, which allows the off-line UPS systems to be cheaper than the other static UPS topologies.

When the line is out of the preset tolerance, the transfer and static switches are off. This mode is known as backup, emergency, or stored energy mode of operation. In order to prevent any power accident, the transfer switch and the static switch disconnect the power between the line and the ac/dc converter and between the line and the dc/ac converter, respectively. At the same time, the dc/ac converter is turned on. It has to meet the power of the load. Until the line is within the preset tolerance again, the critical load is supplied by the battery unit via the dc/ac converter. Transfer time is the most important issue in this mode. It is usually about a quarter line cycle, which is enough for most of the applications such as personal computers.

Simple design, low cost, and small size are the main advantages of the off-line UPS topology. However, the disadvantages of this topology are no output voltage regulation, long switching time, poor performance with nonlinear loads, and lack of the real isolation of the load from the ac line. To overcome the isolation problem, a three-winding transformer is employed in the system as shown in Figure 9.76. This technique has high reliability at a moderate cost as well as limited power conditioning for the output voltage, but this transformer leads to a heavier UPS with a lower efficiency.

As shown in Figure 9.77, a high-frequency transformer with a cycloconverter may be used at the output of the inverter to achieve a compact and lightweight system. Moreover, acoustic noise is reduced in this configuration since the high-frequency transformer carries currents at frequencies close to the switching frequency [5]. To improve the quality of the output voltage, a filter can be used at the output stage of the UPS system. However these modifications increase the cost of off-line UPS systems. These limit off-line UPS system applications to less than 2 kVA [6].

On-Line UPS

This configuration is also known as double-conversion UPS and inverter-preferred UPS. On-line UPS consists of an ac/dc converter (rectifier/charger), a battery unit, a dc/ac converter (inverter), and static (bypass) and the

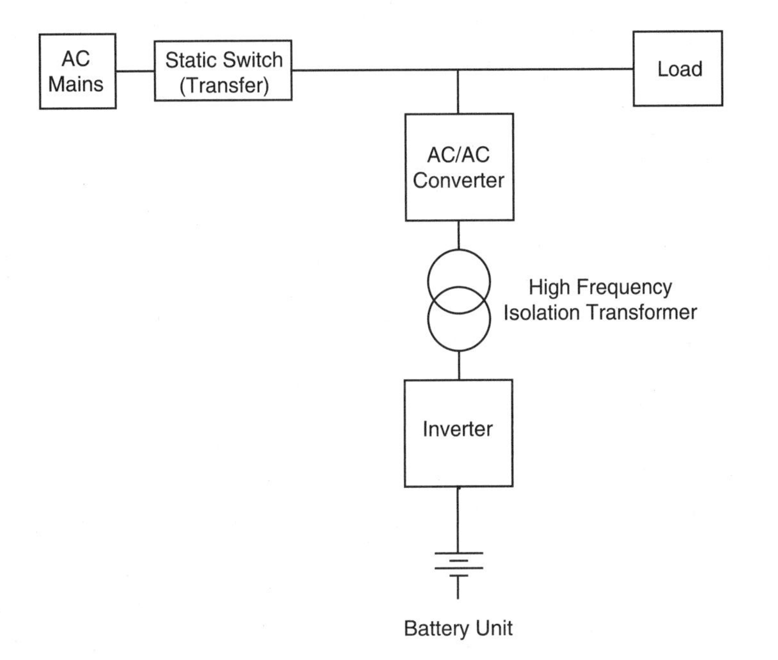

FIGURE 9.77 Block diagram of off-line UPS with high-frequency isolation transformer.

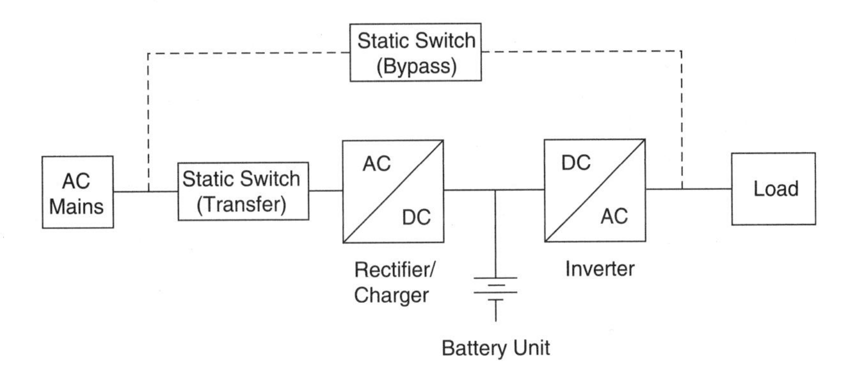

FIGURE 9.78 Block diagram of on-line UPS with bypass switch.

transfer switches as shown in Figure 9.78. There are four operating modes in the on-line UPS systems: normal mode of operation, recovery mode of operation, stored energy mode of operation, and bypass mode of operation.

When the ac line is normal, the transfer and bypass switches are on and off, respectively. The power continuously flows from the dc/ac converter through the ac/dc converter to the critical load. Contrary to the off-line UPS topology, the dc/ac and ac/dc converters are always on in this mode of operation. Very good line conditioning is obtained since double conversion takes place in this configuration.

The ac/dc converter is used as a rectifier and charger in this mode of operation. Therefore, it is rated at 100% of the power demanded by the load and battery unit. Since the dc/ac converter is always on, its power rating is also required to meet 100% of the power demanded by the load. The dc/ac converter must supply the load with the power delivered by the ac/dc converter. Therefore, both converters have the highest power rating which causes high cost in this topology.

When the ac line is out of the preset tolerance, the transfer switch is off to prevent any power accident between the ac line and the load. This mode is called the backup mode of operation. The dc/ac converter continues to deliver the power from the battery to the critical load without any interruption since it is always on. When the ac input voltage is normal, the UPS system returns to the normal operating mode. A phase-locked loop (PLL) is used to make the load voltage in phase with the input voltage.

In case of maintenance of the system or internal malfunction such as over current, the UPS operates in bypass mode. In this mode of operation, the static and bypass switches are off and on, respectively. Fault clearing is also done in this mode. The power is directly transferred through the bypass switch from the ac input to the critical load. Therefore the output frequency should be the same as the ac line frequency.

The main advantage of the on-line UPS system is no transition time from the normal mode to the backup mode operation. This feature makes it more reliable than the off-line UPS topology. Moreover, the output frequency can be regulated or changed 'in this configuration since the dc/ac converter is always on. In addition, double-conversion allows very wide tolerance to the input voltage variations and very good regulation of the output voltage. However, the on-line UPS topology has a low power factor, high THD at the input, and low efficiency. These are the main disadvantages of the on-line UPS topology.

During the ac line outage, the ac/dc converter in the on-line UPS topology is disabled and kept idle. Therefore, traditional on-line UPS topology is not good with nonlinear loads. The load current contains harmonics and is not kept in phase with the output voltage. This causes a low power factor and low battery utilization factor. As a result, the on-line UPS cannot effectively utilize the utility network and local installation. Hence, power installations are oversized on distribution level and eventually the transmission level. To overcome this problem an additional power factor correction (PFC) circuit can be used at the input as shown in Figure 9.79. However, the cost of the UPS system increases with this improvement [7].

The on-line UPS is based on the topology of two power conversion stages. During the normal mode of operation, the power flows through the ac/dc converter and dc/ac converter from the input to the load. Therefore, it has higher power losses and lower efficiency compared to other UPS topologies. However, in terms of performance, power conditioning, and load protection, the on-line UPS system is the most accomplished topology. They are produced with a very wide range of power ratings from a few kVA to several MVA. This is the reason that on-line UPS have a large variety of topologies. Each topology aims to work out different specific problems and the particular choice depends on the particular application.

However, generally, there are two major types of on-line UPS topologies: with bypass switch and without bypass switch. The first topology has a bypass switch or mechanism connected to it. In case of ac/dc converter failure, the critical load will be directly supplied by the AC line through the switch. In the second topology, as the bypass switch is removed, the UPS operates in the on-line mode. Redundancy is a considerable problem in this configuration. This configuration can be realized with a low frequency transformer or a high frequency transformer.

The block diagram of an on-line UPS with low frequency ac link is shown in Figure 9.80. This configuration has transformer isolation at the output which is at low frequency. The weight and volume of the UPS increase since the transformer is heavy and bulky. Therefore, this topology is applicable only in high power ratings

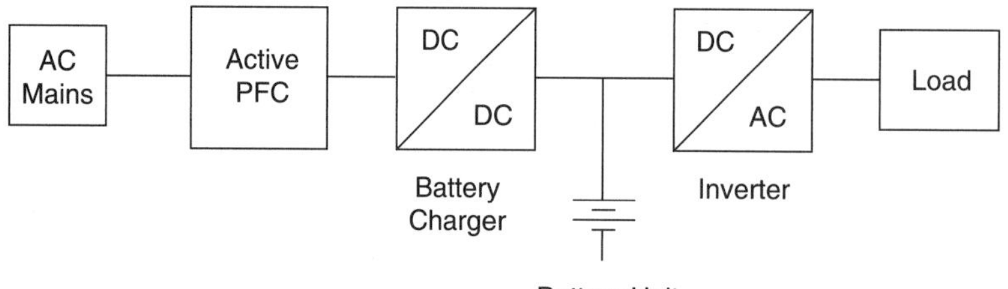

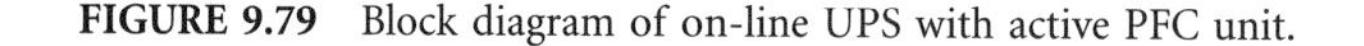

FIGURE 9.79 Block diagram of on-line UPS with active PFC unit.

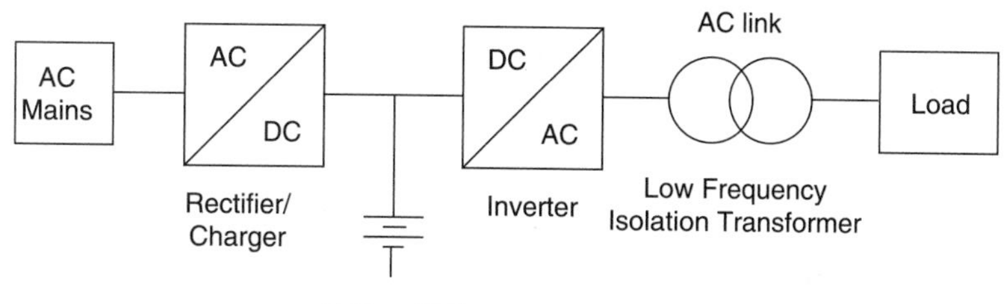

FIGURE 9.80 Block diagram of on-line UPS with low-frequency ac link.

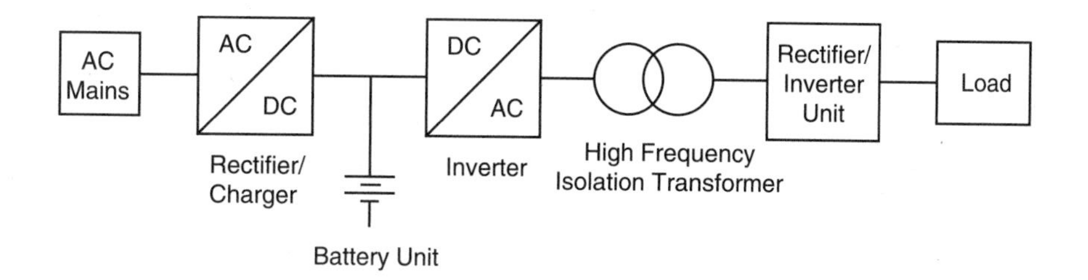

FIGURE 9.81 Block diagram of on-line UPS with high frequency dc link.

which are larger than 20 kVA. The switching frequency is usually limited to less than 2 kHz. Because of low switching frequency, the inductance of the output filter and the transformer produce high acoustic noise in the system. A poor transient response to the load and input voltage changes is another disadvantage of this topology. To solve these problems, the switching frequency of the dc/ac converter can be increased above 20 kHz. However, the size of the isolating transformer cannot be reduced since it is independent of the switching frequency.

To reduce the size and weight of the transformer, a high-frequency ac link can be used in the on-line UPS system as shown in Figure 9.81. This method also reduces the size of the output filter since a high switching frequency is used. The disadvantage of this system is low efficiency because of the cascaded converters.

Line-Interactive UPS

It is also called grid-interactive UPS system. There are three kinds of this UPS in the literature, called line-interactive UPS: single-conversion, in-line, and double-conversion line-interactive UPS systems.

Single-Conversion Topology

The single-conversion topology was introduced in the 1990s. This topology is built on a single-converter structure. The system consists of a static switch, a series inductor, a bidirectional converter, and a battery unit as shown in Figure 9.82. A single-conversion topology can operate either as an on-line or off-line UPS [8]. The series inductor is not required for the off-line operating mode. However, to provide a unity power factor or regulate the output voltage, the single-conversion UPS is usually operated on-line.

When the system is normal, the load is directly supplied by the ac line. The bidirectional converter is in parallel processing mode and charges the battery. It also provides reactive power for the nonlinear load to obtain unity power factor or pure sinusoidal output voltage. The power conditioning function of the bidirectional converter is used only in the single-conversion topology, which is worked as an on-line UPS. This topology has two operating modes: normal mode and backup mode.

During the normal mode of operation, the power flows from the ac line to the critical load through the static transfer switch and the link inductor. The bidirectional converter acts as a parallel active power filter and

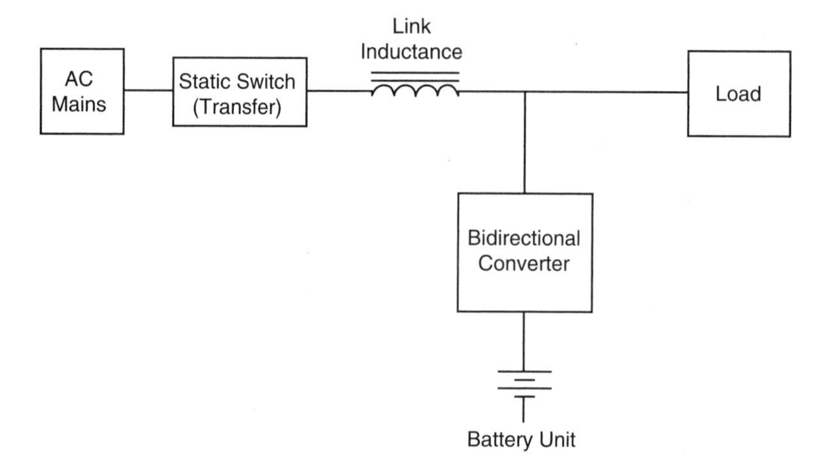

FIGURE 9.82 Block diagram of single-conversion UPS topology.

a current stabilizer. It maintains line current harmonics produced by the nonlinear load and forces the input current to be sinusoidal with unity power factor. In addition, it produces a little active power to control dc bus voltage and operates as a charger to feed the battery unit. Contrary to the conventional on-line UPS topology, it does not cause additional harmonics at the input side. This is an important advantage of the single-conversion line-interactive UPS topology.

When the utility is gone, the static switch is off to prevent back feed from the converter to the ac line. The bidirectional converter acts as an inverter to supply the critical load with a seamless transition. The output voltage can be better regulated with an additional transformer at the output. However, the UPS will be heavy, bulky, and expensive.

The simple design, high reliability, and lower cost compared to the on-line UPS systems are the main advantages of the single-conversion line-interactive UPS topologies. Since they are single-stage, the efficiency is higher than that of the double-conversion topology. As was mentioned before, they also have good harmonic control at the input.

There is no effective isolation between the load and the ac line in this topology. This is the main disadvantage of the system. To overcome this problem, a transformer can be added at the output; however, it will result in additional cost, size, and weight. Since it is not in series with the load, the output voltage cannot be properly regulated. Moreover, the output frequency regulation is not possible in this topology because the ac line directly supplies the load during the normal mode of operation [9].

In-Line Topology

This topology consists of a static switch, a triport high leakage inductance transformer, a bidirectional converter, and a battery unit as shown in Figure 9.83. A voltage-source, voltage-controlled bidirectional converter is used in this topology. It can be operated as a rectifier or inverter. This topology is appropriate to solve a wide range of power quality problems such as dip compensation, harmonic isolation, voltage unbalance, power factor correction, voltage regulation, flicker, and power outages [10].

When the ac line is normal, the load is supplied by the ac line through the static switch and the triport transformer with leakage inductance. The bidirectional converter acts as a series active filter to regulate the input voltage and keep the load voltage in a specific level. The magnitude of the bidirectional converter voltage is kept constant by a suitable control signal. Therefore, it is protected from any sags, swells, and large fluctuations in the ac line voltage. The power factor correction is controlled through the power flow of the system. The bidirectional converter absorbs a lagging or leading VAR from the ac line to obtain pure sinusoidal voltage for the load [11]. It also operates as a charger for charging to the battery unit in the normal mode of operation.

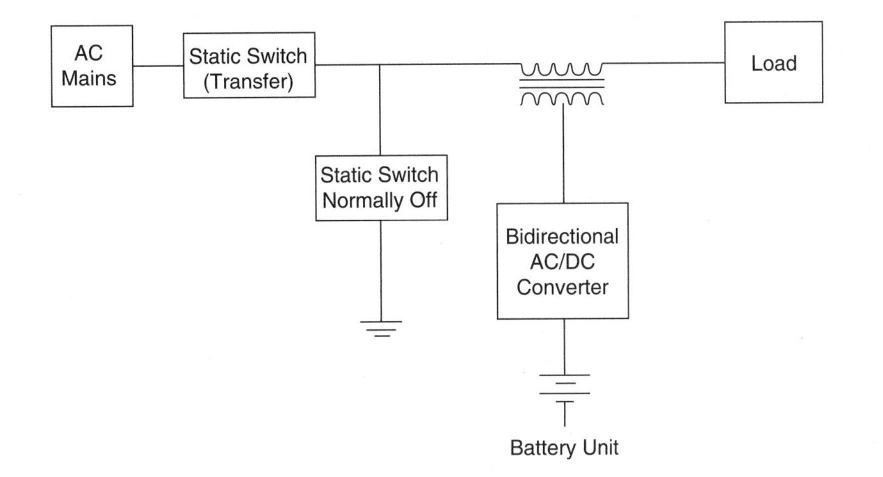

FIGURE 9.83. Block diagram of in-line UPS topology.

When the ac line is out of the preset tolerance, the converter functions as a conventional inverter and operates in the discharge mode. Since the transformer is employed in this topology, the isolation problem between the utility and the load is solved compared to the single-conversion topology. Moreover, this configuration has high efficiency and low cost because of the single power stage. Increased dc bus voltage and reduced harmonic content of the output voltage are also advantages of the in-line UPS configuration [12]. The topology is essentially stable and does not enforce any particular control restrictions.

The main disadvantage of the in-line UPS topology is its low power factor. Moreover, the output voltage of the bidirectional converter is too low to be accepted in case of converter failure. Because of the high leakage inductance of the triport transformer, low voltage problems can occur in this UPS topology [13].

Delta-Conversion Topology

This topology is also called a series-parallel compensated line-interactive UPS system. This configuration can provide the sinusoidal output voltage with low THD even for the nonlinear loads, during the normal or stored energy modes of operation. It can also realize seamless transition from the charging to backup mode and vice versa. At the same time, it can achieve low THD sinusoidal input current with unity power factor at the input side of the system. It can complete these tasks independently of the load nonlinearities or power factor, whereas conventional UPS topologies do not have these features [14].

The delta-conversion UPS topology consists of the auxiliary and the main bidirectional converters, a battery unit, a static switch, and a transformer as shown in Figure 9.84. Two bidirectional converters coupled with a common dc link are used to perform the series and parallel active filtering functions as well as battery charging and feeding the load during the power failures. The battery unit is located in the common dc bus. When a power failure occurs, the static switch provides a fast disconnection between the ac line and the UPS system to avoid any power accidents. The transformer is placed near the input side and provides a series connection between the input and the auxiliary bidirectional converter.

The auxiliary bidirectional converter does not need to provide the power for the critical load. Therefore, it is rated at 20% of the output power of the UPS. The main bidirectional converter is connected in parallel to the load side of the system. In case of power failures, it has to provide the full power for the critical load. Therefore, it is rated at 100% of the output power [15].

The system has two operation modes: normal mode of operation and backup mode of operation. When the ac line is alive, the system operates in active filtering mode. The auxiliary converter acts like a series active filter and regulates voltage fluctuations between the output and the input. It also forces the input current to be in phase with the input voltage and then provides unity power factor for the load. At the same time, it controls the dc link voltage of the system and charges the battery. During the normal operating mode, important

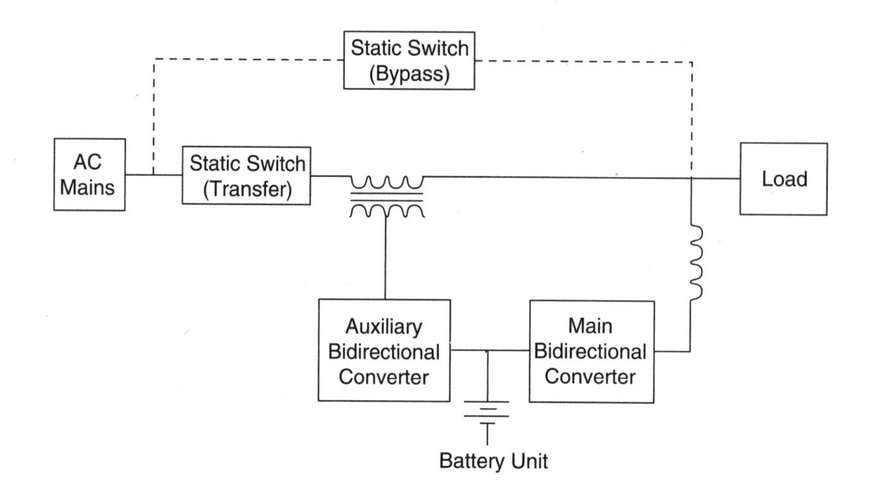

FIGURE 9.84 Block diagram of delta-conversion UPS topology.

portion of the active power (usually up to 85%) directly flows from the ac line to the critical load without any conversion. Series and parallel converters draw only a small part of total active power to realize their active filtering tasks, control the dc bus, and charge the battery unit.

When the ac line is out of the preset tolerance, the static switch is off. The main converter operates as a traditional inverter and supplies load from the battery unit without any interruption.

The main advantages of the delta-conversion UPS topology are high efficiency, low cost compared to the on-line UPS, and applicability in high power rating. Disadvantages of this configuration are a complex control unit and lack of electrical isolation of the load from the ac line.

Rotary UPS

The energy conversion is realized by a motor/generator set in rotary UPS systems. In order to avoid idleness for some components of the power and control circuits, the system can be combined as one machine by the manufacturer. There are two types of rotary UPS systems on the market: flywheel rotary UPS and diesel rotary UPS [16]. A flywheel rotary UPS consists of an ac motor, a flywheel unit, an ac generator, a static switch, and an optional bypass switch as shown in Figure 9.85. The ac motor and ac generator are mechanically coupled through a flywheel unit.

Instead of a battery, the flywheel unit provides energy storage for the UPS in this topology. When the ac line is out of power, the flywheel unit produces dynamic energy for the generator and supplies the critical load through

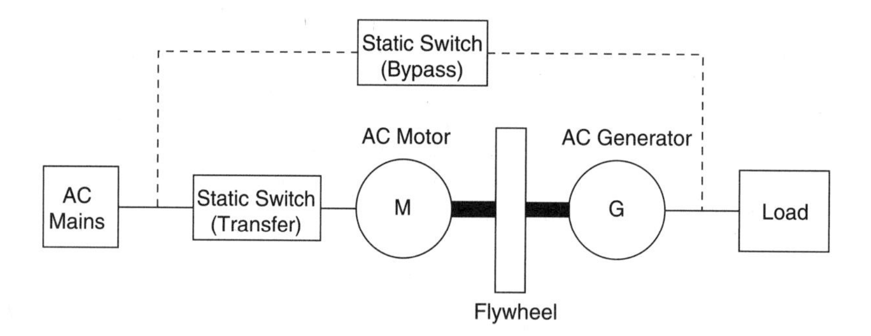

FIGURE 9.85 Block diagram of flywheel rotary UPS topology.

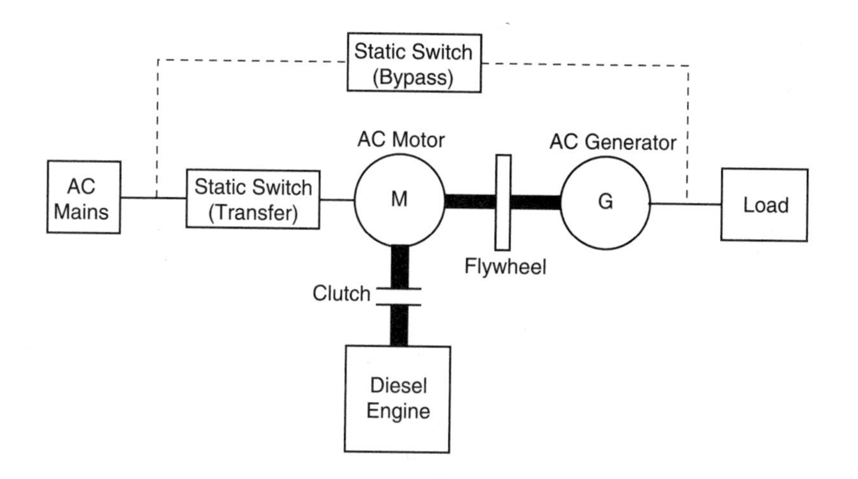

FIGURE 9.86 Block diagram of diesel-flywheel rotary UPS topology.

the generator. Disadvantages of the flywheel UPS are restricted frequency range, restricted speed range, limitation on energy extraction, and huge flywheel dimensions even for a few seconds of backup for the load.

Figure 9.86 shows the diesel rotary UPS schematically. An additional diesel engine is connected to the motor/generator set via the clutch in this system. When the ac line is normal, the motor of the rotary UPS is supplied by the utility. The generator delivers power to the critical load. When the ac line fails, the control unit disconnects the ac motor from the power supply and the diesel engine begins to start up. Because of the start up time, the diesel engine does not provide the power for the load instantly. During this time period, the load is supplied by the rotational energy stored in the flywheel unit. When the diesel engine moves at its rated speed, the clutch is closed and the system operates as a diesel generator set to supply the critical load. Since the system includes a flywheel unit, it may have disadvantages in terms of the diameter and weight of the flywheel. In this topology, the flywheel must be designed to accommodate the diesel engine start time.

The rotary UPS systems are more reliable than the static UPS systems. However, they are of larger size and weight. Yet, they are preferred in high-power applications because of their high transient overload capability, low EMI, and high efficiency.

Hybrid Static/Rotary UPS

Hybrid static/rotary UPS systems have the advantages of both static and rotary UPS systems. They have low output impedance, low THD with nonlinear loads, high reliability, frequency stability, better isolation, and low cost for maintenance.

A hybrid UPS can be considered similar to a static UPS in configuration and operation. In both, an ac/dc converter is used to charge a battery and supply the output dc/ac conversion stage. The inverter stage is realized by the motor/generator set in the rotary UPS which is similar to the dc/ac converter of the static UPS.

The three battery-supported rotary UPS systems are shown in Figure 9.87. Hybrid static/rotary UPS systems based on these three topologies are commercially available.

In Figure 9.87(a), a battery-supported hybrid UPS with dc motor is presented. It consists of an ac/dc converter, a dc motor, an ac generator, a battery unit, and a static switch.

When the ac line is normal, the dc motor is fed from the ac line via the rectifier and drives the generator. The ac generator supplies the load. At the same time, the ac/dc converter charges the battery. When the ac line fails, the battery unit directly supplies the dc motor and the ac generator delivers the power to the critical load. In the case of an internal failure in the UPS system, the static bypass switch is turned on and the ac line directly supplies the critical load. However, there is a transient time for the synchronizing of the ac line and the output voltage [16].

Two other battery-supported hybrid UPS systems with an ac motor are shown in Figure 9.87 as well. The only difference between the two topologies is the drive part of the motor/generator set [17]. Both topologies

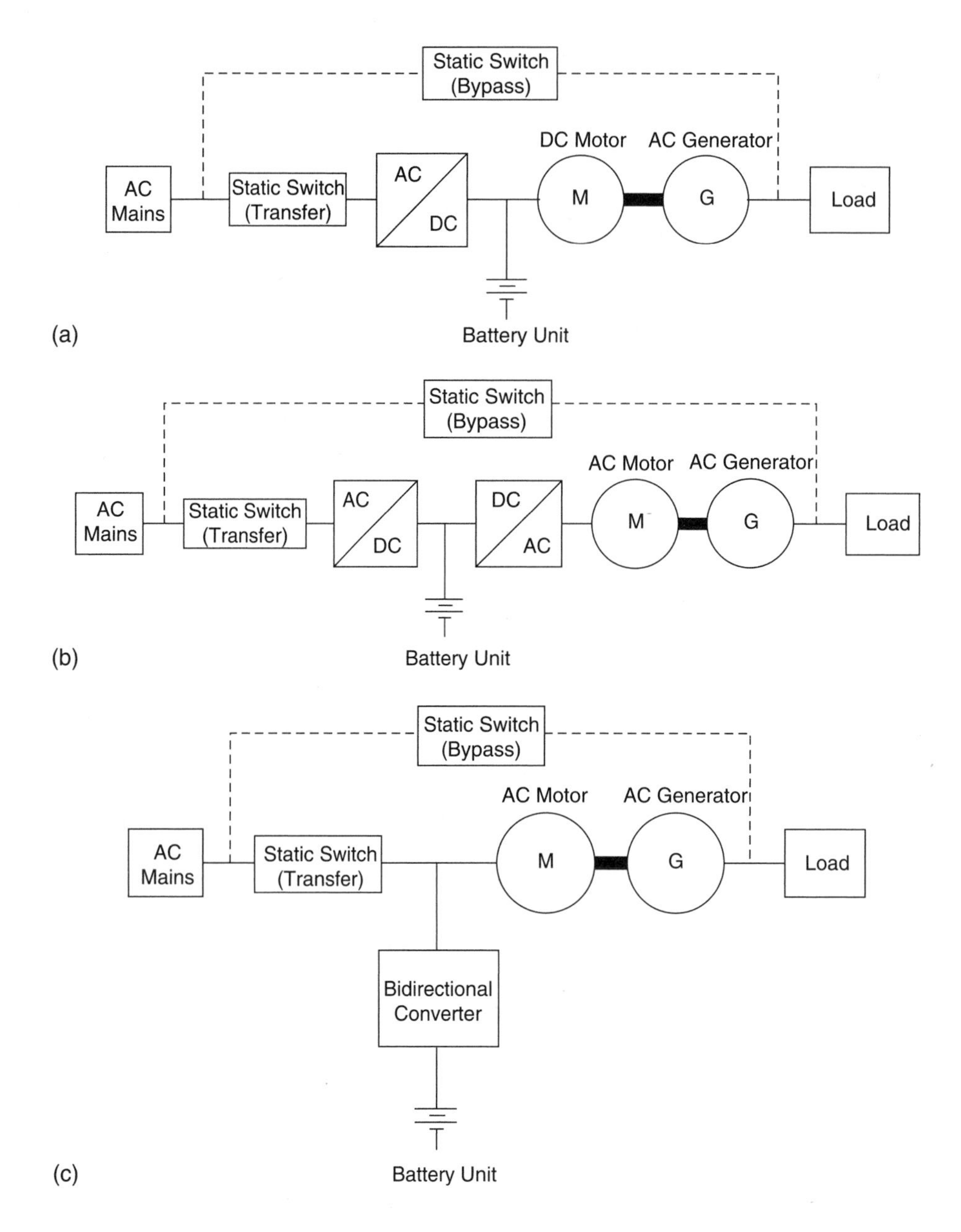

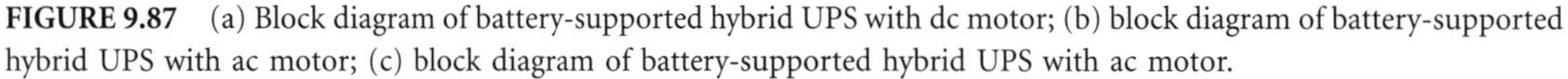

FIGURE 9.87 (a) Block diagram of battery-supported hybrid UPS with dc motor; (b) block diagram of battery-supported hybrid UPS with ac motor; (c) block diagram of battery-supported hybrid UPS with ac motor.

use an ac motor to provide mechanical energy for the generator. To improve the system, manufacturers may integrate the motor and generator as one machine. As a result, the topologies have small size, low weight, and low cost compared to a conventional rotary UPS. Hybrid UPS systems are usually used in very high power applications, i.e., several hundreds kVA.

Applications

UPS systems are used for power conditioning in the systems as well as providing uninterruptible and reliable power for the critical load. They have various applications at present. They are used to meet different standards in many circumstances. Distributed and centralized approaches are two trends in UPS system development today.

Distributed Approach

In a distributed UPS system, critical loads are supplied by many parallel UPS units, which are located in an interconnected secure network. The power flows between the UPS units and the critical load flexibly [18]. There are two distributed UPS schemes in the literature: on-line and line-interactive distributed UPS approaches.

The parallel UPS units are connected between the secure and utility networks in on-line distributed topology as shown in Figure 9.88. dc links of the UPS units provide isolation between the secure and utility networks. Loads are supplied by the ac line through the UPS units during the normal operation mode. UPS units also provide reactive currents for the secure network loads. At the same time, the secure network is isolated from the utility network current harmonics.

When a power failure occurs, the UPS batteries supply the secure network loads without any interruption. If a UPS unit fails, it is rapidly disconnected from the network to prevent power disturbances. There is power sharing between parallel UPS units. Therefore, other UPS units provide the power to the network instead of disconnected UPS unit.

The line-interactive distributed approach is shown in Figure 9.89. There is only one network in this configuration instead of the utility and the secure networks. Therefore, the system has a simpler structure than an online distributed configuration [19]. The secure network is supplied by the ac mains during the normal operating mode. The UPS units only operate during the failure mode. Therefore, the system is cost effective compared to the online distributed approach. Moreover, each UPS unit includes only one inverter which provides bidirectional power flow for the system. Since the UPS unit does not have any rectifier stage, the system is more reliable and more efficient than the online distributed approach [20].

The secure network loads are supplied by the ac line in the normal operation. At the same time, the inverters charge the batteries and provide the reactive power for the secure network loads. The ac line absorbs all load changes with a delaying process (such as 0.5 ms) because of the link inductance. In backup mode, the secure network loads are supplied from the batteries. The system has to provide a seamless transition from the normal mode to the backup mode. To achieve this, the inverter with high switching frequency (10 kHz) can rapidly change the power angle from a negative value to a positive value.

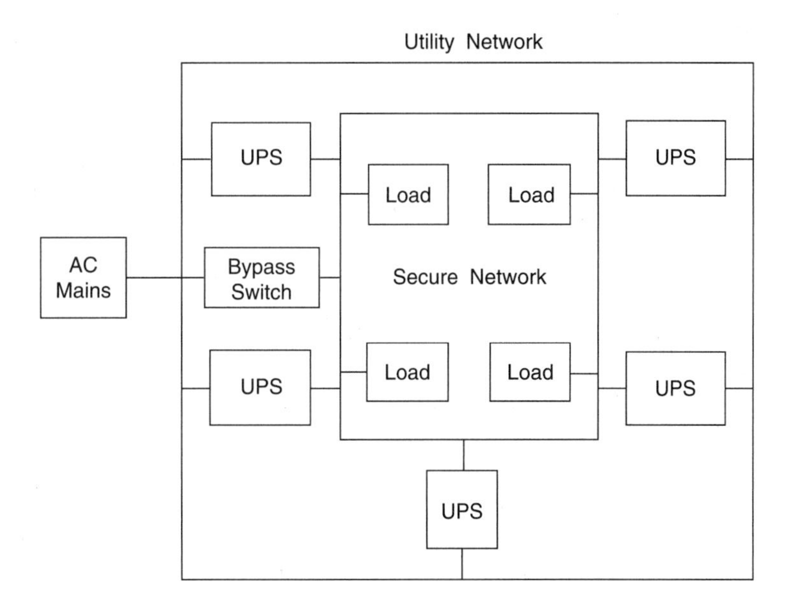

FIGURE 9.88 Block diagram of online-distributed UPS system.

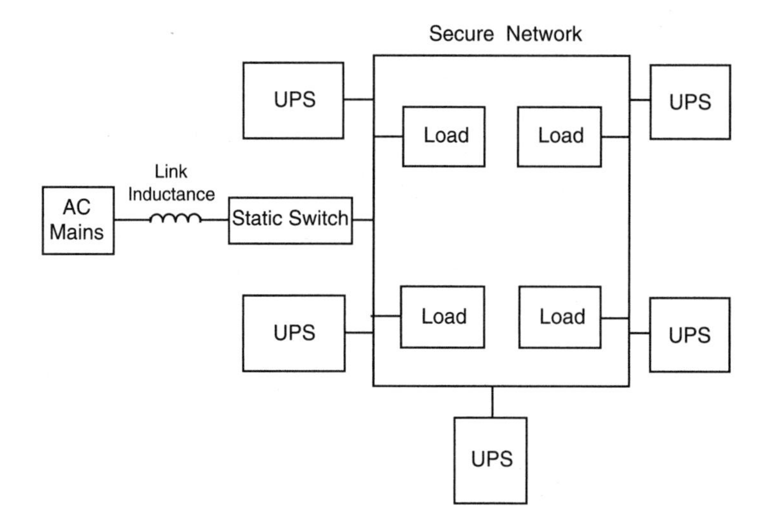

FIGURE 9.89 Block diagram of line interactive-distributed UPS system.

Disadvantages of the distributed UPS systems are difficulty in achieving the load sharing and requirement of very fast digital controllers. In addition, the monitoring of the whole system is difficult.

Centralized Approach

There is only one large UPS unit to supply the secure network loads in the centralized approach as shown in Figure 9.90. The large UPS provides continuous operation for the whole system. Since it requires low maintenance, this approach is more attractive for industrial and utility applications. Disadvantages of the centralized approach include high cost to achieve redundancy and to increase capacity of the load. Furthermore, the system needs special staff for service and maintenance. However, this specially trained group reduces the risk and increases the reliability of the system.

Control Techniques

The most important issue is the control strategy applied to the inverter in UPS systems. The control strategy changes the output voltage THD and dynamic response of the system. In addition, power electronic configurations used in the UPS affect the overall performance of the system.

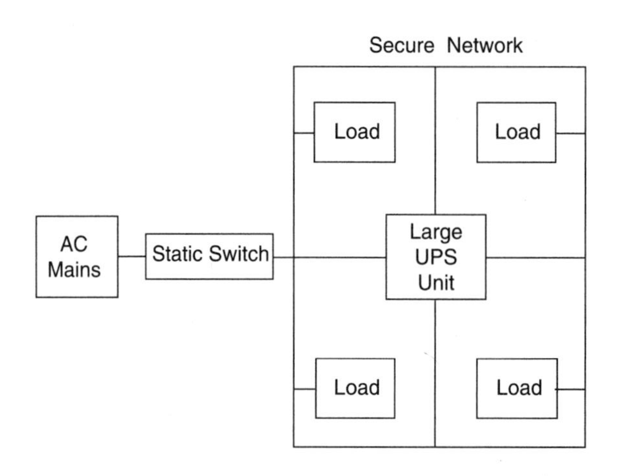

FIGURE 9.90 Block diagram of centralized UPS system.

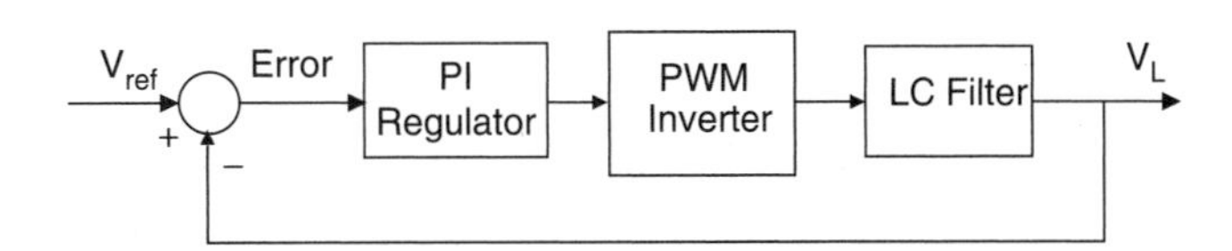

FIGURE 9.91 Block diagram of the single feedback loop control method.

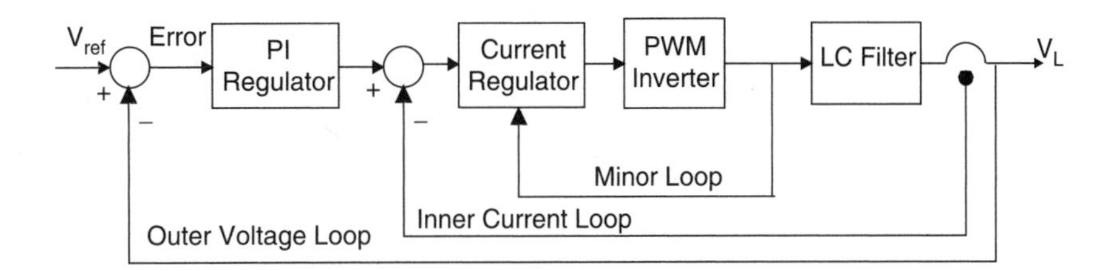

FIGURE 9.92 Block diagram of the multiple feedback loop control method.

Single Feedback Control Loop Method

To achieve regulated output voltage with low THD, the single voltage feedback loop is used in this method as shown in Figure 9.91. Continuous and discontinuous approaches can be used as a feedback controller. The most popular continuous feedback method is sine-triangle comparison or sinusoidal pulse width modulation (SPWM) [21]. This method can be a natural sampling type, average type, or instantaneous type. In the sampling approach, a feedback signal proportional to the peak value of the output voltage is compared with a reference. The error is used to control the PWM modulator. The average type is almost the same except the peak value of the output voltage is converted to an average value. Then, it is compared with a reference signal to find the error. These methods are suitable only at high switching frequencies. In addition, they can only control the amplitude of the output voltage. To obtain better performance in low switching frequencies, digital PWM techniques can be used. Specific harmonics can be minimized by calculating switching angles and also the output filter size can be reduced with digital PWM strategies. To improve the dynamic performance of a UPS inverter, an instantaneous type of PWM can be used. The output voltage is continuously compared with the reference signal in this type of control.

Multiple Feedback Control Loop Method

To obtain better performance with nonlinear loads at low frequencies, multiple control strategies are developed as shown in Figure 9.92. There are three control loops in this method: an outer, an inner, and a minor. The output voltage is used by the outer control as a feedback signal (outer loop). Then, the obtained feedback signal is compared with a reference signal to find the error signal. A PI regulator compensates the error to realize a stable output voltage under steady-state operation. This error is also compared with the current signal from the output filter (inner loop). The sensed current signal from the output of the inverter is used as feedback for the current regulator (minor loop). A minor loop provides fast dynamic responses, achieving better performance with nonlinear or step-changing loads.

Conclusion

Since waveform quality, power reliability, and availability have become important issues for consumers and utilities, the market of uninterruptible power supplies has seen considerable growth in the U.S. and worldwide. Many different UPS topologies have been developed for solving these problems.

References

1. S.B. Bekiarov, *Reduced Parts Uninterruptible Power Supplies*, Ph.D. dissertation, Illinois Institute of Technology, Chicago, IL, May 2004.
2. R. Krishnan and S. Srinivasan, "Topologies for uninterruptible power supplies," in *Proc. IEEE Int. Symp. Ind. Electron.*, Budapest, Hungary, June 1993, pp. 122–127.
3. S.B. Bekiarov and A. Emadi, "Uninterruptible power supplies: classification, operation, dynamics, and control," In *Proc. 17th Ann. IEEE Appl. Power Electron. Conf. Expo.*, Dallas, TX, March 2002, pp. 597–604.
4. S. Karve, "Three of a kind," *IEE Rev.*, vol. 46, no. 2, 2000, pp. 27–31.
5. V. John and N. Mohan, "Standby power supply with high frequency isolation," In *Proc. 10th Ann. IEEE Appl. Power Electron. Conf. Expo.*, Dallas, TX, March 1995, pp. 990–994.
6. A.A. McLennan, "Static UPS technologies," *IEE Colloq. Uninterruptible Power Supplies*, London, U.K., February 1994, pp. 2/1–2/5.
7. R. Caceras, N. Vazquez, C. Aguilar, J. Alvarez, I. Barbi, and J. Arau, "A high performance uninterruptible power supply system with power factor correction," in *Proc. 28th Ann. IEEE Power Electron. Spec. Conf.*, St. Louis, MO, June 1997, pp. 304–309.
8. M.-T. Tsai, C.-E. Lin, W.-I. Tsai, and C.-L. Huang, "Design and implementation of a demand side multifunction battery energy storage system," *IEEE Trans. Ind. Electron.*, vol. 42, no. 6, December 1995, pp. 642–652.
9. J.-C. Wu and H.-L. Jou, "A new UPS scheme provides harmonic suppression and input power factor correction," *IEEE Trans. Ind. Electron.*, vol. 42, no. 6, December 1995, pp. 629–635.
10. A.D. le Roux and H. du Mouton, "A series-shunt compensator with combined UPS operation," in *Proc. IEEE Int. Symp. Ind. Electron.*, Pusan, South Korea, June 2001, pp. 2038–2043.
11. H. Dehbonei, C. Nayar, L. Borle, and M. Malengret, "A solar photovoltaic in-line UPS system using space vector modulation technique," In *Proc. IEEE Power Eng. Soc. Ann. Meet.*, Vancouver, BC, Canada, July 2001, pp. 632–637.
12. G. Joos, "Three-phase static series voltage regulator control algorithms for dynamic sag compensation," in *Proc. IEEE Int. Symp. Ind. Electron.*, Bled, Slovenia, July 1999, pp. 515–520.
13. H.-Y. Chu, H.-L. Jou, L.-C. Wang, and C.-L. Huang, "A steady state model for parameter design of a novel bidirectional UPS," In *Proc. IEEE Int. Symp. Ind. Electron.*, Xian, China, May 1992, pp. 41–44.
14. S.A.O. da Silva, P.F. Donosa-Garcia, P.C. Cortizo, and P.F. Seixas, "A three phase line-interactive UPS system implementation with series-parallel active power line conditioning capabilities," *IEEE Trans. Ind. Appl.*, vol. 38, no. 6, 2002, pp. 1581–1590.
15. A. Nasiri, S.B. Bekiarov, and A. Emadi, "Reduced-parts three-phase series-parallel UPS system with active filter capabilities," in *Proc. IEEE 38th Ind. Appl. Soc. Ann. Meet.*, Salt Lake City, UT, October 2003, pp. 963–969.
16. S.R. Philpott, "Large scale UPS installations," in *Proc. IEEE Int. Conf. Electr. Installation Eng. Eur.*, London, U.K., June 1993, pp. 64–68.
17. W.W. Hung and G.W.A. McDowell, "Hybrid UPS for standby power systems," *Power Eng. J.*, vol. 4, no. 6, 1990, pp. 281–291.
18. Y.D. Liu, Y. Xing, L. Huang, and K. Hirachi, "A novel distributed control scheme for the parallel operation of digital controlled UPS," In *Proc. 29th Ann. IEEE Ind. Electron. Conf.*, Roanoke, VA, November 2003, pp. 668–672.
19. M.C. Chandorkar, D.M. Divan, Y. Hu, and B. Banerjee, "Novel architectures and control for distributed UPS systems," in *Proc. 9th Ann. IEEE Appl. Power Electron. Conf. Expo.*, Orlando, FL, February 1994, pp. 683–689.
20. M.C. Chandorkar, "Control of distributed UPS systems," in *Proc. 25th Ann. IEEE Power Electron. Spec. Conf.*, Taipei, Taiwan, June 1994, pp.197–204.
21. C.D. Manning, "Control of UPS inverters," *IEE Colloq. Uninterruptible Power Supplies*, February 1994, pp. 3/1–3/5.

10

Optoelectronics

Jeff Hecht
Laser Focus World

Laurence S. Watkins
Lucent Technologies

R.A. Becker
Integrated Optical Circuit Consultants

10.1 Lasers

Jeff Hecht

The word *laser* is an acronym for "light amplification by the stimulated emission of radiation," a phrase that covers most, though not all, of the key physical processes inside a laser. Unfortunately, that concise definition may not be very enlightening to the nonspecialist who wants to use a laser and cares less about its internal physics than its external characteristics. From a practical standpoint, a laser can be considered as a source of a narrow beam of monochromatic, coherent light in the visible, infrared, or ultraviolet parts of the spectrum. The power in a continuous beam can range from a fraction of a milliwatt to around 25 kilowatts (kW) in commercial lasers, and up to more than a megawatt in special military lasers. Pulsed lasers can deliver much higher peak powers during a pulse, although the power averaged over intervals while the laser is off and on is comparable to that of continuous lasers.

The range of laser devices is broad. The laser medium, i.e., the material emitting the laser beam, can be a gas, liquid, glass, crystalline solid, or semiconductor crystal, and can range from the size of a grain of salt to big enough to fill the inside of a moderate-sized building. Lasers can take many forms including semiconductor chips, optical fibers, glass or crystalline rods, and gas-filled tubes. Not every laser produces a narrow beam of monochromatic, coherent light. Edge-emitting semiconductor diode lasers, for example, produce beams that spread out over an angle of 20 to 40°, hardly a pencil-thin beam. Liquid dye lasers and solid-state titanium-doped sapphire lasers emit at a broad or narrow range of wavelengths, depending on the optics used with them. Other types emit at a number of spectral lines, producing light that is neither truly monochromatic nor coherent. Table 10.1 summarizes important commercial lasers.

Practically speaking, lasers contain three key elements. First is the laser medium itself, which generates the laser light. Second is the power supply, which delivers energy to the laser medium in the form needed to excite it to emit light. Third is the optical cavity or resonator, which concentrates the light to stimulate the emission of laser radiation. All three elements can take various forms, and although they are not always immediately evident in all types of lasers, their functions are essential. Figure 10.1 shows these elements in a ruby and a helium-neon laser. Figure 10.2 shows two types of semiconductor diode lasers.

TABLE 10.1 Important Commercial Lasers

Wavelength (mm)	Type	Output Type and Power
0.157	Molecular fluorine (F_2)	Pulsed, avg. to a few watts
0.192	ArF excimer	Pulsed, avg. to tens of watts
0.2–0.35	Doubled dye	Pulsed
0.235–0.3	Tripled Ti-sapphire	Pulsed
0.24–0.27	Tripled alexandrite	Pulsed
0.248	KrF excimer	Pulsed, avg. to over 100 W
0.266	Quadrupled Nd	Pulsed, watts
0.275–0.306	Argon ion	CW, 1 W range
0.308	XeCl excimer	Pulsed, to tens of watts
0.32–1.0	Pulsed dye	Pulsed, to tens of watts
0.325	He–Cd	CW, to tens of milliwatts
0.337	Nitrogen	Pulsed, under 1 W avg.
0.35–0.47	Doubled Ti-sapphire	Pulsed
0.351	XeF excimer	Pulsed, to tens of watts
0.355	Tripled Nd	Pulsed, to tens of watts
0.36–0.4	Doubled alexandrite	Pulsed, watts; CW mW
0.37–1.0	CW dye	CW, to a few watts
0.405, 0.44	InGaN diode	CW, milliwatts
0.442	He–Cd	CW, to over 0.1 W
0.45–0.53	Argon	CW, to tens of watts
0.51	Copper vapor	Pulsed, tens of watts
0.520–0.569	Krypon ion	CW, >1 W
0.523	Doubled Nd-YLF	Pulsed, watts
0.532	Doubled Nd-YAG	Pulsed to 50 W, or CW to watts
0.5435	He–Ne	CW, 1 mW range
0.578	Copper vapor	Pulsed, tens of watts
0.594	He–Ne	CW, to several milliwatts
0.612	He–Ne	CW, to several milliwatts
0.6328	He–Ne	CW, to about 50 mW
0.635–0.66	InGaAlP diode	CW, milliwatts to watts
0.647–0.676	Krypton ion	CW, to several watts
0.67	GaInP diode	CW, mW to watts
0.68–1.13	Ti-sapphire	CW, watts
0.694	Ruby	Pulsed, to a few watts
0.72–0.8	Alexandrite	Pulsed, to tens of watts; CW 100 mW
0.75–0.9	GaAlAs diode	CW, to many watts in arrays
0.98	InGaAs diode	CW, to tens of watts
1.02–1.13	Ytterbium-glass (fiber)	CW or pulsed, to kW, peak at 1.07 μm
1.047 or 1.053	Nd-YLF	CW or pulsed, to tens of watts
1.061	Nd-glass	Pulsed, to 100 W
1.064	Nd-YAG	CW or pulsed, to kilowatts
1.15	He–Ne	CW, milliwatts
1.2–1.6	InGaAsP diode	CW, to 100 mW
1.313	Nd-YLF	CW or pulsed, to 0.1 W
1.32	Nd-YAG	Pulsed or CW, to a few watts
1.4–1.6	Color center	CW, under 1 W
1.523	He–Ne	CW, milliwatts
1.54	Erbium glass (bulk)	Pulsed, to 1 W
1.54	Erbium fiber (amplifier)	CW, milliwatts
1.75–2.5	Cobalt-MgF_2	Pulsed, 1 W range
2.3–3.3	Color center	CW, under 1 W
2.6–3.0	HF chemical	CW or pulsed, to hundreds of watts
3.3–29	Lead-salt diode	CW, milliwatt range
3.39	He–Ne	CW, to tens of milliwatts
3.6–4.0	DF chemical	CW or pulsed, to hundreds of watts
5–6	Carbon monoxide	CW, to tens of watts
9–11	Carbon dioxide	CW or pulsed, to tens of kilowatts
40–100	Far-infrared gas	CW, generally under 1 W

Source: Updated from J. Hecht, *The Laser Guidebook*, 2nd ed., New York: McGraw-Hill, 1991. With permission.

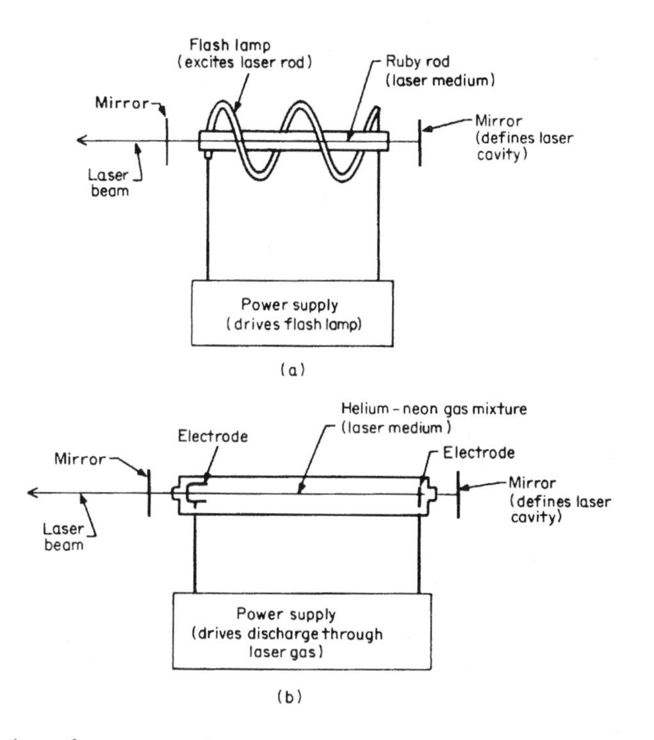

FIGURE 10.1 Simplified views of two common lasers, (a) ruby and (b) helium-neon, showing the basic components that make a laser. (*Source*: J. Hecht, *The Laser Guidebook*, 2nd ed., New York: McGraw-Hill, 1991. With permission.)

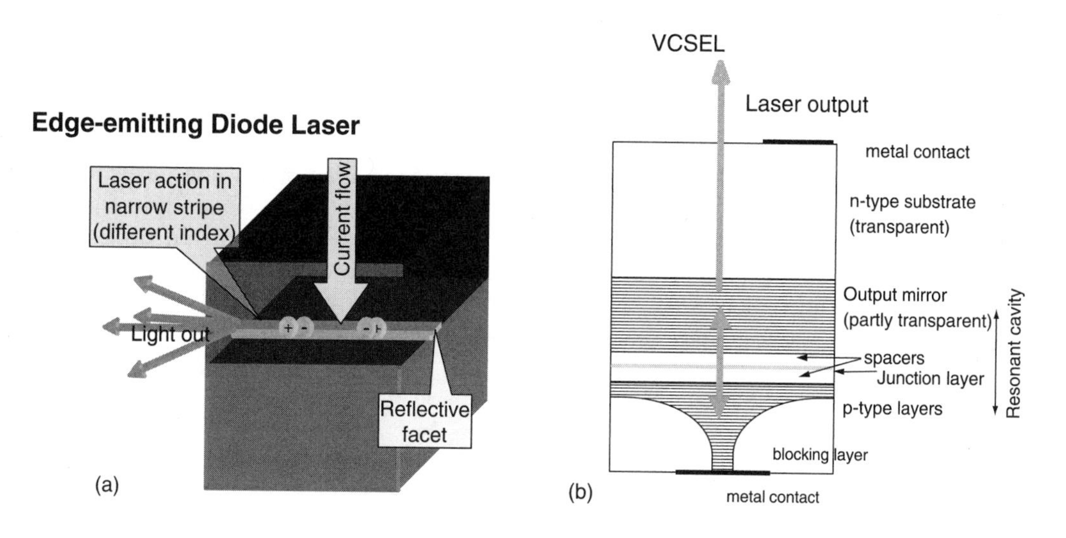

FIGURE 10.2 Simplified views of two common semiconductor diode lasers: (a) an edge-emitter and (b) a vertical-cavity surface-emitting laser (VCSEL). (*Source*: J. Hecht *Understanding Lasers: An Entry Level Guide*, 2nd ed., Piscataway, NJ: IEEE Press/Wiley, 1993. With permission.)

Laser-like devices called optical parametric oscillators have come into increasing use. They are more costly and complex than lasers, but can be tuned across a broad range, with wavelengths from 0.2 to 4 micrometers.

Several general characteristics are common to most lasers that new users may not expect. Like most other light sources, lasers are inefficient in converting input energy into light. Efficiencies range from less than 0.001 to more than 50%, but except for semiconductor and diode-pumped lasers, few types are much above 1% efficient. These low efficiencies can lead to special cooling requirements and duty-cycle limitations, particularly for high-power lasers. In some cases, special equipment may be needed to produce the right conditions for laser operation, such as cryogenic temperatures for lead salt semiconductor lasers. Operating characteristics of individual lasers depend strongly on structural components such as cavity optics, and in many cases a wide range is possible. Packaging can also have a strong impact on laser characteristics and the use of lasers for certain applications. Thus, there are wide ranges of possible characteristics, although single devices will have much more limited ranges of operation.

Differences from Other Light Sources

The basic differences between lasers and other light sources are the characteristics often used to describe a laser: the output beam is narrow, the light is monochromatic, and the emission is coherent. Each of these features is important for certain applications and deserves more explanation.

Most gas or solid-state lasers emit beams with a divergence angle of about a milliradian, meaning that they spread to about 1 m in diameter after traveling a kilometer. (Edge-emitting semiconductor lasers have a much larger beam divergence, but suitable optics can reshape the beam to make it much narrower.) The actual beam divergence depends on the type of laser and the optics used with it. The fact that laser light is contained in a beam serves to concentrate the output power onto a small area. Thus, a modest laser power can produce a high intensity inside the small area of the laser beam; the intensity of light in a 1 mW red diode laser beam is comparable to that of sunlight on a clear day, for example. The beams from high-power lasers, delivering tens of watts or more of continuous power or higher peak powers in pulses, can be concentrated to high enough intensities that they can weld, drill, or cut many materials.

The laser beam's concentrated light delivers energy only where it is focused. For example, a tightly focused laser beam can write a spot on a light-sensitive material without exposing the adjacent area, allowing high-resolution printing. Similarly, the beam from a surgical laser can be focused onto a tiny spot for microsurgery, without heating or damaging surrounding tissue. Lenses can focus the parallel rays in a laser beam to a much smaller spot than they can the diverging rays from a point source, a factor that helps compensate for the limited light-production efficiency of lasers.

Most lasers deliver a beam that contains only a narrow range of wavelengths, and thus the beam can be considered monochromatic for all practical purposes. Conventional light sources, in contrast, emit light over much of the visible and infrared spectrum. For most applications, the range of wavelengths emitted by lasers is narrow enough to make life easier for designers by avoiding the need for achromatic optics and simplifying the task of understanding the interactions between laser beam and target. For some applications in spectroscopy and communications, however, that range of wavelengths is not narrow enough, and special line-narrowing options may be required.

One of the beam's unique properties is its coherence, the property that the light waves it contains are in phase with one another. Strictly speaking, all light sources have a finite coherence length or distance over which the light they produce is in phase. However, for conventional light sources that distance is essentially zero. For many common lasers, it is a fraction of a meter or more, allowing their use for applications requiring coherent light. The most important of these applications is probably holography, although coherence is useful in some types of spectroscopy. Spectroscopy also takes advantage of the fact that laser light can be concentrated in a very narrow range of wavelengths, delivering a much higher power in that narrow range than brighter sources which spread their light across a wide range of wavelengths.

Some types of lasers have two further advantages over other light sources: higher power and longer lifetime. For some high-power semiconductor lasers, lifetime must be traded off against higher power, but for most others the life vs. power trade-off is minimal. The combination of high power and strong directionality makes certain lasers the logical choice to deliver high light intensities to small areas. For some applications, lasers offer longer lifetimes than do other light sources of comparable brightness and cost. In addition, despite their low efficiency, some lasers may be more efficient in converting energy to light than other light sources.

The Laser Industry and Commercial Lasers

There is a big difference between the world of laser research and the world of the commercial laser industry. Unfortunately, many text and reference books fail to differentiate between types of lasers that can be built in the laboratory and those that are readily available commercially. This distinction is a crucial one for laser users.

Laser emission has been obtained from hundreds of materials at many thousands of emission lines in laboratories around the world. Extensive tabulations of these laser lines are available (Weber, 1982), and even today researchers are adding more lines to the list. However, most of these laser lines are of purely academic interest. Many are weak lines close to much stronger lines that dominate the emission in practical lasers. Most of the lasers that have been demonstrated in the laboratory have proved to be cumbersome to operate, low in power, inefficient, and/or simply less practical to use than other types.

Only a few dozen types of lasers have proved to be commercially viable on any significant scale; these are summarized in Table 10.1. Some of these types, notably the ruby and helium-neon lasers, have been around since the beginning of the laser era, although their uses have decreased over the years. The most widely used lasers today are semiconductor diode lasers used in CD players, laser pointers, laser printers, and fiber-optic communication systems. Some lasers are in only limited production, and the economic realities of manufacturing limit the number of different types of lasers that are commercially viable.

There are many possible reasons why certain lasers do not find their way onto the market. Some require exotic operating conditions or laser media, such as high temperatures or highly reactive metal vapors. Some emit only feeble powers. Others have only limited applications, particularly lasers emitting low powers in the far-infrared or in parts of the infrared where the atmosphere is opaque. Some simply cannot compete with materials already on the market.

Defining Terms

Coherence: The condition of light waves that stay in the same phase relative to each other; they must have the same wavelength.

Continuous wave (CW): A laser that emits a steady beam rather than pulses.

Laser medium: The material in a laser that emits light; it may be a gas, solid, or liquid.

Monochromatic: Of a single wavelength or frequency.

Resonator: Mirrors that reflect light back and forth through a laser medium, usually on opposite ends of a rod, tube, or semiconductor wafer. One mirror lets some light escape to form the laser beam.

Solid-state laser: A laser in which light is emitted by atoms in a glass or crystalline matrix. Laser specialists do not consider semiconductor lasers to be solid-state types.

References

J. Hecht, *The Laser Guidebook*, 2nd ed., New York: McGraw-Hill, 1991; this section is excerpted from the introduction.

M.J. Weber, Ed., *CRC Handbook of Laser Science and Technology* (2 vols.), Boca Raton, FL: CRC Press, 1982.

M.J. Weber, Ed., *CRC Handbook of Laser Science and Technology*, Supplement 1, Boca Raton, FL: CRC Press, 1989; other supplements are in preparation.

Further Information

Several excellent introductory college texts are available, which concentrate on laser principles. These include: Anthony E. Siegman, *Lasers*, University Science Books, Mill Valley, CA, 1986, and Orzio Svelto, *Principles of Lasers*, 3rd ed., Plenum, New York, 1989.

Two trade magazines serve the laser field; each publishes an annual directory issue. For further information contact: *Laser Focus World*, PennWell Publishing, 98 Spit Brook Rd., Nashua, NH 03062 (http://www.laserfocusworld.com); or *Photonics Spectra*, Laurin Publishing Co., Berkshire Common, PO Box 1146, Pittsfield, MA 01202 (http://www.photonics.com). Contact the publishers for information.

10.2 Sources and Detectors

Laurence S. Watkins

Properties of Light

The strict definition of *light* is electromagnetic radiation to which the eye is sensitive. Optical devices, however, can operate over a larger range of the electromagnetic spectrum, and so the term usually refers to devices which can operate in some part of the spectrum from the near ultraviolet (UV) through the visible range to the near infrared. Figure 10.3 shows the whole spectrum and delineates these ranges.

Optical radiation is electromagnetic radiation and thus obeys and can be completely described by Maxwell's equations. We will not discuss this analysis here but just review the important properties of light.

Phase Velocity

In isotropic media light propagates as transverse electromagnetic (TEM) waves. The electric and magnetic field vectors are perpendicular to the propagation direction and orthogonal to each other. The velocity of light propagation in a medium (the velocity of planes of constant phase, i.e., wavefronts) is given by

$$v = \frac{c}{\sqrt{\varepsilon\mu}} \tag{10.1}$$

where c is the velocity of light in a vacuum ($c = 299{,}796$ km/sec). The denominator in Equation (10.1) is a term in optics called the refractive index of the medium:

$$n = \sqrt{\varepsilon\mu} \tag{10.2}$$

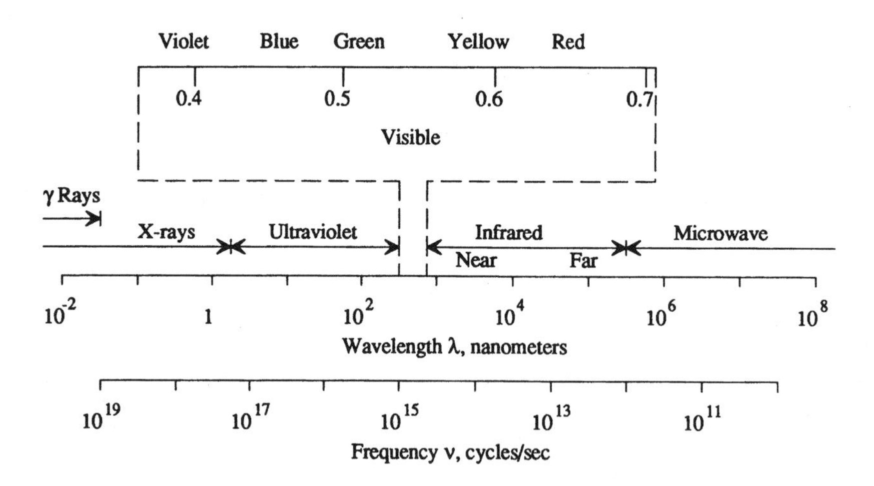

FIGURE 10.3 Electromagnetic spectrum showing visible and optical wavelengths.

where ε is the dielectric constant (permittivity) and μ is the magnetic permeability. The wavelength of light, λ, which is the distance between phase fronts is

$$\lambda = \frac{\lambda_0}{n} = \frac{v}{\upsilon} \tag{10.3}$$

where λ_0 is the wavelength in vacuum and υ is the light frequency. The refractive index varies with wavelength, and this is referred to as the dispersive property of a medium.

Another parameter used to describe light frequency is wave number. This is given by

$$\sigma = \frac{1}{\lambda} \tag{10.4}$$

and is usually expressed in cm^{-1}, giving the number of waves in a 1-cm path.

Group Velocity

When traveling in a medium, the velocity of energy transmission (e.g., a light pulse) is less than c and is given by

$$u = v - \lambda \frac{\mathrm{d}v}{\mathrm{d}\lambda} \tag{10.5}$$

In vacuum the phase and group velocities are the same.

Polarization

Light polarization is defined by the direction of the electric field vector. For isotropic media this direction is perpendicular to the propagation direction. It can exist in a number of states, described as follows.

Unpolarized. The electric field vector has a random and constantly changing direction, and when there are multiple frequencies the vector directions are different for each frequency.

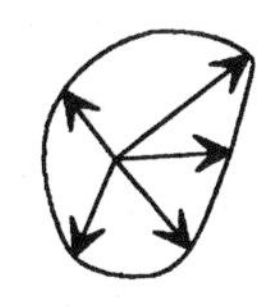

Linear. The electric field vector is confined to one direction.

Elliptical. The electric field vector rotates, either left hand or right hand, at the light frequency. The magnitude of the vector (intensity of the light) traces out an ellipse.

Circular. Circular is the special case of the above where the electric field vector traces out a circle.

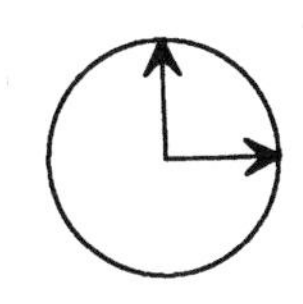

Absorption

Light in traveling through media can be absorbed. This can be represented in two ways. The light flux propagating through a medium can be written as

$$I = I_0 e^{-\alpha x} \tag{10.6}$$

where x is the distance through the medium with incident light flux I_0. α is the absorption coefficient, usually stated in cm^{-1}. An alternative way of describing absorption is to use the imaginary term in the media refractive index. The complex refractive index is

$$\bar{n} = n(1 + ik) \tag{10.7}$$

where k is the attenuation index. α and k are related as

$$\alpha = \frac{4\pi}{\lambda_0} nk \tag{10.8}$$

Coherence

Light can be partially or fully coherent or incoherent, depending on the source and subsequent filtering operations. Common sources of light are incoherent because they consist of many independent radiators. An example of this is the fluorescent lamp in which each excited atom radiates light independently. There is no fixed phase relationship between the waves from these atoms. In a laser the light is generated in a resonant cavity using a light amplifier and the resulting coherent light has well-defined phase fronts and frequency characteristics.

Spatial and Temporal Coherence. Spatial coherence describes the phase front properties of light. A beam from a single-mode laser which has one well-defined phase front is fully spatially coherent. A collection of light waves from a number of light emitters is incoherent because the resulting phase front has a randomly indefinable form. Temporal coherence describes the frequency properties of light. A single-frequency laser output is fully temporally coherent. White light, which contains many frequency components, is incoherent, and a narrow band of frequencies is partially coherent.

Laser Beam Focusing

The radial intensity profile of a collimated single-mode TEM_{00} (Gaussian) beam from a laser is given by

$$I(r) = I_0 \exp\left[2\left(\frac{-r^2}{w_0^2}\right)\right] \tag{10.9}$$

where w_0 is the beam radius ($1/e^2$ intensity). This beam will diverge as it propagates out from the laser, and the half angle of the divergence is given by

$$\theta_{1/2} = \frac{\lambda}{\pi w_0} \tag{10.10}$$

When this beam is focused by a lens the resulting light spot radius is given by

$$w_\text{f} = \frac{\lambda l}{\pi w_\text{d}} \tag{10.11}$$

where l is the distance from the lens to the position of the focused spot and w_d is the beam radius entering the lens. It should be noted that $l \cong f$, the lens focal length, for a collimated beam entering the lens. However, l will be a greater distance than f if the beam is diverging when entering the lens.

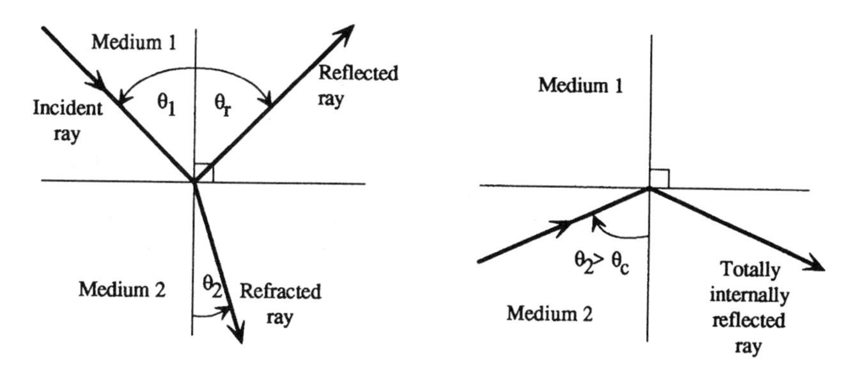

FIGURE 10.4 (a) Diagram of a light ray in medium 1 incident at angle θ_1 on the surface to medium 2. The ray is refracted at angle θ_2. (b) Diagram of the situation when the ray in medium 2 is incident at an angle greater than the critical angle θ_c and totally internally reflected.

Geometric Optics

The wavelength of light can be approximated to zero for many situations. This permits light to be described in terms of light rays which travel in the direction of the wave normal. This branch of optics is referred to geometric optics.

Properties of Light Rays

Refraction. When light travels from one medium into another, it changes propagation velocity (Equation (10.1)). This results in refraction (bending) of the light as shown in Figure 10.4.

The change in propagation direction of the light ray is given by Snell's law:

$$n_1 \sin\theta_1 = n_2 \sin\theta_2 \tag{10.12}$$

where n_1 and n_2 are the refractive indices of media 1 and 2, respectively.

Critical Angle. When a light ray traveling in a medium is incident on a surface of a less dense medium, there is an incidence angle θ_2, where sin $\theta_1 = 1$. This is the critical angle; for light incident at angles greater than θ_2 the light is totally internally reflected as shown in Figure 10.4(b). The critical angle is given by $\theta_c = \sin^{-1}(n_1/n_2)$.

Image Formation with a Lens

Many applications require a lens to focus light or to form an image onto a detector. A well-corrected lens usually consists of a number of lens elements in a mount, and this can be treated as a black box system. The characteristics of this lens are known as the cardinal points. Figure 10.5 shows how a lens is used to form an image from an illuminated object.

The equation which relates the object, image, and lens system is

$$\frac{1}{f} = \frac{1}{s_1} + \frac{1}{s_2} \tag{10.13}$$

The image magnification is given by $M = s_2/s_1$. When the object is very far away s_1 is infinite and the image is formed at the back focal plane.

Incoherent Light

When two or more incoherent light beams are combined, the resulting light flux is the sum of their energies. For coherent light this is not necessarily true and the resulting light intensity depends on the phase relationships between the electric fields of the two beams, as well as the degree of coherence.

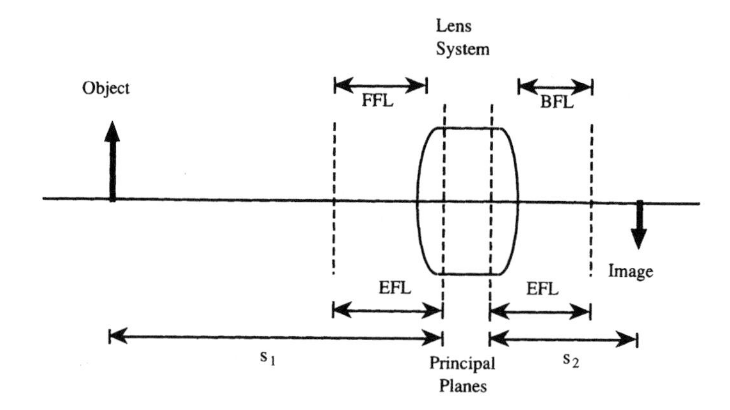

FIGURE 10.5 Schematic of an optical system forming an image of an object. Light rays from the object are captured by the lens which focuses them to form the image. EFL, effective focal length, *f*, of the lens; FFL and BFL, distances from the focal points to the outer lens surface. Principal planes are the positions to which the focal points, object distance, and image distance are measured; in a simple lens they are coincident.

Brightness and Illumination

The flux density of a light beam emitted from a **point source** decreases with the square of distance from it. Light sources are typically extended sources (being larger than point sources). The illumination of a surface from light emitted from an **extended source** can be calculated using Figure 10.6.

The flux incident on a surface element dA from a source element dS is given by

$$dE = \frac{B\,dA\cos\theta\,dS\cos\Psi}{r^2} \qquad (10.14)$$

The constant B is called the luminance or photometric brightness of the source. Its units are candles per square meter (1 stilb $= \pi$ lamberts) and dE is the luminous flux in lumens. The total illumination E of the surface element is calculated by integrating over the source. The illuminance or flux density on the surface is thus:

$$I = \frac{E}{dA} \quad (\text{lumens/cm}^2) \qquad (10.15)$$

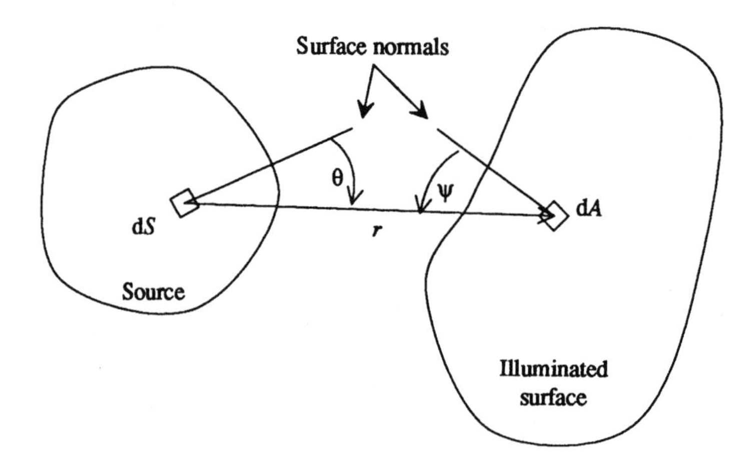

FIGURE 10.6 Surface being illuminated by an extended source. Illumination of surface element dA is calculated by summing the effects of elements dS.

Two methods are commonly used for quantifying light energy, namely, the radiometric unit of watts and the photometric unit of candelas. The candela is an energy unit which is derived from light emission from a blackbody source. The two can be related using the relative visibility curve $V(\lambda)$, which describes the eye's sensitivity to the visible light spectrum, it being maximum near a wavelength of 550 nm. The constant which relates lumens to watts at this wavelength is 685 lm/W. The luminous flux emitted by a source can therefore be written as

$$F = 685 \int V(\lambda)P(\lambda)\mathrm{d}\lambda \quad \text{(lumens)} \tag{10.16}$$

where V is the spectral response of the eye and P is the source radiant intensity in watts.

The source radiance is normally stated as luminance in candle per square centimeter (1 lumen per steradian per square centimeter) or radiance in watts per square centimeter per steradian per nanometer. The lumen is defined as the luminous flux emitted into a solid angle of a steradian by a point source of intensity 1/60th that of a 1-cm^2 blackbody source held at 2042 K temperature (molten platinum).

Thermal Sources

Objects emit and absorb radiation, and as their temperature is increased the amount of radiation emitted increases. In addition, the spectral distribution changes, with proportionally more radiation emitted at shorter wavelengths. A blackbody is defined as a surface which absorbs all radiation incident upon it, and Kirchhoff's law of radiation is given by

$$\frac{W}{a} = \text{constant} = WB \tag{10.17}$$

stating that the ratio of emitted to absorbed radiation is a constant a at a given temperature.

The energy or wavelength distribution for a blackbody is given by Planck's law:

$$\begin{aligned} W &= \frac{c_1}{\lambda^5}\left[\exp\left(\frac{c_2}{\lambda T}\right) - 1\right]^{-1} \text{(watts/cm}^2\text{ area per }\mu\text{m wavelength)} \\ c_1 &= 3.7413 \times 10^4 \\ c_2 &= 1.4380 \times 10^4 \end{aligned} \tag{10.18}$$

T is in degrees Kelvin, λ is in micrometers, and W is the power emitted into a hemisphere direction. Blackbody radiation is incoherent, with atoms or molecules emitting radiation independently. Figure 10.7 is a plot of the blackbody radiation spectrum for a series of temperatures.

Very few materials are true blackbodies; carbon lampblack is one. For this reason a surface emissivity is used which describes the ratio of actual radiation emitted to that from a perfect blackbody. Table 10.2 is a listing of emissivities for some common materials.

Tungsten Filament Lamp

In the standard incandescent lamp a tungsten filament is heated to greater than 2000°C, and it is protected from oxidation and vaporization by an inert gas. In a quartz halogen lamp the envelope is quartz, which allows the filament to run at a higher temperature. This increases the light output and gives a whiter wavelength spectrum with proportionally more visible radiation to infrared.

Standard Light Source—Equivalent Black Blackbody

Because the emissivity of incandescent materials is less than 1, an equivalent source is needed for measurement and calibration purposes. This is formed by using an enclosed space which has a small opening in it. Provided the opening is much smaller than the enclosed area, the radiation from the opening will be nearly equal to that from a blackbody at the same temperature, as long as the interior surface emissivity is >0.5. Blackbody radiation from such a source at the melting point of platinum is defined as 1/60 cd/cm^2.

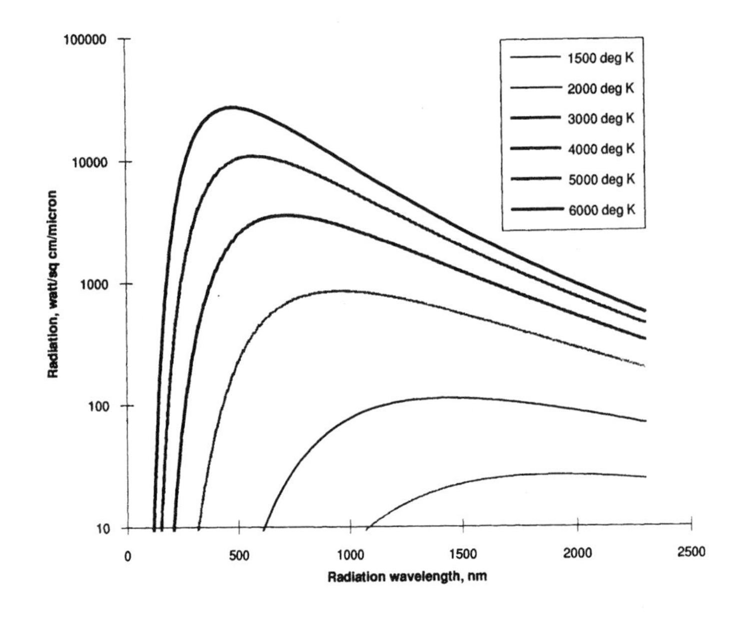

FIGURE 10.7 Plot of blackbody radiation for a series of temperatures. Radiation is in watts into a hemisphere direction from a 1-cm^2 of surface in a 1-μm wavelength band.

TABLE 10.2 Emissivities of Some Common Materials

Material	Temperature (°C)	Emissivity
Tungsten	2000	0.28
Nickel–chromium (80–20)	600	0.87
Lampblack	20–400	0.96
Polished silver	200	0.02
Glass	1000	0.72
Platinum	600	0.1
Graphite	3600	0.8
Aluminum (oxidized)	600	0.16
Carbon filament	1400	0.53

Arc Lamp

A gas can be heated to temperatures of 6000 K or more by generating an electric arc between two electrodes. The actual resulting temperature is dependent on the current flowing through the arc, the gas pressure and its composition, and other factors. This does provide a light source which is close to the temperature of the sun. Using an inert xenon gas results in essentially a white light spectrum. The use of a gas such as mercury gives more light in the UV as well as a number of strong peak light intensities at certain wavelengths. This is due to excitation and fluorescence of the mercury atoms.

Fluorescent Lamp

A fluorescent source is a container (transparent envelope) in which a gas is excited by either a dc discharge or an RF excitation. The excitation causes the electrons of the gas to move to higher energy orbits, raising the atoms to a higher excited state. When the atoms relax to lower states they give off energy, and some of this energy can be light. The wavelength of the light is characteristically related to the energy levels of the excited states of the gas involved. Typically a number of different wavelengths are associated with a particular gas.

Low-pressure lamps have relatively low luminance but provide light with narrow linewidths and stable spectral wavelengths. If only one wavelength is required, then optical filters can be used to isolate it by blocking the unwanted wavelengths.

Higher luminance is achieved by using higher gas pressures. The fluorescent lamp is very efficient since a high proportion of the input electrical energy is converted to light. White light is achieved by coating the inside of the container with various types of phosphor. The gas, for example a mercury–argon mixture, provides UV and violet radiation which excites the phosphor. Since the light is produced by fluorescence and phosphorescence, the spectral content of the light does not follow Planck's radiation law but is characteristic of the coating (e.g., soft white, cool white).

Light-Emitting Diode (LED)

Light can be emitted from a semiconductor material when an electron and hole pair recombine. This is most efficient in a direct gap semiconductor like GaAs and the emitted photons have energy close to the bandgap energy E_g. The wavelength is then given by

$$\lambda \cong \frac{hc}{E_g} \tag{10.19}$$

where h is Planck's constant (6.626×10^{-34} J-s) and c the velocity of light in vacuum. The spectral width of the emission is quite broad, a few hundred nanometers, and is a function of the density of states, transition probabilities, and temperature.

For **light emission** to occur, the conduction band must be populated with many electrons. This is achieved by forward biasing a pn junction to inject electrons and holes into the junction region as shown in Figure 10.8.

Figure 10.9(a) shows the cross section of a surface emitting LED with an integral lens fabricated into the surface. The light from the LED is incoherent and emitted in all directions. The lens and the bottom reflecting surface increase the amount of light transmitted out of the front of the device. The output from the LED is approximately linear with current but does decrease with increasing junction temperature.

Figure 10.9(b) shows an edge emitting LED. Here the light is generated in a waveguide region which confines the light, giving a more directional output beam.

Various wavelengths are available and are obtained by using different bandgap semiconductors. This is done by choosing different binary, ternary, and quaternary compositions. Table 10.3 is a listing of the more common ones.

The output power is usually specified in milliwatts per milliamp current obtained in a given measurement situation, e.g., into a fiber or with a 0.5 numerical aperture large area detector. Other parameters are peak wavelength, wavelength band (usually full width half max), and temperature characteristics.

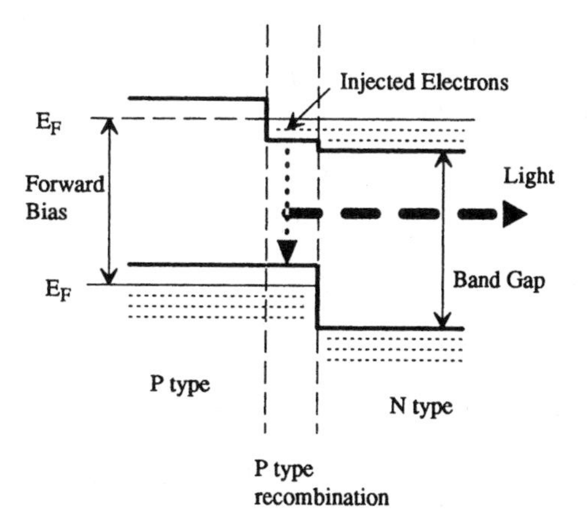

FIGURE 10.8 Band structure of a double heterostructure LED. Forward bias injects holes and electrons into the junction region where they recombine and emit light.

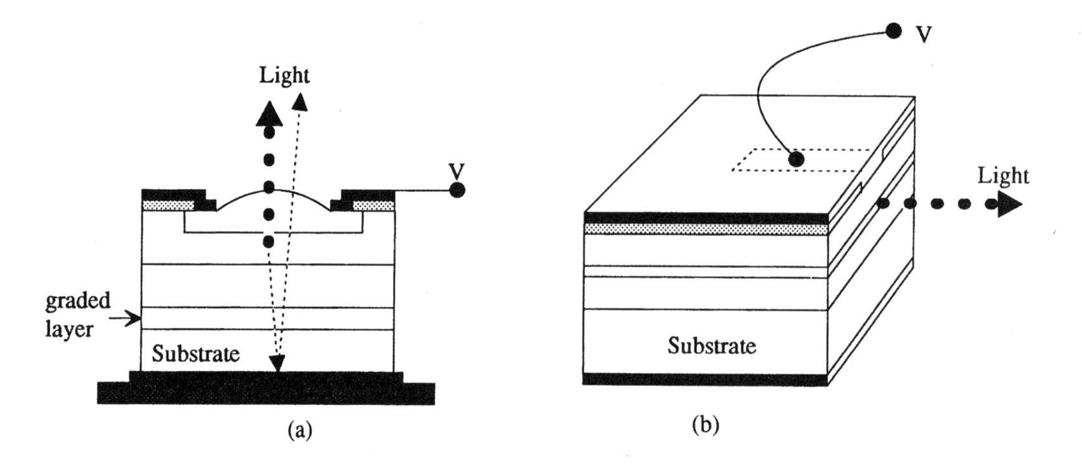

FIGURE 10.9 Cross-sectional diagrams of (a) surface-emitting LED and (b) edge-emitting LED. The light output from the edge emitter is more directional because of confinement by the junction guide region.

TABLE 10.3 Common Light-Emitting Diode Compounds and Wavelengths

Compound	Wavelength (nm)	Color
GaP	565	Green
GaAsP	590	Yellow
GaAsP	632	Orange
GaAsP	649	Red
GaAlAs	850	Near IR
GaAs	940	Near IR
InGaAs	1060	Near IR
InGaAsP	1300	Near IR
InGaAsP	1550	Near IR

LEDs for Fiber Optic Communications

GaAs and InGaAsP LEDs are commonly used as sources for fiber optic communications. Since they are an incoherent source, it is only practical to use them with multimode fiber. Only multimode fiber has a large enough core diameter and numerical aperture (NA) to couple in enough light to be able to propagate any useful distance. Applications for LEDs in fiber optics are for short distance links using glass or plastic fiber at relatively low bandwidths, typically in Mb/sec rather than Gb/sec. Primary applications of these are for low cost datalinks.

The detector can be packaged two ways: first with a fiber pigtail directly attached to the detector package; or a more common package is to have a fiber connector molded in as part of the package so that a connectorized fiber can be plugged into it. Many LEDs for fiber optics are now packaged with electronic drive circuits to form a transmitter module ready to receive standard format data signals.

Detectors, Semiconductor

When light interacts electronically with a medium, by changing the energy of electrons or creating carriers, for example, it interacts in a quantized manner. The light energy can be quantized according to Planck's theory:

$$E = h\upsilon \tag{10.20}$$

where υ is the light frequency and h is Planck's constant. The β energy of each photon is very small; however, it does increase with shorter wavelengths.

Photoconductors

Semiconductors can act as photoconductors, where incident light increases the carrier density, thus increasing the conductivity. There are two basic types: intrinsic and extrinsic. Figure 10.10 shows a simple energy diagram containing conduction and valence bands. Also indicated are the levels which occur with the introduction of donor and acceptor impurities.

Intrinsic photoconduction effect is when a photon with energy $h\upsilon$, which is greater than the bandgap energy, excites an electron from the valence band into the conduction band, creating a hole–electron pair. This increases the conductivity of the material. The spectral response of this type of detector is governed by the bandgap of the semiconductor.

In an extrinsic photoconductor (see Figure 10.10), the photon excites an electron from the valence band into the acceptor level corresponding to the hole of the acceptor atom. The resulting energy $h\upsilon$ is much smaller than the bandgap and is the reason why these detectors have applications for long wavelength infrared sensors. Table 10.4 is a list of commercial photoconductors and their peak wavelength sensitivities.

The doping material in the semiconductor determines the acceptor energy level, and so both the host material and the dopant are named. Since the energy level is quite small it can be populated by a considerable amount by thermal excitation. Thus, for useful detection sensitivity the devices are normally operated at liquid nitrogen and sometimes liquid helium temperatures. The current response, i, of a photoconductor can be written as

$$i = \frac{P\eta\tau_0 ev}{h\upsilon d} \tag{10.21}$$

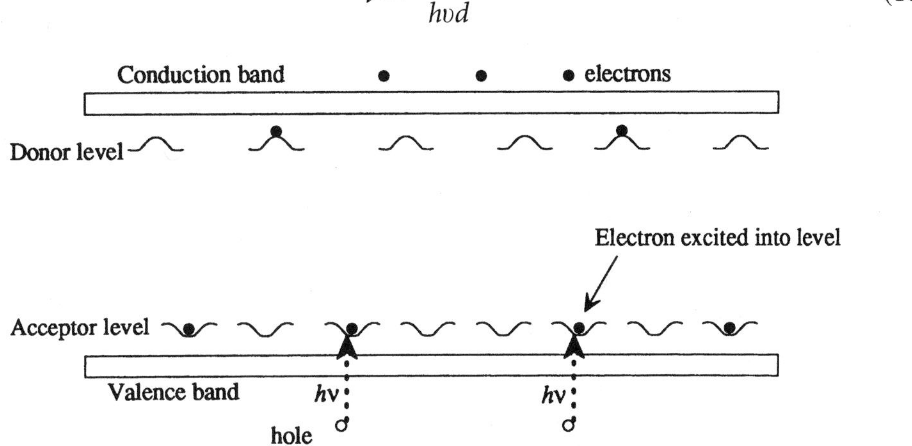

FIGURE 10.10 A simplified energy diagram for a photoconductive semiconductor, showing extrinsic effect of electrons into the acceptor level.

TABLE 10.4 Photoconductor Materials and Their Peak Wavelength Sensitivity

Photoconductor	Peak Wavelength (μm)
PbS	3
PbSe	5
HgCd	4
HgCaTe	10
HgCdTe	11
Si:Ga (4.2 K)	11
Si:As (4.2 K)	20
Si:Sb (4.2 K)	28

where P is the optical power at frequency υ; h is Planck's constant; v is drift velocity $= \mu E$, where μ is mobility and E is electric field; η is quantum efficiency (at frequency υ); τ_0 is lifetime of carriers; and e is charge on electron.

Charge Amplification. For semiconductor photoconductors such as CdS, there can be traps. These are holes, which under the influence of a bias field will be captured for a period of time. This allows electrons to move to the anode instead of recombining with a hole, resulting in a longer period for the conduction increase. This provides a photoconductive gain which is equal to the mean time the hole is trapped divided by the electron transit time in the photoconductor. Gains of 10^4 are typical.

The charge amplification can be written as

$$\frac{\tau_0}{\tau_d} \tag{10.22}$$

where $\tau_d = d/v$, the drift time for a carrier to go across the semiconductor. The response time of this type of sensor is consequently slow, ~10 msec, and the output in quite nonlinear.

Junction Photodiodes

In a simple junction photodiode, a pn-junction is fabricated in a semiconductor material. Figure 10.11 shows the energy diagram of such a device with a reverse voltage bias applied. Incident light with energy greater than the bandgap creates electrons in the p region and holes in the n region. Those that are within the diffusion length of the junction are swept across by the field. The light also creates electron–hole pairs in the junction region, and these are separated by the field. In both cases, an electron charge is contributed to the external circuit. In the case of no bias, the carrier movement creates a voltage with p region being positive. The maximum voltage is equal to the difference in the Fermi levels in the p and n regions and approaches the bandgap energy E_g.

PIN Photodiodes. The carriers that are generated in the junction region experience the highest field and so, being separated rapidly, give the fastest time response. The PIN diode has an extra intrinsic high field layer between the p and n regions, designed to absorb the light. This minimizes the generation of slow carriers and results in a fast response detector.

The signal current generated by incident light power P is

$$i = \frac{Pe\eta}{h\upsilon} + \text{dark current} \tag{10.23}$$

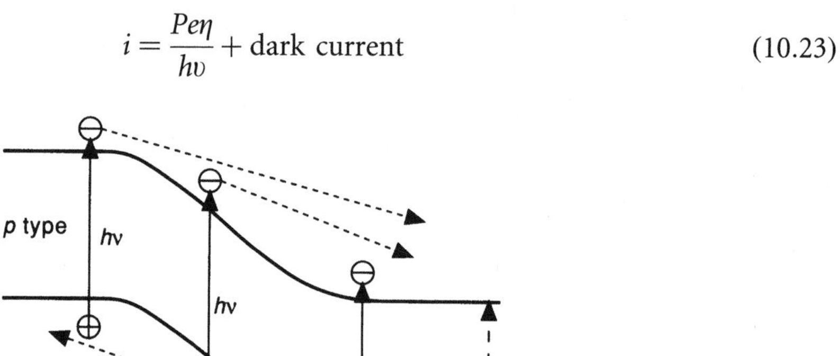

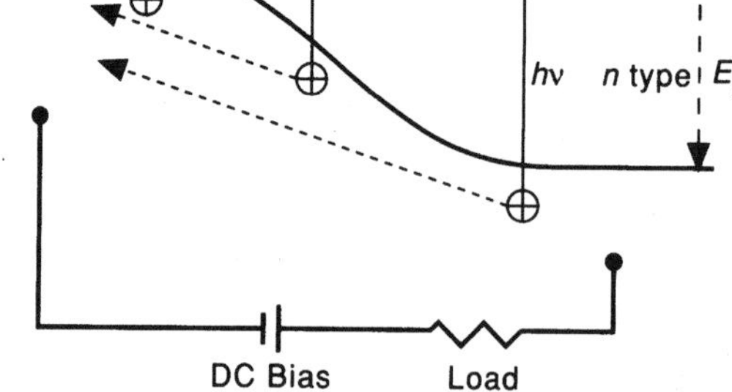

FIGURE 10.11 Energy diagram of a pn-junction photodiode showing the three ways electron–hole pairs are created by absorbing photons and the contribution to current flow in the circuit.

The output current is linear with incident power plus a constant dark current due to thermal generation of carriers; η is the quantum efficiency.

Avalanche Photodiodes

When the reverse bias of a photodiode is increased to near the breakdown voltage, carriers in the depletion region can be accelerated to the point where they will excite electrons from the valence band into the conduction band, creating more carriers. This current multiplication is called avalanche gain, and typical gains of 50 are available. Avalanche diodes are specially designed to have uniform junction regions to handle the high applied fields.

Detectors for Fiber-Optic Communications

A major application for junction photodioldes is detectors for fiber-optic communications. Silicon detectors are typically used for short-wavelength light such as with GaAs sources. InP detectors are used for the 1.3- and 1.5-μm wavelength bands. The specific type and design of a detector are tailored to the fiber-optic application, depending on whether it is low-cost lower-frequency datalinks or higher-cost high-frequency bit-rates in Gb/sec. The detector is packaged either with a fiber pigtail or with a fiber connector receptacle molded as part of the package body.

Fiber-optic detectors can also be packaged with pre-amplifier electronics or complete receiver and communications electronics into a module. For very high frequency response, it is important to minimize the capacitance of the detector and the attached preamplifier circuit.

Solar Cells

Solar cells are large-area pn-junction photodiodes, usually in silicon, which are optimized to convert light to electrical power. They are normally operated in the photovoltaic mode without a reverse voltage bias being applied.

Linear Position Sensors

Large-area photodiodes can be made into single-axis and two-axis position sensors. The single-axis device is a long strip detector, and the two-axis device is normally square. In the single-axis device, the common terminal is in the middle and there are two signal terminals, one at each end. When a light beam is directed onto the detector, the relative output current from each signal terminal depends on how close the beam is to the terminal. The sum of the output currents from both terminals is proportional to the light intensity.

Phototransistors

For bipolar devices, the light generates carriers which inject current into the base of the transistor. This modulates the collector base current, providing a higher output signal. For a field effect device, the light generates carriers that create a gate voltage. PhotoFETs can have very high sensitivities.

SEEDs

A self-electro-optic effect device (SEED) is a multiple quantum well semiconductor optical PIN device and forms the combination of a photodiode and a modulator. It can operate as a photodetector where incident light will generate a photocurrent in a circuit. It can also act as a modulator where the light transmitted through the device is varied by an applied voltage.

Devices are normally connected in pairs to form symmetric SEEDs, as demonstrated in Figure 10.12(a). These can then be operated as optical logic flip-flop devices. They can be set in one of two bistable states by application of incident light beams. The bistable state can be read out by similar light beams which measure the transmitted intensity. The hysteresis curve is shown in Figure 10.12(b). These and similar devices are the emerging building blocks for optical logic and are sometimes referred to as smart pixels.

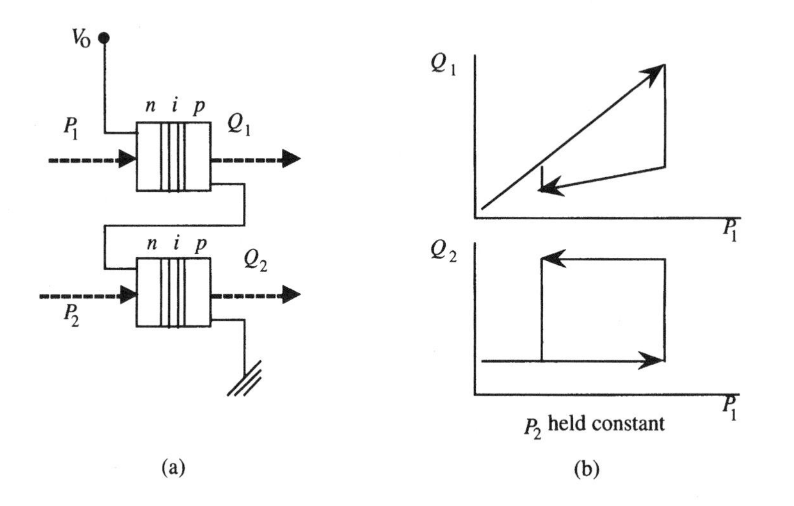

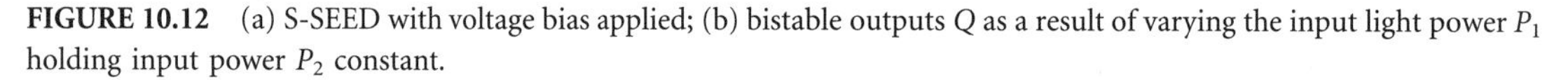
FIGURE 10.12 (a) S-SEED with voltage bias applied; (b) bistable outputs Q as a result of varying the input light power P_1 holding input power P_2 constant.

Detectors, Photoemissive

In the photoemissive effect, light falls onto a surface (photocathode) and the light energy causes electrons to be emitted. These electrons are then collected at a positively biased anode. There is a threshold energy required for the electron to be emitted from the surface. This energy is called the work function, f, and is a property of the surface material. The photon energy $h\upsilon$ must be greater than f, and this determines the longest wavelength sensitivity of the photocathode.

Vacuum Photodiodes

A vacuum photodiode comprises a negatively biased photocathode and a positive anode in a vacuum envelope. Light falling on the cathode causes electrons to be emitted, and these electrons are collected at the anode. Not all photons cause photoelectrons to be emitted, and quantum efficiencies, η, typically run 0.5 to 20%. These devices are not very sensitive; however, they have very good linearity of current to incident light power, P. They are also high-speed devices, with risetime being limited by the transit time fluctuations of electrons arriving at the anode. The photocurrent is given by

$$i = \frac{Pe\eta}{h\upsilon} + \text{dark current} \tag{10.24}$$

This kind of detector exhibits excellent short-term stability. The emissive surface can fatigue with exposure to light but will recover if the illumination is not excessive. Because of these properties, these devices have been used for accurate light measurement, although in many cases semiconductor devices are now supplanting them.

Gas-Filled Tubes

The light sensitivity of vacuum phototubes can be increased by adding 0.1 mm pressure of argon. The photoelectrons under the influence of the anode voltage accelerate and ionize the gas, creating more electrons. Gains of 5 to 10 can be realized. These devices are both low frequency, in the 10-kHz range, and nonlinear and are suitable only for simple light sensors. Semiconductor devices again are displacing these devices for most applications.

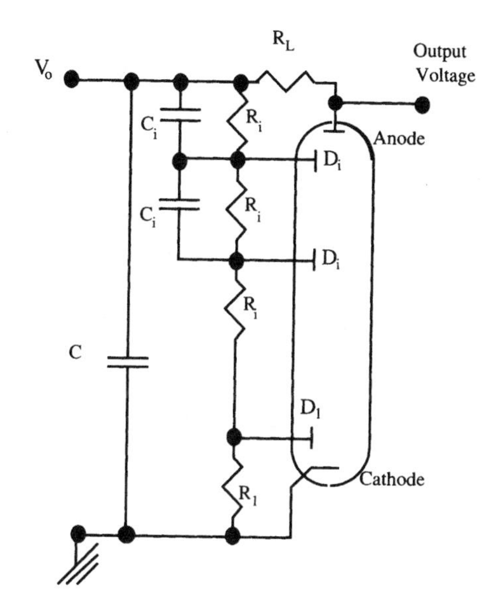

FIGURE 10.13 The basic layout of a photomultiplier tube showing the dynodes and the electrical circuit to bias them.

Photomultiplier Tubes

Photomultiplier tubes are the most sensitive light sensors, especially for visible radiation. Figure 10.13 is a schematic showing the electrical circuit used to bias it and form the output voltage signal. Light is incident on the photocathode, and the resulting photoelectrons are accelerated to a series of dynodes to generate secondary electrons and through this **electron multiplication** amplify the signal. Gains of 10^8 can be achieved with only minor degradation of the linearity and speed of vacuum photodiodes. The spectral response is governed by the emission properties of the photocathode.

There are various types of photomultipliers with different physical arrangements to optimize for a specific application. The high voltage supply ranges from 700 to 3000 V, and the electron multiplication gain is normally adjusted by varying the supply voltage. The linearity of a photomultiplier is very good, typically 3% over 3 decades of light level. Saturation is normally encountered at high anode currents caused by space charge effects at the last dynode where most of the current is generated. The decoupling capacitors, C_1, on the last few dynodes are used for high-frequency response and to prevent saturation from the dynode resistors.

Photon Counting

For the detection of very low light levels and for measuring the statistical properties of light, photon counting can be done using photomultipliers. A pulse of up to 10^8 electrons can be generated for each photoelectron emitted from the cathode, and so the arrival of individual photons can be detected. There is a considerable field of study into the statistical properties of light fields as measured by photon counting statistics.

Imaging Detectors

A natural extension to single photodetectors is to arrange them in arrays, both linear single dimension and two dimensions. Imaging detectors are made from both semiconductors and vacuum phototubes.

Semiconductor Detector Arrays

Detector arrays have been made using either photodiodes or photoconductors. The applications are for visible and infrared imaging devices. For small-sized arrays each detector is individually connected to an electrical lead on the package. This becomes impossible for large arrays, however, and these contain additional electronic switching circuits to provide sequential access to each diode.

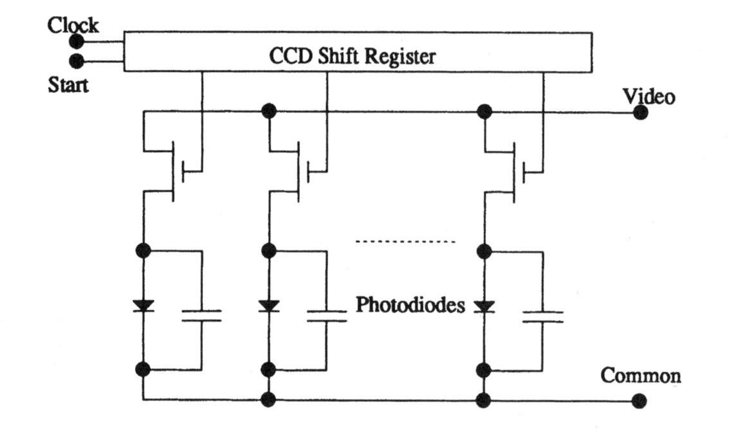

FIGURE 10.14 Schematic diagram of a linear CCD diode array sensor. The CCD shift register sequentially clocks out charge from each photodiode to the video line.

Figure 10.14 shows an example of a **charge-coupled device** (CCD) linear photodiode array. The device consists of a linear array of pn-junction photodiodes. Each diode has capacitance associated with it, and when light falls on the detector the resulting photocurrent charges this capacitance. The charge is thus the time integral of the light intensity falling on the diode. The CCD periodically and sequentially switches the charge to the video line, resulting in a series of pulses. These pulses can be converted to a voltage signal which represents the light pattern incident on the array.

The location of the diodes is accurately defined by the lithographic fabrication process and, being solid state, is also a rugged detector. These devices are thus very suitable for linear or two-dimensional optical image measurement. The devices can be quite sensitive and can have variable sensitivity by adjusting the CCD scan speed since the diode integrates the current until accessed by the CCD switch. The spectral sensitivity is that of the semiconductor photodiode, and the majority of devices now available are silicon. Smaller arrays are becoming more available in many types of semiconductors, however.

Image-Intensifier Tubes

An image-intensifier tube is a vacuum device which consists of a photoemissive surface onto which a light image is projected, an electron accelerator, and a phosphor material to view the image. This device, shown in Figure 10.15, can have a number of applications, for example, brightening a very weak image for night vision or converting an infrared image into a visible one.

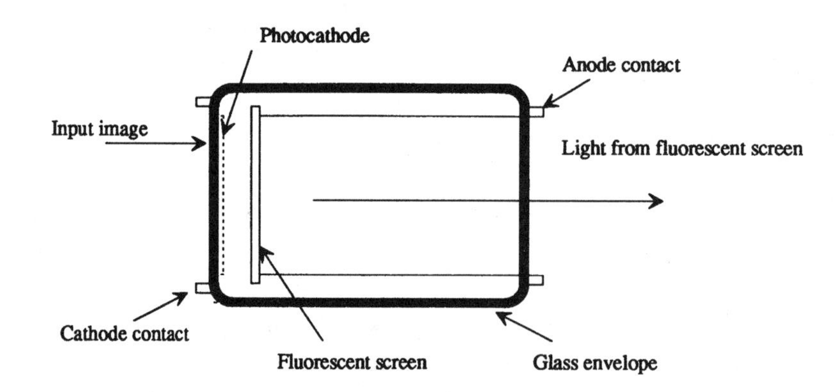

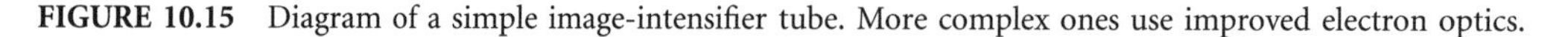

FIGURE 10.15 Diagram of a simple image-intensifier tube. More complex ones use improved electron optics.

Light falling on the cathode causes electrons to be emitted in proportion to the light intensity. These electrons are accelerated and focused by the applied electric field onto the fluorescent screen to form a visible image. Luminance gains of 50 to 100 times can be achieved, and a sequence of devices can be used to magnify the gain even more.

Image Orthicon Tube (TV Camera)

There are two basic types of **television** (TV) camera tubes, the orthicon and the vidicon. The orthicon uses the photoemissive effect. A light image is focused onto the photocathode, and the electrons emitted are attracted toward a positively based target (see Figure 10.16). The target is a wire mesh, and the electrons pass through it to be collected on a glass electron target screen. This also causes secondary electrons to be emitted, and they also collect on the screen. This results in a positive charge image which replicates the light image on the photocathode.

A low-velocity electron beam is raster scanned across the target to neutralize the charge. The surplus electrons return to the electron multiplier and generate a current for the signal output. The output current is thus inversely proportional to the light level at the scanning position of the beam. The orthicon tube is very sensitive because there is both charge accumulation between scans and gain from the electron multiplier.

Vidicon TV Camera Tube

A simple TV camera tube is the vidicon. This is the type used in camcorders and for many video applications where a rugged, simple, and inexpensive camera is required. Figure 10.17 is a schematic of a vidicon tube; the optical image is formed on the surface of a large-area photoconductor, causing corresponding variations in the conductivity. This causes the rear surface to charge toward the bias voltage V_b in relation to the conductivity image. The scanning electron beam periodically recharges the rear side to 0 V, resulting in a recharging current flow in the output. The output signal is a current signal proportional to the light incident at the position of the scanning electron beam.

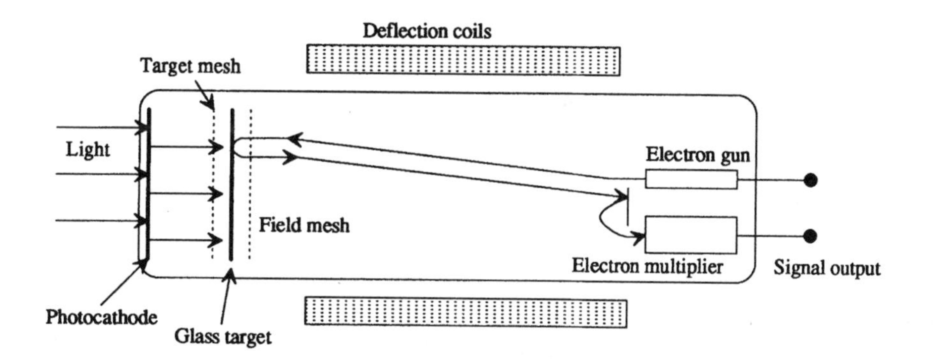

FIGURE 10.16 Schematic diagram of an image orthicon TV camera tube.

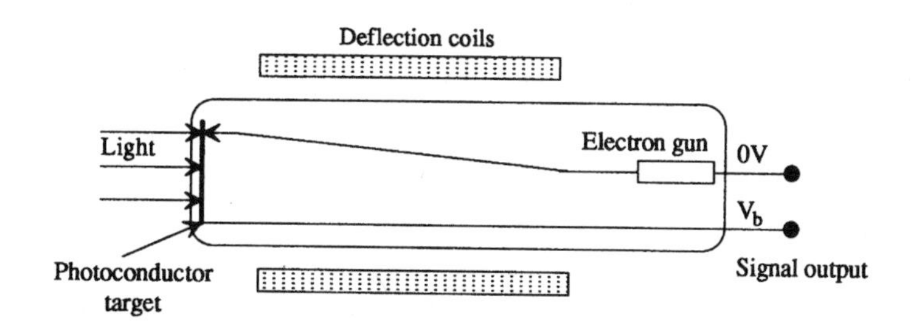

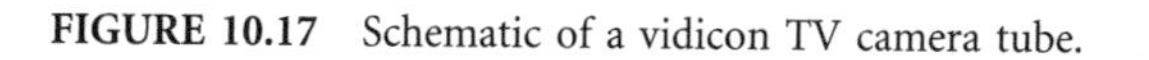

FIGURE 10.17 Schematic of a vidicon TV camera tube.

The primary disadvantages of the vidicon are its longer response time and smaller dynamic range. The recent availability of longer wavelength photoconducting films has resulted in new infrared cameras becoming available.

A recent advance in these types of image sensor is to replace the photoconductor with a dense array of very small semiconductor photodiodes. Photocurrent in the diode charges a capacitor connected to it. The raster scanned electron beam discharges this capacitor in the same way.

Image Dissector Tube

The image dissector tube is a photosensitive device which uses an electron deflection lens to image the electron from the cathode onto a pinhole in front of an electron multiplier. The image can be deflected around in front of the pinhole in a random access manner. The primary application of this kind of device is for tracking purposes.

Noise and Detectivity

Noise

There are two primary sources of noise in photodetectors: Johnson noise due to thermal effects in the resistive components of the device and its circuits, and shot noise or its equivalent, which is due to the quantized nature of electro-optic interactions. In semiconductor devices noise is usually given in terms of noise current,

$$\delta \bar{i}^2 = 2eiM^{2+x}\Delta f + \frac{4kT\Delta f}{R} \tag{10.25}$$

where i includes signal and dark currents, e is electron charge, M is avalanche gain (x depends on avalanche photodetector characteristics), Δf is frequency bandwidth, k is Boltzmann's constant, T is in degrees Kelvin, and R is the total circuit resistance at temperature, T.

For photoconductor devices (including effects of charge amplification) the noise current is given by

$$\delta \bar{i}^2 = \frac{4ei(\tau_0/\tau d)\Delta f}{1 + 4\pi^2 \upsilon^2 \tau_0^2} + \frac{4kT\Delta f}{R} \tag{10.26}$$

The first term is analogous to shot noise but includes the effects of carrier creation and recombination. τ_0 is the carrier lifetime, τ_d is the drift time for a carrier to go across the photoconductor, and υ is the light frequency.

The noise for photoemissive devices is usually written as a noise voltage and is given by

$$\delta_{\bar{v}}^{2} = 2eiG^2 \Delta f R^2 + 4kT\Delta f R \tag{10.27}$$

where G is the current gain for the photomultiplier.

Detectivity

The performance of a detector is often described using the term D^*, detectivity. This term is useful for comparison purposes by normalizing with respect to detector size and/or noise bandwidth. This is written as

$$D^* = \frac{\sqrt{A\Delta f}}{\text{NEP}} \tag{10.28}$$

where NEP is the noise equivalent power (for signal-to-noise ratio equal to 1) and A is detector area. The term $D^*(\lambda)$ is used for quoting the result using a single-wavelength light source and $D^*(T)$ is used for the unfiltered blackbody radiation source.

Defining Terms

Charge-coupled device (CCD): A series of electronic logic cells in a device in which a signal is represented and stored as an electronic charge on a capacitor. The signal is moved from one cell (memory position or register) to an adjacent cell by electronically switching the charge between the capacitors.

Electron multiplication: The phenomenon where a high-energy electron strikes a surface and causes additional electrons to be emitted from the surface. Energy from the incident electron transfers to the other electrons to cause this. The result is electron gain which is proportional to the incident electron energy.

Extended source: A light source with finite size where the source size and shape can be determined from the emitted light characteristics. The light is spatially incoherent.

Light detection: The conversion of light energy into an electrical signal, either current or voltage.

Light emission: The creation or emission of light from a surface or device.

Point source: A light source that is so small that its size and shape cannot be determined from the characteristics of the light emanating from it. The light emitted has a spherical wave front and is spatially coherent.

Television (TV): The process of detecting an image and converting it to a serial electronic representation. A detector raster scans the image, producing a voltage proportional to the light intensity. The time axis represents the distance along the raster scan. Several hundred horizontal scans make up the image starting at the top. The raster scan is repeated to provide a continuing sequence of images.

References

B. Crosignani, P. DiPorto, and M. Bartolotti, *Statistical Properties of Scattered Light*, New York: Academic Press, 1975.

A.L. Lentine et al., "A 2 kbit array of symmetric self-electrooptic effect devices," *IEEE Photon. Technol. Lett.*, vol. 2, no. 1, 1990.

Reticon Corp., subsidiary of EG&G, Inc., Application notes #101.

Further Information

W.J. Smith, *Modern Optical Engineering*, New York: McGraw-Hill, 1966.

M.J. Howes and D.V. Morgan, *Gallium Arsenide Materials, Devices and Circuits*, New York: John Wiley, 1985.

M.K. Baroski, *Fundamentals of Optical Fiber Communications*, New York: Academic Press, 1981.

C.Y. Wyatt, *Electro-Optic System Design for Information Processes*, New York: McGraw-Hill, 1991.

S. Ungar, *Fibre Optics—Theory and Applications*, New York: John Wiley, 1990.

10.3 Circuits

R.A. Becker

In 1969, Stewart Miller of AT&T Bell Laboratories published his landmark article on integrated optics. This article laid the foundation for what has now developed into optoelectronic circuits. In it he described the concepts of planar optical guided-wave devices formed as thin films on various substrates using fabrication techniques similar to those used in the semiconductor integrated circuit (IC) industry. The attributes of these new circuits included small size, weight, power consumption, and mechanical robustness, because all components were integrated on a single substrate. The field of optoelectronic circuits began as a hybrid implementation where optical sources (laser diodes) and detectors have historically been fabricated on separate semiconductor substrates, and waveguide devices, such as modulators and switches, have been fabricated on electro-optic single-crystal oxides such as lithium niobate ($LiNbO_3$). Often, the two dissimilar substrates have been connected using single-mode polarization preserving optical fiber. Now, although the hybrid concept has found widespread commercial applications in telecommunications and sensors, most active research is performed on monolithic implementations, where all devices are fabricated on a common semiconductor substrate. After a discussion of semiconductor, glass, and polymer material systems, we will deal exclusively with the most mature hybrid implementation of optoelectronic circuits based on $LiNbO_3$. Hundreds of thousands of $LiNbO_3$ digital electro-optic modulators are in use today in OC-48 (2.5 Gb/sec) and OC-192 (10 Gb/sec) fiber-optic communication systems.

Because sources and detectors have been covered in previous sections, in this section the devices that are utilized in between, such as modulators and switches, will be discussed.

Integrated Optics

Integrated optics can be defined as the monolithic integration of one or more optical guided-wave structures on a common substrate. These structures can be passive, such as a fixed optical power splitter, or active, such as an optical switch. Active devices are realized by placing metal electrodes in close proximity to the optical waveguides. Applying a voltage to the electrodes changes the velocity of the light within the waveguide. Depending on the waveguide geometry and the electrode placement, a wide variety of technologically useful devices and circuits can be realized.

The technological significance of integrated optics stemmed from its natural compatibility with two other rapidly expanding technologies: fiber optics and semiconductor laser diodes. These technologies have moved in the past 15 years from laboratory curiosities to large-scale commercial ventures. Integrated optic devices typically use laser diode optical sources or diode-pumped yttrium–aluminum–garnet (YAG) lasers, and transmit the modified optical output in a single-mode optical fiber. Integrated optic devices are typically very high speed, compact, and require only moderate control voltages compared to their bulk-optical counterparts.

In integrated optic devices, the optical channel waveguides are formed on a thin, planar, optically polished substrate using photolithographic techniques similar to those used in the semiconductor IC industry. Waveguide routing is accomplished by the mask used in the photolithographic process, similar to the way electrically conductive paths are defined in semiconductor ICs. The photolithographic nature of device fabrication offers the potential of readily scaling the technology to large volumes, as is done in the semiconductor IC industry. For example, the typical device is 0.75 in. $\times$ 0.078 in. in size. Dividing the substrate size by the typical device size and assuming a 50% area usage indicates that one can achieve over 100 devices per 4-in. wafer.

Substrate materials for integrated optics include semiconductors, such as GaAs and InP, glass, polymer coated glass or Si, and $LiNbO_3$. In the past five years, primarily passive glass-based devices have been commercially introduced as replacements for passive all-fiber devices, such as splitters and combiners. In addition, there are slow-speed (millisecond) switches and voltage-controlled variable optical attenuators (VOA) now available, which utilize the thermooptic effect in glass. Glass-based devices are fabricated by either depositing glass waveguiding layers on Si, or through the indiffusion of dopants into glass, which results in a waveguiding layer. Both fabrication approaches are used in commercially available devices.

Recently, low-speed polymer-on-Si switches have been commercially introduced. These also operate via the thermooptic effect. However, since polymers can be engineered with electrooptic properties, high-speed devices have also become available. The primary impediment to market penetration of electrooptic polymer-based devices has been their relatively poor stability, especially at temperatures above 100°C. However, if polymers can be produced with both strong electrooptic properties and enhanced stability at elevated temperatures, they could be the material system of choice for many applications because of their low-cost potential.

The area of semiconductor-based integrated optics has attracted much attention worldwide because it offers the potential of integrating electronic circuitry, optical sources and detectors, and optical waveguides on a single substrate. Being quite promising, the technology has recently become commercialized. Both electro-optic (EO) and electro-absorption (EA) modulators based on both GaAs and InP are available commercially. The EA modulators are typically used in OC-48 or shorter-reach OC-192 fiber-optic systems, whereas the EO modulators are used almost exclusively for longer-reach OC-192 systems. These modulators are usually integrated in a hybrid fashion with the semiconductor optical source, although the EA modulator can be integrated on the same substrate as the laser diode. Technical problems in semiconductor-based integrated optics include lower electrooptic coefficients, higher optical waveguide attenuation, and an incompatibility of the processing steps needed to fabricate the various types of devices on a single substrate. However, considerable attention is being paid to these problems, and improvements are continually occurring.

The devices discussed up to this point can be classified as linear. There is also an emerging class of devices that utilize nonlinear optical effects, such as second harmonic generation (SHG), for example. These devices can be very useful for wavelength conversion and dispersion compensation, and have recently been commercially introduced.

The primary substrate material in integrated optics today is the widely available synthetic crystal, $LiNbO_3$, which has been commercially produced in volume for more than 30 years. This material is transparent to optical wavelengths between 400 nm and 4500 nm, has a hardness similar to glass, and is nontoxic.

$LiNbO_3$-based devices have been commercially available since 1985, and since 1995 have been fielded in a large number of fiber-optic telecommunication systems. The basic $LiNbO_3$ waveguide fabrication technique was developed in 1974, and has been continually refined and improved during subsequent years. The material itself finds wide application in a number of electrical and optical devices because of its excellent optical, electrical, acoustic, and electro- and acousto-optic properties. For example, almost all color television sets manufactured today incorporate a surface-acoustic-wave (SAW) electrical filter based on $LiNbO_3$.

In $LiNbO_3$-based integrated optics, optical waveguides are formed in one of two ways. The first uses photolithographically patterned lines of titanium (Ti), several hundred angstroms thick, on the substrate surface. The titanium is then diffused into the substrate surface at a temperature of about 1000°C for several hours. This process locally raises the refractive index in the regions where titanium has been diffused, forming high-refractive index stripes that will confine and guide light. Because the diffusion is done at exceedingly high temperatures, the waveguide stability is excellent. The waveguide mechanism used is similar to that used in fiber optics, where the higher-index, doped cores guide the light. The exact titanium stripe width, the titanium thickness, and diffusion process are critical parameters in implementing a low-loss single-mode waveguide. Different fabrication recipes are required to optimize the waveguides for operation at the three standard diode laser wavelengths: 800 nm, 1300 nm, and 1500 nm. The second approach uses a technique known as proton exchange. In this approach, a mask is used to define regions of the substrate where hydrogen will be exchanged for lithium, resulting in an increase in the refractive index. This reaction takes place at lower temperatures (200 to 250°C) but has been found to produce stable waveguides if an anneal at 350 to 400°C is performed. Waveguides formed using the proton exchange method support only one polarized mode of propagation, whereas those formed using Ti indiffusion support two. Proton exchange waveguides are also capable of handling much higher optical power densities, especially at the shorter wavelengths, than are those formed by Ti indiffusion. More fabrication detail will be provided later.

Light modulation is realized via the electro-optic effect, i.e., inducing a small change in the waveguide refractive index by applying an electric field within the waveguide. On an atomic scale the applied electric field causes slight changes in the basic crystal unit cell dimensions, which changes the crystal's refractive index. The magnitude of this change depends on the orientation of the applied electric field and the optical polarization. As a result, only certain crystallographic orientations are useful for device fabrication and devices are typically polarization dependent. The electro-optic coefficients of $LiNbO_3$ are among the highest (30.8 pm/V for r_{33}) of any inorganic material, making the material very attractive for integrated optic applications.

Combining the concepts of optical waveguides and electro-optic modulation with the geometric freedom of photolithographic techniques leads to an extremely diverse array of passive and active devices.

Passive components do not require any electric fields and are used for power splitting and combining functions. Two types of passive power division structures have been fabricated: Y-junctions and directional couplers. A single waveguide can be split into two by fabricating a shallow-angle Y-junction as shown in Figure 10.18.

An optical signal entering from the single-waveguide side of the junction is split into two optical signals with the same relative phase but one half the original intensity. Conversely, light incident on the two-waveguide side of the junction will be combined into the single waveguide with a phase and intensity dependent on the original inputs. Directional couplers consist of two or more waveguides fabricated in close

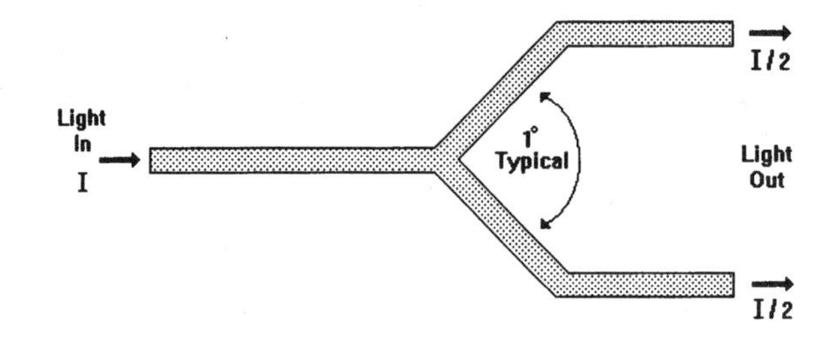

FIGURE 10.18 Passive Y-splitter.

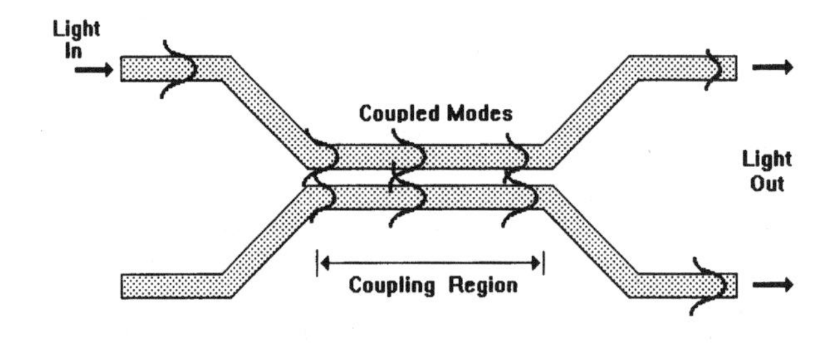

FIGURE 10.19 Directional coupler power splitter.

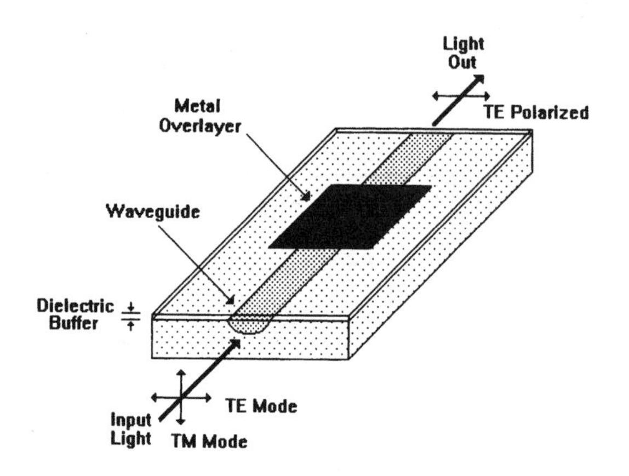

FIGURE 10.20 Thin-film optical polarizer.

proximity to each other so that the optical fields overlap as shown in Figure 10.19. As a result, optical power is transferred between the waveguides. The extent of the power transfer is dependent on the waveguide characteristics, the waveguide spacing, and the interaction length.

A different type of passive component is an optical polarizer which can be made using several different techniques. One such method is the metal-clad, dielectric-buffered waveguide shown in Figure 10.20.

In this passive device, the TM polarization state is coupled into the absorbing metal and is thus attenuated, while the TE polarization is virtually unaffected. Measurements of a 2 mm long polarizer of this type have demonstrated TM attenuations exceeding 50 dB (100,000:1). Polarizers can also be fabricated in others ways. One interesting technique involves the diffusion of hydrogen ions into the $LiNbO_3$. This results in a waveguide that, as discussed earlier, will only support the TE-polarized mode and, thus, is a natural polarizer.

Active components are realized by placing electrodes in close proximity to the waveguide structures. Depending on the substrate crystallographic orientation, the waveguide geometry, and the electrode geometry,

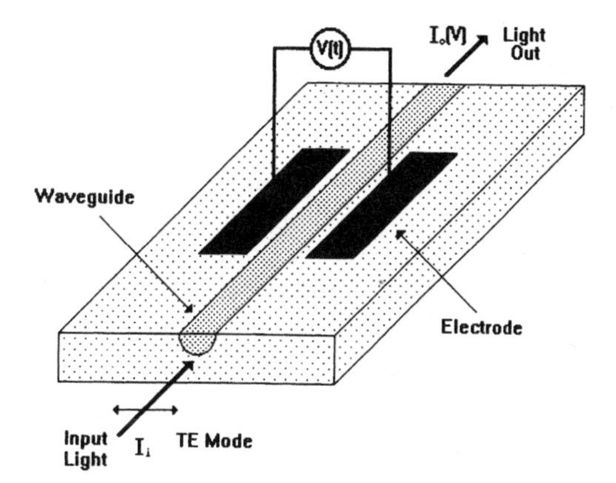

FIGURE 10.21 Electro-optic integrated optic phase modulator.

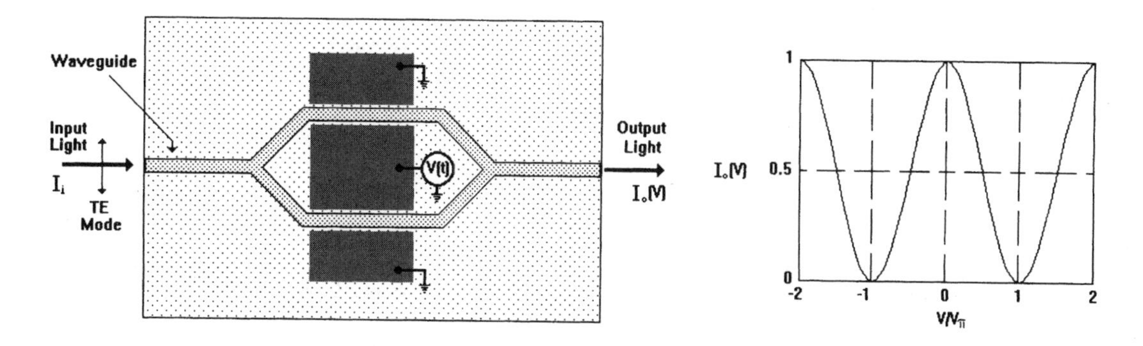

FIGURE 10.22 Mach–Zehnder intensity modulator and transfer function.

a wide variety of components can be demonstrated. The simplest active device is the phase modulator which is a single waveguide with electrodes on either side as shown in Figure 10.21.

Applying a voltage across the electrodes induces an electric field across the waveguide, which changes its refractive index via the electro-optic effect. For 800-nm wavelength operation, a typical phase modulator would be 6 mm long and would induce a π-phase shift for an applied voltage of 4 V. The transfer function (light out vs. voltage in) can be expressed as

$$I_0(V) = I_i \exp(j\omega t + \pi V/V_\pi), \tag{10.29}$$

where V_π is the voltage required to cause a 180° phase shift. Note that there is no change in the intensity of the light. Coherent techniques are used to measure the amount of phase change.

Optical intensity modulators can be fabricated by combining two passive Y-junctions with a phase modulator situated between them. The result, which is shown in Figure 10.22, is a guided-wave implementation of the classic Mach–Zehnder interferometer.

In this device the incoming light is split into two equal components by the first Y-junction. An electrically controlled differential phase shift is then introduced by the phase modulator, and the two optical signals are recombined in the second Y-junction. If the two signals are exactly in phase, then they recombine to excite the lowest-order mode of the output waveguide and the intensity modulator is turned fully on. If instead there exists a π-phase shift between the two signals, then they recombine to form the second mode, which is radiated into the substrate and the modulator is turned fully off. Contrast ratios greater than 25 dB (300:1) are routinely achieved in commercial devices. The transfer function for the Mach–Zehnder modulator can be expressed as

$$I_0(V) = I_i \cos^2(\pi V/2V_\pi + \varphi), \tag{10.30}$$

where V_π is the voltage required to turn the modulator from on to off, and φ is any static phase imbalance between the interferometer arms. This transfer function is shown graphically in Figure 10.22. Note that the modulator shown in Figure 10.22 has push–pull electrodes. This means that when a voltage is applied, the refractive index is changed in opposite directions in the two arms, yielding a twice-as-efficient modulation.

Optical switches can be realized using a number of different waveguide, electrode, and substrate orientations. Two different designs are used in commercially available optical switches: the balanced-bridge and the $\Delta\beta$ directional coupler. The balanced-bridge design is similar to that of the Mach–Zehnder interferometer, except that the Y-junctions have been replaced by 3-dB directional couplers as shown in Figure 10.23.

Similar to the Mach–Zehnder, the first 3-dB coupler splits the incident signal into two signals, ideally of equal intensity. Once again, if a differential phase shift is electro-optically induced between these signals, then when they recombine in the second 3-dB coupler, the ratio of power in the two outputs will be altered. Contrast ratios greater than 20 dB (100:1) are routinely achieved in commercial devices. The transfer function for this switch can be expressed as

$$I_{0a} = I_i \cos^2(\pi V/2V_\pi + \pi/2) \tag{10.31}$$

$$I_{0b} = I_i \sin^2(\pi V/2V_\pi + \pi/2) \tag{10.32}$$

and is graphically depicted in Figure 10.23.

In the other type of switch, the $\Delta\beta$ directional coupler, the electrodes are placed directly over the directional coupler as shown in Figure 10.24.

The applied electric field alters the power transfer between the two adjacent waveguides. Research versions of this switch have demonstrated contrast ratios greater than 40 dB (10,000:1); however, commercial versions typically achieve 20 dB, which is competitive with that achieved with the balanced-bridge switch. The transfer function for the $\Delta\beta$ directional coupler switch can be expressed as

$$I_{0a} = \sin^2 \mathrm{KL}^* \mathrm{sqrt}(1 + (\Delta\beta/2K)^2)/(1 + (\Delta\beta/2K)^2) \tag{10.33}$$

$$I_{0b} = 1 - I_{0a} \tag{10.34}$$

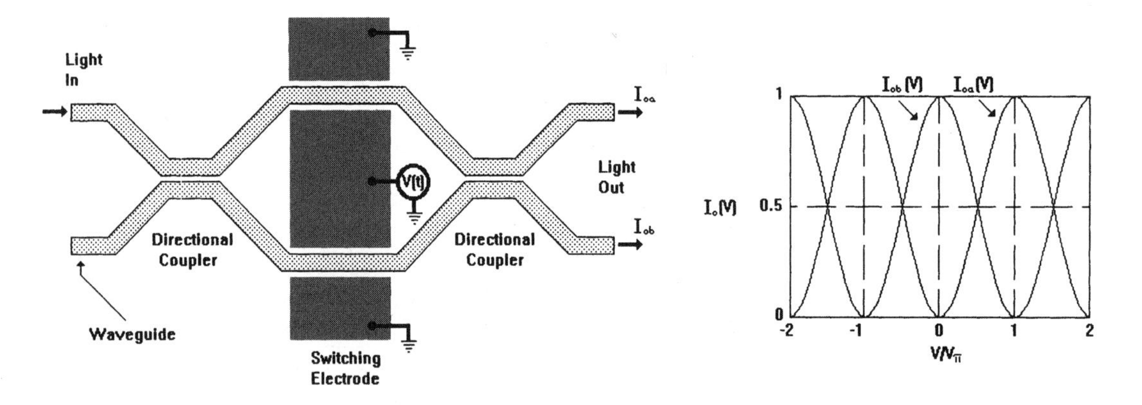

FIGURE 10.23 Balanced-bridge modulator/switch and transfer function.

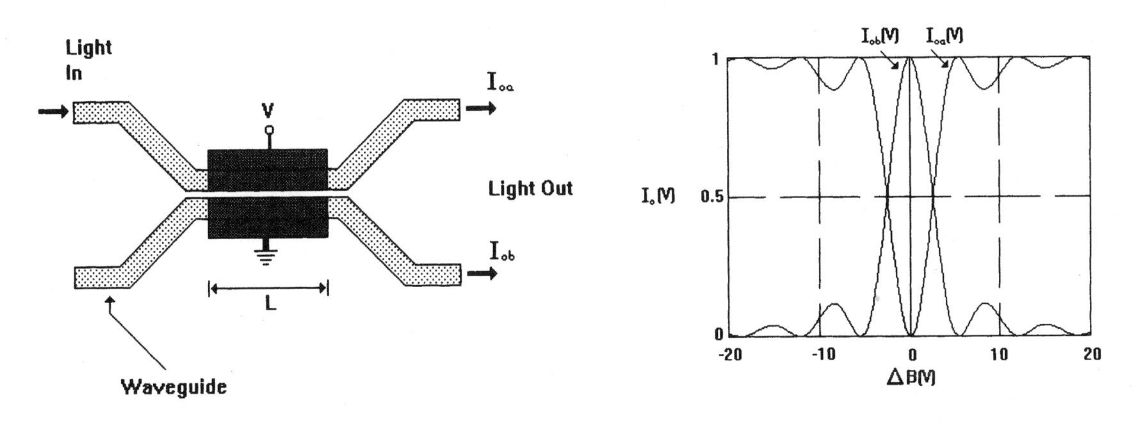

FIGURE 10.24 Directional coupler switch and transfer function.

where K is the coupling constant and $\Delta\beta$ is the voltage-induced change in the propagation constant. This transfer function is depicted in Figure 10.24.

Another type of active component that is commercially available is the polarization controller. This component allows the incoming optical polarization to be continuously adjustable. The device functions as an electrically variable optical waveplate, where both the birefringence and axis orientation can be controlled. The controller is realized by using a three-electrode configuration as shown in Figure 10.25 on a substrate orientation where the TE and TM optical polarizations have almost equal velocities. Typical performance values are TE/TM conversion of greater than 99% with less than 50 V applied.

One of the great strengths of integrated optic technology is the possibility of integrating different types or multiple copies of the same type of device on a single substrate. While this concept is routinely used in the semiconductor IC industry, its application in the optical domain is novel. The scale of integration in integrated optics is quite modest by semiconductor standards. To date, the most complex component demonstrated is a 16×16 optical switch matrix that uses 256 identical 2×2 optical switches. The greatest device diversity on a given substrate is found in fiber gyro applications. Here, components incorporating six phase modulators, two electrically tunable directional couplers, and two passive directional couplers have been demonstrated.

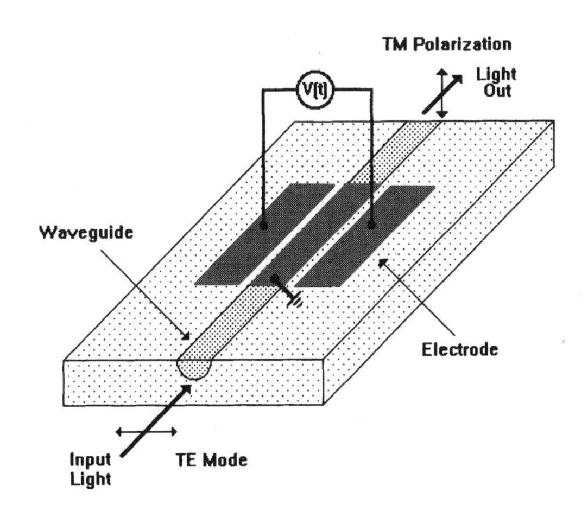

FIGURE 10.25 Guided-wave polarization controller.

Device Fabrication

The fabrication of an integrated optic device uses some of the same techniques as used in the semiconductor IC industry. Device designs are first entered into a computer-aided design (CAD) system for accurate feature placement and dimensional control. This design is then output as a digital file that will control a pattern generation system for fabrication of the chrome masks that are used in device fabrication. A variety of equipment, such as step-and-repeat and E-beam systems, has been developed for the semiconductor IC industry for the generation of chrome masks. These same systems are used today for generation of masks for integrated optic devices.

The waveguides can be fabricated by using either the Ti indiffusion method or the proton exchange method. The first step in fabricating a waveguide device using Ti indiffusion is patterning in titanium. The bare $LiNbO_3$ surface is first cleaned and then coated with photoresist. Next, the coated substrate is exposed using the waveguide-layer chrome mask. The photoresist is then developed. The areas that have been exposed are removed in the development cycle. The patterned substrates are then coated with titanium in a vacuum evaporator. The titanium covers the exposed regions of the substrate as well as the surface of the remaining photoresist. The substrate is soaked in a photoresist solvent. This causes all the residual photoresist (with titanium on top) to be removed, leaving only the titanium that coated the bare regions of the substrate. This process is known as lift-off. Finally, the substrate, which is now patterned with titanium, is placed in a diffusion system. At temperatures above 1000°C the titanium diffuses into the substrate, slightly raising the refractive index of these regions. This process typically takes less than 10 hours. This sequence of steps is depicted in Figure 10.26.

The proton exchange method is depicted in Figure 10.27. Here, a chrome or an oxide masking layer is first deposited on the $LiNbO_3$ substrate. It is patterned using photoresist and etching. Next, the substrate is submerged in hot benzoic acid or some other suitable protonic source. Finally, the masking layer is removed and the substrate is annealed. The regions that have been exposed to the protonic source will have an increased refractive index and will guide light.

If the devices being fabricated are to be active (i.e., voltage controlled), then an electrode fabrication step is also required. This sequence of steps parallels the waveguide fabrication sequence. The only differences are that

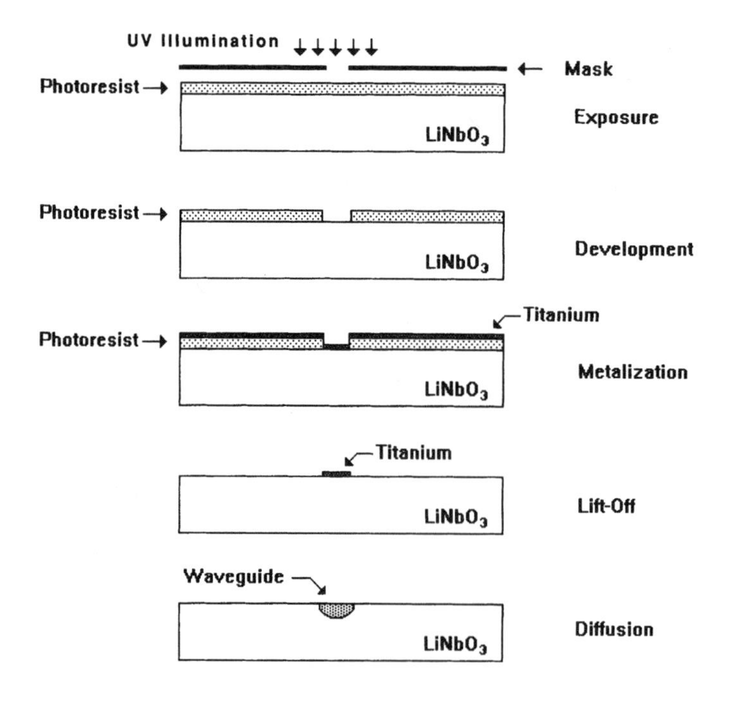

FIGURE 10.26 Ti-indiffused $LiNbO_3$ waveguide fabrication.

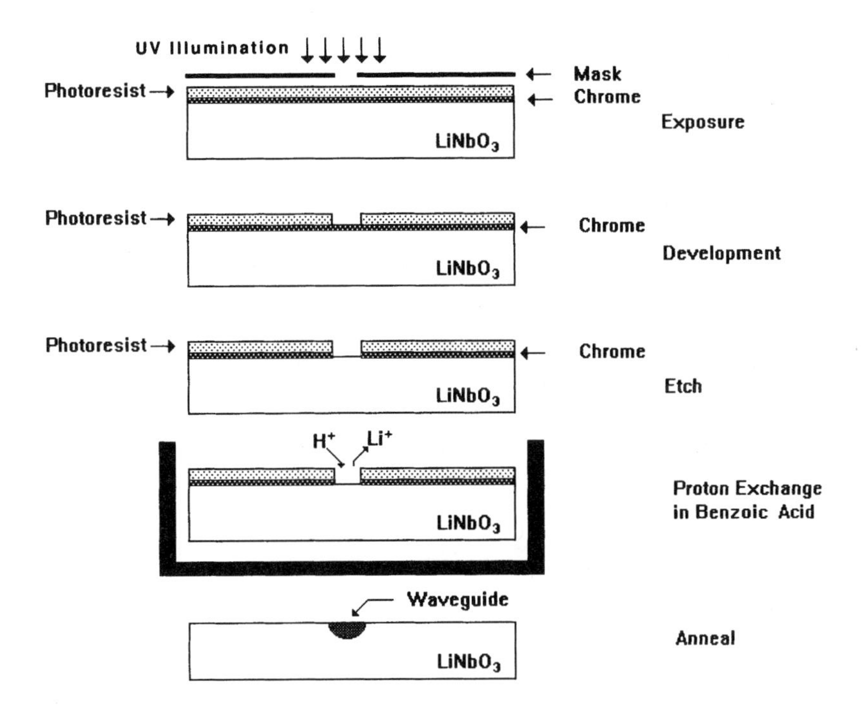

FIGURE 10.27 Proton exchange $LiNbO_3$ waveguide fabrication.

an electrode mask is used and the vacuum-deposited metal used is chrome/gold or chrome/aluminum. This sequence of steps is shown in Figure 10.28.

In order to get the light in and out of the waveguide, the endfaces have to be lapped and polished flat with chip-free knife edges. This is currently accomplished using standard lapping and polishing techniques. After this step, the substrate can be diced into as many devices as were included on the substrate. Finally, the diced parts need to be electrically and optically packaged.

Packaging

To get the light in and out of an integrated optic waveguide requires a tiny optical window to be polished on to the waveguide's end. Currently, the entire endface of the substrate is polished to a sharp, nearly perfect corner, making the whole endface into an optical window. An optical fiber can then be aligned to the waveguide end and attached. Typically, centration of the fiber axis to the waveguide axis must be better than 0.2 μm. Some devices require multiple inputs and outputs. In this case the fibers are prealigned in silicon V-grooves. These V-grooves are fabricated by anisotropic etching of a photolithographically defined pattern on the silicon. The center-to-center spacing of the fiber V-groove array can be made to closely match that of the multiple waveguide inputs and outputs.

Integrated optic devices built on $LiNbO_3$ are inherently single-mode devices. This means that the light is confined in a cross-sectional area of approximately 30 μm^2. The optical mode has a near-field pattern that is 5 to 10 μm across and 3 to 6 μm deep, depending on the wavelength. These mode spot sizes set limits on how light can be coupled in and out. There are a number of methods that can be used to couple the light into $LiNbO_3$ waveguides. These include prism coupling, grating coupling, end-fire coupling with lenses, and end-fire coupling with single-mode optical fibers. In general, most of these techniques are only useful for laboratory purposes. The most practical real-world technique is end-fire coupling with an optical fiber. In this case the optical fiber is aligned to the waveguide end. This is an excellent practical method since

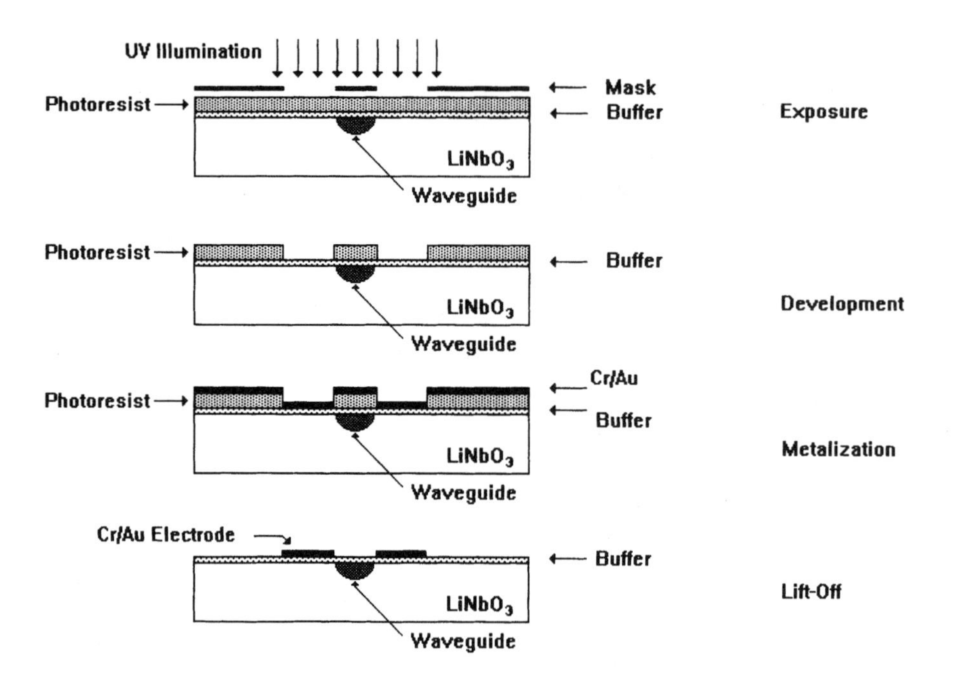

FIGURE 10.28 Electrode fabrication via lift-off.

integrated optic devices are most often used in fiber optic systems. Therefore, the coupling problem is one of aligning and fixing a single-mode fiber to the single-mode $LiNbO_3$ waveguide. The size of the single-mode radiation pattern and its angular divergence set the alignment tolerances. A low-loss connection between a fiber and a $LiNbO_3$ waveguide requires $<1/20$ of a mode spot diameter (0.25 to 0.5 μm) in transverse offset, angular tilt of <2 degrees, and a longitudinal offset of <1 mode spot diameter (5 to 10 μm). These are very stringent alignment requirements, especially if they have to be maintained over a wide operating temperature range.

Another aspect of the problem is that many integrated optic devices require a single, well-defined linearly polarized input. Ordinary single-mode fiber is not suitable in this case. The solution is to use polarization-preserving fiber. This fiber is made such that it will maintain a single linearly polarized input over long distances. The use of polarization preserving fiber, however, adds another requirement to the coupling problem. This requirement is that the fiber must be rotationally aligned about its cylindrical axis so that the linearly polarized light coincides with the desired rotational axis of the $LiNbO_3$ waveguide. The rotational precision needed can be $<0.5°$.

Many $LiNbO_3$ devices, such as fiber gyro components, require multiple input and/or output optical connections. Thus, the packaging must be able to accommodate multiple inputs/outputs and maintain strict alignment for all connections.

The method of end-fire coupling optical fibers to the $LiNbO_3$ waveguide is commonly called pigtailing. This is the only practical packaging method now used for integrated optic devices that operate in a real system and outside the laboratory. The reasons for this are quite logical. The end user installs the device in his system by connecting to the fiber pigtails. The connection can be made with single-mode connectors or by splicing. Flexibility is one of the big advantages of using fiber pigtails.

The typical $LiNbO_3$ device is packaged in a metallic case with optical fiber pigtails connected at both ends. Electrical connections are provided by rf connectors or pins, which are common in the electronics industry. If hermetically sealed packages are desired, then the optical fiber pigtails must be hermetically sealed to the metallic package.

Applications

Many useful systems have been accomplished using $LiNbO_3$-based integrated optic devices. These system applications can be grouped into four broad categories: telecommunications, instrumentation, signal processing, and sensors. In some cases, only a single integrated optic device is used, while in other applications a multifunction component is required.

Optical modulators have been shown to be quite useful in the telecommunications area. Hundreds of thousands of high-speed modulators for digital transmission at OC-48 (2.5 Gb/sec), OC-192 (10 Gb/sec), and OC-768 (40 Gb/sec) are in commercial use. In addition, the ability of the Mach–Zehnder intensity modulator to control intensity with a controlled wavelength change (i.e., chirp) has made its use in long-haul telecommunications systems the preferred choice.

Aside from the telecommunication application, there also is the high-speed analog modulation application. Analog transmission systems using integrated optic devices are particularly attractive as remote antenna links because of their high speed and ability to be driven directly by the received signal without amplification. The use of high-power diode lasers operating at 1300 and 1500 nm and external intensity modulators based on $LiNbO_3$ has found wide application in the cable TV industry. Another demonstrated application of integrated optics in the telecommunications area is in coherent communication systems. These systems require both phase modulators and polarization controllers. Current optical fiber transmission systems rely on intensity modulated data transmission schemes. Coherent communication systems are attractive because of the promise of higher bit rates, wavelength division multiplexing capability, and greater noise immunity. In coherent communication systems the information is coded by varying either the phase or frequency of the optical carrier with a phase modulator. At the receiver, a polarization controller is used to ensure a good signal-to-noise ratio in the heterodyne detection system.

One promising application of integrated optic devices in instrumentation is a high-speed, polarization-independent optical switch for use in optical-time-domain reflectometers (OTDR). OTDRs are used to locate breaks or poor splices in fiber-optic networks. The instruments work by sending out an optical pulse and measuring the backscattered radiation returning to the instrument as a function of time. By employing an optical switch, which will be used to rapidly switch the optical fiber under test from the pulsed light source to the OTDR receiver, the instrument could detect faults closer to the OTDR than currently possible, which is very important in short-haul METRO systems now being installed. This feature is also necessary for local area network (LAN) installations.

Several types of sensors using integrated optic devices have been demonstrated. Two of the most promising are electric/magnetic field or voltage sensors and rotation sensors (fiber-optic gyro). Electric field sensors typically consist of either a Mach–Zehnder intensity modulator or an optical switch that is biased midway between the on and off states. For small modulation depths about this midpoint the induced optical modulation is linear with respect to applied voltage. Linear dynamic ranges in excess of 80 dB have been accomplished. This is larger than that obtained using any other known technology.

Perhaps the most promising near-term application of integrated optic devices in the field of rotation sensing is as a key component in optical fiber gyroscopes. A typical fiber-optic gyro component is shown in Figure 10.29.

The device consists of a polarizer, a Y-junction, and two phase modulators, all integrated on a single substrate. In fiber gyro systems, the integrated optic component replaces individual, fiber-based components that perform the same function. The integrated optic component offers greatly improved performance, at a significant reduction in cost, compared to the fiber-based components. Most fiber-optic gyro development teams have done away with the fiber components and have adopted $LiNbO_3$-based components as the technology of choice.

Summary

Optical circuits have made impressive commercial in-roads over the past five years. Revenues from such products exceed several hundred million dollars a year. As manufacturing and packaging costs are reduced, this amount is expected to substantially increase over the next five years.

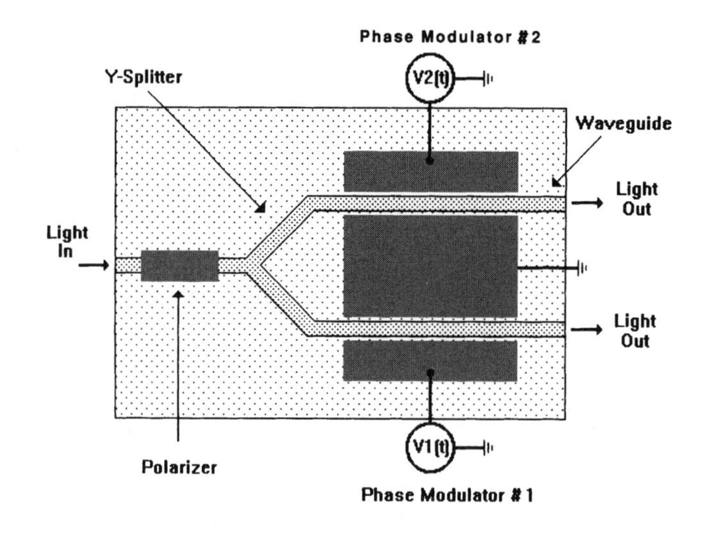

FIGURE 10.29 Fiber optic gyro chip.

Defining Terms

Integrated optics: The monolithic integration of one or more optical guided-wave devices on a common substrate.

Intensity modulator: A modulator that alters only the intensity of the incident light.

Lithium niobate ($LiNbO_3$): A single-crystal oxide that displays electro-optic, acousto-optic, piezoelectric, and pyroelectric properties, which is often the substrate of choice for surface acoustic wave devices and integrate optical devices.

Optical guided-wave device: An optical device that transmits or modifies light while it is confined in a thin-film optical waveguide.

Phase modulator: A modulator that alters only the phase of the incident light.

Polarization controller: A device that alters only the polarization state of the incident light.

References

R. Alferness, "Waveguide electrooptic modulators," *IEEE Trans. Microw. Theory Tech.*, vol. MTT-30, p. 1121, 1982.

R.A. Becker, "Commercially available integrated optics products and services," *SPIE*, vol. 993, p. 246, 1988.

R. Childs and V. O'Byrne, "Predistortion Linearization of Directly Modulated DFB Lasers and External Modulators for AM Video Transmission," *OFC'90 Tech. Dig.*, Paper WHG, 1990, p. 79.

C. Cox, G. Betts, and L. Johnson, "An analytic and experimental comparison of direct and external modulation in analog fiber-optic links," *IEEE Trans. Microw. Theory Tech.*, vol. 38, p. 501, 1990.

T. Findakly and M. Bramson, "High-performance integrated-optical chip for a broad range of fiber-optic gyro applications," *Opt. Lett.*, vol. 15, p. 673, 1990.

P. Granestrand, B. Stoltz, L. Thylen, K. Bergvall, W. Doldisen, H. Heinrich, and D. Hoffmann, "Strictly nonblocking 8×8 integrated optical switch matrix," *Electron. Lett.*, vol. 22, p. 816, 1986.

M. Howerton, C. Bulmer, and W. Burns, "Effect of intrinsic phase mismatch on linear modulator performance of the 1×2 directional coupler and Mach-Zehnder interferometer," *J. Lightw. Tech.*, vol. 8, p. 1177, 1990.

E. J. Lim, S. Matsumoto, and M. M. Fejer, "Noncritical phase matching for guided-wave frequency conversion," *Appl. Phys. Lett.*, vol. 57, p.2294, 1990.

S.E. Miller, "Integrated optics: An introduction," *Bell Syst. Tech. J.*, vol. 48, p. 2059, 1969.

Further Information

Integrated Optical Circuits and Components, Design and Applications, edited by Lynn D. Hutcheson (Marcel Dekker, Inc., New York, 1987) and *Optical Integrated Circuits*, by H. Nishihara, M. Haruna, and T. Suhara (McGraw-Hill Book Company, New York, 1989) provide excellent overviews of the field of integrated and guided-wave optics.

Integrated Optics: Devices and Applications, edited by J. T. Boyd (IEEE Press, New York, 1991) provides an excellent cross section of recent publications in the field.

In addition, the monthly journals *IEEE Journal of Lightwave Technology* and *IEEE Photonic Technology Letters* provide many publications on current research and development on integrated and guided-wave optic devices and systems.

11

D/A and A/D Converters

Fang Lin Luo
Nanyang Technological University

Hong Ye
Nanyang Technological University

Susan A.R. Garrod
Purdue University

11.1 Introduction

Fang Lin Luo and Hong Ye

A digital-to-analog (D/A) converter (or DAC) is the circuitry to convert digital codes (parallel or series) to continuous analog signals. An analog-to-digital (A/D) converter (ADC) is the circuitry to convert continuous analog signals to digital codes (parallel or series) via quantization operation. A digital signal processor (DSP) can be used in analog control systems provided it is equipped with DAC and ADC, as nearly all DSPs are in the world market.

The digital code can be presented in series and/or parallel or combined formats. A series digital code is a pulse-train. The digital number counted is the number of pulses in a certain sampling interval, i.e.:

$$N = fT$$

where N is the digital number counted, f is the frequency of the pulse-train, and T is the sampling interval.

A parallel digital code is an 8-bit binary code. The digital number counted is an 8-bit binary number from the least significant bit (LSB) to the most significant bit (MSB), which corresponds to $0 = (00000000)_B$ to $255 = (11111111)_B$. Some digital numbers are counted by hexadecimal numbers, which correspond to $0 = (00)_H$ to $255 = (FF)_H$. In recent decades, most digital codes take parallel format. In the further description, the digital code usually takes parallel format if not specifically stating series format. The analog signal is usually arranged in the full scale 0 to 10 V (particularly, 9.961 V). The resolution of the corresponding analog signals is better than 0.4% (e.g., 39 mV) for taking parallel digital code. This is sufficiently high accuracy for military and industrial applications. For the D/A and A/D converters with 8-bit binary digital code, the corresponding digital resolution is 1 LSB (or 1/255) and the analog resolution is 0.039 V (full scale 10 V).

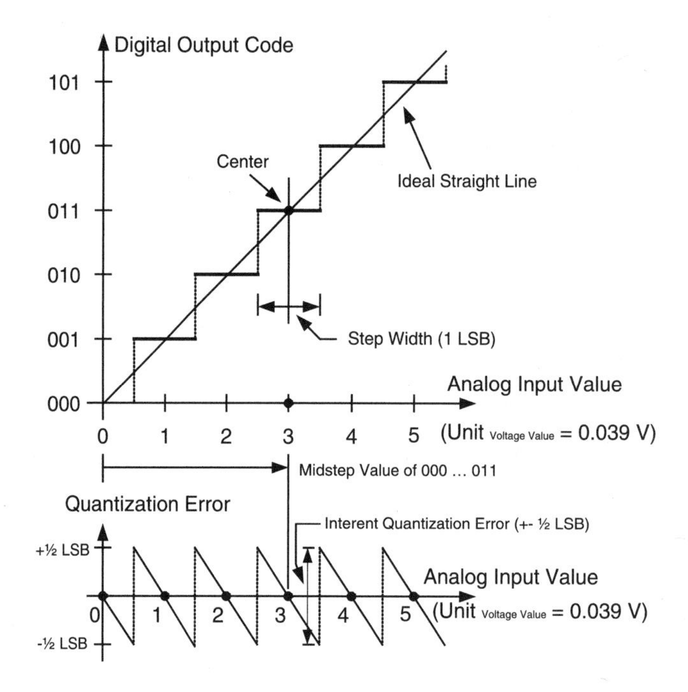

FIGURE 11.1 The ideal transfer function (ADC).

Quantization is the clue of D/A and A/D conversion procession. For example, an ADC uniquely represents all analog inputs within the range by a limited number of digital output codes. An ideal transfer function of an ADC is shown in Figure 11.1. The continuous input analog value is quantized into the discrete analog values with the increment 0.039 V. Since the analog scale is continuous, while the digital codes are discrete, there is a quantization process that introduces an error. As the number of discrete codes increases, the corresponding step width increases and the transfer function approaches an ideal straight line. The steps are designed to have transitions such that the midpoint of each step corresponds to the point on this ideal line. The width of one step is defined as 1 LSB, and this is often used as the reference unit for other quantities in the specification.

Ideal DACs and ADCs have the best quality with highly accurate resolution, quick sampling rate, high conversion speed, and good linearity. They are used for general description to simplify analysis of the fundamental characteristics. Real DACs and ADCs unavoidably have various errors: limited resolution, certain sampling rate, and conversion speed. Error and resolution are the important technical features of all DACs and ADCs. Modern DSPs use a 16 bit or 32 bit central processing unit (CPU); but DSP still has DACs and ADCs with the 8-bit binary digital code. Considering the errors and resolution, this determination is reasonable.

Real DACs and ADCs are unavoidably affected by noise, which is an important technical feature of DAC and ADC operation. Conversion speed is another technical feature of DACs and ADCs. All factors will be analyzed in Section 11.4.

11.2 D/A and A/D Circuits

Fang Lin Luo, Hong Ye, and Susan A.R. Garrod

D/A conversion is the process of converting digital codes into a continuous range of analog signals. A/D conversion is the complementary process of converting a continuous range of analog signals into digital codes. Such conversion processes are necessary to interface real-world systems, which typically monitor continuously varying analog signals with digital systems that process, store, interpret, and manipulate the analog values.

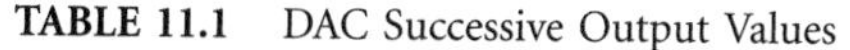

TABLE 11.1 DAC Successive Output Values

Bit	MSB	...	...	...	...	...	...	LSB
Volt	5	2.5	1.25	0.625	0.3125	0.1562	0.0781	0.0390

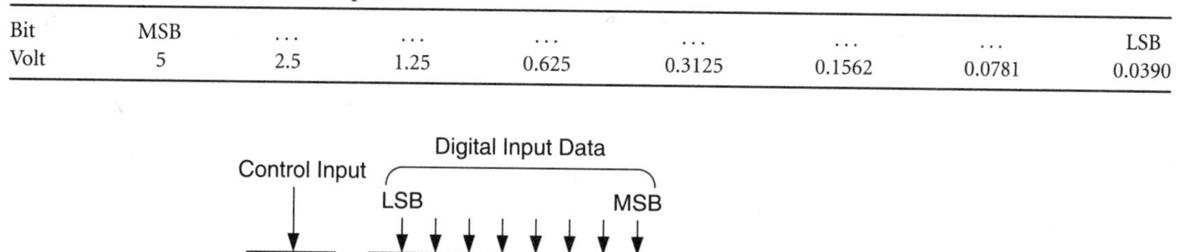

FIGURE 11.2 A typical DAC circuitry.

D/A and A/D applications have evolved from predominately military-driven applications to consumer-oriented applications. Up to the mid-1980s, the military applications determined the design of many D/A and A/D devices. They required very high performance coupled with hermetic packaging, radiation hardening, shock and vibration testing, and military specification and record keeping. Cost was of little concern, and "low-power" applications required approximately 2.8 W. The major applications up to the mid-1980s included military radar warning and guidance systems, digital oscilloscopes, medical imaging, infrared systems, and professional video.

The applications requiring D/A and A/D circuits in the 1990s have different performance criteria from those of earlier years. In particular, low-power and high-speed applications are driving the development of D/A and A/D circuits as the devices are used extensively in battery-operated consumer products. The predominant applications include cellular telephones, handheld camcorders, portable computers, and set-top cable TV boxes. These applications generally have low power and long battery life requirements, and they may have high-speed and high-resolution requirements, as is the case with the set-top cable TV boxes.

DAC is the fundamental circuitry; its successive output values are illustrated in Table 11.1.

A typical DAC circuit is shown in Figure 11.2. The digital input data are from the digital code with 8-bit parallel format, the control input is from DSP itself, and the output analog voltage is the final value of the DAC.

There are five typical A/D converters:

- Counting ADC
- Successive approximation ADC
- Dual slope ADC
- Parallel (flash) ADC
- Sigma-delta ADC

Some A/D converters, such as the counting ADC and successive approximation ADC, can be constructed by using D/A converters. One sample circuit of each typical A/D converter is illustrated in the following figures for reference:

- A simple counting ADC is shown in Figure 11.3.
- A successive approximation ADC is shown in Figure 11.4.
- A dual-slope integrating ADC is shown in Figure 11.5.
- A parallel (flash) ADC is shown in Figure 11.6.
- A sigma-delta ADC (3-bit circuitry) is shown in Figure 11.7.

D/A and A/D conversion circuits are available as integrated circuits (ICs) from many manufacturers. A huge array of ICs exists, consisting of not only the D/A or A/D conversion circuits, but also closely related circuits such as sample-and-hold amplifiers, analog multiplexers, voltage-to-frequency and frequency-to-voltage

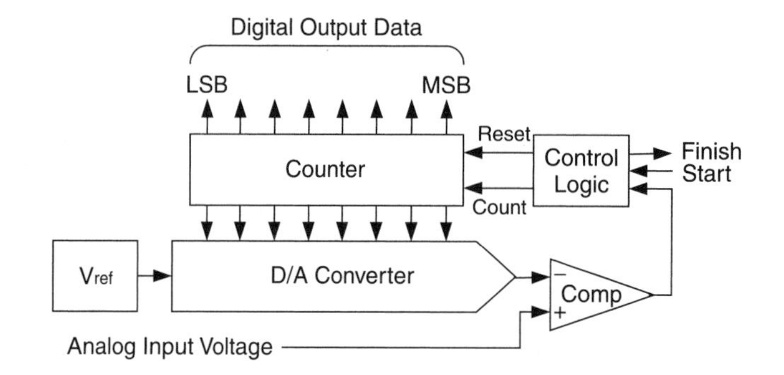

FIGURE 11.3 Simple counting ADC.

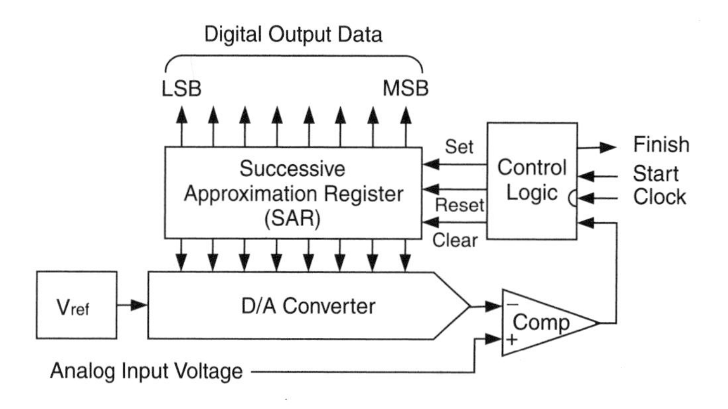

FIGURE 11.4 Successive approximation ADC.

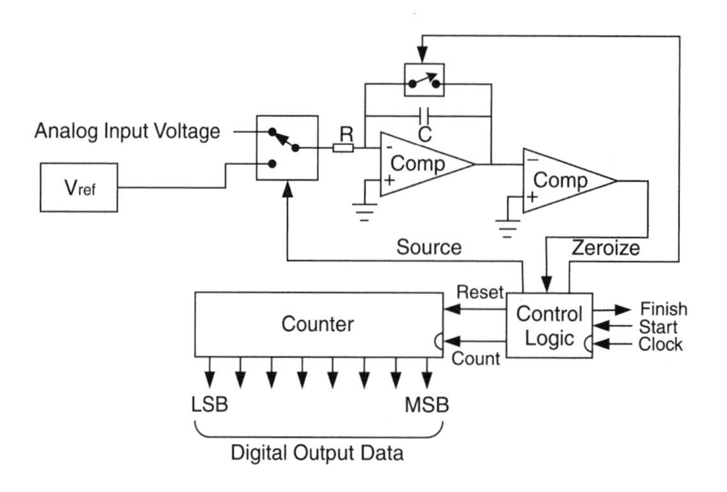

FIGURE 11.5 Dual-slope integrating ADC.

converters, voltage references, calibrators, operation amplifiers, isolation amplifiers, instrumentation amplifiers, active filters, DC/DC converters, analog interfaces to digital signal processing systems, and data acquisition subsystems. Data books from the IC manufacturers contain an enormous amount of information about these devices and their applications to assist the design engineer.

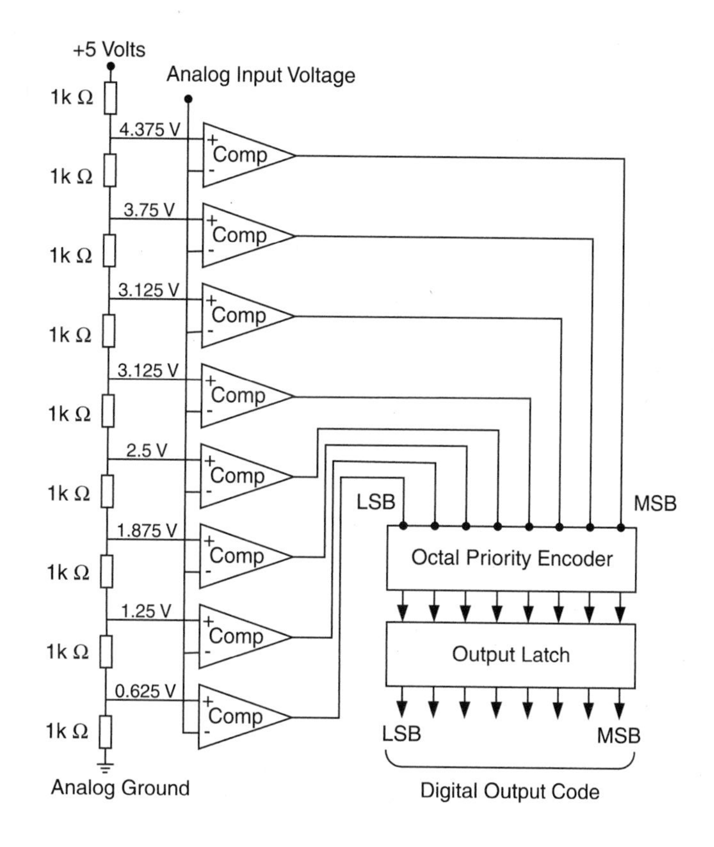

FIGURE 11.6 A parallel (flash) ADC block diagram.

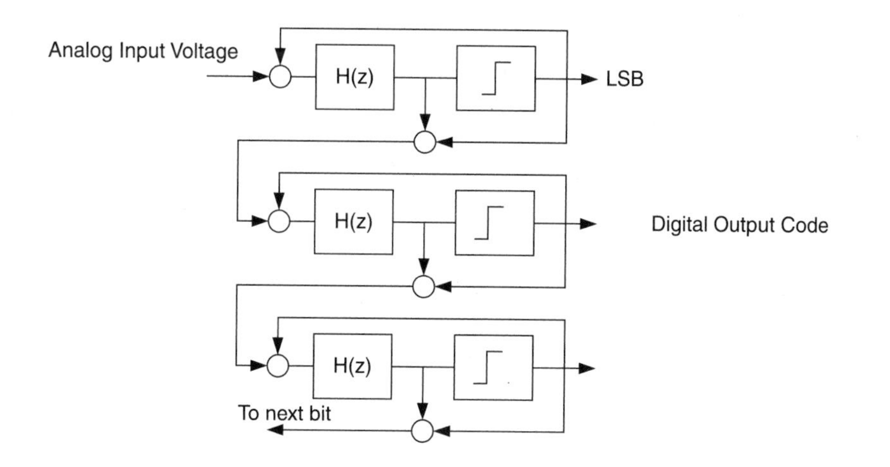

FIGURE 11.7 Sigma-delta ADC (three-bit circuitry).

The ICs discussed in this chapter will be strictly the D/A and A/D conversion circuits. Table 11.2 lists a small sample of the variety of the DACs and ADCs currently available. The ICs usually perform either D/A or A/D conversion. There are serial interface ICs, however, typically for high-performance audio and digital signal processing applications, that perform A/D and D/A processes.

TABLE 11.2 D/A and A/D Integrated Circuits

D/A Converter ICs	Resolution, b	Multiplying vs. Fixed Reference	Setting Time, μsec	Input Data Format
Analog Devices AD558	8	Fixed reference	3	Parallel
Analog Devices AD7524	8	Multiplying	0.400	Parallel
Analog Devices AD390	Quad, 12	Fixed reference	8	Parallel
Analog Devices AD1856	16	Fixed reference	1.5	Series
Burr-Brown DAC729	18	Fixed reference	8	Parallel
Datel DAC-HF8	8	Multiplying	0.025	Parallel
National DAC0800	8	Multiplying	0.1	Parallel
A/D Converter ICs	**Resolution, b**	**Signal Inputs**	**Conversion Speed, μsec**	**Output Data Format**
Analog Devices AD572	12	1	25	Series & parallel
Burr-Brown ADC803	12	1	1.5	Parallel
Burr-Brown ADC701	16	1	1.5	Parallel
National ADC1005B	10	1	50	Parallel
TI, National ADC0808	8	8	100	Parallel
TI, National ADC0834	8	4	32	Series
TI TLC0820	8	1	1	Parallel
TI TLC1540	10	11	21	Series
D/A and A/D Interface ICs	**Resolution, b**	**On-Board Filters**	**Sampling Rate, kHz**	**Data Format**
TI TLC32040	14	Yes	19.2 (programmable)	Series
TI 2914 PCM codec and filter	8	Yes	8	Series

11.3 DAC and ADC Performance Criteria

Susan A.R. Garrod, Fang Lin Luo, and Hong Ye

The major factors that determine the quality of performance of DACs and ADCs are resolution, sampling rate, conversion speed, and linearity (errors).

The resolution of a D/A circuit is the smallest change in the output analog signal. In an A/D system, the resolution is the smallest change in voltage that can be detected by the system and that can produce a change in the digital code. The resolution determines the total number of digital codes or quantization levels that will be recognized or produced by the circuit.

The resolution of a D/A or A/D IC is usually specified in terms of the bits in the digital code or in terms of the LSB of the system. An n-bit code allows for $2n$ quantization levels or $2n - 1$ steps between quantization levels. As the number of bits increases, the step size between quantization levels decreases, therefore increasing the accuracy of the system when a conversion is made between an analog and a digital signal. The system resolution can be specified also as the voltage step size between quantization levels. For A/D circuits, the resolution is the smallest input voltage that is detected by the system.

The conversion speed of a DAC or ADC is determined by the time it takes to perform the conversion process. For DACs, the speed is specified as the settling time. For ADCs, the speed is specified as the conversion time. The settling time for DACs will vary with supply voltage and transition in the digital code; thus, it is specified in the data sheet with the appropriate conditions stated.

ADCs have a maximum sampling rate that limits the speed at which they can perform continuous conversions. The sampling rate is the number of times per second that the analog signal can be sampled and converted into a digital code. For proper A/D conversion, the minimum sampling rate must be at least two times the highest frequency of the analog signal being sampled to satisfy the Nyquist sampling criterion. The conversion speed and other timing factors must be taken into consideration to determine the maximum sampling rate of an ADC. Nyquist ADCs use a sampling rate that is slightly more than twice the highest frequency in the analog signal. Oversampling ADCs use sampling rates of N times this rate, where N typically ranges from 2 to 64.

Both DACs and ADCs require a voltage reference in order to achieve absolute conversion accuracy. Some conversion ICs have internal voltage references, while others accept external voltage references. For high-performance systems, an external precision reference is needed to ensure long-term stability, load regulation, and control over temperature fluctuations. External precision voltage reference ICs can be found in manufacturers' data books.

Measurements accuracy is specified by the converter's linearity. Integral linearity is a measure of linearity over the entire conversion range. It is often defined as the deviation from a straight line drawn between the endpoints and through zero (or the offset value) of the conversion range. Integral linearity is also referred to as relative accuracy. The offset value is the reference level required to establish the zero or midpoint of the conversion range. Differential linearity is the linearity between code transitions and is a measure of the monotonicity of the converter. A converter is said to be monotonic if increasing input values result in increasing output values.

The accuracy and linearity values of a converter are specified in the data sheet in units of the LSB of the code. The linearity can vary with temperature, so the values are often specified at +25°C as well as over the entire temperature range of the device. The working environment temperature range is

- A version in 0°C to +70°C
- B version in −25°C to +85°C
- Storage temperature range in −25°C to +125°C

11.4 Errors, Resolution, Noises, and Conversion Speed

Fang Lin Luo and Hong Ye

The linearity (errors), resolution, noise, and conversion speed are main factors in determining the quality of performance of DACs and ADCs.

Errors

There are five main errors considered below and combining them together, we call them the absolute accuracy error or total error:

1. Quantization error
2. Offset error
3. Gain error
4. Differential nonlinearity error
5. Integral nonlinearity error

Generally, quantization error is ±1 LSB, offset error is up to 1¼ LSB, gain error is −3/4 to +1/2 LSB, differential nonlinearity error is ± 1/2 LSB, and integral nonlinearity error is −1/4 to +1/2 LSB. Usually the total error is 1¼ LSB.

Some errors are usually ignored since they are usually smaller than ±1/4 LSB. They are the aperture error, sampling error, and filtering error.

Resolution and Accuracy

From the above analysis, the resolution and accuracy can be considered ±1 LSB in digital code and ±0.039 V in analog signal.

Noises

There are three main noises considered below:

1. Quantization noise
2. Thermal noise
3. Sampling (switching) noise

The quantization noise is calculated as:

$$V_{Q(rms)} = \frac{V_{LSB}}{\sqrt{12}} \tag{11.1}$$

$$SNR = 20\log\frac{V_{in(rms)}}{V_{Q(rms)}} = 20\log\frac{V_{ref}/2\sqrt{2}}{V_{LSB}/\sqrt{12}} \tag{11.2}$$

Finally:

$$SNR = 6.02N + 1.76\,dB \tag{11.3}$$

The thermal noise happens randomly, and its average value is zero.

The sampling (switching) noise happens periodically, and its average value is zero.

The signal-to-noise ratio (SNR) is used to describe the effect of the noises on ADC and DAC. It is usually in the range 50 dB to 150 dB. For example, the ADC4320, ADC4322, and ADC4325 are complete 16 bit, 1 MHz, 2 MHz, and 500 kHz. Their typical SNR is 91 dB.

Conversion Speed

Conversion speed mainly relies on the sampling frequency and characteristics of the CPU used in the DSP. Updated DSPs have a sampling frequency in the range 500 kHz to 5 MHz, and the corresponding conversion speed is in hundreds of nanoseconds. For example, the conversion speed of ADC4322 is 300 ns.

11.5 D/A Conversion Processes and DAC ICs

Susan A.R. Garrod, Fang Lin Luo, and Hong Ye

Using previous analysis, D/A conversion processes and DAC ICs are described in detail in this section.

D/A Conversion Processes

Digital codes in parallel 8-bit binary format are typically converted to analog voltages by assigning a voltage weight to each bit in the digital code and then summing the voltage weights of the entire code. A general DAC consists of a network of precision resistors, input switches, and level shifters to activate the switches to convert a digital code to an analog current or voltage. D/A ICs that produce an analog current output usually have a faster settling time and better linearity than those that produce a voltage output. When the output current is available, the designer can convert this to a voltage through the selection of an appropriate output amplifier to achieve the necessary response speed for the given application.

DACs commonly have a fixed or variable reference level. The reference level, which determines the switching threshold of the precision switches that form a controlled impedance network, which in turn controls the value of the output signal. Fixed reference DACs produce an output signal that is proportional to the digital input. Multiplying DACs produces an output signal that is proportional to the product of a varying reference level times a digital code.

DACs can produce bipolar, positive, or negative polarity analog signals. A four-quadrant multiplying DAC allows the reference signal and the value of the binary code to have a positive or negative polarity. The four-quadrant multiplying DAC produces bipolar output signals.

DAC ICs

Most DACs are designed for general purpose control applications. Some DACs, however, are designed for special applications, such as video or graphic outputs, high-definition video displays, ultra high-speed signal processing, digital video tape recording, digital attenuators, or high-speed function generators.

DAC ICs often include special features that enable them to be interfaced easily to microprocessors or other systems. Microprocessor control inputs, input latches, buffers, input registers, and compatibility to standard logic families are features that are readily available in DAC ICs. In addition, the ICs usually have laser-trimmed precision resistors to eliminate the need for user trimming to achieve full-scale performance.

11.6 A/D Conversion Processes and ADC ICs

Susan A.R. Garrod, Fang Lin Luo, and Hong Ye

D/A conversion processes and DAC ICs are described in detail in this section.

A/D Conversion Processes

Analog signals can be converted to digital codes by many methods, including integration, successive approximation, parallel (flash) conversion, delta modulation, pulse code modulation, and sigma-delta conversion. Two of the most common A/D conversion processes are successive approximation A/D conversion and parallel or flash A/D conversion. Very high-resolution digital audio or video systems require specialized A/D techniques that often incorporate one of these general techniques as well as specialized A/D conversion processes. Examples of specialized A/D conversion techniques are pulse code modulation (PCM) and sigma-delta conversion. PCM is a common voice-encoding scheme used not only by the audio industry in digital audio recordings but also by the telecommunications industry for voice encoding and multiplexing. Sigma-delta conversion is an oversampling A/D conversion where signals are sampled at very high frequencies. It has very high resolution and low distortion and is being used in the digital audio recording industry.

Successive approximation A/D conversion is a technique that is commonly used in medium- to high-speed data acquisition applications. It is one of the fastest A/D conversion techniques that require a minimum amount of circuitry. The conversion times for successive approximation A/D conversion typically range from 10 to 300 msec for 8-bit systems.

The successive approximation ADC can approximate the analog signal to form an n-bit digital code in n steps. The successive approximation register (SAR) individually compares an analog input voltage to the midpoint of one of n ranges to determine the value of 1 bit. This process is repeated a total of n times, using n ranges, to determine the n bits in the code. The comparison is accomplished as follows: The SAR determines if the analog input is above or below the midpoint and sets the bit of the digital code accordingly. The SAR assigns the bits beginning with the most significant bit. The bit is set to 1 if the analog input is greater than the midpoint voltage, or it is set to 0 if it is less than the midpoint voltage. The SAR then moves to the next bit and sets it to 1 or 0 based on the results of comparing the analog input with the midpoint of the next allowed range. Because the SAR must perform one approximation for each bit in the digital code, an n-bit code requires n approximations. A successive approximation ADC consists of four functional blocks, as shown in Figure 11.4: the SAR, the analog comparator, a DAC, and a clock.

Parallel or flash A/D conversion is used in high-speed applications such as video signal processing, medical imaging, and radar detection systems. A flash ADC simultaneously compares the input analog voltage to $2n - 1$ threshold voltages to produce an n-bit digital code representing the analog voltage. Typical flash ADCs with 8-bit resolution operate at 20 to 100 MHz.

The functional block diagram of a flash ADC is shown in Figure 11.6. The circuitry consists of a precision resistor ladder network, $2n - 1$ analog comparators, and a digital priority encoder. The resistor network establishes threshold voltages for each allowed quantization level. The analog comparators indicate whether or not the input analog voltage is above or below the threshold at each level. The output of the analog comparators is input to the digital priority encoder. The priority encoder produces the final digital output code that is stored in an output latch.

An 8-bit flash ADC requires 255 comparators. The cost of high-resolution A/D comparators escalates as the circuit complexity increases and as the number of analog converters rises by $2n - 1$. As a low-cost alternative, some manufacturers produce modified flash ADCs that perform the A/D conversion in two steps to reduce the amount of circuitry required. These modified flash ADCs are also referred to as half-flash ADCs, since they perform only half of the conversion simultaneously.

ADC ICs

ADC ICs can be classified as general purpose, high-speed, flash, and sampling ADCs. The general purpose ADCs are typically low speed and low cost, with conversion times ranging from 2 ms to 33 ms. A/D conversion techniques used by these devices typically include successive approximation, tracking, and integrating. The general purpose ADCs often have control signals for simplified microprocessor interfacing. These ICs are appropriate for many process control, industrial, and instrumentation applications, as well as for environmental monitoring such as seismology, oceanography, meteorology, and pollution monitoring.

High-speed ADCs have conversion times typically ranging from 300 ns to 3 ms. The higher speed performance of these devices is achieved by using the successive approximation technique, modified flash techniques, and statistically derived A/D conversion techniques. Applications appropriate for these A/D ICs include fast Fourier transform (FFT) analysis, radar digitization, medical instrumentation, and multiplexed data acquisition. Some ICs have been manufactured with an extremely high degree of linearity to be appropriate for specialized applications in digital spectrum analysis, vibration analysis, geological research, sonar digitizing, and medical imaging.

Flash ADCs have conversion times ranging typically from 10 to 50 nsec. Flash A/D conversion techniques enable these ICs to be used in many specialized high-speed data acquisition applications such as TV video digitizing (encoding), radar analysis, transient analysis, high-speed digital oscilloscopes, medical ultrasound imaging, high-energy physics, and robotic vision applications.

Sampling ADCs have a sample-and-hold amplifier circuit built into the IC, eliminating the need for an external one. The throughput of these ADCs ICs ranges typically from 35 kHz to 100 MHz. The speed of the system is dependent on the A/D technique used by the sampling ADC.

ADC ICs produce digital codes in a serial or parallel format, and some ICs offer the designer both formats. The digital outputs are compatible with standard logic families to facilitate interfacing to other digital systems. In addition, some ADC ICs have a built-in analog multiplexer and therefore can accept more than one analog input signal.

Pulse code modulation (PCM) ICs are high-precision ADCs. The PCM IC is often referred to as a PCM codec with encoder and decoder functions. The encoder portion of the codec performs the A/D conversion, and the decoder portion of the codec performs the D/A conversion. The digital code is usually formatted as a serial data stream for ease of interfacing to digital transmission and multiplexing systems.

PCM is a technique where an analog signal is sampled, quantized, and then encoded as a digital word. The PCM IC can include successive approximation techniques or other techniques to accomplish the PCM encoding. In addition, the PCM codec can employ nonlinear data compression techniques such as companding if it is necessary to minimize the number of bits in the output digital code. Companding is a logarithmic technique used to compress a code to fewer bits before transmission. The inverse logarithmic function is then used to expand the code to its original number of bits before converting it to the analog signal. Companding is typically used in telecommunications transmission systems to minimize data transmission rates without degrading the resolution of low-amplitude signals. Two standardized companding techniques are used extensively: *A law* and *μ law.* A law companding is used in Europe, whereas μ law is used predominantly in the U.S. and Japan. Linear PCM conversion is used in high-fidelity audio systems to preserve the integrity of the audio signal throughout the entire analog range.

Digital signal processing (DSP) techniques provide another type of A/D conversion IC. Specialized A/D conversion such as adaptive differential pulse code modulation (ADPCM), sigma-delta modulation, speech subband encoding, adaptive predictive speech encoding, and speech recognition can be accomplished through the use of DSP systems. Some DSP systems require analog front ends that employ traditional PCM codec ICs or DSP interface ICs. These ICs can interface to a digital signal processor for advanced A/D applications. Some manufacturers have incorporated DSP techniques on board the single-chip A/D IC, as in the case of the DSP56ACD16 sigma-delta modulation IC by Motorola.

Integrating ADCs are used for conversions that must take place over a long period of time, e.g., digital voltmeter applications or sensor applications such as thermocouples. The integrating ADC produces a digital code that represents the average of the signal over time. Noise is reduced by means of the signal averaging or integration. Dual-slope integration is accomplished by a counter that advances while an input voltage charges a capacitor in a specified time interval, T. This is compared to another count sequence that advances while a reference voltage discharges across the same capacitor in a time interval, (d T). The ratio d of the charging count value to the discharging count value is proportional to the ratio of the input voltage to the reference voltage. Hence, the integrating converter provides a digital code that is a measure of the input voltage averaged over time. The conversion accuracy is independent of the capacitor and the clock frequency since they affect both the charging and discharging operations. The charging period, T, is selected to be the period of the fundamental frequency to be rejected. The maximum conversion rate is slightly less than $1/(2T)$ conversions per second, making it too slow for high-speed data acquisition applications, but it is appropriate for long-duration applications of slowly varying input signals.

11.7 Grounding and Bypassing on D/A and A/D ICs

Susan A.R. Garrod, Fang Lin Luo, and Hong Ye

DAC and ADC ICs require correct grounding and capacitive bypassing in order to operate according to performance specifications. The digital signals can severely impair analog signals. To combat the electromagnetic interference induced by the digital signals, the analog and digital grounds should be kept separate and should have only one common point on the circuit board. If possible, this common point should be the connection to the power supply.

Bypass capacitors are required at the power connections to the IC, the reference signal inputs, and the analog inputs to minimize noise that is induced by the digital signals. Each manufacturer specifies the recommended bypass capacitor locations and values in the data sheet. The 1-mF tantalum capacitors are commonly recommended, with additional high-frequency power supply decoupling sometimes being recommended through the use of ceramic disc shunt capacitors. The manufacturer's recommendations should be followed to ensure proper performance.

11.8 Selection Criteria for D/A and A/D Converter ICs

Susan A.R. Garrod, Fang Lin Luo, and Hong Ye

Hundreds of DAC and ADC ICs are available, with prices ranging from a few dollars to several hundred dollars each. The selection of the appropriate type of converter is based on the application requirements of the system, the performance requirements, and cost. The following issues should be considered in order to select the appropriate converter:

1. What are the input and output requirements of the system? Specify all signal current and voltage ranges, logic levels, input and output impedances, digital codes, data rates, and data formats.
2. What level of accuracy is required? Determine the resolution needed throughout the analog voltage range, the dynamic response, the degree of linearity, and the number of bits encoding.

3. What speed is required? Determine the maximum analog input frequency for sampling in an A/D system, the number of bits for encoding each analog signal, and the rate of change of input digital codes in a D/A system.
4. What is the operating environment of the system? Obtain information on the temperature range and power supply to select a converter that is accurate over the operating range.

Final selection of DAC and ADC ICs should be made by consulting manufacturers to obtain their technical specifications for the devices. Major manufacturers of DACs and ADCs include Analog Devices, Burr-Brown, DATEL, Maxim, National, Phillips Components, Precision Monolithics, Signetics, Sony, Texas Instruments, Ultra Analog, and Yamaha. Information on contacting these manufacturers and others can be found in an IC Master Catalog.

Defining Terms

Companding: A process designed to minimize the transmission bit rate of a signal by compressing it prior to transmission and expanding it upon reception. It is a rudimentary data compression technique that requires minimal processing.

Delta modulation: An A/D conversion process where the digital output code represents the change, or slope, of the analog input signal rather than the absolute value of the analog input signal. A 1 indicates a rising slope of the input signal. A 0 indicates a falling slope of the input signal. The sampling rate is dependent on the derivative of the signal, since a rapidly changing signal would require a rapid sampling rate for acceptable performance.

Fixed reference DAC: The analog output is proportional to a fixed (nonvarying) reference signal.

Flash A/D: The fastest A/D conversion process available to date, also referred to as parallel A/D conversion. The analog signal is simultaneously evaluated by $2n-1$ comparators to produce an n bit digital code in one step. Because of the large number of comparators required, the circuitry for flash ADCs can be very expensive. This technique is commonly used in digital video systems.

Integrating A/D: The analog input signal is integrated over time to produce a digital signal that represents the area under the curve, or the integral.

Multiplying D/A: A D/A conversion process where the output signal is the product of a digital code multiplied by an analog input reference signal. This allows the analog reference signal to be scaled by a digital code.

Nyquist A/D converters: ADCs that sample analog signals having a maximum frequency that is less than the Nyquist frequency. The Nyquist frequency is defined as one-half of the sampling frequency. If a signal has frequencies above the Nyquist frequency, a distortion called aliasing occurs. To prevent this, an antialiasing filter with a flat passband and very sharp roll-off is required.

Oversampling converters: ADCs that sample frequencies at a rate much higher than the Nyquist frequency. Typical oversampling rates are 32 and 64 times the sampling rate that would be required with the Nyquist converters.

Pulse code modulation (PCM): An A/D conversion process requiring three steps: the analog signal is sampled, quantized, and encoded into a fixed length digital code. This technique is used in many digital voice and audio systems. The reverse process reconstructs an analog signal from the PCM code. The operation is very similar to other A/D techniques, but specific PCM circuits are optimized for the particular voice or audio application.

Sigma-delta A/D conversion: An oversampling A/D conversion process where the analog signal is sampled at rates much higher (typically 64 times) than the sampling rates that would be required with a Nyquist converter. Sigma-delta modulators integrate the analog signal before performing the delta modulation. The integral of the analog signal is encoded rather than the change in the analog signal, as is the case for traditional delta modulation. A digital sample rate reduction filter (also called a digital decimation filter) is used to provide an output sampling rate at twice the

Nyquist frequency of the signal. The overall result of oversampling and digital sample rate reduction is greater resolution and less distortion compared to a Nyquist converter process.

Successive approximation: An A/D conversion process that systematically evaluates the analog signal in *n* steps to produce an *n*-bit digital code. The analog signal is successively compared to determine the digital code, beginning with the determination of the most significant bit of the code.

References

1. Analog Devices, *Analog Devices Data Conversion Products Data Book*, Norwood, Mass.: Analog Devices, Inc., 1989.
2. Burr-Brown, *Burr-Brown Integrated Circuits Data Book*, Tucson, Ariz.: Burr-Brown, 1989.
3. DATEL, *DATEL Data Conversion Catalog*, Mansfield, Mass.: DATEL, Inc., 1988.
4. W. Drachler and M. Bill, "New high-speed, low-power data-acquistion ICs," *Analog Dialogue*, vol. 29, no. 2, Norwood, Mass.: Analog Devices, Inc., 1995, pp. 3–6.
5. S. Garrod and R. Borns, *Digital Logic: Analysis, Application and Design*, Philadelphia, Pa.: Saunders College Publishing, 1991, Chapter 16.
6. J.M. Jacob, *Industrial Control Electronics*, Englewood Cliffs, N.J.: Prentice-Hall, 1989, Chapter 6.
7. B. Keiser and E. Strange, *Digital Telaphony and Network Integration*, 2nd ed., New York: Van Nostrand Reinhold, 1995.
8. Motorola, *Motorola Telecommunications Data Book*, Phoenix, Ariz.: Motorola, Inc., 1989.
9. National Semiconductor, *National Semiconductor Data Acquisition Linear Devices Data Book*, Santa Clara, Calif.: National Semiconductor Corp., 1989.
10. S. Park, *Principles of Sigma-Delta Modulation for Analog-to-Digital Converters*, Phoenix, Ariz.: Motorola, Inc., 1990.
11. Texas Instruments, *Texas Instruments Digital Signal Processing Applications with the TMS320 Family*, Dallas, Texas: Texas Instruments, 1986.
12. Texas Instruments, 1989. *Texas Instruments Linear Circuits Data Acquisition and Conversion Data Book*, Dallas, Texas: Texas Instruments, 1989.
13. Texas Instruments, *Understanding Data Converters: Application Report*, Texas Instruments Incorporated, 1999.
14. Timo Rahkonen, *Error Correction Techniques in High-Speed A/D and D/A Converters/i*, University of Oulu, Finland, 2001.
15. Analog IC Catalog, *ADC4320/ADC4322/ADC4325, Very High Speed 16 bit, Sampling A/D Converters*, Texas Instruments Incorporated, 2001.
16. D. Johns and K. Martin, *Data Converter Fundamentals*, Toronto, Canada: University of Toronto, 1997.

Further Information

Analog Devices, Inc. has edited or published several technical handbooks to assist design engineers with their data acquisition system requirements. These references should be consulted for extensive technical information and depth. The publications include *Analog-Digital Conversion Handbook*, by the engineering staff of Analog Devices, published by Prentice-Hall, Englewood Cliffs, NJ, 1986; *Nonlinear Circuits Handbook*, *Transducer Interfacing Handbook*, and *Synchro and Resolver Conversion*, all published by Analog Devices Inc., Norwood, MA.

Engineering trade journals and design publications often have articles describing recent A/D and D/A circuits and their applications. These publications include *EDN Magazine*, *EE Times*, and *IEEE Spectrum*. Research-related topics are covered in *IEEE Transactions on Circuits and Systems*, and also the *IEEE Transactions on Instrumentation and Measurement*.

12

Digital and Analog Electronic Design Automation

Allen Dewey
Duke University

12.1 Introduction

The field of **design automation** (DA) technology, also commonly called *computer-aided design* (CAD) or *computer-aided engineering* (CAE), involves developing computer programs to conduct portions of product design and manufacturing on behalf of the designer. Competitive pressures to produce more efficiently new generations of products having improved function and performance are motivating the growing importance of DA. The increasing complexities of microelectronic technology, shown in Figure 12.1, illustrate the importance of relegating portions of product development to computer automation (Barbe, 1980).

Advances in microelectronic technology enable over 1 million devices to be manufactured on an **integrated circuit** that is smaller than a postage stamp; yet the ability to exploit this capability remains a challenge. Manual design techniques are unable to keep pace with product design cycle demands and are being replaced by automated design techniques (Saprio, 1986; Dillinger, 1988).

Figure 12.2 summarizes the historical development of DA technology. DA computer programs are often simply called *applications* or *tools*. DA efforts started in the early 1960s as academic research projects and captive industrial programs; these efforts focused on tools for physical and logical design. Followon developments extended logic simulation to more-detailed *circuit* and *device* simulation and more-abstract *functional* simulation. Starting in the mid to late 1970s, new areas of test and synthesis emerged and vendors started offering commercial DA products. Today, the electronic design automation (EDA) industry is an international business with a well-established and expanding technical base (Trimberger, 1990). EDA will be examined by presenting an overview of the following areas:

- Design entry
- Synthesis

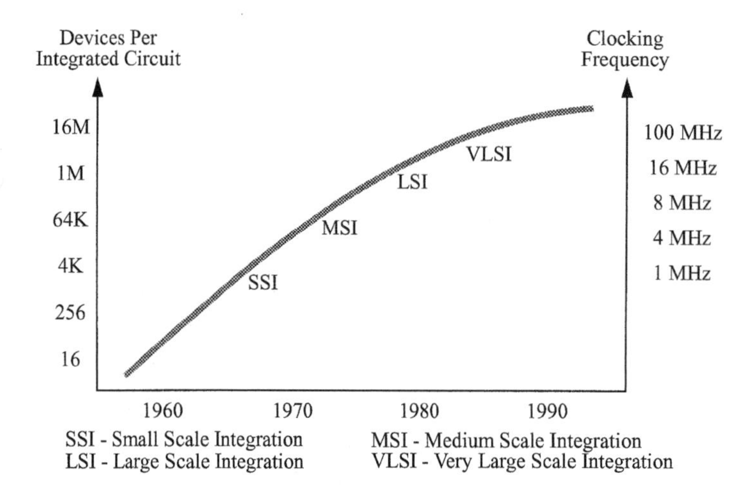

FIGURE 12.1 Microelectronic technology complexity.

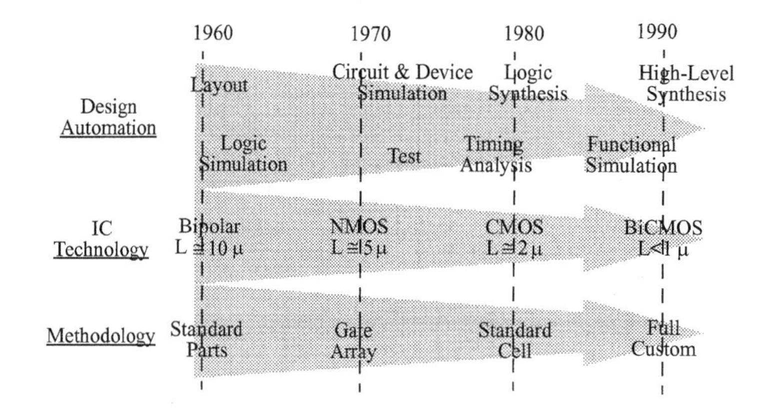

FIGURE 12.2 DA technology development.

- Verification
- Physical design
- Test

12.2 Design Entry

Design entry, also called *design capture*, is the process of communicating with a DA system. In short, design entry is how an engineer "talks" to a DA application and/or system.

Any sort of communication is composed of two elements: language and mechanism. Language provides common semantics; mechanism provides a means by which to convey the common semantics. For example, people communicate via a language, such as English or German, and a mechanism, such as a telephone or electronic mail. For design, a digital system can be described in many ways, involving different perspectives or *abstractions*. An abstraction defines at a particular level of detail the behavior or semantics of a digital system, i.e., how the outputs respond to the inputs. Figure 12.3 illustrates several popular levels of abstractions. Moving from the lower left to the upper right, the level of abstraction generally increases, meaning that physical models are the most detailed and specification models are the least detailed. The trend toward higher levels of design entry abstraction supports the need to address greater levels of complexity (Peterson, 1981).

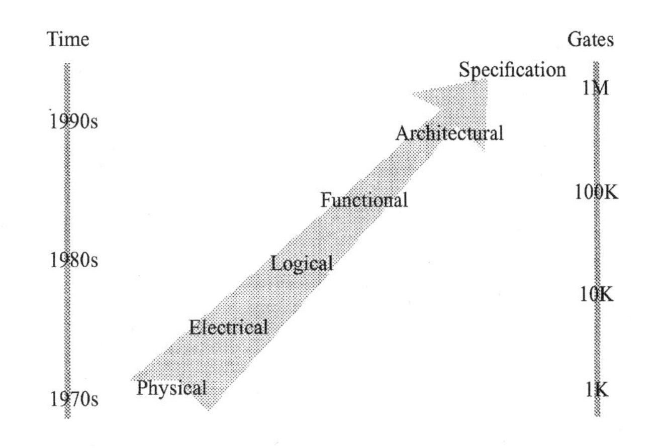

FIGURE 12.3 DA abstractions.

The physical level of abstraction involves geometric information that defines electrical devices and their interconnection. Geometric information includes the shape of objects and how objects are placed relative to each other. For example, Figure 12.4 shows the geometric shapes defining a simple complementary metaloxide semiconductor (**CMOS**) inverter. The shapes denote different materials, such as aluminum and polysilicon, and connections, called *contacts* or **vias**.

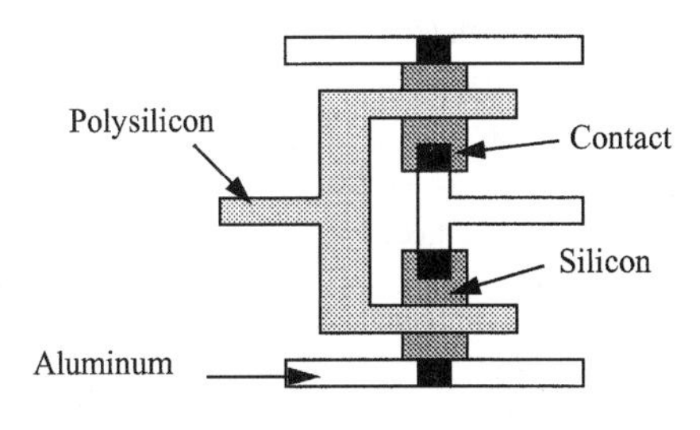

FIGURE 12.4 Physical abstraction.

Design entry mechanisms for physical information involve textual and graphical techniques. With textual techniques, geometric shape and placement are described via an artwork description language, such as Caltech Intermediate Form (CIF) or Electronic Design Intermediate Form (EDIF). With graphical techniques, geometric shape and placement are described by rendering the objects on a display terminal.

The electrical level abstracts physical information into corresponding electrical devices, such as **capacitors**, **transistors**, and **resistors**. Electrical information includes device behavior in terms of terminal current and voltage relationships. Device behavior may also be defined in terms of manufacturing parameters. Figure 12.5 shows the electrical symbols denoting a CMOS inverter.

The logical level abstracts electrical information into corresponding logical elements, such as **and** gates, **or** gates, and inverters. Logical information includes truth table and/or characteristic-switching algebra equations and active-level designations. Figure 12.6 shows the logical symbol for a CMOS inverter. Notice how the amount of information decreases as the level of abstraction increases.

Design entry mechanisms for electrical and logical abstractions are collectively called *schematic capture* techniques. Schematic capture defines hierarchical structures, commonly called **netlists**, of components.

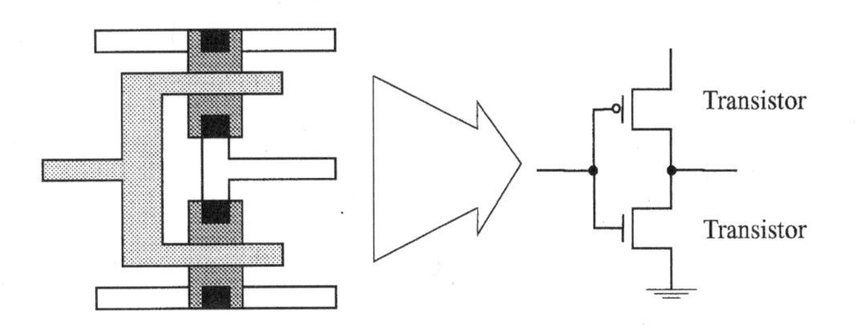

FIGURE 12.5 Electrical abstraction.

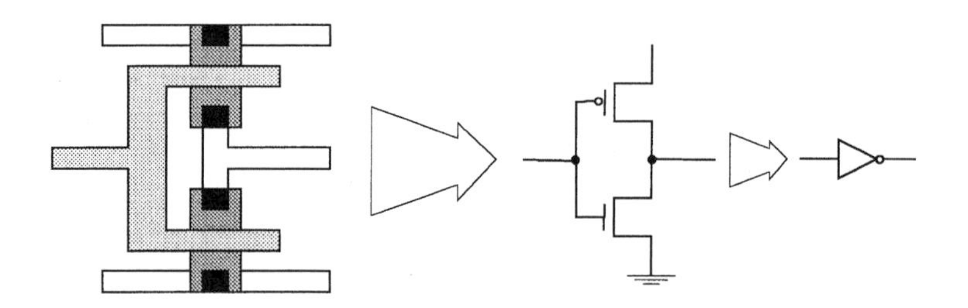

FIGURE 12.6 Logical abstraction.

A designer creates instances of components supplied from a library of predefined components and connects component pins or ports via wires (Douglas-Young, 1988; Pechet, 1991).

The functional level abstracts logical elements into corresponding computational units, such as registers, multiplexers, and arithmetic logic units (ALUs). The architectural level abstracts functional information into computational algorithms or paradigms. Examples of common computational paradigms are listed below:

- State diagrams
- Petri nets
- Control/data flow graphs
- Function tables
- Spreadsheets
- Binary decision diagrams

These higher levels of abstraction support a more expressive, "higher-bandwidth" communication interface between engineers and DA programs. Engineers can focus their creative, cognitive skills on concept and behavior, rather than on the complexities of detailed implementation. Associated design entry mechanisms typically use hardware description languages with a combination of textual and graphic techniques (Birtwistle and Subrahmanyan, 1988).

Figure 12.7 shows an example of a simple state diagram. The state diagram defines three states, denoted by circles. State-to-state transitions are denoted by labeled arcs; state transitions depend on the present state and the input X. The output, Z, per state is given within each state. Since the output is dependent on only the present state, the digital system is classified as a Moore **finite state machine**. If the output is dependent on the present state and input, then the digital system is classified as a Mealy finite state machine.

A hardware description language model written in **VHDL** of the Moore finite state machine is given in Figure 12.8. The VHDL model, called a *design entity*, uses a "**data flow**" description style to describe the state machine (Dewey, 1983, 1992, 1997). The entity statement defines the interface, i.e., the ports. The ports include two input signals, X and CLK, and an output signal Z. The ports are of type BIT, which specifies that

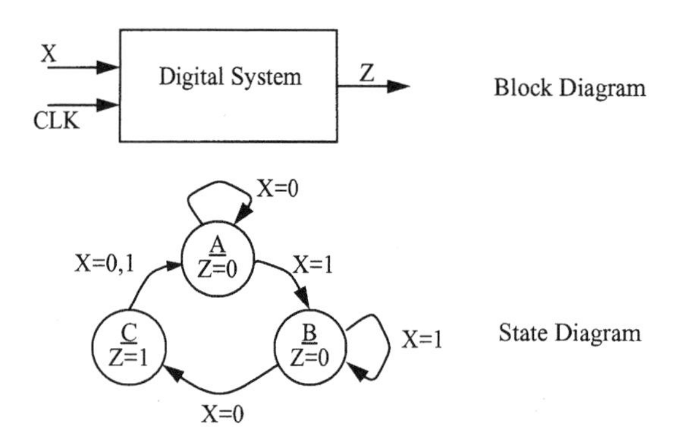

FIGURE 12.7 State diagram.

```
-- entity statement
entity MOORE_MACHINE is
  port (X, CLK : in BIT;  Z : out BIT);
end MOORE_MACHINE;

-- architecture statement
architecture FSM of MOORE_MACHINE is
  type STATE_TYPE is (A, B, C);
  signal STATE : STATE_TYPE := A;
begin
 NEXT_STATE:
 block (CLK='1' and not CLK'STABLE)
 begin
   -- guarded conditional concurrent signal assignment statement
   STATE <= guarded B when (STATE=A and X='1') else
                    C when (STATE=B and X='0') else
                    A when (STATE=C) else
                    STATE;
 end block NEXT_STATE;
 -- unguarded selected concurrent signal assignment statement
 with STATE select
   Z <= '0' when A,
        '0' when B,
        '1' when C;
end FSM;
```

FIGURE 12.8 VHDL model.

the signals may only carry the values 0 or 1. The architecture statement defines the input/output transform via two concurrent signal assignment statements. The internal signal STATE holds the finite state information and is driven by a guarded, conditional concurrent signal assignment statement that executes when the associated block expression:

$$(\mathrm{CLK} = {}'1' \textbf{ and not } \mathrm{CLK}'\mathrm{STABLE})$$

is true, which is only on the rising edge of the signal CLK. STABLE is a predefined attribute of the signal CLK; CLK'STABLE is true if CLK has *not* changed value. Thus, if "**not** CLK'STABLE" is true, meaning that CLK has just changed value, and "CLK='1'," then a rising transition has occurred on CLK. The output signal Z is driven by a nonguarded, selected concurrent signal assignment statement that executes any time STATE changes value.

12.3 Synthesis

Figure 12.9 shows that the **synthesis** task generally follows the design entry task. After describing the desired system via design entry, synthesis DA programs are invoked to assist generating the required detailed design.

Synthesis translates or transforms a design from one level of abstraction to another, more detailed level of abstraction. The more detailed level of abstraction may be only an intermediate step in the entire design process, or it may be the final implementation. Synthesis programs that yield a final implementation are sometimes called **silicon compilers** because the programs generate sufficient detail to proceed directly to silicon fabrication (Ayres, 1983; Gajski, 1988).

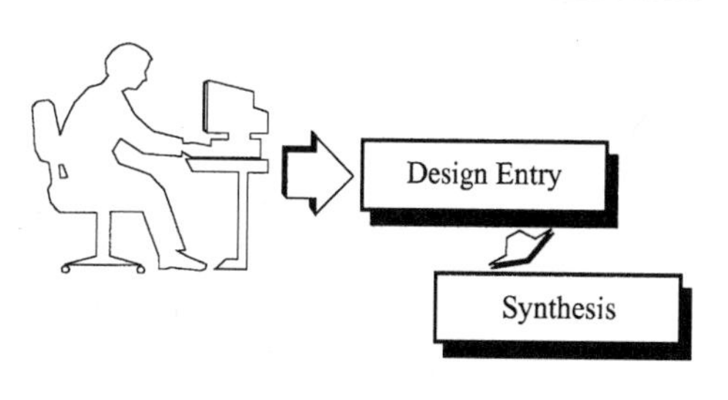

FIGURE 12.9 Design process synthesis.

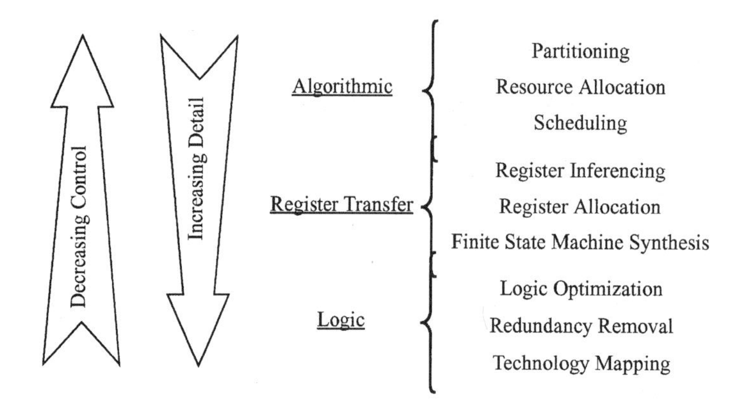

FIGURE 12.10 Taxonomy of synthesis techniques.

Like design abstractions, synthesis techniques can be hierarchically categorized, as shown in Figure 12.10. The higher levels of synthesis offer the advantage of less complexity, but also the disadvantage of less control over the final design.

Algorithmic synthesis, also called *behavioral* synthesis, addresses "multicycle" behavior, which means behavior that spans more than one *control step*. A control step equates to a clock cycle of a synchronous, sequential digital system, i.e., a state in a finite-state machine controller or a microprogram step in a microprogrammed controller. Algorithmic synthesis typically accepts sequential design descriptions that define an input/output transform, but provide little information about the parallelism of the final design (Camposano and Wolfe, 1991; Gajski et al., 1992).

Partitioning decomposes the design description into smaller behaviors. Partitioning is an example of a high-level transformation. High-level transformations include common software programming compiler optimizations, such as loop unrolling, subprogram in-line expansion, constant propagation, and common subexpression elimination.

Resource allocation associates behaviors with hardware computational units, and scheduling determines the order in which behaviors execute. Behaviors that are mutually exclusive can potentially share computational resources. Allocation is performed using a variety of graph clique covering or node coloring algorithms. Allocation and scheduling are interdependent, and different synthesis strategies perform allocation and scheduling different ways. Sometimes scheduling is performed first, followed by allocation; sometimes allocation is performed first, followed by scheduling; and sometimes allocation and scheduling are interleaved.

Scheduling assigns computational units to control steps, thereby determining which behaviors execute in which clock cycles. At one extreme, all computational units can be assigned to a single control step, exploiting maximum concurrency. At the other extreme, computational units can be assigned to individual control steps, exploiting maximum sequentiality. Several popular scheduling algorithms are listed below:

- As-soon-as-possible (ASAP)
- As-late-as-possible (ALAP)
- List scheduling
- Force-directed scheduling
- Control step splitting/merging

ASAP and ALAP scheduling algorithms order computational units based on data dependencies. List scheduling is based on ASAP and ALAP scheduling, but considers additional, more-global constraints, such as maximum number of control steps. Force-directed scheduling computes the probabilities of computational units being assigned to control steps and attempts to evenly distribute computation activity among all control steps. Control step splitting starts with all computational units assigned to one control step and generates a schedule by splitting the computational units into multiple control steps. Control step merging starts with all computational units assigned to individual control steps and generates a schedule by merging or combining units and steps (Paulin and Knight, 1989; Camposano and Wolfe, 1991).

Register transfer synthesis takes as input the results of algorithmic synthesis and addresses "per-cycle" behavior, which means the behavior during one clock cycle. Register transfer synthesis selects logic to realize the hardware computational units generated during algorithmic synthesis, such as realizing an addition operation with a carry–save adder or realizing addition and subtraction operations with an arithmetic logic unit. Data that must be retained across multiple clock cycles are identified, and registers are allocated to hold the data. Finally, finite-state machine synthesis involves state minimization and state assignment. State minimization seeks to eliminate redundant or equivalent states, and state assignment assigns binary encodings for states to minimize combinational logic (Brayton et al., 1992; Sasao, 1993).

Logic synthesis optimizes the logic generated by register transfer synthesis and maps the optimized logic operations onto physical gates supported by the target fabrication technology. Technology mapping considers the foundry cell library and associated electrical restrictions, such as **fan-in/fan-out** limitations.

12.4 Verification

Figure 12.11 shows that the **verification** task generally follows the synthesis task. The verification task checks the correctness of the function and performance of a design to ensure that an intermediate or final design faithfully realizes the initial, desired specification. Three major types of verification are listed below:

- Timing analysis
- Simulation
- Emulation

Timing Analysis

Timing analysis checks that the overall design satisfies operating speed requirements and that individual signals within a design satisfy transition requirements. Common signal transition requirements, also called *timing hazards*, include *rise* and *fall times*, *propagation delays*, *clock periods*, *race conditions*, *glitch detection*, and *setup* and *hold times*. For instance, setup and hold times specify relationships between data and control signals to ensure that memory devices (level-sensitive latches or edge-sensitive flip-flops) correctly and reliably store desired data. The data signal carrying the information to be stored in the memory device must be stable for a period equal to the setup time prior to the control signal transition to ensure that the correct value is sensed by the memory device. Also, the data signal must be stable for a period equal to the hold time after the control signal transition to ensure that the memory device has enough time to store the sensed value.

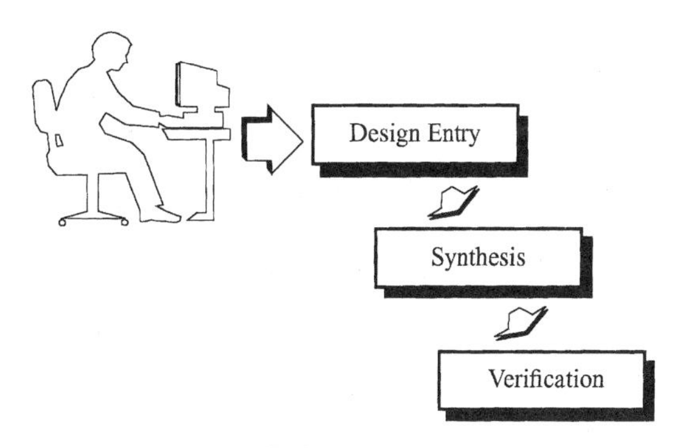

FIGURE 12.11 Design process verification.

Another class of timing transition requirements, commonly called signal integrity checks, include *reflections*, *crosstalk*, **ground bounce**, and *electromagnetic interference*. Signal integrity checks are typically required for high-speed designs operating at clock frequencies above 75 MHz. At such high frequencies, the transmission line behavior of wires must be analyzed. A wire should be properly terminated, i.e., connected, to a port having an impedance matching the wire characteristic impedance to prevent signal reflections. Signal reflections are portions of an emanating signal that "bounce back" from the destination to the source. Signal reflections reduce the power of the emanating signal and can damage the source. Crosstalk refers to unwanted reactive coupling between physically adjacent signals, providing a connection between signals that are supposed to be electrically isolated. Ground bounce is another signal integrity problem. Since all conductive material has a finite impedance, a ground signal network does not in practice offer the exact same electrical potential throughout an entire design. These potential differences are usually negligible because the distributive impedance of the ground signal network is small compared with other finite-component impedances. However, when many signals switch value simultaneously, a substantial current can flow through the ground signal network. High intermittent currents yield proportionately high intermittent potential drops, i.e., ground bounces, which can cause unwanted circuit behavior. Finally, electromagnetic interference refers to signal harmonics radiating from design components and interconnects. This harmonic radiation may interfere with other electronic equipment or may exceed applicable environmental safety regulatory limits (McHaney, 1991).

Timing analysis can be performed dynamically or statically. Dynamic timing analysis exercises the design via simulation or emulation for a period of time with a set of input stimuli and records the timing behavior. Static timing analysis does not exercise the design via simulation or emulation. Rather, static analysis records timing behavior based on the timing behavior, e.g., propagation delay, of the design components and their interconnection.

Static timing analysis techniques are primarily *block oriented* or *path oriented*. Block-oriented timing analysis generates design input (also called primary input) to design output (also called primary output), and propagation delays by analyzing the design "stage-by-stage" and by summing up the individual stage delays. All devices driven by primary inputs constitute stage 1, all devices driven by the outputs of stage 1 constitute stage 2, and so on. Starting with the first stage, all devices associated with a stage are annotated with worst-case delays. A worst-case delay is the propagation delay of the device plus the delay of the last input to arrive at the device, i.e., the signal path with the longest delay leading up to the device inputs. For example, the device labeled "H" in stage 3 in Figure 12.12 is annotated with the worst-case delay of 13, representing the device propagation delay of 4 and the delay of the last input to arrive through devices "B" and "C" of 9 (McWilliams and Widdoes, 1978). When the devices associated with the last stage, i.e., the devices driving the primary outputs, are processed, the accumulated worst-case delays record the longest delay from primary inputs to primary outputs, also call the critical paths. The critical path for each primary output is highlighted in Figure 12.12.

Path-oriented timing analysis generates primary input to primary output propagation delays by traversing all possible signal paths one at a time. Thus, finding the critical path via path-oriented timing analysis is equivalent to finding the longest path through a directed acyclic graph, where devices are graph vertices and interconnections are graph edges (Sasiki et al., 1978).

To account for realistic variances in component timing due to manufacturing tolerances, aging, or environmental effects, timing analysis often provides stochastic or statistical checking capabilities. Statistical timing

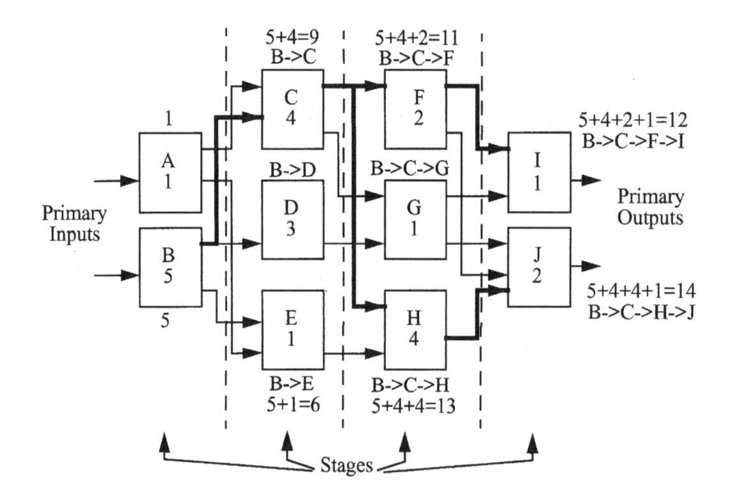

FIGURE 12.12 Block-oriented static timing analysis.

analysis uses random-number generators based on empirically observed probabilistic distributions to determine component timing behavior. Thus, statistical timing analysis describes design performance and the likelihood of the design performance.

Simulation

Simulation exercises a design over a period of time by applying a series of input stimuli and generating the associated output responses. The general event-driven, also called schedule-driven, simulation algorithm is diagrammed in Figure 12.13. An event is a change in signal value. Simulation starts by initializing the design; initial values are assigned to all signals. Initial values include starting values and pending values that constitute future events. Simulation time is advanced to the next pending event(s), signals are updated, and sensitized models are evaluated (Pooch, 1993). The process of evaluating the sensitized models yields new, potentially different, values for signals, i.e., a new set of pending events. These new events are added to the list of pending events, time is advanced to the next pending event(s), and the simulation algorithm repeats. Each pass through the loop in Figure 12.13 of evaluating sensitized models at a particular time step is called a **simulation cycle**. Simulation ends when the design yields no further activity, i.e., when there are no more pending events to process.

Logic simulation is computationally intensive for large, complex designs. As an example, consider simulating 1 sec of a 200K-gate, 20 MHz processor design. By assuming that, on average, only 10% of the total 200K gates are active or sensitized on each processor clock cycle, Equation (12.1) shows that simulating 1 sec of actual processor time equates to 400 billion events:

$$\begin{aligned} 400\text{ billion events} &= (20\text{ million clock cycles})(200\text{K gates})(10\%\text{ activity}) \\ 140\text{ h} &= (400\text{ billion events})\left(\frac{50\text{ instructions}}{\text{event}}\right)\left(\frac{50\text{ million instructions}}{\text{sec}}\right) \end{aligned} \tag{12.1}$$

Assuming that, on average, a simulation program executes 50 computer instructions per event on a computer capable of processing 50 million instructions per second (MIP), Equation (12.1) also shows that processing 400 billion events requires 140 h or just short of 6 days. Figure 12.14 shows how simulation computation generally scales with design complexity.

To address the growing computational demands of simulation, several simulation acceleration techniques have been introduced. Schedule-driven simulation, explained above, can be accelerated by removing layers of interpretation and running a simulation as a native executable image; such an approach is called complied, scheduled-driven simulation.

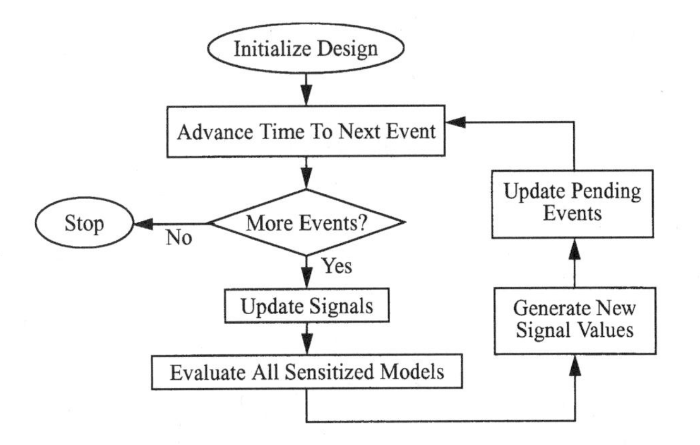

FIGURE 12.13 General event-driven simulation algorithm.

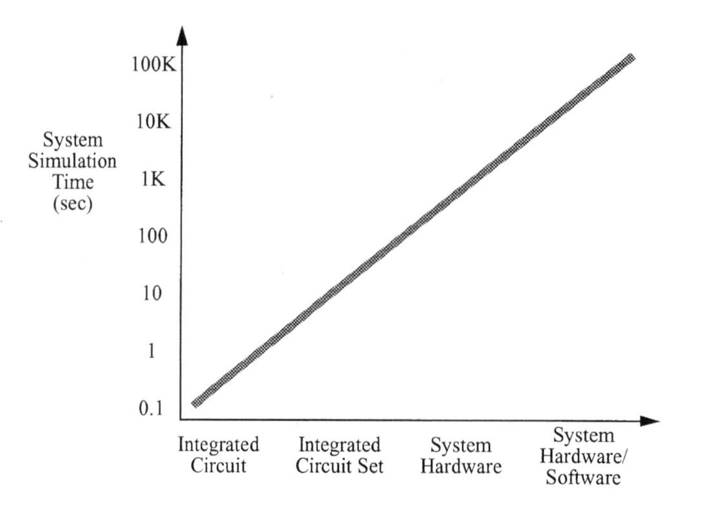

FIGURE 12.14 Simulation requirements.

As an alternative to schedule-driven simulation, *cycle-driven* simulation avoids the overhead of event queue processing by evaluating all devices at regular intervals of time. Cycle-driven simulation is efficient when a design exhibits a high degree of concurrency, i.e., when a large percentage of the devices are active per simulation cycle. Based on the staging of devices, devices are *rank-ordered* to determine the order in which they are evaluated at each time step to ensure the correct causal behavior yielding the proper ordering of events. For functional verification, logic devices are often assigned zero-delay and memory devices are assigned unit-delay. Thus, any number of stages of logic devices may execute between system clock periods.

In another simulation acceleration technique, *message-driven* simulation, also called *parallel* or *distributed* simulation, device execution is divided among several processors and the device simulations communicate event activity via messages. Messages are communicated using conservative or optimistic strategies. Optimistic message-passing strategies, such as *time warp* and *lazy cancellation,* make assumptions about future event activity to advance local device simulation. If the assumptions are correct, the processors operate more independently and better exploit parallel computation. However, if the assumptions are incorrect, then local device simulations may be forced to "roll back" to synchronize local device simulations (Bryant, 1979; Chandy and Misra, 1981).

Schedule-driven, cycle-driven, and message-driven simulations are software-based simulation acceleration techniques. Simulation can also be accelerated by relegating certain simulation activities to dedicated hardware. For example, *hardware modelers* can be attached to software simulators to accelerate the activity of device evaluation. As the name implies, hardware modeling uses actual hardware devices instead of software models to obtain stimulus/response information. Using actual hardware devices reduces the expense of generating and maintaining software models and provides an environment to support application software development. However, it is sometimes difficult for a slave hardware modeler to preserve accurate real-time device operating response characteristics within a master nonreal-time software simulation environment. For example, some hardware devices may not be able to retain state information between invocations, so the hardware modeler must save the history of previous inputs and reapply them to bring the hardware device to the correct state to apply a new input.

Another technique for addressing the growing computational demands of simulation is via simulation engines. A simulation engine can be viewed as an extension of the simulation acceleration technique of hardware modeling. With a hardware modeler, the simulation algorithm executes in software and component evaluation executes in dedicated hardware. With a simulation engine, the simulation algorithm *and* component evaluation execute in dedicated hardware. Simulation engines are typically two to three orders of magnitude faster than software simulation (Takasaki et al., 1989).

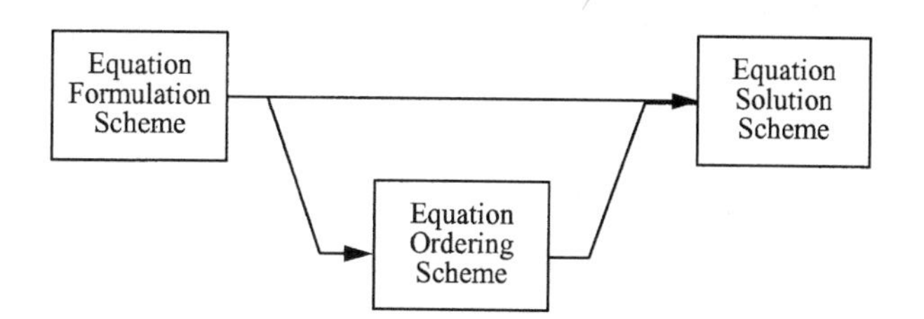

FIGURE 12.15 dc simulation.

TABLE 12.1 Common Circuit Equation Formulation Schemes

Equation Formulation Schemes	Desired Unknowns
Nodal analysis	Node voltages
Modified nodal analysis	Node voltages
	Dependent source currents
	Independent voltage source currents
Sparse tableau analysis	Node voltages
	Branch currents
	Branch voltages
Reduced tableau analysis	Node voltages
	Branch currents
Tree analysis	Tree branch voltages
Link analysis	Link branch currents

Analog Simulation

Analog simulation involves time-domain analyses and frequency-domain analyses, which are generally conducted using some form of direct current (dc) simulation, diagrammed in Figure 12.15. dc simulation determines the quiescent or steady-state operating point for a circuit, specifying **node voltages**, **branch** currents, input/output resistances, element sensitivities, and input/output gains (Chua and Lin, 1975; Nagel, 1975).

Several popular equation formulation schemes are summarized in Table 12.1. Equation formulation schemes generate a set of linear equations denoting relationships between circuit voltages and currents; these relationships are based on the physical principle of the conservation of energy expressed via **Kirchoff's current law** (KCL), **Kirchoff's voltage law** (KVL), and branch constitutive relationships (BCRs). A circuit having N nodes and B branches possesses $2B$ independent variables defining B branch voltages and B branch currents. These variables are governed by $2B$ linearly independent equations composed of N–1 KCL equations, B–N+1 KVL equations, and B BCR equations (Hachtel et al., 1971; Ho et al., 1975).

Equation-ordering schemes augment equation formulation schemes by reorganizing, modifying, and scaling the equations to improve the efficiency and/or accuracy of the subsequent equation solution scheme. More specifically, equation-ordering schemes seek to improve the "diagonal dominance" structure of the coefficient matrix by maximizing the number of "off-diagonal" zeros. Popular equation-ordering schemes include pivoting and row ordering (Markowitz) (Zlatev, 1980).

Finally, equation solution schemes determine the values for the independent variables that comply with the governing equations. There are basically two types of equation solution schemes: explicit and implicit. Explicit solution schemes, such as Gaussian elimination and/or LU factorization, determine independent variable values using closed-form, deterministic techniques. Implicit solution schemes, such as Gauss–Jacobi and Gauss–Seidel, determine independent variable values using iterative, nondeterministic techniques.

Emulation

Emulation, also called *computer-aided prototyping*, verifies a design by realizing the design in "preproduction" hardware and exercising the hardware. The term *preproduction* hardware means nonoptimized hardware providing the correct functional behavior, but not necessarily the correct performance. That is, emulation hardware may be slower, require more area, or dissipate more power than production hardware. At present, preproduction hardware commonly involves some form of **programmable logic devices** (PLDs), typically field-programmable **gate arrays** (FPGAs). PLDs provide generic combinational and sequential digital system logic that can be programmed to realize a wide variety of designs (Walters, 1991).

Emulation offers the advantage of providing prototype hardware early in the design cycle to check for errors or inconsistencies in initial functional specifications. Problems can be isolated and design modifications can be easily accommodated by reprogramming the logic devices. Emulation can support functional verification at computational rates much greater than conventional simulation. However, emulation does not generally support performance verification because, as explained above, prototype hardware typically does not operate at production clock rates.

12.5 Physical Design

Figure 12.16 shows that the physical design task generally follows the verification task. Having validated the function and performance of the detailed design during verification, physical design realizes the detailed design by translating logic into actual hardware. Physical design involves placement, routing, artwork generation, rules checking, and back annotation (Sait and Youseff, 1995).

Placement transforms a logical hierarchy into a physical hierarchy by defining how hardware elements are oriented and arranged relative to each other. Placement determines the overall size, i.e., area, a digital system will occupy. Two popular placement algorithms are *mincut* and *simulated annealing*. Mincut placement techniques group highly connected cells into clusters. Then, the clusters are sorted and arranged according to user-supplied priorities. Simulated annealing conducts a series of trial-and-error experiments by pseudo-randomly moving cells and evaluating the resulting placements, again according to user-supplied priorities.

Routing defines the wires that establish the required port-to-port connections. Routing is often performed in two stages: global and local. Global routing assigns networks to major wiring regions, called tracks; local routing defines the actual wiring for each network within its assigned track. Two common classes of routing algorithms are *channel* and *maze*. Channel routing connects ports abutting the same track. Maze routing, also called switch-box routing, connects ports abutting different channels. Routing considers a variety of metrics, including timing skew, wire length, number of vias, and number of jogs (corners) (Spinks, 1985; Preas et al., 1988).

Rules checking verifies that the final layout of geometric shapes and their orientation complies with logical, electrical, and physical constraints. Logical rules verify that the implementation realizes the desired digital system. Electrical rules verify conformance to loading, noise margins, and fan-in/fan-out connectivity

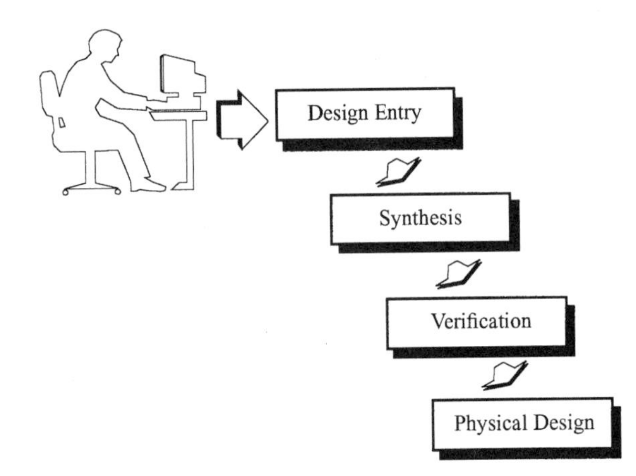

FIGURE 12.16 Design process—physical design.

constraints. Finally, physical rules verify conformance to dimensional, spacing, and alignment constraints (Hollis, 1987).

12.6 Test

Figure 12.17 shows that **test** follows physical design. After physical design, the digital system is manufactured and test checks the resulting hardware for correct function and performance. Thus, the primary objective of test is to detect a faulty device by applying input test stimuli and observing expected results (Buckroyd, 1989; Weyerer and Goldemund, 1992).

The test task is difficult because designs are growing in complexity; more components provide more opportunity for manufacturing defects. Test is also challenged by new microelectronic fabrication processes. These new processes support higher levels of integration that provide fewer access points to probe internal electrical nodes and new failure modes that provide more opportunity for manufacturing defects.

Fault Modeling

What is a fault? A fault is a manufacturing or aging defect that causes a device to operate incorrectly or to fail. A sample listing of common integrated circuit physical faults are given below:

- Wiring faults
- Dielectric faults
- Threshold faults
- Soft faults

Wiring faults are unwanted opens and shorts. Two wires or networks that should be electrically connected, but are not connected constitute an open. Two wires or networks that should not be electrically connected, but are connected constitute a short. Wiring faults can be caused by manufacturing defects, such as metallization and etching problems, or aging defects, such as corrosion and **electromigration**. Dielectric faults are electrical

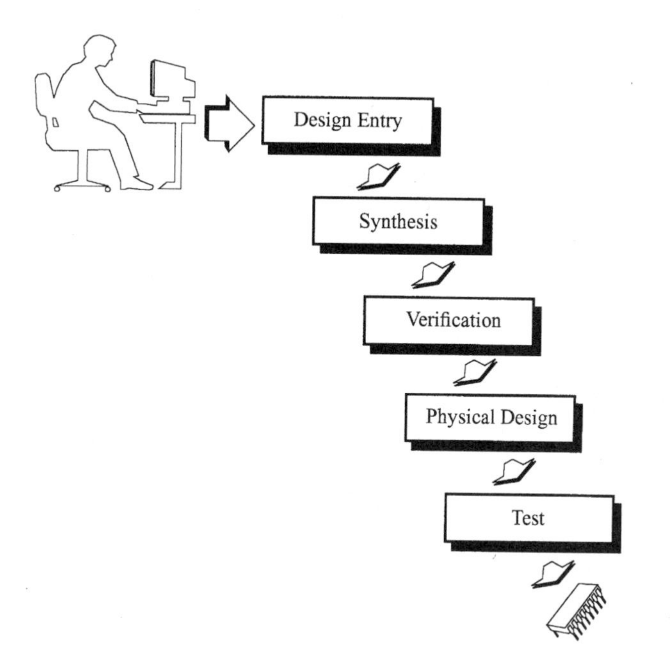

FIGURE 12.17 Design process—test.

isolation defects that can be caused by **masking defects**, chemical impurities, material imperfections, or electrostatic discharge. Threshold faults occur when the turn-on and turn-off voltage potentials of electrical devices exceed allowed ranges. Soft faults occur when radiation exposure temporarily changes electrical charge distributions. Such changes can alter circuit voltage potentials, which can, in turn, change logical values, also called *dropping bits*. Radiation effects are called "soft" faults because the hardware is not permanently damaged (Zobrist, 1993).

To simplify the task of fault testing, the physical faults described above are translated into logical faults. Typically, a single logical fault covers several physical faults. A popular logical fault model is the *single stuck line* (SSL) fault model. The single stuck line fault model supports faults that denote wires permanently set to a logic 0, "stuck-at-0," or a logic 1, "stuck-at-1." Building on the single stuck line fault model, the *multiple stuck line* (MSL) fault model supports faults where multiple wires are stuck-at-0/stuck-at-1. Stuck fault models do not address all physical faults because not all physical faults result in signal lines permanently set to low or high voltages, i.e., stuck-at-0 or stuck-at-1 logic faults. Thus, other fault models have been developed to address specific failure mechanisms. For example, the *bridging* fault model addresses electrical shorts that cause unwanted coupling or spurious feedback loops.

Fault Testing

Once the physical faults that may cause device malfunction have been identified and categorized and how the physical faults relate to logical faults has been determined, the next task is to develop tests to detect these faults. When the tests are generated by a computer program, this activity is called *automatic test program generation* (ATPG). Examples of fault testing techniques are listed below:

- Stuck-at techniques
- Scan techniques
- Signature techniques
- Coding techniques
- Electrical monitoring techniques

Basic stuck-at fault testing techniques address combinational digital systems. Three of the most popular stuck-at fault testing techniques are the D algorithm, the path-oriented decision making (Podem) algorithm, and the Fan algorithm. These algorithms first identify a circuit fault, e.g., stuck-at-0 or stuck-at-1, and then try to generate an input stimulus that detects the fault and makes the fault visible at an output. Detecting a fault is called *fault sensitization* and making a fault visible is called *fault propagation*. To illustrate this process, consider the simple combinational design in Figure 12.18 (Goel, 1981; Fujiwara and Shimono, 1983).

The combinational digital design is defective because a manufacturing defect has caused the output of the top **and** gate to be permanently tied to ground, i.e., stuck-at-0, using a positive logic convention. To sensitize the fault, the inputs A and B should both be set to 1, which should force the top **and** gate output to a 1 for a good circuit. To propagate the fault, the inputs C and D should both be set to 0, which should force the **xor** gate output to 1, again for a good circuit. Thus, if A = 1, B = 1, C = 0, and D = 0 in Figure 12.18, then a good circuit would yield a 1, but the defective circuit yields a 0, which detects the stuck-at-0 fault at the top **and** gate output.

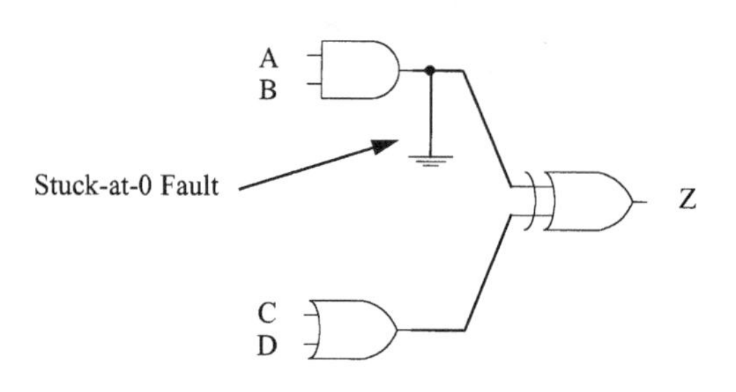

FIGURE 12.18 Combinational logic stuck-at fault testing.

Sequential ATP generation is more difficult than combinational ATPG because exercising or sensitizing a particular circuit path to detect the presence of a possible manufacturing fault may require a *sequence* of input test vectors. One technique for testing sequential digital systems is scan fault testing. Scan fault testing is an example of *design for testability* (DFT) because it modifies or constrains a design in a manner that facilitates fault testing. Scan techniques impose a logic design discipline that connects all state registers into one or more chains to form "scan rings," as shown in Figure 12.19 (Eichelberger and Williams, 1977).

During normal device operation, the scan rings are disabled and the registers serve as conventional memory (state) storage elements. During test operation, the scan rings are enabled and stimulus test vectors are shifted into the memory elements to set the state of the digital system. The digital system is exercised for one clock cycle and then the results are shifted out of the scan ring to record the response.

A variation of scan DFT, called *boundary scan*, has been defined for testing integrated circuits on printed circuit boards (PCBs). Advancements in PCB manufacturing, such as fine-lead components, surface mount assembly, and **multichip modules**, have yielded high-density boards with fewer access points to probe individual pins. These PCBs are difficult to test. As the name implies, boundary scan imposes a design discipline for PCB components to enable the input/output pins of the components to be connected into scan chains. As an example, Figure 12.20 shows a simple PCB containing two integrated circuits configured for boundary scan. Each integrated circuit contains scan registers between its input/output pins and its core logic to enable the PCB test bus to control and observe the behavior of individual integrated circuits (Parker, 1989).

Another DFT technique is signature analysis, also called *built-in self-test* (BIST). Signature testing techniques use additional logic, typically linear feedback shift registers, to generate automatically pseudorandom test vectors. The output responses are compressed into a single vector and compared with a known good vector. If the output response vector does not exactly match the known good vector, then the design is considered faulty. Matching the output response vector and a known good vector does not guarantee correct

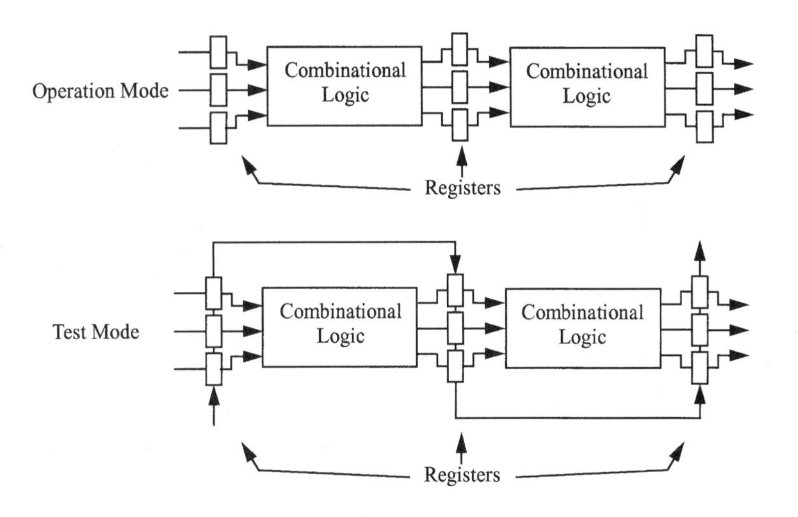

FIGURE 12.19 Scan-based DFT.

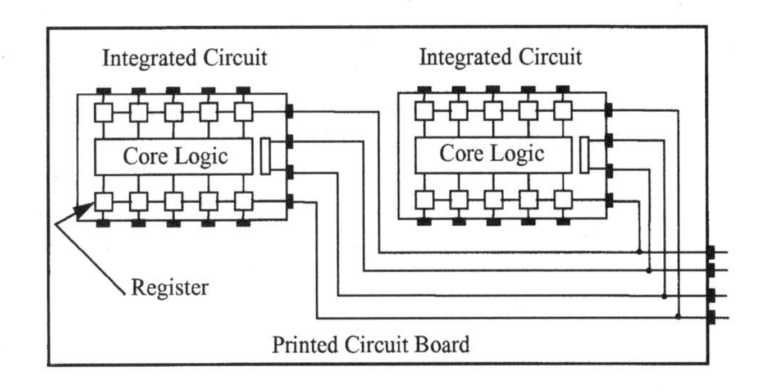

FIGURE 12.20 Boundary scan.

hardware; however, if enough pseudorandom test vectors are exercised, then the chances are acceptably small of obtaining a false positive result. Signature analysis is often used to test memories (Agrawal et al., 1993).

Coding test techniques encode signal information so that errors can be detected and possibly corrected. Although often implemented in software, coding techniques can also be implemented in hardware. For example, a simple coding technique called *parity checking* is often implemented in hardware. Parity checking adds an extra bit to multibit data. The parity bit is set such that the total number of logic 1s in the multibit data *and* parity bit is either an even number (even parity) or an odd number (odd parity). An error has occurred if an even-parity-encoded signal contains an odd number of logic 1s or if an odd-parity-encoded signal contains an even number of logic 1s. Coding techniques are used extensively to detect and correct transmission errors on system buses and networks, storage errors in system memory, and computational errors in processors (Peterson and Weldon, 1972).

Finally, electrical monitoring testing techniques, also called current/voltage testing, rely on the simple observation that an out-of-range current or voltage often indicates a defective or bad part. Possibly a short or open is present causing a particular input/output signal to have the wrong voltage or current. Current testing, or I_{ddq} *testing*, is particularly useful for digital systems using CMOS integrated circuit technology. Normally, CMOS circuits have very low static or quiescent currents. However, physical faults, such as **gate oxide** defects, can increase static current by several orders of magnitude. Such a substantial change in static current is straightforward to detect. The principal advantages of current testing are that the tests are simple and the fault models address detailed transistor-level defects. However, current testing requires that enough time be allotted between input stimuli to allow the circuit to reach a static state, which slows testing down and causes problems with circuits that cannot be tested at scaled clock rates.

12.7 Summary

DA technology offers the potential of serving as a powerful fulcrum in leveraging the skills of a designer against the growing demands of electronic system design and manufacturing. DA programs help to relieve the designer of the burden of tedious, repetitive tasks that can be labor-intensive and error prone.

DA technology can be broken down into several topical areas, such as design entry, synthesis, verification, physical design, and test. Each topical area has developed an extensive body of knowledge and experience.

Design entry defines a desired specification. Synthesis refines the initial design specification into a detailed design ready for implementation. Verification checks that the detailed design faithfully realizes the desired specification. Physical design defines the implementation, i.e., the actual hardware. Finally, test checks that the manufactured part is functionally and parametrically correct.

Defining Terms

Bipolar: Type of semiconductor transistor that involves both minority and majority carrier conduction mechanisms.

BiCMOS: Bipolar/complementary metal-oxide semiconductor. A logic family and form of microelectronic fabrication.

Branch: A circuit element between two nodes. Branch current is the current through the branch. Branch voltage is the potential difference between the nodes. The relationship between the branch current and voltage is defined by the branch constitutive relationship.

Capacitor: Two-terminal electronic device governed by the branch constitutive relationship, charge = capacitance × voltage.

CMOS: Complementary metal-oxide semiconductor. A logic family and form of microelectronic fabrication.

Data Flow: Nonprocedural modeling style in which the textual order that statements are written has no bearing on the order in which they execute.

Design automation: Computer programs that assist engineers in performing digital system development.

Design entry: Area of DA addressing modeling analog and digital electronic systems. Design entry uses a hierarchy of models involving physical, electrical, logical, functional, and architectural abstractions.

Electromigration: Gradual erosion of metal due to excessive currents.

Fan-in/fan-out: Fan-in defines the maximum number of logic elements that may drive another logic element. Fan-out defines the maximum number of logic elements a logic element may drive.

Finite state machine: Sequential digital system. A finite state machine is classified as either Moore and Mealy.

Gate array: Application-specific integrated circuit implementation technique that realizes a digital system by programming the metal interconnect of a prefabricated array of gates.

Gate oxide: Dielectric insulating material between the gate and source/drain terminals of a MOS transistor.

Ground bounce: Transient condition when the potential of a ground network varies appreciably from its uniform static value.

Integrated circuit: Electronic circuit manufactured on a monolithic piece of semiconductor material, typically silicon.

Kirchoff's current law: The amount of current entering a circuit node equals the amount of current leaving a circuit node.

Kirchoff's voltage law: Any closed loop of circuit branch voltages sums to zero.

Masking defects: Defects in masking plate patterns used for integrated circuit lithography that result in errant material composition and/or placement.

Multichip modules: Multiple integrated circuits interconnected on a monolithic substrate.

Netlist: Collection of wires that are electrically connected to each other.

NMOS: N-type metal-oxide semiconductor. A logic family and form of microelectronic fabrication.

Node voltage: Potential of a circuit node relative to ground potential.

Programmable logic devices (PLDs): Generic logic devices that can be programmed to realize specific digital systems. PLDs include programmable logic arrays, programmable array logic, memories, and field-programmable gate arrays.

Resistor: Two-terminal electronic device governed by the branch constitutive relationship, voltage $=$ resistance $\times$ current.

Silicon compilation: Synthesis application that generates final physical design ready for silicon fabrication.

Simulation: Computer program that examines the dynamic semantics of a model of a digital system by applying a series of inputs and generating the corresponding outputs. Major types of simulation include schedule driven, cycle driven, and message driven.

Skew: Timing difference between two events that are supposed to occur simultaneously.

Standard cell: Application-specific integrated circuit implementation technique that realizes a digital system using a library of predefined (standard) logic cells.

Synthesis: Computer program that helps generate a digital/analog system design by transforming a high-level model of abstract behavior into a lower-level model of more-detailed behavior.

Test: Area of EDA that addresses detecting faulty hardware. Test involves stuck-at, scan, signature, coding, and monitoring techniques.

Timing analysis: Verifies timing behavior of electronic system including rise time, fall time, setup time, hold time, glitch detection, clock periods, race conditions, reflections, and crosstalk.

Transistor: Electronic device that enables a small voltage and/or current to control a larger voltage and/or current. For analog systems, transistors serves as amplifiers. For digital systems, transistors serve as switches.

Verification: Area of EDA that addresses validating designs for correct function and expected performance. Verification involves timing analysis, simulation, emulation, and formal proofs.

VHDL: Hardware description language used as an international standard for communicating electronic systems information.

Via: Connection or contact between two materials that are otherwise electrically isolated.

References

Design Automation

D. Barbe, Ed., *Very Large Scale Integration (VLSI)—Fundamentals and Applications*, New York: Springer-Verlag, 1980.

T. Dillinger, *VLSI Engineering*, Englewood Cliffs, NJ: Prentice-Hall, 1988.

S. Sapiro, *Handbook of Design Automation*, Englewood Cliffs, NJ: Prentice-Hall, 1986.

S. Trimberger, *An Introduction to CAD for VLSI*, CA: Domancloud Publishers, 1990.

Design Entry

G. Birtwistle and P. Subrahmanyan, *VLSI Specification, Verification, and Synthesis*, Boston, Mass.: Kluwer Academic Publishers, 1988.

A. Dewey, "VHSIC hardware description language development program," *Proceedings Design Automation Conference*, June, 1983.

A. Dewey, "VHDL: towards a unified view of design," *IEEE Design and Test of Computers*, June, 1992.

A. Dewey, *Analysis and Design of Digital Systems with VHDL*, Boston, Mass.: PWS Publishing, 1997.

J. Douglas-Young, *Complete Guide to Reading Schematic Diagrams*, Englewood Cliffs, N.J.: Prentice-Hall, 1988.

M. Pechet, Ed., *Handbook of Electrical Package Design*, New York: Marcel Dekker, 1981.

J. Peterson, *Petri Net Theory and Modeling of Systems*, Englewood Cliffs, N.J.: Prentice-Hall, 1981.

Synthesis

R. Ayres, *VLSI: Silicon Compilation and the Art of Automatic Microchip Design*, Englewood Cliffs, N.J.: Prentice-Hall, 1983.

Brayton et al., *Logic Minimization Algorithms for VLSI Synthesis*, Boston, Mass.: Kluwer Academic Publishers, 1992.

R. Camposano and W. Wolfe, *High-Level VLSI Synthesis*, Boston, Mass.: Kluwer Academic Publishers, 1991.

D. Gajski, Ed., *Silicon Compilation*, Boston, Mass.: Addison-Wesley, 1988.

D. Gajski, et al., *High-Level Synthesis — Introduction to Chip and System Design*, Boston, Mass.: Kluwer Academic Publishers, 1992.

P. Paulin and J. Knight, "Force-directed scheduling for the behavioral synthesis of ASIC's," *IEEE Design and Test of Computers*, October, 1989.

T. Sasao, Ed., *Logic Synthesis and Optimization*, Boston, Mass.: Kluwer Academic Publishers, 1993.

Verification

R. Bryant, "Simulation on distributed systems," *Proceedings International Conference on Distributed Systems*, 1979.

K. Chandy and J. Misra, "Asynchronous distributed simulation via a sequence of parallel computations," *Communications of the ACM*, April, 1981.

L. Chua and P. Lin, *Computer-Aided Analysis of Electronic Circuits: Algorithms and Computational Techniques*, Englewood Cliffs, N.J.: Prentice-Hall, 1975.

G. Hachtel, R. Brayton, and F. Gustavson, "The sparse tableau approach to network analysis and design," *IEEE Transactions on Circuit Theory*, CT-18, 1971.

W. Hahn and K. Fischer, "High performance computing for digital design simulation", *VLSI85*, New York: Elsevier Science Publishers, 1985.

C. Ho, A. Ruehli, and P. Brennan, "The modified nodal analysis approach to network analysis," *IEEE Transactions on Circuits and Systems*, 1975.

R. McHaney, *Computer Simulation: A Practical Perspective*, New York: Academic Press, 1991.

T. McWilliams and L. Widdoes, "SCALD — structured computer aided logic design," in *Proceedings Design Automation Conference*, June, 1978.

L. Nagel, SPICE2: A Computer Program to Simulate Semiconductor Circuits, Electronic Research Laboratory, ERL-M520, Berkeley, Calif.: University of California, 1975.

U. Pooch, *Discrete Event Simulation: A Practical Approach*, Boca Raton, Fla.: CRC Press, 1993.

T. Sasiki, et al., "Hierarchical design and verification for large digital systems," in *Proceedings Design Automation Conference*, June, 1978.

S. Takasaki, F. Hirose, and A. Yamada, "Logic simulation engines in Japan," *IEEE Design and Test of Computers*, October, 1989.

S. Walters, "Computer-aided prototyping for ASIC-based synthesis," *IEEE Design and Test of Computers*, June, 1991.

Z. Zlatev, "On some pivotal strategies in Gaussian elimination by sparse technique," *SIAM Journal of Numerical Analysis*, vol. 17, no. 1, 1980.

Physical Design

E. Hollis, *Design of VLSI Gate Array Integrated Circuits*, Englewood Cliffs, N.J.: Prentice-Hall, 1987.

B. Preas, B. Lorenzetti, M., and B. Ackland, Eds., *Physical Design Automation of VLSI Systems*, New York: Benjamin Cummings, 1988.

S. Sait and H. Youssef, *VLSI Physical Design Automation: Theory and Practice*, New York: McGraw-Hill, 1995.

B. Spinks, *Introduction to Integrated Circuit Layout*, Englewood Cliffs, N.J.: Prentice-Hall, 1985.

Test

V. Agrawal, C. Kime, and K. Saluja, "A tutorial on built-in self-test," *IEEE Design and Test of Computers*, June, 1993.

A. Buckroyd, *Computer Integrated Testing*, New York: Wiley, 1989.

E. Eichelberger and T. Williams, "A logic design structure for LSI testability," *Proceedings Design Automation Conference*, June, 1977.

H. Fujiwar and T. Shimono, "On the acceleration of test generation algorithms," *IEEE Transactions on Computers*, December, 1983.

P. Goel, "An implicit enumeration algorithm to generate tests for combinational logic circuits," *IEEE Transactions on Computers*, March, 1981.

K. Parker, "The impact of boundary scan on board test," *IEEE Design and Test of Computers*, August, 1989.

W. Peterson and E. Weldon, *Error-Correcting Codes*, Boston, Mass.: The MIT Press, 1972.

M. Weyerer and G. Goldemund, *Testability of Electronic Circuits*, Englewood Cliffs, N.J.: Prentice-Hall, 1992.

G. Zobrist, Ed., *VLSI Fault Modeling and Testing Technologies*, New York: Ablex Publishing Company, 1993.

13

Electronic Data Analysis Using PSPICE and MATLAB

John Okyere Attia
Prairie View A&M University

SPICE is an of the industry standard software for circuit simulation [1]. It can perform dc, ac, transient, Fourier, and Monte Carlo analyses. There are several SPICE-derived simulations packages, including Orcad PSPICE, Meta-software HSPICE, and Intusoft IS-SPICE.

PSPICE contains more features than classical SPICE. Some of the most useful additional features are a post-processor program, PROBE, that can be used for interactive graphical display of simulation results; and an analog behavioral model facility that allows modeling of analog circuit functions by using mathematical equations, tables, and transfer functions [2,3].

MATLAB is primarily a tool for matrix computations [4,5]. MATLAB has a rich set of plotting capabilities. The graphics are integrated in MATLAB. Since MATLAB is also a programming environment, a user can extend the functional capabilities by writing new modules (*m-files*). This chapter shows how the strong features of PSPICE and the powerful functions of MATLAB can be used to perform extensive and complex data analysis of electronic circuits [3,6].

13.1 PSPICE Fundamentals

A general SPICE program consists of the following components: (a) title; (b) element statements; (c) control statements, and (d) end statement. The following two sections will discuss the element and control statements.

TABLE 13.1 Element Name and Corresponding Element

First Letter of Element Name	Circuit Element, Sources, and Subcircuit
B	GaAs MES field-effect transistor
C	Capacitor
D	Diode
E	Voltage-controlled voltage source
F	Current-controlled current source
G	Voltage-controlled current source
H	Current-controlled voltage source
I	Independent current source
J	Junction field-effect transistor
K	Mutual inductors (transformers)
L	Inductor
M	MOS field-effect transistor
Q	Bipolar junction transfer
R	Resistor
S	Voltage-controlled switch
T	Transmission line
V	Independent voltage source
X	Subcircuit

Element Statements

Element statements specify the elements in the circuit. An element statement contains the (a) element name; (b) the circuit nodes to which each element is connected; and (c) the values of the parameters that electrically characterize the element. The element name must begin with a letter of the alphabet that is unique to a circuit element, source, or subcircuit. Table 13.1 shows the beginning alphabet of an element name and the corresponding element.

Circuit nodes are positive integers. The nodes in the circuit need not be numbered sequentially. Node 0 is predefined for ground of a circuit. Element values can be integer, floating point number, and floating point followed by an exponent.

Resistors

The general format for describing resistors is

```
Rname N + N − value*
```

where the name must start with the letter *R*. *N*+ and *N*− are the positive and negative nodes of the resistor, respectively.

Value specifies the value of the resistor. The latter may be positive or negative, but not zero.

Inductors

The general format for describing linear inductors is

```
Lname N + N− value [IC = initial_current]
```

where the inductor name must start with the letter *L*, *N* + and *N*− are positive and negative nodes of the inductor, respectively, *Value* specifies the value of the inductance, and the initial condition for transient analysis is assigned using IC = *initial_current*, to specify the initial current.

Capacitors

The general format for describing linear capacitors is

```
Cname N + N− value [IC = initial_voltage]
```

where the capacitor name must start with the letter *C*, *N+* and *N−* are the positive and negative nodes of the capacitor, respectively, *Value* indicates the value of the capacitance, and the initial condition for transient analysis is assigned using IC = *initial_voltage* on the capacitor.

Independent Voltage Source

The general format for describing independent voltage source is

```
Vname N+ N− [DC value] [AC magnitude phase]
[PULSE V1 V2 td tr tf pw per]
or
[SIN V0 Va freq td df phase]
or
[EXP V1 V2 td1 t1 td2 t2]
or
[PWL t1 V1 t2 V2... tn, Vn]
or
[SFFM V0 Va freq md fs]
```

where the voltage source must start with letter *V*, *N+* and *N−* are the positive and negative nodes of the source, respectively, and sources can be assigned values for dc analysis [*DC value*], ac analysis [*AC magnitude phase*], and transient analysis. Only one of the transient response source options (*PULSE, SIN, EXP, PWL, SFFM*) can be selected for each source. The AC phase angle is in degrees. Discussion on the transient signal generators (*PULSE, SIN, EXP, PWL, SFFM*) can be found in Al-Hashimi [2], Attia [3], or Roberts and Sedra [7].

Independent Current Source

The general format for describing independent current source is

```
Iname N+ N− [DC value] [AC magnitude phase]
[PULSE V1 V2 td tr tf pw per]
or
[SIN V0 Va freq td df phase]
or
[EXP V1 V2 td1 t1 td2 t2]
or
[PWL t1 V1 t2 V2... tn, Vn]
or
[SFFM V0 Va freq md fs]
```

where the current source must start with letter *I*, *N+* and *N−* are the positive and negative nodes of the source, respectively. Current flows from a positive node to the negative node, and independent current sources can be assigned values for dc analysis [*DC value*], ac analysis [*AC magnitude phase*], and transient analysis. Only one of the transient response source options (*PULSE, SIN, EXP, PWL, SFFM*) can be selected for each source. The ac phase angle is in degrees.

13.2 Control Statements

Circuit Title

The circuit title must be the first statement in the SPICE program or circuit netlist. If this is not done, the program will assume that the first statement is the circuit title.

dc Analysis (.DC)

The .DC control statement specifies the values that will be used for dc sweep or dc analysis. The general format for the .DC statement is

```
.DC SOURCE_NAME START-VALUE STOP_VALUE INCREMENT_VALUE
```

where SOURCE_NAME is the name of an independent voltage or current source and START_VALUE, STOP_VALUE and INCREMENT_VALUE represent the starting, ending, and increment values of the source, respectively.

Transient Analysis (.TRAN)

The .TRAN control statement is used to perform transient analysis on a circuit. The general format of the .TRAN statement is

```
.TRAN TSTEP TSTOP <TSTART> <TMAX> <UIC>
```

where the terms inside the angle brackets are optional, *TSTEP* is the printing or plotting increment, *TSTOP* is the final time of the transient analysis, *TSTART* is the starting time for printing out the results of the analysis. If it is omitted, it is assumed to be zero. The transient analyses always start at time zero, *TMAX* is the maximum step size that PSPICE uses for the purposes of computation. If TMAX is omitted, the default is the smallest value of either TSTEP or (TSTOP − TSTART)/50, and *UIC* (use initial conditions) is used to specify the initial conditions of capacitors and inductors.

There are five SPICE-supplied sources that can be used for transient analysis: (1) PULSE (for periodic pulse waveform); (2) EXP (for exponential waveform); (3) PWL (for piece-wise linear waveform); (4) SIN (for a sine wave); and (5) SFFM (for frequency-modulated waveform). The format for specifying the above sources for transient analysis can be obtained from Al-Hashimi [2], Attia [3], or Roberts and Sedra [7].

AC Analysis (.AC)

The .AC control statement is used to perform ac analysis on a circuit. The general format of the .AC statement is

```
.AC FREQ_VAR NP FSTART FSTOP
```

where *FREQ_VAR* is one of three keywords that indicates the frequency variation by decade (DEC), octave (OCT), or linearly LIN), NP is the number of points; its interpretation depends on the keyword (DEC, OCT, or LIN) in the FREQ. DEC - NP is the number of points per decade. OCT - NP is the number of points per octave. LIN - NP is the total number of points spaced evenly from frequency FSTART and ending at FSTOP, *FSTART* is the starting frequency; FSTART cannot be zero, and *FSTOP* is the final or ending frequency.

Printing Command (.PRINT)

The .PRINT control statement is used to print tabular outputs. The general format of the .PRINT statement is

```
.PRINT ANALYSIS_TYPE OUTPUT_VARIABLE
```

TABLE 13.2 Name Types for AC Output Variable

Output Variable	Meaning
V or I	Magnitude of V or I
VR or IR	Real part of complex value V or I
VI or II	Imaginary part of complex number V or I
VM or IM	Magnitude of complex number V or I
VDB or IDB	Decibel value of magnitude, i.e., 20 log 10$\|V\|$ or 20 log10$\|I\|$

where *ANALYSIS_TYPE* can be *DC, AC, TRAN.* Only one analysis type must be specified for .PRINT statement and *OUTPUT-VARIABLE* can be voltages or currents. Up to eight output variables can accompany one .PRINT statement. If more than eight output variables are to be printed, additional .PRINT statements can be used.

The output variable may be node voltages and current through voltage sources. PSPICE allows one to obtain current flowing through passive elements. The voltage output variable has the general form:

```
V(node 1, node 2) or V(node 1) if node 2 is node "0."
```

The current output variable has the general form

```
I (Vname)
```

where *Vname* is an independent-voltage source specified in the circuit netlist.

For PSPICE, the current output variable can also be specified as *I(Rname)*, where Rname is resistance defined in the input circuit. For AC analysis, output voltage and current variables may be specified as real or imaginary magnitude and phase. Table 13.2 shows the name types for AC output variables.

Initial Conditions (.IC, UIC)

The .IC statement is only used when the transient analysis statement, .TRAN, includes the '*UIC*' option. The initial voltage across a capacitor or the initial current flowing through an inductor can be specified as part of capacitor or inductor component statement. For example, for a capacitor we have:

```
Cname N+ N− value IC = initial voltage
```

and for an inductor, we use the statement:

```
Lname N+ N− value IC = initial current
```

It should be noted that the initial conditions on an inductor or capacitor are used provided that .TRAN statement includes the 'UIC' option.

Several SPICE control commands are shown in Table 13.3. Details of using the control statements can be found in Vladimirescu [1], Al-Hashimi [2], and Attia [3].

PSPICE Probe Statement (.PROBE)

Probe is a PSPICE interactive graphics processor that allows the user to display SPICE simulation results in graphical format on a computer monitor. Probe has facilities that allow the user to access any point on a displayed graph and obtain its numerical values. In addition, Probe has many built-in functions that enable a user to compute and display mathematical expression that models aspects of circuit behavior. The general format for specifying the probe statement is

```
.PROBE OUTPUT_VARIABLES
```

where *OUTPUT_VARIABLES* can be node voltages and/or devices currents. If no OUTPUT_VARIABLE is specified, Probe will save all node voltages and device currents.

TABLE 13.3 Other PSPICE Control Statements

Control Statement	Description
*	* in the first column indicates a comment line.
.FOUR	Allows the user to perform Fourier analysis. Fourier components from dc to the *n*th harmonic are calculated.
.MC	Used to vary device parameter and to observe the overall system for variation in circuit parameters.
.NODESET	Used to set the operating point at specified nodes of a circuit during the initial run of a transient analysis.
.OP	Used to obtain the nodal voltages and the current flowing through independent voltage sources.
.SENS	Performs dc sensitivity of circuit element values and variation of model parameters.
.TEMP	Used to change the temperature at which a simulation is performed.
.TF	Used to obtain the small-signal gain, dc input resistance, and dc output resistance.
.WCASE	Causes sensitivity and worst-case analysis to be performed.

13.3 MATLAB Fundamentals

The Colon Symbol

The colon symbol (:) is one of the most important operators in MATLAB. It can be used (a) to create vectors and matrices; (b) to specify sub-matrices and vectors; and (c) to perform iterations.

Creation of Vectors and Matrices

The statement

```
j1 = 1:9
```

will generate a row vector containing the numbers from 1 to 9 with unit increment. Nonunity, positive, or negative increments may be specified.

Specifying Submatrices and Vectors

Individual elements in a matrix can be referenced with subscripts inside parentheses. For example, j2(4) is the fourth element of vector j2. Also, for matrix j3, j3(2, 3) denotes the entry in the second row and third column. Using the colon as one of the subscripts denotes all of the corresponding row or column. For example, j3(:,4) is the fourth column of matrix j3. Also, the statement j3(2,:) is the second row of matrix j3. If the colon exists as the only subscript, such as j3(:), the latter denotes the elements of matrix j3 strung out in a long column vector.

Iterative Uses of Colon Command

The iterative uses of the colon command are discussed in the next section.

FOR Loops

FOR loops allow a statement or group of statements to be repeated a fixed number of times. The general form of a FOR loop is

```
for index = expression
statement group C
end
```

The expression is a matrix and the statement group *C* is repeated as many times as the number of elements in the columns of the expression matrix. The index takes on the elemental values in the matrix expression. Usually, the expression is something like:

```
m:n or m:i:n
```

where *m* is the beginning value, *n* the ending value, and *i* the increment.

TABLE 13.4 Relational Operators

Relational Operator	Meaning
<	Less than
<=	Less than or equal
>	Greater than
>=	Greater than or equal
==	Equal
~=	Not equal

IF Statements

The general form of the *simple IF statement* is

```
if logical expression 1
statement group G1
end
```

In the case of a simple IF statement, if the logical expression *1* is true, the statement group *G1* is executed. However, if the logical expression is false, the statement group G1 is bypassed and the program control jumps to the statement that follows the end statement. Several variations of the IF statement are described in Etter [5] and Attia [6].

IF statements use relational or logical operations to determine what steps to perform in the solution of a problem. The relational operators in MATLAB for comparing two matrices of equal size are shown in Table 13.4.

When any one of the above relational operators is used, a comparison is done between the pairs of corresponding elements.

Graph Functions

MATLAB has built-in functions that allow you to generate x–y and 3-D plots. MATLAB also allows you to give titles to graphs, label the x- and y-axes, and add grids to graphs.

x–y Plots and Annotations

If x and y are vectors of the same length, then the command

```
plot(x, y)
```

plots the element of x (x-axis) versus the elements of y (y-axis).

To plot multiple curves on a single graph, one can use the *plot* command with multiple arguments, such as

```
plot(x1, y1, x2, y2, x3, y3, ..., xn, yn)
```

The variables $x1$, $y1$, $x2$, $y2$, and so on are pairs of vector. Each x–y pair is graphed, generating multiple lines on the plot. The above plot command allows vectors of different lengths to be displayed on the same graph. MATLAB automatically scales the plots.

When a graph is drawn, one can add a grid, title, label, and x- and y-axes to the graph. The commands for grid, title, x-axis label, and y-axis label are *grid* (grid lines), *title* (graph title), *xlabel* (x-axis label), and *ylabel* (y-axis label), respectively.

Logarithmic and Plot3 Functions

Logarithmic and semilogarithmic plots can be generated using the commands *loglog*, *semilogx*, and *semilogy*. Descriptions of these commands follow:

loglog(x,y) — generates a plot of $\log_{10}(x)$ versus $\log_{10}(y)$
semilogx(x, y) — generates a plot of $\log_{10}(x)$ versus linear axis of y
semilogy(x, y) — generates a plot of linear axis of x versus $\log_{10}(y)$

The *plot3* function can be used to do three-dimensional line plots. The function is similar to the two-dimensional plot function. The plot3 function supports the same line size, line style, and color options that are supported by the plot function. The simplest form of the plot3 function is

```
plot(x, y, z)
```

where *x*, *y*, and *z* are equal-sized arrays containing the locations of the data points to be plotted.

Subplot Function

The graph window can be partitioned into multiple windows. The *subplot* command allows the user to split the graph into two subdivisions or four subdivisions. The general form of the subplot command is

```
subplot(i j k)
```

The digits *i* and *j* specify that the graph window is to be split into an *i*-by-*j* grid of smaller windows, arranged in *i* rows and *j* columns. The digit *k* specifies the *k*th window for the current plot. The subwindows are numbered from left to right, top to bottom.

fprintf

The *fprintf* command can be used to print both text and matrix values. The format for printing the matrix can be specified, and line feed can also be specified. The general form of this command is

```
fprintf('text with format specification', matrices)
```

The format specifier, such as *%7.3e*, is used to show where the matrix value should be printed in the text. 7.3e indicates that the value should be printed with an exponential notation of seven digits, three of which should be decimal digits.

Other format specifiers are %c (single character); %d (signed decimal notation); %e (exponential notation); %f (fixed-point notation); and %g (signed decimal number in either %e or %f format, whichever is shorter). The text with format specification should end with *n* to indicate the end of line.

m-files

MATLAB is capable of processing a sequence of commands that are stored in files with extension *m*. MATLAB files with extension *m* are called *m-files*. The latter are ASCII text files and are created with a text editor or word processor. *m*-files can either be scripts or functions. Script and function files contain a sequence of commands. However, function files take arguments and return values.

Script Files

Script files are especially useful for analysis and design problems that require long sequences of MATLAB commands. With a script file written using a text editor or word processor, the file can be invoked by entering the name of the *m*-file without the extension. Statements in a script file operate globally on the workspace data.

Function Files

Function files are *m*-files that are used to create new MATLAB functions. Variables defined and manipulated inside a function file are local to the function, and they do not operate globally on the workspace. However, arguments may be passed into and out of a function file.

The general form of a function file is

```
function variable(s) = function_name (arguments)
% help text in the usage of the function
.
.
end
```

TABLE 13.5 Data Analysis Functions

Function	Description
corrcoef(x)	Determines correlation coefficients.
diff(x)	Computes the differences between elements of an array x. It approximates derivatives.
max(x)	Obtains the largest value of x. If x is a matrix, max(x) returns a row vector containing the maximum elements of each column.
[y, k]=max(x)	Obtains the maximum value of x and corresponding locations (indices) of the first maximum value for each column of x.
mean(x)	Determines the mean or the average value of the elements in the vector. If x is a matrix, mean(x) returns a row vector that contains the mean value of each column.
median(x)	Finds the median value of elements in vector x. If x is a matrix, this function returns a row vector containing the median value of each column.
min(x)	Finds the smallest value of x. If x is a matrix, min(x) returns a row vector containing the minimum values from each column.
[y, k]=min(x)	Obtains the smallest value of x and the corresponding locations(indices) the first minimum value from each column of x.
sort(x)	Sort the rows of matrix x in ascending order.
std(x)	Calculates and returns the standard deviation of x if it is a one-dimensional array. If x is a matrix, a row vector containing the standard deviation of each column is computed and returned.
sum(x)	Calculates and returns the sum of the elements in x. If x is a matrix, this function calculates and returns a row vector that contains the sum of each column.
trapz(x,y)	Trapezoidal integration of the function $y = f(x)$.

Data Analysis Functions

In MATLAB, data analyses are performed on column-oriented matrices. Variables are stored in the individual column cells, and each row represents different observation of each variable. Functions act on the elements in the column. Table 13.5 gives a brief description of various MATLAB functions for performing data analysis.

Curve Fitting (Polyfit, Polyval)

The MATLAB *polyfit* function is used to compute the best fit of a set of data points to a polynomial with a specified degree. The general form of the function is

```
poly_xy = polyfit(x, y, n)
```

where x and y are the data points, n is the nth degree polynomial that will fit the vectors x and y, and *poly_xy* is a polynomial that fits the data in vector y to x in the least-squares sense. poly_*xy* returns $(n+1)$ coefficients in descending powers of x.

Thus, the polynomial fit to vectors x and y is given as

$$poly\_xy(x) = a_1x^n + a_2x^{n-1} + \ldots a_m$$

The degree of the polynomial is n and the number of coefficients $m = n + 1$. The coefficients $(a_1, a_2, \ldots, a_m)$ are returned by the MATLAB *polyfit* function.

Save, Load, and Textread Functions

Save and Load Commands

The *save* command saves data in the MATLAB workspace to disk. The save command can store data either in memory-efficient binary format, called a *MAT-file* or *ASCII file*. The general form of the save command is

```
save filename [List of variables] [options]
```

TABLE 13.6 Save Command Options

Option	Description
-mat	Save data in MAT file format (default)
-ascii	Save data using 8-digit ASCII format
-ascii -double	Save data using 16-digit ASCII format
-ascii -double -tab	Saves data using 16-digit ASCII format with tabs

TABLE 13.7 Load Command Option

Option	Description
-mat	Load data from MAT file (default in file extension is mat)
-ascii	Load data from space-separated file

where *save* (without filename, list of variables, and options) saves all the data in the current workspace to a file named *matlab.mat* in the current directory. If a filename is included in the command line, the data will be saved in file "*filename.mat*." If a list of variables in included, only those variables will be saved.

Options for the save command are shown in Table 13.6.

MAT-files are preferable for data that are generated to be used by MATLAB. MAT-files are platform independent. The files can be written and read by any computer that supports MATLAB. The *ASCII files* are preferable if the data are to be exported or imported to programs other than MATLAB.

The *load* command will load data from MAT-file or ASCII file into the current workspace. The general format of the *load* command is

```
load filename [options]
```

where *load* (by itself without filename and options) will load all the data in file *matlab.mat* into the current workspace and *load filename* will load data from the specified filename.

Options for the load command are shown in Table 13.7.

The Textread Function

The *textread* command can be used to read ASCII files that are formatted into columns of data, where values in each column might be a different type. The general form of the textread command is

```
[a, b, c,...] = textread(filename, format, n)
```

where *filename* is the name of file to open. The filename should be in quotes, as in 'filename', *format* is a string containing a description of the type of data in each column. The format list should be in quotes. Supported functions include %d (read a signed integer value); %u (read an integer value); %f (read a floating point value); and %s (read a whitespace separated string), n is the number of lines to read. If n is missing, the command reads to the end of the file, and $a, b, c\ldots$ are the output arguments. The number of output arguments must match the number of columns that are being read.

13.4 Interfacing SPICE to MATLAB

To exploit the best features of PSPICE and MATLAB, circuit simulation is performed by using PSPICE. The PSPICE results, which are written into a file, *filename.out*, are edited using a text editor or a word processor and the data will be saved as *filename.dat*. The data are read using either MATLAB *textread* or *load* commands. Further processing of the data is performed using MATLAB. The methodology is shown in Figure 13.1.

In the following examples, the methodology described in this section will be used to analyze data obtained from electronic circuits.

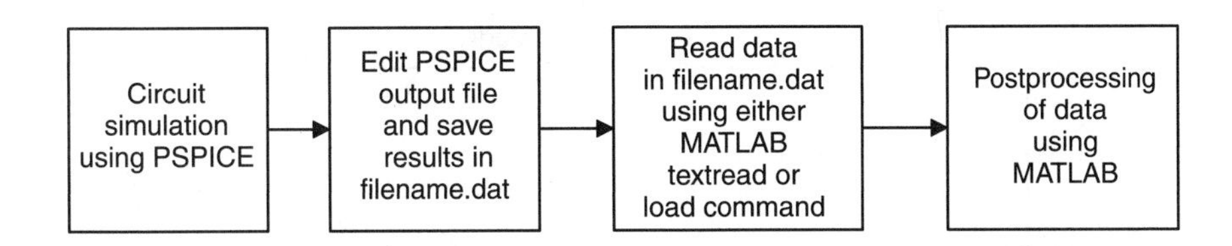

FIGURE 13.1 Flowchart of circuit simulation using PSPICE and postprocessing of PSPICE.

Best-Fit Linear Model of a Diode

A forward-biased diode has the corresponding voltage and current shown in Table 13.8. (a) Draw the equation of best fit for the diode data. (b) For the voltage of 0.64 V, what is the diode current?

The MATLAB script for analyzing the diode data follows:

```
% Example of Best-fit linear model
vt = 25.67e-3;
vd =[0.1 0.2 0.3 0.4 0.5 0.6 0.7];
id= [1.33e-13 1.79e-12 24.02e-12...
0.321e-9 4.31e-9 57.69e-9 7.72e-7];%
lnid= log(id);% Natural log of current
% Determine coefficients
pfit = polyfit (vd, lnid, 1); % Curve fitting
% Linear equation is y = mx + b
b = pfit(2);
m = pfit(1);
ifit = m*vd + b;
% Calculate current when diode voltage is 0.64 V
Ix = m*0.64 + b;
I_64v = exp(Ix);
% Plot v versus ln(i) and best-fit linear model
plot(vd, ifit, 'b', vd, lnid, 'ob')
xlabel ('Voltage, V')
ylabel ('ln(i),A')
title ('Best-Fit Linear Model')
fprintf('Diode current for voltage of 0.64V is %9.3e\n', I_64v)
```

The plot is shown in Figure 13.2. Based on MATLAB, the diode current for voltage of 0.64 V is 1.629e-007 A.

TABLE 13.8 Voltage versus Current of a Diode

Forward-Biased Voltage, V	Forward Current, A
0.1	1.33e-13
0.2	1.79e-12
0.3	24.02e-12
0.4	0.321e-9
0.5	4.31e-9
0.6	57.69e-9
0.7	7.72e-7

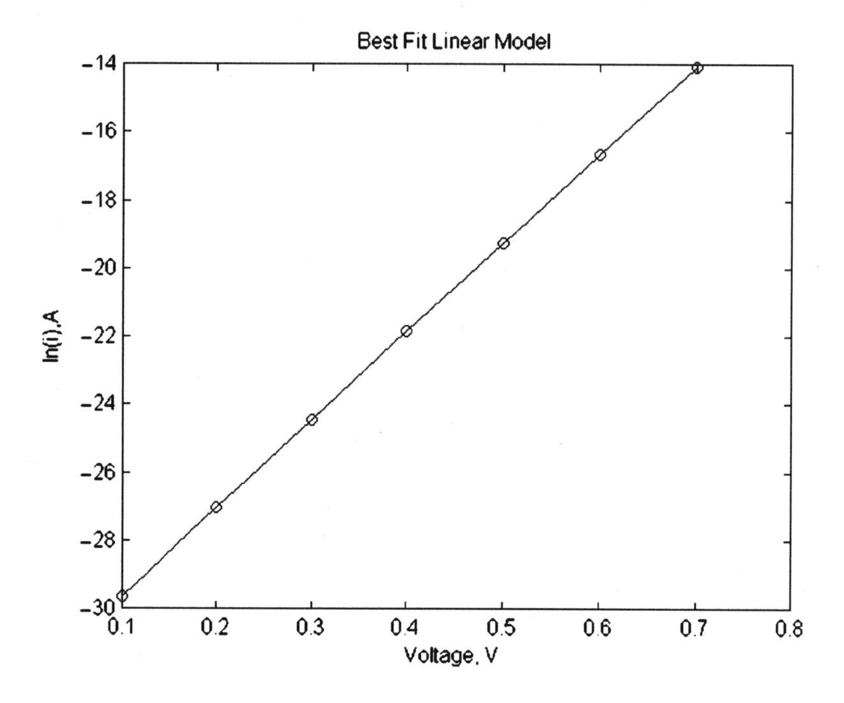

FIGURE 13.2 Best fit linear model of a diode.

Op Amp Circuit with Series-Shunt Feedback Network

Figure 13.3 shows an op amp circuit with series-shunt feedback network. RS = 1 KΩ, RL = 10 KΩ, and R1 = 5 KΩ. Find the gain, V_0/V_S if RF varies from 10 KΩ to 100 KΩ. Plot voltage gain with respect to RF. Assume that the op amp is UA741 and the input voltage VS is sinusoidal waveform with a frequency of 5 KHz and a peak voltage of 1 mV.

PSPICE program for determining the gain follows:

```
* Example - OP AMP CIRCUIT WITH SERIES-SHUNT FEEDBACK
VS    1    0    AC    1E-3 0
RS    1    2    1K
R1    3    0    5K
RL    4    0    10K
VCC   5    0    15V
VEE   6    0    −15V
RF    4    3    RMOD 1
.MODEL  RMOD  RES(R=1)
.STEP RES  RMOD(R)  10.0E3  100.0E3  10.0E3
X1    2    3    5    6    4  UA741;UA741  OP  AMP
* +INPUT; −INPUT; +VCC; −VEE; OUTPUT; CONNECTIONS FOR UA741
** ANALYSIS TO BE DONE
.LIB NOM.LIB
* UA741 OP AMP MODEL IN PSPICE LIBRARY FILE NOM.LIB
** OUTPUT
.AC    LIN    1    5K    5K
.PRINT ACV(4)
.END
```

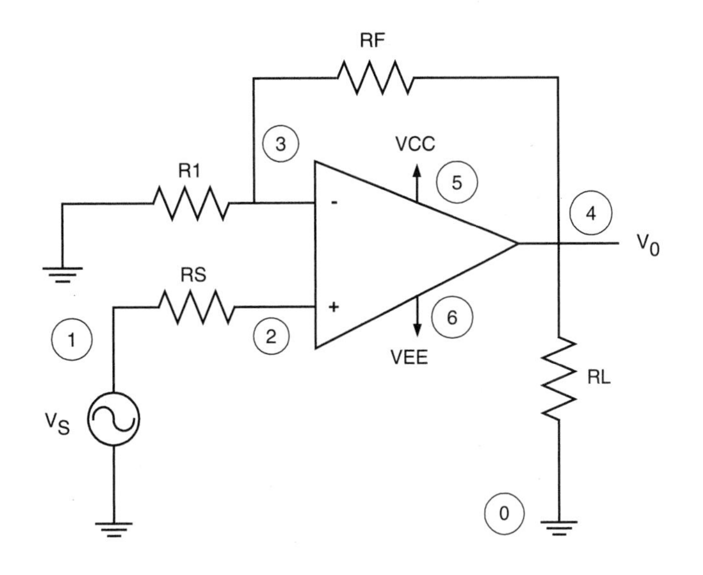

FIGURE 13.3 Op amp circuit with series-shunt feedback network.

PSPICE results are shown in Table 13.9. The data were stored in file *sol_52 ps.dat*. The data from the latter file are used by MATLAB. The MATLAB script for solving the problem follows:

```
% Example
% Load data
load 'sol_52 ps.dat' -ascii;
rf = sol_52ps(:,1);
gain = 1000*sol_52ps(:,2);
% Plot data
plot(rf, gain, rf, gain,'ob')
xlabel('Feedback Resistance, Ohms')
ylabel('Voltage Gain')
title('Voltage Gain vs. Feedback Resistance')
```

The gain versus feedback resistance is shown in Figure 13.4.

TABLE 13.9 Output Voltage versus Feedback Resistance

Feedback Resistance, RF	Output Voltage
10.0E03	3.000E-03
20.0E03	4.998E-03
30.0E03	6.996E-03
40.0E03	8.991E-03
50.0E03	1.098E-02
60.0E03	1.297E-02
70.0E03	1.496E-02
80.0E03	1.694E-02
90.0E03	1.892E-02
100.0E03	2.089E-02

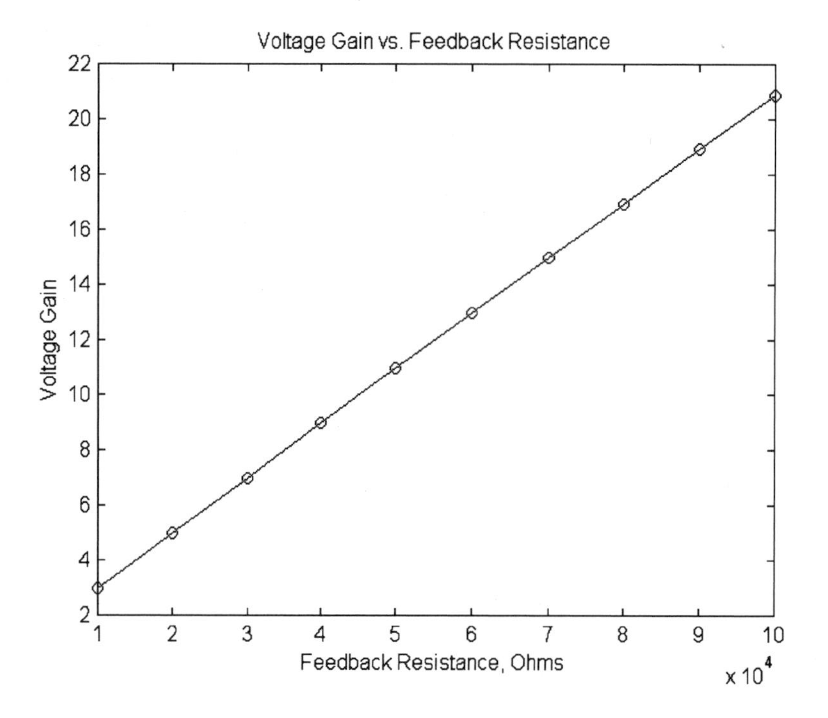

FIGURE 13.4 Voltage gain versus feedback resistance of an op amp circuit.

13.5 Conclusions

SPICE is an industry-standard software for circuit simulation. MATLAB is primarily a tool for matrix computations. It has numerous functions for data processing and analysis. PSPICE can be used to perform dc, ac, transient, Fourier, temperature and Monte Carlo analysis of electronic circuits with device models and subsystem subcircuits. MATLAB can then be used to perform calculation of device parameters, curve fitting, statistical analysis, and two- and three-dimensional plots. The strong features of PSPICE and powerful functions of MATLAB, when used in conjunction, allow extensive and complex data analysis of electronic circuits.

References

1. A. Vladimirescu, *The SPICE Book*, New York: John Wiley & Sons, 1994.
2. B. Al-Hashimi, *The Art of Simulation Using PSPICE, Analog, and Digital*, Boca Raton, FL: CRC Press, 1994.
3. J.O. Attia, *PSPICE and MATLAB for Electronics: An Integrated Approach*, Boca Raton, FL: CRC Press, 2002.
4. S.J. Chapman, *MATLAB Programming for Engineers*, 2nd ed., Pacific Grove, CA: Brook, Cole Thompson, Learning, 2002.
5. D.M. Etter, *Engineering Problem Solving with MATLAB*, 2nd ed., Upper Saddle River, NJ: Prentice-Hall, 1997.
6. J.O. Attia, *Electronics and Circuit Analysis Using MATLAB*, Boca Raton, FL: CRC Press, 1999.
7. G.W. Roberts and A.S. Sedra, *SPICE for Microelectronic Circuits*, Fort Worth, TX: Saunders College Publishing, 1992.

II

Electromagnetics

14

Electromagnetic Fields

Banmali S. Rawat
University of Nevada

Moncef B. Tayahi
University of Nevada

The subject of electromagnetic fields is the study of fields created due to moving and/or stationary electric charges and their applications. The study of electromagnetic fields is important for understanding the basic principles and design of electrical analog and digital devices and circuits. These digital and analog systems are expanding the use of GHz range spectrums in communication technology in order to increase bandwidth and speed. Electromagnetic fields can be divided into three groups (i) steady or static electric fields (ii) static magnetic fields, and (iii) time varying fields.

14.1 Static Electric Fields

The steady electric fields are produced by a moving or stationary electric charge. The most common mathematical tool used for analyzing electric and magnetic fields is vector algebra. The application of vector algebra makes the analysis very simple and less time consuming. Most of the electric and magnetic fields laws first investigated by scientists were laws derived from experiment and were known by their discoverers. These laws included Coulomb's law, Faraday's law, Ohm's law, and Ampere's law. Finally, James Clerk Maxwell gave a mathematical formulation to these laws in the form of Maxwell's four equations. By solving these four Maxwell's equations of electromagnetic fields, the electric and magnetic fields for any configuration can be obtained with appropriate boundary conditions. The focus of this chapter is a brief explanation and analysis of these basic laws, as well as associated laws which led to Maxwell's four equations. In addition, some important applications are discussed.

Experimental Law of Coulomb and Electric Field Intensity

A French army officer, Charles Coulomb, experimented with electric charges and concluded that a force exists between two electric charges. This force is proportional to the charges and inversely proportional to the square of the distance between them, provided the distance is very large compared to the size of the charges. His analysis showed similarity with Newton's gravitational law. The mathematical form of the law is given as [1]:

$$F = k\frac{Q_1 Q_2}{R^2} \tag{14.1}$$

where Q_1 and Q_2 represent positive or negative charges (C), R is the distance (m) between the charges, and k is the constant of proportionality. In SI units $k = 1/4\pi\varepsilon_0$ with $\varepsilon_0 = 1 \times 10^{-9}/36\pi = 8.854 \times 10^{-12}$ F/m. Coulomb's law in free space or vacuum is written as

$$F = \frac{Q_1 Q_2}{4\pi\varepsilon_0 R^2} N \tag{14.2}$$

In vector notations Equation (14.2) is written as

$$\mathbf{F} = \frac{Q_1 Q_2}{4\pi\varepsilon_0 R^2}\mathbf{a}_\mathrm{R}\, N \tag{14.3}$$

where $\mathbf{a}_\mathrm{R}$ is the unit vector in the direction of R given by

$$\mathbf{a}_\mathrm{R} = \mathbf{R}/|\mathbf{R}| = \mathbf{R}/R \tag{14.4}$$

This force is known as electric field and now the electric field intensity can be explained as the force on a unit positive charge as

$$\mathbf{E} = \mathbf{F}/Q_\mathrm{t}\,\mathrm{V/m} \tag{14.5}$$

where Q_t is the test charge. The units of electric field intensity are N/C or equivalently, V/m. Now the electric field intensity due to a charge Q on a test charge Q_t at a distance R can be written as

$$\mathbf{E} = \mathbf{F}/Q_\mathrm{t} = \frac{QQ_t}{4\pi\varepsilon_0 R^2 Q_t}\mathbf{a}_\mathrm{R} = \frac{Q}{4\pi\varepsilon_0 R^2}\mathbf{a}_\mathrm{R} \tag{14.6}$$

This electric field intensity vector is in the direction of test charge from charge Q and is directed radially outward along unit vector $\mathbf{a}_\mathrm{R}$.

Concept of Differential Charge Elements and Electric Field Intensity Due to Various Charge Distributions

In the natural world all the charge distributions are not as point forms but may be as line, surface or volume charge distributions. In order to determine the electric field intensity due to these charge distributions using Coulomb's law, as discussed in the last section, the concept of differential charge element has to be introduced. Any charge distribution can be assumed to be made of an infinite number of differential charge elements separated by an infinitely small distance. The electric field intensity at any point due to this particular charge distribution is the summation of field intensities due to all these differential charge elements. In other words, we can state that Coulomb's law is linear. If the distance between two adjacent charge elements approaches zero, and the number of charges is assumed to be infinity, the summation can be replaced by integration. This differential charge elements concept is the basis for finding electric field intensity due to different charge distributions. This concept is shown in Figure 14.1.

The electric field intensity at point P is the summation of the field intensities due to differential charge elements 1, 2, ... n as given by

$$\mathbf{E} = \mathrm{d}\mathbf{E}_1 + \mathrm{d}\mathbf{E}_2 + \cdots + \mathrm{d}\mathbf{E}_n \tag{14.7}$$

where $\mathrm{d}\mathbf{E}_n = \frac{\mathrm{d}Q_n}{4\pi\epsilon_0 R_n^2}\mathbf{a}_{Rn,\ n=1,2,\ldots n}$.

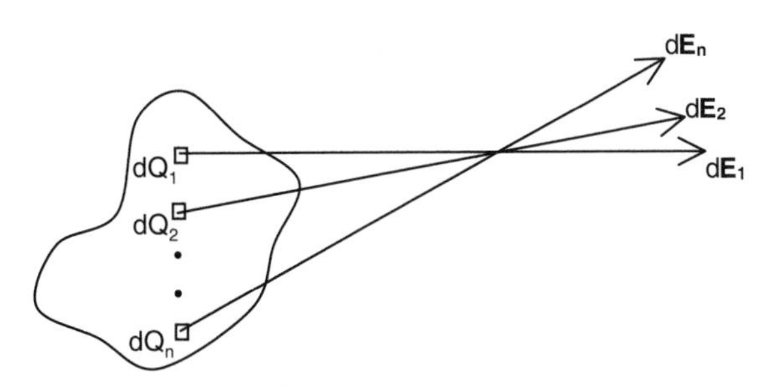

FIGURE 14.1 Electric field intensity due to differential charge elements.

The concept of differential charges can be easily applied to line, surface and volume charge distributions. They have been expressed as follows:

Infinite line charge: $\mathbf{E} = \dfrac{\rho_l}{2\pi\varepsilon_0 r}\mathbf{a}_r \mathrm{V/m}$, radially outward along r; ρ_l is the line charge density per unit length, r is the distance of the point from the line and ε_0 is the permittivity of free space.

Infinite surface charge: $\mathbf{E} = \dfrac{\rho_s}{2\varepsilon_0}\mathbf{a}_n \mathrm{V/m}$, normal to the surface; ρ_s is the surface charge density per unit surface area.

Volume charge: $\mathbf{E} = \int\limits_{vol} \dfrac{\rho_v dv}{4\pi\varepsilon_0 R^2}\mathbf{a}_R$ V/m; ρ_v is the volume charge density and R is distance of the point where electric field intensity is to be determined.

All these electric field intensities are determined using a particular coordinate system. The rectangular coordinate system is simplest and also easy to implement. However, where there is cylindrical and spherical symmetry, it is easier to use cylindrical or spherical coordinate systems, respectively. In cylindrical symmetry there is symmetry about a line which can be considered as one of the axes of the rectangular coordinate system. In the case of spherical coordinate system there is symmetry about a point which can be considered as the origin of the rectangular coordinate system. In practice, a coaxial cable can be analyzed using the concept of line charge while a capacitor or strip line can be analyzed on the basis of surface or sheet charge. The formula developed for an infinite surface charge can be used for a parallel capacitor under the condition that the area of plates is very large compared to the separation between the plates so that the uniform charge distribution is valid.

Concept of Electric Flux Density and Gauss's Law

In 1837, Michael Faraday conducted an experiment with two concentric charged spheres where he showed an induced charge on the outer sphere due to a charged inner sphere. He concluded that this was possible only if some sort of "displacement" took place from inner to outer spherical conductor. This displacement is known as "displacement", "displacement flux", or simply "the electric flux." The induced charge on the outer conductor is opposite and is directly proportional to the charge on the inner conductor. Also, the induced charge is independent of the medium between two conductors. In SI units the relationship between the charge on inner conductor and electric flux is given by

$$\Psi = Q \tag{14.8}$$

where Ψ is the electric flux in coulomb and Q is the charge on inner conductor. We can now define the electric flux density vector in radial direction on the surface of the spherical conductor of radius "a" as

$$\mathbf{D} = \frac{\Psi}{4\pi a^2}\mathbf{a}_r = \mathbf{a}_r\ \mathrm{C/m^2} \tag{14.9}$$

For concentric spheres, as shown in Figure 14.2, where a and b are inner and outer radii, respectively, the flux density is given by

$$\mathbf{D}|_{r=a} = \frac{Q}{4\pi a^2}\mathbf{a}_r \ \mathrm{C/m^2} \quad \text{and} \qquad (14.10)$$
$$\mathbf{D}|_{r=b} = \frac{Q}{4\pi b^2}\mathbf{a}_r \ \mathrm{C/m^2}$$

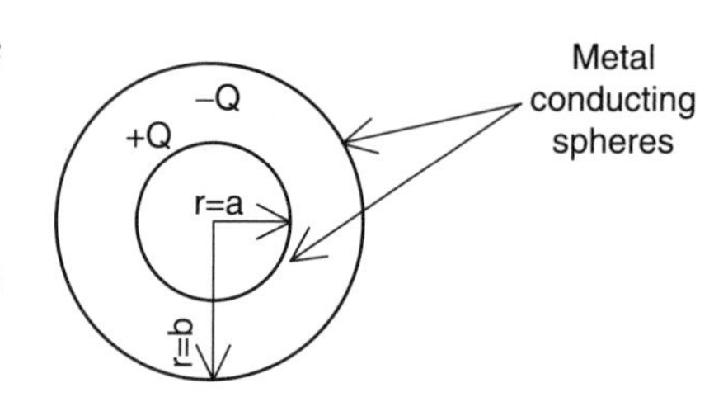

FIGURE 14.2 Electric flux between two conducting spheres.

The fundamental difference between surface charge density and flux density is that surface charge density can only be defined on a conducting surface while flux density can be defined on any surface, even on an imaginary surface. In the case of a point charge, the electric flux density is given by

$$\mathbf{D} = \frac{Q}{4\pi r^2}\mathbf{a}_r \ \mathrm{C/m^2} \qquad (14.11)$$

From Equation (14.6) and Equation (14.11) the relationship between **D** and **E** is obtained as

$$\mathbf{D} = \varepsilon_0 \mathbf{E} \quad \text{(in free space only)} \qquad (14.12)$$

Again, in practice, the medium surrounding a charge distribution is generally a dielectric medium. Thus, it is important to understand the effect on the electric field due to these dielectric materials. The most commonly used dielectric materials in electrical circuits are: quartz, Teflon, rubber, glass, duroid, and alumina. The positive and negative charges separated by atomic distances in these dielectrics are known as bound charges, as they are not free to move. These bound charges form the dipoles in the dielectric materials [2]. In the presence of an external electric field these dipoles rotate so as to align with the external field. The force of rotation is dependent on the charge value and distance between the charges. The quantity dipole moment **p** is defined as

$$\mathbf{p} = Q\mathbf{d} \ \mathrm{Cm} \qquad (14.13)$$

where Q is the absolute value of the charge and **d** is the distance vector from negative to positive charge. In the case of a dielectric volume containing a large number of dipoles, the polarization vector **P** is the dipole moments per unit volume given as

$$\mathbf{P} = \lim_{\Delta v \to 0} \Sigma_i \mathbf{p}_i / \Delta v \qquad (14.14)$$

If there is no electric field applied, the dipoles are oriented in a random fashion canceling their dipole moments and the material is said to be nonpolarized. The polarization in the direction of an applied electric field is the direction of polarization vector **P**. Thus, for the dielectrics the flux density is related to **E** and **P** as

$$\mathbf{D} = \varepsilon_0 \mathbf{E} + \mathbf{P} \ \mathrm{C/m^2} \qquad (14.15)$$

The polarization vector **P** is further related to electric field through electric susceptibility of the material, χ_e as

$$\mathbf{P} = \varepsilon_0 \chi_e \mathbf{E} \qquad (14.16)$$

Now from Equation (14.15) and Equation (14.16) we obtain:

$$\mathbf{D} = \varepsilon_0 (1 + \chi_e)\mathbf{E} \qquad (14.17)$$

The term $(1+\chi_e) = \varepsilon_r$ is known as the relative permittivity or dielectric constant of the material which relates to **D** and **E** as

$$\mathbf{D} = \varepsilon_0\, \varepsilon_r\, \mathbf{E} = \varepsilon\, \mathbf{E} \tag{14.18}$$

where $\varepsilon = \varepsilon_0\, \varepsilon_r$ is the permittivity of the material.

From Equation (14.18) it is evident that **D** and **E** can be conveniently interchanged. Also for isotropic dielectrics **D** and **E** have linear relationship, i.e., each component of **D** is a function of only that component of **E**. However, in the case of anisotropic dielectrics, each component of **D** is a function of every component of **E**. The relationship $\mathbf{D} = \varepsilon\, \mathbf{E}$ still holds, but now ε is a tensor given in rectangular coordinates as

$$\acute{\varepsilon} = \begin{bmatrix} \varepsilon_{xx} & \varepsilon_{xy} & \varepsilon_{xz} \\ \varepsilon_{yx} & \varepsilon_{yy} & \varepsilon_{yz} \\ \varepsilon_{zx} & \varepsilon_{zy} & \varepsilon_{zz} \end{bmatrix} \tag{14.19}$$

Anisotropic materials are particularly useful in microstrip directional couplers where they provide very high directivity and coupling coefficients [3].

Gauss's Law: Based on Faraday's experiment, Gauss, one of the greatest mathematicians, developed a general form of the law which can be stated as "if a charge is enclosed within a surface, the electric flux passing through the closed surface is equal to the total charge enclosed by that surface." Faraday's experiment and laws applied only to concentric spherical surfaces but Gauss's law applied to any surface. Gauss's main contribution was not merely the wider applicability of his law but its mathematical formulation:

$$\Psi = \oint_s \mathbf{D} \cdot d\mathbf{S} = \text{charge enclosed} = Q \tag{14.20}$$

where **D** is the flux density on the surface "s."

The charge enclosed can be in the form of a point charge, line charge, surface charge or volume charge. For volume charge distribution, Gauss's law can be written as

$$\Psi = \oint_s \mathbf{D} \cdot d\mathbf{S} = Q = \int_{vol} \rho_v dv \tag{14.21}$$

where ψ is the flux crossing through the surface "s," also known as the "Gaussian surface," and ρ_v is the volume charge distribution enclosed by this surface.

In order to get a solution using Gauss's law it is important that the surface is symmetrical, where the flux density **D** is constant on the surface. Because of this there is no need at all to do integration and the total charge enclosed is simply the multiplication of the constant flux density on the surface and the total surface area. Thus, the solution of symmetrical charge distribution problems becomes simpler using Gauss's law compared to Coulomb's law. If we apply Gauss's law to a differential volume element with the volume approaching to zero, the flux density **D** can be almost constant on an approximate Gaussian surface, but not on an asymmetrical surface. This particular approximation leads to the application of a divergence theorem to electrical charge distribution problems resulting from Maxwell's equation:

$$\text{Div}\,\mathbf{D} = \nabla \cdot \mathbf{D} = \rho_v = \frac{\partial D_x}{\partial x} + \frac{\partial D_y}{\partial y} + \frac{\partial D_z}{\partial z} \quad \text{(in rectangular coordinates)} \tag{14.22}$$

Now from Equation (14.21) and Equation (14.22) we obtain a divergence theorem for electrical charge distribution as

$$\oint_s \mathbf{D} \cdot d\mathbf{S} = \int_{vol} \rho_v dv = \int_{vol} \nabla \cdot \mathbf{D}\, dv \tag{14.23}$$

Using the divergence theorem, the surface integral is converted to a volume integral to solve electrical charge distribution problems. It is important to note that while Coulomb's law is applicable to any charge distribution, Gauss's law provides simpler solutions to symmetrical charge distribution problems only.

Energy and Potential

If we want to move a charge placed in an electric field, work is required as the charge is moved against the field. This work is also known as the energy expended. However, if the charge is moved in the direction of the field, no work is required. As charge Q is moved through distance $\mathbf{dL}$ in an electric field $\mathbf{E}$, the force on Q due to electric field is

$$\mathbf{F} = Q\mathbf{E} \tag{14.24}$$

The component of this force in the direction $\mathbf{dL}$ against which the external force has to be applied is given as

$$F_{\mathrm{L}} = \mathbf{F} \cdot \mathbf{a}_{\mathrm{L}} = Q\mathbf{E} \cdot \mathbf{a}_{\mathrm{L}} \tag{14.25}$$

where $\mathbf{a}_{\mathrm{L}}$ is a unit vector in the direction of $\mathbf{dL}$. The external force to move the charge must be equal and opposite to the force due to the field, i.e., the applied force is

$$F_{\mathrm{applied}} = -F_{\mathrm{L}} = -Q\mathbf{E} \cdot \mathbf{a}_{\mathrm{L}} \tag{14.26}$$

Now the energy expended or the work done in moving charge Q a distance dL through the field $\mathbf{E}$ is given by

$$\mathrm{dW} = \mathrm{F}_{\mathrm{applied}}\, dL = -Q\mathbf{E} \cdot \mathbf{a}_{\mathrm{L}}\, dL = -Q\mathbf{E} \cdot \mathbf{dL} \tag{14.27}$$

From Equation (14.27), it is evident that if $\mathbf{E}$ and $\mathbf{dL}$ are at 90°, i.e., if the charge is moved in a normal direction with respect to the direction of electric field, the work done is zero. Now the work done or energy expended to move the charge through a finite distance is obtained by integrating Equation (14.27) as

$$W = -Q \int_{\mathrm{initial}}^{\mathrm{final}} \mathbf{E} \cdot \mathbf{dL} \text{ joule} \tag{14.28}$$

It is interesting to note that the work done is independent of the path along which the charge is moved from initial to final position. This concept of energy has been very well utilized in the description of potential energy associated with electrons in the orbits of an atom. Here it is assumed that each electron has been moved from infinity to the orbits of the atom and is held in this orbit. The kinetic energy in moving these electrons is transformed to potential energy. Also, since the potential at infinity is assumed to be zero, the energy associated with each electron is increasingly negative at lower energy levels or at inner orbits of the atom.

Potential Difference and Potential

Now we are in a position to define the potential difference V in terms of the work done by the external source in moving a positive unit charge from initial to final position in an electric field:

$$V = -\int_{\mathrm{initial}}^{\mathrm{final}} \mathbf{E} \cdot \mathbf{dL} \text{ joule}/C \text{ or volts } (V) \tag{14.29}$$

In terms of initial and final positions as "b" and "a," Equation (14.28) can be written as

$$V = -\int_{b}^{a} \mathbf{E} \cdot \mathbf{dL}\ V \tag{14.30}$$

For a point charge Q placed at the origin of the coordinate system, the potential difference between points A and B at radial distances r_A and r_B is obtained as

$$V_{AB} = -\int_B^A \mathbf{E}.d\mathbf{L} = -\int_{r_B}^{r_A} \frac{Q}{4\pi\varepsilon_0 r^2} dr = \frac{Q}{4\pi\varepsilon_0}\left(\frac{1}{r_A} - \frac{1}{r_B}\right) V \qquad (14.31)$$

In practice we prefer to define potential in absolute terms as simply "potential." It means the initial point potential is considered a reference potential which may be zero or some constant value. The most commonly used references with zero potential are: ground, infinity, and the outer conductor of a coaxial line. The ground or surface of the earth is considered as an infinite conductor with zero potential. The reference, infinity, is useful for theoretical problems and for applications where it is not possible to use a ground as a reference. For a reference point potential to be used as a constant, the potential at some point has to be known. It is important to note that the potential is a scalar field and does not involve any unit vector. Surfaces having the same potential are known as equipotential surfaces, i.e., the potential difference between two points on an equipotential surface is always zero.

Another important parameter associated with potentials is the "potential gradient," which is the maximum rate of change of a potential in any direction. It is related to the electric field intensity as

$$\mathbf{E} = -\text{grad } V = -\nabla V \qquad (14.32)$$

It is evident that the gradient of V is a vector quantity and can be written in rectangular, cylindrical, and spherical coordinate systems as

$$\mathbf{E} = -\nabla V = -\left(\frac{\partial V}{\partial x}\mathbf{a}_x + \frac{\partial V}{\partial y}\mathbf{a}_y + \frac{\partial V}{\partial z}\mathbf{a}_z\right) \quad \text{rectangular} \qquad (14.33)$$

$$\mathbf{E} = -\nabla V = -\left(\frac{\partial V}{\partial \rho}\mathbf{a}_\rho + \frac{1}{\rho}\frac{\partial V}{\partial \phi}\mathbf{a}_\varphi + \frac{\partial V}{\partial z}\mathbf{a}_z\right) \quad \text{cylindrical} \qquad (14.34)$$

$$\mathbf{E} = -\nabla V = -\left(\frac{\partial V}{\partial r}\mathbf{a}_r + \frac{1}{r}\frac{\partial V}{\partial \theta}\mathbf{a}_\theta + \frac{1}{r\sin\theta}\frac{\partial V}{\partial \phi}\mathbf{a}_\varphi\right) \quad \text{spherical} \qquad (14.35)$$

In actual practice the *potential* and the *potential gradient* are the terms used as measurable parameters in place of the term *electric field intensity*. One of the most important applications of the potential concept is in the analysis of dipoles. An electric dipole, or simply a dipole, consists of two point charges of equal magnitude and opposite sign and separated by a small distance. Dipole analysis compares the distance from the dipole to the point where potential or electric field is to be determined. The dipole concept is very important in understanding the behavior of dielectric materials and analysis of capacitors. The dipole configuration is shown in Figure 14.3.

The potential and electric field of a dipole are given by

$$V = \frac{Qd\cos\theta}{4\pi\varepsilon_0 r^2}$$
$$\mathbf{E} = \frac{Qd}{4\pi\varepsilon_0 r^3}(2\cos\theta\,\mathbf{a}_r + \sin\theta\,\mathbf{a}_\theta) \qquad (14.36)$$

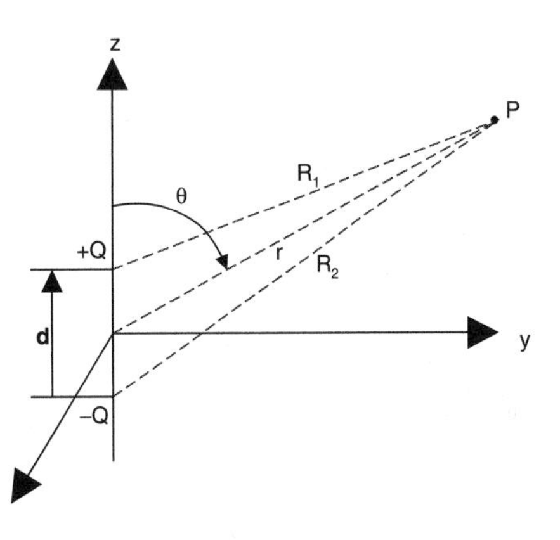

FIGURE 14.3 Dipole configuration.

where r is the distance of the field point from dipole center. The vector distance between two equal point charges is measured from negative to positive charge. It is evident that the V and **E** fields of dipoles fall at a faster rate compared to the fields of a point charge.

14.2 Static Magnetic Fields and Basic Laws

Static magnetic fields are produced by a permanent magnet or an electric field linearly changing with time or a direct current. It is important to note that an electric field is produced by a stationary or a moving charge. A magnetic field, however, is produced only by a moving charge. Also, a moving charge produces a current which in turn produces a magnetic field. Similar to an electric field, a magnetic field produces a force on another magnetic field or current element.

Biot–Savart Law

In some respects a magnetic field is similar to an electric field but in other respects it is completely different. In order to understand the concept of a magnetic field, we start by examining a differential current element. A differential current element is a very small, approaching zero, section of a filamentary conductor. A filamentary conductor is a cylindrical, current-carrying conductor with a radius approaching zero. According to the Biot–Savart law, the magnitude of the magnetic field intensity at a point due to a differential current element is proportional to the product of current, the magnitude of differential length, and the sine of the smaller angle between the current element and a line joining the current element to the point where field is to be determined. It is also inversely proportional to the square of the distance from the differential element to the field point. Using the constant of proportionality $1/4\pi$ in SI units, the law can be written as

$$\mathrm{d}\mathbf{H} = (I\mathrm{d}\mathbf{L} \times \mathbf{a}_{\mathrm{R}})/4\pi R^2 = (I\mathrm{d}\mathbf{L} \times \mathbf{R})/4\pi R^3 \tag{14.37}$$

where I is the current through differential current element of length $\mathrm{d}\mathbf{L}$, R is the distance from differential element to the field point P, and $\mathbf{a}_{\mathrm{R}}$ is the unit vector in the direction of vector **R**. It is important to note that the current element is part of a closed conducting circuit. This was shown when we considered dc or non-time varying fields requiring a closed circuit. On closer examination of the Biot–Savart law it is evident that it is analogous to Coulomb's law for differential electric charges except with respect to the direction of the field. Since a closed circuit for dc analysis is necessary, the more practical magnetic field equation is the integral form of Equation (14.37) as

$$\mathbf{H} = \oint (I\mathrm{d}\mathbf{L} \times \mathbf{a}_{\mathrm{R}})/4\pi \mathrm{R}^2 \ \mathrm{A/m} \tag{14.38}$$

For an infinitely long filamentary current-carrying conductor (I), the magnetic field intensity is found along the concentric circles around the conductor, and is given by

$$\mathbf{H} = \frac{I}{2\pi\rho}\mathbf{a}_{\varphi} \ \mathrm{A/m} \tag{14.39}$$

For distributed current sources of surface current density **K** and volume current density **J**, Equation (14.38) is modified as

$$\mathbf{H} = \int_{\mathrm{S}} (\mathbf{K} \times \mathbf{a}_{\mathrm{R}}\,\mathrm{d}S)/4\pi \mathrm{R}^2 \quad \text{and} \quad \mathbf{H} = \int_{\mathrm{vol}} (\mathbf{J} \times \mathbf{a}_{\mathrm{R}}\,\mathrm{d}v)/4\pi \mathrm{R}^2 \tag{14.40}$$

These equations can be used to determine magnetic field intensities due to filamentary, surface, and volume current distributions.

Ampere's Circuital Law

As we have seen in the case of electric field intensities, some symmetrical charge distribution problems can be easily analyzed using Gauss's law. Similarly, Ampere's circuital law in magnetic fields can be considered as analogous to Gauss's law, and can be applied to certain current distribution problems with greater ease. According to Ampere's circuital law, the line integral of **H** around a closed path is equal to the direct current enclosed by that path, and is written as

$$\oint \mathbf{H} \cdot d\mathbf{L} = I_{\text{enclosed}} \tag{14.41}$$

For solving problems, we determine the enclosed path known as the Amperian path around the current distribution just like a Gaussian surface in the electric field case. This method can be used for determining the magnetic field intensity due to a coaxial line, current sheet, solenoid, and toroid. The magnetic field intensity due to a current sheet with current density **K** A/m is obtained as

$$\mathbf{H} = \frac{1}{2}\mathbf{K} \times \mathbf{a}_N \tag{14.42}$$

If there are two current sheets carrying currents in opposite directions, the magnetic field intensities are

$$\mathbf{H} = \mathbf{K} \times \mathbf{a}_N \quad \text{(point is between the two sheets)}; \quad \mathbf{H} = 0 \quad \text{(point is outside)} \tag{14.43}$$

This approach is very useful for determining magnetic field intensity due to a coaxial line or a strip line. If we apply Ampere's circuital law to the perimeter of a differential surface which encloses the current **J**, we obtain a very important result:

$$\nabla \times \mathbf{H} = \text{curl}\,\mathbf{H} = \mathbf{J} \tag{14.44}$$

which is actually the point or differential form of Ampere's circuital law. Similarly for static electric field we can write

$$\nabla \times \mathbf{E} = 0 \tag{14.45}$$

Concept of Magnetic Flux Density and Gauss's Law for Magnetic Fields

The magnetic flux density **B** is similar to electric flux density **D** and is related to magnetic field intensity **H** in a similar manner as **D** is related to **E** as

$$\mathbf{B} = \mu_0\,\mathbf{H} \text{ (free space)} \quad \text{or} \quad \mathbf{B} = \mu\,\mathbf{H} \text{ (for other material) Wb/m}^2 \text{ or tesla (T)} \tag{14.46}$$

where μ is the permeability of the material which is $\mu_0 = 4\pi \times 10^{-7}$ H/m for free space. Now we can define the total magnetic flux crossing through a surface S as

$$\Phi = \int_S \mathbf{B} \cdot d\mathbf{S} \text{ Wb} \tag{14.47}$$

By comparing the magnetic flux passing through a closed surface to the electric flux case or Gauss's law for magnetic fields is

$$\Phi = \oint_S \mathbf{B} \cdot d\mathbf{S} = 0 \tag{14.48}$$

Since the magnetic field or magnetic flux due to a filamentary conductor is found along concentric circles around the conductor, the flux crossing through a closed surface is zero. This clearly shows that unlike electric flux lines, magnetic flux lines do not cross the surface and are continuous lines around the conductor. This also suggests that there are not isolated magnetic poles like electric charges, which is a fundamental difference between electric and magnetic fields. Thus, Gauss's law for a magnetic field becomes Equation (14.48), or in differential or point form using divergence theorem, it is written as

$$\nabla \cdot \mathbf{B} = 0 \tag{14.49}$$

These four important equations discussed so far are known as Maxwell's equations for static or non-time varying fields, and can be summarized in differential and integral forms as

$$\nabla \cdot \mathbf{D} = \rho_v; \ \nabla \times \mathbf{E} = 0; \ \nabla \times \mathbf{H} = \mathbf{J}; \ \nabla \cdot \mathbf{B} = 0 \quad \text{(differential form)} \tag{14.50a}$$

$$\oint_S \mathbf{D} \cdot d\mathbf{S} = Q = \int_{vol} \rho_v dv; \ \oint \mathbf{E} \cdot d\mathbf{L} = 0; \ \oint \mathbf{H} \cdot d\mathbf{L} = I_{encl} = \int_S \mathbf{J} \cdot d\mathbf{S};$$
$$\oint_S \mathbf{B} \cdot d\mathbf{S} = 0 \quad \text{(integral form)} \tag{14.50b}$$

Concept of Magnetic Potential

In order to draw parallelism between electric and magnetic fields, we can define magnetic scalar potential similar to electric potential as

$$\mathbf{H} = -\nabla V_m \tag{14.51}$$

The main difficulty with this scalar magnetic potential is that, unlike electric potential, it is not constant. We have seen that once the reference has been defined, the electric potential at any point in the electric field is constant. It does not depend on the path along which the charge is moved, but in the case of scalar magnetic potential it depends on the specific path selected. It means the scalar magnetic potential at any point in the field is changing according to the path selected. Because of this, the scalar magnetic potential does not have any practical significance. It is the vector magnetic potential **A**, which is extremely useful in analyzing electromagnetic radiation from antennae, transmission lines, and waveguides, and is related to **B** as

$$\mathbf{B} = \nabla \times \mathbf{A} \ \text{Wb/m}^2 \tag{14.52}$$

From Equation (14.52) we can see that the units of vector magnetic potential **A** are Wb/m. The vector magnetic potential **A** for a filamentary current conductor, surface current, and volume current, respectively, is obtained from

$$\mathbf{A} = \oint \frac{\mu I}{4\pi R} d\mathbf{L}; \quad \mathbf{A} = \int_S \frac{\mu}{4\pi R} \mathbf{K}\, dS; \quad \mathbf{A} = \int_{vol} \frac{\mu}{4\pi R} \mathbf{J}\, dv \tag{14.53}$$

As in the electric potential reference for theoretical problems, the reference for vector magnetic potential is taken at infinity. Similarly, the vector magnetic potential due to a differential current element that is part of a closed circuit is given as

$$d\mathbf{A} = \frac{\mu I}{4\pi R} d\mathbf{L} \tag{14.54}$$

Magnetic Forces

If a moving charge is placed in a magnetic field, it experiences a force as in an electric field. For a differential charge element dQ moving with a velocity **v** and placed in a magnetic field **B**, the differential force is

$$\mathrm{d}\mathbf{F} = \mathrm{d}Q\ \mathbf{v} \times \mathbf{B} \tag{14.55}$$

In terms of volume charge density ρ_v, the differential charge $\mathrm{d}Q = \rho_v\ \mathrm{d}v$, and the force on this differential charge from Equation (14.55) can be written as

$$\mathrm{d}\mathbf{F} = \rho_v\ \mathrm{d}v\ \mathbf{v} \times \mathbf{B} = \mathbf{J} \times \mathbf{B}\ \mathrm{d}v \tag{14.56}$$

The force on a differential current element (for surface current and filamentary current) is obtained as

$$\mathrm{d}\mathbf{F} = \mathbf{K} \times \mathbf{B}\ \mathrm{d}S \quad \text{and} \quad \mathrm{d}\mathbf{F} = \mathrm{I}\ \mathrm{d}\mathbf{L} \times \mathbf{B} \tag{14.57}$$

In all the above equations it is assumed that the differential current element is part of a closed circuit. Now from Equation (14.56) and Equation (14.57) the force on a complete path in a volume or on a surface or on a closed filamentary path is obtained through integration as

$$\mathbf{F} = \int_{\mathrm{vol}} \mathbf{J} \times \mathbf{B}\ \mathrm{d}v \quad \text{or} \quad \mathbf{F} = \int_{S} \mathbf{K} \times \mathbf{B}\ \mathrm{d}S \tag{14.58a}$$

$$\mathbf{F} = \oint \mathrm{I}\ \mathrm{d}\mathbf{L} \times \mathbf{B} = -\mathrm{I} \oint \mathbf{B} \times \mathrm{d}\mathbf{L} \tag{14.58b}$$

The force on a straight filamentary conductor of length L placed in a magnetic field **B** is obtained as

$$\mathbf{F} = \mathrm{I}\mathbf{L} \times \mathbf{B} \tag{14.59}$$

and magnitude of this force is given by

$$\mathrm{F} = \mathrm{BIL}\sin\theta \tag{14.60}$$

where θ is the angle between the direction of current flow and direction of magnetic flux density.

The concept of magnetic forces is useful in electric motor and generators and magnetic forklift trucks that upload metallic objects. It is important to note that electric and magnetic fields vary from one material to another and follow certain boundary conditions. These boundary conditions are important in the study of materials and electric/magnetic circuit components. Electric circuit components like resistors, capacitors, and inductors are analyzed on the basis of electric and magnetic fields, and boundary conditions of different materials in these fields.

14.3 Time-Varying Electromagnetic Fields and Maxwell's Equations

Ampere's circuital law (or simply Ampere's law), Gauss's law, and Faraday's law form the four Maxwell's equations of electromagnetic fields. The main contribution of Maxwell is the mathematical formulation of these laws. These four equations have become the heart of the study of electromagnetic fields for both the static fields and time-varying fields. So far, for the sake of simplicity, our focus has been the static, or steady electric, and magnetic fields. In practice, we have to solve many problems related to time-varying fields. We have already discussed the first two laws in the previous sections and now attention is given to Faraday's law so that these four equations can be discussed in greater depth. According to Faraday's law, a time-varying magnetic field produces an electric field. It means if a conducting coil is placed in a magnetic field that is changing with time, a voltage is induced across the terminals of the coil resulting in current flow through the coil. The induced voltage is also known as electromotive force (emf), and Faraday's law can be mathematically written as

$$\text{induced voltage or emf} = -\mathrm{d}\Phi/\mathrm{d}t \tag{14.61}$$

where the negative sign indicates that the magnetic flux due to induced voltage opposes the applied magnetic flux, which is also known as "Lenz's law," and Φ is the time varying magnetic flux. It is evident that a steady or static magnetic field cannot produce induced voltage. If there are N number of turns in the coil, the induced voltage is

$$\text{induced voltage or emf} = -N\,\mathrm{d}\Phi/\mathrm{d}t \tag{14.62}$$

Using the definition of voltage for emf with electric field **E**, and writing Φ in terms of magnetic flux density **B**, the emf is obtained as

$$\text{emf} = \oint \mathbf{E}\cdot\mathrm{d}\mathbf{L} = -\frac{\mathrm{d}}{\mathrm{d}t}\int_S \mathbf{B}\cdot\mathrm{d}\mathbf{S} \tag{14.63}$$

Since this emf can be obtained by varying the magnetic field in any direction, by varying the closed path, or by changing both, the emf can be written as

$$\text{emf} = \oint \mathbf{E}\cdot\mathrm{d}\mathbf{L} = -\int_S \frac{\partial}{\partial t}\mathbf{B}\cdot\mathrm{d}\mathbf{S} \tag{14.64}$$

which is the integral form of Faraday's law and the differential form is written as

$$\nabla\times\mathbf{E} = -\frac{\partial}{\partial t}\mathbf{B} \tag{14.65}$$

Again from Equation (14.65) it is evident that for static magnetic fields $\nabla\times\mathbf{E} = 0$, i.e., a steady magnetic field cannot produce an electric field. Now we can write the four Maxwell's equations as

Law	Integral form	Point form	
Faraday	$\oint \mathbf{E}\cdot\mathrm{d}\mathbf{L} = -\int_S \frac{\partial}{\partial t}\mathbf{B}\cdot\mathrm{d}\mathbf{S}$	$\nabla\times\mathbf{E} = -\frac{\partial}{\partial t}\mathbf{B}$	(14.66a)
Ampere	$\oint \mathbf{H}\cdot\mathrm{d}\mathbf{L} = \int_S \mathbf{J}\cdot\mathrm{d}\mathbf{S} + \frac{\mathrm{d}}{\mathrm{d}t}\int_S \mathbf{D}\cdot\mathrm{d}\mathbf{S}$	$\nabla\times\mathbf{H} = \mathbf{J} + \frac{\partial}{\partial t}\mathbf{D}$	(14.66b)
Gauss (electric field)	$\oint_S \mathbf{D}\cdot\mathrm{d}\mathbf{S} = \int_{vol} \rho_v\,\mathrm{d}v$	$\nabla\cdot\mathbf{D} = \rho_v$	(14.66c)
Gauss (magnetic field)	$\oint_S \mathbf{B}\cdot\mathrm{d}\mathbf{S} = 0$	$\nabla\cdot\mathbf{B} = 0$	(14.66d)

If we consider sinusoidal time variation of electric and magnetic fields, and replace the time derivative $\frac{\partial}{\partial t}$ by $j\omega$, Maxwell's equations in Equation (14.66) are modified as

$$\oint \mathbf{E} \cdot \mathrm{d}\mathbf{L} = -j\omega \oint_S \mathbf{B} \cdot \mathrm{d}\mathbf{S} \quad \nabla \times \mathbf{E} = -j\omega \mathbf{B} \tag{14.67a}$$

$$\oint \mathbf{H} \cdot \mathrm{d}\mathbf{L} = \int_S J \cdot \mathrm{d}\mathbf{S} + j\omega \int_S \mathbf{D} \cdot \mathrm{d}\mathbf{S} \quad \nabla \times \mathbf{H} = \mathbf{J} + j\omega \mathbf{D} \tag{14.67b}$$

$$\int_S \mathbf{D} \cdot \mathrm{d}\mathbf{S} = \int_{\mathrm{vol}} \rho_{\mathrm{v}} \, \mathrm{d}v \quad \nabla \cdot \mathbf{D} = \rho_{\mathrm{v}} \tag{14.67c}$$

$$\oint_S \mathbf{B} \cdot \mathrm{d}\mathbf{S} = 0 \quad \nabla \cdot \mathbf{B} = 0 \tag{14.67d}$$

In addition to these four Maxwell's equations, we also need two other relationships to determine electric and magnetic fields in a medium. According to Kong [4], these relationships are known as constitutive relationships, written as

$$\mathbf{D} = \varepsilon \mathbf{E} \quad \text{and} \quad \mathbf{B} = \mu \mathbf{H} \text{ for isotropic medium} \tag{14.68}$$

where ε and μ are permittivity and permeability, respectively. Maxwell's equations are very useful tools to determine electric and magnetic fields in any medium with proper boundary conditions.

References

1. W.H. Hayt, Jr. and J.A. Buck, *Engineering Elecromagnetics*, 6th ed., New York: McGraw-Hill, 2001.
2. C.R. Paul, *Electromagneics for Engineers*, New York: John Wiley & Sons, 2004.
3. L. Yu and B. Rawat, "Quasi-Static analysis of three line microstrip symmetrical coupler on anisotropic substrates," *IEEE Trans. Microwave Theory and Techniques*, vol. 39, no. 8, pp. 1433–1437, 1991.
4. J.A. Kong, *Electromagnetic Wave Theory*, New York: Wiley-Interscience, 1990.

15
Magnetism and Magnetic Fields

Geoffrey Bate
Consultant in Information Storage Technology
Mark H. Kryder
Carnegie Mellon University

15.1 Magnetism

Geoffrey Bate

Static Magnetic Fields

To understand the phenomenon of magnetism we must also consider electricity and vice versa. A stationary electric charge produces, at a point a fixed distance from the charge, a static (i.e., time-invariant) electric field. A moving electric charge, i.e., a current, produces at the same point a time-dependent electric field and a magnetic field, d**H**, whose magnitude is constant if the electric current, I, represented by the moving electric charge, is constant.

Fields from Constant Currents

Figure 15.1 shows that the direction of the magnetic field is perpendicular both to the current I and to the line, **R**, from the element d**L** of the current to a point, P, where the magnetic field, d**H**, is being calculated or measured:

$$\mathrm{d}\mathbf{H} = I\ \mathrm{d}\mathbf{L} \times \mathbf{R}/4\pi R^3 \quad \mathrm{A/m} \text{ when } I \text{ is in amps and } \mathrm{d}\mathbf{L} \text{ and } R \text{ are in meters}$$

If the thumb of the right hand points in the direction of the current, then the fingers of the hand curl in the direction of the magnetic field. Thus, the stream lines of H, i.e., the lines representing at any point the direction of the H field, will be an infinite set of circles having the current as center. The magnitude of the field $H_\phi = I/2\pi R$ A/m. The line integral of H about any closed path around the current is $\oint H \cdot \mathrm{d}L = I$.

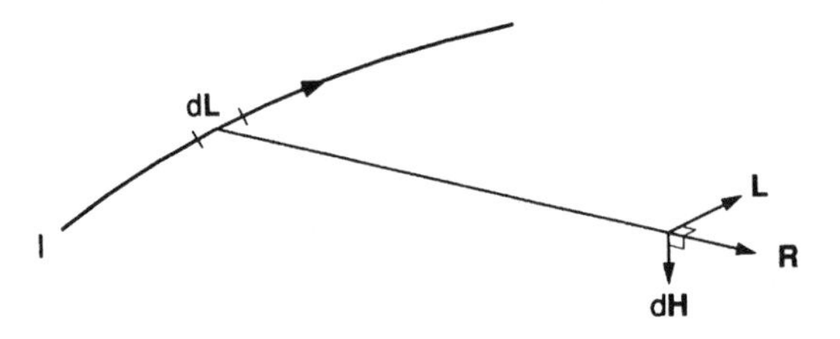

FIGURE 15.1 A current I flowing through a small segment **dL** of a wire produces at a distance **R** a magnetic field whose direction **dH** is perpendicular both to **R** and **dL**.

This relationship (known as Ampère's circuital law) allows one to find formulas for the magnetic field strength for a variety of symmetrical coil geometries, e.g.:

1. At a radius, ρ, between the conductors of a coaxial cable:

$$H_\phi = I/2\pi\rho \text{ A/m}$$

2. Between two infinite current sheets in which the current, **K**, flows in opposite directions:

$$\mathbf{H} = \mathbf{K} \times \mathbf{a_n}$$

   where $\mathbf{a_n}$ is the unit vector normal to the current sheets.

3. Inside an infinitely long, straight solenoid of diameter d, having N turns closely wound:

$$\mathbf{H} = NI/d \text{ A/m}$$

4. Well inside a toroid of radius ρ, having N closely wound turns:

$$\mathbf{H} = NI/2\pi\rho \cdot \mathbf{a}_\phi \text{ A/m}$$

Applying Stokes' theorem to Ampère's circuital law we find the point form of the latter:

$$\nabla \times \mathbf{H} = \mathbf{J}$$

where **J** is the current density in amps per square meter.

Time-Dependent Electric and Magnetic Fields

A constant current I produces a constant magnetic field **H** which, in turn, polarizes the medium containing **H**. While we cannot obtain isolated magnetic poles, it is possible to separate the "poles" by a small distance to create a magnetic dipole (i.e., to *polarize* the medium), and the dipole moment (the product of the pole strength and the separation of the poles) per unit volume is defined as the *magnetization M*. The units are emu/cc in the cgs system and amps per meter in the SI system of units. Because it is usually easier to determine the mass of a sample than to determine its volume, we also have a magnetization per unit mass, σ, whose units are emu/g or Am^2/kg. The conversion factors between cgs and SI units in magnetism are shown in Table 15.1. The effects of the static and time-varying currents may be summarized as follows:

Static	$[\mathbf{I}]_o \rightarrow [\mathbf{H}]_o \rightarrow [\mathbf{M}]_o$
	↙ **motion**
Time-varying	$[\mathbf{I}]_t \rightarrow [\mathbf{H}]_t \rightarrow [\mathbf{M}]_t$

where the suffixes "o" and "t" signify *static* and *time-dependent*, respectively.

TABLE 15.1 Units in Magnetism

Quality	Symbol	cgs Units	× Factor	=	SI units
		$B = H + 4\pi M$			$B = \mu_o(H + M)$
Magnetic flux density	B	gauss (G)	$\times 10^{-4}$	=	tesla (T), Wb/m^2
Magnetic flux	Φ	maxwell (Mx) G cm^2	$\times 10^{-8}$	=	webers (Wb)
Magnetic potential difference (magnetomotive force)	U	gilbert (Gb)	$\times\ 10/4\pi$	=	ampere (A)
Magnetic field strength	H	oersted (Oe)	$\times 10^3/4\pi$	=	A/m
Magnetization (per volume)	M	emu/cc	$\times 10^3$	=	A·m
Magnetization (per mass)	σ	emu/g	$\times 1$	=	A·m^2/kg
Magnetic moment	M	emu	$\times 10^{-3}$	=	A·m^2
Susceptibility (volume)	χ	dimensionless	$\times 4\pi$	=	dimensionless
Susceptibility (mass)	κ	dimensionless	$\times 4\pi$	=	dimensionless
Permeability (vacuum)	μ_o	dimensionless	$\times 4\pi.10^{-7}$	=	Wb/A·m
Permeability (material)	μ	dimensionless	$\times 4\pi.10^{-7}$	=	Wb/A·m
Bohr magneton	μ_B	$= 0.927 \times 10^{-20}$ erg/Oe	$\times 10^{-3}$	=	Am^2
Demagnetizing factor	N	dimensionless	$\times 1/4\pi$	=	dimensionless

Magnetic Flux Density

In the case of electric fields there is in addition to **E** an electric flux density field **D**, the lines of which begin on positive charges and end on negative charges. **D** is measured in coulombs per square meter and is associated with the electric field **E** (V/m) by the relation $\mathbf{D} = \epsilon_r \epsilon_o \mathbf{E}$ where ϵ_o is the *permittivity* of free space ($\epsilon_o = 8.854 \times 10^{-12}$ F/m) and ϵ_r is the (dimensionless) dielectric constant.

For magnetic fields there is a magnetic flux density $\mathbf{B}$(Wb/m^2) $= \mu_r\mu_o\mathbf{H}$, where μ_o is the *permeability* of free space ($\mu_o = 4\pi \times 10^{-7}$ H/m) and μ_r is the (dimensionless) permeability. In contrast to the lines of the **D** field, lines of **B** are closed, having no beginning or ending. This is not surprising when we remember that while isolated positive and negative charges exist, no magnetic monopole has yet been discovered.

Relative Permeabilities

The range of the relative permeabilities covers about six orders of magnitude (Table 15.2) whereas the range of dielectric constants is only three orders of magnitude.

Forces on a Moving Charge

A charged particle, q, traveling with a velocity v and subjected to a magnetic field experiences a force:

$$\mathbf{F} = q\mathbf{v} \times \mathbf{B}$$

This equation reveals how the Hall effect can be used to determine whether the majority current carriers in a sample of a semiconductor are (negatively charged) electrons flowing, say, in the negative direction or (positively charged) holes flowing in the positive direction. The (transverse) force (Figure 15.2) will be in the same direction in either case, but the *sign* of the charge transported to the voltage probe will be positive for holes and negative for electrons.

In general, when both electric and magnetic fields are present, the force experienced by the carriers is given by

$$\mathbf{F} = q\ (\mathbf{E} + \mathbf{v} \times \mathbf{B})$$

The Hall effect is the basis of widely used and sensitive instruments for measuring the intensity of magnetic fields over a range of 10^{-5} to 2×10^6 A/m.

TABLE 15.2 Relative Permeability, μ_r, of Some Diamagnetic, Paramagnetic, and Ferromagnetic Materials

Material	μ_r	M_s, A/m^2
Diamagnetics		
Bismuth	0.999833	
Mercury	0.999968	
Silver	0.9999736	
Lead	0.9999831	
Copper	0.9999906	
Water	0.9999912	
Paraffin wax	0.99999942	
Paramagnetics		
Oxygen (s.t.p.)	1.000002	
Air	1.00000037	
Aluminum	1.000021	
Tungsten	1.00008	
Platinum	1.0003	
Manganese	1.001	
Ferromagnetics		
Purified iron: 99.96% Fe	280,000	2.158
Motor-grade iron: 99.6% Fe	5,000	2.12
Permalloy: 78.5% Ni, 21.5% Fe	70,000	2.00
Supermalloy: 79% Ni, 15% Fe, 5% Mo, 0.5% Mn	1,000,000	0.79
Permendur: 49% Fe, 49% Ca, 2% V	5,000	2.36
Ferrimagnetics		
Manganese–zinc ferrite	750	0.34
	1,200	0.36
Nickel–zinc ferrite	650	0.29

Source: F. Brailsford, *Physical Principles of Magnetism*, London: Van Nostrand, 1966. With permission.

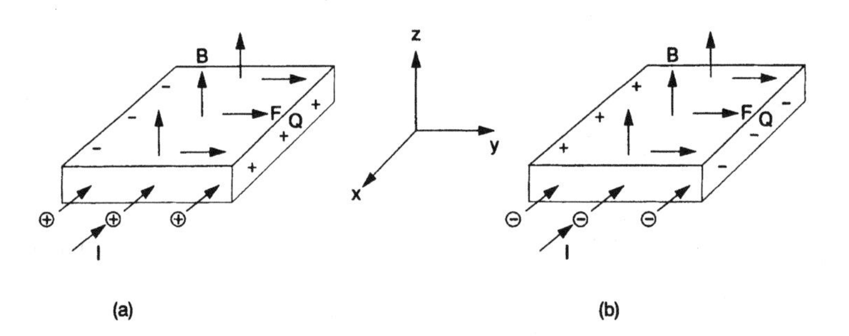

FIGURE 15.2 Hall effect. A magnetic field **B** applied to a block of semiconducting material through which a current I is flowing exerts a force $\mathbf{F} = \mathbf{V} \times \mathbf{B}$ on the current carriers (electrons or holes) and produces an electric charge on the right face of the block. The charge is positive if the carriers are holes and negative if the carriers are electrons.

Time-Varying Magnetic Fields

In 1831, 11 years after Oersted demonstrated that a current produced a magnetic field which could deflect a compass needle, Faraday succeeded in showing the converse effect—that a magnetic field could produce a current. The reason for the delay between the two discoveries was that it is only when a magnetic field is changing that an emf is produced:

$$\text{emf} = -\frac{d\Phi}{dt} \text{ V}$$

where $\Phi = BS$ = flux density (in gauss) × area S. The time-changing flux, $d\Phi/dt$, can happen as a result of:

1. A changing magnetic field within a stationary circuit
2. A circuit moving through a steady magnetic field
3. A combination of 1 and 2

The electrical circuit may have N turns and then:

$$\text{emf} = -N\frac{d\Phi}{dt}$$

We can write emf = $\mathbf{E}\cdot d\mathbf{L}$ and in the presence of changing magnetic fields or a moving electrical circuit $\mathbf{E}\cdot d\mathbf{L}$ is no longer required to be equal to 0 as it was for stationary fields and circuits.

Maxwell's Equations

Because the flux Φ can be written $\int \mathbf{B}\cdot ds$ we have emf = $\mathbf{E}\cdot d\mathbf{L} = -d/dt\,\mathbf{B}\cdot ds$, and by using Stokes' theorem:

$$(\nabla \times \mathbf{E})\cdot ds = -d\mathbf{B}/dt\ ds$$

or

$$\nabla \times \mathbf{E} = -d\mathbf{B}/dt$$

That is, a spatially changing *electric field* produces a time-changing *magnetic field*. This is one of Maxwell's equations linking electric and magnetic fields.

By a similar argument it can be shown that:

$$\nabla \times \mathbf{H} = \mathbf{J} + d\mathbf{D}/dt$$

This is another of Maxwell's equations and shows a spatially changing *magnetic field* produces a time-changing *electric field*. The latter $d\mathbf{D}/dt$ can be treated as an electric current which flows through a dielectric, e.g., in a capacitor, when an alternating potential is applied across the plates. This current is called the *displacement current* to distinguish it from the conduction current which flows in conductors. The *conduction current* involves the movement of electrons from one electrode to the other through the conductor (usually a metal). The *displacement current* involves no translation of electrons or holes but rather an alternating polarization throughout the dielectric material which is between the plates of the capacitor.

From the last two equations we see a key conclusion of Maxwell: that in electromagnetic fields a time-varying magnetic field produces a spatially varying electric field and a time-varying electric field produces a spatially varying magnetic field.

Maxwell's equations in point form, then, are

$$\nabla \times \mathbf{E} = d\mathbf{B}/dt$$

$$\nabla \times \mathbf{H} = \mathbf{J} + d\mathbf{D}/dt$$

$$\nabla \cdot \mathbf{D} = \rho_v$$

$$\nabla \cdot \mathbf{B} = 0$$

These equations are supported by the following auxiliary equations:

$$\mathbf{D} = \epsilon\mathbf{E}\ (\text{displacement} = \text{permittivity} \times \text{electric field intensity})$$

$$\mathbf{B} = \mu\mathbf{H} \text{ (flux density = permeability } \times \text{ magnetic field intensity)}$$

$$\mathbf{J} = \sigma\mathbf{E} \text{ (current density = conductivity } \times \text{ electric field strength)}$$

$$\mathbf{J} = \rho_v\mathbf{V} \text{ (current density = volume charge density } \times \text{ carrier velocity)}$$

$$\mathbf{D} = \epsilon_o \mathrm{E} + \mathrm{P} \text{ (displacement as function of electric field and polarization)}$$

$$\mathbf{B} = \mu_o(\mathbf{H} + \mathbf{M}) \text{ (magnetic flux density as function of magnetic field strength and magnetization)}$$

$$\mathbf{P} = \chi_e\epsilon_o\mathbf{E} \text{ (polarization = electric susceptibility permittivity of free space electrical field strength)}$$

$$\mathbf{M} = \chi_m\mu_o\mathbf{H} \text{ (magnetization = magnetic susceptibility } \times \text{ permeability of free space magnetic field strength)}$$

The last two equations relate, respectively, the electric polarization **P** to the displacement $\mathbf{D} = \epsilon_o\mathbf{E}$ and the magnetic moment **M** to the flux density $\mathbf{B} = \mu_o\mathbf{H}$. They apply only to "linear" materials, i.e., those for which **P** is linearly related to **E** and **M** to **H**. For magnetic materials we can say that nonlinear materials are usually of greater practical interest.

Dia- and Paramagnetism

The phenomenon of magnetism arises ultimately from moving electrical charges (electrons). The movement may be orbital around the nucleus or the other degree of freedom possessed by electrons which, by analogy with the motion of the planets, is referred to as *spin*. In technologically important materials, i.e., ferromagnetics and ferrimagnetics, spin is more important than orbital motion. Each arrow in Figure 15.3 represents the *total spin* of an atom.

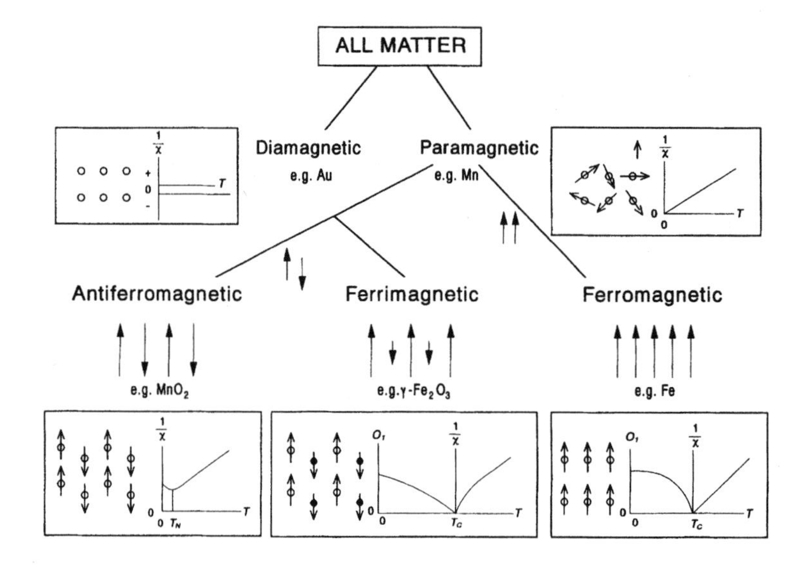

FIGURE 15.3 All matter consists of diamagnetic material (atoms having no permanent magnetic dipole moment) or paramagnetic material (atoms having magnetic dipole moment). Paramagnetic materials may be further divided into ferromagnetics, ferrimagnetics, and antiferromagnetics.

An atom may have a permanent magnetic moment, in which case it is referred to as belonging to a paramagnetic material, or the atom may be magnetized only when in the presence of a magnetic field, in which case it is called *diamagnetic.* Diamagnetics are magnetized in the *opposite direction* to that of the applied magnetic field, i.e., they display *negative* susceptibility (a measure of the induced magnetization per unit of applied magnetic field). Paramagnetics are magnetized in the *same* direction as the applied magnetic field, i.e., they have *positive* susceptibility. *All* atoms are diamagnetic by virtue of their having electrons. *Some* atoms are also paramagnetic as well, but in this case they are called *paramagnetics* since paramagnetism is roughly a hundred times stronger than diamagnetism and overwhelms it. Faraday discovered that paramagnetics are attracted by a magnetic field and move toward the region of maximum field, whereas diamagnetics are repelled and move toward a field minimum.

The total magnetization of both paramagnetic and diamagnetic materials is zero in the absence of an applied field, i.e., they have zero **remanence**. Atomic paramagnetism is a necessary condition but not a sufficient condition for ferro- or ferrimagnetism, i.e., for materials having useful magnetic properties.

Ferromagnetism and Ferrimagnetism

To develop technologically useful materials, we need an additional force that ensures that the spins of the outermost (or almost outermost) electrons are mutually parallel. Slater showed that in iron, cobalt, and nickel this could happen if the distance apart of the atoms (D) was more than 1.5 times the diameter of the 3d electron shell (d). (These are the electrons, *near* the outside of atoms of iron, cobalt, and nickel, that are responsible for the strong paramagnetic moment of the atoms. Paramagnetism of the atoms is an essential prerequisite for ferro- or ferrimagnetism in a material.)

Slater's result suggested that, of these metals, iron, cobalt, nickel, and gadolinium should be ferromagnetic at room temperature, while chromium and manganese should not be ferromagnetic. This is in accordance with experiment. Gadolinium, one of the rare earth elements, is only weakly ferromagnetic in a cool room. Chromium and manganese in the elemental form narrowly miss being ferromagnetic. However, when manganese is alloyed with copper and aluminum ($Cu_{61}Mn_{24}Al_{15}$) to form what is known as a Heusler alloy [Crangle, 1962], it becomes ferromagnetic. The radius of the 3*d* electrons has not been changed by alloying, but the atomic spacing has been increased by a factor of 1.53/1.47. This small change is sufficient to make the difference between positive exchange, parallel spins, and ferromagnetism and negative exchange, antiparallel spins, and antiferromagnetism.

For all ferromagnetic materials there exists a temperature (the **Curie temperature**) above which the thermal disordering forces are stronger than the exchange forces that cause the atomic spins to be parallel. From Table 15.3 we see that in order of descending Curie temperature we have Co, Fe, Ni, Gd. From Figure 15.4 we find that this is also the order of descending values of the exchange integral, suggesting that high positive values of the exchange integral are indicative of high Curie temperatures rather than high magnetic intensity in ferromagnetic materials.

Negative values of exchange result in an antiparallel arrangement of the spins of adjacent atoms and in antiferromagnetic materials (Figure 15.3). Until five years ago, it was true to say that antiferromagnetism had

TABLE 15.3 The Occurrence of Ferromagnetism

	Cr	Mn	Fe	Co	Ni	Gd
Atomic number	24	25	26	27	28	64
Atomic spacing/diameter	1.30	1.47	1.63	1.82	1.97	1.57
Ferromagnetic moment/mass (Am^2/kg)						
At 293 K	—	—	217.75	161	54.39	0
At 0 K	—	—	221.89	162.5	57.50	250
Curie point, Θ_c K	—	—	1043	1400	631	289
Néel temp., Θ_n K	475	100	—	—	—	—

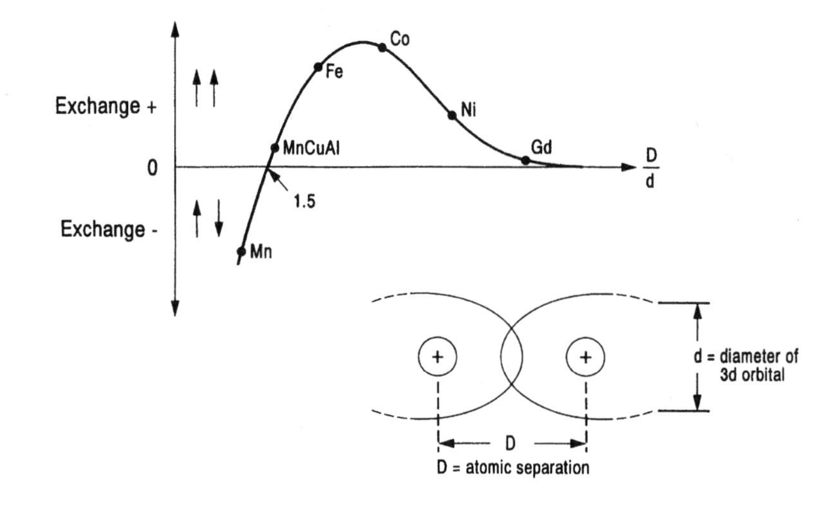

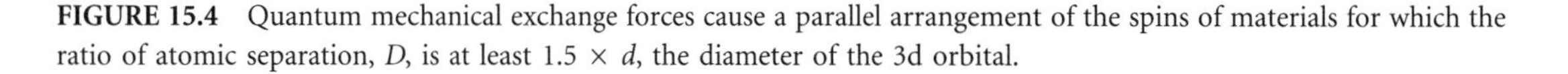

FIGURE 15.4 Quantum mechanical exchange forces cause a parallel arrangement of the spins of materials for which the ratio of atomic separation, *D*, is at least 1.5 × *d*, the diameter of the 3d orbital.

no practical application. Thin films on antiferromagnetic materials are now used to provide the bias field which is used to linearize the response of some magnetoresistive reading heads in magnetic disk drives. Ferrimagnetism, also illustrated in Figure 15.3, is much more widely used. It can be produced as soft, i.e., low **coercivity**, ferrites for use in magnetic recording and reading heads or in the core of transformers operating at frequencies up to tens of megahertz. High-coercivity, single-domain particles (which are discussed later) are used in very large quantities to make magnetic recording tapes and flexible disks γ-Fe_2O_3 and cobalt-impregnated iron oxides and to make barium ferrite, the most widely used material for permanent magnets.

Intrinsic Magnetic Properties

Intrinsic magnetic properties are those properties that depend on the type of atoms and their composition and crystal structure, but not on the previous history of a particular sample. Examples of intrinsic magnetic properties are the *saturation* magnetization, Curie temperature, magnetocrystallic anisotropy, and magnetostriction.

Extrinsic magnetic properties depend on type, composition, and structure, but they also depend on the previous history of the sample, e.g., heat treatment. Examples of extrinsic magnetic properties include the technologically important properties of *remanent* magnetization, coercivity, and permeability. These properties can be substantially altered by heat treatment, quenching, cold-working the sample, or otherwise changing the size of the magnetic particle.

A ferromagnetic or ferrimagnetic material, on being heated, suffers a reduction of its magnetization (per unit mass, i.e., σ, and per unit volume, M). The slope of the curve of M_s vs. T increases with increasing temperature as shown in Figure 15.5. This figure represents the conflict between the ordering tendency of the exchange interaction and the disordering effect of increasing temperature. At the Curie temperature, the order no longer exists and we have a paramagnetic material. The change from ferromagnetic or ferrimagnetic materials to paramagnetic is completely reversible on reducing the temperature to its initial value. Curie temperatures are always lower than melting points.

A single crystal of iron has the body-centered structure at room temperature. If the magnetization as a function of applied magnetic field is measured, the shape of the curve is found to depend on the direction of the field. This phenomenon is *magnetocrystalline anisotropy.* Iron has body-centered structure at room temperature, and the "easy" directions of magnetization are those directions parallel to the cube edges [100],

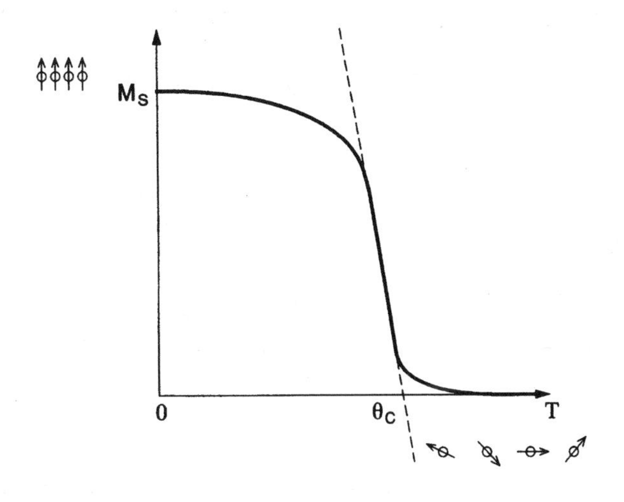

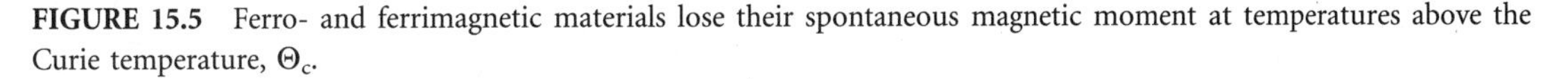

FIGURE 15.5 Ferro- and ferrimagnetic materials lose their spontaneous magnetic moment at temperatures above the Curie temperature, Θ_c.

[010], and [001] or, collectively. The hard direction of magnetization for iron is the body diagonal [111]. At higher temperatures, the anisotropy becomes smaller and disappears above 300°C.

Nickel crystals (face-centered cubic) have an easy direction of [111] and a hard direction of [100]. Cobalt has the hexagonal close-packed (HCP) structure and the hexagonal axis is the easy direction at room temperature.

Magnetocrystalline anisotropy plays a very important part in determining the coercivity of ferro- or ferrimagnetic materials, i.e., the field value at which the direction of magnetization is reversed.

Many magnetic materials change dimensions on becoming magnetized: the phenomenon is known as *magnetostriction* and can be positive, i.e., length increases, or negative. Magnetostriction plays an important role in determining the preferred direction of magnetization of *soft*, i.e., low H_c, films such as those of alloys of nickel and iron, known as *permalloy*.

The origin of both magnetocrystalline anisotropy and magnetostriction is *spin–orbit* coupling. The magnitude of the magnetization of the film is controlled by the electron spin as usual, but the preferred direction of that magnetization with respect to the crystal lattice is determined by the electron orbits which are large enough to interact with the atomic structure of the film.

Extrinsic Magnetic Properties

Extrinsic magnetic properties are those properties that depend not only on the shape and size of the sample, but also on the shape and size of the magnetic constituents of the sample. For example, if we measure the hysteresis loop like the one shown in Figure 15.6 on a disk-shaped sample punched from a magnetic recording tape, the result will depend not only on the diameter and thickness of the disk coating but also on the distribution of shapes and sizes of the magnetic particles within the disk. They display hysteresis individually and collectively. For a soft magnetic material, i.e., one that might be used to make the laminations of a transformer, the dependence of magnetization, M, on the applied magnetic field, H, is also complex. Having once left a point described by the coordinates (H_1, M_1), it is not immediately clear how one might return to that point.

Alloys of nickel and iron, in which the nickel content is the greater, can be capable of a reversal of magnetization by the application of a magnetic field, H, which is weaker than the earth's magnetic field (0.5°Oe, 40°A/m) by a factor of five. (To avoid confusion caused by the geomagnetic field it would be necessary to screen the sample, for example, by surrounding it by a shield of equally soft material or by measuring the earth's field and applying a field which is equal in magnitude but opposite in direction to the earth's field in order to cancel its effects.)

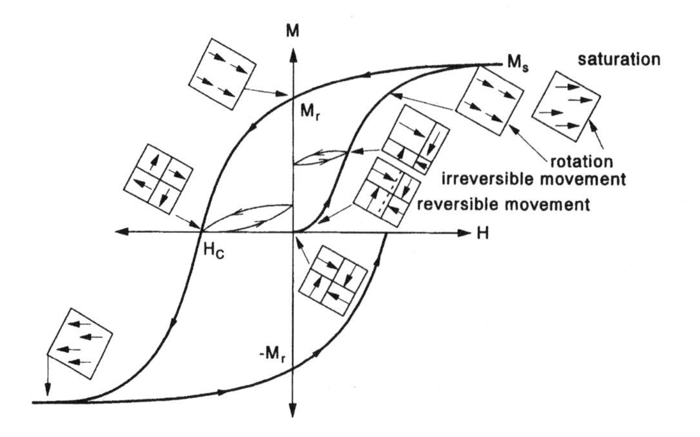

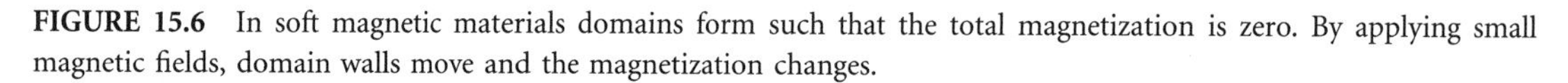

FIGURE 15.6 In soft magnetic materials domains form such that the total magnetization is zero. By applying small magnetic fields, domain walls move and the magnetization changes.

The magnetization of the sample in zero field may be macroscopically zero, but locally the material may be magnetized virtually to the saturation state. As shown in Figure 15.6, which shows a greatly simplified domain structure, the net magnetization at the center of the loop is zero because the magnetization of the four "domains" cancels in pairs. A domain is a region (not necessarily square or even of a regular shape, although the shape often is regular in Ni–Fe thin films or sheets) over which the magnetization is constant in magnitude and direction. Thus, the sample in Figure 15.6 consists of four domains, initially separated from each other by "domain walls." If a magnetic field is applied in the direction of $+H$, that domain will grow whose direction of magnetization is closest to the field direction and the domains will shrink if their magnetization is opposed to the field. For small applied fields, the movement of the walls is reversible, i.e., on reducing the applied field to zero, the original domain configuration will be obtained. Beyond a certain field the movement of the walls is irreversible, and eventually near the knee of the magnetization curve all the domain walls have been swept away by the applied field. The sample is not yet in the saturated state since the direction of M is not quite the same as the direction of the applied field. However, a small increase in the strength of the applied field finally achieves the saturated state by rotating the magnetization of the whole sample into the field direction.

On removing the applied field, the sample does not retrace the magnetization curve, and when the applied field is zero, we can see that a considerable amount of magnetization remains, M_r. Appropriately, this is referred to as the remanent state, and M_r is the *remanent magnetization*. By reversing the original direction of the applied field, domains reappear and the magnetization is eventually reduced to zero at the *coercive field*, H_c. It should be noticed that, at H_c, although the net magnetization is clearly zero, the individual domains may be magnetized in directions that are different from those at the starting point. Figure 15.6 shows an incomplete hysteresis loop. If the field H were increased beyond $+H_c$ the loop would be completed.

The differences between ideally magnetically soft materials (used in transformers and magnetic read/write heads) and magnetically hard materials (used in permanent magnets and in recording tapes and disks) are as follows:

$$\begin{array}{llll} \text{Magnetically soft materials:} & H_c 0; & M_r 0; & M_s \text{ high value} \\ \text{Magnetically hard materials:} & H_c \text{ high value}; & M_r/M_s (\text{squareness}); & M_s \text{ high value} \end{array}$$

Examples are given in Table 15.4.

It is noticeable that the differences between hard and soft magnetic materials are confined to the extrinsic properties, M_r, H_c, and permeability, μ. The latter is related to M and H as follows:

$$\mathbf{B}(\text{G}) = \mathbf{H}\ (\text{Oe}) + 4\pi\mathbf{M}\ (\text{emu/cc})$$

$$\mu = \mathbf{B}/\mathbf{H} = 1 + 4\pi\kappa\ (\text{cgs units})$$

TABLE 15.4 "Hard" and "Soft" Magnetic Materials

	High M_s	Low H_c	Low M_r	High μ
Soft				
Fe	1700 emu/cc	1 Oe	< 500	20,000
80 Ni 20 Fe	660	0.1	< 300	50,000
Mn Zn ferrite	400	0.02	< 200	5,000
$Co_{70}Fe_5Si_{15}B_{10}$	530	0.1	< 250	10,000
	High M_s	**High H_c**	**High M_r**	**T_c**
Hard				
Particles				
γ-Fe_2O_3	400	250–450	200–300	115–126
CrO_2	400	450–600	300	120
Fe	870–1100	1,100–1,500	435–550	768
$BaO.6Fe_2O_3$	238–370	800–3,000	143–260	320
Alloys				
$SmCo_5$	875	40,000	690	720
Sm_2Co_{17}	1,000	17,000	875	920
$Fe_{14}BNd_2$	1,020	12,000	980	310

or

$$\begin{aligned}\mathbf{B}\,(\text{Wb/m})^2 &= \mu_0(\mathbf{H}\ \text{A/m} + \mathbf{M}\ \text{A/m}) \\ &= \mu_0\mu_1\mathbf{H} = \mu\mathbf{H}\ (\text{SI units})\end{aligned}$$

Domain walls form in order to minimize the magnetic energy of the sample. The magnetic energy is $\mu H^2/8\pi$ cgs units or $1/2\mu\mathbf{H}^2$ J/m^3 (SI units) and clearly depends on **H**, the magnetic field emanating from the sample. In the *initial* domain configuration shown in Figure 15.6, there is no net magnetization of the sample and thus no substantial **H** exists outside the sample and the magnetostatic energy is zero. Thus, the establishment of domains *reduces* the energy associated with **H** but it *increases* the energy needed to establish domain walls within the sample. A compromise is reached in which domain walls are formed until the establishment of one more wall would *increase*, rather than decrease, the total magnetic energy of the sample.

The wall energy depends on the area of the wall, i.e., L^2, while the energy associated with the external magnetic field depends on L^3, the volume of the sample. Clearly, as the size of single particles becomes small, terms in L^2 are more important than terms in L^3, and so for small magnetic particles, the formation of domain walls may not be energetically feasible and a *single-domain particle* results. These are found in the particles of iron oxide, cobalt-modified iron oxide, chromium dioxide, iron, or barium ferrite, which are used to make magnetic recording tapes, and in barium ferrite, samarium cobalt, and neodymium iron boron, which are used to make powerful permanent magnets. In the latter cases, the very high coercivities are caused by domain walls being pinned at grain boundaries between the main phase grains and finely precipitated secondary phases. This is an example of *nucleation-controlled coercivity.*

The amount of available energy that can be stored in a permanent magnet is the area of the largest rectangle that can be drawn in the second quadrant of the B vs. H hysteresis loop. The *energy product* has grown remarkably by a factor of about 50 since 1900 (Strnat, 1986). We see from the graph in Figure 15.7 of *intrinsic* coercive force, i.e., the coercive force obtained from the graph of M vs. H (in contrast to the smaller coercive force obtained by plotting B vs. H), that increases in H_c (rather than increases in M_r) have been responsible for almost all the improvement in the energy product.

The key attributes of technologically important magnetic materials are:

1. Large, spontaneous atomic magnetic moments
2. Large, positive exchange integrals
3. Magnetic anisotropy and heterogeneity which are small for soft magnetic materials and large for hard magnetic materials

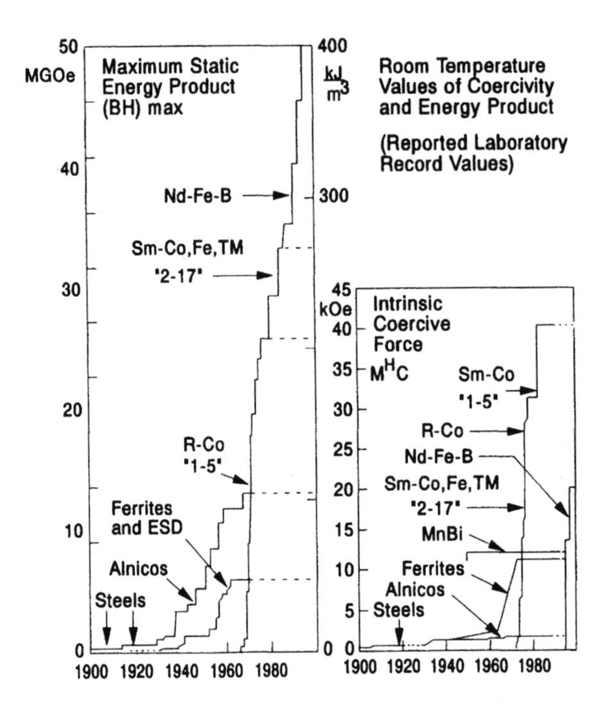

FIGURE 15.7 The development of magnetic materials for permanent magnets showing the increase in energy product and in intrinsic coercivity as a function of time.

In single-domain materials, the magnetic particles are so small that reversal of the magnetization can only occur by rotation of the magnetization vector. This rotation can be resisted by combinations of three anisotropies: crystalline anisotropy, shape anisotropy, and magnetoelastic anisotropy (which depends on the magnetostrictive properties of the material).

Crystalline Anisotropy

Crystalline anisotropy arises from the existence of easy and hard directions of magnetization within the crystal structure of the material. For example, in iron the {100} directions are easy directions while the {111} directions are hard. In nickel crystals the reverse is true. In cobalt the hexagonal {00.1} directions are easy and the {10.0} directions in the basal plane are hard.

Hard and easy directions in crystalline materials come about as a result of spin–orbit coupling. The spin, as usual, determines the occurrence of ferro- or ferrimagnetization, while the orbital motion of the electrons (3d in the case of Fe, Co, and Ni) responds to the structure of the crystal lattice.

The maximum value of coercivity resulting from crystalline anisotropy is given by $H_c = 2K_1/M_s$, where K_1 is the first magnetocrystalline anisotropy constant and M_s is the saturation magnetization.

Shape Anisotropy

A spherical particle has no shape anisotropy, i.e., all directions are equally easy (or hard). For particles (having low crystalline anisotropy) of any other shape, the longest dimension is the easy direction and the shortest dimension is the hardest direction of magnetization. Thus, a needle-shaped (acicular) particle will tend to be magnetized along the long dimension, whereas a particle in the form of a disk will have the axis of the disk as its hard direction, while any direction in the plane of the disk will be equally easy (assuming that shape is the dominant anisotropy).

For an acicular particle, the maximum value of the particle's switching field is (Stoner and Wohlfarth, 1948):

$$H_c = (N_b - N_a)\ M_s$$

where N_b is the demagnetizing factor in the shorter dimension and N_a is the factor for the long axis of the particle. When the ratio $b/a \to \infty$, then:

$$\left.\begin{array}{l} N_a \to 0 \\ N_b \to 2\pi \end{array}\right\} N_a + N_b + N_c \equiv 4\pi$$

and H_c for iron > 10,000 Oe (7.95×10^5 A/m), higher than has been achieved in the laboratory for single-domain iron particles. Particles of iron having $H_c \leqslant 2000$ Oe are widely used in high-quality audio and video tapes.

The reason for the discrepancy is that the simplest single-domain model makes the assumption that the spins on all the atoms in a particle rotate in the same direction and at the same time, i.e., are coherent. This seems to be improbable since switching may begin at different places in the single-domain particle at the same time. Jacobs and Bean (1955) proposed an incoherent mode, *fanning,* in which different segments on a longitudinal chain of atoms rotate in opposite directions. Shtrikman and Treves (1959) introduced another incoherent mode, *buckling*. These incoherent modes of magnetization reversal within single-domain particles not only predicted values of coercivity closer to the observed values, but also they could explain why the observed coercivity values for single-domain particles increased with decreasing particle size (Bate, 1980).

Shape anisotropy also plays an important role in determining the magnetization direction in thin magnetic films. It, of course, favors magnetization in the film plane.

Magnetoelastic Anisotropy

Spin–orbit coupling is also responsible for magnetostriction (the increase or decrease of the dimensions of a body on becoming magnetized or demagnetized). The magnetostriction coefficient λ_s = fractional change of a dimension of the body. It can be positive or negative, and it varies with changes in the direction and magnitude of the applied stress (or internal stress) and of the applied magnetic field. It is highly sensitive to composition, to structure, and to the previous history of the sample. The maximum coercivity is given by the formula $H_c = 3\lambda_s T/M_s$ where λ_s is the saturation magnetostriction coefficient, T is the tension, and M_s the saturation magnetization. Magnetostriction has been put to practical use in the generation of sonar waves for the detection of schools of fish or submarines.

For samples made of single-domain particles, the maximum coercivity for three ferromagnetic metals and one ferrimagnetic oxide (widely used in magnetic recording) is calculated using the preceding formula. Table 15.5 shows maximum coercivity (Oe) for single-domain particles (coherent rotation).

The assumption is made that all the spins rotate so that they remain parallel at all times. This is known as *coherent rotation*. In the case of γ-Fe_2O_3, an incoherent mode of reversal probably occurs since the maximum observed coercivity is only 350 Oe. Several incoherent modes have been proposed, e.g., chain-of-spheres fanning (Jacobs and Bean, 1955), curling (Shtrikman and Treves, 1959). Their characteristics

TABLE 15.5 Maximum Coercivity (Oe) for Single-Domain Particles (Coherent Rotation)

	Iron	Cobalt	Nickel	γ-Fe_2O_3
Crystalline	250	3000	70	230
Strain	300	300	2000	<10
Shape (10:1)	5300	4400	1550	2450

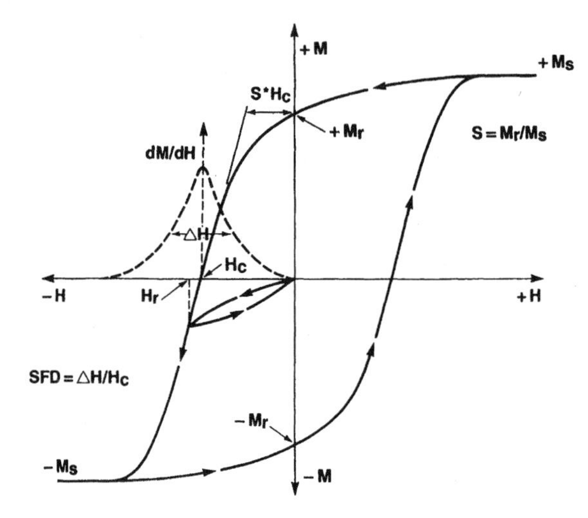

FIGURE 15.8 The switching field distribution (SFD) for a magnetic recording medium can be obtained in two ways: (1) SFD = $\Delta H/H_c$ or (2) SFD = $1 - S^*$.

and differences are discussed by Bate (1980). The coercivity of particles of γ-Fe_2O_3 is increased (in order to make recording tapes of extended frequency response) by precipitating cobalt hydroxide on the surface of the particles. After gentle warming, the cobalt is incorporated on the surface of the particles and increases the coercivity to 650 Oe (51 · 73 kA/m).

Figure 15.8 illustrates two additional extrinsic magnetic properties of importance. They are the remanence coercivity, H_r, and the switching field distribution (SFD). These are of particular importance in magnetic particles used in magnetic tapes (audio, video, or data) or magnetic disks. The coercivity, H_c, of a magnetic material is the value of the magnetic field (the major loop) at which $M = 0$. However, if the applied field is allowed to go to zero, a small magnetization remains. It is necessary to increase the applied field from $H_c > H_r$ (Figure 15.8) to achieve $M_r = 0$. H_r is the *remanence coercivity* and is more relevant than is H_c in discussing the writing process in magnetic recording since it corresponds to the center of the remanent magnetization transition on the recording medium.

Particles, for example, in a magnetic recording do not all reverse their magnetization at the same field; there is a distribution of switching fields, which can be found by differentiating M with respect to H around the point H_c. The result is shown as the broken curve on Figure 15.8, where the SFD = $\Delta H/H_c$, ΔH being the width at half the maximum of the curve. Typically, SFD = 0.2 to 0.3 for a high-quality particulate medium and smaller than this for thin-film recording media.

Figure 15.8 also shows the construction used to find the parameter S^*. The quantity $1 - S^*$ is found to be very close to $\Delta H/H_c$, and it is quicker to evaluate. Either of these parameters can be used to determine the distribution of switching fields, small values of which are required to achieve high recording densities.

Amorphous Magnetic Materials

Before 1960, all the known ferro- and ferrimagnetic materials were crystalline. Because the occurrence of these magnetic states is known to depend on short-range interaction between atoms, there is no reason why amorphous materials (which have *only* short-range order) should not have useful magnetic properties. This was found to be true in 1960, when thin, amorphous ribbons of $Au_{81}Si_{19}$ were made by rapidly cooling the molten alloy through the melting point and the lower glass transition temperature. Because there is always at least one crystalline phase more stable than the amorphous state, the problem is to invent a production method that yields the amorphous phase rather than the crystalline one.

Most methods involve cooling the molten mixture so rapidly that there is insufficient time for crystals to form. Cooling rates of 10^5 to 10^6 degrees per second are needed and can be achieved in several ways:

1. Pouring the molten mixture from a silica crucible onto the edge of a rapidly rotating copper wheel. This yields a ribbon of the amorphous alloy, typically 1 mm wide and 25 μm thick.
2. Depositing a thin film from a metal vapor or a solution of metal ions.
3. Irradiating a thin sample of the metal with high-energy particles.

Once an amorphous alloy is formed, it will remain indefinitely in the *glassy* state at room temperature. The problem is that only *thin* films are obtained, and large areas are required to make, for example, the core of a transformer.

There are three main groups of amorphous films:

1. Metal-metalloid alloys, e.g., $Au_{81}Si_{19}$
2. Late transition–early transition metal alloys, e.g., $Ni_{60}Nb_{40}$
3. Simple metal alloys, e.g., $Cu_{65}Al_{35}$

When normal metals freeze, crystallization begins at a fixed temperature, the *liquidus*, T_e. In amorphous alloys *configurational freezing* occurs at a lower temperature, the *glass temperature*, T_g, which is not as well defined as T_e. There is an abrupt increase in the time required for the rearrangement of the atoms, from 10^{-12} sec for liquids to 10^5 sec (a day) for glasses. Not surprisingly, this increase in atomic rearrangement time is associated with an abrupt increase in viscosity, from 10^{-2} poise for liquids, e.g., water or mercury, to 10^{15} poise for glasses.

The principal difference between magnetic glasses and ferromagnetic alloys is that the glasses are completely isotropic (all directions of magnetization are very easy directions), and consequently, considering only the magnetic properties, soft amorphous alloys are almost ideally suited for use in the core of power transformers or magnetic recording heads, where almost zero remanence and coercivity are desired at frequencies up to megahertz. The limit on their performance seems to be the magnetic anisotropy which arises from strains generated during the manufacturing process.

When an amorphous material is required to store energy (as in a permanent magnet) or information (as in magnetic bubbles or thermomagneto-optic films), it must have magnetic anisotropy. This is generally produced by applying a magnetic field at high temperatures to the amorphous material. The field and temperature must be high enough to allow a local rearrangement of atoms to take place in order to create the desired degree of magnetocrystalline anisotropy.

Amorphous materials will apparently play increasingly important roles as magnetic materials. To accelerate their use, we need to have answers to the questions "What governs the formation of amorphous materials?" and "What is the origin of their anisotropy and magnetostriction?".

Defining Terms

Coercivity, H_c (Oe, A/m): The property of a magnetized body enabling it to resist reversal of its magnetization.

Compensation temperature, T_c (°C, K): The temperature at which the magnetization of a material comprising ferromagnetic atoms (e.g., Fe, Co, Ni) and rare earth atoms (e.g., Gd, Tb) becomes zero because the magnetization of the sublattice of ferromagnetic atoms is canceled by the opposing magnetization of the rare earth sublattice.

Curie temperature, Θ_c (°C, K): The temperature at which the spontaneous magnetization of a ferromagnetic or ferrimagnetic body becomes zero.

Remanence, M_r (emu/cc, A/m): The property of a magnetized body enabling it to retain its magnetization.

References

G. Bate, in *Recording Materials in Ferromagnetic Materials,* vol. 2, Amsterdam: North-Holland, 1980, pp. 381–507.

G. Bate, *J. Magnetism and Magnetic Materials,* vol. 100, 1991, pp. 413–424.

F. Brailsford, *Physical Principles of Magnetism*, London: Van Nostrand, 1966.

J. Crangle, "Ferromagnetism and antiferromagnetism in non-ferrous metals and alloys," *Metall. Rev.*, 1962, pp. 133–174.

I.S. Jacobs and C.P. Bean, *Phys. Rev.*, vol. 100, p. 1060, 1955.

K. Moorjani and J.M.D. Coey, *Magnetic Glasses: Methods and Phenomena, Their Application in Science and Technology*, vol. 6, Amsterdam: Elsevier, 1984.

S. Shtrikman and D. Treves, *J. Phys. Radium*, vol. 20, p. 286, 1959.

J.C. Slater, *Phys. Rev.*, vol. 36, p. 57, 1930.

E.C. Stoner and E.P. Wohlfarth, *Phil. Trans. Roy. Soc.*, vol. A240, p. 599, 1948.

K.J. Strnat, *Proceedings of Symposium on Soft and Hard Magnetic Materials with Applications*, vol. 8617–005, Metals Park, OH: American Society of Metals, 1986.

Further Information

A substantial fraction of the papers published in English on the technologically important aspects of magnetism appear in the *IEEE Transactions on Magnetics* or in the *Journal of Magnetism and Magnetic Materials.*

The two major annual conferences are Intermag (proceedings published in the *IEEE Transactions on Magnetics*) and the Magnetism and Magnetic Materials Conference, MMM (proceedings published in the American Physical Society's *Journal of Applied Physics*).

15.2 Magnetic Recording

Mark H. Kryder

Magnetic recording is used in a wide variety of applications and formats, ranging from relatively low-density, low-cost floppy disk drives and audio recorders to high-density videocassette recorders, digital audio tape recorders, computer tape drives, rigid disk drives, and instrumentation recorders. The storage density of this technology has been advancing at a very rapid pace. With a storage density exceeding 1 Gbit/in.^{2}, magnetic recording media today can store the equivalent of about 50,000 pages of text on one square inch. This is more than 500,000 times the storage density on the RAMAC, which was introduced in 1957 by IBM as the first disk drive for storage of digital information. The original Seagate 5.25-in. magnetic disk drive, introduced in 1980, stored just 5 Mbytes. Today, instead of storing megabytes, 5.25-in. drives store tens of gigabytes, and drives as small as 2.5 in. store over a gigabyte.

This astounding rate of progress shows no sign of slowing. Fundamental limits to magnetic recording density are still several orders of magnitude away, and recent product announcements and laboratory demonstrations indicate the industry is accelerating the rate of progress rather than approaching practical limits. Recently IBM demonstrated the feasability of storing information at a density of 3 Gbit/in.^{2} (Tsang et al., 1996). Similar advances can also be expected in audio and video recording.

Fundamentals of Magnetic Recording

Although magnetic recording is practiced in a wide variety of formats and serves a wide variety of applications, the fundamental principles by which it operates are similar in all cases. The fundamental magnetic recording configuration is illustrated in Figure 15.9. The recording head consists of a toroidally shaped core of soft magnetic material with a few turns of conductor around it. The magnetic medium below the head could be either tape or disk, and the substrate could be either flexible (for tape and floppy disks) or rigid (for rigid disks). To record on the medium, current is applied to the coil around the core of the head, causing the high-permeability magnetic core to magnetize. Because of the gap in the recording head, magnetic flux emanates from the head and penetrates the medium. If the field produced by the head is sufficient to overcome the **coercive force** of the medium, the medium will be magnetized by the head field. Thus, a representation of the current waveform applied to the head is stored in the magnetization pattern in the medium.

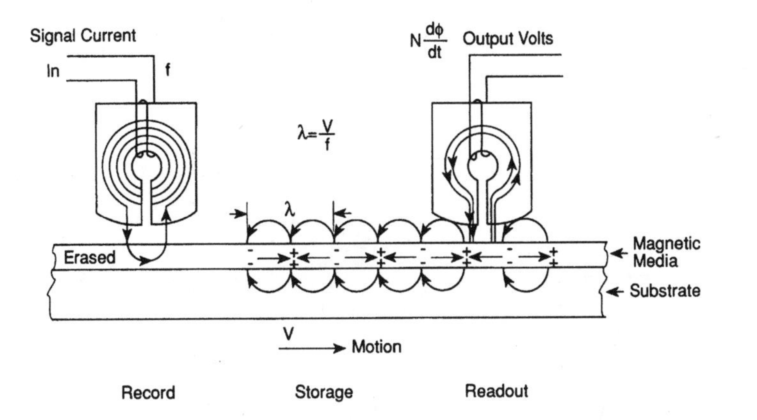

FIGURE 15.9 The fundamental magnetic recording configuration.

Readout of previously recorded information is typically accomplished by using the head to sense the magnetic stray fields produced by the recorded patterns in the medium. The recorded patterns in the medium cause magnetic stray fields to emanate from the medium and to flow through the core of the head. Thus, if the medium is moved with respect to the head, the flux passing through the coil around the head will change in a manner which is representative of the recorded magnetization pattern in the medium. By Faraday's law of induction, a voltage representative of the recorded information is thus induced in the coil.

The Recording Process

During recording the head is used to produce large magnetic fields which magnetize the medium. It was shown by Karlqvist (1954) that, in the case where the track width and length of the poles along the gap are both large compared to the gap length, the fields produced by a recording head could be described by:

$$H_x = \frac{NI}{\pi g}\left[\tan^{-1}\left(\frac{x+g/2}{y}\right) - \tan^{-1}\left(\frac{x-g/2}{y}\right)\right] \tag{15.1a}$$

$$H_y = \frac{NI}{2\pi g}\ln\left[\frac{(x+g/2)^2+y^2}{(x-g/2)^2+y^2}\right] \tag{15.1b}$$

where H_x and H_y are the longitudinal and perpendicular components of field, as indicated by the coordinates in Figure 15.10, N is the number of turns on the head, I is the current driving the head, and g is the gap width of the head. In this approximation, the contours of equal longitudinal field are described by circles which intersect the gap corners as shown in Figure 15.10.

In digital or saturation recording, the recording head is driven with sufficiently large currents that a portion of the recording medium is driven into saturation. However, because of the gradient in the head fields, other portions of the medium see fields less than those required for saturation. This is illustrated in Figure 15.10 where the contours for three different longitudinal fields are drawn. In this figure H_{cr} is the **remanence coercivity** or the field required to produce zero **remanent magnetization** in the medium after it was saturated in the opposite direction, and H_1 and H_2 are fields which would produce negative and positive remanent magnetization, respectively. Note that the head field gradient is the sharpest near the pole tips of the head. This means that smaller head-to-medium spacing and thinner medium both lead to narrower transitions being recorded.

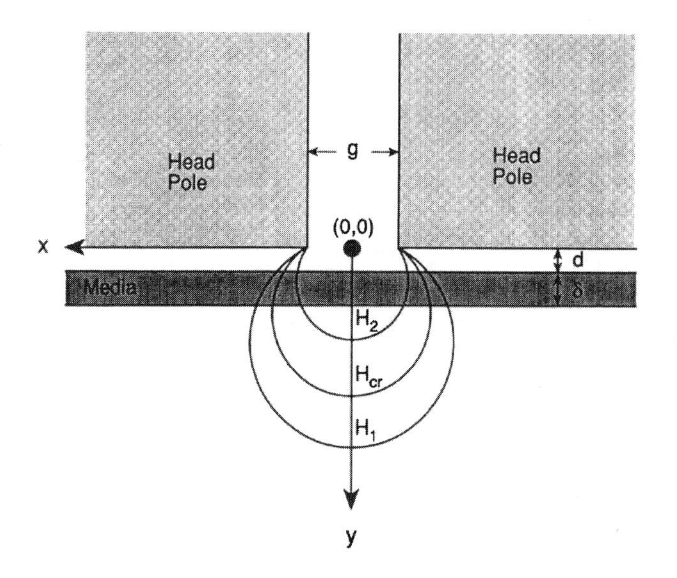

FIGURE 15.10 The constant longitudinal field contours in the gap region of a recording head.

Modeling the recording process involves convolving the head field contours with the very nonlinear and hysteretic magnetic properties of the recording medium. A typical *M–H* hysteresis loop for a longitudinal magnetic recording medium is shown in Figure 15.11. Whether the medium has positive or negative magnetization depends upon not only the magnetic field applied but the past history of the magnetization. If the medium was previously saturated at $-M_s$, then when the magnetic field H is reduced to zero, the remanent magnetization will be $-M_r$; however, if it was previously saturated at $+M_s$, then the remanent magnetization would be $+M_r$. Similarly, if the medium was initially saturated to $-M_s$, then magnetized by a field $+H_1$, and finally allowed to go to a remanent state, the magnetization would go to value M_1. This hysteretic behavior is the basis for the use of the medium for long-term storage of information but makes the recording process highly nonlinear.

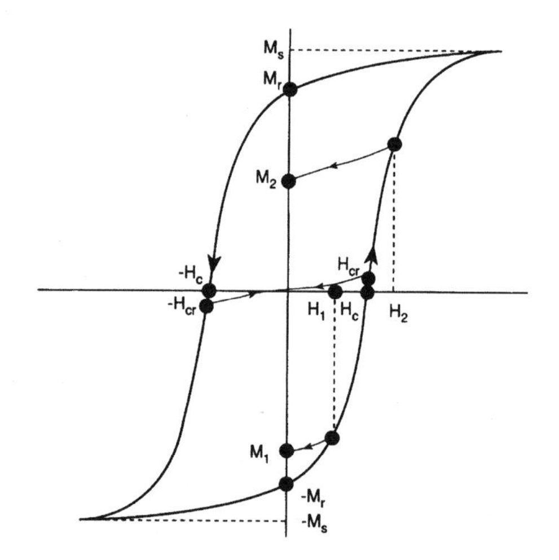

FIGURE 15.11 A remanent *M–H* hysteresis loop for a longitudinal recording medium.

An additional complicating factor in determining the actual recorded pattern is the **demagnetizing field** of the medium itself. As shown in Figure 15.9, transitions in the recorded magnetization direction produce effective magnetostatic charge given by

$$\rho_M = -\nabla \cdot \vec{M} \tag{15.2}$$

which in turn results in demagnetizing fields. The demagnetizing fields outside the medium are what is sensed by the head during readback, but demagnetizing fields also exist inside the medium and act to alter the total field seen by the medium during the recording process from that of the head field alone.

Taking into account the head field gradients, the nonlinear *M–H* loop characteristics of the medium and the demagnetizing fields, Williams and Comstock (1971) developed a model for the recording process. This model predicts the width of a recorded transition, in a material with a square hysteresis loop, to be

$$a = \sqrt{\frac{M_r \delta d}{\pi H_c}} \tag{15.3}$$

where δ is the medium thickness, d is the head-medium spacing, M_r is the remanent magnetization of the medium, and H_c is the coercivity of the medium. That the transition widens with the product $M_r \cdot \delta$ is a result of the fact that the demagnetizing fields increase linearly with this quantity. Similarly, the transition narrows as H_c is increased, because with high coercivity, the medium can resist the transition broadening due to the demagnetizing fields. The increase in transition width with d is due to the fact that a poorer head field gradient is obtained with larger head-medium spacing.

The nonlinearities of the recording process can be largely removed by a technique referred to as ac bias recording. This is frequently used in analog recording in audio and video recorders. In this technique, a high-frequency ac bias is added to the signal to be recorded. This ac bias signal is ramped from a value much larger than the coercivity of the medium to zero. This removes the hysteretic behavior of the medium and causes it to assume a magnetization state which represents the minimum energy state determined by the amount of field produced by the signal to be recorded.

The Readback Process

As opposed to the recording process, the readback process can usually be modeled as a linear process. This is because the changes in magnetization which occur in either the head or the medium during readback are typically small.

The most common way to model the readback process is to use the principle of reciprocity, which states that the flux produced by the head through a cross-section of an element of the medium, normalized by the number of ampere turns of current driving the head, is equal to the flux produced in the head by the element of medium, normalized by the equivalent current required to produce the magnetization of that element. For a magnetic recording head, which produces the longitudinal field $H_x(x,y)$ when driven by NI ampere turns of current, this principle leads to the following expression for the voltage induced in the head by a recording medium with magnetization $M(x,y)$ and moving with velocity v relative to the head:

$$e = \frac{\mu_o W v}{I} \int_{d}^{d+\delta} \int_{-\infty}^{\infty} \frac{\partial M_x(x-\bar{x}, y)}{\partial \bar{x}} H_x(x,y) \mathrm{d}x\mathrm{d}y \tag{15.4}$$

where W is the track width of the head.

This expression shows that the readback voltage induced in the recording head is linearly dependent upon the magnitude of the magnetization in the recording medium being sensed and the relative head-to-medium velocity. The linearity of the readback process ensures that analog recordings such as those recorded on audio or video tapes are faithfully reproduced.

Magnetic Recording Media

A wide variety of magnetic recording media are available today. Different applications require different media, but furthermore, in many cases the same application will be able to utilize a variety of different competitive media.

Just a decade ago, essentially all recording media consisted of fine acicular magnetic particles embedded in a polymer and coated onto either flexible substrates such as mylar for floppy disks and tapes or onto rigid aluminum-alloy substrates for rigid disks. Today, although such particulate media are still widely used for tape and floppy disks, thin-film media have almost entirely taken over the rigid disk business, and metal-evaporated thin-film media has been introduced into the tape marketplace. Furthermore, many new particle types have been introduced.

The most common particulate recording media today are γ-Fe_2O_3, Co surface-modified γ-Fe_2O_3, CrO_2, and metal particle media. All of these particles are acicular in shape with aspect ratios on the order of 5 or 10 to 1. The particles are sufficiently small that it is energetically most favorable for them to remain in a single-domain saturated state. Because of demagnetizing effects caused by the acicular shape, the magnetization prefers to align along the long axis of the particle.

As was noted in the discussion of Equation (15.3), to achieve higher recording densities requires media with higher coercivity. The coercivity of a particle is determined by the field required to cause the magnetization to switch by 180°. If the magnetization remained in a single-domain state during the switching process, then the coercivity should be given by

$$H_c = (N_a - N_b)M_s \tag{15.5}$$

where N_a and N_b are the demagnetizing factors in the directions transverse and parallel to the particle axis, respectively. In practice the coercivity is measured to be less than this. This has been explained as being a result of the fact the magnetization does not remain uniform during the switching process, but switches inhomogeneously (White, 1984). In addition to the effects which the shape anisotropy of the particles has on the coercivity, crystalline anisotropy can also be used to control coercivity.

The coercivity of the medium which is made from the particles is determined by the distribution of coercivities of the particles from which it is made, their orientation in the medium relative to the fields from the head, and their interactions among each other. The coercivities of a variety of particulate recording media are summarized in Table 15.6.

Although coercivity is indeed an important parameter for magnetic recording media, it is by no means the only one. Particle size affects the medium noise because, at any time, the head is sensing a fixed volume of the medium. Because the particles are quantized and there are statistical variations in their switching behavior, the medium power signal-to-noise ratio varies linearly with the number of particles contained in that volume. To reduce particulate medium noise, it is therefore generally desirable to use small particles.

There is a limit, however, to how small particles may be made and still remain stable. When the thermal energy kT is comparable in magnitude to the energy required to switch a particle, $M \cdot H_c$, the particle becomes unstable and may switch because of thermal excitation. This phenomenon is known as **superparamagnetism** and can lead to decay of recorded magnetization patterns over time.

The remanent magnetization of a medium is important because it directly affects the signal level during readback as shown by Equation (15.4). The remanent magnetizations of several particulate media are listed in Table 15.6. Obtaining high remanent magnetization in particulate media requires the use of particles with high **saturation magnetization** and a high-volume packing fraction of particles in the polymer binder. Obtaining a high-volume packing fraction of particles in the binder, however, can lead to nonuniform distributions of particles and agglomerates of many particles, which switch together, also causing noise during readback.

TABLE 15.6 Magnetic Material, Saturation Remanence M_r (∞), Coercivity H_c, Switching-Field Distribution Δh_r, and Number of Particles per Unit Volume, N, of Various Particulate Magnetic Recording Media

Application	Material	M_r (∞), kA/m (emu/cm^3)	H_c, kA/m (4π Oe)	Δ h	N, $10^3/\mu m^3$
Reel-to-reel audio tape	γ-Fe_2O_3	100–120	23–28	0.30–0.35	0.3
Audio tape IEC I	γ-Fe_2O_3	120–140	27–32	0.25–0.35	0.6
Audio tape IEC II	CrO_2	120–140	38–42	0.25–0.35	1.4
	γ-Fe_2O_3 + Co	120–140	45–52	0.25–0.35	0.6
Audio tape IEC IV	Fe	230–260	80–95	0.30–0.37	3
Professional video tape	γ-Fe_2O_3	75	24	0.4	0.1
	CrO_2	110	42	0.3	1.5
	γ-Fe_2O_3 + Co	90	52	0.35	1
Home video tape	CrO_2	110	45–50	0.35	2
	γ-Fe_2O_3 + Co	105	52–57	0.35	1
	Fe	220	110–120	0.38	4
Instrumentation tape	γ-Fe_2O_3	90	27	0.35	0.6
	γ-Fe_2O_3 + Co	105	56	0.50	0.8
Computer tape	γ-Fe_2O_3	87	23	0.30	0.16
	CrO_2	120	40	0.29	1.4
Flexible disk	γ-Fe_2O_3	56	27	0.34	0.3
	γ-Fe_2O_3 + Co	60	50	0.34	0.5
Computer disk	γ-Fe_2O_3	56	26–30	0.30	0.3
	γ-Fe_2O_3 + Co	60	44–55	0.30	0.5

Source: E. Köster and T.C. Arnoldussen, "Recording media," in *Magnetic Recording*, C.D. Mee and E.D. Daniel, Eds., New York: McGraw-Hill, 1987. With permission.

Generally, then, to obtain good high-density particulate recording media it is desired to have adequate coercivity (to achieve the required recording density), small particles (for low noise), with a very narrow switching field distribution (to obtain a narrow transition), to have them oriented along the direction of recording (to obtain a large remanence), and to have them uniformly dispersed (to obtain low modulation noise), with high packing density (to obtain large signals).

Thin-film recording media generally have excellent magnetic properties for high-density recording. Because they are nearly 100% dense (voids at the grain boundaries reduce the density somewhat), they can be made to have the highest possible magnetization. Because of their high magnetization, they can be made extremely thin and still provide adequate signal during readback. This helps narrow the recorded transition since the head field gradient is sharper for thinner media, as was discussed in reference to Figure 15.10.

Thin-film media can also be made extremely smooth. To achieve the smallest possible head-to-medium spacing and therefore the sharpest head field gradient and the least spacing loss, smooth media are required.

The coercivity of thin-film media can also be made very high. In volume production today are media with coercivities of 160 kA/m; however, media with coercivities to 250 kA/m have been made and appear promising (Velu and Lambeth, 1992). Such high coercivities are adequate to achieve more than an order of magnitude higher recording density than today.

Numerical models indicate that noise in thin-film media increases when the grains in polycrystalline films are strongly exchange coupled (Zhu and Bertram, 1988). Exchange coupled films tend to exhibit zigzag transitions, which produce considerable jitter in the transition position relative to the location where the record current in the head goes through zero. A variety of experimental studies have indicated that the introduction of nonmagnetic elements which segregate to the grain boundaries and careful control of the sputtering conditions to achieve a porous microstructure at the grain boundaries reduce such transition jitter (Chen and Yamashita, 1988).

Magnetic Recording Heads

Early recording heads consisted of toroids of magnetically soft ferrites, such as NiZn–ferrite and MnZn–ferrite, with a few turns of wire around them. For high-density recording applications, however, ferrite can no longer be used, because the saturation magnetization of ferrite is limited to about 400 kA/m. Saturation of the pole tips of a ferrite head begins to occur when the deep gap field in the head approaches one-half the saturation magnetization of the ferrite. Because the fields seen by a medium are one-half to one-quarter the deep gap field, media with coercivities above about 80 kA/m cannot be reliably written with a ferrite head. High-density thin-film disk media, metal particle media, and metal evaporated media, therefore, cannot be written with a ferrite head.

Magnetically soft alloys of metals such as Permalloy (NiFe) and Sendust (FeAlSi) have saturation magnetizations on the order of 800 kA/m, about twice that of ferrites, but because they are metallic may suffer from eddy current losses when operated at high frequencies. To overcome the limitations imposed by eddy currents, they are used in layers thinner than a skin depth at their operating frequency. To prevent saturation of the ferrite heads, the high magnetization metals are applied to the pole faces of the ferrite, making a so-called metal-in-gap or MiG recording head, as shown in Figure 15.12. Since the corners of the pole faces are the first parts of a ferrite head to saturate, the high magnetization metals enable these MiG heads to be operated to nearly twice the field to which a ferrite head can be operated. Because the layer of metal is thin, it can furthermore be less than a skin depth, and eddy current losses do not limit performance at high frequencies.

Yet another solution to the saturation problem of ferrite heads is to use thin-film heads. Thin-film heads are made of Permalloy and are therefore metallic, but the films are made sufficiently thin that they are thinner than the skin depth and, consequently, the heads operate well at high frequencies. A diagram of a thin-film head is shown in Figure 15.13. It consists of a bottom yoke of Permalloy, some insulating layers, a spiral conductor, and a top yoke of Permalloy, which is joined to the bottom yoke at the back gap but separated from it by a thin insulator at the recording gap. These thin-film heads are made using photolithography and microfabrication techniques similar to those used in the manufacture of semiconductor devices. The thin pole tips of these heads actually sharpen the head field function and, consequently, the pulse shape produced by an isolated transition, although at the expense of some undershoot, as illustrated in Figure 15.14.

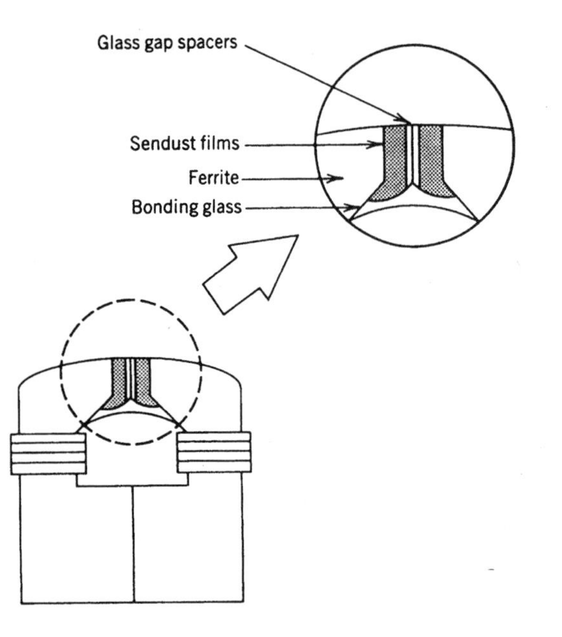

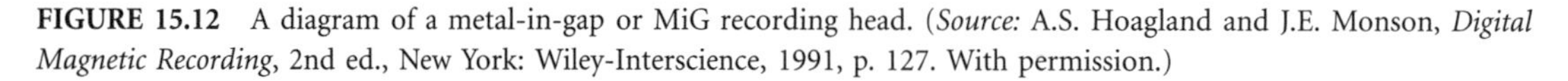
FIGURE 15.12 A diagram of a metal-in-gap or MiG recording head. (*Source:* A.S. Hoagland and J.E. Monson, *Digital Magnetic Recording*, 2nd ed., New York: Wiley-Interscience, 1991, p. 127. With permission.)

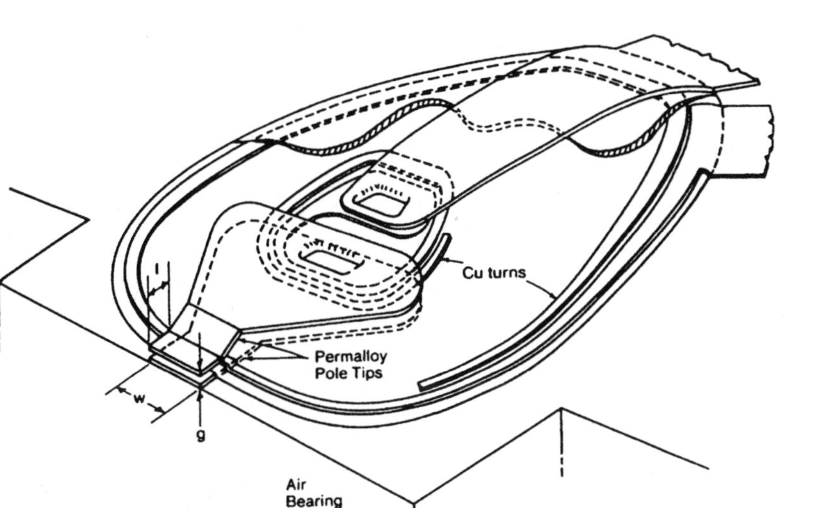

FIGURE 15.13 A thin-film head. (*Source:* R.M. White, Ed., *Introduction to Magnetic Recording*, New York: IEEE Press, p. 28. © 1985 IEEE.)

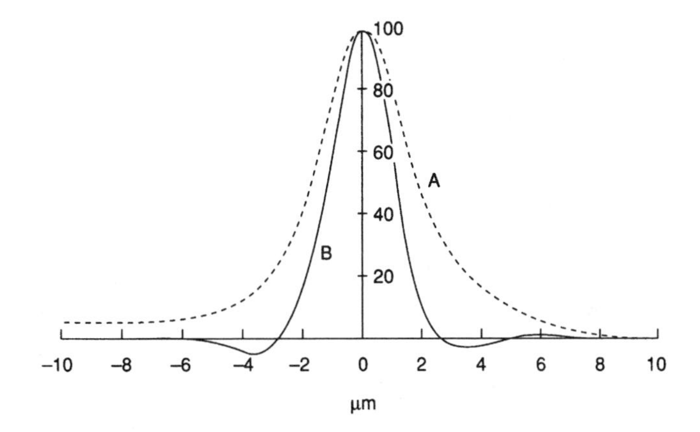

FIGURE 15.14 Pulse shapes for (curve A) long- and (curve B) short-pole heads normalized for equal amplitude. (*Source:* E.P. Valstyn and L.F. Shew, "Performance of single-turn film heads," *IEEE Trans. Magnet.*, vol. MAG-9, no. 3, p. 317. © 1973 IEEE.)

Because thin-film heads are made by photolithographic techniques, they can be made extremely small and to have low inductance. This, too, helps extend the frequency of operation.

A relatively new head which is now being used for readback of information in high-density recording is the magnetoresistive (MR) head. MR heads are based on the phenomenon of **magnetoresistance**, in which the electrical resistance of a magnetic material is dependent upon the direction of magnetization in the material relative to the direction of current flow. An unshielded MR head is depicted in Figure 15.15. Current flows in one end of the head and out the other. The resistivity of Permalloy from which the head is made varies as

$$\rho = \rho_o + \Delta\rho \cos^2\theta \tag{15.6}$$

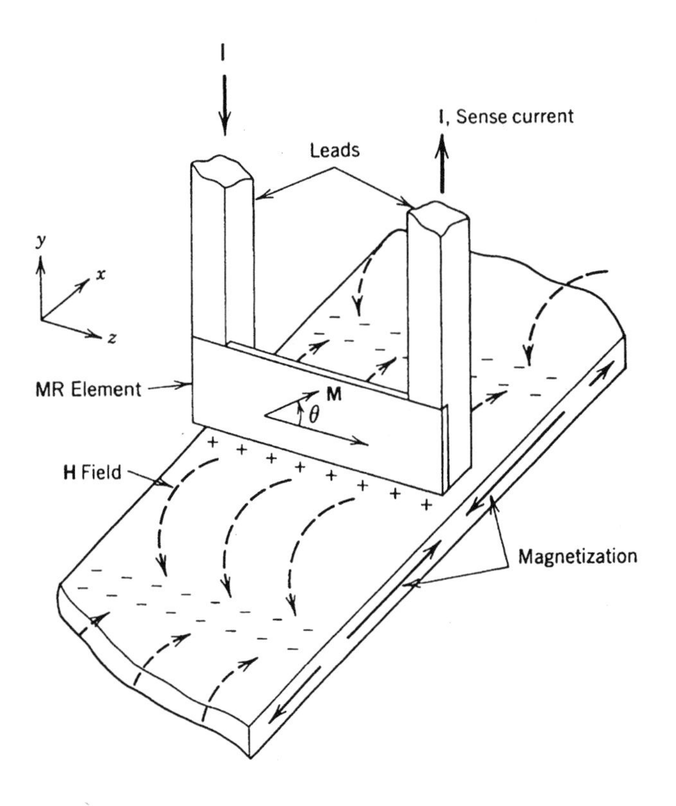

FIGURE 15.15 An unshielded magnetoresistive element. (*Source:* A.S. Hoagland and J.E. Monson, *Digital Magnetic Recording*, 2nd ed., New York: Wiley-Interscience, 1991, p. 131. With permission.)

where θ is the angle which the magnetization in the Permalloy makes relative to the direction of current flow, ρ_o is the isotropic resistivity, and $\Delta\rho$ is the magnetoresistivity. When the recording medium with a changing magnetization pattern moves under the MR head, the stray fields from the medium cause a change in the direction of magnetization and, consequently, a change in resistance in the head. With a constant current source driving the head, the head will therefore exhibit a change in voltage across its terminals.

Magnetoresistive heads are typically more sensitive than inductive heads and therefore produce larger signal amplitudes during readback. The increased sensitivity and the fact that the read head is independent of the write head can be used to make a write/read head combination in which the write head writes a wider track than the read head senses. Thus, adjacent track interference is reduced during the readback process.

Another advantage of the MR head is that it senses magnetic flux ϕ, not the time rate of change of flux $d(\phi)/dt$ as an inductive head does. Consequently, whereas the inductive head output voltage is dependent upon the head-to-medium velocity as was shown by Equation (15.4), the output voltage of an MR head is independent of velocity.

Conclusions

Magnetic recording today is used in a wide variety of formats for a large number of applications. Formats range from tape, which has the highest volumetric packing density and lowest cost per bit stored, to rigid disks, which provide fast access to a large volume of data. Applications include computer data storage, audio and video recording, and collecting data from scientific instruments.

The technology has increased storage density by more than a factor of 1,000,000 over the past 35 years since it was first used in a disk format for computer data storage; however, fundamental limits, set by superparamagnetism, are estimated yet to be a factor of more than 1000 from where we are today. Furthermore, recent product announcements and developments in research labs suggest that the rate of progress is likely to accelerate. Storage densities of over 5 Gbit/in.^{2} are likely by the end of this decade and densities of 100 Gbit/in.^{2} appear likely in the early twenty-first century.

Defining Terms

Coercive force or coercivity: The magnetic field required to reduce the mean magnetization of a sample to zero after it was saturated in the opposite direction.

Demagnetizing field: The magnetic field produced by divergences in the magnetization of a magnetic sample.

Magnetoresistance: The resistance change produced in a magnetic sample when its magnetization is changed.

Remanence coercivity: The magnetic field required to produce zero remanent magnetization in a material after the material was saturated in the opposite direction.

Remanent magnetization: The magnetic moment per unit volume of a material in zero field.

Saturation magnetization: The magnetic moment per unit volume of a material when the magnetization in the sample is aligned (saturated) by a large magnetic field.

Superparamagnetism: A form of magnetism in which the spins in small particles are exchange coupled but may be collectively switched by thermal energy.

References

T. Chen and T. Yamashita, "Physical origin of limits in the performance of thin-film longitudinal recording media," *IEEE Trans. Magnet.*, vol. MAG-24, p. 2700, 1988.

A.S. Hoagland and J.E. Monson, *Digital Magnetic Recording*, New York: John Wiley & Sons, 1991.

O. Karlqvist, "Calculation of the magnetic field in the ferromagnetic layer of a magnetic drum," *Trans. Roy. Inst. Technol., Stockholm*, No. 86, 1954. Reprinted in R.M. White, Ed., *Introduction to Magnetic Recording*, New York: IEEE Press, 1985.

E. Köster and T.C. Arnoldussen, "Recording media," in *Magnetic Recording*, C.D. Mee and E.D. Daniel, Eds., New York: McGraw-Hill, 1987.

M.-M. Tsang, H. Santini, T. Mccown, J. Lo, and R. Lee, "3 Gbit/in.^{2} recording demonstration with dual element heads of thin film drives," *IEEE Trans Magnet.*, MAG-32, p. 7, 1996.

E.P. Valstyn and L.F. Shew, "Performance of single-turn film heads," *IEEE Trans. Magnet.*, vol. MAG-9, p. 317, 1973.

E. Velu and D. Lambeth, "High Density Recording on SmCo/Cr Thin Film Media," Paper KA-01, Intermag Conference, St. Louis, April 1992; to be published in *IEEE Trans. Magnet.*, vol. MAG-28, 1992.

R.M. White, *Introduction to Magnetic Recording*, New York: IEEE Press, p. 14, 1984.

M.L. Williams and R.L. Comstock, "An analytical model of the write process in digital magnetic recording," *AIP Conf. Proc.*, part 1, no. 5, 1971, pp. 738–742.

J-G. Zhu and H.N. Bertram, "Recording and transition noise simulations in thin film media," *IEEE Trans. Magnet.*, vol. MAG-24, p. 2706, 1988.

Further Information

There are several books which provide additional information on magnetic and magneto-optic recording. They include the following:

R.M. White, *Introduction to Magnetic Recording*, New York: IEEE Press, 1984.

C.D. Mee and E.D. Daniel, *Magnetic Recording*, New York: McGraw-Hill, 1987.

A.S. Hoagland and J.E. Monson, *Digital Magnetic Recording*, New York: John Wiley & Sons, 1991.

16

Wave Propagation

Matthew N.O. Sadiku
Prairie View A&M University

Kenneth Demarest
University of Kansas

16.1 Space Propagation

Matthew N.O. Sadiku

This section summarizes the basic principles of electromagnetic (EM) **wave propagation** in space. The principles essentially state how the characteristics of the Earth and the atmosphere affect the propagation of EM waves. Understanding such principles is of practical interest to communication system engineers. Engineers cannot competently apply formulas or models for communication system design without an adequate knowledge of the propagation issue.

Propagation of an EM wave may be regarded as a means of transferring energy or information from one point (a transmitter) to another (a receiver). EM wave propagation is achieved through guided structures such as transmission lines and waveguides or through space. Wave propagation through waveguides and microstrip lines will be treated in Section "Waveguides". In this section, our major focus is on EM wave propagation in space and the power resident in the wave.

For a clear understanding of the phenomenon of EM wave propagation, it is expedient to break the discussion of propagation effects into categories represented by four broad frequency intervals (Collin, 1985):

- Very low frequencies (VLF), 3 to 30 kHz
- Low-frequency (LF) band, 30 to 300 kHz
- High-frequency (HF) band, 3 to 30 MHz
- Above 50 MHz

In the first range, a wave propagates as in a waveguide, using the Earth's surface and the ionosphere as boundaries. Attenuation is comparatively low, and hence VLF propagation is useful for long-distance worldwide telegraphy and submarine communication. In the second frequency range, the availability of increased bandwidth makes standard AM broadcasting possible. Propagation in this band is by means of surface waves due to the presence of the ground. The third range is useful for long-range broadcasting services via sky wave reflection and refraction by the ionosphere. Basic problems in this band include fluctuations in the ionosphere and a limited usable frequency range. Frequencies above 50 MHz allow for line-of-sight space wave propagation, FM radio and TV channels, radar and navigation systems, and so on. In this band, due consideration must be given to reflection from the ground, refraction by the troposphere, scattering by atmospheric hydrometeors, and **multipath** effects of buildings, hills, trees, etc.

EM wave propagation can be described by two complementary models. The physicist attempts a theoretical model based on universal laws, which extends the field of application more widely than currently known. The engineer prefers an empirical model based on measurements that can be used immediately. This section presents complementary standpoints by discussing theoretical factors affecting wave propagation and the semiempirical rules allowing handy engineering calculations. First, we consider wave propagation in idealistic simple media, with no obstacles. We later consider the more realistic case of wave propagation around the Earth, as influenced by its curvature and by atmospheric conditions.

Propagation in Simple Media

The conventional propagation models, on which the basic calculation of radio links is based, result directly from Maxwell's equations:

$$\nabla \cdot \mathbf{D} = \rho_v \tag{16.1}$$

$$\nabla \cdot \mathbf{B} = 0 \tag{16.2}$$

$$\nabla \times \mathbf{E} = -\frac{\partial \mathbf{B}}{\partial t} \tag{16.3}$$

$$\nabla \times \mathbf{H} = \mathbf{J} + \frac{\partial \mathbf{D}}{\partial t} \tag{16.4}$$

In these equations, **E** is electric field strength in volts per meter, **H** is magnetic field strength in amperes per meter, **D** is electric flux density in coulombs per square meter, **B** is magnetic flux density in webers per square meter, **J** is conduction current density in amperes per square meter, and ρ_v is electric charge density in coulombs per cubic meter. These equations go hand in hand with the constitutive equations for the medium:

$$\mathbf{D} = \epsilon \mathbf{E} \tag{16.5}$$

$$\mathbf{B} = \mu \mathbf{H} \tag{16.6}$$

$$\mathbf{J} = \sigma \mathbf{E} \tag{16.7}$$

where $\epsilon = \epsilon_o \epsilon_r$, $\mu = \mu_o \mu_r$, and σ are the permittivity, the permeability, and the conductivity of the medium, respectively.

Consider the general case of a lossy medium which is charge-free ($\rho_v = 0$). Assuming time-harmonic fields and suppressing the time factor $e^{j\omega t}$, Equation (16.1) to Equation (16.7) can be manipulated to yield Helmholtz's wave equations:

$$\nabla^2 \mathbf{E} - \gamma^2 \mathbf{E} = 0 \tag{16.8}$$

$$\nabla^2 \mathbf{H} - \gamma^2 \mathbf{H} = 0 \tag{16.9}$$

where $\gamma = \alpha + j\beta$ is the **propagation constant,** α is the *attenuation constant* in nepers per meter or decibels per meter, and β is the *phase constant* in radians per meter. Constants α and β are given by

$$\alpha = \omega\sqrt{\frac{\mu\epsilon}{2}\left[\sqrt{1+\left(\frac{\sigma}{\omega\epsilon}\right)^2}-1\right]} \tag{16.10}$$

$$\beta = \omega\sqrt{\frac{\mu\epsilon}{2}\left[\sqrt{1+\left(\frac{\sigma}{\omega\epsilon}\right)^2}+1\right]} \tag{16.11}$$

where $\omega = 2\pi f$ is the frequency of the wave. The wavelength λ and wave velocity u are given in terms of β as

$$\lambda = \frac{2\pi}{\beta} \tag{16.12}$$

$$u = \frac{\omega}{\beta} = f\lambda \tag{16.13}$$

Without loss of generality, if we assume that wave propagates in the z-direction and the wave is polarized in the x-direction, solving the wave Equation (16.8) and Equation (16.9) results in

$$\mathrm{E}(z,t) = E\, e_0^{-\alpha z}\cos(\omega t - \beta z)\mathrm{a}_x \tag{16.14}$$

$$\mathrm{H}(z,t) = \frac{E_0}{|\eta|}e^{-\alpha z}\cos(\omega t - \beta z - \theta_\eta)\mathrm{a}_y \tag{16.15}$$

where $\eta = |\eta|\angle\theta_\eta$ is the *intrinsic impedance* of the medium and is given by

$$|\eta| = \frac{\sqrt{\mu/\epsilon}}{\sqrt[4]{\left[1+\left(\frac{\sigma}{\omega\epsilon}\right)\right]^{1/4}}}, \quad \tan 2\theta_\eta = \frac{\sigma}{\omega\epsilon}, \quad 0 \leqslant \theta_\eta \leqslant 45° \tag{16.16}$$

Equation (16.14) and Equation (16.15) show that as the EM wave travels in the medium, its amplitude is attenuated according to $e^{-\alpha z}$, as illustrated in Figure 16.1. The distance δ through which the wave amplitude is reduced by a factor of e^{-1} (about 37%) is called the *skin depth* or *penetration depth* of the medium, i.e.:

$$\delta = \frac{1}{\alpha} \tag{16.17}$$

The power density of the EM wave is obtained from the Poynting vector:

$$\mathbf{P} = \mathbf{E} \times \mathbf{H} \tag{16.18}$$

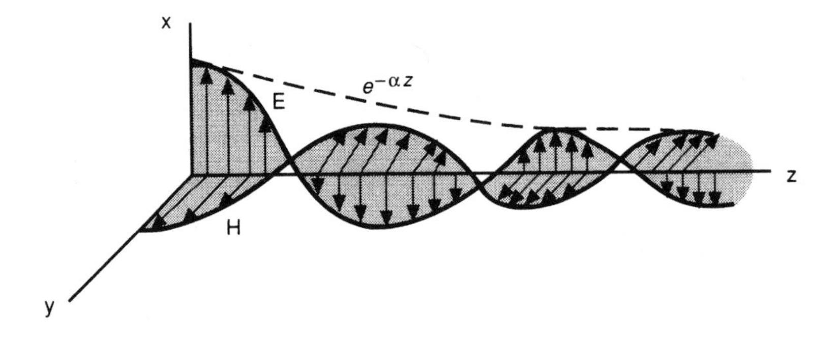

FIGURE 16.1 The magnetic and electric field components of a plane wave in a lossy medium.

with the time-average value of

$$\begin{aligned} P_{\text{ave}} &= \frac{1}{2}\text{Re}\,(\mathbf{E}\times\mathbf{H}^*) \\ &= \frac{E_o^2}{2|\eta|}e^{-2\alpha z}\cos\theta_\eta a_z \end{aligned} \tag{16.19}$$

It should be noted from Equation (16.14) and Equation (16.15) that **E** and **H** are everywhere perpendicular to each other and also to the direction of wave propagation. Thus, the wave described by Equation (16.14) and Equation (16.15) is said to be *plane-polarized*, implying that the electric field is always parallel to the same plane (the *xz*-plane in this case) and is perpendicular to the direction of propagation. Also, as mentioned earlier, the wave decays as it travels in the *z*-direction because of loss. This loss is expressed in the *complex relative permittivity* of the medium:

$$\epsilon_c = \epsilon_r' - j\epsilon_r'' = \epsilon_r\left(1 - j\frac{\sigma}{\omega\epsilon}\right) \tag{16.20}$$

and measured by the *loss tangent,* defined by

$$\tan\delta = \frac{\epsilon_r''}{\epsilon_r'} = \frac{\sigma}{\omega\epsilon} \tag{16.21}$$

The imaginary part $\epsilon_r'' = \sigma/\omega\epsilon_0$ corresponds to the losses in the medium. The refractive index of the medium n is given by

$$n = \sqrt{\epsilon_c} \tag{16.22}$$

Having considered the general case of wave propagation through a lossy medium, we now consider wave propagation in other types of media. A medium is said to be a good conductor if the loss tangent is large ($\sigma >> \omega\epsilon$) or a lossless or good dielectric if the loss tangent is very small ($\sigma << \omega\epsilon$). Thus, the characteristics of wave propagation through other types of media can be obtained as special cases of wave propagation in a lossy medium as follows:

1. Good conductors: $\sigma >> \omega\epsilon$, $\epsilon = \epsilon_o$, $\mu = \mu_o\mu_r$
2. Good dielectric: $\sigma << \omega\epsilon$, $\epsilon = \epsilon_o\epsilon_r$, $\mu = \mu_o\mu_r$
3. Free space: $\sigma = 0$, $\epsilon = \epsilon_o$, $\mu = \mu_o$

TABLE 16.1 Attenuation Constant, Phase Constant, and Intrinsic Impedance for Different Media

	Lossy Medium	Good Conductor $\sigma/\omega\epsilon >> 1$	Good Dielectric $\sigma/\omega\epsilon << 1$	Free Space
Attenuation constant α	$\omega\sqrt{\frac{\mu\epsilon}{2}\left[\sqrt{1+\left(\frac{\sigma}{\omega\epsilon}\right)^2}-1\right]}$	$\sqrt{\frac{\omega\mu\sigma}{2}}$	$\simeq 0$	0
Phase constant β	$\omega\sqrt{\frac{\mu\epsilon}{2}\left[\sqrt{1+\left(\frac{\sigma}{\omega\epsilon}\right)^2}+1\right]}$	$\sqrt{\frac{\omega\mu\sigma}{2}}$	$\omega\sqrt{\mu\epsilon}$	$\omega\sqrt{\mu_o\epsilon_o}$
Intrinsic impedance η	$\sqrt{\frac{j\omega\mu}{\sigma+j\omega\varepsilon}}$	$\sqrt{\frac{\omega\mu}{2\sigma}}(1+j)$	$\sqrt{\frac{\mu}{\epsilon}}$	377

where $\epsilon_o = 8.854\times10^{-12}$ F/m is the free-space permittivity, and $\mu_o = 4\pi\times10^{-7}$ H/m is the free-space permeability.

The conditions for each medium type are merely substituted in Equation (16.10) to Equation (16.21) to obtain the wave properties for that medium. The formulas for calculating attenuation constant, phase constant, and intrinsic impedance for different media are summarized in Table 16.1.

The classical model of a wave propagation presented in this subsection helps us understand some basic concepts of EM wave propagation and the various parameters that play a part in determining the motion of a wave from the transmitter to the receiver. We now apply the ideas to the particular case of wave propagation in the atmosphere.

Propagation in the Atmosphere

Wave propagation hardly occurs under the idealized conditions assumed in the previous subsection. For most communication links, the analysis must be modified to account for the presence of the Earth, the ionosphere, and atmospheric precipitates such as fog, raindrops, snow, and hail. This will be done in this subsection.

The major regions of the Earth's atmosphere that are of importance in radio wave propagation are the troposphere and the ionosphere. At radar frequencies (approximately 100 MHz to 300 GHz), the troposphere is by far the most important. It is the lower atmosphere consisting of a nonionized region extending from the Earth's surface up to about 15 km. The ionosphere is the Earth's upper atmosphere in the altitude region from 50 km to one Earth radius (6370 km). Sufficient ionization exists in this region to influence wave propagation.

Wave propagation over the surface of the Earth may assume one of the following three principal modes:

1. Surface wave propagation along the surface of the Earth
2. Space wave propagation through the lower atmosphere
3. Sky wave propagation by reflection from the upper atmosphere

These modes are portrayed in Figure 16.2. The sky wave is directed toward the ionosphere, which bends the propagation path back toward the Earth under certain conditions in a limited frequency range (0 to 50 MHz approximately). The surface wave is directed along the surface over which the wave is propagated. The space wave consists of the direct wave and the reflected wave. The direct wave travels from the

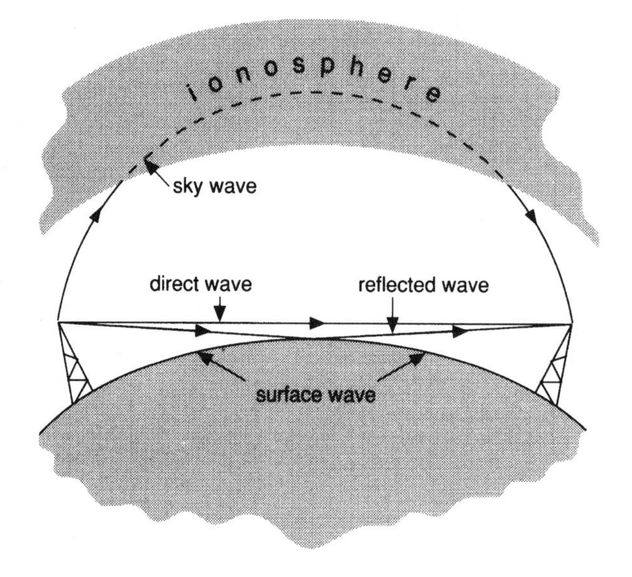

FIGURE 16.2 Modes of wave propagation.

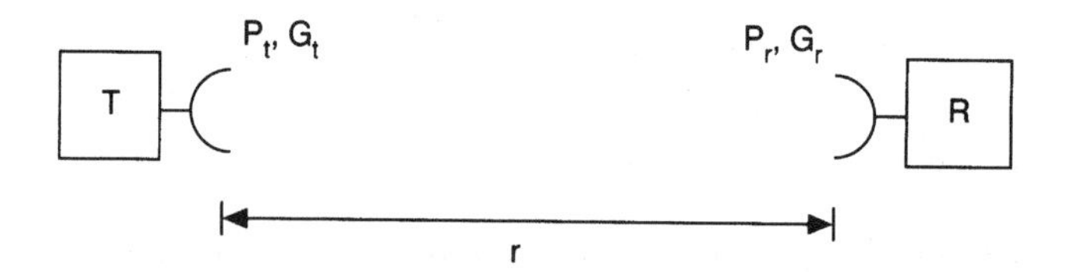

FIGURE 16.3 Transmitting and receiving antennas in free space.

transmitter to the receiver in nearly a straight path, while the reflected wave is due to ground reflection. The space wave obeys the optical laws in that direct and reflected wave components contribute to the total wave. Although the sky and surface waves are important in many applications, we will only consider space waves in this section.

Figure 16.3 depicts the electromagnetic energy transmission between two antennas in space. As the wave radiates from the transmitting antenna and propagates in space, its power density decreases, as expressed ideally in Equation (16.19). Assuming that the antennas are in free space, the power received by the receiving antenna is given by the *Friis transmission equation* (Liu and Fang, 1988):

$$P_{\mathrm{r}} = G_{\mathrm{r}}G_{\mathrm{t}}\left(\frac{\lambda}{4\pi r}\right)^2 P_{\mathrm{t}} \tag{16.23}$$

where the subscripts t and r, respectively, refer to transmitting and receiving antennas. In Equation (16.23), P is the power in watts, G is the antenna gain (dimensionless), r is the distance between the antennas in meters, and λ is the wavelength in meters. The Friis equation relates the power received by one antenna to the power transmitted by the other provided that the two antennas are separated by $r > 2d^2/\lambda$, where d is the largest dimension of either antenna. Thus, the Friis equation applies only when the two antennas are in the far-field of each other. In case the propagation path is not in free space, a correction factor F is included to account for the effect of the medium. This factor, known as the **propagation factor**, is simply the ratio of the electric field intensity E_m in the medium to the electric field intensity E_o in free space, i.e.:

$$F = \frac{E_m}{E_o} \tag{16.24}$$

The magnitude of F is always less than unity since E_m is always less than E_o. Thus, for a lossy medium, Equation (16.23) becomes

$$P_r = G_r G_t \left(\frac{\lambda}{4\pi r}\right)^2 P_t |F|^2 \tag{16.25}$$

For practical reasons, Equation (16.23) and Equation (16.25) are commonly expressed in logarithmic form. If all terms are expressed in decibels (dB), Equation (16.25) can be written in logarithmic form as

$$P_r = P_t + G_r + G_t - L_o - L_m \tag{16.26}$$

where P is power in decibels referred to 1 W (or simply dBW), G is gain in decibels, L_o is free-space loss in decibels, and L_m is loss in decibels due to the medium.

The free-space loss is obtained from standard monograph or directly from

$$L_o = 20 \log \left(\frac{4\pi r}{\lambda}\right) \tag{16.27}$$

while the loss due to the medium is given by

$$L_m = -20 \log |F| \tag{16.28}$$

Our major concern in the rest of the section is to determine L_o and L_m for two important cases of space propagation that differ considerably from the free-space conditions.

Effect of the Earth

The phenomenon of multipath propagation causes significant departures from free-space conditions. The term *multipath* denotes the possibility of EM wave propagation along various paths from the transmitter to the receiver. In multipath propagation of an EM wave over the Earth's surface, two such paths exist: a direct path and a path via reflection and diffractions from the interface between the atmosphere and the Earth. A simplified geometry of the multipath situation is shown in Figure 16.4. The reflected and diffracted component is commonly separated into two parts, one specular (or coherent) and the other diffuse (or incoherent), that can be separately analyzed. The specular component is well defined in terms of its amplitude, phase, and incident direction. Its main characteristic is its conformance to Snell's law for reflection, which requires that the angles of incidence and reflection be equal and coplanar. It is a plane wave and, as such, is uniquely specified by its direction. The diffuse component, however, arises out of the random nature of the scattering surface and, as such, is nondeterministic. It is not a plane wave and does not obey Snell's law for reflection. It does not come from a given direction but from a continuum.

The loss factor F that accounts for the departures from free-space conditions is given by

$$F = 1 + \Gamma \rho_s D S(\theta) e^{-j\Delta} \tag{16.29}$$

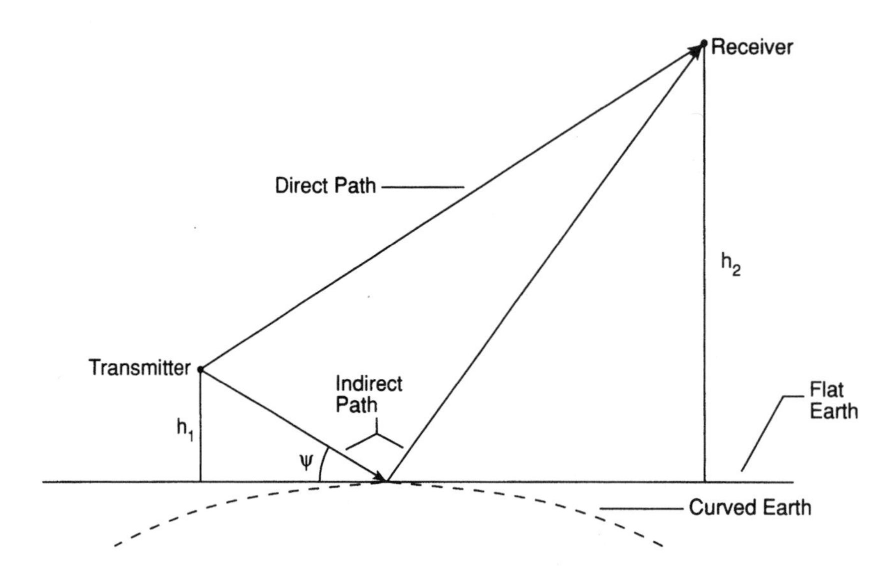

FIGURE 16.4 Multipath geometry.

where Γ is the Fresnel reflection coefficient, ρ_s is the roughness coefficient, D is the divergence factor, $S(\theta)$ is the shadowing function, and Δ is the phase angle corresponding to the path difference. We now account for each of these terms.

The Fresnel reflection coefficient Γ accounts for the electrical properties of the Earth's surface. Because the Earth is a lossy medium, the value of the reflection coefficient depends on the complex relative permittivity ϵ_c of the surface, the grazing angle Ψ and the wave polarization. It is given by

$$\Gamma = \frac{\sin\Psi - z}{\sin\Psi + z} \tag{16.30}$$

where

$$z = \sqrt{\epsilon_c - \cos^2\Psi} \quad \text{for horizontal polarization} \tag{16.31}$$

$$z = \frac{\sqrt{\epsilon_c - \cos^2\Psi}}{\epsilon_c} \quad \text{for vertical polarization} \tag{16.32}$$

$$\epsilon_c = \epsilon_r - j\frac{\sigma}{\omega\epsilon_o} = \epsilon_r - j\,60\sigma\lambda \tag{16.33}$$

ϵ_r and σ are the dielectric constant and conductivity of the surface; ω and λ are the frequency and wavelength of the incident wave; and Ψ is the grazing angle. It is apparent that $0<|\Gamma|<1$.

To account for the spreading (or divergence) of the reflected rays because of the Earth's curvature, we introduce the divergence factor D. The curvature has a tendency to spread out the reflected energy more than a corresponding flat surface. The divergence factor is defined as the ratio of the reflected field from curved surface to the reflected field from flat surface (Kerr, 1951). Using the geometry of Figure 16.5, D is given by

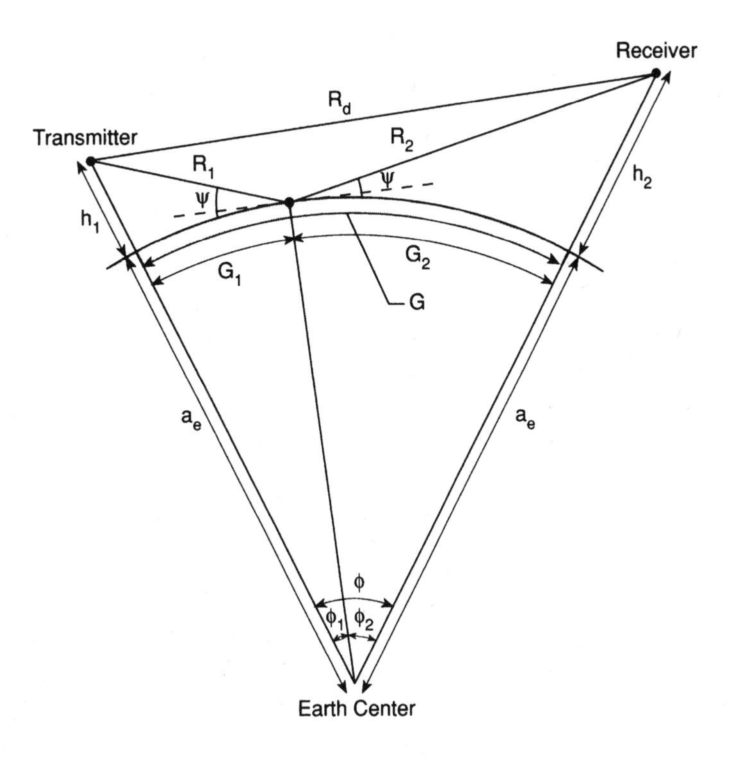

FIGURE 16.5 Geometry of spherical Earth reflection.

$$D \simeq \left(1 + \frac{2G_1G_2}{a_eG \sin \psi}\right)^{-1/2} \tag{16.34}$$

where $G = G_1+G_2$ is the total ground range and $a_e = 6370$ km is the effective Earth radius. Given the transmitter height h_1, the receiver height h_2, and the total ground range G, we can determine G_1, G_2, and Ψ. If we define

$$P = \frac{2}{\sqrt{3}}\left[a_e(h_1 + h_2) + \frac{G^2}{4}\right]^{1/2} \tag{16.35}$$

$$\alpha = \cos^{-1}\left[\frac{2a_e(h_1 - h_2)G}{p^3}\right] \tag{16.36}$$

and assume $h_1 \leq h_2$, $G_1 \leq G_2$, using small angle approximation yields (Blake, 1986):

$$G_1 = \frac{G}{2} + p\cos\left(\frac{\pi + \alpha}{3}\right) \tag{16.37}$$

$$G = G_2 - G_1 \tag{16.38}$$

$$\phi_i = \frac{G_i}{a_e}, \quad i = 1, 2 \tag{16.39}$$

$$R_i = \left[h_i^2 + 4a_e(a_e + h_i)\sin^2(\phi_i/2)\right]^{1/2}, \quad i = 1, 2 \tag{16.40}$$

The grazing angle is given by

$$\Psi = \sin^{-1}\left[\frac{2a_e h_1 + h_1^2 - R_1^2}{2a_e R_1}\right] \tag{16.41}$$

or

$$\Psi = \sin^{-1}\left[\frac{2a_e h_1 + h_1^2 - R_1^2}{2(a_e + h_1)R_1}\right] - \phi_1 \tag{16.42}$$

Although D varies from 0 to 1, in practice D is a significant factor at low grazing angle Ψ.

The phase angle corresponding to the path difference between direct and reflected waves is given by

$$\Delta = \frac{2\pi}{\lambda}(R_1 + R_2 - R_d) \tag{16.43}$$

The roughness coefficient ρ_s takes care of the fact that the Earth's surface is not sufficiently smooth to produce specular (mirrorlike) reflection except at a very low grazing angle. The Earth's surface has a height distribution that is random in nature. The randomness arises out of the hills, structures, vegetation, and ocean waves. It is found that the distribution of the heights of the Earth's surface is usually the Gaussian or normal distribution of probability theory. If σ_h is the standard deviation of the normal distribution of heights, we define the roughness parameters

$$g = \frac{\sigma_h \sin\Psi}{\lambda} \tag{16.44}$$

If $g < 1/8$, specular reflection is dominant; if $g > 1/8$, diffuse scattering results. This criterion, known as *Rayleigh criterion*, should only be used as a guideline since the dividing line between a specular and diffuse reflection or between a smooth and a rough surface is not well defined (Beckman and Spizzichino, 1963). The roughness is taken into account by the roughness coefficient ($0 < \rho_s < 1$), which is the ratio of the field strength after reflection with roughness taken into account to that which would be received if the surface were smooth. The roughness coefficient is given by

$$\rho_s = \exp\left[-2(2\pi g)^2\right] \tag{16.45}$$

The shadowing function $S(\theta)$ is important at a low grazing angle. It considers the effect of geometric shadowing—the fact that the incident wave cannot illuminate parts of the Earth's surface shadowed by higher parts. In a geometric approach, where diffraction and multiple scattering effects are neglected, the reflecting surface will consist of well-defined zones of illumination and shadow. As there will be no field on a shadowed portion of the surface, the analysis should include only the illuminated portions of the surface. The phenomenon of shadowing a stationary surface was first investigated by Beckman in 1965 and subsequently refined by Smith (1967) and others. A pictorial representation of rough surfaces illuminated at angle of

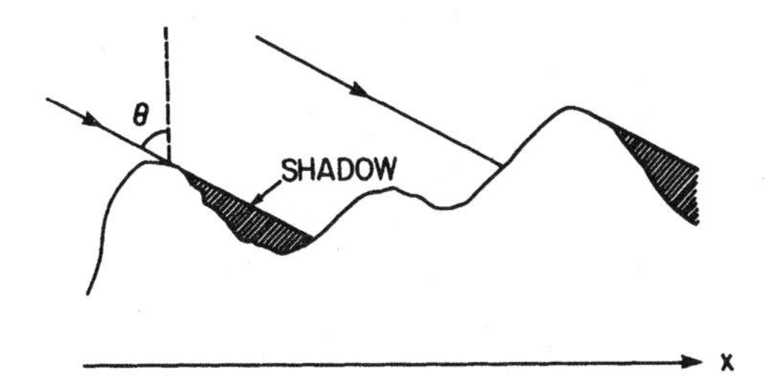

FIGURE 16.6 Rough surface illuminated at an angle of incidence θ.

incidence θ ($= 90° - \Psi$) is shown in Figure 16.6. It is evident from the figure that the shadowing function $S(\theta)$ is equal to unity when $\theta = 0$ and zero when $\theta = \pi/2$. According to Smith (1967):

$$S(\theta) \simeq \frac{\left[1 - \frac{1}{2}\,\mathrm{erfc}\,(a)\right]}{1 + 2B} \tag{16.46}$$

where erfc(x) is the complementary error function:

$$\mathrm{erfc}\,(x) = 1 - \mathrm{erf}\,(x) = \frac{2}{\sqrt{\pi}} \int_x^{\infty} e^{-t^2}\, dt \tag{16.47}$$

and

$$B = \frac{1}{4a}\left[\frac{1}{\sqrt{\pi}} e^{a^2} - a\,\mathrm{erfc}\,(a)\right] \tag{16.48}$$

$$a = \frac{\cot\theta}{2s} \tag{16.49}$$

$$s = \frac{\sigma_h}{\sigma_l} = \text{rms surface slope} \tag{16.50}$$

In Equation (16.50) σ_h is the rms roughness height and σ_l is the correlation length. Alternative models for $S(\theta)$ are available in the literature. Using Equation (16.30) to Equation (16.50), the loss factor in Equation (16.29) can be calculated. Thus,

$$L_o = 20 \log\left(\frac{4\pi R_d}{\lambda}\right) \tag{16.51}$$

$$L_m = -20 \log\left[1 + \Gamma \rho_s D s\,(\theta) e^{-j\Delta}\right] \tag{16.52}$$

Effect of Atmospheric Hydrometeors

The effect of atmospheric hydrometeors on satellite–Earth propagation is of major concern at microwave frequencies. The problem of scattering of electromagnetic waves by atmospheric hydrometeors has attracted much interest since the late 1940s. The main hydrometeors that exist for long duration and have the greatest interaction with microwaves are rain and snow. At frequencies above 10 GHz, rain has been recognized as the most fundamental obstacle on the Earth–space path. Rain has been known to cause attenuation, phase difference, and depolarization of radio waves. For analog signals, the effect of rain is more significant above 10 GHz, while for digital signals, rain effects can be significant down to 3 GHz. Attenuation of microwaves because of precipitation becomes severe owing to increased scattering and beam energy absorption by raindrops, thus impairing terrestrial as well as Earth–satellite communication links. Cross-polarization distortion due to rain has also engaged the attention of researchers. This is of particular interest when frequency reuse employing signals with orthogonal polarizations is used for doubling the capacity of a communication system. A thorough review of the interaction of microwaves with hydrometeors has been given by Oguchi (1983).

The loss due to a rain-filled medium is given by

$$L_{\mathrm{m}} = \gamma(R)\ \ell_{\mathrm{e}}(R)\, p\,(R) \tag{16.53}$$

where γ is attenuation per unit length at rain rate R, ℓ is the equivalent path length at rain rate R, and $p(R)$ is the probability in percentage of rainfall rate R.

Attenuation is a function of the cumulative rain-rate distribution, drop-size distribution, refractive index of water, temperature, and other variables. A rigorous calculation of $\gamma(R)$ incorporating raindrop-size distribution, velocity of raindrops, and refractive index of water can be found in Sadiku (1992). For practical engineering purposes, what is needed is a simple formula relating attenuation to rain parameters. Such is found in the aR^{b} empirical relationship, which has been used to calculate rain attenuation directly (Collin, 1985), i.e.:

$$\gamma(R) = aR^{b}\ \mathrm{dB/km} \tag{16.54}$$

where R is the rain rate and a and b are constants. At 0°C, the values of a and b are related to frequency f in gigahertz as follows:

$$a = G_{a} f^{\mathrm{Ea}} \tag{16.55}$$

where $G_{\mathrm{a}} = 6.39 \times 10^{-5}$, $E_{\mathrm{a}} = 2.03$, for $f < 2.9$ GHz; $G_{\mathrm{a}} = 4.21 \times 10^{-5}$, $E_{\mathrm{a}} = 2.42$, for 2.9 GHz $\leq f \leq$ 54 GHz; $G_{\mathrm{a}} = 4.09 \times 10^{-2}$, $E_{\mathrm{a}} = 0.699$, for 54 GHz $\leq f <$ 100 GHz; $G_{\mathrm{a}} = 3.38$, $E_{\mathrm{a}} = -0.151$, for 180 GHz $< f$; and

$$b = G_{b} f^{\mathrm{Eb}} \tag{16.56}$$

where $G_{\mathrm{b}} = 0.851$, $E_{\mathrm{b}} = 0.158$, for $f <$ 8.5 GHz; $G_{\mathrm{b}} = 1.41$, $E_{\mathrm{b}} = -0.0779$, for 8.5 GHz $\leq f <$ 25 GHz; $G_{\mathrm{b}} = 2.63$, $E_{\mathrm{b}} = -0.272$, for 25 GHz $\leq f <$ 164 GHz; $G_{\mathrm{b}} = 0.616$, $E_{\mathrm{b}} = 0.0126$, for 164 GHz $\leq f$.

The effective length $\ell_{\mathrm{e}}(R)$ through the medium is needed since rain intensity is not uniform over the path. Its actual value depends on the particular area of interest and therefore has a number of representations (Liu and Fang, 1988). Based on data collected in western Europe and eastern North America, the effective path length has been approximated as (Hyde, 1984):

$$\ell_{e}(R) = [0.00741\mathrm{R}^{0.766} + (0.232 - 0.00018\mathrm{R})\sin\theta]^{-1} \tag{16.57}$$

where θ is the elevation angle.

The cumulative probability in percentage of rainfall rate R is given by (Hyde, 1984):

$$p(R) = \frac{M}{87.66}[0.03\beta e^{-0.03R} + 0.2(1-\beta)(e^{-0.258R} + 1.86e^{-1.63R})] \quad (16.58)$$

where M is the mean annual rainfall accumulation in millimeters and β is the Rice–Holmberg thunderstorm ratio.

The effect of other hydrometeors such as water vapor, fog, hail, snow, and ice is governed by similar fundamental principles as the effect of rain (Collin, 1985). In most cases, however, their effects are at least an order of magnitude less than the effect of rain.

Other Effects

Besides hydrometeors, the atmosphere has the composition given in Table 16.2. While attenuation of EM waves by hydrometeors may result from both absorption and scattering, gases act only as absorbers. Although some of these gases do not absorb microwaves, some possess permanent electric and/or magnetic dipole moments and play some part in microwave absorption. For example, nitrogen molecules do not possess permanent electric or magnetic dipole moments and therefore play no part in microwave absorption. Oxygen has a small magnetic moment, which enables it to display weak absorption lines in the centimeter and millimeter wave regions. Water vapor is a molecular gas with a permanent electric dipole moment. It is more responsive to excitation by an EM field than is oxygen.

TABLE 16.2 Composition of Dry Atmosphere from Sea Level to about 90 km

Constituent	Percent by Volume	Percent by Weight
Nitrogen	78.088	75.527
Oxygen	20.949	23.143
Argon	0.93	1.282
Carbon dioxide	0.03	0.0456
Neon	1.8×10^{-3}	1.25×10^{-3}
Helium	5.24×10^{-4}	7.24×10^{-5}
Methane	1.4×10^{-4}	7.75×10^{-5}
Krypton	1.14×10^{-4}	3.30×10^{-4}
Nitrous oxide	5×10^{-5}	7.60×10^{-5}
Xenon	8.6×10^{-6}	3.90×10^{-5}
Hydrogen	5×10^{-5}	3.48×10^{-6}

Source: D.C. Livingston, *The Physics of Microwave Propagation*, Englewood Cliffs, NJ: Prentice-Hall, 1970, p. 11. With permission.

Defining Terms

Multipath: Propagation of electromagnetic waves along various paths from the transmitter to the receiver.

Propagation constant: The negative of the partial logarithmic derivative, with respect to the distance in the direction of the wave normal, of the phasor quantity describing a traveling wave in a homogeneous medium.

Propagation factor: The ratio of the electric field intensity in a medium to its value if the propagation took place in free space.

Wave propagation: The transfer of energy by electromagnetic radiation.

References

P. Beckman and A. Spizzichino, *The Scattering of Electromagnetic Waves from Random Surfaces*, New York: Macmillan, 1963.

L.V. Blake, *Radar Range-Performance Analysis*, Norwood, MA: Artech House, 1986, pp. 253–271.

R.E. Collin, *Antennas and Radiowave Propagation*, New York: McGraw-Hill, 1985, pp. 339–456.

G. Hyde, "Microwave propagation," in *Antenna Engineering Handbook*, 2nd ed., R.C. Johnson and H. Jasik, Eds., New York: McGraw-Hill, 1984, pp. 45.1–45.17.

D.E. Kerr, *Propagation of Short Radio Waves*, New York: McGraw-Hill (republished by Peter Peregrinus, London, 1987), 1951, pp. 396–444.

C.H. Liu and D.J. Fang, "Propagation," in *Antenna Handbook: Theory, Applications, and Design*, Y.T. Lo and S.W. Lee, Eds., New York: Van Nostrand Reinhold, 1988, pp. 29.1–29.56.

T. Oguchi, "Electromagnetic wave propagation and scattering in rain and other hydrometeors," *Proc. IEEE*, vol. 71, pp. 1029–1078, 1983.
M.N.O. Sadiku, *Numerical Techniques in Electromagnetics*, Boca Raton, FL: CRC Press, 1992, pp. 96–116.
B.G. Smith, "Geometrical shadowing of a random rough surface," *IEEE Trans. Ant. Prog.*, vol. 15, pp. 668–671, 1967.

Further Information

There are several sources of information dealing with the theory and practice of wave propagation in space. Some of these are in the reference section. Journals such as *Radio Science, IEE Proceedings Part H,* and *IEEE Transactions on Antennas and Propagation* are devoted to EM wave propagation. *Radio Science* is available from the American Geophysical Union, 2000 Florida Avenue NW, Washington, DC 20009; *IEE Proceedings Part H* from the IEE Publishing Department, Michael Faraday House, 6 Hills Way, Stevenage, Herts SG1 2AY, U.K.; and *IEEE Transactions on Antennas and Propagation* from IEEE, 445 Hoes Lane, P.O. Box 1331, Piscataway, NJ 08855-1331.

Other mechanisms that can affect EM wave propagation in space, not discussed in this section, include clouds, dust, and the ionosphere. The effect of the ionosphere is discussed in detail in standard texts.

16.2 Waveguides

Kenneth Demarest

Waveguide Modes

Any structure that guides electromagnetic waves can be considered to be a *waveguide*. Most often, however, this term refers to closed metal cylinders that maintain the same cross-sectional dimensions over long distances. Such a structure is shown in Figure 16.7, which consists of a metal cylinder that is filled with a dielectric. When they are filled with low-loss dielectrics (such as air), waveguides typically exhibit lower losses than transmission lines, which make them useful for transporting RF energy over relatively long distances. They are most often used for frequencies ranging from 1 to 150 GHz.

Every type of waveguide has an infinite number of distinct electromagnetic field configurations that can exist inside it. Each of these configurations is called a *mode*. The characteristics of these modes depend upon the cross-sectional dimensions of the conducting cylinder, the type dielectric material inside the waveguide, and the frequency of operation. Waveguide modes are typically classed according to the nature of the electric and magnetic field components that are directed parallel to the waveguide axis E_z H_z. These components are called the longitudinal components of the fields. Several types of modes are possible in waveguides.

TE modes: Transverse-electric modes, sometimes called *H* modes. These modes have $E_z = 0$ at all points within the waveguide, which means that the electric field vector is always perpendicular (i.e., transverse) to the waveguide axis. These modes are always possible in waveguides with uniform dielectrics.

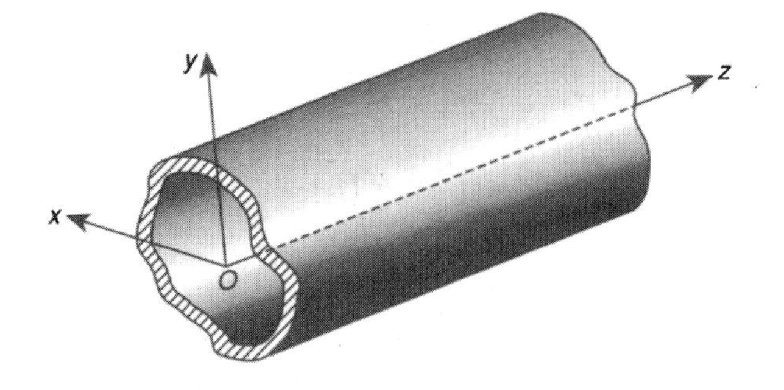

FIGURE 16.7 A uniform waveguide with arbitrary cross-section.

TM modes: Transverse-magnetic modes, sometimes called E modes. These modes have $H_z = 0$ at all points within the waveguide, which means that the magnetic field vector is perpendicular to the waveguide axis. Like TE modes, they are always possible in waveguides with uniform dielectrics.

EH modes: These are hybrid modes in which neither E_z nor H_z are zero, but the characteristics of the transverse fields are controlled more by E_z than H_z. These modes are often possible in waveguides with inhomogeneous dielectrics.

HE modes: These are hybrid modes, in which neither E_z nor H_z are zero, but the characteristics of the transverse fields are controlled more by H_z than E_z. Like EH modes, these modes are often possible in waveguides with inhomogeneous dielectrics.

TEM modes: Transverse-electromagnetic modes, often called transmission line modes. These modes can exist only when a second conductor exists within the waveguide, such as a center conductor on a coaxial cable. Since these modes cannot exist in single, closed conductor structures, they are *not* waveguide modes.

Waveguide modes are most easily determined by first computing the longitudinal field components, E_z and H_z, that can be supported by a waveguide. From these, the fields transverse (i.e., perpendicular) to the waveguide axis can easily be found simply by taking spatial derivatives of the longitudinal fields (Collin, 1992).

When the waveguide properties are constant along the z-axis, E_z and H_z vary in the longitudinal direction as $E_z, H_z \propto \exp(\omega t - \gamma z)$, where $\omega = 2\pi f$ is the radian frequency of operation and γ is a complex number of the form:

$$\gamma = \alpha + j\beta \tag{16.59}$$

The parameters γ, α, and β are called the *propagation*, *attenuation*, and *phase constants*, respectively, and $j = \sqrt{-1}$. When there are no metal or dielectric losses, γ is always either purely real or imaginary. When γ is real, E_z and H_z have constant phase and decay exponentially with increasing z. When γ is imaginary, E_z and H_z vary in phase with increasing z but do not decay in amplitude. When this latter condition occurs, the fields are said to be *propagating*.

When the dielectric is uniform (i.e., homogeneous), E_z and H_z satisfy the scalar wave equation at all points within the waveguide (Demarest, 1998):

$$\nabla_t^2 E_z + h^2 E_z = 0 \tag{16.60}$$

and

$$\nabla_t^2 H_z + h^2 H_z = 0 \tag{16.61}$$

where

$$h^2 = \omega^2 \mu\varepsilon + \gamma^2 = k^2 + \gamma^2 \tag{16.62}$$

Here, μ and ε are the permeability and permittivity of the dielectric media, respectively, and $k = \omega\sqrt{\mu\varepsilon}$ is the wavenumber of the dielectric. The operator ∇_t^2 is called the transverse Laplacian operator. In Cartesian coordinates $\nabla_t^2 = \frac{\partial^2}{\partial x^2} + \frac{\partial^2}{\partial y^2}$.

Most of the properties of the allowed modes in practical waveguides can be found by first assuming that the metal walls are perfectly conducting. Under this condition, $E_z = 0$ and $\frac{\partial H_z}{\partial p} = 0$ at the waveguide walls, where p is the direction perpendicular to the waveguide wall. When these conditions are imposed upon the general solutions of Equation (16.60) and Equation (16.61), it is found that only certain values of h are allowed. These values are called the *modal eigenvalues* and are determined by the cross-sectional shape of the waveguide. Using Equation (16.62), the propagation constant γ for each mode varies with frequency according to

$$\gamma = \alpha + j\beta = h\sqrt{1-(f/f_c)^2} \tag{16.63}$$

where

$$f_c = \frac{h}{2\pi\sqrt{\mu\varepsilon}} \tag{16.64}$$

The parameter f_c has units Hz and is called the *model cut-off frequency.* According to Equation (16.63), when $f > f_c$, the propagation constant γ is imaginary and, thus, the mode is propagating. However, when $f < f_c$, γ is real; this means that the fields decay exponentially with increasing values of z. Modes operating below their cut-off frequency are not able to transport energy over large distances and are called *evanescent modes.*

The dominant mode of a waveguide is the one with the lowest cut-off frequency. Although higher order modes are often useful for a variety of specialized uses of waveguides, signal distortion is usually minimized when a waveguide is operated in the frequency range where only the dominant mode exists. This range of frequencies is called the *dominant range* of the waveguide.

The distance over which the fields of propagating modes repeat themselves is called the guide wavelength λ_g. From Equation (16.63), it can be shown that λ_g always varies with frequency according to

$$\lambda_g = \frac{\lambda_o}{\sqrt{1-(f_c/f)^2}} \tag{16.65}$$

where $\lambda_o = (f\sqrt{\mu\varepsilon})^{-1}$ is the wavelength of a plane wave of the same frequency in an infinite sample of the waveguide dielectric. For $f >> f_c$, $\lambda_g \approx \lambda_o$. Also, $\lambda_g \rightarrow \infty$ as $f \rightarrow f_c$, which is one reason why it is usually undesirable to operate a waveguide mode near modal cut-off frequencies.

Although waveguide modes are not plane waves, the ratios of the magnitudes of their transverse electric and magnetic fields are constant throughout the cross-section of a waveguide, just as for plane waves. This ratio is called the *modal wave impedance* and has the following values for TE and TM modes:

$$Z_{TE} = \frac{E_T}{H_T} = \frac{j\omega\mu}{\gamma} \tag{16.66}$$

and

$$Z_{TM} = \frac{E_T}{H_T} = \frac{\gamma}{j\omega\varepsilon} \tag{16.67}$$

where E_T and H_T are the magnitudes of the transverse electric and magnetic fields, respectively. In the limit as $f \rightarrow \infty$, Z_{TE} and Z_{TM} approach $\sqrt{\mu/\varepsilon}$, which is the intrinsic impedance of the dielectric medium. However, as $f \rightarrow f_c$, $Z_{TE} \rightarrow \infty$ and $Z_{TM} \rightarrow 0$, which means that the transverse electric fields are dominant in TE modes near cut-off and vice versa for TM modes.

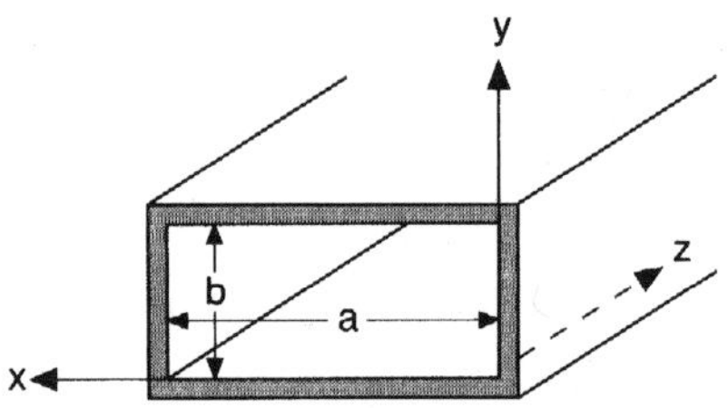

FIGURE 16.8 A rectangular waveguide.

Rectangular Waveguides

A rectangular waveguide is shown in Figure 16.8. The conducting walls are formed such that the inner surfaces form a rectangular cross-section, with dimensions a and b along the x and y coordinate axes, respectively. If the walls are perfectly conducting and the dielectric material is lossless, the field components for the TE_{mn} modes are given by

$$E_x = H_o\left(\frac{j\omega\mu}{h_{mn}^2}\right)\left(\frac{n\pi}{b}\right)\cos\left(\frac{m\pi}{a}x\right)\sin\left(\frac{n\pi}{b}y\right)\exp\left(j\omega t - j\beta_{mn}z\right) \tag{16.68a}$$

$$E_y = -H_o\left(\frac{j\omega\mu}{h_{mn}^2}\right)\left(\frac{m\pi}{a}\right)\sin\left(\frac{m\pi}{a}x\right)\cos\left(\frac{n\pi}{b}y\right)\exp\left(j\omega t - j\beta_{mn}z\right) \tag{16.68b}$$

$$E_z = 0 \tag{16.68c}$$

$$H_x = H_o\left(\frac{\gamma_{mn}}{h_{mn}^2}\right)\left(\frac{m\pi}{a}\right)\sin\left(\frac{m\pi}{a}x\right)\cos\left(\frac{n\pi}{b}y\right)\exp\left(j\omega t - j\beta_{mn}z\right) \tag{16.68d}$$

$$H_y = H_o\left(\frac{\gamma_{mn}}{h_{mn}^2}\right)\left(\frac{n\pi}{b}\right)\cos\left(\frac{m\pi}{a}x\right)\sin\left(\frac{n\pi}{b}y\right)\exp\left(j\omega t - j\beta_{mn}z\right) \tag{16.68e}$$

$$H_z = H_o\cos\left(\frac{m\pi}{a}x\right)\cos\left(\frac{n\pi}{b}y\right)\exp\left(j\omega t - j\beta_{mn}z\right) \tag{16.68f}$$

where

$$h_{mn} = \left(\frac{m\pi}{a}\right)^2 + \left(\frac{n\pi}{b}\right)^2 = 2\pi f_{c_{mn}}\sqrt{\mu\varepsilon} \tag{16.69}$$

For the TE_{mn} modes, m and n can be any positive integer value, including zero, so long as both are not zero.

The field components for the TM_{mn} modes are

$$E_x = -E_o\left(\frac{\gamma_{mn}}{h_{mn}^2}\right)\left(\frac{m\pi}{a}\right)\cos\left(\frac{m\pi}{a}x\right)\sin\left(\frac{n\pi}{b}y\right)\exp\left(j\omega t - j\beta_{mn}z\right) \tag{16.70a}$$

$$E_y = -E_o\left(\frac{\gamma_{mn}}{h_{mn}^2}\right)\left(\frac{n\pi}{b}\right)\sin\left(\frac{m\pi}{a}x\right)\cos\left(\frac{n\pi}{b}y\right)\exp\left(j\omega t - j\beta_{mn}z\right) \tag{16.70b}$$

$$E_z = E_o\sin\left(\frac{m\pi}{a}x\right)\sin\left(\frac{n\pi}{b}y\right)\exp\left(j\omega t - j\beta_{mn}z\right) \tag{16.70c}$$

$$H_x = E_o\left(\frac{j\omega\varepsilon}{h_{mn}^2}\right)\left(\frac{n\pi}{b}\right)\sin\left(\frac{m\pi}{a}x\right)\cos\left(\frac{n\pi}{b}y\right)\exp\left(j\omega t - j\beta_{mn}z\right) \tag{16.70d}$$

$$H_y = -E_o\left(\frac{j\omega\varepsilon}{h_{mn}^2}\right)\left(\frac{m\pi}{a}\right)\cos\left(\frac{m\pi}{a}x\right)\sin\left(\frac{n\pi}{b}y\right) \times \exp\left(j\omega t - j\beta_{mn}z\right) \quad (16.70e)$$

$$H_z = 0 \quad (16.70f)$$

where the values of h_{mn} and $f_{c_{mn}}$ are given by Equation (16.69). For the TM_{mn} modes, m and n can be any positive integer value except zero.

The dominant mode in a rectangular waveguide is TE_{10}, which has a cut-off frequency:

$$f_{c_{10}} = \frac{1}{2a\sqrt{\mu\varepsilon}} = \frac{c}{2a} \quad (16.71)$$

where c is the speed of light in the dielectric media. The modal field patterns for this mode are shown in Figure 16.9. Table 16.3 shows the cut-off frequencies of the lowest order rectangular waveguide modes (as referenced to the cut-off frequency of the dominant mode) when $a/b = 2.1$. The modal field patterns of several lower order modes are shown in Figure 16.10.

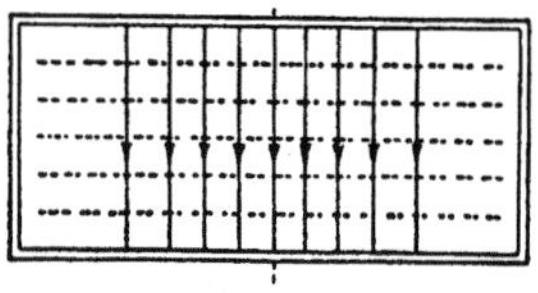

FIGURE 16.9 Field configuration for the TE_{10} (dominant) mode of a rectangular waveguide. Solid lines, **E**; dashed lines, **H**. (*Source*: Adapted from Marcuvitz, 1986.)

TABLE 16.3 Cut-off Frequencies of the Lowest Order Rectangular Waveguide Modes (referenced to the cut-off frequency of the dominant mode) for a Rectangular Waveguide with $a/b = 2.1$

$f_c/f_{c_{10}}$	Modes
1.0	TE_{10}
2.0	TE_{20}
2.1	TE_{01}
2.326	TE_{11}, TM_{11}
2.9	TE_{21}, TM_{21}
3.0	TE_{30}
3.662	TE_{31}, TM_{31}
4.0	TE_{40}

Circular Waveguides

A circular waveguide with inner radius a is shown in Figure 16.11. Here the axis of the waveguide is aligned with the z-axis of a circular-cylindrical coordinate system, where ρ and ϕ are the radial and azimuthal coordinates, respectively. If the walls are perfectly conducting and the dielectric material is lossless, the equations for the TE_{nm} modes are

$$E_\rho = H_o \frac{j\omega\mu n}{h_{nm}^2\rho} J_n(h_{nm}\rho)\sin(n\phi)\exp\left(j\omega t - j\beta_{nm}z\right) \quad (16.72a)$$

$$E_\phi = H_o \frac{j\omega\mu}{h_{nm}} J_n'(h_{nm}\rho)\cos(n\phi)\exp\left(j\omega t - j\beta_{nm}z\right) \quad (16.72b)$$

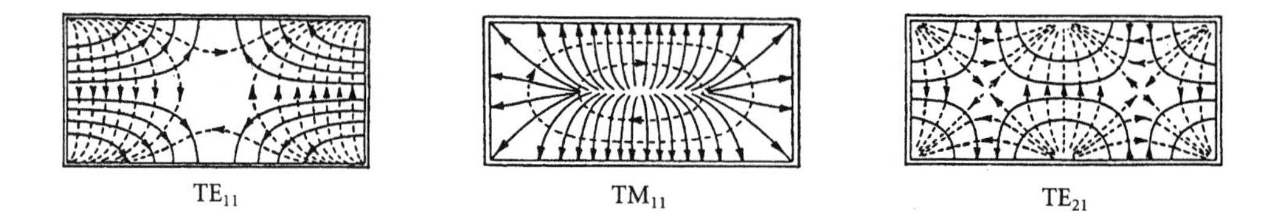

FIGURE 16.10 Field configurations for the TE_{11}, TM_{11}, and the TE_{21} modes. Solid lines, E; dashed lines, H. (*Source*: Adapted from Marcuvitz, 1984.)

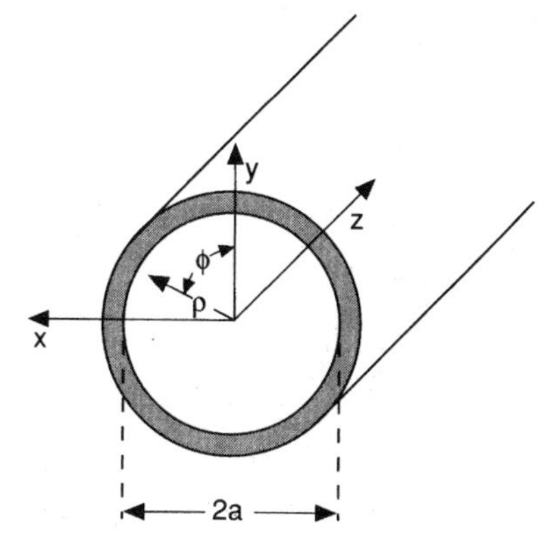

FIGURE 16.11 A circular waveguide.

$$E_z = 0 \tag{16.72c}$$

$$H_\rho = -H_o \frac{\gamma_{nm}}{h_{nm}} J'_n(h_{nm}\rho)\cos(n\phi)\exp(j\omega t - j\beta_{nm}z) \tag{16.72d}$$

$$H_\phi = H_o \frac{\gamma_{nm} n}{h^2_{nm}\rho} J_n(h_{nm}\rho)\sin(n\phi)\exp(j\omega t - j\beta_{nm}z) \tag{16.72e}$$

$$H_z = H_o J_n(h_{nm}\rho)\cos(n\phi)\exp(j\omega t - j\beta_{nm}z) \tag{16.72f}$$

where n is any positive valued integer, including zero, and $J_n(x)$ and $J'_n(x)$ are the regular Bessel function of order n and its first derivative, respectively. The allowed values of the modal eigenvalues h_{nm} satisfy

$$J'_n(h_{nm}a) = 0 \tag{16.73}$$

where m signifies the root β number of Equation (16.73). By convention, $1 < m < \infty$, where m=1 indicates the smallest root.

The equations that define the TM_{nm} modes in circular waveguides are

$$E_\rho = -E_o \frac{\gamma_{nm}}{h_{nm}} J'_n(h_{nm}\rho)\cos(n\phi)\exp(j\omega t - j\beta_{nm}z) \tag{16.74a}$$

$$E_\phi = E_o \frac{\gamma_{nm} n}{h^2_{nm}\rho} J_n(h_{nm}\rho)\sin(n\phi)\exp(j\omega t - j\beta_{nm}z) \tag{16.74b}$$

$$E_z = E_o J_n(h_{nm}\rho)\cos(n\phi)\exp(j\omega t - j\beta_{nm}z) \tag{16.74c}$$

$$H_\rho = -E_o \frac{j\omega\varepsilon n}{h_{nm}^2 \rho} J_n(h_{nm}\rho) \sin(n\phi) \times \exp(j\omega t - j\beta_{nm} z) \quad (16.74d)$$

$$H_\phi = -E_o \frac{j\omega\varepsilon}{h_{nm}} J_n'(h_{nm}\rho) \cos(n\phi) \times \exp(j\omega t - j\beta_{nm} z) \quad (16.74e)$$

$$H_z = 0 \quad (16.74f)$$

where n is any positive valued integer, including zero. For the TM_{nm} modes, the values of the modal eigenvalues are solutions of

$$J_n(h_{nm} a) = 0 \quad (16.75)$$

where m signifies the root number of Equation (16.75). As in the case of the TE modes, $1 < m < \infty$.

The dominant mode in a circular waveguide is the TE_{11} mode, which has a cut-off frequency given by

$$f_{c_{11}} = \frac{0.293}{a\sqrt{\mu\varepsilon}} \quad (16.76)$$

The configuration of the electric and magnetic fields of this mode are shown in Figure 16.12. Table 16.4 shows the cut-off frequencies of the lowest order modes for circular waveguides, referenced to the cut-off frequency of the dominant mode. The modal field patterns several lower order modes are shown in Figure 16.13.

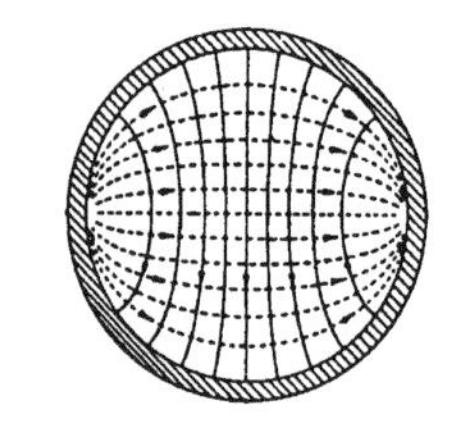

FIGURE 16.12 Field configuration for the TE_{11} (dominant) mode of a circular waveguide. Solid lines, **E**; dashed lines, **H**. (*Source*: Adapted from Marcuvitz, 1986.)

TABLE 16.4 Cut-off Frequencies of the Lowest Order Circular Waveguide Modes, Referenced to the Cut-off Frequency of the Dominant Mode

$f_c/f_{c_{11}}$	Modes
1.0	TE_{11}
1.307	TM_{01}
1.66	TE_{21}
2.083	TE_{01}, TM_{11}
2.283	TE_{31}
2.791	TM_{21}
2.89	TE_{41}
3.0	TE_{12}

Commercially Available Waveguides

The dimensions of standard rectangular waveguides are given in Table 16.5. In this table, "WR-xxx" is the Electronic Industry Association designation for "waveguide, rectangular," and the numbers that follow are the inside width dimension, in inches.

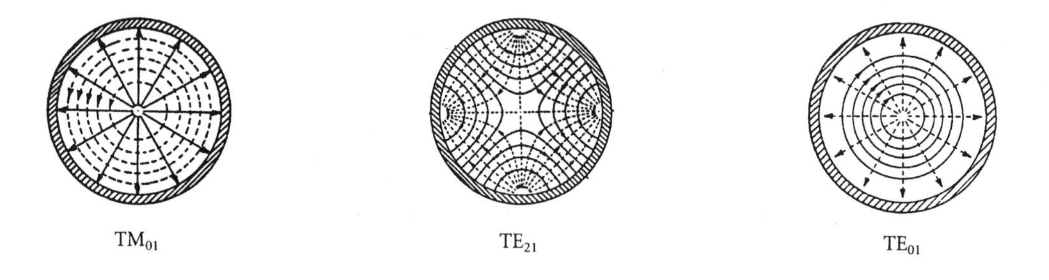

FIGURE 16.13 Field configurations for the TM_{01}, TE_{21}, and TE_{01} circular waveguide modes. Solid lines, E; dashed lines, **H**. (*Source*: Adapted from Marcuvitz, 1986.)

TABLE 16.5 Standard Rectangular Waveguide Data

WR	Inside Dimensions Width × Height (cm)	Cut-off Frequency (GHz)	Recommended Range (GHz)
WR-2300	58.42 × 29.21	0.257	0.32–0.49
WR-2100	53.34 × 26.67	0.281	0.35–0.53
WR-1800	45.72 × 22.86	0.328	0.41–0.62
WR-1500	38.10 × 19.05	0.394	0.49–0.75
WR-1150	29.21 × 14.61	0.514	0.64–0.98
WR-975	24.77 × 12.38	0.606	0.76–1.15
WR-770	19.55 × 9.779	0.767	0.96–1.46
WR-650	16.51 × 8.255	0.909	1.14–1.73
WR-510	12.95 × 6.477	1.158	1.45–2.2
WR-430	10.92 × 5.461	1.373	1.72–2.61
WR-340	8.636 × 4.318	1.737	2.17–3.3
WR-284	7.214 × 3.404	2.079	2.60–3.95
WR-229	5.817 × 2.908	2.597	3.22–4.90
WR-187	4.755 × 2.215	3.155	3.94–5.99
WR-159	4.039 × 2.019	3.714	4.64–7.05
WR-137	3.485 × 1.580	4.304	5.38–8.17
WR-112	2.850 × 1.262	5.263	6.57–9.99
WR-90	2.286 × 1.016	6.562	8.20–12.50
WR-75	1.905 × 0.953	7.874	9.84–15.00
WR-62	1.580 × 0.790	9.494	1.90–18.00
WR-51	1.295 × 0.648	11.583	14.50–22.00
WR-42	1.067 × 0.432	14.058	17.60–26.70
WR-34	0.864 × 0.432	17.361	21.70–33.00
WR-28	0.711 × 0.356	21.097	26.140–40.00
WR-22	0.569 × 0.284	26.363	32.90–50.10
WR-19	0.478 × 0.239	31.381	39.20–59.60
WR-15	0.376 × 0.188	39.894	49.80–75.80
WR-12	0.310 × 0.155	48.387	60.50–91.90
WR-10	0.254 × 0.127	59. 055	73.80–112.00
WR-8	0.203 × 0.406	73.892	92.20–140.00
WR-7	0.165 × 0.084	90.909	114.00–173.00
WR-5	0.130 × 0.066	115.385	145.00–220.00
WR-4	0.109 × 0.056	137.615	172.00–261.00
WR-3	0.086 × 0.043	174.419	217.00–333.00

In addition to rectangular and circular waveguides, there are several other waveguide types commonly used in microwave applications. Among these are elliptical waveguides and ridge waveguides. The modes of elliptical waveguides can be expressed in terms of Mathieu functions (Kretzschmar, 1970) and are similar to those of circular waveguides, but are less perturbed by minor twists and bends of the waveguide. This property makes them attractive for coupling to antennas.

Single and double ridge waveguides are shown in Figure 16.14. The modes of these waveguides bear similarities to those of rectangular guides, but can only be derived numerically (Montgomery, 1971). Ridge waveguides are useful because their dominant ranges exceed those of rectangular waveguides. However, this range increase is obtained at the expense of higher losses and lower power-handling capabilities. The dimensions and performance parameters of standard double ridge waveguides are shown in Table 16.6. In this table, the "D" in "WRD" stands for double ridge, and D24 means a bandwidth ratio of 1.24:1, etc.

Waveguides are also available in a number of construction types, including rigid, semirigid, and flexible. In applications where it is not necessary for the waveguide to bend, rigid construction is always the best, since it exhibits the lowest loss. In general, the more flexible the waveguide construction is, the higher the loss.

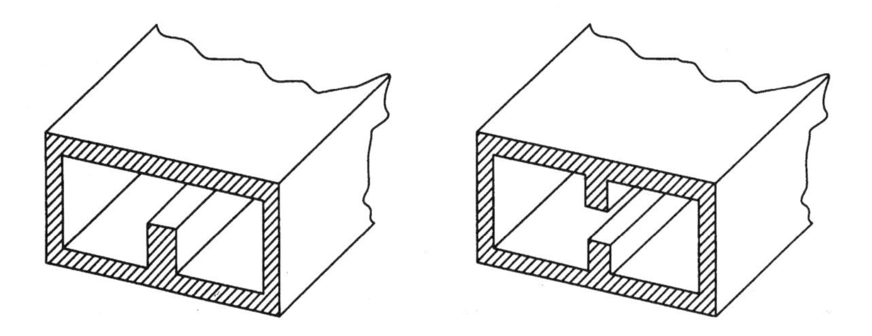

FIGURE 16.14 Single- and double-ridged waveguides

TABLE 16.6 Standard Double-Ridge Waveguide Data

WRD	Inside Dimensions Width × Height (cm)	Cutoff Frequency (GHz)	Recommended Range (GHz)
WRD-250	4.20×1.82	2.093	2.60–7.80
WR-350 D24	3.76×1.75	2.915	3.50–8.20
WR-475 D24	2.77×1.29	3.961	4.75–11.0
WRD-500 D36	1.91×0.820	4.222	5.00–18.00
WRD-650	1.83×0.815	5.348	6.50–18.00
WRD-750 D24	1.76×0.815	6.239	7.50–18.00
WRD-110 D24	1.20×0.556	9.363	11.00–26.50
WRD-180 D24	0.732×0.340	14.995	18.00–40.00

Waveguide Losses

There are two mechanisms that cause losses in waveguides: dielectric losses and metal losses. In both cases, these losses cause the amplitudes of the propagating modes decay as $\exp(-\alpha z)$, where α is the attenuation constant, measured in units of Nepers/m. Typically, the attenuation constant is considered as the sum of two components, $\alpha = \alpha_{die} + \alpha_{met}$, where α_{die} and α_{met} are the dielectric and metal attenuation constants, respectively.

The attenuation constant α_{die} can be found directly from Equation (16.63) simply by generalizing the dielectric wavenumber k to include the effect of the dielectric conductivity σ. For a lossy dielectric, the wavenumber is given by $k^2 = \omega^2\mu\varepsilon\left(1 + \frac{\sigma}{j\omega\varepsilon}\right)$. Thus, from Equation (16.62) and Equation (16.63), the attenuation constant α_{die} due to dielectric losses is given by

$$\alpha_{die} = \text{Re}\left[\sqrt{h^2 - \omega^2\mu\varepsilon\left(1 + \frac{\sigma}{j\omega\varepsilon}\right)}\right] \tag{16.77}$$

where "Re" stands for "real part of," and the allowed values of h are given by Equation (16.69) for rectangular modes and Equation (16.73) and Equation (16.75) for circular modes.

The metal loss constant α_{met} is usually obtained by assuming that the wall conductivity is high enough to have only a negligible effect on the transverse properties of the modal field patterns. Using this assumption, the power loss in the walls per unit distance along the waveguide can then be calculated to obtain α_{met} (Marcuvitz, 1986). Figure 16.15 shows the metal attenuation constants for several circular waveguide modes, each normalized to the resistivity R_s of the walls, where $R_s = \sqrt{\pi f \mu / \sigma}$ and where μ and σ are the permeability and conductivity of the metal walls, respectively. As can be seen from this figure, the TE_{0m} modes exhibit particularly low loss at

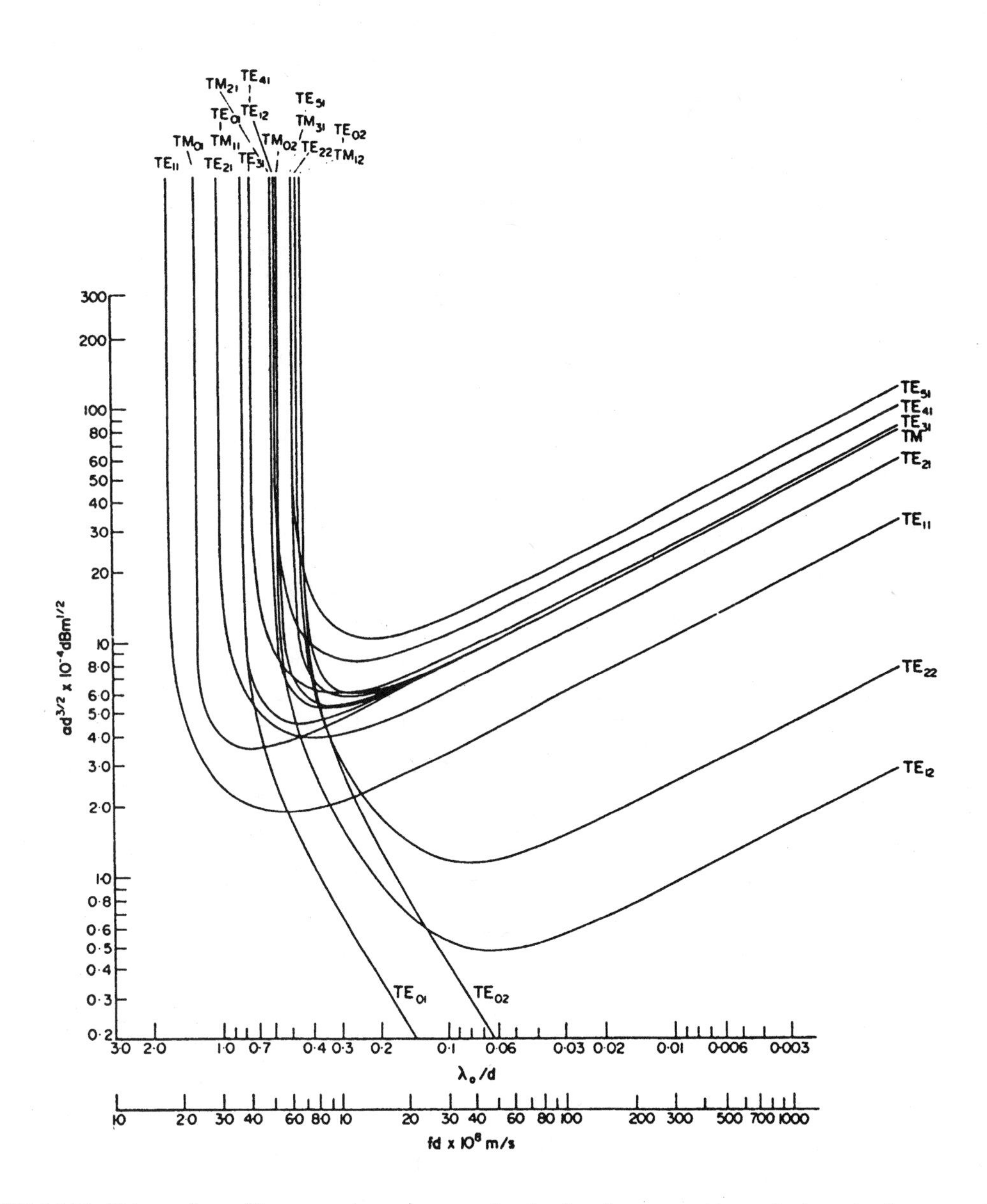

FIGURE 16.15 Values of metallic attenuation constant α for the first few waveguide modes in a circular waveguide of diameter d, plotted against normalized wavelength. (*Source*: Baden Fuller, 1979. With permission.)

frequencies significantly above their cut-off frequencies, making them useful for transporting microwave energy over large distances.

Figure 16.16 shows a comparison of the losses of various waveguide and transmission line types as a function of frequency over their operating bandwidths. This figure clearly shows the superior loss characteristics of waveguides over transmission lines.

Mode Launching

Signals can be coupled into or out of waveguides through apertures or couplers. In either case, it is important to ensure that the desired modes are excited and that reflections back to the source are minimized. This is

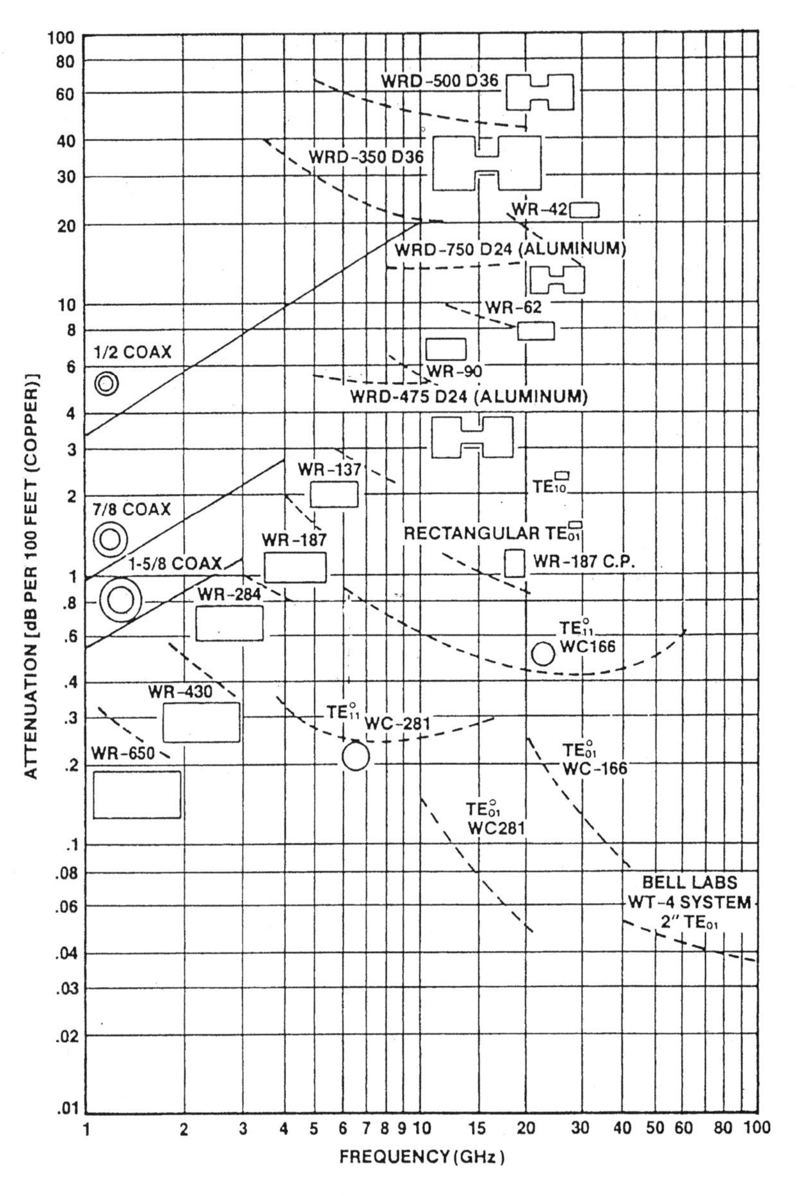

FIGURE 16.16 Attenuation vs. frequency for a variety of waveguides. (*Source*: T. Anderson, 1982. With permission.)

achieved by using launching (or coupling) structures that allow strong coupling between the desired modes on both structures.

Figure 16.17 shows a mode launching structure for coaxial cable to rectangular waveguide transitions. This structure provides good coupling between the TEM (transmission line) mode on the coaxial cable and the TE_{10} mode in the waveguide because the antenna probe excites a strong transverse electric field in the center of the waveguide, directed between the broad walls.

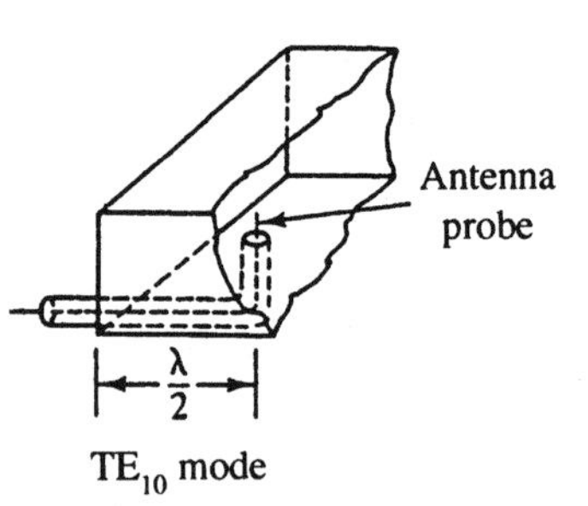

FIGURE 16.17 Coaxial to rectangular waveguide transition that couples the transmission line mode to the dominant waveguide mode.

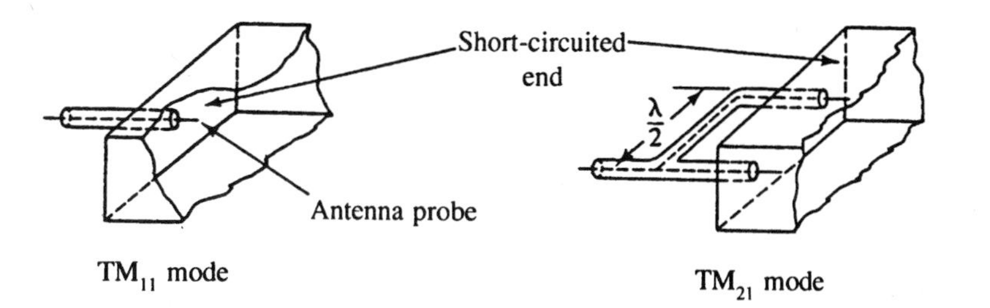

FIGURE 16.18 Coaxial to rectangular waveguide transitions that couple the transmission line mode to the TM_{11} and TM_{21} waveguide modes.

The distance between the probe and the short circuit back wall is chosen to be approximately $\lambda/4$, which allows the TE_{10} mode launched in this direction to reflect from the short circuit and arrive in phase with the mode launched toward the right.

Launching structures can also be devised to launch higher-order modes. Mode launchers that couple the transmission line mode on a coaxial cable to the TM_{11} and TM_{21} waveguide mode are shown in Figure 16.18.

References

K. Demarest, *Engineering Electromagnetics*, Upper Saddle River, NJ: Prentice-Hall, pp. 508–546, 1998.

J. Montgomery, "On the complete eigenvalue solution of ridged waveguide," *IEEE Trans. Microwave Theory Tech.*, vol. MTT-19, no. 6, pp. 457–555, 1971.

J. Kretzschmar, "Wave propagation in hollow conducting elliptical waveguides," *IEEE Trans. Microwave Theory Tech.*, vol. MTT-18, no. 9, pp. 547–554, 1970.

S.Y. Liao, *Microwave Devices and Circuits*, 3rd ed., Englewood Cliffs, NJ: Prentice-Hall, 1990.

A.J. Baden Fuller, *Microwaves*, 2nd ed., New York: Pergamon Press, 1979.

N. Marcuvitz, *Waveguide Handbook*, 2nd ed., London: Peter Peregrinus Ltd., 1986.

R.E. Collin, *Foundations for Microwave Engineering*, 2nd ed., New York: Wiley–IEEE Press, 1992.

T.N. Anderson, "State of the Waveguide Art," *Microwave J.*, vol. 25, no. 12, pp. 22–48, 1982.

Further Information

There are many textbooks and handbooks that cover the subject of waveguides in great detail. In addition to the references cited above, others include:

T.K. Ishii, *Handbook of Microwave Technology*, San Diego, CA: Academic Press, 1995.

H. Carpentier, B.L. Smith, *Microwave Engineering Handbook*, New York: Van Nostrand Reinhold, 1993.

J. Whitaker, *RF Transmission Systems Handbook*, Boca Raton, FL: CRC Press, 002.

L. Lewin, *Theory of Waveguides*, New York: John Wiley & Sons, 1975.

R.E. Collin, *Field Theory of Guided Waves*, 2nd ed., Piscataway, NJ: Wiley–IEEE Press, 1991.

F. Gardiol, *Introduction to Microwaves*, Dedham, MA: Artech House, 1984.

S. Ramo, J. Whinnery, T. Van Duzer, *Fields and Waves in Communication Electronics*, New York: John Wiley & Sons, 1994.

17

Antennas

Nicholas J. Kolias
Raytheon Company

Richard C. Compton
DV Wireless Group

J. Patrick Fitch
Lawrence Livermore National Laboratory

James C. Wiltse
Georgia Tech Research Institute

17.1 Wire

Nicholas J. Kolias and Richard C. Compton

Antennas have been widely used in communication systems since the early 1900s. Over this span of time scientists and engineers have developed a vast number of different antennas. The radiative properties of each of these antennas are described by an antenna pattern. This is a plot as a function of direction of the power P_r per unit solid angle Ω radiated by the antenna. The antenna pattern, also called the **radiation pattern,** is usually plotted in spherical coordinates θ and φ. Often two orthogonal cross-sections are plotted: one where the E-field lies in the plane of the slice (called the E-plane) and one where the H-field lies in the plane of the slice (called the H-plane).

Short Dipole

Antenna patterns for a short dipole are plotted in Figure 17.1. In these plots the radial distance from the origin to the curve is proportional to the radiated power. Antenna plots are usually either on linear scales or decibel scales (10 log power).

The antenna pattern for a short dipole may be determined by first calculating the vector potential $\mathbf{A}$ (Harrington, 1961; Lorrain and Corson, 1970; Balanis, 1982; Collin, 1985). Using Collin's notation, the vector potential in spherical coordinates is given by

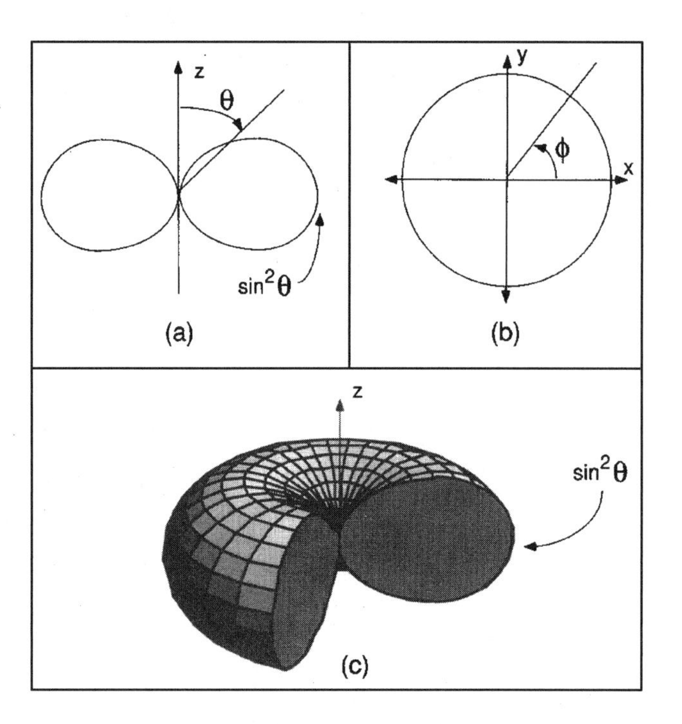

FIGURE 17.1 Radiation pattern for a short dipole of length dl (dl << λ_0). These are plots of power density on linear scales. (a) E-plane; (b) H-plane; (c) three-dimensional view with cutout.

$$\mathbf{A} = \mu_0 I\, \mathrm{d}l \frac{\mathrm{e}^{-jk_0 r}}{4\pi r}(\mathbf{a}_r \cos\theta - \mathbf{a}_\theta \sin\theta) \tag{17.1}$$

where $k_0 = 2\pi/\lambda_0$, and I is the current, assumed uniform, in the short dipole of length dl (dl << λ_0). Here, the assumed time dependence $e^{j\omega t}$ has not been explicitly shown. The electric and magnetic fields may then be determined using

$$\mathbf{E} = -j\omega\mathbf{A} + \frac{\nabla\nabla\cdot\mathbf{A}}{j\omega\mu_0\varepsilon_0} \qquad \mathbf{H} = \frac{1}{\mu_0}\nabla\times\mathbf{A} \tag{17.2}$$

The radiated fields are obtained by calculating these fields in the so-called *far-field region* where $r >> \lambda$. Doing this for the short dipole yields

$$\begin{aligned} \mathbf{E} &= jZ_0 I\mathrm{d}l\, k_0 \sin\theta \frac{\mathrm{e}^{-jk_0 r}}{4\pi r}\mathbf{a}_\theta \\ \mathbf{H} &= jI\mathrm{d}l\, k_0 \sin\theta \frac{e^{-jk_0 r}}{4\pi r}\mathbf{a}_\varphi \end{aligned} \tag{17.3}$$

where $Z_0 = \sqrt{\mu_0/\varepsilon_0}$. The average radiated power per unit solid angle Ω can then be found to be

$$\frac{\Delta P_r(\theta,\varphi)}{\Delta\Omega} = \frac{1}{2}r^2 \Re e\{\mathbf{E}\times\mathbf{H}^* \cdot \mathbf{a}_r\} = |I|^2 Z_0 (\mathrm{d}l)^2 k_0^2 \frac{\sin^2\theta}{32\pi^2} \tag{17.4}$$

Directivity

The **directivity** $D(\theta,\varphi)$ and **gain** $G(\theta,\varphi)$ of an antenna are defined as

$$D(\theta,\varphi) = \frac{\text{radiated power per solid angle}}{\text{total radiated power}/4\pi} = \frac{\Delta P_r(\theta,\varphi)/\Delta\Omega}{P_r/4\pi}$$
$$G(\theta,\varphi) = \frac{\text{radiated power per solid angle}}{\text{total input power}/4\pi} = \frac{\Delta P_r(\theta,\varphi)/\Delta\Omega}{P_{in}/4\pi} \tag{17.5}$$

Antenna efficiency, η, is given by

$$\eta \equiv \frac{P_r}{P_{in}} = \frac{G(\theta,\varphi)}{D(\theta,\varphi)} \tag{17.6}$$

For many antennas $\eta \approx 1$ and so the words *gain* and *directivity* can be used interchangeably. For the short dipole:

$$D(\theta,\varphi) = \frac{3}{2}\sin^2\theta \tag{17.7}$$

The maximum directivity of the short dipole is 3/2. This single number is often abbreviated as the antenna directivity. By comparison, for an imaginary isotropic antenna which radiates equally in all directions, $D(\theta,\psi) = 1$. The product of the maximum directivity with the total radiated power is called the *effective isotropic radiated power* (EIRP). It is the total radiated power that would be required for an isotropic radiator to produce the same signal as the original antenna in the direction of maximum directivity.

Magnetic Dipole

A small loop of current produces a *magnetic dipole.* The far fields for the magnetic dipole are dual to those of the electric dipole. They have the same angular dependence as the fields of the electric dipole, but the polarization orientations of **E** and **H** are interchanged:

$$\mathbf{H} = -Mk_0^2 \sin\theta \frac{e^{-jk_0 r}}{4\pi r}\mathbf{a}_\theta$$
$$\mathbf{E} = MZ_0 k_0^2 \sin\theta \frac{e^{-jk_0 r}}{4\pi r}\mathbf{a}_\varphi \tag{17.8}$$

where $M = \pi r_0^2 I$ for a loop with radius r_0 and uniform current I.

Input Impedance

At a given frequency the impedance at the feedpoint of an antenna can be represented as $Z_a = R_a + jX_a$. The real part of Z_a (known as the input resistance) corresponds to radiated fields plus losses, while the imaginary part (known as the input reactance) arises from stored evanescent fields. The radiation resistance is obtained from $R_a = 2P_r/|I|^2$ where P_r is the total radiated power and I is the input current at the antenna terminals. For electrically small electric and magnetic dipoles with uniform currents:

$$R_a = 80\pi^2\left(\frac{dl}{\lambda_0}\right)^2 \quad \text{electric dipole}$$
$$R_a = 320\pi^6\left(\frac{r_0}{\lambda_0}\right)^4 \quad \text{magnetic dipole} \tag{17.9}$$

The reactive component of Z_a can be determined from $X_a = 4\omega(W_m - W_e)/|I|^2$ where W_m is the average magnetic energy and W_e is the average electric energy stored in the near-zone evanescent fields. The reflection coefficient, Γ, of the antenna is just

$$\Gamma = \frac{Z_a - Z_0}{Z_a + Z_0} \tag{17.10}$$

where Z_0 is the characteristic impedance of the system used to measure the reflection coefficient.

Arbitrary Wire Antennas

An arbitrary wire antenna can be considered as a sum of small current dipole elements. The vector potential for each of these elements can be determined in the same way as for the short dipole. The total vector potential is then the sum over all these infinitesimal contributions and the resulting E in the far field can be found to be

$$\mathbf{E}(r) = jk_0Z_0\frac{e^{-jk_0r}}{4\pi r}\int_C [(\mathbf{a}_r \cdot \mathbf{a})\mathbf{a}_r - \mathbf{a}]I(l')e^{jk_0\mathbf{a}_r\cdot r'}dl' \tag{17.11}$$

where the integral is over the contour C of the wire, $\mathbf{a}$ is a unit vector tangential to the wire, and $\mathbf{r}'$ is the radial vector to the infinitesimal current element.

Resonant Half-Wavelength Antenna

The resonant half-wavelength antenna (commonly called the half-wave dipole) is used widely in antenna systems. Factors contributing to its popularity are its well-understood radiation pattern, its simple construction, its high efficiency, and its capability for easy impedance matching.

The electric and magnetic fields for the half-wave dipole can be calculated by substituting its current distribution, $I = I_0 \cos(k_0z)$, into Equation (17.11) to obtain

$$\mathbf{E} = jZ_0I_0\frac{\cos\left(\frac{\pi}{2}\cos\theta\right)}{\sin\theta}\frac{e^{-jk_0r}}{2\pi r}\mathbf{a}_\theta$$
$$\mathbf{H} = jI_0\frac{\cos\left(\frac{\pi}{2}\cos\theta\right)}{\sin\theta}\frac{e^{-jk_0r}}{2\pi r}\mathbf{a}_\varphi \tag{17.12}$$

The total radiated power, P_r, can be determined from the electric and magnetic fields by integrating the expression 1/2 ($\Re e\{\mathbf{E}\times\mathbf{H}^* \cdot \mathbf{a}_r\}$ over a surface of radius r. Carrying out this integration yields $P_r = 36.565\ |I_0|^2$. The radiation resistance of the half-wave dipole can then be determined from

$$R_a = \frac{2P_r}{|I_0|^2} \approx 73\ \Omega \tag{17.13}$$

This radiation resistance is considerably higher than the radiation resistance of a short dipole. For example, if we have a dipole of length 0.01λ, its radiation resistance will be approximately 0.08 Ω (from Equation (17.9)).

This resistance is probably comparable to the ohmic resistance of the dipole, thereby resulting in a low efficiency. The half-wave dipole, having a much higher radiation resistance, will have much higher efficiency. The higher resistance of the half-wave dipole also makes impedance matching easier.

End Loading

At many frequencies of interest, for example, the broadcast band, a half-wavelength becomes unreasonably long. Figure 17.2 shows a way of increasing the effective length of the dipole without making it longer. Here, additional wires have been added to the ends of the dipoles. These wires increase the end capacitance of the dipole, thereby increasing the effective electrical length.

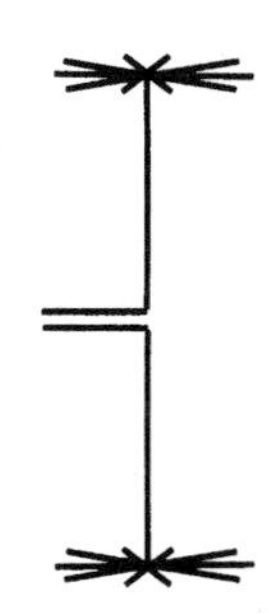

FIGURE 17.2 Using end loading to increase the effective electrical length of an electric dipole.

Monopole Antennas

A monopole antenna consists of half of a dipole placed above and perpendicular to a conducting ground plane. A monopole of length L behaves like a dipole of length $2L$ except that the radiation only takes place above the conducting plane, resulting in double the directivity and half the radiation resistance. The quarter-wave monopole ($L = \lambda/4$) thus approximates the half-wave dipole and is often used for mobile antennas, where the conducting plane might be the body of an automobile or the case of a handset.

Arrays of Wire Antennas

Often it is advantageous to have several antennas operating together in an **array**. Arrays of antennas can be made to produce highly directional radiation patterns. Also, small antennas can be used in an array to obtain the level of performance of a large antenna at a fraction of the area. Arrays of dipoles are often used as base station antennas for mobile communication systems.

The radiation pattern of an array depends on the number and type of antennas used, the spacing in the array, and the relative phase and magnitude of the excitation currents. The ability to control the phase of the exciting currents in each element of the array allows one to electronically scan the main radiated beam. An array that varies the phases of the exciting currents to scan the radiation pattern through space is called an electronically scanned **phased array**. Phased arrays are used extensively in radar applications.

Analysis of General Arrays

To obtain analytical expressions for the radiation fields due to an array one must first look at the fields produced by a single array element. For an isolated radiating element positioned as in Figure 17.3, the electric field at a far-field point P is given by

$$\mathbf{E}_i = a_i \mathbf{K}_i(\theta, \varphi) e^{j[k_0(\mathbf{R}_i \cdot \mathbf{i}_p) - \alpha_i]} \tag{17.14}$$

where $\mathbf{K}_i(\theta, \varphi)$ is the electric field pattern of the individual element, $a_i e^{-j\alpha_i}$ is the excitation of the individual element, $\mathbf{R}_i$ is the position vector from the phase reference point to the element, $\mathbf{i}_p$ is a unit vector pointing toward the far-field point P, and k_0 is the free space wave vector.

Now, for an array of N of these arbitrary radiating elements the total E-field at position P is given by the vector sum:

$$\mathbf{E}_{tot} = \sum_{i=0}^{N-1} \mathbf{E}_i = \sum_{i=0}^{N-1} \mathbf{a}_i \mathbf{K}_i(\theta, \varphi) e^{j[k_0(\mathbf{R}_i \cdot \mathbf{i}_p) - \alpha_i]} \tag{17.15}$$

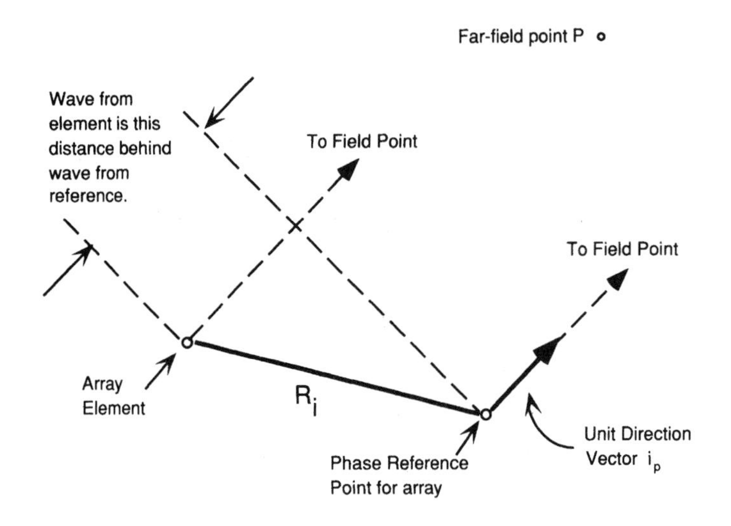

FIGURE 17.3 Diagram for determining the far field due to radiation from a single array element. (*Source: Reference Data for Radio Engineers,* Indianapolis: Howard W. Sams & Co., 1975, chap. 27–22. With permission.)

This equation may be used to calculate the total field for an array of antennas where the mutual coupling between the array elements can be neglected. For most practical antennas, however, there is mutual coupling, and the individual patterns will change when the element is placed in the array. Thus, Equation (17.15) should be used with care.

Arrays of Identical Elements

If all the radiating elements of an array are identical, then $\mathbf{K}_i(\theta, \varphi)$ will be the same for each element and Equation (17.15) can be rewritten as

$$\mathbf{E}_{\text{tot}} = \mathbf{K}(\theta, \varphi) \sum_{i=0}^{N-1} a_i e^{j[k_0(\mathbf{R}_i \cdot \mathbf{i}_\text{p}) - \alpha_i]} \tag{17.16}$$

This can also be written as

$$\mathbf{E}_{\text{tot}} = \mathbf{K}(\theta, \varphi) f(\theta, \varphi) \quad \text{where} \quad f(\theta, \varphi) = \sum_{i=0}^{N-1} a_i e^{j[k_0(\mathbf{R}_i \cdot \mathbf{i}_\text{p}) - \alpha_i]} \tag{17.17}$$

The function $f(\theta, \varphi)$ is normally called the array factor or the array polynomial. Thus, one can find $\mathbf{E}_{\text{tot}}$ by just multiplying the individual element's electric field pattern, $\mathbf{K}(\theta, \varphi)$, by the array factor, $f(\theta, \varphi)$. This process is often referred to as pattern multiplication.

The average radiated power per unit solid angle is proportional to the square of $\mathbf{E}_{\text{tot}}$. Thus, for an array of identical elements:

$$\frac{\Delta P_\text{r}(\theta, \varphi)}{\Delta \Omega} \sim |\mathbf{K}(\theta, \varphi)|^2 |f(\theta, \varphi)|^2 \tag{17.18}$$

Equally Spaced Linear Arrays

An important special case occurs when the array elements are identical and are arranged on a straight line with equal element spacing, d, as shown in Figure 17.4. If a linear phase progression, α, is assumed for the excitation

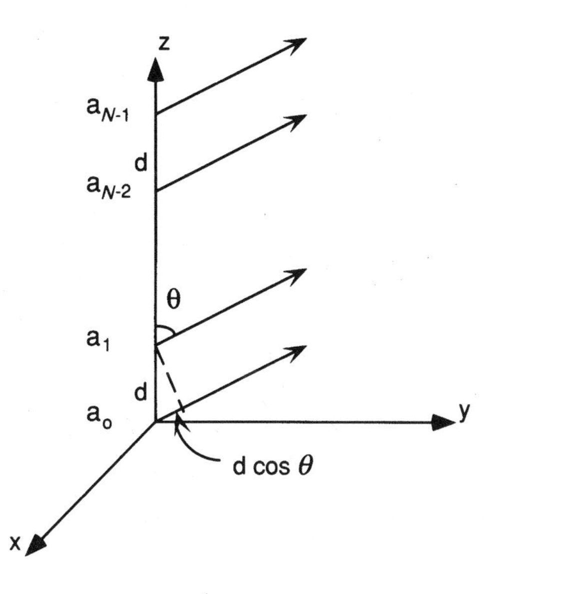

FIGURE 17.4 A linear array of equally spaced elements.

currents of the elements, then the total field at position *P* in Figure 17.4 will be

$$\begin{aligned} \mathbf{E}_{\text{tot}} &= \mathbf{K}(\theta, \varphi) \sum_{n=0}^{N-1} a_n e^{jn(k_0 d \cos\theta - \alpha)} \\ &= \mathbf{K}(\theta, \varphi) \sum_{n=0}^{N-1} a_n e^{jn\psi} = \mathbf{K}(\theta, \varphi) f(\psi) \end{aligned} \tag{17.19}$$

where $\psi = k_0 \, d \, \cos\, \theta - \alpha$.

Broadside Arrays

Suppose that, in the linear array of Figure 17.4, all the excitation currents are equal in magnitude and phase ($a_0 = a_1 = \ldots = a_{N-1}$ and $\alpha = 0$). The array factor, $f(\psi)$, then becomes

$$f(\psi) = a_0 \sum_{n=0}^{N-1} e^{jn\psi} = a_0 \frac{1 - e^{jN\psi}}{1 - e^{j\psi}} \tag{17.20}$$

This can be simplified to obtain the normalized form:

$$f'(\psi) = \left| \frac{f(\psi)}{a_0 N} \right| = \left| \frac{\sin \frac{N\psi}{2}}{N \sin \frac{\psi}{2}} \right| \tag{17.21}$$

Note that $f'(\psi)$ is maximum when $\psi = 0$. For our case, with $\alpha = 0$, we have $\psi = k_0 d \cos\theta$. Thus, $f'(\psi)$ will be maximized when $\theta = \pi/2$. This direction is perpendicular to the axis of the array (see Figure 17.4), and so the resulting array is called a broadside array.

Phased Arrays

By adjusting the phase of the elements of the array it is possible to vary the direction of the maximum of the array's radiation pattern. For arrays where all the excitation currents are equal in magnitude but not necessarily phase, the array factor is a maximum when $\psi = 0$. From the definition of ψ, one can see that at the pattern maximum:

$$k_0 d \cos\theta = \alpha$$

Thus, the direction of the array factor maximum is given by

$$\theta = \cos^{-1}\left(\frac{\alpha}{k_0 d}\right) \tag{17.21b}$$

Note that if one is able to control the phase delay, α, the direction of the maximum can be scanned without physically moving the antenna.

Planar (2D) Arrays

Suppose there are M linear arrays, all identical to the one pictured in Figure. 17.4, lying in the yz-plane with element spacing d in both the y and the z direction. Using the origin as the phase reference point, the array factor can be determined to be

$$f(\theta, \varphi) = \sum_{n=0}^{N-1} \sum_{m=0}^{M-1} a_{mn} e^{[jn(k_0 d \cos\theta - \alpha_z) + jm(k_0 d \sin\theta \sin\varphi - \alpha_y)]} \tag{17.22}$$

where α_y and α_z are the phase differences between the adjacent elements in the y and z directions, respectively. The formula can be derived by considering the 2D array to be a 1D array of subarrays, where each subarray has an antenna pattern given by Equation (17.19).

If all the elements of the 2D array have excitation currents equal in magnitude and phase (all the a_{mn} are equal and $\alpha_z = \alpha_y = 0$), then the array will be a broadside array and will have a normalized array factor given by

$$f'(\theta, \varphi) = \frac{\sin\left(\frac{Nk_0 d}{2}\cos\theta\right)}{N\sin\left(\frac{k_0 d}{2}\cos\theta\right)} \frac{\sin\left(\frac{Mk_0 d}{2}\sin\theta\sin\varphi\right)}{M\sin\left(\frac{k_0 d}{2}\sin\theta\sin\varphi\right)} \tag{17.23}$$

Yagi-Uda Arrays

The Yagi-Uda array is very popular in the HF–VHF–UHF frequency range and is often used as a home TV antenna. The Yagi-Uda array avoids the problem of needing to control the feeding currents to all of the array elements by driving only one element. The other elements in the Yagi-Uda array are excited by near-field coupling from the driven element.

The basic three-element Yagi-Uda array is shown in Figure. 17.5. The array consists of a driven antenna of length l_1, a reflector element of length l_2, and a director element of length l_3. Typically, the director element is shorter than the driven element by 5% or more, while the reflector element is longer than

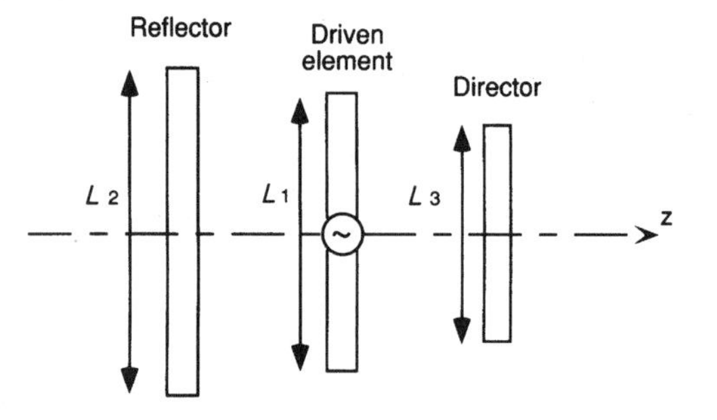

FIGURE 17.5 Three-element Yagi-Uda antenna. (*Source:* Shintaro Uda and Yasuto Mushiake, *Yagi-Uda Antenna*, Sendai, Japan: Sasaki Printing and Publishing Company, 1954, p. 100. With permission.)

the driven element by 5% or more (Stutzman and Thiele, 1981). The radiation pattern for the array in Figure 17.5 will have a maximum in the $+z$ direction.

One can increase the gain of the Yagi–Uda array by adding additional director elements. Adding additional reflector elements, however, has little effect because the field behind the first reflector element is small.

Yagi-Uda arrays typically have directivities between 10 and 100, depending on the number of directors (Ramo et al., 1984). TV antennas usually have several directors.

Log-Periodic Dipole Arrays

Another variation of wire antenna arrays is the log-periodic dipole array. The log-periodic is popular in applications that require a broadband, frequency-independent antenna. An antenna will be independent of frequency if its dimensions, when measured in wavelengths, remain constant for all frequencies. If, however, an antenna is designed so that its characteristic dimensions are periodic with the logarithm of the frequency, and if the characteristic dimensions do not vary too much over a period of time, then the antenna will be essentially frequency independent. This is the basis for the log-periodic dipole array, shown in Figure 17.6.

In Figure 17.6, the ratio of successive element positions equals the ratio of successive dipole lengths. This ratio is often called the scaling factor of the log-periodic array and is denoted by

$$\tau = \frac{Z_{n+1}}{Z_n} = \frac{L_{n+1}}{L_n} \tag{17.24}$$

Also note that there is a mechanical phase reversal between successive elements in the array caused by the crossing over of the interconnecting feed lines. This phase reversal is necessary to obtain the proper phasing between adjacent array elements.

To get an idea of the operating range of the log-periodic antenna, note that for a given frequency within the operating range of the antenna, there will be one dipole in the array that is half-wave resonant or is nearly so. This half-wave resonant dipole and its immediate neighbors represent the active region of the log-periodic array. As the operating frequency changes, the active region shifts to a different part of the log-periodic. Hence, the frequency range for the log-periodic array is roughly given by the frequencies at which the longest and shortest dipoles in the array are half-wave resonant (wavelengths such that $2L_N < \lambda < 2L_1$) (Stutzman and Thiele, 1981).

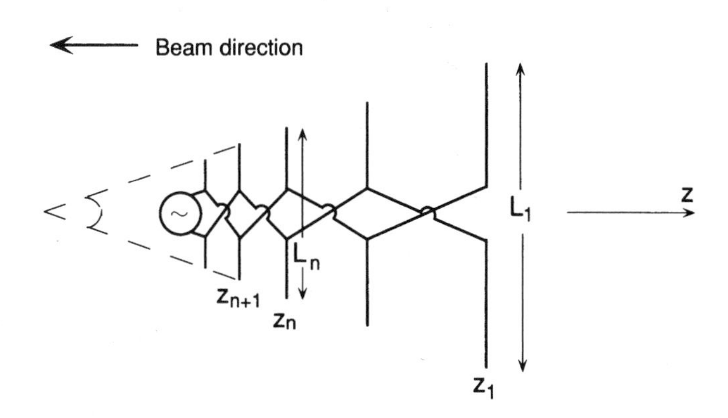

FIGURE 17.6 The log-periodic dipole array. (*Source:* D.G. Isbell, "Log periodic dipole arrays," *IRE Transactions on Antennas and Propagation,* vol. AP-8, p. 262, 1960. With permission.)

Defining Terms

Antenna gain: The ratio of the actual radiated power per solid angle to the radiated power per solid angle that would result if the total input power were radiated isotropically.

Array: Several antennas arranged together in space and interconnected to produce a desired radiation pattern.

Directivity: The ratio of the actual radiated power per solid angle to the radiated power per solid angle that would result if the radiated power was radiated isotropically. Oftentimes the word *directivity* is used to refer to the maximum directivity.

Phased array: An array in which the phases of the exciting currents are varied to scan the radiation pattern through space.

Radiation pattern: A plot as a function of direction of the power per unit solid angle radiated in a given polarization by an antenna. The terms *radiation pattern* and *antenna pattern* can be used interchangeably.

References

C.A. Balanis, *Antenna Theory Analysis and Design*, New York: Harper and Row, 1982.

R. Carrel, "The design of log-periodic dipole antennas," *IRE International Convention Record* (part 1), 1961, pp. 61–75.

R.E. Collin, *Antennas and Radiowave Propagation*, New York: McGraw-Hill, 1985.

R.F. Harrington, *Time Harmonic Electromagnetic Fields*, New York: McGraw-Hill, 1961.

D.E. Isbell, "Log periodic dipole arrays," *IRE Transactions on Antennas and Propagation*, vol. AP-8, pp. 260–267, 1960.

P. Lorrain and D.R. Corson, *Electromagnetic Fields and Waves*, San Francisco, CA: W.H. Freeman, 1970.

S. Ramo, J.R. Whinnery, and T. Van Duzer, *Fields and Waves in Communication Electronics*, New York: John Wiley & Sons, 1984.

W.L. Stutzman and G.A. Thiele, *Antenna Theory and Design*, New York: John Wiley & Sons, 1981.

S. Uda and Y. Mushiake, *Yagi–Uda Antenna*, Sendai, Japan: Sasaki Printing and Publishing Company, 1954.

Further Information

For general-interest articles on antennas the reader is directed to the *IEEE Antennas and Propagation Magazine.* In addition to providing up-to-date articles on current issues in the antenna field, this magazine also provides easy-to-read tutorials. For the latest research advances in the antenna field the reader is referred to the *IEEE Transactions on Antennas and Propagation.* In addition, a number of very good textbooks are devoted to antennas. The books by Collin and by Stutzman and Thiele were especially useful in the preparation of this section.

17.2 Aperture

J. Patrick Fitch

The main purpose of an **antenna** is to control a wavefront at the boundary between two media: a source (or receiver) and the medium of propagation. The source can be a fiber, cable, waveguide, or other transmission line. The medium of propagation may be air, vacuum, water, concrete, metal, or tissue, depending on the application. Antenna aperture design is used in acoustic, optic, and electromagnetic systems for imaging, communications, radar, and spectroscopy applications.

There are many classes of antennas: wire, horn, slot, notch, reflector, lens, and **array** to name a few (see Figure 17.7). Within each class is a variety of subclasses. For instance, the horn antenna can be pyramidal or conical. The horn can also have flaring in only one direction (sectoral horn), asymmetric components, shaped

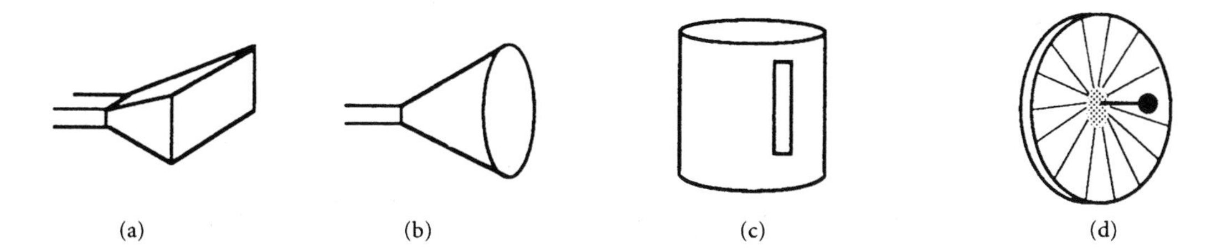

FIGURE 17.7 Examples of several types of antennas: (a) pyramidal horn, (b) conical horn, (c) axial slot on a cylinder, and (d) parabolic reflector.

edges, or a compound design of sectoral and pyramidal combined. For all antennas, the relevant design and analysis will depend on antenna aperture size and shape, the center wavelength, λ, and the distance from the aperture to a point of interest (the range, R). This section covers discrete **oscillators**, arrays of oscillators, synthetic apertures, geometric design, Fourier analysis, and parameters of some typical antennas. The emphasis is on microwave-type designs.

The Oscillator or Discrete Radiator

The basic building block for antenna analysis is a linear conductor. Movement of electrons (current) in the conductor induces an electromagnetic field. When the electron motion is oscillatory—e.g., a dipole with periodic electron motion—the induced electric field, E, is proportional to $\cos(\omega t - kx + \phi)$, where ω is radian frequency of oscillation, t is time, k is wave number, x is distance from the oscillator, and ϕ is the phase associated with this oscillator (relative to the time and spatial coordinate origins). When the analysis is restricted to a fixed position x, the electric field can be expressed as

$$E(t) = A\cos(\omega t + \phi) \tag{17.25}$$

where the phase term ϕ now includes the kx term, and all of the constants of proportionality are included in the amplitude A. Basically, the assumption is that oscillating currents produce oscillating fields. The description of a receiving antenna is analogous: an oscillating field induces a periodic current in the conductor.

The field from a pair of oscillators separated in phase by δ radians is

$$E_\delta(t) = A_1\cos(\omega t + \phi) + A_2\cos(\omega t + \phi + \delta) \tag{17.26}$$

Using phasor notation, $\tilde{E}_\delta$, the cosines are converted to complex exponentials and the radial frequency term, ωt, is suppressed:

$$\tilde{E}_\delta(t) = A_1 e^{i\phi} + A_2 e^{i(\phi+\delta)} \tag{17.27}$$

The amplitude of the sinusoidal modulation $E_\delta(t)$ can be calculated as $|\tilde{E}_\delta|$. The intensity is

$$I = |\tilde{E}_\delta|^2 = |A_1|^2 + |A_2|^2 + 2A_1A_2\cos(\delta) \tag{17.28}$$

When the oscillators are of the same amplitude, $A = A_1 = A_2$, then:

$$\begin{aligned} E_\delta(t) &= A\cos(\omega t + \phi) + A\cos(\omega t + \phi + \delta) \\ &= 2A\cos\left(\frac{\delta}{2}\right)\cos\left(\omega t + \phi + \frac{\delta}{2}\right) \end{aligned} \tag{17.29}$$

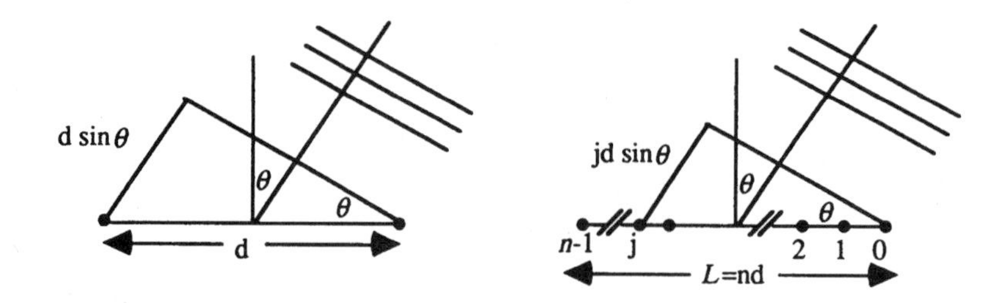

FIGURE 17.8 A two-element and an n-element array with equal spacing between elements. The propagation length difference between elements is $d \sin\theta$, which corresponds to a phase difference of $kd \sin\theta$, where k is the wave number $2\pi/\lambda$. The length L corresponds to a continuous aperture of length nd with the sample positions beginning $d/2$ from the ends.

For a series of n equal amplitude oscillators with equal phase spacing:

$$E_{n\delta}(t) = \sum_{j=0}^{n-1} A\cos(\omega t + \phi + j\delta) \tag{17.30}$$

By using phasor arithmetic the intensity is given as

$$I_{n\delta}(t) = |\tilde{E}_{n\delta}|^2 = \left| Ae^{i\phi}\sum_{j=0}^{n-1} e^{ij\delta}\right|^2 = A^2\left|\frac{1-e^{in\delta}}{1-e^{i\delta}}\right|^2 = I_0\frac{1-\cos(n\delta)}{1-\cos(\delta)} = I_0\frac{\sin^2(n\delta/2)}{\sin^2(\delta/2)} \tag{17.31}$$

where $I_0 = n^{-2}$ to normalize the intensity pattern at $\delta = 0$.

For an incoming plane wave which is tilted at an angle θ from the normal, the relative phase difference between two oscillators is $kd \sin\theta$, where d is the distance between oscillators and k is the wavenumber $2\pi/\lambda$ (see Figure 17.8). For three evenly spaced oscillators, the phase difference between the end oscillators is $2kd \sin\theta$. In general, the end-to-end phase difference for n evenly spaced oscillators is $(n-1)kd \sin\theta$. This formulation is identical to the phase representation in Equation (17.30) with $\delta = kd \sin\theta$. Therefore, the intensity as a function of incidence angle θ for an evenly spaced array of n elements is

$$I_{nL}(\theta) = I_0\frac{\sin^2\left(\frac{1}{2}knd \sin\theta\right)}{\sin^2\left(\frac{1}{2}kd \sin\theta\right)} = I_0\frac{\sin^2\left(\frac{1}{2}kL \sin\theta\right)}{\sin^2\left(\frac{1}{2n}kL \sin\theta\right)} = I_0\frac{\sin^2\left(\frac{\pi L}{\lambda}\sin\theta\right)}{\sin^2\left(\frac{\pi L}{n\lambda}\sin\theta\right)} \tag{17.32}$$

where $L = nd$ corresponds to the physical dimension (length) of the aperture of oscillators. The zeros of this function occur at $kL \sin\theta = 2m\pi$, for any nonzero integer m. Equivalently, the zeros occur when $\sin\theta = m\lambda/L$. When the element spacing d is less than a wavelength, the number of zeros for $0 < \theta < \pi 2$ is given by the largest integer M such that $M \leq L/\lambda$. Therefore, the ratio of wavelength to largest dimension, λ/L, determines both the location (in θ space) and the number of zeros in the intensity pattern when $d \leq \lambda$. The number of oscillators controls the amplitude of the side lobes.

For $n = 1$, the intensity is constant—i.e., independent of angle. For $\lambda > L$, both the numerator and denominator of Equation (17.32) have no zeros and as the length of an array shortens (relative to a wavelength), the intensity pattern converges to a constant ($n = 1$ case). As shown in Figure 17.9, a separation of $\lambda/4$ has an intensity rolloff less than 1 dB over $\pi/2$ radians (a $\lambda/2$ separation rolls off 3 dB). This implies that placing antenna elements closer than $\lambda/4$ does not significantly change the intensity pattern. Many microwave antennas exploit this and use a mesh or parallel wire (for polarization sensitivity) design rather than covering

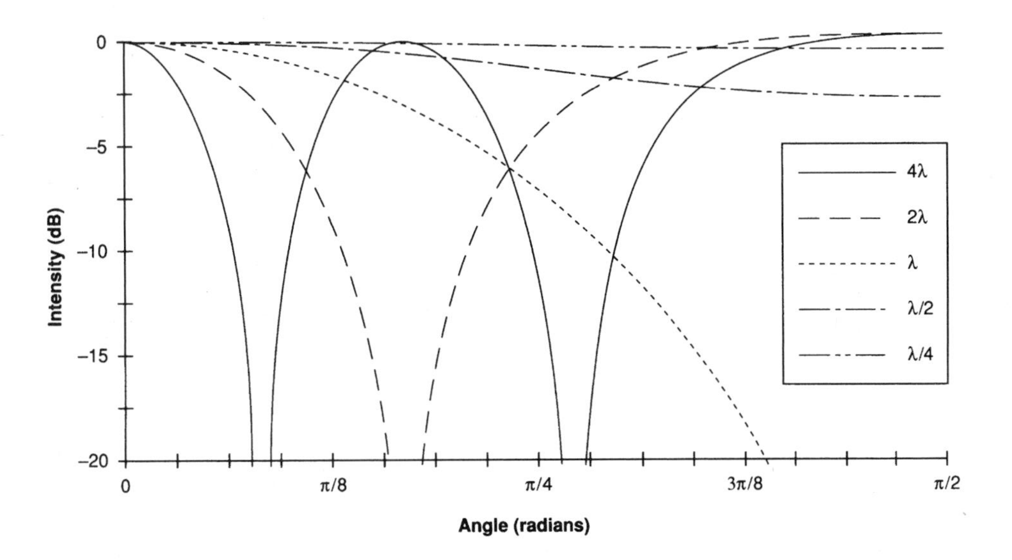

FIGURE 17.9 Normalized intensity pattern in decibels (10 log(I)) for a two-element antenna with spacing 4λ, 2λ, λ, $\lambda/2$, and $\lambda/4$ between the elements.

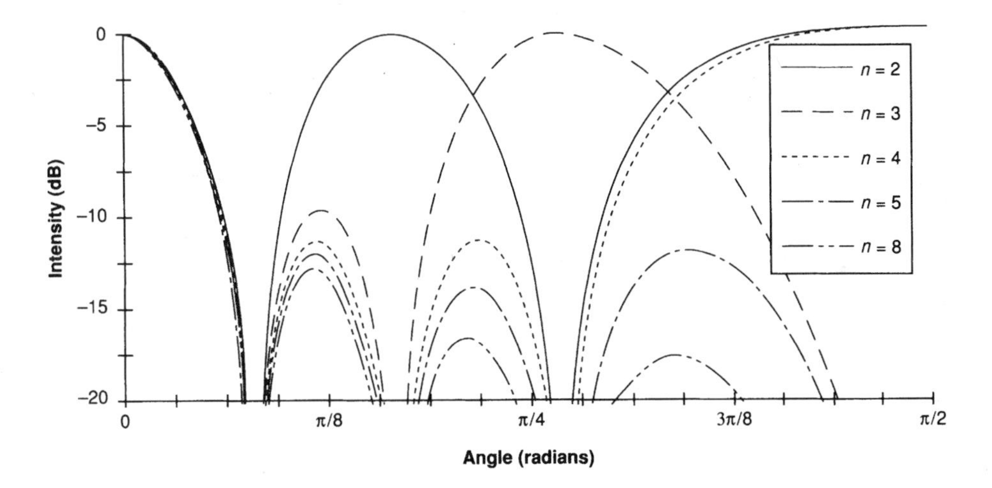

FIGURE 17.10 Normalized intensity pattern in decibels (10 log(I)) for a length 4λ array with 2, 3, 4, 5, and 8 elements.

the entire aperture with conductor. This reduces both weight and sensitivity to wind loading. Note that the analysis has not accounted for phase variations from position errors in the element placement where the required accuracy is typically better than $\lambda/10$.

For $L >> \lambda$, $\sin\theta \approx \theta$, which implies that the first zero is at $\theta = \lambda/L$. The location of the first zero is known as the Rayleigh resolution criteria. That is, two plane waves separated by at least λ/L radians can be discriminated. For imaging applications, this corresponds roughly to the smallest detectable feature size. As shown in Figure 17.10, the first zero occurs at approximately $\lambda/L = 0.25$ radians (the Rayleigh resolution). Note that there is no side lobe suppression until $d \leq \lambda$, when the location of the zeros becomes fixed. Having more than eight array elements (separation of less than a quarter wavelength) only moderately reduces the height of the maximum side lobe.

Synthetic Apertures

In applications such as air- and space-based radar, size and weight constraints prohibit the use of very large antennas. For instance, if the L-band (23.5-cm wavelength) radar imaging system on the Seasat satellite (800-km altitude, launched in 1978) had a minimum resolution specification of 23.5 m, then, using the Rayleigh resolution criteria, the aperture would need to be 8 km long. In order to attain the desired resolution, an aperture is "synthesized" from data collected with a physically small (10 m) antenna traversing an 8-km flight path. Basically, by using a stable oscillator on the spacecraft, both amplitude and phase are recorded, which allows postprocessing algorithms to combine the individual echoes in a manner analogous to an antenna array. From an antenna perspective, an individual scattering element produces a different round trip propagation path based on the position of the physical antenna—a synthetic antenna array. Using the geometry described in Figure 17.11, the phase is

$$\phi(x) = \frac{2\pi}{\lambda}2R(x) = \frac{2\pi}{\lambda}2\sqrt{x^2 + y^2 + z^2} \tag{17.33}$$

It is convenient to assume a straight-line flight path along the x-axis, a planar earth (x, y plane), and a constant velocity, v, with range and cross-range components $v_r(x)$ and $v_c(x)$, respectively. In many radar applications the broad side distance to the center of the footprint, R, is much larger than the size of the footprint. This allows the distance $R(x)$ to be expanded about R resulting in

$$\phi(t) = \frac{2\pi}{\lambda}2R(vt) = 2\pi\left\{\frac{2R}{\lambda} + \frac{2v_r}{\lambda}t + \frac{v_c^2}{\lambda R}t^2\right\} \tag{17.34}$$

The first term in Equation (17.34) is a constant phase offset corresponding to the center of beam range bin and can be ignored from a resolution viewpoint. The second term, $2v_r/\lambda$, is the Doppler frequency shift due to the relative (radial) velocity between the antenna and scattering element. The third term represents a quadratic correction of the linear flight path to approximate the constant range sphere from a scattering element. It is worth noting that synthetic aperture systems do not require the assumptions used here, but accurate position and motion compensation is required.

For an antenna with cross range dimension D and a scattering element at range R, the largest synthetic aperture that can be formed is of dimension $\lambda R/D$ (the width of the footprint). Because this data collection scenario is for round-trip propagation, the phase shift at each collecting location is twice the shift at the edges of a single physical antenna. Therefore, at a range R, the synthetic aperture resolution is

$$\frac{\lambda R}{D_{SA}} = \frac{\lambda R}{2\lambda R/D} = \frac{D}{2} \tag{17.35}$$

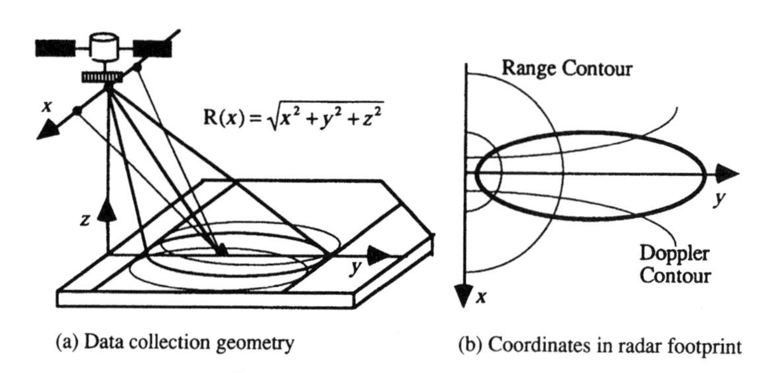

FIGURE 17.11 Synthetic aperture radar geometry and nearly orthogonal partitioning of the footprint by range (circular) and Doppler frequency (hyperbolic) contours.

The standard radar interpretation for synthetic apertures is that information coded in the Doppler frequency shift can be decoded to produce high-resolution images. It is worth noting that the synthetic aperture can be formed even with no motion (zero Doppler shift). For the no-motion case the antenna array interpretation is appropriate. This approach has been used for acoustic signal processing in nondestructive evaluation systems as well as wave migration codes for seismic signal processing. When there is motion, the Doppler term in the expansion of the range dominates the phase shift and therefore becomes the useful metric for predicting resolution.

Geometric Designs

The phase difference in a linear array was caused by the spatial separation and allowed the discrimination of plane waves arriving at different angles. Desired phase patterns can be determined by using analytic geometry to position the elements. For example, if coherent superposition across a wave front is desired, the wavefront can be directed (reflected, refracted, or diffracted) to the receiver in phase. For a planar wavefront, this corresponds to a constant path length from any point on the reference plane to the receiver. Using the geometry in Figure 17.12, the sum of the two lengths $(x, R + h)$ to (x, y) and (x, y) to $(0, h)$ must be a constant independent of x—which is $R + 2h$ for this geometry. This constraint on the length is

$$R + h - y + \sqrt{x^2 + (h - y)^2} = R + 2h \quad \text{or} \quad x^2 = 4hy \tag{17.36}$$

This is the equation for a parabola. Losses would be minimized if the wave front were specularly reflected to the transceiver. Specular reflection occurs when the angles between the normal vector **N** [or equivalently the tangent vector $\mathbf{T}=(x,f'(x))] = (1, x/2h)$ and the vectors $\mathbf{A} = (0, -1)$ and $\mathbf{B} = (-x, h - y)$ are equal. This is the same as equality of the inner products of the normalized vectors, which is shown by

$$\hat{\mathbf{T}} \cdot \hat{\mathbf{A}} = \frac{(2h, x)}{\sqrt{x^2 + 4h^2}} \cdot (0, -1) = \frac{-x}{\sqrt{x^2 + 4h^2}} \tag{17.37}$$

$$\hat{\mathbf{T}} \cdot \hat{\mathbf{B}} = \frac{(2h, x)}{\sqrt{x^2 + 4h^2}} \cdot \frac{(-x, h - y)}{\sqrt{x^2 + (h - y)^2}} = \frac{-x(x^2 + 4h^2)}{(x^2 + 4h^2)^{3/2}} = \frac{-x}{\sqrt{x^2 + 4h^2}} \tag{17.38}$$

The constant path length and high gain make the parabolic antenna popular at many wavelengths including microwave and visible. More than one reflecting surface is allowed in the design. The surfaces are typically conical sections and may be designed to reduce a particular distortion or to provide better functionality. Compound designs often allow the active elements to be more accessible and eliminate long transmission lines. A two-bounce reflector with a parabolic primary and a hyperbolic secondary is known as a Cassegrain system.

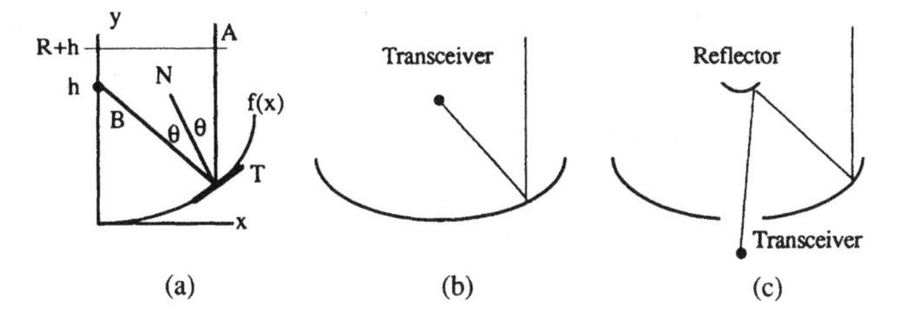

FIGURE 17.12 Parabolic reflector systems: (a) geometry for determining the function with a constant path length and specular reflection, (b) single-bounce parabolic reflector, (c) two-bounce reflector with a parabolic primary and hyperbolic secondary (Cassegrain).

In all reflector systems it is important to account for the blockage ("shadow" of the feed, secondary reflector, and support structures) as well as the spillover (radiation propagating past the intended reflecting surface).

Continuous Current Distributions (Fourier Transform)

Ideally, antennas would be designed using solutions to Maxwell's equations. Unfortunately, in most cases exact analytic and numerical solutions to Maxwell's equations are difficult to obtain. Under certain conditions, approximations can be introduced that allow solution to the wave equations. Approximating spherical wavefronts as quadratics has been shown for the synthetic aperture application and is valid when the propagation distance is greater than $(\pi L^2/4\lambda)^{1/3}$, where L is the aperture size. In general, this is known as the **Fresnel** or **near-field** approximation. When the propagation distance is at least $2L^2/\lambda$, the angular radiation pattern can be approximated as independent of distance from the aperture. This pattern is known as the normalized **far-field** or **Fraunhofer** distribution, $E(\theta)$, and is related to the normalized current distributed across an antenna aperture, $i(x)$, by a Fourier transform:

$$E(u) = \int i(x')e^{i2\pi ux'}dx' \tag{17.39}$$

where $u = \sin\theta$ and $x' = x/\lambda$.

Applying the Fraunhofer approximation to a line source of length L:

$$E_L(u = \sin\theta) = \int_{-L/2\lambda}^{L/2\lambda} e^{i2\pi ux'}dx' = \frac{\sin\left(\frac{\pi L}{\lambda}u\right)}{\frac{\pi L}{\lambda}u} = \frac{\sin\left(\frac{\pi L}{\lambda}\sin\theta\right)}{\frac{\pi L}{\lambda}\sin\theta} \tag{17.40}$$

which is Equation (17.32) when $n >> L/\lambda$. As with discrete arrays, the ratio L/λ is the important design parameter: $\sin\theta = \lambda/L$ is the first zero (no zeros for $\lambda > L$) and the number of zeros is the largest integer M such that $M \leq L/\lambda$.

In two dimensions, a rectangular aperture with uniform current distribution produces

$$E_R(u_1, u_2) = \frac{\sin\left(\frac{\pi}{\lambda}u_1L_1\right)}{\frac{\pi}{\lambda}u_1L_1}\frac{\sin\left(\frac{\pi}{\lambda}u_2L_2\right)}{\frac{\pi}{\lambda}u_2L_2} \text{ and } I_R(u_1, u_2) = |E_L(u_1)|^2|E_L(u_2)|^2 \tag{17.41}$$

The field and intensity given in Equation (17.41) are normalized. In practice, the field is proportional to the aperture area and inversely proportional to the wavelength and propagation distance.

The normalized far-field intensity distribution for a uniform current on a circular aperture is a circularly symmetric function given by

$$I_C(u) = \left[\frac{2J_1\left(\frac{\pi}{\lambda}uL\right)}{\frac{\pi}{\lambda}uL}\right]^2 \tag{17.42}$$

where J_1 is the Bessel function of the first kind, order one. This far-field intensity is called the Airy pattern. As with the rectangular aperture, the far-field intensity is proportional to the square of the area and inversely proportional to the square of the wavelength and the propagation distance. The first zero (Rayleigh resolution criteria) of the Airy pattern occurs for $uL/\lambda = 1.22$ or $\sin\theta = 1.22\lambda/L$. As with linear and rectangular apertures, the resolution scales with λ/L.

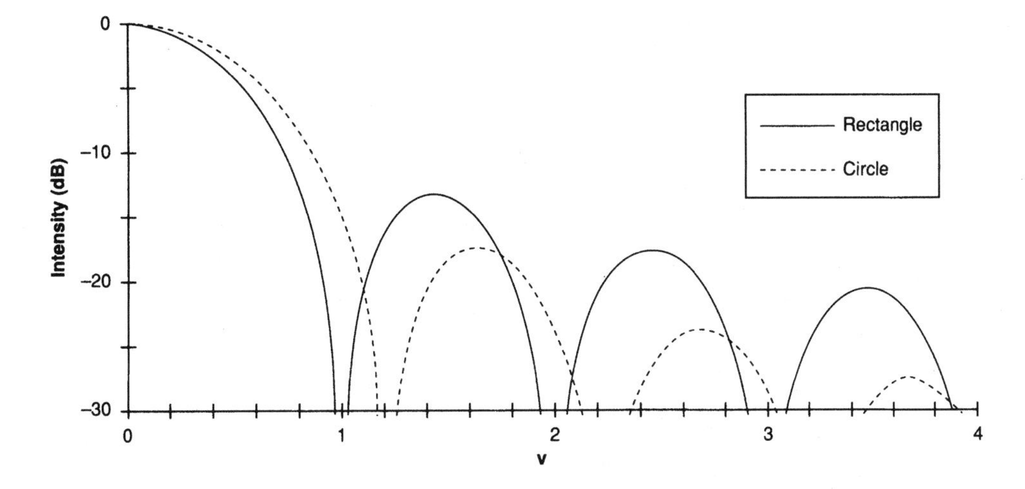

FIGURE 17.13 Normalized intensity pattern in decibels ($10 \log[I(v = uL/\lambda)]$) for a rectangular and a circular antenna aperture with uniform current distributions.

TABLE 17.1 Directivity and Gain of Some Higher Frequency Antennas

Antenna Type	Directivity[1]	Gain[1]
Uniform rectangular aperture	$\frac{4\pi}{\lambda^2} L_x L_y$	$\frac{4\pi}{\lambda^2} L_x L_y$
Large square aperture	$12.6\left(\frac{L}{\lambda}\right)^2$	$7.7\left(\frac{L}{\lambda}\right)^2$
Large circular aperture (parabolic reflector)	$9.87\left(\frac{D}{\lambda}\right)^2$	$7\left(\frac{D}{\lambda}\right)^2$
Pyramidal horn	$\left(\frac{4\pi}{\lambda^2}\right) L_x L_y$	$0.5\left(\frac{4\pi}{\lambda^2}\right) L_x L_y$

[1]Directivity and gain are relative to a half-wave dipole.

Figure 17.13 shows a slice through the normalized far-field intensity of both a rectangular aperture and a circular aperture. The linearity of the Fourier transform allows apertures to be represented as the superposition of subapertures. The primary reflector, the obscurations from the support structures, and the secondary reflector of a Cassegrain-type antenna can be modeled. Numerical evaluation of the Fourier transform permits straightforward calculation of the intensity patterns, even for nonuniform current distributions.

Antenna Parameters

Direct solutions to Maxwell's equations or solutions dependent on approximations provide the analytic tools for designing antennas. Ultimately, the analysis must be confirmed with experiment. Increasingly sensitive radar and other antenna applications have resulted in much more attention to edge effects (from the primary aperture, secondary, and/or support structures). The geometric theory of diffraction as well as direct Maxwell solvers are making important contributions.

With the diversity of possible antenna designs, a collection of design rules of thumb are useful. The **directivity** and **gain** for a few popular antenna designs are given in Table 17.1. Directivity is the ratio of the maximum to average radiation intensity. The gain is defined as the ratio of the maximum radiation intensity from the subject antenna to the maximum radiation intensity from a reference antenna with the same power input. The directivity, D, and gain, G, of an antenna can be expressed as

$$D = \left(\frac{4\pi}{\lambda^2}\right)A_{\mathrm{em}} \quad \text{and} \quad G = \left(\frac{4\pi}{\lambda^2}\right)A_{\mathrm{e}} \tag{17.43}$$

where A_{em} is the maximum effective aperture and A_{e} is the actual effective aperture of the antenna. Because of losses in the system, $A_{\mathrm{e}} = kA_{\mathrm{em}}$, where k is the radiation efficiency factor. The gain equals the directivity when there are no losses ($k = 1$), but is less than the directivity if there are any losses in the antenna ($k < 1$), that is, $G = kD$.

As an example, consider the parabolic reflector antenna where efficiency degradation includes:

- Ohmic losses are small ($k = 1$)
- Aperture taper efficiency ($k = 0.975$)
- Spillover (feed) efficiency ($k = 0.8$)
- Phase errors in aperture field ($k = 0.996$ to 1)
- Antenna blockage efficiency ($k = 0.99$)
- Spar blockage efficiency ($k = 0.994$)

Each antenna system requires a customized analysis of the system losses in order to accurately model performance.

Defining Terms

Antenna: A physical device for transmitting or receiving propagating waves.

Aperture antenna: An antenna with a physical opening, hole, or slit. Contrast with a wire antenna.

Array antenna: An antenna system performing as a single aperture but composed of antenna subsystems.

Directivity: The ratio of the maximum to average radiation intensity.

Fraunhofer or far field: The propagation region where the normalized angular radiation pattern is independent of distance from the source. This typically occurs when the distance from the source is at least $2L^2/\lambda$, where L is the largest dimension of the antenna.

Fresnel or near field: The propagation region where the normalized radiation pattern can be calculated using quadratic approximations to the spherical Huygens' wavelet surfaces. The pattern can depend on distance from the source and is usually valid for distances greater than $(\pi/4\lambda)^{1/3}L^{2/3}$, where L is the largest dimension of the antenna.

Gain: The ratio of the maximum radiation intensity from the subject antenna to the maximum radiation intensity from a reference antenna with the same power input. Typical references are a lossless isotropic source and a lossless half-wave dipole.

Oscillator: A physical device that uses the periodic motion within the material to create propagating waves. In electromagnetics, an oscillator can be a conductor with a periodic current distribution.

Reactive near field: The region close to an antenna where the reactive components of the electromagnetic fields from charges on the antenna structure are very large compared to the radiating fields. Considered negligible at distances greater than a wavelength from the source (decay as the square or cube of distance). Reactive field is important at antenna edges and for electrically small antennas.

References

R. Feynman, R.B. Leighton, and M.L. Sands, *The Feynman Lectures on Physics*, Reading, MA: Addison-Wesley, 1989.

J.P. Fitch, *Synthetic Aperture Radar*, New York: Springer-Verlag, 1988.

J.W. Goodman, *Introduction to Fourier Optics*, New York: McGraw-Hill, 1968.

H. Jasik, *Antenna Engineering Handbook*, New York: McGraw-Hill, 1961.

R.W.P. King and G.S. Smith, *Antennas in Matter*, Cambridge, Mass.: MIT Press, 1981.

J.D. Krause, *Antennas*, New York: McGraw-Hill, 1950.
Y.T. Lo and S.W. Lee, *Antenna Handbook*, New York: Van Nostrand Reinhold, 1988.
A.W. Rudge, K. Milne, A.D. Olver, and P. Knight, *The Handbook of Antenna Design*, London: Peter Peregrinus, 1982.
M. Skolnik, *Radar Handbook*, New York: McGraw-Hill, 1990.
B.D. Steinberg, *Principles of Aperture & Array System Design*, New York: John Wiley & Sons, 1976.

Further Information

The monthly *IEEE Transactions on Antennas and Propagation* as well as the proceedings of the annual *IEEE Antennas and Propagation International Symposium* provide information about recent developments in this field. Other publications of interest include the *IEEE Transactions on Microwave Theory and Techniques* and the *IEEE Transactions on Aerospace and Electronic Systems.*

Readers may also be interested in the "IEEE Standard Test Procedures for Antennas," The Institute for Electrical and Electronics Engineers, Inc., ANSI IEEE Std. 149–1979, 1979.

17.3 The Fresnel Zone Plate Antenna

James C. Wiltse

Introduction

The Fresnel zone plate is a quasi-optical component that provides the functions of a lens (or a reflective antenna when the zone plate is backed by a mirror). The zone plate has the advantages of simplicity of design and construction, planar configuration, low loss, low weight, and low cost while giving performance comparable to a lens (sometimes better). The zone plate produces lens-like focusing and imaging of electromagnetic waves by means of diffraction, rather than refraction. Thus, the field is often referred to as diffraction optics. The zone plate transforms a normally-incident plane wave into a converging wave, concentrating the radiation field to a small focal region. Much of the material that follows has been adapted from the author's articles in the list of references.

There are two categories of zone plates in practice. One has been used in optical systems for decades; this is the configuration employing focal lengths (F) much larger than the diameter (D) of the plate (the small angle or high F-number case). The other is the case employing focal lengths that are comparable to the diameter (F/D near unity, often in the range between 0.3 and 2.5), referred to as the large-angle or "fast" configuration. The latter case has seen extensive investigations at microwave and millimeter wavelengths [1–8] and a few studies at terahertz frequencies [9] between 210 GHz and 1 THz. The phase-correcting zone plate is most often a stepped planar lens (Figure 17.14) made from low-loss materials such as polystyrene, Teflon, polyethylene, TPX, polycarbonate, foamed polystyrene, low density polytetrafluroethylene (PTFE), foamed polyethylene, quartz, sapphire, ceramics, and semiconductor materials such as high resistivity silicon. These materials range in dielectric constant from approximately 1.1 to 12. It has been shown that materials having dielectric constants below 5 are preferable because they have lower surface and internal reflections [6]. There are nonplanar zone plate versions, such as spherical or paraboloidal types, but those are relatively rare and will not be discussed here. Figure 17.14 illustrates a stepped, quarter-wave corrected zone plate. An alternative approach would be to use rings of different dielectric constants (including air) to produce the stepped phase corrections [2,3].

At microwave or millimeter wavelengths the steps, which are fractions of the operating wavelength, are of convenient dimensions for machining or molding of dielectrics. At terahertz frequencies, where wavelengths are on the order of a millimeter or less, the steps are inconveniently small. To cope with this, very low dielectric constant materials are recommended [9]. The other major difficulty in the terahertz region is keeping losses low.

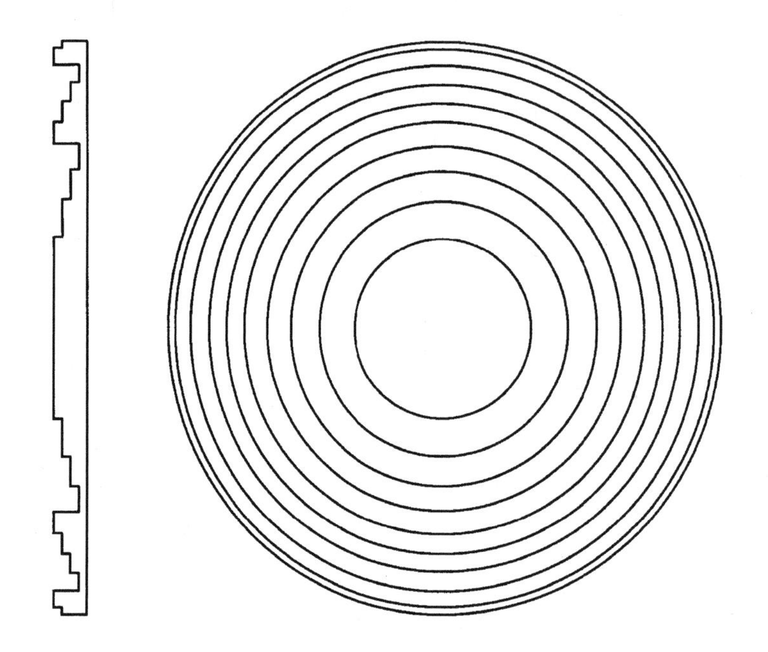

FIGURE 17.14 Quarter-wave corrected Fresnel zone plate antenna. (*Source:* Ref. [9].)

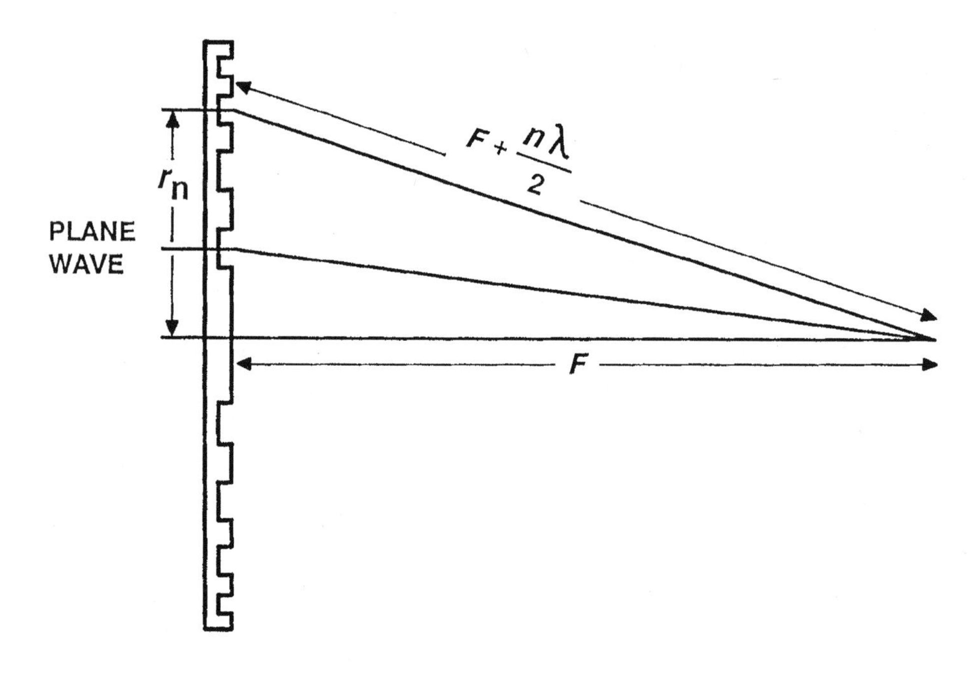

FIGURE 17.15 Ray geometry for half-wave correction. (*Source:* Ref. [9].)

Design Considerations

The typical geometry for a zone plate antenna is shown in Figure 17.15 and Figure 17.16. From Figure 17.16, the equation for the radii of the Fresnel zones is given by

$$r_n = [2nF\lambda/P + (n\lambda/P)^2]^{1/2} \tag{17.44}$$

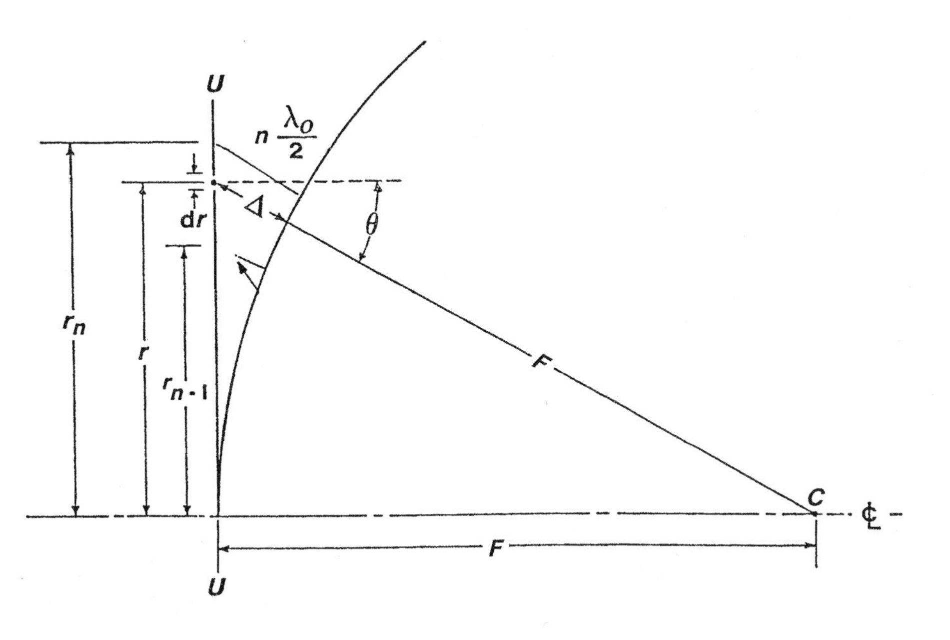

FIGURE 17.16 Geometry of the nth zone. (*Source:* Ref. [9].)

where n is the zone number, F is the focal length, λ is the free-space wavelength, and P is the number of steps to implement the phase correction (e.g., $P = 2$ is half-wave, $P = 4$ is quarter-wave, $P = 8$ is eighth-wave). If we set $r_n = D/2$ we can find the value of n at the outer edge of the zone plate, thus giving us the number of zones. Equation (17.44) is said to define interferometric zone plates, which are free of spherical aberration. Note that the radii of the zones are independent of the dielectric constant, ε. Typical millimeter-wave zone plates have an F/D near unity, as illustrated in Figure 17.16. Defining θ as the angle between the optical axis and the ray from the focal point to the outer edge of the zone plate, we find that θ ranges from 11.3° for $F/D = 2.5$ to 45° for $F/D = 0.5$.

For the grooved zone plate, the expression for the depth of the grooves is

$$d = \lambda/P(\sqrt{\varepsilon} - 1) \tag{17.45}$$

where the parameters are as defined above. Notice that d is independent of the focal length and the diameter of the zone plate. It is obvious that the depth of the grooves is tending toward very small physical dimensions as frequency increases toward the terahertz region.

Feed Considerations

The typical geometry for an FZP and feed is shown in Figure 17.15 and Figure 17.16. Usually the feed device is a horn antenna, although dipole feeds have been used. The corrugated horn gives equal beamwidths in both principal planes, which is ideal for the circular symmetry case. If the amplitude distribution produced across the zone plate were uniform the resulting gain is maximum, but the sidelobe level is not minimized. The feedhorn should have a beam pattern wide enough to cover the zone plate with an amplitude taper that is several dB down at the edge of the plate. The amplitude taper reduces sidelobes, but also decreases the gain and broadens the beamwidth. If we assume a transmitting configuration, with a feed at the focal point, there are two factors that produce a taper across the zone plate aperture. One is the radiation pattern from the feedhorn, and the other is that the radiation from the focus travels a greater distance (R_{max}) to the edge than to the center of the zone plate (F). The zone plate is normally in the far-field of the feedhorn, and the path length dependence is inverse-squared. This type of dependence $[F/R_{max}]^2$ and the calculated angles are given in

TABLE 17.2 Calculated Angles as a Function of *F/D*

F/D	$\theta°$	$2\theta°$	$[F/R_{max}]^2$ (dB)
2.5	11.31	22.62	
2.0	14.036	28.072	−0.263
1.5	18.435	36.87	−0.46
1.0	26.565	53.130	−0.97
0.5	45	90	−3.0
1/3	56.31	112.62	−5.13
00.2	68.199	136.397	−8.6

Table 17.2 for various values of *F*/*D*. It can be seen that this path length dependence does not have a significant effect on the edge illumination of the zone plate until *F*/*D* becomes very small.

In addition to this path length dependence we would like some amplitude taper due to the feedhorn pattern. We could design a feed pattern for which the first nulls would fall at the edge of the zone plate. This would improve efficiency because all the main-beam energy from the feed pattern would be included, and there would be low spill-over loss. If the feedhorn (with circular aperture) were uniformly illuminated, the horn main beam would contain 83.3% of the total radiated power. Since the horn aperture undoubtedly has some amplitude taper, it will have more energy in the horn main beam, and less in the horn pattern sidelobes. Thus the horn beam efficiency may be over 90%. By contrast, for a uniformly illuminated feedhorn, only 47.6% of the total radiated horn power is included in the 3dB beamwidth. If the horn pattern illuminated the zone plate with a 3 dB edge taper, approximately half of the radiated power would be lost, although this has been recommended by other authors in a previous article. A preferred horn beam pattern would provide a truncated Gaussian distribution across the zone plate aperture [10]. High-quality scalar feedhorns (corrugated horns) often produce a pattern which is nearly Gaussian. Some recently developed multimode horns produce Gaussian beams. While an untruncated Gaussian aperture distribution has the desirable property that its Fourier transform is also Gaussian (far-field pattern with no sidelobes), real antenna patterns have sidelobes resulting from the truncation of the aperture distribution. In addition, a zone plate pattern is affected by nonconstant phase across the aperture; nonetheless, the far-field pattern is primarily determined by the amplitude distribution in the aperture. Goldsmith [10] has carried out an analysis to show that the aperture efficiency for a circular aperture has a maximum value of 81.5% for a Gaussian distribution with an edge taper of 10.9 dB. Aperture efficiency is defined as the product of spillover efficiency and the taper efficiency (resulting from the effect of the taper compared to uniform illumination). The first sidelobe level peak is at approximately –24 dB below the main beam compared to the value of –17.6 dB for uniform illumination. In an actual application, there are still the questions of how much gain is needed or how narrow the main beam should be, and given the conditions above, the answer is determined by how large the aperture must be. The gain is related to the efficiency of the zone plate configuration.

Efficiency

The overall efficiency of the antenna system contains not only the aperture efficiency mentioned above (relating to the feed and the aperture distribution) for the transmitting situation, but also the diffraction efficiency of a zone plate lens compared to a true lens, as well as the ohmic transmission loss in the lens. The diffraction efficiency is determined by comparing the maximum on-axis, or focused, intensity (due to a plane wave illumination of the FZP) to that of an ideal lens, obviously relating to the receiving situation. Generally, the diffraction efficiency of the zone plate is less than for a true lens, but the ohmic loss of the zone plate is less and in many cases the combination of these two efficiency factors is better (higher) for the zone plate [11].

The diffraction efficiency of the FZP compared to a lens has already been analyzed by numerous investigators, based on simplified plane-wave analysis [1]. These results will be described first. Other recent analyses that give different efficiency values are discussed in Ref. [1], but as yet the analyses have no experimental verification.

From the simplified analysis, the diffraction efficiency is given by

$$\frac{\sin^2(\phi/2)}{(\phi/2)^2}$$

where φ is the increment of phase correction. A plot of this function is shown in Figure 17.17. For a zone plate with half-wave phase correction ($\varphi = \pi$) the efficiency is only 40.5%; for a quarter-wave plate ($\varphi = 90°$) it is 81%, and for an eighth-wave plate ($\varphi = 45°$) it is 95% (0.22 dB loss). When an eighth-wave or smaller correction was used, the FZP with its lower attenuation loss has been analyzed to be better than a standard lens for a number of cases considered between 35 and 200 GHz [11]. In two recent papers, the authors have reported on a planar quarter-wave zone plate utilizing four dielectrics (one of which is air), rather than a grooved dielectric, and have measured exceptionally good efficiencies, as high as 56.7% aperture efficiency [2,12]. This is comparable to efficiencies for excellent parabolic reflector antennas. The authors did not measure diffraction efficiencies for their cases.

A new zone plate structure has recently been proposed which offers higher efficiency, lower phase errors, and better phase correction [13,14]. In the usual grooved FZP, the grooves are milled out so that the bottoms of the grooves are parallel to the back of the zone plate. However, in the so-called stepped conical zone plate (SCZP), the bottoms of the zones are tilted so that they approximate the curve of a lens. For circular symmetry the tilted subzones form a section of a cone. The result is that when rays are traced through the SCZP, the path lengths show smaller errors and are better-corrected. For this configuration, the simplified analysis shows an improvement for the half-wave correction from 40.5 to 95.2%. For the quarter-wave case the efficiency is improved from 81 to 99%. The tilted or tapered conical subzones also offer broader bandwidth capability.

At microwave and millimeter-wave frequencies the wavelengths of interest may range from 10 cm (3 GHz) to 3 mm (100 GHz). Fractional values (half-wave, quarter-wave) of these dimensions can easily be machined into dielectrics like polystyrene by conventional cutters on vertical milling machines or lathes. However, diffraction optics or binary optics for wavelengths around 1 μm require different approaches. Possible dielectric materials are formed by semiconductor growth techniques or laser machining. These optical diffractive elements employ phase correction to much smaller increments than quarter-wave or eighth-wave, and very complex analysis has been developed to explain the behavior of axially symmetric diffractive elements.

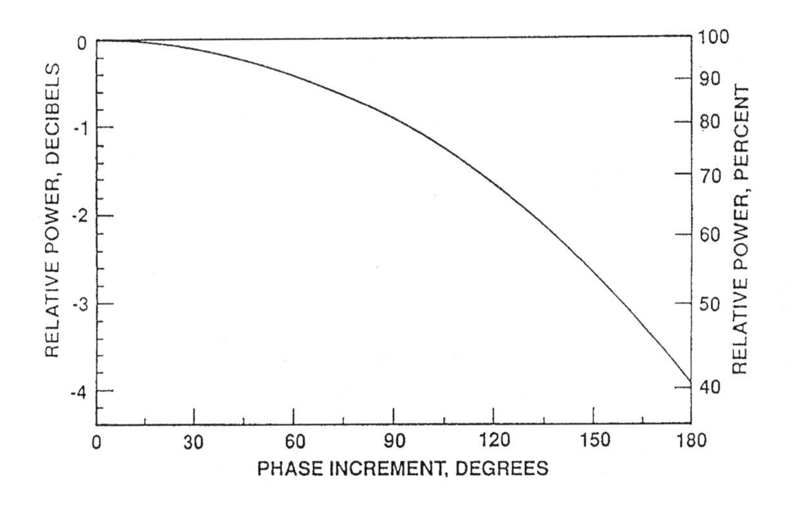

FIGURE 17.17 Relative power of subzoned plates. (*Source:* Ref. [1].)

Loss Considerations

The overall efficiency of the FZP also depends on the attenuation or ohmic loss of the lens. Examples of such losses were calculated for the comparison between a true lens and a half-wave corrected FZP of 20 cm diameter and 20 cm focal length at frequencies from 35 to 200 GHz [11]. The extra attenuation for polystyrene material ranges from about 1/4 to 2.3 dB at these frequencies. Loss will also be reduced if the FZP center is open (has air dielectric), which is one possible design option [1,2,12]. A low dielectric constant is desirable to reduce lens or FZP surface and/or multiple internal reflection losses. An analysis carried out by van Houten and Herben showed that these reflection losses are low if the dielectric constant is below about 5, which is true for polystyrene, Teflon, polyethylene, polycarbonate, TPX, and quartz [6].

Far-Field Patterns

The beamwidth of an FZP is essentially the same, perhaps 1 to 3% wider, as that of a true lens, for the same illumination conditions. The usual way to obtain the antenna far-field or Fraunhofer pattern is to apply the Kirchoff–Fresnel integral knowing the phase and amplitude distribution across the aperture. When the distributions are not constant, the far-field integral (Fourier transform) usually cannot be solved in closed form and computer numerical solutions are required [3,7,8]. However, we can see the improvement in the far-field pattern resulting from reduced phase errors by using the curves given by Baggen and Herben in Ref. [8]. Figure 5 of that reference shows plots of the far-field patterns for three types of zone plates. One is the Soret type with alternate absorbing zones (not phase-correcting) and it has the lowest gain and the highest average sidelobes. The phase-correcting versions have improved gain and sidelobes. For half-wave correction ($P = 2$) the on-axis gain is about 6 dB higher and the average sidelobes are better. Going to quarter-wave correction ($P = 4$) gives 2 dB more on-axis gain and greatly reduced principal sidelobes and average sidelobe level. This illustrates the improvement due to reducing the phase error.

Multiple Frequency Bands

FZP antennas are normally designed for a particular center frequency and band of operation. The response of the zone plate is frequency dependent, however, and it is possible to design the FZP so that it will function well at two or three widely spaced bands [3,15]. This permits the development of high-gain, narrow beamwidth patterns at two or more bands, or one can operate in a given band with narrow beamwidth and high gain, but suppress other bands to reduce interference or jamming, or for signal hiding purposes. It turns out, for example, that a quarter-wave corrected FZP designed for one frequency will normally operate as a half-wave corrected zone plate at twice the design frequency. Although the half-wave zone plate would have a 3 dB lower gain due to the poorer phase correction, doubling the frequency for a given aperture size would add a gain increment of about 6 dB (for a comparable amplitude distribution), so there would be an overall gain of about 3 dB at the doubled frequency.

Ref. [15] gives the analysis and derives the conditions for such FZPs, while additional theory is given in the earlier Refs. [3,4]. Examples include a design which behaves as a sixth-wave ($P = 6$) plate at 30 GHz and a half-wave plate at 90 GHz, a quarter-wave design at 30 GHz, and a half-wave plate at 60 GHz, all having high gain. A sample response curve is included which shows relative gain as a function of frequency ratio for two zone plates with $F = 50\lambda$; a quarter-period plate has 24 zones, while a half-period FZP has 12 zones.

Frequency Behavior

The response of the zone plate is frequency dependent, and the intensity at the focal point exhibits a periodic behavior with frequency. As frequency changes, the focal point moves along the axis of the zone plate. An increase in frequency moves the focus away from the lens, and a decrease does the reverse, with a decrease in signal level in either direction as the focus moves along the axis from the original position, reaching half-power levels at 1.5 or 0.5 times the design frequency. The effect may be used for "focal isolation" and frequency filtering.

Equation (17.44) may be rearranged to solve for F as a function of frequency:

$$F = \frac{P r_{\mathrm{n}}^2}{2n\lambda} - \frac{n\lambda}{2P}$$

where $\lambda = c/f$ and $f =$ frequency. So:

$$F = \frac{Pfr_n^2}{2nc} - \frac{nc}{2Pf} \tag{17.46}$$

When numerical examples are substituted, the first term is larger than the second term. As frequency increases, the first term increases and the second (smaller) term decreases. Thus the response is not linear (but may be nearly linear in some cases). When the frequency increases, F also increases, and when frequency decreases, F decreases.

Summary Comments

Zone plate lens antennas offer several advantages over standard lenses and certain reflective antennas. Usually zone plates give comparable performance, while offering advantages of lower weight, volume, and cost, as well as simplicity of fabrication. Zone plate antennas have seen application in several fields, including satellite communications, radar, radiometry, point-to-point communications and missile terminal guidance.

References

1. J.C. Wiltse, "Large-angle zone plate antennas," *Proc. SPIE*, vol. 5104, Orlando, FL, 2003, pp. 45–56.
2. H.D. Hristov and M.H.A.J. Herben, "Millimeter-wave Fresnel zone plate lens and antenna," *IEEE Trans. Microwave Theory Tech.*, vol. 43, pp. 2779–2785, 1995.
3. D.N. Black and J.C. Wiltse, "Millimeter-wave characteristics of phase correcting Fresnel zone plates," *IEEE Trans. Microwave Theory Tech.*, Mtt-35, 1122–1129, 1987.
4. H.D. Hristov, *Fresnel Zones in Wireless Links, Zone Plate Lenses and Antennas*, Boston, MA: Artech House, 2000.
5. I. Minin and O. Minin, *Diffractional Optics of Millimetre Waves*, Bristol, U.K.: Institute of Physics, 2004.
6. J.M. van Houten and M.H.A.J. Herben, "Analysis of a phase-correcting Fresnel-zone plate antenna with dielectric/transparent zones", *J. Electromagn. Waves Appl.*, vol. 8, no. 7, pp. 847–858, 1994.
7. L.C. Baggen and M.H.A.J. Herben, "Calculating the radiation pattern of Fresnel-zone plate antenna: a comparison between UTD/GTD and PO," *Electromagnetics*, vol. 15, pp. 321–345, 1995.
8. L.C. Baggen and M.H.A.J. Herben, "Design procedure for a Fresnel-zone plate antenna," *Int. J. Infrared Millim. Waves*, vol. 14, no. 6, pp. 1341–1352, 1993.
9. J.C. Wiltse, "Diffraction optics for terahertz waves," *Proc. SPIE*, vol. 5411, Orlando, FL, 2004.
10. P.F. Goldsmith, "Radiation patterns of circular apertures with Gaussian illumination," *Int. J. Infrared Millim. Waves*, vol. 8, pp. 771–781, 1987.
11. J.C. Wiltse, "High efficiency, high gain Fresnel zone plate antennas," *Proc. SPIE*, vol. 3375, Orlando, FL, 1998, pp. 286–290.
12. H.D. Hristov and M.H.A.J. Herben, "Quarter-wave Fresnel zone planar lens and antenna," *IEEE Microwave Guided Wave Lett.*, vol. 5, pp. 249–251, 1995.
13. J.C. Wiltse, "The stepped conical zone plate antenna," *Proc. SPIE*, vol. 4386, Orlando, FL, 2001, pp. 85–92.
14. J.C. Wiltse, "Bandwidth characteristics for the stepped conical zoned antenna," *Proc. SPIE*, vol. 4732, Orlando, 2002, pp. 59–68.
15. J.C. Wiltse, "Dual-band Fresnel zone plate antennas," *Proc. SPIE*, vol. 3062, Orlando, FL, 1997, pp. 181–185.

18 Microwave Devices

Michael B. Steer
North Carolina State University

Robert J. Trew
North Carolina State University

18.1 Passive Microwave Devices

Michael B. Steer

Wavelengths in air at microwave and millimeter-wave frequencies range from 1 m at 300 MHz to 1 mm at 300 GHz and are comparable to the physical dimensions of fabricated electrical components. For this reason circuit components commonly used at lower frequencies, such as resistors, capacitors, and inductors, are not readily available above 10 GHz. The available microwave frequency lumped elements have dimensions of around 1 mm. The relationship between the wavelength and physical dimensions enables new classes of distributed components to be constructed that have no analogy at lower frequencies. Components are realized by disturbing the field structure on a transmission line, resulting in energy storage and thus reactive effects. Electric (E) field disturbances have a capacitive effect and the magnetic (H) field disturbances appear inductive. Microwave components are fabricated in waveguide, coaxial lines, and strip lines. The majority of circuits are constructed using strip lines as the cost is relatively low and they are highly reproducible due to the photolithographic techniques used. Fabrication of waveguide components requires precision machining but they can tolerate higher power levels and are more easily realized at millimeter-wave frequencies (30 to 300 GHz) than either coaxial or microstrip components.

Characterization of Passive Elements

Passive microwave elements are defined in terms of their reflection and transmission properties for an incident wave of electric field or voltage. Scattering (S) parameters are based on traveling waves and so naturally describe these properties. As well they are the only ones that can be measured directly at microwave frequencies. S parameters are defined in terms of root power waves which in turn are defined using forward and backward traveling voltage waves. Consider the N port network of Figure 18.1 where the nth port has a reference transmission line of characteristic impedance Z_{0n} and of infinitesimal length. The transmission line at the nth port serves to separate the forward and backward traveling voltage (V_n^+ and V_n^-) and current

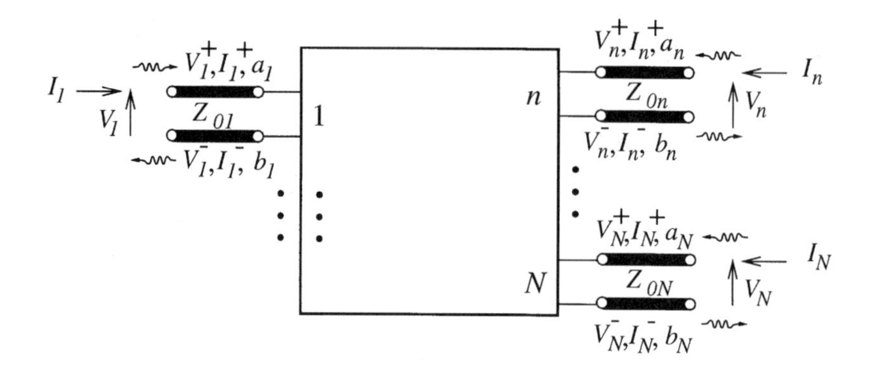

FIGURE 18.1 *N* port network with reference transmission lines used in defining *S* parameters.

(I_n^+ and I_n^-) waves. The reference characteristic impedance matrix, $\mathbf{Z}_0$ is a diagonal matrix, $\mathbf{Z}_0 = \text{diag}(\mathbf{Z}_{01} \ldots \mathbf{Z}_{0n} \ldots \mathbf{Z}_{0N})$, and the root power waves at the nth port, a_n and b_n, are defined by

$$a_n = V_n^+ / \sqrt{\mathbf{Z}_{0n}} \quad \text{and} \quad b_n = V_n^- \sqrt{\mathbf{Z}_{0n}} \tag{18.1}$$

In matrix form:

$$a = \mathbf{Z}_0^{-1/2} V^+ = V_0^{1/2} V^+, \quad b = \mathbf{Z}_0^{-1/2} V^- = \mathbf{Y}_0^{1/2} V^-, \tag{18.2}$$

$$V^+ = \mathbf{Z}_0^{1/2} a = \mathbf{Y}_0^{-1/2} a \quad \text{and} \quad V^- = \mathbf{Z}_0^{1/2} b = \mathbf{Y}_0^{-1/2} b \tag{18.3}$$

where

$$a = [a_1 \ldots a_n \ldots a_N]^{\mathrm{T}}, \quad b = [b_1 \ldots b_n \ldots b_N]^{\mathrm{T}}, \tag{18.4}$$

$$V^+ = [V_1^+ \ldots V_n^+ \ldots V_N^+]^{\mathrm{T}}, \quad V^- = [V_1^- \ldots V_n^- \ldots V_N^-]^{\mathrm{T}}. \tag{18.5}$$

and the characteristic admittance matrix $\mathbf{Y}_0$ and $\mathbf{Z}_0^{-1}$. Now S parameters can be formally defined:

$$\boldsymbol{b} = \boldsymbol{Sa} \tag{18.6}$$

Thus, $\mathbf{Y}_0^{1/2}\boldsymbol{V}^- = \mathbf{S}\mathbf{Y}_0^{1/2}\boldsymbol{V}^+$ and so $\boldsymbol{V}^- = \mathbf{Y}_0^{-1/2}\mathbf{S}\mathbf{Y}_0^{1/2}\boldsymbol{V}^+$. This reduces to $\boldsymbol{V}^- = \boldsymbol{S}\boldsymbol{V}^+$ when all of the reference transmission lines have the same characteristic impedance.

S parameters can be related to other network parameters after first considering the relationship of total port voltage $V = [V_1 \ldots V_n \ldots V_N]^{\mathrm{T}}$ and current $I = [I_1 \ldots I_n \ldots I_N]^{\mathrm{T}}$ to forward and backward voltage and current waves:

$$V = V^+ + V^- \quad \text{and} \quad I = I^+ + I^- \tag{18.7}$$

where $\boldsymbol{I}^+ = \mathbf{Y}_0\boldsymbol{V}^+ = \mathbf{Y}_0^{1/2\boldsymbol{a}}$ and $\boldsymbol{I}^- = \mathbf{Y}_0\boldsymbol{V}^+ = \mathbf{Y}_0^{1/2}\boldsymbol{b}$. The development of the relationship between S parameters and other network parameters is illustrated by considering Y parameters defined by

$$\boldsymbol{I} = \boldsymbol{YV} \tag{18.8}$$

Using traveling waves this becomes

$$I^+ + I^- = Y(V^+ + V^-) \tag{18.9}$$

$$\mathbf{Y}_0(V^+ - V^-) = Y(V^+ + V^-) \tag{18.10}$$

$$\mathbf{Y}_0(1 - \mathbf{Y}_0^{-1/2}S\mathbf{Y}_0^{1/2})V^+ = Y(1 + \mathbf{Y}_0^{-1/2}S\mathbf{Y}_0^{1/2})V^+ \tag{18.11}$$

$$Y = \mathbf{Y}_0\left(1 - \mathbf{Y}_0^{-1/2}S\mathbf{Y}_0^{1/2}\right)\left(1 + \mathbf{Y}_0^{-1/2}S\mathbf{Y}_0^{1/2}\right)^{-1} \tag{18.12}$$

Alternatively (18.10) can be rearranged as

$$(\mathbf{Y}_0 + Y)V^- = (\mathbf{Y}_0 - Y)V^+ \tag{18.13}$$

$$V^- = (\mathbf{Y}_0 + Y)^{-1}(\mathbf{Y}_0 - Y)V^+ \tag{18.14}$$

$$\mathbf{Y}_0^{-1/2}\mathrm{b} = (\mathbf{Y}_0 + Y)^{-1}(\mathbf{Y}_0 - Y)\mathbf{Y}_0^{-1/2}\mathrm{a} \tag{18.15}$$

Comparing this to the definition of *S* parameters, (18.6), leads to

$$S = \mathbf{Y}_0^{1/2}(\mathbf{Y}_0 + Y)^{-1}(\mathbf{Y}_0 - Y)\mathbf{Y}_0^{-1/2} \tag{18.16}$$

For the usual case where all of the reference transmission lines have the same characteristic impedance $\mathbf{Z}_0 = 1/\mathbf{Y}_0$, $\boldsymbol{Y} = \mathbf{Y}_0(\mathbf{1}-\boldsymbol{S})(\mathbf{1}+\boldsymbol{S})^{-1}$ and $\boldsymbol{S} = (\mathbf{Y}_0+\boldsymbol{Y})^{-1}(\mathbf{Y}_0-\boldsymbol{Y})$.

The most common situation involving conversion to and from *S* parameters is for a two port with both ports having a common reference characteristic impedance $\mathbf{Z}_0$. Table 18.1 lists the most common conversions. *S* parameters require that the reference impedances be specified. If they are not it is assumed that it is 50 Ω. They are commonly plotted on Smith Charts—polar plots with lines of constant resistance and reactance (Vendelin *et al.*, 1990).

In Figure 18.2(a) a travelling voltage wave with phasor V_1^+ is incident at port 1 of a two-port passive element. A voltage V_1^- is reflected and V_2^- is transmitted. V_2^- is then reflected by Z_L to produce V_2^+. V_2^+ is zero if $Z_L = \mathbf{Z}_0$. The input voltage reflection coefficient

$$\Gamma_1 = V_1^-/V_1^+ = s_{11} + s_{12}s_{21}/(1 - s_{22}\Gamma_L)$$

transmission coefficient

$$T = V_2^-/V_1^+$$

and the load reflection coefficient

$$\Gamma_L = (Z_L - \mathbf{Z}_0)/(Z_L + \mathbf{Z}_0)$$

TABLE 18.1 Two-Port *S* Parameter Conversion Chart for Impedance, *Z*, Admittance, *Y*, and Hybrid, *H*, Parameters

	S	In Terms of *S*
Z	$z'_{11} = z_{11}/\mathbf{Z}_0 \quad z'_{12} = z_{12}/\mathbf{Z}_0$	$z'_{21} = z_{21}/\mathbf{Z}_0 \quad z'_{22} = z_{22}/\mathbf{Z}_0$
	$\delta = (Z'_{11} + 1)(Z'_{22} + 1) - Z'_{12}Z'_{21}$	$\delta = (1 - S_{11})(1 - S_{22}) - S_{12}S_{21}$
	$S_{11} = [(Z'_{11} - 1)(Z'_{22} + 1) - Z'_{12}Z'_{21}]/\delta$	$Z'_{11} = [(1 + S_{11})(1 - S_{22}) + S_{12}S_{21}]/\delta$
	$S_{12} = 2Z'_{12}/\delta$	$Z'_{12} = 2S_{12}/\delta$
	$S_{21} = 2Z'_{21}/\delta$	$Z'_{21} = 2S_{21}/\delta$
	$S_{22} = [(Z'_{11} + 1)(Z'_{22} - 1) - Z'_{12}Z'_{21}]/\delta$	$Z'_{22} = [(1 - S_{11})(1 + S_{22}) + S_{12}S_{21}]/\delta$
Y	$Y'_{11} = Y_{11}\mathbf{Z}_0 \quad Y'_{12} = Y_{12}\mathbf{Z}_0$	$Y'_{21} = Y'_{21}\mathbf{Z}_0 \quad Y'_{22} = Y_{22}\mathbf{Z}_0$
	$\delta = (1 + Y'_{11})(1 + Y'_{22}) = Y'_{12}Y'_{21}$	$\delta = (1 + S_{11})(1 + S_{22}) - S_{12}S_{21}$
	$S_{11} = [(1 - Y'_{11})(1 + Y'_{22}) + Y'_{12}Y'_{21}]/\delta$	$Y'_{11} = [(1 - S_{11})(1 + S_{22}) + S_{12}S_{21}]/\delta$
	$S_{12} = -2Y'_{12}/\delta$	$Y'_{12} = -2S_{12}/\delta$
	$S_{21} = -2Y'_{21}/\delta$	$Y'_{21} = -2S_{21}/\delta$
	$S_{22} = [(1 + Y'_{11})(1 - Y'_{22}) + Y'_{12}Y'_{21}]/\delta$	$Y'_{22} = [(1 + S_{11})(1 - S_{22}) + S_{12}S_{21}]/\delta$
H	$H'_{11} = H_{11}/\mathbf{Z}_0 \quad H'_{12} = H_{12}$	$H'_{21} = H_{21} \quad H'_{22} = H_{22}\mathbf{Z}_0$
	$\delta = (1 + H'_{11})(1 + H'_{22}) - H'_{12}H'_{21}$	$\delta = (1 - S_{11})(1 + S_{22}) + S_{12}S_{21}$
	$S_{11} = [(H'_{11} - 1)(H'_{22} + 1) - H'_{12}H'_{21}]/\delta$	$H'_{11} = [(1 + S_{11})(1 + S_{22}) - S_{12}S_{21}]/\delta$
	$S_{12} = 2H'_{12}/\delta$	$H'_{12} = 2S_{12}/\delta$
	$S_{21} = -2H'_{21}/\delta$	$H'_{21} = -2S'_{21}/\delta$
	$S_{22} = [(1 + H'_{11})(1 - H'_{22}) + H'_{12}H'_{21}]/\delta$	$H'_{22} = [(1 - S_{11})(1 - S_{22}) - S_{12}S_{21}]/\delta$

Note: The Z', Y', and H' parameters are normalized to $\mathbf{Z}_0$.

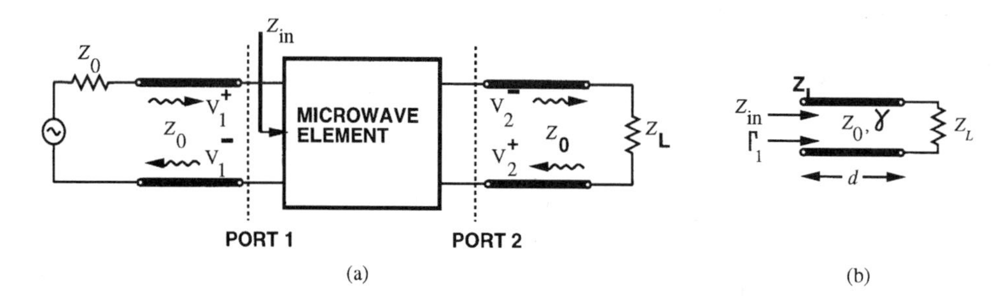

FIGURE 18.2 Incident, reflected and transmitted traveling voltage waves at (a) a passive microwave element and (b) a transmission line.

More convenient measures of reflection and transmission performance are the **return loss** and **insertion loss** as they are relative measures of power in transmitted and reflected signals. In decibels:

$$\text{RETURN LOSS} = -20 \log \Gamma_1 \text{ (dB)} \qquad \text{INSERTION LOSS} = -20 \log T \text{ (dB)}$$

The input impedance at port 1, Z_{in}, is related to Γ by

$$Z_{in} = Z_0(1 + \Gamma_1/1 - \Gamma_1)$$

The reflection characteristics are also described by the voltage standing wave ratio (VSWR), a quantity that can be measured using relatively simple equipment. The VSWR is the ratio of the maximum voltage amplitude on the imput transmission line ($|V_1^+| + |V_1^-|$) to the minimum voltage amplitude ($|V_1^+| - |V_1^-|$). Thus:

$$\text{VSWR} = (1 + |\Gamma_1|)/(1 - |\Gamma_1|)$$

Most passive devices, with the notable exception of ferrite devices, are reciprocal and so $S_{pq}=S_{qp}$. *A loss-less passive device also satisfies the unitary condition:* $\Sigma_p|S_{pq}|^2 = 1$, *which is a statement of power* conservation indicating that all power is either reflected or transmitted.

Most microwave circuits are designed to minimize the reflected energy and maximize transmission at least over the frequency range of operation. Thus, the return loss is high and the VSWR ≈ 1 for well-designed circuits.

A terminated transmission line such as that in Figure 18.2(b) has an input impedance:

$$Z_{\text{in}} = Z_0 \frac{Z_L + jZ_0 \tanh \gamma d}{Z_0 + jZ_L \tanh \gamma d}$$

Thus, a short section ($\gamma_d << 1$) of a short circuited ($Z_L = 0$) transmission line looks like an inductor and a capacitor if it is open circuited ($Z_L = \infty$). When the line is a half wavelength long, an open circuit is presented at the input to the line if the other end is short circuited.

Transmission Line Sections

The simplest microwave circuit element is a uniform section of transmission line which can be used to introduce a time delay or a frequency-dependent phase shift. Other line segments for interconnections include bends, corners, twists, and transitions between lines of different dimensions (see Figure 18.3). The dimensions and shapes are designed to minimize reflections and so maximize return loss and minimize insertion loss.

Discontinuities

The waveguide discontinuities shown in Figure 18.4(a)-(f) illustrate most clearly the use of E and H field disturbances to realize capacitive and inductive components. An E-plane discontinuity (Figure 18.4(a)) can be modeled approximately by a frequency-dependent capacitor. H-plane discontinuities (Figure 18.4(b) and (c)) resemble inductors as does the circular iris of Figure 18.4(d). The resonant waveguide iris of Figure 18.4(e) disturbs both the E and H fields and can be modeled by a parallel LC resonant circuit near the frequency of resonance. Posts in waveguide are used both as reactive elements (Figure 18.4(f)) and to mount active devices (Figure 18.4(g)). The equivalent circuits of microstrip discontinuities (Figure 18.4(k)–(o)) are again modeled by capacitive elements if the E field is interrupted and by inductive elements if the H field (or current) is

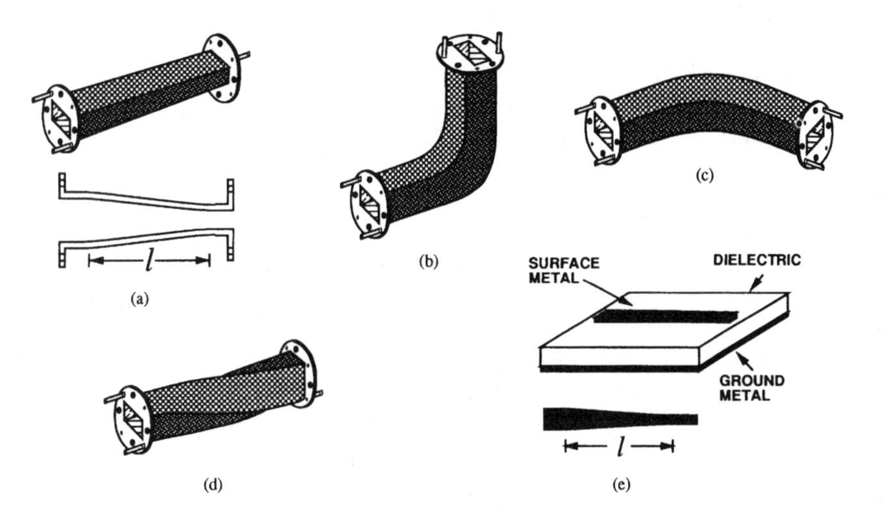

FIGURE 18.3 Sections of transmission lines used for interconnecting components: (a) waveguide tapered section, (b) waveguide E-plane bend, (c) waveguide H-plane bend, (d) waveguide twist, and (e) microstrip taper.

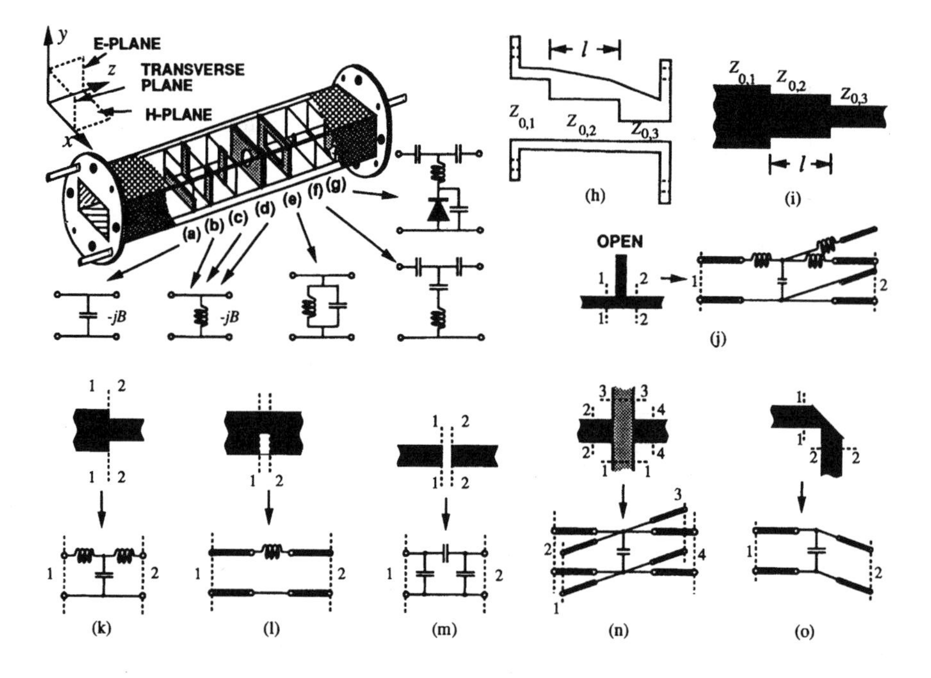

FIGURE 18.4 Discontinuities. Waveguide discontinuities: (a) capacitive *E*-plane discontinuity, (b) inductive *H*-plane discontinuity, (c) symmetrical inductive *H*-plane discontinuity, (d) inductive post discontinuity, (e) resonant window discontinuity, (f) capacitive post discontinuity, (g) diode post mount, and (h) quarter-wave impedance transformer. Microstrip discontinuities: (i) quarter-wave impedance transformer, (j) open microstrip stub, (k) step, (l) notch, (m) gap, (n) crossover, and (o) bend.

interrupted. The stub shown in Figure 18.4(j) presents a short circuit to the through transmission line when the length of the stub is $\lambda_g/4$. When the stubs are electrically short ($<<\lambda_g/4$) they introduce shunt capacitances in the through transmission line.

Impedance Transformers

Impedance transformers are used to interface two sections of line with different **characteristic impedances.** The smoothest transition and the one with the broadest bandwidth is a tapered line as shown in Figure 18.3(a) and (e). This element tends to be very long and so step terminations called quarter-wave impedance transformers (see Figure 18.4(h) and (i)) are sometimes used although their bandwidth is relatively small centered on the frequency at which $l = \lambda_g/4$. Ideally, $Z_{0,2} = \sqrt{Z_{0,1}Z_{0,3}}$.

Terminations

In a termination, power is absorbed by a length of lossy material at the end of a shorted piece of transmission line (Figure 18.5 (a) and (c)). This type of termination is called a matched load as power is absorbed and reflections are very small irrespective of the characteristic impedance of the transmission line. This is generally preferred as the characteristic impedance of transmission lines varies with frequency, particularly so for waveguides. When the characteristic impedance of a line does not vary much with frequency, as is the case with a coaxial line, a simpler smaller termination can be realized by placing a resistor to ground (Figure 18.5(b)).

Attenuators

Attenuators reduce the level of a signal traveling along a transmission line. The basic construction is to make the line lossy but with a characteristic impedance approximating that of the connecting lines so as to reduce

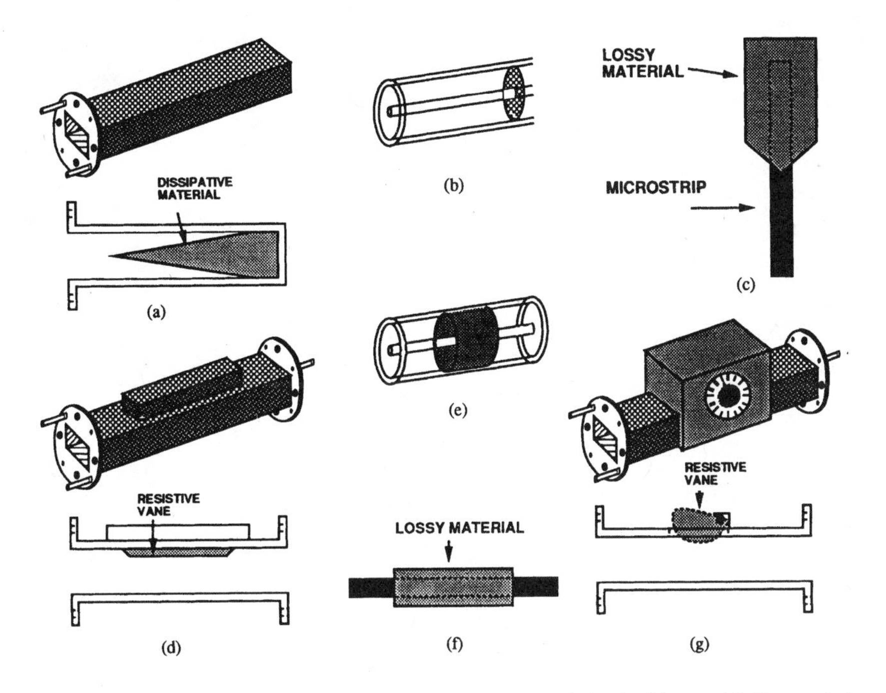

FIGURE 18.5 Terminations and attenuators: (a) waveguide matched load, (b) coaxial line resistive termination, (c) microstrip matched load, (d) waveguide fixed attenuator, (e) coaxial fixed attenuator, (f) microstrip attenuator, and (g) waveguide variable attenuator.

reflections. The line is made lossy by introducing a resistive vane in the case of a waveguide (Figure 18.5(d)), replacing part of the outer conductor of a coaxial line by resistive material (Figure 18.5(e)), or covering the line by resistive material in the case of a microstrip line (Figure 18.5(f)). If the amount of lossy material introduced into the transmission line is controlled, a variable attenuator is achieved (e.g., Figure 18.5(d)).

Microwave Resonators

In a lumped element resonant circuit, stored energy is transferred between an inductor which stores magnetic energy and a capacitor which stores electric energy, and back again every period. Microwave resonators function the same way, exchanging energy stored in electric and magnetic forms but with the energy stored spatially. Resonators are described in terms of their quality factor:

$$Q = 2\pi f_0 \left(\frac{\text{maximum energy stored in the resonator at } f_0}{\text{power lost in the cavity}} \right) \tag{18.17}$$

where f_0 is the resonant frequency. The Q is reduced and thus the resonator bandwidth is increased by the power lost due to coupling to the external circuit so that the loaded Q:

$$\begin{aligned} Q_L &= 2\pi f_0 \left(\frac{\text{maximum energy stored in the resonator at } f_0}{\text{power lost in the cavity and to the external circuit}} \right) \\ &= \frac{1}{1/Q + 1/Q_{ext}} \end{aligned} \tag{18.18}$$

where Q_{ext} is called the external Q. Q_L accounts for the power extracted from the resonant circuit and is typically large. For the simple response shown in Figure 18.6(a) the half power (3 dB) bandwidth is f_0/Q_L.

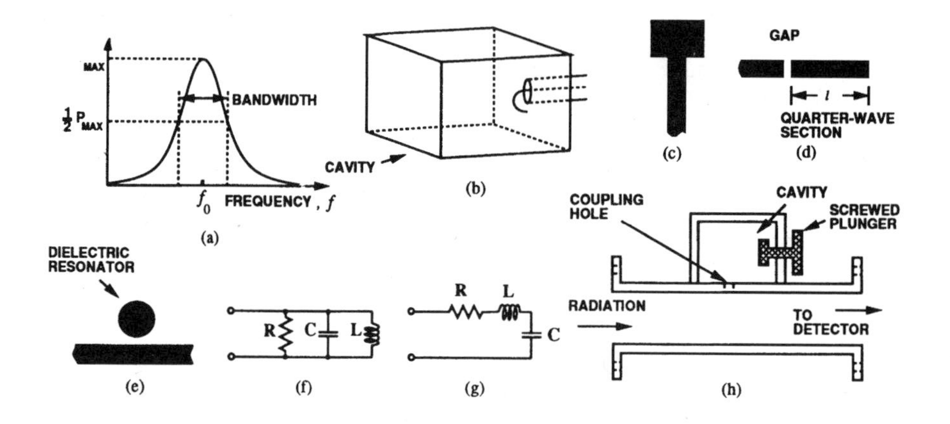

FIGURE 18.6 Microwave resonators: (a) resonator response, (b) rectangular cavity resonator, (c) microstrip patch resonator, (d) microstrip gap-coupled reflection resonator, (e) transmission dielectric transmission resonator in microstrip, (f) parallel equivalent circuits, (g) series equivalent circuits, and (h) waveguide wavemeter.

Near resonance the response of a microwave resonator is very similar to the resonance response of a parallel or series *R*, *L*, *C* resonant circuit (Figure 18.6(f) and (g)). These equivalent circuits can be used over a narrow frequency range.

Several types of resonators are shown in Figure 18.6. Figure 18.6(b) is a rectangular cavity resonator coupled to an external coaxial line by a small coupling loop. Figure 18.6(c) is a microstrip patch reflection resonator. This resonator has large coupling to the external circuit. The coupling can be reduced and photolithographically controlled by introducing a gap as shown in Figure 18.6(d) for a microstrip gap-coupled transmission line reflection resonator. The *Q* of a resonator can be dramatically increased by using a high dielectric constant material as shown in Figure 18.6(e) for a dielectric transmission resonator in microstrip. One simple application of a cavity resonator is the waveguide wavemeter (Figure 18.6(h)). Here, the resonant frequency of a rectangular cavity is varied by changing the physical dimensions of the cavity with a null of the detector indicating that the frequency corresponds to the resonant cavity frequency.

Tuning Elements

In rectangular waveguide the basic adjustable tuning element is the sliding short shown in Figure 18.7(a). Varying the position of the short will change resonance frequencies of cavities. It can be combined with hybrid tees to achieve a variety of tuning functions. The post in Figure 18.4(f) can be replaced by a screw to obtain a screw tuner which is commonly used in waveguide filters. Sliding short circuits can be used in coaxial lines and in conjunction with branching elements to obtain stub tuners. Coaxial slug tuners are also used to provide adjustable matching at the input and output of active circuits. The slug is movable and changes the characteristic impedance of the transmission line. It is more difficult to achieve variable tuning in passive microstrip circuits. One solution is to

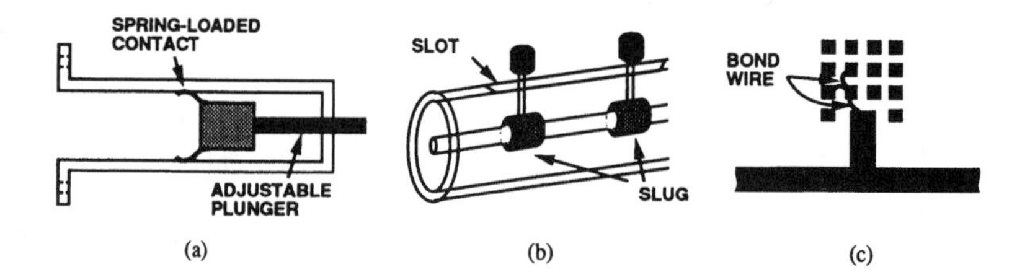

FIGURE 18.7 Tuning elements: (a) waveguide sliding short circuit, (b) coaxial line slug tuner, (c) microstrip stub with tuning pads.

provide a number of pads as shown in Figure 18.7(c) which, in this case, can be bonded to the stub to obtain an adjustable stub length. Variable amounts of phase shift can be inserted by using a variable length of line called a line stretcher, or by a line with a variable propagation constant. One type of waveguide variable phase shifter is similar to the variable attenuator of Figure 18.5(d) with the resistive material replaced by a low-loss dielectric.

Hybrid Circuits and Directional Couplers

Hybrid circuits are multiport components which preferentially route a signal incident at one port to the other ports. This property is called directivity. One type of hybrid is called a directional coupler, the schematic of which is shown in Figure 18.8(a). Here the signal incident at port 1 is coupled to ports 2 and 3 while very little is coupled to port 4. Similarly, a signal incident at port 2 is coupled to ports 1 and 4 but very little power appears at port 3. The feature that distinguishes a directional coupler from other types of hybrids is that the power at the output ports (here ports 2 and 3) is different. The performance of a directional coupler is specified by three parameters:

$$\begin{aligned} \text{coupling factor} &= \mathrm{P}_1/\mathrm{P}_3 \\ \text{directivity} &= \mathrm{P}_3/\mathrm{P}_4 \\ \text{isolation} &= \mathrm{P}_1/\mathrm{P}_4 \end{aligned} \tag{18.19}$$

Microstrip and waveguide realizations of directional couplers are shown in Figure 18.8(b) and (c) where the microstrip coupler couples in the backward direction and the waveguide coupler couples in the forward direction. The powers at the output ports of the hybrids shown in Figure 18.9 are equal and so the hybrids serve to split a signal into half as well as having directional sensitivity.

Filters

Filters are combinations of microwave passive elements designed to have a specified frequency response. Typically, a topology of a filter is chosen based on established lumped element filter design theory. Then computer-aided design techniques are used to optimize the response of the circuit to the desired response.

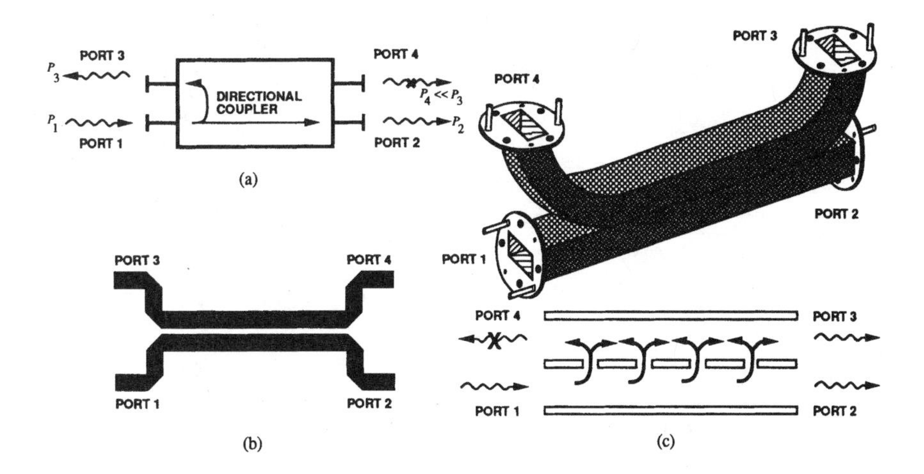

FIGURE 18.8 Directional couplers: (a) schematic, (b) backward-coupling microstrip directional coupler, (c) forward-coupling waveguide directional coupler.

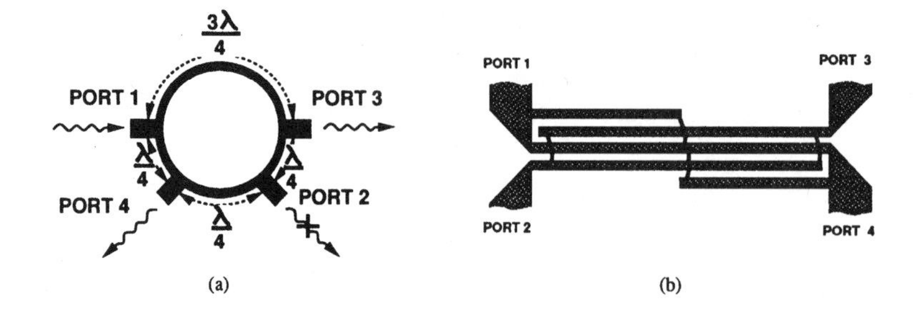

FIGURE 18.9 Microstrip hybrids: (a) rat race hybrid and (b) Lange coupler.

Ferrite Components

Ferrite components are nonreciprocal in that the insertion loss for a wave traveling from port A to port B is not the same as that from port B to port A.

Circulators and Isolators

The most important type of ferrite component is a circulator (Figure 18.10(a) and (b)). The essential element of a circulator is a piece of ferrite which when magnetized becomes nonreciprocal, preferring progression of electromagnetic fields in one circular direction. An ideal circulator has the scattering matrix:

$$[S] = \begin{bmatrix} 0 & 0 & S_{13} \\ S_{21} & 0 & 0 \\ 0 & S_{32} & 0 \end{bmatrix} \qquad (29.20)$$

In addition to the insertion and return losses, the performance of a circulator is described by its isolation which is its insertion loss in the undesired direction. An isolator is just a three-port circulator with one of the ports terminated in a matched load as shown in the microstrip realization of Figure 18.10(c). It is used in a

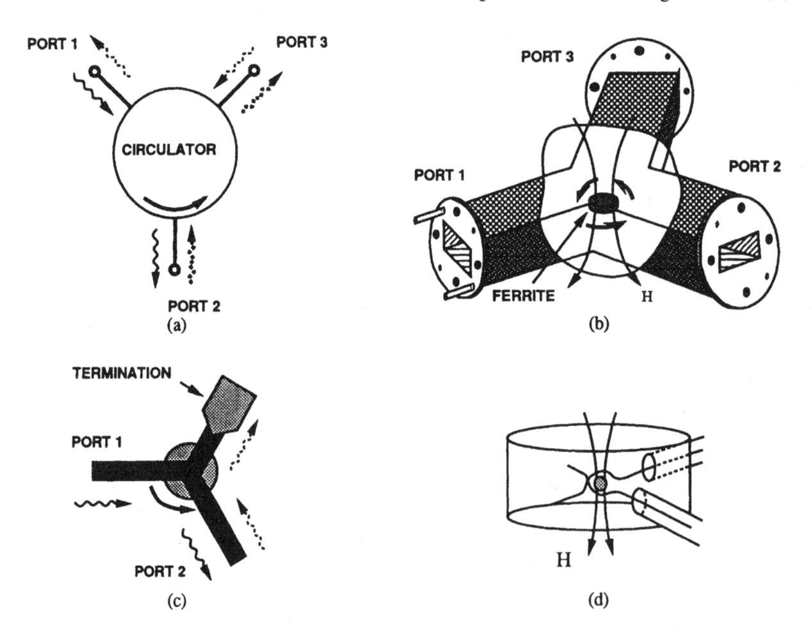

FIGURE 18.10 Ferrite components: (a) schematic of a circulator, (b) a waveguide circulator, (c) a microstrip isolator, and (d) a YIG tuned bandpass filter.

transmission line to pass power in one direction but not in the reverse direction. It is commonly used to protect the output of equipment from high reflected signals. The heart of isolators and circulators is the nonreciprocal element. Electronic versions have been developed for MMICs. A four-port version is called a duplexer and is used in radar systems and to separate the received and transmitted signals in a transceiver.

YIG Tuned Resonator

A magnetized yttrium iron garnet (YIG) sphere, shown in Figure 18.10(d), provides coupling between two lines over a very narrow bandwidth. The center frequency of this bandpass filter can be adjusted by varying the magnetizing field.

Passive Semiconductor Devices

A semiconductor diode can be modeled by a voltage-dependent resistor and capacitor in shunt. Thus an applied dc voltage can be used to change the value of a passive circuit element. Diodes optimized to produce a voltage variable capacitor are called varactors. In detector circuits a diode's voltage variable resistance is used to achieve rectification and, through design, produce a dc voltage proportional to the power of an incident microwave signal. A controllable variable resistance is used in a PIN diode to realize an electronic switch.

Defining Terms

Characteristic impedance: Ratio of the voltage and current on a transmission line when there are no reflections.

Insertion loss: Power lost when a signal passes through a device.

Reference impedance: Impedance to which scattering parameters are referenced.

Return loss: Power lost upon reflection from a device.

Voltage standing wave ratio (VSWR): Ratio of the maximum voltage amplitude on a line to the minimum voltage ampitude.

Reference

G.D. Vendelin, A.M. Pavio, and U.L. Rohde, *Microwave Circuit Design Using Linear and Nonlinear Techniques*, New York: Wiley, 1990.

Further Information

The following books provide good overviews of passive microwave components: *Microwave Engineering Passive Circuits* by P.A. Rizzi, Prentice-Hall, Englewood Cliffs, NJ, 1988; *Microwave Devices and Circuits* by S.Y. Liao, 3rd ed., Prentice-Hall, Englewood Cliffs, NJ, 1990; *Microwave Theory, Components and Devices* by J.A. Seeger, Prentice-Hall, Englewood Cliffs, NJ, 1986; *Microwave Technology* by E. Pehl, Artech House, Dedham, Mass., 1985; *Microwave Engineering and Systems Applications* by E.A. Wolff and R. Kaul, Wiley, New York, 1988; and *Microwave Engineering* by T.K. Ishii, 2nd ed., Harcourt Brace Jovanovich, Orlando, FL, 1989. *Microwave Circuit Design Using Linear and Nonlinear Techniques* by G.D. Vendelin, A.M. Pavio, and U.L. Rohde, Wiley, New York, 1990, provides a comprehensive treatment of computer-aided design techniques for both passive and active microwave circuits. *Microwave Transistor Amplifiers: Analysis and Design,* 2nd ed., by G. Gonzalez, Prentice-Hall, Englewood Cliffs, NJ, 1996.

The monthly journals *IEEE Transactions on Microwave Theory Techniques, IEEE Microwave and Guided Wave Letters,* and *IEEE Transactions on Antennas and Propagation* publish articles on modeling and design of microwave passive circuit components. Articles in the first two journals are more circuit and component oriented while the third focuses on field theoretic analysis. These are published by The Institute of Electrical and Electronics Engineers, Inc. For subscription or ordering contact: IEEE Service Center, 445 Hoes Lane, P.O. Box 1331, Piscataway, NJ 08855-1331.

Articles can also be found in the biweekly magazine *Electronics Letters* and the bimonthly magazine *IEE Proceedings Part H—Microwave, Optics and Antennas*. Both are published by the Institute of Electrical Engineers and subscription inquiries should be sent to IEE Publication Sales, P.O. Box 96, Stenage, Herts SG1 2SD, United Kingdom. Telephone number +44(0)438 313311.

The *International Journal of Microwave and Millimeter-Wave Computer-Aided Engineering* is a quarterly journal devoted to the computer-aided design aspects of microwave circuits and has articles on component modeling and computer-aided design techniques. It has a large number of review-type articles. For subscription information contact John Wiley & Sons, Periodicals Division, P.O. Box 7247–8491, Philadelphia, PA 19170-8491.

18.2 Active Microwave Devices

Robert J. Trew

Active devices that can supply gain at microwave frequencies can be fabricated from a variety of semiconductor materials. The availability of such devices permits a wide variety of system components to be designed and fabricated. Systems are generally constructed from components such as filters, amplifiers, oscillators, mixers, phase shifters, switches, etc. Active devices are primarily required for the oscillator and amplifier components. For these functions, devices that can supply current, voltage, or power gain at the frequency of interest are embedded in circuits that are designed to provide the device with the proper environment to create the desired response. The operation of the component is dictated, therefore, by both the capabilities of the active device and its embedding circuit.

It is common to fabricate microwave integrated circuits using both hybrid and monolithic techniques. In the hybrid approach, discrete active devices are mounted in radio frequency (RF) circuits that can be fabricated from waveguides or transmission lines fabricated using coaxial, microstrip, stripline, coplanar waveguide, or other such media. Monolithic circuits are fabricated with both the active device and the RF circuit fabricated in the same semiconductor chip. Interconnection lines and the embedding RF circuit are generally fabricated using microstrip or coplanar waveguide transmission lines.

Active microwave devices can be fabricated as **two-terminal devices** (diodes) or **three-terminal devices** (transistors). Generally, three-terminal devices are preferred for most applications since the third terminal provides a convenient means to control the RF performance of the device. The third terminal allows for inherent isolation between the input and output RF circuit. Amplifiers and oscillators can easily be designed by providing circuits with proper stabilization or feedback characteristics. Amplifiers and oscillators can also be designed using two-terminal devices (diodes), but input/output isolation is more difficult to achieve since only one RF port is available. In this case it is generally necessary to use RF isolators or circulators.

The most commonly used two-terminal active devices consist of Gunn, tunnel, and IMPATT diodes. These devices can be designed to provide useful gain from low gigahertz frequencies to high millimeter-wave frequencies. Three-terminal devices consist of bipolar (BJT), heterojunction bipolar (HBT), and field-effect transistors (MESFETs and HEMTs). These devices can also be operated from UHF to millimeter-wave frequencies.

Semiconductor Material Properties

Active device operation is strongly dependent upon the charge transport characteristics of the semiconductor materials from which the device is fabricated. Semiconductor materials can be grown in single crystals with very high purity. The electrical conductivity of the crystal can be precisely controlled by introduction of minute quantities of dopant impurities. When these impurities are electrically activated, they permit precise values of current flow through the crystals to be controlled by potentials applied to contacts, placed upon the crystals. By clever positioning of the metal contacts, various types of semiconductor devices are fabricated. In this section we will briefly discuss the important material characteristics.

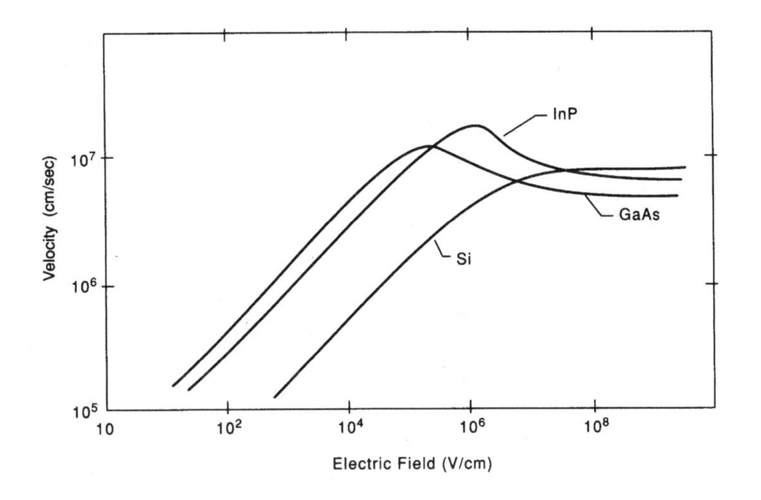

FIGURE 18.11 Electron velocity versus electric field for several semiconductors. This figure shows the electron velocity in several common semiconductors as a funcation of electric field strength. At low electric field the electron velocity is ohmic, as indicated by the linear characteristic. At higher electric field strength the electron velocity saturates and becomes nonlinear. Compound semiconductors such as GaAs and InP have highly nonlinear behavior at large electric fields.

Semiconductor material parameters of interest for device fabrication consist of those involved in charge transport through the crystal, as well as thermal and mechanical properties of the semiconductor. The charge transport properties describe the ease with which free charge can flow through the material. For example, the velocity–electric field characteristics for several commonly used semiconductors are shown in Figure 18.11. At low values of electric field, the charge transport is ohmic and the charge velocity is directly proportional to the magnitude of the electric field. The proportionality constant is called the mobility and has units of $cm^2/V\,s$. Above a critical value for the electric field, the charge velocity saturates and either becomes constant (e.g., Si) or decreases with increasing field (e.g., GaAs). Both of these behaviors have implications for device fabrication, especially for devices intended for high-frequency operation. Generally, a high velocity is desired since current is directly proportional to velocity. Also, a low value for the saturation electric field is desirable since this implies a high-charge mobility. High mobility implies low resistivity and, therefore, low values for parasitic and access resistances for semiconductor devices.

The decreasing electron velocity with electric field characteristic for compound semiconductors such as GaAs and InP makes possible active two-terminal devices called transferred electron devices (TEDs) or Gunn diodes. The negative slope of the velocity versus electric field characteristic implies a decreasing current with increasing voltage. That is, the device has a negative resistance. When a properly sized piece of these materials is biased and placed in a resonant cavity, the device will be unstable up to very high frequencies. By proper selection of embedding impedances oscillators or amplifiers can be constructed.

Other semiconductor materials parameters of interest include thermal, dielectric constant, energy bandgap, electric breakdown characteristics, and minority carrier lifetime. The thermal conductivity of the material is important because it describes how easily heat can be extracted from the device. The thermal conductivity has units of W/cm K. Generally, high thermal conductivity is desirable. Compound semiconductors, such as GaAs and InP, have relatively poor thermal conductivity compared to elemental semiconductors such as Si. Materials such as SiC have excellent thermal conductivity and have uses in high-power electronic devices. The dielectric constant is important since it affects the size of the semiconductor device. The larger the dielectric constant, the smaller the device. Electric breakdown characteristics are important since breakdown limits the magnitudes of the dc and RF voltages that can be applied to the device. This is turn limits the RF power that can be handled by the device. The electric breakdown for the material is generally described by the critical value of electric field that produces avalanche ionization. Minority carrier lifetime is important for bipolar

TABLE 18.2 Material Parameters for Several Semiconductors

Semiconductor	E_g (eV)	ε_r	κ (W/cm K) @300 K	E_c(V/cm)	$\tau_{minority}$ (s)
Si	1.12	11.9	1.5	3×10^5	2.5×10^{-3}
GaAs	1.42	12.5	0.54	4×10^5	$\sim10^{-8}$
InP	1.34	12.4	0.67	4.5×10^5	$\sim10^{-8}$
α-SiC	2.86	10.0	4	$(1–5)\times10^6$	$\sim(1–10)\times10^{-9}$
β-SiC	2.2	9.7	4	$(1–5)\times10^6$	$\sim(1–10)\times10^{-9}$

devices, such as pn-junction diodes, rectifiers, and bipolar junction transistors (BJTs). A low value for minority carrier lifetime is desirable for devices such as diode temperature sensors and switches where low reverse bias leakage current is desirable. A long minority carrier lifetime is desirable for devices such as bipolar transistors. For materials such as Si and SiC, the minority carrier lifetime can be varied by controlled impurity doping. A comparison of some of the important material parameters for several common semiconductors is presented in Table 18.2.

Two-Terminal Active Microwave Devices

The IMPATT diode, transferred electron device, and tunnel diode are the most commonly used two-terminal devices. These devices can operate from the low microwave through high millimeter-wave frequencies. They were the first semiconductor devices that could provide useful RF power levels at microwave and millimeter-wave frequencies. The three devices are similar in that they are fabricated from blocks of semiconductors and require two electrodes (anode and cathode) for supplying dc bias. The same electrodes are used for the RF port, and since only two electrodes are available, the devices must be operated as a **one-port network.** This is generally accomplished by mounting the semiconductor in a pin-type package. The package can then be positioned in an RF circuit or resonant cavity and the top and bottom pins on the package used as the dc and RF electrical contacts. This arrangement works quite well and packaged devices can be operated up to about 90 to 100 GHz. For higher-frequency operation, the devices are generally mounted directly into circuits using microstrip or some other similar technology.

All three devices operate as negative immittance components. That is, their active characteristics can be described as either a negative resistance or a negative conductance. Which description to use is determined by the physical operating principles of the particular device.

Tunnel Diodes

Tunnel diodes (Sze, 1981) generate active characteristics by a mechanism involving the physical tunneling of electrons between energy bands in highly doped semiconductors. For example, if a pn-junction diode is heavily doped, the conduction and valence bands will be located in close proximity and **charge carriers** can tunnel through the electrostatic barrier separating the p-type and n-type regions, rather than be thermionically emitted over the barrier as generally occurs in this type of diode. When the diode is biased (either forward or reverse bias) current immediately flows and junction conduction is basically ohmic. In the forward bias direction, conduction occurs until the applied bias forces the conduction and valence bands to separate. The tunnel current then decreases and normal junction conduction occurs. In the forward bias region where the tunnel current is decreasing with increasing bias voltage, a negative immittance characteristic is generated. The immittance is called "N-type" because the I–V characteristic "looks like" the letter N. This type of active element is short-circuit stable and is described by a negative conductance in shunt with a capacitance. Tunnel diodes are limited in operation frequency by the time it takes for charge carriers to tunnel through the junction. Since this time is very short (on the order of 10^{-12} sec) operation frequency can be very high, approaching 1000 GHz. Tunnel diodes have been operated at hundreds of gigahertz, limited by practical packaging and parasitic impedance considerations. The RF power available from a tunnel diode is limited (hundreds of milliwatts level) since the maximum RF voltage swing that can be applied across the junction is limited by the forward turn-on characteristics of the device (typically 0.6 to 0.9 V). Increased RF power can only be obtained by increasing device area to increase RF current, but device area is limited by operation frequency according to an inverse scaling law. Tunnel diodes have moderate dc-to-RF conversion efficiency (<10%), very low **noise figures**, and are useful in low-noise systems applications, such as microwave and millimeter-wave receivers.

Transferred Electron Devices

Transferred electron devices (i.e., Gunn diodes) (Bosch and Engelmann, 1975) also have N-type active characteristics and can be modeled as a negative conductance in parallel with a capacitance. Device operation, however, is based upon a fundamentally different principle. The negative conductance derives from the complex conduction band structure of certain compound semiconductors, such as GaAs and InP. In these direct bandgap materials the central (or Γ) conduction band is in close energy-momentum proximity to secondary, higher-order conduction bands (i.e., the X and L valleys). The electron effective mass is determined by the shape of the conduction bands, and the effective mass is "light" in the Γ valley but "heavy" in the higher-order X and L valleys. When the crystal is biased, current flow is initially due to electrons in the light effective mass Γ valley and conduction is ohmic. However, as the bias field is increased, an increasing proportion of the free electrons are transferred into the X and L valleys where the electrons have heavier effective mass. The increased effective mass slows down the electrons, with a corresponding decrease in conduction current through the crystal. The net result is that the crystal displays a region of applied bias voltages where current decreases with increasing voltage. That is, a negative conductance is generated. The device is unstable and, when placed in an RF circuit or resonant cavity, oscillators or amplifiers can be fabricated. The device is not actually a diode since no pn or Schottky junction is used. The phenomenon is a characteristic of the bulk material and the special structure of the conduction bands in certain compound semiconductors. Most semiconductors do not have the conduction band structure necessary for the transferred electron effect. The term *Gunn diode* is actually a misnomer since the device is not a diode. TEDs are widely used in oscillators from the microwave through high millimeter-wave frequency bands. They have good RF output power capability (milliwatts to watts level), moderate efficiency (<20%), and excellent noise and bandwidth capability. Octave band tunable oscillators are easily fabricated using devices such as YIG (yttrium iron garnet) resonators or varactors as the tuning element. Most commercially available solid-state sources for 60- to 100-GHz operation generally use InP TEDs.

IMPATT Diodes

Impact avalanche transit time (IMPATT) diodes (Bhartia and Bahl, 1984) are fabricated from pn or Schottky junctions. A typical pn junction device is shown in Figure 18.12. For optimum RF performance the diode is separated, by use of specially designed layers of controlled impurity doping, into avalanche and drift regions. In operation the diode is reverse biased into avalanche breakdown. Due to the very sensitive I–V characteristic, it is best to bias the diode using a constant current source in which the magnitude of the current is limited. When the diode is placed in a microwave resonant circuit, RF voltage fluctuations in the bias circuit grow and are forced into a narrow frequency range by the impedance characteristics of the resonant circuit. Because of the avalanche process the RF current across the avalanche region lags the RF voltage by 90°. This inductive delay is not sufficient, by itself, to produce active characteristics. However, when the 90° phase shift is added to that arising from

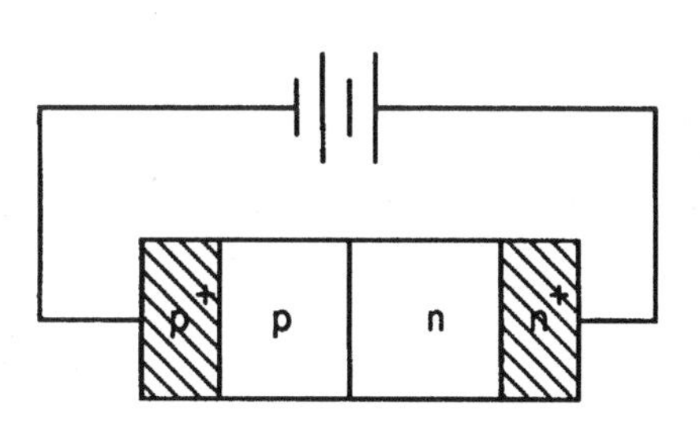

FIGURE 18.12 Diagram showing the structure for a typical pn-junction IMPATT diode. This particular diode is called a double-drift device because avalanche breakdown occurs at the pn-junction, which is located in the middle of the device. When operated in breakdown, electrons would travel through the n-type region towards the positive terminal of the bias source and holes would travel through the p-type region towards the negative terminal of the source. The diode, therefore, operates as two diodes connected in a back-to-back configuration. The frequency capability of the device is directly proportional to the width of the n and p regions.

an additional inductive delay caused by the transit time of the carriers drifting through the remainder of the diode external to the avalanche region, a phase shift between the RF voltage and current greater than 90° is obtained. A Fourier analysis of the resulting waveforms reveals a device impedance with a negative real part. That is, the device is active and can be used to generate or amplify RF signals. The device impedance has an "S-type" active characteristic and the device equivalent circuit consists of a negative resistance in series with an inductor. The device has significant pn-junction capacitance that must be considered, and a complete equivalent circuit would include the device capacitance in parallel with the series negative resistance-inductance elements. For optimum performance the drift region is designed so that the electric field throughout the RF cycle is sufficiently high to produce velocity saturation for the charge carriers. In order to achieve this, it is common to design complex structures consisting of alternating layers of highly doped and lightly doped semiconductor regions. These structures are called "high-low," "low-high-low," or "Read" diodes, after the man who first proposed their use. They can also be fabricated in a back-to-back arrangement to form double-drift structures. These devices are particularly attractive for millimeter-wave applications. IMPATT diodes can be fabricated from most semiconductors, but are generally fabricated from Si or GaAs. The devices are capable of good RF output power (mW to W) and good dc-to-RF conversion efficiency (~10 to 20%). They operate well into the millimeter-wave region and have been operated as high as 340 GHz. They have moderate bandwidth capability, but have relatively poor noise performance due to the impact ionization process.

Although the two-terminal active devices are used in many electronic systems, the one-port characteristic can introduce significant complexity into circuit design. Isolators and circulators are generally required, and these components are often large and bulky. They are often fabricated from magnetic materials, which can introduce thermal sensitivities. For these reasons three-terminal devices have replaced two-terminal devices in many practical applications. Generally, if two-terminal and three-terminal devices with comparable capability are available, the three-terminal device offers a more attractive design solution and will be selected. Two-terminal devices are generally only used when a comparable three-terminal device is not available. For this reason IMPATT and TED devices are used in millimeter-wave applications, where they retain an advantage in providing good RF power. Tunnel diodes are not often used, except in a few special applications where their low-noise and wide-bandwidth performance can be used to advantage.

Three-Terminal Active Microwave Devices

The high-frequency performance of three-terminal semiconductor devices has improved dramatically during the past two decades. Twenty years ago transistors that could provide useful **gain** at frequencies above 10 GHz were a laboratory curiosity. Today, such devices are readily available, and state-of-the-art transistors operate well above 100 GHz. This dramatic improvement has been achieved by advances in semiconductor growth technology, coupled with improved device design and fabrication techniques. Semiconductor materials technology continues to improve and new device structures that offer improved high-frequency performance are continually being reported.

In this section we will discuss the two most commonly employed transistors for microwave applications, the metal-semiconductor field-effect transistor (MESFET) (Liechti, 1976) and the bipolar transistor (BJT) (Cooke, 1971). These two transistors are commonly employed in practical microwave systems as amplifiers, oscillators, and gain blocks. The transistors have replaced many two-terminal devices due to their improved performance and ease of use. Transistors are readily integrated into both hybrid and monolithic integrated circuit environments (MICs). This, in turn, has resulted in significantly reduced size, weight, and dc power consumption, as well as increased reliability and mean time to failure for systems that use these components. Transistors are easily biased and the **two-port network** configuration leads naturally to inherent separation between input and output networks.

Field-Effect Transistors

A cross-sectional view of a microwave MESFET is shown in Figure 18.13. The device is conceptionally very simple. The MESFET has two ohmic contacts (the source and drain) separated by some distance, usually in

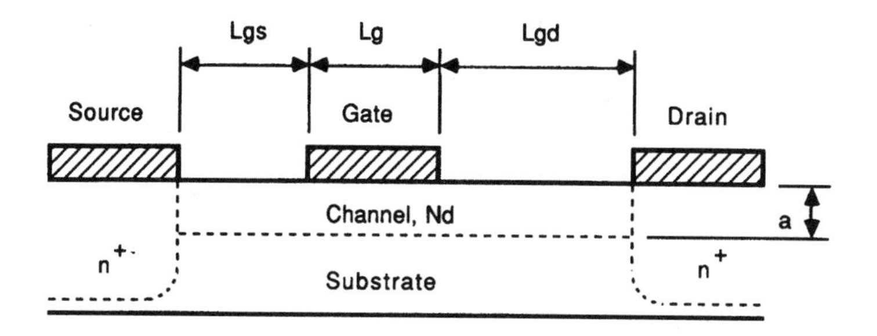

FIGURE 18.13 Cross-sectional view of a microwave MESFET. The cross-hatched areas indicate metal electrodes placed upon the semiconductor to provide for electrical connections. The areas indicated as n^+ are highly doped, highly conducting regions to reduce ohmic access resistances. The channel contains the region of current flow and the substrate is highly resistive and nonconducting so that the current flow is confined to the channel.

the range of 3 to 10 μm. A rectifying Schottky contact (the gate) is located between the two ohmic contacts. Typically, the gate length is on the order of 0.1 to 2 μm for modern microwave devices. The width of the device scales with frequency and typically ranges from about 1 to 10 μm for power microwave devices to 50 μm for millimeter-wave devices. All three contacts are located on the surface of a thin conducting layer (the channel) which is located on top of a high-resistivity, nonconductive substrate to form the device. The channel region is typically very thin (on the order of 0.1 to 0.3 μm) and is fabricated by epitaxial growth or ion implantation. In operation, the drain contact is biased at a specified potential (positive drain potential for an n-channel device) and the source is grounded. The flow of current through the conducting channel is controlled by negative dc and superimposed RF potentials applied to the gate, which modulate the channel current and provide RF gain. The current flow is composed of only one type of charge carrier (generally electrons) and the device is termed *unipolar.* The MESFET can be fabricated from a variety of semiconductors, but is generally fabricated from GaAs. MESFETs fabricated from Si do not work at high frequencies as well as those fabricated from GaAs due to lower electron mobility in Si (e.g., $\mu_n \sim 6000\ \text{cm}^2/\text{V s}$ for GaAs and $1450\ \text{cm}^2/\text{V s}$ for Si). The lower electron mobility in Si produces high source resistance, which seriously degrades the high-frequency gain possible from the device.

MESFETs can be optimized for small-signal, low-noise operation or for large-signal, RF power applications. Generally, low-noise operation requires short gate lengths, relatively narrow gate widths, and highly doped channels. Power devices generally have longer gate lengths, much wider gate widths, and lower doped channels. Low-noise devices can be fabricated that operate with good gain ($\sim$10 dB) and low noise figure ($<$3 dB) to above 100 GHz. Power devices can provide RF power levels on the order of watts (W) up to over 20 GHz.

The current gain of the MESFET is indicated by the f_T of the device, sometimes called the gain-bandwidth product. This parameter is defined as the frequency at which the short-circuited current gain is reduced to unity and can be expressed as

$$f_{\mathrm{T}} = \frac{g_{\mathrm{m}}}{2\pi C_{\mathrm{gs}}} \tag{18.21}$$

where g_m is the device transconductance (a measure of gain capability) and C_{gs} is the gate source capacitance. High f_T is desirable and this is achieved with highly doped channels and low capacitance gates. The RF power gain is also of interest and this performance can be indicated by the unilateral power gain defined as

$$U = \frac{1}{4}\left(\frac{f_{\mathrm{T}}}{f}\right)^2 \frac{R_{\mathrm{ds}}}{R_{\mathrm{g}}} \tag{18.22}$$

where U is the unilateral power gain, f is the operating frequency, R_{ds} is the drain-source resistance, and R_g is the gate resistance. As this expression indicates, large power gain requires a high f_T, and a large R_{ds}/R_g ratio. The highest frequency at which the device could be expected to produce power gain can be defined from the frequency, f, at which U goes to zero. This frequency is called the maximum frequency of oscillation, or f_{max}, and is defined as

$$f_{max} = \frac{f_T}{2}\sqrt{\frac{R_{ds}}{R_g}} \tag{18.23}$$

A different form of field-effect transistor can be fabricated by inserting a highly doped, wider-bandgap semiconductor between the conducting channel and the gate electrode (Drummond et al., 1986). The conducting channel is then fabricated from undoped semiconductor. The discontinuity in energy bandgaps between the two semiconductors, if properly designed, results in free charge transfer from the highly doped, wide-bandgap semiconductor into the undoped, lower-bandgap channel semiconductor. The charge accumulates at the interface and creates a two-dimensional electron gas (2DEG). The sheet charge is essentially two-dimensional and allows current to flow between the source and drain electrodes. The amount of charge in the 2DEG can be controlled by the potential applied to the gate electrode. In this manner the current flow through the device can be modulated by the gate and gain results. Since the charge flows at the interface between the two materials, but is confined in the undoped channel semiconductor, very little impurity scattering occurs and extremely high charge carrier mobility results. The device, therefore, has very high transconductance and is capable of very high frequency operation and very low noise figure operation. This type of device is called a high electron mobility transistor (HEMT). HEMTs can be fabricated from material systems such as AlGaAs/GaAs or AlInAs/GaInAs/InP. The latter material system produces devices that have f_T's above 300 GHz and have produced noise figures of about 1 dB at 100 GHz.

Bipolar Transistors

A cross-sectional view of a bipolar transistor is shown in Figure 18.14. The bipolar transistor consists of back-to-back pn junctions arranged in a sandwich structure. The three regions are designated the emitter, base, and collector. This type of device differs from the field-effect transistors in that both electrons and holes are involved in the current transport process (thus the designation *bipolar*). Two structures are possible: pnp or npn, depending upon the conductivity type common to both pn junctions. Generally, for microwave applications the npn structure is used since device operation is controlled by electron flow. In general, electron transport is faster than that for holes, and npn transistors are capable of superior high-frequency performance compared to comparable pnp transistors. In operation, the base-emitter pn-junction is forward biased and the collector-base pn-junction is reverse biased. When an RF signal is applied to the base-emitter junction the junction allows a current to be injected into the base region. The current in the base region consists of minority charge carriers (i.e., carriers with the opposite polarity compared to the base material—electrons for an npn transistor). These charge carriers then diffuse across the base region to the base-collector junction, where they are swept across the junction by the large reverse bias electric field. The reverse bias electric field in the base-collector region is generally made sufficiently large that the carriers travel at their saturation velocity. The transit time of the charge carriers across this region is small, except for millimeter-wave transistors where the base-collector region transit time can be a significant fraction of the total time required for a charge carrier to travel from the emitter through the collector. The operation of the transistor is primarily controlled by the ability of the minority charge carriers to diffuse across the base region. For this reason microwave transistors are designed with narrow base regions in order to minimize the time required for the carriers to travel through this region. The base region transit time is generally the limiting factor in determining the high-frequency capability of the transistor. The gain of the transistor is also significantly affected by minority carrier behavior in the base region. The density of minority carriers is significantly smaller than the density of majority carriers (majority carrier density is approximately equal to

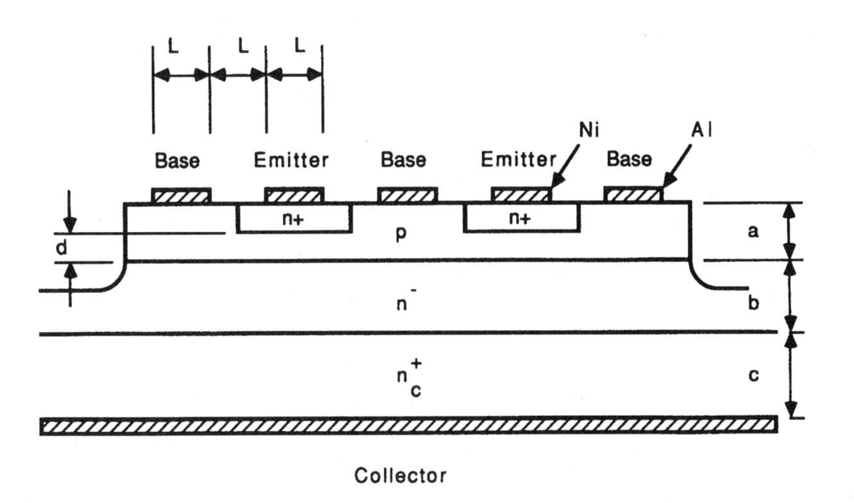

FIGURE 18.14 Cross-sectional view of a microwave bipolar transistor (BJT). The cross-hatched areas indicate metal electrodes. The electrode pattern on the surface is interdigitated with the base electrodes connected together at one end and the emitter electrodes connected together at the other end. Due to the interdigitated structure, there will always be one more base electrode than the number of emitter electrodes. The n^+, p, n^-, and $n_c{}^+$ designations indicate the impurity doping type and relative concentration level. This device has the collector electrode on the bottom of the device.

the impurity doping density) for typical operating conditions and the probability that the minority charge will recombine with a majority carrier is high. If recombination occurs, the minority charge cannot reach the base-collector junction but appears as base current. This, in turn, reduces the current gain capability of the transistor. Narrow base regions reduce the semiconductor volume where recombination can occur and, therefore, result in increased **gain**. Modern microwave transistors typically have base regions on the order of 0.1 to 0.25 μm.

The frequency response of a bipolar transistor can be determined by an analysis of the total time it takes for a charge carrier to travel from the emitter through the collector. The total time can be expressed as

$$\tau_{ec} = \tau_e + \tau_b + \tau_c + \tau_c' \tag{18.24}$$

where τ_{ec} is the total emitter–collector transit time, τ_e is the base-emitter junction capacitance charging time, τ_b is the base region transit time, τ_c is the base-collector junction capacitance charging time, and τ_c' is the base-collector region transit time. The total emitter-base time is related to the gain-bandwidth capability of the transistor according to the relation:

$$f_T = \frac{1}{2\pi\tau_{ec}} \tag{18.25}$$

Since the bipolar transistor has three terminals, it can be operated in various configurations, depending upon the electrode selected as the common terminal. The two most commonly employed are the common emitter (CE) and the common base (CB) configurations, although the common collector (CC) configuration can also be used. Small-signal amplifiers generally use the CE configuration and power amplifiers often use the CB configuration.

The current gain for a bipolar transistor is shown in Figure 18.15. The current gains of the transistor operated in the CE and CB configurations are called β and α, respectively. As indicated in the figure, the CE current gain β is much larger than the CB current gain α, which is limited to values less than unity. For modern microwave transistors $\alpha_o \sim 0.98$ to 0.99 and $\beta_o \sim 50$ to 60.

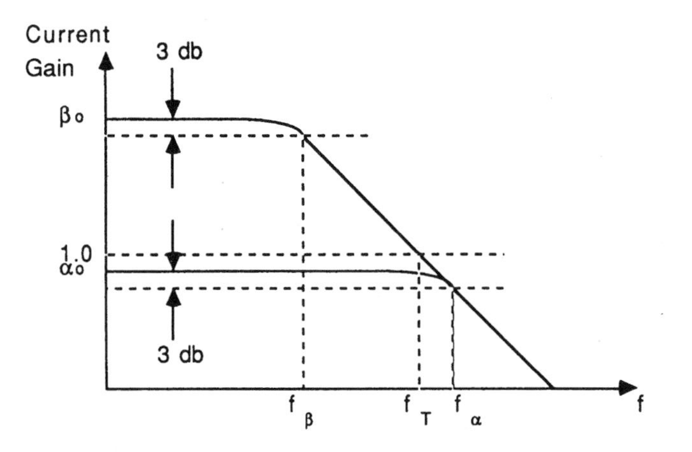

FIGURE 18.15 Current gains versus frequency for bipolar transistors. The common-emitter and common-base current gains are designated as α and β, respectively. The subscript "o" indicates the dc value. The gains decrease with frequency above a certain value. The frequencies where the gains are reduced by 3 dB from their dc values are indicated as the CE and CB cutoff frequencies, f_β and f_α, respectively. The frequency at which the CE current gain is reduced to unity is defined as the gain-bandwidth product f_T. Note that the CB current gain is restricted to values less than unity and that the CE current gain has values that significantly exceed unity.

A measure of the RF power gain for the transistor is indicated by the unilateral power gain, which can be expressed as

$$U \cong \frac{\alpha_o}{16\pi^2 r_b C_c f^2 \left(\tau_{ec} + \frac{r_e C_c}{a_o} \right)} \tag{18.26}$$

where U is the power gain, α_o is the dc CB current gain, r_b is the base resistance, C_c is the collector capacitance, τ_{ec} is the total emitter-to-collector transit time, and r_e is the emitter resistance. The frequency at which U is reduced to unity (f_{max}) is the maximum frequency at which the device will have active characteristics. This frequency is

$$f_{max} = \left[\frac{f_T}{8\pi r_b C_c} \right]^{1/2} \tag{18.27}$$

In order to maximize the high-frequency performance of a transistor, it is necessary to design the device so that it has high current gain (f_T), low base resistance (r_b), and low collector capacitance (C_c).

Bipolar transistors operating to about 20 GHz are generally fabricated from Si. These devices are easily fabricated and low cost. They are useful in moderate gain and low to high RF power applications. The have relatively high noise figure that varies from about 1 dB at 1 GHz to about 4 to 5 dB at 10 GHz.

An improved high-frequency bipolar transistor can be fabricated using heterostructures of compound semiconductors, such as AlGaAs/GaAs (Kroemer, 1982). These devices have their emitters fabricated from a wide-bandgap semiconductor (such as AlGaAs) and the remainder of the device fabricated from the lower-bandgap semiconductor (GaAs). The wide-bandgap emitter results in improved charge injection efficiency across the base-emitter junction into the base region and much improved RF performance. While the operation of standard Si bipolar transistors is limited to frequencies less than about 40 GHz, the heterojunction bipolar transistors (HBTs) can operate in excess of 100 GHz. They are useful in both low-noise and high RF power applications. The heterostructure concept has recently been applied in Si-based devices using heterostructures using SiGe/Si compounds. These devices show consider promise for high-frequency applications and the transistors have demonstrated RF performance comparable to that obtained from the AlGaAs/GaAs HBTs.

Comparison of Bipolar Transistor and MESFET Noise Figures

In low-noise applications, GaAs MESFETs are generally preferred to Si bipolar transistors. The MESFET demonstrates a lower noise figure than the bipolar transistor throughout the microwave frequency range, and the advantage increases with frequency. This advantage is demonstrated by a comparison of the expressions for the minimum noise figure for the two devices. The bipolar transistor has a minimum noise figure that can be expressed as

$$F_{\mathrm{min}} \cong 1 + bf^2\left[1 + \sqrt{1 + \frac{2}{bf^2}}\right] \tag{18.28}$$

where F_{min} is the noise figure and

$$b = \frac{40 I_{\mathrm{c}} r_{\mathrm{b}}}{f_{\mathrm{T}}^2} \tag{18.29}$$

where I_{c} is the collector current and the other terms are as previously defined. The minimum noise figure for the MESFET is

$$F_{\mathrm{min}} \cong 1 + mf \tag{18.30}$$

where

$$m = \frac{2.5}{f_{\mathrm{T}}}\sqrt{g_{\mathrm{m}}(R_{\mathrm{g}} + R_{\mathrm{s}})} \tag{18.31}$$

where g_{m} is the MESFET **transconductance,** R_{g} is the gate resistance, and R_{s} is the source resistance.

Comparing these expressions shows that the minimum noise figure increases with frequency quadratically for bipolar transistors and linearly for MESFETs. Therefore, as operating frequency increases, the MESFET demonstrates increasingly superior noise figure performance as compared to Si bipolar transistors.

Conclusions

Various active solid-state devices that are useful at microwave and millimeter-wave frequencies have been discussed. Both two-terminal and three-terminal devices were included. The most commonly used two-terminal devices are tunnel diodes, transferred-electron devices, and IMPATT diodes. Three-terminal devices consist of various forms of field-effect transistors and bipolar transistors. Recent advances employ heterostructures using combinations of different semiconductors to produce devices with improved RF performance, especially for high-frequency applications. Both two-terminal and three-terminal devices can provide useful gain at frequencies in excess of 100 GHz. Further improvements are likely as fabrication technology continues to improve.

Defining Terms

Active device: A device that can convert energy from a dc bias source to a signal at an RF frequency. Active devices are required in oscillators and amplifiers.

Charge carriers: Units of electrical charge that when moving produce current flow. In a semiconductor two types of charge carriers exist: electrons and holes. Electrons carry unit negative charge and have an effective mass that is determined by the shape of the conduction band in energy-momentum space. The effective mass of an electron in a semiconductor is generally significantly less than an electron in free space. Holes have unit positive charge. Holes have an effective mass that is determined by the shape of the valence band in energy-momentum space. The effective mass of a hole is generally significantly larger than that for an electron. For this reason electrons generally move much faster than holes when an electric field is applied to the semiconductor.

Gain: A measure of the ability of a network to increase the energy level of a signal. Gain is generally measured in decibels. For voltage or current gain: $G(\text{dB}) = 20 \log(S_{out}/S_{in})$, where S is the RF voltage or current out of and into the network. For power gain $G(\text{dB}) = 10 \log(P_{out}/P_{in})$. If the network has net loss, the gain will be negative.

Noise figure: A measure of the noise added by a network to an RF signal passing through it. Noise figure can be defined in terms of signal-to-noise ratios at the input and output ports of the network. Noise figure is generally measured in decibels and can be defined as $F(\text{dB}){=}10 \log[(S/N)_{in}/(S/N)_{out}]$.

One-port network: An electrical network that has only one RF port. This port must be used as both the input and output to the network. Two-terminal devices result in one-port networks.

Three-terminal device: An electronic device that has three contacts, such as a transistor.

Transconductance: A measure of the gain capability of a transistor. It is defined as the change in output current as a function of a change in input voltage.

Two-port network: An electrical network that has separate RF ports for the input and output. Three-terminal devices can be configured into two-port networks.

Two-terminal device: An electronic device, such as a diode, that has two contacts. The contacts are usually termed the cathode and anode.

References

P.B. Bhartia and I.J. Bahl, *Millimeter Wave Engineering and Applications*, New York: Wiley-Interscience, 1984.

B.G. Bosch and R.W. Engelmann, *Gunn-Effect Electronics*, New York: Halsted Press, 1975.

H.F. Cooke, "Microwave transistors: Theory and design," *Proc. IEEE*, vol. 59, pp. 1163–1181, Aug. 1971.

T.J. Drummond, W.T. Masselink, and H. Morkoc, "Modulation-doped GaAs/AlGaAs heterojunction field-effect transistors: MODFET's," *Proc. IEEE*, vol. 74, pp. 773–822, June 1986.

H. Kroemer, "Heterostructure bipolar transistors and integrated circuits," *Proc. IEEE*, vol. 70, pp. 13–25, Jan. 1982.

C.A. Liechti, "Microwave field-effect transistors—1976," *IEEE Trans. Microwave Theory and Tech.*, vol. MTT-24, pp. 128–149, June 1976.

S.M. Sze, *Physics of Semiconductor Devices*, 2nd ed., New York: Wiley-Interscience, 1981.

Further Information

Additional details on the various devices discussed in this chapter can be found in the following books:

I. Bahl and P. Bhartia, *Microwave Solid State Circuit Design*, New York: Wiley-Interscience, 1988.

M. Shur, *Physics of Semiconductor Devices*, Englewood Cliffs, NJ: Prentice-Hall, 1990.

S.M. Sze, *High-Speed Semiconductor Devices*, New York: Wiley-Interscience, 1990.

S. Tiwari, *Compound Semiconductor Device Physics*, San Diego, CA: Academic Press, 1992.

S. Wang, *Fundamentals of Semiconductor Theory and Device Physics*, Englewood Cliffs, NJ: Prentice-Hall, 1989.

19

Compatibility

Leland H. Hemming
McDonnell Douglas Helicopter Systems

Ken Kaiser
Kettering University

Halit Eren
Curtin University of Technology

Bert Wong
Western Australia Telecommunications Research Institute

Martin A. Uman
University of Florida

19.1 Grounding, Shielding, and Filtering

Leland H. Hemming and Ken Kaiser

Electromagnetic interference (EMI) is defined to exist when undesirable voltages or currents are present to influence adversely the performance of an electronic circuit or system. Interference can be within the system (intrasystem), or it can be between systems (intersystem). The system is the equipment or circuit over which one exercises design or management control.

The cause of an EMI problem is an unplanned coupling between a source and a receptor by means of a transmission path. Transmission paths may be conducted or radiated. Conducted interference occurs by means of metallic paths. Radiated interference occurs by means of near- and far-field coupling. These different paths are illustrated in Figure 19.1.

The control of EMI is best achieved by applying good interference control principles during the design process. These involve the selection of signal levels, impedance levels, frequencies, and circuit configurations that minimize conducted and radiated interference. In addition, signal levels should be selected to be as low as possible, while being consistent with the required signal-to-noise ratio. Impedance levels should be chosen to minimize undesirable capacitive and inductive coupling.

The frequency spectral content should be designed for the specific needs of the circuit, minimizing interference by constraining signals to desired paths, eliminating undesired paths, and separating signals from interference. Interference control is also achieved by physically separating leads carrying currents from different sources.

For optimum control, the three major methods of EMI suppression—grounding, shielding, and filtering—should be incorporated early in the design process. The control of EMI is first achieved by proper grounding, then by good shielding design, and finally by filtering.

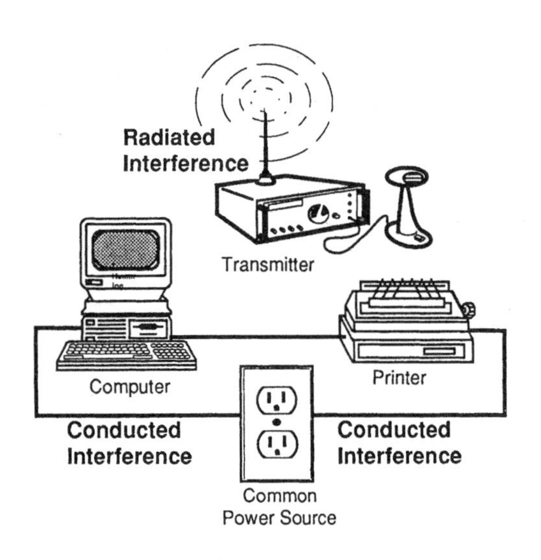

FIGURE 19.1 Electromagnetic interference is caused by uncontrolled conductive paths and radiated near/far fields.

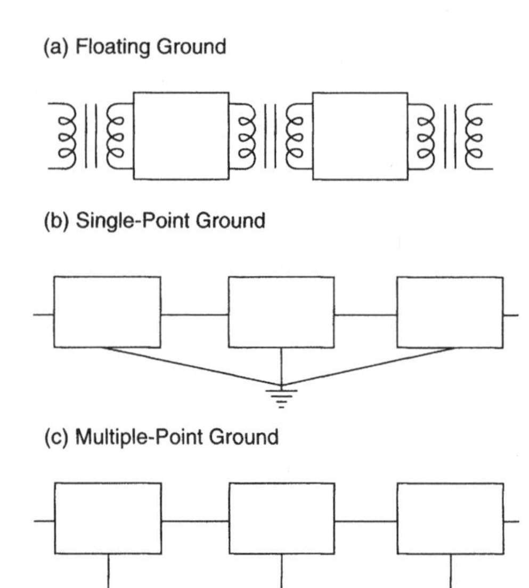

FIGURE 19.2 The type of ground system used must be selected carefully.

Grounding is the process of electrically establishing a low impedance path between two or more points in a system. An ideal ground plane is a zero potential, zero impedance body that can be used as reference for all signals in the system. Associated with grounding is *bonding*, which is the establishment of a low impedance path between two metal surfaces.

Shielding is the process of confining radiated energy to the bounds of a specific volume or preventing radiated energy from reaching a specific volume. *Filtering* is the process of eliminating conducted interference by controlling the spectral content of the conducted path. Filtering is the last step in the EMI design process.

Grounding

Grounding Principles

The three fundamental grounding techniques—floating, single-point, and multiple-point—are illustrated in Figure 19.2.

Floating grounds are used to isolate circuits or equipment from a common ground plane. Static charges are a hazard with this type of ground. Dangerous voltages may develop or a noise-producing discharge might occur. Generally, bleeder resistors are used to control the static problem. Floating grounds are useful only at low frequencies where capacitive coupling paths are negligible.

The single-point ground is a single physical point in a circuit. By connecting all grounds to a common point, no interference will be produced in the equipment because the configuration does not result in potential differences across the equipment. At high frequencies care must be taken to prevent capacitive coupling, which will result in interference.

A multipoint ground system exists when each ground connection is made directly to the ground plane at the closest available point on it, thus minimizing ground lead lengths. A large conductive body is chosen for the ground. Care must be taken to avoid ground loops.

Circuit grounding design is dependent on the function of each type of circuit. In unbalanced systems, care must be taken to reduce the potential of common mode noise. Differential devices are commonly used to suppress this form of noise. The use of high circuit impedances should be minimized. Where it cannot be avoided, all interconnecting leads should be shielded, with the shield well grounded. Power supply grounding must be done properly to minimize load inducted noise on a power supply bus. When electromechanical relays are used in a system, it is best that they be provided with their own power supplies.

Cable shield grounding must be designed based upon the frequency range, impedance levels (whether balanced or unbalanced) and operating voltage and/or current. Cross-talk between cables is a major problem and must be carefully considered during the design process.

Building facility grounds must be provided for electrical faults, signal, and lightning. The fault protection (green wire) subsystem is for the protection of personnel and equipment from the hazards of electrical power faults and static charge buildup. The lightning protection system consists of air terminals (lightning rods), heavy duty down-conductors, and ground rods. The **signal reference subsystem** provides a ground for signal circuits to control static charges and noise and to establish a common reference between signals and loads. Earth grounds may consist of vertical rods, horizontal grids or radials, plates, or incidental electrodes such as utility pipes or buried tanks. The latter must be constructed and tested to meet the design requirements of the facility.

Grounding Design Guidelines

The following design guidelines represent good practice but should be applied subject to the detailed design objectives of the system.

Fundamental Concepts

- Use single-point grounding for circuit dimensions less than 0.03 λ (wavelength) and multipoint grounding for dimensions greater than 0.15 λ.
- The type of grounding for circuit dimensions between 0.03 and 0.15 λ depends on the physical arrangement of the ground leads as well as the conducted emission and conducted susceptibility limits of the circuits to be grounded. Hybrid grounds may be needed for circuits that must handle a broad portion of the frequency spectrum.
- Apply floating ground isolation techniques (i.e., transformers) if ground loop problems occur.
- Keep all ground leads as short as possible.
- Design ground reference planes so that they have high electrical conductivity and can be maintained easily to retain good conductivity.

Safety Considerations

- Connect test equipment grounds directly to the grounds of the equipment being tested.
- Make certain the ground connections can handle fault currents that might flow unexpectedly.

Circuit Grounding

- Maintain separate circuit ground systems for signal returns, signal shield returns, power system returns, and chassis or case grounds. These returns then can be tied together at a single ground reference point.
- For circuits that produce large, abrupt current variations, provide a separate grounding system, or provide a separate return lead to the ground to reduce transient coupling into other circuits.
- Isolate the grounds of low-level circuits from all other grounds.
- Where signal and power leads must cross, make the crossing so that the wires are perpendicular to each other.
- Use balanced differential circuitry to minimize the effects of ground circuit interference.
- For circuits whose maximum dimension is significantly less than $\lambda/4$, use tightly twisted wires (either shielded or unshielded, depending on the application) that are single-point grounded to minimize equipment susceptibility.

Cable Grounding

- Avoid pigtails when terminating cable shields.
- When coaxial cable is needed for signal transmission, use the shield as the signal return and ground at the generator end for low-frequency circuits. Use multipoint grounding of the shield for high-frequency circuits.
- Provide multiple shields for low-level transmission lines. Single-point grounding of each shield is recommended.

Shielding

The control of near- and far-field coupling (radiation) is accomplished using shielding techniques. The first step in the design of a shield is to determine what undesired field level may exist at a point with no shielding and what the tolerable field level is. The difference between the two then is the needed **shielding effectiveness.**

This section discusses the shielding effectiveness of various solid and nonsolid materials and their application to various shielding situations. **Penetrations** and their design are discussed so that the required shielding effectiveness is maintained. Finally, common shielding effectiveness testing methods are reviewed.

Enclosure Theory

The attenuation provided by a shield results from three loss mechanisms as illustrated in Figure 19.3:

1. Incident energy is reflected (R) by the surface of the shield because of the impedance discontinuity of the air–metal boundary. This mechanism does not require a particular material thickness but simply an impedance discontinuity.
2. Energy that does cross the boundary (not reflected) is attenuated (A) in passing through the shield.
3. The energy that reaches the opposite face of the shield encounters another air–metal boundary and thus some of it is reflected (B) back into the shield. This term is only significant when $A < 15$ dB and is generally neglected because the barrier thickness is generally great enough to exceed the 15-dB loss rule of thumb.

Thus:

$$S = R + A + B\,\mathrm{dB} \tag{19.1}$$

Absorption loss is independent of the type of wave (electric/magnetic) and is given by

$$A = 1.314(f\mu_r\sigma_r)^{1/2}d\,\mathrm{dB} \tag{19.2}$$

where d is shield thickness in centimeters, μ_r is relative permeability, f is frequency in Hz, and σ_r is conductivity of metal relative to that of copper. Typical absorption loss is provided in Table 19.1.

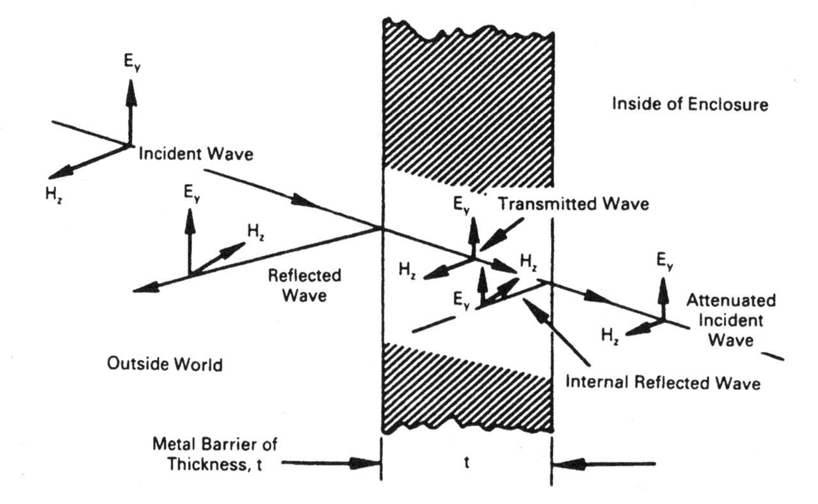

FIGURE 19.3 Shielding effectiveness is the result of three loss mechanisms.

TABLE 19.1 Absorption Loss as a Function of Type of Material and Frequency (Loss Shown Is at 150 kHz)

Metal	Relative Conductivity	Relative Permeability	Absorption Loss *A*, dB/mm
Silver	1.05	1	52
Copper—annealed	1.00	1	51
Copper—hard drawn	0.97	1	50
Gold	0.70	1	42
Aluminum	0.61	1	40
Magnesium	0.38	1	31
Zinc	0.29	1	28
Brass	0.26	1	26
Cadmium	0.23	1	24
Nickel	0.20	1	23
Phosphor–bronze	0.18	1	22
Iron	0.17	1,000	650
Tin	0.15	1	20
Steel, SAE1045	0.10	1,000	500
Beryllium	0.10	1	16
Lead	0.08	1	14
Hypernik	0.06	80,000	3,500[1]
Monel	0.04	1	10
Mu-metal	0.03	80,000	2,500[1]
Permalloy	0.03	80,000	2,500[1]
Steel, stainless	0.02	1,000	220[1]

[1]Assuming that material is not saturated.
Source: MIL-HB-419A.

Reflection loss is a function of the intrinsic impedance of the metal boundary with respect to the wave impedance, and therefore, three conditions exist: near-field magnetic, near-field electric, and plane wave.

The relationship for low-impedance (magnetic field) source is

$$R = 20\log_{10}\left\{\left[1.173(\mu_r/f\sigma_r)^{1/2}/D\right] + 0.0535D(f\sigma_r/\mu_r)^{1/2} + 0.354\,\text{dB}\right\} \tag{19.3}$$

where D is distance to source in meters. For a plane wave source the reflection loss is

$$R = 168 - 10\log_{10}(f\mu_r/\sigma_r)\,\text{dB} \tag{19.4}$$

For a high-impedance (electric field) source the reflection loss R is

$$R = 362 - 20\log_{10}[(\mu_r f^3/\sigma_r)^{1/2}D]\,\text{dB} \tag{19.5}$$

Figure 19.4 illustrates the shielding effectiveness of a variety of common materials versus various thicknesses for a source distance of 1 m. This is the shielding effectiveness of a six-sided enclosure. To be useful, the enclosure must be penetrated for various services or devices. This is illustrated in Figure 19.5(a) for small enclosures and Figure 19.5(b) for room-sized enclosures.

Shielding Penetrations

Total shielding effectiveness of an enclosure is a function of the basic shield and all of the leakages associated with the penetrations in the enclosure. The latter includes seams, doors, vents, control shafts, piping, filters, windows, screens, and fasteners.

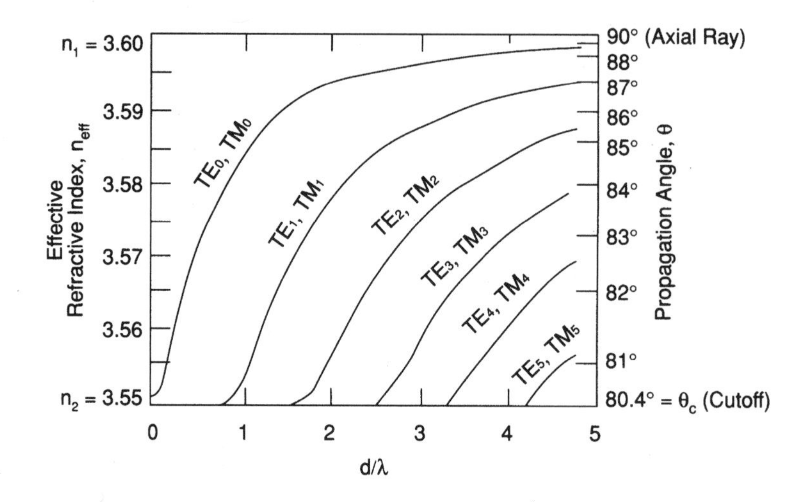

FIGURE 19.4 The shielding effectiveness of common sheet metals, 1-m separation: (a) 26-gage steel; (b) 3-oz copper foil; (c) 0.030-in. aluminum sheet; (d) 0.003-in. Permalloy; (e) a common specification for shielded enclosures.

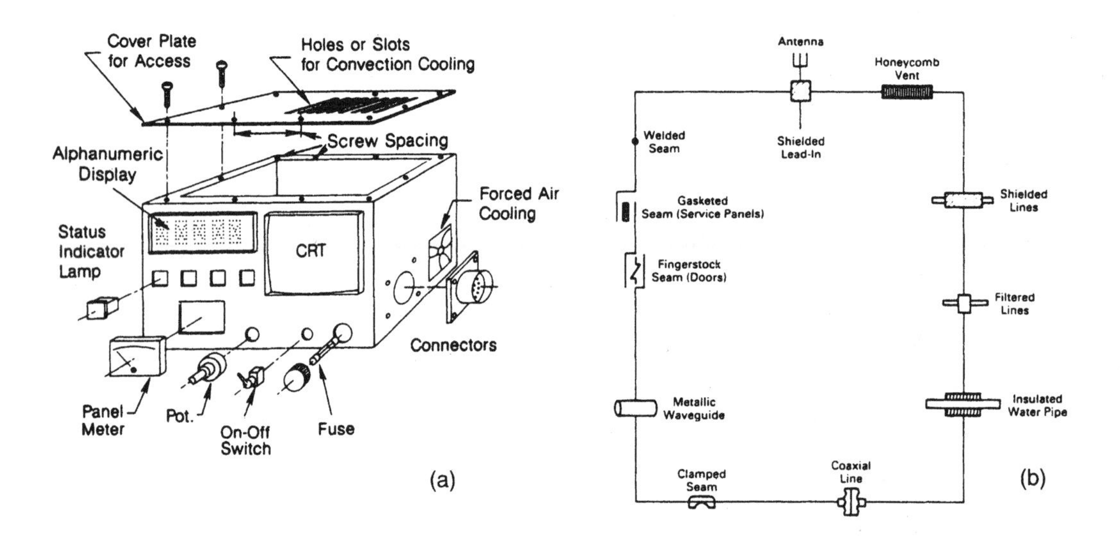

FIGURE 19.5 Penetrations in small (a) and large (b) enclosures.

The design of the seams is a function of the type of enclosure and the level and nature of the shielding effectiveness required. For small instruments, computers, and similar equipment, the typical shielding required is on the order of 60 dB for electric and plane wave shielding. EMI gaskets are commonly used to seal the openings in sheet metal construction. In some high-performance applications the shielding is achieved using very tight-fitting machined housings. Examples are IF strips and large dynamic range log amplifier circuits. Various methods of sealing joints are illustrated in Figure 19.6. EMI gasketing methods are shown in Figure 19.7. For large room-sized enclosures, the performance requirements typically range from 60 to 120 dB. Conductive EMI shielding tape is used in the 60-dB realm, clamped seams for 80 to 100 dB, and continuous welded seams for 120-dB performance. These are illustrated in Figure 19.8.

A good electromagnetic shielded door design must meet a variety of physical and electrical requirements. Figure 19.9 illustrates a number of ways this is accomplished.

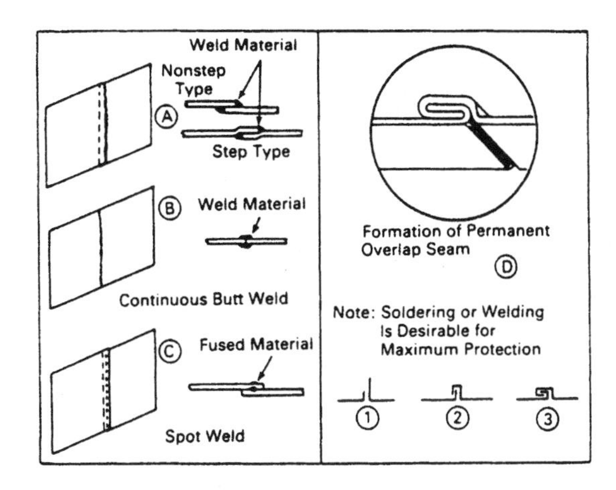

FIGURE 19.6 Methods of sealing enclosure seams.

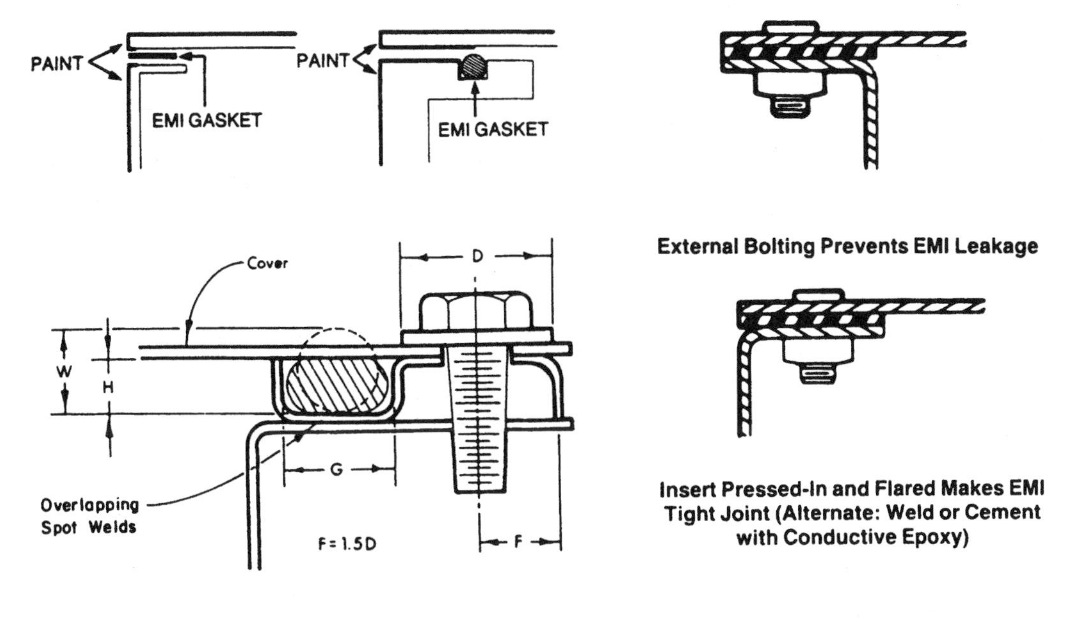

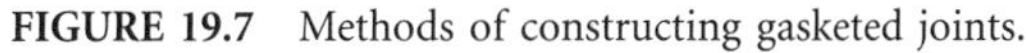

FIGURE 19.7 Methods of constructing gasketed joints.

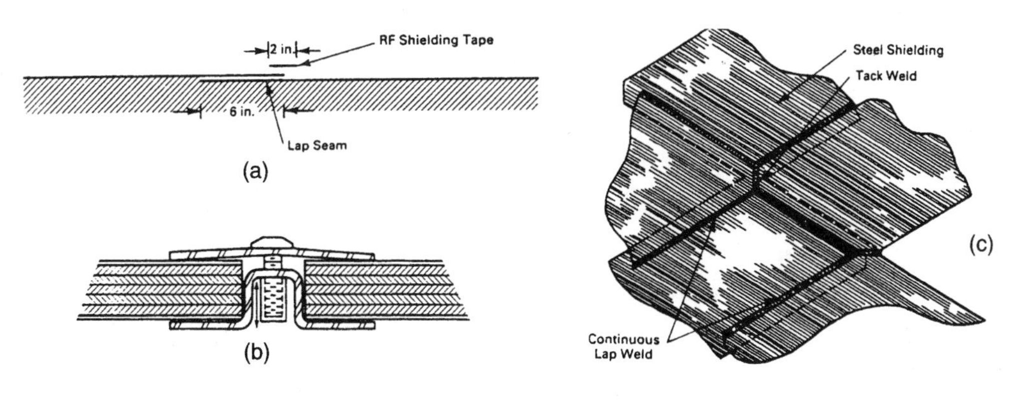

FIGURE 19.8 Most common seams in large enclosures. (a) Foil and shielding tape; (b) clamped; (c) welded.

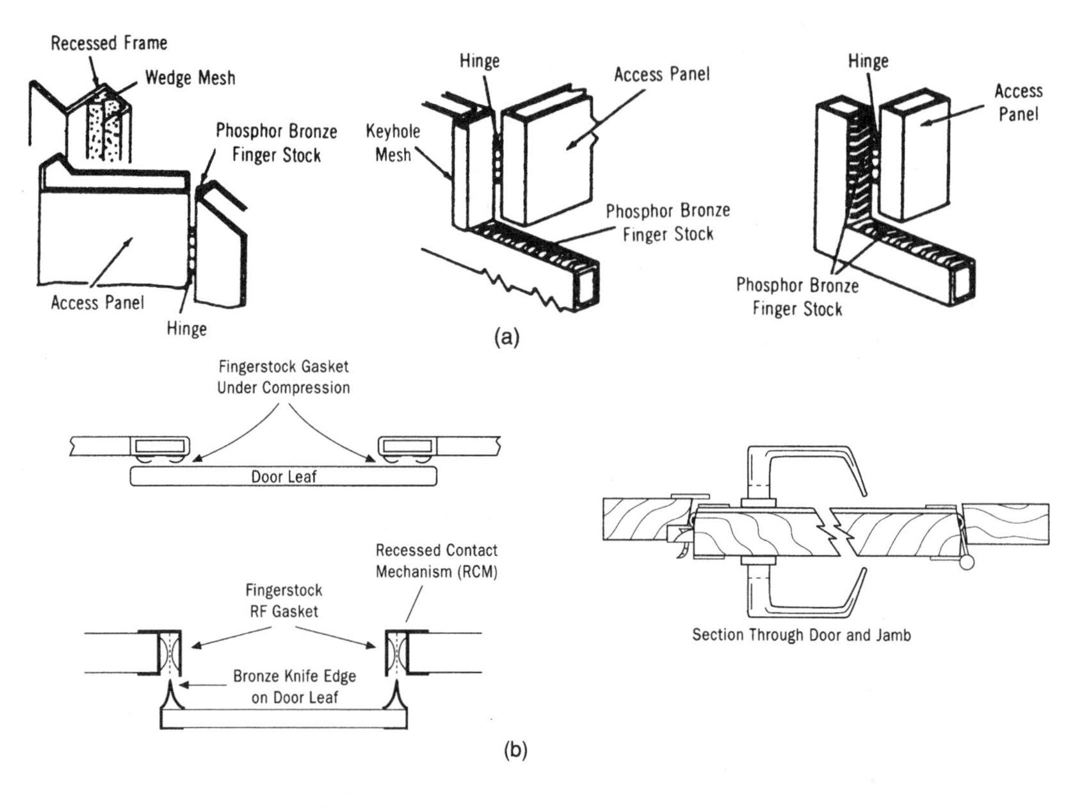

FIGURE 19.9 Methods of sealing seams in RF enclosure small (a) and large (b) doors.

For electronic equipment, a variety of penetrations must be made to make the shielded volume functional. These include control shafts, windows, lights, filters, and displays. Careful design is required to maintain the required shielding integrity.

Shield Testing

The most common specification used for shield evaluation is the procedure given in MIL-STD-285. This consists of establishing a reference level without the shield and then enclosing the receiver within the shield and determining the difference. The ratio is the shielding effectiveness. This applies regardless of materials used in the construction of the shield. Care must be taken in evaluating the results since the measured value is a function of a variety of factors, not all of which are definable.

Summary of Good Shielding Practice

Shielding Effectiveness

- Good conductors, such as copper and aluminum, should be used for electric field shields to obtain high reflection loss. A shielding material thick enough to support itself usually provides good electric shielding at all frequencies.
- Magnetic materials, such as iron and special high-permeability alloys, should be used for magnetic field shields to obtain high absorbtion loss.
- In the plane wave region, the sealing of all apertures is critical to good shielding practice.

Multiple Shields

- Multiple shields are quite useful where high degrees of shielding effectiveness are required.

Shield Seams

- All openings or discontinuities should be addressed in the design process to ensure achievement of the required shielding effectiveness. Shield material should be selected not only from a shielding requirement, but also from electrochemical corrosion and strength considerations.
- Whenever system design permits, use continuously overlapping welded seams. Obtain intimate contact between mating surfaces over as much of the seam as possible.
- Surfaces to be mated must be clean and free from nonconducting finishes, unless the bonding process positively and effectively cuts through the finish. When electromagnetic compatibility (**EMC**) and finish specifications conflict, the finishing requirements must be modified.

Case Construction

- Case material should have good shielding properties.
- Seams should be welded or overlapped.
- Panels and cover plates should be attached using conductive gasket material with closely spaced fasteners.
- Mating surfaces should be cleaned just before assembly to ensure good electrical contact and to minimize corrosion.
- A variety of special devices are available for sealing around doors, vents, and windows.
- Internal interference generating circuits must be isolated both electrically and physically. Electrical isolation is achieved by circuit design; physical isolation may be achieved by proper shielding.
- For components external to the case, use EMI boots on toggle switches, EMI rotary shaft seals on rotary shafts, and screening and shielding on meters and other indicator faces.

Cable Shields

- Cabling that penetrates a case should be shielded and the shield should be terminated in a peripheral bond at the point of entry. This peripheral bond should be made to the connector or adaptor shell.

Filtering

An electrical filter is a combination of lumped or distributed circuit elements arranged so that it has a frequency characteristic that passes some frequencies and blocks others.

Filters provide an effective means for the reduction and suppression of electromagnetic interference as they control the spectral content of signal paths. The application of filtering requires careful consideration of an extensive list of factors, including insertion loss, impedance, power handling capability, signal distortion, tunability, cost, weight, size, and rejection of undesired signals. Often they are used as stopgap measures, but if suppression techniques are used early in the design process, then the complexity and cost of interference fixes can be minimized. There are many textbooks on filtering, which should be used for specific applications.

The types of filters are classified according to the band of frequencies to be transmitted or attenuated. The basic types illustrated in Figure 19.10 include low-pass, high-pass, bandpass, and bandstop (reject).

Filters can be composed of lumped, distributed, or dissipative elements; the type used is mainly a function of frequency.

Filtering Guidance

- It is best to filter at the interference source.
- Suppress all spurious signals.
- Design nonsusceptible circuits.
- Ensure that all filter elements interface properly with other EMC elements, i.e., proper mounting of a filter in a shielded enclosure.

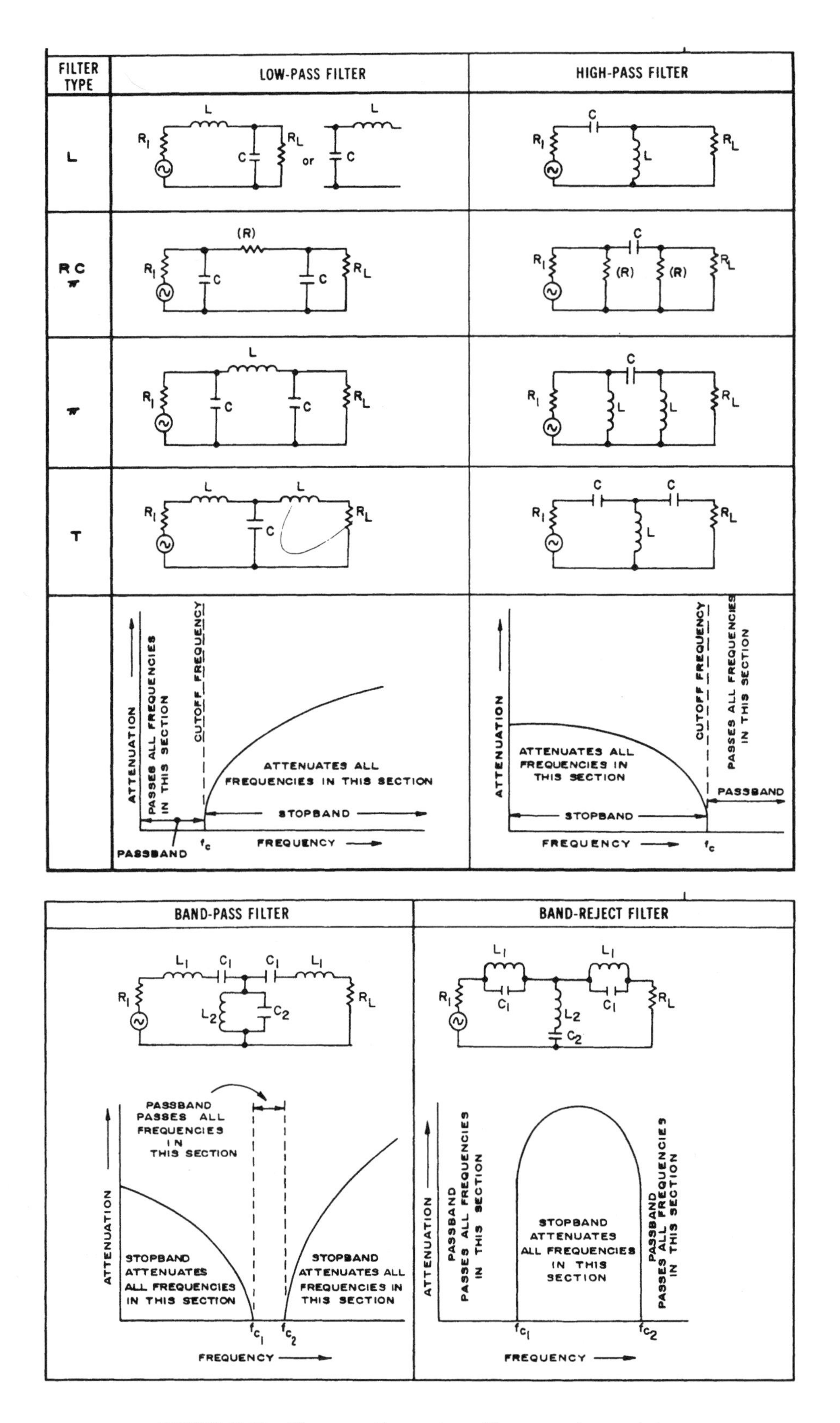

FIGURE 19.10 Filters provide a variety of frequency characteristics.

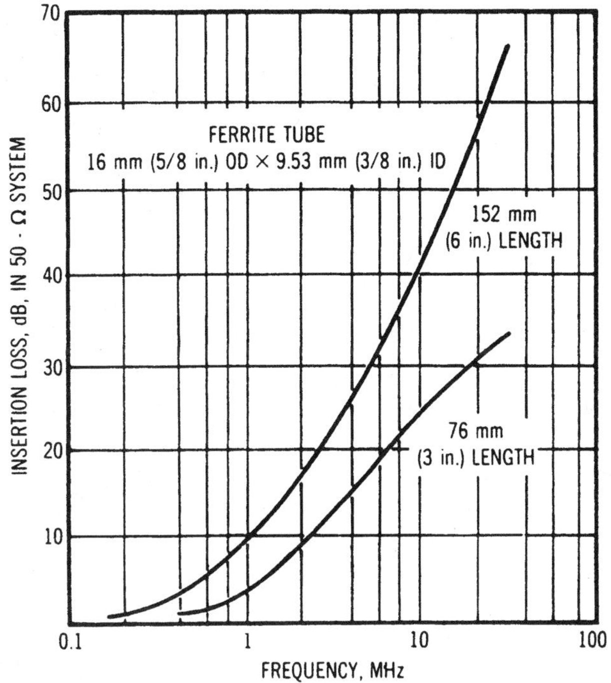

FIGURE 19.11 Ferrite provides a flexible means of achieving a low-pass filter with good high-frequency loss characteristics.

Filter Design

Filters using lumped and distributive elements generally are reflective, in that the various component combinations are designed for high series impedance and low shunt impedance in the stopband while providing low series impedance and high shunt impedance in the passband.

The impedance mismatches associated with the use of reflective filters can result in an increase of interference. In such cases, the use of dissipative elements is found to be useful. A broad range of ferrite components are available in the form of beads, tubes, connector shells, and pins. A very effective method of low-pass filtering is to form the ferrite into a coaxial geometry, the properties of which are proportional to the length of the ferrite, as shown in Figure 19.11.

Application of filtering takes many forms. A common problem is transient suppression as illustrated in Figure 19.12. All sources of transient interference should be treated at the source.

Power line filtering is recommended to eliminate conducted interference from reaching the powerline and adjacent equipment. Active filtering is very useful in that it can be built in as part of the circuit design and can be effective in passing only the design signals. A variety of noise blankers, cancelers, and limiter circuits are available for active cancellation of interference.

Special Filter Types

A variety of special-purpose filters are used in the design of electronic equipment. Transmitters require a variety of filters to achieve a noise-free output.

Receive preselectors play a useful role in interference rejection. Both distributed (cavity) and lumped element components are used.

IF filters control the selectivity of a receiving system and use a variety of mechanical and electrical filtering components.

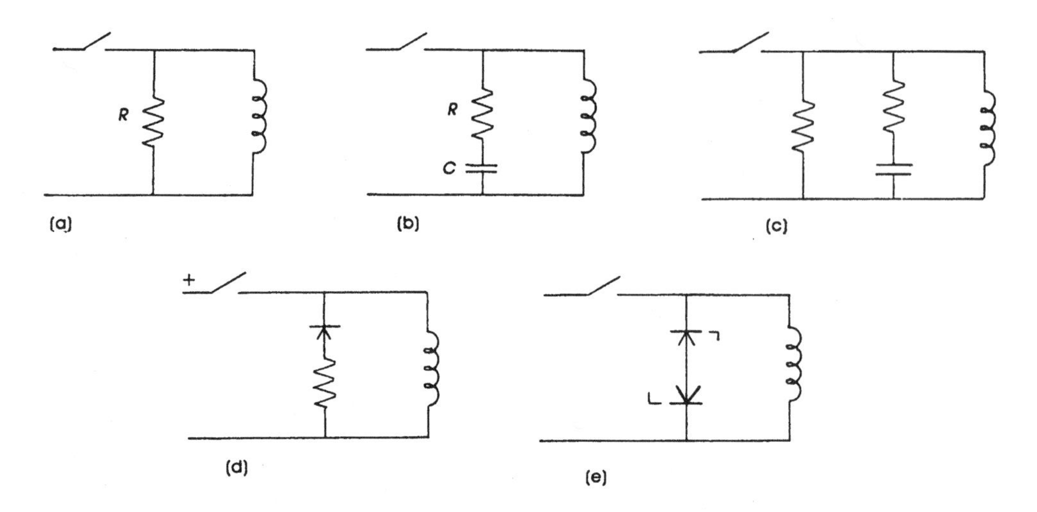

FIGURE 19.12 Transient responses are controlled using simple filters at the source. (a) Resistance damping (b) capacitance suppression; (c) RC suppression; (d) diode suppression; (e) back-to-back diode suppression.

Testing

The general requirements for electromagnetic filters are detailed in MIL-F-15733, MIL-F-18327, and MIL-F-25880. Insertion loss is measured in accordance with MIL-STD-220.

Defining Terms

Earth electrode system: A network of electrically interconnected rods, plates, mats, or grids, installed for the purpose of establishing a low-resistance contact with earth. The design objective for resistance to earth of this subsystem should not exceed 10 Ω.

Electromagnetic compatibility (EMC): The capability of equipment or systems to be operated in their intended operational environment at designed levels of efficiency without causing or receiving degradation owing to unintentional electromagnetic interference. Electromagnetic compatibility is the result of an engineering planning process applied during the life cycle of the equipment. The process involves careful considerations of frequency allocation, design, procurement, production, site selection, installation, operation, and maintenance.

Electromagnetic pulse (EMP): A large, impulsive-type electromagnetic wave generated by nuclear or chemical explosions.

Field strength: A general term that means the magnitude of the electric field vector (in volts per meter) or the magnitude of the magnetic field vector (in ampere-turns per meter). As used in the field of EMC/EMI, the term *field strength* shall be applied only to measurements made in the far field and shall be abbreviated as FS. For measurements made in the near field, the term *electric field strength* (EFS) or *magnetic field strength* (MFS) shall be used, according to whether the resultant electric or magnetic field, respectively, is measured.

Penetration: The passage through a partition or wall of an equipment or enclosure by a wire, cable, pipe, or other conductive object.

Radio frequency interference (RFI): Synonymous with *electromagnetic interference.*

Shielding effectiveness: A measure of the reduction or attenuation in the electromagnetic field strength at a point in space caused by the insertion of a shield between the source and that point.

Signal reference subsystem: This subsystem provides the reference points for all signal grounding to control static charges, noise, and interference. It may consist of any one or a combination of the lower frequency network, higher frequency network, or hybrid signal reference network.

TEMPEST: A code word (not an acronym) that encompasses the government/industrial program for controlling the emissions from systems processing classified data. Individual equipment may be *TEM-PESTed* or commercial equipment may be placed in shielded enclosures.

References

AFSC Design Handbook, DH1–4, Electromagnetic Compatibility, 4th ed., U.S. Air Force, Wright-Patterson Air Force Base, OH, January 1991.

R.F. Ficchi, Ed., *Practical Design for Electromagnetic Compatibility*, Hayden, 1971.

E.R. Freeman, *Electromagnetic Compatibility Design Guide for Avionics and Related Ground Support Equipment*, Norwood, MA: Artech House, 1982.

L.H. Hemming, *Architectural Electromagnetic Shielding Handbook*, New York: IEEE Press, 1991.

K. Kaiser, *Electromagnetic Compatibility Handbook*, Boca Raton, FL: CRC Press, 2005.

B. Keiser, *Principles of Electromagnetic Compatibility*, 3rd ed., Norwood, MA: Artech House, 1987.

Y.J. Lubkin, *Filter Systems and Design: Electrical, Microwave, and Digital*, Reading, MA: Addison-Wesley, 1970.

MIL-HDBK-419A, Grounding, Bonding, and Shielding of Electronic Equipment and Facilities, U.S. Department of Defense, Washington, DC, 1990.

R. Morrison, *Grounding and Shielding Techniques in Instrumentation*, New York: John Wiley, 1986.

R. Morrison and W. H. Lewis, *Grounding and Shielding Techniques in Facilities*, New York: John Wiley, 1990.

T. Rikitake, *Magnetic and Electromagnetic Shielding*, Amsterdam: D. Reidel, 1987.

N.O.N. Violetto, *Electromagnetic Compatibility Handbook*, New York: Van Nostrand Reinhold, 1987.

D.R.J. White, *Shielding Design, Methodology and Procedures*, Springfield, VA: Interference Control Technologies, 1986.

D.R.J. White, *A Handbook on Electromagnetic Shielding Materials and Performance*, Springfield VA: Interference Control Technologies, 1975.

Further Information

The annual publication *Interference Technology Engineers' Master (Item)*, published by R&B Enterprises, West Conshohocken, Pa., covers all aspects of EMI including an extensive product directory.

The periodical *IEEE Transactions on Electromagnetic Compatibility*, which is published by The Institute of Electrical and Electronics Engineers, Inc., provides theory and practice in the EMI field.

The periodical *emf-emi control* published bimonthly by EEC Press, Gainesville, VA, is an excellent source of practical EMI information.

The periodical *Compliance Engineering*, published quarterly by Compliance Engineering, Inc., is a good source of information on EMC regulations and rules.

19.2 Spectrum, Specifications, and Measurement Techniques

Halit Eren and Bert Wong

The electromagnetic energy, created by moving or oscillating charges, propagates in air with a velocity of 300,000 km/sec. Electromagnetic energy may be generated intentionally (e.g., by communication systems) or unintentionally (e.g., by power lines). Generally, all electrical and electronic devices create some form of unintentional electromagnetic energy due to inductive and capacitive properties of the circuits. In some cases, this energy interferes with the operation of other devices, termed as the electromagnetic interference (EMI). Minimization or elimination of the EMI from devices and systems is a concern for electromagnetic compatibility (EMC). The EMC may be defined as the capability of two or more devices to operate simultaneously without mutual interference.

Because of the increasing number of man-made EMI generated devices and systems around the globe, allowable limits as well as measurement techniques on the EMC are set and regulated by authorities at national and international levels. The problems or the potential problems that can be caused due to emission and susceptibility of devices and systems have led to numerous standards and regulations worldwide. The Federal Communication Commission (FCC), the U.S. Military, and the International Electrotechnical Commission (IEC) are few examples of the organizations that develop and maintain standards on EMC.

Electromagnetic Spectrum

The electromagnetic spectrum is a continuum of all electromagnetic waves arranged according to frequency and wavelength. The frequency spectrum of electromagnetic energy can span from dc to gamma ray (10^{21} Hz) and beyond. Table 19.2 gives the approximate wavelengths, frequencies and the radiated energy levels for a limited range of electromagnetic spectrum. Figure 19.13 illustrates the typical frequency spectrum chart over a fraction of Hertz to 6×10^{22} Hz.

The frequency spectrum of interest for EMC purposes ranges from extreme low frequency (ELF) of a few Hertz well into microwave bands (40 GHz and beyond). Low frequencies are the primary concern of biological research, industrial equipment, and low frequency communication devices, while high frequencies are concern for most analogue and digital electronic equipment, and devices used by the military and in the health industry.

TABLE 19.2 Approximate Wavelengths, Frequencies, and Radiated Energy Levels

Electromagnetic Spectrum			
Region	Wavelength (cm)	Frequency (Hz)	Energy (eV)
Radio	>10	$<3\times10^9$	$<10^{-5}$
Microwave	10–0.01	3×10^9–3×10^{12}	10^{-5}–0.01
Infrared	0.01–7×10^{-5}	3×10^{12}–4.3×10^{14}	0.01–2
Visible	7×10^{-5}–4×10^{-5}	4.3×10^{14}–7.5×10^{14}	2–3
Ultraviolet	4×10^{-5}–10^{-7}	7.5×10^{14}–3×10^{17}	3–10^3
X-rays	10^{-7}–10^{-9}	3×10^{17}–3×10^{19}	10^3–10^5
Gamma rays	$<10^{-9}$	$>3\times10^{19}$	$>10^5$

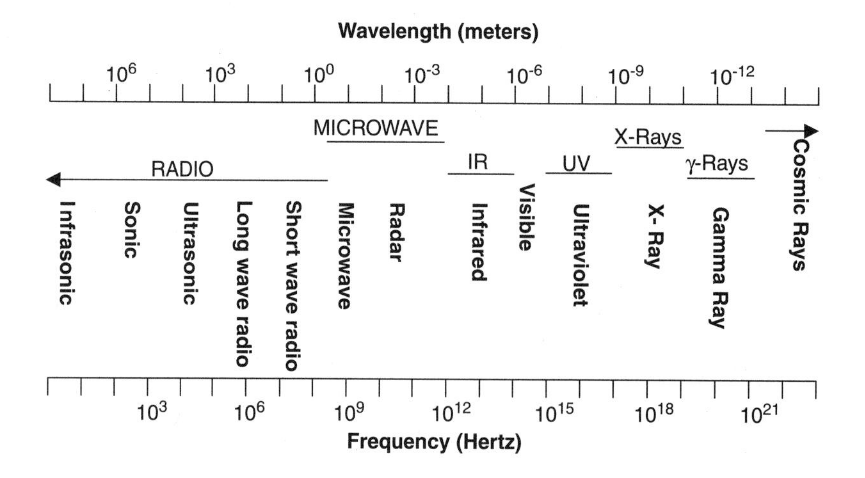

FIGURE 19.13 The frequency spectrum chart.

Spectrum Specifications

Spectrum specifications that are allocated by the FCC are generally accepted as the norm by other organizations and countries. Although there are many other organizations concerned with the allocation of spectrum, the FCC will be examined in detail here. The FCC is an independent United States Government agency, directly responsible to Congress. The Office of Engineering and Technology (OET) advises the Commission concerning engineering matters. The main mission of the OET is to manage the spectrum and create new opportunities for competitive technologies and services.

The FCC sets limits on the amount of electromagnetic radiation allowed to be emitted from commercial electronic equipment. Any electronic device capable of emitting radio frequency (RF) energy by radiation or conduction is defined as a RF device and is required to comply with the standards. The relevant limits and some measurement techniques, along with equipment authorization procedures, are given in the *Code of Federal Regulations* (CFR), Title 47 on telecommunications. The FCC rules are located in Parts 2, 5, 15, and 18 of the Title; these are:

Part 2 — Frequency allocations and radio treaty matters; general rules and regulations
Part 5 — Experimental radio service (other than broadcast)
Part 15 — RF devices
Part 18 — Industrial, scientific, and medical equipment

Title 47 — Part 15 contains three categories of equipment: (1) *incidental radiators* (such as dc motors, mechanical light switches, etc.) that are not subject to FCC Part 15 emission control requirements; (2) *unintentional radiators* that are RF devices that intentionally generate RF energy for use within the device, but they are not intended to emit RF energy outside the device by radiation or induction; and (3) *intentional radiators* that are RF devices that intentionally radiate RF energy by radiation or by induction.

In Title 47, five different types of equipment authorization procedures are listed: type acceptance, type approval, notification, certification, and verification. Restrictions are placed on the marketing and sale of all RF devices until the appropriate equipment authorization criteria are met.

Devices and systems that require allocation of the frequency spectrum fall under either type acceptance or type approval equipment authorization procedures. These are the RF devices that usually radiate high powers as in the radio and television broadcast transmitters. RF devices not within the allocated part of the RF spectrum would require either certification or verification equipment authorization. Some receivers, such as pagers, require notification and equipment authorization as well. It is recommended that the CFR, Title 47, be consulted for the compliance if the device in question intentionally radiates energy in the listed bands. Requirement for these types of devices are detailed in Part 15 of the CFR and, summarized in Table 19.3.

Electromagnetic Compatibility

Electromagnetic compatibility is the ability of a device or system to function satisfactorily without introducing intolerable disturbances on other devices or systems that are operating simultaneously in the same environment. The EMC has two main components: (1) EMI, which is the ability of a device or a system to operate without emitting too much electromagnetic energy that may cause harm to other devices; and (2) electromagnetic susceptibility (EMS), which is the ability of a device or system to reject interference and operate without being interfered by others. Susceptibility implies immunity, that is, a nonsusceptible device is immune to interference. Mathematically, the EMC can be expressed as

$$\mathrm{EMC} = \mathrm{EMI} + \mathrm{EMS}$$

Most practical EMC problems fall into four main key areas: emission, disturbances, RF interface, and electrostatic discharges.

TABLE 19.3 *Code of Federal Regulations*, Title 47, Requirements for Operations at Various Frequency Bands

Part	Operational Bands
Part 15.217	Operation in the band 160–190 kHz
Part 15.219	Operation in the band 510–1705 kHz
Part 15.221	Operation in the band 525–1705 kHz
Part 15.223	Operation in the band 1.705–10 MHz
Part 15.225	Operation in the band 13.553–13.567 MHz
Part 15.227	Operation in the band 26.96–27.28 MHz
Part 15.229	Operation in the band 40.66–40.70 MHz
Part 15.231	Periodic operation in the band 40.66–40.70 MHz and above 70 MHz
Part 15.233	Operation within the bands 43.71–44.49 MHz, 46.60–46.98 MHz, 48.75–49.51 MHz, and 49.66–50.0 MHz
Part 15.235	Operation within the band 49.82–49.90 MHz
Part 15.237	Operation within the bands 72.0–73.0 MHz, 74.6–74.8 MHz, and 75.2–76.0 MHz
Part 15.239	Operation in the band 88–108 MHz
Part 15.241	Operation in the band 174–216 MHz
Part 15.242	Operation in the band 174–216 MHz and 470–668 MHz
Part 15.243	Operation in the band 890–940 MHz
Part 15.245	Operation in the bands 902–928 MHz, 2435–2465 MHz, 5785–5815 MHz, 10,500–10,550 MHz, and 24,075–24,175 MHz
Part 15.247	Operation within the bands 902–928 MHz, 2400–2483.5 MHz, and 5725–5850 MHz
Part 15.249	Operation within the bands 902–928 MHz, 2400–2483.5 MHz, 5725–5875 MHz, and 24.0–24.25 GHz
Part 15.251	Operation within the bands 2.9–3.26 GHz, 3.267–3.332 GHz, 3.339–3.3458 GHz, and 3.358–3.6 GHz
Part 15.253	Operation within the bands 46.7–46.9 GHz and 76.0–77.0 GHz
Part 15.255	Operation within the bands 57–64 GHz
Part 15.321	Specific requirements for asynchronous devices operating in the 1910–1920 MHz and 2390–2400 MHz bands
Part 15.323	Specific requirements for asynchronous devices operating in the 1920–1930 subband

Emission refers to electromagnetic energy originating within a device or a system that can interfere with other equipment or systems. Emissions are best addressed at the design stages of equipment. Strategies include suitable printed circuit board design techniques, filtering of unwanted frequencies, using shielded cables and enclosures, and so on.

Disturbances are caused in the power distribution systems and the mains that take many different forms, such as short transients, frequency and waveform variations, fast transients, RF on the power lines, etc. Some of the solutions can include proper grounding, use of power filters, and installation of transient protectors.

Electrostatic Discharge (ESD) refers to the sudden discharge of electrical charges that occur after a gradual buildup. Static discharges are in the form of electric current, and almost instantaneous. Extremely fast discharge results in emission of high frequency electromagnetic energy that spans well into the ultra high frequency (UHF) ranges. For example, a 1-nsec discharge can cause electromagnetic waves with frequencies over 300 MHz. Because of intense electromagnetic fields, the ESD is known to severely upset many devices and systems located at distances of 5 to 10 m. Solutions to ESD problems may include transient protection, high-frequency filtering, proper cable and enclosure shielding.

Radio Frequency Interference (RFI) is a very common problem. The RFI interference problem is expected to grow due to a wide proliferation of handheld communication devices worldwide. Also, wider application of wireless local area networks (LANs) is likely to intensify this problem still further. Solutions to the RFI may include high frequency filtering, using shielded cables, effective shielding of the enclosures, and so on. Analogue devices are known to demonstrate a particular vulnerability to RFI. The RF interface can be viewed to take place in two forms, these being *near-field* and *far-field*.

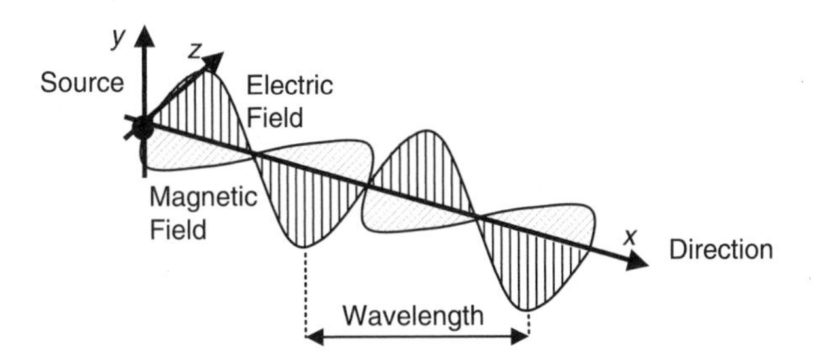

FIGURE 19.14 Propagation of plane waves.

Near-field (also known as electric and magnetic fields) are located less than about 1/6 wavelength from the source. In this region, the reactive energy dominates and it can be divided into electric fields and magnetic fields, both dependent on the source and circuit impedance.

Far-field (also known as the plane wave field) is located greater than about 1/6 wavelength from the source. Plane waves have electric field and magnetic field components propagating as shown in Figure 19.14. Once this takes place, the propagation is independent of the source and has constant wave impedance (ratio of electric field intensity to magnetic field intensity) of 377 Ω in free space.

Determination of EMC

A common EMI problem is gathering information about the characteristics of the electromagnetic waves and organizing data. For this purpose, the *source–path–receptor* model (Figure 19.15) is the most popular method. The three elements of this method are:

1. *Source* is the emitter of energy
2. *Receptor* is the device affected by the energy emitted by the source
3. *Coupling path* is the link (media) between the source and the receptor

For EMI to take place all three elements must exist at the same time. The objectives of the EMC standards and regulations are to minimize the unwanted generation of energy, block the energy from reaching the receptor, and to minimize the susceptibility or increase the immunity of the receptor.

In the determination of the EMC, the key metric is the field intensity measured in volt/meter (V/m). In its simplified form, the intensity may be expressed as a function of transmitter power, and distance from the emitting source, as

$$E = 5.5\sqrt{PA}/d\,(\mathrm{V/m})$$

where P is the transmitter power in watts, A is the antenna gain, and d is the distance from the source.

As an example, the field intensity at a distance of 1 m from a 1 W transmitter with antenna gain of 1 is about 5.5 V/m, while the electric field from a 10 kW broadcast station at a 1-km distance can be calculated to

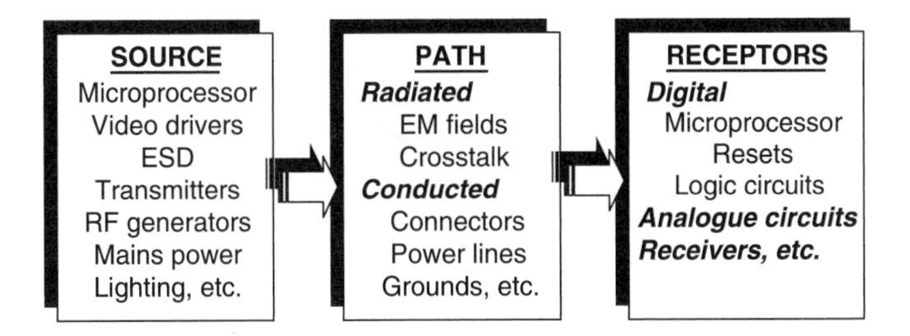

FIGURE 19.15 Three elements of EMC and examples.

be 0.55 V/m. This example shows that the distance of the receptor from the offending source is very significant.

Shielding: Many systems require shielding to meet emission and immunity requirements. Shielding involves two independent mechanisms, reflection and absorption. In reflection, an electromagnetic wave bounces off the surface. In absorption, the electromagnetic wave penetrates the material and is absorbed as it passes through.

Although much more complex, the effectiveness of shielding can be expressed simply by

$$\mathrm{SE}\ (\mathrm{dB}) = R\ (\mathrm{dB}) + A\ (\mathrm{dB})$$

where *SE* is the total shielding effectiveness in dB, *R* and *A* are the reflections and absorptions, respectively.

Reflection is the key mechanism for high frequency shielding (emission, RFI, ESD), while absorption is the main mechanism for low frequency shielding.

EMC Standards, Regulations, and Organizations

Electromagnetic radiation is a form of energy that can propagate through a medium. It can be generated intentionally or unintentionally and can interface with electronic systems, instruments, and other working equipment, thus degrading their performance. Therefore standards and regulations are set by many organizations to control the EMC problems. The related standards and regulations can be divided into three major groups; EMC in general, emission, and immunity. The application areas of the specific groups may differ greatly, as listed below:

(a) Domestic radio and television receivers
(b) Domestic appliances and household electronic equipment
(c) Mobile radio equipment
(d) Mobile radio and commercial radiotelephone equipment
(e) Radio and television broadcast transmitters
(f) Information technology equipment
(g) Telecommunications networks and equipment
(h) Educational electronic equipment
(i) Medical and scientific equipment
(j) Aeronautical and marine radio equipment
(k) Industrial manufacturing equipment
(l) Lights and fluorescent lamps, etc.

Each one of these specific groups may further be divided into application areas, e.g.:

- Vehicles
- Earth moving machinery
- Space systems
- Agricultural and forestry machinery
- Lifts, escalators, conveyors, etc.

Some EMC standards and regulatory organizations specialize with a particular group of applications while others have much wider interests. In general, the organization responsible for EMC can be divided into three main groups: (1) international organizations, (2) regional organizations, and (3) national and professional organizations.

International Standards Organizations

There are a number of international organizations that are involved in EMC standards, some of which include the International Electrotechnic Commission (EIC), International Standards Organization (ISO), International Telecommunication Union (ITU), and International Organization of Legal Metrology (OIML). Another organization of importance is the International Special Committee on Radio Interference (CISPR), which was

TABLE 19.4 Examples of Recent EMC-Related Standards from the IEC and CISPR

Standard Number	Related Title
IEC 61204-3-3:2000	Low voltage supplies, dc output
IEC 61000-1-2:2001	General, metrology for safety
IEC 61000-3-3:2002	Limitations of voltage fluctuations and flicker
IEC 60974-10-10:2002	Arc welding equipment
CISPR 61000-6-3:1996	Emission standards for residential, commercial, and light industrial environments
CISPR/TR 16-3:2000	Specifications for radio disturbance and immunity measuring apparatus and methods

established in 1934 by a group of international organizations to address radio interference. CISPR is a nongovernmental group composed of National Committees of the IEC. The operations of the IEC and ISO will briefly be explained next.

The IEC specifies the general conditions or rules necessary for achieving electromagnetic compatibility. The IEC 61000 series deals with terminology, descriptions of electromagnetic phenomena, measurement and testing techniques, and guidelines on installation and mitigation. The IEC standards consist of nine parts as follows:

Part 1: General
- General considerations (introduction, fundamental principles, safety)
- Definitions, terminology

Part 2: Environment
- Description of the environment
- Classification of the environment
- Compatibility levels

Part 3: Limits
- Emission limits
- Immunity limits

Part 4: Testing and measurement techniques
- Measurement techniques and testing techniques

Part 5: Installation and mitigation guidelines
- Installation guidelines
- Mitigation methods and devices

Part 6: Generic standards
Part 7: Open
Part 8: Open
Part 9: Miscellaneous

Among many others, some of the recent publications of the IEC and CISPR are given in Table 19.4.

The ISO forms committees and teams from national delegations nominated by the member institutes of the countries concerned with the EMC. The ISO rules state that the member institutions are expected to take account of the views of a broad range of interested parties and other stakeholders so that acceptance of the proposal becomes a consolidated and national consensus.

Some of the recent publications of the ISO on the EMC are given in Table 19.5.

Regional Standards Organizations

The main regional standards organizations concerned with the EMC are from the European Union (EU). Recently, European nations have introduced the European Standard Organization (ESO) to unify access to information on standards and the routes into the standardization process. The ESO plays a significant role in aligning commercial, judicial, and financial objectives within the community. There are 25 countries that participate in the EU who have agreed to the common regulatory requirements placed on commercial products.

TABLE 19.5 Examples of Recent EMC-Related Standards from the ISO

Standard Number	Related Title
ISO 7637-1-1:2002	Road vehicles, electric disturbance
ISO 13766:1999	Earth moving machinery
ISO 14302:2002	Space systems

TABLE 19.6 Examples of Recent EMC-Related Standards from the EU Organizations

Standard Number	Related Title
EN 50065-2-1:2003	Signaling on low-voltage electrical installations
EN 50083-2:2001	Cabled networks for television signals
EN 50370-2:2003	Product family standards for machine tools
EN 55020:2002	Sound and television broadcast receivers, associated equipment
EN 60730-2-9:2002	Automatic electrical controls for household and similar use
EN 60947-5-7:2003	Low voltage switchgear and controlgear
EN 61000-6-2:2001	Generic standards—immunity for industrial environment
EN 62052-11:2003	AC electricity metering equipment
EN 13309:2000	Construction machinery (from CEN)
EN 300 386 V.1.2.1	EMC radio spectrum matters (from ETSI)
ETS 300 673 V.1.2.1	Radio equipment and systems

These agreements, called directives, are listed in the *Official Journal of the European Council (OJEC)*. In a recent publication (31st March, 2004), the OJEC states that it is desirable to harmonize standards of all EU countries. To this end, the European Committee of Standardization (CEN), The European Committee of Electrotechnical Standardization (CENELEC), and the European Telecommunications Standards Institute (ETSI) are recognized as the main bodies for adoption and harmonization of standards in accordance with the general guidelines.

The EU approves standards on EMC with the prefix EN (from European Norm). The development of EMC standards is an ongoing process in all countries but Europeans face an additional problem of harmonization of practice common to all member countries. For example, the CENELEC 50000 series of standards deals with terminology and descriptions of standards, and measurement and testing techniques in particular applications, which have been adopted across the board.

The European organizations have many publications on the EMC standards generated from CENELEC, CEN, and ETSI, some of which are listed in Table 19.6.

There are other regional organizations such as the Association of Southeast Asian Nations (ASEAN), but they are not as effective as the European organizations on EMC matters.

National and Professional Organizations

Almost every nation has its own standards organization concerned with EMC. In the U.S., the FCC and the military (MIL) are the two regulating bodies for the allocation of spectrums. In the U.S., there are also other organizations such as the American National Standards Institute (ANSI), the Institute of Electrical and Electronics Engineering (IEEE), and others who are directly involved in EMC standards. In Europe, each country has its own EMC governing body as well as its own standards. *Deutsches Institut für Normung* (DIN) in Germany, British Standards Institute (BSI) in Britain, and *Ente Nazionale Italiano di Unificazione* (UNI) in Italy are among these. Organizations such as the Standards Council of Canada (SCC), Standards Australia (SAA), and the China Electronics Standardization Institute (CESI) are other organizations that are scattered around the globe. Here, only few examples will be provided.

The Australian EMC/EMI standards closely follow CISPR models. Since 1992, Australia and New Zealand have adopted joint national standards that carry the prefix AS/NZS. The newer EMC standards are generally similar to those of CISPR or CENELEC, with some deviations. In Australia, empowered by the Radiocommunications Act of 1992, the Spectrum Management Agency (SMA) had been developing EMC guidelines in conjunction with other agencies such as the Civil Aviation Authority as well as the local industry. Some of the recent Australian/New Zealand publications on the EMC standards are given Table 19.7.

The Canadian EMC/EMI standards are controlled by the Standards Council of Canada (SCC). A particular branch of the SCC, the Canadian Electrical Code, is dedicated to harmonized electrically related standards. Some of the recent Canadian publications on the EMC standards are listed in Table 19.8.

The U.S. EMC/EMI standards are developed by various government, private, and professional organizations such as the American National Standards Institute (ANSI), the Institute of Electrical and Electronics Engineering (IEEE), and several other bodies. ANSI, for example, is a private, nonprofit organization that administers and coordinates U.S. voluntary standardization and conformity assessment systems. It has approximately 1000 companies, organizations, government agencies, institutional, and international members. The IEEE, however, is a professional organization with visions of promoting the engineering processes of creating, developing, integrating, sharing, and applying knowledge about electronic and information technologies and sciences. ANSI and IEEE are active in developing and maintaining EMC standards, some of which are listed in Table 19.9.

The U.S. Military has its own standards and requirements related to EMI and susceptibility, as described in the MIL-STD-461C documents. Electromagnetic interference is defined as the radiated and conducted energy emitted from any device. Electromagnetic susceptibility is defined as the amount of radiated or conducted energy that the device can withstand without degrading its performance.

The standards are broken down into 17 segments defined by a two-letter suffix code and followed by three numbers ranging from 101 to 999 in the requirement name. The letter codes are: *conducted emissions* (CE), *conducted susceptibility* (CS), *radiated emission* (RE), and *radiated susceptibility* (RS). Table 19.10 is a list of descriptions of the different emission and susceptibility requirements for a particular branch or type of application.

TABLE 19.7 Examples of Recent EMC-Related Standards from Australia and New Zealand

Standard Number	Related Title
AS/NZS 1044:1995	Limits and disturbances of electric motor operated and thermal appliances for household and similar purpose apparatus (EN55014 and CISPR 14)
AS/NZS 2557:1999	Limits and measurements of radio interface of vehicles, motor boats, etc. (CISPR 12)
AS/NZ 4251.1:1999	Generic mission standard — residential, commercial, and light industry (from EN50081.1 and IEC 61000-6-3)

TABLE 19.8 Examples of Recent EMC-Related Standards from Canada

Standard Number	Related Title
CAN/CSA C108.1.1:1999	Electromagnetic interference measuring instrument—CISPR type
CAN/CSA C108.1.1:1999	Electromagnetic interference measuring instrument—CISPR type
CAN/CS-CEI/IEC 61000-4-9: 2001	Testing and measurement techniques

TABLE 19.9 Examples of Recent EMC-Related Standards from the US Organizations

Standard Number	Related Title
ANSI C37.32:2002	American national standards for switchgear
ANSI C63.4:2000	Methods of measurement of radio-noise emissions from low-voltage electrical and electronic equipment, range 9 kHz to 40 GHz
ANSI C63.12:1999	American national standard for recommended practice for EMC limits
IEEE 644:2001	IEEE standard procedure for measurement of power frequency electric and magnetic fields from ac power lines
IEEE C37.41:2000	IEEE standard design for high-voltage fuses, etc.
IEEE C63.4:2000	Methods of measurement or radio-noise emission from low-voltage electric and electronic equipment in the range of 9 kHz to 40 GHz
IEEE C63.14:1998	Dictionary for EMC, electromagnetic pulse (EPM), and electrostatic discharge (ESD)

TABLE 19.10 Classification of Requirements for Different Platforms and Installations

CE101	Conducted emissions, power leads, 30 Hz to 10 kHz
CE102	Conducted emissions, power leads, 10 kHz to 10 MHz
CE106	Conducted emissions, antenna terminal, 10 kHz to 40 GHz
CS101	Conducted susceptibility, power leads, 30 Hz to 50 kHz
CS103	Conducted susceptibility, antenna port, intermodulation, 15 kHz to 10 MHz
CS104	Conducted susceptibility, antenna port, rejection of signals, 30 Hz to 20 GHz
CS105	Conducted susceptibility, antenna port, cross modulation, 30 Hz to 20 GHz
CS109	Conducted susceptibility, structure current, 60 Hz to 100 kHz
CS114	Conducted susceptibility, bulk cable injection, 10 kHz to 400 MHz
CS115	Conducted susceptibility, bulk cable injection, impulse excitation
CS116	Conducted susceptibility, damped sinusoidal transients, cables, and power leads, 10 kHz to 100 MHz
RE101	Radiated emission, magnetic field, 30 Hz to 100 kHz
RE102	Radiated emission, electric field, 10 kHz to 18 GHz
RE103	Radiated emission, antenna spurious and harmonic outputs, 10 kHz to 40 GHz
RS101	Radiated susceptibility, magnetic field, 30 Hz to 100 kHz
RS103	Radiated susceptibility, electric field, 10 kHz to 40 GHz
RS105	Radiated susceptibility, transient electromagnetic field

TABLE 19.11 RE102 Applicability and Frequency Band of Testing

Applied For	Frequency Band
Ground	2 MHz to 18 GHz
Ships, surface	10 kHz to 18 GHz
Submarine	10 kHz to 1 GHz
Aircraft (Army)	10 kHz to 18 GHz
Aircraft (Air Force and Navy)	2 MHz to 18 GHz

TABLE 19.12 Examples of Recent EMC Related Standards from the U.S. Military

Standard Number	Related Title
MIL-HDBK-235/1B:2000	Electromagnetic environment considerations for design and procurement of electrical and electronic equipment, subsystems, and systems
MIL-HDBK-241B	Design guide for EMI reduction in power supplies
MIL-HDBK-1857	Grounding, bonding, and shielding design practices
MIL-STD-220B	Method of insertion loss measurement
MIL-STD-464A:2002	Electromagnetic environmental effects for systems
MIL-STD-469B:1996	Radar engineering interface requirements, EMC metric
MIL-STD-1541A	EMC requirements for space systems
MIL-STD-1818A	Electromagnetic effects requirements for systems

Testing requirements of equipment in each segment are different. For example, from Table 19.6, RE102 encompasses electric fields in the frequency bands of the testing requirements, typically from 10 kHz to 18 GHz, depending on the clock frequency of the device or types of platforms (described in Table 19.11). Up to 30 MHz, only the vertical polarization of the electric field will be measured and compared with the limits. Above 30 MHz, horizontal and vertical field components must be measured and again compared with the limits. The device is in conformance with RE102 if the electric field intensity in the appropriate frequency band is less than the prescribed limit.

Some U.S. military standards are given in Table 19.12.

Measurement Techniques

In many applications, the measurement of electrical and magnetic fields to determine the intensity of radiated and conducted coupling is essential. The maximum stray field strengths that cause no EMI are incorporated in the standards and regulations explained above. The procedure for determining emission and susceptibility levels is well established. There are many private and government organizations around the globe that are qualified to conduct measurements. Each country may adopt its own measurement guidelines.

Measurement of electromagnetic radiation is regulated by various standards, such as CISPR 22 of the IEC, FCC Part 15 of the U.S., ICES 003 of Canada, VCCI-V series of Japan, EN 55022 and IEC/CISPR 22 of the EU, AS 3548 of Australia, CNS 13438 of China, etc. Measurements are for conducted emission, radiated emission, ESD and transient immunity, surge immunity, and RF immunity, and are necessary for: (1) diagnostic purposes, (2) precompliance requirements, (3) full compliance requirements, and (4) production purposes.

Electromagnetic radiation has two perpendicularly propagating components: the electric field and the magnetic field. The electric field, E, is normally expressed in volts per meter (V/m). The magnetic field, H, is expressed in amperes per meter (A/m). A source may generate electric and magnetic fields. In the near-field the ratio of E and H varies greatly, approaching infinity for an electric field source or zero for a magnetic field source. For example, near a magnetic field source, the electric field is weak, but becomes stronger at distances larger than 1/6th if the wavelength, the ratio of E and H approaches to a constant 377 Ω. This is also applicable for an electric field source.

Electromagnetic emission measurements can be conducted for near-field and far-field. In the near-field regions, E and H, the fields must be measured separately. The far-field strengths are normally measured in terms of E fields. Alternatively, the far-field strength may be specified in terms of *power density*, expressed in watts per square meter. The power density denotes the amount of radiated power passing through each square meter of a surface perpendicular to the direction away from the source. The *peak-power density* is known to be equal to $E \times H$, and, for a sinusoidal source, the *average power density* is half of this value. For sinusoidal sources, each frequency component must be considered separately. The total average power can be determined as the sum of the average powers for all frequencies. It follows that the total average power is $P = E^2/377$ W.

An important factor that merits attention during the far-field measurements is *polarization*. The propagating electromagnetic fields may be oriented at many different angles with respect to the surface of the Earth. The direction of the electric field is called the ***polarization*** of the wave, which may be vertical, horizontal, or something in between. The wave may be elliptically polarized, which may result from two waves that are not exactly in phase. If the waves are equal in magnitude and exactly 90° out of phase, the wave is circularly polarized. To account for all these cases, the field must be checked separately for vertically and horizontally polarized waves.

In general, conducted emissions and radiated emissions measurements are taken below and above the 30 MHz mark, respectively. Although this mark is not absolute, spurious measurements can occur. At higher frequencies, above 30 MHz, resonance becomes dominant in typical cable lengths, disturbing conducted emission measurements. In the case of radiated emissions measurements, it is easier to capture the far-field characteristics as the equipment under test (EUT) becomes an effective radiator. At lower frequencies, below 30 MHz, near-field probes are required to target hot-spots, as capturing such complex radiation field patterns can be very involved and difficult.

There are three important considerations in electromagnetic field measurement: the measurement environment, the antenna and backup equipment, and software (illustrated in Figure 19.16).

Measurement Environment

A major difficulty in electromagnetic field measurement is the repeatability of results. Electromagnetic fields are affected by materials in the vicinity; therefore the measurement environment must carefully be defined and maintained for repeatability.

Measurements can be conducted in large outdoor areas, called *open-field sites*. Assuming that the conductivity, permittivity, and permeability of the Earth are constant, every open-field site would have the same effect on electromagnetic fields radiated from the equipment under test (EUT). Once compensated for the reflections by using ground planes, the radiated energy must be measured in all directions from the EUT, at various angles of inclination of the receiving antenna. However, open-field measurements are prone to interference from other sources emitting electromagnetic energy. To avoid interference from stray sources, tests can be conducted inside shielded enclosures.

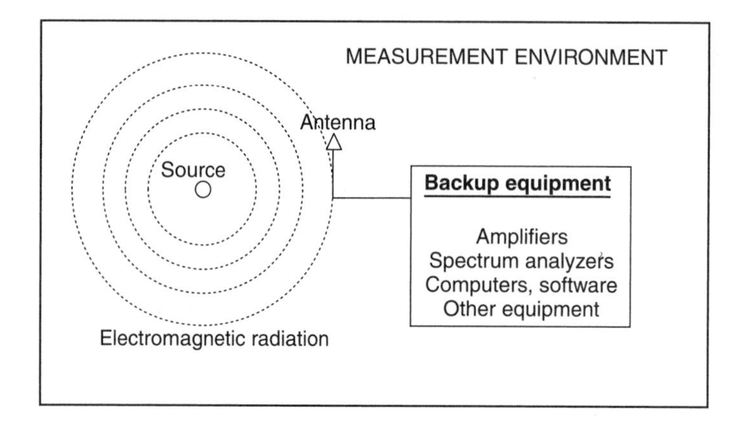

FIGURE 19.16 An electromagnetic measurement setup.

An ideal test environment is a shielded enclosure lined with material that does not reflect electromagnetic waves. Such an enclosure is called an *anechoic chamber.* Another type of test chamber is the *transverse electromagnetic cell* (TEM), which consists of an enlarged section of waveguide. TEMs are suitable for testing small devices at relatively low frequencies. Shielded rooms also are used in measurement.

Antennas

Antennas are simple conductors that respond to electric fields and provide voltage and current at the terminals. There are many different types, such as dipole antennas, biconical antennas, log-periodic antennas, spiral antennas, etc. Most antennas are linearly polarized since they respond to only one polarization component of the propagating wave. That is, if the antenna is oriented horizontally, only the horizontally polarized component of the wave will affect it. The correct use of antennas in electromagnetic measurements is very important since antenna factors, matching, balancing, and such play important roles in the accuracy of the results.

Backup Equipment and Software

Backup equipment for electromagnetic measurements mainly consists of amplifiers, tracking generators, spectrum analyzers, quasi-peak detectors, near-field probes, field strength meters, computers, and appropriate software support. All the necessary components are commercially available from many companies. However, correct selection of equipment is necessary as they have different power handling capacities and frequencies of operation, and other characteristics. Here, only the spectrum analyzer and quasi-peak detectors will briefly be explained.

The spectrum analyzer: A standard EMC test measurement generally requires a measuring receiver or spectrum analyzer. Spectrum analyzers are useful devices to quickly and reliably scan a range of frequencies during diagnostic tests. They are considerably cheaper than measuring receivers but lack dynamic range and are prone to overload. A typical spectrum analyzer has the following features: frequency range 9 kHz to 22 GHz, maximum resolution 30 Hz, noise level -102 to -127 dBm.

The performance of a spectrum analyzer can be facilitated with a preselector, also known as a preamplifier, connected to the front end. A preselector has input protection, preamplification, and a sweep-tuned filter locked to the local oscillator of the spectrum analyzer. Some of the more sophisticated spectrum analyzers come with a built-in tracking generator or with an option to incorporate one into the system. A tracking generator is basically a signal generator in which its output frequency tracks the measurement frequency and sweeps at the same rate. The measurement uncertainty is typically ± 1 dB over the range 100 kHz to 1 GHz.

The quasi-peak detector is a peak detector with weighted charge and discharge times. A peak detector functions by responding to the peak signal value and discharges at a fairly rapid rate, and usually applies to

TABLE 19.13 The CISPR 16 Quasi-Peak Definition

	Frequency Bands			
	A	B	C	D
	9 kHz–150 kHz	150 kHz–30 MHz	30 MHz–300 MHz	300 MHz–1 GHz
Quasi-Peak				
6-dB bandwidth (kHz)	0.2	9	120	
Charge time constant (msec)	45	1	1	
Discharge time constant (msec)	500	160	550	
Pre-detector overload factor (dB)	24	30	43.5	

military standards. Another detector commonly used in RF measurements is the average detector that averages the measured signal. CISPR derivative EN55022 standards entail average detection on conducted emissions.

As expected, an interference signal measured at a given frequency relies on the bandwidth of the receiver and response of its detector. As an example, commercially based CISPR emissions standards as in CISPR 16 define four measurement bands between 9 kHz and 1 GHz for quasi-peak detection. A summary is given in Table 19.13.

Other Instruments are now available to fulfill the functionality of the spectrum analyzer. Instrumentation manufacturers allow for add-on options to some of their models. For instance, there are oscilloscopes that are able to convert the measured signal into its frequency representation. These devices are handy for quick diagnostic purposes but have limitations in terms of stability, bandwidth, and sensitivity.

Most manufacturers offer complete test systems. A typical set has the following features:

- Antenna and cable test set with spectrum analyzer
- Frequency range 9 kHz to 3 GHz
- Broadband antenna and preamplifier
- E field and H field probes
- Interfering signal feature allows identification and type of interference
- RS-232 port to interface with printers and computers
- Analysis software, Smith chart tools, etc.

Conclusions

Because of the increasing number of man-made EMI generated devices and systems, the problems caused by the intentional and unintentional emission of electromagnetic energy have led numerous organizations worldwide to set standards and regulations. Bodies such as the FCC set limits on the amount of electromagnetic emissions, at certain frequency bands, from commercial electronic equipment. EMC/EMI can be analyzed by the *source–path–receptor* model and carefully conducted measurements can reveal information on the emission levels and susceptibility of receptors. There are well-established test procedures and a diverse range of equipment to conduct measurements.

Defining Terms

Conducted emission: An RF current propagated through an electrical conductor.

Emission: Electric energy emanating from a source.

Far-field: The region where the ratio of the electric to the magnetic field is approximately equal to 377 Ω.

Field strength: An amount of electric or magnetic field measured in the far-field region and expressed in volt/meter or amps/meter.

Immunity: Ability of equipment to reject interference.

Near-field: Any location less than 1/6th wavelength from the source.

Polarization: The direction of electric field of an electromagnetic wave.

Power density: Radiated power per unit of cross-sectional area.

Radiated emission: An electromagnetic field propagated through space.
Susceptibility: Vulnerability of electronic equipment to external sources of interference.

Abbreviations

ANSI: American National Standards Institute
ASEAN: Association of Southeast Asian Nations
BSI: The British Standards Institute
CFR: Code of Federal Regulations
CE: Conducted emissions
CEN: The European Committee of Standardization
CENELEC: The European Committee of Electrotechnical Standardization
CESI: The China Electronics Standardization Institute
CISPR: International Special Committee on Radio Interference
CS: Conducted susceptibility
CSS: Standards Council of Canada
DIN: Deutsches Institut für Normung, Germany
ELF: Extreme low frequency
EMC: Electromagnetic compatibility, the capability of two or more devices to operate simultaneously without mutual interference
EMI: Electromagnetic interference
EN: European Norms
EPM: Electromagnetic pulse
ESD: Electrostatic discharge that often follows a buildup of static charge
ETSI: European Telecommunications Standards Institute
EU: European Union
EUT: Equipment under test
FCC: The Federal Communications Commission
IEC: International Electrotechnical Commission
IEEE: The Institute of Electrical and Electronics Engineering
ISO: International Standards Organization
ITE: Information technology equipment
ITU: International Telecommunication Union
LISN: Line impedance stabilization network
OET: Office of Engineering and Technology
OIML: International Organization of Legal Metrology
OJEC Official Journal of the European Council
RE: Radiated emission
RF: Radio frequency
RFI: Radio frequency interface
RS: Radiated susceptibility
SAA: Standards Australia
SMA: Spectrum Management Agency
TEM: Transverse electromagnetic cell
UHF: Ultrahigh frequency
UNI: Ente Nazionale Italiano di Unificazione, Italy
MIL: The U.S. Military

Bibliography

American National Standards Institute, ANSI (http://www.ansi.org/).
British Standards Institution, BSI (http://www.bsi-global.com/).

H. Eren, *Electronic Portable Instruments—Design and Applications*, Boca Raton, Fla.: CRC Press, 2004.
IEEE Xplore (http://ieeexplore.ieee.org/).
International Electrotechnical Commission, IEC (http://www.iec.org/).
International Organization for Standardization, ISO (http://www.iso.org/).
National Standards Services Networks, NSSN (http://www.nssn.org/).
National Information Standards Organization, NISO (http://www.niso.org/).
Standards Council of Canada, SCC (http://www.scc.ca/).
Techstreet Thomson, standards catalogs (http://www.techstreet.com/).
J.G. Webster, Ed., *The Measurements, Instrumentation and Sensors Handbook*, Boca Raton, Fla.: CRC and IEEE Press, 1999.
World Standards Services Networks, WSSN (http://www.wssn.net/).

19.3 Lightning

Martin A. Uman

An understanding of lightning and of the electric and magnetic fields produced by lightning is critical to an understanding of lightning-induced effects on communication and electric power systems. This section begins with an overview of the terminology and physics of lightning; then statistics on lightning occurrences are given. Next, the characteristics of the electric and magnetic fields resulting from lightning discharges and currents are examined and the models used to describe that relationship are discussed. The section ends with a discussion of the coupling of the electric and magnetic fields from lightning to overhead wires.

Terminology and Physics

Lightning is a transient, high-current electric spark whose length is measured in kilometers. Lightning discharges can occur within a cloud, between clouds, from cloud to air, and from cloud to ground. All discharges except the latter are known as cloud discharges. The usual cloud-to-ground lightning is initiated in the cloud, has a duration of about half a second, and carries to ground some 20 to 30 Coulombs of negative cloud charge. A less frequent type of cloud-to-ground discharge, accounting for less than 10% of all cloud-to-ground lightning, also begins in the cloud but lowers positive cloud charge. An even less frequent type of cloud-to-ground lightning is initiated in an upward direction from tall man-made structures such as TV towers or tall geographical features such as mountain tops. A complete lightning discharge of any type is called a *flash*. The usual negative cloud-to-ground lightning flash starts in the cloud when a so-called preliminary breakdown, a particular type of electric discharge in the cloud, occurs. This process is followed by an electrical discharge, termed the *stepped leader*, which propagates toward the ground in a series of luminous steps having tens of meters length. In progressing toward the ground, the negatively charged stepped leader branches in a downward direction. When one or more leader branches approach within 100 m or so of the ground, after 10 to 20 msec of stepped leader travel at an average speed of 10^5 to 10^6 m/sec, the electric field at the ground (or at objects on the ground) increases above the critical breakdown field of the surrounding air and one or more upward-going discharges are initiated, starting the attachment process. After traveling a few tens of meters, one of the upward-going discharges, which is essentially at ground potential, contacts the tip of one branch of the stepped leader, which is at a high negative potential, probably some tens of megavolts. From that point, ground potential propagates upward, discharging to ground some or all of the negative charge previously deposited along the channel by the stepped leader. This upward propagating potential discontinuity is called the *return stroke*. Its front is a region of high electric field that causes increased ionization, current, temperature, and pressure as it travels up the 5-km or more length of the leader channel. That trip is made in about 100 μsec at an initial speed on the order of one third to one half the speed of light, the speed decreasing with height. The current at ground associated with the negative first return stroke has a peak of typically 30 kA achieved in a few microseconds, has a maximum current derivative of about 10^{11} A/sec and falls to half of peak value in some tens of microseconds. The cessation of the first return stroke current may or may not end the flash. If more cloud charge is made available to the first stroke channel by in-cloud discharges, another leader-return stroke sequence may ensue, typically after tens of milliseconds. Preceding and initiating a subsequent return stroke is a negatively charged continuous leader called a *dart leader*. The dart leader typically propagates down the residual channel of the previous stroke, generally ignoring the first stroke branches, although in about 50% of cloud-to-ground flashes there is at least one dart leader which transforms to a stepped leader on the downward trip, creating a new path to ground. There are typically three or four leader-return stroke sequences per negative cloud-to-ground flash, but ten or more is not uncommon.

Of the many different processes that occur during the various phases of a negative cloud-to-ground lightning (e.g., the in-cloud K processes, in-cloud J processes, and cloud-to-ground M components that occur between strokes and after the final stroke and that are not discussed here), the electric and magnetic fields associated with the return stroke described above generally are the largest and hence the most significant in inducing unwanted voltages in communication and electric power systems. This is the case because the currents in all other lightning processes are generally smaller than return stroke currents and the ground strike point of the return stroke can be much closer to objects on the ground than are in-cloud discharges. Cloud discharges exhibit currents similar to those of the in-cloud processes occurring in ground discharges and hence produce similar relatively small fields at or near ground level.

Positive flashes to ground, those initiated in the cloud and conducting positive charge to earth, generally contain only one return stroke, which is preceded by a "pulsating" leader rather than the stepped leader characteristically preceding negative first strokes and is generally followed by a period of continuous current flow. Positive flashes contain a greater percentage of very large return stroke currents, in the 100- to 300-kA range, than do negative flashes. Positive flashes may represent half of all flashes to ground in winter storms, which produce few total flashes, and typically represent 1 to 20% of the overall flashes in summer storms, that percentage increasing with increasing latitude.

A complete discussion of all aspects of the physics of negative and positive lightning flashes is found in Rakov and Uman (2003).

Lightning Occurrence Statistics

Lightning flash density is defined as the number of lightning flashes per unit time per unit area and is usually measured in units of lightning flashes, either cloud or cloud-to-ground or both, per square kilometer per year. The two most common techniques for directly measuring flash density are (1) the use of so-called flash counters, relatively crude devices which trigger on electric fields above a value of the order of 1 kV/m in a frequency band centered in the hundreds of hertz to kilohertz range, of which two models have been used extensively, the CIGRE 10-kHz and the CIGRE 500-Hz, and (2) the use of networks of wideband magnetic direction finders, networks of wideband time-of-arrival detectors, and networks combining the two technologies, such networks now covering the U.S., Canada, Japan, Korea, Taiwan, most of Europe, and parts of many other countries. The average flash density varies considerably with geographical location, generally increasing with decreasing latitude. Typical ground flash densities are 1 to 5 km^{-2} yr^{-1}, with the world's highest being 30 to 50 km^{-2} yr^{-1}. Significant variations in flash density are observed with changes in local meteorological conditions within distances of the order of 10 km, for example, perpendicular to and inland from the Florida coastline. The first ground flash density map for the U.S. obtained from the National Lightning Detection Network (NLDN) of 114 wideband magnetic direction finders, containing over 13 million ground flashes that occurred in 1989, is given by Orville (1991). A ten-year ground flash density map (1989 to 1998) for the US is published in Figure 2.11 of Rakov and Uman (2003). There are on average about 25 million cloud-to-ground flashes occurring annually over the contiguous US. Flash densities in the U.S. are maximum in Florida with 10 to 15 km^{-2} yr^{-1} and minimum along portions of the Pacific coast, which has essentially no lightning.

An extensively measured parameter used to describe lightning activity worldwide is the thunderday or isokeraunic level, T_D, the number of days per year that thunder is heard at a given location. This parameter has been recorded by weather station observers worldwide for many decades, whereas the relatively accurate direct measurement of flash density by networks of sensors (discussed above) has been possible only recently.

Commonly used relations to convert thunderday level to ground flash density N_g are of the form:

$$N_g = aT_D^b \text{ km}^{-2} \text{ yr}^{-1} \tag{19.6}$$

where the value of a is near, but usually less than 0.1 and the value of b is near, but usually greater than one. It should be noted that Equation (19.6) may be relatively inaccurate for a given location or year in that the data to which the formula is fit are highly variable. The literature contains more than ten different values of a and b determined in different studies.

Finally, from Earth-orbiting satellite measurements, it has been estimated that there are about 50 to 100 total flashes, cloud and cloud-to-ground, per second over the whole Earth. This number corresponds to an average global total flash density of 3 to 6 km^{-2} yr^{-1}. Rakov and Uman (2003) review the available satellite observations and in their Figure 2.12 give a flash density map for the world based on eight years of data from two satellites.

Electric and Magnetic Fields

For the usual negative return stroke, measurements of the vertical component of the electric field and the two horizontal components of the magnetic field at ground level using wideband systems with upper frequency 3-dB points in the 1- to 20-MHz range are well documented in the literature. Measured vertical electric field and horizontal magnetic field waveshapes are shown in Figure 19.17. Sketches of typical electric and magnetic fields are given in Figure 19.18 for lightning in the 1- to 5-km range and in Figure 19.19 for lightning at 10, 15, 50, and 200 km. Measured vertical and horizontal electric fields near ground are shown in Figure 19.20. The mean value of the initial peak vertical electric field, normalized to 100 km by assuming an inverse distance dependence, is about 7 V/m for negative first strokes and about 4 V/m for negative subsequent strokes.

The return stroke vertical electric field rise to peak comprises two distinguishable parts, evident in Figure 19.17: a slow front immediately followed by a fast transition to peak. For first strokes the slow front has a duration of a few microseconds and rises to typically half the peak amplitude, while for subsequent strokes the same slow front lasts less than 1 μsec and rises only to typically 20% of the peak. The mean 10 to 90% fast transition time is about 200 nsec regardless of stroke order for strokes observed over saltwater, where there is minimal distortion of the waveform due to propagation. The waveforms in Figure 19.17 have suffered distortion in propagation over land.

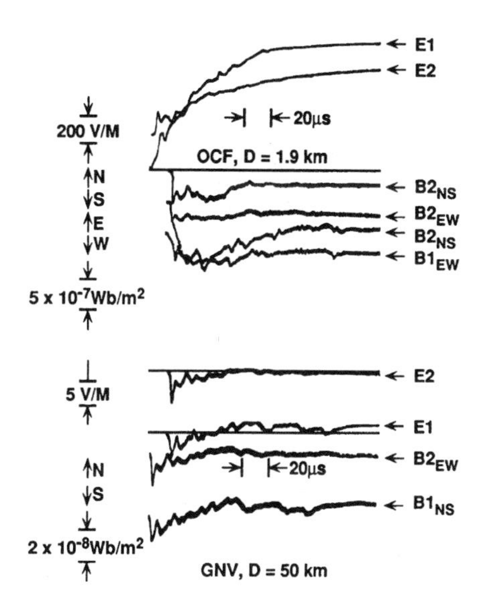

FIGURE 19.17 Simultaneously measured return stroke vertical electric field (E) and two horizontal magnetic flux densities (B_{EW} and B_{NS}) as observed about 2 and 50 km from a two-stroke flash, the first stroke being designated "one," the second "two." (*Source:* Adapted from Y.T. Lin et al., *J. Geophys. Res.*, vol. 84, pp. 6307–6314, 1979. With permission.)

After the initial field peak, the waveshapes of the vertical and horizontal magnetic fields for close lightning exhibit a valley followed by a hump in the case of the magnetic field and by a ramp in the case of the electric field, as is evident from Figure 19.17 to Figure 19.20. Relative to the amplitude of the initial peak, the hump and the ramp decrease with increasing distance to the return stroke. For distances of 25 km or greater, the ramp in the electric field is no longer significant, and for distances of 50 km or more, and for times of the order of 100 μsec, the waveshapes of the electric and magnetic fields are nearly identical, exhibiting a zero crossing and polarity reversal at some tens of microseconds. For very close distances, of the order of tens to hundreds of meters, triggered lightning (lightning initiated by firing small rockets trailing grounded wires toward thunderstorms) exhibits a return stroke electric field waveshape that is more or less a step function with a

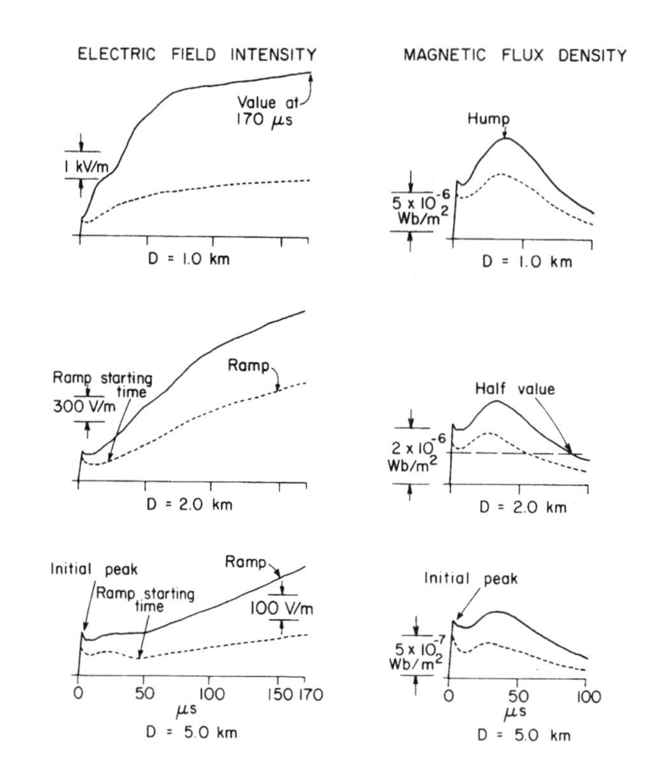

FIGURE 19.18 Drawings of typical return stroke electric fields and magnetic flux densities at 1, 2, and 5 km with definition of pertinent characteristic features. Solid lines represent first strokes, dotted subsequent strokes. (*Source:* Adapted from Y.T. Lin et al., *J. Geophys. Res.*, vol. 84, pp. 6307–6314, 1979. With permission.)

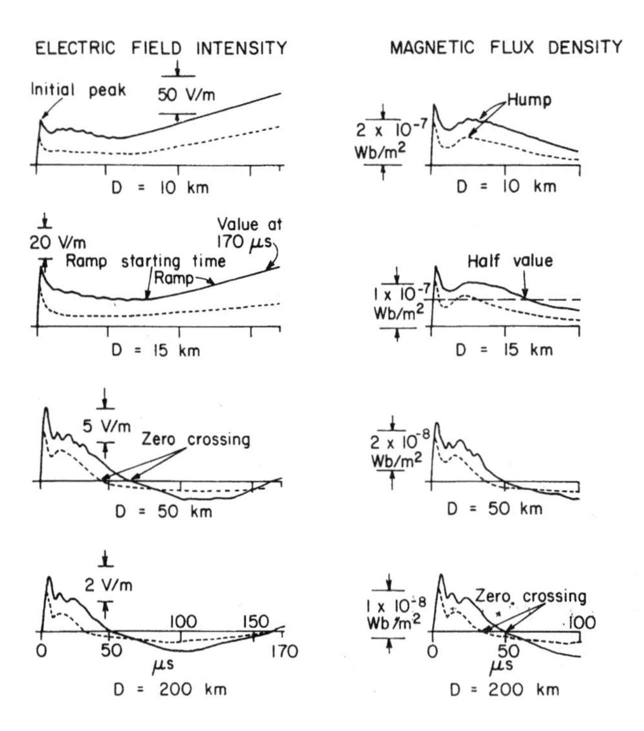

FIGURE 19.19 Drawings of typical return stroke electric fields and magnetic flux densities at 10, 15, 50, and 200 km; a continuation of Figure 19.20. (*Source:* Adapted from Y.T. Lin et al., *J. Geophys. Res.*, vol. 84, pp. 6307–6314, 1979. With permission.)

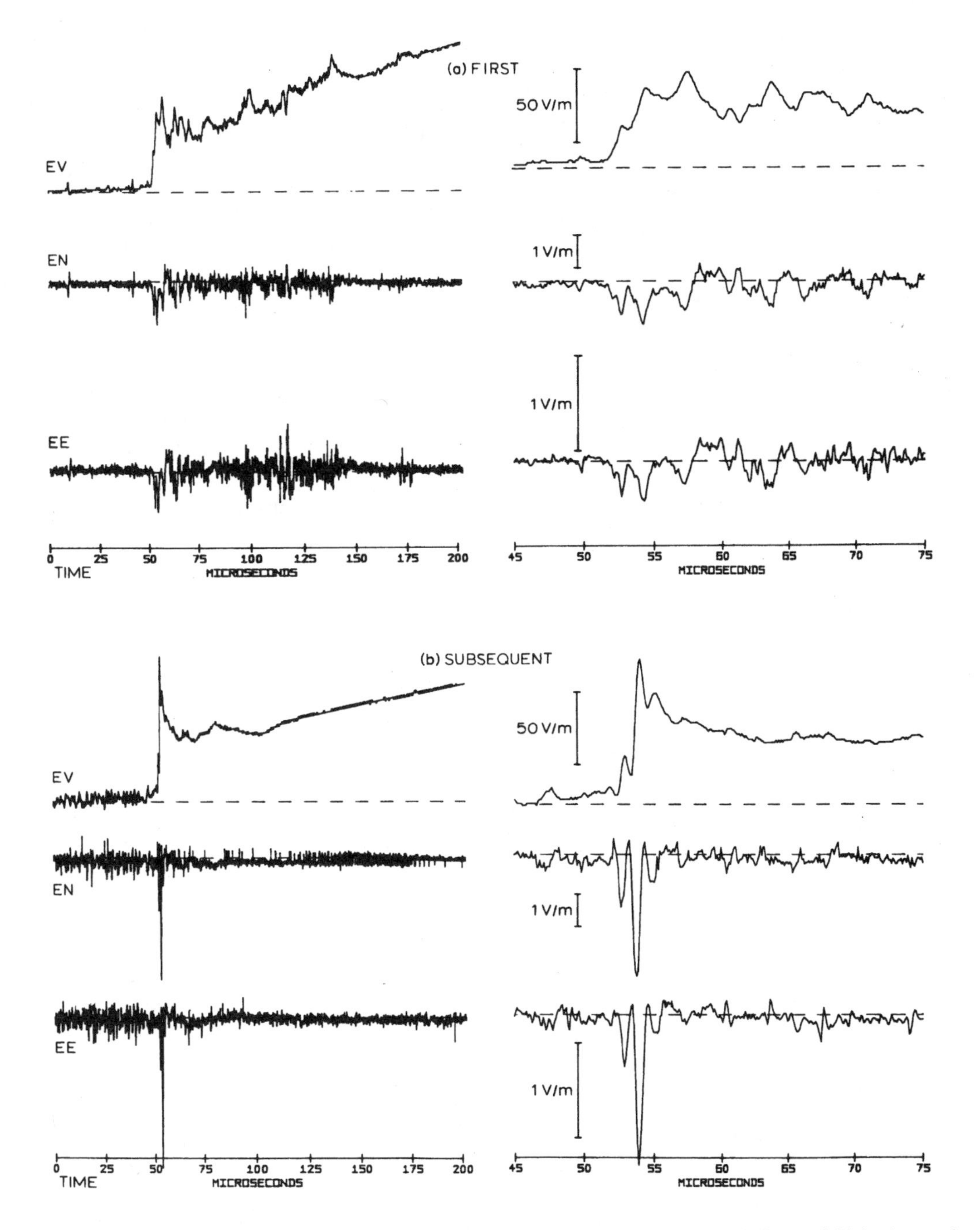

FIGURE 19.20 Measured horizontal electric field components (EN and EE) and vertical electric field (EV) 1 m above ground for a first stroke (a) and a subsequent stroke (b) at a distance of 7 km presented on two time scales. (*Source:* Adapted from E.M. Thomson et al., *J. Geophys. Res.*, vol. 93, pp. 2429–2441, 1988. With permission.)

risetime on the order of a microsecond and a magnetic field waveshape that is similar to the return stroke current waveshape at ground level. Very close natural return stroke fields are likely similar.

For positive return strokes, there are more very large peak currents at the channel base, in the 100-kA range, than for return strokes lowering negative charge to ground, although the median value for both positives and negatives is not much different (Berger et al., 1975). This observation is supported by measurements of the initial peak magnetic field from positive and negative return strokes made with magnetic direction-finding networks worldwide.

The horizontal component of the electric field has not been as well studied or characterized as the vertical. For the case of a finite-conducting earth and lightning beyond a few kilometers, Thomson et al. (1988) give wideband measurements of the three perpendicular components of the electric field about 1 m above ground level. An example is shown in Figure 19.20. The horizontal field waveshapes are more impulsive and vary on a faster time scale than their associated vertical electric field waveshapes. In fact, the horizontal field appears to be a crude derivative of the vertical. The peak amplitudes of the horizontal electric fields are on the order of 30 times smaller than those of the vertical fields for ground conductivities of the order of 10^{-2} mho/m, this ratio being roughly proportional to the square root of the ground conductivity. The horizontal field, although considerably smaller for distant lightning, can be as important as the vertical electric field in inducing voltages on an overhead horizontal wire because of the greater horizontal extent of the wire relative to its height above ground, a fact well established by recent research, whereas in the earlier literature on power line coupling, for example, only the vertical field was considered to be important.

The so-called wavetilt formula, given in Equation (19.7), models the ratio, in the frequency domain, of the horizontal to vertical electric field of a plane wave at the surface of Earth with conductivity σ and permittivity $\varepsilon_r\varepsilon_0$ for the case of grazing incidence and is certainly applicable to lightning return strokes occurring beyond a few kilometers, probably beyond a few hundred meters:

$$\frac{E_H(\omega)}{E_V(\omega)} = \frac{1}{\sqrt{\varepsilon_r + \dfrac{\sigma}{j\omega\varepsilon_0}}} \tag{19.7}$$

For closer lightning, the Cooray–Rubinstein formula (Cooray, 1992; Rubinstein, 1996) should be used

$$E_r(r, z, j\omega) = E_{rp}(r, z, j\omega) - H_{\psi p}(r, 0, j\omega) \times \frac{c\mu_0}{\sqrt{\varepsilon_r + \sigma/j\omega\varepsilon_0}} \tag{19.8}$$

where μ_0 is the permeability of free space, $E_{rp}(r, z, j\omega)$ and $H_{\phi p}(r, 0, j\omega)$ are the Fourier transforms of the horizontal electric field at height z above ground and the azimuthal magnetic field at ground level, respectively, both computed for the case of a perfectly (subscript "p") conducting ground. The second term is equal to zero for $\sigma \rightarrow \infty$ and becomes increasingly important as σ decreases. A generalization of the Cooray–Rubinstein formula has been offered by Wait (1997).

Apparently no horizontal or vertical electric fields very close to natural lightning, at distances of tens of meters, necessary to the understanding of the voltages induced by very close lightning, have been published (such very close fields have been measured for triggered lightning), although such close fields have been calculated by Diendorfer (1990) and by Rubinstein et al. (1990) using different return stroke models. These two sets of calculated fields are to be considered model-dependent estimates. Although there is disagreement between the two studies as to the waveshape of fields and the significance of the influence of the finite ground conductivity at small distances, both studies yield horizontal field amplitudes at the height of a typical power distribution line comparable to the amplitude of the vertical field. Note that no return stroke model used to date (see next section) takes proper account of the attachment process referred to earlier and hence probably none accurately models the fields at very early times. Further, the leader fields preceding the return stroke field change are not taken into account in the existing models, although such fields at very close range are clearly important since it is the leader charge near ground that the return stroke discharges to ground, and hence the leader and return stroke electrostatic field changes are of equivalent magnitude very close to the ground strike point (Rubinstein et al., 1995; Crawford et al., 2001).

Modeling the Return Stroke

General

Rakov and Uman (1998) review all types of return stroke models and rank the so-called "engineering" models in terms of their validity and usefulness. For the engineering models, if the current at the channel base is specified (e.g., from measurement) along with the model parameters, the channel current can be calculated as a function of height and time, and thereafter the electric and magnetic fields computed from the channel current. It is commonly assumed in these models that the lightning channel is perfectly straight and vertical and that the ground is a perfect conductor. The remote electric and magnetic fields are calculated from Equation (19.9) to Equation (19.15):

$$\overline{E} = \overline{E}_{\text{ele}} + \overline{E}_{\text{ind}} + \overline{E}_{\text{rad}} \tag{19.9}$$

$$\overline{E}_{\text{ele}} = \frac{1}{4\pi\epsilon_0} \int_{-h}^{h} \left\{ \frac{2\cos\theta'\hat{a}_R + \sin\theta'\hat{a}_{\theta'}}{R^3} \int_0^t i\left(|z'|, \tau - \frac{R}{c}\right)\mathrm{d}t \right\} \mathrm{d}z' \tag{19.10}$$

$$\overline{E}_{\text{ind}} = \frac{1}{4\pi\epsilon_0} \int_{-h}^{h} \frac{2\cos\theta'\hat{a}_R + \sin\theta'\hat{a}_{\theta'}}{cR^2} i\left(|z'|, t - \frac{R}{c}\right)\mathrm{d}z' \tag{19.11}$$

$$\overline{E}_{\text{rad}} = \frac{1}{4\pi\epsilon_0} \int_{-h}^{h} \frac{1}{c^2 R} \frac{\partial i\left(|z'|, t - \frac{R}{c}\right)}{\partial t} \hat{a}_{\theta'}\mathrm{d}z' \tag{19.12}$$

$$\overline{B} = \overline{B}_{\text{ind}} + \overline{B}_{\text{rad}} \tag{19.13}$$

$$\overline{B}_{\text{ind}} = \frac{\mu_0}{4\pi} \int_{-h}^{h} \frac{\sin\theta'}{R^2} i\left(|z'|, t - \frac{R}{c}\right)\hat{a}_{\phi}\mathrm{d}z' \tag{19.14}$$

$$\overline{B}_{\text{rad}} = \frac{\mu_0}{4\pi} \int_{-h}^{h} \frac{\sin\theta'}{cR} \frac{\partial i\left(|z'|, t - \frac{R}{c}\right)}{\partial t} \hat{a}_{\phi}\mathrm{d}z' \tag{19.15}$$

where $i(z', t)$ is the current along the channel obtained from one of the return stroke current models mentioned above, and the geometry by which the above equations are to be interpreted is shown in Figure 19.21. Note that the spatial integral includes the image current below the perfectly conducting ground plane to take account of reflections from the earth's surface. The three terms on the right-hand side of Equation (19.9) (expanded in Equation (19.10) to Equation (19.12)) are called, from left to right, the electrostatic, induction, and radiation terms. Similarly, the two terms on the right-hand side of Equation (19.13) (expanded in Equation (19.14) and Equation (19.15)) are termed the induction and radiation terms.

For large distances from the lightning channel, the radiation part of the electric and magnetic fields is dominant due to its $1/R$ dependence (as compared to the $1/R^2$ and $1/R^3$ dependencies of the induction and electrostatic terms, respectively). By a similar argument, for close distances, the dominant terms will be the electrostatic term in the case of the electric field and the induction term for the magnetic field. It can be readily shown from Equation (19.9) to Equation (19.15) and the preceding discussion that for any individual lightning return stroke model, the waveshapes of the vertical electric field and the horizontal magnetic field are almost identical for great distances, and this fact is also evident in the experimental data (see Figure 19.19). Moreover, it can be shown that for great distances, the ratio of the electric field intensity E to the magnetic flux density B is the speed of light c.

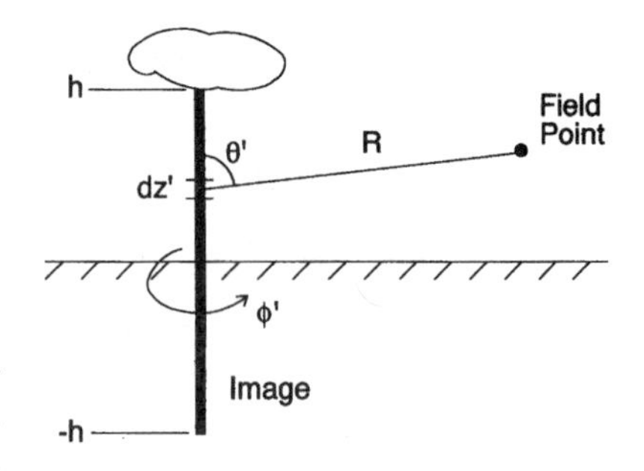

FIGURE 19.21 The geometry for the calculation of the fields using Equation (19.9) to Equation (19.15).

A brief examination of return stroke current models follows. We discuss here only the transmission line (TL) model and its modified versions (modified transmission line linear—MTLL, modified transmission line exponential—MTLE) and the DU model of Diendorfer and Uman (1990) along with its modified version. More details of all models are found in Rakov and Uman (2003), Thottappillil and Uman (1993), Nucci et al. (1990), Diendorfer and Uman (1990), and Thottappillil et al. (1991), including fields calculated from the various models. By assessing all published attempts at the validation of the various return stroke models, Rakov and Uman (1998) conclude that for modeling the initial microsecond or so of the return stroke (up to the initial field peak) the TL model is preferred (see also Schoene et al., 2003, for a later validation); and, for tens of microseconds of the return stroke, the ranking of the engineering models, with best first, is MTLL, DU, MTLE, BG, and TL.

The Transmission Line (TL) Model and Its Modified Versions

In the transmission line model it is assumed that the current waveform at the ground travels undistorted up the lightning channel at a constant speed v. Mathematically, this current is represented by

$$\begin{aligned} &i(z', t) = i(0, t - z'/v) \quad z' < vt \\ &i(z', t) = 0 \quad z' > vt \end{aligned} \tag{19.16}$$

No charge is removed by the transmission line return stroke current along the channel since the charge entering the bottom of any section of the channel leaves the top when the current reaches it. All the charge is therefore transferred from the bottom to the top of the channel, an unrealistic situation given our knowledge of lightning physics. Nevertheless, the TL model appears to be entirely adequate to model the fields for the first microsecond or so (Schoene et al., 2003).

Willett et al. (1989) have presented return stroke current, field, and speed data from triggered lightning in an attempt to validate the TL model. Using these data, Rakov et al. (1992) have shown, at least for strokes in triggered lightning, that return stroke peak current can be derived from return stroke peak field by the expression $I = 1.5 - 0.037DE$ where the peak current I is in kA and is negative, the distance D is in km, and the peak electric field E is in V/m and is positive.

In the modified transmission line models, the upward propagating current given by Equation (19.16) is multiplied by an attenuation factor, $(1-z'/H)$ for the MTLL where H is the channel height, or $e^{-z'/\lambda}$ for the MTLE where λ is the height constant. Details are found in Rakov and Uman (1998).

The Diendorfer–Uman (DU) Model and a Modification of It

In the DU model (Diendorfer and Uman, 1990), the channel current above ground is assumed to discharge the leader by way of two independent processes: (1) the discharge of the highly ionized core of the leader channel, termed as the breakdown discharge process, with a time constant of 1 μsec or less, and (2) the discharge of the corona envelope with a larger time constant. In both cases, the discharge at a height z' starts when the return stroke front, assumed to travel up at a constant speed v, arrives at z'. The liberated currents are assumed to flow to the ground at the speed of light.

For a single current component at ground $i(0,t)$, Diendorfer and Uman (1990) show that the current as a function of height and time is

$$i(z', t) = i(0, t_{\mathrm{m}}) - i(0, z'/v*) \exp(-t_{\mathrm{e}}/\tau) \tag{19.17}$$

where $t_{\mathrm{m}} = (t + z'/v)$, $t_{\mathrm{e}} = (t - z'/v)$, $v* = v(1 + v/c)$, and τ is the discharge time constant. An analytical generalization of the DU model which allows for the return stroke speed and the downward current speed to be arbitrary functions of height has been presented by Thottappillil et al. (1991).

Lightning–Overhead Wire Interactions

General

Lightning interactions with overhead wires such as power distribution lines are a major source of electromagnetic compatibility problems, resulting in inferior power quality, power outages, and damaged electronics. Only a small fraction of all the cloud-to-ground lightning flashes directly strikes overhead lines, making induced overvoltages a significant source of power disturbances. This section begins with a discussion of the appropriate transmission line figures. Then, examples of measured lightning-induced voltages on overhead lines as well as calculated voltages are presented.

Transmission-Line Equations

To estimate voltages induced on distribution lines by nearby strikes, three basic coupling models have been used: the model of Rusck (1958, 1977), the model of Chowdhuri and Gross (1967), and the model of Agrawal et al. (1980). All of these models are based on transmission-line theory. The model of Agrawal et al. (1980) and its equivalent formulations (Taylor et al., 1965; Rachidi, 1993; Nucci and Rachidi, 1995) can be considered to be accurate within the limits of the transmission-line theory (Nucci and Rachidi, 1995), but in both the Rusck and the Chowdhuri–Gross models, some source terms have been omitted (Nucci et al., 1995a,b; Cooray, 1994; Cooray and Scuka, 1998). However, for the case of the electromagnetic field radiated from a vertical lightning channel, the Rusck model is equivalent to the model of Agrawal et al. (Cooray, 1994). Besides his detailed model, Rusck also proposed a simplified analytical formula for the peak-induced voltage on an infinitely long overhead line above a perfectly conducting ground as a function of the return-stroke peak current. A discussion of the validity of the transmission-line theory vs. the more sophisticated scattering theory is found in Tesche (1992).

The derivation of the time-domain coupling equations is conceptually simple. Maxwell's equations are first integrated over closed cylindrical surfaces and along closed rectangular paths. The resulting integral equations, which are in terms of electric and magnetic fields, are then recast in terms of voltages and currents. One version of the transmission-line equations, due to Agrawal et al. (1980), follows:

$$\frac{\partial V^{\mathrm{s}}(x,t)}{\partial t} + Z_{\mathrm{g}} * I(x,t) + L\frac{\partial I(x,t)}{\partial t} = E_{\mathrm{x}}^{\mathrm{i}}(x, z = h, t) \tag{19.18}$$

$$\frac{\partial I(x,t)}{\partial x} + C\frac{\partial V^{\mathrm{s}}(x,t)}{\partial t} = 0 \tag{19.19}$$

$$V^{\mathrm{t}} = V^{\mathrm{i}} + V^{\mathrm{s}} = -\int_{0}^{h} E_{\mathrm{z}}^{\mathrm{i}}(x,z,t)\mathrm{d}z + V^{\mathrm{s}} \tag{19.20}$$

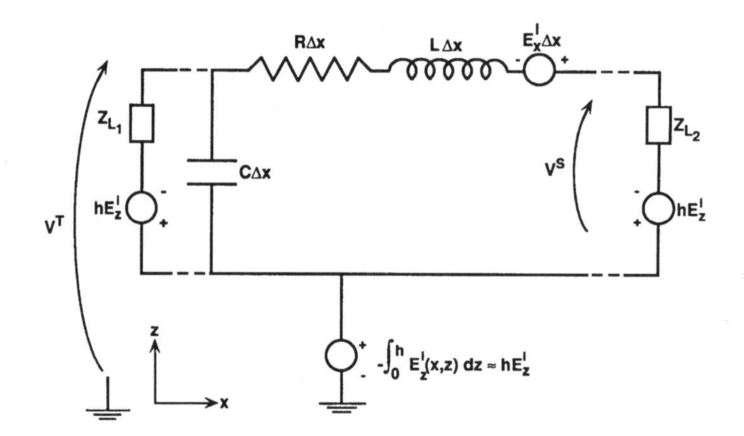

FIGURE 19.22 Equivalent circuit model obtained from Equation (19.18) to Equation (19.20).

where the superscript s identifies the "scattered" quantities, the superscript i identifies the "incident" quantities, the superscript t identifies the total measurable quantities, and the asterisk is the convolution operator.

In these equations, the only source along the horizontal portion of the line is the horizontal component of the incident electric field. At the line terminations, the boundary condition and the termination current, I, are used to determine the end voltage. At those vertically oriented terminations, the vertical electric fields drive currents through the terminations into the line. The total voltage, $V^t(x, t)$, at the line terminations must equal $I_T{*}Z_T$ at all times, where Z_T is the termination impedance. Equation (19.18) to Equation (19.20) can be represented by the circuit model in Figure 19.22.

Two basic assumptions are used to arrive at Equation (19.18) to Equation (19.20): (1) The response of the power line (scattered voltages and currents) to the impinging EM wave (incident field) is quasi TEM (i.e., the scattered fields can be approximated as transverse electromagnetic). This allows us to define a "static" voltage along the line and to relate the line current and the scattered magnetic flux by an inductance, as well as the line scattered voltage and charge by a capacitance. (2) The transverse dimensions of the line system are small compared to the minimum wavelength, λ_{min}, of the excitation wave, and the height of the line is much larger than the diameter of the wire.

Measured and Calculated Lightning-Induced Voltages on Overhead Wires

Coupling models have been tested by means of natural lightning (Yokoyama et al., 1983; Master et al., 1984; Eriksson, 1987; De la Rosa et al., 1988) and triggered lightning (Rubinstein et al., 1989; Georgiadis et al., 1992; Barker et al., 1996). Laboratory tests have also been performed in a more controlled environment using NEMP (nuclear electromagnetic pulse) simulators (e.g., Gerrieri et al., 1995) and also using reduced scale models (e.g., Ishii et al., 1994, 1999; Piantini and Janiszewski, 1992; Nucci et al., 1998). Nucci (1995) presented a survey of some of these tests.

Calculations of the voltages induced on overhead lines under various conditions have been given, for example, by Master and Uman (1984), Diendorfer (1990), Chowdhuri (1989a,b, 1990), Georgiadis et al. (1992), Nucci et al. (1993a,b), Ishii et al. (1994), Rachidi et al. (1997a,b), Michishita et al. (1996, 1997), Rachidi et al. (1999), and Paolone et al. (2004).

The strategy in the various experiments designed to test a coupling model is similar. Measure the lightning electric and magnetic fields in the vicinity of an instrumented overhead line (or compute them from a return stroke model) while simultaneously measuring the voltages induced on the line, the fields then being used as inputs to a computer program written to solve Equation (19.18) to Equation (19.20) and the computer-calculated voltage waveforms being compared with the measured voltage waveforms. The following discussion illustrates the types of voltage waveforms induced on overhead wires by lightning beyond a few kilometers and the degree of agreement that has been obtained in the coupling-model calculations. Examples of voltages

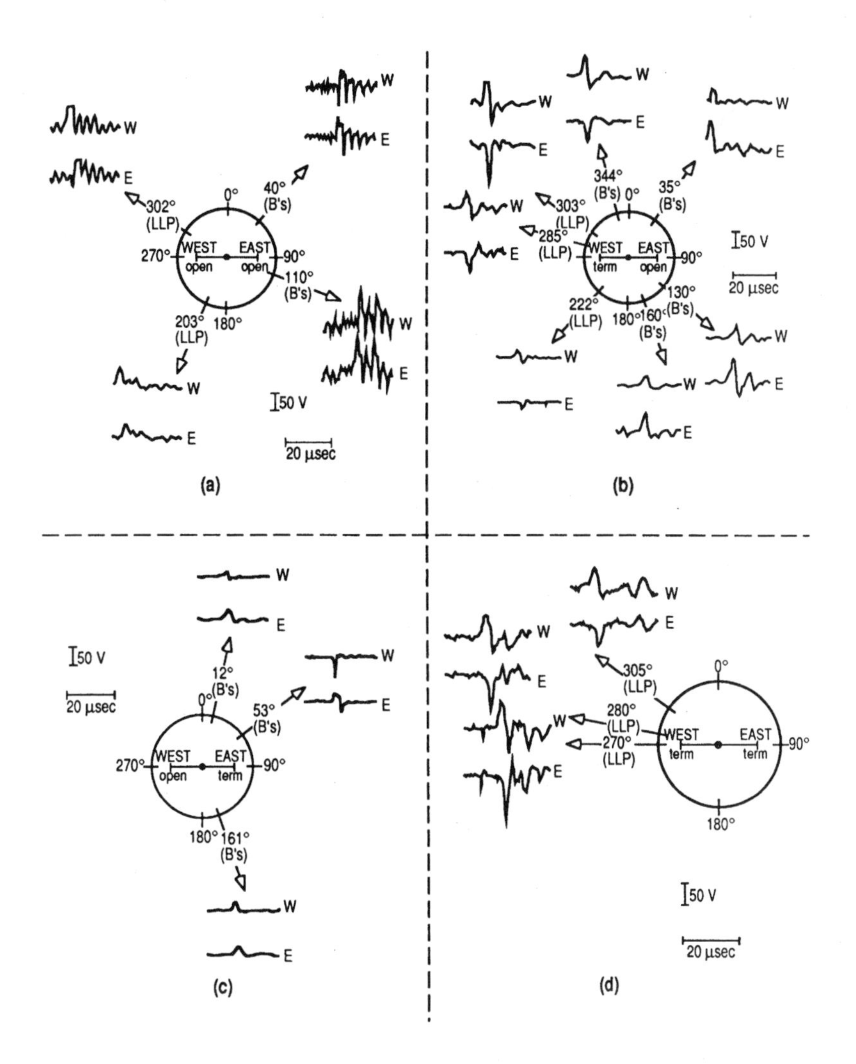

FIGURE 19.23 Examples of simultaneously measured lightning-induced voltages at the east end (E) and west end (W) of a 450-m line. Each line end is either open or terminated in its characteristic impedance, as noted. Directions to the lightning are determined from the ratio of the horizontal magnetic flux densities (*Bs*) or from a commercial lightning location system (LLP). (*Source:* Adapted from N. Georgiadis et al., "Lightning-induced voltages at both ends of a 450-meter distribution line," *IEEE Trans. EMC*, vol. 34, pp. 451–460, 1992. ©1992 IEEE. With permission.)

induced on a 450 m overhead line about 10 m above the ground are shown in Figure 19.23. Each line end was either terminated in its characteristic impedance or open-circuited (four different cases), and voltages were measured simultaneously at each end. Figure 19.24 and Figure 19.25 contain specific examples of measured and calculated voltage waveforms at each line end as well as the measured vertical electric field and calculated horizontal electric field via Equation (19.7). It is clear from Figure 19.23 to Figure 19.25 that the induced voltage polarities and waveshapes are strongly dependent on the angle to the lightning and on the line end terminations. It is apparent also from Georgiadis et al. (1992) that while the measured and calculated voltage waveshapes are in good agreement, the measured voltage amplitudes are, on average, a factor of three smaller than calculated voltages. This amplitude discrepancy is probably due to the fact that the fields reaching the power line were shielded by trees along the line whereas the fields measured were in an open area and hence were unshielded.

Barker et al. (1996) measured voltages on a 10 m high test distribution line when lightning was triggered (artificially initiated) using the rocket-and-wire technique at a distance of 145 m from the line. Stroke current,

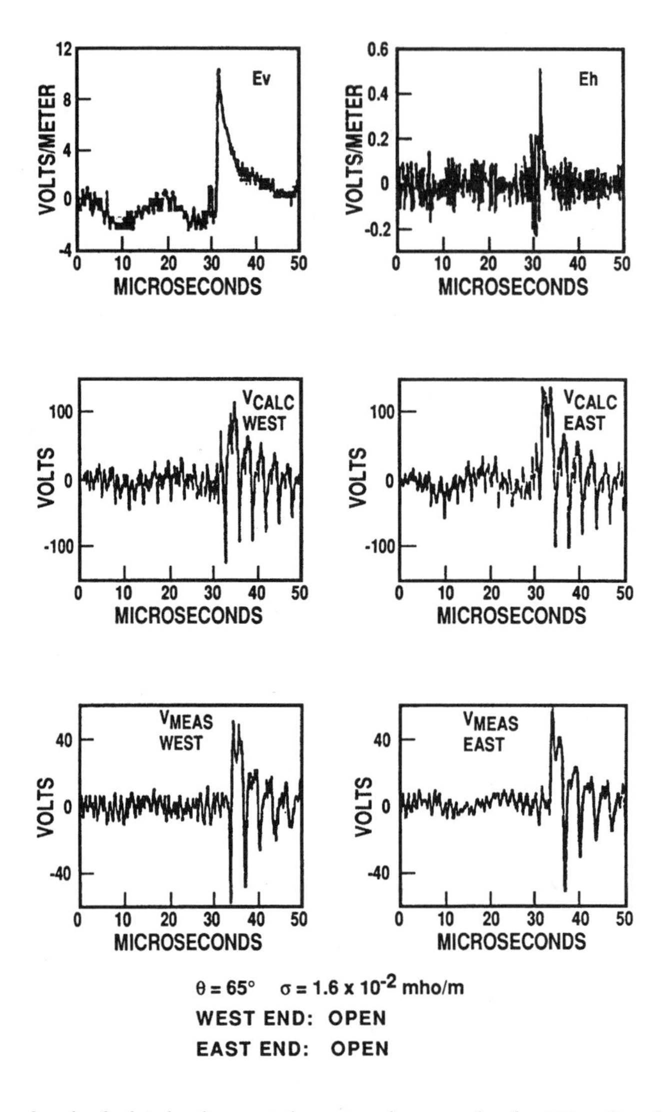

FIGURE 19.24 Measured and calculated voltages at the east and west ends of a 450-m line for both line ends open. Although the direction to the lightning as determined from the ratio of the two magnetic flux density (B) components was 40°, as shown in Figure 19.25(a), the best calculated fit to the data was found for 65° as shown, the angular error apparently being caused by variation in the magnitudes of the magnetic flux density components due to nearby conductors as determined from comparing azimuths computed from the Bs and from a commercial lightning location system (LLP). (*Source:* Adapted from N. Georgiadis et al., "Lightning-induced voltages at both ends of a 450-meter distribution line," *IEEE Trans. EMC*, vol. 34, pp. 451–460, 1992. ©1992 IEEE. With permission.)

line voltage, and nearby electric and magnetic fields vs. time were recorded with submicrosecond time resolution. For 63 strokes having peak currents between a few kA and 44 kA, peak-induced voltages at the center of the line relative to a grounded wire about 2 m below ranged between 8 and 100 kV, the peak values being linearly correlated with a correlation coefficient of 0.75. The waveshapes of the induced voltage were unipolar with a median width at half-peak value of about 4 μsec, whereas the typical width of the current was nearly an order of magnitude larger. According to Barker et al. (1996), the induced voltage waveform resembles the derivative of the vertical electric field, the theory of Agrawal et al. (1980) and other equivalent approaches accounting well for the observations. The peak induced voltages were roughly 60% larger than predicted by the simplified formula of Rusck (1958, 1977) discussed above.

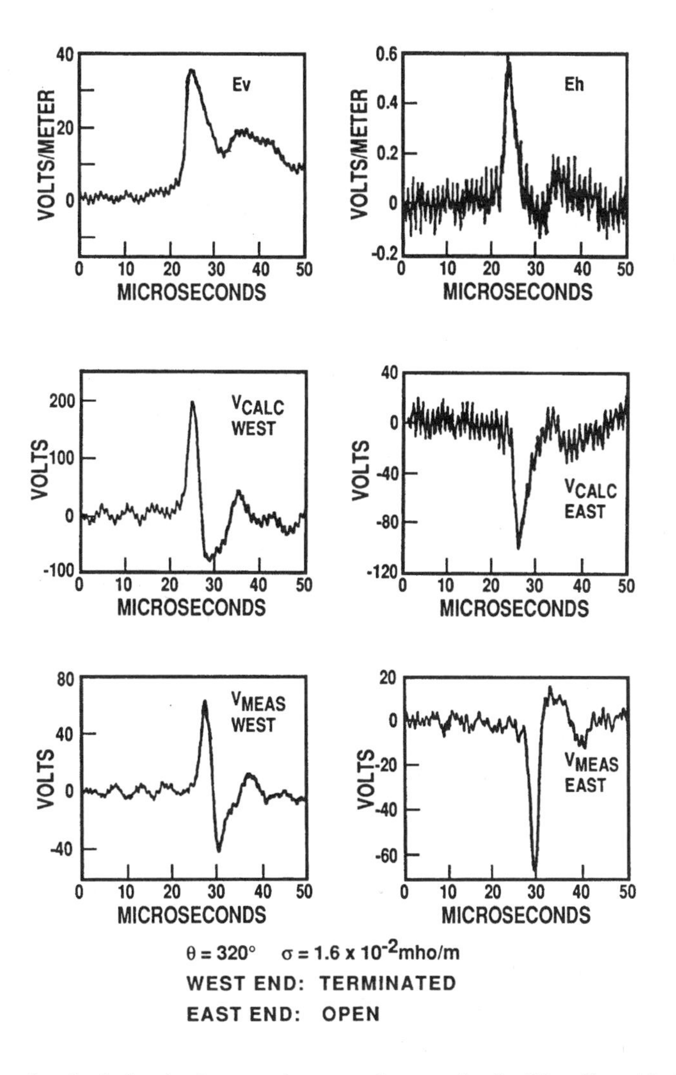

FIGURE 19.25 Measured and calculated voltages at the east and west ends of a 450-m line with the west end terminated and the east end open. The azimuth was determined from LLP data and is shown in Figure 19.25(b). (*Source:* Adapted from N. Georgiadis et al., "Lightning-induced voltages at both ends of a 450-meter distribution line," *IEEE Trans. EMC*, vol. 34, pp. 451–460, 1992. ©1992 IEEE. With permission.)

Defining Terms

Attachment process: A process that occurs when one or more stepped leader branches approach within 100 m or so of the ground and the electric field at the ground increases above the critical breakdown field of the surrounding air. At that time one or more upward-going discharges are initiated. After traveling a few tens of meters, one of the upward discharges, which is essentially at ground potential, contacts the tip of one branch of the stepped leader, which is at a high potential, completing the leader path to ground.

Dart leader: A continuously moving leader lowering charge preceding a return stroke subsequent to the first. A dart leader typically propagates down the residual channel of the previous stroke.

Flash: A complete lightning discharge of any type.

Preliminary breakdown: An electrical discharge in the cloud that initiates a cloud-to-ground flash.

Return stroke: The upward propagating high-current, bright, potential discontinuity following the leader that discharges to the ground some or all of the charge previously deposited along the channel by the leader.

Stepped leader: A discharge following the preliminary breakdown that propagates from cloud towards the ground through virgin air in a series of intermittent luminous steps with an average speed of 10^5 to 10^6 m/sec. Negatively charged leaders clearly step, while positively charged leaders are more pulsating than stepped.

Triggered (artificially initiated) lightning: Lightning to ground that is induced to occur from natural thunderclouds by firing upward rockets trailing grounded wires. Triggered lightning does not have a first stroke as in natural lightning but exhibits return strokes similar to the subsequent strokes in natural cloud-to-ground flashes.

References

A.K. Agrawal, H.J. Price, and S.H. Gurbaxani, "Transient response of multiconductor transmission lines excited by a non-uniform electromagnetic field," *IEEE Trans. EMC*, vol. EMC-22, pp. 119–129, 1980.

P. Barker, T. Short, A. Eybert-Berard, and J. Berlandis, "Induced voltage measurements on an experimental distribution line during nearby rocket triggered lightning flashes," *IEEE Trans. Power Delivery*, vol. 11, pp. 980–995, 1996.

K. Berger, R.B. Anderson, and H. Kroninger, "Parameters of lightning flashes," *Electra*, vol. 80, 23–37, 1975.

P. Chowdhuri, and E.T.B. Gross, "Voltage surges induced on overhead lines by lightning strokes," *Proc. IEE*, vol. 114, pp. 1899–907, 1967.

P. Chowdhuri, "Parametric effects on the induced voltages on overhead lines by lightning strokes to nearby ground," *IEEE Trans. Power Delivery*, vol. 4, pp. 1185–94, 1989a.

P. Chowdhuri, "Analysis of lightning-induced voltages on overhead lines," *IEEE Trans. Power Delivery*, vol. 4, pp. 479–92, 1989b.

P. Chowdhuri, "Lightning-induced voltages on multiconductor overhead lines," *IEEE Trans. Power Delivery*, vol. 5, pp. 658–66, 1990.

V. Cooray, "Horizontal fields generated by return strokes," *Radio Sci.*, vol. 27, pp. 529–37, 1992.

V. Cooray, "Calculating lightning-induced voltages in power lines: a comparison of two coupling models," *IEEE Trans. EMC*, vol. 36, pp. 170–182, 1994.

V. Cooray and V. Scuka, "Lightning-induced overvoltages in power lines: validity of various approximations made in overvoltage calculations," *IEEE Trans. Electromagn. Compat.*, vol. 40, pp. 355–63, 1998.

D.E. Crawford, V.A. Rakov, M.A. Uman, G.H. Schnetzer, K.J. Rambo, and M.V. Stapleton, "The close lightning electromagnetic environment: dart-leader electric field change versus distance," *J. Geophys. Res.*, vol. 106, 14,909–14,917, 2001.

F. De la Rosa, R. Valdivia, H. Perez, and J. Loza, J., "Discussion about the inducing effects of lightning in an experimental power distribution line in Mexico," *IEEE Trans. Power Delivery*, vol. 3, pp. 1080–1089, 1988.

G. Diendorfer, "Induced voltage on an overhead line due to nearby lightning," *IEEE Trans. Electromag. Compat.*, vol. 32, pp. 292–299, 1990.

G. Diendorfer and M. A. Uman, "An improved return stroke model with specified channel-base current," *J. Geophys. Res.*, vol. 95, 13,621–13,644, 1990.

A.J. Eriksson, "The incidence of lightning strikes to power lines," *IEEE Trans. Power Delivery*, vol. 2, pp. 859–70, 1987.

N. Georgiadis, M. Rubinstein, M.A. Uman, P.J. Medelius, and E.M. Thomson, "Lightning-induced voltages at both ends of a 450-meter distribution line," *IEEE Trans. EMC*, vol. 34, pp. 451–460, 1992.

S. Guerrieri, F. Rachidi, M. Ianoz, P. Zweiacker, and C.A. Nucci, "A time-domain approach to evaluate induced voltages on tree-shaped electrical networks by external electromagnetic fields," *Zurich EMC Symp. Electromagn. Compat.*, 1995.

M. Ishii, K. Michishita, Y. Hongo, and S. Oguma, "Lightning-induced voltage on an overhead wire dependent on ground conductivity," *IEEE Trans. Power Delivery*, vol. 9, pp. 109–118, 1994.

M. Ishii, K. Michishita, and Y. Hongo, "Experimental study of lightning-induced voltage on an overhead wire over lossy ground," *IEEE Trans. Electromagn. Compat.*, vol. 41, 39–45, 1999.

Y.T. Lin, M.A. Uman, J.A. Tiller, R.D. Brantley, W.H. Beasley, E.P. Krider, and C.D. Weidman, "Characterization of lightning return stroke electric and magnetic fields from simultaneous two-station measurements," *J. Geophys. Res.*, vol. 84, pp. 6307–6314, 1979.

M.J. Master and M.A. Uman, "Lightning induced voltage on power lines: Theory," *IEEE Trans. Power Apparatus Syst.*, vol. 103, pp. 2502–2518, 1984.

M.J. Master, M.A. Uman, W.H. Beasley, and M. Darveniza, "Lightning induced voltages on power lines: Experiment," *IEEE Trans. Power Apparatus Syst.*, vol. 103, pp. 2519–2529, 1984.

K. Michishita, M. Ishii, and Y. Hongo, "Induced voltage on an overhead wire associated with inclined return-stroke channel-model experiment on finitely conductive ground," *IEEE Trans. Electromagn. Compat.*, vol. 38, pp. 508–513, 1996.

K. Michishita, M. Ishii, and Y. Hongo, "Lightning-induced voltage on an overhead wire influenced by a branch line," *IEEE Trans. Power Delivery*, vol. 12, pp. 296–301, 1997.

C.A. Nucci, G. Diendorfer, M.A. Uman, F. Rachidi, M. Ianoz, and C. Mazzetti, "Lightning return stroke current models with specified channel-base current: a review and comparison," *J. Geophys. Res.*, vol. 95, p. 20, 395–20,408, 1990.

C.A. Nucci, F. Rachidi, M.V. Ianoz, and C. Mazzetti, "Lightning-induced voltages on overhead lines," *IEEE Trans. Electromagn. Compat.*, vol. 35, pp. 75–86, 1993a.

C.A. Nucci, F. Rachidi, M.V. Ianoz, and C. Mazzetti, "Corrections to lightning-induced voltages on overhead lines," *IEEE Trans. Electromagn. Compat.*, vol. 35, p. 488, 1993b.

C.A. Nucci and F. Rachidi, "On the contribution of the electromagnetic field components in field-to-transmission line interaction," *IEEE Trans. EMC*, vol. 37, pp. 505–508, 1995.

C.A. Nucci, F. Rachidi, M. Ianoz, and C. Mazzetti, "Comparison of two coupling models for lightning-induced overvoltage calculations," *IEEE Trans. Power Delivery*, vol. 10, pp. 330–338, 1995a.

C.A. Nucci, M. Ianoz, R. Rachidi, M. Rubinstein, F.M. Tesche, M.A. Uman, M.A., and C, Mazzetti, "Modelling of lightning-induced voltages on overhead lines: recent developments," *Elektrotechnik und Informationstechnik*, vol. 112, no. 6, pp. 290–296, 1995b.

C.A. Nucci, "Lightning-induced voltages on overhead power lines, Part II: coupling models for the evaluation of the induced voltages," *Electra*, vol. 162, pp. 121–145, 1995.

C.A. Nucci, A. Borghetti, A. Piantini, and J.M. Janiszewski, "Lightning-induced voltages on distribution overhead lines: comparison between experimental results from a reduced-scale model and most recent approaches," *Proc. 24th Int. Conf. Lightning*, Birmingham, pp. 314–320, 1998.

R.E. Orville, "Annual summary — lightning ground flash density in the contiguous United States—1989," *Mon. Weather Rev.*, vol. 119, pp. 573–577, 1991.

M. Paolone, C.A. Nucci, E. Petrache, and F. Rachidi, "Mitigation of lightning-induced overvoltages in medium voltage distribution lines by means of periodical grounding of shielding wires and of surge arresters: modeling and experimental validation," *IEEE Trans. Power Delivery*, vol. 19, pp. 423–431, 2004.

A. Piantini, and J.M. Janiszewski, "An experimental study of lightning induced voltages by means of a scale model," *Proc. 21st Int. Conf. Lightning Protection*, Berlin, pp. 195–199, 1992.

F. Rachidi, "Formulation of the field-to-transmission line coupling Figures in terms of magnetic excitation field," *IEEE Trans. EMC*, vol. 35, pp. 404–407, 1993.

F. Rachidi, C.A. Nucci, M. Ianoz, and C. Mazzetti, "Response of multiconductor power lines to nearby lightning return stroke electromagnetic fields," *IEEE Trans. Power Delivery*, vol. 12, pp. 1404–1411, 1997a.

F. Rachidi, M. Rubinstein, S. Guerrieri, and C.A. Nucci, "Voltages induced on overhead lines by dart leaders and subsequent return strokes in natural and rocket-triggered lightning," *IEEE Trans. Electromagn. Compat.*, vol. 39, pp. 160–166, 1997b.

F. Rachidi, C.A. Nucci, and M. Ianoz, "Transient analysis of multiconductor lines above a lossy ground," *IEEE Trans. Power Delivery*, vol. 14, pp. 294–302, 1999.

V.A. Rakov, R. Thottappillil, and M.A. Uman, "On the empirical formula of Willett, et al., relating lightning return stroke peak current and peak electric field," *J. Geophys. Res.*, vol. 97, p. 11, 527–11,533, 1992.

V.A. Rakov and M.A. Uman, "Review and evaluation of lightning return stroke models, including some aspects of their application," *IEEE Trans. EMC*, vol. 40, pp. 403–426 1998.

M. Rubinstein, A. Tzeng, M.A. Uman, P.J. Medelius, and E.M. Thomson, "An experimental test of a theory of lightning induced voltages on an overhead wire," *IEEE Trans. Electromagn. Compat.*, vol. 31, pp. 376–383, 1989.

M. Rubinstein, M.A. Uman, E.M. Thomson, and P.J. Medelius, "Voltages induced on a test distribution line by artificially initiated lightning at close range: measurement and theory," *Proc. 20th Int. Conference on Lightning Protection*, Interlaken, Switzerland, September 24–28, 1990.

M. Rubinstein, M.A. Uman, P.J. Medelius, and E.M. Thomson, "Measurements of the voltage induced on an overhead power line 20 m from triggered lightning," *IEEE Trans. EMC*, vol. 36, pp. 134–140, 1994.

M. Rubinstein, F. Rachidi, M.A. Uman, R. Thottappillil, V.A. Rakov, and C.A. Nucci, "Characterization of vertical electric fields 500 m and 30 m from triggered lightning," *J. Geophys. Res.*, vol. 100, pp. 8863–8872, 1995.

M. Rubinstein, "An approximate formula for the calculation of the horizontal electric field from lightning at close, intermediate, and long range," *IEEE Trans. Electromagn. Compat.*, vol. 38, pp. 531–535, 1996.

S. Rusck, "Induced lightning overvoltages on power transmission lines with special reference to the overvoltage protection of low voltage networks," *Trans. Roy. Inst. Tech., (K. Tek. Högsk. Handl.), Stockholm*, vol. 120, 1958.

S. Rusck, "Protection of distribution systems," in *Lightning, Vol. 2: Lightning Protection,* R.H. Golde, Ed., New York: Academic, pp. 747–772, 1977.

J. Schoene, M.A. Uman, V.A. Rakov, K.J. Rambo, J. Jerauld, and G. Schnetzer, "Test of the transmission line model and the traveling current source model with triggered lightning return strokes at very close range," *J. Geophys. Res.*, vol. 108, no. D23, p. 4737 (ACL 10–1 through 10–14), doi:10.1029/2003JD003683, 2003.

C.D. Taylor, R.S. Satterwhite, and C.W. Harrison, "The response of a terminated two-wire transmission line excited by a non-uniform electromagnetic field," *IEEE Trans. Antennas Propag.*, vol. AP-13, pp. 987–989, 1965.

F.M. Tesche, "Comparison of the transmission line and scattering models for computing the HEMP response of overhead cables," *IEEE Trans. Electrom. Comp.*, vol. 34, pp. 93–99, 1992.

E.M. Thomson, P. Medelius, M. Rubinstein, M.A. Uman, J. Johnson, and J. Stone, "Horizontal electric fields from lightning return strokes," *J. Geophys. Res.*, vol. 93, pp. 2429–2441, 1988.

R. Thottappillil and M.A. Uman, "Comparison of return stroke models," *J. Geophys. Res.*, vol. 98, pp. 22,903–22,914, 1993.

R. Thottappillil, D.K. McLain, G. Diendorfer, and M.A. Uman, "Extension of the Diendorfer–Uman lightning return stroke model to the case of a variable upward return stroke speed and a variable downward discharge current speed," *J. Geophys. Res.*, vol. 96, p. 17,143–17,150, 1991.

J.R. Wait, "Concerning the horizontal electric field of lightning," *IEEE Trans. Electromagn. Compat.*, vol. 39, p. 186, 1997.

J.E. Willett, J.C. Bailey, V.P. Idone, A. Eybert-Berard, and L. Barret, "Submicrosecond intercomparison of radiation fields and currents in triggered lightning return strokes based on the transmission-line model," *J. Geophys. Res.*, vol. 94, pp. 13,275–13,286, 1989.

S. Yokoyama, K. Miyake, H. Mitani, and A. Takanishi, "Simultaneous measurement of lightning induced voltages with associated stroke currents," *IEEE Trans. Power Apparatus Syst.*, vol. 102, pp. 2420–2429. 1983.

Further Information

For more details on the material presented here, see *Lightning Physics and Effects* (Cambridge University Press, Cambridge, 2003) by V.A. Rakov and M.A. Uman; *The Lightning Discharge* (Academic Press, San Diego, 1987; Dover paperback, 2001) by M.A. Uman; and the review article "Natural and Artificially Initiated Lightning" (*Science*, vol. 246, 457–464, 1989) by M.A. Uman and E.P. Krider.

20 Radar

Melvin L. Belcher
Georgia Tech Research Institute

Josh T. Nessmith
Georgia Tech Research Institute

Samuel O. Piper
Georgia Tech Research Institute

James C. Wiltse
Georgia Tech Research Institute

20.1 Pulse Radar

Melvin L. Belcher and Josh T. Nessmith

Overview of Pulsed Radars

Basic Concept of Pulse Radar Operation

The basic operation of a pulse radar is depicted in Figure 20.1. The radar transmits pulses superimposed on a radio frequency (RF) carrier and then receives returns (reflections) from desired and undesired scatterers. Scatterers corresponding to desired targets may include space, airborne, and sea- or surface-based vehicles. They can also include the Earth's surface and the atmosphere in remote sensing applications. Returns from undesired scatterers are denoted as clutter. Clutter sources include the Earth's surface, natural and man-made discrete objects, the sea, and volumetric atmospheric phenomena such as rain and birds. Short-range/low-altitude radar operation is often constrained by clutter since a multitude of undesired returns masks returns from targets of interest such as aircraft. Conversely, volumetric atmospheric phenomena may be considered as targets for weather radar systems. The magnitude of the clutter returns exceeds those from small targets by multiple orders of magnitudes so a combination of angle, range, and Doppler filtering is required to extract returns from targets of interest.

The range, azimuth angle, elevation angle, and range rate can be directly measured from a return to estimate target metrics, position, and velocity to support tracking. Signature data to support noncooperative target identification or environmental remote sensing can be extracted by measuring the amplitude, phase, and polarization of the return.

Pulse radar affords a great deal of design and operational flexibility. Pulse duration, pulse rate, and pulse bandwidth can be tailored to specific applications to provide optimal performance. Modern computer-controlled multiple-function radars exploit this capability by choosing the best waveform from a repertoire for a given operational mode and interference environment automatically. Pulsed operation also enables the extreme ratio between transmitter power and receiver sensitivity, exceeding 10^{20} in some systems, necessitated by the detection of small targets at long ranges.

FIGURE 20.1 Pulse radar.

Radar Applications

The breadth of pulse radar applications is summarized in Table 20.1 in terms of operating frequencies. Radar applications can also be grouped into search, track, and signature measurement applications. Search radars are used for surveillance tracking but have relatively coarse metric accuracy. The search functions favor broad beamwidths and low bandwidths in order to efficiently search over a large spatial volume. As indicated in Table 20.1, search is preferably performed in the lower frequency bands. The antenna pattern is typically narrow in azimuth and has a cosecant pattern in elevation to provide acceptable coverage from the horizon to the zenith.

Tracking radars are typically characterized by a narrow beamwidth and moderate bandwidth in order to provide accurate range and angle measurements on a given target. The antenna pattern is typically characterized as a pencil beam with approximately the same dimensions in azimuth and elevation. Track is usually conducted at the higher frequency bands in order to minimize the beamwidth for a given antenna aperture area as well as provide enhanced range and Doppler resolution. After each return from a target is received, the range and angle are measured and input into a track filter. Track filtering smoothes the data to refine the estimate of target position and velocity. It also predicts the target's flight path to provide range gating and antenna pointing control to the radar system.

Signature measurement applications include remote sensing of the environment as well as the measurement of target characteristics. In some applications, synthetic aperture radar (SAR) imaging is conducted from aircraft or satellites to characterize land usage over broad areas. Moving targets that present changing aspects to the radar can be imaged from airborne or ground-based radars via inverse synthetic aperture radar (ISAR)

TABLE 20.1 Radar Bands

Band	Frequency Range	Principal Applications
HF	3–30 MHz	Over-the-horizon radar
VHF	30–300 MHz	Long-range search
UHF	300–1000 MHz	Long-range surveillance
L	1000–2000 MHz	Long-range surveillance
S	2000–4000 MHz	Surveillance
		Long-range weather characterization
		Terminal air traffic control
C	4000–8000 MHz	Fire control
		Instrumentation tracking
X	8–12 GHz	Fire control
		Air-to-air missile seeker
		Marine radar
		Airborne weather characterization
Ku	12–18 GHz	Short-range fire control
		Remote sensing
Ka	27–40 GHz	Remote sensing
		Weapon guidance
V	40–75 GHz	Remote sensing
		Weapon guidance
W	75–110 GHz	Remote sensing
		Weapon guidance

techniques. As defined in the subsection "Resolution and Accuracy," cross-range resolution improves with increasing antenna extent. SAR/ISAR effectively substitutes an extended observation interval over which coherent returns are collected from different target aspect angles for a large antenna structure that would not be physically realizable in many instances.

In general, characterization performance improves with increasing frequency because of the associated improvement in range, range rate, and cross-range resolution. However, phenomenological characterization to support environmental remote sensing may require data collected across a broad swath of frequencies. For a given sensitivity, as defined by the product of transmit power, antenna gain, and receive antenna aperture area divided by the product of system noise temperature and composite loss factor, the cost of a radar system generally increases with carrier frequency.

A multiple-function *phased array* radar generally integrates these functions to some degree. Its design is typically driven by the necessity to maintain a given surveillance rate while supporting precision track of a specified set of targets. Its operational frequency is generally a compromise between the lower frequency of the search radar and the higher frequency desired for the tracking radar. The degree of signature measurement implemented to support such functions as noncooperative target identification depends on the resolution capability of the radar as well as the operational user requirements. Multiple-function radar design represents a compromise among these different requirements. However, implementation constraints, multiple-target handling requirements, and reaction time requirements often dictate the use of phased array radar systems integrating search, track, and characterization functions.

Critical Subsystem Design and Technology

The major subsystems making up a pulse radar system are depicted in Figure 20.2. The associated interaction between function and technology is summarized in this subsection.

Antenna

The radar antenna provides spatial directivity to the transmitted EM wave and intercepts the scattering of that wave from a target while attenuating interference signals from undesired angles of arrival. Most radar antennas may be categorized as mechanically scanning or electronically scanning. Mechanically scanned reflector antennas are used in applications where rapid beam scanning is not required. Electronic scanning antennas include phased arrays and frequency scanned antennas. Phased array beams can be steered to any point in their field-of-view, typically within 10 to 100 μsec, depending on the latency of the beam steering subsystem and the switching time of the phase shifters. Phased arrays are desirable in multiple function radars since they can interleave search operations with multiple target tracks.

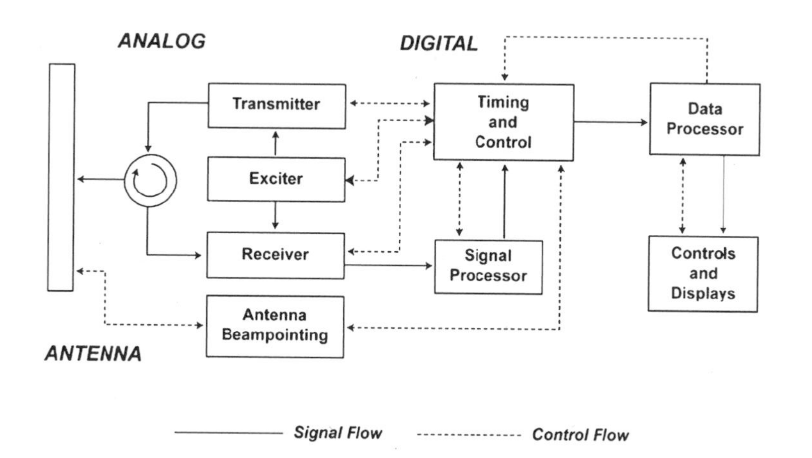

FIGURE 20.2 Radar system architecture.

There is a Fourier transform relationship between the antenna illumination function and the far-field antenna pattern analogous to spectral analysis. Hence, tapering the illumination to concentrate power near the center of the antenna suppresses spatial sidelobes while reducing the effective antenna aperture area. The phase and amplitude control of the antenna illumination determine the achievable sidelobe suppression and angle measurement accuracy.

Perturbations in the illumination due to the mechanical and electrical sources distort the illumination function and constrain performance in these areas. Mechanical illumination error sources include antenna shape deformation due to sag and thermal effects as well as manufacturing defects. Electrical illumination error is of particular concern in phased arrays where sources include beam steering computational error and phase shifter quantization. Control of both the mechanical and electrical perturbation errors is the key to both low sidelobes and highly accurate angle measurements. Control denotes that either tolerance is closely held and maintained or that there must be some means for monitoring and correction. Phased arrays are attractive for low sidelobe applications since they can provide element-level phase and amplitude control.

Transmitter

The transmitter function is to amplify waveforms to a power level sufficient for target detection and estimation. There is a general trend away from tube-based transmitters toward solid-state transmitters. In particular, solid-state transmit/receive modules appear attractive for constructing phased array radar systems. In this case, each radiating element is driven by a module that contains a solid-state transmitter, phase shifter, low-noise amplifier, and associated control components. Active electronically scanned arrays built from such modules appear to offer significant reliability advantages over radar systems driven from a single transmitter. In addition to the redundancy inherent in a distributed transmitter, the solid-state components and supporting low-voltage power supplies are more reliable than tubes or their attendant high-voltage power supplies. In general, solid-state transmitters offer less peak power than tube-based transmitters but can provide equivalent radar performance by operating at higher transmit duty cycles and enabling reduced RF losses. There is a strong trend in developmental radars toward the use of solid-state transmitters due to production base economics as well as performance considerations. Table 20.2 summarizes transmitter technologies.

Receiver and Exciter

This subsystem contains the precision timing and frequency reference source or sources used to derive the master oscillator and local oscillator reference frequencies. These reference frequencies are used to down convert received signals in a multiple-stage superheterodyne architecture to accommodate signal amplification and interference rejection. Filtering is conducted at the carrier and intermediate frequencies in processing to reject interference outside the operating band of the radar. The receiver front end is typically protected from overload during transmission through the combination of a circulator and a transmit/receive switch.

The exciter generates the waveforms for subsequent transmission. As in signal processing, the trend is toward programmable digital signal synthesis because of the associated flexibility and performance stability. Rejection of strong clutter via Doppler filtering in order to detect small low-altitude targets also demands very stable frequency reference sources as spurious sidebands of the clutter can mask the target returns.

Signal and Data Processing

Digital processing is generally divided between two processing subsystems, i.e., signals and data, according to the algorithm structure, and throughput demands. Signal processing includes pulse compression, Doppler filtering, and detection threshold estimation and testing. Data processing includes track filtering, user interface support, and such specialized functions as electronic protection and built-in test (BIT), as well as the resource management process required to control the radar system.

The signal processor is often optimized to perform the repetitive complex multiply-and-add operations associated with the fast Fourier transform (FFT). FFT processing is used for implementing *pulse compression* via fast convolution and for Doppler filtering. Pulse compression consists of matched filtering on receive to an intrapulse modulation imposed on the transmitted pulse. As delineated subsequently, the imposed intrapulse

TABLE 20.2 Median Target RCS (m^2)

Technology	Mode of Operation	Maximum Frequency (GHz)	Demonstrated Peak/ Average Power (kW)	Typical Gain (dB)	Typical Bandwidth
Thermionic					
Magnetron	Oscillator	95	1 MW/500 W @ X-band	n/a	Fixed-10%
Helix traveling wave tube (TWT)	Amplifier	95	4 kW/400 W @ X-band	40–60	Octave/multioctave
Ring-loop TWT	Amplifier	18	8 kW/200 W @ X-band	40–60	5–15%
Coupled-cavity TWT	Amplifier	95	120 kW/36 kW @ X-band	40–60	5–15%
Extended interaction oscillator (EIO)	Oscillator	280	1.4 kW/140 W @ 95 GHz	n/a	0.2% (elec.) 4% (mech.)
Extended interaction					
Klystron (EIK)	Amplifier	280	1.2 kW/120 W @ 95 GHz	40–50	0.5–1%
Klystron	Amplifier	36	50 kW/5 kW @ X-band	30–60	0.1–2% (inst.) 1–10% (mech.)
Crossed-field amplifier (CFA)	Amplifier	18	500 kW/1 kW @ X-band	10–20	5–15%
Solid-state					
Silicon BJT	Amplifier	5	300 W/30 W @ 1 GHz	5–10	10–25%
GaAs FET	Amplifier	40	15 W/5W @ X-band	5–10	5–30%
Impatt diode	Oscillator	170	30 W/10 W @ X-band	n/a	Fixed-5%

Source: Tracy V. Wallace, Georgia Tech Research Institute, Atlanta, GA.

bandwidth determines the range resolution of the pulse while the modulation format determines the suppression of the waveform matched-filter response outside the nominal range resolution extent. Fast convolution consists of taking the FFT of the digitized receiver output, multiplying it by the stored FFT of the desired filter function and then taking the inverse FFT of the resulting product. Fast convolution results in significant computational saving over performing the time-domain convolution of returns with the filter function corresponding to the matched filter. Similarly, Doppler filtering can be conducted by performing an N-point FFT across the equivalent range-sampling point of N successive pulses presupposing coherent transmission and reception. The signal processor output can be characterized in terms of range gates and Doppler filters corresponding approximately to the range and Doppler resolution, respectively.

In contrast, the radar data processor typically consists of a general-purpose computer with an operating system and application software suite supporting real-time operation. Fielded radar data processors range from microcomputers to mainframe computers, depending on the requirements of the radar system. The heterogonous algorithmic mixture of radar signal/data processing typically results in significant processing inefficiency so benchmarking is essential. The timing and control subsystem typically functions as the two-way interface between the data processor and the other radar subsystems. The increasing inclusion of BIT and built-in calibration capability in timing and control subsystem designs promises to result in significant improvement in fielded system performance and availability. The trend is toward increasing use of commercial off-the-shelf digital processing elements and software development/maintenance tools for radar applications and tighter integration of the signal and data processing functions. Radar signal/data processing is anticipated to benefit from increased usage of open system architectures.

Radar Performance Prediction

Radar Line-of-Sight

With the exception of over-the-horizon (OTH) radar systems, which exploit either sky-wave bounce or ground-wave propagation modes and sporadic ducting effects at higher frequencies, surface and airborne platform radar operation is limited to the refraction-constrained line of sight. Atmospheric refraction effects can be closely approximated by setting the Earth's radius to 4/3 its nominal value in estimating

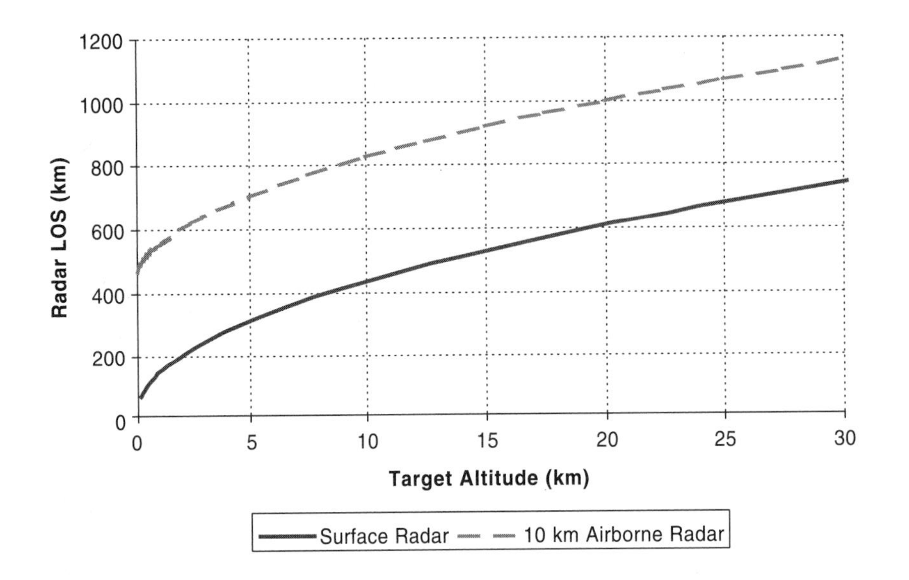

FIGURE 20.3 Maximum line-of-sight range for surface-based radar, an airborne surveillance radar, and a space-based radar.

horizon-limited range. The resulting line-of-sight range is depicted in Figure 20.3 for a surface-based radar and a notional 10-km altitude airborne radar.

As evident in the plot, airborne surveillance radar systems offer significant advantages in the detection of low-altitude targets that would otherwise be masked by Earth's curvature and terrain features from surface-based radars. However, efficient clutter rejection techniques must be used in order to detect targets since surface clutter returns will be present at almost all ranges of interest.

Radar Range Equation

The radar range equation is commonly used to estimate radar system performance, given that line-of-sight conditions are satisfied. This formulation essentially computes the signal-to-noise ratio (S/N) at the output of the radar signal processor. In turn, S/N is used to provide estimates of radar detection and position measurement performance as described in subsequent subsections. S/N can be calculated in terms of the number of pulses coherently integrated over a single coherent processing interval (CPI) using the radar range equation such that:

$$S/N = \frac{PGAT_pN_p\sigma}{(4\pi)^2R^4L_tL_{rn}L_{sp}kT_s} \tag{20.1}$$

where P is peak transmitter power output, G is directivity of the transmit antenna, A is effective aperture area of the receive antenna in meters squared, T_p is pulse duration, σ is *radar cross-section* in square meters, N_p is the number of coherently integrated pulses within the coherent processing interval, R is range to target in meters, L_t is system ohmic and nonohmic transmit losses, L_{rn} is system nonohmic receive losses, L_{sp} is signal processing losses, k is Boltzman's constant (1.38×10^{-23}K), and T_s is system noise temperature, including receive ohmic losses (Kelvin). S/N losses are derived relative to matched filter performance.

At X band and above it is prudent to include propagation loss due to atmospheric absorption (Blake, 1986). This form of the radar range equation is applicable to radar systems using pulse compression or pulse Doppler waveforms as well as the unmodulated single-pulse case. In many applications, average power is a better measure of system performance than peak power since it indicates the S/N improvement achievable with pulse integration over a given interval of time. Hence, the radar range equation can be modified such that:

$$S/N = \frac{P_a G A T_c \sigma}{(4\pi)^2 R^4 L_t L_{rn} L_{sp} k T_s} \tag{20.2}$$

where P_a is average transmitter power and T_c is the coherent processing interval (CPI).

The portion of time over which the transmitter is in operation is referred to as the radar duty cycle. The average transmitter power is the product of duty cycle and peak transmitter power. Duty cycle ranges from less than 1% for typical *noncoherent* pulse radars to somewhat less than 50% for high pulse repetition frequency (PRF) pulse Doppler or interrupted continuous wave radar systems. The CPI is the period over which returns are collected for *coherent* processing functions such as pulse integration and Doppler filtering. The CPI can be estimated as the product of the number of coherently integrated pulses and the interval between pulses. Noncoherent pulse integration is less efficient and alters the statistical character of the signal and interference.

Antenna Directivity and Aperture Area

The directivity of the antenna, G, is

$$G = \frac{4\pi A \eta}{\lambda^2} \tag{20.3}$$

where η is aperture efficiency and λ is radar carrier wavelength. Aperture inefficiency is due to the antenna illumination factor.

The common form of the radar range equation uses power gain rather than directivity. Antenna gain is equal to the directivity divided by the antenna losses. In the design and analysis of modern radars, directivity is a more convenient measure of performance because it permits designs with distributed active elements, such as solid-state phased arrays, to be assessed to permit direct comparison with passive antenna systems. Beamwidth and directivity are inversely related, as beamwidth is proportion to the RF carrier wavelength divided by antenna extent. A useful approximation summarizing these relationships is

$$G \approx \frac{30{,}000}{\theta_{az}\theta_{el}} \tag{20.4}$$

where θ_{az} and θ_{el} are the radar azimuth and elevation beamwidths, respectively, in degrees.

Radar Cross-Section

In practice, the *radar cross-section* (RCS) of a realistic target must be considered a random variable with an associated correlation interval. Targets are composed of multiple interacting scatters so that the composite return varies in magnitude with the constructive and destructive interference of the contributing returns. The target RCS is typically estimated as the mean or median of the target RCS distribution. The associated correlation interval indicates the rate at which the target RCS varies over time. RCS fluctuation degrades single-look target detection performance at moderate to high probability of detection but is mitigated by a search strategy that enables collecting several independent returns on any given target. The radar coherent processing interval is generally chosen sufficiently short that the RCS does not significantly change over that period.

The median RCS of typical targets is given in Table 20.3. The composite RCS measured by a radar system may be composed of multiple individual targets in the case of closely spaced targets such as a bird flock.

TABLE 20.3 Median Target RCS (m^2)

Carrier Frequency (GHz)	1–2	3	5	10	17
Aircraft (nose/tail avg.)					
Small propeller	2	3	2.5	—	—
Small jet (Lear)	1	1.5	1	1.2	—
T38-twin jet, F5	2	2–3	2	1–2/6	—
T39-Sabreliner	2.5	—	10/8	9	—
F4, large fighter	5–8/5	4–20/10	4	4	—
737, DC9, MD80	10	10	10	10	10
727, 707, DC8-type	22–40/15	40	30	30	—
DC-10-type, 747	70	70	70	70	—
Ryan drone	—	—	—	2/1	—
Standing man (180 lb)	0.3	0.5	0.6	0.7	0.7
Automobiles	100	100	100	100	100
Ships-incoming ($\times 10^4$ m^2)					
4 K tons	1.6	2.3	3.0	4.0	5.4
16 K tons	13	18	24	32	43
Birds					
Sea birds	0.002	0.001–0.004	0.004	—	—
Sparrow, starling, etc.	0.001	0.001	0.001	0.001	0.001

Slash marks indicate different set.

Source: F.E. Nathanson, *Radar Design Principles,* 2nd ed., New York: McGraw-Hill, 1991. With permission.

Loss and System Temperature Estimation

Sources of *S/N* loss include ohmic and nonohmic (mismatch) loss in the antenna and other radio frequency components, propagation effects, signal processing deviations from matched filter operation, detection thresholding, and search losses. Scan loss in phased array radars is due to the combined effects of the decrease in projected antenna area and element mismatch with increasing scan angle.

Search operations impose additional losses due to target position uncertainty. Because the target position is unknown before detection, the beam, range gate, and Doppler filter will not be centered on the target return. Hence, straddling loss will occur as the target effectively straddles adjacent resolution cells in range and Doppler. Beamshape loss is a consequence of the radar beam not being pointed directly at the target so that there is a resultant loss in both transmit and receive antenna gain. Other search losses include detection threshold loss associated with radar system adaptation to interference (Nathanson, 1991).

System noise temperature estimation corresponds to assessing the system thermal noise floor referenced to the antenna output. Assuming the receiver hardware is at ambient temperature, the system noise temperature can be estimated as

$$T_s = T_a + 290(L_{ro}F - 1) \tag{20.5}$$

where T_a is the antenna noise temperature, L_{ro} is receive ohmic losses, and F is the receiver noise figure.

In phased array radars, the thermodynamic temperature of the antenna receive beamformer may be significantly higher than ambient, so a more complete analysis is required. The antenna noise temperature is determined by the external noise received by the antenna from solar, atmospheric, Earth surface, and other sources. In addition, active electronically scanned arrays often utilize several levels of active amplification within the beamformer so the analysis must consider the composite system noise temperature imposed by spatial combining, amplifier characteristics, and intervening losses.

Table 20.4 provides typical loss and noise temperature budgets for several major radar classes. In general, loss increases with the complexity of the radar hardware between the transmitter/receiver and the antenna radiator.

TABLE 20.4 Typical Microwave Loss and System Temperature Budgets

	Mechanically Scanned Reflector Antenna	Electronically Scanned Slotted Array	Solid-State Phased Array
Nominal losses			
Transmit loss, $L_{t\ (dB)}$	1	1.5	0.5
Nonohmic receiver loss, $L_{r\ (dB)}$	0.5	0.5	0.1
Signal processing loss, $L_{sp\ (dB)}$	1.4	1.4	1.4
Scan loss (dB)	N/A	−30 log(cos [scan angle])	−30 log(cos [scan angle])
Search losses, L_{DS}			
Beam shape (dB)	3	3	3
Range gate straddle (dB)	0.5	0.5	0.5
Doppler filter straddle (dB)	0.5	0.5	0.5
Detection thresholding (dB)	1	1	1
System noise temperature (Kelvin)	500	600	400

Reflector antennas and active phased arrays impose relatively low loss, while passive array antennas impose relatively high loss.

Resolution and Accuracy

The fundamental resolution capabilities of a radar system are summarized in Table 20.5. In general, there is a trade-off between mainlobe resolution corresponding to the nominal range, Doppler, and angle resolution, and effective dynamic range corresponding to suppression of sidelobe components. This is evident in the use of weighting to suppress Doppler sidebands and angle sidelobes at the expense of broadening the mainlobe and S/N loss.

Cross range denotes either of the two dimensions orthogonal to the radar line of sight. Cross-range resolution in real-aperture antenna systems is closely approximated by the product of target range and radar beamwidth in radians. Attainment of the nominal ISAR/SAR cross-range resolution generally requires complex signal processing to generate a focused image, including correction for scatterer change in range over the CPI.

The best accuracy performance occurs for the case of thermal noise-limited error. The resulting accuracy is the resolution of the radar divided by the square root of the S/N and an appropriate monopulse or interpolation factor. In this formulation, the single-pulse S/N has been multiplied by the number of pulses integrated within the CPI as indicated in Equation (20.1) and Equation (20.2).

TABLE 20.5 Resolution and Accuracy

Dimension	Nominal Resolution	Noise-Limited Accuracy
Angle	$\frac{\alpha\lambda}{d}$	$\frac{\alpha\lambda}{dK_m\sqrt{2S/N}}$
Range	$\frac{\alpha C}{2B}$	$\frac{\alpha C}{2BK_i\sqrt{2S/N}}$
Doppler	$\frac{\alpha}{\mathrm{CPI}}$	$\frac{\alpha}{\mathrm{CPI}K_i\sqrt{2S/N}}$
SAR/ISAR	$\frac{\alpha\lambda}{2\Delta\theta}$	$\frac{\alpha\lambda}{2\Delta\theta K_i\sqrt{2S/N}}$

α, taper broadening factor, typically ranging from 0.89 (unweighted) to 1.3 (Hamming); d, antenna extent in azimuth/elevation; B, waveform bandwidth; K_m, monopulse slope factor, typically on the order of 1.5; K_i, interpolation factor, typically on the order of 1.8; $\Delta\theta$, line-of-sight rotation of target relative to radar over CPI.

In practice, accuracy is also constrained by environmental effects, target characteristics, and instrumentation error as well as the available S/N. Environmental effects include multipath and refraction. Target glint is characterized by an apparent wandering of the target position because of coherent interference effects associated with the composite return from the individual scattering centers on the target. Instrumentation error is minimized with alignment and calibration but may significantly constrain track filter performance as a result of the relatively long correlation interval of some error sources.

Radar Range Equation for Search and Track

The radar range equation can be modified to directly address performance in the two primary radar missions: search and track.

Search performance is basically determined by the capability of the radar system to detect a target of specific RCS at a given maximum detection range while scanning a given solid angle extent within a specified period of time. S/N can be set equal to the minimum value required for a given detection performance, $S/N|r$, while R can be set to the maximum required target detection range, $R_{\max}$. Manipulation of the radar range equation results in the following expression:

$$\frac{P_a A}{L_t L_r L_{sp} L_{os} T_s} \geqslant \left(\frac{S}{N}\right)_r \frac{R_{\max}^4 \Omega}{\sigma T_{fs}} \cdot 16 \tag{20.6}$$

where Ω is the solid angle over which search must be performed (steradians), T_{fs} is the time allowed to search Ω by operational requirements, and L_{os} is the composite incremental loss associated with search.

The left-hand side of the equation contains radar design parameters, while the right-hand side is determined by target characteristics and operational requirements. The right-hand side of the equation is evaluated to determine radar requirements. The left-hand side of the equation is evaluated to determine if the radar design meets the requirements.

The track radar range equation is conditioned on noise-limited angle accuracy as this measure stresses radar capabilities significantly more than range accuracy in almost all cases of interest. The operational requirement is to maintain a given data rate track providing a specified single-measurement angle accuracy for a given number of targets with specified RCS and range. Antenna beamwidth, which is proportional to the radar carrier wavelength divided by antenna extent, impacts track performance since the degree of S/N required for a given measurement accuracy decreases as the beamwidth decreases. Track performance requirements can be bounded as

$$\frac{P_a A^3}{\lambda^4 L_t L_r L_{sp} T_s} k_m^2 \eta^2 \geqslant 5k \frac{r N_t R^4}{\sigma \sigma_\theta^2} \tag{20.7}$$

where r is the single-target track rate, N_t is the number of targets under track in different beams, σ_θ is the required angle accuracy standard deviation (radians), and σ is the RCS. In general, a phased array radar antenna is required to support multiple target tracking when $N_t > 1$.

Incremental search losses are suppressed during single-target-per-beam tracking. The beam is pointed as closely as possible to the target to suppress beamshape loss. The tracking loop centers the range gate and Doppler filter on the return. Detection thresholding loss can be minimal since the track range window is small, although the presence of multiple targets generally mandates continual detection processing.

Radar Waveforms

Pulse Compression

Typical pulse radar waveforms are summarized in Table 20.6. In most cases, the signal processor is designed to closely approximate a matched filter. As indicated in Table 20.5, the range and Doppler resolution of any match-filtered waveform are inversely proportional to the waveform bandwidth and duration, respectively. Pulse compression, using modulated waveforms, is attractive since S/N is proportional to pulse duration rather than bandwidth in matched filter implementations. Ideally, the intrapulse modulation is chosen to

TABLE 20.6 Selected Waveform Characteristics

	Comments	Time Bandwidth Product	Range Sidelobes (dB)	S/N Loss (dB)	Range/Doppler Coupling	ECM/EMIR Robustness
Unmodulated	No pulse compression	~1	Not applicable	0	No	Poor
Linear frequency modulation	Linearly swept over bandwidth	>10	Unweighted: −13.5 Weighted: −<−40[1]	00.7–1.4	Yes	Poor
Nonlinear FM	Multiple variants	Waveform specific	Waveform specific	0	Waveform specific	Fair
Barker	N-bit biphase	≤13 (N)	−20 log (N)	0	No	Fair
LRS	N-bit biphase	~N; > 64/pulse[a]	~ −10 log (N)	0	No	Good
Frank	N-bit polyphase (N = integer^2)	~N	~ −10 log (π^{2N})	0	Limited	Good
Frequency coding	N subpulses noncoincidental in time and frequency	~N^2	Waveform specific ◆ Periodic ◆ Pseudorandom	0.7–1.40 / 0	Waveform specific	Good

[a]Constraint due to typical techonology limitations rather than fundamental waveform characteristics.

attain adequate range resolution and range sidelobe suppression performance while the pulse duration is chosen to provide the required sensitivity. Pulse compression waveforms are characterized as having a time bandwidth product (TBP) significantly greater than unity, in contrast to an unmodulated pulse, which has a TBP of approximately unity.

Pulse Repetition Frequency

The radar system pulse repetition frequency (PRF) determines its ability to unambiguously measure target range and range rate in a single CPI as well as determining the inherent clutter rejection capabilities of the radar system. In order to obtain an unambiguous measurement of target range, the interval between radar pulses (1/PRF) must be greater than the time required for a single pulse to propagate to a target at a given range and back. The maximum unambiguous range is then given by $C/(2/\text{PRF})$ where C is the velocity of electromagnetic propagation.

Returns from moving targets and clutter sources are offset from the radar carrier frequency by the associated Doppler frequency. As a function of range rate, V_r, the Doppler frequency, f_D, is given by $f_D = 2 \cdot V_r/\lambda$. A coherent pulse train samples the returns Doppler modulation at the PRF. Most radar systems employ parallel sampling in the in-phase and quadrature baseband channels so that the effective sampling rate is twice the PRF. The targets return is folded in frequency if the PRF is less than the target Doppler.

Clutter returns are primarily from stationary or near-stationary surfaces such as terrain. In contrast, targets of interest often have a significant range rate relative to the radar clutter. Doppler filtering can suppress returns from clutter. With the exception of frequency ambiguity, the Doppler filtering techniques used to implement pulse Doppler filtering are quite similar to those described for CW radar in Section 20.2. Ambiguous measurements can be resolved over multiple CPIs by using a sequence of slightly different PRFs and correlating detections among the CPIs.

Detection and Search

Detection processing consists of comparing the amplitude of each range gate/Doppler filter output with a threshold. A detection is reported if the amplitude exceeds that threshold. A false alarm occurs when noise or other interference produces an output of sufficient magnitude to exceed the detection threshold. As the detection threshold is decreased, the detection probability and the false alarm probability increase. S/N must be increased to enhance detection probability while maintaining a constant false alarm probability.

As noted in the subsection Radar Cross-Section, RCS fluctuation effects must be considered in assessing detection performance. The Swerling models which use chi-squared probability density functions (PDFs) of two and four degrees of freedom (DOF) are commonly used for this purpose. The Swerling 1 and 2 models are based on the two DOF PDF and can be derived by modeling the target as an ensemble of independent scatterers of comparable magnitude. This model is considered representative of complex targets such as aircraft. The Swerling 3 and 4 models use the four DOF PDF and correspond to a target with a single dominant scatterer and an ensemble of lesser scatterers. Missiles are sometimes represented by Swerling 2 and 4 models. The Swerling 1 and 3 models presuppose slow fluctuation such that the target RCS is constant from pulse to pulse within a scan. In contrast, the RCS of Swerling 2 and 4 targets is modeled as independent on a pulse to pulse basis.

Single-pulse detection probabilities for nonfluctuating, Swerling 1/2, and Swerling 3/4 targets are depicted in Figure 20.4. This curve is based on a typical false alarm number corresponding approximately to a false alarm probability of 10^{-6}. The difference in S/N required for a given detection probability for a fluctuating target relative to the nonfluctuating case is termed as the fluctuation loss.

The detection curves presented here and in most other references presuppose noise-limited operation. In many cases, the composite interference present at the radar system output will be dominated by clutter returns or electromagnetic interference such as that imposed by hostile electronic countermeasures. The standard

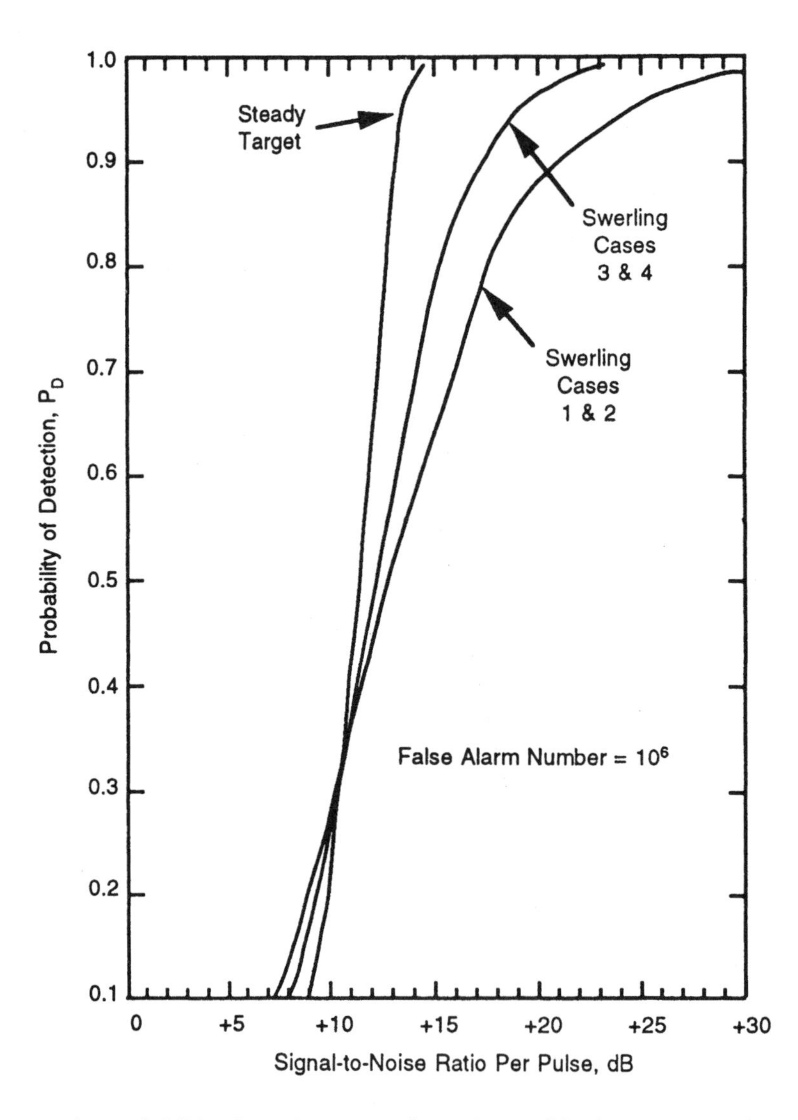

FIGURE 20.4 Detection probabilities for various target fluctuation models. (*Source:* F.E. Nathanson, *Radar Design Principles*, 2nd ed., New York: McGraw-Hill, 1991, p. 91. With permission.)

textbook detection curves cannot be applied in these situations unless the composite interference is statistically similar to thermal noise with a Gaussian PDF and a white power spectral density. The presence of non-Gaussian interference is generally characterized by an elevated false alarm probability, as the PDF tails are relatively greater. Adaptive detection threshold estimation techniques are often required to search for targets in environments characterized by such interference (Nathanson, 1991).

Estimation and Tracking

Measurement Error Sources

Radars measure target range and angle position, and potentially, Doppler frequency. Angle measurement performance is emphasized here since the corresponding cross-range error dominates range error for most practical applications. Target returns are generally smoothed in a tracking filter, but tracking performance is ultimately determined by the measurement accuracy and associated error characteristics of the subject radar system. Radar measurement error can be characterized as indicated in Table 20.7.

The radar design and the alignment and calibration process development must consider the characteristics and interaction of these error components. Integration of automated techniques to support alignment and calibration is an area of strong effort in modern radar design that can lead to significant performance improvement in fielded systems.

As indicated previously, angle measurement generally is the limiting factor in measurement accuracy. Target azimuth and elevation position is primarily measured by a monopulse technique in modern radars though early systems used sequential lobing and conical scanning. Specialized monopulse tracking radars utilizing reflectors have achieved instrumentation and S/N angle residual systematic error as low as 50 μrad. Phased array antennas have achieved a random error of less than 60 μrad, but the composite systematic residual errors remain to be measured. The limitations are primarily in the tolerance on the phase and amplitude of the antenna illumination function.

Figure 20.5 shows the monopulse beam patterns. The first is the received sum pattern that is generated by a feed that provides the energy from the reflector or phased array antenna through two ports in equal amounts and summed in phase in a monopulse comparator shown in Figure 20.6. The second is the difference pattern generated by providing the energy through the same two ports in equal amounts but taken out with a phase difference of p radians, giving a null at the center. A target located at the center of the same beam would receive a strong signal from the sum pattern with which the target could be detected and ranged. The received difference pattern would produce a null return, indicating the target was at the center of the beam. If the target were off the null, the signal output or difference voltage would be almost linear and proportional to the distance off the center (off-axis), as shown in the figure. This output of the monopulse processor is the real part of the dot product of the complex sums and the difference signals divided by the absolute magnitude of the sum signal squared, i.e.:

$$e_{\mathrm{d}} = \mathrm{Re}\left[\frac{\Sigma \cdot \Delta}{|\Sigma|^2}\right] \tag{20.8}$$

TABLE 20.7 Radar Measurement Error

Random errors	Those errors that cannot be predicted except on a statistical basis. The magnitude of the random error can be termed the *precision* and is an indication of the repeatability of a measurement.
Bias errors	A systematic error whether due to instrumentation or propagation conditions. A nonzero mean value of a random error.
Systematic error	An error whose quantity can be measured and reduced by calibration.
Residual systematic error	Those errors remaining after measurement and calibration. A function of the systematic and random errors in the calibration process.
Accuracy	The magnitude of the rms value of the residual systematic and random errors.

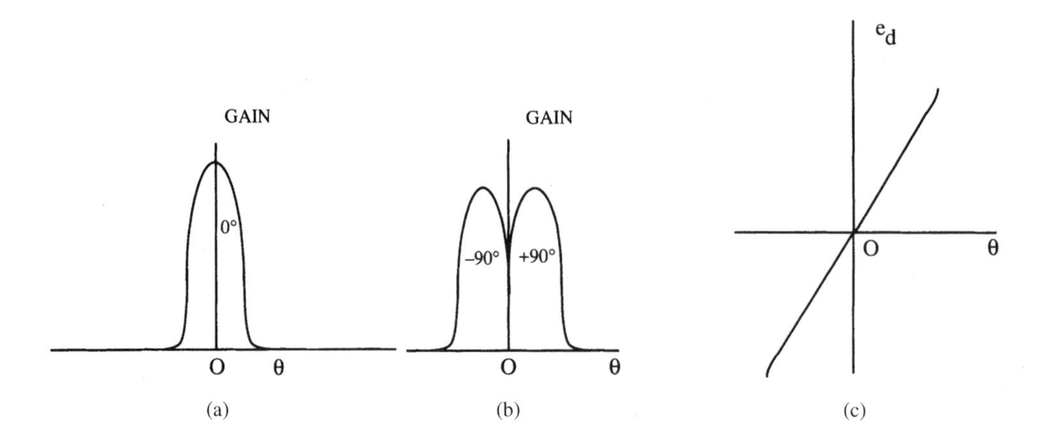

FIGURE 20.5 Monopulse beam patterns and difference voltage: (a) sum (*S*); (b) difference (*D*); (c) difference voltage.

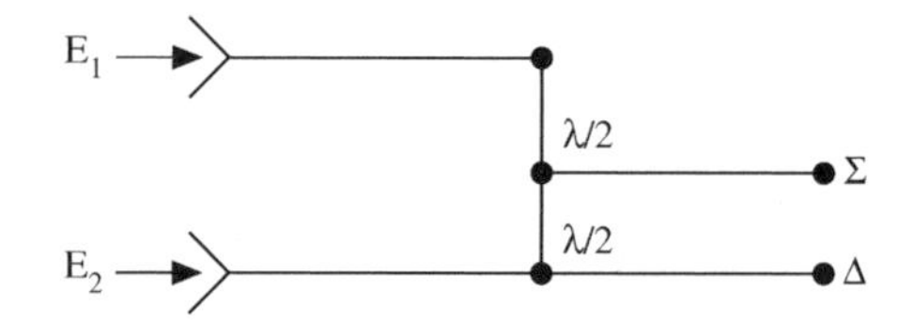

FIGURE 20.6 Monopulse comparator.

The random instrumentation measurement errors in the angle estimator are caused by phase and amplitude errors of the antenna illumination function. In reflector systems, such errors occur because of the position of the feedhorn, differences in electrical length between the feed and the monopulse comparator, mechanical precision of the reflector, and its mechanical rotation. In phased array radars, these errors are a function of the phase shifters, time-delay units, and combiners between the antenna elements and the monopulse comparator as well as the precision of the array. Although these errors are random, they may have correlation intervals considerably longer than the white noise considered in the thermal-noise random error and may depend upon the flight path of the target. For a target headed radially from or toward the radar, the correlation period of angle-measurement instrumental errors is essentially the tracking period. For crossing targets, the correlation interval may be pulse to pulse.

As in the estimate of range, the propagation effects of refraction and multipath also enter into the tracking error. The bias error in range and elevation angle by refraction can be estimated as

$$\Delta R = 0.007\, N_s \text{cosecant}\, E_0 \quad (\text{m}) \tag{20.9}$$

$$\Delta E_0 = N_s \cot E_0 \quad (\mu\text{rad})$$

where N_s is the surface refractivity and E_0 is the elevation angle (Barton and Ward, 1984).

One can calculate the average error in multipath. However, one cannot correct for it as in refraction since the direction of the error cannot be known in advance unless there are controlled conditions such as in a carefully controlled experiment. Hence, the general approach is to design the antenna sidelobes to be as low as feasible and accept the multipath error that occurs when tracking close to the horizon. There has been considerable research to find means to reduce the impact, including using very wide bandwidths to separate the direct path from the multipath return as well as specialized track filtering techniques that accommodate multipath effects.

Tracking Filter Performance

Target tracking based on processing returns from multiple CPIs generally provides a target position and velocity estimate of greater accuracy than the single-CPI measurement accuracy delineated in Table 20.5. In principle, the error variance of the estimated target position with the target moving at a constant velocity is approximately $4/n \cdot \sigma_{\mathrm{m}}^2$, where n is the number of independent measurements processed by the track filter and σ_{m} is the single measurement accuracy. In practice, the variance reduction factor afforded by a track filter is often limited to about one order of magnitude because of the reasons summarized in the following paragraphs.

Track filtering generally provides smoothing and prediction of target position and velocity via a recursive prediction–correction process. The filter predicts the targets position at the time of the next measurement based on the current smoothed estimates of position, velocity, and possibly acceleration. The subsequent difference between the measured position at this time and the predicted position is used to update the smoothed estimates. The update process incorporates a weighting vector that determines the relative significance given the track filter prediction versus the new measurement in updating the smoothed estimate.

Target model fidelity and adaptivity are fundamental issues in track filter mechanization. Independent one-dimensional tracking loops may be implemented to control pulse-to-pulse range gate positioning and antenna pointing. The performance of one-dimensional polynomial algorithms, such as the alpha-beta filter, to track targets from one pulse to the next and provide modest smoothing is generally adequate to maintain track. However, one-dimensional closed-loop tracking ignores knowledge of the equations of motion governing the target so that their smoothing and long-term prediction performance is relatively poor for targets with known equations of motion. In addition, simple one-dimensional tracking-loop filters do not incorporate any adaptivity or measure of estimation quality.

Kalman filtering addresses these shortcomings at the cost of significantly greater computational complexity and potentially increased maneuver response latency. Target equations of motion are modeled explicitly such that the position, velocity, and potentially higher-order derivatives of each measurement dimension are estimated by the track filter as a state vector. The error associated with the estimated state vector is modeled via a covariance matrix that is also updated with each iteration of the track filter. The covariance matrix determines the weight vector used to update the smoothed state vector in order to incorporate such factors as measurement *S*/*N* and dynamic target maneuvering.

Smoothing performance is constrained by the degree of a priori knowledge of the target's kinematic motion characteristics. For example, Kalman filtering can achieve significantly better error reduction against ballistic or orbital targets than against maneuvering aircraft. In the former case, the equations of motion are explicitly known, while the latter case imposes motion model error because of the presence of unpredictable pilot or guidance system commands. Similar considerations apply to the fidelity of the track filters model of radar measurement error. Failure to consider the impact of correlated measurement errors may result in underestimating track error when designing the system. A number of innovative track filtering techniques have been formulated to address these issues.

Many modern tracking problems are driven by the presence of multiple targets that impose a need for assigning measurements to specific tracks as well as accommodating unresolved returns from closely spaced targets. Existing radars generally employ some variant of the nearest-neighbor algorithm where a measurement is uniquely assigned to the track with a predicted position minimizing the normalized track filter update error. More sophisticated techniques assign measurements to multiple tracks if they cannot clearly be resolved or make the assignment on the basis of several contiguous update measurements.

Defining Terms

Coherent: Integration where magnitude and phase of received signals are preserved in summation.
Noncoherent: Integration where only the magnitude of received signals is summed.

Phased array: Antenna composed of an aperture of individual radiating elements. Beam scanning is implemented by imposing a phase taper across the aperture to collimate signals received from a given angle of arrival.

Pulse compression: The processing of a wideband, coded signal pulse, of initially long time duration and low-range resolution, to result in an output pulse of time duration corresponding to the reciprocal of the bandwidth.

Radar cross-section (RCS): Measure of the reflective strength of a radar target; usually represented by the symbol σ, measured in square meters, and defined as 4π times the ratio of the power per unit solid angle scattered in a specified direction of the power unit area in a plane wave incident on the scatterer from a specified direction.

References

D.K. Barton, and H.R. Ward, *Handbook of Radar Measurement*, Dedham, MA: Artech, 1984.
L.V. Blake, *Radar Range-Performance Analysis*, Dedham, MA: Artech, 1986.
J.L. Eaves and E.K. Reedy, Eds., *Principles of Modern Radar*, New York: Van Nostrand, 1987.
G.V. Morris, *Airborne Pulsed Doppler Radar*, Dedham, MA: Artech, 1988.
F.E. Nathanson, *Radar Design Principles*, 2nd ed., New York: McGraw-Hill, 1991.

Further Information

M.I. Skolnik, Ed., *Radar Handbook*, 2nd ed., New York: McGraw-Hill, 1990.
IEEE Standard Radar Definitions, IEEE Standard 686-1990, April 20, 1990.

20.2 Continuous Wave Radar

Samuel O. Piper and James C. Wiltse

Continuous wave (CW) radar employs a transmitter which is on all or most of the time. Unmodulated CW radar is very simple and is able to detect the Doppler-frequency shift in the return signal from a target which has a component of motion toward or away from the transmitter. While such a radar cannot measure range, it is used widely in applications such as police radars, motion detectors, burglar alarms, proximity fuzes for projectiles or missiles, illuminators for semiactive missile guidance systems (such as the Hawk surface-to-air missile), and scatterometers (used to measure the scattering properties of targets or clutter such as terrain surfaces) (Saunders, 1990; Ulaby and Elachi, 1990; Nathanson, 1991; Komarov, 2003; Pace, 2003).

Modulated versions include frequency-modulated (FMCW), interrupted frequency-modulated (IFMCW), and phase-modulated. Typical waveforms are indicated in Figure 20.7. Such systems are used in altimeters, Doppler navigators, proximity fuzes, over-the-horizon radar, and active seekers for terminal guidance of air-to-surface missiles. The term *continuous* is often used to indicate a relatively long waveform (as contrasted to pulse radar using short pulses) or a radar with a high duty cycle (for instance, 50% or greater, as contrasted with the typical duty cycle of less than 10% for the usual pulse radar). As an example of a long waveform, planetary radars may transmit for up to 10 h and are thus considered to be CW (Freiley et al., 1992). Another example is interrupted CW (or pulse-Doppler) radar, where the transmitter is pulsed at a high rate for 10 to 60% of the total time (Nathanson, 1991). All of these modulated CW radars are able to measure range.

The first portion of this section discusses concepts, principles of operation, and limitations. The latter portion describes various applications. In general, CW radars have several potential advantages over pulse radars. Advantages include simplicity and the facts that the transmitter leakage is used as the local oscillator,

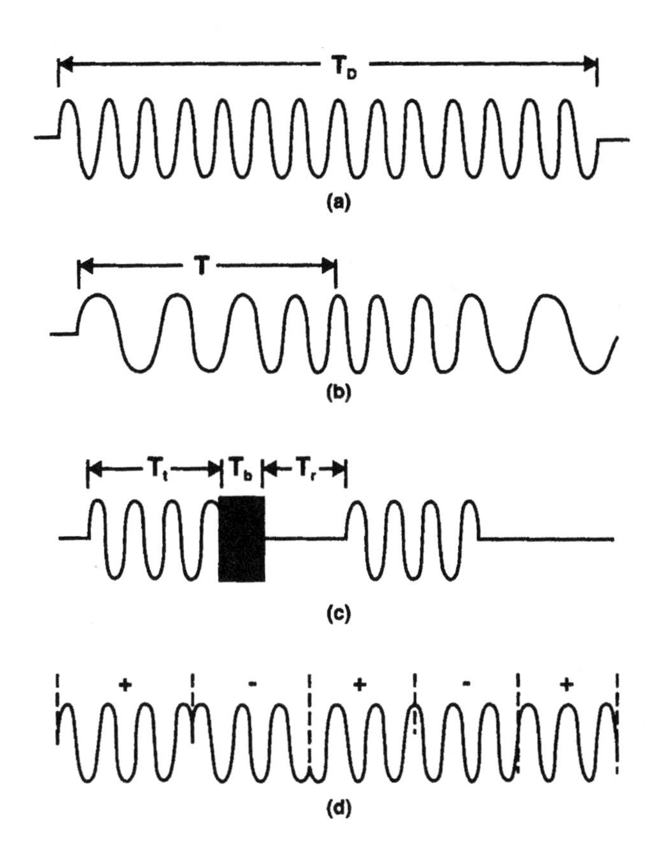

FIGURE 20.7 Waveforms for the general class of CW radar: (a) continuous sine wave CW; (b) frequency modulated CW; (c) interrupted CW; (d) binary phase-coded CW. (*Source*: F.E. Nathanson, *Radar Design Principles*, New York: McGraw-Hill, 1991, p. 450. With permission.)

transmitter spectral spread is minimal (not true for wide-deviation FMCW), and peak power is the same as (or only a little greater than) the average power. This latter situation means that the radar is less detectable by simple intercepting equipment.

The largest disadvantage for CW radars is the need to provide antenna isolation (reduce spillover) so that the transmitted signal does not interfere with the receiver. In a pulse radar, the transmitter is off before the receiver is enabled (by means of a duplexer and/or receiver-protector switch). Isolation is frequently obtained in the CW case by employing two antennas, one for transmit and one for reception. When this is done, there is also a reduction of close-in clutter return from rain or terrain. A second disadvantage is the existence of noise sidebands on the transmitter signal which reduce sensitivity because the Doppler frequencies are relatively close to the carrier. This is considered in more detail below.

CW Doppler Radar

If a sine wave signal is transmitted, the return from a moving target would be Doppler-shifted in frequency by an amount given by the following equation:

$$f_D = \frac{2v_r f_T}{c} = \text{Doppler frequency} \tag{20.10}$$

where f_T = transmitted frequency; c = velocity of propagation, 3×10^8 m/sec; and v_r = radial component of velocity between radar and target.

TABLE 20.8 Doppler Frequencies for Several Transmitted Frequencies and Various Relative Speeds (1 m/sec = 2.237 mph)

Microwave Frequency — f_T (GHz)	Relative Speed			
	1 m/sec (Hz)	300 m/sec (kHz)	1 mph (Hz)	600 mph (kHz)
3	20	6	8.9	5.4
10	67	20	30	17.9
35	233	70	104	63
95	633	190	283	170

Using Equation (20.10) the Doppler frequencies have been calculated for several speeds and are given in Table 20.8.

As may be seen, the Doppler frequencies at 10 GHz (X band) range from 30 Hz to about 18 kHz for a speed range between 1 and 600 mph. The spectral width of these Doppler frequencies will depend on target fluctuation and acceleration, antenna scanning effects, frequency variation in oscillators or components (for example, due to microphonism from vibrations), but most significantly, by the spectrum of the transmitter, which inevitably will have noise sidebands that extend much higher than these Doppler frequencies, probably by orders of magnitude. At higher microwave frequencies, the Doppler frequencies are also higher and more widely spread. In addition, the spectra of higher frequency transmitters are also wider, and, in fact, the transmitter noise-sideband problem is usually worse at higher frequencies, particularly at millimeter wavelengths (i.e., above 30 GHz). These characteristics may necessitate frequency stabilization or phaselocking of transmitters to improve the spectra.

Simplified block diagrams for CW Doppler radars are shown in Figure 20.8. The transmitter is a single-frequency source, and leakage (or coupling) of a small amount of transmitter power serves as a local oscillator signal in the mixer. This is called homodyning. The transmitted signal will produce a Doppler-shifted return from a moving target. In the case of scatterometer measurements, where, for example, terrain reflectivity is to be measured, the relative motion may be produced by moving the radar (perhaps on a vehicle) with respect to the stationary target (Wiltse et al., 1957). The return signal is collected by the antenna and then also fed to the

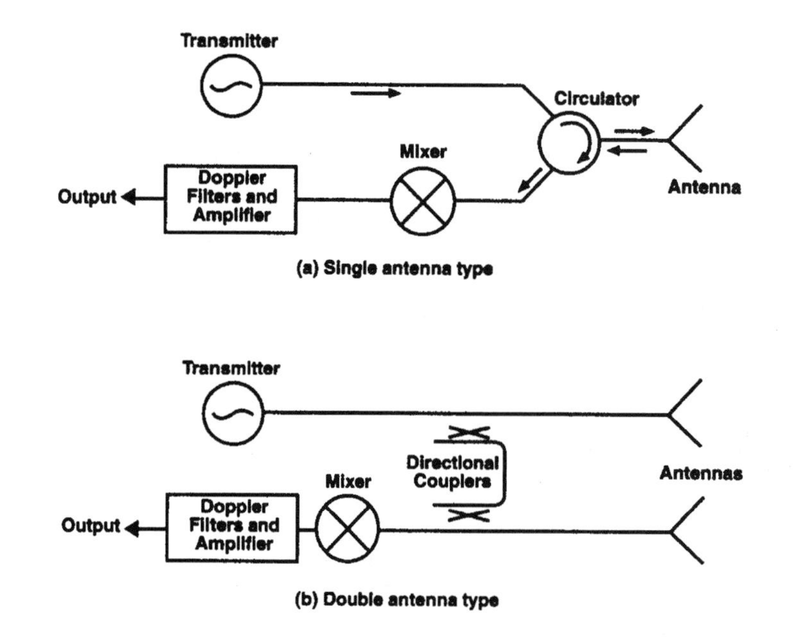

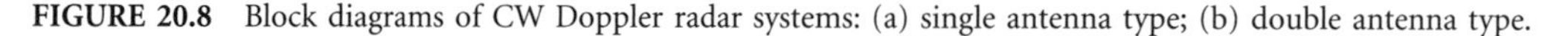

FIGURE 20.8 Block diagrams of CW Doppler radar systems: (a) single antenna type; (b) double antenna type.

mixer. After mixing with the transmitter leakage, a difference frequency will be produced which is the Doppler shift. As indicated in Table 20.8, this difference is apt to range from low audio to over 100 kHz, depending on relative speeds and choice of microwave frequency. The Doppler amplifier and filters are chosen based on the information to be obtained, and this determines the amplifier bandwidth and gain, as well as the filter bandwidth and spacing. The transmitter leakage may include reflections from the antenna or nearby clutter in front of the antenna, as well as mutual coupling between antennas in the two-antenna case.

The detection range for such a radar can be obtained from the following (Nathanson, 1991):

$$R^4 = \frac{\bar{P}_T G_T L_T A_e L_R L_p L_a L_s \sigma_T}{(4\pi)^2 k T_s b (S/N)} \tag{20.11}$$

where R = the detection range of the desired target

$\bar{P}_T$ = the average power

G_T = the transmit power gain of the antenna with respect to an omnidirectional radiator

L_T = the losses between the transmitter output and free space including power dividers, waveguide or coax, radomes, and any other losses not included in A_e

A_e = the effective aperture of the antenna, which is equal to the projected area in the direction of the target times the efficiency

L_R = the receive antenna losses defined in a manner similar to the transmit losses

L_p = the beamshape and scanning and pattern factor losses

L_a = the two-way-pattern propagation losses of the medium; often expressed as $\exp(-2\alpha R)$, where α is the attenuation constant of the medium and the factor 2 is for a two-way path

L_s = signal-processing losses that occur for virtually every waveform and implementation

σ_T = the radar cross section of the object that is being detected

k = Boltzmann's constant (1.38×10^{-23} W s/K)

T_s = system noise temperature

b = Doppler filter or speedgate bandwidth

S/N = signal-to-noise ratio

S_{min} = the minimum detectable target-signal power that, with a given probability of success, the radar can be said to detect, acquire, or track in the presence of its own thermal noise or some external interference. Since all these factors (including the target return itself) are generally noiselike, the criterion for a detection can be described only by some form of probability distribution with an associated probability of detection P_D and a probability that, in the absence of a target signal, one or more noise or interference samples will be mistaken for the target of interest.

While the Doppler filter should be a matched filter, it is usually wider because it must include the target spectral width. There is usually some compensation for the loss in detectability by the use of postdetection filtering or integration. The *S/N* ratio for a CW radar must be at least 6 dB, compared with the value of 13 dB required for pulse radars when detecting steady targets (Nathanson, 1991, p. 449).

The Doppler system discussed above has a maximum detection range based on signal strength and other factors, but it cannot measure range. The rate of change in signal strength as a function of range has sometimes been used in fuzes to estimate range closure and firing point, but this is a relative measure.

FMCW Radar

The most common technique for determining target range is the use of frequency modulation. Typical modulation waveforms include sinusoidal, linear sawtooth, or triangular, as illustrated in Figure 20.9. For a linear sawtooth, a frequency increasing with time may be transmitted. Upon being reflected from a stationary point target, the same linear frequency change is reflected back to the receiver, except it has a time delay

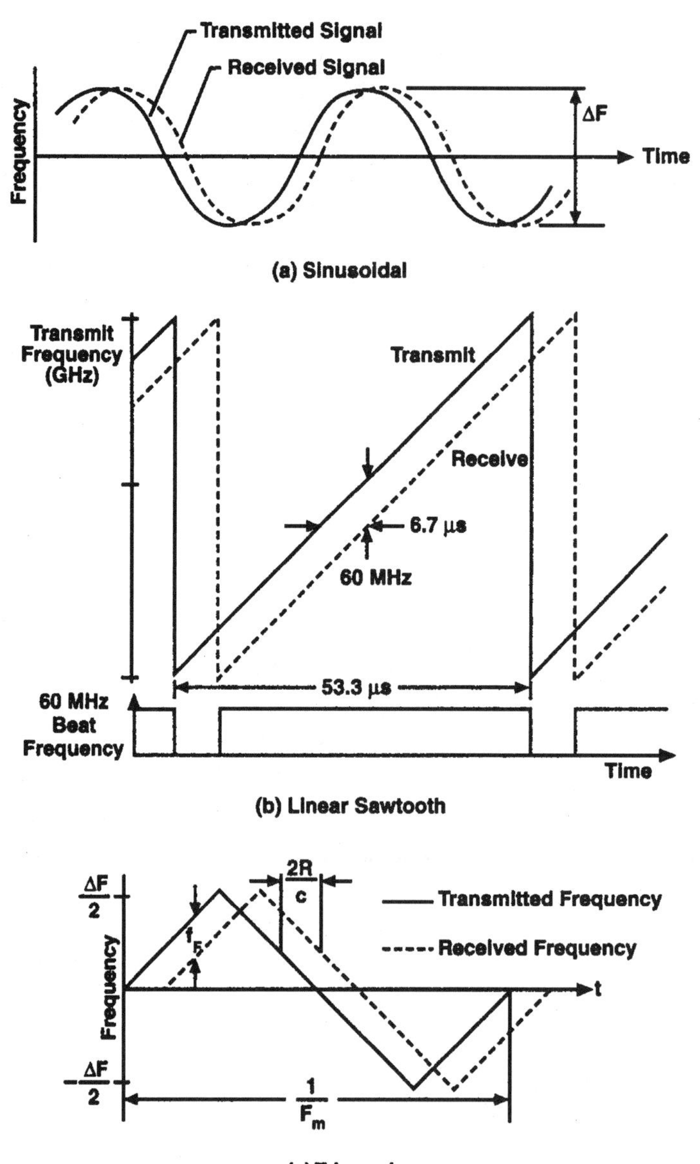

FIGURE 20.9 Frequency vs. time waveforms for FMCW radar: (a) sinusoidal, (b) linear sawtooth, (c) triangular modulations.

which is related to the range to the target. The time is $T = (2R)/c$ where R is the range. The received signal is mixed with the transmit signal and the difference or beat frequency (F_b) is obtained. (The sum frequency is much higher and is rejected by filtering.) For a stationary target the beat frequency for a sawtooth waveform is given by

$$F_b = \frac{2R}{c} \cdot \Delta F \cdot F_m \tag{20.12}$$

and for a triangle wave the beat frequency is

$$F_{\mathrm{b}} = \frac{4R}{c} \cdot \Delta F \cdot F_{\mathrm{m}} \tag{20.13}$$

where $\Delta F =$ frequency deviation and $F_{\mathrm{m}} =$ modulation rate.

The beat frequency is constant except near the turnaround region of the sawtooth, but, of course, it is different for targets at different ranges. (If it is desired to have a constant intermediate frequency for different ranges, which is a convenience in receiver design, then the modulation rate or the frequency deviation must be adjusted.) Multiple targets at a variety of ranges will produce multiple-frequency outputs from the mixer and are handled in the receiver by a bank of range-bin filters as may be provided by digital signal processing using fast Fourier transform (FFT) techniques.

If the target is moving with a component of velocity toward (or away from) the radar, then there will be a Doppler frequency component added to (or subtracted from) the difference frequency (F_{b}), and the Doppler will be slightly higher at the upper end of the sweep range than at the lower end. This will introduce an uncertainty or ambiguity in the measurement of range, which may or may not be significant depending on the parameters chosen and the application. For example, if the Doppler frequency is low (as in an altimeter) and the difference frequency is high, the error in range measurement may be tolerable. For the symmetrical triangular waveform, a Doppler less than F_{b} averages out, since it is higher on one-half of a cycle and lower on the other half. With a sawtooth modulation, only a decrease or increase is noted since the frequencies produced in the transient during a rapid flyback are out of the receiver passband. Exact analyses of triangular, sawtooth, dual triangular, dual sawtooth, and combinations of these with noise have been carried out by Tozzi (1972). Specific design parameters are given later in this chapter for an application utilizing sawtooth modulation in a missile terminal guidance seeker.

For the case of sinusoidal frequency modulation the spectrum consists of a series of lines spaced away from the carrier by the modulation frequency or its harmonics. The amplitudes of the carrier and these sidebands are proportional to the values of the Bessel functions of the first kind (J_n, $n = 0, 1, 2, 3, \ldots$), whose argument is a function of the modulation frequency and range. By choosing a particular modulation frequency, the values of the Bessel functions and thus the characteristics of the spectral components can be influenced. For instance, the signal variation with range at selected ranges can be optimized, which is important in fuzes. A short-range dependence that produces a rapid increase in signal, greater than that corresponding to the normal range variation, is beneficial in producing well-defined firing signals. This can be accomplished by proper choice of modulation frequency and filtering to obtain the signal spectral components corresponding to the appropriate order of the Bessel function. In a similar fashion, spillover and reflections from close-in objects can be reduced by filtering to pass only certain harmonics of the modulation frequency (F_{m}). Receiving only frequencies near $3F_{\mathrm{m}}$ results in considerable spillover rejection, but at a penalty of 4 to 10 dB in signal-to-noise ratio (Nathanson, 1991).

For the sinusoidal modulation case, Doppler frequency contributions complicate the analysis considerably. For details of this analysis the reader is referred to Saunders (1990) or Nathanson (1991).

Interrupted Frequency-Modulated CW (IFMCW)

To improve isolation during reception, the IFMCW format involves preventing transmission for a portion of the time during the frequency change. Thus, there are frequency gaps, or interruptions, as illustrated in Figure 20.10. This shows a case where the transmit time equals the round-trip propagation time, followed by an equal time for reception. This duty factor of 0.5 for the waveform reduces the average transmitted power by 3 dB relative to using an uninterrupted transmitter. However, the improvement in the isolation should reduce the system noise by more than 3 dB, thus improving the signal-to-noise ratio (Piper, 1987). For operation at short range, Piper points out that a high-speed switch is required (Piper, 1987). He also points out that the ratio of frequency deviation to beat frequency should be an even integer and that the minimum ratio is typically 6, which produces an out-of-band loss of 0.8 dB.

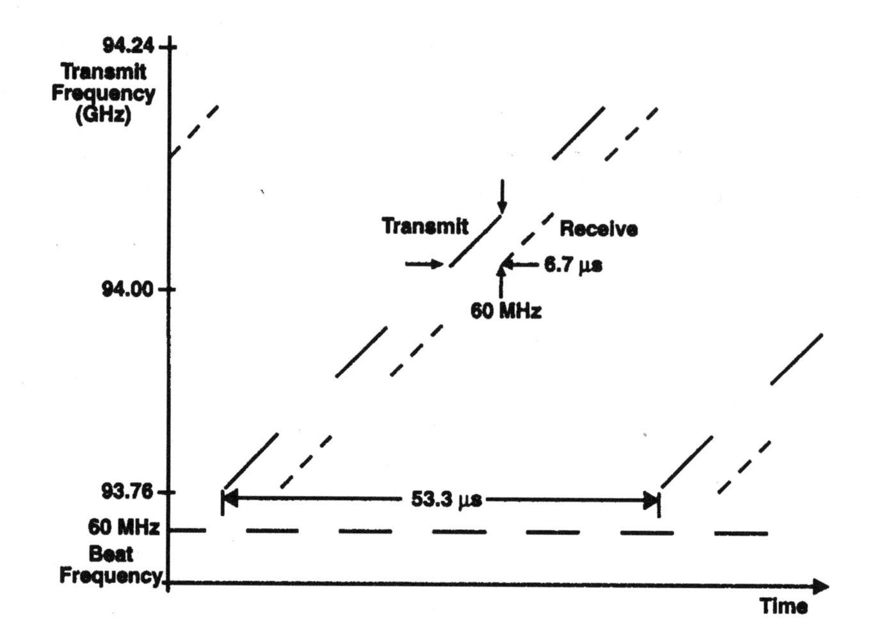

FIGURE 20.10 Interrupted FMCW waveform. (*Source:* S.O. Piper, "MMW seekers," in *Principles and Applications of Millimeter Wave Radar*, N. Currie and C.E. Brown, Eds., Norwood, MA: Artech House, 1987, p. 683. With permission.)

IFMCW may be compared with pulse compression radar if both use a wide bandwidth. Pulse compression employs a "long" pulse (i.e., relatively long for a pulse radar) with a large frequency deviation or "chirp." A long pulse is often used when a transmitter is peak-power limited, because the longer pulse produces more energy and gives more range to targets. The frequency deviation is controlled in a predetermined way (frequently a linear sweep) so that a matched filter can be used in the receiver. The large time-bandwidth product permits the received pulse to be compressed in time to a short pulse in order to make an accurate range measurement. A linear-sawtooth IFMCW having similar pulse length, frequency deviation, and pulse repetition rate would thus appear similar, although arrived at from different points of view.

Applications

Space does not permit giving a full description of the many applications mentioned at the beginning of this chapter, but several will be discussed.

Radar Proximity Fuzes

Projectiles or missiles designed to be aimed at ships or surface land targets often need a height-of-burst (HOB) sensor (or target detection device) to fire or fuze the warhead at a height of a few meters. There are two primary generic methods of sensing or measuring height to generate the warhead fire signal. The most obvious, and potentially the most accurate, is to measure target round trip propagation delay employing conventional radar ranging techniques. The second method employs a simple CW Doppler radar or variation thereof, with loop gain calibrated in a manner that permits sensing the desired burst height by measurement of target return signal amplitude and/or rate of change. Often the mission requirements do not justify the complexity and cost of overcoming the short range eclipsing challenges associated with the radar ranging approach. Viable candidates are thus narrowed down to variations on the CW Doppler fuze.

In its simplest form, the CW Doppler fuze consists of a fractional watt RF oscillator, homodyne detector, Doppler amplifier, Doppler envelope detector, and threshold circuit. When the Doppler envelope amplitude derived from the returned signal reaches the preset threshold, a fire signal is generated. The height at which the fire signal occurs depends on the radar loop gain, threshold level, and target reflectivity. Fuze gain is designed to produce the desired height of burst under nominal trajectory angle and target reflectivity conditions, which may have large fluctuations due to glint effects, and deviations from the desired height due to antenna gain variations with angle, target reflectivity, and fuze gain tolerances are accepted. A loop gain change of 6 dB, (2 to 1 in voltage), whether due to a change in target reflection coefficient, antenna gain, or whatever, will result in a 2 to 1 HOB change.

HOB sensitivity to loop gain factors can be reduced by utilizing the slope of the increasing return signal, or so-called rate-of-rise. Deriving HOB solely from the rate-of-rise has the disadvantage of rendering the fuze sensitive to fluctuating signal levels such as might result from a scintillating target. The use of logarithmic amplifiers decreases the HOB sensitivity to the reflectivity range. An early (excessively high) fire signal can occur if the slope of the signal fluctuations equals the rate-of-rise threshold of the fuze. In practice a compromise is generally made in which Doppler envelope amplitude and rate-of-rise contribute in some proportion of HOB.

Another method sometimes employed to reduce HOB sensitivity to fuze loop gain factors and angle of fall is the use of FM sinusoidal modulation of suitable deviation to produce a range correlation function comprising the zero order of a Bessel function of the first kind. The subject of sinusoidal modulation is quite complex, but has been treated in detail by Saunders (1990, pp. 14.22–14.46 and 14.41). The most important aspects of fuze design have to do with practical problems such as low cost, small size, ability to stand very high-g accelerations, long life in storage, and countermeasures susceptibility.

Police Radars

Down-the-road police radars, which are of the CW Doppler type, operate at 10.525 (X Band), 24.150 (K Band), or in the 33.4 to 36.0 GHz (Ka band) range, frequencies approved in the United States by the Federal Communications Commission. Antenna half-power beamwidths are typically in the 0.21 to 0.31 radian (12° to 18°) range. The sensitivity is usually good enough to provide a range exceeding 800 meters. Target size has a dynamic range of 30 dB (from smallest cars or motorcycles to large trucks). This means that a large target can be seen well outside the antenna 3-dB point at a range exceeding the range of a smaller target near the center of the beam. Thus, there can be uncertainty about which vehicle is the target. Fisher (1992) has given a discussion of a number of the limitations of these systems, but despite these factors probably tens of thousands have been built.

The transmitter is typically a Gunn oscillator in the 30 to 100 mW power range, and antenna gain is usually around 20 to 24 dB, employing circular polarization. The designs typically have three amplifier gains for detection of short, medium, or maximum range targets, plus a squelch circuit so that sudden spurious signals will not be counted. Provision is made for calibration to assure the accuracy of the readings. Speed resolution is about 1 mph. The moving police radar system uses stationary (ground) clutter to derive patrol car speed. Then closing speed minus patrol car speed yields target speed.

The limitations mentioned about deciding which vehicle is the correct target have led to the development of laser police radars, which utilize much narrower beamwidth, making target identification much more accurate. Of course, the use of microwave and laser radars has spawned the development of automotive radar detectors, which are also in wide use.

Altimeters

A very detailed discussion of FMCW altimeters has been given by Saunders (1990, pp. 14.34–14.36), in which he has described commercial products built by Bendix and Collins. The parameters will be summarized below and if more information is needed, the reader may want to turn to other references (Bendix Corp., 1982; Saunders, 1990; Maoz et al., 1991; Stratahos, 2000). In his material, Saunders gives a general overview of altimeters, all of which use wide-deviation FM at a low modulation frequency. He discusses the limitations on

TABLE 20.9 Parameters for Two Commercial Altimeters

Manufacturer and Model	Modulation Frequency	Frequency Deviation	Prime Power	Weight (pounds)	Radiated Power
Bendix ALA-52A	150 Hz	130 MHz	30 W	11[a]	
Collins ALT-55	100 kHz	100 MHz		8	350 mW

[a]Not including antenna and indicator.

narrowing the antenna pattern, which must be wide enough to accommodate attitude changes of the aircraft. Triangular modulation is used, since for this waveform the Doppler averages out, and dual antennas are employed. There may be a step error or quantization in height (which could be a problem at low altitudes), due to the limitation of counting zero crossings. A difference of one zero crossing (i.e., 0.5 Hz) corresponds to 0.75 m for a frequency deviation of 100 MHz. Irregularities are not often seen, however, since meter response is slow. Also, if terrain is rough, there will be actual physical altitude fluctuations. Table 20.9 shows some of the altimeters' parameters.

These altimeters are not acceptable for military aircraft because their relatively wide-open front ends make them potentially vulnerable to electronic attack (EA) or electronic countermeasures (ECM). A French design has some advantages in this respect by using a variable frequency deviation, a difference frequency that is essentially constant with altitude, and a narrowband front-end amplifier (Saunders, 1990).

Doppler Navigators

These systems are mainly sinusoidally modulated FMCW radars employing four separate downward looking beams aimed at about 15° off the vertical. Because commercial airlines have shifted to nonradar forms of navigation, these units are designed principally for helicopters. Saunders (1990) cites a particular example of a commercial unit operating at 13.3 GHz, employing a Gunn oscillator as the transmitter, with an output power of 50 mW, and utilizing a 30-kHz modulation frequency. A single microstrip antenna is used. A low-altitude piece of equipment (below 15,000 ft), the unit weighs less than 12 lb. A second unit cited has an output power of 300 mW, dual antennas, dual modulation frequencies, and an altitude capability of 40,000 ft.

Phase Modulated CW Radar

An example of a phase modulated CW radar is a developmental obstacle avoidance radar described by Honeywell. This 35 GHz radar is derived from a 4.3 GHz high duty cycle biphase modulated CW covert radar altimeter waveform for helicopter low-level collision avoidance (Proctor, 1997). This obstacle avoidance radar system, developed by NASA Ames and Honeywell, has a 20° vertical by 50° horizontal field of view (FOV). Figure 20.11 shows the radar on a helicopter.

This radar successfully detected metal high tension towers and cables, water towers and small radio towers, wood telephone poles, vehicles, boats, fences, leafless trees, and ground return at less than 4° grazing angle (Raymer and Weingartner, 1994).

The radar requirements include 914.4 m (3000 ft) maximum range in 16 mm/h rainfall rate and for −20 dB surface reflectivity for 5° grazing angle. The maximum velocity is 150 knots and the false alarm rate is 5%. Low cost and low probability of intercept and detection were also required (Becker and Almsted, 1995). Table 20.10 lists the radar parameters.

As described in Table 20.11, the 32 nsec biphase code chip corresponds to 4.8 m (16 ft) range resolution. This will require a 31.3 MHz phase modulator and an analog-to-digital converter (ADC) sample rate of at least 62.5 MHz, along with 62.5 MHz memory. A code sequence of at least 190 chips is required for 0.9 km maximum range. For a maximal length code length of 190 the range sidelobe levels will be approximately 23 dB. The radar receiver must perform a 190-element correlation. For a brute force correlator this will require 190 multiplies every 16 nsec or 11.9×10^9 multiplies per second.

FIGURE 20.11 Honeywell 35 GHz biphase modulated CW obstacle avoidance radar (http://ccf.arc.nasa.gov/dx/basket/pix/RASCAL.jpg). (Photo: Dominic Hart.)

TABLE 20.10 Honeywell Phase Modulated CW Radar Parameters

Parameter	Value
RF center frequency	35 GHz
Transmit power	35 mW
Receiver noise figure	6 dB
Antenna gain	34 dBi
Beamwidth	3°
Antenna sidelobes	25 dB
Azimuth field of view	±45°
Elevation field of view	±10°
Biphase code chip	32 nsec or 16 ft
Codes transmitted	1, 5, 7, 11, and 13 bit
Receiver bandwidth	25 kHz

TABLE 20.11 Honeywell Phase Modulated CW Radar Waveform

Biphase Modulation Parameter	Performance	Requirements
32 nsec chip length (1120 35 GHz cycles per chip)	4.8 m (16 ft) range resolution	31.3 MHz bandwidth phase modulator 62.5 MHz ADC sample rate—8 to 10 bit (48 to 60 dB) dynamic range (ADC technology limitation) 62.5 MHz memory
190 code length (190 code length × 32 nsec chip length ~6 μsec code repetition interval)	0.9 km (3 kft) unambiguous range 23 dB range sidelobe levels	190 element correlator 190 multiplies every 16 nsec 11.9×10^9 multiplies per second

TABLE 20.12 Phase Modulated CW Challenges

To Improve Performance	Benefit	Requirements
Decrease chip length (analogous to compressed pulse length of pulse radar)	• Finer range resolution	• Wider bandwidth phase modulator • Faster ADC sample rate; reduces dynamic range (ADC technology limitation) • Faster memory
Increase code length Code length × chip length (analogous to pulse repetition interval of pulse radar)	• Longer unambiguous range • Lower range sidelobe levels	• Longer correlator • More operations

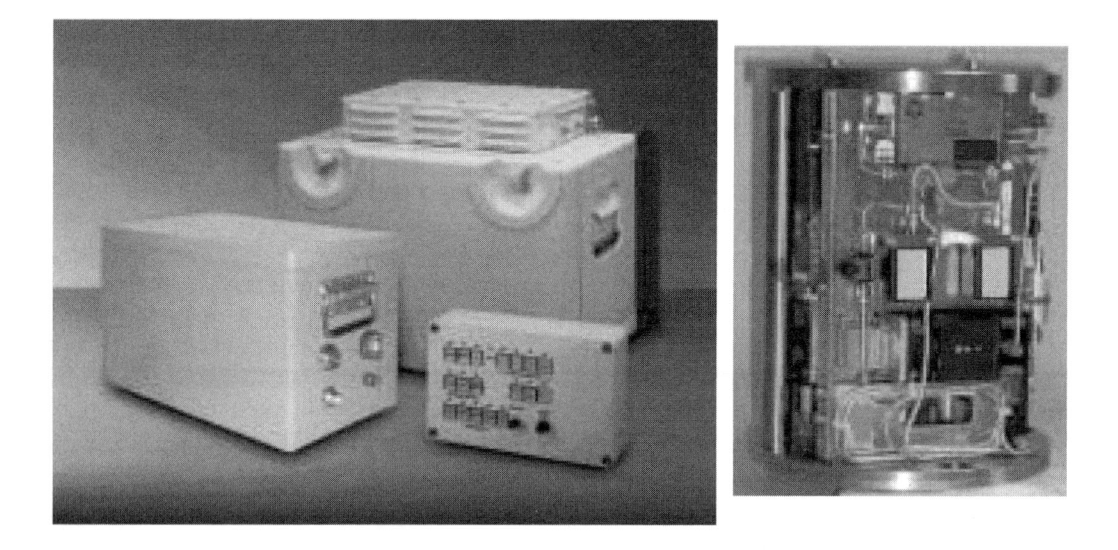

FIGURE 20.12 PILOT FMCW radar transceiver, signal processor, and remote control panel on left and transceiver interior on right (http://products.saab.se/PDBWeb/GetFile.aspx?pathtype = ProductFiles&filetype = Files&id = 1118).

This example illustrates the challenges for fine-resolution phase-modulated CW radar. As described in Table 20.12 in order to achieve finer range resolution shorter chip length is required which will require a corresponding increase in the phase modulator bandwidth and the ADC sample rate along with the memory. Longer unambiguous range or lower range sidelobe levels will require longer code sequences requiring a longer correlator with more operations.

PILOT FMCW Radars

The PILOT radar from Saab Bofors Dynamics was first developed in the 1980s. It is a navigation and detection radar for use on ships and submarines that employs the FMCW waveform to achieve low probability of intercept (LPI) performance. Table 20.13 lists the parameters of the PILOT radar and Figure 20.12 shows the PILOT FMCW radar units. The PILOT Mk3 version includes frequency agility over the 400 MHz bandwidth from 9.1 to 9.5 GHz to enhance its LPI performance and a reflected power canceller (RPC) to permit the system to operate with standard single navigation radar antennas. The PILOT transmit power of 1 W gives it the same average power and

TABLE 20.13 PILOT Mk3 FMCW Radar Parameter Summary

	PILOT
RF center frequency	9.3 GHz, X(I)-Band
Frequency agility bandwidth	400 MHz
Features	Reflected Power Canceller
Output power	1, 10, 100, and 1,000 mW
Receiver noise figure	5 dB
Instrumented range	2.4, 6, 24 nautical miles
Range cell size	2.4, 6, 24 meters

detection range of pulsed navigation radars, but it enjoys an advantage relative to simple intercept receivers and antiradiation missile (ARM) receivers that are not matched to the FMCW waveform. In addition the operator can reduce the transmit power in 10-dB increments down to 1 mW for operation at shorter ranges and against larger targets (http://products.saab.se/PDBWeb/GetFile.aspx?pathtype = ProductFiles&filetype = Files&id = 1118).

Frequency Shift Keying CW Radar

CW radar can also use frequency modulation that steps back and forth between two frequencies. Eaton VORAD radars use this frequency shift keying (FSK) CW waveform. One version operated at 24.725-GHz RF center frequency dwelling approximately 5 msec at each of two frequencies separated by 500 kHz, which yields 300-m maximum unambiguous range. The radar had 0.5-mW transmit power, and 4.0° azimuth and 5.5° elevation beamwidths (Will, 1995).

The Eaton VORAD model EVT-300 Collision Warning System operates at 24.725 GHz with 3 mW (typical) transmitted RF power. The operating range is from 0.9 to 107 m (3 to 350 ft) with 0.5 to 120 mph host vehicle speeds and 0.25 to 100 mph vehicle closing rates. Drivers' warnings are issued via beeper and LEDs. The "SmartCruise" add-on offers the adaptive cruise control (ACC) function (http://truck.eaton.com/na/spec_products/product_features_specs/vorad/service.htm).

Millimeter-Wave Seeker for Terminal Guidance Missile

Terminal guidance for short-range (less than 2 km) air-to-surface missiles has seen extensive development in the past two decades. Targets such as tanks are frequently immersed in a clutter background which may give a radar return that is comparable to that of the target. To reduce the clutter return in the antenna footprint, the antenna beamwidth is reduced by going to millimeter wavelengths. For a variety of reasons the choice is usually a frequency near 35 or 90 GHz. Antenna beamwidth is inversely proportional to frequency, so in order to get a reduced beamwidth we would normally choose 90 GHz; however, more deleterious effects at 90 GHz due to atmospheric absorption and scattering can modify that choice. Despite small beamwidths, the clutter is a significant problem, and in most cases signal-to-clutter is a more limiting condition than signal-to-noise in determining range performance. Piper (1987) analyzed the situation for 35- and 90-GHz pulse radar seekers and compared them with a 90-GHz FMCW seeker. His FMCW results are summarized below.

In his approach to the problem, Piper gives a summary of the advantages and disadvantages of a pulse system compared to the FMCW approach. One difficulty for the FMCW can be emphasized here. That is the need for a highly linear sweep, and, because of the desire for the wide bandwidth, this requirement is accentuated. Direct digital chirp synthesizers (DDCS) offer high fidelity linear FM sweeps over wide bandwidths. The STEL-2375B DDCS and the STEK-9949 DDCS from ITT Microwave offer 400-MHz bandwidth with 0.23-Hz frequency steps (http://www.ittmicrowave.com/ddsModules.asp?subsection = 0, 6 June 2004 and http://www.ittmicrowave.com/ddsModules.asp?subsection = 1, 6 June 2004). The wide bandwidth is desired in order to average the clutter return and to smooth the glint effects. In particular, glint occurs from a complex target because of the vector addition of coherent signals scattered back to the receiver from various reflecting surfaces. At some angles the vectors may add in phase (constructively) and at others they may cancel, and the effect is specifically dependent on wavelength. For a narrowband system, glint may provide a very large signal change over a small variation of angle, but, of course, at another wavelength it would be different. Thus, very wide bandwidth is desirable from this smoothing point of view, and typical numbers used in millimeter-wave radars are in the 450- to 650-MHz range. Piper chose 480 MHz.

Another tradeoff involves the choice of FM waveform. Here, the use of a triangular waveform is undesirable because the Doppler frequency averages out and Doppler compensation is then required. Thus, the sawtooth version is chosen, but because of the large frequency deviation desired, the difficulty of linearizing the frequency sweep is made greater. In fact many components must be extremely wideband, and this generally increases cost and may adversely affect performance. However, the difference frequency (F_b) and the intermediate frequency (F_{IF}) will be higher and thus further from the carrier, so the phase noise will be lower. After discussing the other tradeoffs, Piper chose 60 MHz for the beat frequency. A modern FMCW system is likely to use a lower beat frequency to reduce the ADC sample rate requirement for digital signal processing.

With a linear FM/CW waveform, the inverse of the frequency deviation provides the theoretical time resolution, which is 2.1 nsec for 480 MHz (or range resolution of 0.3 m). For an RF sweep linearity of 300 kHz, the range resolution is actually 5 m at the 1000-m nominal search range. (The system has a mechanically scanned antenna.) An average transmitting power of 25 mW was chosen, which was equal to the average power of the 5-W peak IMPATT assumed for the pulse system. The antenna diameter was 15 cm. For a target radar cross section of 20 m^2 and assumed weather conditions, the signal-to-clutter and signal-to-noise ratios were calculated and plotted for ranges out to 2 km and for clear weather or 4 mm per hour rainfall. The results show that for 1 km range the target-to-clutter ratios are higher for the FMCW case than the pulse system in clear weather or in rain, and target-to-clutter is the determining factor.

Summary Comments

From this brief review it is clear that there are many uses for CW radars, and various types (such as fuzes) have been produced in large quantities. Because of their relative simplicity, today there are continuing trends toward the use of digital processing and integrated circuits. In fact, this is exemplified in articles describing FMCW radars built on single microwave integrated circuit chips (Maoz et al., 1991; Chang et al., 1995; Haydl et al., 1999; Menzel, 1999).

Defining Terms

Doppler-frequency shift: The observed frequency change between the transmitted and received signal produced by motion along a line between the transmitter/receiver and the target. The frequency increases if the two are closing and decreases if they are receding.

Missile terminal guidance seeker: Located in the nose of a missile, a small radar with short-range capability which scans the area ahead of the missile and guides it during the terminal phase toward a target such as a tank.

Pulse Doppler: A coherent radar, usually having high pulse repetition rate and duty cycle and capable of measuring the Doppler frequency from a moving target. Has good clutter suppression and thus can see a moving target despite background reflections.

References

R.C. Becker and L. Almsted, Honeywell, "Flight test evaluation of a 35 GHz forward looking altimeter for terrain avoidance," *IEEE AES Systems Magazine*, pp. 19–22, 1995.

Bendix Corporation, Service Manual for ALA-52A Altimeter; Design Summary for the ALA-52A, Bendix Corporation, Ft. Lauderdale, FL, May 1982.

K.W. Chang, H. Wang, G. Shreve, J.G. Harrison, M. Core. A. Paxton. M. Yu, C.H. Chen, and G.S. Dow, "Forward-looking automotive radar using a W-band single-chip transceiver," *IEEE Trans. Microwave Theory Tech.*, vol. 43, 1995, pp. 1659–1668.

Collins (Rockwell International), ALT-55 Radio Altimeter System; Instruction Book, Cedar Rapids, IA, October 1984.

P.D. Fisher, "Improving on police radar," *IEEE Spectrum*, vol. 29, pp. 38–43, 1992.

A.J. Freiley, B.L. Conroy, D.J. Hoppe, and A.M. Bhanji, "Design concepts of a 1-MW CW X-band transmit/receive system for planetary radar," *IEEE Trans. Microwave Theory Tech.*, vol. 40, pp. 1047–1055, 1992.

W.H. Haydl, et al., "Single-chip coplanar 94 GHz FMCW radar sensors," *IEEE Microwave Guided Wave Lett.*, vol. 9, pp. 73–75, February 1999.

Igor V. Komarov, Sergey M. Smolskiy, and David K. Barton (translator), *Fundamentals of Short-Range FM Radar*, Artech House, September 2003.

B. Maoz, L.R. Reynolds, A. Oki, and M. Kumar, "FM-CW radar on a single GaAs/AlGaAs HBT MMIC chip," *IEEE Microwave Millim.-Wave Monolithic Circ. Symp. Dig.*, pp. 3–6, June 1991.

W. Menzel, D. Pilz, and R. Lererer, "A 77-GHz FM/CW radar front-end with a low-profile low-loss printed antenna," *IEEE Trans. Microwave Theory Tech.*, 47, 2237–2241, December 1999.

F.E. Nathanson, *Radar Design Principles*, New York: McGraw-Hill, 1991, pp. 448–467.

Phillip E. Pace, *Detecting and Classifying Low Probability of Intercept Radar*, Artech House; Book and CD-ROM edition, November, 2003.

S.O. Piper, "MMW seekers," in *Principles and Applications of Millimeter Wave Radar*, N.C. Currie and C.E. Brown, Eds., Norwood, MA: Artech House, 1987, chap. 14.

P. Proctor, "Low-Level Collision Avoidance Tested," *Aviat. Week Space Tech.*, vol. 144, p. 58, 1997.

K. Raymer and T. Weingartner, "Advanced Terrain Data Processor," *Proceedings of the IEEE/AIAA 13th Digital Avionics Systems Conference (DASC)*, 31 October to 3 November 1994, pp. 636–639.

W.K. Saunders, "CW and FM radar," in *Radar Handbook*, M. I. Skolnik, Ed., New York: McGraw-Hill, 1990, chap. 14.

G.E. Stratakos, P. Bourgas, and K. Gotsis, "A low cost, high accuracy radar altimeter," *Microwave J.*, 43, 120–128, February 2000.

L.M. Tozzi, Resolution in Frequency-Modulated Radars, Ph.D. thesis, University of Maryland, College Park, MD, 1972.

F.T. Ulaby and C. Elachi, *Radar Polarimetry for Geoscience Applications*, Norwood, MA: Artech House, 1990, pp. 193–200.

J.D. Will, "Monopulse Doppler radar for vehicle applications," *Proceedings of the Intelligent Vehicles '95 Symposium*, 25–26 September 1995, p. 42.

J.C. Wiltse, S.P. Schlesinger, and C.M. Johnson, "Back-scattering characteristics of the sea in the region from 10 to 50 GHz," *Proc. IRE*, vol. 45, pp. 220–228, 1957.

Further Information

For a general treatment, including analysis of clutter effects, Nathanson's (1991) book is very good and generally easy to read. For extensive detail and specific numbers in various actual cases, Saunders (1990) gives good coverage. The treatment of millimeter-wave seekers by Piper (1987) is comprehensive and easy to read.

21

Lightwave

Samuel O. Agbo
California Polytecnic State University

Gerd Keiser
Photonics Comm Solutions, Inc.

21.1 Lightwave Waveguides

Samuel O. Agbo

Lightwave waveguides fall into two broad categories: dielectric slab waveguides and optical fibers. As illustrated in Figure 21.1, slab waveguides generally consist of a middle layer (the film) of **refractive index** n_1 and lower and upper layers of refractive indices n_2 and n_3, respectively.

Optical fibers are slender glass or plastic cylinders with annular cross-sections. The core has a refractive index, n_1, which is greater than the refractive index, n_2, of the annular region (the cladding). Light propagation is confined to the core by total internal reflection, even when the fiber is bent into curves and loops. Optical fibers fall into two main categories: step-index and graded-index (GRIN) fibers. For step-index fibers, the refractive index is constant within the core. For GRIN fibers, the refractive index is a function of radius r given by

$$n(r) = \begin{cases} n_1\left[1 - 2\Delta\left(\dfrac{r}{a}\right)^{\alpha}\right]^{1/2}; & r<a \\ n_1(1-2\Delta)^{1/2} = n_2; & a<r \end{cases} \tag{21.1}$$

In Equation (21.1), Δ is the **relative refractive index difference,** a is the core radius, and α defines the type of graded-index profile. For triangular, parabolic, and step-index profiles, α is, respectively, 1, 2, and ∞. Figure 21.2 shows the raypaths in step-index and graded-index fibers and the cylindrical coordinate system used in the analysis of lightwave propagation through fibers. Because rays propagating within the core in a GRIN fiber undergo progressive refraction, the raypaths are curved (sinusoidal in the case of parabolic profile).

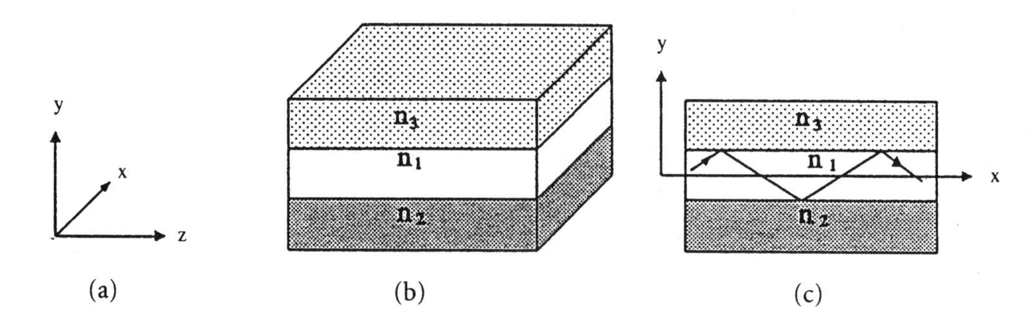

FIGURE 21.1 Dielectric slab waveguide: (a) the Cartesian coordinates used in analysis of slab waveguides; (b) the slab waveguide; (c) light guiding in a slab waveguide.

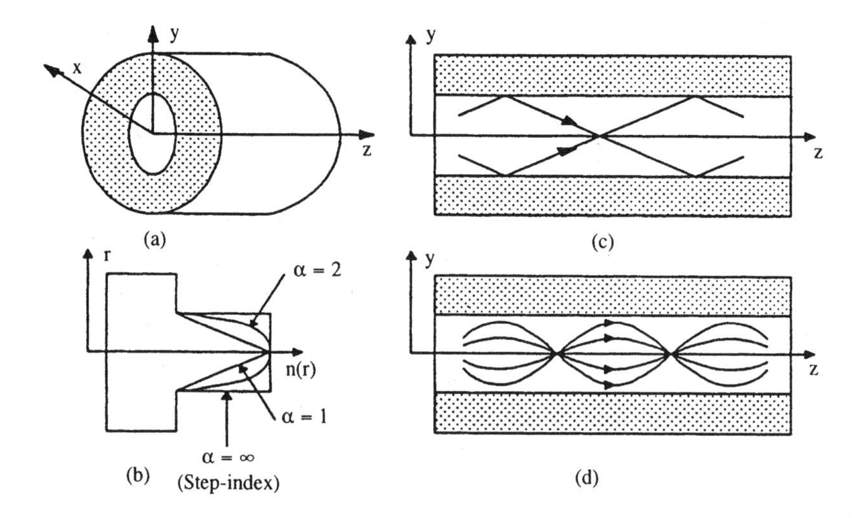

FIGURE 21.2 The optical fiber: (a) the cylindrical coordinate system used in analysis of optical fibers; (b) some graded-index profiles; (c) raypaths in step-index fiber; (d) raypaths in graded-index fiber.

Ray Theory

Consider Figure 21.3, which shows possible raypaths for light coupled from air (refractive index n_0) into the film of a slab waveguide or the core of a step-index fiber. At each interface, the transmitted raypath is governed by Snell's law. As θ_0 (the acceptance angle from air into the waveguide) decreases, the angle of incidence θ_i increases until it equals the critical angle, θ_c, making θ_0 equal to the maximum acceptance angle, θ_a. According to ray theory, all rays with acceptance angles less than θ_a propagate in the waveguide by total internal reflections. Hence, the numerical aperture (NA) for the waveguide, a measure of its light-gathering ability, is given by

$$\mathrm{NA} = n_0 \sin\theta_\mathrm{a} = n_1 \sin\left(\frac{\pi}{2} - \theta_\mathrm{c}\right) \tag{21.2}$$

By Snell's law, $\sin\theta_\mathrm{c} = n_2/n_1$. Hence:

$$\mathrm{NA} = [n_1^2 - n_2^2]^{1/2} \tag{21.3}$$

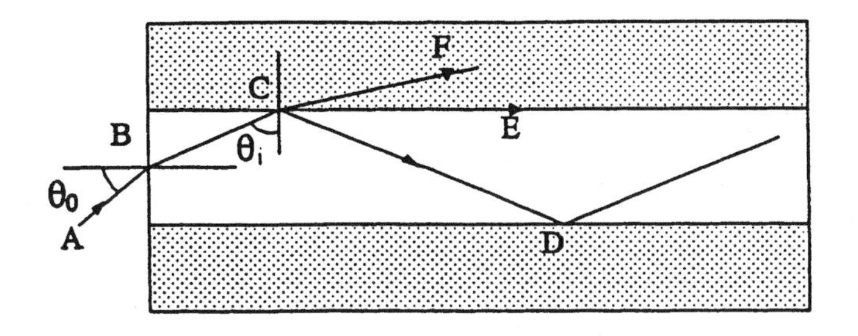

FIGURE 21.3 Possible raypaths for light coupled from air into a slab waveguide or a step-index fiber.

For step-index fibers, the preceding analysis applies to **meridional rays**. Skew (nonmeridional) rays have larger maximum acceptance angles, θ_{as}, given by

$$\sin\theta_{as} = \frac{\text{NA}}{\cos\gamma} \tag{21.4}$$

where NA is the numerical aperture for meridional rays and γ is the angle between the core radius and the projection of the ray onto a plane normal to the fiber axis.

Wave Equation for Dielectric Materials

Only certain discrete angles, instead of all acceptance angles less than the maximum acceptance angle, lead to guided propagation in lightwave waveguides. Hence, ray theory is inadequate, and wave theory is necessary, for analysis of light propagation in optical waveguides.

For lightwave propagation in an unbounded dielectric medium, the assumption of a linear, homogeneous, charge-free, and nonconducting medium is appropriate. Assuming also sinusoidal time dependence of the fields, the applicable Maxwell's equations are

$$\nabla \times \mathbf{E} = -j\omega\mu\mathbf{H} \tag{21.5a}$$

$$\nabla \times \mathbf{H} = j\omega\epsilon\mathbf{E} \tag{21.5b}$$

$$\nabla \times \mathbf{E} = 0 \tag{21.5c}$$

$$\nabla \times \mathbf{H} = 0 \tag{21.5d}$$

The resulting wave equations are

$$\nabla^2\mathbf{E} - \gamma^2\mathbf{E} = 0 \tag{21.6a}$$

$$\nabla^2\mathbf{H} - \gamma^2\mathbf{H} = 0 \tag{21.6b}$$

where

$$\gamma^2 = \omega^2\mu\epsilon = (j\kappa)^2 \tag{21.7}$$

and

$$\kappa = n\kappa_0 = \omega\sqrt{\mu\epsilon} = \frac{\omega}{v} \tag{21.8}$$

In Equation (21.8), κ is the phase propagation constant and n is the refractive index for the medium, while κ_0 is the phase propagation constant for free space. The velocity of propagation in the medium is $v = 1/\sqrt{\mu\epsilon}$.

Modes in Slab Waveguides

Consider a plane wave polarized in the y direction and propagating in z direction in an unbounded dielectric medium in the Cartesian coordinates. The vector wave Equation (21.6) lead to the scalar equations:

$$\frac{\partial^2 E_y}{\partial z} - \partial^2 E_y = 0 \tag{21.9a}$$

$$\frac{\partial^2 H_x}{\partial z} - \partial^2 H_x = 0 \tag{21.9b}$$

The solutions are

$$E_y = A\mathrm{e}^{j(\omega t - \kappa z)} \tag{21.10a}$$

$$H_x = \frac{-E_y}{\eta} = \frac{A}{\eta}\mathrm{e}^{j(\omega t - \kappa z)} \tag{21.10b}$$

where A is a constant and $\eta = \sqrt{\mu\epsilon}$ is the intrinsic impedance of the medium.

Because the film is bounded by the upper and lower layers, the rays follow the zigzag paths as shown in Figure 21.3. The upward and downward traveling waves interfere to create a standing wave pattern. Within the film, the fields transverse to the z axis, which have even and odd symmetry about the x axis, are given, respectively, by

$$E_y = A\cos(hy)\mathrm{e}^{j(\omega t - \beta z)} \tag{21.11a}$$

$$E_y = A\sin(hy)\mathrm{e}^{j(\omega t - \beta z)} \tag{21.11b}$$

where β and h are the components of κ parallel to and normal to the z axis, respectively. The fields in the upper and lower layers are evanescent fields decaying rapidly with attenuation factors α_3 and α_2, respectively, and are given by

$$E_y = A_3\mathrm{e}^{-\alpha_3\left(y - \frac{d}{2}\right)}\mathrm{e}^{j(\omega t - \beta z)} \tag{21.12a}$$

$$E_y = A_2\mathrm{e}^{-\alpha_2\left(y + \frac{d}{2}\right)}\mathrm{e}^{j(\omega t - \beta z)} \tag{21.12b}$$

Only waves with raypaths for which the total phase change for a complete (up and down) zigzag path is an integral multiple of 2π undergo constructive interference, resulting in guided modes. Waves with raypaths not satisfying this mode condition interfere destructively and die out rapidly. In terms of a raypath with an angle of incidence $\theta_i = \theta$ in Figure 21.3, the mode conditions (Haus, 1984) for fields transverse to the z axis and with even and odd symmetry about the x axis are given, respectively, by

$$\tan\left(\frac{hd}{2}\right) = \frac{1}{n_1 \cos\theta}[n_1^2 \sin^2\theta - n_2^2]^{1/2} \tag{21.13a}$$

$$\tan\left(\frac{hd}{2} - \frac{\pi}{2}\right) = \frac{1}{n_1 \cos\theta}[n_1^2 \sin^2\theta - n_2^2]^{1/2} \tag{21.13b}$$

where $h = \kappa \cos\theta = (2\pi n_1/\lambda) \cos\theta$ and λ is the free space wavelength.

Equation (21.13a) and Equation (21.13b) are transcendental, have multiple solutions, and are better solved graphically. Let $(d/\lambda)_0$ denote the smallest value of d/λ, the film thickness normalized with respect to the wavelength, satisfying Equation (21.13a) and Equation (21.13b). Other solutions for both even and odd modes are given by

$$\left(\frac{d}{\lambda}\right)_{\mathrm{m}} = \left(\frac{d}{\lambda}\right)_0 + \frac{m}{2n_1 \cos\theta} \tag{21.14}$$

where m is a nonnegative integer denoting the order of the mode.

Figure 21.4 (Palais, 1992) shows a **mode chart** for a symmetrical slab waveguide obtained by solving Equation (21.13a) and Equation (21.13b). For the $\mathrm{TE_m}$ modes, the E field is transverse to the direction (z) of propagation, while the H field lies in a plane parallel to the z axis. For the $\mathrm{TM_m}$ modes, the reverse is the case. The highest-order mode that can propagate has a value m given by the integer part of

$$m = \frac{2d}{\lambda}[n_1^2 - n_2^2]^{1/2} \tag{21.15}$$

To obtain a single-mode waveguide, d/λ should be smaller than the value required for $m = 1$, so that only the $m = 0$ mode is supported. To obtain a multimode waveguide, d/λ should be large enough to support many modes.

Shown in Figure 21.5 are transverse mode patterns for the electric field in a symmetrical slab waveguide. These are graphical illustrations of the fields given by Equation (21.11) and Equation (21.12). Note that, for

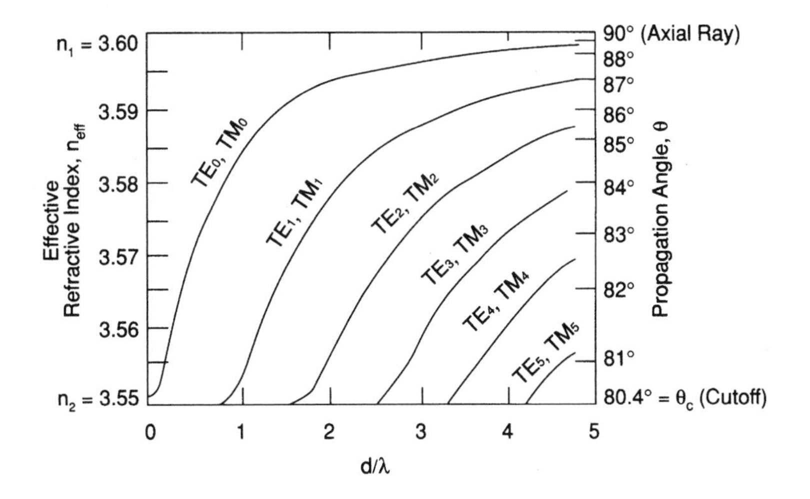

FIGURE 21.4 Mode chart for the symmetric slab waveguide with $n_1 = 3.6$, $n_2 = 3.55$. (*Source:* J.C. Palais, *Fiber Optic Communications*, Englewood Cliffs, NJ: Prentice-Hall, 1992, p. 84. With permission.)

TE_m, the field has m zeros in the film, and the evanescent field penetrates more deeply into the upper and lower layers for high-order modes.

For asymmetric slab waveguides, the equations and their solutions are more complex than those for symmetric slab waveguides. Shown in Figure 21.6 (Palais, 1992) is the mode chart for the asymmetric slab waveguide. Note that the TE_m and TM_m modes in this case have different propagation constants and do not overlap. By contrast, for the symmetric case, TE_m and TM_m modes are degenerate, having the same propagation constant and forming effectively one mode for each value of m.

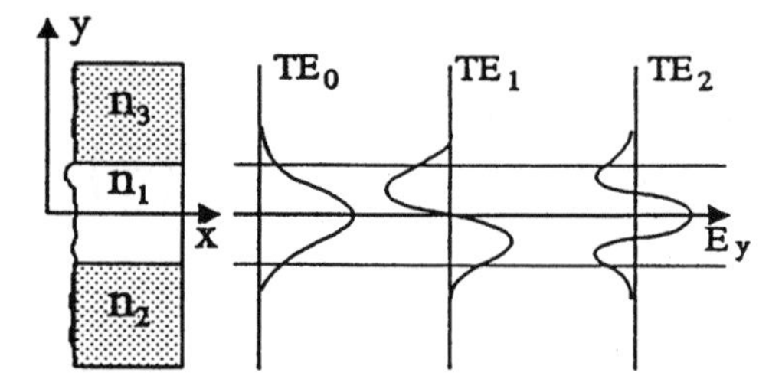

FIGURE 21.5 Transverse mode field patterns in the symmetric slab waveguide.

Figure 21.7 shows typical mode patterns in the asymmetric slab waveguide. Note that the asymmetry causes the evanescent fields to have unequal amplitudes at the two boundaries and to decay at different rates in the two outer layers.

The preceding analysis of slab waveguides is in many ways similar to, and constitutes a good introduction to, the more complex analysis of cylindrical (optical) fibers. Unlike slab waveguides, cylindrical waveguides are bounded in two dimensions rather than one. Consequently, skew rays exist in optical fibers, in addition to the meridional rays found in slab waveguides. In addition to transverse modes similar to those found in slab waveguides, the skew rays give rise to hybrid modes in optical fibers.

Fields in Cylindrical Fibers

Let ψ represent E_z or H_z and β be the component of κ in z direction. In the cylindrical coordinates of Figure 21.2, with wave propagation along the z axis, the wave equation (Equation (21.6)) correspond to the scalar equation:

$$\frac{\partial^2\psi}{\partial r^2}+\frac{1}{r}\frac{\partial\psi}{\partial r}+\frac{1}{r^2}\frac{\partial^2\psi}{\partial\Phi^2}+(\kappa^2-\beta^2)\psi=0 \tag{21.16}$$

The general solution to the preceding equation is

$$\psi(r)=C_1J_\ell(\mathrm{hr})+C_2Y_\ell(\mathrm{hr});\ \kappa^2>\beta^2 \tag{21.17a}$$

$$\psi(r)=C_1I_\ell(\mathrm{qr})+C_2K_\ell(\mathrm{qr});\ \kappa^2<\beta^2 \tag{21.17b}$$

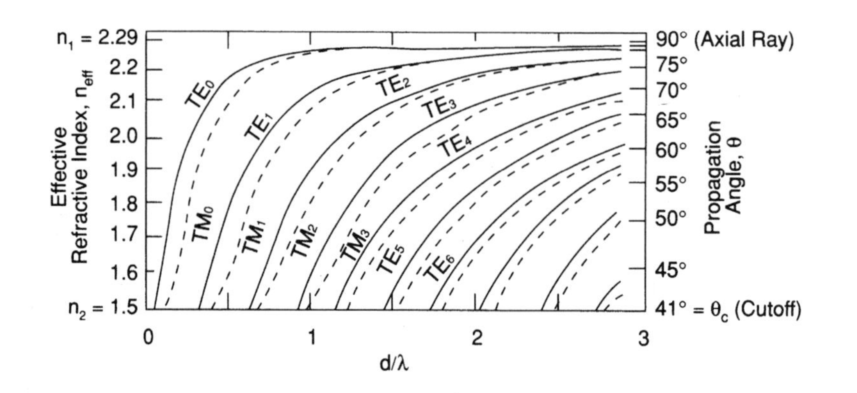

FIGURE 21.6 Mode chart for the asymmetric slab waveguide with $n_1 = 2.29$, $n_2 = 1.5$, and $n_3 = 1.0$. (*Source:* J.C. Palais, *Fiber Optic Communications*, Englewood Cliffs, NJ: Prentice-Hall, 1992, p. 88. With permission.)

In Equation (21.17a) and Equation (21.17b), J_ℓ and Y_ℓ are Bessel functions of the first kind and second kind, respectively, of order ℓ; I_ℓ and K_ℓ are modified Bessel functions of the first kind and second kind, respectively, of order ℓ; C_1 and C_2 are constants; $h^2=\kappa^2-\beta^2$ and $q^2=\beta^2-\kappa^2$.

E_z and H_z in a fiber core are given by Equation (21.17a) or Equation (21.17b), depending on the sign of $\kappa^2-\beta^2$. For guided propagation in the core, this sign is negative to ensure that the field is evanescent in the cladding. One of the coefficients vanishes because of asymptotic behavior of the respective Bessel functions in the core or cladding. Thus, with A_1 and A_2 as arbitrary constants, the fields in the core and cladding are given, respectively, by

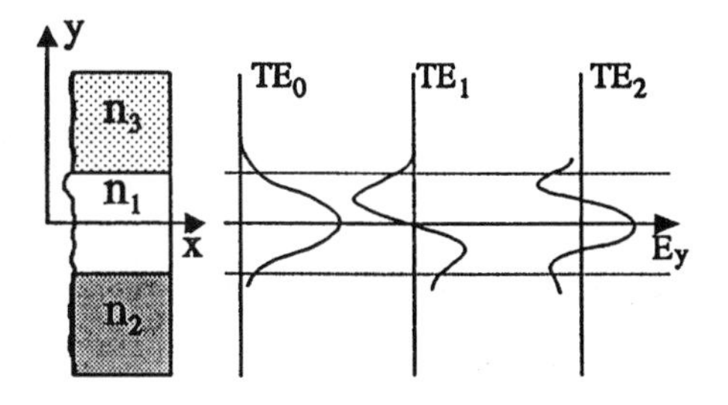

FIGURE 21.7 Transverse mode field patterns in the asymmetric slab waveguide.

$$\psi(r) = A_1 J_\ell(\mathrm{hr}) \tag{21.18a}$$

$$\psi(r) = A_2 \kappa_\ell(\mathrm{hr}) \tag{21.18b}$$

Because of the cylindrical symmetry:

$$\psi(r,t) = \psi(r,\phi)\mathrm{e}^{j(\omega t-\beta z)} \tag{21.19}$$

Thus, the usual approach is to solve for E_z and H_z and then express E_r, E_ϕ, H_r, and H_ϕ in terms of E_z and H_z.

Modes in Step-Index Fibers

Derivation of the exact modal field relations for optical fibers is complex. Fortunately, fibers used in optical communication satisfy the weekly guiding approximation in which the relative index difference, ∇, is much less than unity. In this approximation, application of the requirement for continuity of transverse and tangential electric field components at the core-cladding interface (at $r=a$) to Equation (21.18a) and Equation (21.18b) results in the following eigenvalue equation (Snyder, 1969):

$$haJ_{\ell\pm1}\frac{(ha)}{J_\ell(ha)} = \pm\frac{qa\kappa_{\ell\pm1}(qa)}{\kappa_\ell(qa)} \tag{21.20}$$

Let the normalized frequency V be defined as

$$V = a(q^2+h^2)^{1/2} = a\kappa_0(n_1^2-n_2^2)^{1/2} = \frac{2\pi}{\lambda}a(\mathrm{NA}) \tag{21.21}$$

Solving Equation (21.20) allows β to be calculated as a function of V. Guided modes propagating within the core correspond to $n_{2w}\kappa_0 \le \beta \le n_1\kappa$. The normalized frequency V corresponding to $\beta=n_1\kappa$ is the cut-off frequency for the mode.

As with planar waveguides, TE ($E_z=0$) and TM ($H_z=0$) modes corresponding to meridional rays exist in the fiber. They are denoted by EH or HE modes, depending on which component, E or H, is stronger in the plane transverse to the direction or propagation. Because the cylindrical fiber is bounded in two dimensions rather than one, two integers, ℓ and m, are needed to specify the modes, unlike one integer, m, required for planar waveguides. The exact modes, $\mathrm{TE}_{\ell m}$, $\mathrm{TM}_{\ell m}$, $\mathrm{EH}_{\ell m}$, and $\mathrm{HE}_{\ell m}$, may be given by two linearly polarized

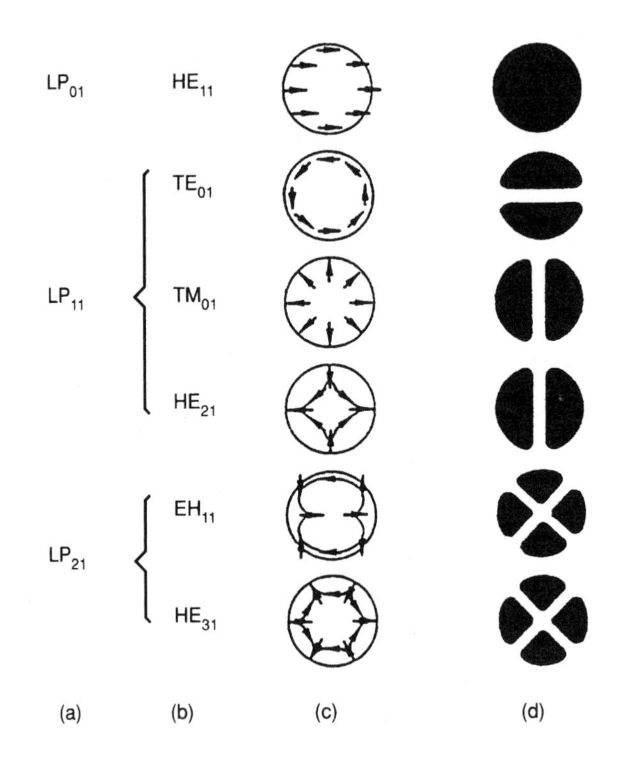

FIGURE 21.8 Transverse electric field patterns and field intensity distributions for the three lowest LP modes in a step-index fiber: (a) mode designations; (b) electric field patterns; (c) intensity distribution. (*Source:* J.M. Senior, *Optical Fiber Communications: Principles and Practice*, Englewood Cliffs, NJ: Prentice-Hall, 1985, p. 36. With permission.)

modes, $LP_{\ell m}$. The subscript ℓ is now such that $LP_{\ell m}$ corresponds to $HE_{\ell+1,m}$, $EH_{\ell-1,m}$, $TE_{\ell-1,m}$, and $TM_{\ell-1,m}$. In general, there are 2ℓ field maxima around the fiber core circumference and m field maxima along a radius vector. Figure 21.8 illustrates the correspondence between the exact modes and the LP modes and their field configurations for the three lowest LP modes.

Figure 21.9 gives the mode chart for step-index fiber on a plot of the refractive index, β/κ_0, against the normalized frequency. Note that for a single-mode (LP_{01} or HE_{11}) fiber, $V<2.405$. The number of modes supported as a function of V is given by

$$N = \frac{V^2}{2} \tag{21.22}$$

Modes in Graded-Index Fibers

A rigorous modal analysis for optical fibers based on the solution of Maxwell's equations is possible only for step-index fiber. For graded-index fibers, approximate methods are used. The most widely used approximation is the Wenzel–Kramers–Brillouin (WKB) method (Marcuse, 1982). This method gives good modal solutions for graded-index fiber with arbitrary profiles, when the refractive index does not change appreciably over distances comparable to the guided wavelength (Yariv, 1991). In this method, the transverse components of the fields are expressed as

$$E_t = \psi(r)e^{j1\phi}e^{j(\omega t-\beta z)} \tag{21.23}$$

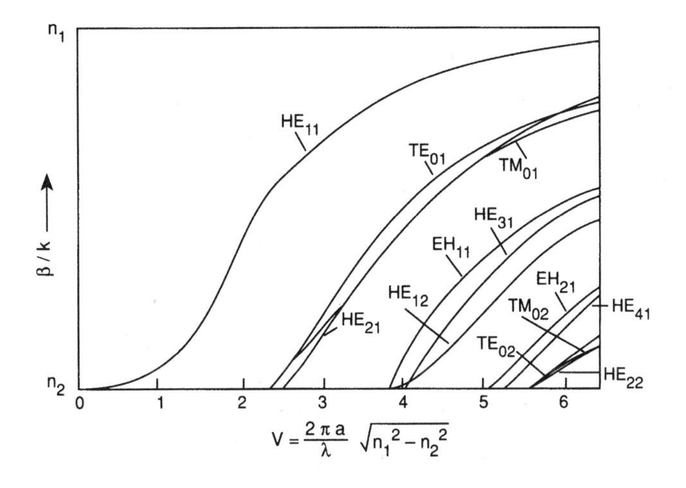

FIGURE 21.9 Mode chart for step-index fibers: $b = (\beta/\kappa_0 - n_2)/(n_1 - n_2)$ is the normalized propagation constant. (*Source:* D.B. Keck, *Fundamentals of Optical Fiber Communications,* M.K. Barnoski, Ed., New York: Academic Press, 1981, p. 13. With permission.)

$$H_t = \frac{\beta}{\omega\mu} E_t \tag{21.24}$$

In Equation (21.23), ℓ is an integer. Equation (21.16), the scalar wave equation in cylindrical coordinates, can now be written with $\kappa = n(r)\ \kappa_0$ as

$$\left[\frac{d^2}{\mathrm{d}r^2} + \frac{1}{2}\frac{d}{\mathrm{d}r} + p^2(r)\right]\psi(r) = 0 \tag{21.25}$$

where

$$p^2(r) = n^2(r)\kappa_0^2 - \frac{\ell^2}{r^2} - \beta^2 \tag{21.26}$$

Let r_1 and r_2 be roots of $p^2(r) = 0$ such that $r_1 < r_2$. A ray propagating in the core does not necessarily reach the core-cladding interface or the fiber axis. In general, it is confined to an annular cylinder bounded by the two caustic surfaces defined by r_1 and r_2. As illustrated in Figure 21.10, the field is oscillatory within this annular cylinder and evanescent outside it. The fields obtained as solutions to Equation (21.25) are

$$\psi(r) = \frac{A}{[rp(r)]^{1/2}} \exp\left[-\int_r^{r_1} |p(r)|\mathrm{d}r\right]; r < r_1 \tag{21.27a}$$

$$\psi(r) = \frac{B}{[rp(r)]^{1/2}} \sin\left[\int_{r_1}^{r} p(r)\mathrm{d}r + \frac{\pi}{4}\right]; r_1 < r \tag{21.27b}$$

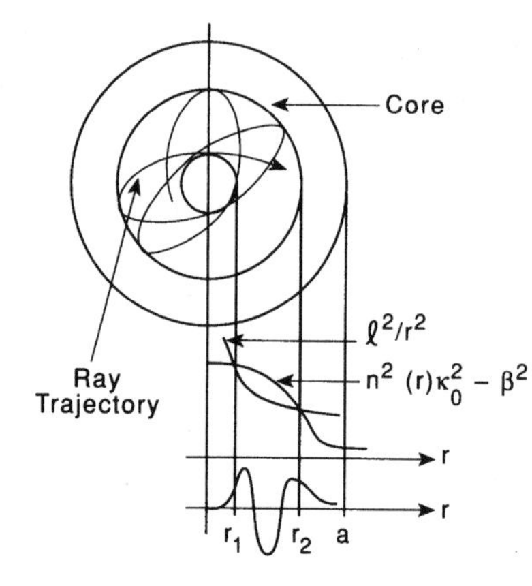

FIGURE 21.10 End view of a skew ray in a graded-index fiber, its graphical solution in the WKB method, and the resulting field that is oscillatory between r_1 and r_2 and evanescent outside that region.

$$\psi(r) = \frac{C}{[rp(r)]^{1/2}} \sin\left[\int_r^{r_2} p(r)\mathrm{d}r + \frac{\pi}{4}\right]; r < r_2 \tag{21.27c}$$

$$\psi(r) = \frac{D}{[rp(r)]^{1/2}} \exp\left[-\int_{r_2}^{r} |p(r)|\mathrm{d}r\right]; r_2 < r \tag{21.27d}$$

Equation (21.27b) and Equation (21.27c) represent fields in the same region. Equating them leads to the mode condition:

$$\int_{r_1}^{r_2}\left[n^2(r)\kappa_0^2 - \frac{\ell^2}{r^2} - \beta^2\right]^{1/2} \mathrm{d}r = (2m+1)\frac{\pi}{2} \tag{21.28}$$

In Equation (21.28), ℓ and m are the integers denoting the modes. A closed analytical solution of this equation for β is possible only for a few simple graded-index profiles. For other cases, numerical or approximate methods are used. It can be shown (Marcuse, 1982) that for fiber of graded index profile α, the number of modes supported N_g, and the normalized frequency V (and hence the core radius), for single mode operations are given, respectively, by

$$N_g = \left(\frac{\alpha}{2+\alpha}\right)\left(\frac{V^2}{2}\right) \tag{21.29}$$

$$V = 2.405\left(\frac{2+\alpha}{\alpha}\right)^{\frac{1}{2}} \tag{21.30}$$

For a parabolic ($a = 2$) index profile Equation (21.29) gives $N_g = \frac{V^*}{4}$, which is half the corresponding number of modes for step index fiber, and Equation (21.30) gives $V \leqslant 2.405\sqrt{2}$. Thus, compared with step index fiber, graded index fiber will have larger core radios for single mode operation, and for the same core radius, will support a fewer number of modes.

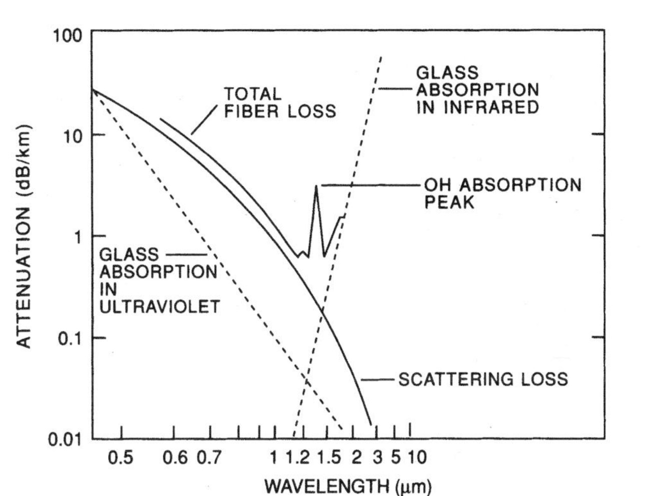

FIGURE 21.11 Attenuation of a germanium-doped low-loss silica glass fiber. (*Source:* H. Osanai et al., "Effects of dopants on transmission loss of low-OH content optical fibers," *Electronic Letters*, vol. 12, no. 21, p. 550, 1976. With permission.)

Attenuation

The assumption of a nonconducting medium for dielectric waveguides led to solutions to the wave equation with no attenuation component. In practice, various mechanisms give rise to losses in lightwave waveguides. These mechanisms contribute a loss factor of $e^{-\alpha z}$ to Equation (21.10) and comparable field expressions, where α is the attenuation coefficient. The attenuation due to these mechanisms and the resulting total attenuation as a function of wavelength is shown in Figure 21.11 (Osanai et al., 1976). Note that the range of wavelengths (0.8 to 1.6 μm) in which communication fibers are usually operated corresponds to a region of low overall attenuation. Brief discussions follow of the mechanisms responsible for the various types of attenuation shown in Figure 21.11.

Intrinsic Absorption

Intrinsic absorption is a natural property of glass. In the ultraviolet region, it is due to strong electronic and molecular transition bands. In the infrared region, it is caused by thermal vibration of chemical bonds.

Extrinsic Absorption

Extrinsic absorption is caused by metal (Cu, Fe, Ni, Mn, Co, V, Cr) ion impurities and hydroxyl (OH) ion impurity. Metal ion absorption involves electron transition from lower to higher energy states. OH absorption is caused by thermal vibration of the hydroxyl ion. Extrinsic absorption is strong in the range of normal fiber operation. Thus, it is important that impurity level be limited.

Rayleigh Scattering

Rayleigh scattering is caused by localized variations in refractive index in the dielectric medium, which are small relative to the optic wavelength. It is strong in the ultraviolet region. It increases with decreasing wavelength, being proportional to λ^{-4}. It contributes a loss factor of $\exp(-\alpha_R z)$. The Rayleigh scattering coefficient, α_R, is given by

$$\alpha_R = \left(\frac{8\pi^3}{3\lambda^4}\right)(\delta n^2)^2 \delta V \tag{21.31}$$

where δn^2 is the mean-square fluctuation in refractive index and V is the volume associated with this index difference.

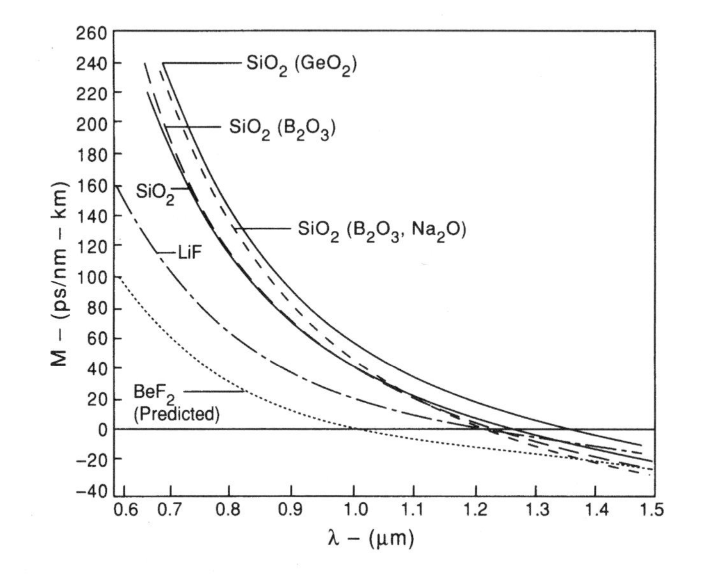

FIGURE 21.12 Material dispersion as a function of wavelength for silica and several solids. (*Source:* S.H. Wemple, "Material dispersion in optical fibers," *Applied Optics*, vol. 18, no. 1, p. 33, 1979. With permission.)

Mie Scattering

Mie scattering is caused by inhomogeneities in the medium, with dimensions comparable to the guided wavelength. It is independent of wavelength.

Dispersion and Pulse Spreading

Dispersion refers to the variation of velocity with frequency or wavelength. Dispersion causes pulse spreading, but other nonwavelength-dependent mechanisms also contribute to pulse spreading in optical waveguides. The mechanisms responsible for pulse spreading in optical waveguides include material dispersion, waveguide dispersion, and multimode pulse spreading.

Material Dispersion

In material dispersion, the velocity variation is caused by some property of the medium. In glass, it is caused by the wavelength dependence of refractive index. For a given pulse, the resulting pulse spread per unit length is the difference between the travel times of the slowest and fastest wavelengths in the pulse. It is given by

$$\Delta\tau = \frac{-\lambda}{c} n'' \Delta\lambda = -M\Delta\lambda \tag{21.32}$$

In Equation (21.32), n'' is the second derivative of the refractive index with respect to λ, $M = (\lambda/c)n''$ is the material dispersion, and $\Delta\lambda$ is the **linewidth** of the pulse. Figure 21.12 shows the wavelength dependence of material dispersion (Wemple, 1979). Note that for silica, zero dispersion occurs around 1.3 μm, and material dispersion is small in the wavelength range of small fiber attenuation.

Waveguide Dispersion

The effective refractive index for any mode varies with wavelength for a fixed film thickness, for a slab waveguide, or a fixed core radius, for an optical fiber. This variation causes pulse spreading, which is termed *waveguide dispersion*. The resulting pulse spread is given by

$$\Delta\tau = \frac{-\lambda}{c} n''_{\text{eff}} \Delta\lambda = -M_{\text{G}} \Delta\lambda \tag{21.33}$$

where $M_{\text{G}} = (\lambda/c)n''_{\text{eff}}$ is the waveguide dispersion.

Polarization Mode Dispersion

The HE_{11} propagating in a single-mode fiber actually consists of two orthogonally polarized modes, but the two modes have the same effective refractive index and propagation velocity except in birefringent fibers. Birefringent fibers have asymmetric cores or asymmetric refractive index distribution in the core, which result in different refractive indices and group velocities for the orthogonally polarized modes. The different group velocities result in a group delay of one mode relative to the other, known as polarization mode dispersion. Birefringent fibers are polarization preserving and are required for several applications, including coherent optical detection and fiber optic gyroscopes. In high birefringence fibers, polarization dispersion can exceed 1 nsec/km. However, in low birefringence fibers, polarization mode dispersion is negligible relative to other pulse spreading mechanisms (Payne et al., 1982).

Multimode Pulse Spreading

In a multimode waveguide, different modes travel different path lengths. This results in different travel times and, hence, in pulse spreading. Because this pulse spreading is not wavelength dependent, it is not usually referred to as dispersion. Multimode pulse spreads are given, respectively, for a slab waveguide, a step-index fiber, and a parabolic graded-index fiber by the following equations:

$$\Delta\tau_{\text{mod}} = \frac{n_1(n_1 - n_2)}{cn_2} \quad \text{(slab waveguide)} \tag{21.34}$$

$$\Delta\tau_{\text{mod}} = \frac{n_1\Delta}{c} \quad \text{(step-index fiber)} \tag{21.35}$$

$$\Delta\tau_{\text{mod}} = \frac{n_1\Delta^2}{8c} \quad \text{(GRIN fiber)} \tag{21.36}$$

Total Pulse Spread

Total pulse spread is the overall effect of material dispersion, waveguide dispersion, and multimode pulse spread. It is given by

$$\Delta\tau_{\text{T}}^2 = \Delta\tau_{\text{mod}}^2 + \Delta\tau_{\text{dis}}^2 \tag{21.37}$$

where

$$\Delta\tau_{\text{dis}} = \text{total dispersion} = -(M + M_{\text{G}})\Delta\lambda$$

In a multimode waveguide, multimode pulse spread dominates, and dispersion can often be ignored. In a single-mode waveguide, only material and waveguide dispersion exist; material dispersion dominates, and waveguide dispersion can often be ignored.

Total pulse spread imposes an upper limit on the bandwidth of an optical fiber. This upper limit is equal to $1/(2\Delta\tau_{\text{T}})Hz$.

Defining Terms

Linewidth: The range of wavelengths emitted by a source or present in a pulse.

Meridional ray: A ray that is contained in a plane passing through the fiber axis.

Mode chart: A graphical illustration of the variation of effective refractive index (or, equivalently, propagation angle θ) with normalized thickness d/λ for a slab waveguide or normalized frequency V for an optical fiber.

Refractive index: The ratio of the velocity of light in free space to the velocity of light in a given medium.

Relative refractive index difference: The ratio $(n_1^2 - n_2^2)/2n_1^2 \approx (n_1 - n_2)/n_1$, where $n_1 > n_2$ and n_1 and n_2 are refractive indices.

References

H.A. Haus, *Waves and Fields in Optoelectronics*, Englewood Cliffs, NJ: Prentice-Hall, 1984.

D.B. Keck, "Optical fiber waveguides," in *Fundamentals of Optical Fiber Communications*, 2nd ed., M.K. Barnoski, Ed., New York: Academic Press, 1981.

D. Marcuse, *Light Transmission Optics*, 2nd ed., New York: Van Nostrand Reinhold, 1982.

H. Osanai et al., "Effects of dopants on transmission loss of low-OH-content optical fibers," *Electron. Lett.*, vol. 12, no. 21, 1976.

J.C. Palais, *Fiber Optic Communications*, Englewood Cliffs, NJ: Prentice-Hall, 1992.

D.N. Payne, A.J. Barlow, and J.J.R. Hansen, "Development of low-and-high birefringence optical fibers," *IEEE J. Quantum Electron.*, vol. QE-18, no. 4, pp. 477–487, 1982.

J.M. Senior, *Optical Fiber Communications: Principles and Practice*, Englewood Cliffs, NJ: Prentice-Hall, 1985.

J.M. Snyder, "Asymptotic expressions for eigenfunctions and eigenvalues of a dielectric or optical waveguide," *Trans. IEEE Microwave Theory Tech.*, vol. MTT-17, 1130–1138, 1969.

S.H. Wemple, "Material dispersion in optical fibers," *Appl. Optics*, vol. 18, no. 1, p. 33, 1979.

A. Yariv, *Optical Electronics*, 4th ed., Philadelphia, PA: Saunders College Publishing, 1991.

Further Information

IEEE Journal of Lightwave Technology, a bimonthly publication of the IEEE, New York.

IEEE Lightwave Telecommunications Systems, a quarterly magazine of the IEEE, New York.

Applied Optics, a biweekly publication of the Optical Society of America, 2010 Massachusetts Avenue NW, Washington, D.C. 20036.

D. Macruse, *Theory of Optical Waveguides*, 2nd ed., Boston, Mass.: Academic Press, 1991.

21.2 Optical Fibers and Cables

Gerd Keiser

Introduction

Optical fiber communication links and networks use light as a signal carrier and optical fibers as the transmission media. The optical transmission system converts voice, video, or data from an electrical format into a coded pulse stream of light using a suitable semiconductor light source. Optical fibers then carry this stream of light pulses to a regenerating or receiving station. At the destination the light pulses are converted to electric signals, decoded, and transformed back into the form of the original information. Optical fiber links are used for telecommunications, data transmission, military applications, industrial controls, medical applications, and CATV.

Optical Fiber Applications

The physical properties of optical fibers give them a number of inherent cost and operational advantages over copper wires, and make them highly attractive for communication links. Included in these advantages are the following:

- *Long transmission distance.* Optical fibers have lower transmission losses compared to copper wires. This means that data can be sent over longer distances, thereby reducing the number of intermediate repeaters needed for these spans. This reduction in equipment and components decreases system cost and complexity.
- *Large information capacity.* Optical fibers have wider bandwidths than copper wires, which means that more information can be sent over a single physical line. This property results in a decrease in the number of physical lines needed for sending a certain amount of information.
- *Small size and low weight.* The low weight and the small dimensions of fibers offer a distinct advantage over heavy, bulky wire cables in crowded underground city ducts or in ceiling-mounted cable trays. This is of particular importance in aircraft, satellites, and ships where small, lightweight cables are advantageous, and in tactical military applications where large amounts of cable must be unreeled and retrieved rapidly.
- *Immunity to electrical interference.* An especially important feature of optical fibers relates to the fact that they consist of dielectric materials, which means they do not conduct electricity. This makes optical fibers immune to the electromagnetic interference effects seen in copper wires, such as inductive pickup from other adjacent signal-carrying wires or coupling of electrical noise into the line from any type of nearby equipment.
- *Enhanced safety.* Optical fibers do not have the problems of ground loops, sparks, and potentially high voltages inherent in copper lines. However, precautions with respect to laser light emissions need to be observed to prevent possible eye damage.
- *Increased signal security.* An optical fiber offers a high degree of data security, since the optical signal is well confined within the fiber and any signal emissions are absorbed by an opaque coating around the fiber. This is in contrast to copper wires where electrical signals often can be tapped off easily. This makes optical fibers attractive in applications where information security is important, such as financial, legal, government, and military systems.

Optical fibers can be applied to interconnections ranging from localized links within an equipment rack to links that span continents or oceans. Communication networks may be divided into the following four broad categories:

- *Local area networks* (LANs) interconnect users in a localized area such as a room, a department, a building, an office or factory complex, or a campus. The word *campus* refers to any group of buildings within reasonable walking distance of each other. For example, it could be the collocated buildings of a corporation, a large medical facility, or a university complex. LANs usually are owned, used, and operated by a single organization.
- *Metropolitan area networks* (MANs) span a larger area than a LAN. This can range from interconnections between buildings covering several blocks within a city or could encompass an entire city and the metropolitan area surrounding it. MANs are owned and operated by many organizations. Typically MAN fiber optic implementations are referred to as *metro applications*.
- *Access networks* provide a high-speed, high-capacity means of interconnecting the metro resources with communication entities located in LANs and at customer premises.
- *Wide area networks* (WANs) span a large geographical area. The links can range from connections between switching facilities in neighboring cities to long-haul terrestrial or undersea transmission lines running across a country or between countries. WANs invariably are owned and operated by many transmission service providers.

The basic function of an optical fiber link is to transport a signal from some piece of electronic equipment (e.g., a computer, telephone, or video device) at one location to corresponding equipment at another location

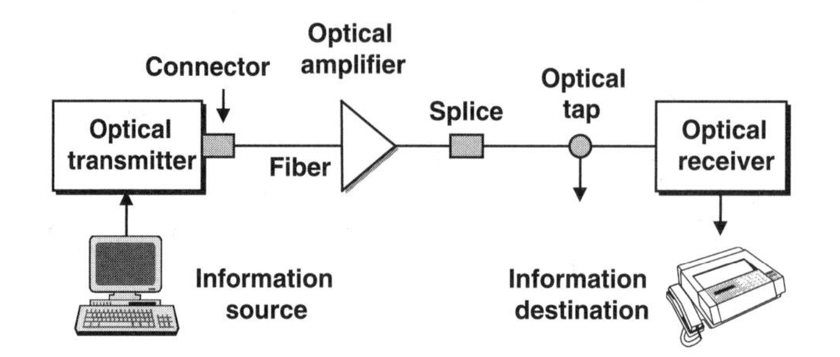

FIGURE 21.13 The key sections of an optical fiber communications link.

with a high degree of reliability and accuracy. Figure 21.13 shows the key sections of an optical fiber communications link, which are:

- *Transmitter.* The transmitter consists of a semiconductor light source and associated electronic circuitry. The source is about the size of a grain of salt in order to couple its light emission efficiently into a fiber. The electronics set the source operating point, control the light output stability, and vary the optical output in proportion to an electrically formatted information input signal.
- *Optical fiber.* The optical fiber is placed inside a cable that offers mechanical and environmental protection. A variety of fiber types exist and there are many different cable configurations depending on whether the cable is to be installed inside a building, in underground pipes, outside on poles, or underwater.
- *Receiver.* Inside the receiver is a photodiode that detects the weakened and distorted optical signal emerging from the end of an optical fiber and converts it to an electrical signal. The receiver also contains amplification devices and circuitry to restore signal fidelity.
- *Passive devices.* Passive devices are optical components that require no electronic control for their operation. Among these are optical connectors for attaching cables, splices for joining one fiber to another, optical isolators that prevent unwanted light from flowing in a backward direction, optical filters that select only a narrow spectrum of desired light, and couplers that are used to tap off a certain percentage of light, usually for performance monitoring purposes.
- *Optical amplifiers.* After an optical signal has traveled a certain distance along a fiber, it becomes weakened due to power loss along the fiber. At that point the optical signal needs to get a power boost. Traditionally the optical signal was converted to an electrical signal, amplified electrically, and then converted back to an optical signal. The invention of an optical amplifier that boosts the power level completely in the optical domain circumvented these transmission bottlenecks.
- *Active components.* Lasers and optical amplifiers fall into the category of active devices, which require an electronic control for their operation. Not shown in Figure 21.13 are numerous other diverse active optical components. These include light signal modulators, tunable (wavelength-selectable) optical filters, variable optical attenuators, and optical switches.

Electromagnetic Spectrum

To get an understanding of the distinction between electrical and optical communication systems and what the advantages are of lightwave technology, let us examine the spectrum of electromagnetic (EM) radiation shown in Figure 21.14.

All telecommunication systems use some form of electromagnetic energy to transmit signals from one device to another. Electromagnetic energy is a combination of electrical and magnetic fields, and includes power, radio waves, microwaves, infrared light, visible light, ultraviolet light, x-rays, and gamma rays. Each of these makes up a portion (or band) of the electromagnetic spectrum. The fundamental nature of all radiation

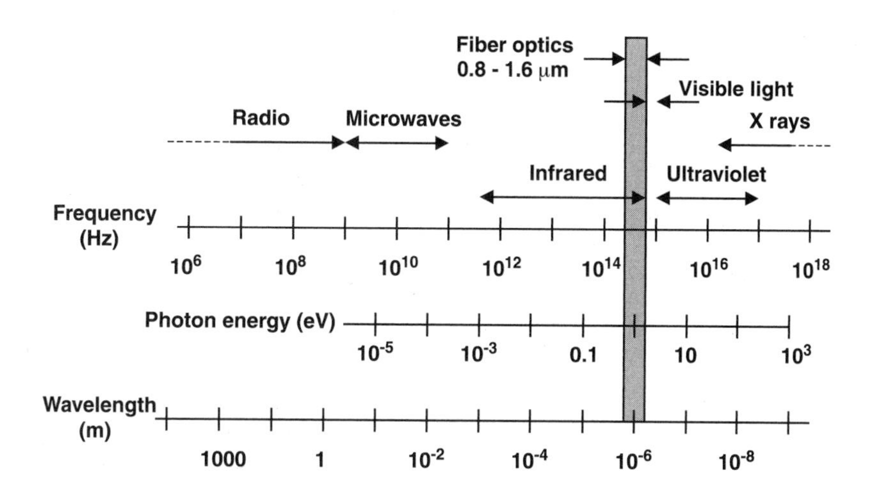

FIGURE 21.14 The spectrum of electromagnetic (EM) radiation.

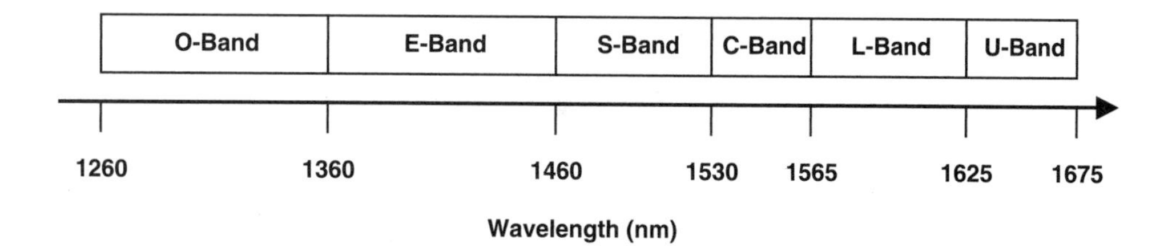

FIGURE 21.15 Spectral band designations for optical communications.

within this spectrum is the same in that it can be viewed as electromagnetic waves that travel at the speed of light, which is about $c = 300{,}000$ km/sec in a vacuum. Note that the speed of light in a material is less than c.

The physical properties of radiation in different parts of the spectrum can be measured in several interrelated ways. These are the length of one period of the electromagnetic wave, the energy contained in the wave, or the oscillating frequency of the wave. Whereas electrical signal transmission tends to use frequency to designate the signal operating bands, optical communications generally uses *wavelength* to designate the spectral operating region and *photon energy* or *optical power* when discussing topics such as signal strength or electro-optical component performance.

The optical spectrum ranges from about 5 nm (ultraviolet) to 1 mm (far infrared), the visible region being the 400 to 700 nm band. Optical fiber communications use the spectral band ranging from 800 to 1675 nm.

The International Telecommunications Union (ITU) has designated six spectral bands for use in intermediate-range and long-distance optical fiber communications within the 1260 to 1675 nm region. These band designations arose from the physical characteristics of optical fibers and the performance behavior of optical amplifiers in multiple-wavelength systems. As shown in Figure 21.15, the regions are known by the letters O, E, S, C, L, and U, defined as follows:

1. Original band (O band): 1260 to 1360 nm
2. Extended band (E band): 1360 to 1460 nm
3. Short band (S band): 1460 to 1530 nm
4. Conventional band (C band): 1530 to 1565 nm
5. Long band (L band): 1565 to 1625 nm
6. Ultra-long band (U band): 1625 to 1675 nm

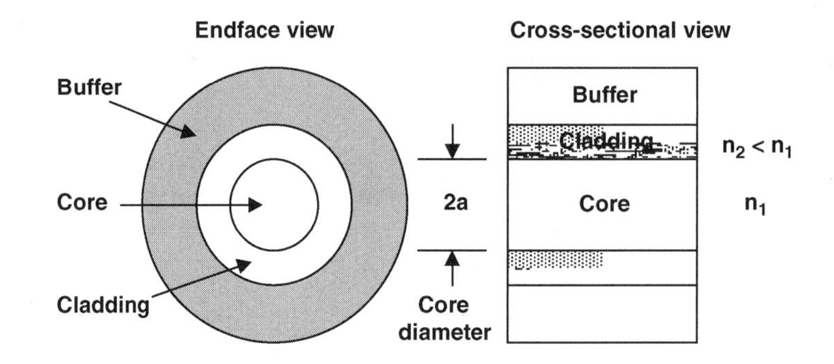

FIGURE 21.16 The endface and a longitudinal cross-section of a standard optical fiber.

Optical Fiber Characteristics

Figure 21.16 shows the endface and a longitudinal cross-section of a standard optical fiber, which consists of a cylindrical glass core surrounded by a glass cladding. The *core* has a refractive index n_1 and the *cladding* has a refractive index n_2. Surrounding these two layers is a polymer *buffer coating* that protects the fiber from mechanical and environmental effects. Traditionally the core radius is designated by the letter a. In almost all cases, for telecommunication fibers the core and cladding are made of silica glass (SiO_2).

The refractive index of pure silica varies with wavelength ranging from 1.453 at 850 nm to 1.445 at 1550 nm. By adding certain impurities such as germanium or boron to the silica during the fiber manufacturing process the index can be changed slightly, usually as an increase in the core index. This is done so that the refractive index n_2 of the cladding is slightly smaller than the index of the core (i.e., $n_2 < n_1$), which is the condition required for light traveling in the core to be totally internally reflected at the boundary with the cladding. The difference in the core and cladding indices also determine how light signals behave as they travel along a fiber. Typically the index differences range from 0.2 to 3.0% depending on the desired behavior of the resulting fiber.

Light Propagation in Fibers

To get an understanding of how light travels along a fiber, let us first examine the case when the core diameter is much larger than the wavelength of the light. For such a case one can consider a simple geometric optics approach using the concept of light rays. Figure 21.17 shows a light ray entering the fiber core from a medium of refractive index n_0, which is less than the index n_1 of the core. The ray meets the core endface at an angle θ_0 with respect to the fiber axis and is refracted into the core where it now makes an angle θ with respect to the fiber axis. The behavior of a light ray at the interface between these two dielectric media can be described by Snell's law, given by

$$n_0 \cos\theta_0 = n_1 \cos\theta_1$$

or equivalently as

$$n_0 \sin\phi_0 = n_1 \sin\phi_1$$

where $\phi = \theta - 90°$ is the angle with respect to a line perpendicular to the axis. Inside the core the ray strikes the core-cladding interface at a normal angle ϕ_1. If the light ray strikes this interface at such an angle that it is totally internally reflected, then the ray follows a zigzag path along the fiber core.

Critical Angle. Now suppose that the angle θ_0 is the largest entrance angle for which total internal reflection can occur at the core-cladding interface. Then rays outside of the acceptance cone shown in Figure 21.17, such as the ray given by the dashed line, will refract out of the core and be lost in the cladding.

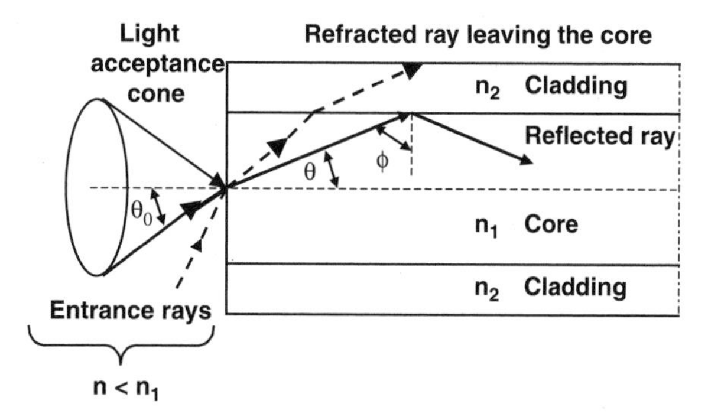

FIGURE 21.17 Illustration of the concept of light rays.

This condition defines a *critical angle* ϕ_c, which is the smallest angle ϕ that supports total internal reflection at the core–cladding interface.

Referring to Figure 21.17 from Snell's law, the minimum angle $\phi = \phi_{\min}$ that supports total internal reflection is given by $\phi_c = \phi_{\min} = \arcsin(n_2/n_1)$. Rays striking the core–cladding interface at angles less than $\phi_{\min}$ will refract out of the core and be lost in the cladding. Now suppose the medium outside of the fiber is air for which $n = 1.00$. By applying Snell's law to the air–fiber interface boundary, the condition for the critical angle can be related to the maximum entrance angle $\theta_{0,\max}$ through the relationship:

$$\sin \theta_{0,\max} = n_1 \sin \theta_c = (n_1^2 - n_2^2)^{1/2}$$

where $\theta_c = \pi/2 - \phi_c$. Thus, those rays having entrance angles θ_0 less than $\theta_{0,\max}$ will be totally internally reflected at the core–cladding interface.

Optical Fiber Modes. Although it is not directly obvious from the ray picture shown in Figure 21.17, only a finite set of rays at certain discrete angles greater than or equal to the critical angle ϕ_c are capable of propagating along a fiber. These angles are related to a set of electromagnetic wave patterns or field distributions called *modes* that can propagate along a fiber. When the fiber core diameter is on the order of 8 to 10 μm, which is only a few times the value of the wavelength, then only the one single *fundamental ray* that travels straight along the axis is allowed to propagate in a fiber. Such a fiber is referred to as a *single-mode fiber.* The operational characteristics of single-mode fibers cannot be explained by a ray picture, but instead need to be analyzed in terms of the *fundamental mode* using the electromagnetic wave theory. Fibers with larger core diameters (e.g., greater than or equal to 50 μm) support many propagating rays or modes and are known as *multimode fibers.* A number of performance characteristics of multimode fibers can be explained by the ray theory whereas other attributes (such as the optical coupling concept) need to be described by the wave theory.

Figure 21.18 shows the field patterns of the three lowest-order *transverse electric* (TE) modes as seen in a cross-sectional view of an optical fiber. They are labeled as the TE_0, TE_1, and TE_2 modes and illustrate three of many possible power distribution patterns in the fiber core. The subscript refers to the *order* of the mode, which is equal to the number of zero crossings within the guide. In single-mode fibers only the lowest-order or *fundamental mode* (TE_0) will be guided along the fiber core.

As the plots in Figure 21.18 show, the power distributions are not confined completely to the core, but instead they extend partially into the cladding. The fields vary harmonically within the core guiding region of index n_1 and decay exponentially outside of this region (in the cladding). For low-order modes the fields are concentrated tightly near the axis of the fiber with little penetration into the cladding. On the other hand, for higher-order modes the fields are distributed more towards the edges of the core and penetrate further into the cladding region. The characteristic that the power of a mode extends partially into the cladding is important to applications such as coupling of power from one fiber to another.

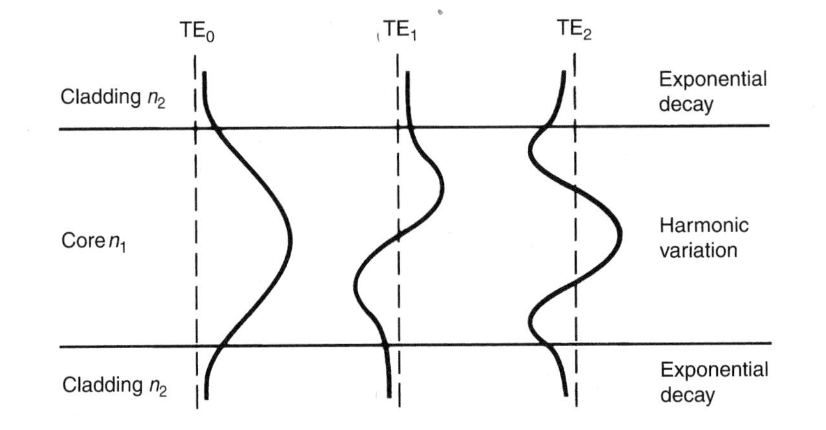

FIGURE 21.18 The field patterns of the three lowest-order transverse electric (TE) modes of an optical fiber.

Variations of Fiber Types

Variations in the material composition of the core and the cladding give rise to the two basic fiber types shown in Figure 21.19a. In the first case, the refractive index of the core is uniform throughout and undergoes an abrupt change (or step) at the cladding boundary. This is called a *step-index fiber.* In the second case, the core refractive index varies as a function of the radial distance from the center of the fiber. This defines a *graded-index fiber.* More complex structures of the cladding index profile allow fiber designers to tailor the signal dispersion characteristics of the fiber. Figure. 21.19b shows two of many different possible configurations.

Table 21.1 lists typical core, cladding, and buffer coating sizes of optical fibers for use in telecommunications, in a metro network, or in a LAN. The outer diameter of the buffer coating can be either 250 or 500 μm. Single-mode fibers are used for long-distance communication and for transmissions at very high data rates. The larger-core multimode fibers typically are used for local area network applications.

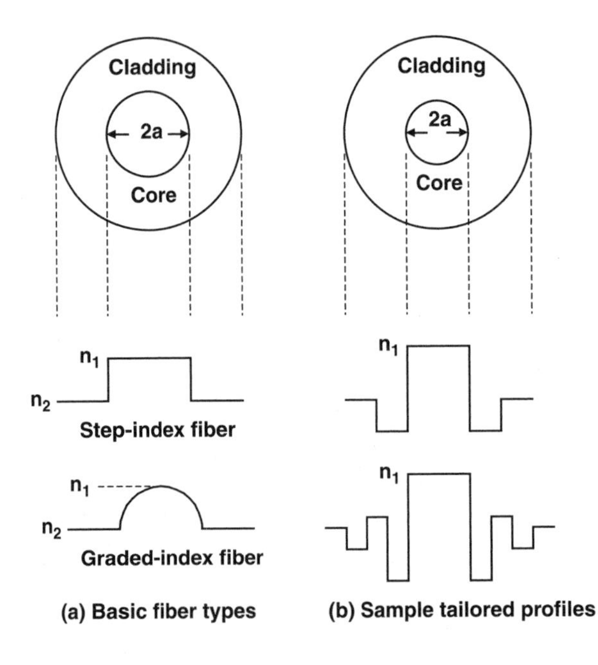

FIGURE 21.19 Illustration of different fiber types: (a) simple refractive index profiles define step- and graded-index fibers; (b) complex cladding-index profiles tailor the signal dispersion characteristics of a fiber.

TABLE 21.1 Typical Core, Cladding, and Buffer Coating Sizes of Optical Fibers

Fiber Type	Core Diameter (μm)	Cladding Diameter (μm)	Buffer Outer Diameter (μm)	Application
Single-mode	7 to 10	125	250 or 500	Telecommunications
Multimode	50.0	125	250 or 500	LAN or MAN
Multimode	62.5	125	250 or 500	LAN

The critical angle also defines a parameter called the *numerical aperture* (NA), which is used to describe the light acceptance or gathering capability of fibers that have a core size much larger than a wavelength. This parameter defines the size of the acceptance cone shown in Figure 21.17. The NA is a dimensionless quantity less than unity, with values ranging from 0.14 to 0.50.

The critical angle condition on the entrance angle defines the NA of a step-index fiber. This is given by

$$\mathrm{NA} = n\sin\theta_{0,\max} = n_1 \sin\theta_c = (n_1^2 - n_2^2)^{1/2} \approx n_1\sqrt{2\Delta}$$

where the parameter Δ is called the *core–cladding index difference* or simply the *index difference*. It is defined through the equation $n_2 = n_1(1-\Delta)$. Typical values of Δ range from 1 to 3% for multimode fibers and from 0.2 to 1.0% for single-mode fibers. Thus, since Δ is much less than 1, the approximation on the right-hand side of the above equation is valid. Since the NA is related to the maximum acceptance angle, it may be used to describe the light acceptance or gathering capability of a multimode fiber and to calculate the source-to-fiber optical power coupling efficiencies.

Optical Fiber Attenuation

Light traveling in a fiber loses power over distance, mainly because of absorption and scattering mechanisms in the fiber. The fiber loss is referred to as *signal attenuation* or simply *attenuation*. Attenuation is an important property of an optical fiber because, together with signal distortion mechanisms, it determines the maximum transmission distance possible between a transmitter and a receiver (or an amplifier) before the signal power needs to be interpreted or boosted to an appropriate level above the signal noise for high-fidelity reception. The degree of the attenuation depends on the wavelength of the light and on the fiber material.

Figure 21.20 shows an approximate attenuation-versus-wavelength curve for a typical silica fiber. The loss of power is measured in decibels and the loss within a cable is described in terms of decibels per kilometer (dB/km). Early optical fibers had a large attenuation spike between 900 and 1200 nm due to the fourth-order absorption peak from water molecules. Another spike from the third-order water absorption occurs between 1350 and 1480 nm for commonly fabricated fibers. Because of such absorption peaks, three transmission windows were defined initially. The *first window* ranges from 800 to 900 nm, the *second window* is centered at 1310 nm, and the *third window* ranges from 1480 to 1600 nm. However, since the attenuation of highly pure fibers with low water content (known as *full-spectrum fibers*) makes the designation of these windows obsolete, the concept of operational spectral bands arose for the 1260 to 1675 nm region.

Light power also can be lost as a result of fiber bending. Fibers can be subject to two types of bends: (a) *macroscopic bends* that have radii which are large compared with the fiber diameter, for example, such as those occurring when a fiber cable turns a corner or when it is coiled within an equipment rack, and (b) random *microscopic bends* of the fiber axis that can arise when fibers are incorporated into cables. Since the microscopic bending loss is determined in the manufacturing process, the user has little control over the degree of loss resulting from them. In general cable fabrication processes keep these values very low, which is included in published cable-loss specifications.

For slight bends, the excess optical power loss due to macroscopic bending is extremely small and is essentially unobservable. As the radius of curvature decreases, the loss increases exponentially until at a certain critical bend radius the curvature loss becomes observable. If the bend radius is made a bit smaller once this threshold has been reached, the losses suddenly become extremely large. Bending losses depend on wavelength.

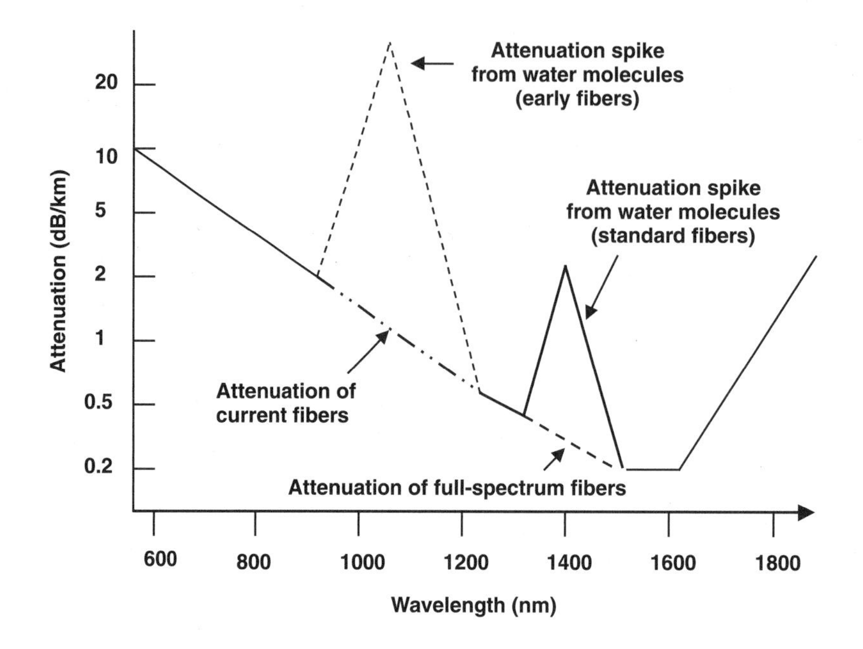

FIGURE 21.20 A typical attenuation-versus-wavelength curve for a silica fiber.

For a given bend radius, the loss is progressively greater at longer wavelengths. In some cases, near the critical bend radius there can be a large difference in losses between operation at 1310 and 1550 nm. As a rule of thumb, it is best not to make the bend radius of a standard telecommunications fiber be less than 2.5 cm.

Since often fibers need to be bent into very tight loops within component packages, special fibers that are immune to bending losses have been developed for such applications. This class of specialty fibers is described later.

Fiber Information Capacity

The information-carrying capacity of the fiber is limited by various distortion mechanisms in the fiber, such as signal dispersion factors and nonlinear effects. The three main dispersion categories are modal, chromatic, and polarization mode dispersions. These distortion mechanisms cause optical signal pulses to broaden as they travel along a fiber. If optical pulses travel sufficiently far in a fiber they will eventually overlap with neighboring pulses, thereby creating errors in the output since they become indistinguishable to the receiver. Nonlinear effects occur when there are high power densities (optical power per cross-sectional area) in a fiber. Their impact on signal fidelity includes shifting of power between wavelength channels, appearances of spurious signals at other wavelengths, and decreases in signal strength.

Modal dispersion arises from the different path lengths associated with various modes (as represented by light rays at different angles). It appears only in multimode fibers, since in a single-mode fiber there is only one mode. If many rays traveling at different angles are launched into a fiber at the same time in a given light pulse, then they will arrive at the fiber end at slightly different times. This causes the pulse to spread out and is the basis of modal dispersion.

In a graded-index fiber the index of refraction is lower near the core–cladding interface than at the center of the core. Therefore in such a fiber the rays that strike this interface at a steeper angle will travel slightly faster as they approach the cladding than those arriving at a smaller angle. Thereby the various rays tend to keep up with each other to some degree. Consequently the graded-index fiber exhibits less pulse spreading than a step-index fiber where all rays travel at the same speed.

The index of refraction of silica varies with wavelength; for example, it ranges from 1.453 at 850 nm to 1.445 at 1550 nm. In addition, a light pulse from an optical source is not a sharp line but instead contains a certain slice of wavelength spectrum. For example, a laser diode source may emit pulses that have a 1-nm spectral width. Consequently different wavelengths within an optical pulse travel at slightly different speeds through the fiber. Therefore each wavelength will arrive at the fiber end at a slightly different time, which leads to pulse spreading. This factor is called *chromatic dispersion*, which often is referred to simply as *dispersion*. It is a fixed quantity at a specific wavelength and is measured in units of picoseconds per kilometer of fiber per nanometer of optical source spectral width or "psec/(km nm)." For example, a single-mode fiber might have a chromatic dispersion value of $D_{CD} = 2$ psec/(km nm) at 1550 nm.

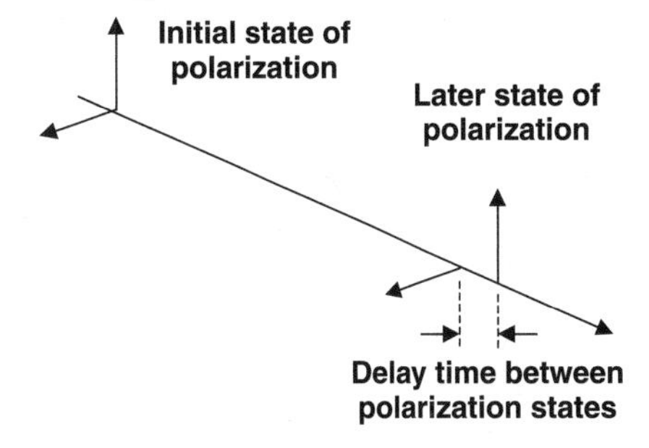

FIGURE 21.21 Signal energy at a given wavelength occupies two orthogonal polarization states.

Polarization mode dispersion (PMD) results from the fact that light-signal energy at a given wavelength in a single-mode fiber actually occupies two orthogonal polarization states or modes. Figure 21.21 shows this condition. At the start of the fiber the two polarization states are aligned. However, fiber material is not perfectly uniform throughout its length. In particular, the refractive index is not perfectly uniform across any given cross-sectional area. This condition is known as the *birefringence* of the material. Consequently each polarization mode will encounter a slightly different refractive index, so that each will travel at a slightly different velocity and the polarization orientation will rotate with distance. The resulting difference in propagation times between the two orthogonal polarization modes will result in pulse spreading. This is the basis of polarization mode dispersion. PMD is not a fixed quantity but fluctuates with time due to factors such as temperature variations and stress changes on the fiber. It varies as the square root of distance and thus is specified as a mean value in units of psec/$\sqrt{\text{km}}$. A typical value is $D_{PMD} = 0.05\,\text{psec}/\sqrt{\text{km}}$.

If t_{mod}, t_{CD}, and t_{PMD} are the modal, chromatic, and polarization mode dispersion times, respectively, then the total dispersion t_T can be calculated by the relationship:

$$t_T = \sqrt{(t_{mod})^2 + (t_{CD})^2 + (t_{PMD})^2}$$

Note that $t_{mod} = 0$ for single-mode fibers. As a rule of thumb, the information-carrying capacity over a certain length of fiber then is determined by specifying that the pulse spreading cannot be more than 10% of the pulse width at a designated data rate.

Optical Fiber Standards

The Telecommunications Sector of the International Telecommunication Union (ITU-T) and the Telecommunications and Electronics Industry Associations (TIA/EIA) are the main organizations that have published standards for both multimode and single-mode optical fibers used in telecommunications. The recommended bounds on fiber parameters (e.g., attenuation, cutoff wavelength, and chromatic dispersion) designated in these standards assure the users of product capability and consistency. In addition the standards allow fiber manufacturers to have a reasonable degree of flexibility to improve and develop new products.

Multimode fibers are used in LAN environments, storage area networks, and central-office connections, where the distances between buildings are typically 2 km or less. The two principal multimode fiber types for these applications have either 50- or 62.5-μm core diameters and both have 125-μm cladding diameters. To meet the demands for short-reach low-cost transmission of high-speed Ethernet signals, a 50-μm multimode fiber is available for 10 Gb/sec operation at 850 nm over distance up to 300 m. The standards document TIA/EIA-568 lists the specifications for such fiber. The ITU-T Recommendation G.651 describes other multimode fiber specifications for LAN applications using 850-nm optical sources.

The ITU-T has published a series of recommendations for single-mode fibers. The characteristics of these fibers are summarized in Table 21.2 and more details are given in the following listing.

ITU-T G.652A/B: This recommendation deals with the single-mode fiber that was installed widely in telecommunication networks starting in the early 1990s. It has a Ge-doped silica core which has a diameter

TABLE 21.2 ITU-T Recommendations for Single-Mode Fibers

ITU-T Rec. No.	Description
G.652A/B	Standard single-mode fiber
G.652C/D	Low-water-peak fiber for CWDM applications
G.653	Dispersion-shifted fiber (made obsolete by NZDSF)
G.654	Submarine applications (1500 nm cutoff wavelength)
G.655A/B	Nonzero dispersion-shifted fiber (NZDSF)

between 5 and 8 μm. Since early applications used 1310-nm laser sources, this fiber was optimized to have a zero-dispersion value at 1310 nm. The G.652B subclass meets stricter polarization mode dispersion tolerances than the G.652A subclass.

ITU-T G.652C/D: *Low-water-peak fiber* for CWDM applications are created by reducing the water ion concentration of the G.652 material in order to eliminate the attenuation spike in the 1360 to1460 nm E band. The main use of this fiber is for low-cost short-reach coarse wavelength division multiplexing (CWDM) applications ranging from the O band through the L band. In CWDM the wavelength channels are sufficiently spaced (20 nm apart) so that minimum wavelength stability control is needed for the optical sources. The G.652D subclass meets stricter polarization mode dispersion tolerances than the G.652C subclass.

ITU-T G.653: *Dispersion-shifted fiber* (DSF) was developed for use with 1550-nm lasers. In this fiber type the zero-dispersion point is shifted to 1550 nm where the fiber attenuation is about half of that at 1310 nm. Although this fiber allows a high-speed data stream of a single-wavelength channel to maintain its fidelity over long distances, it presents dispersion-related problems in dense wavelength division multiplexing (DWDM) applications where many wavelength channels are packed with spacings of 0.8 nm or less into one or more of the operational bands. As a result, this fiber type became obsolete with the introduction of G.655 NZDSF.

ITU-T G.654: This specification deals with *cutoff-wavelength-shifted fiber* that is designed for long-distance high-power signal transmission. Since it has a high cutoff wavelength of 1500 nm, this fiber is restricted to operation at 1550 nm. It typically is used only in submarine applications.

ITU-T G.655A/B: *Nonzero dispersion-shifted fiber* (NZDSF) was introduced in the mid-1990s for DWDM applications. Its principal characteristic is that it has a nonzero dispersion value over the entire C band, which is the spectral operating region for erbium-doped optical fiber amplifiers. The G.655A fiber is intended for DWDM use in long-haul telecommunications applications. The G.655B subclass was introduced in October 2000 to extend DWDM applications into the S band. Its principal characteristic is that it has a nonzero dispersion value over the entire S band and the C band. This makes it very useful for metropolitan telecommunications applications.

Specialty Fibers

Whereas telecommunication fibers, such as those described above, are designed to transmit light over long distances with minimal change in the signal, *specialty fibers* are used to manipulate the light signal. Specialty fibers interact with light and are custom-designed for specific applications such as optical signal amplification, wavelength selection, wavelength conversion, and sensing of physical parameters.

A number of both passive and active optical devices use specialty fibers to direct, modify, or strengthen an optical signal as it travels through the device. Among these optical devices are light transmitters, optical signal modulators, optical receivers, wavelength multiplexers, couplers, splitters, optical amplifiers, optical switches, wavelength add/drop modules, and light attenuators. Table 21.3 gives a summary of some specialty fibers and their applications.

Cables

Cabling of optical fibers involves enclosing them within some type of protective structure. The cable structure will vary greatly depending on whether the cable is to pulled or blown into underground or intrabuilding

TABLE 21.3 Summary of Some Specialty Fiber and Their Applications

Specialty Fiber Type	Application
Erbium-doped fiber	Gain medium for optical fiber amplifiers
Photosensitive fibers	Fabrication of fiber Bragg gratings
Bend-insensitive fibers	Tightly looped connections in device packages
High-loss attenuating fiber	Termination of open optical fiber ends
Polarization preserving fibers	Pump lasers, polarization-sensitive devices, sensors
High-index fibers	Fused couplers, short-λ sources, DWDM devices
Holey (photonic crystal) fibers	Switches; dispersion compensation

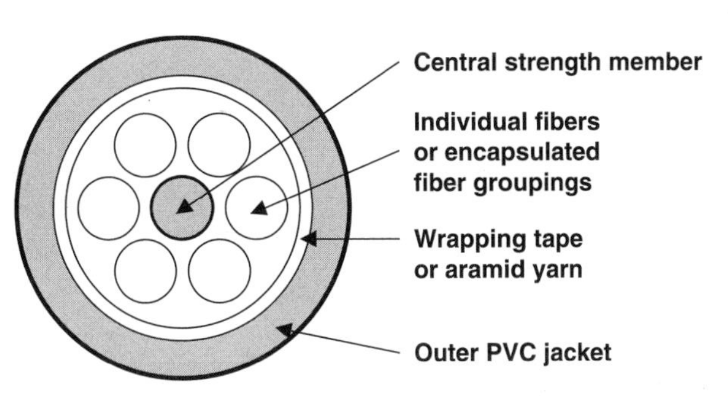

FIGURE 21.22 Schematic of a generic cable configuration.

tubes (called *ducts*), buried directly in the ground, installed on outdoor poles, or placed underwater. Different cable configurations are required for each type of application, but certain fundamental cable design principles will apply in every case. The objectives of cable manufacturers have been that the optical fiber cables should be installable with the same type of equipment, installation techniques, and precautions as those used for conventional wire cables. This requires special cable designs because of the unique properties of optical fibers such as their strength, dielectric (nonmetallic) nature, small size, and low weight.

Cable Materials and Structures

Figure 21.22 shows a generic cable configuration. To prevent excessive stretching, the cabling process usually includes the incorporation of *strength members* into the cable design. This is especially important in the design of aerial cables that can experience severe stresses due to factors such as wind forces or ice loading. The strength member can be strong nylon yarns, steel wires, or fiberglass rods. A commonly used yarn is Kevlar®, which is a soft but tough yellow synthetic nylon material belonging to a generic yarn family known as *aramids*.

Individual fibers or modules consisting of fiber groupings are wound loosely around the central member. Groupings of up to 12 fibers nominally are encapsulated by a miniature protective tube or by a wrapping tape. A cable wrapping tape and optionally another strength member then encapsulate a multiple of these fiber groupings.

Surrounding all of this is a tough polymer *jacket* that provides crush resistance and handles any tensile stresses applied to the cable so that the fibers inside are not damaged. The jacket also protects the fibers inside against abrasion, moisture, oil, solvents, and other contaminants. The jacket type defines the application characteristics; for example, heavy-duty cables for direct burial and aerial use have thicker, tougher jackets than light-duty cables for indoor use.

An important factor for using a cable in a building is the *flammability rating*. The National Electrical Code (NEC) in the United States establishes flame ratings for cables, while on a global scale the Underwriters Laboratories (UL) has developed cable test procedures. For example, the NEC requires that all cables which

TABLE 21.4 Standard Optical Fiber Jacket and Encapsulation-Tube Color Identifications

Fiber number	Color	Fiber number	Color
1	Blue	13	Blue/black tracer
2	Orange	14	Orange/yellow tracer
3	Green	15	Green/black tracer
4	Brown	16	Brown/black tracer
5	Slate (gray)	17	Slate/black tracer
6	White	18	White/black tracer
7	Red	19	Red/black tracer
8	Black	20	Black/yellow tracer
9	Yellow	21	Yellow/black tracer
10	Violet	22	Violet/black tracer
11	Rose (pink)	23	Rose/black tracer
12	Aqua	24	Aqua/black tracer

run through plenums (the air-handling space between walls, under floors, and above drop ceilings) must either be placed in fireproof conduits or be constructed of low-smoke and fire-retardant materials.

Fiber and Jacket Color Coding

If there is more than one fiber in a protective fiber encapsulation tube, then each fiber is designated by a separate and distinct jacket color. The ANSI/TIA/EIA-598-A standard, *Optical Fiber Cable Color Coding*, prescribes a common set of fiber colors. Since nominally there are up to 12 fiber strands in a single tube, strands 1 to 12 are uniquely color-coded as listed in Table 21.4. If there are more than 12 fibers within an individual tube, then strands 13 through 24 repeat the same fundamental color code as those for strands 1 through 12 with the addition of a black or yellow dashed or solid tracer line, as noted in Table 21.4. For cables having more than one encapsulation tube, the tubes also are color-coded in the same manner as the fibers, that is, tube 1 is blue, tube 2 is orange, and so on.

Ribbon cables follow the same color-coding scheme. Thus, one of the outside fibers would have a blue jacket, the next fiber would be orange, and so on, until the other outer edge is reached where, for a 12-fiber ribbon, the fiber would be aqua (light blue).

Indoor Cable Designs

Indoor cables can be used for interconnecting instruments, for distributing signals among office users, for connections to printers or servers, and for short patch cords in telecommunication equipment racks. The three main indoor types are described here:

1. *Interconnect cable.* Interconnect cables are designed for light-duty low-fiber count indoor applications such as fiber-to-the-desk links, patch cords, and point-to-point runs in conduits and trays. The cable is flexible, compact, and lightweight. Fiber optic *patch cords*, also known as *jumper cables*, are short lengths (usually less than 2 m) of simplex or duplex cable with connectors on both ends. They are used to connect lightwave test equipment to a fiber patch panel or to interconnect optical transmission modules within an equipment rack.
2. *Breakout or fanout cable.* Up to 12 fibers can be stranded around a central strength member to form what is called a *breakout* or *fanout* cable. Breakout cables are designed specifically for low to medium fiber count applications where it is necessary to protect individual jacketed fibers. The breakout cable facilitates easy installation of an independent connector to each fiber. With such a cable configuration, running (breaking out) the individually terminated fibers to separate pieces of equipment can be achieved easily.
3. *Distribution cable.* Individual or small groupings of fibers can be stranded around a nonconducting central strength member to form what is called a *distribution cable.* This cable can be used for a wide range of intra- and interbuilding network applications for sending data, voice, and video signals. If groupings of fibers are desired, they can be wound around a smaller strength member and held together

with a cable wrapping tape. Distribution cables are designed for use in intrabuilding trays, conduits, backbone premise pathways, and drop ceilings. A main feature is that they enable groupings of fibers within the cable to be branched (distributed) to various locations.

For indoor cables the NEC requires that the cables be marked correctly and installed properly in accordance with their intended use. The NEC identifies three indoor cable types for different building regions: plenum, riser, and general-purpose areas. A basic difference in the cable types is the material used for the outer protective cable jacket. NEC Article 770 addresses the flammability and smoke emission requirements of indoor fiber optic cables. These requirements vary based on the particular application as described here:

- *Plenum cables*: A plenum is the empty space within walls, under floors, or above drop ceilings used for air flow, or it can form part of an air distribution system used for heating or air conditioning. Plenum-rated cables are UL-certified by the UL-910 *plenum fire test method* as having adequate fire resistance and low smoke-producing characteristics for installations in these spaces without the use of a conduit.
- *Riser cables*: A riser is an opening, shaft, or duct that runs vertically between one or more floors. Riser cables can be used in these vertical passages. Riser-rated cables are UL-certified by the UL-1666 *riser fire test method* as having adequate fire resistance for installation without conduit in areas such as elevator shafts and wiring closets. Plenum cables may be substituted for riser cables, but not vice versa.
- *General-purpose cables*: A general-purpose area refers to all other regions on the same floor that are not plenum or riser spaces. General-purpose cables can be installed in horizontal, single-floor connections, for example, to connect from a wall jack to a computer. However, they cannot be used in riser or plenum applications without being placed in fireproof conduits. To qualify as a general-purpose cable, it must pass the UL 1581 *vertical-tray fire test*. Plenum or riser cables may be substituted for general-purpose cables.

Outside Plant Cables

Outside plant cable installations include aerial, duct, direct-burial, underwater, and tactical military applications. Many different designs and sizes of outdoor cables are available depending on the physical environment in which the cable will be used and the particular application.

Aerial cable. An aerial cable is intended for mounting outside between buildings or on poles or towers. The two main designs that are being used are the self-supporting and the facility-supporting cable structures. The *self-supporting cable* contains an internal strength member that allows the cable to be strung between poles without implementing any additional support to the cable. For the *facility-supporting cable*; first a separate wire or strength member is strung between the poles, then the cable is lashed or clipped to this member. Three common self-supporting aerial cable structures are known as OPGW, ADSS, and Figure 21.20.

In addition to housing the optical fibers, the *optical ground wire* (OPGW) cable structure contains a steel or aluminum tube that is designed to carry the ground current of an electrical system. The metal structure acts as the strength member of the cable. OPGW cables with up to 144 fibers are available.

The *all-dielectric self-supporting* (ADSS) cable uses only dielectric materials, such as aramid yarns and glass-reinforced polymers, for strength and protection of the fibers. An ADSS cable typically contains 288 fibers in a loose-tube stranded-cable-core structure.

A popular aerial cable is known as a Figure 21.20 cable because of its shape. A key feature is the factory-attached *messenger*, which is a support member used in aerial installations. The built-in messenger runs along the entire length of the cable and is an all-dielectric material or a high-tension steel cable. This self-supporting structure allows the cable to be installed easily and quickly on low-voltage utility or railway poles. In some cases the steel messenger is placed at the center of a self-supporting aerial cable to reduce stresses on the cable from wind or ice loading. This configuration usually is for short distances between poles or for short distances between adjacent buildings.

Alternatively a cable for an aerial application need not contain a built-in messenger. Instead a separate steel messenger is first strung between poles and the optical cable is then lashed to this messenger. This lashing method supports the cable at short intervals between poles instead of just at the poles themselves, thereby reducing stress along the length of the cable.

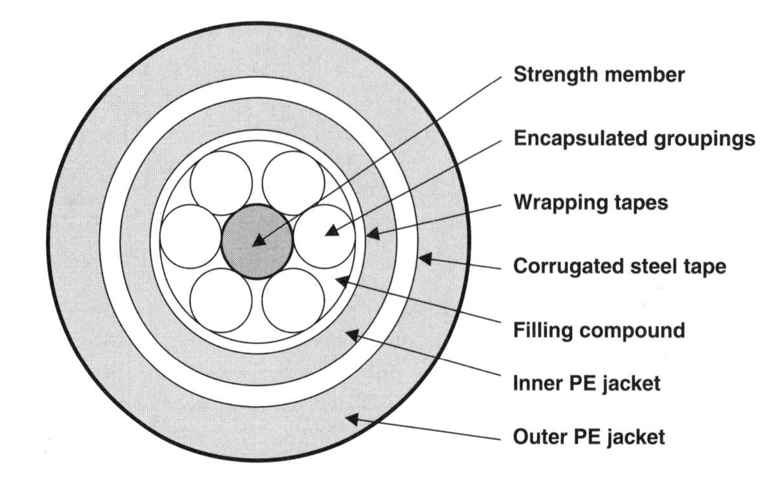

FIGURE 21.23 Schematic of an armored cable for direct-burial or underground-duct applications.

Armored Cable. An armored cable for direct-burial or underground-duct applications has one or more layers of steel-wire or steel-sheath protective armoring below a layer of polyethylene jacketing as shown in Figure 21.23. This not only provides additional strength to the cable but also protects it from gnawing animals such as squirrels or burrowing rodents. For example, in the United States the plains pocket gopher will destroy unprotected cable that is buried less than 6 feet deep.

Underwater Cable. Underwater cable, also known as *submarine cable*, can be used in rivers, lakes, and ocean environments. Since such cables normally are exposed to high water pressures, they have much more stringent requirements than underground cables. For example, they have various water-blocking layers and a heavier armor jacket. Cables that run under the ocean have further layers of armoring and contain copper wires to provide electrical power for submersed optical amplifiers or regenerators.

In addition to these designs, there are numerous other cable configurations for specific applications, such as telecommunication ribbon cables, air-drop cables for emergency and military operations, small and extremely flexible cables designed for air-blown installation in ducts, and highly rugged cables for military, mining, and petrochemical environments. Additional details on optical fibers, cables, installation methods, and equipment can be found in the books by Chomycz and by Keiser.

Further Reading

1. BiCSI, *Telecommunications Cabling Installation*, New York: McGraw-Hill, 2001.
2. Chomycz, B., *Fiber Optic Installer's Field Manual*, New York: McGraw-Hill, 2000.
3. Keiser, G., *Optical Communication Essentials*, New York: McGraw-Hill, 2003.
4. Keiser, G., *Optical Fiber Communications*, 3rd ed., Burr Ridge, IL: McGraw-Hill, 2000.
5. ANSI/TIA/EIA-598-A, *Optical Fiber Cable Color Coding*, 1995.
6. *Canadian Electrical Code*, "Optical Fiber Cables," Canadian Standards Association, 2002.
7. ITU-T Recommendations G.651 through G.655 for multimode and single-mode fibers.
8. *NFPA 70 — National Electrical Code*, Article 770, "Optical Fiber Cables and Raceways," National Fire Protection Association, 2002.

22

Solid-State Circuits

Ian D. Robertson
University of Leeds

Inder J. Bahl
M/A-COM, Inc.

22.1 Introduction

Solid-state circuits are extensively used in such applications as radar, communication, navigation, electronic warfare (EW), smart weapons, consumer electronics, and microwave instruments and equipment. This chapter briefly describes the performance status of amplifiers, oscillators, multipliers, mixers, and microwave control circuits.

High-frequency circuits can be broadly categorized into three technologies:

1. High-density active circuits using established analogue integrated circuit design techniques
2. Planar transmission line circuits including active and passive components and interconnections
3. Intricate assemblies using active devices mounted in rectangular waveguides or quasi-optical systems

As a result of rapid development of high-performance deep submicron silicon technologies, the first technology is able to operate up to the tens of GHz range and is widely adopted for commercial products in the wireless communications arena. These circuits almost always employ silicon technology and are referred to as radio frequency integrated circuits (RFICs). Single-chip transceivers have become commonly available and the rapid development of this technology in the late 1990s and early 2000s has enabled a wireless connectivity revolution and partly displaced established microwave design techniques (transmission lines, Smith charts, electromagnetism, etc.) in the lower microwave bands.[1] This type of solid-state circuit is not described in detail in this chapter, but further information can be found elsewhere in the handbook.

At higher frequencies, especially above 20 GHz, and for many specialist high-performance components it is still necessary to employ traditional transmission-line design techniques, and this leads to the second technology type in which the active devices are closely integrated with transmission-line circuitry [2–4]; the hybrid microwave integrated circuit (HMIC) consists of an interconnect pattern and distributed circuit components printed on a suitable substrate, with active and lumped circuit components (in packaged or chip form) attached individually to the substrate by the use of soldering and wire bonding. Alternatively, in the monolithic microwave integrated circuit (MMIC), all interconnections and components, both active and passive, are fabricated simultaneously on a semiinsulating semiconductor substrate (usually gallium arsenide, GaAs) using deposition and etching processes, thereby eliminating discrete components and wire bond interconnects. An exhaustive review of the history of HMIC and MMIC technology can be found in Ref. [5]. The term MMIC is popularly used for monolithic circuits operating in both the microwave (1 to 30 GHz) and millimeter wave (30 to 300 GHz) regions of the frequency spectrum. Major advantages of MMICs include low cost, small size, low weight, circuit design flexibility, broadband performance, elimination of circuit tweaking, high-volume manufacturing capability, package simplification, improved reproducibility, improved reliability, and multifunction performance on a single chip.

In the region above 100 GHz, two-terminal devices such as Schottky, Gunn, tunnel, impact avalanche and transit time (IMPATT), and varactor diodes start to offer better performance than transistors, and planar transmission lines often suffer from too much loss. This leads to the requirement for the third type of technology in which very specialist devices are assembled into rectangular waveguide and similar transmission media to achieve the best possible performance. Micromachining techniques are increasingly used to fabricate these circuits with better precision than can be obtained from a mechanical workshop. These techniques are widely used for submillimeter-wave and terahertz solid-state circuits. In passing, we note that the distinction between these two terms is the subject of some debate but the IEEE standard definition of submillimeter waves is the 300 GHz to 3 THz frequency range. At such high frequencies even rectangular waveguides have high loss and there is a great deal of research activity on quasi-optical techniques where the "circuit" treats the signal as a wave manipulated by antennas, lenses, holograms, mirrors, etc.

22.2 Device Technologies

In the vast majority of cases up to ~100 GHz, RFIC or MMIC technology is used nowadays to realize solid-state circuits. Some exceptions are where extremely high-power or low-noise performance is required and HMIC realization is used because of the lower transmission-line losses. A wide range of technologies has evolved and the different active devices that can be used are as follows:

GaAs MESFET
GaAs HEMT
GaAs HBT
Silicon bipolar, CMOS, BiCMOS
Silicon-germanium HBT, HMOST
Indium phosphide HEMT or HBT
Gallium nitride (GaN) and silicon carbide (SiC) based devices
Two terminal devices; Gunn, Schottky, IMPATT, tunneling devices.

The GaAs MESFET (metal-semiconductor field effect transistor) was the first microwave transistor and is still used extensively in many applications. It is easily fabricated using ion implantation for high-volume applications and has good noise figure and quite high output power. GaAs foundries typically offer 0.5 μm gate length MESFET processes, which are useful for circuits operating to 20 GHz. Figure 22.1 shows the typical layout of a four-fingered MESFET.

The GaAs-based pseudomorphic high electron mobility transistor (pHEMT) has a similar layout but the high mobility afforded by the channel design offers a significant increase in performance compared to the MESFET. With gate lengths as short as 0.05 μm, circuits operating to over 100 GHz have been reported. For higher performance the indium concentration in the channel must be increased beyond that which a GaAs substrate can accommodate; lattice-matched HEMTs on indium phosphide substrates have produced excellent high-frequency performance and MMIC amplifiers operating at over 200 GHz have been reported using InP-based HEMTs.

The GaAs heterojunction bipolar transistor (HBT) has an advantage over the HEMT in that it is a vertical structure with critical device dimensions which are defined by the material growth and doping rather than by lithography. An HBT with an emitter width of 1 or 2 microns can offer good microwave performance. Many manufacturers are now offering HBT foundry processes, most often in the GaAs/GaInP material system, because there has been a move away from the use of aluminum due to reliability problems. GaAs HBTs have found widespread application in mobile handset power amplifiers due to their good power efficiency and high linearity.

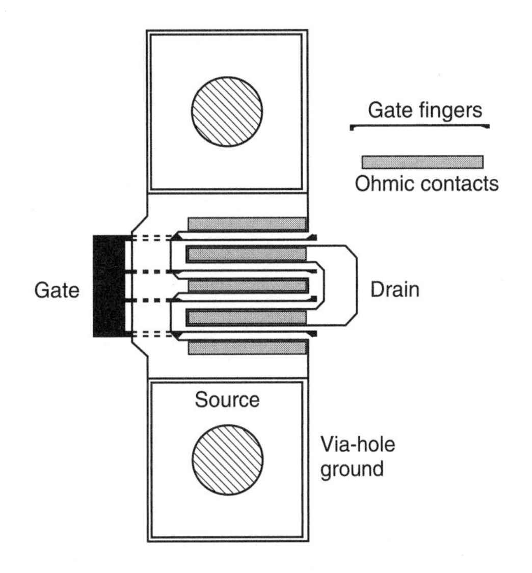

FIGURE 22.1 Layout of a 4×75 pi-gate MESFET.

Silicon bipolar technology has advanced tremendously in recent years and the silicon-germanium HBT has been reported to have *f*Ts of over 300 GHz. CMOS technology has also advanced dramatically and 50 nm devices have achieved *f*Ts of over 200 GHz. The major advantage of silicon technology is that 300 mm silicon wafers are readily available, whilst GaAs MMIC fabrication is limited to 150 mm at the time of writing, and that silicon microwave circuits can readily be integrated with digital circuits, i.e., memory, DSP, and microprocessors. However, silicon suffers from a substrate loss problem that makes it difficult to integrate active devices and low-loss transmission lines. This makes it challenging to realize millimeter-wave circuits, but already at the time of writing circuits operating at 77 GHz have been reported. Crucially, the highly optimized silicon devices push the capability of the material to the limit and the breakdown voltage of silicon devices does not match that of GaAs devices. GaAs pHEMTs can have 20 V breakdown voltage compared with only 2 to 3 V for some SiGe HBTs, therefore GaAs can achieve higher powers. Gallium nitride and silicon carbide technologies have even higher breakdown voltages due to the wide bandgap materials. A discrete GaN HEMT device has been reported with >200 W output power at 2.1 GHz with operation from a 50-V supply [6].

InP devices achieve the highest levels of performance, but in manufacture the fragile substrate restricts wafer sizes to even less than those of GaAs. The unique compatibility of InP HBTs and HEMTs with lasers and photodetectors required for long-haul optical communications has led to tremendous interest in ultrahigh-speed digital circuits such as dividers, multiplexers, and demultiplexers operating to 100 Gb/sec [7].

22.3 Amplifiers

There are six key microwave amplifier design topologies, as illustrated in Figure 22.2. The reactively matched amplifier (a) uses purely reactive matching networks at the input and output of the transistor; either lumped inductors and capacitors or transmission lines can be used. The technique gives moderate bandwidth and good noise and power performance. However, because of the transistor inherent instability and gain roll-off, wideband design is difficult. Figure 22.3 shows a 20 to 40 GHz reactively matched GaAs MMIC amplifier and it can clearly be seen how the active devices, passive components, and transmission-line matching elements are fabricated on a single chip with this technology. The lossy match amplifier uses resistors within its matching networks to enable flat gain to be achieved over a broad bandwidth [8]. The most typical topology is to employ resistors in series with high-impedance stubs on both the input and output, as shown in Figure 22.2. (b) At low frequencies the stubs have little electrical length, and the resistors load the transistor and lower its gain. At high frequencies the resistors have little effect on the transistor because of the inductive effect of the stubs. Hence, the matching networks can introduce a positive gain slope to compensate the transistor's gain roll-off. The feedback amplifier (c) uses a resistance (of the order of hundreds of ohms) from the drain to the gate. This has the effect of stabilizing the device and can make the input and output impedances much closer to the desired 50 Ω. Closely related, the popular Darlington pair configuration (d) ensures a high gain-bandwidth product, and the resistor

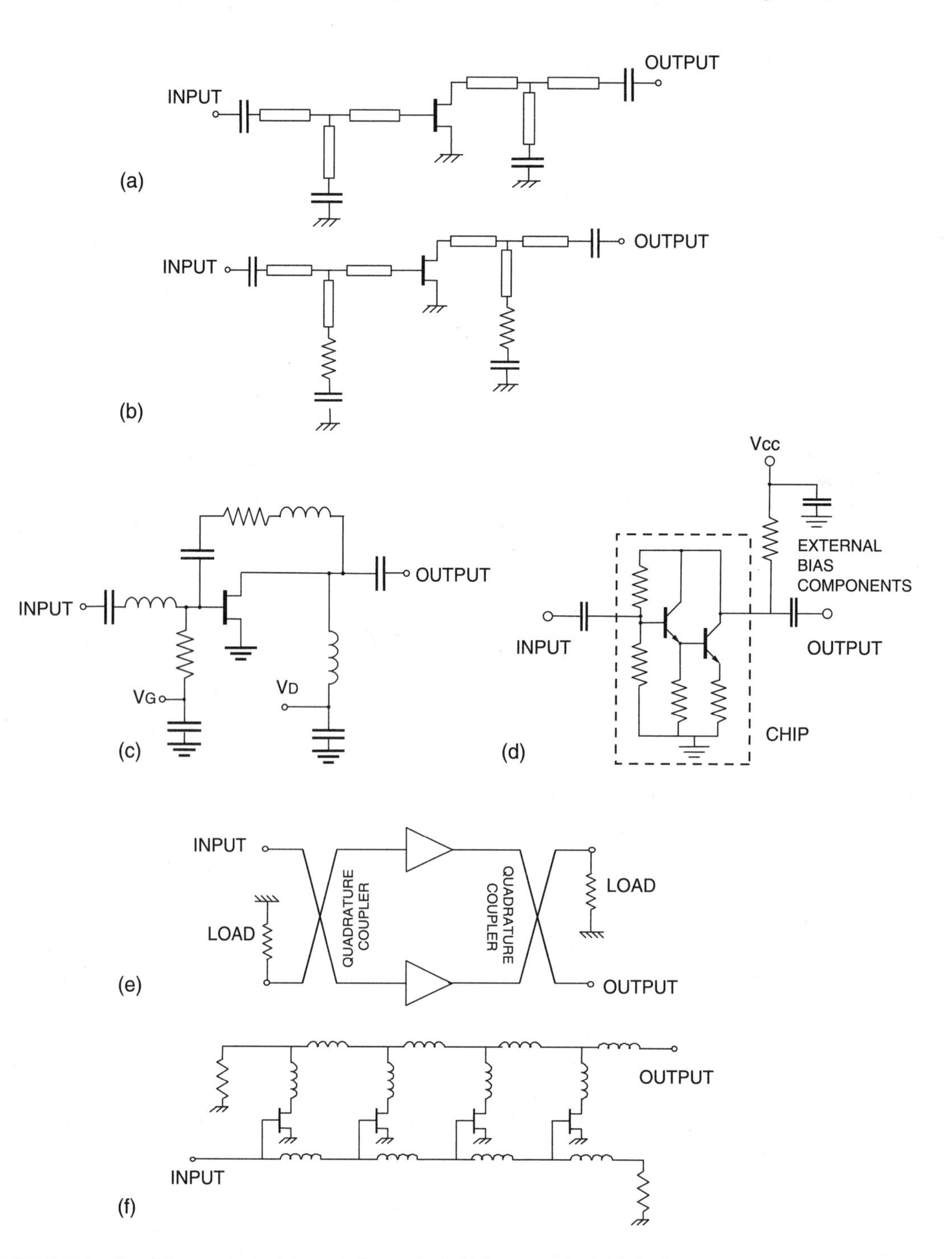

FIGURE 22.2 Amplifier topologies (a) reactively matched, (b) lossy matched, (c) feedback, (d) Darlington pair feedback, (e) balanced, (f) distributed.

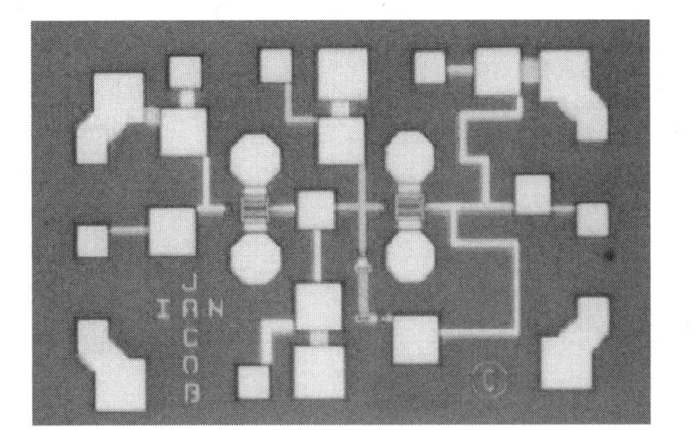

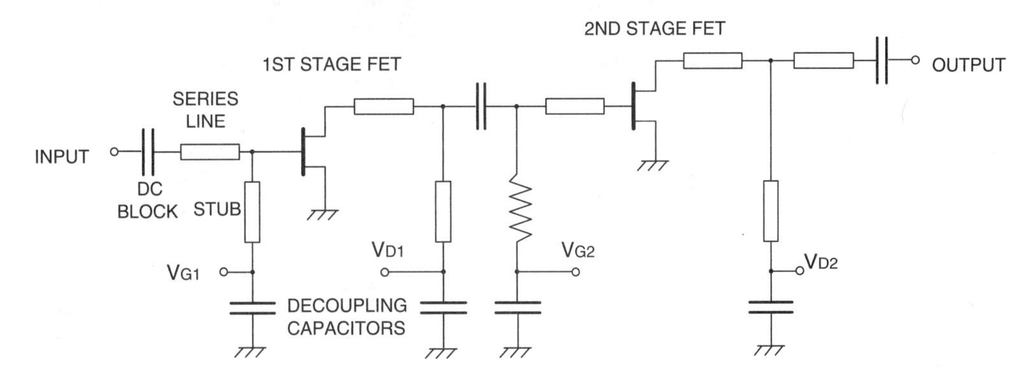

FIGURE 22.3 pHEMT GaAs MMIC 20 to 40 GHz two-stage amplifier using transmission-line matching elements: (a) photograph, (b) circuit diagram.

feedback ensures a smooth broadband response with good input and output matches. Emitter resistors are employed for stabilization of the operating point and for series feedback. This amplifier topology is capable of achieving more than dc to 40 GHz bandwidth; with HBTs (e) the balanced amplifier is a generic approach for improving matching and increasing output power. By operating an identical pair of amplifiers fed in quadrature using directional couplers, the reflected signals are dumped into the load terminations. This is especially valuable for the reactively matched amplifier for controlling VSWR, when noise or power matching is needed. Figure 22.4 shows the photograph of a 60 GHz SiGe HBT balanced low-noise amplifier (LNA) which achieved 14 dB gain and 6.3 dB noise figure [9]. In the distributed (or traveling-wave) amplifier (f) the problem of achieving a broadband match to the transistor input and output impedance is

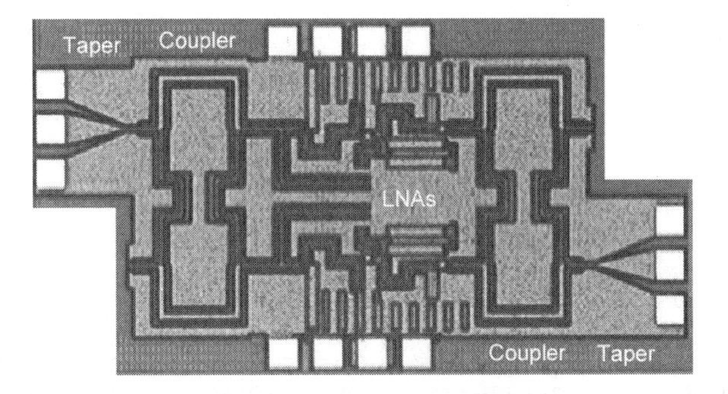

FIGURE 22.4 Photograph of a 60 GHz SiGe HBT balanced LNA [9]. (© IEEE 2004, used with permission.)

TABLE 22.1 Broadband MMIC Distributed Amplifier Performance

Device	Bandwidth (GHz)	RF power (dBm)	Gain (dB)	Ref.
GaAs MESFET	30	22	7	[10]
GaAs pHEMT	46	16.5	14	[11]
InP HEMT	157	—	5	[12]
SiGe HBT	50	5	6	[13]
Si CMOS	27	10	6	[14]
InP DHBT	75	19	6	[15]
GaN HEMT	8	35 to 37.8	12	[16] (tapered)

overcome by incorporating the input and output capacitances of a number of transistors into artificial transmission-line structures. Table 22.1 compares the performance of various distributed amplifiers in terms of bandwidth and output power capability. All these operate down to low frequencies—as low as 30 kHz if suitable bias networks are designed.

Monolithic power amplifier design involves power combining as many devices as is practical in order to achieve increased power. A single large device is impractical on an MMIC because of the very low impedance level; the cluster matching technique has emerged as the optimum means of integrating the matching network into the splitting and combining manifolds [17,18]. The 50-W MMIC amplifier [19] shown in Figure 22.5 illustrates very clearly the output manifold. The insertion loss of this output manifold imposes a practical limit on the number of devices that can be combined, both for economical reasons (wasteful use of expensive chip space) and because the efficiency drops quickly as the combining loss increases, causing a thermal problem. Table 22.2 compares some MMICs that have achieved the highest powers for various frequency ranges. The challenge in obtaining high power at millimeter-wave frequencies and over multioctave bandwidths is evident from this comparison. To further increase the achievable power it is necessary to power combine multiple MMICs. Depending on the frequency this might be done at the package level or by using very low loss combining in waveguide, or even quasi-optically as discussed later.

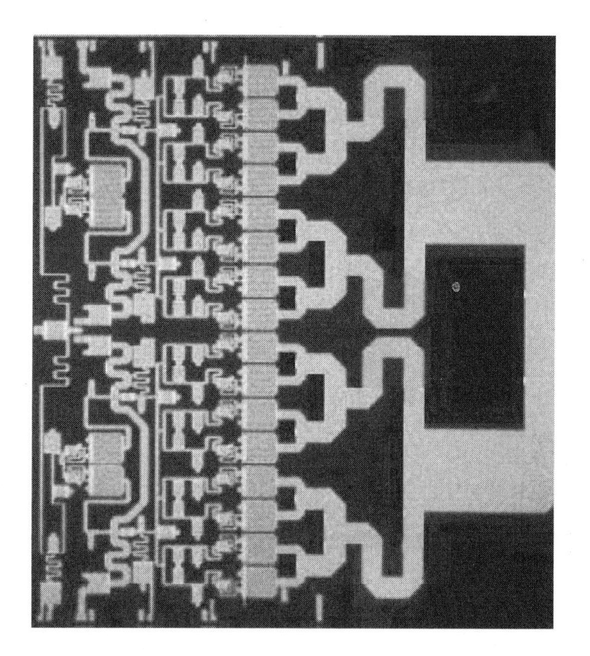

FIGURE 22.5 Filtronic 50 W 2.1 to 2.2 GHz power amplifier MMIC [19]. (© IEEE 2004, used with permission.)

TABLE 22.2 Some Typical GaAs MMIC Power Amplifier Performance Parameters (2004)

Company	Frequency (GHz)	Power	No. Stages	Gain (dB)	PAE (%)	Device
Filtronic [19]	2.1 to 2.2	50 W P1dB	2	21	50	pHEMT
M/A-COM MA08509D	8 to 11	10 W sat	3	22	32	MESFET
Agilent AMMC-6440	38 to 43	1 W sat	3	15	31	pHEMT
TRW [20]	95	427 mW sat	2	15	19	InP HEMT
Agilent AMMC-5024	30 kHz to 40	180 mW P1dB	9 section DA	16	~ 15	pHEMT

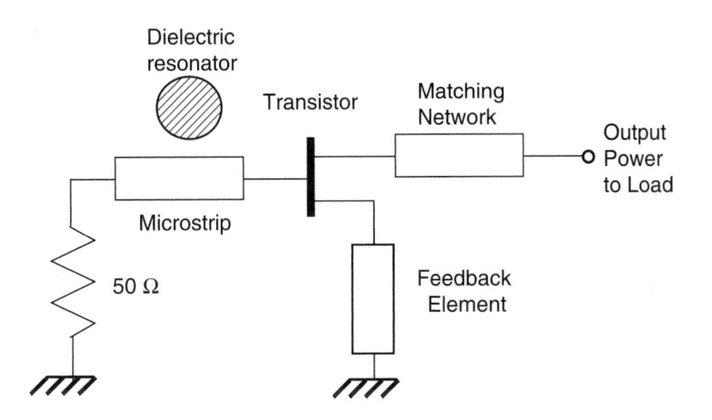

FIGURE 22.6 Basic configuration of a dielectric resonator oscillator.

22.4 Oscillators and Multipliers

Solid-state oscillators represent the basic microwave energy source and have the advantages of light weight and small size compared to microwave tubes. As shown in Figure 22.6, a typical microwave oscillator consists of a MESFET as an active device (a diode can also be used) and a passive frequency-determining resonant element, such as a microstrip, surface acoustic wave (SAW), cavity resonator, or dielectric resonator for fixed tuned oscillators, and a varactor or a yttrium iron garnet (YIG) sphere for tunable oscillators. These oscillators have the capability of temperature stabilization and phase locking. Dielectric resonator oscillators provide stable operation from 1 to 100 GHz as fixed frequency sources. In addition to their good frequency stability, they are simple in design, have high efficiency, and are compatible with MMIC technology. Gunn and IMPATT oscillators provide higher power levels and cover microwave and millimeter wavebands. The transistor oscillators using MESFETs, HEMTS, and HBTs provide highly cost-effective, miniature, reliable, and low-noise sources for use up to the mm-wave frequency range. Compared to a GaAs MESFET oscillator, a BJT or a HBT oscillator typically has 6 to 10 dB lower phase noise very close to the carrier. Figure 22.6 shows the basic configuration of a dielectric resonator oscillator which is commonly used at microwave frequencies. The feedback element is used to make the active device unstable, the matching network allows transfer of maximum power to the load, and the dielectric resonator provides frequency stability.

In recent years the synthesized oscillator has become a key component in wireless transceivers, and a great deal of research has been published on the design of integer-N and fractional-N phase locked loops. The fractional-N method can give low phase noise and small frequency step size, but to reduce fractional spurs requires advanced techniques employing Σ-Δ modulators and dithering [21,22].

Microwave frequency multipliers are used to generate microwave power at levels above those obtainable with fundamental frequency oscillators. Several different nonlinear phenomena can be used to achieve frequency multiplication, e.g., nonlinear reactance in varactors and step-recovery diodes, nonlinear resistance in Schottky barrier diodes, and three-terminal devices (BJT, MESFET, HEMT, HBT). Figure 22.7 shows the power available from carefully optimized Schottky varactor diode multipliers in the submillimeter range [23]. At the higher end of the THz range, quantum cascade lasers are being investigated by many researchers [24]. These, however, have the opposite power/frequency slope, leading to what has been termed the "terahertz gap" where neither electronic nor photonic solutions have yet generated sufficient power from compact sources.

22.5 Mixers and Modulators

Mixers convert (heterodyne) the input frequency to a new frequency, where filtering and gain is easier to implement. A mixer converting the RF frequency to a low intermediate frequency (IF) is called a down-converter, whereas a mixer converting an input IF signal to the RF frequency is called an up-converter. A mixer

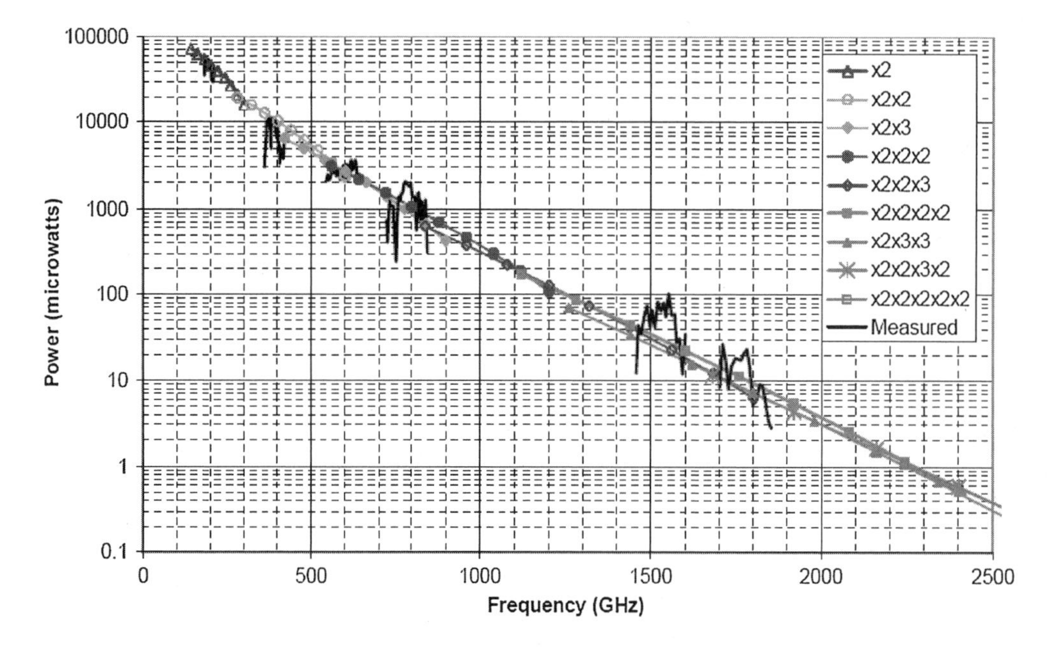

FIGURE 22.7 Output power of Schottky diode frequency multipliers in the submillimeter wave range [23]. (© IEEE 2004, used with permission.)

is basically a multiplier that requires two signals and uses any solid-state device that exhibits nonlinear properties. Mixing is achieved by applying an RF and a high-power local oscillator signal to a nonlinear element, which can be a diode or a transistor.

There are many types of mixers such as single-ended, balanced, double-balanced, and even double-double-balanced [25]. Mixers have for many years used diodes but can also be realized using the nonlinearities associated with transistors that provide conversion gain. The most commonly used mixer configuration in the microwave frequency band is the double-balanced mixer having better isolation between the ports and better spurious response. However, the single-ended and single-balanced mixers place lower power requirements on the local oscillator and have lower conversion loss and are often preferred for submillimeter wave applications where LO power is limited. Subharmonic mixing (where the local oscillator frequency is lower than that needed in conventional mixers) has been extensively used at mm-wave frequencies. This technique is quite useful when reliable stable local oscillators are either unavailable or prohibitively expensive at high frequencies.

Figure 22.8 shows the two most common mixer circuits; the double-balanced diode ring mixer and the Gilbert cell. The former is common in MIC and MMIC applications and gives reasonable conversion loss with good power handling (measured in terms of the third-order intercept point). In MMIC applications the baluns, shown as wirewound transformers, are realized with planar forms of the Marchand balun [26–28]. At lower frequencies, and especially in RFICs, the Gilbert cell mixer is almost always used for its compact size and conversion gain. Often a whole transceiver will use balanced signals in order to ease RF grounding issues (due to the virtual ground) and eliminate unwanted common-mode signals such as substrate noise and bounce from digital circuitry; then, the baluns may be realized off-chip in the package.

In many wireless systems, and sometimes in radar transceivers, the direct conversion transceiver concept has been adopted. Here, to enable a single-chip solution, the IF sections are removed and the mixers are then required to translate the baseband information signal to and from the carrier frequency directly. For modern digital communications, complex modulation is required (QPSK, QAM, etc.) and the vector modulator is widely employed. As shown in Figure 22.9(a), this consists of a pair of mixers fed with quadrature LO signals. The baseband I and Q signals modulate each arm independently and the two outputs are summed to give the quadrature modulated RF output signal. On the receiver side the input RF signal is down-converted directly to

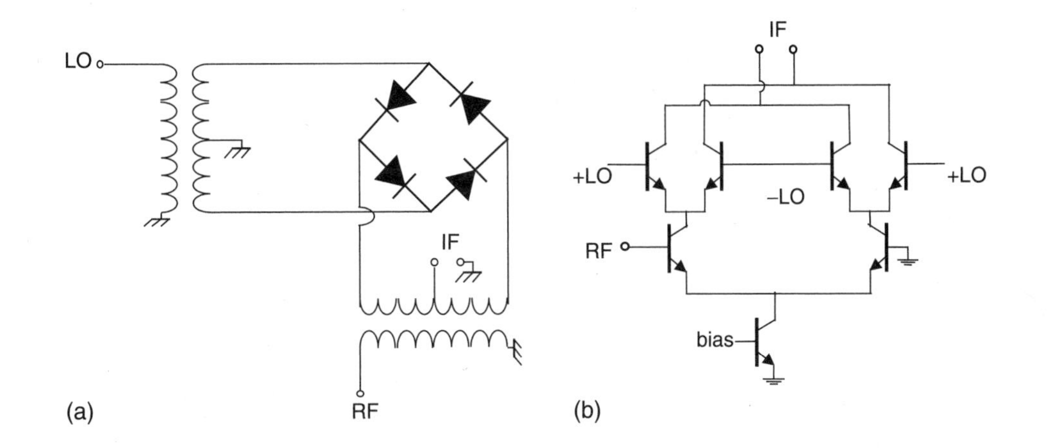

FIGURE 22.8 The two most common basic mixer configurations: (a) double-balanced diode ring mixer and (b) double-balanced Gilbert cell mixer (with bipolar transistors).

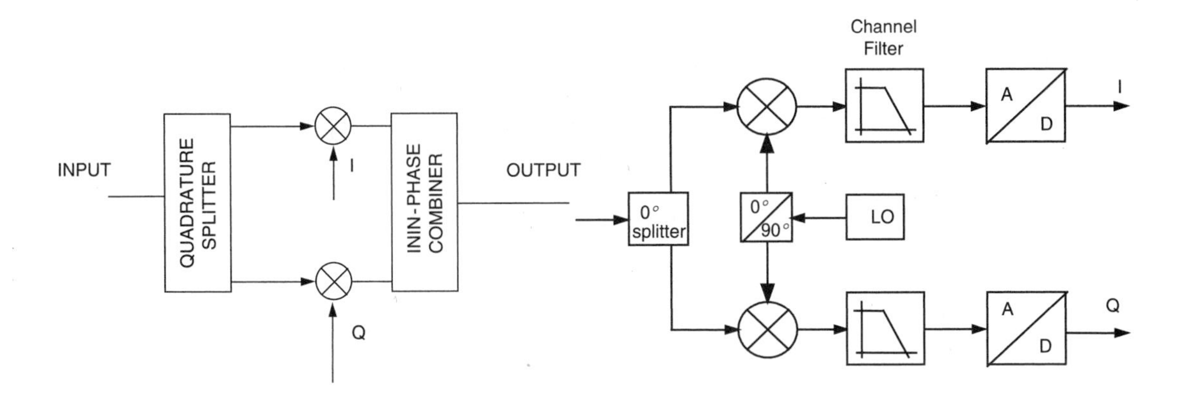

FIGURE 22.9 Block diagram of (a) IQ vector modulator and (b) demodulator with analogue-to-digital converters.

baseband, as shown in Figure 22.9(b). The A/D converters are shown to emphasize how modern RF design spans a wide topic area from electromagnetism to digital signal processing (DSP). Converting directly to baseband does mean that special attention must be paid to the mixer design to minimize unwanted signal leakage, second-order intermodulation and phase noise. A number of researchers have proposed means of addressing these issues [29].

22.6 Control Circuits

MMICs are extensively used for switching functions, and for amplitude and phase control. The best-known application is in adaptive beam-formers for phased-array antennas. Other applications include switching for redundancy or for antenna diversity, TX–RX switching for radar and communications, automatic gain control, and direct carrier amplitude/phase modulation for communications.

Switches

FET switches are common because the control electrode (gate) is isolated from the drain and source electrodes, no dc-blocking capacitors are needed and negligible dc power is consumed. The FET switch is a three-terminal device in which the gate bias voltage is used to control the drain-to-source resistance of the channel: At zero bias the resistance is at a minimum, and at pinch-off the channel resistance is very high. However, the source/drain contact and access resistances, as well as the intrinsic resistance of the channel itself, mean that the ON-state resistance is never zero. A large gate-periphery device results in a reduced ON-state resistance, but also in larger intrinsic drain-to-source capacitance which limits the OFF-state isolation of the FET switch at high frequencies. The physical parameters of the device, such as channel geometry, gate length, doping, and pinch-off voltage, can be optimized to produce switches with improved performance. Selective ion implantation can be used to enable the use of high quality switch FETs on the same chip as standard amplifying FETs.

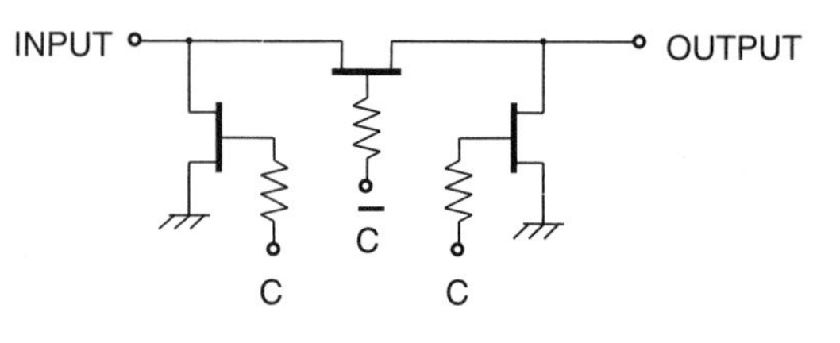

FIGURE 22.10 Circuit diagram of a reflective SPST FET switch.

The performance of a single FET switch is generally rather poor, and instead a combination of series and shunt FETs is often necessary to provide low insertion loss and high isolation. Figure 22.10 shows the topology of a typical reflective SPST switch, where C and $\bar{C}$ are complementary control voltages. Figure 22.11 shows the topology of a SPDT switch. Since switches are passive elements they are capable of operating to surprising high power levels, and MMIC switches capable of handling more than 100 W RF power have been reported [30].

Variable Attenuators

Attenuators can be designed for either digital or analogue control. Digitally controlled components have the obvious advantage of compatibility with the processor controlling the system, and binary-weighted attenuation values are generally used (e.g., 1, 2, 4, 8, and 16 dB). Analogue control has the advantage of simplicity and provides continuous gain control. In the switched attenuator type [31], two single-pole double-throw (SPDT) switches are used to switch the signal between a through line and an attenuator, as shown in Figure 22.12(a). This technique is best suited to larger attenuation values, and the attenuator arm can use a T- or π-type resistor network. The switched scaled-FET technique uses different sizes of FETs in two paths connected between the input and output ports, as shown in Figure 22.12(b). The FETs are used as switchable low-value resistances using the fact that the ON-state resistance depends on the gate peripheries

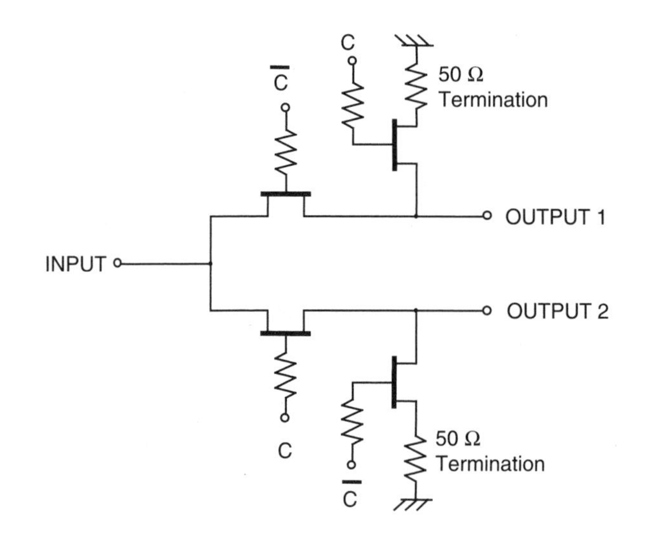

FIGURE 22.11 Circuit diagram of an absorptive SPDT FET switch.

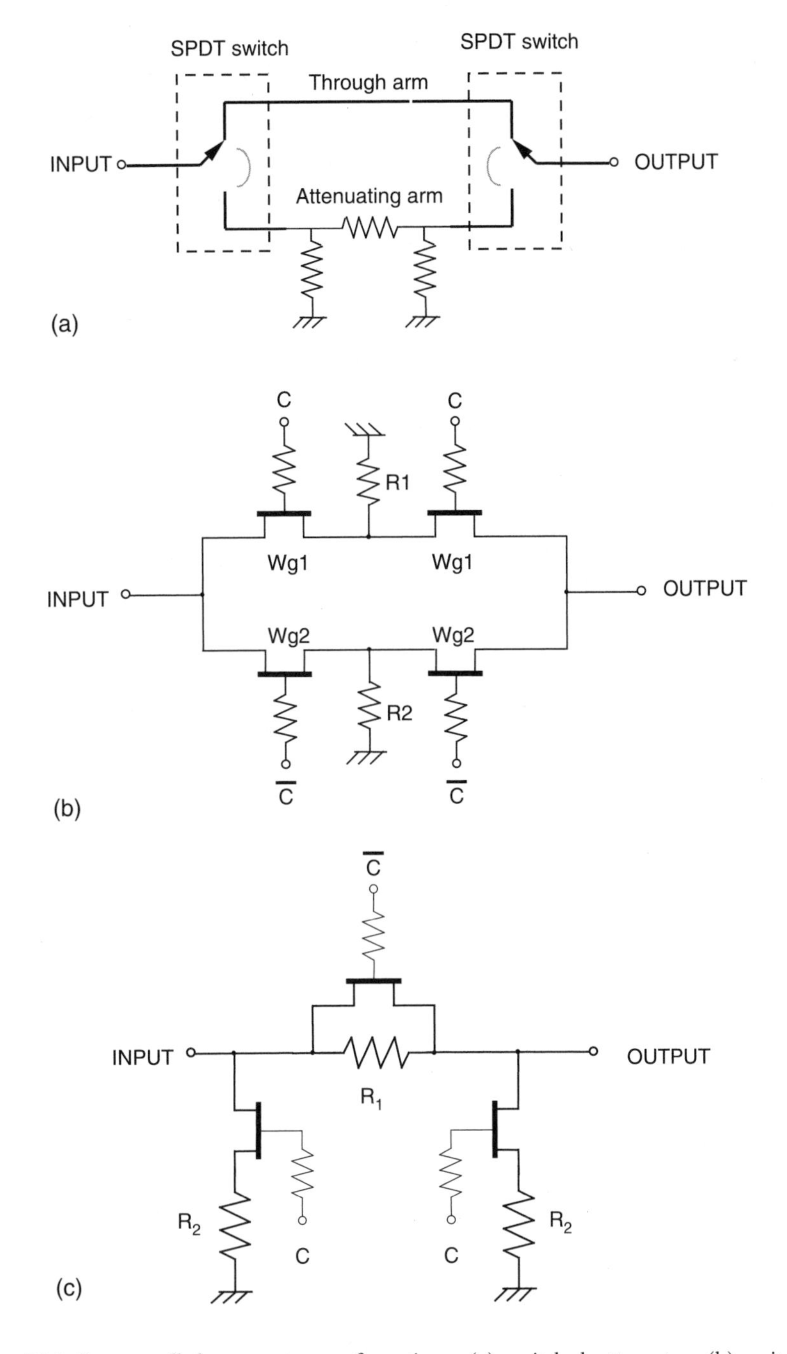

FIGURE 22.12 Digitally controlled attenuator configurations: (a) switched attenuator, (b) switched scaled FET, (c) switched resistor.

W_{g1} and W_{g2}. The switched resistor method [32] uses T- or π-attenuators in which FET switches are used to switch the resistors in or out of the circuit, as shown in Figure 22.12(c).

An analogue attenuator can be realized using a reflection-type topology in which variable resistance devices are used to control the magnitude of the reflection coefficient of the terminations [33], or, rather than employing an attenuator, a variable gain amplifier can be employed for continuous gain control [34,35].

Phase Shifters

MMIC phase shifters can be implemented with either analogue or digital control. In the case of analogue control most phase shifters are based on the continuous voltage control of a varactor diode's capacitance. Dual-gate FETs have also been used as analogue phase shifters but the performance is limited. Digital phase shifters can have a number of topologies, but the general principle is either that the signal is switched between two networks which have a fixed phase difference, or the phase shift of a single network is controlled by switching elements in or out of the network. A digital phase shifter will generally use binary weighting (e.g., with separate networks for 22.5, 45, 90, and 180°) and generally a different technique is optimum for each of these individual "bits."

In order to achieve constant insertion loss and good port matches almost all analogue phase shifters employ the reflection-type phase shifter topology [36,37], as shown in Figure 22.13. The signal from the output port is that reflected from the two varactor diodes. If the varactors are ideal variable capacitances then the magnitude of their reflection coefficient is fixed at unity, but the phase varies from zero (no capacitance) to 180 (maximum capacitance,virtually short circuit). Thus, the phase of the output signal can be varied continuously using the bias voltage on the varactor diodes. In practice, the phase shift range is limited and is determined by the capacitance tuning ratio of the diodes.

The reflection-type topology can also be used for digitally controlled phase shifters. In this case the varactor diodes are replaced with FET switches (for a 0/180° phase shifter bit) or with switched short-circuited transmission-lines. In the latter case, a number of diode or FET switches are placed along the transmission lines, as shown in Figure 22.14(a), and the one that is turned ON determines the point at which the line is short circuited [38]. This switched-line approach can also be used more directly; the signal is simply routed through one of two transmission lines of different lengths [39,40], as shown in Figure 22.14(b). However, it should be noted that strictly speaking this becomes a time-shifter, and that the phase shift frequency response would not be flat. The length of the transmission lines precludes the full monolithic integration of the switched-line topology at the lower microwave frequencies. Loaded-line phase shifters are more compact and use a single transmission line which is periodically loaded with switchable reactive elements [41]. This method has been used extensively [42,43], but can be difficult to use for large values of phase shift because it is difficult to keep the line's impedance constant.

The switched-filter phase shifter shown in Figure 22.14(c) employs SPDT switches to switch the signal between a high-pass and a low-pass filter [44]. The filters have equal amplitude responses over the range of interest, but the low-pass filter introduces phase-lag whereas the high-pass filter introduces phase-lead. Using lumped elements for the filters makes this technique suitable for low-frequency applications. In order to further reduce the size, it has been shown that FET switches can be directly integrated into the filters, and the FET parasitic capacitances can be absorbed into the filter network [45].

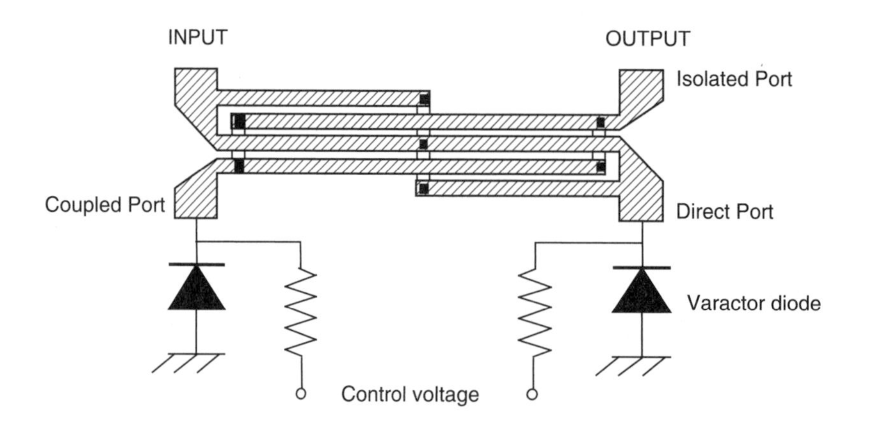

FIGURE 22.13 Analogue reflection-type phase shifter.

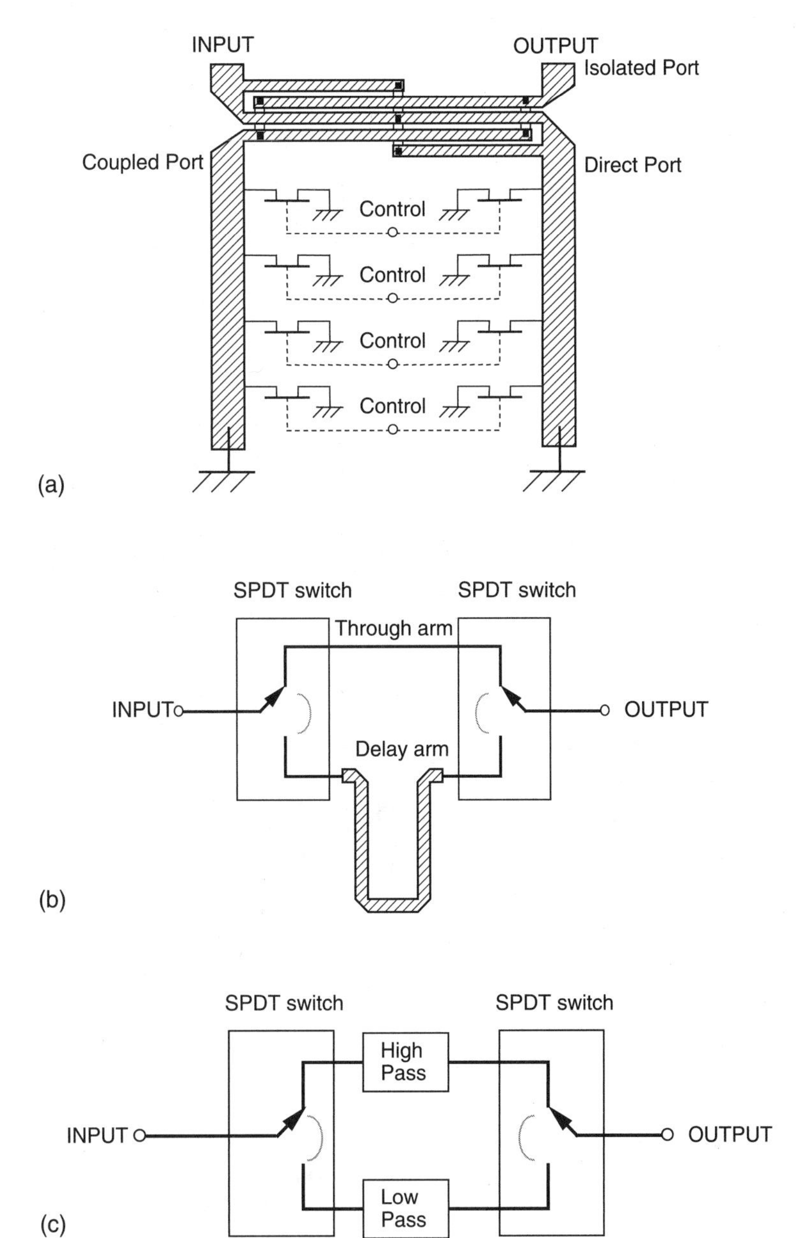

FIGURE 22.14 Digital phase shifters: (a) reflection type, (b) switched line, (c) switched filter.

Most of the topologies described so far can be used for a number of different phase shift bits. There is another class of digital phase shifter where the phase shift value is fixed intrinsically. For example, a 90° bit can be realized using a switched-coupler approach [46] and a 180-bit can be realized by switching between the two outputs of a balun which can be active [47] or passive [48].

22.7 Integrated Antennas

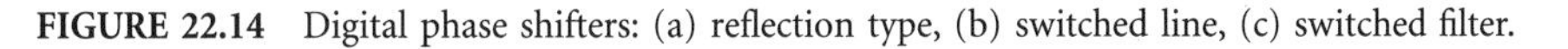

Above 30 GHz, the on-chip antenna becomes an effective means of coupling the signal into or out of the MMIC. A major advantage of such an approach is the simplified packaging, with reduced bond-wire parasitics and variability in production. However, the integrated antenna really yields the most benefits when integrated

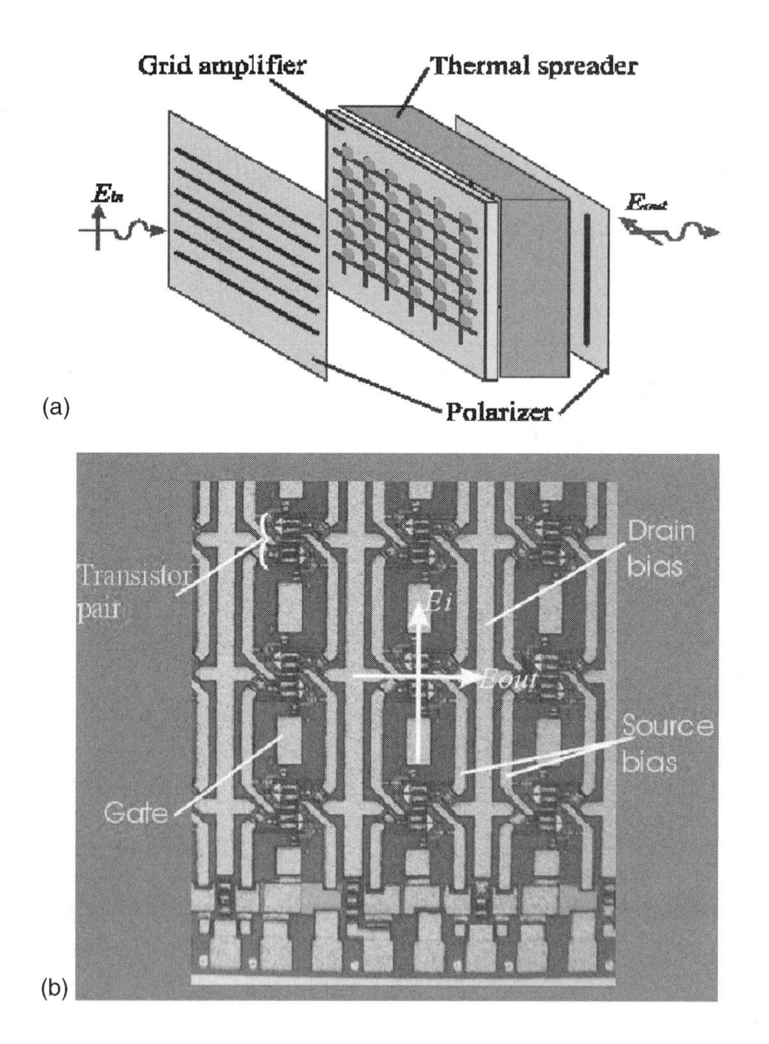

FIGURE 22.15 Grid amplifer: (a) concept, showing polarizers, (b) detail of the MMIC grid amplifier circuits [49]. (© IEEE 2004, used with permission.)

into arrays for imaging, beam steering, and spatial power combining. The grid amplifier [49] shown in Figure 22.15 is an important technique for spatial power combining, in which the input and output waves are in orthogonal polarizations in order to achieve isolation, and a large MMIC contains the amplifier array. Spatial power combining is very important since the combining loss can be greatly reduced compared to planar transmission lines, making it practical to combine the output power of hundreds of devices.

22.8 MEMS Technology and Micromachining

Micromachined silicon components, using selective crystallographic etching techniques, have been widely developed for high-volume commercial markets, such as, air-bag sensors, displays, disk drives, and print-heads. These miniature components are classified as microelectromechanical systems (MEMS) or microsystems when moving parts are incorporated. For microwave circuits, MEMS technology has the important feature of being able to realize moving parts for switching, tuning, and steering [50]. Figure 22.16 shows a switchable capacitor, for example, used as part of a reconfigurable impedance tuner [51]. The same fabrication techniques can be used to construct transmission lines and passive components using air as the main dielectric, leading to low loss. Figure 22.17 shows a low-loss 72 GHz power combining network which

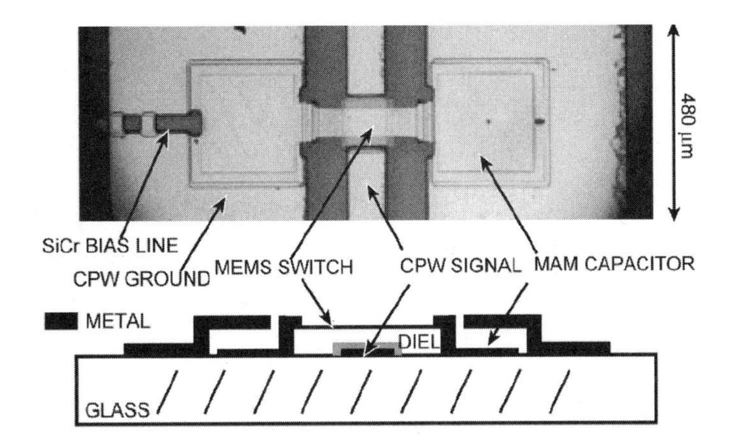

FIGURE 22.16 RF MEMS switched capacitor [51]. (© IEEE 2004, used with permission.)

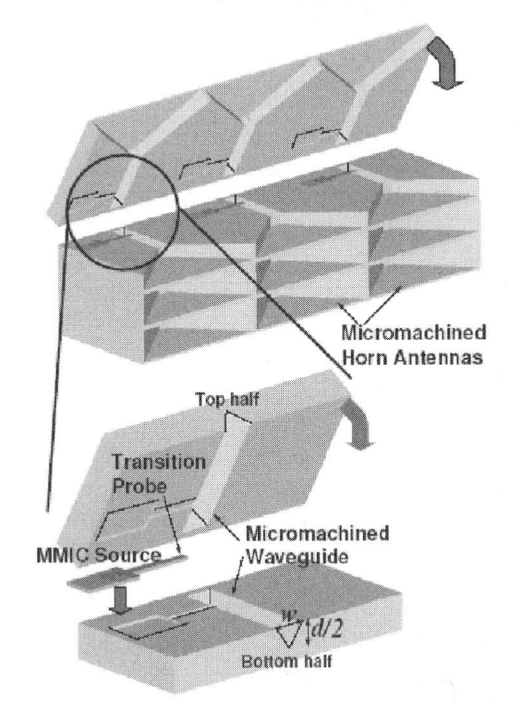

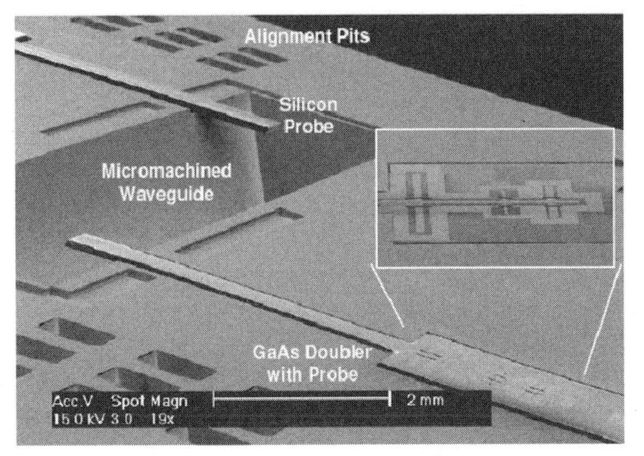

FIGURE 22.17 Micromachined millimeter-wave module for power combining [52]. (© IEEE 2004, used with permission.)

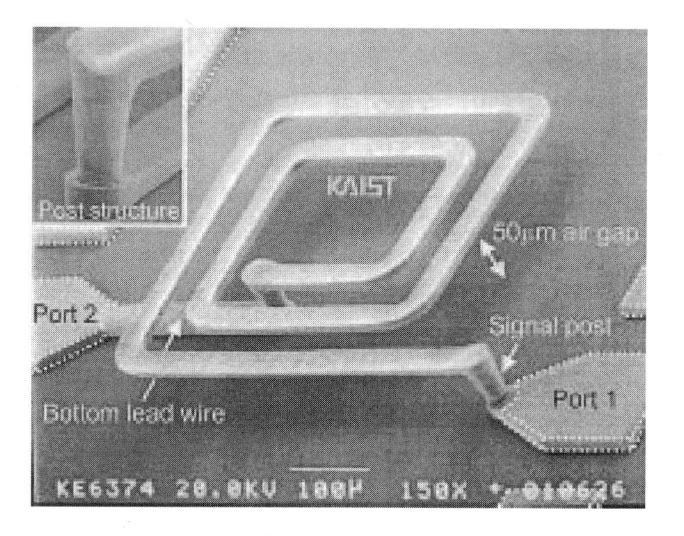

FIGURE 22.18 Airbridge inductor fabricated by micromachining [53]. (© IEEE 2004, used with permission.)

was fabricated using micromachined waveguides [52]. Micromachining has also been extensively applied to the realization of high-Q inductors on silicon substrates. Here, the challenge is to isolate the inductor from the low resistivity substrate, especially for CMOS technology where the substrate resistivity is kept low, to minimize coupling and bounce in digital circuits. Figure 22.18 shows an airbridge inductor that has an inductance of 1.32 nH and achieved a Q of 70 at 6 GHz [53].

22.9 Summary and Future Trends

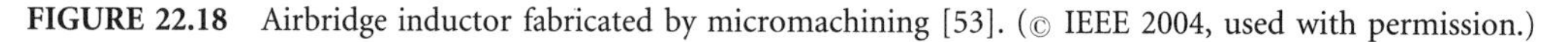

In this chapter we have briefly described the performance of microwave and millimeter wave solid-state circuits. Silicon technology has advanced dramatically and circuits operating to over 70 GHz have been realized in full production processes. Up to 10- or even 20-GHz RFIC design techniques make use of established analogue IC design techniques, rather than the traditional transmission line and Smith chart methods. GaAs, InP, and GaN technologies have also advanced rapidly and find special applications where high-power, low-noise, and millimeter-wave operation are required. For optimum product design a range of technologies may be required and the system-in-package approach has evolved as a high technology solution with concepts such as chip-scale packaging being employed to realize truly 3D modules. The exploitation of the millimeter wave and terahertz frequency ranges continues to advance rapidly and many advanced concepts such as spatial power combining, quasi-optical techniques and MEMS technology require full usage of the principles of electromagnetism. Fortunately, electromagnetic design software and powerful computers have developed in parallel to assist the engineer with these challenges. These extremely high-frequency circuits enable a wide range of new applications to be developed in communications, security, medicine, sensing, and imaging.

Defining Terms

Amplifier: Active two-port device with signal of higher amplitude than the input signal while retaining the essential signal characteristics of the input signal.

Attenuator: Two-port device with output signal of lower amplitude than the input signal while retaining the essential signal characteristics of the input signal.

Bandwidth (BW): A measure of the frequency range over which the circuit performs to specified parameters such as gain, noise figure, power output, etc.

Computer aided design (CAD): A design tool that constitutes circuit simulators and optimization programs/algorithms to aid in the design of microwave circuits to meet the specified performance goals.

Gain: The ratio of the output signal to the input signal of an amplifier.

Heterojunction: A junction between two dissimilar semiconductor materials having different bandgaps. Often used in high-performance device design.

Hybrid microwave integrated circuit (HMIC): A planar assembly that combines different circuit functions formed by strip or microstrip transmission lines printed on a dielectric substrate and incorporating discrete semiconductor solid-state devices and passive distributed or lumped circuit elements, interconnected with wire bonds.

Micromachining: The fabrication of miniature mechanical structures using photolithography and microelectronic fabrication methods (etching, etc.).

Mixer: A three-port device in which an input signal is multiplied by a local oscillator signal in order to achieve frequency translation. An up-converter is a mixer in which the output signal frequency is above the input signal. A down-converter is a mixer in which the output signal frequency is below the input signal.

Monolithic microwave integrated circuit (MMIC): An MMIC is formed by fabricating all active and passive circuit elements or components and interconnections on to or into the surface of a semiinsulating semiconducting substrate by deposition and etching schemes such as epitaxy, ion implantation, sputtering, evaporation and/or diffusion, and utilizing photolithographic processes for pattern definition, thus eliminating the need for internal wire bond interconnects.

Multiplier: A two-port device in which the output signal is a harmonic of the input signal.

Noise figure: The noise figure of any linear two-port circuit is defined as the signal-to-room temperature thermal noise ratio at the input divided by the signal-to-noise ratio at the output.

Oscillator: An active one-port device which produces a nominally frequency stable constant amplitude signal.

PIN diode: A two-port semiconductor device in which a *p*-doped contact is isolated from an *n*-doped contact by an intrinsic region forming an anisotropic junction.

Phase shifter: A circuit that provides a shift in the phase of the output signal with respect to a reference value.

Return loss: Ratio of reflected power to input power of a signal port.

RFIC: An ill-defined term used to describe integrated circuits operating at high frequency for radio systems.

Smith chart: A special graph used for transmission-line calculations. Essentially a polar plot of reflection coefficient with constant resistance and reactance contours superimposed.

Switch: A circuit designed to close or open one or more transmission paths for the microwave signals.

Varactor diode: A diode optimized for use as a voltage-variable capacitance device.

Vector modulator: A circuit which can control the amplitude and phase of an input signal. Often the control signal is formed of a the baseband information signal in Cartesian format ("*I* and *Q*").

Voltage standing wave ratio (VSWR): Ratio of maximum voltage amplitude to the minimum voltage amplitude at the specified port.

References

1. J. Sevenhans, F. Op't Eynde, and P. Reusens, "The silicon radio decade," *IEEE Trans. Microwave Theory Tech.*, vol. 50, no. 1, pp. 235–244, 2002.
2. I.J. Bahl and P. Bhartia, *Microwave Solid State Circuit Design*, 2nd ed., New York: Wiley, 2003.
3. D. Fisher and I.J. Bahl, *Gallium Arsenide IC Applications Handbook*, San Diego, Calif.: Academic Press, 1995.
4. I.D. Robertson and S. Lucyszyn, Eds., "RFIC and MMIC design and technology," *IEE*, 2001.
5. E.C. Niehenke, R.A. Pucel, and I.J. Bahl, "Microwave and millimeter-wave integrated circuits," *IEEE Trans. Microwave Theory Tech.*, vol. 50, no. 3, pp. 846–857, 2002.
6. T. Kikkawa et al., "An over 200-W output power GaN HEMT push-pull amplifier with high reliability," *IEEE MTT-S*, pp. 1347–1350, 2004.

7. K. Sano et al., "InP-based optical system ICs operating at 40 Gbit/s and beyond," *IEEE RFIC Symp. Dig.*, 2004.
8. K.B. Niclas, "On the design and performance of lossy match GaAs MESFET amplifiers," *IEEE Trans. Microwave Theory Tech.*, pp. 1900–1906, 1982.
9. B.A. Floyd, "V-band and W-band SiGe bipolar low-noise amplifiers and voltage-controlled oscillators," *IEEE RFIC Symposium*, 2004.
10. Agilent HMMC-5027 data sheet, www.agilent.com.
11. Northrop Grumman Space Technology, AUH232 MMIC data sheet.
12. B. Agarwal et al., "112-GHz, 157-GHz, and 180-GHz InP HEMT traveling-wave amplifiers," *IEEE Trans. Microwave Theory Tech.*, vol. 46, no. 12, pp. 2553–2559, 1998.
13. J. Aguirre and C. Plett, "A 0.1–50 GHz SiGe HBT distributed amplifier employing constant-k m-derived sections," *IEEE MTT-S*, pp. 923–926, 2003.
14. R.E. Amaya, N.G. Tarr, and C. Plett, "A 27 GHz fully integrated CMOS distributed amplifier using coplanar waveguides," *IEEE RFIC Symp. Dig.*, 2004.
15. Y. Wei et al., "75 GHz 80 mW InP DHBT power amplifier," *IMS Dig.*, pp. 919–921, 2003.
16. B.M. Green et al., "High power broadband AlGaN/GaN HEMT MMIC's on SiC substrates," *IEEE Trans. MTT*, vol. 49, no. 12, pp. 2486–2493, 2001.
17. J.M. Schellenberg and H. Yamasaki, "A new approach to FET power amplifiers," *Microwave J.*, pp. 51–66, March 1982.
18. D. Pavlidis et al., "A new specifically monolithic approach to microwave power amplifiers," *IEEE Microwave Millim.-Wave Monolithic Circ. Symp.*, pp. 54–58, 1983.
19. M. Akkul et al., "50 Watt MMIC power amplifier design for 2 GHz applications," *IEEE MTT-S*, pp. 1355–1358, 2004.
20. Y.C. Chen et al., "A 95-GHz InP HEMT MMIC amplifier with 427-mW power output," *IEEE MWGWL*, pp. 399–401, November 1998.
21. J. Craninckx and M. Steyaert, *Wireless CMOS Frequency Synthesizer Design*, Dordrecht, The Ntehrlands: Kluwer, 1998, ISBN 0-7923-8138-6.
22. B. De Muer and M.S.J. Steyaert, "A CMOS monolithic ΣΔ-controlled fractional-N frequency synthesizer for DCS-1800," *IEEE J. Solid State Circ.*, vol. 37, no. 7, pp. 835–844, 2002.
23. J. Ward et al., "Capability of THz sources based on Schottky diode frequency multiplier chains," *IEEE MTT-S*, pp. 1587–1560, 2004.
24. G. Davies and E. Linfield, "Terahertz quantum cascade lasers," *IEEE MTT-S*, 2004.
25. S. Maas, *Microwave Mixers*, Boston, Mass.: Artech House, 1988.
26. N. Marchand, "Transmission line conversion transformers," *Electronics*, vol. 17, no. 12, pp. 142–145, 1944.
27. M.C. Tsai, "A new compact wideband balun," *IEEE Microwave Millim.-Wave Monolithic Circ. Symp.*, pp. 123–125, June 1993.
28. R. Schwindt and C. Nguyen, "Computer-aided analysis and design of a planar multilayer Marchand balun," *IEEE Trans. Microwave Theory Tech.*, MTT-42, pp. 1429–1434, 1994.
29. D. Manstretta, M. Brandolini, and F. Svelto, "Second-order intermodulation mechanisms in CMOS downconverters," *IEEE J. Solid-State Circ.*, vol. 38, no. 3, pp. 394–406, 2003.
30. P. Katzin et al., "High speed, 100+ W RF switches using GaAs MMICs," *IEEE Trans. Microwave Theory Tech.*, vol. 40, no. 11, pp. 1989–1996, 1992.
31. A.K. Anderson and J.S. Joshi, "Wideband constant phase digital attenuators for space applications," *Microwave Eng. Eur.*, February 1994.
32. B. Bedard and B. Maoz, "Fast GaAs MMIC attenuator has 5-bit resolution," *Microwaves RF*, pp. 71–76, October 1991.
33. L.M. Devlin and B.J. Minnis, "A versatile vector modulator design for MMIC," *IEEE MTT-S Int. Symp. Dig.*, pp. 519–522, 1990.
34. K. Snow, J. Komiak, and D. Bates, "Wideband variable gain amplifiers in GaAs," *IEEE Microwave Millim.-Wave Monolithic Circ. Symp.*, pp. 133–137, 1988.

35. R. Naster et al., "An L-band variable gain amplifier in GaAs MMIC with binary step control," *IEEE GaAs IC Symp.*, pp. 235–237, 1987.
36. R.N. Hardin, E.J. Downey, and J. Munushian, "Electronically-variable phase shifters utilizing variable capacitance diodes," *Proc. IRE*, 48, pp. 944–945, 1960.
37. D.E. Dawson et al., "An analog X-band phase shifter," *IEEE Microwave Millim.-Wave Monolithic Circ. Symp. Dig.*, pp. 6–10, 1984.
38. K. Wilson et al., "A novel MMIC X-band phase shifter," *IEEE Trans. Microwave Theory Tech.*, MTT-33, pp. 1572–1578, 1985.
39. P. Bauhahn et al., "30 GHz multibit monolithic phase shifters," *IEEE Microwave Millim.-Wave Monolithic Circ. Symp. Dig.*, pp. 4–7, 1985.
40. V.E. Dunn et al., "MMIC phase shifters and amplifiers for millimeter-wavelength active arrays," *IEEE MTT-S Symp. Dig.*, pp. 127–130, 1989.
41. H.N. Dawirs and W.G. Swarner, "A very fast, voltage-controlled, microwave phase shifter," *Microwave J.*, pp. 99–107, 1962.
42. Y. Ayasli et al., "A monolithic single chip X-band four-bit phase shifter," *IEEE Trans. Microwave Theory Tech.*, vol. MTT-30, no. 12, pp. 2201–2206, 1982.
43. A.J. Slobodnik, R.T. Webster, and G.A. Roberts, "A monolithic GaAs 36 GHz four-bit phase shifter," *Microwave J.*, pp. 106–111, 1993.
44. L.M. Devlin, "Digitally controlled, 6 bit, MMIC phase shifter for SAR applications," in *22nd European Microwave Conf. Dig.*, Espoo, pp. 225–230, 1992.
45. Y. Ayasli et al., "Wideband monolithic phase shifter," *IEEE Trans. Microwave Theory Tech.*, vol. MTT-32, no. 12, pp. 1710–1714, 1984.
46. P. Miller and J.S. Joshi, "MMIC phase shifters for space applications," in *ESA Proc. Int. Workshop on Monolithic Microwave Integr. Circ. Space Appl.*, ESTEC, Noordwijk, March 1990.
47. Walters P.C. and Fikart J.L., "A fully integrated 5-bit phase shifter for phased array applications," *IEEE Int. Symp. Dig. MMICs Commun. Syst.*, King's College London, September 1992.
48. D.C. Boire, J.E. Degenford, and M. Cohn, "A 4.5 to 18 GHz phase shifter,' *IEEE MTT-S Symp. Dig.*, 601–604, 1985.
49. C.T. Cheung et al., "V-band transmission and reflection grid amplifier packaged in waveguide," *IEEE MTT-S*, pp. 1863–1866, 2003.
50. G.M. Rebeiz, *RF MEMS: Theory, Design, and Technology*, New York: Wiley, February 2003.
51. T. Vähä-Heikkilä1, J. Varis, J. Tuovinen, and G.M. Rebeiz, "A reconfigurable 6–20 GHz RF MEMS impedance tuner," *IEEE MTT-S*, pp. 729–732, 2004.
52. Y. Lee, J.R. East, and L.P.B. Katehi, "Micromachined millimeter-wave module for power combining," *IEEE MTT-S*, pp. 349–352, 2004.
53. J.B. Yoon et al., "CMOS-compatible surface-micromachined suspended-spiral inductors for multi-GHz silicon RF ICs," *IEEE Electron Device Lett.*, vol. 23, no. 10, pp. 591–593, 2002.

Further Information

The *IEEE Transactions on Microwave Theory and Techniques* and the *IEEE Microwave and Wireless Component Letters* routinely publish articles on the design and performance of solid state circuits.

The *International Journal of Microwave and Millimeter-Wave Computer-Aided Engineering* (John Wiley & Sons) specializes in aspects of modeling and simulation of microwave and millimeter-wave solid-state circuits.

The IEEE RFIC Symposium (previously called the Microwave and Millimeter-Wave Monolithic Circuits Symposium and held every year since 1982) is one of the leading international conferences in this area and includes comprehensive information on the design and performance of monolithic microwave and millimeter-wave solid-state circuits.

Books included in the references of this chapter discuss thoroughly the design, circuit implementation, and performance of solid-state circuits.

23 Computational Electromagnetics

Matthew N.O. Sadiku
Prairie View A&M University

Sudarshan Rao Nelatury
The Pennsylvania State University

Until the 1940s, most electromagnetic (EM) problems were solved using the classical methods of separation of variables and integral equations. Besides the fact that a high degree of ingenuity, experience, and effort were required to apply those methods, only a narrow range of practical problems could be investigated due to the complex geometries defining the problems. While theory and experiment remain the two conventional pillars of science and engineering, modeling and simulation represent the third pillar that complements them.

Computational electromagnetics (CEM) is the theory and practice of solving EM field problems on digital computers. It offers the key to comprehensive solutions of Maxwell's equations. CEM techniques can be used to model electromagnetic interaction phenomena in circuits, devices, and systems.

Commonly used numerical methods for solving EM problems include the method of moments, the finite difference method, the finite element method, the Monte Carlo method, and the method of lines. In this chapter, only the first three will be presented because of their popularity and efficiency. Some commercial codes developed for CEM will also be discussed.

23.1 Moment Method

The method of moments (MOM) is firmly established as one of the most powerful numerical tools used to solve EM problems. Stated in a line, it is a technique by which one can convert an inhomogeneous functional equation into a matrix equation that can be solved by known techniques. The error in the conversion process can be reduced to zero or made small by making proper choices in its implementation. The seminal idea was first proposed by Galerkin, a Russian engineer, around 1920. A detailed treatment of the method, as applied to EM problems, has been given by Harrington. The literature on MOM is enormous. In this section, we shall first briefly explain the underlying concept and then furnish three numerical examples.

The Method

Subsequent to formulating an EM problem, one encounters the task of having to solve equations of the form:

$$\mathcal{L}f = g \tag{23.1}$$

where $\mathcal{L}$ is a linear operator, which could be differential, integral, or integro-differential, g is the known excitation or source function, and f is the unknown quantity or response to be determined. We might have a generalized view of Equation (23.1) by thinking that $\mathcal{L}$ could be operating on a set of functions called the domain of $\mathcal{L}$ denoted by $D(\mathcal{L})$, resulting in another set called the range of $\mathcal{L}$ denoted by $R(\mathcal{L})$. Thus, we view f as one member of $D(\mathcal{L})$ and g as one of $R(\mathcal{L})$. Further, we topologize these two sets by equipping them with a suitable inner product. The inner product of two functions $u(x)$ and $v(x)$, denoted by $\langle u, v\rangle$ is usually defined as

$$\langle u, v\rangle = \int_{\Omega} u(x)v^*(x)\mathrm{d}x \tag{23.2}$$

where the asterisk stands for the complex conjugation. The inner product is a scalar and gives a degree of resemblance between u and v. It may be interpreted as the "projection" of u in the direction of v. Here, Ω is used to indicate the domain over which u and v are defined. We require that the inner product satisfy the following properties:

$$\langle u, v\rangle = \langle v, u\rangle^* \tag{23.3}$$

$$\langle \alpha u + \beta v, f\rangle = \alpha\langle u, f\rangle + \beta\langle v, f\rangle \tag{23.4}$$

$$\langle u, u\rangle = \; ||u||^2 \geqslant 0 \tag{23.5}$$

where α and β are arbitrary scalars, $||u||$ is called the norm of u and is zero if and only if $u = 0$.

Now to solve for f in Equation (23.1), MOM starts with the possibility of viewing f as a linear combination of a set of basis functions $\{f_n\}$:

$$f = \sum_n \alpha_n f_n \tag{23.6}$$

where αs are constants to be determined. Theoretically, in a typical case, the dimension of $D(\mathcal{L})$ is infinite and thus n runs from 1 to ∞ in the above summation. However, for practical purposes if we use only N basis functions, the equality becomes approximate. Substituting Equation (23.6) in Equation (23.1) with finite N, and using the fact that $\mathcal{L}$ is linear, we get:

$$\sum_{n=1}^{N} \alpha_n \mathcal{L}f_n \cong g \tag{23.7}$$

Assuming R_N to be the residual error in the above approximation, we might write Equation (23.7) as

$$\sum_{n=1}^{N} \alpha_n \mathcal{L} f_n = g_N = g + R_N \tag{23.8}$$

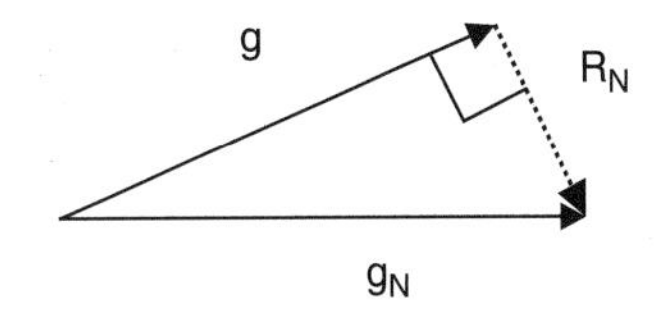

FIGURE 23.1 Vector analogy for the approximation $g_N = g + R_N$. We note that R_N is minimum if and only if it is orthogonal to g. We choose αs such that R_N, obtained when g and g_N are projected to the subspace spanned by $\{w_m\}$, is minimized.

We note that g and R_N are members of $R(\mathcal{L})$. As we want the best approximation, we wish to minimize R_N using a suitable criterion. From approximation theory and from vector analogy, we know that this is possible when g and R_N are orthogonal as illustrated in Figure 23.1.

Just as f_n spans $D(\mathcal{L})$, we could choose a set of functions that span $R(\mathcal{L})$ called *weighting functions* denoted by $\{w_m\}$, and let R_N be orthogonal to each member of $\{w_m\}$ ensuring, hopefully, its orthogonality to g. Again, we might need infinite number of ws in theory, but we just take only N in practice, thereby limiting ourselves to a subspace $W \subseteq R(\mathcal{L})$. For finite N, this produces

$$\langle R_N,\ w_m \rangle = 0 \text{ for } m = 1, 2, \dots N \tag{23.9}$$

With Equation (23.8) in the above:

$$\left\langle \sum_{n=1}^{N} \alpha_n \mathcal{L} f_n - g, w_m \right\rangle = 0 \text{ for } m = 1, 2, \dots N \tag{23.10}$$

Or equivalently:

$$\sum_{n=1}^{N} \langle \mathcal{L} f_n,\ w_m \rangle \alpha_n = \langle g,\ w_m \rangle \text{ for } m = 1, 2, \dots N \tag{23.11}$$

which might be cast into a matrix equation of the form:

$$[l_{mn}][\alpha_n] = [g_m] \tag{23.12}$$

where

$$[l_{mn}] = \begin{bmatrix} \langle \mathcal{L} f_1, w_1 \rangle & \langle \mathcal{L} f_2, w_1 \rangle & \cdots & \langle \mathcal{L} f_N, w_1 \rangle \\ \langle \mathcal{L} f_1, w_2 \rangle & \langle \mathcal{L} f_2, w_2 \rangle & \cdots & \langle \mathcal{L} f_1, w_2 \rangle \\ \vdots & \vdots & \cdots & \vdots \\ \langle \mathcal{L} f_1, w_N \rangle & \langle \mathcal{L} f_2, w_N \rangle & \cdots & \langle \mathcal{L} f_N, w_N \rangle \end{bmatrix} \tag{23.13}$$

$$[\alpha_n] = \begin{bmatrix} \alpha_1 \\ \alpha_2 \\ \vdots \\ \alpha_N \end{bmatrix}, \quad [g_m] = \begin{bmatrix} g_1 \\ g_2 \\ \vdots \\ g_N \end{bmatrix} = \begin{bmatrix} \langle g, w_1 \rangle \\ \langle g, w_2 \rangle \\ \vdots \\ \langle g, w_N \rangle \end{bmatrix} \tag{23.14}$$

If the matrix $[l_{mn}]$ in Equation (23.13) is nonsingular, it has an inverse $[l_{mn}]^{-1}$ that can be found using known techniques like Gauss elimination. Then, $[\alpha_n]$ would be obtained as

$$[\alpha_n] = [l_{mn}]^{-1}[g_m] \tag{23.15}$$

This helps us to find f. With the notation $[\tilde{f}] = [f_1\ f_2 \cdots f_N]$, we can write

$$f = [\tilde{f}][l_{mn}]^{-1}[g_m] \tag{23.16}$$

The moment method attempts to equate the orthogonal projection of $R(\mathcal{L})$ on to the space W spanned by $\{w_m\}$ to that spanned by g_N. The weighted error is thus minimized in this technique. There is an infinite number of possible sets of basis and weighting functions. Crucial to the effective working of the MOM is the choice of these sets an engineer is supposed to make. This could be specific to the problem at hand, but there are generic guiding rules to maximize success in obtaining accurate results with the least computer time and storage. First of all, they should form a linearly independent set. In choosing $\{f_n\}$, the assumption of Equation (23.6) should be reasonably good. This choice is dictated by the boundary conditions and the differentiability conditions in case of differentials in the operator $\mathcal{L}$. In practice, the basis sets may be put into two general classes. The first consists of subdomain functions, which have finite support along the domain of their definition. By this we mean the spatial surface of the electromagnetic structure involved in the problem. The second class consists of entire domain functions with the support over the entire domain, as the name suggests. These are chosen with some prior knowledge of the response to be found, whereas, the subdomain class does not demand any prior knowledge. Examples of subdomain basis are pulse, triangular or sinusoidal functions defined on a smaller interval, but the union of their supports covers the original domain. Next, $\{w_m\}$ should be in $R(\mathcal{L})$. They ought to form a basis and as N gets sufficiently large, they must be able to represent any arbitrary member in $R(\mathcal{L})$. Hence, if they cannot approximate g in $R(\mathcal{L})$ to a high degree of accuracy the result would be a far cry from the true solution. In particular, if $R(\mathcal{L})\subseteq D(\mathcal{L})$, we might conveniently choose $\{w_m\} = \{f_m\}$. The technique then goes by the name of *Galerkin's method* for this specific choice, but how do we know if $R(\mathcal{L})\subseteq D(\mathcal{L})$ and if Galerkin's method is applicable? To understand this we consider what is termed the adjoint operator of $\mathcal{L}$. We define the adjoint of $\mathcal{L}$ as $\mathcal{L}^a$ when

$$\langle \mathcal{L}u, v\rangle = \langle u, \mathcal{L}^a v\rangle \ \forall u \in D(\mathcal{L}), v \in D(\mathcal{L}^a) \tag{23.17}$$

It is known that

$$D(\mathcal{L}^a) = N(\mathcal{L}^a) \oplus \overline{R}(\mathcal{L}) \tag{23.18}$$

where $D(\mathcal{L}^a)$ is the domain of $\mathcal{L}^a$, $N(\mathcal{L}^a)$ is the null space of $\mathcal{L}^a$, $\overline{R}(\mathcal{L})$ is the closure of $R(\mathcal{L})$, and $\oplus$ is the direct sum. Also $N(\mathcal{L}^a)$ and $\overline{R}(\mathcal{L})$ intersect only at the identity point.

As $R(\mathcal{L})\subset D(\mathcal{L}^a)$, and since it is easier to find the latter than the former, we question if $\{w_m\}$ are in $D(\mathcal{L}^a)$. If they are not in $D(\mathcal{L}^a)$, they cannot be in $R(\mathcal{L})$. The dual argument settles the question of $\{f_n\}$. Thus Galerkin's method is applicable when $\mathcal{L} = \mathcal{L}^a$, in which case the operator is said to be self-adjoint, leading to $D(\mathcal{L}) = D(\mathcal{L}^a)$. An alternative choice is to have impulses for $\{w_m\}$, in which case the method goes by the name *collocation technique* or *point-matching*. This is slightly simpler than Galerkin's method but slower in convergence. In most electromagnetic problems $D(\mathcal{L}) \neq D(\mathcal{L}^a)$ and the use of Galerkin's method should be viewed with "a pinch of salt." For instance, take the classic example of finding the impedance of a dipole antenna, which involves solving *Pocklington's* integral equation with a plane wave excitation. If the piecewise sinusoidal expansion functions are used, they are zero at the ends of the wire, therefore, they satisfy the boundary conditions that current should go to zero at the ends. However, the function g in this case is supposed to be constant, which cannot be made up by the sinusoidal expansion functions. If they cannot span the range $R(\mathcal{L})$, they are not candidates which qualify as weighting functions. So, Galerkin's method is unsuited for solving Pocklington's integral equation, in principle. Whereas, *Hellen's* integral equation is in terms of the vector potential, which is finite at the ends, so Galerkin's method might be used justifiably.

In the next section we shall give some numerical examples to describe the method. First we take a boundary value problem from differential equations. Second, we take the problem of electromagnetic scattering from a line source in the presence of a conducting strip of finite width. Third, we take an electrostatic problem of finding capacitance of a square plate. All three, being introductory in nature, help the reader gain insight into the implementation of MOM.

An Example from Differential Equations

(a) Suppose we are given the differential equation:

$$-\frac{\mathrm{d}^2 f(x)}{\mathrm{d}x^2} = 6x - 2 \quad \text{with} \quad f(0) = f(1) = 0. \tag{23.19}$$

Let us find the response function $f(x)$ using Galerkin's method.

(b) We shall also repeat the problem for

$$-\frac{\mathrm{d}^2 f(x)}{\mathrm{d}x^2} = -12x^2 + 4 \quad \text{with} \quad f(0) = f(1) = 0 \tag{23.20}$$

using the collocation technique.

Solution: (a) Here, we recognize that $\mathcal{L} = -\frac{\mathrm{d}^2}{\mathrm{d}x^2}$. Let us attempt the Galerkin's method and choose the basis $f_n = x - x^{n+1}$ and write

$$f(x) \cong \sum_{n=1}^{N} \alpha_n f_n \tag{23.21}$$

The elements of l_{mn} can be found as

$$l_{mn} = \langle \mathcal{L} f_n, w_m \rangle = \int_0^1 \left[-\frac{\mathrm{d}^2 (x - x^{n+1})}{\mathrm{d}x^2} \right] (x - x^{n+1}) \mathrm{d}x = \frac{mn}{m + n + 1} \tag{23.22}$$

Likewise, the elements of g_m can be determined:

$$g_m = \langle g, w_m \rangle = \int_0^1 (6x - 2)(x - x^{n+1}) \mathrm{d}x = \frac{m(m+1)}{(m+2)(m+3)} \tag{23.23}$$

Knowing l_{mn} and g_m we can calculate αs and hence plot $f(x)$. For $N = 1$, we find that $l_{11} = 1/3$, $g_1 = 1/6$, $\alpha_1 = 1/2$. Then $f(x) = \frac{x(1-x)}{2}$. Similarly for $N = 2$, we can verify that $f(x) = x^2(1 - x) = f_{\text{exact}}$. These are plotted in Figure 23.2.

We shall now continue the next part of this example.

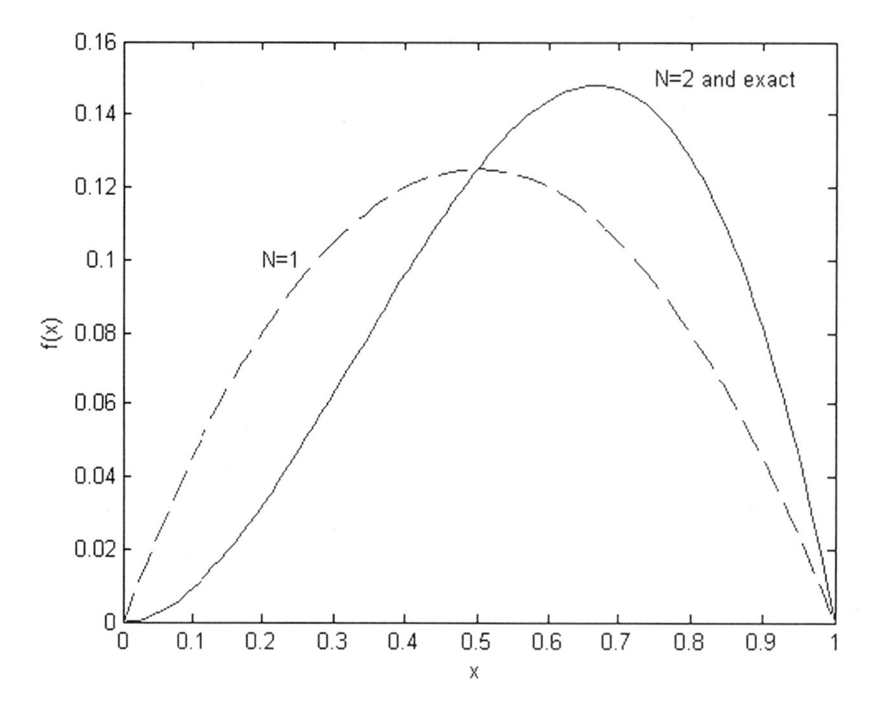

FIGURE 23.2 Numerical and exact solutions for Example 1a.

(b) Note that $g(x)$ has changed. In point-matching technique, we select for the weighting functions $\{w_m\} = \{\delta(x - x_m)\}$. Following the same lines we arrive at:

$$l_{mn} = \langle \mathcal{L} f_n, w_m \rangle = \int_0^1 \left[-\frac{\mathrm{d}^2(x - x^{n+1})}{\mathrm{d}x^2} \right] \delta(x - x_m)\mathrm{d}x = n(n+1)\left(\frac{m}{N}\right)^{n-1} \tag{23.24}$$

and

$$g_m = \langle g, w_m \rangle = \int_0^1 (-12x^2 + 4)\delta(x - x_m)\mathrm{d}x = 4 - 12\left(\frac{m}{N}\right)^2 \tag{23.25}$$

With these values for l_{mn} and g_m, for $N = 1$, 2, and 3, respectively, we get: $f(x) = 4x(x-1), x(2 - 5x + 3x^2)$ and $x(1 - 2x + x^3)$ These are plotted in Figure 23.3.

Radiation of a Line Source above a Two-Dimensional Strip of Finite Width

In this section we consider the radiation of an infinite line source above a two-dimensional conducting strip of finite width as shown in Figure 23.4. The line is along the z-axis and the strip of width w is h m below the line symmetrically located from $x' = -w/2$ to $w/2$. Using MOM, we solve for the current induced on the strip and then obtain the radiation pattern due to the original line current and the induced current. The results are compared to those obtained using physical optics and also method images.

The electric field radiated by a line source of electric current I_e flowing in the z-direction in the absence of the strip is given by

$$E_z^{\mathrm{i}} = -\frac{\beta^2 I_{\mathrm{e}}}{4\omega\varepsilon} H_0^{(2)}(\beta\rho) \tag{23.26}$$

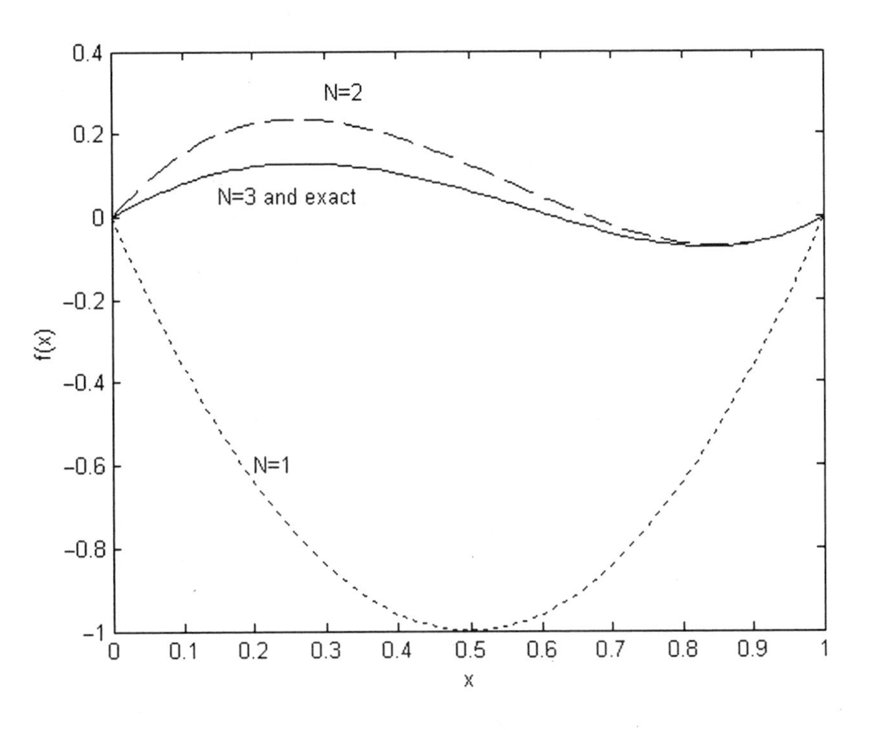

FIGURE 23.3 Numerical and exact solutions for Example 1b.

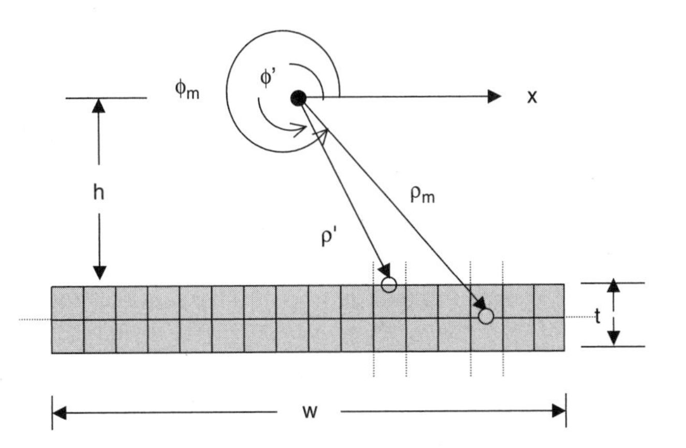

FIGURE 23.4 Geometry of a line source above a two-dimensional conducting strip of finite width.

where β is the phase constant, ω is the angular frequency, ε is the permittivity, ρ is the cylindrical coordinate, $H_0^{(2)}(\beta\rho)$ is the *Hankel* function of the second kind and zero order.

This field is incident on and interacts with the strip and induces a surface current on the strip of density, say $J_z(x')$, which in turn gives rise to scattered field:

$$E_z^s(\rho) = -\frac{\beta^2}{4\omega\varepsilon}\int_{-w/2}^{w/2} J_z(x')H_0^{(2)}(\beta|\boldsymbol{\rho} - \boldsymbol{\rho}'|)\mathrm{d}x' \tag{23.27}$$

The total field is the sum of incident and scattered field:

$$E_z^{\mathrm{t}} = E_z^{\mathrm{i}} + E_z^{\mathrm{s}} \tag{23.28}$$

To find the scattered component we need to know the induced current. This can be posed as a problem of integral equation as follows. Since the only information we know is that field vanishes on the strip, we evaluate Equation (23.28) on the strip itself arriving at

$$-\frac{\beta^2 I_{\mathrm{e}}}{4\omega\varepsilon} H_0^{(2)}(\beta\rho_m)\Bigg|_{\rho_m \in \mathrm{strip}} = \frac{\beta^2}{4\omega\varepsilon} \int\limits_{-w/2}^{w/2} J_z(x')H_0^{(2)}(\beta|\boldsymbol{\rho}_m - \boldsymbol{\rho}'|)\mathrm{d}x'\Bigg|_{\rho_m \in \mathrm{strip}} \tag{23.29}$$

For convenience letting I_{e} to be unity, the above can be rewritten as

$$H_0^{(2)}(\beta\rho_m) = -\int\limits_{-w/2}^{w/2} J_z(x')H_0^{(2)}(\beta|\boldsymbol{\rho}_m - \boldsymbol{\rho}'|)\mathrm{d}x' \tag{23.30}$$

An integral equation of the above form is called *Fredholm's Integral equation* of the first kind. To solve this, we use MOM as outlined below.

Let us first divide the domain of integration $-w/2 < x' < w/2$ into N equal subdivisions. We choose pulse basis functions defined below to compose the unknown $J_z(x')$:

$$J_z(x') \approx \sum_{n=1}^{N} \alpha_n f_n(x') \tag{23.31}$$

$$f_n(x') = \begin{cases} 1, & x'_{n-1} \leqslant x' \leqslant x'_n \\ 0 & \text{otherwise} \end{cases} \tag{23.32}$$

where αs are constants yet to be determined. Substituting the above in Equation (23.31), the latter can be written as a linear combination of several integrals evaluated over the various subdivisions:

$$H_0^{(2)}(\beta\rho_m) = -\sum_{n=1}^{N} \alpha_n \int\limits_{x'_{n=1}}^{x'_n} H_0^{(2)}(\beta|\boldsymbol{\rho}_m - \boldsymbol{\rho}'|)\mathrm{d}x' \tag{23.33}$$

By choosing impulses for the set of weighting functions, we can use the point-matching technique. This amounts to enforcing Equation (23.33) for the values of ρ_m at N equidistant points on the strip collocated at the centers of the subsections. In doing this, whenever the argument of the Hankel function is zero, the integral cannot be evaluated numerically. So to resolve this problem we consider the thickness of the strip to be nonzero and place the matching points along the line passing midway through the thickness, thus slightly shifting them as shown in Figure 23.4. Here, the circle on the middle line of the strip shows the field point and the one on the top surface of the strip represents the source point. With this arrangement, we can rewrite Equation (23.33) in matrix notation as

$$[V_m] = [Z_{mn}][\alpha_n] \tag{23.34}$$

where

$$V_m = H_0^{(2)}(\beta \rho_m) \tag{23.35}$$

$$Z_{mn} = -\int_{x'_{n-1}}^{x'_n} H_0^{(2)}(\beta|\boldsymbol{\rho}_m - \boldsymbol{\rho}'|)\mathrm{d}x' \tag{23.36}$$

The evaluation of Z_{mn} requires some care. The diagonal terms can be obtained by writing the asymptotic expressions of Hankel function for the small arguments. For the off-diagonal terms Hankel functions can be crudely thought of as essentially constant. With these ideas, we can express:

$$Z_{mn} \approx \begin{cases} -\Delta x_n\left[1 - j\frac{2}{\pi}\ln\left(\frac{1.781\beta\Delta x_n}{4e}\right)\right], & \text{for } m = n \\ -\Delta x_n H_0^{(2)}(\beta|\boldsymbol{\rho}_m - \boldsymbol{\rho}'_n|) & m \neq n \end{cases} \tag{23.37}$$

where

$$\Delta x_n = x_{n+1} - x_n \tag{23.38}$$

Once we calculate these matrices, we can get α's as

$$[\alpha_n] = [Z_{mn}]^{-1}[V_m] \tag{23.39}$$

and the current density can be obtained using Equation (23.31).

We now provide some simulations. Suppose the parameters of the strip are $w = 2\lambda$, $t = 0.01\lambda$, $h = 0.5\lambda$. Current density obtained with $N = 100$, and point matching technique is plotted in Figure 23.5. Also, for comparison we compute the current density obtained using physical optics (PO) as

$$J_z^{\mathrm{PO}} \approx 2\hat{n} \times \mathbf{H}^{\mathrm{i}} \tag{23.40}$$

where $\hat{n}$ is the unit normal to the strip and $\mathbf{H}^{\mathrm{i}}$ is the incident magnetic field vector. Expressing $\mathbf{H}^{\mathrm{i}}$ in terms of the incident electric field we get

$$J_z^{\mathrm{PO}} \approx -jI_{\mathrm{e}}\frac{\beta y_m}{2\rho_m}H_1^{(2)}(\beta\rho_m) \tag{23.41}$$

This is shown by a dotted line in Figure 23.5.

If we know the current induced on the strip, it is then straightforward to compute the far-field radiated by both line current and the induced current. For brevity, we avoid a few mathematical steps and directly give the total far field as

$$E_z^{\mathrm{total}} \approx -\frac{\beta^2}{4\omega\varepsilon}\sqrt{\frac{2j}{\pi\beta\rho}}\mathrm{e}^{-j\beta\rho}\left[1 + \int_{-w/2}^{w/2} J_z(x')\mathrm{e}^{j\beta\rho'\cos(\phi-\phi')}\mathrm{d}x'\right] \tag{23.42}$$

which might be simplified to give the normalized field:

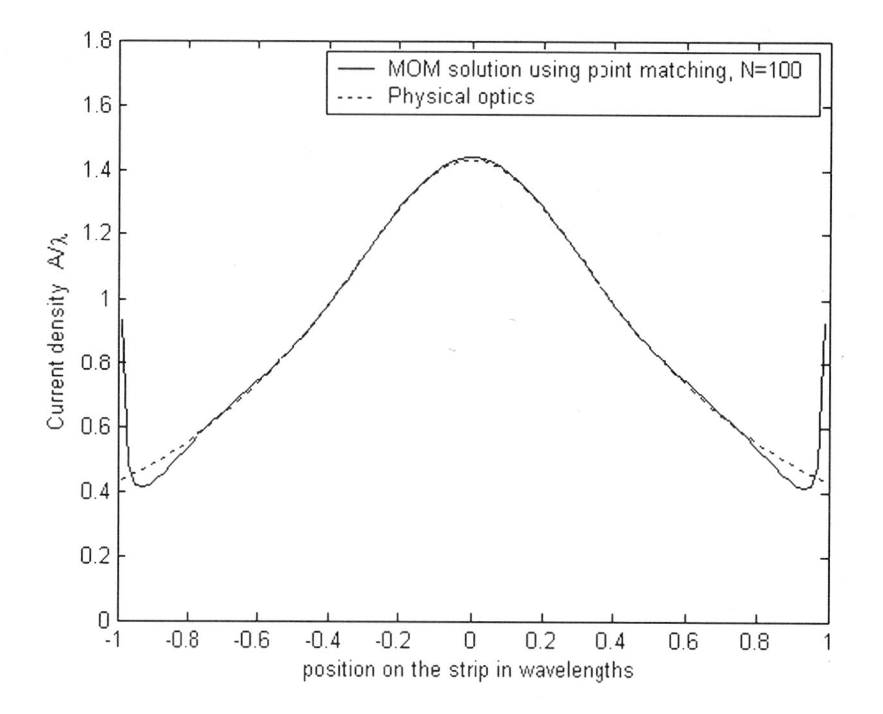

FIGURE 23.5 Current density on the finite conducting strip due to the infinite line current above.

$$E_z^{\text{total}}(\text{normalized}) \approx \left[1 + \int_{-w/2}^{w/2} J_z(x')e^{j\beta\rho'\cos(\phi-\phi')}\text{d}x' \right] \tag{23.43}$$

For the strip parameters $w=2\lambda$, $t=0.01\lambda$, $h=0.5\lambda$, and using point matching as previously, we give the normalized filed amplitude pattern in Figure 23.6.

Numerical Evaluation of Capacitance of a Parallel Plate Capacitor of Finite Plate Area

It is known from basic physics that the capacitance of a parallel plate capacitor of plate area A, separation d, and with a dielectric constant ε is

$$C = \varepsilon A/d \tag{23.44}$$

This formula is indeed simple, but not accurate. We need to consider the fringing effects of the fields at the ends. Thus, a more accurate model of the practical capacitor with finite plate dimensions is to have the ideal capacitor of value from Equation (23.44) in parallel with the fringing capacitor. To find the total capacitance we can use the MOM in the following way.

Consider a square conducting plate of side $2a$ lying in the xy plane around the origin as shown in Figure 23.7. Let $\rho(x, y)$ be the surface charge density on the plate. The potential at any point in space is given by

$$\phi(x,y,z) = \int_{-a}^{a} \text{d}x' \int_{-a}^{a} \frac{\rho(x',y')}{4\pi\varepsilon R}\text{d}y' \tag{23.45}$$

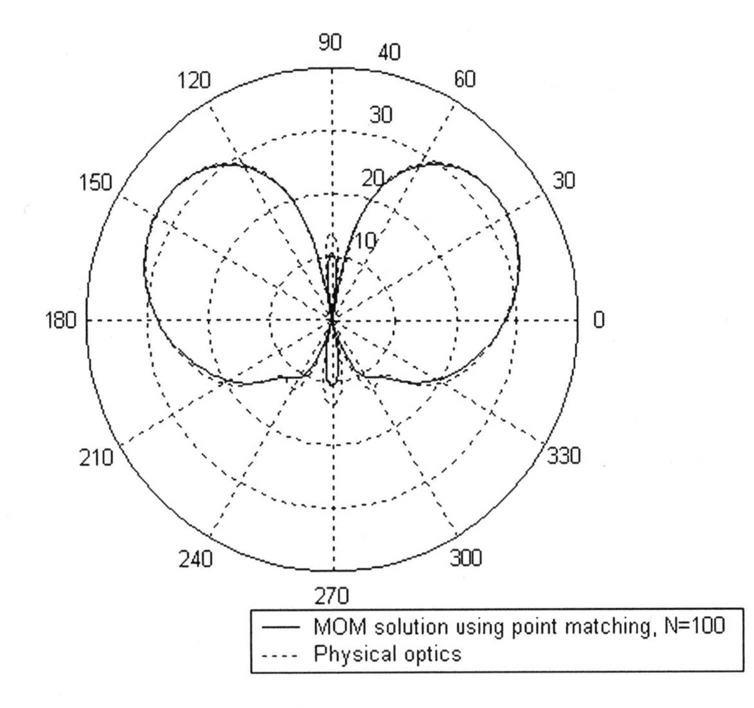

FIGURE 23.6 Normalized amplitude radiation pattern (on a log scale) of the line source placed $h = 0.5\lambda$ above a finite strip of width $w = 2\lambda$. Results using MOM and PO are shown.

where

$$R = \sqrt{(x - x')^2 + (y - y')^2 + z^2} \tag{23.46}$$

We can easily set up an integral equation for this by placing the field point on the plate itself where the potential is known as V volts. Thus, we get

$$V = \int_{-a}^{a} \mathrm{d}x' \int_{-a}^{a} \frac{\rho(x', y')}{4\pi\varepsilon\sqrt{(x - x')^2 + (y - y')^2}} \mathrm{d}y' \tag{23.47}$$

In Equation (23.47), (x, y) lies on the plate. If we know the charge density $\rho(x', y')$ we can determine the capacitance. Application of MOM starts with the choice of basis function set to express the required charge density. Let us pick pulse expansion functions f_n, whose value is unity on each of the plate subsections of area Δs_n shown in Figure 23.7. With these members, we can write the charge density as

$$\rho(x', y') \approx \sum_{n=1}^{N} \alpha_n f_n \tag{23.48}$$

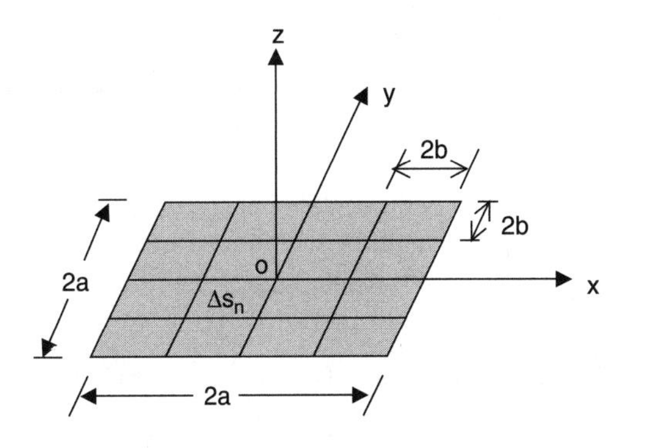

FIGURE 23.7 Square conducting plate and subsections.

Substituting Equation (23.48) in Equation (23.47), and enforcing the same at the mid-points of the subsections, which is the same as saying that we choose impulses for the weighting function set, we get

$$V = \sum_{n=1}^{N} l_{mn}\alpha_n, \quad \text{for} \quad m = 1, 2, \ldots, N \tag{23.49}$$

where

$$l_{mn} = \int_{\Delta x_n} dx' \int_{\Delta y_n} \frac{1}{4\pi\varepsilon\sqrt{(x_m - x')^2 + (y_m - y')^2}} dy' \tag{23.50}$$

Equation (23.49) for various values of m can be expressed in matrix notation thus:

$$[l_{mn}][\alpha_n] = [g_m] \tag{23.51}$$

where

$$[g_m] = \begin{bmatrix} V \\ V \\ \vdots \\ V \end{bmatrix} \tag{23.52}$$

The integral in Equation (23.50) in practice can be evaluated approximately and we get

$$l_{mn} \approx \begin{cases} \dfrac{b^2}{\pi\varepsilon\sqrt{(x_m - x_n)^2 + (y_m - y_n)^2}}, & \text{for} \quad m \neq n \\ \frac{2b}{\pi\varepsilon} 0.8814 & m = n \end{cases} \tag{23.53}$$

With the above approximation, a solution $[\alpha_n] = [l_{mn}]^{-1}[g_m]$ to the set of equations in Equation (23.51) allows us to compute the charge and hence the capacitance as

$$C \approx \frac{1}{V} \sum_{n=1}^{N} \alpha_n \Delta s_n \tag{23.54}$$

Table 23.1 shows the value of the capacitance in picorfarads for a plate of side 2 m and for different values of N.

The three examples presented above are intended to give insight into the underlying concept and implementation of the MOM. One might appreciate the generality of MOM in solving a wide class of problems. The method is limited by the number of linear equations to be solved. Especially when the structures under consideration are electrically large, we need to employ techniques such as conjugate gradient method.

TABLE 23.1 Capacitance of a Square Plate of Side 2 Meters Found via MOM Using Pulse Expansion and Point Matching Technique

N	1	4	9	16	25	36	49	64	81	100
C, pF	31.5155	35.6542	37.3215	38.1871	38.7151	39.0694	39.3229	39.5129	39.6604	39.778

23.2 Finite Difference Method

Many problems in engineering electromagnetics may be formulated in terms of partial differential equations. The static fields are governed by Laplace and Poisson equations, which respectively, for the electric potential are of the form:

$$\nabla^2 \phi = 0 \tag{23.55}$$

$$\nabla^2 \phi = -\frac{\rho}{\varepsilon} \tag{23.56}$$

Another example is the dynamic fields in a waveguide satisfy the Helmholtz equation, which for a wave function ψ is of the form:

$$\left(\nabla^2 + k^2\right)\psi = 0 \tag{23.57}$$

In radiation or scattering problems, the right-hand side of Equation (23.57) would be nonzero. The operator $\mathcal{L}$ in the preceding section can now be related to the operators ∇^2 and $(\nabla^2 + k^2)$ in the equations above. Only in selected cases, one is able to obtain the exact analytic solutions, but when the geometries involved are arbitrary, numerical methods are the preferred choice. Similar to MOM, the method of finite differences also converts the operator equation plus the boundary conditions into a system of simultaneous equations. Basically, the importance of the finite difference method (FDM) lies in the ease with which many logically complicated operations and functions may be discretized.

The FDM was first developed by A. Thom in 1920 under the name "the method of squares" to solve nonlinear hydrodynamic equations. Since then, the method is used for solving problems occurring in various engineering disciplines including electromagnetics. The next subsection offers the basic steps in the FDM followed by three examples of its application.

Finite Differences

The first step in solving an electromagnetic boundary value problem using finite differences is to obtain difference equations. Given a function $f(x)$ shown in Figure 23.8 one can approximate the derivative, the slope of the tangent drawn at P, using the values at A$(x_o-h, f(x_o-h))$, P $(x_o\ f(x_o))$, and B $(x_o + h,\ f(x_o + h))$ in one of the three possible ways

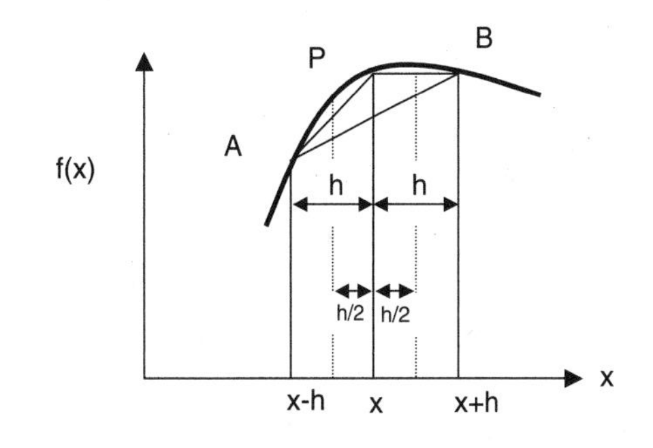

FIGURE 23.8 Approximating the derivative of $f(x)$ at P using forward, backward, and central differences.

$$f'(x) \approx \frac{\Delta f}{h} = \frac{f(x+h) - f(x)}{h} \tag{23.58}$$

$$f'(x) \approx \frac{\nabla f}{h} = \frac{f(x) - f(x-h)}{h} \tag{23.59}$$

$$f'(x) \approx \frac{\delta f}{h} = \frac{f(x+h/2) - f(x-h/2)}{h} \tag{23.60}$$

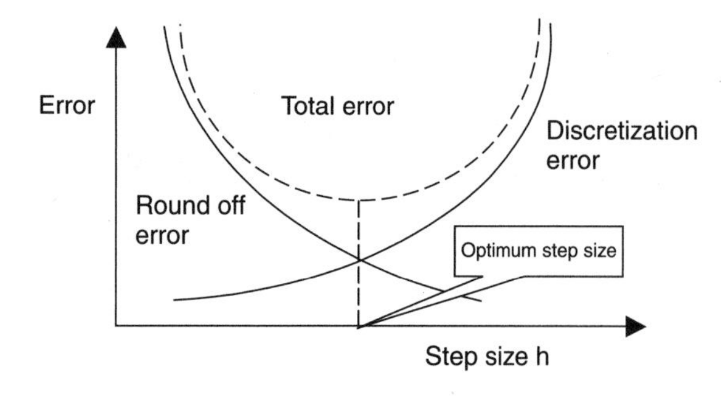

FIGURE 23.9 Trade-off between discretization error and round-off error.

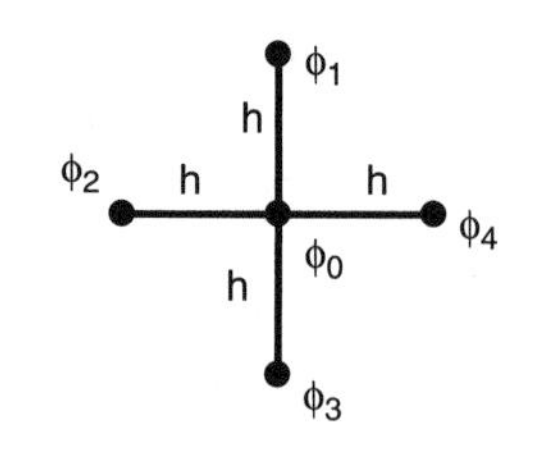

FIGURE 23.10 The Laplacean of a two-dimensional function ϕ is approximated by its discrete values as in Equation (23.62).

The symbols Δ, ∇, and δ respectively represent forward, backward (in the present context), and central difference operators. From Figure 23.8, we understand that the slope of the tangent at P is approximately equal to that of the chords PB, AB, and PA. Taylor series expansion of $f(x)$ around A and B reveals that the forward, backward, and central differences are of orders, h, h, and h^2, respectively. Thus, as the step size h becomes smaller and smaller, we expect that truncation error reduces, but the round off errors occurring in the calculations grows bigger and bigger. This is suggested in Figure 23.9. So, one has to make a trade-off in the choice of step size.

Using central difference, the second derivative at P may be expressed as

$$f''(x) \approx \frac{\delta f'}{h} = \frac{f(x+h) - 2f(x) + f(x-h)}{h^2} \tag{23.61}$$

The above may be extended for a two-dimensional scalar Laplace operator acting on a function $\phi(x, y)$ discretized (as shown in Figure 23.10) as

$$\nabla^2\phi = \frac{\partial^2\phi}{\partial x^2} + \frac{\partial^2\phi}{\partial y^2} \approx \frac{\phi_1 + \phi_2 + \phi_3 + \phi_4 - 4\phi_0}{h^2} \tag{23.62}$$

In certain applications, the boundary of the domain of interest might be irregular in shape and it is hard to discretize the region unless one uses a fine grid or a nonuniform grid. The Laplacean of ϕ for a nonuniform grid such as the one in Figure 23.11 would then be

$$\begin{aligned}\nabla^2\phi &= \frac{\partial^2\phi}{\partial x^2} + \frac{\partial^2\phi}{\partial y^2} \\ &\approx \frac{2}{h^2}\left[\frac{\phi_1}{\alpha_2(\alpha_2+\alpha_4)} + \frac{\phi_2}{\alpha_1(\alpha_1+\alpha_3)} + \frac{\phi_3}{\alpha_4(\alpha_2+\alpha_4)} + \frac{\phi_4}{\alpha_3(\alpha_1+\alpha_3)} - \frac{\phi_0}{\alpha_1\alpha_3} + \frac{\phi_0}{\alpha_2\alpha_4}\right]\end{aligned} \tag{23.63}$$

Further, if the region of interest has multiple dielectrics or if the permittivity varies from point to point, we need to modify the preceding approximations. For instance, consider the interface between two dielectrics as shown in Figure 23.12. Gauss's law states that

$$\oint_c \mathbf{D}.\mathbf{dI} = \oint_c \varepsilon\mathbf{E}.\mathbf{dI} = \text{charge enclosed per unit length} = 0 \tag{23.64}$$

Since the electric field can be thought of as the negative normal derivative of potential, we get

$$\phi_0=\frac{1}{4}\left(\frac{2\varepsilon_1}{\varepsilon_1+\varepsilon_2}\phi_1+\phi_2+\frac{2\varepsilon_2}{\varepsilon_1+\varepsilon_2}\phi_3+\phi_4\right) \quad (23.65)$$

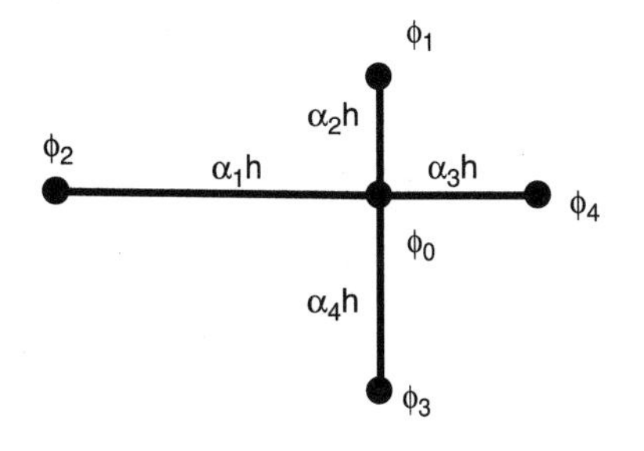

FIGURE 23.11 A nonuniform grid used in Equation (23.63).

With the finite difference formulas as derived above, we can now outline the method of solving a Laplace equation. The steps involved are: (1) Divide the domain of interest into a suitably fine grid. (2) Apply the relevant finite difference formula at each node of the grid to obtain a system of simultaneous equations in terms of the discrete values of the unknown. In this process some nodes called *fixed nodes* that fall on the boundary have known values set by the boundary conditions, the remaining nodes where the function is unknown are called *free nodes.* (3) Solve the system of equations either directly or iteratively for the unknowns.

Let us first take a simple problem of solving Laplace Equation (23.55) in a rectangular region with a homogenous dielectric subject to the boundary conditions shown in Figure 23.13.

Application of finite differences with a grid of size 40×30 gives the potential, which is sketched in Figure 23.14.

Next let us consider a cylindrical region with multiple dielectrics shown in Figure 23.15. The geometry in the problem has an inner conductor at $\rho=a$, and an outer conductor at $\rho=d$. The space in between is filled with three different layers of thickness $b-a$, $c-b$, $d-c$ with permittivities ε_1, ε_2, and ε_3, respectively. For the numerical values $a=2$, $b=6$, $c=10$, $d=14$; $\varepsilon_1=14$, $\varepsilon_2=4$, $\varepsilon_3=10$, and for $\Phi=0$ at $\rho=$a, and $\Phi=10$ at $\rho=d$, over a grid size of 60×60, FDM furnishes the solution shown in Figure 23.16.

The idea behind finite differences is extended to time-varying fields in the finite-difference time-domain technique discussed next.

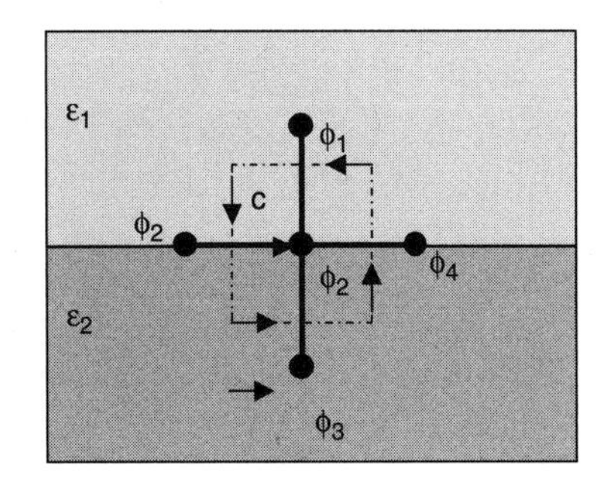

FIGURE 23.12 Interface between two dielectrics of permittivities ε_1 and ε_2.

The Finite-Difference Time-Domain Method (FDTD)

This was first introduced by Yee in 1966 and later developed by Taflove and others. In this method, essentially the Maxwell's curl equations are rewritten using finite difference notation. The principal idea in this method is to employ time and space centered approximations for the derivatives occurring in the Maxwell's equations with a second-order accuracy. In an isotropic medium, Maxwell's equations can be written as

$$\nabla\times\mathbf{E}=-\mu\frac{\partial\mathbf{H}}{\partial t} \quad (23.66a)$$

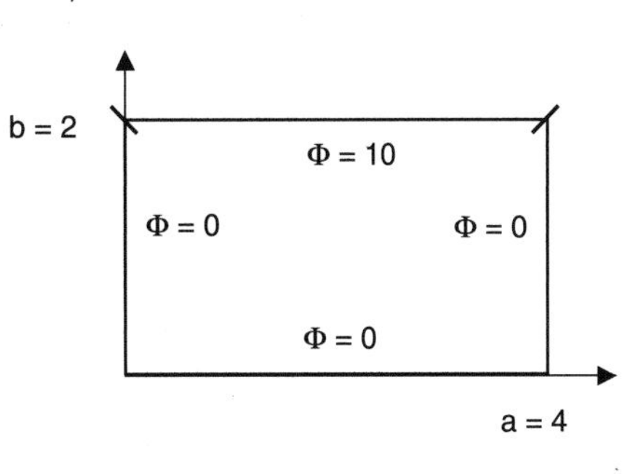

FIGURE 23.13 A rectangular trough with the prescribed boundary conditions.

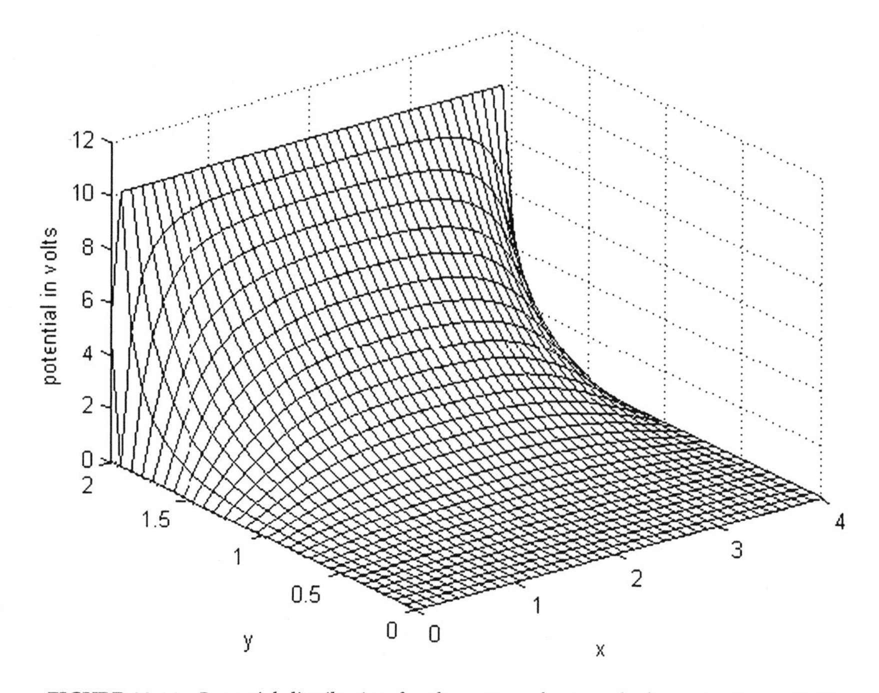

FIGURE 23.14 Potential distribution for the rectangular trough shown in Figure 23.13.

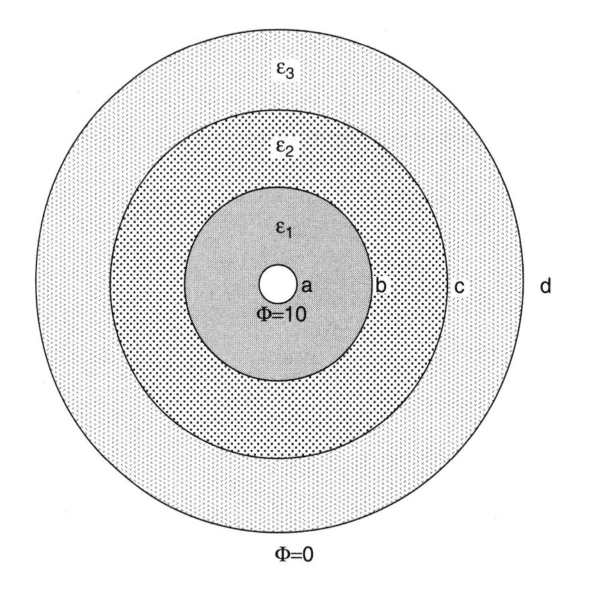

FIGURE 23.15 A cylindrical region with multiple dielectrics.

$$\nabla \times \mathbf{H} = \sigma \mathbf{E} + \varepsilon \frac{\partial \mathbf{E}}{\partial t} \tag{23.66b}$$

Written component-wise, these correspond to six scalar equations. Following Yee's notation, we define a grid point in the solution region as

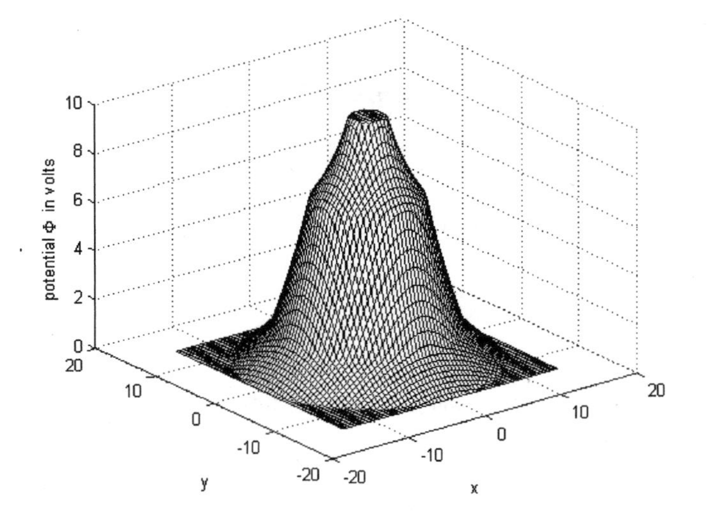

FIGURE 23.16 Potential distribution for the cylindrical region shown in Figure 23.15.

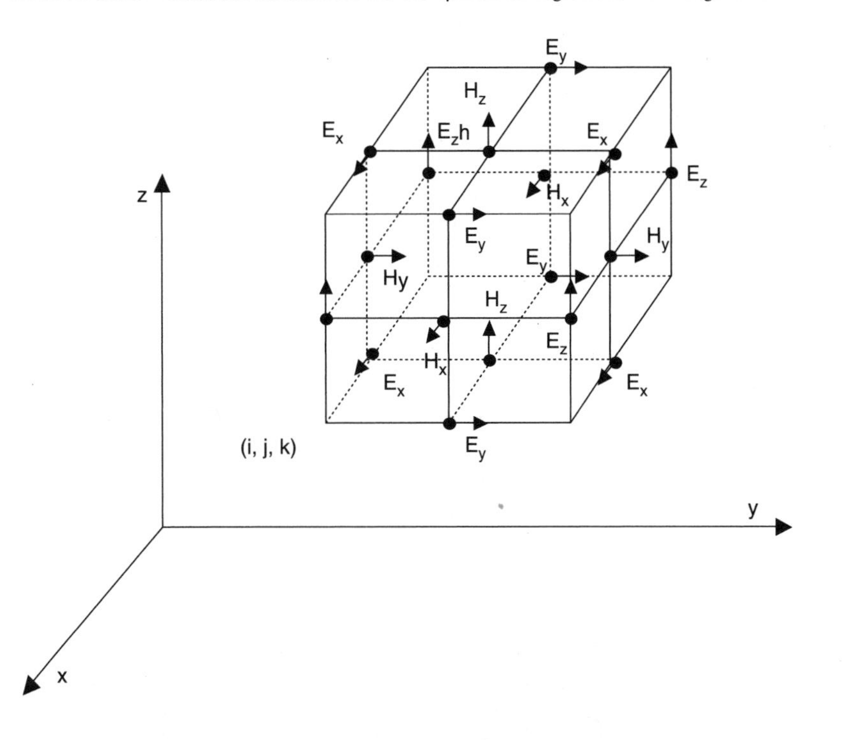

FIGURE 23.17 Positions of the field components in the Yee cell.

$$(i, j, k) \equiv (i\Delta x, j\Delta y, k\Delta z) \tag{23.67}$$

and a function F of space and time at nth time instant as

$$F^n(i, j, k) = F(i\delta, j\delta, k\delta, n\Delta t) \tag{23.68}$$

where $\delta = \Delta x = \Delta y = \Delta z$ is the space increment, Δt is the time increment, while i, j, k and n are integers.

Using central difference approximations for space and time derivatives that are second-order accurate:

$$\frac{\partial F^n(i,j,k)}{\partial x} = \frac{F^n(i+1/2,j,k) - F^n(i-1/2,j,k)}{\delta} + O(\delta^2) \tag{23.69}$$

$$\frac{\partial F^n(i,j,k)}{\partial t} = \frac{F^{n+1/2}(i,j,k) - F^{n-1/2}(i,j,k)}{\Delta t} + O(\Delta t^2) \tag{23.70}$$

In applying Equation (23.69) to all the space derivatives in Equation (23.66), Yee positions the components of **E** and **H** about a unit cell of lattice as shown in Figure 23.17. To incorporate Equation (23.70), the components of **E** and **H** are evaluated at alternate half-time steps. Thus, we obtain the explicit finite difference approximations of Equation (23.66) as

$$\begin{aligned} H_x^{n+1/2}(i,j+1/2,k+1/2) &= H_x^{n-1/2}(i,j+1/2,k+1/2) + \frac{\Delta t}{\mu(i,j+1/2,k+1/2)\delta} \\ &\times \begin{bmatrix} E_y^n(i,j+1/2,k+1) - E_y^n(i,j+1/2,k) \\ +E_z^n(i,j,k+1/2) - E_z^n(i,j+1,k+1/2) \end{bmatrix} \end{aligned} \tag{23.71a}$$

$$\begin{aligned} H_y^{n+1/2}(i+1/2,j,k+1/2) &= H_y^{n-1/2}(i+1/2,j,k+1/2) \\ &+ \frac{\Delta t}{\mu(i+1/2,j,k+1/2)\delta} \begin{bmatrix} E_z^n(i+1,j,k+1/2) - E_z^n(i,j,k+1/2) \\ +E_x^n(i+1/2,j,k) - E_x^n(i+1/2,j,k+1) \end{bmatrix} \end{aligned} \tag{23.71b}$$

$$\begin{aligned} H_z^{n+1/2}(i+1/2,j+1/2,k) &= H_z^{n-1/2}(i+1/2,j+1/2,k) \\ &+ \frac{\Delta t}{\mu(i+1/2,j+1/2,k)\delta} \begin{bmatrix} E_x^n(i+1/2,j+1,k) - E_x^n(i+1/2,j,k) \\ +E_y^n(i,j+1/2,k) - E_y^n(i+1,j+1/2,k) \end{bmatrix} \end{aligned} \tag{23.71c}$$

$$\begin{aligned} E_x^{n+1}(i+1/2,j,k) &= \left(1 - \frac{\sigma(i+1/2,j,k)\Delta t}{\varepsilon(i+1/2,j,k)}\right) E_x^n(i+1/2,j,k) \\ &+ \frac{\Delta t}{\varepsilon(i+1/2,j,k)\delta} \begin{bmatrix} H_z^{n+1/2}(i+1/2,j+1/2,k) - H_z^{n+1/2}(i+1/2,j-1/2,k) \\ +H_y^{n+1/2}(i+1/2,j,k-1/2) - H_y^{n+1/2}(i+1/2,j,k+1/2) \end{bmatrix} \end{aligned} \tag{23.71d}$$

$$\begin{aligned} E_y^{n+1}(i,j+1/2,k) &= \left(1 - \frac{\sigma(i,j+1/2,k)\Delta t}{\varepsilon(i,j+1/2,k)}\right) E_y^n(i,j+1/2,k) \\ &+ \frac{\Delta t}{\varepsilon(i,j+1/2,k)\delta} \begin{bmatrix} H_z^{n+1/2}(i,j+1/2,k+1/2) - H_x^{n+1/2}(i,j+1/2,k-1/2) \\ +H_z^{n+1/2}(i-1/2,j+1/2,k) - H_z^{n+1/2}(i+1/2,j+1/2,k) \end{bmatrix} \end{aligned} \tag{23.71e}$$

$$E_z^{n+1}(i,j,k+1/2) = \left(1 - \frac{\sigma(i,j,k+1/2)\Delta t}{\varepsilon(i,j,k+1/2)}\right) E_z^n(i,j,k+1/2) + \frac{\Delta t}{\varepsilon(i,j,k+1/2)\delta}\begin{bmatrix} H_y^{n+1/2}(i+1/2,j,k+1/2) - H_y^{n+1/2}(i-1/2,j,k+1/2) \\ +H_x^{n+1/2}(i,j-1/2,k+1/2) - H_x^{n+1/2}(i,j+1/2,k+1/2) \end{bmatrix} \tag{23.71f}$$

Notice from Equation (23.71) that the components of **E** and **H** are interlaced within the unit cell and are evaluated at alternate half-time steps. All the field components are present in a quarter of a unit cell. In translating the system of Equation (23.71) into a computer code one must make sure that, within the same time loop, one type of field components is calculated first and the results obtained are then used in calculating another type.

To ensure accuracy of the computed results, the spatial increment δ must be small compared to the wavelength (usually $\leq \lambda/10$) or the minimum dimension of the conducting body. This amounts to having ten or more cells per wavelength. To ensure the stability of the FDTD scheme, the time increment must satisfy:

$$u_{\max}\Delta t \leqslant \left[\frac{1}{\Delta x^2} + \frac{1}{\Delta y^2} + \frac{1}{\Delta z^2}\right]^{-1/2} \tag{23.72}$$

where $u_{\max}$ is the maximum phase velocity within the model. Equation (23.72) is called the *Courant stability condition.* Since we are using a cubic cell with $\Delta x = \Delta y = \Delta z = \delta$, Equation (23.72) becomes

$$\frac{u_{\max}\Delta t}{\delta} \leqslant \frac{1}{\sqrt{n}} \tag{23.73}$$

where n is the number of space dimensions. For practical reasons, it is preferable to choose a ratio of time increment to spatial increment as large as possible, yet satisfying Equation (23.73).

A basic difficulty encountered in applying the FDTD method to problems involving open or unbounded geometries is that the domain in which the field exists is infinite. Since no computer can store unlimited data, we should devise a scheme to confine our solution to a finite region. In other words, artificial boundaries must be simulated so that a numerical illusion is created. The conditions under which this can happen are called *radiation conditions, absorbing boundary conditions,* or *lattice truncation conditions.*

The FDTD method has several advantages. It is conceptually simple. It can treat even inhomogeneous conducting or dielectric structures of complex geometries because the medium parameters can be assigned to each lattice point. It also permits computation of the solution in the frequency domain from a single time-domain simulation via Fourier transform techniques. A disadvantage of the method is that it necessitates modeling of an object as well as its surroundings. Thus, the program execution time may be excessive. Its accuracy is at least one order of magnitude less than that of the moment method. Despite these limitations, the FDTD method has been successfully applied to solve scattering and other problems including aperture penetration, antenna/radiation problems, microwave circuits, eigenvalue problems, and EM absorption in human tissues.

23.3 Finite Element Method

Like the finite difference method and the moment method, the finite element method is useful in solving electromagnetic problems. Application of the finite difference and moment methods becomes difficult with irregularly shaped boundaries. Such problems can be handled more easily using the finite element method.

The finite element analysis of any problem involves basically four steps: (a) discretizing the solution region into a finite number of subregions called *elements,* (b) deriving governing equations for a typical element,

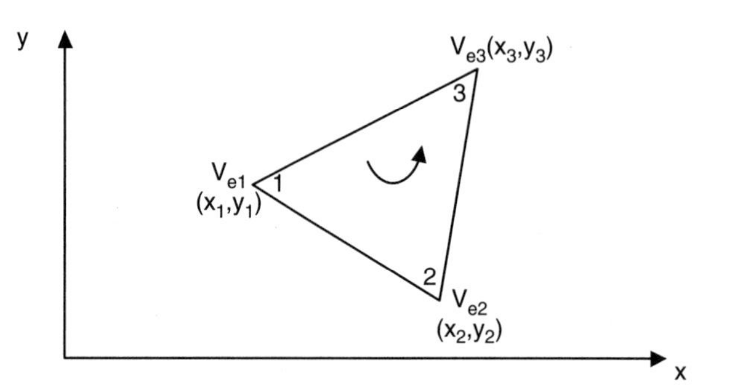

FIGURE 23.18 Typical triangular element; the local node numbering 1, 2, 3 must be counterclockwise as indicated by the arrow.

(c) assembling of all elements in the solution region, and (d) solving the system of equations obtained. These steps are best shown with an example.

Suppose we desire to solve a Laplace equation ($\nabla^2 V = 0$) in a two-dimensional domain. The domain is first divided into triangular elements. This is usually done either by hand or by an automatic mesh generator. We seek an approximation for the potential V_e within an element and then interrelate the potential distribution in various elements such that the potential is continuous across interelement boundaries. The approximate solution for the entire region is

$$V(x,y) \simeq \sum_{e=1}^{N} V_e(x,y) \tag{23.74}$$

where N is the number of triangular elements into which the region has been divided. The most common form of approximation for V_e within an element is polynomial approximation, namely:

$$V_e(x,y) = a + bx + cy \tag{23.75}$$

Consider a typical triangular element shown in Figure 23.18. The potential V_{e1}, V_{e2}, and V_{e3} at nodes 1, 2, and 3 respectively, are obtained from Equation (23.75) as

$$\begin{bmatrix} V_{e1} \\ V_{e2} \\ V_{e3} \end{bmatrix} = \begin{bmatrix} 1 & x_1 & y_1 \\ 1 & x_2 & y_2 \\ 1 & x_3 & y_3 \end{bmatrix} \begin{bmatrix} a \\ b \\ c \end{bmatrix} \tag{23.76}$$

Solving for coefficients a, b, and c and substituting for them in Equation (23.75) gives

$$V_e = \sum_{i=1}^{3} \alpha_i(x,y) V_{ei} \tag{23.77}$$

where

$$\alpha_1 = \frac{1}{2A}\left[(x_2 y_3 - x_3 y_2) + (y_2 - y_3)x + (x_3 - x_2)y\right] \tag{23.78}$$

$$\alpha_2 = \frac{1}{2A}\left[(x_3y_1 - x_1y_3) + (y_3 - y_1)x + (x_1 - x_3)y\right] \tag{23.79}$$

$$\alpha_3 = \frac{1}{2A}\left[(x_1y_2 - x_2y_1) + (y_1 - y_2)x + (x_2 - x_1)y\right] \tag{23.80}$$

$$A = \frac{1}{2}\left[(x_2 - x_1)(y_3 - y_1) - (x_3 - x_1)(y_2 - y_1)\right] \tag{23.81}$$

The interpolation functions α_i are called element shape functions.

The functional corresponding to Laplace's equation is given by

$$W_e = \frac{1}{2}\int \varepsilon|E_e|^2 dS = \frac{1}{2}\int \varepsilon|\nabla V_e|^2 dS \tag{23.82}$$

but

$$\nabla V_e = \sum_{i=1}^{3} V_{ei}\nabla\alpha_i$$

$$W_e = \frac{1}{2}\sum_{i=1}^{3}\sum_{j=1}^{3}\varepsilon V_{ei}\left[\int \nabla\alpha_i\nabla\alpha_j dS\right]V_{ej} \tag{23.83}$$

If we define the term in brackets as

$$C_{ij}^{(e)} = \int \nabla\alpha_i\nabla\alpha_j dS \tag{23.84}$$

we may write Equation (23.84) in matrix form as

$$W_e = \frac{1}{2}\varepsilon[V_e]^t[C^{(e)}][V_e] \tag{23.85}$$

where the superscript t denotes the transpose of the matrix:

$$[V_e] = \begin{bmatrix} V_{e1} \\ V_{e2} \\ V_{e3} \end{bmatrix} \tag{23.86a}$$

and

$$[C^{(e)}] = \begin{bmatrix} C_{11}^{(e)} & C_{12}^{(e)} & C_{13}^{(e)} \\ C_{21}^{(e)} & C_{22}^{(e)} & C_{23}^{(e)} \\ C_{31}^{(e)} & C_{32}^{(e)} & C_{33}^{(e)} \end{bmatrix} \tag{23.86b}$$

The matrix $[C^{(e)}]$ is usually called the *element coefficient matrix* (or "stiffness matrix" in structural analysis). The matrix element $C_{ij}^{(e)}$ of the coefficient matrix may be regarded as signifying the coupling between nodes i and j; its value is obtained from Equation (23.84).

Having considered a typical element, the next step is to assemble all such elements in the solution region. The energy associated with the assemblage of elements is

$$W = \sum_{e=1}^{N} W_e = \frac{1}{2}\varepsilon[V]^t[C][V] \tag{23.87}$$

where

$$[V] = \begin{bmatrix} V_1 \\ V_2 \\ V_3 \\ \vdots \\ V_n \end{bmatrix} \tag{23.88}$$

n is the number of nodes, N is the number of elements, and $[C]$ is called the over-all or *global coefficient matrix* which is the assemblage of individual element coefficient matrices. Notice that to obtain Equation (23.87), we have assumed that the whole solution is homogeneous so that ε is constant. For an inhomogeneous solution region, the region is discretized such that each finite element is homogeneous. In this case, Equation (23.85) still holds but Equation (23.87) does not apply since $\varepsilon = \varepsilon_o\varepsilon_r$ varies from element to element. To apply Equation (23.87), we may replace ε by ε_o and multiply the integrand in Equation (23.84) by ε_r.

The last step is to solve the resulting equations. It can be shown that Laplace's (or Poisson's) equation is satisfied when the total energy in the solution region is minimum. Thus, we require that the partial derivatives of W with respect to each nodal value of the potential be zero, i.e.:

$$\frac{\partial W}{\partial V_k} = 0, \ k = 1, 2, \cdots, n \tag{23.89}$$

By writing Equation (23.89) for all nodes, we obtain a set of simultaneous equations. These equations can be solved iteratively as

$$V_k = -\frac{1}{C_{kk}} \sum_{i=1, i\neq k}^{n} V_i C_{ki} \tag{23.90}$$

Thus far, we have applied the finite element method to solve Laplace's equation. To solve the two-dimensional Poisson's equation:

$$\nabla^2 V = -\frac{\rho_v}{\varepsilon} \tag{23.91}$$

we take essentially the same steps taken for Laplace's equation except that we must include the source term. We approximate the potential distribution $V_e(x, y)$ and the source term ρ_{se} as

$$V_e = \sum_{i=1}^{3} \alpha_i(x, y) V_{ei} \tag{23.92}$$

$$\rho_{se} = \sum_{i=1}^{3} \alpha_i(x, y) \rho_{ei} \tag{23.93}$$

where the coefficients V_{ei} and ρ_{ei}, respectively, represent the values of V and ρ_s at vertex i of element e.

The energy functional associated with the Poisson's equation is

$$F(V_e) = \frac{1}{2} \int_S \left[\varepsilon |\nabla V_e|^2 - 2\rho_{se} V_e \right] dS \tag{23.94}$$

$F(V_e)$ represents the total energy per length within element e. Substitution of Equation (23.92) and Equation (23.93) into Equation (23.94) yields

$$F(V_e) = \frac{1}{2} \sum_{i=1}^{3} \sum_{j=1}^{3} \varepsilon V_{ei} \left[\int \nabla\alpha_i \nabla\alpha_j dS \right] V_{ej} - \sum_{i=1}^{3} \sum_{j=1}^{3} V_{ei} \left[\int \alpha_i \alpha_j dS \right] \rho_{ej} \tag{23.95}$$

This can be written in matrix form as

$$F(V_e) = \frac{1}{2} \varepsilon [V_e]^t [C^{(e)}][V_e] - [V_e]^t [T^{(e)}][\rho_e] \tag{23.96}$$

where $[C^{(e)}]$ is already defined in Equation (23.86b) and

$$T_{ij}^{(e)} = \int \alpha_i \alpha_j dS = \begin{cases} A/12, & i \neq j \\ A/6, & i = j \end{cases} \tag{23.97}$$

and A is the area of the triangular element.

Equation (23.95) can be applied to every element in the solution region. All the resulting equations can be assembled into one global discretized functional. We obtain the discretized functional for the whole solution region (with N elements and n nodes) as the sum of the functionals for the individual elements, i.e., from Equation (23.95):

$$F(V) = \sum_{e=1}^{N} F(V_e) = \frac{1}{2} \varepsilon [V]^t [C][V] - [V]^t [T][\rho] \tag{23.98}$$

In Equation (23.98), the column matrix $[V]$ consists of the values of V_{ei} while the column matrix $[\rho]$ contains n values of the source function ρ_s at the nodes. The functional in Equation (23.98) is now minimized by differentiating with respect to V_{ei} and setting the result equal to zero. The resulting equations can be solved either by the iteration method or band matrix method.

Similarly, the FEM can be applied to the wave equation. A typical wave equation is the inhomogeneous scalar Helmholtz's equation:

$$\nabla^2 \Phi + k^2 \Phi = g \tag{23.99}$$

where Φ is the potential (for waveguide problem, $\Phi = H_z$ for TE mode or E_z for TM mode) to be determined, g is the source function, and $k = \omega\sqrt{\mu\varepsilon}$ is the wave number of the medium. The solution of Equation (23.99) is equivalent to satisfying the boundary conditions and minimizing the functional:

$$I(\Phi) = \frac{1}{2}\int_S \left[|\nabla\Phi|^2 - k^2\Phi^2 + 2\Phi g\right]\mathrm{d}S \tag{23.100}$$

If other than the natural boundary conditions (i.e., Dirichlet or homogeneous Neumann conditions) must be satisfied, appropriate terms must be added to the functional. We now express potential Φ and source function g in terms of the shape functions α_i over a triangular element as

$$\Phi_e(x, y) = \sum_{i=1}^{3} \alpha_i \Phi_{ei} \tag{23.101}$$

$$g_e(x, y) = \sum_{i=1}^{3} \alpha_i g_{ei} \tag{23.102}$$

where Φ_{ei} and g_{ei} are respectively the prescribed values of Φ and g at nodal point i of element e.

Substituting Equation (23.101) and Equation (23.102) into Equation (23.100) gives

$$I(\Phi_e) = \frac{1}{2}\sum_{i=1}^{3}\sum_{j=1}^{3} \Phi_{ei}\Phi_{ej} \int\int \nabla\alpha_i \nabla\alpha_j \mathrm{d}S - \frac{k^2}{2}\sum_{i=1}^{3}\sum_{j=1}^{3} \Phi_{ei}\Phi_{ej} \int\int \alpha_i\alpha_j \mathrm{d}S + \sum_{i=1}^{3}\sum_{j=1}^{3} \Phi_{ei} g_{ej} \int\int \alpha_i\alpha_j \mathrm{d}S \tag{23.103}$$

or

$$I(\Phi_e) = \frac{1}{2}[\Phi_e]^t[C^{(e)}][\Phi_e] - \frac{k^2}{2}[\Phi_e]^t[T^{(e)}][\Phi_e] + [\Phi_e]^t[T^{(e)}][G_e] \tag{23.104}$$

where $[\Phi_e] = [\Phi_{e1}, \Phi_{e2}, \Phi_{e3}]^t$, $[G_e] = [g_{e1}, g_{e2}, g_{e3}]$ and $[C^{(e)}]$ and $[T^{(e)}]$ are defined in Equation (23.84) and Equation (23.97), respectively. Equation (23.104) derived for a single element can be applied for all N elements in the solution region. Thus:

$$I(\Phi) = \sum_{e=1}^{N} I(\Phi_e) \tag{23.105}$$

From Equation (23.104) and Equation (23.105), I(Φ) can be expressed in matrix form as

$$I(\Phi) = \frac{1}{2}[\Phi]^t[C][\Phi] - \frac{k^2}{2}[\Phi]^t[T][\Phi] + [\Phi]^t[T][G] \tag{23.106}$$

where

$$[\Phi] = [\Phi_1, \Phi_2, \cdots, \Phi_N]^t \tag{23.107a}$$

$$[G] = [g_1, g_2, \cdots, g_N]^t \tag{23.107b}$$

$[C]$, and $[T]$ are global matrices consisting of local matrices $[C^{(e)}]$ and $[T^{(e)}]$, respectively.

Consider the special case in which the source function $g = 0$. We now minimize $I(\Phi)$, i.e.,

$$\frac{\partial I}{\partial \Phi_i} = 0, \; i = 1, 2, \cdots, N \tag{23.108}$$

This leads to a linear matrix equation of order N:

$$[C][\Phi] = k^2[T][\Phi] \tag{23.109}$$

Solution of the algebraic eigenvalue problem in Equation (23.109) furnishes eigenvalues and eigenvectors which form good approximations to the eigenvalues and eigenfunctions of the Helmholtz problem, i.e., the cutoff wavelengths and field distribution patterns of the various modes possible in a given waveguide.

The finite element approximation we have used so far has been the linear type. A higher order element is one in which the shape functions or interpolation polynomial is of order 2 or more. The accuracy of a finite element solution can be improved by using finer mesh or using higher order elements or both. In general, fewer higher order elements are needed to achieve the same degree of accuracy in the final results. The higher order elements are particularly useful when the gradient of the field variable is expected to vary rapidly. They have been applied with great success in solving EM-related problems.

The finite element techniques developed thus far are for two-dimensional elements. The idea can be extended to three-dimensional elements. We first divide the solution region into tetrahedral or hexahedral (rectangular prism) elements. One would expect three-dimensional problems to require a large total number of elements to achieve an accurate result and demand a large storage capacity and computation time.

23.4 Commercial Packages for CEM

Numerical modeling and simulation have revolutionized all aspects of engineering design. As a result, several software packages that have been developed to aid designing and modeling. Widely used software packages for CEM include N*umerical Electromagnetics Code* (NEC), and its PC version (MININEC) based on the method of moments and developed at Lawrence Livermore National Laboratory, *High Frequency Structure Simulator* (HFSS) based on finite element method and developed by Ansoft, *Microwave Office* based on the method of moments and developed by Applied Wave Research, *FEMLAB* (which stands for finite element modeling laboratory) from Comsol, *Sonnet* developed by Sonnet, and *IE3D* from Zeland Software. These software applications put powerful tools and techniques, previously available only to full-time theorists, into the hands of engineers not formally trained in CEM. The best method or software package to use depends on the particular problem one is trying to solve. Substantial progress has been made in CEM and further gains have been derived from the availability of these powerful commerical packages. Still, the available modeling tools sometimes fall short of the user's need.

References

M.N.O. Sadiku, *Numerical Techniques in Electromagnetics*, 2nd ed., Boca Raton, FL: CRC Press, 2001.

P.P. Silvester and G. Pelosi, *Finite Elements for Wave Electromagnetics*, New York: IEEE Press, 1994.

R.F. Harrington, *Field Computation by Moment Methods*, Malabar, FL: Robert E. Krieger, 1968.

A.F. Peterson, S.L. Ray, and R. Mittra, *Computational Methods for Electromagnetics*, New York: IEEE Press, 1998.

K.S. Kunz and R.J. Luebbers, *The Finite Difference Time Domain Method for Electromagnetics*, Boca Raton, FL: CRC Press, 1993.

A. Taflove, *Computational Electrodynamics: the Finite-Difference Time-Domain Method*, Boston, MA: Artech House, 1995.

J. Jin, *The Finite Element Method in Electromagnetics*, New York: Wiley, 1993.

C.A. Balanis, *Advanced Engineering Electromagnetics*, New York: Wiley, 1989.

III

Electrical Effects and Devices

24

Electroacoustic Transducers

Peter H. Rogers
Georgia Institute of Technology

Charles H. Sherman
Image Acoustics

Mark B. Moffett
Antion Corporation

24.1 Introduction

Electroacoustics is concerned with the transduction of acoustical to electrical energy and vice versa. Transducers that convert acoustical signals into electrical signals are referred to as "microphones" or "hydrophones", depending on whether the acoustic medium is air or water, respectively. Transducers that convert electrical signals into acoustical waves are referred to as loudspeakers (or earphones) in air and projectors in water.

24.2 Transduction Mechanisms

Piezoelectricity

Certain crystals produce charge on their surfaces when strained or conversely become strained when placed in an electric field. Important piezoelectric crystals include quartz, ADP, lithium sulfate, Rochelle salt, and tourmaline. Lithium sulfate and tourmaline are volume expanders; that is, their volume changes when subjected to an electric field in the proper direction. Such crystals can detect hydrostatic pressure directly. Crystals that are not volume expanders must have one or more surfaces shielded from the pressure field in order to convert the pressure to a uniaxial strain that can be detected. Tourmaline is relatively insensitive and used primarily in blast gauges, while quartz is used principally in high-Q ultrasonic transducers.

It has been discovered that certain polymers, notably polyvinylidene fluoride, can be made piezoelectric during manufacture by stretching. Such piezoelectric polymers are finding use in directional microphones and ultrasonic hydrophones.

Electrostriction

Certain materials, such as lead titanate, lead zirconate titanate (PZT), barium titanate, and lead magnesium niobate (PMN), are ferroelectric and exhibit strong electrostriction. Since barium titanate and PZT have large remanent polarization, they can be permanently polarized, which linearizes the quadratic electrostrictive effect and makes them equivalent to piezoelectric materials for most practical purposes. These materials are used mainly as ceramics, but some have recently become available as single crystals of sufficient size for transducer use. They are capable of producing very large forces, and are used extensively as sources and receivers for underwater sound. PZT and barium titanate have only a small volume sensitivity; hence, they must have one or more surfaces shielded in order to detect sound efficiently. Piezoelectric ceramics have extraordinarily high dielectric coefficients and hence high capacitance, and they are thus capable of driving long cables without preamplifiers.

Magnetostriction

Some ferromagnetic materials display a significant magnetostrictive strain when subjected to a magnetic field. The effect is analogous in many respects to electrostriction, but the magnetic materials do not have large remanent magnetization, and thus require a bias field or dc current for linear operation. Important magnetostrictive metals and alloys include nickel and permendur. At one time, magnetostrictive transducers were used extensively in active sonars but have now been largely replaced by ceramic transducers. Magnetostrictive transducers are rugged and reliable but inefficient and configurationally awkward. Certain rare-earth iron alloys such as terbium–dysprosium–iron (Terfenol-D) possess extremely large magnetostriction (as much as 100 times that of nickel). They have relatively low eddy current losses but require large bias fields and have yet to find significant applications. Gallium–iron alloys (Galfenol) have somewhat less magnetostriction, but they possess high tensile strength and are machinable. Metallic glasses have also been considered for magnetostrictive transducers.

Electrodynamic (Moving Coil)

Electrodynamic transducers exploit the force on a current-carrying conductor in a magnetic field and, conversely, the voltage generated by a conductor moving in a magnetic field. Moving-coil transducers dominate the loudspeaker field and are the most numerous of all the transducer types. Prototypes of high-power underwater projectors have been constructed using superconducting magnets. Electrodynamic microphones, particularly the directional ribbon microphones, are also common.

Electrostatic

Electrostatic sources utilize the force of attraction between charged capacitor plates. The force is independent of the sign of the voltage, and so a bias voltage is necessary for linear operation. Because the forces are relatively weak, a large area is needed to obtain significant acoustic output. The effect is reciprocal, with the change in the separation of the plates (i.e., the capacitance) produced by an incident acoustic pressure generating a voltage. The impedance of a condenser microphone, however, is high, and so a preamplifier located close to the transducer is required. Condenser microphones are frequency-independent and extremely sensitive. The change in capacitance induced by an acoustic field can also be detected by making the capacitor a part of a bridge circuit or, alternatively, a part of an oscillator circuit. The acoustic signal will then appear as either an amplitude or frequency modulation of an ac carrier. The charge storage properties of electrets have been exploited to produce electrostatic microphones that do not require a maintained bias voltage. Micromachined silicon electrostatic transducers are used as miniature microphones and accelerometers.

Magnetic (Variable Reluctance)

Magnetic transducers utilize the force of attraction between magnetic poles and, reciprocally, the voltages produced when the reluctance of a magnetic circuit is changed. They are analogous in many ways to electrostatic transducers. Magnetic speakers are used in telephone receivers.

Parametric Sources and Receivers

The nonlinear interaction of sound waves can be used to produce highly directional sound sources with no side lobes and small physical apertures. Despite of their inherent inefficiency, useful source levels can be achieved and such parametric sources have found a number of underwater applications. Parametric receivers have also been investigated but practical applications have yet to be found.

Carbon Microphones

Carbon microphones utilize a change in electrical resistance with pressure and have been used extensively in telephones.

Hydraulic

Nonreversible, low-frequency, high-power underwater projectors can be constructed utilizing hydraulic forces to move large pistons. The piston motion is controlled by modulating the hydraulic pressure with a spool valve driven by a PZT electromechanical transducer.

Fiber Optic

An acoustic field acting on an optical fiber will change the optical path length by changing the length and index of refraction of the fiber. Extremely sensitive hydrophones and microphones can be made by using a fiber exposed to an acoustic field as one leg of an optical interferometer. Path length changes of the order of 10^{-6} optical wavelengths can be detected. The principal advantages of such sensors are their configurational flexibility, their sensitivity, and their suitability for use with fiber optic cables. Fiber optic sensors which utilize amplitude modulation of the light (microbend transducers) are also being developed.

24.3 Receiving and Transmitting Responses

A microphone or hydrophone is characterized by its free-field voltage sensitivity, M, which is defined (usually in units of dB re 1 V/μPa) as 20 times the logarithm of the ratio of the output voltage, E, to the free-field amplitude of an incident plane acoustic wave. That is, for an incident wave which, in the absence of the transducer, is given by

$$P = P_0 \cos(\mathbf{k} \cdot \mathbf{R} - \omega t) \tag{24.1}$$

M is defined by

$$M = 20 \log|E/P_0| \tag{24.2}$$

where the wave vector, $\mathbf{k}$, has magnitude ω/c and points in the direction of propagation. (Here, $\omega = 2\pi f$ is the angular frequency and c is the speed of sound.) In general, M is a function of frequency and the orientation of the transducer with respect to $\mathbf{k}$ (i.e., the direction of the incident wave). However, for most purposes it is customary to let M be the sensitivity for an incident wave on the maximum response axis (MRA). M is then a function of frequency but not orientation, and the directivity is a separate, but equally important, characteristic that will be discussed in Section "Directivity". It is usually desirable for a microphone or hydrophone to have a frequency-independent free-field voltage sensitivity over the broadest possible range of frequencies to assure fidelity of the output electrical signal. For example, for piezoelectric transducers, this means the frequency range below resonance.

A loudspeaker or projector is characterized in a similar manner by its transmitting current response, S. In the far field of the transducer the acoustic pressure is a spherically spreading wave which can be expressed as

$$P(R) = A_0 \cos(kR - \omega t)/R \tag{24.3}$$

along the direction of maximum response. The amplitude A_0, which is the product of the range R and the amplitude of the pressure measured at R, characterizes the strength of the acoustic source, whose peak source level is defined as 20 times the logarithm of A_0 (e.g., in units of dB re 1 μPa m). Thus, the transmitting current response is given (e.g., in units of μPa m/amp) by

$$S = 20 \log|A_0/I| \tag{24.4}$$

where I is the amplitude of the driving current and S is a function of the frequency ω. For high-fidelity sound reproduction S should be as flat as possible over the broadest possible bandwidth. For some purposes, however, such as ultrasonic cleaning or long-range underwater acoustic propagation, fidelity is unnecessary and high-Q resonant transducers are employed to produce high-intensity sound over a narrow bandwidth.

24.4 Reciprocity

Most conventional transducers are reversible; that is, they can be used as either sources or receivers of sound (a carbon microphone and a fiber optic hydrophone are examples of transducers which are not reversible). A transducer is said to be linear if the input and output variables are linearly proportional (hot-wire microphones and unbiased magnetostrictive transducers are examples of nonlinear transducers). A transducer is said to be passive if the only source of energy is the input electrical or acoustical signal (a microphone with a built-in preamplifier is an example of a nonpassive transducer). Most transducers that are linear, passive, and reversible exhibit a remarkable property called reciprocity. For a reciprocal transducer of any kind (moving coil, piezoelectric, magnetostrictive, electrostatic, magnetic, etc.), the free-field voltage sensitivity is proportional to the transmitting current response and is independent of the geometry and construction of the transducer. That is:

$$M - S = 20 \log|J(\omega)| = 20 \log(4\pi/\rho_0 \omega) \tag{24.5}$$

where J is known as the spherical-wave reciprocity factor, and where ρ_0 is the density of the medium. (It should be noted that J is often defined as $4\pi d_0/\rho_0\omega$, where d_0 is a reference distance, usually equal to 1 m. This factor is used when S is defined in terms of the pressure extrapolated to 1 m, rather than the pressure–distance product. Numerically the results are the same, but the dimensions of J and S differ from those presented above.) Reciprocity has a number of important consequences:

1. The receiving and transmitting beam patterns of a reciprocal transducer are identical.
2. A reciprocal transducer cannot be simultaneously flat as a receiver and transmitter because of the frequency-dependence of J.
3. Equation (24.5) provides the basis for the three-transducer reciprocity calibration technique whereby an absolute calibration of a hydrophone or microphone can be obtained from purely electrical measurements.

24.5 Canonical Equations and Electroacoustic Coupling

Simple electroacoustic transducers can be characterized by a pair of linear equations that relate the four variables: voltage, E, current, I, velocity of the moving surface, V, and the external force on the surface, F. One of the most useful of these equation pairs, especially for magnetic transducers where the internal force is proportional to current, is

$$E = Z_e I + T_{em} V \tag{24.6}$$

$$F = T_{me} I + Z_m V \tag{24.7}$$

In these equations, T_{em} and T_{me} are transduction coefficients, Z_e is the blocked ($V = 0$) electrical impedance, and Z_m is the open-circuit ($I = 0$) mechanical impedance. For reciprocal transducers $T_{em} = \pm T_{me}$. For example, for an electrodynamic transducer with an axially moving coil in a radial magnetic field, B:

$$T_{em} = -T_{me} = BL \tag{24.8}$$

where L is the length of the wire in the coil.

Transduction coefficients are a measure of electromechanical coupling that are useful for comparing transducers of the same type. The electromechanical coupling coefficient (or factor), κ, is a dimensionless measure that can be used to compare all types of transducers, because its values are always between zero and unity. The physical meaning of κ^2 can be best understood from its definition as the ratio of converted energy to total input energy. For example, in a loudspeaker or projector the input energy is electrical and the fraction of it converted to mechanical energy equals κ^2. Two specific examples are:

$$\text{Moving coil transducer:} \quad \kappa^2 = (BL)^2 / K_m^E L_b \tag{24.9}$$

where K_m^E is the short circuit mechanical stiffness and L_b is the blocked inductance:

$$\text{Piezoelectric transducer:} \quad \kappa^2 = d_{33}^2 / \varepsilon_{33}^T s_{33}^E \tag{24.10}$$

where d_{33} is the piezoelectric strain constant, ε_{33}^T is the permittivity at constant stress, and s_{33}^E is the short-circuit compliance coefficient. Note that κ^2 is a ratio of stored, not dissipated, energies, and so is independent of the electroacoustic efficiency, which is defined as the ratio of output power to total input power. For example, quartz crystal transducers can be very efficient, but have low coupling, while moving-coil loudspeakers are usually inefficient devices with high coupling factors.

If a piston transducer is placed in an acoustic field such that the average pressure over the surface of the piston is P_B, then $F = P_B A$, where A is the area of the piston. For a receiver $I = 0$,

$$E = (T_{em} A / Z_m) P_B \tag{24.11}$$

If the transducer is small compared with an acoustic wavelength, $P_B \approx P_0$, and so the free-field voltage sensitivity is given by

$$M \approx 20 \log |T_{em} A / Z_m| \tag{24.12}$$

while the transmitting current response is

$$S \approx 20 \log |\rho_0 \omega T_{em} A / 4\pi Z_m| \tag{24.13}$$

From these simple considerations a number of principles of practical transducer design can be deduced. The mechanical impedance Z_m is in general given by

$$Z_m = R_m + j\omega M + K_m/j\omega \tag{24.14}$$

where R_m is the mechanical resistance, M the effective mass, and K_m an effective spring constant. For a moving-coil transducer (Equation (24.8)) T_{em} is independent of frequency; hence from Equation (24.13) we derive the fundamental tenet of loudspeaker design, that a moving coil loudspeaker will have a flat transmitting current response above resonance (i.e., where $Z_m \approx j\omega M$). Accordingly, moving coil loudspeakers are designed to have the lowest possible resonance frequency (by means of a high compliance since the output is inversely proportional to the mass). Similarly, it can be shown that an acoustically small condenser microphone or piezoelectric hydrophone will have a flat free-field voltage sensitivity below resonance.

An interesting and important consequence of electromechanical coupling is the effect of the motion of the transducer on the electrical impedance. In the absence of external forces ($F = 0$), Equation (24.6) and Equation (24.7) yield

$$E = (Z_e - T_{em}T_{me}/Z_m)I \tag{24.15}$$

That is, the electrical impedance has a "motional" component given by $-T_{em}T_{me}/Z_m$. The motional component can be quite significant near resonance where Z_m is small. This effect is the basis of crystal-controlled oscillators.

24.6 Radiation Impedance

An oscillating surface produces a pressure, p, in the medium and a reaction force, F_R, on the surface given by

$$F_R = \iint p dS = -Z_R V \tag{24.16}$$

where Z_R is the radiation impedance. For projector operation, where F_R is the only external force, we can thus rewrite Equation (24.7) as

$$0 = T_{me}I + (Z_R + Z_m)V \tag{24.17}$$

For an acoustically small baffled circular piston of radius a:

$$Z_R = \pi a^4 \rho_0 \omega^2/2c + j(8/3)\omega\rho_0 a^3 \tag{24.18}$$

The radiation impedance has a mass-like reactance with an equivalent radiation mass of $(8/3)\rho_0 a^3$ and a small resistive component proportional to ω^2 responsible for the radiated power. A transducer will thus have a lower resonance frequency when operated underwater than when operated in air or vacuum. The total radiated power of the piston transducer is given by

$$\pi = ReZ_r|V|^2 = \left(\pi a^4 \rho_0 \omega^2/2c\right)|V|^2 \tag{24.19}$$

Most transducers are displacement limited, and so for a direct-radiating transducer, V in Equation (24.19) is limited. To obtain the most output power the piston should have the largest possible surface area. Alternatively, the driver can be placed at the apex of a horn. For a conical horn, the fluid velocity at the end of the horn (where the radius is a_e) will be reduced to $V(a/a_e)$ but the radiating piston will now have an effective radius of a_e so the radiated power will increase by a factor of $(a_e/a)^2$. For high-power operation at a single frequency, the driver can be placed at the end of a quarter-wave resonator.

24.7 Directivity

It is often desirable for transducers to be directional. Directional sound sources are needed in diagnostic and therapeutic medical ultrasonics, for acoustic depth sounders, and to reduce the power requirements and reverberation in active sonars, etc. Directional microphones are useful to reduce unwanted noise (e.g., to pick up the voice of a speaker and not the audience); directional hydrophones or hydrophone arrays increase signal-to-noise and aid in target localization. The directional characteristics of a transducer are embodied in the normalized directivity function, $D(\theta, \phi)$, which has a value of unity in the direction of maximum response, i.e., $D(0, 0) = 1$. The three-dimensional pressure field is then expressed as $P(R, \theta, \phi) = P(R)D(\theta, \phi)$ with $P(R)$ given by Equation (24.3). The directivity factor, D_f, is

$$D_\mathrm{f} = \frac{4\pi}{\int_0^{2\pi}\int_0^{\pi} D^2(\theta, \phi)\,\sin\theta \mathrm{d}\theta \mathrm{d}\phi} \tag{24.20}$$

and the directivity index is

$$DI = 10 \log D_\mathrm{f} \tag{24.21}$$

One way to achieve directionality is to make the radiating surface large. A baffled circular piston has a directivity function given by

$$D = 2J_1(ka \sin\theta)/(ka \sin\theta) \tag{24.22}$$

where $k = \omega/c$. D equals unity for $\theta = 0$ and 1/2 when $ka \sin\theta = 2.2$. For small values of ka, D is near unity for all angles.

Some transducers respond to the gradient of the acoustic pressure rather than pressure, for example, the ribbon microphone which works by detecting the motion of a thin conducting strip orthogonal to a magnetic field. Such transducers have a directivity function which is dipole in nature, i.e.:

$$D = \cos\theta \tag{24.23}$$

Note that a ribbon microphone will have flat receiving sensitivity when its impedance is mass controlled. By combining a dipole receiver with a monopole receiver one obtains a unidirectional cardioid receiver with

$$D = (1 + \cos\theta)/2 \tag{24.24}$$

Defining Terms

Electroacoustics: Concerned with the transduction of acoustical to electrical energy and vice versa.
Microphones: Devices that convert acoustical signals into electrical signals.

References

D. Berlincourt, "Piezoelectric crystals and ceramics," in *Ultrasonic Transducer Materials*, O.E. Mattiat, Ed., New York: Plenum Press, 1971.

D.A. Berlincourt, D.R. Curran, and H. Jaffe, "Piezoelectric and piezomagnetic materials and their function in transducers," in *Physical Acoustics*, vol. 1, Part A, W.P. Mason, Ed., New York: Academic Press, 1964.

R.J. Bobber, "New types of transducer," in *Underwater Acoustics and Signal Processing*, L. Bjorno, Ed., Dordrecht, The Netherlands: D. Riedel, 1981.

R.J. Bobber, *Underwater Electroacoustic Measurements*, Washington, DC: Government Printing Office, 1969.
J.V. Bouyoucos, "Hydroacoustic transduction," *J. Acoust. Soc. Am.*, 57, 1341, 1975.
J.A. Bucaro, H.D. Dardy, and E.F. Carome, "Fiber optic hydrophone," *J. Acoust. Soc. Am.*, 62, 1302, 1977.
A. Caronti, R. Carotenuto, and M. Pappalardo, "Electromechanical coupling factor of capacitive micromachined ultrasonic transducers", *J. Acoust. Soc. Am.*, 113, 279, 2003.
A.E. Clark, J.B. Restorff, M. Wun-Fogle, T.A. Lograsso, and D.L. Schlagel, "Magnetostrictive properties of b.c.c. Fe–Ga and Fe–Ga–Al alloys," *IEEE Trans. Magn.*, 36, 3238, 2000.
A.E. Clark, "Magnetostrictive rare earth-Fe_2 compounds," *Ferromagnetic Materials*, vol. 1, Amsterdam: North-Holland, 1980, p. 531.
F.V. Hunt, *Electroacoustics*, Cambridge: Harvard University Press, and New York: Wiley, 1954.
W.P. Mason, *Electromechanical Transducers and Wave Filters*, 2nd ed., New York: D. Van Nostrand, 1948.
N.W. McLachlan, *Loud Speakers — Theory Performance Testing and Design*, Mineola, NY: Dover, 1960.
S.W. Meeks and R.W. Timme, "Rare earth iron magnetostrictive underwater sound transducer," *J. Acoust. Soc. Am.*, 62, 1158, 1977.
M.B. Moffett, M.D. Jevnager, S.S. Gilardi, and J.M. Powers, "Biased lead zirconate titanate as a high-power transduction material," *J. Acoust. Soc. Am.*, 105, 2248, 1999.
M.B. Moffett, A.E. Clark, M. Wun-Fogle, J.F. Lindberg, J.P. Teter, and E.A. McLaughlin, "Characterization of Terfenol-D for magnetostrictive transducers," *J. Acoust. Soc. Am.*, 89, 1448, 1991.
M.B. Moffett and R.H. Mellen, "Model for parametric acoustic sources," *J. Acoust. Soc. Am.*, 61, 325, 1977.
W.Y. Pan, W.Y. Gu, D.J. Taylor, and L.E. Cross, "Large piezoelectric effect induced by direct current bias in PMN-PT relaxor ferroelectric ceramics," *Jpn. J. Appl. Phys.*, 28, 653, 1989.
D. Ricketts, "Electroacoustic sensitivity of piezoelectric polymer cylinders," *J. Acoust. Soc. Am.*, 68, 1025, 1980.
G.M. Sessler and J.E. West, "Applications," in *Electrets*, G.M. Sessler, Ed., New York: Springer, 1980.
C.H. Sherman and J.L. Butler, *Transducers and Arrays for Underwater Sound*, Berlin: Springer, to be published in 2005.
D. Stansfield, *Underwater Electroacoustic Transducers*, Bath, U.K.: Bath University Press and Institute of Acoustics, 1991.
S. Trolier-McKinstry, L.E. Cross, and Y. Yamashita, Eds., *Piezoelectric Single Crystals and Their Application*, Pennsylvania State University and Toshiba Corp., Pennsylvania State University Press, 2004.
R.S. Woollett, "Basic problems caused by depth and size constraints in low-frequency underwater transducers," *J. Acoust. Soc. Am.*, 68, 1031, 1980.
R.S. Woollett, "Procedures for comparing hydrophone noise with minimum water noise," *J. Acoust. Soc. Am.*, 54, 1376, 1973.
R.S. Woollett, "Power limitations of sonic transducers," *IEEE Trans. Sonics Ultrason.*, SU-15, 218, 1968.
R.S. Woollett, "Effective coupling factor of single-degree-of-freedom transducers," *J. Acoust. Soc. Am.*, 40, 1112, 1966.

Further Information

IEEE Transactions on Acoustics, Speech, and Signal Processing
IEEE Transactions on Ultrasonics, Ferroelectrics, and Frequency Control

25

Ferroelectric and Piezoelectric Materials

K.F. Etzold
IBM T.J. Watson Research Center

25.1 Introduction

Piezoelectric materials have been used extensively in resonator, actuator, and ultrasonic receiver applications, while **ferroelectric** materials have recently received much attention for their potential use in nonvolatile (NV) memory applications. We will discuss the basic concepts in the use of these materials, highlight their applications, and describe the constraints limiting their uses. This chapter emphasizes properties which need to be understood for the effective use of these materials. Among the properties discussed are **hysteresis** and **domains**.

Ferroelectric and piezoelectric materials derive their properties from a combination of structural and electrical properties. As the name implies, both types of materials have electric attributes. A large number of materials, which are ferroelectric, are also piezoelectric. However, the converse is not true. Pyroelectricity is closely related to ferroelectric and piezoelectric properties via the symmetry properties of the crystals.

Examples of the classes of materials that are technologically important are given in Table 25.1. It is apparent that many materials exhibit electric phenomena which can be attributed to ferroelectric, piezoelectric, and **electret** properties. It is also clear that vastly different materials (organic and inorganic) can exhibit ferroelectricity or piezoelectricity, and many have actually been commercially exploited for these properties.

As shown in Table 25.1, there are two dominant classes of bulk ferroelectric materials: ceramics and organics. Both classes have important applications of their piezoelectric properties. To exploit the ferroelectric property for memory applications, recently a large effort has been devoted to producing thin films of the ceramic **PZT** (lead [Pb] zirconate titanate) on various substrates for silicon-based memory chips for nonvolatile storage. In these devices, data is retained in the absence of external power as positive and negative **polarization**. Other applications of thin films are resonators and surface acoustic wave (SAW) devices for signal filtering and the use of the nonlinear properties for phase shifters on electrically steered antennas. Organic materials have not been used for their ferroelectric properties. Liquid crystals in display applications are used for their ability to rotate the plane of polarization of light and not their ferroelectric attribute.

It should be noted that the prefix *ferro* refers to the permanent nature of the electric polarization in analogy with the magnetization in the magnetic case. It does not imply the presence of iron, even though the root of the word means iron. The root of the word *piezo* means pressure; hence the original meaning of the word

TABLE 25.1 Ferroelectric, Piezoelectric, and Electrostrictive Materials

Type	Material Class	Example	Applications
Electret	Organic	Waxes	No recent
Electret	Organic	Fluorine based	Microphones
Ferroelectric	Organic	PVF2	No known
Ferroelectric	Organic	Liquid crystals	Displays
Ferroelectric	Ceramic	PZT thin film	NV memory
Piezoelectric	Organic	PVF2	Transducer
Piezoelectric	Ceramic	PZT	Transducer
Piezoelectric	Ceramic	PLZT	Optical
Piezoelectric	Single crystal	Quartz	Frequency control
Piezoelectric	Single crystal	$LiNbO_3$	SAW devices
Electrostrictive	Ceramic	PMN	Actuators

piezoelectric implied "pressure electricity"—the generation of an electric field from applied pressure. This definition ignores the fact that these materials are reversible, allowing the generation of mechanical motion by applying a field.

25.2 Mechanical Characteristics

When materials are acted on by forces (stresses) the resulting deformations are called strains. An example of a strain due to a force to the material is the change of dimension parallel and perpendicular to the applied force. It is useful to introduce the coordinate system and the numbering conventions, which are used when discussing these materials. Subscripts 1, 2, and 3 refer to the x, y, and z directions, respectively. Displacements have single indices associated with their direction. If the material has a preferred axis, such as the poling direction in PZT, the axis is designated the z or 3 axis. Stresses and strains require double indices such as xx or xy. To make the notation less cluttered and confusing, contracted notation has been defined. The following mnemonic rule is used to reduce the double index to a single index:

$$\begin{array}{ccc} 1 & 6 & 5 \\ xx & xy & xz \\ & 2 & 4 \\ & yy & yz \\ & & 3 \\ & & zz \end{array}$$

This rule can be thought of as a matrix with the diagonal elements having repeated indices in the expected order, then continuing the count in a counterclockwise direction. Note that $xy = yx$, etc., so that subscript 6 applies equally to xy and yx.

Any mechanical object is governed by the well-known relationship between stress and strain:

$$\mathbf{S} = \mathbf{sT} \tag{25.1}$$

where **S** is the strain (relative elongation), **T** is the stress (force per unit area), and **s** contains the coefficients connecting the two. All quantities are tensors; **S** and **T** are second rank, and **s** is fourth rank. Note, however, that usually contracted notation is used so that the full complement of subscripts is not visible. PZT converts electrical fields into mechanical displacements and vice versa. The connection between the two is via the d and g coefficients. The d coefficients give the displacement when a field is applied (transmitter), while the g coefficients give the field across the device when a stress is applied (receiver). The electrical effects are added to

TABLE 25.2 Properties of Well-Known PZT Formulations (Based on the Original Navy Designations)

	Units	PZT4	PZT5A	PZT5H	PZT8
ε_{33}	—	1300	1700	3400	1000
d_{33}	10^{-2} Å/V	289	374	593	225
d_{13}	10^{-2} Å/V	−123	−171	−274	−97
d_{15}	10^{-2} Å/V	496	584	741	330
g_{33}	10^{-3} Vm/N	26.1	24.8	19.7	25.4
k_{33}	—	70	0.705	0.752	0.64
T_{Θ}	°C	328	365	193	300
Q	—	500	75	65	1000
p	G/cm³	7.5	7.75	7.5	7.6
Application		High signal	Medium signal	Receiver	Highest signal

the basic Equation (25.1) such that:

$$\mathbf{S} = \mathbf{sT} + \mathbf{dE} \tag{25.2}$$

where **E** is the electric field and **d** is the tensor which contains the coupling coefficients. The latter parameters are reported in Table 25.2 for representative materials. One can write the matrix equation (Equation (25.2)) explicitly as Equation (25.3) as shown below:

$$\begin{bmatrix} S_1 \\ S_2 \\ S_3 \\ S_4 \\ S_5 \\ S_6 \end{bmatrix} = \begin{bmatrix} s_{11} & s_{12} & s_{13} & & & \\ s_{12} & s_{11} & s_{13} & & 0 & \\ s_{13} & s_{13} & s_{33} & & & \\ & & & s_{44} & & \\ & 0 & & & s_{44} & \\ & & & & & 2(s_{11}-s_{12}) \end{bmatrix} \begin{bmatrix} T_1 \\ T_2 \\ T_3 \\ T_4 \\ T_5 \\ T_6 \end{bmatrix} + \begin{bmatrix} 0 & 0 & d_{13} \\ 0 & 0 & d_{13} \\ 0 & 0 & d_{33} \\ 0 & d_{15} & 0 \\ d_{15} & 0 & 0 \\ 0 & 0 & 0 \end{bmatrix} \begin{bmatrix} E_1 \\ E_2 \\ E_3 \end{bmatrix} \tag{25.3}$$

Note that **T** and **E** are shown as column vectors for typographical reasons; they are in fact row vectors. This equation shows explicitly the stress-strain relation and the effect of the electromechanical conversion.

A similar equation applies when the material is used as a receiver:

$$\mathbf{E} = -\mathbf{gT} + (\boldsymbol{\varepsilon}^{\mathrm{T}})^{-1}\mathbf{D} \tag{25.4}$$

where $\boldsymbol{\varepsilon}^{\mathrm{T}}$ is the transpose of $\boldsymbol{\varepsilon}$ and **D** the electric displacement. The matrices are not fully populated for all materials. Whether a coefficient is nonzero depends on the crystalline symmetry. For PZT, a ceramic which is given a preferred direction by the poling operation (the z axis), only d_{33}, d_{13}, and d_{15} are nonzero. Also, again by symmetry, $d_{13} = d_{23}$ and $d_{15} = d_{25}$.

Applications

Historically the material which was used earliest for its piezoelectric properties was single-crystal quartz. Sonar devices were built by Langevin using quartz transducers, but the most important application of piezoelectric quartz was, and still is, frequency control. Crystal oscillators are today at the heart of every clock that does not derive its frequency reference from the ac power line. They are also used in every color television set and personal computer. In these applications, at least one (or more) quartz crystals controls frequency or time. This explains the label "quartz" appearing on many clocks and watches. The use of quartz resonators for frequency control relies on another unique property. Not only is the material piezoelectric (which allows one to excite mechanical vibrations), but the material has also a very high mechanical "Q" or quality factor ($Q > 100{,}000$). The actual value depends on the mounting details, whether the crystal is in a vacuum, and other details. Compare this value to a Q for PZT between 65 and 1000. The Q factor is a measure of the rate of decay and thus the mechanical losses of an excitation with no external drive. A high Q leads to a very sharp resonance and thus tight frequency control. For frequency control it has been possible to find orientations of cuts of quartz, which reduce the influence of temperature on the vibration frequency.

Ceramic materials of the PZT family have also found increasingly important applications. The piezoelectric but not the ferroelectric property of these materials is made use of in transducer applications. PZT has a very high efficiency (electric energy to mechanical energy coupling factor k) and can generate high-amplitude ultrasonic waves in water or solids. The coupling factor is defined by

$$k^2 = \frac{\text{energy stored mechanically}}{\text{total energy stored}} \tag{25.5}$$

Typical values of k_{33} are 0.7 for PZT 4 and 0.09 for quartz, showing that PZT is a much more efficient transducer material than quartz. Note that the energy is a scalar; the subscripts are assigned by finding the energy conversion coefficient for a specific vibrational mode and field direction, and selecting the subscripts accordingly. Thus, k_{33} refers to the coupling factor for a longitudinal mode driven by a longitudinal field.

Probably the most important applications of PZT today are based on ultrasonic echo ranging. Sonar uses the conversion of electrical signals to mechanical displacement as well as the reverse transducer property, which is to generate electrical signals in response to a stress wave. Medical diagnostic ultrasound, sonar, and nondestructive testing systems devices rely on the same properties. Actuators have also been built but a major obstacle is the small displacement which can conveniently be generated. Even then, the required voltages are typically hundreds of volts and the displacements are only a few hundred angstroms. For PZT the strain in the z-direction due to an applied field in the z-direction is (no stress, $\mathbf{T} = 0$):

$$s_3 = d_{33} E_3 \tag{25.6}$$

or

$$s_3 = \frac{\Delta d}{d} = d_{33} \frac{V}{d} \tag{25.7}$$

where s is the strain, E the electric field, and V the potential; d_{33} is the coupling coefficient which connects the two. Thus:

$$\Delta d = d_{33} V \tag{25.8}$$

Note that this expression is independent of the thickness d of the material but this is true only when the applied field is parallel to the displacement. Let the applied voltage be 100 V and let us use PZT8 for which d_{33} is 225 (from Table 25.2). Hence, $\Delta d = 225$ Å or 2.25 Å/V, a small displacement indeed. We also note that Equation (25.6) is a special case of Equation (25.2) with the stress equal to zero. This is the situation when an actuator is used in a force-free environment, for example, as a mirror driver. This arrangement results in the maximum displacement. Any forces which tend to oppose the free motion of the PZT will subtract from the available displacement with the reduction given by the normal stress–strain relation (Equation (25.1)).

It is possible to obtain larger displacements with mechanisms that exhibit mechanical gain, such as laminated strips (bimorph, similar to bimetallic strips). The motion then is typically up to about 1 mm, but at a cost of a reduced available force. An example of such an application is the video head translating device to provide tracking in VCRs.

There is another class of ceramic bulk materials which recently has become important. **PMN** (Pb, magnesium niobate, typically doped with $\approx$10% lead titanate) is an **electrostrictive** material which has seen applications where the absence of hysteresis is important. For example, deformable mirrors require repositioning of the reflecting surface to a defined location regardless of whether the old position was above or below the original position.

Electrostrictive materials exhibit a strain which is quadratic as a function of the applied field. Producing a displacement requires an internal polarization. Because the latter polarization is induced by the applied field and is not permanent, as it is in the ferroelectric materials, electrostrictive materials have essentially no hysteresis. Unlike PZT, electrostrictive materials are not reversible; PZT will change shape on application of an electric potential and generate an electric field when a strain is induced. Electrostrictive materials only change shape on application of a field and, therefore, cannot be used as receivers. PZT has inherently large hysteresis because of the domain nature of the polarization.

Organic electrets have important applications in self-polarized condenser (or capacitor) microphones where the required electric bias field in the gap is generated by the diaphragm material rather than by an external power supply.

Structure of Ferroelectric and Piezoelectric Materials

Ferroelectric materials have, as their basic building block, atomic groups which have an associated electric field, either as a result of their structure or as result of distortion of the charge clouds making up the groups. In the first case, the field arises from an asymmetric placement of the individual ions in the group (these groupings are called unit cells). In the second case, the electronic cloud is moved with respect to the ionic core. If the group is distorted permanently, then a permanent electric field can be associated with each group. We can think of these distorted groups as represented by electric dipoles, defined as two equal but opposite charges, which are separated by a small distance. Electric dipoles are similar to magnetic dipoles which have the familiar north and south poles. The external manifestation of a magnetic dipole is a magnetic field and that of an electric dipole is an electric field.

Figure 25.1(a) represents a hypothetical slab of material in which the dipoles are perfectly arranged. In actual materials the atoms are not as uniformly arranged, but, nevertheless, from this model there would be a very strong field emanating from the surface of the crystal. The common observation, however, is that the fields are either absent or weak. This effective charge neutrality arises from the fact that there are free, mobile charges available, which can be attracted to the surfaces. The polarity of the mobile charges is opposite to the charge of the free dipole end. The added charges on the two surfaces generate their own field, equal and opposite to the field due to the internal dipoles. Thus the effect of the internal field is canceled and the external field is zero, as if no charges were present at all (Figure 25.1(b)). This name given to this phenomenon is charge compensation.

In ferroelectric materials a crystalline asymmetry exists, which allows electric dipoles to form. In their absence the dipoles are absent and the internal field disappears. Consider an imaginary horizontal line drawn through the middle of a dipole. We can see readily that the dipole is not symmetric about that line. The asymmetry thus requires that there be no center of inversion when the material is in the ferroelectric state.

All ferroelectric and piezoelectric materials have phase transitions at which the material changes crystalline symmetry. For example, in PZT there is a change from tetragonal or rhombohedral symmetry to cubic as the

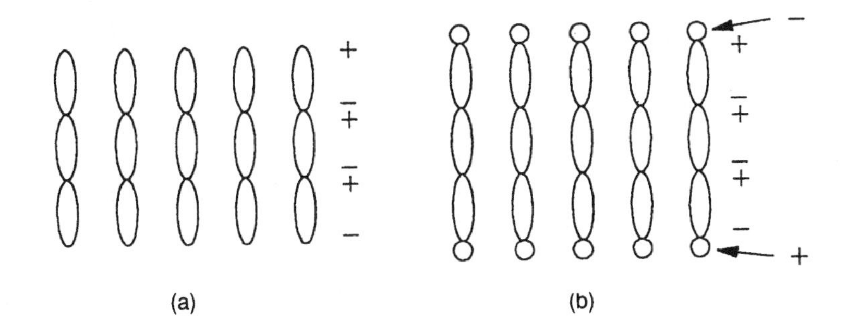

FIGURE 25.1 Charge configurations in ferroelectric model materials: (a) uncompensated and (b) compensated dipole arrays.

temperature is increased. The temperature at which the material changes **crystalline phases** is called the **Curie temperature,** T_{Θ}. For typical PZT compositions the Curie temperature is between 250 and 450°C. This temperature is affected by composition and doping.

A consequence of a phase transition is that a rearrangement of the lattice takes place when the material is cooled through the transition. Intuitively we would expect that the entire crystal assumes the same orientation throughout as we pass through the transition. By orientation we mean the direction of the preferred axis (say the tetragonal axis). Experimentally it is found, however, that the material breaks up into smaller regions in which the preferred direction, and thus the polarization, is uniform. Note that cubic materials have no preferred direction. In tetragonal crystals the polarization points along the *c* axis (the longer axis) whereas in rhombohedral lattices the polarization is along the body diagonal. The volume in which the preferred axis is pointing in the same direction is called a domain and the boundary between the regions is called a domain wall. The energy of the multidomain state is slightly lower than the single-domain state and is thus the preferred configuration. The direction of the polarization changes by either 90° or 180° as we pass from one uniform region to another. Thus the domains are called 90° and 180° domains. Whether an individual crystallite or grain consists of a single domain depends on the size of the crystallite and external parameters such as strain gradients, impurities, etc. It is also possible that a domain extends beyond the grain boundary and encompasses two or more grains of the crystal.

Real materials consist of large numbers of unit cells, and the manifestation of the individual charged groups is an internal and an external electric field when the material is stressed. Internal and external refer to inside and outside of the material. The interaction of an external electric field with a charged group causes a displacement of certain atoms in the group. The macroscopic manifestation of this is a displacement of the surfaces of the material. This motion is called the piezoelectric effect, the conversion of an applied field into a corresponding displacement.

25.3 Ferroelectric Materials

PZT ($PbZr_xTi_{(1-x)}O_3$) is an example of a ceramic material which is ferroelectric. We will use PZT as a prototype system for many of the ferroelectric attributes to be discussed. The concepts, of course, have general validity. The structure of this material is ABO_3 where A is lead and B is one or the other atoms, Ti or Zr. This material consists of many randomly oriented crystallites which vary in size between approximately 10 nm and several microns depending on the details of the preparation. The crystalline symmetry of the material is determined by the magnitude of the parameter *x*. The material changes from rhombohedral to tetragonal symmetry when $x > 0.48$. This transition is almost independent of temperature. The line that divides the two phases is called a **morphotropic phase boundary** (change of symmetry as a function of composition only). Commercial materials are made with $x \approx 0.48$, where the *d* and *g* sensitivity of the material is maximum. It is clear from Table 25.2 that there are other parameters which can be influenced as well. Doping the material with donors or acceptors often changes the properties dramatically. Thus, niobium is important to obtain higher sensitivity and resistivity, and to lower the Curie temperature. As the Curie temperature is lowered the *g* and *d* coefficients increase making the material more sensitive. The tradeoff is the increased temperature dependence of the parameters.

PZT typically is a p-type conductor and niobium will significantly decrease the conductivity because of the electron which Nb^{5+} contributes to the lattice. The Nb ion substitutes for the **B-site** ion Ti^{4+} or Zr^{4+}. The resistance to depolarization (the "hardness" of the material) is affected by iron doping. Hardness is a definition giving the relative resistance to depolarization. It should not be confused with mechanical hardness. Many other dopants and admixtures have been used, often in very exotic combinations to affect aging, sensitivity, etc.

The designations used in Table 25.2 reflect very few of the many combinations which have been developed. The PZT designation types were originally designed by the U.S. Navy to reflect certain property combinations. The property types can be obtained with different combinations of compositions and dopants. The examples given in the table are representative of typical PZT materials, but today essentially all applications have their

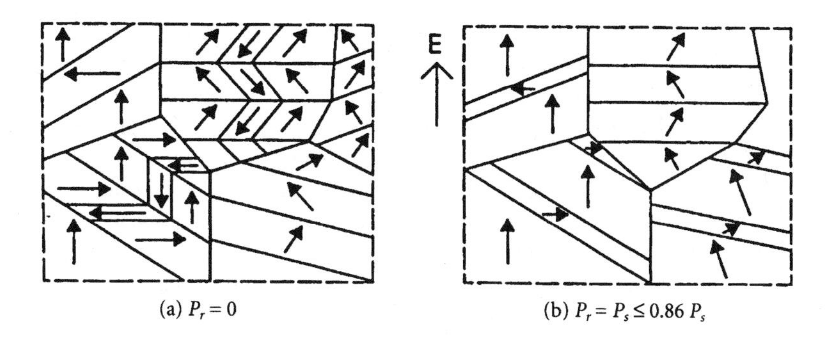

FIGURE 25.2 Domains in PZT: as prepared (a) and poled (b).

own custom formulation. The name PZT has become generic for the lead zirconate titanates and does not reflect Navy or proprietary designations.

When PZT ceramic material is prepared, the crystallites and domains are randomly oriented, and therefore the material does not exhibit any piezoelectric behavior; (see Figure 25.2(a)). The random nature of the displacements for the individual crystallites causes the net displacement to average to zero when an external field is applied. The tetragonal axis has three equivalent directions 90° apart and the material can be poled by reorienting the polarization of the domains into a direction nearest to the applied field. When a sufficiently high field is applied, some but not all of the domains will be rotated toward the electric field through the allowed angle 90° or 180°. If the field is raised further, eventually all domains will be oriented as close as possible to the direction of the field. Note however, that the polarization will not point exactly in the direction of the field (Figure 25.2(b)). At this point, no further domain motion is possible and the material is saturated. As the field is reduced, the majority of domains retain the orientation they had with the field on leaving the material in an oriented state which now has a net polarization. Poling is accomplished for commercial PZT by raising the temperature to about 150°C in oil (which has the effect of lowering the **coercive field**, E_c) and applying a field of about 30 to 60 kV/cm for several minutes. The temperature is then lowered, but it is not necessary to keep the field on during cooling because the domains will not spontaneously rerandomize.

Electrical Characteristics

Before considering the dielectric properties, we will consider the equivalent circuit for a slab of ferroelectric material. In Figure 25.3 the circuit shows a mechanical (acoustic) component and the static or clamped capacity C_o (and the dielectric loss R_d), which are connected in parallel. The acoustic components are due to their motional or mechanical equivalents, the compliance (capacity, C), and the mass (inductance, L). There will be mechanical losses which are indicated in the mechanical branch by R_M. The electrical branch has the clamped capacity C_o and a dielectric loss (R_d) distinct from the mechanical losses. This configuration will have a resonance which is usually assumed to correspond to the mechanical thickness mode but can represent other

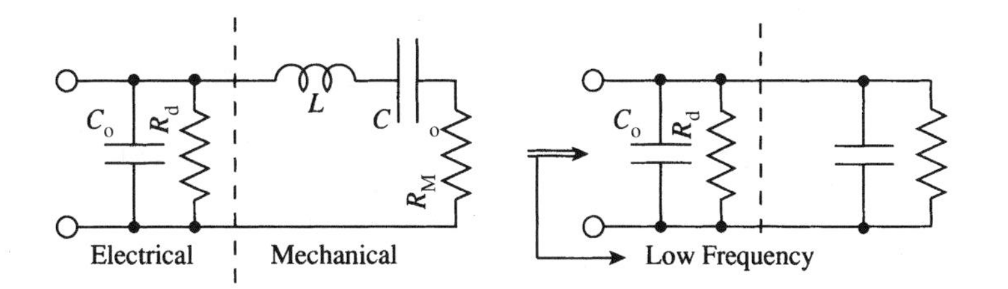

FIGURE 25.3 Equivalent circuit for a piezoelectric resonator. The reduction of the equivalent circuit at low frequencies is shown on the right.

modes as well. This simple model does not show the many other modes a slab (or rod) of material will have. Thus, transverse, plate, and flexural modes are present. Each can be represented by its own combination of L, C, and R_M. The presence of a large number of modes often causes difficulties in characterizing the material since some parameters must be measured either away from the resonances or from clean, nonoverlapping resonances. For instance, the clamped capacity (or clamped dielectric constant) of a material is measured at high frequencies where there are usually a large number of modes present. For an accurate measurement these must be avoided and often a low-frequency measurement is made, in which the material is physically clamped to prevent motion. This yields the static, nonmechanical capacity, C_o. The circuit can be approximated at low frequencies by ignoring the inductor, and redefining R and C. Thus, the coupling constant can be extracted from the value of C and C_o. From the previous definition of k (Equation (25.5)) we find:

$$k^2 = \frac{CV^2/2}{(C + C_o)V^2/2} = \frac{1}{\frac{C_o}{C} + 1} \tag{25.9}$$

It requires a charge to rotate or flip a domain. Thus, there is charge flow associated with the rearrangement of the polarization in the ferroelectric material. If a slow bipolar signal is applied to a ferroelectric material, its hysteresis loop is traced out and the charge in the circuit can be measured using the Sawyer Tower circuit (Figure 25.4). In some cases, the drive signal to the material is not repetitive and only a single cycle is used. In that case the starting point and the end point do not have the same polarization value because of relaxation effects and the hysteresis curve will not close on itself.

The charge flow through the sample is due to the rearrangement of the polarization vectors in the domains (the polarization) and contributions from the static capacity and losses (C_o and R_d in Figure 25.3). The charge is integrated by the measuring capacitor which is in series with the sample. The measuring capacitor is sufficiently large to avoid a significant voltage loss. The polarization is plotted on an X–Y scope or plotter against the applied voltage and therefore the applied field.

Ferroelectric and piezoelectric materials are lossy. This will distort the shape of the hysteresis loop and can even lead to incorrect identification of materials as ferroelectric when they merely have nonlinear conduction characteristics. A resistive component (from R_d in Figure 25.3) will introduce a phase shift in the polarization signal. Thus, the display has an elliptical component, which looks like the beginning of the opening of a hysteresis loop. However, if the horizontal signal has the same phase shift, the influence of this lossy component is eliminated because it is in effect subtracted. Obtaining the exact match is the function of the optional phase shifter, and in the original Sawyer Tower circuit a bridge was constructed, which had a second measuring capacitor in the comparison arm (identical to the one in series with the sample). The phase was then matched with adjustable high-voltage components matching C_o and R_d.

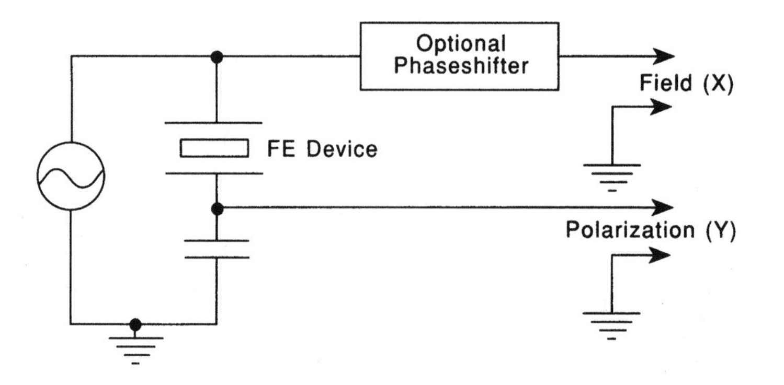

FIGURE 25.4 Sawyer Tower circuit.

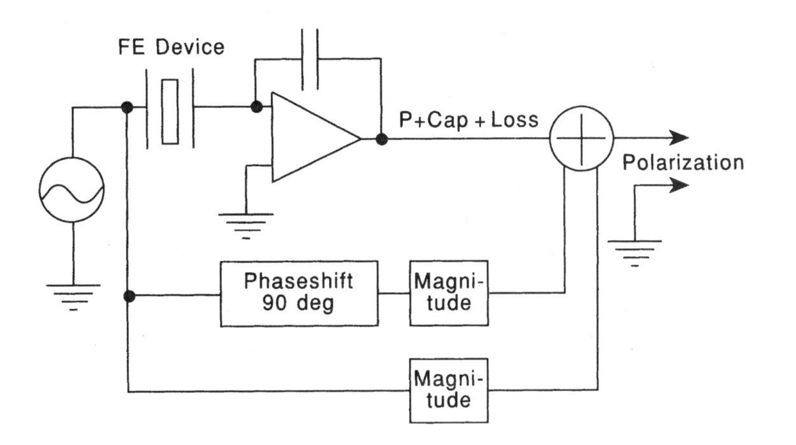

FIGURE 25.5 Modern hysteresis circuit. An op amp is used to integrate the charge; loss and static capacitance compensation are included schematically and are digital in contemporary testers.

This design is inconvenient to implement and modern hysteresis loopers have the capability to shift the reference phase either electronically or digitally to compensate for the loss and static components. A contemporary version, which has compensation and no voltage loss across the integrating capacitor, is shown in Figure 25.5. The op-amp integrator provides a virtual ground at the input, reducing the voltage loss to negligible values. The output from this circuit is the sum of the polarization and the capacitive and loss components. These contributions can be canceled using a purely real (resistive) and a purely imaginary (capacitive, 90° phaseshift) compensation component proportional to the drive across the sample. Both need to be scaled (magnitude adjustments) to match them to the device being measured and then have to be subtracted (adding negatively) from the output of the op amp. The remainder is the polarization. The hysteresis for typical ferroelectrics is frequency dependent and traditionally the reported values of the polarization are measured at 50 or 60 Hz.

The improved, digital version of the Sawyer Tower circuit allows the cancellation of C_o and R_d and the losses, thus determining the ferroelectrically active component. A representative plot with compensation is shown in Figure 25.6. The ability to separate the contributions is important in the development of materials for ferroelectric or piezoelectric applications. It is far easier to judge the shape and squareness of the loop when the inactive components are cancelled. Also, by calibrating the "magnitude controls", the value of the inactive components can be read off directly. In typical measurements, the resonance is far above the test frequency used, so ignoring the inductance in the equivalent circuit is justified. There are commercial testers on the market; examples are: the Precision LC™ by Radiant Technologies (U.S.) and the TF Analyzer 2000™ by aixACCT (Germany). The latter has specialized adapter heads for various thin film measurements. Newer instruments now also have the capability to make measurements at higher frequencies, which is important for memory devices.

The measurement of the dielectric constant and the losses is usually very straightforward. A slab with a circular or other well-defined cross section is prepared, electrodes are applied, and the capacity and loss are measured (usually as a function of frequency). The dielectric constant is found from

$$C = \varepsilon_o \varepsilon \frac{A}{t} \tag{25.10}$$

where A is the area of the device and t the thickness. In this definition (also used in Table 25.2), ε is the relative dielectric constant and ε_o is the permittivity of vacuum. Until recently, the dielectric constant, like the polarization, was measured at 50 or 60 Hz (typical powerline frequencies) primarily due to instrumentation constraints. Today the dielectric parameters are typically specified at 1 kHz, which is possible because

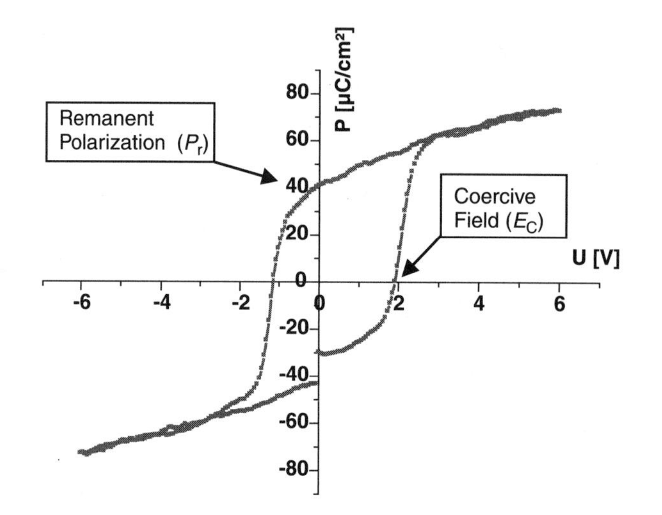

FIGURE 25.6 Hysteresis curve for typical 40/60 PZT material targeted for memory applications. The trace is from a 300 μm × 300 μm sample showing the sensitivity of a contemporary tester. The curve is not closed because only a single, slow sweep was taken. Many PZT materials display offsets from the origin and have asymmetries with respect to the origin, some of which is visible on this sample. The figure also shows how the remanent polarization (P_r) and the coercive field (E_C) are defined.

impedance analyzers with high-frequency capability are readily available. Even higher frequencies such as 1 MHz are often used when evaluating PZT thin films.

Low-frequency anomalies caused by interface layers are not included in the equivalent circuit (Figure 25.3). These layers will cause both the resistive and reactive components to increase at low frequencies producing readings which are not representative of the intrinsic dielectric properties.

A piezoelectric component often has a very simple geometric shape, especially when it is prepared for measurement purposes. There will be mechanical resonances associated with the major dimensions of a sample piece. The resonance spectrum will be more or less complicated, depending on the shape of a sample piece. If the object has a simple shape, then some of the resonances will be well separated from each other and can be associated with specific vibrations and dimensions (modes). Each of these resonances has an electrical equivalent, and inspection of the equivalent circuit shows that there will be a resonance (minimum impedance) and an antiresonance (maximum impedance). Thus, an impedance plot can be used to determine the frequencies and also the coupling constants and mechanical parameters for the various modes.

25.4 Ferroelectric and High Epsilon Thin Films

While PZT and other ferroelectric (FE) bulk materials have had major commercial importance, thin films prepared from these materials have only recently been the focus of significant research efforts. In this section the material properties and process issues will be discussed. Because of the potentially large payoff, major efforts have been directed at developing the technologies for depositing thin films of ferroelectric and nonferroelectric but high epsilon (high dielectric constant) thin films.

A recent trend has been the ever increasing density of dynamic random access memory (DRAM). The storage capacitor in these devices is becoming a major limiting factor because the dielectric has to be very thin in order to achieve the desired capacitance values to yield, in turn, a sufficient signal for the storage cell. It is often also desirable to have nonvolatile operation (no data loss on power loss). These two desires have, probably more than anything else, driven the development of high epsilon and FE thin films. Of course, these are not the only applications of FE films. Table 25.3 lists the applications of FE (nonvolatile, NV) and high epsilon films (volatile), and highlights which of the properties are important for their use. It is seen that the

TABLE 25.3 Materials Properties and Applications Areas

	Ferroelectric	Epsilon	Polarization	Coercive Field	Leakage	Aging	Electro-Optical	Electro-Mechanical
NV RAM	X		X	X	X	X		
DRAM		X			X	X		
Actuator				X				X
Saw devices				X				X
Filters (FBAR)								X
TFT display	X				X	X	X	
Optical Modulator	X				X	X	X	
Phase shifter	X	X						

memory application is very demanding. Satisfying all these requirements simultaneously has produced significant challenges in the manufacture of these films. At the present time (2004) some progress has been made in producing nonvolatile memory devices using perovskites such as PZT but the devices are for niche applications such as smart cards. The true mass application, as a memory element in computers large or small, has not come to pass. In this arena flash memory is still the dominant technology.

There has been a success in the acoustic application of piezoelectric materials for bandpass filters in the form of a mechanical filter called FBAR (film bulk acoustic resonator). This is a filter device for cell phones etc., operating typically at 2.4 GHz. Traditional LC filters are bulky and have low Q at these frequencies. The FBAR is small and has good band shape characteristics. Interestingly the transducer is made of AlN, a piezoelectric material which has not seen significant other applications. This material has the desirable property of a high velocity of sound, which eases some of the scaling issues. In passing we note that other group III nitrides are also piezoelectric, but have lower coupling factors. The other desirable property is that the material itself is very compatible with silicon wafer processing, unlike PZT.

Perhaps the least understood and to some extent still unsolved problem is that of fatigue. In nonvolatile memory applications the polarization represents the memory state of the material (positive $\equiv$ bit 1; negative $\equiv$ bit 0). In use the device can be switched at the clock rate, say 100 MHz. Thus, for a lifetime of five years, the material must withstand $\simeq 10^{16}$ polarization reversals or large field reversals. Typical materials for ferroelectric memory applications are PZTs with the ratio of zirconium to titanium adjusted to yield the maximum dielectric constant and polarization. This maximum will be near the morphotropic phase boundary for PZT. Small quantities of other materials can be added, such as lanthanum or niobium to modify optical or switching characteristics. The Sol-Gel method discussed below is particularly suitable for incorporating these additives. Devices made from materials at the current state-of-the-art lose a significant portion of their polarization after 10^{10} to 10^{12} cycles, rendering them useless for their intended memory use because of the associated signal loss. This is a topic of intensive investigation and no superior material besides PZT material has emerged which might be suitable for memory use.

High epsilon nonferroelectric materials are of great interest for DRAM applications. As an example, major efforts are extant to produce thin films of mixtures of barium and strontium titanate (BST). Dielectric constants of 600 and above have been achieved (compared to 4 to 6 for silicon oxides and nitrides).

In applications for FE films, significant opportunities also exist for electro-optical modulators for fiber-optic devices and light valves for displays. Another large scale application is actuators and sensors. For the latter the electro-mechanical conversion property is used and large values of d_{33} (the conversion coefficient) are desirable. A highly desirable application of field dependent dielectric properties is being currently investigated. Steerable array antennas and tunable devices thus become feasible if suitable materials and configurations can be found. However, economically the importance of all other applications is and probably will be in the foreseeable future, less significant than that of memory devices.

Integration of ferroelectric or nonferroelectric materials with silicon devices and substrates has proved to be very challenging. Contacts and control of the crystallinity and crystal size, and the stack structure of the capacitor device are the principal issues. In both volatile and nonvolatile memory cells the dielectric material tends to interact with the silicon substrate. Thus an appropriate barrier layer must be incorporated while at the same time obtaining a suitable substrate on which to grow the dielectric films. A typical device structure starts with an oxide layer (SiO_x) on the silicon substrate followed by a thin titanium layer which prevents diffusion of the final substrate and contact layer, e.g., platinum (the actual growth substrate).

Significant differences have been observed in the quality of the films depending on the nature of the substrate. The quality can be described by intrinsic parameters such as the crystallinity (i.e., the degree to which noncrystalline phases are present). The uniformity of the orientation of the crystallites also seems to play a role in determining the electrical properties of the films. In the extreme case of perfect alignment of the crystallites of the film and the formation of large single crystal areas, an epitaxial film is obtained. These films tend to have the best electrical properties. In addition to amorphous material, other crystalline but nonferroelectric phases can be present. One example is the pyrochlore phase in PZT. These phases often form incidentally to the growth process of the desired film and usually degrade one or more of the desired properties of the film (for instance the dielectric constant). The pyrochlore and other oxide materials can accumulate between the electrode and the desired PZT or BST layer. The interface layer is then electrically in series with the desired dielectric layer and degrades its properties. The apparent reduction of the dielectric constant which is often observed in these films as the thickness is reduced can be attributed to the presence of these low dielectric constant layers.

There are many growth methods for these films. Table 25.4 lists the most important techniques along with some of the critical parameters. Wet methods use metal organic compounds in liquid form. In the Sol-Gel process the liquid is spun on to the substrate. The wafer is then heated, typically to a lower, intermediate temperature (around 300°C). This spin-on and heat process is repeated until the desired thickness is reached. At this temperature only an amorphous film forms. The wafer is then heated to between 500 and 700°C, usually in oxygen, and the actual crystal growth takes place. Instead of simple long term heating (order of hours) rapid thermal annealing (RTA) is also used. In this process the sample is only briefly exposed to the elevated temperature, usually by a scanning infrared beam. It is in the transition between the low decomposition temperature and the firing temperature that the pyrochlore tends to form. At the higher temperatures the more volatile components have a tendency to evaporate, thus producing a chemically unbalanced compound which also has a great propensity to form one or more of the pyrochlore phases. In the case of PZT, 5 to 10% excess lead is usually incorporated, which helps to form the desired perovskite material and compensates for the loss. In preparing Sol-Gel films it is generally easy to prepare the compatible liquid compounds of the major constituents and the dopants. The composition is then readily adjusted by appropriately changing the ratio of the constituents. Very fine quality films have been prepared by this method, including epitaxial films.

The current semiconductor technology is tending toward dry processing. Thus, despite of the advantages of the Sol-Gel method, other methods using physical vapor deposition (PVD) are being investigated. These methods use energetic beams or plasma to move the constituent materials from the target to the heated substrate. The compound then forms *in situ* on the heated wafer ($\simeq$500°C). Even then, however, a subsequent anneal is typically required. With PVD methods it is much more difficult to change the composition since now

TABLE 25.4 Deposition Methods for PZT and Perovskites

	Process Type	Rate (nm/min)	Substrate Temperature	Anneal Temperature	Target/Source
Wet	Sol-Gel	100 nm/coat	RT	450–750	Metal organic
Wet	MOD	300 nm/coat	RT	500–750	Metal organic
Dry	RF sputter	0.5–5	RT–700	500–700	Metals and oxides
Dry	Magnetron sputter	5–30	RT–700	500–700	Metals and oxides
Dry	Ion beam sputter	2–10	RT–700	500–700	Metals and oxides
Dry	Laser sputter	5–100	RT–700	500–700	Oxide
Dry	MOCVD	5–100	400–800	500–700	MO vapor and carrier gas

RT, room temperature

the oxide or metal ratios of the target have to be changed or dopants have to be added. This involves the fabrication of a new target for each composition ratio. MOCVD is an exception here; the ratio is adjusted by regulating the carrier gas flow. However, the equipment is very expensive and the processing temperatures tend to be high (up to 800°C, uncomfortably high for semiconductor device processing). The laser sputtering method is very attractive and it has produced very fine films. The disadvantage is that the films are defined by the plume which forms when the laser beam is directed at the source. This produces only small areas of good films and scanning methods need to be developed to cover full size silicon wafers. Debris is also a significant issue in laser deposition. However, it is a convenient method to produce films quickly and with a small investment. In the long run MOCVD or Sol-Gel will probably evolve as the method of choice for realistic DRAM or FRAM devices with state-of-the-art densities.

Defining Terms

A-site: Many ferroelectric materials are oxides with a chemical formula ABO_3. The A-site is the crystalline location of the A atom.

B-site: Analogous to the definition of the A-site.

Coercive field: When a ferroelectric material is cycled through the hysteresis loop the coercive field is the electric field value at which the polarization is zero. A material has a negative and a positive coercive field, and these are usually, but not always, equal in magnitude to each other.

Crystalline phase: In crystalline materials the constituent atoms are arranged in regular geometric ways; for instance in the cubic phase the atoms occupy the corners of a cube (edge dimensions $\approx$2 to 15 Å for typical oxides).

Curie temperature: The temperature at which a material spontaneously changes its crystalline phase or symmetry. Ferroelectric materials are often cubic above the Curie temperature and tetragonal or rhombohedral below.

Domain: Domains are portions of a material in which the polarization is uniform in magnitude and direction. A domain can be smaller, larger, or equal in size to a crystalline grain.

Electret: A material which is similar to ferroelectrics but in which charges are macroscopically separated and thus are not structural. In some cases the net charge in the electrets is not zero, for instance when an implantation process was used to embed the charge.

Electrostriction: The change in size of a nonpolarized, dielectric material when it is placed in an electric field.

Ferroelectric: A material with permanent charge dipoles which arise from asymmetries in the crystal structure. The electric field due to these dipoles can be observed external to the material when certain conditions are satisfied (ordered material and no charge on the surfaces).

Hysteresis: When the electric field is raised across a ferroelectric material the polarization lags behind. When the field is cycled across the material the hysteresis loop is traced out by the polarization.

Morphotropic phase boundary (MPB): Materials which have a MPB assume a different crystalline phase depending on the composition of the material. The MPB is sharp (a few percent in composition) and separates the phases of a material. It is approximately independent of temperature in PZT.

Piezoelectric: A material which exhibits an external electric field when a stress is applied to the material and a charge flow proportional to the strain is observed when a closed circuit is attached to electrodes on the surface of the material.

PLZT: A PZT material with a lanthanum doping or admixture (up to approximately 15% concentration). The lanthanum occupies the A-site.

PMN: Generic name for electrostrictive materials of the lead (Pb) magnesium niobate family.

Polarization: The polarization is the amount of charge associated with the dipolar or free charge in a ferroelectric or an electret, respectively. For dipoles the direction of the polarization is the direction of the dipole. The polarization is equal to the external charge which must be supplied to the material to produce a polarized state from a random state (twice that amount is necessary to reverse the polarization). The statement is rigorously true if all movable charges in the material are reoriented (i.e., if complete saturation can be achieved).

PVF2: An organic polymer which can be ferroelectric. The name is an abbreviation for polyvinyledene difluoride.

PZT: Generic name for piezoelectric materials of the lead (Pb) zirconate (Zr) titanate (Ti) family.

Remanent polarization: The residual or remanent polarization of a material after an applied field is reduced to zero. If the material was saturated, the remanent value is usually referred to as the polarization, although even at smaller fields a (smaller) polarization remains.

References

J.C. Burfoot and G.W. Taylor, *Polar Dielectrics and their Applications*, Berkeley, Calif.: University of California Press, 1979.

H. Diamant, K. Drenck, and R. Pepinsky, *Rev. Sci. Instrum.*, 28, 30, 1957.

T. Hueter and R. Bolt, *Sonics*, New York: Wiley, 1954.

B. Jaffe, W. Cook, and H. Jaffe, *Piezoelectric Ceramics*, London: Academic Press, 1971.

M.E. Lines and A.M. Glass, *Principles and Applications of Ferroelectric Materials*, Oxford: Clarendon Press, 1977.

R.A. Roy and K.F. Etzold, "Ferroelectric film synthesis, past and present: a select review," *Mater. Res. Soc. Symp. Proc.*, 200, 141, 1990.

C.B. Sawyer and C.H. Tower, *Phys. Rev.*, 35, 269, 1930.

Z. Surowiak, J. Brodacki, and H. Zajosz, *Rev. Sci. Instrum.*, 49, 1351, 1978.

Further Information

IEEE Transactions on Ultrasonics, Ferroelectrics, and Frequency Control (UFFC).

IEEE Proceedings of International Symposium on the Application of Ferroelectrics (ISAF) (these symposia are held at irregular intervals).

Materials Research Society, Symposium Proceedings (this society holds symposia on ferroelectric materials, typically as part of the fall meeting).

K.-H. Hellwege, Ed., *Landolt-Bornstein: Numerical Data and Functional Relationships in Science and Technology*, New Series, Gruppe III, vols. 11 and 18, Berlin: Springer-Verlag, 1979 and 1984 (these volumes have elastic and other data on piezoelectric materials).

American Institute of Physics Handbook, 3rd ed., New York: McGraw-Hill, 1972.

26

Electrostriction

R. Yimnirun
Chiang Mai University

V. Sundar
Dentsply Prosthetics – Ceramco

Robert E. Newnham
The Pennsylvania State University

26.1 Introduction

Electrostriction is the basic electromechanical coupling mechanism in centric crystals and amorphous solids. It has been recognized as the primary electromechanical coupling in centric materials since early in the twentieth century (Cady, 1929). Electrostriction is the quadratic coupling between the strain developed in a material and the electric field applied, and it exists in all insulating materials. Piezoelectricity is a better-known linear coupling mechanism that exists only in materials without a center of symmetry.

Electrostriction is a second-order property that is tunable and nonlinear. Electrostrictive materials exhibit a reproducible, nonhysteretic, and tunable strain response to electric fields, which gives them an advantage over piezoelectrics in micropositioning applications. While most electrostrictive actuator materials are perovskite ceramics, recently there has been much interest in large electrostriction effects in such polymer materials as polyvinylidene fluoride (PVDF) and its copolymer with trifluoroethylene (P(VDF-TrFE)) (Zhang et al., 1998).

This chapter discusses the three electrostrictive effects and their applications. A discussion of the sizes of these effects and typical electrostrictive coefficients is followed by an examination of lead magnesium niobate (PMN) as a prototype electrostrictive material. The electromechanical properties of some common electrostrictive materials are also compared. A few common criteria used to select relaxor ferroelectrics for electrostrictive applications are also outlined.

26.2 Defining Equations

Electrostriction is defined as the quadratic coupling between strain (x) and electric field (E), or between strain and polarization (P). It is a fourth-rank tensor defined by the following relationship:

$$x_{ij} = M_{ijmn} E_m E_n \tag{26.1}$$

where x_{ij} is the strain tensor, E_m and E_n components of the electric field vector, and M_{ijmn} the fourth-rank field-related electrostriction tensor. The M coefficients are defined in units of m^2/V^2.

Ferroelectrics and related materials often exhibit nonlinear dielectric properties with changing electric fields. To better express the quadratic nature of electrostriction, it is useful to define a polarization-related electrostriction coefficient Q_{ijmn}, as

$$x_{ij} = Q_{ijmn} P_m P_n \tag{26.2}$$

Q coefficients are defined in units of m^4/C^2. The M and Q coefficients are equivalent. Conversions between the two coefficients are carried out using the field-polarization relationships:

$$P_m = \eta_{mn} E_n, \quad \text{and} \quad E_n = \chi_{mn} P_m \tag{26.3}$$

where η_{mn} is the dielectric susceptibility tensor and χ_{mn} is the inverse dielectric susceptibility tensor.

Electrostriction is not a simple phenomenon but manifests itself as three thermodynamically related effects (Newnham et al., 1997). The first is the well-known variation of strain with polarization, called the direct effect $d^2x_{ij}/dE_k dE_l = M_{ijkl}$). The second is the stress (X_{kl}) dependence of the dielectric stiffness χ_{mn}, or the reciprocal dielectric susceptibility, called the first converse effect ($d\chi_{mn}/dX_{kl} = M_{mnkl}$). The third effect is the polarization dependence of the piezoelectric voltage coefficient g_{jkl}, called the second converse effect ($dg_{jkl}/dP_i = \chi_{mk}\chi_{nl}M_{ijmn}$). These effects are of importance because they offer independent and equivalent techniques for electrostriction measurements, which include the strain gage, the capacitance dilatometer, and the laser interferometer (Yimnirun et al., 2002, 2003).

Piezoelectricity and Electrostriction

Piezoelectricity is a third-rank tensor property found only in acentric materials and is absent in most materials. The noncentrosymmetric point groups generally exhibit piezoelectric effects that are larger than the electrostrictive effects and obscure them. The electrostriction coefficients M_{ijkl} or Q_{ijkl} constitute fourth-rank tensors which, like the elastic constants, are found in all insulating materials, regardless of symmetry.

Electrostriction is the origin of piezoelectricity in ferroelectric materials, in both conventional ceramic ferroelectrics such as $BaTiO_3$ as well as in organic polymer ferroelectrics such as PVDF copolymers (Furukawa and Seo, 1990). In a ferroelectric material that exhibits both spontaneous and induced polarizations, P^s_i and P'_l, the strains arising from spontaneous polarizations, piezoelectricity, and electrostriction may be formulated as

$$x_{ij} = Q_{ijkl}P^s_k P^s_l + 2Q_{ijkl}P^s_k P'_l + Q_{ijkl}P'_k P'_l \tag{26.4}$$

In the paraelectric state, we may express the strain as $x_{ij} = Q_{ijkl}P_k P_l$, so that $dx_{ij}/dP_k = g_{ijk} = 2Q_{ijkl}P_l$. Converting to the commonly used d_{ijk} coefficients:

$$d_{ijk} = \chi_{mk}\, g_{ijm} = 2\chi_{mk} Q_{ijmn} P_n \tag{26.5}$$

This origin of piezoelectricity in electrostriction provides us an avenue into nonlinearity. In this case, it is the ability to tune the piezoelectric coefficient and the dielectric behavior of a transducer. The piezoelectric coefficient varies with the polarization induced in the material, and may be controlled by an applied electric field. The electrostrictive element may be tuned from an inactive to a highly active state. The electrical impedance of the element may be tuned by exploiting the dependence of permittivity on the biasing field for these materials, and the saturation of polarization under high fields (Newnham, 1990).

Electrostriction and Compliance Matrices

The fourth-rank electrostriction tensor is similar to the elastic compliance tensor, but is not identical. Compliance is a more symmetric fourth-rank tensor than is electrostriction. For compliance, in the most general case:

$$s_{ijkl} = s_{jikl} = s_{ijlk} = s_{jilk} = s_{klij} = s_{lkij} = s_{klji} = s_{lkij} \tag{26.6}$$

but for electrostriction

$$M_{ijkl} = M_{jikl} = M_{ijlk} = M_{jilk} \neq M_{klij} = M_{lkij} = M_{klji} = M_{lkij} \qquad (26.7)$$

This means that for most point groups the number of independent electrostriction coefficients exceeds those for elasticity. M and Q coefficients may also be defined in a matrix (Voigt) notation. The electrostriction and elastic compliance matrices for point groups 6/mmm and ∞/mm are compared below:

$$\begin{bmatrix} s_{11} & s_{12} & s_{13} & 0 & 0 & 0 \\ s_{12} & s_{11} & s_{13} & 0 & 0 & 0 \\ s_{13} & s_{13} & s_{11} & 0 & 0 & 0 \\ 0 & 0 & 0 & s_{44} & 0 & 0 \\ 0 & 0 & 0 & 0 & s_{44} & 0 \\ 0 & 0 & 0 & 0 & 0 & 2(s_{44}-s_{12}) \end{bmatrix} \begin{bmatrix} M_{11} & M_{12} & M_{13} & 0 & 0 & 0 \\ M_{12} & M_{11} & M_{13} & 0 & 0 & 0 \\ M_{31} & M_{31} & M_{33} & 0 & 0 & 0 \\ 0 & 0 & 0 & M_{44} & 0 & 0 \\ 0 & 0 & 0 & 0 & M_{44} & 0 \\ 0 & 0 & 0 & 0 & 0 & (M_{11}-M_{12}) \end{bmatrix}$$

Compliance coefficients s_{13} and s_{31} are equal, but M_{13} and M_{31} are not. The difference arises from an energy argument which requires the elastic constant matrix to be symmetric.

It is possible to define sixth-rank and higher-order electrostriction coupling coefficients. The electrostriction tensor can also be treated as a complex quantity, similar to the dielectric and the piezoelectric tensors. The imaginary part of the electrostriction is also a fourth-rank tensor. Our discussion is confined to the real part of the quadratic electrostriction tensor.

Magnitudes and Signs of Electrostrictive Coefficients

The values of M coefficients range from about 10^{-24} m^2/V^2 in low-permittivity materials to 10^{-16} m^2/V^2 in high-permittivity actuator materials made from relaxor ferroelectrics such as PMN–lead titanate (PMN–PT) compositions. Large strains of the order of strains in ferroelectric piezoelectric materials such as lead zirconate titanate (PZT) may be induced in these materials. Q values vary in an opposite way to M values. Q ranges from 10^{-3} m^4/C^2 in relaxor ferroelectrics to greater than 1 m^4/C^2 in low-permittivity materials. Since the strain is directly proportional to the square of the induced polarization, it is also proportional to the square of the dielectric permittivity. This implies that materials with large dielectric permittivities, like relaxor ferroelectrics, can produce large strains despite having small Q coefficients.

As a consequence of the quadratic nature of the electrostriction effect, the sign of the strain produced in the material is independent of the polarity of the field. This is in contrast with linear piezoelectricity where reversing the direction of the field causes a change in the sign of the strain. The sign of the electrostrictive strain depends only on the sign of the electrostriction coefficient. In most oxide ceramics, the longitudinal electrostriction coefficients are positive. The transverse coefficients are negative as expected from Poisson ratio effects. Another consequence is that electrostrictive strain occurs at twice the frequency of an applied ac field. In acentric materials, where both piezoelectric and electrostrictive strains may be observed, this fact is very useful in separating the strains arising from piezoelectricity and from electrostriction.

26.3 PMN–PT: A Prototype Electrostrictive Material

Most commercial applications of electrostriction involve high-permittivity materials such as relaxor ferroelectrics. PMN ($Pb(Mg_{1/3}Nb_{2/3})O_3$) relaxor ferroelectric compounds were first synthesized more than 40 years ago. Since then, the PMN system has been well characterized in both single-crystal and ceramic forms, and may be considered the prototype ferroelectric electrostrictor (Jang et al., 1980). Lead titanate ($PbTiO_3$, PT) and other materials are commonly added to PMN to shift T_{max} or increase the maximum dielectric constant. The addition of PT to PMN gives rise to a range of compositions, the PMN–PT system, that have a higher Curie range and superior electromechanical coupling coefficients. The addition of other oxide compounds, mostly other ferroelectrics, is a widely used method to tailor the electromechanical properties of electrostrictors (Voss et al., 1983). Some properties of the PMN–PT system are presented here.

Based on dielectric constant vs. temperature plots, the electromechanical behavior of a relaxor ferroelectric may be divided into three regimes (Figure 26.1). At temperatures less than T_d, the depolarization temperature, the relaxor material is macropolar, exhibits a stable remanent polarization, and behaves as a piezoelectric. T_{max} is the temperature at which the maximum dielectric constant is observed. Between T_d and T_{max}, the material possesses nanometer-scale microdomains that strongly influence the electromechanical behavior. Large dielectric permittivities and large electrostrictive strains arising from micro-macrodomain reorientation are observed. Above T_{max}, the material is a "true electrostrictor" in that it is paraelectric and exhibits nonhysteretic, quadratic strain-field behavior. Since macroscale domains are absent, no remanent strain is observed. Improved reproducibility in strain and low-loss behavior are achieved.

Figure 26.2 illustrates the quadratic dependence of the transverse strain on the induced polarization for ceramic 0.9PMN–0.1PT. Figure 26.3(a) and (b) show the longitudinal strain as a function of the applied

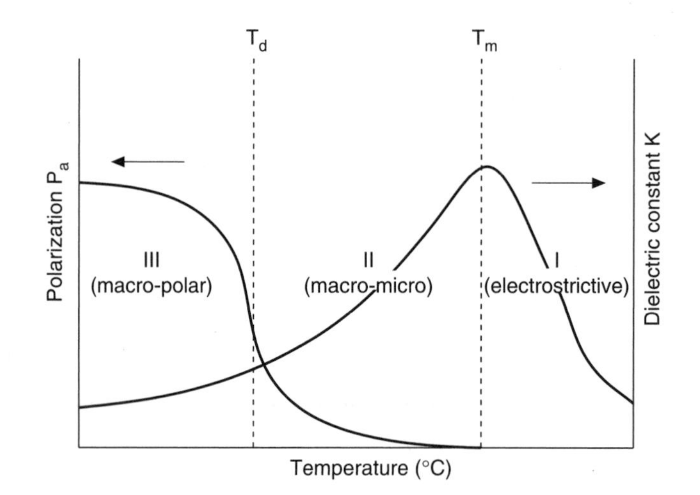

FIGURE 26.1 Polarization and dielectric behavior of a relaxor ferroelectric as a function of temperature, showing the three temperature regimes.

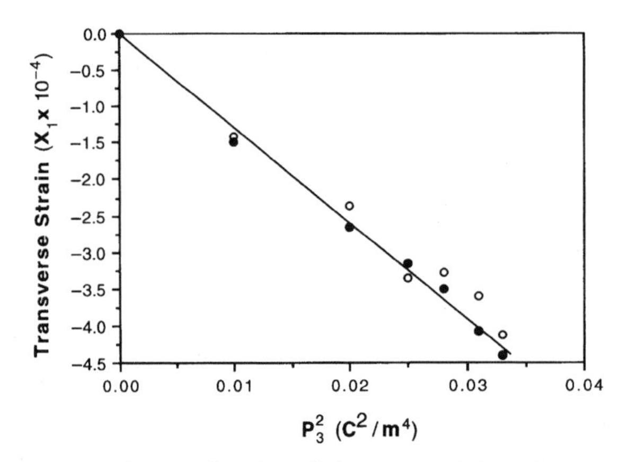

FIGURE 26.2 Transverse strain as a function of the square of the polarization in ceramic 0.9PMN–0.1PT, at *RT*. The quadratic ($x = QP^2$) nature of electrostriction is illustrated. Shaded circles indicate strain measured while increasing polarization and unshaded circles indicate decreasing polarization.

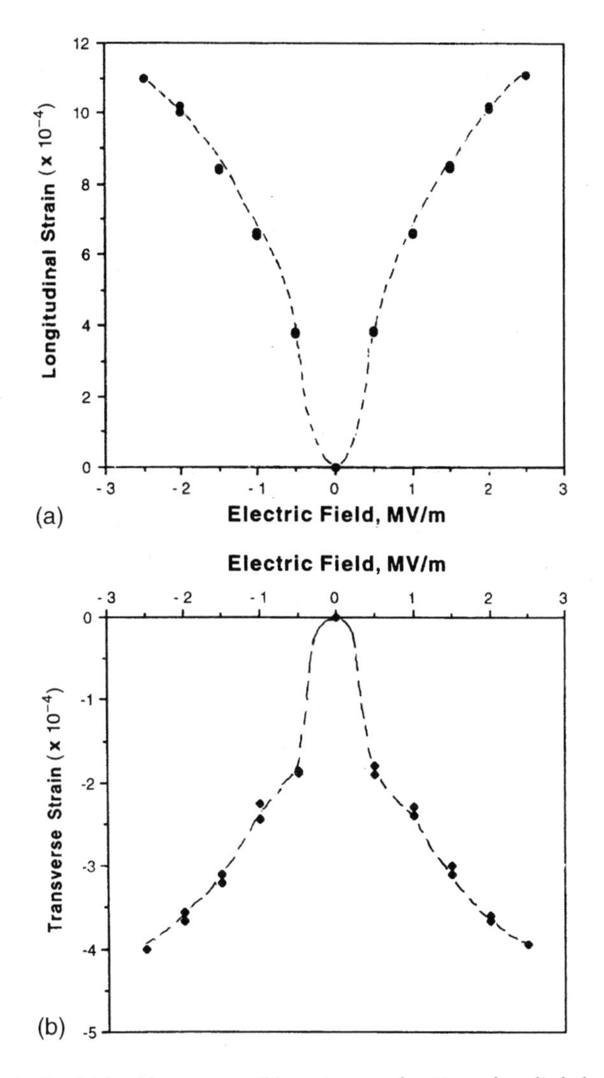

FIGURE 26.3 Longitudinal (a) and transverse (b) strains as a function of applied electric field in 0.9PMN–0.1PT, at *RT*. x is not quadratic with E except at low fields.

electric field for the same composition. The strain-field plots are not quadratic, and illustrate essentially anhysteretic nature of electrostrictive strain. The transverse strain is negative, as expected.

The averaged longitudinal and transverse electrostriction coefficients have been measured for poled ceramic PMN to be $Q_{33} \sim 2.3 \times 10^{-2}$ m^4/C^2, $Q_{13} \sim -0.64 \times 10^{-2}$ m^4/C^2. The corresponding field-related coefficients are $M_{33} \sim 1.50 \times 10^{-16}$ m^2/V^2 and $M_{13} \sim -4.19 \times 10^{-17}$ m^2/V^2. Induced strains of the order of 10^{-4} may be achieved with moderate electric fields of ~40 kV/cm. These strains are much larger than thermal expansion strains, and are in fact equivalent to thermal expansion strains induced by a temperature change of ~1000°C. M_{33} values for some other common ferroelectrics, P(VDF–TrFE) copolymers, and nonferroelectric low-permittivity dielectrics are listed in Table 26.1.

The mechanical quality factor Q_M for PMN is 8100 (at a field of ~200 kV/m) compared with 300 for poled barium titanate or 75 for poled PZT 5-A (Nomura and Uchino, 1983) The induced piezoelectric coefficients d_{33} and d_{31} can vary with field (Figure 26.4). The maxima in the induced piezoelectric coefficients for PMN as a function of biasing electric field are at $E \sim 1.2$ MV/m, with $d_{33} = 240$ pC/N and $-d_{31} = 72$ pC/N. $Pb(Mg_{0.3}Nb_{0.6}Ti_{0.1})O_3$ is a very active composition, with a maximum $d_{33} = 1300$ pC/N at a biasing field of 3.7 kV/cm.

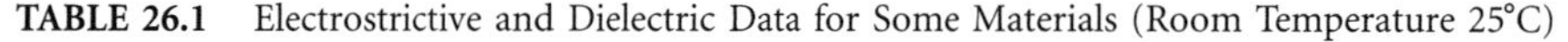

TABLE 26.1 Electrostrictive and Dielectric Data for Some Materials (Room Temperature 25°C)

Composition	$M_{33} \times 10^{-17}$ (m^2/V^2)	Dielectric Constant (K)	Ref.
$Pb(Mg_{1/3}Nb_{2/3})O_3$ (PMN)	15.04	9140	Nomura and Uchino, 1983
$(Pb_{1-x}La_{2x/3})(Zr_{1-y}Ti_y)O_3$ (PLZT 11/65/35)	1.52	5250	Landolt-Bornstein, 1979, 1984
$BaTiO_3$ (poled)	1.41	1900	Nomura and Uchino, 1983
$PbTiO_3$	1.65	1960	Landolt-Bornstein, 1979, 1984
$SrTiO_3$	5.61×10^{-2}	247	Landolt-Bornstein, 1979, 1984
PVDF/TrFE copolymer (70/30)	−43	12	Elhami et al., 1995
PVDF/TrFE copolymer (50/50) (e^--irradiated)	-1.4×10^{-1}	45	Zhang et al., 1998
NaCl (single crystal) (M_{11})	5.65×10^{-4}	5.9	Yimnirun, 2001
CaF_2 (single crystal) (M_{11})	0.66×10^{-4}	6.8	Yimnirun, 2001
Al_2O_3 (ceramic)	0.57×10^{-4}	10	Yimnirun, 2001
MgO (ceramic)	0.67×10^{-4}	8.1	Yimnirun, 2001

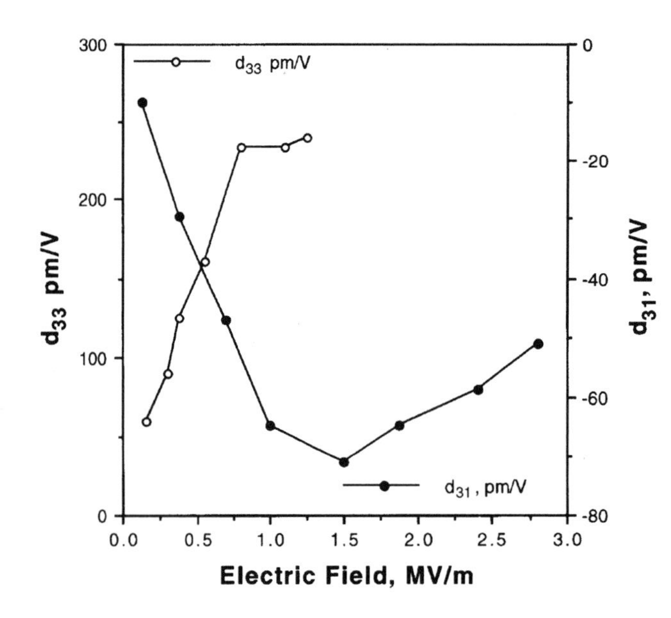

FIGURE 26.4 Induced piezoelectric coefficients d_{33} and $-d_{31}$ as a function of applied biasing field for ceramic PMN, 18°C.

26.4 Applications

The advantages that electrostrictors have over other actuator materials include low hysteresis of the strain-field response, no remanent strain (walk off), reduced aging and creep effects, a high response speed (<10 msec), and strain values ($>0.03\%$) achievable at realizable electric fields. Displacement ranges of several tens of microns may be achieved with ± 0.01 μ reproducibility. Most actuator applications of electrostrictors as servotransducers and micropositioning devices take advantage of these characteristics (Uchino, 1986).

In selecting electrostrictive relaxor ferroelectrics for actuator and sensor applications, the following criteria are commonly used. A large dielectric constant and field stability in the K vs. E relations are useful in achieving large electrostrictive strains. These criteria also lead to large induced polarizations and large induced piezoelectric coefficients through the second converse effect.

Broad dielectric transitions allow for a large operating temperature range. In the case of relaxors, this implies a large difference between T_{max} and T_d. Minimal E–P hysteresis and no remanent polarization are useful in achieving a low-loss material that is not susceptible to joule heating effects. These factors are listed in Table 26.2.

TABLE 26.2 Selection Criteria for Relaxor Ferroelectrics for Electrostrictive Devices

Desirable Properties	Material Behavior
• Large strain, induced polarization, and induced piezoelectricity	• Large dielectric constants
• Large operating temperature range	• $T_{max}-T_d$ is large
• Broad dielectric transition	• Broad dielectric transition
• Low-loss, low-joule heating, minimal hysteresis, no remanent polarization	• Operation in paraelectric regime ($T < T_{max}$)

Mechanical applications range from stacked actuators through inchworms, microangle adjusting devices, and oil pressure servovalves (Uchino, 1996). Multilayer actuators produce large displacements and high forces at low drive voltages. The linear change in capacitance with applied stress of an electrostrictor can be used as a capacitive stress gauge. Electrostrictors may also be used in field-tunable piezoelectric transducers (Sundar and Newnham, 1992). Recently, electrostrictive materials have been integrated into ultrasonic motors and novel flextensional transducers (Uchino, 1996; Meyer et al., 2001). Electrostrictors have also been integrated into "smart" optical systems such as bistable optical devices, interferometric dilatometers, and deformable mirrors. Electrostrictive correction of optical aberrations is a significant tool in active optics. Electrostrictors also find applications in "very smart" systems such as sensor-actuator active vibration-suppression elements (Uchino and Giniewicz, 2003). A shape memory effect arising from inverse hysteretic behavior (Meyer and Newnham, 2000) and electrostriction in PZT family antiferroelectrics is also of interest (Uchino, 1985).

26.5 Summary

Electrostriction is a fundamental electromechanical coupling effect. In ceramics with large dielectric constants and in some polymers, large electrostrictive strains may be induced that are comparable in magnitude with piezoelectric strains in actuator materials such as PZT. The converse electrostrictive effect, which is the change in dielectric susceptibility with applied stress, facilitates the use of the electrostrictor as a stress gauge. The second converse effect may be used to tune the piezoelectric coefficients of the material as a function of the applied field. Electrostrictive materials offer tunable nonlinear properties that are suitable for application in very smart systems.

Defining Terms

Elastic compliance: A fourth-rank tensor (s_{ijkl}) relating the stress (X) applied on a material and the strain (x) developed in it, $x_{ij} = s_{ijkl} X^{kl}$. Its inverse is the elastic stiffness tensor (c_{ijkl}).

Electrostriction: The quadratic coupling between strain and applied field or induced polarization. Conversely, it is the linear coupling between dielectric susceptibility and applied stress. It is present in all insulating materials.

Ferroelectricity: The phenomenon by which a material exhibits a permanent spontaneous polarization that can be reoriented (switched) between two or more equilibrium positions by the application of a realistic electric field (i.e., less than the breakdown field of the material).

Perovskite: A crystal structure with the formula ABO_3, with A atoms at the corners of a cubic unit cell, B atoms at the body-center position, and O atoms at the centers of the faces. Many oxide perovskites are used as transducers, capacitors, and thermistors.

Piezoelectricity: The induction of polarization when stress is applied in acentric materials. The converse effect is the linear coupling between applied electric field and induced strain.

Relaxor ferroelectric: Relaxor ferroelectric materials exhibit a diffuse phase transition between paraelectric and ferroelectric phases, and a frequency dependence of the dielectric properties.

Smart and very smart systems: A system that can sense a change in its environment, and tune its response suitably to the stimulus. A system that is only smart can sense a change in its environment and react to it.

References

W.G. Cady, *International Critical Tables*, vol. 6, p. 207, 1929.

K. Elhami, B. Gauthier-Manuel, J.F. Manceau, and F. Bastien, *J. Appl. Phys.*, vol. 77, p. 3987, 1995.

T. Furukawa and N. Seo, *Jpn. J. Appl. Phys.*, vol. 29, p. 675, 1990.

S.J. Jang, K. Uchino, S. Nomura, and L.E. Cross, *Ferroelectrics*, vol. 27, p. 31, 1980.

Landolt-Bornstein, *Numerical Data and Functional Relationships in Science and Technology*, New Series, Gruppe III, vols. 11 and 18, Berlin: Springer, 1979, 1984.

R.J. Meyer Jr. and R.E. Newnham, *J. Intell. Mater. Syst. Struct.*, vol. 11, p. 199, 2000.

R.J. Meyer Jr., A. Dogan, C. Yoon, S.M. Pilgrim, and R.E. Newnham, *Sensor Actuat. A-Phys.*, vol. 87, p. 157, 2001.

R.E. Newnham, *Chemistry of Electronic Ceramic Materials*, in *Proc. Int. Conf.*, Jackson, WY, 1990; NIST Special Publication vol. 804, p. 39, 1991.

R.E. Newnham, V. Sundar, R. Yimnirun, J. Su, and Q.M. Zhang, *J. Phys. Chem. B*, vol. 101, p. 10141, 1997.

S. Nomura and K. Uchino, *Ferroelectrics*, vol. 50, p. 197, 1983.

V. Sundar and R.E. Newnham, *Ferroelectrics*, vol. 135, p. 431, 1992.

K. Uchino, *Jpn. J. Appl. Phys.*, vol. 24, p. 460, 1985.

K. Uchino, *Ceramic Bull.*, vol. 65, p. 647, 1986.

K. Uchino, *MRS Bull.*, vol. 18, p. 42, 1993.

K. Uchino, *Piezoelectric Actuators and Ultrasonic Motors*, Dordrecht, The Netherlands: Kluwer, 1996, Chaps. 7–9.

K. Uchino and J. Giniewicz, *Micromechatronics*, New York: Marcel Dekker, 2003, Chaps. 8–11.

D.J. Voss, S.L. Swartz, and T.R. Shrout, *Ferroelectrics*, vol. 50, p. 1245, 1983.

R. Yimnirun, Ph.D. thesis, The Pennsylvania State University, University Park, 2001.

R. Yimnirun, P.J. Moses, R.E. Newnham, and R.J. Meyer Jr., *J. Electroceram.*, vol. 8, p. 87, 2002.

R. Yimnirun, P.J. Moses, R.E. Newnham, and R.J. Meyer Jr., *Rev. Sci. Instrum.*, vol. 74, p. 3429, 2003.

Q.M. Zhang, V. Bharti, and X. Zhao, *Science*, vol. 280, p. 2101, 1998.

Further Information

IEEE Proceedings of the International Symposium on the Applications of Ferroelectrics (ISAF).

IEEE Transactions on Ultrasonics, Ferroelectrics, and Frequency Control (UFFC).

American Institute of Physics Handbook, 3rd ed., New York: McGraw-Hill, 1972.

M.E. Lines and A.M. Glass, *Principles and Applications of Ferroelectric Materials*, Oxford: Clarendon Press, 1977.

27

The Hall Effect

Alexander C. Ehrlich
U.S. Naval Research Laboratory

27.1 Introduction

The Hall effect is a phenomenon that arises when an electric current and magnetic field are simultaneously imposed on a conducting material. Specifically, in a flat plate conductor, if a current density, J_x, is applied in the x direction and (a component of) a magnetic field, B_z, in the z direction, then the resulting electric field, E_y, transverse to J_x and B_z is known as the Hall electric field E_{H} (see Figure 27.1) and is given by

$$E_y = RJ_xB_z \tag{27.1}$$

where R is known as the Hall coefficient. The Hall coefficient can be related to the electronic structure and properties of the **conduction bands** in metals and semiconductors and historically has probably been the most important single parameter in the characterization of the latter. Some authors choose to discuss the Hall effect in terms of the Hall angle, ϕ, shown in Figure 27.1, which is the angle between the net electric field and the imposed current. Thus,

$$\tan\phi = E_{\mathrm{H}}/E_x \tag{27.2}$$

For the vast majority of Hall effect studies that have been carried out, the origin of E_{H} is the Lorentz force, F_{L}, that is exerted on a charged particle as it moves in a magnetic field. For an electron of charge e with velocity v, F_{L} is proportional to the vector product of **v** and **B**; that is:

$$F_{\mathrm{L}} = e\,\mathbf{v}\,\mathbf{x}\,\mathbf{B} \tag{27.3}$$

In these circumstances a semiclassical description of the phenomenon is usually adequate. This description combines the classical Boltzmann transport equation with the Fermi–Dirac distribution function for the charge carriers (electrons or holes) (Ziman, 1960), and this is the point of view that will be taken in this chapter. Examples of Hall effect that cannot be treated semiclassically are the spontaneous (or extraordinary) Hall effect that occurs in ferromagnetic conductors (Berger and Bergmann, 1980), the quantum Hall effect (Prange and Girvin, 1990), and the Hall effect that arises in conjuction with hopping conductivity (Emin, 1977).

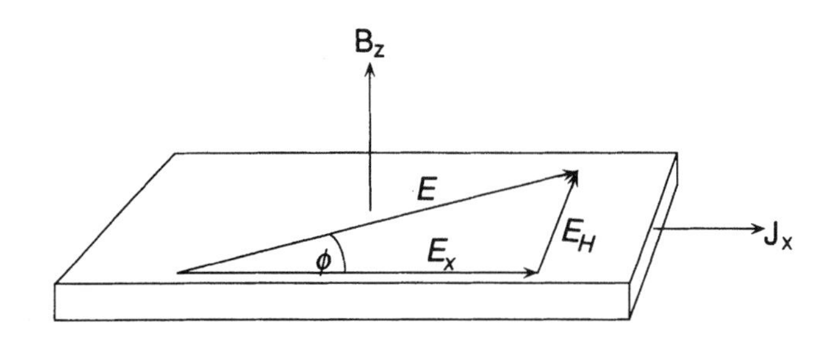

FIGURE 27.1 Typical Hall effect experimental arrangement in a flat plate conductor with current J_x and magnetic field B_z. The Hall electric field $E_H = E_y$ in this geometry arises because of the Lorentz force on the conducting charges and is of just such a magnitude that in combination with the Lorentz force there is no net current in the y direction. The angle ϕ between the current and net electric field is called the Hall angle.

In addition to its use as an important tool in the study of the nature of electrically conducting materials, the Hall effect has a number of direct practical applications. For example, the sensor in some commercial devices for measuring the magnitude and orientation of magnetic fields is a Hall sensor. The spontaneous Hall effect has been used as a nondestructive method for exploring the presence of defects in steel structures. The quantum Hall effect has been used to refine our knowledge of the magnitudes of certain fundamental constants such as the ratio of e^2/h where h is Planck's constant.

27.2 Theoretical Background

The Boltzmann equation for an electron gas in a homogeneous, isothermal material that is subject to constant electric and magnetic fields is (Ziman, 1960):

$$e[\mathbf{E} + \mathbf{vXB}]\left(\frac{1}{\hbar}\right)\nabla_k f(\mathbf{k}) - \left(\frac{\partial f}{\partial t}\right)_s = 0 \tag{27.4}$$

Here, **k** is the quantum mechanical wave vector, $\hbar$ is Planck's constant divided by 2π, t is the time, f is the electron distribution function, and s is meant to indicate that the time derivative of f is a consequence of scattering of the electrons. In static equilibrium (E = 0, B = 0) f is equal to f_0 and f_0 is the Fermi–Dirac distribution function:

$$f_0 = \frac{1}{e^{(\mathscr{E}(\mathbf{k})-\zeta)/KT} + 1} \tag{27.5}$$

where $\mathscr{E}(\mathbf{k})$ is the energy, ζ is the chemical potential, K is Boltzmann's constant, and T is the temperature. Each term in Equation (27.4) represents a time rate of change of f and in dynamic equilibrium their sum has to be zero. The last term represents the effect of collisions of the electrons with any obstructions to their free movement such as lattice vibrations, crystallographic imperfections, and impurities. These collisions are usually assumed to be representable by a relaxation time, $\tau(\mathbf{k})$; that is:

$$\left(\frac{\partial f}{\partial t}\right)_c = \frac{-(f - f_0)}{\tau(\mathbf{k})} = \frac{(\partial f_0/\partial \mathscr{E})g(\mathbf{k})}{\tau(\mathbf{k})} \tag{27.6}$$

where $f - f_0$ is written as $(\partial f_0/\partial \varepsilon)g(\mathbf{k})$, which is essentially the first term in an expansion of the deviation of f from its equilibrium value, f_0. Equation (27.6) and Equation (27.4) can be combined to give

$$e[\mathbf{E} + \mathbf{vXB}]\frac{1}{\hbar}\nabla_{\mathbf{k}} f(\mathbf{k}) = \frac{(\partial f_0/\partial \mathscr{E})g(\mathbf{k})}{\tau(\mathbf{k})} \tag{27.7}$$

If Equation (27.7) can be solved for $g(\mathbf{k})$, then expressions can be obtained for both the E_H and the magnetoresistance (the electrical resistance in the presence of a magnetic field). Solutions can in fact be developed that are linear in the applied electric field (the regime where Ohm's law holds) for two physical situations: (i) when $\omega_c\tau << 1$ (Hurd, 1972, p. 69) and (ii) when $\omega_c\tau >> 1$ (Lifshitz et al., 1956; Hurd, 1972) where $\omega_c = Be/m$ is the cyclotron frequency. Situation (ii) means the electron is able to complete many cyclotron orbits under the influence of **B** in the time between scatterings and is called the high (magnetic) field limit. Conversely, situation (i) is obtained when the electron is scattered in a short time compared to the time necessary to complete one cyclotron orbit and is known as the low field limit. In effect, the solution to Equation (27.7) is obtained by expanding $g(\boldsymbol{k})$ in a power series in $\omega_c\tau$ or $1/\omega_c\tau$ for (i) and (ii), respectively. Given $g(\mathbf{k})$ the current vector, J_l $(l = x,y,z)$ can be calculated from (Blatt, 1957):

$$J_l = \left(\frac{e}{4\pi^3}\right)\int v_l(\mathbf{k})g(\mathbf{k})(\partial f_0/\partial \mathscr{E})d^3k \tag{27.8}$$

where $v_l(\mathbf{k})$ is the velocity of the electron with wave vector **k.** Every term in the series defining J_l is linear in the applied electric field, **E,** so that the conductivity tensor $\sigma_{\rm lm}$ is readily obtained from $J_l = \sigma_{lm}E_m$ (Hurd, 1972, p. 9). This matrix equation can be inverted to give $E_l = \rho_{\rm lm}J_{\rm m}$. For the same geometry used in defining Equation (27.1):

$$E_y = E_{\rm H} = \rho_{yx}J_x \tag{27.9}$$

where ρ_{21} is a component of the resistivity tensor sometimes called the Hall resistivity. Comparing Equation (27.1) and Equation (27.9) it is clear that the B dependence of $E_{\rm H}$ is contained in ρ_{12}. However, nothing in the derivation of ρ_{12} excludes the possibility of terms to the second or higher powers in B. Although these are usually small, this is one of the reasons that experimentally one usually obtains R from the measured transverse voltage by reversing magnetic fields and averaging the measured $E_{\rm H}$ by calculating $(1/2)[E_{\rm H}(\mathbf{B}) - E_{\rm H}(-\mathbf{B})]$. This eliminates the second-order term in B and in fact all even power terms contributing to the $E_{\rm H}$. Using the Onsager relation (Smith and Jensen, 1989, p. 60) $\rho_{12}(\mathbf{B}) = \rho_{21}(-\mathbf{B})$, it is also easy to show that in terms of the Hall resistivity:

$$R = \frac{1}{2}\frac{1}{B}[\rho_{12}(\mathbf{B}) + \rho_{21}(\mathbf{B})] \tag{27.10}$$

Strictly speaking, in a single crystal the electric field resulting from an applied electric current and magnetic field, both of arbitrary direction relative to crystal axes and each other, cannot be fully described in terms of a second-order resistivity tensor (Hurd, 1972, p. 71). However, Equation (27.1), Equation (27.9), and Equation (27.10) do define the Hall coefficient in terms of a second-order resistivity tensor for a polycrystalline (assumed isotropic) sample or for a cubic single crystal or for a lower symmetry crystal when the applied fields are oriented along major symmetry directions. In real-world applications the Hall effect is always treated in this manner.

27.3 Relation to the Electronic Structure—(i) $\omega_c\tau \ll 1$

General expressions for R in terms of the parameters that describe the electronic structure can be obtained using Equation (27.7) to Equation (27.10) and have been given by Blatt (1957) for the case of crystals having cubic symmetry. An even more general treatment has been given by McClure(1956). Here, the discussion of specific results will be restricted to the free electron model wherein the material is assumed to have one or more conducting bands, each of which has a quadratic dispersion relationship connecting E and **k**; that is:

$$\mathscr{E}_i = \frac{\hbar^2 k_i^2}{2m_i} \tag{27.11}$$

where the subscript specifies the band number and m_i, the **effective mass** for each band. These masses need not be equal nor the same as the free electron mass. In effect, some of the features lost in the free electron approximation are recovered by allowing the masses to vary. The **relaxation times**, τ_i, will also be taken to be isotropic (not **k** dependent) within each band but can be different from band to band. Although extreme, these approximations are often qualitatively correct, particularly in polycrystalline materials, which are macroscopically isotropic. Further, in semiconductors these results will be strictly applicable only if τ_i is energy independent as well as isotropic.

For a single spherical band, R_H is a direct measure of the number of current carriers and turns out to be given by (Blatt, 1957):

$$R_H = \frac{1}{ne} \tag{27.12}$$

where n is the number of conduction carriers/volume. R_H depends on the sign of the charge of the current carriers being negative for electrons and positive for **holes**. This identification of the carrier sign is itself a matter of great importance, particularly in semiconductor physics. If more than one band is involved in electrical conduction, then by imposing the boundary condition required for the geometry of Figure 27.1 that the total current in the y direction from all bands must vanish, $J_y = 0$, it is easy to show that (Wilson, 1958):

$$R_H = (1/\sigma)^2 \sum [\sigma_i^2 R_i] \tag{27.13}$$

where R_i and σ_i are the Hall coefficient and electrical conductivity, respectively, for the ith band ($\sigma_i = n_i e^2 \tau_i / m_i$), $\sigma = \Sigma \sigma_i$ is the total conductivity of the material, and the summation is taken over all bands. Using Equation (27.12), Equation (27.13) can also be written:

$$R_H = \frac{1}{e n_{\text{eff}}} = \frac{1}{e} \sum \left[\frac{1}{n_i} \left(\frac{\sigma_i}{\sigma} \right)^2 \right] \tag{27.14}$$

where n_{eff} is the effective or apparent number of electrons determined by a Hall effect experiment. (Note that some workers prefer representing Equation (27.13) and Equation (27.14) in terms of the current carrier mobility for each band, μ_i, defined by $\sigma_i = n_i e \mu_i$.)

The most commonly used version of Equation (27.14) is the so-called two-band model, which assumes that there are two spherical bands with one composed of electrons and the other of holes. Equation (27.14) then takes the form:

$$R_H = \frac{1}{e} \left[\frac{1}{n_e} \left(\frac{\sigma_e}{\sigma} \right)^2 - \frac{1}{n_h} \left(\frac{\sigma_h}{\sigma} \right)^2 \right] \tag{27.15}$$

From Equation (27.14) or Equation (27.15) it is clear that the Hall effect is dominated by the most highly conducting band. Although for fundamental reasons it is often the case that $n_e = n_h$ (a so-called compensated material), R_H would rarely vanish since the conductivities of the two bands would rarely be identical. It is also clear from any of Equation (27.12), Equation (27.14), or Equation (27.15) that, in general, the Hall effect in semiconductors will be orders of magnitude larger than that in metals.

27.4 Relation to the Electronic Structure—(ii) $\omega_c \tau \gg 1$

The high field limit can be achieved in metals only in pure, crystalographically well-ordered materials and at low temperatures, which circumstances limit the electron scattering rate from impurities, crystallographic imperfections, and lattice vibrations, respectively. In semiconductors, the much longer relaxation time and smaller effective mass of the electrons makes it much easier to achieve the high field limit. In this limit the result analogous to Equation (27.15) is (Blatt, 1968, p. 290):

$$R_H = \frac{1}{e}\frac{1}{n_e - n_h} \tag{27.16}$$

Note that the individual band conductivities do not enter in Equation (27.16). Equation (27.16) is valid provided the cyclotron orbits of the electrons are closed for the particular direction of **B** used. It is not necessary that the bands be spherical or the τ's isotropic. Also, for more than two bands R_H depends only on the net difference between the number of electrons and the number of holes. For the case where $n_e = n_h$, in general, the lowest order dependence of the Hall electric field on B is B^2 and there is no simple relationship of R_H to the number of current carriers. For the special case of the two-band model, however, R_H is a constant and is of the same form as Equation (27.15) (Fawcett, 1964).

Metals can have geometrically complicated Fermi surfaces wherein the Fermi surface contacts the Brillouin zone boundary as well as encloses the center of the zone. This leads to the possibility of open electron orbits in place of the closed cyclotron orbits for certain orientations of **B**. In these circumstances R can have a variety of dependencies on the magnitude of B and in single crystals will generally be dependent on the exact orientation of **B** relative to the crystalline axes (Fawcett, 1964; Hurd, 1972, p. 51). R will not, however, have any simple relationship to the number of current carriers in the material.

Semiconductors have too few electrons to have open orbits but can manifest complicated behavior of their Hall coefficient as a function of the magnitude of B. This occurs because of the relative ease with which one can pass from the low field limit to the high field limit and even on to the so-called quantum limit with currently attainable magnetic fields. (The latter has not been discussed here.) In general, these different regimes of B will not occur at the same magnitude of B for all the bands in a given semiconductor, further complicating the dependence of R on B.

Defining Terms

Conducting band: The band in which the electrons primarily responsible for the electric current are found.

Effective mass: An electron in a lattice responds differently to applied fields than would a free electron or a classical particle. One can, however, often describe a particular response using classical equations by defining an effective mass whose value differs from the actual mass. For the same material the effective mass may be different for different phenomena; e.g., electrical conductivity and cyclotron resonance.

Electron band: A range or band of energies in which there is a continuum (rather than a discrete set as in, for example, the hydrogen atom) of allowed quantum mechanical states partially or fully occupied by electrons. It is the continuous nature of these states that permits them to respond almost classically to an applied electric field.

Hole or hole state: When a conducting band, which can hold two electrons/unit cell, is more than half full, the remaining unfilled states are called *holes*. Such a band responds to electric and magnetic fields as if it contained positively charged carriers equal in number to the number of holes in the band.

Relaxation time: The time for a distribution of particles, out of equilibrium by a measure Φ, to return exponentially toward equilibrium to a measure Φ/e out of equilibrium when the disequilibrating fields are removed (e is the natural logarithm base).

References

L. Berger and G. Bergmann, in *The Hall Effect and Its Applications*, C.L. Chien and C.R. Westlake, Eds., New York: Plenum Press, 1980, p. 55.

F.L. Blatt, in *Solid State Physics*, vol. 4, F. Seitz and D. Turnbull, Eds., New York: Academic Press, 1957, p. 199.

F.L. Blatt, *Physics of Electronic Conduction in Solids,* New York: McGraw-Hill, 1968, p. 290. See also N.W. Ashcroft and N.D. Mermin in *Solid State Physics,* New York: Holt, Rinehart and Winston, 1976, p. 236.

D. Emin, *Phil. Mag.*, vol. 35, p. 1189, 1977.

E. Fawcett, *Adv. Phys.*, vol. 13, p. 139, 1964.

C.M. Hurd, *The Hall Effect in Metals and Alloys,* New York: Plenum Press, 1972, p. 69.

I.M. Lifshitz, M.I. Azbel, and M.I. Kaganov, *Zh. Eksp. Teor. Fiz.*, vol. 31, p. 63, 1956 (*Soviet Phys. JETP* [Engl. Trans.], vol. 4, p. 41, 1956).

J.W. McClure, *Phys. Rev.*, vol. 101, p. 1642, 1956.

R.E. Prange and S.M. Girvin, Eds., *The Quantum Hall Effect,* New York: Springer-Verlag, 1990.

H. Smith and H.H. Jensen, *Transport Phenomena,* Oxford: Oxford University Press, 1989, p. 60.

A.H. Wilson, *The Theory of Metals,* London: Cambridge University Press, 1958, p. 212.

J.M. Ziman, *Electrons and Phonons,* London: Oxford University Press, 1960. See also N.W. Ashcroft and N.D. Mermin in *Solid State Physics,* New York: Holt, Rinehart and Winston, 1976, chapters 12 and 16.

Further Information

In addition to the texts and review article cited in the references, an older but still valid article by J. P. Jan, in *Solid State Physics* (edited by F. Seitz and D. Turnbull, New York: Academic Press, 1957, p. 1) can provide a background in the various thermomagnetic and galvanomagnetic properties in metals. A parallel background for semiconductors can be found in the monograph by E. H. Putley, *The Hall Effect and Related Phenomena* (Boston, MA: Butterworths, 1960).

Examples of applications of the Hall effect can be found in the book *Hall Generators and Magnetoresistors,* by H. H. Wieder, edited by H. J. Goldsmid (London: Pion Limited, 1971).

An index to the most recent work on or using any aspect of the Hall effect reported in the major technical journals can be found in *Physics Abstracts* (Science Abstracts Series A).

28
Superconductivity

Kevin A. Delin
Jet Propulsion Laboratory

Terry P. Orlando
Massachusetts Institute of Technology

28.1 Introduction

The fundamental ideal behind all of a superconductor's unique properties is that **superconductivity** is a quantum mechanical phenomenon on a macroscopic scale created when the motions of individual electrons are correlated. According to the theory developed by John Bardeen, Leon Cooper, and Robert Schrieffer (BCS theory), this correlation takes place when two electrons couple to form a Cooper pair. For our purposes, we may therefore consider the electrical charge carriers in a superconductor to be Cooper pairs (or more colloquially, superelectrons) with a mass m^* and charge q^* twice those of normal electrons. The average distance between the two electrons in a Cooper pair is known as the coherence length, ξ. Both the coherence length and the binding energy of two electrons in a Cooper pair, 2Δ, depend upon the particular superconducting material. Typically, the coherence length is many times larger than the interatomic spacing of a solid, and so we should not think of Cooper pairs as tightly bound electron molecules. Instead, there are many other electrons between those of a specific Cooper pair allowing for the paired electrons to change partners on a time scale of $h/(2\Delta)$, where h is Planck's constant.

If we prevent the Cooper pairs from forming by ensuring that all the electrons are at an energy greater than the binding energy, we can destroy the superconducting phenomenon. This can be accomplished, for example, with thermal energy. In fact, according to the BCS theory, the critical temperature, T_c, associated with this energy is

$$\frac{2\Delta}{k_B T_c} \approx 3.5 \tag{28.1}$$

where k_B is Boltzmann's constant. For low critical temperature (conventional) superconductors, 2Δ is typically on the order of 1 meV, and we see that these materials must be kept below temperatures of about 10 K to exhibit their unique behavior. Superconductors with high critical temperature, in contrast, will superconduct up to temperatures of about 100 K, which is attractive from a practical view because the materials can be cooled cheaply using liquid nitrogen. A second way of increasing the energy of the electrons is electrically driving them. In other words, if the critical current density, J_c, of a superconductor is exceeded, the electrons have sufficient kinetic energy to prevent the formation of Cooper pairs. The necessary kinetic energy can also be generated through the induced currents created by an external magnetic field. As a result, if a superconductor is placed in a magnetic field larger than its critical field, H_c, it will return to its normal metallic state.

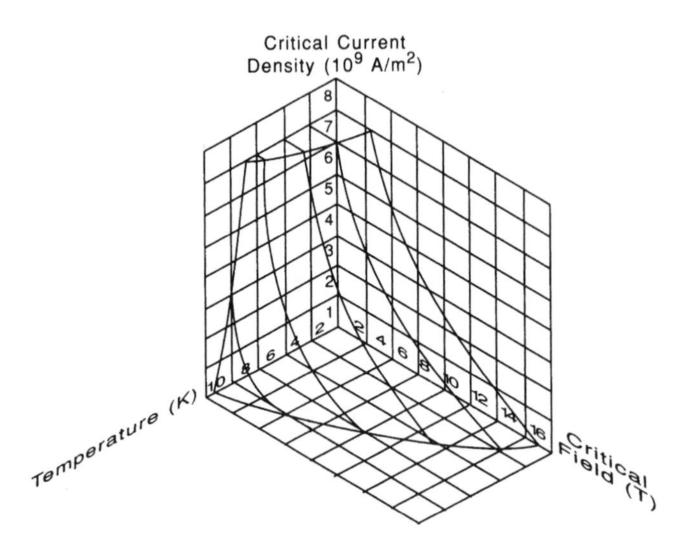

FIGURE 28.1 The phase space for the superconducting alloy niobium–titanium. The material is superconducting inside the volume of phase space indicated. (*Source:* T.P. Orlando and K.A. Delin, *Foundations of Applied Superconductivity*, Reading, MA: Addison-Wesley, 1991, p. 110. With permission. [As adapted from Wilson, 1983.])

To summarize, superconductors must be maintained under the appropriate temperature, electrical current density, and magnetic field conditions to exhibit their special properties. An example of this phase space is shown in Figure 28.1.

28.2 General Electromagnetic Properties

The hallmark electromagnetic properties of a superconductor are its ability to carry a static current without any resistance and its ability to exclude a static magnetic flux from its interior. It is this second property, known as the Meissner effect that distinguishes a superconductor from merely being a perfect conductor (which conserves the magnetic flux in its interior). Although superconductivity is a manifestly quantum mechanical phenomenon, a useful classical model can be constructed around these two properties. In this section we outline the rationale for this classical model, which is useful in engineering applications such as waveguides and high-field magnets.

The zero's dc resistance criterion implies that the superelectrons move unimpeded. The electromagnetic energy density, w, stored in a superconductor is therefore:

$$w = \frac{1}{2}\varepsilon\mathbf{E}^2 + \frac{1}{2}\mu_o\mathbf{H}^2 + \frac{n^*}{2}m^*\mathbf{v}_s^2 \tag{28.2}$$

where the first two terms are the familiar electric and magnetic energy densities, respectively. (Our electromagnetic notation is standard: ε is the permittivity, μ_o is the permeability, $\mathbf{E}$ is the electric field, and the magnetic flux density, $\mathbf{B}$, is related to the magnetic field, $\mathbf{H}$, via the constitutive law $\mathbf{B} = \mu_o\,\mathbf{H}$.) The last term represents the kinetic energy associated with the undamped superelectrons' motion (n^* and $\mathbf{v}_s$ are the superelectrons' density and velocity, respectively). Because the supercurrent density, $\mathbf{J}_s$, is related to the superelectron velocity by $\mathbf{J}_s = n^*q^*\mathbf{v}_s$, the kinetic energy term can be rewritten:

$$n^*\left(\frac{1}{2}m^*\mathbf{v}_s^2\right) = \frac{1}{2}\Lambda\mathbf{J}_s^2 \tag{28.3}$$

where Λ is defined as

$$\Lambda = \frac{m^*}{n^*(q^*)^2} \tag{28.4}$$

Assuming that all the charge carriers are superelectrons, there is no power dissipation inside the superconductor, and so Poynting's theorem over a volume V may be written:

$$-\int_V \nabla \cdot (\mathbf{E} \times \mathbf{H}) \mathrm{dv} = \int_V \frac{\partial w}{\partial t} \mathrm{dv} \tag{28.5}$$

where the left side of the expression is the power flowing into the region. By taking the time derivative of the energy density and appealing to Faraday's and Ampère's laws to find the time derivatives of the field quantities, we find that the only way for Poynting's theorem to be satisfied is if

$$\mathbf{E} = \frac{\partial}{\partial t}(\Lambda \mathbf{J}_s) \tag{28.6}$$

This relation, known as the *first London equation* (after the London brothers, Heinz and Fritz), is thus necessary if the superelectrons have no resistance to their motion.

Equation (28.6) also reveals that the superelectrons' inertia creates a lag between their motion and that of the electric field. As a result, a superconductor can support a time-varying voltage drop across itself. The impedance associated with the supercurrent, therefore, is an inductor, and it will be useful to think of Λ as an inductance created by the correlated motion of the superelectrons.

If the first London equation is substituted into Faraday's law, $\nabla \times \mathbf{E} = -(\partial \mathbf{B}/\partial t)$, and integrated with respect to time, the *second London equation* results:

$$\nabla \times (\Lambda \mathbf{J}_s) = -\mathbf{B} \tag{28.7}$$

where the constant of integration has been defined to be zero. This choice is made so that the second London equation is consistent with the Meissner effect, as we now demonstrate. Taking the curl of the quasistatic form of Ampère's law, $\nabla \times \mathbf{H} = \mathbf{J}_s$, results in the expression $\nabla^2 \mathbf{B} = -\mu_o \nabla \times \mathbf{J}_s$, where a vector identity, $\nabla \times \nabla \times \mathbf{C} = \nabla (\nabla \cdot \mathbf{C}) - \nabla^2 \mathbf{C}$; the constitutive relation, $\mathbf{B} = \mu_o \mathbf{H}$; and Gauss's law, $\nabla \cdot \mathbf{B} = 0$, have been used. By now appealing to the second London equation, we obtain the vector Helmholtz equation:

$$\nabla^2 \mathbf{B} - \frac{1}{\lambda^2} \mathbf{B} = 0 \tag{28.8}$$

where the penetration depth is defined as

$$\lambda \equiv \sqrt{\frac{\Lambda}{\mu_o}} = \sqrt{\frac{m^*}{n^*(q^*)^2 \mu_o}} \tag{28.9}$$

From Equation (28.8) we find that a flux density applied parallel to the surface of a semiinfinite superconductor will decay away exponentially from the surface on a spatial length scale of order λ. In other words, a bulk superconductor will exclude an applied flux as predicted by the Meissner effect.

The London equations reveal that there is a characteristic length λ over which electromagnetic fields can change inside a superconductor. This penetration depth is different from the more familiar skin depth of electromagnetic theory, the latter being a frequency-dependent quantity. Indeed, the penetration depth at zero temperature is a distinct material property of a particular superconductor.

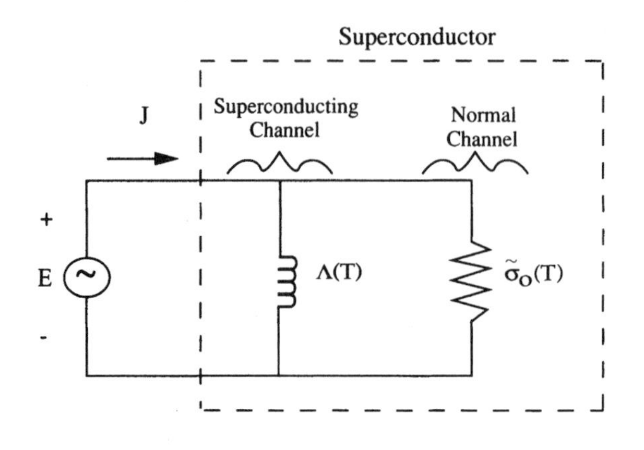

FIGURE 28.2 A lumped element model of a superconductor.

Notice that λ is sensitive to the number of correlated electrons (the superelectrons) in the material. As previously discussed, this number is a function of temperature, and so only at $T = 0$ do *all* the electrons that usually conduct ohmically participate in the Cooper pairing. For intermediate temperatures, $0 > T > T_c$, there are actually two sets of interpenetrating electron fluids: the uncorrelated electrons providing ohmic conduction and the correlated ones creating supercurrents. This two-fluid model is a useful way to build temperature effects into the London relations.

Under the two-fluid model, the electrical current density, **J**, is carried by both the uncorrelated (normal) electrons and the superelectrons: $\mathbf{J} = \mathbf{J}_n + \mathbf{J}_s$, where $\mathbf{J}_n$ is the normal current density. The two channels are modeled in a circuit, as shown in Figure 28.2, by a parallel combination of a resistor (representing the ohmic channel) and an inductor (representing the superconducting channel). To a good approximation, the respective temperature dependences of the conductor and inductor are

$$\tilde{\sigma}_o(T) = \sigma_o(T_c)\left(\frac{T}{T_c}\right)^4 \quad \text{for } T \leqslant T_c \tag{28.10}$$

$$\Lambda(T) = \Lambda(0)\left(\frac{1}{1-(T/T_c)^4}\right) \quad \text{for } T \leqslant T_c \tag{28.11}$$

where σ_o is the dc conductance of the normal channel. (Strictly speaking, the normal channel should also contain an inductance representing the inertia of the normal electrons, but typically such an inductor contributes negligibly to the overall electrical response.) Since the temperature-dependent penetration depth is defined as $\lambda(T) = \sqrt{\Lambda(T)/\mu_o}$, the effective conductance of a superconductor in the sinusoidal steady state is

$$\sigma = \tilde{\sigma}_o + \frac{1}{j\omega\mu_o\lambda^2} \tag{28.12}$$

where the explicit temperature dependence notation has been suppressed.

Most of the important physics associated with the classical model is embedded in Equation (28.12). As is clear from the lumped element model, the relative importance of the normal and superconducting channels is a function not only of temperature but also of frequency. The familiar *L/R* time constant, here equal to $\Lambda\tilde{\sigma}_o$, delineates the frequency regimes where most of the total current is carried by $\mathbf{J}_n$ (if $\omega\Lambda\tilde{\sigma}_o >> 1$) or $\mathbf{J}_s$ (if $\omega\Lambda\tilde{\sigma}_o << 1$). This same result can also be obtained by comparing the skin depth associated with the normal channel, $\delta = \sqrt{2/(\omega\mu_o\tilde{\sigma}_o)}$, to the penetration depth to see which channel provides more field screening. In addition, it is straightforward to use Equation (28.12) to rederive Poynting's theorem for systems that involve superconducting materials:

$$-\int_V \nabla\cdot(\mathbf{E}\times\mathbf{H})\mathrm{d}\mathbf{v} = \frac{\mathrm{d}}{\mathrm{d}t}\int_V\left(\frac{1}{2}\varepsilon\mathbf{E}^2 + \frac{1}{2}\mu_o\mathbf{H}^2 + \frac{1}{2}\Lambda(T)\mathbf{J}_\mathrm{s}^2\right)\mathrm{d}\mathbf{v} + \int_V \frac{1}{\tilde{\sigma}_o(T)}\mathbf{J}_\mathrm{n}^2\mathrm{d}\mathbf{v} \tag{28.13}$$

Using this expression, it is possible to apply the usual electromagnetic analysis to find the inductance (L_o), capacitance (C_o), and resistance (R_o) per unit length along a parallel plate transmission line. The results of such analysis for typical cases are summarized in Table 28.1.

TABLE 28.1 Lumped Circuit Element Parameters per Unit Length for Typical Transverse Electromagnetic Parallel Plate Waveguides[a]

Transmission Line Geometry	L_o	C_o	R_o
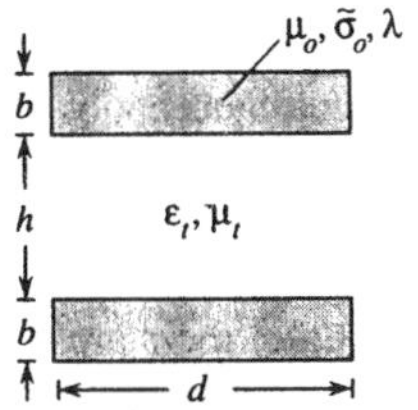 Two identical, thin ($\lambda \gg b$) superconducting plates	$\frac{\mu_t h}{d} + \frac{2\mu_0\lambda^2}{db}$	$\frac{\varepsilon_t d}{h}$	$\frac{8}{db\tilde{\sigma}_o}\left(\frac{\lambda}{\delta}\right)^4$
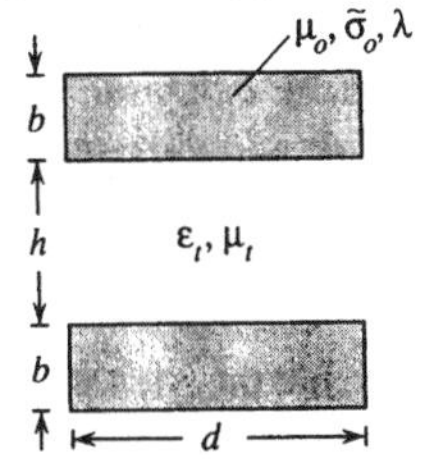 Two identical, thick ($\lambda \ll b$) superconducting plates	$\frac{\mu_t h}{d} + \frac{2\mu_0\lambda}{d}$	$\frac{\varepsilon_t d}{h}$	$\frac{4}{d\delta\tilde{\sigma}_o}\left(\frac{\lambda}{\delta}\right)^3$
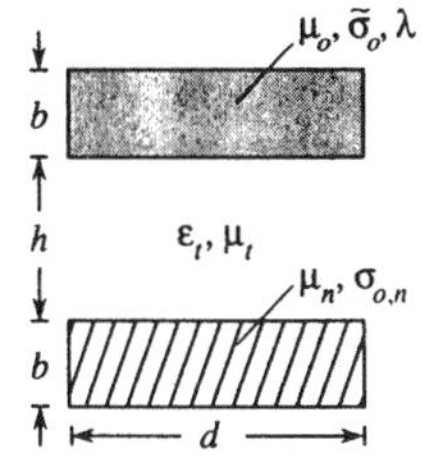 One thick ($\lambda \ll b$) superconducting plate and one thick ($\lambda \ll b$) ohmic plate	$\frac{\mu_t h}{d} + \frac{\mu_0\lambda}{d} + \frac{\mu_\mathrm{n}\delta_\mathrm{n}}{2d}$	$\frac{\varepsilon_t d}{h}$	$\frac{1}{d\delta_\mathrm{n}\sigma_{\mathrm{o,n}}}$

[a]The subscript n refers to parameters associated with a normal (ohmic) plate. Using these expressions, line input impedance, attenuation, and wave velocity can be calculated.

Source: T.P. Orlando and K.A. Delin, *Foundations of Applied Superconductivity,* Reading, MA: Addison-Wesley, p. 171. With permission.

28.3 Superconducting Electronics

The macroscopic quantum nature of superconductivity can be usefully exploited to create a new type of electronic device. Because all the superelectrons exhibit correlated motion, the usual wave-particle duality normally associated with a single quantum particle can now be applied to the entire ensemble of superelectrons. Thus, there is a spatiotemporal phase associated with the ensemble that characterizes the supercurrent flowing in the material.

Naturally, if the overall electron correlation is broken, this phase is lost and the material is no longer a superconductor. There is a broad class of structures, however, known as *weak links*, where the correlation is merely perturbed locally in space rather than outright destroyed. Colloquially, we say that the phase "slips" across the weak link to acknowledge the perturbation.

The unusual properties of this phase slippage were first investigated by Brian Josephson and constitute the central principles behind superconducting electronics. Josephson found that the phase slippage could be defined as the difference between the macroscopic phases on either side of the weak link. This phase difference, denoted as ϕ, determined the supercurrent, i_s, through voltage, v, across the weak link according to the Josephson equations:

$$i_s = I_c \sin \phi \tag{28.14}$$

$$v = \frac{\Phi_o}{2\pi} \frac{\partial \phi}{\partial t} \tag{28.15}$$

where I_c is the critical (maximum) current of the junction and Φ_o is the quantum unit of flux. (The flux quantum has a precise definition in terms of Planck's constant, h, and the electron charge, e: $\Phi_o \equiv h/(2e) \approx 2.068 \times 10^{-15}$ Wb). As in the previous section, the correlated motion of the electrons, here represented by the superelectron phase, manifests itself through an inductance. This is straightforwardly demonstrated by taking the time derivative of Equation (28.14) and combining this expression with Equation (28.15). Although the resulting inductance is nonlinear (it depends on cos ϕ), its relative scale is determined by

$$L_j = \frac{\Phi_o}{2\pi I_c} \tag{28.16}$$

a useful quantity for making engineering estimates.

A common weak link, known as the Josephson tunnel junction, is made by separating two superconducting films with a very thin (typically 20 Å) insulating layer. Such a structure is conveniently analyzed using the resistively and capacitively shunted junction (RCSJ) model shown in Figure 28.3. Under the RCSJ model, an

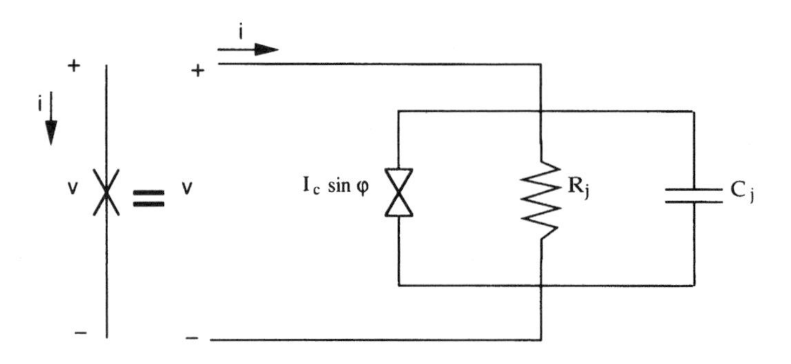

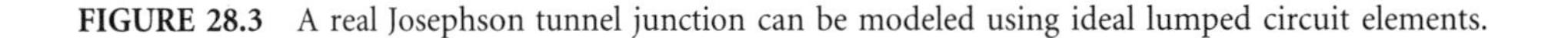

FIGURE 28.3 A real Josephson tunnel junction can be modeled using ideal lumped circuit elements.

ideal lumped junction (described by Equation (28.14) and Equation (28.15)) and a resistor R_j represent how the weak link structure influences the respective phases of the super and normal electrons, and a capacitor C_j represents the physical capacitance of the sandwich structure. If the ideal lumped junction portion of the circuit is treated as an inductor-like element, many Josephson tunnel junction properties can be calculated with the familiar circuit time constants associated with the model. For example, the quality factor Q of the RCSJ circuit can be expressed as

$$Q^2 = \frac{R_j C_j}{L_j / R_j} = \frac{2\pi I_c R_j^2 C_j}{\Phi_o} \equiv \beta \tag{28.17}$$

where β is known as the Stewart–McCumber parameter. Clearly, if $\beta >> 1$, the ideal lumped junction element is underdamped in that the capacitor readily charges up, dominates the overall response of the circuit, and therefore creates a hysteretic *i–v* curve as shown in Figure 28.4(a). In the case when the bias current is raised from zero, no time-averaged voltage is created until the critical current is exceeded. At this point the junction switches to the voltage $2\Delta/e$ with a time constant $\sqrt{L_j C_j}$. Once the junction has latched into the voltage state, however, the bias current must be lowered to zero before it can again be steered through the superconducting path. Conversely, $\beta << 1$ implies that the L_j/R_j time constant dominates the circuit response, so that the capacitor does not charge up and the *i–v* curve is not hysteretic (Figure 28.4(b)).

Just as the correlated motion of the superelectrons creates the frequency-independent Meissner effect in a bulk superconductor through Faraday's law, so too the macroscopic quantum nature of superconductivity allows the possibility of a device whose output voltage is a function of a static magnetic field. If two weak links are connected in parallel, the lumped version of Faraday's law gives the voltage across the second weak link as $v_2 = v_1 + (d\Phi/dt)$, where Φ is the total flux threading the loop between the links. Substituting Equation (28.15) and integrating with respect to time yields

$$\phi_2 - \phi_1 = (2\pi\,\Phi)/\Phi_o \tag{28.18}$$

showing that the spatial change in the phase of the macroscopic wavefunction is proportional to the local magnetic flux. The structure described is known as a *superconducting quantum interference device (SQUID)* and can be used as a highly sensitive magnetometer by biasing it with current and measuring the resulting voltage as a function of magnetic flux. Such SQUID structures have also been proposed for quantum bits in quantum computing. From this discussion, it is apparent that a duality exists in how fields interact with the

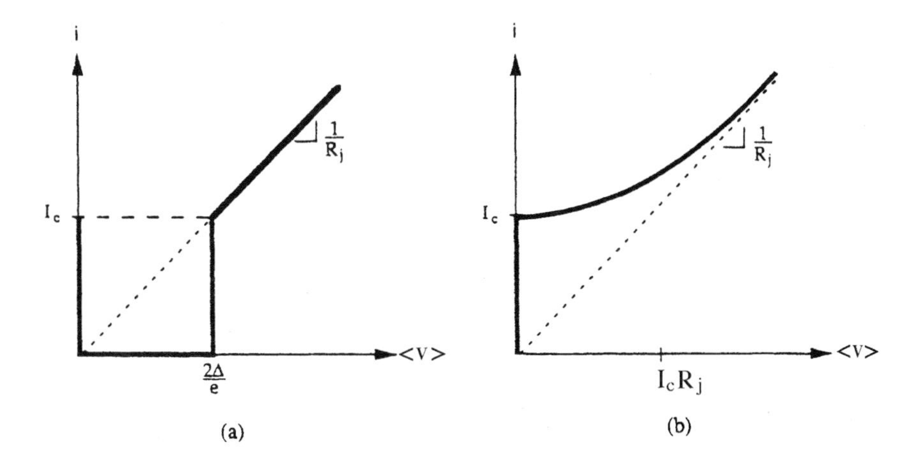

FIGURE 28.4 The *i–v* curves for a Josephson junction: (a) $\beta >> 1$, (b) $\beta << 1$.

macroscopic phase: electric fields are coupled to its rate of change in time, and magnetic fields are coupled to its rate of change in space.

28.4 Types of Superconductors

The macroscopic quantum nature of superconductivity also affects the general electromagnetic properties previously discussed. This is most clearly illustrated by the interplay of the characteristic lengths ξ, representing the scale of quantum correlations, and λ, representing the scale of electromagnetic screening. Consider the scenario where a magnetic field, H, is applied parallel to the surface of a semiinfinite superconductor. The correlations of the electrons in the superconductor must lower the overall energy of the system or else the material would not be superconducting in the first place. Because the critical magnetic field H_c destroys all the correlations, it is convenient to define the energy density gained by the system in the superconducting state as $(1/2)\mu_o H_c^2$. The electrons in a Cooper pair are separated on a length scale of ξ however, the correlations cannot be fully achieved until a distance roughly ξ from the boundary of the superconductor. There is thus an energy per unit area, $(1/2)\mu_o H_c^2 \xi$, that is lost because of the presence of the boundary. Now consider the effects of the applied magnetic field on this system. It costs the superconductor energy to maintain the Meissner effect, $\mathbf{B} = 0$, in its bulk; in fact, the energy density required is $(1/2)\mu_o H^2$. However, since the field can penetrate the superconductor a distance roughly λ, the system need not expend an energy per unit area of $(1/2)\mu_o H^2 \lambda$ to screen this volume. To summarize, more than a distance ξ from the boundary, the energy of the material is lowered (because it is superconducting), and more than a distance λ from the boundary the energy of the material is raised (to shield the applied field).

Now, if $\lambda > \xi$, the region of superconducting material greater than λ from the boundary but less than ξ will be higher in energy than that in the bulk of the material. Thus, the surface energy of the boundary is positive and therefore costs the total system some energy. This class of superconductors is known as type I. Most elemental superconductors, such as, aluminum, tin, and lead, are type I. In addition to having $\lambda > \xi$, type I superconductors are generally characterized by low critical temperatures ($\sim$5 K) and critical fields ($\sim$0.05 T). Typical type I superconductors and their properties are listed in Table 28.2.

Conversely, if $\lambda > \xi$, the surface energy associated with the boundary is negative and lowers the total system energy. It is therefore thermodynamically favorable for a normal superconducting interface to form inside these type II materials. Consequently, this class of superconductors does not exhibit the simple Meissner effect as do type I materials. Instead, there are now two critical fields: for applied fields below the lower critical field, H_{c1}, a type II superconductor is in the Meissner state, and for applied fields greater than the upper critical field, H_{c2}, superconductivity is destroyed. The three critical fields are related to each other by $H_c \approx \sqrt{H_{c1} H_{c2}}$.

In the range $H_{c1} > H > H_{c2}$, a type II superconductor is said to be in the vortex state because now the applied field can enter the bulk superconductor. Because flux exists in the material, however, the superconductivity is destroyed locally, creating normal regions. Recall that for type II materials the boundary

TABLE 28.2 Material Parameters for Type I Superconductors*

Material	T_c (K)	λ_o (nm)	ξ_o (nm)	Δ_o (meV)	$\mu_o H_{co}$ (mT)
Al	1.18	50	1600	0.18	10.5
In	3.41	65	360	0.54	23.0
Sn	3.72	50	230	0.59	30.5
Pb	7.20	40	90	1.35	80.0
Nb	9.25	85	40	1.50	198.0

*The penetration depth λ_o is given at zero temperature, as are the coherence length ξ_o, the thermodynamic critical field H_{co}, and the energy gap Δ_o.

Source: R.J. Donnelly, "Cryogenics", in *Physics Vade Mecum,* H. L. Anderson, Ed., New York: American Institute of Physics, 1981. With permission.

between the normal and superconducting regions lowers the overall energy of the system. Therefore, the flux in the superconductor creates as many normal-superconducting interfaces as possible without violating quantum criteria. The net result is that flux enters a type II superconductor in quantized bundles of magnitude Φ_o known as *vortices* or *fluxons* (the former name derives from the fact that current flows around each quantized bundle in the same manner as a fluid vortex circulates around a drain). The central portion of a vortex, known as the core, is a normal region with an approximate radius ξ. If a defect-free superconductor is placed in a magnetic field, the individual vortices, whose cores essentially follow the local average field lines, form an ordered triangular array, or flux lattice. As the applied field is raised beyond H_{c1} (where the first vortex enters the superconductor), the distance between adjacent vortex cores decreases to maintain the appropriate flux density in the material. Finally, the upper critical field is reached when the normal cores overlap and the material is no longer superconducting. Indeed, a precise calculation of H_{c2} using the phenomenological theory developed by Vitaly Ginzburg and Lev Landau yields:

$$H_{c2} = \frac{\Phi_o}{2\pi\mu_o\xi^2} \tag{28.19}$$

which verifies out simple picture. The values of typical type II material parameters are listed in Table 28.3 and Table 28.4.

TABLE 28.3 Material Parameters for Conventional Type II Superconductors*

Material	T_c (K)	λ_{GL} (0)(nm)	ξ_{GL} (0)(nm)	Δ_o (meV)	$\mu_o H_{c2,o}$ (T)
Pb–In	7.0	150	30	1.2	0.2
Pb–Bi	8.3	200	20	1.7	0.5
Nb–Ti	9.5	300	4	1.5	13
Nb–N	16	200	5	2.4	15
$PbMo_6S_8$	15	200	2	2.4	60
V_3Ga	15	90	2–3	2.3	23
V_3Si	16	60	3	2.3	20
Nb_3Sn	18	65	3	3.4	23
Nb_3Ge	23	90	3	3.7	38

*The values are only representative because the parameters for alloys and compounds depend on how the material is fabricated. The penetration depth $\lambda_{GL}(0)$ is given as the coefficient of the Ginzburg–Landau temperature dependence as $\lambda_{GL}(T) = \lambda_{GL}(0)(1 - T/T_c)^{-1/2}$; likewise for the coherence length where $\xi_{GL}(T) = \xi_{GL}(0)(1-T/T_c)^{-1/2}$. The upper critical field $H_{c2,o}$ is given at zero temperature as well as the energy gap Δ_o.

Source: R.J. Donnelly, "Cryogenics," in *Physics Vade Mecum*, H.L. Anderson, Ed., New York: American Institute of Physics, 1981. With permission.

TABLE 28.4 Type II (Nonconventional and High-Temperature Superconductors)

Material	T_c (K)
$Ba_{1-x}K_xBi\ O_3$	30
Rb_3C_{60}	33
MgB_2	39
$YBa_2Cu_3O_7$	95
$Bi_2Sr_2CaCu_2O_8$	85
$Bi_2Sr_2Ca_2Cu_3O_{10}$	110
$TlBa_2Ca_2Cu_3O_{10}$	125
$HgBa_2Ca_2\ Cu_3O_8$	131

See the NIST WebHTS Database at http://www. ceramics. nist.gov/srd/hts/htsquery.htm for more information.

Type II superconductors are of great technical importance because typical H_{c2} values are at least an order of magnitude greater than the typical H_c values of type I materials. It is therefore possible to use type II materials to make high-field magnet wire. Unfortunately, when current is applied to the wire, there is a Lorentz-like force on the vortices, causing them to move. Because the moving vortices carry flux, their motion creates a static voltage drop along the superconducting wire according to Faraday's law. As a result, the wire no longer has a zero dc resistance, even though the material is still superconducting. To fix this problem, type II superconductors are usually fabricated with intentional defects, such as impurities or grain boundaries, in their crystalline structure to pin the vortices and prevent vortex motion. The pinning is created because the defect locally weakens the superconductivity in the material, and it is thus energetically favorable for the normal core of the vortex to overlap the nonsuperconducting region in the material. Critical current densities usually quoted for practical type II materials really represent the depinning critical current density where the Lorentz-like force can overcome the pinning force. (The depinning critical current density should not be confused with the depairing critical current density, which represents the current when the Cooper pairs have enough kinetic energy to overcome their correlation. The depinning critical current density is typically an order of magnitude less than the depairing critical current density, the latter of which represents the theoretical maximum for $\mathbf{J}_c$.)

By careful manufacturing, it is possible to make superconducting wire with tremendous amounts of current-carrying capacity. For example, standard copper wire used in homes will carry about 10^7 A/m^2, whereas a practical type II superconductor like niobium-titanium can carry current densities of 10^{10} A/m^2 or higher even in fields of several teslas. This property, more than a zero dc resistance, is what makes superconducting wire so desirable.

Defining Terms

Superconductivity: A state of matter whereby the correlation of conduction electrons allows a static current to pass without resistance and a static magnetic flux to be excluded from the bulk of the material.

References

W. Buckel and R. Kleiner, *Superconductivity: Fundamentals and Applications*, 2nd ed., New York: John Wiley & Sons, 2004.

K.A. Delin and A.W. Kleinsasser, "Stationary properties of high-critical-temperature proximity effect Josephson junctions," *Supercond. Sci. Technol.*, 9, 227, 1996.

R.J. Donnelly, "Cryogenics," in *Physics Vade Mecum*, H.L. Anderson, Ed., New York: American Institute of Physics, 1981.

S. Foner, and B.B. Schwartz, *Superconducting Machines and Devices*, New York: Plenum Press, 1974.

S. Foner and B.B. Schwartz, *Superconducting Materials Science*, New York: Plenum Press, 1981.

T.P. Orlando and K.A. Delin, *Foundations of Applied Superconductivity*, Reading, MA: Addison-Wesley, 1991.

S.T. Ruggiero and D.A. Rudman, *Superconducting Devices*, Boston, MA: Academic Press, 1990.

B.B. Schwartz and S. Foner, *Superconducting Applications: SQUIDs and Machines*, New York: Plenum Press, 1977.

T. Van Duzer and C.W. Turner, *Principles of Superconductive Devices and Circuits*, 2nd ed., Englewood Cliffs, NJ: Prentice Hall, 1999.

M.N. Wilson, *Superconducting Magnets*, Oxford, U.K.: Oxford University Press, 1983.

Further Information

Every two years an Applied Superconductivity Conference is held devoted to practical technological issues. The proceedings of these conferences have been published every other year from 1977 to 1991 in the *IEEE Transactions on Magnetics.*

In 1991 the *IEEE Transactions on Applied Superconductivity* began publication. This quarterly journal focuses on both the science and the technology of superconductors and their applications, including materials issues, analog and digital circuits, and power systems. The proceedings of the Applied Superconductivity Conference now appear in this journal.

29

Dielectrics and Insulators

R. Bartnikas
Institut de Recherche d'Hydro-Québec

29.1 Introduction

Dielectrics are materials that are used primarily to isolate components electrically from each other or ground, or to act as capacitive elements in devices, circuits, and systems. Their insulating properties are directly attributable to their large energy gap between the highest filled valence band and the conduction band. The number of electrons in the conduction band is extremely low, because the energy gap of a dielectric (5 to 7 eV) is sufficiently large to maintain most of the electrons trapped in the lower band. As a consequence, a dielectric subjected to an electric field, will evince only an extremely small conduction or loss current. This current will be caused by the finite number of free electrons available in addition to other free charge carriers (ions) associated usually with contamination by electrolytic impurities, as well as dipole orientation losses arising with polar molecules under ac conditions. Often the two latter effects will tend to obscure the miniscule contribution of the relatively few free electrons available. Unlike solids and liquids, vacuum and gases (in their nonionized state) approach the conditions of a perfect insulator, i.e., they exhibit virtually no detectable loss or leakage current.

Two fundamental parameters characterizing a dielectric material are its conductivity σ and the value of the real permittivity or dielectric constant ε'. By definition, σ is equal to the ratio of the leakage current density J_l to the applied electric field E:

$$\sigma = \frac{J_l}{E} \tag{29.1}$$

Since J_l is in A cm^{-2} and E in V cm^{-1}, the corresponding units of σ are in S cm^{-1} or Ω^{-1} cm^{-1}. Alternatively, when only mobile charge carriers of charge e and mobility μ, in cm^2 V^{-1} sec^{-1}, with a concentration of n per cm^3 are involved, the conductivity may be expressed as

$$\sigma = e\mu n \tag{29.2}$$

The conductivity is usually determined in terms of the measured insulation resistance R in Ω; it is then given by $\sigma = d/RA$, where d is the insulation thickness in cm and A the surface area in cm^2. Most practical insulating materials have conductivities ranging from 10^{-6} to 10^{-20} S cm^{-1}. Often dielectrics may be classified in terms of their resistivity value ρ, which by definition is equal to the reciprocal of σ.

The real value of the permittivity or dielectric constant ε' is determined from the ratio:

$$\varepsilon' = \frac{C}{C_0} \tag{29.3}$$

where C represents the measured capacitance in F and C_0 is the equivalent capacitance in vacuo, which is calculated for the same specimen geometry from $C_0 = \varepsilon_o A/d$; here, ε_0 denotes the permittivity in vacuo and is equal to 8.854×10^{-14} F cm^{-1} (8.854×10^{-12} F m^{-1} in SI units) or more conveniently to unity in the Gaussian CGS system. In practice, the value of ε_0 in free space is essentially the same as that for a gas (e.g., for air, $\varepsilon_0 = 1.000536$). The majority of liquid and solid dielectric materials presently in use have dielectric constants extending from approximately 2 to 10.

29.2 Dielectric Losses

Under ac conditions *dielectric losses* arise mainly from the movement of free charge carriers (electrons and ions), space charge polarization, and dipole orientation [1]. Ionic, space charge, and dipole losses are temperature- and frequency-dependent, a dependency which is reflected in the measured values of σ and ε'. This necessitates the introduction of a complex permittivity ε defined by

$$\varepsilon = \varepsilon' - j\varepsilon'' \tag{29.4}$$

where ε'' is the imaginary value of the permittivity, which is equal to σ/ω. Note that the conductivity σ determined under ac conditions may include the contributions of the dipole orientation, space charge, and ionic polarization losses in addition to that of the drift of free charge carriers (ions and electrons) which determine its dc value.

The complex permittivity, ε, is equal to the ratio of the dielectric displacement vector $\bar{\mathbf{D}}$ to the electric field vector $\bar{\mathbf{E}}$, i.e., $\varepsilon = \bar{\mathbf{D}}/\bar{\mathbf{E}}$. Since under ac conditions the appearance of a loss or leakage current is manifest as a phase angle difference δ between the $\bar{\mathbf{D}}$ and $\bar{\mathbf{E}}$ vectors, then in complex notation $\bar{\mathbf{D}}$ and $\bar{\mathbf{E}}$ may be expressed as $\bar{\mathbf{D}}_0 \exp[j\omega t - \delta]$ and $\bar{\mathbf{E}}_0 \exp[j\omega t]$, respectively, where ω is the radial frequency term, t the time, and D_0 and E_0 the respective magnitudes of the two vectors. From the relationship between $\bar{\mathbf{D}}$ and $\bar{\mathbf{E}}$, it follows that

$$\varepsilon' = \frac{D_0}{E_0} \cos \delta \tag{29.5}$$

and

$$\varepsilon'' = \frac{D_0}{E_0} \sin \delta \tag{29.6}$$

It is customary under ac conditions to assess the magnitude of loss of a given material in terms of the value of its *dissipation factor*, $\tan \delta$; it is apparent from Equation (29.5) and Equation (29.6), that:

$$\tan \delta = \frac{\varepsilon''}{\varepsilon'} = \frac{\sigma}{\omega \varepsilon'} \tag{29.7}$$

Examination of Equation (29.7) suggests that the behavior of a dielectric material may also be described by means of an equivalent electrical circuit. It is most commonplace and expedient to use a parallel circuit representation, consisting of a capacitance C in parallel with a large resistance R as delineated in Figure 29.1. Here, C represents the capacitance and R the resistance of the dielectric. For an applied voltage V across the dielectric, the leakage current is $I_l = V/R$ and the displacement current is $I_C = j\omega CV$; since $\tan\delta = I_l/I_C$, then:

$$\tan\delta = \frac{1}{\omega RC} \tag{29.8}$$

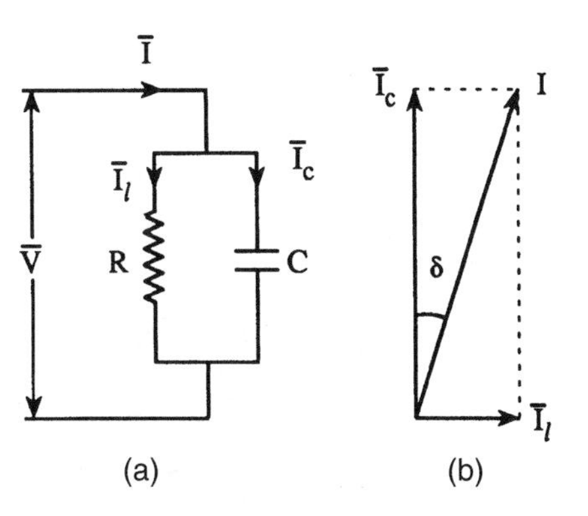

FIGURE 29.1 (a) Parallel equivalent RC circuit and (b) corresponding phasor diagram.

It is to be emphasized that in Equation (29.8), the quantities R and C are functions of temperature, frequency, and voltage. The equivalence between Equation (29.7) and Equation (29.8) becomes more palpable if I_l and I_C are expressed as $\omega\varepsilon'' C_o V$ and $j\omega\varepsilon'' C_o V$, respectively.

Every loss mechanism will exhibit its own characteristic tan δ loss peak, centered at a particular absorption frequency ω_0 for a given test temperature. The loss behavior will be contingent upon the molecular structure of the material, its thickness and homogeneity, and the temperature, frequency, and electric field range over which the measurements are performed [1]. For example, dipole orientation losses will be manifested only if the material contains permanent molecular or side-link dipoles; a considerable overlap may occur between the permanent dipole and ionic relaxation regions. Ionic relaxation losses occur in dielectric structures where ions are able to execute short-range jumps between two or more equilibrium positions. Interfacial or space charge polarization will arise with insulations of multilayered structures where the conductivity and permittivity are different for the individual strata or where one dielectric phase is interspersed in the matrix of another dielectric. Space charge traps also occur at crystalline-amorphous interfaces, crystal defects, and oxidation and localized C–H dipole sites in polymers. Alternatively, space charge losses will occur with mobile charge carriers whose movement becomes limited at the electrodes. This type of mechanism takes place often in thin-film dielectrics and exhibits a pronounced thickness effect. If the various losses are considered schematically on a logarithmic frequency scale at a given temperature, then the tan δ and ε' values will appear as functions of frequency as delineated schematically in Figure 29.2. For many materials the dipole and ionic relaxation losses tend to predominate over the frequency range extending from about 0.5 to 300 MHz, depending upon the molecular structure of the dielectric and temperature. For example, the absorption peak of an oil may occur at 1 MHz, while that of a much lower viscosity fluid such as water may appear at approximately 100 MHz. There is considerable overlap between the dipole and ionic relaxation losses, because the ionic jump distances are ordinarily of the same order of magnitude as the radii of the permanent dipoles. Space charge polarization losses manifest themselves normally over the low-frequency region extending from 10^{-6} Hz to 1 MHz and are characterized by very broad and intense peaks; this behavior is apparent from Equation (29.7), which indicates that even small conductivities may lead to very large tan δ values at very low frequencies. The nonrelaxation-type electronic conduction losses are readily perceptible over the low-frequency spectrum and decrease monotonically with frequency.

The dielectric loss behavior may be phenomenologically described by the Pellat–Debye equations, relating the imaginary and real values of the permittivity to the relaxation time, τ, of the loss process (i.e., the frequency at which the ε'' peak appears: $f_0 = 1/2\,\pi\tau$), the low-frequency or static value of the real permittivity, ε_s, and the high- or optical-frequency value of the real permittivity, ε_∞. Thus, for a loss process characterized by a single relaxation time:

$$\varepsilon' = \varepsilon_\infty + \frac{\varepsilon_s - \varepsilon_\infty}{1 + \omega^2\tau^2} \tag{29.9}$$

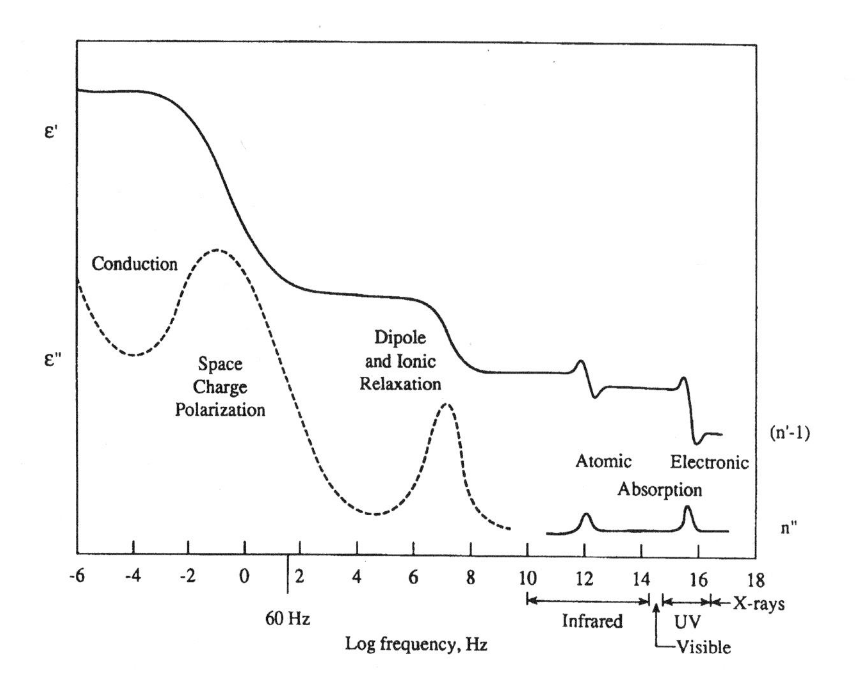

FIGURE 29.2 Schematic representations of different absorption regions [1].

and

$$\varepsilon'' = \frac{(\varepsilon_s - \varepsilon_\infty)\omega\tau}{1 + \omega^2\tau^2} \tag{29.10}$$

In practice, Equation (29.9) and Equation (29.10) are modified due to a distribution in the relaxation times either because several different loss processes are involved or as a result of interaction or cooperative movement between the discrete dipoles or the trapped and detrapped charge carriers in their own particular environment. Since the relaxation processes are thermally activated, an increase in temperature will cause a displacement of the loss peak to higher frequencies. In the case of ionic and dipole relaxation, the relaxation time may be described by the relation:

$$\tau = \frac{h}{kT}\exp\left[\frac{\Delta H}{RT}\right]\exp\left[-\frac{\Delta S}{R}\right] \tag{29.11}$$

where h is the Planck constant (6.624×10^{-34} J s^{-1}), k the Boltzmann constant (1.38×10^{-23} J K^{-1}), ΔH the activation energy of the relaxation process, R the universal gas constant (8.314×103 J K^{-1} kmol^{-1}), and ΔS the entropy of activation. For the ionic relaxation process, τ may alternatively be taken as equal to $1/2\Gamma$, where Γ denotes the ion jump probability between two equilibrium positions. Also for dipole orientation in liquids, τ may be approximately equated to the Debye term $\eta/4\pi r^3 T$, where η represents the macroscopic viscosity of the liquid and r is the dipole radius [2]. With interfacial or space charge polarization, which may arise due to a pile-up of charges at the interface of two contiguous dielectrics of different conductivity and permittivity, Equation (29.10) must be rewritten as [3]

$$\varepsilon'' = \varepsilon_\infty\left[\frac{\tau}{\omega\tau_1\tau_2} + \frac{K\omega\tau}{1 + \omega^2\tau^2}\right] \tag{29.12}$$

where the Wagner absorption factor K is given by

$$K = \frac{(\tau_1 + \tau_2 - \tau)\tau - \tau_1\tau_2}{\tau_1\tau_2} \tag{29.13}$$

where τ_1 and τ_2 are the relaxation times of the two contiguous layers or strata of respective thicknesses d_1 and d_2; τ is the overall relaxation time of the two-layer combination and is defined by $\tau = (\varepsilon_1' d_2 + \varepsilon_2' d_1)/(\sigma_1 d_2 + \sigma_2 d_1)$ where ε_1', ε_2', σ_1, and σ_2 are the respective real permittivity and conductivity parameters of the two discrete layers. Note that since ε_1' and ε_2' are temperature- and frequency-dependent and σ_1 and σ_2 are, in addition, voltage-dependent, the values of τ and ε'' will in turn also be influenced by these three variables. Space charge processes involving electrons are more effectively analyzed, using dc measurement techniques. If retrapping of electrons in polymers is neglected, then the decay current as a function of time t, arising from detrapped electrons, assumes the form [4]

$$i(t) = \frac{kT}{\nu t} n(E) \tag{29.14}$$

where $n(E)$ is the trap density and ν is the attempt jump frequency of the electrons. The electron current displays the usual t^{-1} dependence and the plot of $i(t)t$ vs. $kT \ln(\nu t)$ yields the distribution of trap depths. Equation (29.14) represents an approximation, which underestimates the current associated with the shallow traps and overcompensates for the current due to the deep traps. The mobility of the free charge carriers is determined by the depth of the traps, the field resulting from the trapped charges, and the temperature. At elevated temperatures and low space charge fields, the mobility is proportional to $\exp[-\Delta H/kT]$, and at low temperatures to $(T)^{1/4}$ [5]. A high trapped charge density will create intense fields, which will in turn exert a controlling influence on the mobility and the charge distribution profile. In polymers, shallow traps are of the order of 0.5 to 0.9 eV and deep traps are ca. 1.0 to 1.5 eV, while the activation energies of dipole orientation and ionic conduction in solid and liquid dielectrics fall within the same range. It is known that most charge trapping in the volume occurs in the vicinity of the electrodes; this can now be confirmed by measurement, using methods based on pressure wave propagation, thermal diffusion, and voltage pulse application [6]. In the first procedure, an elastic wave traveling at the speed of sound is transmitted along the dielectric specimen. This causes displacement of charges within the specimen, thereby creating an electrical signal which is readily detected at the measuring electrodes. In the thermal diffusion technique, variation of the temperature at one end of the specimen gives rise to nonuniform diffusion of heat along the dielectric, thereby leading again to a displacement of charges along its path and hence an associated electrical signal. The third method, which involves a rapid application of voltage pulses across the dielectric specimen, is commonly referred to as the pulsed electro-acoustic (PEA) method. It is, perhaps due to its simplicity, the most popular of the three test procedures. The applied electrical pulses produce pressure waves within the dielectric bulk of charge density $\rho(x)$ and surface charge density of σ_1 and σ_2 at the two respective electrodes. A portion of the transmitted pressure waves traversing the low potential electrode is detected by means of an acoustical transducer which provides a proportional electrical signal output. Hence, the first detected pressure wave is comprised of three components, namely that due to the low or ground potential electrode [7]:

$$p_0(t) = k_1\left[\sigma_1 e_\mathrm{p}(t_\mathrm{delay}) - \frac{1}{2}\varepsilon_0\varepsilon_\mathrm{r} e_\mathrm{p}^2(t - t_\mathrm{delay})\right] \tag{29.15}$$

the volume of the dielectric:

$$p_\rho(t) = k_1 \int_0^d \rho(x) e_\mathrm{p}\left(t - \frac{x}{c_2} - t_\mathrm{delay}\right) dx \tag{29.16}$$

and the high potential electrode:

$$p_\mathrm{d}(t) = k_1 k' \left[\sigma_2 e_\mathrm{p}\left(t - \frac{d}{c_2} - t_\mathrm{delay}\right) + \frac{1}{2}\varepsilon_0 \varepsilon_\mathrm{r} e_\mathrm{p}^2\left(t - \frac{d}{2} - t_\mathrm{delay}\right)\right] \tag{29.17}$$

where c_2 denotes the acoustical velocity and e_p the electrical field; d is the thickness of the specimen; and k_1, k_2, and k_3 (with $k_1 = k_2$ and $k' = k_3/k_1$) are constants determined by the acoustical impedances of the two electrodes and the specimen, respectively. The received electrical signals are correlated with the acoustical wave by means of a deconvolution procedure to obtain the profile of the trapped charge $\rho(x)$. Errors in the measurement are primarily due to electrode surface charge effects and difficulties in distinguishing between the polarization of the polar dipoles and that of the trapped charges [8]. Moreover, while the three methods for determining the charge density distribution have a common theoretical basis, the agreement between the charge density distributions derived, utilizing the three methods, may in some cases be rather tenuous [9].

Temperature influences the real value of the permittivity or dielectric constant, ε', insofar as it affects the density of the dielectric material. As the density diminishes with temperature, ε' falls with temperature in accordance with the Clausius–Mossotti equation:

$$[P] = \frac{(\varepsilon' - 1)\, M}{(\varepsilon' + 2)\, d_0} \tag{29.18}$$

where $[P]$ represents the polarization per mole, M the molar mass, d_0 the density at a given temperature, and $\varepsilon' = \varepsilon_\mathrm{s}$. Equation (29.18) is equally valid if the substitution $\varepsilon' = (n')^2$ is made; here n' is the real value of the index of refraction. In fact, the latter provides a direct connection with the dielectric behavior at optical frequencies. In analogy with the complex permittivity, the index of refraction is also a complex quantity, and its imaginary value n'' exhibits a loss peak at the absorption frequencies; however, in contrast to the ε' value, which can only fall with frequency, the real index of refraction, n', exhibits an inflection-like behavior at the absorption frequency. This is illustrated schematically in Figure 29.2 which depicts the kn' or n'' and $n'-1$ values as a function of frequency over the optical frequency regime. The absorption in the infrared results from atomic resonance that arises from a displacement and vibration of atoms relative to each other, while an electronic resonance absorption effect occurs over the ultraviolet frequencies as a consequence of the electrons being forced to execute vibrations at the frequency of the external field.

The characterization of dielectric materials must be carried out in order to determine their properties for various applications over different parts of the electromagnetic frequency spectrum. There are many techniques and methods available for this purpose, which are too numerous and detailed to attempt to present here even in a cursory manner. However, Figure 29.3 portrays schematically the different test methods that are commonly used to carry out the characterization over the different frequencies up to and including the optical regime [10]. A direct relationship exists between the time and frequency domain test methods via the Laplace transforms.

The frequency response of dielectrics at the more elevated frequencies is primarily of interest in the electrical communications field. In contradistinction for electrical power generation, transmission, and distribution, it is the low-frequency spectrum that constitutes the area of application. Also, the use of higher voltages in the electrical power area necessarily requires detailed knowledge of how the electrical losses vary as a function of the electrical field. Since most electrical power apparatus operates at a fixed frequency of 50 or 60 Hz, the main variable apart from the temperature is the applied or operating voltage. At power frequencies the dipole losses are generally very small and invariant with voltage up to the saturation fields, which exceed substantially the operating fields, being in the order of 10^7 kV cm^{-1} or more. However, both the space charge polarization and

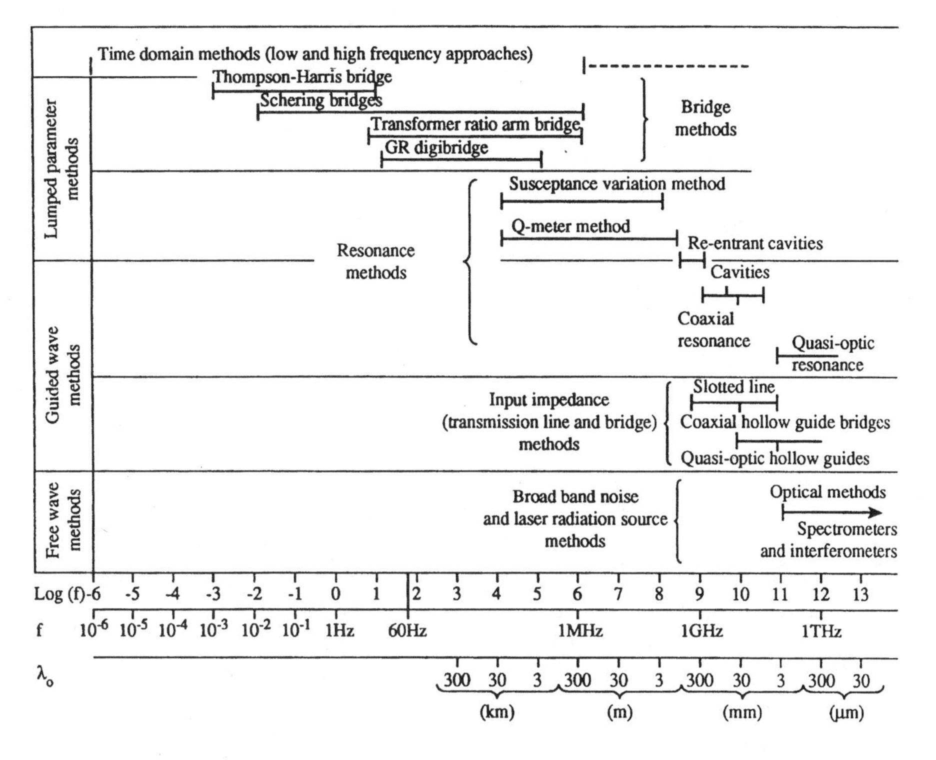

FIGURE 29.3 Frequency range of various dielectric test methods [10].

ionic losses are highly field-dependent. As the electrical field is increased, ions of opposite sign are increasingly segregated; this hinders their recombination and, in effect, enhances the ion charge carrier concentration. As the dissociation rate of the ionic impurities is further augmented by temperature increases, combined rises in temperature and field may lead to appreciable dielectric loss. Thus, for example, for a thin liquid film bounded by two solids, tan δ increases with voltage until at some upper voltage value, the physical boundaries begin to finally limit the amplitude of the ion excursions, at which point tan δ commences a downward trend with voltage (Böning–Garton effect). The interfacial or space charge polarization losses may evince a rather intricate field dependence, depending upon the manner in which the discrete conductivities of the contiguous media change with applied voltage and temperature (as is apparent from the nature of Equation (29.12) and Equation (29.13)). The exact frequency value at which the space charge loss exhibits its maximum is contingent upon the value of the relaxation time τ. Figure 29.4 depicts typical tan δ versus applied voltage characteristics for an oil-impregnated paper-insulated cable model at two different temperatures, in which the loss behavior is primarily governed by ionic conduction and space charge effects. The monotonically rising dissipation factor with increasing applied voltage at room temperature is indicative of the predominating ionic loss mechanism, while at 85°C, an incipient decrease in tan δ is suggestive of space charge effects.

29.3 Dielectric Breakdown

As the voltage is increased across a dielectric material, a point is ultimately reached beyond which the insulation will no longer be capable of sustaining any further rise in voltage and breakdown will ensue, causing a short to develop between the electrodes. If the dielectric consists of a gas or liquid medium, the breakdown will be self-healing in the sense that the gas or liquid will support anew a reapplication of voltage until another breakdown recurs. In a solid dielectric, however, the initial breakdown will result in the formation of a permanent conductive channel, which cannot support a reapplication of voltage. The dielectric breakdown processes are distinctly different for the three states of matter.

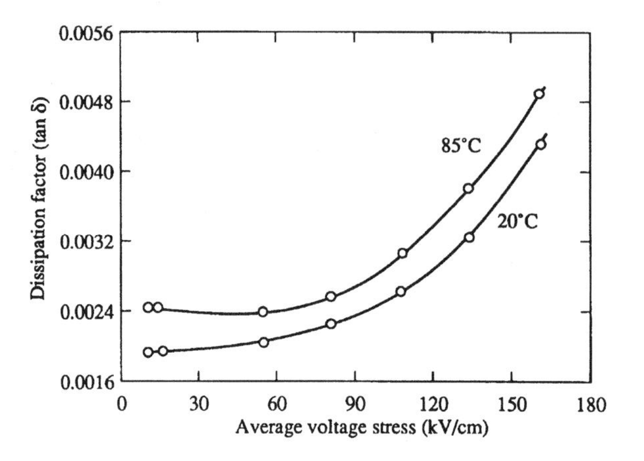

FIGURE 29.4 Loss characteristics of mineral oil-impregnated paper.

In the case of solid dielectrics the breakdown is dependent not only upon the molecular structure and morphology of the solid but also upon extraneous variables, such as the geometry of the material, the temperature, and the ambient environment. Since breakdown often occurs along some fault of the material, the breakdown voltage displays a readily perceptible decrease with area and thickness of the specimen due to increased incidence of faults over larger volumes. This is indeed part of the reason why thin-film inorganic dielectrics, which are normally evaluated using small-diameter dot counter electrodes, exhibit exceptionally high dielectric strengths. With large organic dielectric specimens, recessed electrodes are used to minimize electrode edge effects, leading to greatly elevated breakdown strengths in the order of 10^6 to 10^7 kV cm^{-1}, a range of values considered to represent the ultimate breakdown strength of the material or its intrinsic breakdown strength; as the intrinsic breakdown occurs in approximately 10^{-8} to 10^{-6} sec, an electronic mechanism is implicated.

The breakdown strength under dc and impulse conditions tends to exceed that at ac fields, thereby suggesting the ac breakdown process may be partially of a thermal nature. An additional factor that may lower the ac breakdown strength is that associated with the occurrence of partial discharges either in void inclusions or at the electrode edges; this leads to breakdown values very much less than the intrinsic value. In practice, the breakdowns are generally of an extrinsic nature, and the intrinsic values are useful conceptually in so far as they provide an idea of an upper value that can be attained only under ideal conditions. The intrinsic breakdown theories were essentially developed for crystalline dielectrics, for which it was assumed that a very small number of thermally activated electrons can be thermally excited to move from the valence to the conduction band and that under the influence of an external field they will be impelled to move in the direction of the field, colliding with the lattice of the crystalline dielectric and dissipating their energy by phonon interactions [11]. Accordingly, breakdown is said to occur when the average rate of energy gain by the electrons, $A(E, T, T_e, \xi)$, exceeds that lost in collisions with the lattice, $B(T, T_e, \xi)$. Hence, the breakdown criterion can be stated as

$$A(E, T, T_e, \xi) = B(T, T_e, \xi) \tag{29.19}$$

where E is the applied field, T the lattice temperature, T_e the electron temperature, and ξ an energy distribution constant. Thus, in qualitative terms, as the temperature is increased gradually, the breakdown voltage rises because the interaction between the electrons and the lattice is enhanced as a result of the increased thermal vibrations of the lattice. Ultimately, a critical temperature is attained where the electron–electron interactions surpass in importance those between the electrons and the lattice, and the breakdown strength commences a monotonic decline with temperature; this behavior is borne out in NaCl crystals, as is apparent from Figure 29.5 [12]. However, with amorphous or partially crystalline polymers, as, for example, with polyethylene, the maximum in breakdown strength is seen to be absent and only a decrease is observed [13]; as the crystalline content is increased in amorphous-crystalline solids, the breakdown strength is reduced.

It is generally believed that lamellar crystalline regions play a predominant role in determining the electrical characteristics of polyethylene [14].

The electron avalanche concept has also been applied to explain breakdown in solids, in particular to account for the observed decrease in breakdown strength with insulation thickness. Since breakdown due to electron avalanches involves the formation of space charge, space charges will tend to modify the conditions for breakdown. Any destabilization of the trapping and detrapping process, such as may be caused by a perturbation of the electrical field, will initiate the breakdown event [5]. The detrapping of mobile charge carriers will be accompanied by photon emission and formation of the plasma breakdown channel, resulting in the dissipation of polarization energy. If dipole interaction is neglected, the polarization energy due to a trapped charge is of the order of 5χ eV, where χ is the dielectric susceptibility. The release of the polarization energy will be accompanied by electrical tree growth in and melting of the polymer.

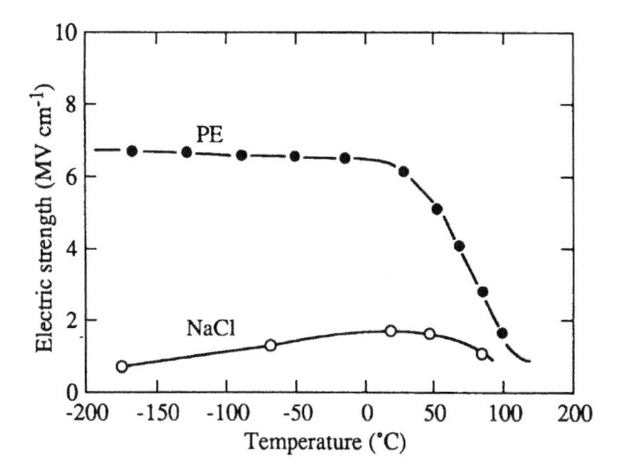

FIGURE 29.5 Dielectric breakdown characteristic of sodium chloride (after von Hippel and Lee [12]) and polyethylene (after Oakes [13]).

The electrical breakdown process in gases is relatively well understood and is explained in terms of the avalanche theory. A free electron, occurring in a gas due to cosmic radiation, will be accelerated in a field and upon collision with neutral molecules in its trajectory will eject, if its energy is sufficient, other electrons that will in turn undergo additional collisions resulting in a production of more free electrons. If the electric field is sufficiently high, the number of free electrons will increase exponentially along the collision route until ultimately an electron avalanche will form. As the fast-moving electrons in the gap disappear into the anode, they leave behind the slower-moving ions, which gradually drift to the cathode where they liberate further electrons with a probability γ. When the height of the positive ion avalanche becomes sufficiently large to lead to a regeneration of a starting electron, the discharge mechanism becomes self-sustaining and a spark bridges the two electrodes. The condition for the Townsend breakdown in a short gap is given by

$$\gamma[\exp(\alpha d) - 1] = 1 \tag{29.20}$$

where d is the distance between the electrodes and α represents the number of ionizing impacts per electron per unit distance. The value of γ is also enhanced by photoemission at the cathode and photon radiation in the gas volume by the metastable and excited gas atoms or molecules. In fact, in large gaps the breakdown is governed by streamer formation in which photon emission from the avalanches plays a dominant role. Breakdown characteristics of gases are represented graphically in terms of the Paschen curves, which are plots of the breakdown voltage as a function of the product of gas pressure p and the electrode separation d. Each gas is characterized by a well-defined minimum breakdown voltage at one particular value of the pd product as depicted in Figure 29.6 [15]. In short gaps, the breakdown may assume either the form of a spark discharge (constricted breakdown channel) or that of a glow or pseudoglow discharge (a highly diffused breakdown channel encompassing usually the entire intervening electrode area) [16].

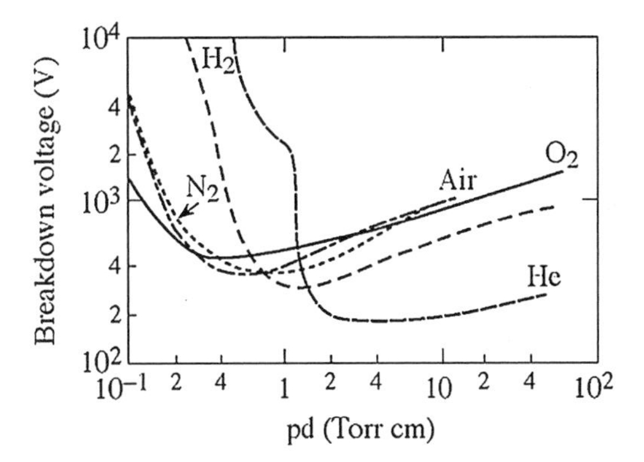

FIGURE 29.6 Paschen's characteristics of breakdown voltages for air, N_2, O_2, He, and H_2 (after Schönhuber [15]).

The breakdown process in liquids is perhaps the least understood due to a lack of a satisfactory theory on the liquid state. The avalanche theory has been applied with limited success to explain the breakdown in liquids, by assuming that electrons injected from an electrode surface exchange energy with the atoms or molecules of the liquid, ultimately causing the atoms and molecules to ionize thus precipitating breakdown. Investigations, utilizing electro-optical techniques, have demonstrated that breakdown involves streamers with tree- or bush-like structures that propagate from the electrodes [17]. High-pressure cavities, which are the precursors to breakdown streamer development, are initiated by electrical field enhancement sites that occur at asperities or protrusions on electrode surfaces. These microcavities undergo partial discharge as they expand until their ultimate collapse when their internal pressure falls to the ambient hydrostatic pressure within the dielectric liquids [18]. If the external field attains its breakdown value, then the cavities develop into breakdown streamers. The partial discharges within these prebreakdown cavities occur in the form of pulse bursts (c.f. Figure 29.7), whose duration corresponds to the time interval between the initiation and collapse of the cavities [19]. The negative streamers, emerging from the cathode, form due to electron emission, while positive streamers originating at the anode are due to free electrons in the liquid itself. The breakdown of liquids is noticeably affected by electrolytic impurities as well as water and oxygen content; also, macroscopic particles may form bridges between the electrodes along which electrons may hop with relative ease, resulting in a lower breakdown. As in solids, there is a volume effect and breakdown strength decreases with thickness; a slight increase in breakdown voltage is also observed with viscosity.

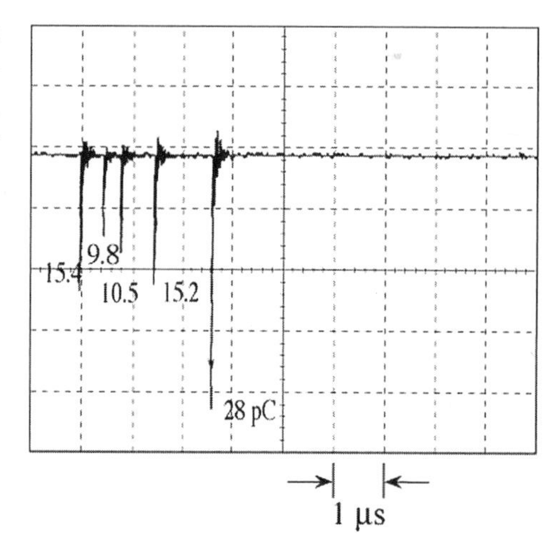

FIGURE 29.7 Partial discharge pulse burst within a prebreakdown streamer cavity in a 13.0 cSt transformer oil in a 30 mm point-to-plane gap at 35 kV rms with the charge transfers associated with each discrete pulse indicated in pC [19].

Thermal breakdown may occur at localized hot spots where the rate of heat generated exceeds that dissipated by the surrounding medium. The temperature at such hot spots continues to rise until it becomes sufficiently high to induce fusion and vaporization, causing eventually the development of a channel along which breakdown ensues between the opposite electrodes. Since a finite amount of time is required for the heat buildup to occur to lead to the thermal instability, thermally induced breakdown is contingent upon the time of the alternating voltage application and is thus implicated as the leading cause of breakdown in many solid and solid-liquid insulating systems under long-term operating conditions. However, under some circumstances thermal instability may develop over a very short time; for example, some materials have been found to undergo thermal breakdown when subjected to very short repetitive voltage pulses. In low-loss solid dielectrics, such as polyethylene, the occurrence of thermal breakdown is highly improbable under low operating temperatures, while glasses with significant ionic content are more likely to fail thermally, particularly at higher frequencies.

The criterion for thermal breakdown under ac conditions may be stated as

$$KA\Delta T/\ell = \omega\varepsilon' E^2 \tan \delta \tag{29.21}$$

where the left-hand side represents the heat transfer in $J\,s^{-1}$ along a length ℓ (cm) of sectional area A (cm^2) of the dielectric surface in the direction of the temperature gradient due to the temperature difference ΔT, in °C, such that the units of the thermal conductivity constant K are in J °C^{-1} cm^{-1} sec^{-1}. The right-hand side of Equation (29.21) is equal to the dielectric loss dissipated in the dielectric in J sec^{-1}, where E is the external field, ε' the real value of the permittivity, and tan δ the dissipation factor at the radial frequency ω.

Other causes of extrinsic breakdown are associated with particular defects in the dielectric or with the environmental conditions under which the dielectric material is employed. For example, some solid dielectrics may contain gas-filled cavities that are inherent with the porous structure of the dielectric or that may be inadvertently introduced either during the manufacturing process or created under load cycling. If the operating electrical field is sufficiently elevated to cause the gas within the cavities to undergo discharge, the dielectric will be subjected to both physical and chemical degradation by the partial discharges; should the discharge process be sustained over a sufficiently long period, breakdown may eventually ensue.

With overhead line insulators or bushings of electrical equipment, breakdown may occur along the surface rather than in the bulk of the material. Insulator surfaces consisting of porcelain, glass, or polymeric materials (usually elastomers) may become contaminated by either industrial pollutants or salt spray near coastal areas, leading to surface tracking and breakdown below the normal flashover voltage. Surface tracking is enhanced in the presence of moisture, which increases the surface conductivity, particularly in the presence of ionic contaminants [20]. The latter is measured in S or Ω^{-1} and must be distinguished from the volume conductivity whose units are stated in S cm^{-1} or Ω^{-1} cm^{-1}. Surface tracking may be prevented by cleaning the surface and by applying silicone greases. When insulators are employed under a vacuum environment, charge accumulation will occur on the surface because the charged surface will no longer be able to discharge due to the finite breakdown strength of an ambient gas. Space charge will thus play a dominant role in the surface breakdown mechanism [21].

29.4 Insulation Aging

All insulating materials will undergo varying degrees of aging or deterioration under normal operating conditions. The rate of aging will be contingent upon the magnitude of the electrical, thermal, and mechanical stresses to which the material is subjected; it will also be influenced by the composition and molecular structure of the material itself as well as the chemical, physical, and radiation environment under which the material must perform. The useful life of an insulating system will thus be determined by a given set and subset of aging variables. For example, the subset of variables in the voltage stress variable are the average and maximum values of the applied voltage, its frequency, and the recurrence rate of superposed impulse or transient voltage surges. For the thermal stress, the upper and lower ambient temperatures, the temperature gradient in the insulation, and the maximum permissible operating temperature constitute the subvariable set. Also, the character of the mechanical stress will differ, depending upon whether torsion, compression, or tension and bending are involved. Furthermore, the aging rate will be differently affected if all stresses (electrical, thermal, and mechanical) act simultaneously, separately, or in some predetermined sequence. The influence exerted on the aging rate by the environment will depend on whether the insulation system will be subjected to corrosive chemicals, petroleum fluids, water or high humidity, air or oxygen, or ultraviolet and nuclear radiation. Organic insulations, in particular, may experience chemical degradation in the presence of oxygen. For example, polyethylene under temperature cycle will undergo both physical and chemical changes. These effects will be particularly acute at the emergency operating temperatures (90 to 130°C); at these temperatures partial or complete melting of the polymer will occur and the increased diffusion rate will permit the oxygen to migrate to a greater depth into the polymer. Ultimately, the antioxidant will be consumed, resulting in an embrittlement of polymer and in extreme cases in the formation of macroscopic cracks. Subjection of the polymer to many overload cycles will be accompanied by repeated melting and recrystallization of the polymer—a process that will inevitably cause the formation of cavities, which, when subjected to sufficiently high voltages, will undergo discharge.

There is a general consensus that electrically induced aging involves the mechanisms of treeing, partial discharge, and dielectric heating. Dielectric heating failures are more characteristic of lossy insulations or when highly conductive contaminants are involved. In the treeing mechanism, a distinction must be made between electrical and water trees [1]. The former refers to growth, resembling a tree, that occurs under dry conditions in the presence of an electric field; its branches or channels are hollow. In contrast, water trees require the presence of moisture, and their branches consist of fine filamentary channels, joining small cavities, all of which contain water; when placed in a dry environment, they eventually disappear. In a translucent dielectric, water trees are invisible and are rendered visible only when stained with a die, whereas electrical trees once formed remain readily discernible. Water trees are intrinsic to solid polymeric insulation, while electrical trees may occur in both solid and impregnated insulating systems. The actual failure path, when the breakdown current is limited to prevent destruction of evidence, even in liquids, invariably consists of an electrical treelike structure.

Both electrical and water trees tend to propagate from electrical stress enhancement points, with water trees requiring appreciably lower stresses for their inception (ca. 2 kV mm^{-1}). Whereas the electrical tree mechanism usually involves partial discharges, the occurrence of the latter is conspicuously absent in water trees [22]. The fact that water trees may bridge the opposite electrodes without precipitating failure infers a nonconductive nature of the branches. Yet water trees are implicated in the overall degradation and aging process of solid dielectrics, because of their presence in areas of failure as well as because of the often-observed phenomenon of an electrical tree emerging from some point along a water tree and causing abrupt failure [1].

The presence of voids or cavities within a solid or solid–liquid insulating system may lead to eventual failure, with the proviso that the electrical field is sufficiently elevated to induce them to undergo either continuous or regularly recurring, though at times intermittent, discharge. While it is well recognized that some insulating systems are more resistant to partial discharges than others, it has not been possible thus far to arrive at any definite relationship between the partial discharge intensity and the life expectancy of insulating materials. Electrical trees may readily ensue from charge injection sites adjacent to discharging cavity inclusions. It has been suggested that the time required for electrical tree initiation may be defined quasi-empirically as [23]:

$$t = \frac{N_c}{2bfn_o[\exp(\alpha l - 1)]} \tag{29.22}$$

where N_c denotes the critical number of impact ionizations necessary for tree channel formation, l is the avalanche or newly formed tree length, n_o is the number of initiating electrons, and b is a constant dependent on the frequency, f. The deterioration rate due to partial discharges is proportional to the power dissipated P by the discharges, which may be expressed as [1]:

$$P = \sum_{i=1}^{l} \sum_{j=1}^{m} n_{ij} \Delta Q_{ij} V_{ij} \tag{29.23}$$

where n_{ij} is the recurrence rate of the jth discharge pulse in the ith cavity and ΔQ_{ij} is its corresponding charge transfer at the instantaneous voltage V_{ij}. Under ac conditions, the discharges will tend to recur regularly in each cycle due to the capacitative voltage division across the void. Under dc conditions, the discharge rate will be controlled by the time constant required to recharge the cavity following a previous discharge. The physical damage arising from discharges consists of surface erosion and pitting, and is caused by the ion and electron bombardment incident on the void's walls at a given discharge site. Chemical degradation results from molecular chain scission due to particle impact on the surface and the reactions between the ambient ionized gases and the gases released due to the molecular chain scission processes. The final chemical composition of the reaction products is generally varied, depending primarily on the molecular structure of the dielectric materials involved and the composition of the ambient ionized gases; one discharge degradation product common to many polymers exposed to discharges in air is that of oxalic acid. Oxalic acid, as a result of its elevated conductivity, when deposited upon the cavity's walls may change the nature of the discharge (e.g., from a spark to a pseudoglow or glow type) or may even ultimately extinguish the discharge (i.e., replace the discharge loss by an I^2R-type loss along the cavity's walls). There are various procedures available for detecting, locating, and identifying partial discharge sources in electrical apparatus and cables [24]. These may employ pulse reflectometry methods, capacitive and inductive probes, acoustical transducers as well as ultraviolet detectors. Also partial discharge pulse pattern recognition and classification techniques, utilizing principally neural networks, fuzzy logic, and other intelligent machine procedures, have been applied with considerable success [25].

A number of aging models have been propounded to predict insulation aging under different types of stress. However, there are essentially only two models whose usefulness has been substantiated in practice and which have, therefore, gained wide acceptance. One of these is Dakin's classical thermal degradation model, which is based upon the approach used in chemical reaction rate theory [26]:

$$t = A\exp[\Delta H/kT] \tag{29.24}$$

where t represents the time to breakdown, A is a preexponential constant, ΔH is the activation energy of the aging process, T is the absolute temperature, and k is the Boltzmann constant ($1.38\ 3\times10^{-23}$ J K^{-1}). If the log t versus $1/T$ plot represents a straight line obtained when the insulation aging is accelerated at various temperatures above the operating temperature, extrapolation of the line to the operating temperature may yield a rudimentary estimate of the service life that can be anticipated from the insulation system when operated under normal temperature and load conditions. Deviations from straight-line behavior are indicative of more than one thermal aging mechanism; for example, a polymeric insulation will exhibit such behavior when thermally stressed beyond its melting or phase transition temperature.

Another extremely useful classical model, applicable to accelerated aging studies under electrical stress, is the so-called inverse power law relationship given by [26]:

$$t = BE^{-n} \tag{29.25}$$

where t is the time to breakdown under an electric stress E, B is a constant, and n is an exponent parameter. The relationship is essentially empirical in nature, and its proof of validity rests primarily on experiment and observation. It is found that a single type of electrically induced aging process will generally result in a straight-line relationship between log t and log E. Consequently, aging data obtained at higher electrical stresses with correspondingly shorter times to breakdown when extrapolated to longer times at stresses in the vicinity of the operating stress may thus conceivably yield a rudimentary service life estimate under the operating stress. The slope of the line determines the exponent n, which constitutes an approximate indicator of the type of aging involved. For example, with polymer insulation water treeing ordinarily results in n values less than 4, while under conditions that may involve discharges and electrical trees, the n values may approach 10 or greater. When accelerated aging tests are carried out to determine life expectancy estimates of newly developed electrical insulating systems, it is advantageous to perform those tests under simultaneous electrical, thermal, and mechanical stresses in order to simulate as much as possible their normal operating conditions. This permits the use of lower stress factor magnitudes to achieve reasonably acceptable test times, while also ensuring that the induced degradation mechanisms approximate more closely those that are likely to occur in service.

29.5 Dielectric Materials

Dielectric materials comprise a variety of solids, liquids, and gases. The breakdown strength generally increases with the density of the material so that dielectric solids as well as liquids tend to have higher dielectric strengths than gases. The same tendency is observed also with the dielectric loss and permittivity; for example, the dielectric losses in gases are virtually too small to be measurable and their dielectric constant in most practical applications can be considered as unity. In this section, we shall describe a number of the more common insulating materials in use.

Gases

The 60-Hz breakdown strength of a 1 cm gap of air at 25°C under atmospheric pressure is 31.7 kV cm^{-1}. Although this is a relatively low value, air is a most useful insulating medium for large electrode separations as is the case for overhead transmission lines. The only dielectric losses in the overhead lines are those due to corona discharges at the line conductor surfaces and leakage losses over the insulator surfaces. In addition, the highly reduced capacitance between the conductors of the lines ensures a small capacitance per unit length, thus rendering overhead lines an efficient means for transmitting large amounts of power over long distances. Nitrogen, which has a slightly higher breakdown strength (33.4 kV cm^{-1}) than air, has been used when compressed in low-loss gas-insulated capacitors. However, as air, which was also used in its compressed form in circuit breakers, it has been replaced by sulfur hexafluoride, SF_6. The breakdown strength of SF_6 is *ca.* 79.3 kV cm^{-1}. Since the outside metal casing of metal-clad circuit breakers is ordinarily grounded, it has become common practice to have much of the substation interconnected bus and interrupting switching equipment insulated with SF_6. To achieve higher breakdown strengths, the SF_6 gas is compressed at pressures on the order of 6 atm. The use of SF_6 in substation equipment has resulted in the saving of space and the elimination of line insulator pollution problems; however, its extensive usage has also posed some environmental concerns in regard to its fluorine content.

The breakdown strength of gases is little affected with increasing frequency until the period of the wave becomes comparable to the transit time of the ions and, finally, electrons across the gap. At this point, a substantial reduction in the breakdown strength is observed.

Insulating Liquids

Insulating liquids are rarely used by themselves and are intended for use mainly as impregnants with cellulose or synthetic papers. The 60-Hz breakdown strength of practical insulating liquids exceeds that of gases and for a 1 cm gap separation, it is of the order of about 100 kV cm^{-1}. However, since the breakdown strength increases with decreasing gap length, and as the oils are normally evaluated using a gap separation of 0.254 cm, the breakdown strengths normally cited range from approximately 138 to 240 kV cm^{-1}. The breakdown values are more influenced by the moisture and particle contents of the fluids than by their molecular structure (cf. Table 29.1).

Mineral oils have been extensively used in high-voltage electrical apparatus since the turn of the century. They constitute a category of hydrocarbon liquids that are obtained by refining petroleum crudes. Their composition consists of paraffinic, naphthenic, and aromatic constituents, and is dependent upon the source of the crude as well as the refining procedure followed. The inclusion of the aromatic constituents is desirable because of their gas absorption and oxidation characteristics. Mineral oils used for cable and transformer applications have low polar molecule contents, and are characterized by dielectric constants extending from about 2.10 to 2.25 with dissipation factors generally between 2×10^{-5} and 6×10^{-5} at room temperature,

TABLE 29.1 Electrical Properties of a Number of Representative Insulating Liquids

Liquid	Viscosity cSt (37.8°C)	Dielectric Constant (at 60 Hz, 25°C)	Dissipation Factor (at 60 Hz, 100°C)	Breakdown Strength (kV cm^{-1})
Capacitor oil	21	2.2	0.001	>118
Pipe cable oil	170	2.15	0.001	>118
Self-contained cable oil	49.7	2.3	0.001	>118
Heavy cable oil	2365	2.23	0.001	>118
Transformer oil	9.75	2.25	0.001	>128
Alkyl benzene	6.0	2.1	0.0004	>138
Polybutene pipe cable oil	110 (SUS)	2.14 (at 1 MHz)	0.0003	>138
Polybutene capacitor oil	2200 (SUS at 100°C)	2.22 (at 1 MHz)	0.0005	>138
Silicone fluid	50	2.7	0.00015	>138
Castor oil	98 (100°C)	3.74	0.06	>138
$C_8F_{16}O$ fluorocarbon	0.64	1.86	<0.0005	>138

After [2] and [27].

depending upon their viscosity and molecular weight. Their dissipation factors increase appreciably at higher temperatures when the viscosities are reduced. Oils may deteriorate in service due to oxidation and moisture absorption. Filtering and treatment with Fullers' earth may improve their properties, but special care must be taken to ensure that the treatment process does not remove the aromatic constituents which are essential to maintaining the gas-absorption characteristics of the oil.

Alkyl benzenes are used as impregnants in high-voltage cables, often as substitutes of the low-viscosity mineral oils in self-contained oil-filled cables [22]. They consist of alkyl chains attached to a benzene ring, having the general formula $C_6H_5(CH_2)_nCH_3$; however, branched alkyl benzenes are also employed. Their electrical properties are comparable to those of mineral oils, and they exhibit good gas inhibition characteristics. Because of their detergent character, alkyl benzenes tend to be more susceptible to contamination than mineral oils.

Polybutenes are synthetic oils that are derived from the polymerization of olefins. Their long molecular chain structures, with isobutene as the base unit, have methyl group side chains with molecular weights in the range from 300 to 1350. Their electrical properties are comparable to those of mineral oils; due to their low cost, they have been employed as pipe cable filling oils. Higher viscosity polybutenes have been also used as capacitor impregnants. Mixtures of polybutenes and alkyl benzenes have been utilized to obtain higher ac breakdown strength with impregnated paper systems. They are also compatible and miscible with mineral oils.

Since the discontinued use of the nonflammable polychlorinated biphenyls (PCBs), a number of unsaturated synthetic liquids have been developed for application to capacitors where, due to high stresses, the evolved gases may readily undergo partial discharge. Most of these new synthetic capacitor fluids are thus gas-absorbing low-molecular-weight derivatives of benzene, with permittivities ranging from 2.66 to 5.25 at room temperature (compared to 3.5 for PCBs). None of these fluids has the nonflammable characteristics of the PCBs; however, they do have high boiling points [2].

Halogenated aliphatic hydrocarbons are derived by replacing the hydrogen by either chlorine or fluorine or both; they may also contain nitrogen and oxygen in their molecular structure. Their dielectric constants range from 1.8 to 3.0, the higher value reflecting some polarity due to molecular asymmetry as a result of branching. They have superior thermal properties to mineral oils, and are highly flame-resistant. Fluorocarbons have been used in large power transformers, where both flammability and heat removal are of prime concern.

Silicone liquids consist of polymeric chains of silicon atoms alternating with oxygen atoms, with methyl side groups. For electrical applications, polydimethylsiloxane fluids are used, primarily for transformers as substitutes for the PCBs, due to their inherently high flash and flammability points and reduced environmental concerns. They have lower tan δ values than mineral oils, but somewhat higher dielectric constants because of their moderately polar nature. The viscosity of silicone fluids exhibits relatively little change with temperature, which is attributed to the ease of rotation about the Si–O–Si bond, thereby overcoming close packing of molecules and reducing intermolecular forces.

There are a large number of organic esters, but only a few are suitable for electrical applications. Their properties are adversely affected by hydrolysis, oxidation, and water content. Because of their reduced dielectric losses at elevated frequencies, they have been used in high-frequency capacitors. Castor oil has found specialized application in energy storage capacitors due to its exceptional resistance to partial discharges. The dielectric constants of esters are substantially higher than those for mineral oils.

Solid Insulating Materials

Solid insulating materials may be classified into two main categories: organic and inorganic [1]. There are an extremely large number of solid insulants available, but in this section only the more commonly representative solid insulants will be considered.

Inorganic Solids

Below are described a number of the more prevalent inorganic dielectrics in use; their electrical and physical properties are listed in Table 29.2.

TABLE 29.2 Electrical and Physical Properties of Some Common Solid Insulating Materials

Material	Specific Gravity	Maximum Operating Temperature (°C)	Dielectric Constant 20°C			Dissipation Factor 20°C			AC Dielectric Strength (kV cm^{-1})
			60 Hz	1 kHz	1 MHz	60 Hz	1 kHz	1 MHz	
Alumina (Al_2O_3)	3.1–3.9	1950	8.5	8.5	8.5	1×10^{-3}	1×10^{-3}	1×10^{-3}	98–157
Porcelain (mullite)	2.3–2.5	1000	8.2	8.2	8.2	1.4×10^{-3}	5.7×10^{-4}	2×10^{-4}	94–157
Steatite $3MgO \cdot 4SiO_2 \cdot H_2O$	2.7–2.9	1000–1100	5.5	5.0	5.0	1.3×10^{-3}	4.5×10^{-4}	3.7×10^{-4}	200
Magnesium oxide (MgO)	3.57	<2800	9.65	9.65	9.69	$<3\times10^{-4}$	$<3\times10^{-4}$	$<3\times10^{-4}$	>2000
Glass (soda lime)	2.47	110–460	6.25	6.16	6.00	5.0×10^{-3}	4.2×10^{-3}	2.7×10^{-3}	4500
Mica ($KA_{l2}(OH)_2 Si_3AlO_{10}$)	2.7–3.1	550	6.9	6.9	5.4	1.5×10^{-3}	2.0×10^{-4}	3.5×10^{-4}	3000–8200
SiO_2 film	—	<900	—	3.9	—	—	7×10^{-4}	—	1000–10,000
Si_3N_4	—	<1000	—	12.7	—	—	$<1\times10^{-4}$	—	1000–10,000
Ta_2O_5	8.2	<1800	—	28	—	—	1×10^{-2}	—	—
HfO_2	—	4700°F	—	35	—	—	1×10^{-2}	—	—
Low-density PE	(density 0.910–0.925 g cm^{-3})	70	2.3	2.3	2.3	2×10^{-4}	2×10^{-4}	2×10^{-4}	181–276
Medium-density PE	(density: 0.926–0.940 g cm^{-3})	70	2.3	2.3	2.3	2×10^{-4}	2×10^{-4}	2×10^{-4}	197–295
High-density PE	(density: 0.941–0.965 g cm^{-3})	70	2.35	2.35	2.35	2×10^{-4}	2×10^{-4}	2×10^{-4}	177–197
XLPE	(density: 0.92 g cm^{-3})	90	2.3	—	2.28	3×10^{-4}	—	4×10^{-4}	217
EPR	0.86	300–350°F	—	3.0–3.5	—	4×10^{-3}	—	—	354–413
Polypropylene	0.90	128–186	2.22–2.28	2.22–2.28	2.22–2.28	$2–3\times10^{-4}$	$2.5–3.0\times10^{-4}$	4.6×10^{-4}	295–314
PTFE	2.13–2.20	<327	2.0	2.0	2.0	$<2\times10^{-4}$	$<2\times10^{-4}$	$<2\times10^{-4}$	189
Glass-reinforced polyester premix	1.8–2.3	265	5.3–7.3	—	5.0–6.4	$1–4\times10^{-2}$	—	$0.8–2.2\times10^{-2}$	90.6–158
Thermoplastic polyester	1.31–1.58	250	3.3–3.8 (100 Hz)	—	—	$1.5–2.0\times10^{-3}$	—	—	232–295
Polyimide polyester	1.43–1.49	480°F	—	3.4 (100 kHz)	—	—	$1–5\times10^{-3}$ (100 kHz)	—	220
Polycarbonate	1.20	215	3.17	—	2.96	9×10^{-4}	—	1×10^{-2}	157
Epoxy (with mineral filler)	1.6–1.9	200 (decomposition temperature)	4.4–5.6	4.2–4.9	4.1–4.6	$1.1–8.3\times10^{-2}$	$0.19–1.4\times10^{-1}$	$0.13–1.4\times10^{-1}$	98.4–158
Epoxy (with silica filler)	1.6–2.0	200 (decomposition temperature)	3.2–4.5	3.2–4.0	3.0–3.8	$0.8–3.0\times10^{-2}$	$0.8–3.0\times10^{-2}$	$2–4\times10^{-2}$	158–217
Silicone rubber	1.1–1.5	700°F	3.3–4.0	—	3.1–3.7	$1.5–3.0\times10^{-2}$	—	$3.0–5.0\times10^{-3}$	158–197

After [1] and [27].

Alumina (Al_2O_3) is produced by heating aluminum hydroxide or oxyhydroxide; it is widely used as a filler for ceramic insulators. Further heating yields the corundum structure which, in its sapphire form, is used for dielectric substrates in microcircuit applications.

Barium titanate ($BaTiO_3$) is an extraordinary dielectric in that at temperatures below 120°C it behaves as a ferroelectric. That is, the electric displacement is both a function of the field as well as its previous history. Because of spontaneous polarization of the crystal, a dielectric hysteresis loop is generated. The dielectric constant is different in the x- and z-axis of the crystal (e.g., at 20°C, $\varepsilon' > 4000$ perpendicular to the z-axis and $\varepsilon' < 300$ in the x-axis direction).

Porcelain is a multiphase ceramic material that is obtained by heating aluminum silicates until a mullite ($3Al_2O_3 \cdot 2SiO_2$) phase is formed. Since mullite is porous, its surface must be glazed with a high-melting-point glass to render it smooth and impervious, and thus applicable for use in overhead line insulators. For high-frequency applications, low-loss single-phase ceramics, such as steatite ($3MgO \cdot 4SiO_2 \cdot H_2O$), are preferred.

Magnesium oxide (MgO) is a common inorganic insulating material, which due to its relatively high thermal conductivity is utilized for insulating heating elements in ovens. The resistance wire elements are placed concentrically within stainless steel tubes, with magnesium oxide packed around them to provide the insulation.

Electrical-grade glasses consist principally of SiO_2, B_2O_3, and P_2O_5 structures that are relatively open to permit ionic diffusion and migration. Consequently, glasses tend to be relatively lossy at high temperatures, though at low temperatures they are suitable for use in overhead line insulators and in transformer, capacitor, and circuit breaker bushings. At high temperatures, their main application lies with incandescent and fluorescent lamps as well as electronic tube envelopes.

Most of the mica used in electrical applications is of the muscovite [$KAl_2(OH)_2Si_3AlO_{10}$] type. Mica is a layer-type dielectric, and mica films are obtained by the splitting of mica blocks. The extended two-dimensionally layered strata prevent the formation of conductive pathways across the mica, resulting in a high dielectric strength. It has excellent thermal stability and due to its inorganic nature is highly resistant to partial discharges. It is used in sheet, plate, and tape form in rotating machine and transformer coils. For example, a mica-epoxy composite is employed in stator bar insulation of rotating machines.

In metal-oxide-silicon (MOS) devices, the semiconductor surface is passivated by thermally growing a silicon dioxide, SiO_2, film (about 5000 Å) with the semiconductor silicon wafer exposed to oxygen ambient at 1200°C. The resulting SiO_2 dielectric film has good adhesion properties, but due to its open glassy structure is not impervious to ionic impurities (primarily sodium). Accordingly, a denser film structure of silicon nitride, Si_3N_4, is formed in a reaction between silane and ammonia, and is pyrolytically deposited on the SiO_2 layer. The thin film of Si_3N_4 is characterized by extremely low losses, and its relatively closed structure does not provide any latitude for free sodium movement, thereby providing complete passivation of the semiconductor device. The high dielectric strength of the double film layer of SiO_2 and Si_3N_4 renders it dielectrically effective in field-effect transistor (FET) applications.

Low loss fused silica (SiO_2) is a dielectric, which constitutes the base material of choice for optical fiber communication cables [22]. The optical fibers consist of a silica inner core surrounded by a cladding that is covered by a protective coating. The light wave guiding characteristics of the silica optical fiber are determined by the index of refraction of the inner core, which is usually enhanced by the use of dopants such as phosphorous pentoxide (P_2O_5) or germanium oxide (GeO_2), and that of the cladding surrounding the core, which is lowered slightly below that of the core by the addition of chlorine, fluorine, or boron oxide (B_2O_3). The optical fibers are designed to operate within the infrared wavelength regions, extending from 0.80 to 1.55 μm, and are thus removed from the electronic and atomic absorption regions (c.f. Figure 29.2).

In integrated circuit devices, a number of materials are suitable for thin-film capacitor applications. In addition to Al_2O_3, tantalum pentoxide, Ta_2O_5, has been extensively utilized. It is characterized by high-temperature stability and is resistant to acids with the exception of hydrofluoric acid (HF). The high dielectric constant material hafnia (HfO_2) has also been used in thin-film capacitors.

Organic Solids

Solid organic dielectrics consist of large polymer molecules, which generally have molecular weights in excess of 600. Primarily, with the exception of paper which consists of cellulose that is comprised of a series of glucose units, organic dielectric materials are synthetically derived.

Polyethylene (PE) is perhaps one of the most common solid dielectrics, and is extensively used as a solid dielectric extruded insulant in power and communication cables [22]. Linear PE is classified as a low- (0.910 to 0.925), medium- (0.926 to 0.940), or high- (0.941 to 0.965) density polymer (cf. Table 29.2). Since PE is essentially a long-chain hydrocarbon material in which the repeat unit is $–CH_2–CH_2–$, a low-density PE necessarily implies a high degree of branching. Decreased branching increases the crystallinity as molecules undergo internal folding, which leads to improved stiffness, tear strength, hardness, and chemical resistance. Cross linking of PE produces a thermosetting polymer with a superior temperature rating, improved tensile strength, and an enhanced resistance to partial discharges. Most of the PE used on extruded cables is of the cross-linked polyethylene (XLPE) type.

Ethylene–propylene rubber (EPR) is an amorphous elastomer, which is synthesized from ethylene and propylene. As an extrudent on cables its composition has filler contents up to 50%, comprising primarily clay, with smaller amounts of added silicate and carbon black. The dielectric losses are appreciably enhanced by the fillers and, consequently, EPR is not suitable for extra-high-voltage applications, with its use being usually confined to lower and intermediate voltages ($\leq$138 kV) and also where high cable flexibility due to its rubber-like properties may be additionally desired.

Polypropylene has a structure related to that of ethylene with one added methyl group. It is a thermoplastic material having properties similar to high-density PE, though due to its lower density, it has also a lower dielectric constant. It has many electrical applications both in bulk form as in molded and extruded insulations as well as in film form in taped capacitor, transformer, and cable insulations.

Polytetrafluoroethylene (PTFE) or Teflon®* is a fully fluorinated version of PE, having a repeat unit of $[–CF_2–CF_2–]$. It is characterized by a low dielectric constant, extremely low losses, and has excellent temperature stability and is resistive to chemical degradation. It has been extensively used in specialized applications on insulators, wires and cables, transformers, motors, and generators. Its relatively high cost is attributable to both the higher cost of the fluorinated monomers as well as the specialized fabrication techniques required.

Polyesters are obtained most commonly by reacting a dialcohol with a diester; they may be either thermosetting or thermoplastic. The former are usually employed in glass laminates and glass fiber-reinforced moldings, while thermoplastic polyesters are used for injection-molding applications. They are used in small and large electrical apparatus as well as in electronic applications.

Polyimides (kaptons®*), as nylons®* (polyamides), have nitrogen in their molecular structure. They constitute a class of high-temperature thermoplastics that may be exposed to continuous operation at 480°F. When glass-reinforced, they may be exposed to temperatures as high as 700°F; they are used in molded, extruded wire, and film form.

Polycarbonates are thermoplastics that are closely related to polyesters. They are primarily employed in the insulation of electrically powered tools and in the casings of electrical appliances. Polycarbonates may be either compression- or injection-molded and extruded as films or sheets.

Epoxy resins are prepared from an epoxide monomer. The first step involves a reaction between two co-monomers; in the subsequent step, the prepolymer is cured by means of a cross-linking agent. Epoxy resins are characterized by low shrinkage and high mechanical strength; they may be reinforced with glass fibers and mixed with mica flakes. Epoxy resins have many applications such as, for example, for insulation of bars in the stators of rotating machines, solid-type transformers, and spacers for compressed-gas-insulated busbars and cables.

Silicone rubber is classified as an organic-inorganic elastomer, which is obtained from the polymerization of organic siloxanes. They are composed of dimethyl-siloxane repeat units, $(CH_3)_2SiO–$, with the side groups being methyl units. Fillers are added to obtain the desired silicone rubber compounds; cross–linking is carried

*Registered trademark, E.I. DuPont de Nemours Co., Inc., Wilmington, DE.

TABLE 29.3 Electrical Properties of Taped Solid–Liquid Insulations

Tape	Impregnating Liquid	Average Voltage Stress (kV cm^{-1})	tan δ at Room Temperature	tan δ at Operating Temperature
Kraft paper	Mineral oil	180	3.8×10^{-3} at 23°C	5.7×10^{-3} at 85°C
Kraft paper	Silicone liquid	180	2.7×10^{-3} at 23°C	3.1×10^{-3} at 85°C
Paper–polypropylene–paper (PPP)	Dodecyl benzene	180	9.8×10^{-4} at 18°C	9.9×10^{-4} at 100°C
Kraft paper	Polybutene	180	2.0×10^{-3} at 25°C	2.0×10^{-3} at 85°C

out with peroxides. Since no softeners and plasticizers are required, silicone rubbers are resistant to embrittlement and may be employed for low-temperature applications down to −120°F. Continuous operation is possible up to 500°F, with intermittent usage as high as 700°F.

Solid–Liquid Insulating Systems

Impregnated-paper insulation constitutes one of the earliest insulating systems employed in electrical power apparatus and cables [22]. Although in some applications alternate solid- or compressed-gas insulating systems are now being used, the impregnated-paper system still constitutes one of the most reliable insulating systems available. Proper impregnation of the paper results in a cavity-free insulating system, thereby eliminating the occurrence of partial discharges that may lead to deterioration and breakdown of the insulating system. The cellulose structure of paper has a finite acidity content as well as a residual colloidal or bound water, which is held by hydrogen bonds. Consequently, impregnated cellulose base papers are characterized by somewhat more elevated tan δ values on the order of 2×10^{-3} at 30 kV cm^{-1}. The liquid impregnants employed are either mineral oils or synthetic fluids. Since the dielectric constant of these fluids is normally about 2.2 and that of dried cellulose may extend from 6.5 to 10, the resulting dielectric constant of the impregnated paper is approximately between 3.1 and 3.5.

Lower-density cellulose papers have slightly lower dielectric losses, but the dielectric breakdown strength is also reduced. The converse is true for impregnated systems, utilizing higher-density papers. The general chemical formula of cellulose paper is $C_{12}H_{20}O_{10}$. If the paper is heated beyond 200°C, the chemical structure of the paper breaks down even in the absence of external oxygen, since the latter is readily available from within the cellulose molecule. To avert this process from occurring, cellulose papers are ordinarily not used beyond 100°C.

In an attempt to reduce the dielectric losses in solid–liquid systems, cellulose papers have been substituted in some applications by synthetic papers (c.f. Table 29.3). For example in extra-high-voltage cables, cellulose paper-polypropylene composite tapes have been employed. A partial paper content in the composite tapes is necessary both to retain some of the impregnation capability of a porous cellulose paper medium and to maintain the relative ease of cellulose-to-cellulose tape sliding capability upon bending. In transformers the synthetic nylon® or polyamide paper (nomex®) is used both in film and board form. It may be continuously operated at temperatures up to 220°C.

Defining Terms

Conductivity σ: Represents the ratio of the leakage current density to the applied electric field intensity. In general its ac and dc values differ because the mechanisms for establishing the leakage current in the two cases are not necessarily identical.

Dielectric: A material in which nearly all or a large portion of the energy required for its charging can be recovered when the external electric field is removed.

Dielectric constant ε': A quantity that determines the amount of electrostatic energy, which can be stored per unit volume per unit potential gradient. It is a real quantity and in Gaussian-CGS units is numerically equal to the ratio of the measured capacitance of the specimen, C, to the equivalent geometrical capacitance in vacuo, C_o. It is also commonly referred to as the real value of the permittivity. Note that when the SI system of units is employed, the ratio C/C_o defines the real value of the relative permittivity ε_r'.

Dielectric loss: The rate at which the electrical energy supplied to a dielectric material by an alternating electrical field is changed to heat.

Dielectric strength: Represents the value of the externally applied electric field at which breakdown or failure of the dielectric takes place. Unless a completely uniform field gradient can be assured across the dielectric specimen, the resulting breakdown value will be a function of the specimen thickness and the test electrode geometry; this value will be substantially below that of the intrinsic breakdown strength.

Dissipation factor (tan δ): Equal to the tangent of the loss angle δ, which is the phase angle between the external electric field vector and the resulting displacement vector $\overline{D}$. It is numerically equal to the ratio of the imaginary permittivity ε'' to the real permittivity ε'; alternatively, it is defined by the ratio of the leakage current to the displacement (charging or capacitive) current.

References

1. R. Bartnikas and R.M. Eichhorn, Eds., *Engineering Dielectrics, Vol. II A, Electrical Properties of Solid Insulating Materials: Molecular Structure and Electrical Behavior*, STP 783, Philadelphia/West Conshohocken: ASTM, 1983.
2. R. Bartnikas, Ed., *Engineering, Dielectrics, Vol. III, Electrical Insulating Liquids*, Monograph 2, Philadelphia/West Conshohocken: ASTM, 1994.
3. A. von Hippel, *Dielectrics and Waves*, New York: Wiley, 1956.
4. P.K. Watson, *IEEE Trans. Dielect. Electr. Insul.*, 2, 915–924, 1995.
5. C. LeGressus and G. Blaise, *IEEE Trans. Electr. Insul.*, 27, 472–481, 1992.
6. S. Holé, T. Ditchi, and J. Lewiner, *IEEE Trans. Dielect. Electr. Insul.*, 10, 670–677, 2003.
7. T. Maeno, T. Futami, H. Kushibe, T. Takada, and C.M. Cooke, *IEEE Trans. Electr. Insul.*, 23, 433–439, 1988.
8. H.J. Wintle, *IEEE Trans. Electr. Insul.*, 25, 27–44, 1990.
9. D.K. Das-Gupta, J.S. Hornsby, G.M. Hornsby, and G.M. Sessler, *J. Phys. D: Appl. Phys.*, 29, 3113–3116, 1996.
10. R. Bartnikas, Ed., *Engineering Dielectrics, Vol. II B, Measurement Techniques*, STP 926, Philadelphia/West Conshohocken: ASTM, 1987.
11. J.K. Nelson, *Breakdown Strength of Solids*, in Ref. [1] Chapter 5.
12. A. von Hippel and G.M. Lee. *Phys. Rev.*, 59, 824–826, 1941.
13. W.G. Oakes, *Proc. IEE*, 90(I), 37–43, 1949.
14. T.J. Lewis, *IEEE Trans. Dielect. Electr. Insul.*, 9, 717–729, 2002.
15. M.J. Schönhuber, *IEEE Trans. Power Apparatus Syst.*, PAS-88, 100–107, 1969.
16. R. Bartnikas and E.J. McMahon, Eds., *Engineering Dielectrics, Vol. I, Corona Measurement and Interpretation*, STP 669, Philadelphia/West Conshohocken: ASTM, 1979.
17. E.O. Forster, *Electrical Breakdown in Dielectric Liquids*, Ref. [2], Chapter 3.
18. F. Aitken, F. McCluskey, and A. Denat, *J. Fluid Mech.*, 327, 373–392, 1996.
19. M. Pompili, R. Bartnikas, and C. Mazzetti, *IEEE Trans. Dielect. Electr. Insul.*, 12, 395–403, 2005.
20. K.N. Mathes, *Surface Failure Measurements*, in Ref. [10] Chapter 4.
21. H.C. Miller, *IEEE Trans. Electr. Insul.*, 28, 512–527, 1993.
22. R. Bartnikas and K.D. Srivastava, Eds., *Power and Communication Cables*, New York: Wiley/IEEE Press, 2003.
23. J.C. Fothergill, L.A. Dissado, and P.J.J. Sweeny, *IEEE Trans. Dielect. Electr. Insul.*, 1, 474–486, 1994.
24. R. Bartnikas, *IEEE Trans. Dielect. Electr. Insul.*, 9, 668–670, 2002.
25. N. C. Sahoo, M.M.A. Salama, and R. Bartnikas, *IEEE Trans. Dielect. Electr. Insul.*, 12, 248–264, 2005.
26. W.D. Wilkens, *Statistical Methods for the Evaluation of Electrical Insulating Systems*, in Ref. [10] Chapter 7.
27. Encyclopedia Issue, *Insulation Circuits*, June/July 1972.

IV

Mathematics, Symbols, and Physical Constants

Ronald J. Tallarida
Temple University

THE GREAT ACHIEVEMENTS in engineering deeply affect the lives of all of us and also serve to remind us of the importance of mathematics. Interest in mathematics has grown steadily with these engineering achievements and with concomitant advances in pure physical science. Whereas scholars in nonscientific fields, and even in such fields as botany, medicine, geology, etc., can communicate most of the problems and results in nonmathematical language, this is virtually impossible in present-day engineering and physics. Yet it is interesting to note that until the beginning of the twentieth century, engineers regarded calculus as something of a mystery. Modern students of engineering now study calculus, as well as differential equations, complex variables, vector analysis, orthogonal functions, and a variety of other topics in applied analysis. The study of systems has ushered in matrix algebra and, indeed, most engineering students now take linear algebra as a core topic early in their mathematical education.

This section contains concise summaries of relevant topics in applied engineering mathematics and certain key formulas, that is, those formulas that are most often needed in the formulation and solution of engineering problems. Whereas even inexpensive electronic calculators contain tabular material (e.g., tables of trigonometric and logarithmic functions) that used to be needed in this kind of handbook, most calculators do not give symbolic results. Hence, we have included formulas along with brief summaries that guide their use. In many cases we have added numerical examples, as in the discussions of matrices, their inverses, and their use in the solutions of linear systems. A table of derivatives is included, as well as key applications of the derivative in the solution of problems in maxima and minima, related rates, analysis of curvature, and finding approximate roots by numerical methods. A list of infinite series, along with the interval of convergence of each, is also included.

Of the two branches of calculus, integral calculus is richer in its applications, as well as in its theoretical content. Though the theory is not emphasized here, important applications such as finding areas, lengths, volumes, centroids, and the work done by a nonconstant force are included. Both cylindrical and spherical polar coordinates are discussed, and a table of integrals is included. Vector analysis is summarized in a separate section and includes a summary of the algebraic formulas involving dot and cross multiplication, frequently needed in the study of fields, as well as the important theorems of Stokes and Gauss. The part on special functions includes the gamma function, hyperbolic functions, Fourier series, orthogonal functions, and both Laplace and z-transforms. The Laplace transform provides a basis for the solution of differential equations and is fundamental to all concepts and definitions underlying analytical tools for describing feedback control systems. The z-transform, not discussed in most applied mathematics books, is most useful in the analysis of discrete signals as, for example, when a computer receives data sampled at some prespecified time interval. The Bessel functions, also called cylindrical functions, arise in many physical applications, such as the heat transfer in a "long" cylinder, whereas the other orthogonal functions discussed—Legendre, Hermite, and Laguerre polynomials—are needed in quantum mechanics and many other subjects (e.g., solid-state electronics) that use concepts of modern physics.

The world of mathematics, even applied mathematics, is vast. Even the best mathematicians cannot keep up with more than a small piece of this world. The topics included in this section, however, have withstood the test of time and, thus, are truly *core* for the modern engineer.

This section also incorporates tables of physical constants and symbols widely used by engineers. While not exhaustive, the constants, conversion factors, and symbols provided will enable the reader to accommodate a majority of the needs that arise in design, test, and manufacturing functions.

Mathematics, Symbols, and Physical Constants

Greek Alphabet

Greek Letter	Greek Name	English Equivalent	Greek Letter	Greek Name	English Equivalent
Α α	Alpha	a	Ν ν	Nu	n
Β β	Beta	b	Ξ ξ	Xi	x
Γ γ	Gamma	g	Ο ο	Omicron	ŏ
Δ δ	Delta	d	Π π	Pi	P
Ε ε	Epsilon	ĕ	Ρ ρ	Rho	r
Ζ ζ	Zeta	z	Σ σ	Sigma	s
Η η	Eta	ē	Τ τ	Tau	t
Θ θ ϑ	Theta	th	Υ υ	Upsilon	u
Ι ι	Iota	i	Φ ϕ φ	Phi	ph
Κ κ	Kappa	k	Χ χ	Chi	ch
Λ λ	Lambda	l	Ψ ψ	Psi	ps
Μ μ	Mu	m	Ω ω	Omega	ō

International System of Units (SI)

The International System of units (SI) was adopted by the 11th General Conference on Weights and Measures (CGPM) in 1960. It is a coherent system of units built form seven *SI base units,* one for each of the seven dimensionally independent base quantities: they are the meter, kilogram, second, ampere, kelvin, mole, and candela, for the dimensions length, mass, time, electric current, thermodynamic temperature, amount of substance, and luminous intensity, respectively. The definitions of the SI base units are given below. The *SI derived units* are expressed as products of powers of the base units, analogous to the corresponding relations between physical quantities but with numerical factors equal to unity.

In the International System there is only one SI unit for each physical quantity. This is either the appropriate SI base unit itself or the appropriate SI derived unit. However, any of the approved decimal prefixes, called *SI prefixes,* may be used to construct decimal multiples or submultiples of SI units.

It is recommended that only SI units be used in science and technology (with SI prefixes where appropriate). Where there are special reasons for making an exception to this rule, it is recommended always to define the units used in terms of SI units. This section is based on information supplied by IUPAC.

Definitions of SI Base Units

Meter: The meter is the length of path traveled by light in vacuum during a time interval of 1/299,792,458 of a second (17th CGPM, 1983).

Kilogram: The kilogram is the unit of mass; it is equal to the mass of the international prototype of the kilogram (3rd CGPM, 1901).

Second: The second is the duration of 9,192,631,770 periods of the radiation corresponding to the transition between the two hyperfine levels of the ground state of the cesium-133 atom (13th CGPM, 1967).

Ampere: The ampere is that constant current which, if maintained in two straight parallel conductors of infinite length, of negligible circular cross-section, and placed 1 m apart in vacuum, would produce between these conductors a force equal to 2×10^{-7} newton per meter of length (9th CGPM, 1948).

Kelvin: The kelvin, unit of thermodynamic temperature, is the fraction 1/273.16 of the thermodynamic temperature of the triple point of water (13th CGPM, 1967).

Mole: The mole is the amount of substance of a system which contains as many elementary entities as there are atoms in 0.012 kg of carbon-12. When the mole is used, the elementary entities must be specified and may be atoms, molecules, ions, electrons, or other particles or specified groups of such particles (14th CGPM, 1971).

Examples of the use of the mole:

1 mol of H_2 contains about 6.022×10^{23} H_2 molecules, or 12.044×10^{23} H atoms.
1 mol of HgCl has a mass of 236.04 g.
1 mol of Hg_2Cl_2 has a mass of 472.08 g.
1 mol of Hg_2^{2+} has a mass of 401.18 g and a charge of 192.97 kC.
1 mol of $Fe_{0.91}S$ has a mass of 82.88 g.
1 mol of e^- has a mass of 548.60 μg and a charge of −96.49 kC.
1 mol of photons whose frequency is 10^{14} Hz has energy of about 39.90 kJ.

Candela: The candela is the luminous intensity in a given direction of a source that emits monochromatic radiation of frequency 540×10^{12} hertz and that has a radiant intensity in that direction of (1/683) watt per steradian (16th CGPM, 1979).

Names and Symbols for the SI Base Units

Physical Quantity	Name of SI Unit	Symbol for SI Unit
Length	meter	m
Mass	kilogram	kg
Time	second	s
Electric current	ampere	A
Thermodynamic temperature	kelvin	K
Amount of substance	mole	mol
Luminous intensity	candela	cd

SI Derived Units with Special Names and Symbols

Physical Quantity	Name of SI Unit	Symbol for SI Unit	Expression in Terms of SI Base Units	
Frequency[1]	hertz	Hz	s^{-1}	
Force	newton	N	$m\ kg\ s^{-2}$	
Pressure, stress	pascal	Pa	$N\ m^{-2}$	$= m^{-1}\ kg\ s^{-2}$
Energy, work, heat	joule	J	N m	$= m^2\ kg\ s^{-2}$
Power, radiant flux	watt	W	$J\ s^{-1}$	$= m^2\ kg\ s^{-3}$
Electric charge	coulomb	C	A s	
Electric potential, electromotive force	volt	V	$J\ C^{-1}$	$= m^2\ kg\ s^{-3}\ A^{-1}$
Electric resistance	ohm	Ω	$V\ A^{-1}$	$= m^2\ kg\ s^{-3}\ A^{-2}$
Electric conductance	siemens	S	Ω^{-1}	$= m^{-2}\ kg^{-1}\ s^3\ A^2$
Electric capacitance	farad	F	$C\ V^{-1}$	$= m^{-2}\ kg^{-1}\ s^4\ A^2$
Magnetic flux density	tesla	T	$V\ s\ m^{-2}$	$= kg\ s^{-2}\ A^{-1}$
Magnetic flux	weber	Wb	V s	$= m^2\ kg\ s^{-2}\ A^{-1}$
Inductance	henry	H	$V\ A^{-1}\ s$	$= m^2\ kg\ s^{-2}\ A^{-2}$
Celsius temperature[2]	degree Celsius	°C	K	

(continued)

SI Derived Units with Special Names and Symbols (continued)

Physical Quantity	Name of SI Unit	Symbol for SI Unit	Expression in Terms of SI Base Units
Luminous flux	lumen	lm	cd sr
Illuminance	lux	lx	cd sr m^{-2}
Activity (radioactive)	becquerel	Bq	s^{-1}
Absorbed dose (of radiation)	gray	Gy	J kg^{-1} = m^2 s^{-2}
Dose equivalent (dose equivalent index)	sievert	Sv	J kg^{-1} = m^2 s^{-2}
Plane angle	radian	rad	1 = m m^{-1}
Solid angle	steradian	sr	1 = m^2 m^{-2}

[1]For radial (circular) frequency and for angular velocity the unit rad s^{-1}, or simply s^{-1}, should be used, and this may not be simplified to Hz. The unit Hz should be used only for frequency in the sense of cycles per second.

[2]The Celsius temperature θ is defined by the equation:

$$\theta/°C = T/K - 273.15$$

The SI unit of Celsius temperature interval is the degree Celsius, °C, which is equal to the kelvin, K. °C should be treated as a single symbol, with no space between the ° sign and the letter C. (The symbol °K and the symbol ° should no longer be used.)

Units in Use Together with the SI

These units are not part of the SI, but it is recognized that they will continue to be used in appropriate contexts. SI prefixes may be attached to some of these units, such as milliliter, ml; millibar, mbar; megaelectronvolt, MeV; kilotonne, ktonne.

Physical Quantity	Name of Unit	Symbol for Unit	Value in SI Units
Time	minute	min	60 s
Time	hour	h	3600 s
Time	day	d	86,400 s
Plane angle	degree	°	(π/180) rad
Plane angle	minute	′	(π/10,800) rad
Plane angle	second	″	(π/648,000) rad
Length	ångstrom[1]	Å	10^{-10} m
Area	barn	b	10^{-28} m^2
Volume	liter	l, L	dm^3 = 10^{-3} m^3
Mass	tonne	t	Mg = 10^3 kg
Pressure	bar[1]	bar	10^5 Pa = 10^5 N m^{-2}
Energy	electronvolt[2]	eV (= $e \times$ V)	≈1.60218 × 10^{-19} J
Mass	unified atomic mass unit[2,3]	u (= $m_a(^{12}C)/12$)	≈1.66054 × 10^{-27} kg

[1]The ångstrom and the bar are approved by CIPM for "temporary use with SI units," until CIPM makes a further recommendation. However, they should not be introduced where they are not used at present.

[2]The values of these units in terms of the corresponding SI units are not exact, since they depend on the values of the physical constants e (for the electronvolt) and N_a (for the unified atomic mass unit), which are determined by experiment.

[3]The unified atomic mass unit is also sometimes called the dalton, with symbol Da, although the name and symbol have not been approved by CGPM.

Conversion Constants and Multipliers

Recommended Decimal Multiples and Submultiples

Multiples and Submultiples	Prefixes	Symbols	Multiples and Submultiples	Prefixes	Symbols
10^{18}	exa	E	10^{-1}	deci	d
10^{15}	peta	P	10^{-2}	centi	c
10^{12}	tera	T	10^{-3}	milli	m
10^{9}	giga	G	10^{-6}	micro	μ (Greek mu)
10^{6}	mega	M	10^{-9}	nano	n
10^{3}	kilo	k	10^{-12}	pico	p
10^{2}	hecto	h	10^{-15}	femto	f
10	deca	da	10^{-18}	atto	a

Conversion Factors—Metric to English

To Obtain	Multiply	By
Inches	centimeters	0.3937007874
Feet	meters	3.280839895
Yards	meters	1.093613298
Miles	kilometers	0.6213711922
Ounces	grams	$3.527396195 \times 10^{-2}$
Pounds	kilogram	2.204622622
Gallons (U.S. liquid)	liters	0.2641720524
Fluid ounces	milliliters (cc)	$3.381402270 \times 10^{-2}$
Square inches	square centimeters	0.155003100
Square feet	square meters	10.76391042
Square yards	square meters	1.195990046
Cubic inches	milliliters (cc)	$6.102374409 \times 10^{-2}$
Cubic feet	cubic meters	35.31466672
Cubic yards	cubic meters	1.307950619

Conversion Factors—English to Metric*

To Obtain	Multiply	By
Microns	mils	**25.4**
Centimeters	inches	**2.54**
Meters	feet	**0.3048**
Meters	yards	**0.9144**
Kilometers	miles	**1.609344**
Grams	ounces	28.34952313
Kilograms	pounds	**0.45359237**
Liters	gallons (U.S. liquid)	**3.785411784**
Millimeters (cc)	fluid ounces	29.57352956
Square centimeters	square inches	**6.4516**
Square meters	square feet	**0.09290304**
Square meters	square yards	**0.83612736**
Milliliters (cc)	cubic inches	**16.387064**
Cubic meters	cubic feet	$2.831684659 \times 10^{-2}$
Cubic meters	cubic yards	0.764554858

*Boldface numbers are exact; others are given to ten significant figures where so indicated by the multiplier factor.

Conversion Factors—General*

To Obtain	Multiply	By
Atmospheres	feet of water @ 4°C	2.950×10^{-2}
Atmospheres	inches of mercury @ 0°C	3.342×10^{-2}
Atmospheres	pounds per square inch	6.804×10^{-2}
BTU	foot-pounds	1.285×10^{-3}
BTU	joules	9.480×10^{-4}
Cubic feet	cords	**128**
Degree (angle)	radians	57.2958
Ergs	foot-pounds	1.356×10^{7}
Feet	miles	**5280**
Feet of water @ 4°C	atmospheres	33.90
Foot-pounds	horsepower-hours	1.98×10^{6}
Foot-pounds	kilowatt-hours	2.655×10^{6}
Foot-pounds per min	horsepower	3.3×10^{4}
Horsepower	foot-pounds per sec	1.818×10^{-3}
Inches of mercury @ 0°C	pounds per square inch	2.036
Joules	BTU	1054.8
Joules	foot-pounds	1.35582
Kilowatts	BTU per min	1.758×10^{-2}
Kilowatts	foot-pounds per min	2.26×10^{-5}
Kilowatts	horsepower	0.745712
Knots	miles per hour	0.86897624
Miles	feet	1.894×10^{-4}
Nautical miles	miles	0.86897624
Radians	degrees	1.745×10^{-2}
Square feet	acres	**43,560**
Watts	BTU per min	17.5796

*Boldface numbers are exact; others are given to ten significant figures where so indicated by the multiplier factor.

Temperature Factors

$$°F = 9/5\,(°C) + 32$$

$$\text{Fahrenheit temperature} = 1.8\,(\text{temperature in kelvins}) - 459.67$$

$$°C = 5/9\,[(°F) - 32)]$$

$$\text{Celsius temperature} = \text{temperature in kelvins} - 273.15$$

$$\text{Fahrenheit temperature} = 1.8\,(\text{Celsius temperature}) + 32$$

Conversion of Temperatures

From	To	
°Celsius	°Fahrenheit	$t_F = (t_C \times 1.8) + 32$
	Kelvin	$T_K = t_C + 273.15$
	°Rankine	$T_R = (t_C + 273.15) \times 18$
°Fahrenheit	°Celsius	$t_C = \frac{t_F - 32}{1.8}$
	Kelvin	$T_k = \frac{t_F - 32}{1.8} + 273.15$
	°Rankine	$T_R = t_F + 459.67$
Kelvin	°Celsius	$t_C = T_K - 273.15$
	°Rankine	$T_R = T_K \times 1.8$
°Rankine	Kelvin	$T_K = \frac{T_R}{1.8}$
	°Fahrenheit	$t_F = T_R - 459.67$

Physical Constants

General

Equatorial radius of the Earth = 6378.388 km = 3963.34 miles (statute)
Polar radius of the Earth, 6356.912 km = 3949.99 miles (statute)
1 degree of latitude at 40° = 69 miles
1 international nautical mile = 1.15078 miles (statute) = 1852 m = 6076.115 ft
Mean density of the earth = 5.522 g/cm^3 = 344.7 lb/ft^3
Constant of gravitation (6.673 ± 0.003) × 10^{-8} cm^3 gm^{-1} s^{-2}
Acceleration due to gravity at sea level, latitude 45° = 980.6194 cm/s^2 = 32.1726 ft/s^2
Length of seconds pendulum at sea level, latitude 45° = 99.3575 cm = 39.1171 in.
1 knot (international) = 101.269 ft/min = 1.6878 ft/s = 1.1508 miles (statute)/h
1 micron = 10^{-4} cm
1 ångstrom = 10^{-8} cm
Mass of hydrogen atom = (1.67339±0.0031) × 10^{-24} g
Density of mercury at 0°C = 13.5955 g/ml
Density of water at 3.98°C = 1.000000 g/ml
Density, maximum, of water, at 3.98°C = 0.999973 g/cm^3
Density of dry air at 0°C, 760 mm = 1.2929 g/l
Velocity of sound in dry air at 0°C = 331.36 m/s – 1087.1 ft/s
Velocity of light in vacuum = (2.997925±0.000002) × 10^{10} cm/s
Heat of fusion of water 0°C = 79.71 cal/g
Heat of vaporization of water 100°C = 539.55 cal/g
Electrochemical equivalent of silver 0.001118 g/s international amp
Absolute wavelength of red cadmium light in air at 15°C, 760 mm pressure = 6438.4696 Å
Wavelength of orange-red line of krypton 86 = 6057.802 Å

π Constants

π = 3.14159 26535 89793 23846 26433 83279 50288 41971 69399 37511
$1/\pi$ = 0.31830 98861 83790 67153 77675 26745 02872 40689 19291 48091
π^2 = 9.8690 44010 89358 61883 44909 99876 15113 53136 99407 24079
$\log_e \pi$ = 1.14472 98858 49400 17414 34273 51353 05871 16472 94812 91531
$\log_{10} \pi$ = 0.49714 98726 94133 85435 12682 88290 89887 36516 78324 38044
$\log_{10} \sqrt{2\pi}$ = 0.39908 99341 79057 52478 25035 91507 69595 02099 34102 92128

Constants Involving e

e = 2.71828 18284 59045 23536 02874 71352 66249 77572 47093 69996
$1/e$ = 0.36787 94411 71442 32159 55237 70161 46086 74458 11131 03177
e^2 = 7.38905 60989 30650 22723 04274 60575 00781 31803 15570 55185
$M = \log_{10} e$ = 0.43429 44819 03251 82765 11289 18916 60508 22943 97005 80367
$1/M \cdot = \log_e 10$ = 2.30258 50929 94045 68401 79914 54684 36420 67011 01488 62877
$\log_{10} M$ = 9.63778 43113 00536 78912 29674 98645 –10

Numerical Constants

$\sqrt{2}$ = 1.41421 35623 73095 04880 16887 24209 69807 85696 71875 37695
$3\sqrt{2}$ = 1.25992 10498 94873 16476 72106 07278 22835 05702 51464 70151
$\log_e 2$ = 0.69314 71805 59945 30941 72321 21458 17656 80755 00134 36026
$\log_{10} 2$ = 0.30102 99956 63981 19521 37388 94724 49302 67881 89881 46211

$\sqrt{3}$ = 1.73205 08075 68877 29352 74463 41505 87236 69428 05253 81039

$\sqrt[3]{3}$ = 1.44224 95703 07408 38232 16383 10780 10958 83918 69253 49935

$\log_e 3$ = 1.09861 22886 68109 69139 52452 36922 52570 46474 90557 82275

$\log_{10} 3$ = 0.47712 12547 19662 43729 50279 03255 11530 92001 28864 19070

Symbols and Terminology for Physical and Chemical Quantities

Name	*Symbol*	*Definition*	*SI Unit*
		Classical Mechanics	
Mass	m		kg
Reduced mass	μ	$\mu = m_1m_2/(m_1 + m_2)$	kg
Density, mass density	ρ	$\rho = M/V$	kg m^{-3}
Relative density	d	$d = \rho/\rho^\theta$	1
Surface density	ρ_A, ρ_S	$\rho_A = m/A$	kg m^{-2}
Momentum	p	$p = mv$	kg m s^{-1}
Angular momentum, action	L	$l = r$ ¥ p	J s
Moment of inertia	I, J	$I = \Sigma m_i r_i^2$	kg m^2
Force	F	$F = d\mathbf{p}/dt = ma$	N
Torque, moment of a force	T, (M)	$T = r \times \mathbf{F}$	N m
Energy	E		J
Potential energy	E_p, V, Φ	E_p=Fds	J
Kinetic energy	E_k, T, K	$e_k = (1/2)mv^2$	J
Work	W, w	$w = F$ds	J
Hamilton function	H	$H(q, p) = T(q, p) + V(q)$	J
Lagrange function	L	$L(q, \dot{q})T(q, \dot{q}) - V(q)$	J
Pressure	p, P	$p = F/A$	Pa, N m^{-2}
Surface tension	γ, σ	$\gamma = dW/dA$	N m^{-1}, J m^{-2}
Weight	G, (W, P)	$G = mg$	N
Gravitational constant	G	$F = Gm_1m_2/r^2$	N m^2 kg^{-2}
Normal stress	σ	$\sigma = F/A$	Pa
Shear stress	τ	$\tau = F/A$	Pa
Linear strain, relative elongation	ε, e	$\varepsilon = \Delta l/l$	1
Modulus of elasticity, Young's modulus	E	$E = \sigma/\varepsilon$	Pa
Shear strain	γ	$\gamma = \Delta x/d$	1
Shear modulus	G	$G = \tau/\gamma$	Pa
Volume strain, bulk strain	θ	$\theta = \Delta V/V_0$	1
Bulk modulus, compression modulus	K	$K = -V_0(dp/dV)$	Pa
Viscosity, dynamic viscosity	η, μ	$\tau_{x,z} = \eta(dv_x/dz)$	Pa s
Fluidity	ϕ	$\phi = 1/\eta$	m kg^{-1} s
Kinematic viscosity	v	$v = \eta/\rho$	m^2 s^{-1}
Friction coefficient	μ, (f)	$F_{frict} = \mu F_{norm}$	1
Power	P	$P = dW/dt$	W
Sound energy flux	P, P_a	$P = dE/dt$	W
Acoustic factors			
Reflection factor	ρ	$\rho = P_t/P_0$	1
Acoustic absorption factor	α_a, (α)	$\alpha_a = 1 - \rho$	1
Transmission factor	τ	$\tau = P_{tr}/P_0$	1
Dissipation factor	δ	$\delta = \alpha_a - \tau$	1

(continued)

Symbols and Terminology for Physical and Chemical Quantities (continued)

Name	*Symbol*	*Definition*	*SI Unit*
		Electricity and Magnetism	
Quantity of electricity, electric charge	Q		C
Charge density	ρ	$\rho = Q/V$	C m^{-3}
Surface charge density	σ	$\sigma = Q/A$	C m^{-2}
Electric potential	V, ϕ	$V = \mathrm{d}W/\mathrm{d}Q$	V, J C^{-1}
Electric potential difference	$U, \Delta V, \Delta\phi$	$U = V_2 - V_1$	V
Electromotive force	E	$E = (F/Q)\mathrm{d}s$	V
Electric field strength	$\mathbf{E}$	$\mathbf{E} = \mathbf{F}/Q = -\mathrm{grad}\ V$	V m^{-1}
Electric flux	Ψ	$\Psi = \mathbf{D}\mathrm{d}\mathbf{A}$	C
Electric displacement	$\mathbf{D}$	$\mathbf{D} = \varepsilon\mathbf{E}$	C m^{-2}
Capacitance	C	$C = Q/U$	F, C V^{-1}
Permittivity	ε	$D = \varepsilon E$	F m^{-1}
Permittivity of vacuum	ε_0	$\varepsilon_0 = \mu_0{}^{-1} c_0{}^{-2}$	F m^{-1}
Relative permittivity	ε_r	$\varepsilon_r = \varepsilon/\varepsilon_0$	1
Dielectric polarization (dipole moment per volume)	$\mathbf{P}$	$\mathbf{P} = \mathbf{D} - \varepsilon_0\mathbf{E}$	C m^{-2}
Electric susceptibility	χ_e	$\chi_e = \varepsilon_r - 1$	1
Electric dipole moment	$\mathbf{p}, \mu$	$\mathbf{p} = Q\mathbf{r}$	C m
Electric current	I	$I = \mathrm{d}Q/\mathrm{d}t$	A
Electric current density	$\mathbf{j}, \mathbf{J}$	$I = \mathbf{j}\mathrm{d}x\mathbf{A}$	A m^{-2}
Magnetic flux density, magnetic induction	$\mathbf{B}$	$\mathbf{F} = Qv \times \mathbf{B}$	T
Magnetic flux	Φ	$\Phi = \mathbf{B}\mathrm{d}A$	Wb
Magnetic field strength	$\mathbf{H}$	$\mathbf{B} = \mu\mathbf{H}$	A M^{-1}
Permeability	μ	$\mathbf{B} = \mu\mathbf{H}$	N A^{-2}, H m^{-1}
Permeability of vacuum	μ_0		H m^{-1}
Relative permeability	μ_r	$\mu_r = \mu/\mu_0$	1
Magnetization (magnetic dipole moment per volume)	$\mathbf{M}$	$\mathbf{M} = \mathbf{B}/\mu_0 - \mathbf{H}$	A m^{-1}
Magnetic susceptibility	$\chi, \kappa, (\chi_m)$	$\chi = \mu_r - 1$	1
Molar magnetic susceptibility	χ_m	$\chi_m = V_m\chi$	m^3 mol^{-1}
Magnetic dipole moment	$\mathbf{m}, \mu$	$E_p = -\mathbf{m} \cdot \mathbf{B}$	A m^2, J T^{-1}
Electrical resistance	R	$\mathbf{P} = \mathbf{Y}/\mathbf{I}$	Ω
Conductance	G	$G = 1/R$	S
Loss angle	δ	$\delta = (\pi/2) + \phi_I - \phi_U$	1, rad
Reactance	X	$X = (U/I)\sin\delta$	Ω
Impedance (complex impedance)	Z	$Z = R + \mathrm{i}X$	Ω
Admittance (complex admittance)	Y	$Y = 1/Z$	S
Susceptance	B	$Y = G + iB$	S
Resistivity	ρ	$\rho = E/j$	Ω m
Conductivity	κ, γ, σ	$\kappa = 1/\rho$	S m^{-1}
Self-inductance	L	$E = -L(\mathrm{d}I/\mathrm{d}t)$	H
Mutual inductance	M, L_{12}	$E_1 = L_{12}(\mathrm{D}i_2/\mathrm{d}t)$	H
Magnetic vector potential	$\mathbf{A}$	$\mathbf{B} = \nabla \times \mathbf{A}$	Wb m^{-1}
Poynting vector	$\mathbf{S}$	$\mathbf{S} = \mathbf{E} \times \mathbf{H}$	W m^{-2}
		Electromagnetic Radiation	
Wavelength	λ		m
Speed of light			m s^{-1}
in vacuum	c_0		
in a medium	c	$c = c_0/n$	

(continued)

Symbols and Terminology for Physical and Chemical Quantities (continued)

Name	*Symbol*	*Definition*	*SI Unit*
Electromagnetic Radiation			
Wavenumber in vacuum	V	$V = V/c_0 = 1/n\lambda$	m^{-1}
Wavenumber (in a medium)	σ	$\sigma = 1/\lambda$	m^{-1}
Frequency	v	$v = c/\lambda$	Hz
Circular frequency, pulsatance	ω	$\omega = 2\pi v$	s^{-1}, rad s^{-1}
Refractive index	n	$n = c_0/c$	1
Planck constant	h		J s
Planck constant/2π	$\hbar$	$\hbar = h/2\pi$	J s
Radiant energy	Q, W		J
Radiant energy density	ρ, w	$\rho = Q/V$	J m^{-3}
Spectral radiant energy density			
in terms of frequency	ρ_v, w_v	$\rho_v = \delta\rho/dv$	J m^{-3} Hz^{-1}
in terms of wavenumber	$\rho_{\bar{v}}$, $w_{\bar{v}}$	$\rho_{\bar{v}} = d\rho/d\bar{v}$	J m^{-2}
in terms of wavelength	ρ_λ, w_λ	$\rho_\lambda = \delta\rho/d\lambda$	J m^{-4}
Einstein transition probabilities			
Spontaneous emission	A_{nm}	$dN_n/dt = -A_{nm}N_n$	s^{-1}
Stimulated emission	B_{nm}	$dn_n/dt = -\rho\bar{v}(\bar{V}_{nm}) \times B_{nm}N_n$	s kg^{-1}
Radiant power, radiant energy per time	Φ, P	$\Phi = dQ/dt$	W
Radiant intensity	I	$I = d\Phi/d\Omega$	W sr^{-1}
Radiant exitance (emitted radiant flux)	M	$M = d\Phi/dA_{source}$	W m^{-2}
Irradiance (radiant flux received)	E, (I)	$E = d\Phi/\delta A$	W m^{-2}
Emittance	ε	$\varepsilon = M/M_{bb}$	1
Stefan–Boltzmann constant	σ	$M_{bb} = \sigma T^4$	W m^{-2} K^{-4}
First radiation constant	c_1	$c_1 = 2\pi hc_0^2$	W m^2
Second radiation constant	c_2	$c_2 = hc_0/k$	K m
Transmittance, transmission factor	τ, T	$\tau = \Phi_{tr}/\Phi_0$	1
Absorptance, absorption factor	α	$\alpha = \phi_{abs}/\phi_0$	1
Reflectance, reflection factor	ρ	$\rho = \phi_{refl}/\Phi_0$	1
(Decadic) absorbance	A	$A = lg(1 - \alpha_i)$	1
Napierian absorbance	B	$B = ln(1 - \alpha_i)$	1
Absorption coefficient			
(Linear) decadic	a, K	$a = A/l$	m^{-1}
(Linear) napierian	α	$\alpha = B/l$	m^{-1}
Molar (decadic)	ε	$\varepsilon = a/c = A/cl$	m^2 mol^{-1}
Molar napierian	κ	$\kappa = \alpha/c = B/cl$	m^2 mol^{-1}
Absorption index	k	$k = \alpha/4\pi\bar{v}$	1
Complex refractive index	$\hat{n}$	$\hat{n} = n + ik$	1
Molar refraction	R, R_m	$R = \frac{(n^2-1)}{(n^2+2)} V_m$	m^3 mol^{-1}
Angle of optical rotation	α		1, rad
Solid State			
Lattice vector	$\mathbf{R}$, $\mathbf{R}_0$		m
Fundamental translation vectors for the crystal lattice	$\mathbf{a}_1$; $\mathbf{a}_2$; $\mathbf{a}_3$, $\mathbf{a}$; $\mathbf{b}$; $\mathbf{c}$	$R = n_1\mathbf{a}_1 + n_2\mathbf{a}_2 + n_3\mathbf{a}_3$	m
(Circular) reciprocal lattice vector	$\mathbf{G}$	$G \cdot R = 2\pi m$	m^{-1}

(continued)

Symbols and Terminology for Physical and Chemical Quantities (continued)

Name	Symbol	Definition	SI Unit
Solid State			
(Circular) fundamental translation vectors for the reciprocal lattice	$\mathbf{b}_1; \mathbf{b}_2; \mathbf{b}_3, \mathbf{a}^*; \mathbf{b}^*; \mathbf{c}^*$	$\mathbf{a}_i \cdot \mathbf{b}_k = 2\pi\delta_{ik}$	m^{-1}
Lattice plane spacing	d		m
Bragg angle	θ	$n\lambda = 2d \sin \theta$	1, rad
Order of reflection	n		1
Order parameters			
Short range	σ		1
Long range	s		1
Burgers vector	b		m
Particle position vector	r, R_j		m
Equilibrium position vector of an ion	R_o		m
Displacement vector of an ion	$\mathbf{u}$	$\mathbf{u} = \mathbf{R} - \mathbf{R}_0$	m
Debye–Waller factor	B, D		1
Debye circular wavenumber	q_D		m^{-1}
Debye circular frequency	ω_D		s^{-1}
Grüneisen parameter	γ, Γ	$\gamma = \alpha V/\kappa C_V$	1
Madelung constant	$\alpha, \mathcal{M}$	$E_{coul} = \frac{\alpha N_A z_+ z_- e^2}{4\pi\varepsilon_0 R_0}$	1
Density of states	N_E	$N_E = dN(E)/dE$	$J^{-1}\ m^{-3}$
(Spectral) density of vibrational modes	N_ω, g	$N_\omega = dN(\omega)/d\omega$	$s\ m^{-3}$
Resistivity tensor	ρ_{ik}	$E = \rho \cdot j$	Ω m
Conductivity tensor	σ_{ik}	$\sigma = \rho^{-1}$	$S\ m^{-1}$
Thermal conductivity tensor	λ_{ik}	$J_q = -\lambda \cdot \text{grad}\ T$	$W\ m^{-1}\ K^{-1}$
Residual resistivity	ρ_R		Ω m
Relaxation time	τ	$\tau = l/v_F$	s
Lorenz coefficient	L	$L = \lambda/\sigma T$	$V^2\ K^{-2}$
Hall coefficient	A_H, R_H	$\mathbf{E} = \rho \cdot \mathbf{j} + R_H(\mathbf{B} \times \mathbf{j})$	$m^3\ C^{-1}$
Thermoelectric force	E		V
Peltier coefficient	Π		V
Thomson coefficient	$\mu,(\tau)$		$V\ K^{-1}$
Work function	Φ	$\Phi = E_\infty - E_F$	J
Number density, number concentration	$n, (p)$		m^{-3}
Gap energy	E_γ		J
Donor ionization energy	E_δ		J
Acceptor ionization energy	E_α		J
Fermi energy	E_Φ, ε_F		J
Circular wave vector, propagation vector	$\boldsymbol{k}, \boldsymbol{q}$	$k = 2\pi/\lambda$	m^{-1}
Bloch function	$u_k(\boldsymbol{r})$	$\psi(\boldsymbol{r}) = u_k(\boldsymbol{r}) \exp(\mathbf{ik} \cdot \mathbf{r})$	$m^{-3/2}$
Charge density of electrons	ρ	$\rho(\mathbf{r}) = -e\psi^*(\mathbf{r})\psi(\mathbf{r})$	$C\ m^{-3}$
Effective mass	m^*		kg
Mobility	μ	$\mu = v_{drift}/E$	$m^2\ V^{-1}\ s^{-1}$
Mobility ratio	b	$b = \mu_n/\mu_p$	1
Diffusion coefficient	D	$dN/dt = -DA(dn/dx)$	$m^2\ s^{-1}$
Diffusion length	L	$L = \sqrt{D\tau}$	m
Characteristic (Weiss) temperature	ϕ, ϕ_W		K
Curie temperature	T_C		K
Néel temperature	T_N		K

Credits

Material in Section IV was reprinted from the following sources:

D. R. Lide, Ed., *CRC Handbook of Chemistry and Physics,* 76th ed., Boca Raton, FL: CRC Press, 1992: International System of Units (SI), conversion constants and multipliers (conversion of temperatures), symbols and terminology for physical and chemical quantities, fundamental physical constants, classification of electromagnetic radiation.

D. Zwillinger, Ed., *CRC Standard Mathematical Tables and Formulae,* 30th ed., Boca Raton, FL: CRC Press, 1996: Greek alphabet, conversion constants and multipliers (recommended decimal multiples and submultiples, metric to English, English to metric, general, temperature factors), physical constants, series expansion.

Probability for Electrical and Computer Engineers

Charles W. Therrien

The Algebra of Events

The study of probability is based upon experiments that have uncertain outcomes. Collections of these outcomes comprise *events* and the collection of all possible outcomes of the experiment comprise what is called the *sample space*, denoted by S. Outcomes are members of the sample space and events of interest are represented as *sets* of outcomes (see Figure IV.1).

The algebra $\mathcal{A}$ that deals with representing events is the usual set algebra. If A is an event, then A^c (the *complement* of A) represents the event that "A did not occur." The complement of the sample space is the *null event*, $\varnothing = S^c$. The event that *both* event A_1 and event A_2 have occurred is the intersection, written as "$A_1 \cdot A_2$" or "$A_1 A_2$" while the event that *either* A_1 or A_2 *or both* have occurred is the union, written as "$A_1 + A_2$."[1]

Table IV.1 lists the two postulates that define the algebra $\mathcal{A}$, while Table IV.2 lists seven axioms that define properties of its operations. Together these tables can be used to show all of the properties of the algebra of events. Table IV.3 lists some additional useful relations that can be derived from the axioms and the postulates.

Since the events "$A_1 + A_2$" and "$A_1 A_2$" are included in the algebra, it follows by induction that for any finite number of events $A_1 + A_2 + \cdots + A_N$ and $A_1 \cdot A_2 \cdots A_N$ are also included in the algebra. Since problems often involve the union or intersection of an *infinite* number of events, however, the algebra of events must be defined to include these infinite intersections and unions. This extension to infinite unions and intersections is known as a sigma algebra.

A set of events that satisfies the two conditions:

1. $A_i A_j = \varnothing \neq$ for $\neq i \neq j$
2. $A_1 + A_2 + A_3 + \cdots = S$

is known as a *partition* and is important for the solution of problems in probability. The events of a partition are said to be *mutually exclusive* and *collectively exhaustive*. The most fundamental partition is the set outcomes defining the random experiment, which comprise the sample space by definition.

Probability

Probability measures the likelihood of occurrence of events represented on a scale of 0 to 1. We often estimate probability by measuring the *relative frequency* of an event, which is defined as

$$\text{relative frequency} = \frac{\text{number of occurrences of the event}}{\text{number of repetitions of the experiment}}$$

(for a large number of repetitions). Probability can be defined formally by the following axioms:

(I) The probability of any event is nonnegative:

$$\Pr[A] \geqslant 0 \qquad \text{(IV.1)}$$

(II) The probability of the universal event (i.e., the entire sample space) is 1:

$$\Pr[S] = 1 \qquad \text{(IV.2)}$$

[1]Some authors use $\cap$ and $\cup$ rather than $\cdot$ and $+$, respectively.

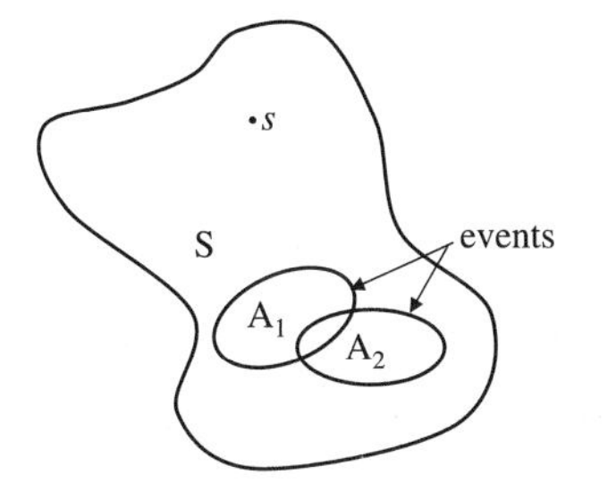

FIGURE IV.1 Abstract representation of the sample space S with outcome s and sets A_1 and A_2 representing events.

(III) If A_1 and A_2 are mutually exclusive, i.e., $A_1A_2 = \varnothing$, then

$$\Pr[A_1 + A_2] = \Pr[A_1] + \Pr[A_2] \qquad \text{(IV.3)}$$

(IV) If $\{A_i\}$ represent a countably infinite set of mutually exclusive events, then

$$\Pr[A_1 + A_2 + A_3 + \cdots] = \sum_{i=1}^{\infty} \Pr[A_i] \quad (\text{ if } A_iA_j = \varnothing \quad i \neq j) \qquad \text{(IV.4)}$$

Note that although the additivity of probability for any finite set of disjoint events follows from (III), the property has to be stated explicitly for an infinite set in (IV). These axioms and the algebra of events can be used to show a number of other important properties which are summarized in Table IV.4. The last item in the table is an especially important formula since it uses probabilistic information about

TABLE IV.1 Postulates for an Algebra of Events

1.	*If* $A \in \mathcal{A}$ *then* $A^c \in \mathcal{A}$
2.	*If* $A_1 \in \mathcal{A}$ *and* $A_2 \in \mathcal{A}$ *then* $A_1 + A_2 \in \mathcal{A}$

TABLE IV.2 Axioms of Operations on Events

$A_1A_1^c = \varnothing$	Mutual exclusion
$A_1S = A_1$	Inclusion
$(A_1^c)^c = A_1$	Double complement
$A_1 + A_2 = A_2 + A_1$	Commutative law
$A_1 + (A_2 + A_3) = (A_1 + A_2) + A_3$	Associative law
$A_1(A_2 + A_3) = A_1A_2 + A_1A_3$	Distributive law
$(A_1A_2)^c = A_1^c + A_2^c$	DeMorgan's law

TABLE IV.3 Additional Identities in the Algebra of Events

$S^c = \varnothing$	
$A_1 + \varnothing = A_1$	Inclusion
$A_1A_2 = A_2A_1$	Commutative law
$A_1(A_2A_3) = (A_1A_2)A_3$	Associative law
$A_1 + (A_2A_3) = (A_1 + A_2)(A_1 + A_3)$	Distributive law
$(A_1 + A_2)^c = A_1^c A_2^c$	DeMorgan's law

TABLE IV.4 Some Corollaries Derived from the Axioms of Probability

$\Pr[A^c] = 1 - \Pr[A]$
$0 \leq \Pr[A] \leq 1$
If $A_1 \subseteq A_2$ then $\Pr[A_1] \leq \Pr[A_2]$
$\Pr[\varnothing] = 0$
If $A_1A_2 = \varnothing -$ then $= \Pr[A_1A_2] = 0$
$\Pr[A_1 + A_2] = \Pr[A_1] + \Pr[A_2] - \Pr[A_1A_2]$

individual events to compute the probability of the union of two events. The term $\Pr[A_1A_2]$ is referred to as the *joint probability* of the two events. This last equation shows that the probabilities of two events add as in Equation (IV.3) only if their joint probability is 0. The joint probability is 0 when the two events have no intersection ($A_1A_2 = \varnothing$).

Two events are said to be statistically *independent* if and only if

$$\Pr[A_1A_2] = \Pr[A_1]\cdot\Pr[A_2] \quad \text{(independent events)} \tag{IV.5}$$

This definition is not derived from the earlier properties of probability. An argument to give this definition intuitive meaning can be found in Ref. [1]. Independence occurs in problems where two events are not influenced by one another and Equation (IV.5) simplifies such problems considerably.

A final important result deals with partitions. *A partition* is a finite or countably infinite set of events $A_1, A_2, A_3, \ldots$ that satisfy the two conditions:

$$A_iA_j = \varnothing \text{ for } i \neq j$$

$$A_1 + A_2 + A_3 + \cdots = S$$

The events in a partition satisfy the relation:

$$\sum_i \Pr[A_i] = 1 \tag{IV.6}$$

Further, if B is *any* other event, then

$$\Pr[B] = \sum_i \Pr[A_iB] \tag{IV.7}$$

The latter result is referred to as the *principle of total probability* and is frequently used in solving problems. The principle is illustrated by a Venn diagram in Figure IV.2. The rectangle represents the sample space and other events are defined therein. The event B is seen to be comprised of all of the pieces

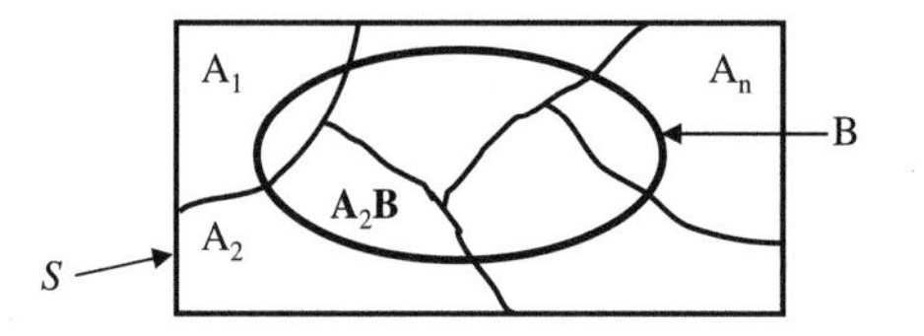

FIGURE IV.2 Venn diagram illustrating the principle of total probability.

that represent intersections or overlap of event B with the events A_i. This is the graphical interpretation of Equation (IV.7).

An Example

Simon's Surplus Warehouse has large barrels of mixed electronic components (parts) that you can buy by the handful or by the pound. You are not allowed to select parts individually. Based on your previous experience, you have determined that in one barrel, 29% of the parts are bad (faulted), 3% are bad resistors, 12% are good resistors, 5% are bad capacitors, and 32% are diodes. You decide to assign probabilities based on these percentages. Let us define the following events:

Event	Symbol
Bad (faulted) component	B
Good component	G
Resistor	R
Capacitor	C
Diode	D

A Venn diagram representing this situation is shown below along with probabilities of various events as given:

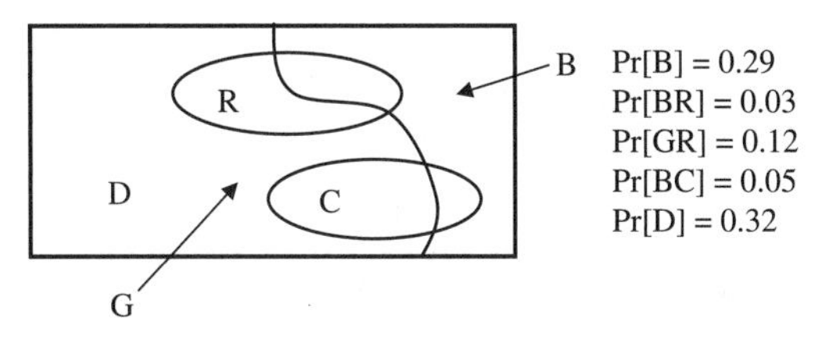

Note that since any component must be a resistor, capacitor, or diode, the region labeled D in the diagram represents everything in the sample space which is not included in R or C.

We can answer a number of questions.

1. What is the probability that a component is a resistor (either good *or* bad)?
   Since the events B and G form a partition of the sample space, we can use the principle of total probability Equation (IV.7) to write:

$$\Pr[R] = \Pr[GR] + \Pr[BR] = 0.12 + 0.03 = 0.15$$

2. Are bad parts and resistors independent?
   We know that $\Pr[BR] = 0.03$ and we can compute:

$$\Pr[B] \cdot \Pr[R] = (0.29)(0.15) = 0.0435$$

   Since $\Pr[BR] \neq \Pr[B] \cdot \Pr[R]$, the events are *not* independent.
3. You have no use for either bad parts or resistors. What is the probability that a part is either bad and/or a resistor?

Using the formula from Table IV.4 and the previous result we can write:

$$\Pr[B + R] = \Pr[B] + \Pr[R] - \Pr[BR] = 0.29 + 0.15 - 0.03 = 0.41$$

4. What is the probability that a part is useful to you?
   Let U represent the event that the part is useful. Then (see Table IV.4):

$$\Pr[U] = 1 - \Pr[U^c] = 1 - 0.41 = 0.59$$

5. What is the probability of a bad diode?
   Observe that the events R, C, and D form a partition, since a component has to be one and only one type of part. Then using Equation (IV.7) we write:

$$\Pr[B] = \Pr[BR] + \Pr[BC] + \Pr[BD]$$

Substituting the known numerical values and solving yields

$$0.29 = 0.03 + 0.05 + \Pr[BD] \text{ or } \Pr[BD] = 0.21$$

Conditional Probability and Bayes' Rule

The *conditional* probability of an event A_1 given that an event A_2 has occurred is defined by

$$\Pr[A_1|A_2] = \frac{\Pr[A_1A_2]}{\Pr[A_2]} \tag{IV.8}$$

($\Pr[A_1|A_2]$ is read "probability of A_1 *given* A_2.") As an illustration, let us compute the probability that a component in the previous example is bad given that it is a resistor:

$$\Pr[B|R] = \frac{\Pr[BR]}{\Pr[R]} = \frac{0.03}{0.15} = 0.2$$

(The value for $\Pr[R]$ was computed in question 1 of the example.) Frequently the statement of a problem is in terms of conditional probability rather than joint probability, so Equation (IV.8) is used in the form:

$$\Pr[A_1A_2] = \Pr[A_1|A_2] \cdot \Pr[A_2] = \Pr[A_2|A_1] \cdot \Pr[A_1] \tag{IV.9}$$

(The last expression follows because $\Pr[A_1A_2]$ and $\Pr[A_2A_1]$ are the same thing.) Using this result, the principle of total probability Equation (IV.7) can be rewritten as

$$\Pr[B] = \sum_j \Pr[B|A_j] \Pr[A_j] \tag{IV.10}$$

where B is any event and $\{A_j\}$ is a set of events that forms a partition.

Now, consider any one of the events A_i in the partition. It follows from Equation (IV.9) that

$$\Pr[A_i|B] = \frac{\Pr[B|A_i] \cdot \Pr[A_i]}{\Pr[B]}$$

Then substituting in Equation (IV.10) yields:

$$\Pr[A_i|B] = \frac{\Pr[B|A_i] \cdot \Pr[A_i]}{\sum_j \Pr[B|A_j]\Pr[A_j]} \tag{IV.11}$$

This result is known as *Baye's theorem* or *Bayes' rule.* It is used in a number of problems that commonly arise in electrical engineering. We illustrate and end this section with an example from the field of communications.

Communication Example

The transmission of bits over a binary communication channel is represented in the drawing below:

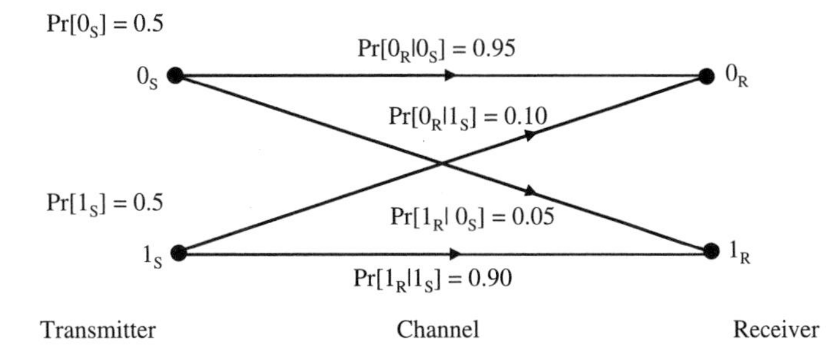

where we use notation like 0_S, 0_R ... to denote events "0 sent," "0 received," etc. When a 0 is transmitted, it is correctly received with probability 0.95 or incorrectly received with probability 0.05. That is, $\Pr[0_R|0_S] = 0.95$ and $\Pr[1_R|0_S] = 0.05$. When a 1 is transmitted, it is correctly received with probability 0.90 and incorrectly received with probability 0.10. The probabilities of sending a 0 or a 1 are denoted by $\Pr[0_S]$ and $\Pr[1_S]$. It is desired to compute the *probability of error* for the system.

This is an application of the principle of total probability. The two events 0_S and 1_S are mutually exclusive and collectively exhaustive and thus form a partition. Take the event B to be the event that an error occurs. It follows from Equation (IV.10) that

$$\begin{aligned}\Pr[\text{error}] &= \Pr[\text{error}|0_S]\Pr[0_S] + \Pr[\text{error}|1_S]\Pr[1_S] \\ &= \Pr[1_R|0_S]\Pr[0_S] + \Pr[0_R|1_S]\Pr[1_S] \\ &= (0.05)(0.5) + (0.10)(0.5) = 0.075\end{aligned}$$

Next, given that an error has occurred, let us compute the probability that a 1 was sent or a 0 was sent. This is an application of Bayes' rule. For a 1, Equation (IV.11) becomes

$$\Pr[1_S|\text{error}] = \frac{\Pr[\text{error}|1_S]\Pr[1_S]}{\Pr[\text{error}|1_S]\Pr[1_S] + \Pr[\text{error}|0_S]\Pr[0_S]}$$

Substituting the numerical values then yields:

$$\Pr[1_S|\text{error}] = \frac{(0.10)(0.5)}{(0.10)(0.5) + (0.05)(0.5)} \approx 0.667$$

For a 0, a similar analysis applies:

$$\begin{aligned}\Pr[0_S|\text{error}] &= \frac{\Pr[\text{error}|0_S]\,\Pr[0_S]}{\Pr[\text{error}|1_S]\,\Pr[1_S] + \Pr[\text{error}|0_S]\,\Pr[0_S]} \\ &= \frac{(0.05)(0.5)}{(0.10)(0.5) + (0.05)(0.5)} \approx 0.333\end{aligned}$$

The two resulting probabilities sum to 1 because 0_S and 1_S form a partition for the experiment.

Reference

1. C. W. Therrien and M. Tummala, *Probability for Electrical and Computer Engineers.* Boca Raton, FL: CRC Press, 2004.

Indexes

Author Index

A

B

C

D

E

F

G

H

K

L

M

N

O

P

R

S

T

U

W

Y

Z

Subject Index

A

Page on which a term is defined is indicated in bold.

B

C

D

E

F

G

H

I

J

K

L

M

N

O

P

Q

R

S

T

U

V

W

Y

Z